Calculus

Single Variable

Brian E. Blank
Steven G. Krantz
Washington University In St. Louis

Debut Edition

Key College Publishing
Innovators in Higher Education
www.keycollege.com
in cooperation with
Springer

Brian E. Blank and Steven G. Krantz
Department of Mathematics
Washington University in St. Louis
St. Louis, MO 63130

Key College Publishing was founded in 1999 as a division of Key Curriculum Press® in cooperation with Springer-Verlag New York, Inc. We publish innovative texts and courseware for the undergraduate curriculum in mathematics and statistics as well as mathematics and statistics education. For more information, visit us at www.keycollege.com.

Key College Publishing
1150 65th Street
Emeryville, CA 94608
(510) 595-7000
info@keycollege.com
www.keycollege.com

Development Editor: Allyndreth Cassidy
Production Director: McKinley Williams
Production Coordinator: Ken Wischmeyer
Editorial Production Project Managers: Beth Masse and Laura Ryan
Project Manager: Eric Houts
Copyeditor: Tara Joffe
Proofreader: Andrea Fox
Indexer: Victoria Baker
Text Designer: Suzanne Montazer
Composition, Illustration: Interactive Composition Corporation
Cover Designer: Jensen Barnes
Cover Photo Credit: St. Louis Arch, Missouri, USA: GettyImages/Charles Thatcher
Printer: RR Donnelley

Editorial Director: Richard J. Bonacci
General Manager: Mike Simpson
Publisher: Steven Rasmussen

Library of Congress Cataloging-in-Publication Data

Blank, Brian E., 1953-
Calculus, single variable / Brian E. Blank, Steven G. Krantz.
p. cm.
Includes index.
ISBN 1-931914-59-1 (pbk.)
1. Calculus. 2. Variables (Mathematics) I. Krantz, Steven G. (Steven George), 1951-II. Title.

QA303.2.B468 2005
515—dc22

2004050701

Printed in China
10 9 8 7 6 5 4 3 2 1 09 08 07 06 05

A pebble for Louis.
BEB

This book is for Hypatia, the sweetheart of my life.
SGK

Contents

Preface *ix*
Features *xi*
Supplements *xv*
Acknowledgments *xvii*
About the Authors *xix*

1 Basics 1

Preview 1
1.1 Number Systems 2
1.2 Planar Coordinates and Graphing in the Plane 11
1.3 Lines and Their Slopes 21
1.4 Functions and Their Graphs 33
1.5 Combining Functions 45
1.6 Trigonometry 58
Summary of Key Topics 68
Genesis & Development 71

2 Limits 75

Preview 75
2.1 The Concept of Limit 77
2.2 Limit Theorems 87
2.3 Continuity 98
2.4 Infinite Limits and Asymptotes 112
2.5 Limits of Sequences 121
2.6 Exponential Functions 131
Summary of Key Topics 144
Genesis & Development 147

3 The Derivative 151

Preview 151
3.1 Rates of Change and Tangent Lines 152
3.2 The Definition of the Derivative 165
3.3 Rules for Differentiation 175
3.4 Differentiation of Some Basic Functions 185
3.5 The Chain Rule 193
3.6 Inverse Functions and the Natural Logarithm 201
3.7 Higher Derivatives 215
3.8 Implicit Differentiation and Related Rates 222
3.9 Differentials and Approximation of Functions 233
Summary of Key Topics 241
Genesis & Development 245

4 Applications of the Derivative 249

Preview 249
4.1 The Derivative and Graphing 250
4.2 Maxima and Minima of Functions 258
4.3 Applied Maximum-Minimum Problems 265
4.4 Concavity 277
4.5 Graphing Functions 286
4.6 L'Hôpital's Rule 298
4.7 The Newton-Raphson Method 305
4.8 Antidifferentiation and Applications 315
4.9 Enrichment: Applications to Economics 326
Summary of Key Topics 338
Genesis & Development 342

5 The Integral 347

Preview 347
5.1 Introduction to Integration—The Area Problem 348
5.2 The Riemann Integral 357
5.3 Rules for Integration 369
5.4 The Fundamental Theorem of Calculus 377
5.5 Integration by Substitution 386
5.6 More on the Calculation of Area 393
5.7 Numerical Techniques of Integration 399
Summary of Key Topics 411
Genesis & Development 414

6 Differential Equations and Transcendental Functions 419

Preview 419

6.1 First Order Differential Equations 420

6.2 A Calculus Approach to the Logarithm 430

6.3 The Exponential Function 442

6.4 Logarithms and Powers with Arbitrary Bases 450

6.5 Applications of the Exponential Function 463

6.6 Inverse Trigonometric Functions 477

6.7 Enrichment: The Hyperbolic Functions 488

Summary of Key Topics 501

Genesis & Development 505

7 Techniques of Integration 509

Preview 509

7.1 Integration by Parts 510

7.2 Partial Fractions—Linear Factors 517

7.3 Powers and Products of Trigonometric Functions 529

7.4 Integrals Involving Quadratic Expressions 537

7.5 Partial Fractions—Irreducible Quadratic Factors 546

Summary of Key Topics 554

Genesis & Development 557

8 Applications of the Integral 561

Preview 561

8.1 Volumes 562

8.2 Arc Length and Surface Area 575

8.3 The Average Value of a Function 583

8.4 Center of Mass 590

8.5 Work 596

8.6 Improper Integrals—Unbounded Integrands 603

8.7 Improper Integrals—Unbounded Intervals 609

Summary of Key Topics 616

Genesis & Development 619

9 Infinite Series 621

Preview 621

9.1 Series 622

9.2 Determining Convergence 630

9.3 Series with Nonnegative Terms—The Integral Test 634

9.4 Series with Nonnegative Terms—The Comparison Test 642

9.5 Alternating Series 647
9.6 The Ratio and Root Tests 653
Summary of Key Topics 658
Genesis & Development 660

10 Taylor Series 663

Preview 663
10.1 Introduction to Power Series 664
10.2 Operations on Power Series 672
10.3 Taylor Polynomials 684
10.4 Estimating the Error Term—The Rate of Convergence of Taylor's Expansion 693
10.5 Taylor Series 697
Summary of Key Topics 705
Genesis & Development 707

Appendix Answers to Selected Exercises *711*
Index *759*

Preface

Calculus is one of the milestones of human thought. Every well-educated person should be acquainted with the basic ideas of the subject. In today's technological world, in which more and more ideas are being quantified, knowledge of calculus has become essential to a broader cross section of the population.

Starting in the late 1980s, a vigorous discussion began about the approaches to and the methods of teaching calculus. This debut edition of *Calculus,* published in two volumes, *Single Variable* and *Multivariable,* offers the best in current calculus teaching. We have worked hard to properly assess, with realism and purpose, the calculus market as it actually exists and to address the needs of today's students, bringing together time-tested, as well as innovative, pedagogy and exposition.

The Changing Face of the Student

More than ever, today's students compose a highly heterogeneous group. Calculus students come from a wide variety of disciplines—some study the subject because it is required, others because it will widen their career options. Mathematics majors often go into law, medicine, genome research, the technology sector, and many other professions. As the teaching and learning of calculus is rethought, instructors must keep their students' backgrounds and futures in mind.

Instructors must also remember that an increasingly larger number of college and university students have already seen quite a bit of calculus in high school. Instructors must build on what students already know, refining their mathematical skills and expanding their conceptual horizons.

The goal is to empower students, enhance their critical thinking skills, and give them the intellectual equipment to proceed successfully in whatever major or discipline they ultimately choose to study. This text is intended to be a cornerstone of that process.

The Changing Role of the Textbook

Many resources are available to instructors and students today, from Web sites to interactive tutorials. The calculus textbook must be a tool that instructors can use to augment and bolster their lectures, classroom activities, and resources. It must enhance the classroom experience and speak compellingly to the students who are actively engaged in the class.

A calculus book must tell the truth. It must be carefully written in the accepted language of mathematics, but it also must be credible—for students and instructors alike—and it must be readable. The textbook should include useful and fascinating applications. It should acquaint students with the history of the subject and with a sense of what mathematics is all about. While teaching technique, it should also teach ideas. It should drill students in basic methods and teach them how to discover and build their own concepts in a scientific subject. In today's world, it is particularly important that a calculus book illustrate ideas using modeling and numerical calculation. We have made every effort to ensure that this text is such a calculus book.

Calculus is designed to increase the student's role as an independent thinker, whether as a potential mathematician, scientist, or practitioner in another analytical field. The intent of this book is to make it natural for students to succeed in their calculus course as well as in future courses. We believe that a good calculus book is a crucial stepping-stone in the foundational education of a student in the twenty-first century.

Brian E. Blank
Steven G. Krantz

Features

Although *Calculus: Single Variable* and *Calculus: Multivariable* have many features that appeal to those teaching mathematics and science majors, it was written to offer a large segment of the calculus market a textbook that provides the necessary tools for students to succeed in their study and to appreciate the subject. Instructors teaching calculus to students with various backgrounds and educational goals will be able to shape their course to fit particular needs.

Students must understand concepts as well as calculations—they must be able to reason through word problems as well as complete drill problems. To strike this balance, the following features are included in the text.

Pedagogy

- The writing style is clear and readable. Students should not have to struggle with the exposition as they learn calculus.
- Motivation for important topics is crisp and clean, enabling students to get to key examples quickly and efficiently.
- All essential ideas are showcased with examples, offering a seamless link between concepts and applications.
- Concepts are reinforced by graphical interpretations. The large number of figures helps students visualize the concepts. The text also presents numerical examples when they will aid student's understanding.
- Material that is not required for subsequent sections is denoted by an asterisk (*). Instructors can choose whether to include this material in their courses.
- At the end of each section, before the exercises, students can immediately reinforce the concepts learned by answering the Quick Quiz questions.

Exercises, Examples, and Applications

- Examples are carefully tied to the Problems for Practice exercises, allowing students to immediately practice and master the needed skills.
- Each example is presented as a problem with a clearly stated task. The authors have taken great care to explain the steps of the solutions to these problems. For example, when an equation is obtained by using a formula labeled by a number, that number is placed over the equals sign.
- Applications, both large and small, abound. Instructors may pick and choose those that best suit the course being taught. Applications to chemistry, biology, medicine, public policy, finance, economics, and other social sciences augment classic applications in physics and engineering.
- Each end-of-section exercise set comprises three types of exercises: Problems for Practice, Further Theory and Practice, and Calculator/Computer Exercises.
- The Problems for Practice develop essential computational skills. They are organized by type and include more exercises than will typically be assigned. Students can use the unassigned exercises for extra practice, as needed.
- The Further Theory and Practice exercises are mixed in nature. In general, these exercises cannot be done by following a worked example. Some exercises in this section extend the theory discussed in the text. Some Further Theory and Practice exercises fill in details of proofs. There are far more Further Theory and Practice exercises than will be assigned in a typical course. Instructors may choose the ones suitable for their courses.

Content

- Transcendental functions (trigonometric, logarithmic, and exponential) receive a thorough, but accessible, treatment. Chapter 1 reviews trigonometric functions. Chapter 2 introduces exponential functions. The natural logarithm is introduced in Chapter 3. The order of these chapters is suitable for an "early transcendentals" course.
- Differential equations are treated throughout the text. Chapter 6 uses first order differential equations to provide a second look at the exponential and logarithmic functions. Differential equations that involve the other transcendental functions are also discussed.
- Sequences, which are introduced in Chapter 1, are treated throughout the text. Chapter 2 defines and studies limits of sequences. This treatment takes advantage of students' prior familiarity with the topic. Although that knowledge, gained through the infinite decimal expansions that students learned before calculus, is informal, it makes for a good intuitive base. Sequences therefore help make the general concept of limit more concrete.

The early and recurring discussion of sequences results in a more comfortable, expeditious treatment of infinite series. Mastering the notion of infinite sequence in advance better prepares students to understand the sophisticated idea on which the infinite series rests.

- When introduced, integration techniques (substitution, integration by parts, partial fractions, and the use of trigonometric identities) are discussed in considerable detail. These topics provide students with useful practice in the important mathematical techniques of substitution and algebraic manipulation. The development of partial fractions has been spread over two sections, allowing instructors greater flexibility in what they choose to cover.
- The chapter on applications of integration introduces and develops concepts from probability theory.
- The development of power series has been given its own chapter. Taylor polynomials are introduced before Taylor series.
- Stokes's, Green's, and the Divergence theorems are thoroughly covered in *Calculus: Multivariable*.

Technology

- Computer modeling and numerical calculations are included both in the text and in the exercises. Screen shots from the popular computer algebra system (CAS), Maple™, are shown throughout so that students learn to recognize CAS interfaces.
- The use of a CAS or graphing calculator provides students with another avenue for exploring calculus.
- Each end-of-section exercise set concludes with several problems that are intended to be solved with CAS or a graphing calculator. We use Maple in our calculus courses, but the exercises are written so that any CAS or graphing calculator can be used.
- Technology complements and enhances traditional mathematical skills; it does not eliminate mathematical techniques.
- Throughout the text, the mathematical notation used is compatible with modern calculators and computer algebra systems. Rather than a bewildering array of brackets for algebraic groupings, parentheses are employed. The argument of every function appears within parentheses. For example, we write $\sin(x)$ and not $\sin x$.

Chapter Structure and Elements

- Each chapter starts with a preview of the topics that will be covered. This short initial discussion gives an overview and provides motivation for the chapter.

Each chapter contains a section that summarizes the chapter's important formulas, theorems, definitions, and general concepts.

- Occasionally within the text, A Look Back/A Look Forward boxes remind students of concepts that were learned earlier in the text or offer previews of still-to-come material.
- Boxed features called Insights highlight further information about concepts. They occur both in margins and in the text; arrows indicate where they flow within the text. The Insights are remarks directed to the student. Sometimes an Insight clears up a point, sometimes it answers a question that arises frequently in lectures. The term *Insight* does not reflect any deep understanding of calculus that the authors claim to have; the remarks are simply the result of day-to-day teaching experience. Naturally, at the chalkboard, instructors will offer their own insights.
- Prior to exercise sets, students test their understanding with the Quick Quiz.
- Each chapter ends with a novel feature, *Genesis & Development*. These sections give history and perspective on key topics in the evolution of the subject. There are occasional references to these sections in the text, but by and large they are intended to be supplementary. Instructors can assign these sections as reading if they wish, but there is no core material that requires any of the *Genesis & Developments* to be covered.

Supplements

For the Instructor

Instructor Resources
Single Variable (1-931914-34-6) and *Multivariable* (1-931914-70-2)
Each volume of the *Instructor Resources* is structured similarly and is designed to provide you with a tool that guides your teaching of *Calculus*. Each volume is devoted to teaching suggestions, such as sample syllabi, lecture outlines, and course pacing. Sample midterms and finals are also included.

Instructor Solutions Manual
Single Variable (1-597570-21-4) and *Multivariable* (1-597570-22-2)
These volumes contain full solutions to all text exercises.

Test Generator (1-931914-73-7)
One comprehensive test generator CD-ROM is available for all *Calculus* chapters. You may build your own tests by choosing from a wide variety of both static and algorithmically generated problems. The problems are written in a style to match the prose of the text. Question types include multiple choice, true/false, and short answer. Tests can be administered with paper and pencil, or they can be delivered online through your school's local network.

You can also keep track of your courses with the software's built-in management. Tests administered over a network can be automatically graded and entered into a grade book. Each of your sections will be monitored separately.

Web site: www.keycollege.com/online
A portion of this course-specific Web site is devoted to instructors using *Calculus: Single Variable* and *Calculus: Multivariable*. If your school uses WebCT, you will find content that is portable into that system. Other resources include PowerPoint® presentations, extra worked examples, and modifiable sample syllabi. Each problem from the exercise sets has been keyed so that homework can be assigned and completed online. You will also be able to access all portions of the student Web site.

For the Student

Student Study and Solutions Companion
Single Variable (1-931914-71-0) and *Multivariable* (1-931914-72-9)
The *Student Study and Solutions Companions* are published in two volumes to correlate with each text volume. Students will find the same valuable resources in each *Companion*. Study hints and additional worked examples are provided for each chapter. Fully worked solutions are given for each text problem that has its answer at the end of the textbook. The *Single Variable* volume also includes algebra review. Both volumes have a section with common formulas and integral tables.

Web site: www.keycollege.com/online
To enter this site, students will use their unique access code, which is packaged with every new text. Once registered, students will be able to complete homework assignments online. They will also be able to access Maple keystrokes that correlate to the book's Computer/Calculator Exercises. Links are also provided to online calculus resources.

Access Code Package (1-931914-74-5)
Students who purchase used books will also need to purchase the Access Code Package to get their unique access code to register for www.keycollege.com/online. If the package is not available through the school's bookstore, students may call toll free 888-877-7240 to order the Access Code Package.

Acknowledgments

Over the years of its development, *Calculus: Single Variable* and *Calculus: Multivariable* have profited from the comments of our colleagues around the country. We are thankful for the reviews at all stages of development. We would particularly like to acknowledge the following people:

David Calvis, Baldwin-Wallace College
Gunnar Carlsson, Stanford University
Chi Keung Cheung, Boston College
Dennis DeTurck, University of Pennsylvania
Bruce Edwards, University of Florida
Saber Elaydi, Trinity University
David Ellis, San Francisco State University
Salvatrice Keating, Eastern Connecticut State University
Jerrold Marsden, California Institute of Technology
Jack Mealy, Austin College
Harold Parks, Oregon State University
Ronald Taylor, Berry College

From Brian E. Blank

I would like to thank William Moser, Kohur Gowrisankaran, Jeremy Hayhurst, and William Hoffman for their important contributions to the early conceptual periods of this project and Mike Simpson for his strategic insights at the critical marketing stage of the published text.

From Steven G. Krantz

In my career, I have benefited from my association with many editors and publishing professionals. Rich Jones was the first to encourage me to write a calculus book. He provided both direction and guidance and taught me how to write. Barbara Holland worked closely with me to develop the text and taught me to listen to and learn from reviewers. Jim Harrison provided me with encouragement, guidance, and my first computer. Seth Howell carefully checked all of the mathematics for correctness. Richard Bonacci and Allyndreth Cassidy at Key College Publishing made it all come together and turned this nascent project into a finished book. Eric Houts, Beth Masse, and Laura Ryan, master production editors, lovingly shepherded the project through every stage of copy editing and composition. No project of this magnitude can be created without the combined efforts of many people. I am grateful to them all—both named and unnamed.

About the Authors

Brian E. Blank and Steven G. Krantz have a combined experience of more than 50 years teaching calculus. They are both award-winning teachers and highly respected writers. Their extensive experience in consulting for a variety of professions enables them to bring to this project diverse and motivational applications as well as realistic and practical uses of the computer.

Brian E. Blank received his B.Sc. degree from McGill University in 1975 and Ph.D. from Cornell University in 1980. He has taught calculus at The University of Texas, the University of Maryland, and Washington University in St. Louis.

Steven G. Krantz received his B.A. degree from the University of California at Santa Cruz in 1971 and his Ph.D. from Princeton University in 1974. He has been on the faculties of the University of California at Los Angeles, Pennsylvania State University, and Princeton University. He is former chair at Washington University in St. Louis. Krantz is holder of the Chauvenet Prize, the Beckenbach Book Award, and the Kemper Prize. He has written more than 120 research papers and more than 45 books, including *How to Teach Mathematics* and *Mathematical Apocrypha*.

1

Basics

PREVIEW

Many of the variables encountered in everyday life are subject to continuous change. When traveling along a straight line from one point to another, we must pass through all points in between. If a tire deflates from 28 psi (pounds per square inch) to 10 psi, then at some time in between it was $19\frac{1}{3}$ psi. Everyone reading this text was, at some time, π years old.

Often, continuous variables depend on other continuous variables through precise mathematical laws. The area A of a circle is related to its radius r by the formula $A = \pi r^2$. When a body that is initially at rest moves under constant acceleration g, the distance s that it travels in time t is given by $s = gt^2/2$. Because r determines A and t determines s, we say that A is a *function* of r and s is a *function* of t. The functional relationships between continuous variables are among the main concerns of calculus. Calculus provides a tool for understanding how continuous variables change and for investigating the functional relationships between continuous variables.

Before beginning a study of calculus, it is important to understand clearly what is meant by a "function" and what is meant by a "variable." We must also have a way to interpret a functional relationship graphically. These topics are reviewed in this chapter. Although some of the topics will be familiar to you from previous studies, it is likely that you will learn some new ideas as well.

1.1 Number Systems

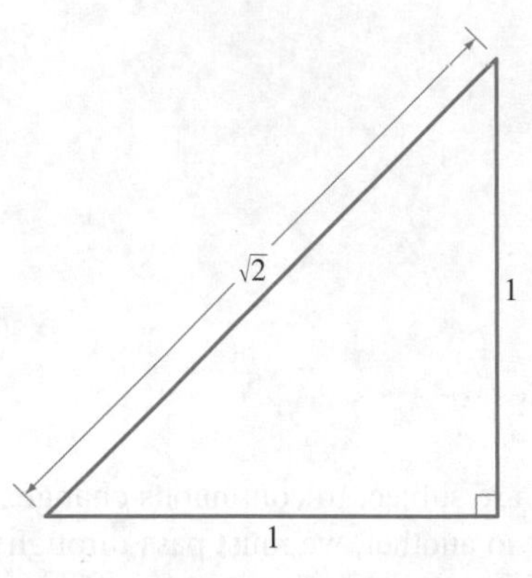

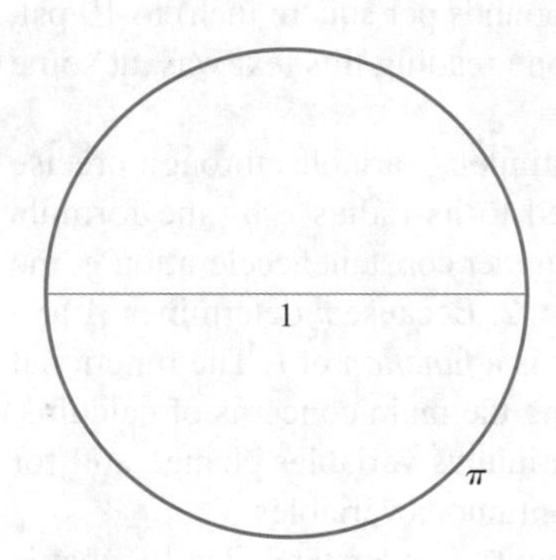

Figure 1
Two numbers that are *not* rational: the hypotenuse of a right triangle with base and height 1 and the circumference of the circle of diameter 1

The simplest number system is the set of *natural numbers* 0, 1, 2, 3, ..., denoted by the letter $\mathbb{N}$. Multiplication and addition are operations in $\mathbb{N}$, meaning that the sum or product of two natural numbers is a natural number. Subtraction, however, may not make sense if we have *only* the natural numbers at our disposal. For example, $3-5$ has no meaning in $\mathbb{N}$. Therefore, we must consider the larger number system $\mathbb{Z}$, the *integers,* consisting of the numbers $\ldots, -3, -2, -1, 0, 1, 2, 3, \ldots$. The set of *positive* integers, $\{1, 2, 3, \ldots\}$, is denoted by $\mathbb{Z}^+$.

Although addition, multiplication, and subtraction make sense in the integers, division may not. For instance, the expression $3/5$ does not represent an integer. So we pass to the larger number system $\mathbb{Q}$, which consists of all expressions of the form a/b where a and b are integers and b is not zero. This is the system of *rational numbers*.

In general, the numbers that we encounter in everyday life—prices, speed limits, weights, temperatures, interest rates, and so on—are rational numbers. However, there are also numbers that are not rational. As a simple example, the length of a diagonal of a square of side length 1 is $\sqrt{2}$, which is *not* a rational number (see Figure 1 and Exercise 57). As another example, the number π, which is defined as the circumference of a circle of diameter 1, is not a rational number (see Figure 1).

When studying calculus, it is important to remember that the rational numbers are not appropriate for the mathematical process of *taking limits*. For instance, although the sequence of rational numbers

$$3, \frac{31}{10}, \frac{314}{100}, \frac{3141}{1000}, \frac{31415}{10000}, \ldots$$

seems to tend to, or approach, the decimal representation of π, $3.141592653\ldots$, π is not a rational number—it cannot be represented as a quotient of integers.

At present, when we use expressions like "tend to" and "approach a limit" or "limits," we are speaking intuitively. Later, when we deal with infinite decimal expansions, we will also do so intuitively. Chapter 2 makes the notion of limit more precise. Chapter 9 discusses in detail the ideas connected with infinite decimal expansions.

Because calculus involves the systematic use of various kinds of limiting processes, we work not with rational numbers but with the larger *real number system* $\mathbb{R}$, which consists of all three types of infinite decimal expansions.

1. *Decimal expansions that, after a finite number of digits, contain only zeros* This type of decimal expansion is called *terminating*. The number $2.657000\ldots$ is an example. We would ordinarily write this number as 2.657 and would recognize it as the rational number $2657/1000$. Any other real number with a decimal expansion that is terminating can be displayed as a rational number in the same way.

2. *Nonterminating expansions that, after a finite number of terms, repeat a single block of terms endlessly* An example is the number $x = 8.347626262\ldots$. Notice that, after the digit 7, the block 62 is repeated endlessly. To understand this number, we write

$$\begin{aligned} 100000x &= 834762.626262\ldots \\ 1000x &= 8347.626262\ldots. \end{aligned}$$

Subtracting the second equation from the first gives $99000x = 826415$, or $x = 826415/99000$. We can adapt this procedure to demonstrate that any number of this type is rational. Conversely, we can use long division to convert every rational number to a decimal expansion that either terminates or repeats.

3. *Nonterminating, nonrepeating decimal expansions* Numbers of this type cannot be rational and are therefore called *irrational*. As was mentioned earlier, the numbers $\sqrt{2}$ and π are irrational. The first several digits of their decimal expansions are $1.41421356\ldots$ and $3.14159265\ldots$, respectively.

Because any decimal expansion must be one of these three types, every real number is either a rational number or an irrational number. To aid in thinking and to display answers, it helps to exhibit real numbers in a diagram, such as a *number line*. Begin with an infinite line with equally spaced points representing the integers. Subdivide the line to represent fractions (see Figure 2). With sufficiently accurate tools, it is possible to determine which point corresponds to any *terminating* decimal expansion. This still leaves many unlabeled points, which are limits of sequences of the points that have already been labeled. In other words, the unlabeled points correspond to the nonterminating decimal expansions. In this way, we can think of each point of the line as corresponding to a real number and vice versa.

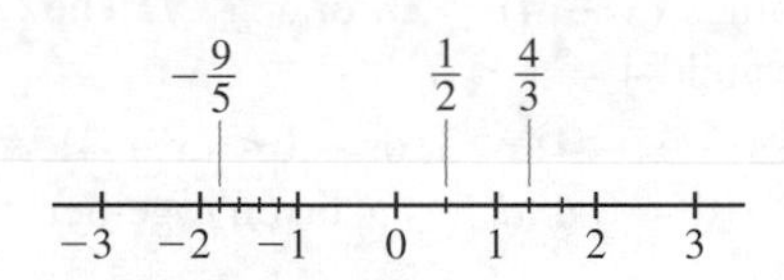

Figure 2

Sets of Real Numbers

The most common way to describe a set of real numbers is with the notation $\{x : P(x)\}$, which is read "the set of all x that satisfy the property $P(x)$." If x belongs to a set S, we say that x is an element, or a member, of S, and we write $x \in S$.

Example 1 Sketch the sets $S = \{s : 1 < s < 4\}$ and $T = \{t : -2 \le t < 3\}$.

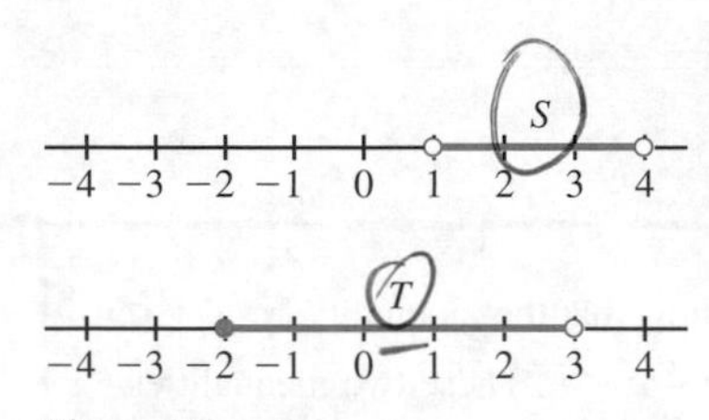

Figure 3

Solution The sets are sketched in Figure 3. A solid dot indicates that -2 is included in T, whereas hollow dots show that 1 and 4 are missing from S and that 3 is missing from T. ■

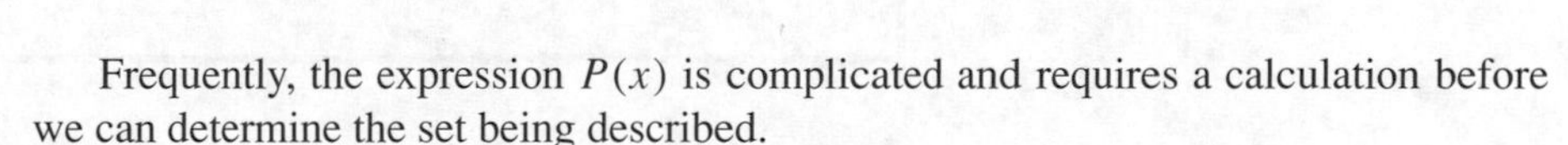

Frequently, the expression $P(x)$ is complicated and requires a calculation before we can determine the set being described.

Example 2 Sketch the set $U = \{u : 2u + 5 > 3u - 9\}$.

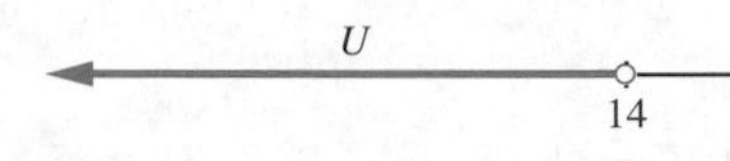

Figure 4

Solution We must first simplify the inequality $2u + 5 > 3u - 9$. Add 9 and subtract $2u$ from both sides to obtain $14 > u$. Therefore, $U = \{u : u < 14\}$. The sketch is in Figure 4. ■

Definition If x is a real number, then the symbol $|x|$, called the *absolute value* of x, represents the linear distance from x to 0 on the number line. If $x > 0$, then $|x| = x$. If $x < 0$, then $|x| = -x$. If $x = 0$, then $|x| = 0$. The three possibilities for $|x|$ are shown in Figure 5.

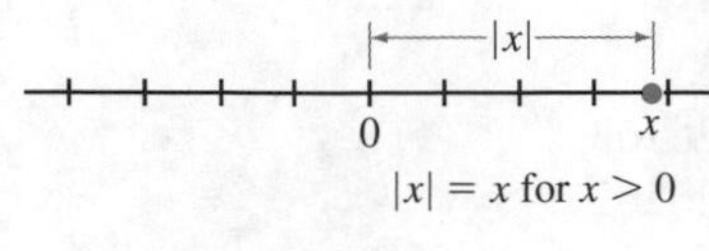

0

$|0| = 0$

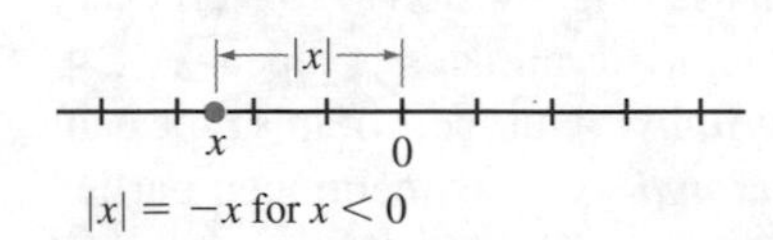

Figure 5

The absolute value arises frequently. For example, both x and $-x$ are square roots of x^2. Because the expression $\sqrt{x^2}$ denotes the *nonnegative* square root of x^2, we may write

$$\sqrt{x^2} = |x|$$

without having to worry about the sign of x. The following example illustrates another convenient use of absolute value.

Example 3 Sketch the set $V = \{x : |x - 4| \leq 3\}$.

Solution We must solve the inequality $|x - 4| \leq 3$.

Case 1 If $x - 4 > 0$, then $|x - 4| \leq 3$ becomes $(x - 4) \leq 3$, or $x \leq 7$. The conditions $x - 4 > 0$ *and* $x \leq 7$ taken together yield $4 < x \leq 7$.

Case 2 If $x - 4 < 0$, then $|x - 4| \leq 3$ becomes $-(x - 4) \leq 3$, or $-3 \leq (x - 4)$, which is equivalent to $1 \leq x$. The conditions $x - 4 < 0$ *and* $1 \leq x$ taken together yield $1 \leq x < 4$.

Case 3 If $x - 4 = 0$, then $|x - 4| \leq 3$ becomes $0 \leq 3$, which is certainly true.

To summarize,

$$V = \{x : 4 < x \leq 7 \text{ or } 1 \leq x < 4 \text{ or } x = 4\}.$$

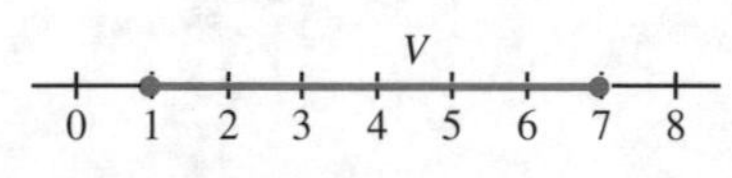

Figure 6

We can write this more concisely as $V = \{x : 1 \leq x \leq 7\}$. A sketch of V is given in Figure 6. ■

inSIGHT

We can generalize the analysis used in Example 3 to show that the inequality $|x - c| \leq r$ is equivalent to the *two* inequalities $-r \leq x - c$ and $x - c \leq r$. These two inequalities can also be expressed as $c - r \leq x \leq c + r$.

Definition The *distance* between two real numbers x and y is either $x - y$ or $y - x$, whichever is nonnegative. A convenient way to say this is that the distance between x and y is $|x - y|$.

The definition of distance helps describe the set $V = \{x : |x - 4| \leq 3\}$ from Example 3 as the set of numbers x with a distance from 4 that is less than or equal to 3. This interpretation reinforces the finding that V is the set of x such that $1 \leq x \leq 7$.

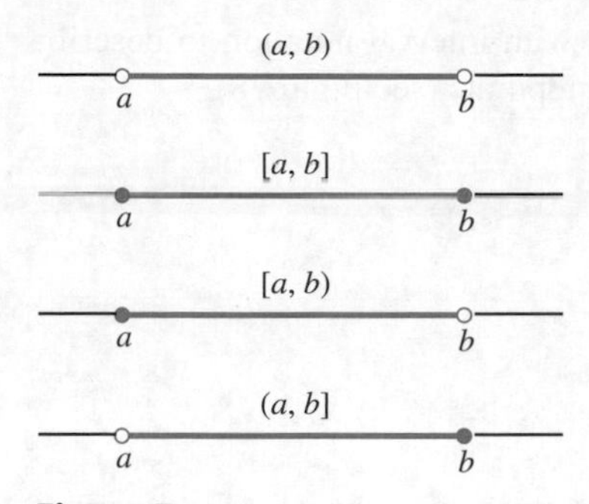

Figure 7

Intervals

The most frequently encountered sets of real numbers are *intervals*. The sets S, T, U, and V from Examples 1, 2, and 3 are examples of intervals. If a and b are real numbers with $a \leq b$, then we can produce the following four types of *bounded intervals* (see also Figure 7).

$$\begin{aligned}
(a, b) &= \{x : a < x < b\} \\
[a, b] &= \{x : a \leq x \leq b\} \\
[a, b) &= \{x : a \leq x < b\} \\
(a, b] &= \{x : a < x \leq b\}
\end{aligned}$$

Intervals of the form (a, b) are called *open*, whereas those of the form $[a, b]$ are called *closed*. The other two types are called either *half open* or *half closed* (these two terms are interchangeable). The variable a is the *left endpoint*, and b is the *right endpoint* of the interval. The interval $[a, a]$ is simply the set $\{a\}$, which means it has only the number a in it. Notice that an interval such as (a, a) does not contain any point. When a set has no elements, it is called the *empty set* and is denoted by the symbol $\emptyset$.

If a and b are the left and right endpoints, respectively, of an interval I, then the *midpoint* of I is $m = (a + b)/2$. It is easy to verify that m is equally distant from each of the endpoints:

$$|m - a| = \left|\left(\frac{a+b}{2}\right) - a\right| = \frac{b-a}{2} = \left|b - \left(\frac{a+b}{2}\right)\right| = |b - m|.$$

Example 4 Write the interval $(3, 11)$ in the form $\{x : |x - c| < r\}$.

Solution Since $I = \{x : |x - c| < r\}$ is the set of points with a distance from c that is less than r, we can deduce that c is the midpoint of I. The midpoint of the interval $(3, 11)$ is $(3+11)/2$, or 7. Therefore, $c = 7$. The distance r between c and the endpoints 3 and 11 is 4. In conclusion, we can write the interval $(3, 11)$ as $\{x : |x - 7| < 4\}$. ■

Example 5 Write the set $W = \{x : |3x - 18| \leq 6\}$ as closed interval $[a, b]$.

Solution We first multiply each side of $|3x - 18| \leq 6$ by $1/3$ to obtain

$$\frac{1}{3}|3x - 18| \leq 2, \text{ or}$$

$$\left|\frac{1}{3}(3x - 18)\right| \leq 2, \text{ or}$$

$$|x - 6| \leq 2.$$

(Note: Inequalities are *preserved* under multiplication by positive numbers and *reversed* under multiplication by negative numbers.) The inequality $|x - 6| \leq 2$ is shorthand for the two inequalities $-2 \leq x - 6 \leq 2$. By adding 6 to each side, we obtain $4 \leq x \leq 8$. Thus, W is the closed interval $[4, 8]$. ■

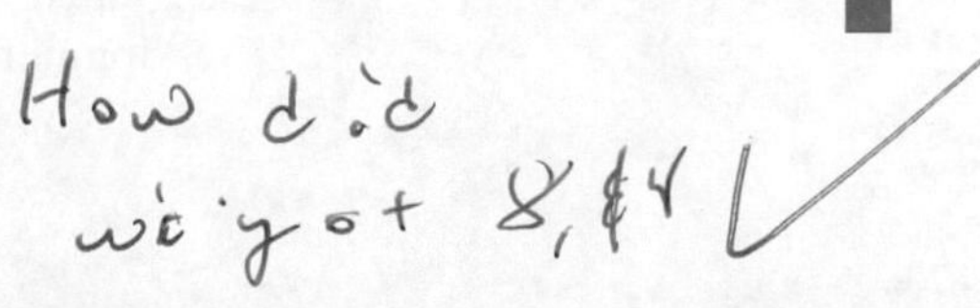

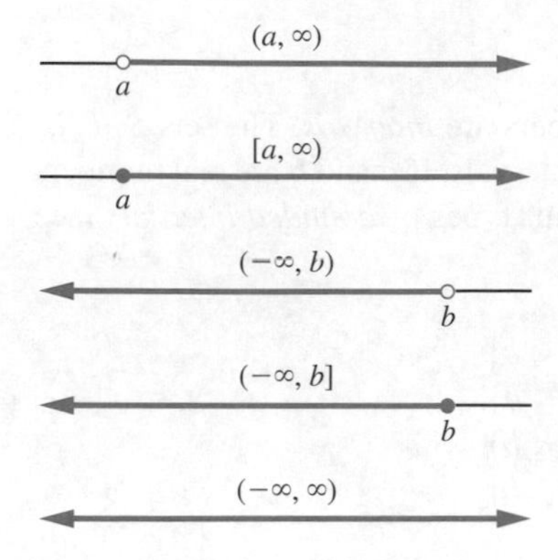

Figure 8

We can use the symbols ∞ and $-\infty$ together with interval notation to describe some intervals with only one endpoint or with no endpoints (see Figure 8).

$$\begin{aligned}
(a, \infty) &= \{x : a < x\} \\
[a, \infty) &= \{x : a \le x\} \\
(-\infty, b) &= \{x : x < b\} \\
(-\infty, b] &= \{x : x \le b\} \\
(-\infty, \infty) &= \{x : -\infty < x < \infty\}
\end{aligned}$$

Each of these intervals is an *unbounded interval,* as opposed to the bounded intervals described earlier.

inSIGHT

> The symbols ∞ and $-\infty$ do not represent real numbers. They are simply a convenient notational device. We never write $[-\infty, a)$ or any other expression suggesting that $+\infty$ or $-\infty$ is an element of a set of real numbers.

Example 6 Solve the inequality

$$\frac{3x-2}{|x+1|} > 1.$$

Solution Observe that x cannot equal -1. For other values of x, we clear the denominator by multiplying both sides of the equation by $|x+1|$: This quantity is never negative, so the multiplication *preserves* the inequality. The result is $3x - 2 > |x+1|$. We divide our analysis of this inequality into two cases, according to the sign of the expression inside the absolute value.

If $x > -1$, then $x + 1 > 0$ and $|x+1| = x + 1$. The inequality $3x - 2 > |x+1|$ becomes $3x - 2 > (x+1)$, which simplifies to $x > 3/2$. The points that satisfy both $x > -1$ and $x > 3/2$ make up the interval $(3/2, \infty)$.

In the second case, when $x < -1$, the expression $x + 1$ is negative and, as a consequence, $|x+1| = -(x+1)$. The inequality $3x - 2 > |x+1|$ becomes $3x - 2 > -(x+1)$, which simplifies to $x > 1/4$. There are, however, no points that satisfy both $x < -1$ and $x > 1/4$. Therefore, the set of points satisfying the given inequality is simply the interval $(3/2, \infty)$ that we found in the first case. ■

The Triangle Inequality

It is not difficult to see that $|a+b| = |a| + |b|$ when a and b have the same sign (see Exercise 53). If a and b have opposite signs, however, then cancellation occurs in the sum $a + b$ but not in the sum $|a| + |b|$. In this case, $|a+b| < |a| + |b|$. In summary, for all real numbers a and b,

$$|a+b| \le |a| + |b|.$$

This relationship, called the *Triangle Inequality,* is a useful tool for estimating the magnitude of a sum in terms of the magnitudes of its summands.

Example 7 Suppose the distance of a to b is less than 3 and the distance of b to 2 is less than 6. Estimate the distance of a from 2.

Solution We have $|a - 2| = |(a - b) + (b - 2)|$. By the Triangle Inequality,

$$|a - 2| \leq |a - b| + |b - 2| < 3 + 6 = 9.$$

Therefore, the distance of a from the number 2 is not greater than 9. ■

The Triangle Inequality can extend to three or more summands:

$$|a + b + c| \leq |a| + |b| + |c|.$$

Approximation

Calculus deals with many infinite processes, such as the decimal expansion of an irrational number. Although we can think abstractly of irrational numbers such as π and $\sqrt{2}$, we cannot write all the digits of their decimal expansions. If a calculator returns the value 3.14159265359 for π, then it is really displaying an *approximation* of π. A small *error* has resulted. In general, the type of error that results when an infinite process is terminated is known as a *truncation error*.

In numerical work, we always have the equation to find the true value of a number:

$$\textit{True value} = \textit{approximate value} + \textit{error}.$$

If a is an approximation of the number x, then $|x-a|$ is the *absolute error* and $|x-a|/|x|$ is the *relative error*. Relative error is often of greater concern than absolute error. For example, an absolute error of $1000 in the national budget is not serious because it is a very small relative error. The same absolute error in your personal budget, however, would be far more significant. When we calculate an approximation, we must ensure that the error term is acceptably small.

Definition If $|x - a|$ is less than or equal to $5 \times 10^{-(q+1)}$, then *a approximates* (or *agrees with*) *x to q decimal places.*

Example 8 Does 22/7 approximate π to as many decimal places as does 3.14? Does 1.418 approximate $\sqrt{2}$ to two decimal places? Use your calculator.

Solution Because $|\pi - 22/7| = 1.26\ldots \times 10^{-3}$, we have

$$5 \times 10^{-(3+1)} = 5 \times 10^{-4} < \left|\pi - \frac{22}{7}\right| < 5 \times 10^{-3} = 5 \times 10^{-(2+1)}.$$

Therefore, 22/7 approximates π to two, but not three, decimal places. Similarly, $\pi - 3.14 = 1.59\ldots \times 10^{-3}$. The number 3.14, like 22/7, approximates π to two, but not three, decimal places.

Since $|\sqrt{2} - 1.418| = 3.78\ldots \times 10^{-3} < 5 \times 10^{-(2+1)}$, we conclude that 1.418 does approximate $\sqrt{2}$ to two decimal places. ■

Even if two numbers a and b agree to q decimal places, we cannot be certain that the results of rounding a and b to q decimal places will be the same. In Example 8, we see that $\sqrt{2}$ and 1.418 agree to two decimal places. Rounding these numbers to two decimal places, however, results in 1.41 and 1.42, respectively.

Floating Point Representations

When working with a calculator or computer software, there are several sources of error. The errors that we now consider are caused by the way real numbers are stored in calculators or computers.

The *floating point decimal representation* of a nonzero real number x is

$$x = \pm 0.a_1 a_2 a_3 a_4 \ldots \times 10^p$$

where p is an integer and $a_1, a_2, \ldots$ are natural numbers from 0 to 9 with $a_1 \neq 0$. The decimal point is thought of as "floating" to the left of the first nonzero digit. A number with a floating point decimal representation of $\pm 0.a_1 a_2 \ldots a_k \times 10^p$ is said to have k *significant digits*.

Since the memory of a calculator or computer is finite, it must limit the number of significant digits with which it works. This can lead to a type of error known as *roundoff error*. It is important to understand that a computation can reduce the number of significant digits. Such a *loss of significant digits*, or, more simply, a *loss of significance*, can occur when two very nearly equal numbers are subtracted. For example, 0.124 and 0.123 have three significant digits, whereas their difference, 0.1×10^{-2}, has only one significant digit. The next example illustrates why loss of significance is sometimes called *catastrophic cancellation*.

Example 9 What is the outcome of evaluating

$$x = 0.412 \times 0.300 - 0.617 \times 0.200$$

on a calculator that uses only three significant digits? What relative error results?

inSIGHT

Throughout the study of calculus, it is important to guard against loss of significance even when using a calculator with ten or more significant digits.

Solution The product 0.412×0.300 to three significant digits is 0.124; 0.617×0.200 equals 0.123 to three significant digits. Therefore, on a three-digit calculator, x evaluates to $0.124 - 0.123 = 0.1 \times 10^{-2}$. Of course, x really equals $0.1236 - 0.1234 = 0.2 \times 10^{-3}$, as you can verify with your own calculator. The relative error is therefore

$$\frac{|0.2 \times 10^{-3} - 0.1 \times 10^{-2}|}{|0.2 \times 10^{-3}|} = 4,$$

or 400%. ■

Computer Algebra Systems

From time to time, this book refers to software packages known as *computer algebra systems*. A computer algebra system not only performs arithmetic operations with specific numbers, it can also carry out those operations symbolically. When the "factor" command of a computer algebra system is applied to $x^6 - y^6$, for example, the result is

$$(x-y)(x+y)(x^2+xy+y^2)(x^2-xy+y^2).$$

Several calculators now offer some of the capabilities of computer algebra systems.

quickquiz

1. What is an irrational number?
2. How do you recognize the decimal expansion of a rational number?
3. What interval does the set $\{x : |x-3| < 5\}$ represent?
4. Write the interval $(-1, 2)$ in the form $\{x : |x-c| < r\}$.

EXERCISES

Problems for Practice

In Exercises 1–6, convert the decimal to a rational fraction. (Ellipses are included in some exercises to indicate repetition.)

1. 2.13
2. 0.00034
3. 0.232323...
4. 2.222...
5. 5.001001001...
6. 15.7231231231...

In Exercises 7–12, use long division to convert the rational fraction to a (possibly nonterminating) decimal with a repeating block. Identify the repeating block.

7. 1/40
8. 25/8
9. 5/3
10. 2/7
11. 18/25
12. 1/17

In Exercises 13–18, write the set using interval notation of the form (a, b), $[a, b]$, $[a, b)$, and so on.

13. $\{x : 1 \le x \le 3\}$
14. $\{x : |x-2| < 5\}$
15. $\{t : t > 1\}$
16. $\{u : |u-4| \ge 6\}$
17. $\{y : |y+4| \le 10\}$
18. $\{s : |s-2| > 8\}$

In Exercises 19–24, sketch the set on a real number line.

19. $\{x : 2x-5 < x+4\}$
20. $\{s : |s-2| < 1\}$
21. $\{t : (t-5)^2 < 9/4\}$
22. $\{y : 7y+4 \ge 2y+1\}$
23. $\{x : |3x+9| \le 15\}$
24. $\{w : |2w-12| \ge 1\}$

In Exercises 25–28, write the interval in the form $\{x : |x-c| < r\}$ or $\{x : |x-c| \le r\}$.

25. $[-1, 3]$
26. $[3, 4\sqrt{2}]$
27. $(-\pi, \pi+2)$
28. $(\pi - \sqrt{2}, \pi)$

Further Theory and Practice

29. Explain why the sum and product of two rational numbers are always rational.

30. Give an example of two irrational numbers with a rational product; give an example of two irrational numbers with a rational sum.

31. Two commonly used approximations of π are 22/7 and 3.14. Explain why these are inexact approximations.

32. Find the smallest interval of the form $\{x : |x-\pi| \le r\}$ that contains both 22/7 and 3.14. Is r rational or irrational?

33. A chemistry experiment, if followed exactly, results in the production of 34.5 g of a compound. To allow for a small amount of experimental error, the lab instructor will accept, without penalty, any reported measurement within 1% of the correct mass. In what interval must a measurement lie if a penalty is to be avoided? Describe this interval as a set of the form $\{x : |x-c| \le \epsilon\}$.

34. A door designed to be 885 mm wide and 1475 mm high is to fit in a 895 mm × 1485 mm rectangular frame. The clearance between the door and the frame can be no greater than 7 mm on any side for an acceptable seal to result. Under anticipated temperature ranges, the door will expand by at most 0.2% in height and width. The frame does not expand significantly under these temperature ranges. At all times, even when the door has expanded, there must be a 1 mm clearance between the door and the frame on each of the four sides. The door does not have to be manufactured precisely to specifications, but there are limitations. In what interval must the door width lie? The height?

In Exercises 35–43, sketch the set on a real number line.

35. $\{x : x > -2 \text{ and } x^2 < 9\}$

36. $\{s : |s+3| < |2s+7|\}$

37. $\{y : y - \sqrt{7} < 3y + 4 \leq 4y + \sqrt{2}\}$

38. $\{t : |t^2 + 6t| \leq 10\}$

39. $\{x : |x^2 - 5| \geq 4\}$

40. $\{s : |s+5| < 4 \text{ and } |s-2| \leq 8\}$

41. $\{x : x + 1 \geq 2x + 5 > 3x + 8\}$

42. $\{t : (t-4)^2 < (t-2)^2 \text{ or } |t+1| \leq 4\}$

43. $\{x : |x^2 + x| > x^2 - x\}$

Describe each set in Exercises 44–47 using interval notation and the notation $\{x : P(x)\}$.

44. The set of points with a distance from 2 that does not exceed 4

45. The set of numbers that are equidistant from 3 and −9

46. The set of numbers with a square that lies strictly between 2 and 10

47. The set of all numbers with a distance less than 2 from 3 and with a square that does not exceed 8

In Exercises 48–51, if the set is given with absolute value signs, then write it without absolute value signs. If it is given without absolute value signs, then write it using absolute value signs.

48. $\{x : x^2 - 9 < |x+1|\}$

49. $\{s : |s-4| > |2s+9|\}$

50. $\{t : t^2 - 3t + 1 < 4t^2 - 5t - 4\}$

51. $\{w : w/(w+1) < 3w + 2\}$

52. Suppose $p(x) = x^2 + Bx + C$ has two real roots, r_1 and r_2 with $r_1 < r_2$. Write the set $\{x : p(x) \leq 0\}$ as an interval.

53. Show that equality holds in the Triangle Inequality when both summands are nonpositive.

54. Use the Triangle Inequality to prove that

$$|a| - |b| \leq |a + b|$$

for all $a, b \in \mathbb{R}$.

55. Prove that there is no smallest positive real number.

56. At first glance, it may appear that the number $y = 0.\overline{9} = 0.999\ldots$ is just a little smaller than 1. In fact, $y = 1$. Let $x = 1 - y$. Explain why it is true that $x < 10^{-n}$ for every positive n. Deduce that $0 \leq x < b$ for every positive b. Explain why this implies that $y = 1$.

57. About 2400 years ago, the followers of Pythagoras discovered that if x is a positive number such that $x^2 = 2$, then x is irrational. Complete the following outline to obtain a proof of this fact.

a. Suppose that $x = a/b$ where a and b are integers with no common factors.

b. Conclude that $2 = x^2 = a^2/b^2$.

c. Conclude that $2b^2 = a^2$.

d. Conclude that 2 divides a evenly, with no remainder. Therefore, $a = 2\alpha$ for some integer α.

e. Conclude that $b^2 = 2\alpha^2$.

f. Conclude that 2 divides b evenly, with no remainder.

g. Notice that parts d and f contradict part a.

58. Explain the error in the following reasoning: Let $x = (\pi + 3)/2$. Then $2x = \pi + 3$ and $2x(\pi - 3) = \pi^2 - 9$. It follows that $x^2 + 2\pi x - 6x = x^2 + \pi^2 - 9$ and $x^2 - 6x + 9 = x^2 - 2\pi x + \pi^2$. Each side is a perfect square: $(x-3)^2 = (x-\pi)^2$. Therefore, $x - 3 = x - \pi$ and $3 = \pi$.

Calculator/Computer Exercises

In Exercises 59–62, determine the interval that y must lie in to agree with x to q decimal places.

59. $x = 0.449, q = 3$

60. $x = 24, q = 2$

61. $x = 0.999 \times 10^{-5}, q = 3$

62. $x = 0.213462 \times 10^{-1}, q = 5$

63. Write a number with four significant digits that agrees with $x = 3.996$ to three decimal places but that differs from x in each digit.

64. Let $x = 0.4449$. Round this number *directly* to three, to two, and to one significant digit(s). Now *successively* round x to three, to two, and to one significant digit(s).

The mathematical phrasing of this phenomenon is that *rounding is not transitive*. Describe this idea in your own words.

65. Let

$$\alpha = 1728148040 - 140634693\sqrt{151}$$

and

$$\beta = 1728148040 + 140634693\sqrt{151}.$$

A direct computation (with integers) shows that $\alpha \cdot \beta = 1$. Calculate decimal representations of α and β. Use them to compute the product $\alpha \cdot \beta$. Note: To 20 significant digits, $140634693\sqrt{151}$ equals

$$1728148039.9999999997.$$

Calculators that do not record enough significant digits will give 0 for the value of α.

66. Find two numbers $0.a_1a_2a_3a_4a_5$ and $0.b_1b_2b_3b_4b_5$ with $a_1 = b_1$ that agree when rounded to four significant digits but that do not agree when rounded directly to one significant digit.

The term a_nx^n is said to be the *leading term* of the polynomial

$$a_nx^n + a_{n-1}x^{n-1} + \cdots + a_1x + a_0,$$

where $a_n \neq 0$. The other terms are referred to as *lower-order terms*. In calculus, for large values of $|x|$, neglecting all the lower-order terms of a polynomial introduces only a small relative error; this relative error tends to zero as $|x|$ tends to infinity. In Exercises 67–70, present two tables, one for $x > 0$ and one for $x < 0$, to support this assertion. For example, for $p(x) = 3x^4 + 100x^3 - 1000$, the table supports the assertion that the relative error $|p(x) - 3x^4|/|p(x)|$ tends to zero as x tends to positive infinity.

x	**Relative Error**
10^3	3.2258×10^{-2}
10^4	3.3223×10^{-3}
10^5	3.3322×10^{-4}
10^6	3.3332×10^{-5}
10^8	3.3333×10^{-7}

67. $x + 10^{10}$

68. $x^2 + 12345x$

69. $-3x^4 + 10^5x^3$

70. $|x|^5 + 1000x^4$

1.2 Planar Coordinates and Graphing in the Plane

A set such as $\{a, b\}$ does not have an implied ordering of its elements. We do not, for example, interpret a as the first element of $\{a, b\}$ and b as the second. When an ordering of a and b is required, we use the *ordered pair* notation (a, b). We say that a is the *first coordinate* and b the *second coordinate* of the ordered pair (a, b). Although $\{a, b\}$ and $\{b, a\}$ are exactly the same set, (a, b) and (b, a) are different ordered pairs (if $a \neq b$). The set of all ordered pairs of real numbers is called the *Cartesian plane* and is denoted by $\mathbb{R}^2$:

$$\mathbb{R}^2 = \{(a, b) : a, b \in \mathbb{R}\}.$$

Just as we can describe a point on a line with a real number, so too can we describe a point in a plane with an ordered pair of real numbers. Begin with an intersecting horizontal and vertical line. Put scales of real numbers on each line so that the zero of each scale is at the point of intersection. The point of intersection of the two lines is the *origin* and is represented by the symbol 0. The vertical and horizontal lines are called *axes*. It is convenient to name each axis by using a variable. Commonly, the horizontal axis is named the x-axis and the vertical axis is named the y-axis. The Cartesian plane is then called the *xy-plane*. The two axes divide the plane into four regions that are known as *quadrants,* which are ordered as in Figure 1.

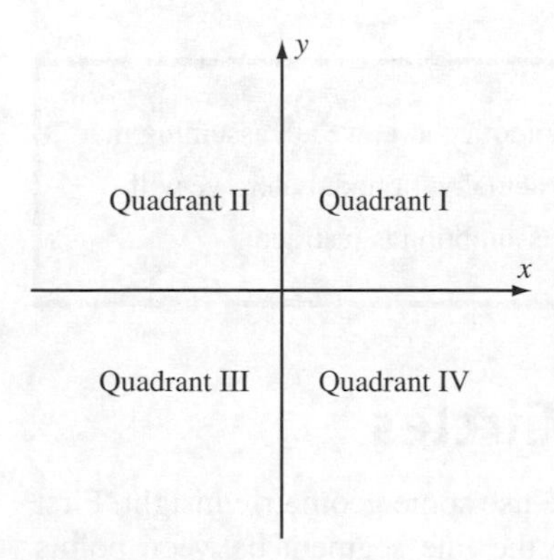

Figure 1
The xy-plane

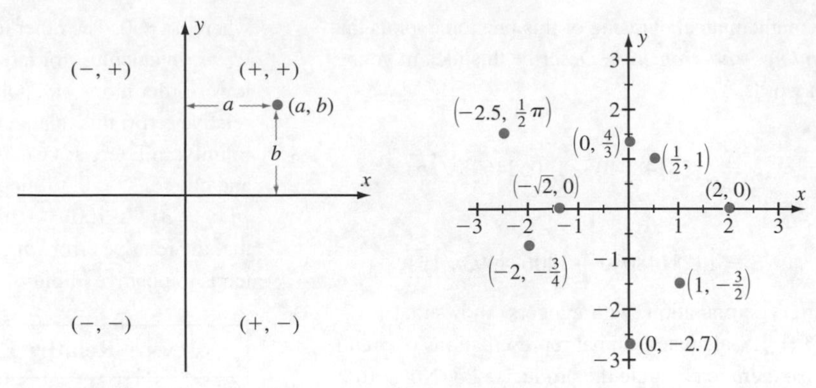

Figure 2

Figure 3

Each point P is determined by its displacement from the axes. Here the word "displacement" is interpreted in the *signed* sense: Positive displacement is to the right or upward, and negative displacement is to the left or downward. The displacement a of the point P from the y-axis is said to be the *x-coordinate,* or *abscissa,* of P. The displacement b of the point P from the x-axis is said to be the *y-coordinate,* or *ordinate,* of P. We represent the point P by the *ordered pair* (a, b), as shown in Figure 2. The signs of the x- and y-coordinates in each of the four quadrants are also illustrated in Figure 2. Figure 3 exhibits several points and their coordinates.

It is easy to plot single points on the xy-plane. Our job, however, is to plot a *set* of points with coordinates that satisfy some given equation in the variables x and y. This set is often referred to as the *locus* of the equation, or the *curve* defined by the equation. The equation is called the *Cartesian equation* of the locus of points.

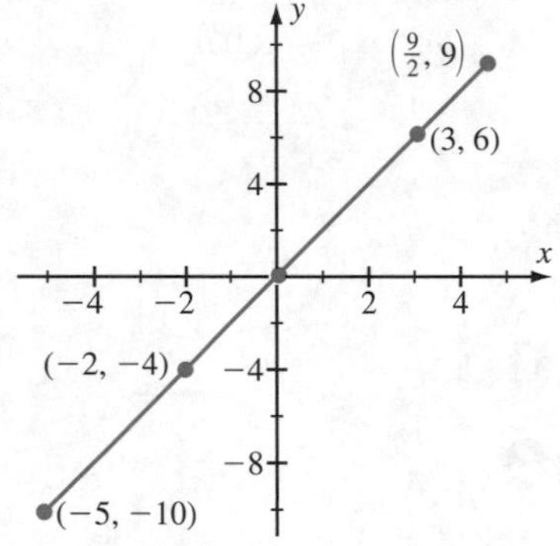

Figure 4

Example 1 Sketch or graph all points (x, y) in the plane that satisfy the equation $y = 2x$.

Solution For now, we do this by plotting some points and connecting them. Notice that the points $(0, 0)$, $(-5, -10)$, $(3, 6)$, $(-2, -4)$, $(9/2, 9)$ all satisfy the equation $y = 2x$. Plotting these points suggests that the collection of all points satisfying $y = 2x$ is a straight line (see Figure 4). ■

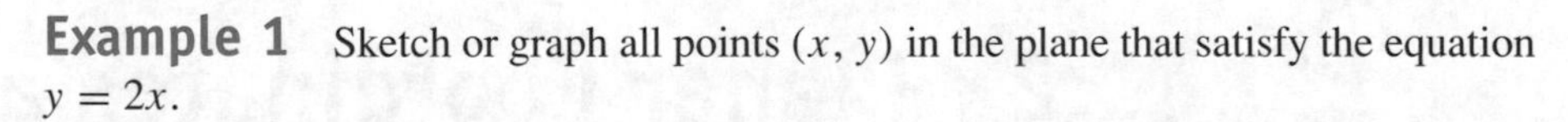

When we plot a few points and connect them with a smooth curve, we are assuming that the curve is well behaved between the points we have actually plotted. Later, we will learn how calculus provides ways to know when this assumption is justified.

The Distance Formula and Circles

We can learn more from Cartesian coordinates if we use some geometric insight. First we review the *distance formula.* Let $\overline{P_1P_2}$ denote the line segment between points $P_1 = (a_1, b_1)$ and $P_2 = (a_2, b_2)$ in the plane. The shortest distance between P_1 and P_2

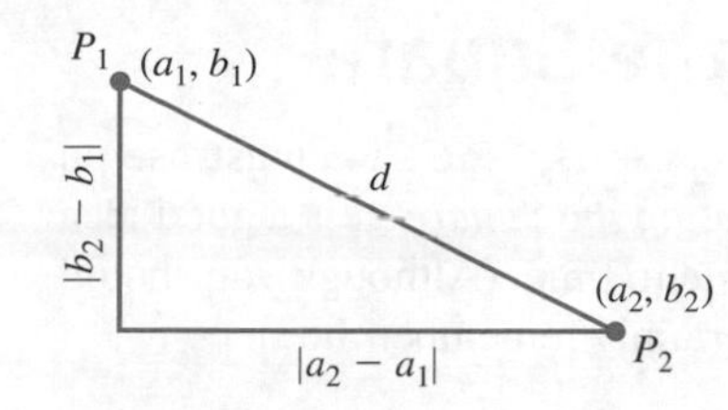

Figure 5
$d = \sqrt{(a_2 - a_1)^2 + (b_2 - b_1)^2}$

is the length of line segment $\overline{P_1P_2}$, denoted by $|\overline{P_1P_2}|$. We can calculate this distance by realizing line segment $\overline{P_1P_2}$ as the hypotenuse d of a right triangle (see Figure 5). By the Pythagorean Theorem,

$$d^2 = |a_2 - a_1|^2 + |b_2 - b_1|^2,$$

or

$$|\overline{P_1P_2}| = \sqrt{(a_2 - a_1)^2 + (b_2 - b_1)^2}.$$

Because the quantities under the square root sign are squared, we may omit the absolute value signs.

Example 2 Calculate the distance between (2, 6) and (−4, 8).

Solution The requested distance is

$$d = \sqrt{(2 - (-4))^2 + (6 - 8)^2} = \sqrt{6^2 + (-2)^2} = \sqrt{40} = 2\sqrt{10}.$$ ■

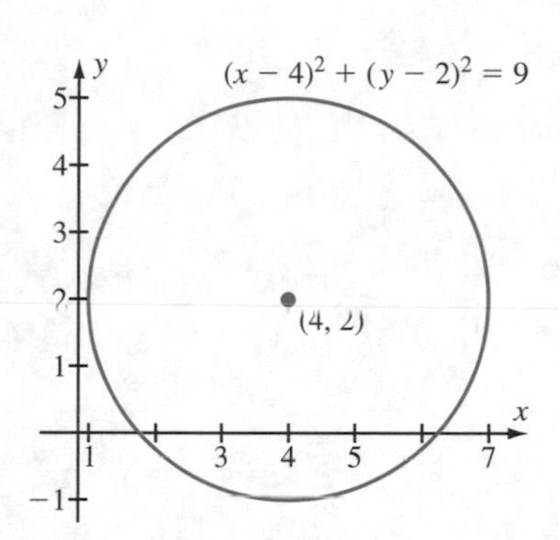

Figure 6

Example 3 Graph the set of points satisfying $(x - 4)^2 + (y - 2)^2 = 9$.

Solution We rewrite the equation as

$$\sqrt{(x - 4)^2 + (y - 2)^2} = 3.$$

According to this equation, 3 is the distance between the point (4, 2) and any point (x, y) that satisfies the equation. The collection of such points is a circle of radius 3 and center (4, 2). The locus is exhibited in Figure 6. ■

The Equation of a Circle

Example 3 illustrates a general principle: An equation of the form

$$(x - h)^2 + (y - k)^2 = r^2, \quad \textbf{(1.1)}$$

$r > 0$, has a graph that is a circle of radius r and center (h, k). Conversely, every circle is described by an equation of this form. We call this equation the *standard form* for a circle. The circle centered at the origin with radius 1 is called the *unit circle*. It is the locus of the equation

$$x^2 + y^2 = 1.$$

Example 4 Write the equation of the circle with center (8, −3/5) and radius 6.

Solution We apply formula (1.1) to obtain

$$(x - 8)^2 + \left(y - \left(-\frac{3}{5}\right)\right)^2 = 6^2,$$

or $(x - 8)^2 + (y + 3/5)^2 = 36$. ■

The Method of Completing the Square

To work efficiently with expressions of the form $Ax^2 + Bx + C$, we must use an algebraic procedure known as the *Method of Completing the Square*. It is a good idea to review this procedure as it will be needed from time to time. (Although you should not memorize the following formulas, you should certainly remember the steps.)

Completing the Square

Step 1 Starting from $Ax^2 + Bx + C$, group the terms that involve x. From this group, factor out the coefficient A of x^2:

$$Ax^2 + Bx + C = A\left(x^2 + \frac{B}{A}x\right) + C.$$

Step 2 To the group of terms that involve x, add the square of half the coefficient of x. Subtract an equal amount from the constant. In doing so, do not forget to account for the coefficient A:

$$Ax^2 + Bx + C = A\left(x^2 + \frac{B}{A}x + \left(\frac{B}{2A}\right)^2\right) + C - A\left(\frac{B}{2A}\right)^2.$$

Step 3 Identify the square:

$$Ax^2 + Bx + C = A\left(x + \frac{B}{2A}\right)^2 + C - A\left(\frac{B}{2A}\right)^2.$$

Example 5 Apply the Method of Completing the Square to $3x^2 + 18x + 16$ and to $5 - 4x - 4x^2$.

Solution We have

$$\begin{aligned} 3x^2 + 18x + 16 &= 3(x^2 + 6x) + 16 && \text{Step 1} \\ &= 3\left(x^2 + 6x + \left(\frac{6}{2}\right)^2\right) + 16 - 3\left(\frac{6}{2}\right)^2 && \text{Step 2} \\ &= 3\left(x + \frac{6}{2}\right)^2 - 11 && \text{Step 3} \end{aligned}$$

and

$$\begin{aligned} 5 - 4x - 4x^2 &= -4(x^2 + x) + 5 && \text{Step 1} \\ &= -4\left(x^2 + x + \left(\frac{1}{2}\right)^2\right) + 5 + 4\left(\frac{1}{2}\right)^2 && \text{Step 2} \\ &= -4\left(x + \frac{1}{2}\right)^2 + 6. && \text{Step 3} \end{aligned}$$

■

Example 6 The locus of the equation $2x^2 + 2y^2 + 8x - 4y + 6 = 0$ is a circle. Convert this equation to the standard form for a circle and determine the center and radius.

Solution We complete the square as follows:

$$2\left(x^2 + \frac{8}{2}x\right) + 2\left(y^2 + \frac{-4}{2}y\right) + 6 = 0 \qquad \text{Step 1}$$

$$2\left(x^2 + 4x + \left(\frac{4}{2}\right)^2\right) + 2\left(y^2 - 2y + \left(\frac{-2}{2}\right)^2\right) + 6 - 2\left(\frac{4}{2}\right)^2 - 2\left(\frac{-2}{2}\right)^2 = 0. \qquad \text{Step 2}$$

Thus, our equation becomes

$$2(x+2)^2 + 2(y-1)^2 + 6 - 8 - 2 = 0, \qquad \text{Step 3}$$

or

$$(x+2)^2 + (y-1)^2 = 2.$$

We write this as

$$(x - (-2))^2 + (y-1)^2 = (\sqrt{2})^2.$$

Therefore, our equation describes the circle with center $(-2, 1)$ and radius $\sqrt{2}$. ■

inSIGHT

It is wrong to suppose that every quadratic equation that we encounter can be put in the form of a circle. Notice that for the method in Example 6 to work, it is necessary that *both* x and y appear to the second power and that both of these squared terms have the same coefficient. Also, no xy term ever appears in the equation of a circle.

Parabolas, Ellipses, and Hyperbolas

Parabolas, ellipses, and hyperbolas are curves that, like circles, can be realized by intersecting a cone with a plane (in three-dimensional space). Such curves are known as *conic sections*. We will study the geometric properties of these curves in greater detail later. For now, it is important simply to recognize the Cartesian equations of these curves when they arise. When discussing a conic section, we often refer to an axis of symmetry. For example, we say that a line L is an *axis of symmetry* for a curve $\mathcal{C}$ if the reflection of $\mathcal{C}$ through L is the same set of points as $\mathcal{C}$ itself (see Figure 7).

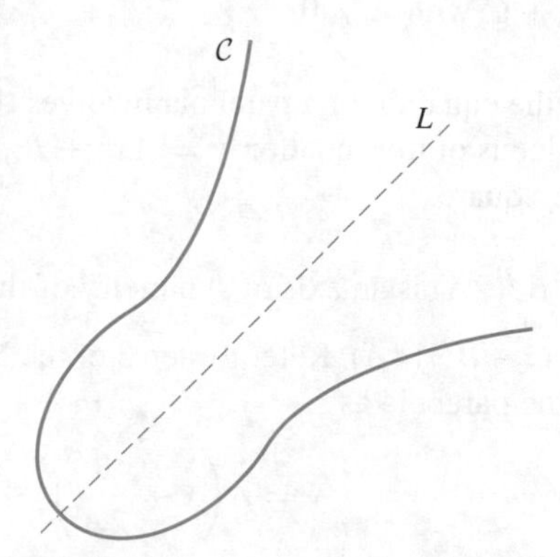

Figure 7
L is an axis of symmetry for $\mathcal{C}$.

Definition Let a and b be positive numbers. An *ellipse* with vertical and horizontal axes of symmetry (see Figure 8) is the locus of an equation of the form

$$\frac{(x-h)^2}{a^2}+\frac{(y-k)^2}{b^2}=1.$$

A *hyperbola* with vertical and horizontal axes of symmetry (see Figures 9 and 10) is the locus of an equation of the form

$$\frac{(x-h)^2}{a^2}-\frac{(y-k)^2}{b^2}=1$$

or

$$\frac{(y-k)^2}{b^2}-\frac{(x-h)^2}{a^2}=1.$$

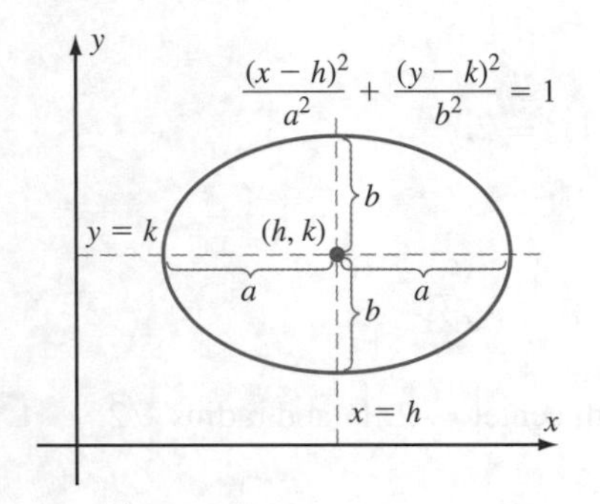

Figure 8

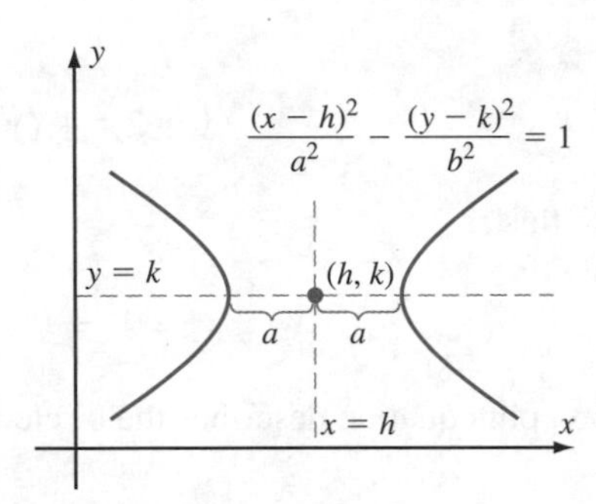

Figure 9

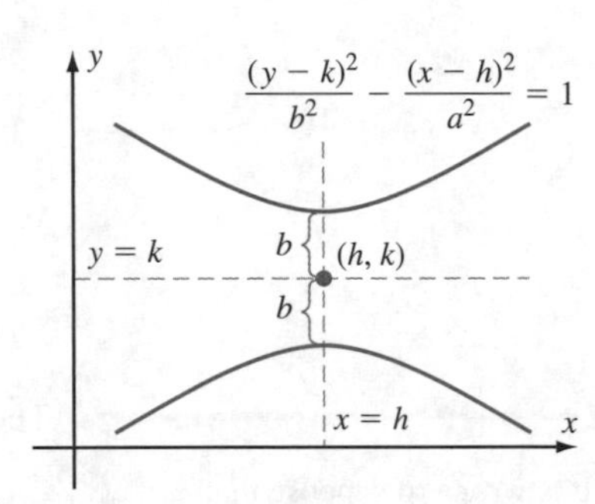

Figure 10

Notice that the Cartesian equations of ellipses and hyperbolas, like those of circles, involve the squares of both variables x and y. The Cartesian equation of an ellipse with vertical and horizontal axes is similar to that of a circle, except that the coefficients of $(x-h)^2$ and $(y-k)^2$ are unequal. Likewise, the equation of a hyperbola is similar except that the coefficients of $(x-h)^2$ and $(y-k)^2$ have different signs.

Definition A *parabola* (with a vertical axis of symmetry) is the locus of an equation of the form $y = Ax^2 + Bx + C$ with $A \neq 0$.

Notice that the equation of a parabola involves the square of *only one* variable. To understand the locus of the equation $y = Ax^2 + Bx + C$, we must use the Method of Completing the Square.

Theorem 1 The line $x = -B/(2A)$ is an axis of symmetry of the parabola $y = Ax^2 + Bx + C$.

Proof Let $\gamma = C - B^2/(4A)$. Refer to step 3 of the Method of Completing the Square. We may write the parabola as

$$y = A\left(x+\frac{B}{2A}\right)^2+\gamma.$$

Using this form, we can easily verify that both of the points

$$\left(-\frac{B}{2A}+t,\, At^2+\gamma\right)$$

and

$$\left(-\frac{B}{2A}-t,\, At^2+\gamma\right)$$

are on the graph of $y = Ax^2 + Bx + C$. Because these points have the same ordinate and the same distance $|t|$ from the vertical line $x = -B/(2A)$, we deduce that $x = -B/(2A)$ is an axis of symmetry. ■

Definition The intersection of the parabola with its axis of symmetry is called the *vertex* of the parabola.

If A is positive, then the parabola $y = Ax^2 + Bx + C$ opens upward. In this case, the *minimum* value of y is the ordinate of the vertex. It occurs when $x = -B/(2A)$ (see Figure 11a). If A is negative, then the parabola opens downward. In this case, the *maximum* value of y is the ordinate of the vertex. It occurs when $x = -B/(2A)$ (see Figure 11b).

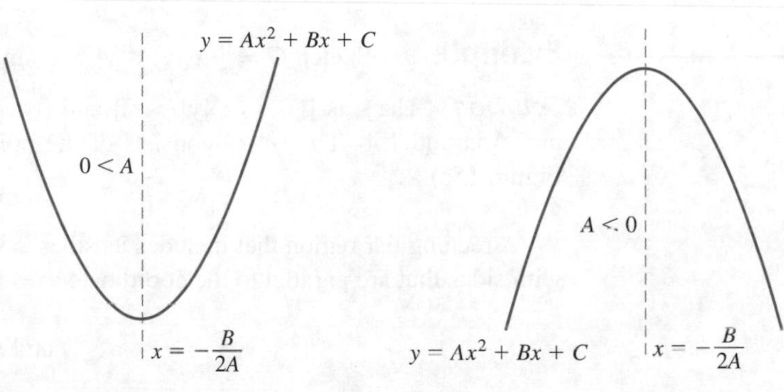

Figure 11a **Figure 11b**

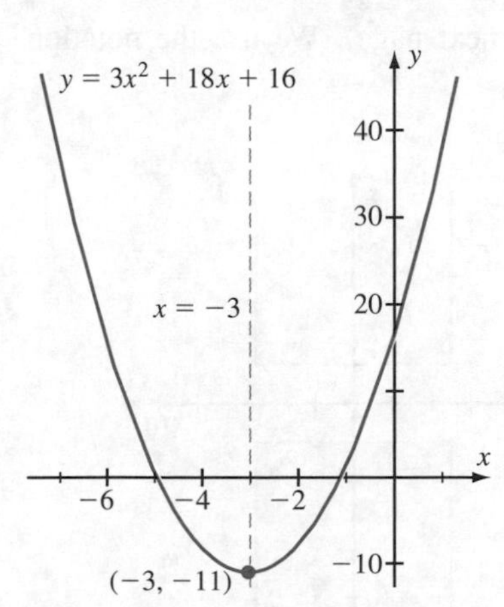

Figure 12

Example 7 What are the axis of symmetry and vertex of the parabola $y = 3x^2 + 18x + 16$?

Solution We complete the square of $3x^2 + 18x + 16$:

$$3x^2 + 18x + 16 = 3\left(x+\frac{6}{2}\right)^2 - 11.$$

Thus,

$$y = 3\left(x+\frac{6}{2}\right)^2 - 11 = 3(x-(-3))^2 - 11.$$

We conclude that $x = -3$ is the axis of symmetry and $(-3, -11)$ is the vertex. The graph of this parabola is given in Figure 12. ■

Regions in the Plane

It is sometimes useful to graph *regions* in the plane. These regions are usually described by inequalities rather than equalities.

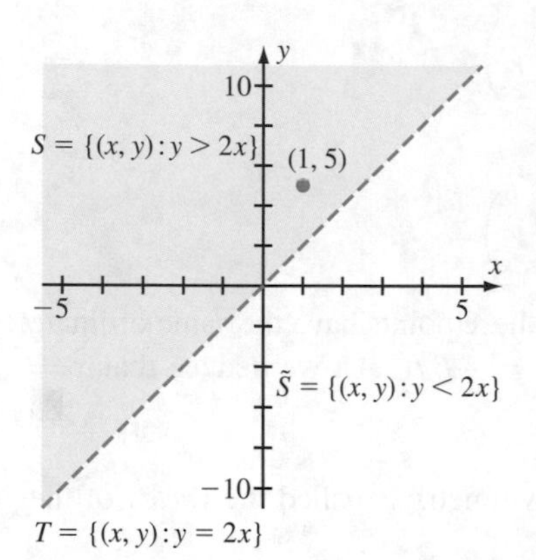

Figure 13

Example 8 Sketch $S = \{(x, y) : y > 2x\}$.

Solution We first use a dotted line to sketch $T = \{(x, y) : y = 2x\}$ to mark the boundary of our region. The set T divides the plane into two regions: $\tilde{S} = \{(x, y) : y < 2x\}$ and $S = \{(x, y) : y > 2x\}$. We pick an arbitrary test point $(1, 5)$ from the upper region. Since $(1, 5)$ is in S and in the upper region in Figure 13, the upper region must be S. ■

inSIGHT

In Example 8, a dotted line bounds the region. Doing so shows that the boundary is not included in the region being sketched. When the boundary is included, we use a solid line. The region $\{(x, y) : y \geq 2x\}$, for example, includes its boundary (see Figure 14).

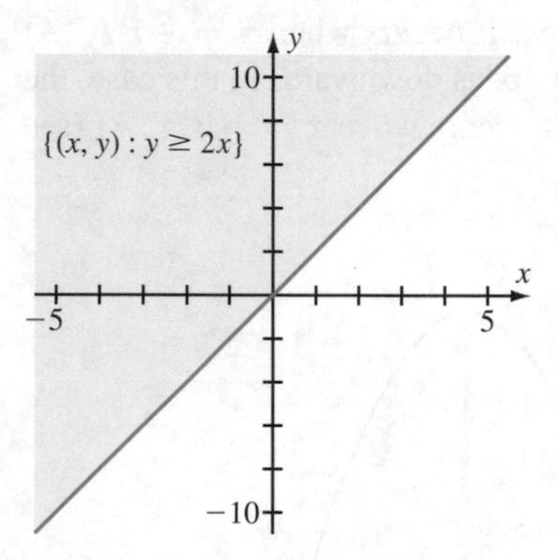

Figure 14

Example 9 Sketch $G = \{(x, y) : |y| > 2 \text{ and } |x| \leq 5\}$.

Solution The sets $\{(x, y) : |y| > 2\}$ and $\{(x, y) : |x| \leq 5\}$ are illustrated in Figures 15a and 15b. The set G consists of the points common to these two sets (see Figure 15c). ■

A rectangular region that includes its sides is said to be *closed*. A closed rectangle with sides that are parallel to the coordinate axes has the form

$$\{(x, y) : a \leq x \leq b \text{ and } c \leq y \leq d\}$$

for some constants a, b, c, and d (see Figure 16, next page). We use the notation $[a, b] \times [c, d]$ for this rectangle.

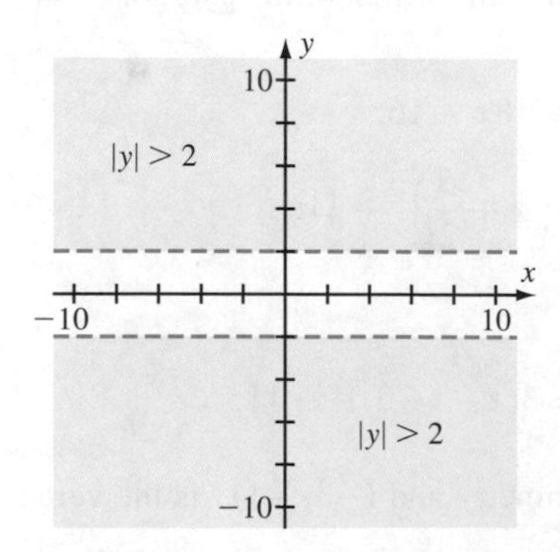

Figure 15a
$\{(x, y) : |y| > 2\}$

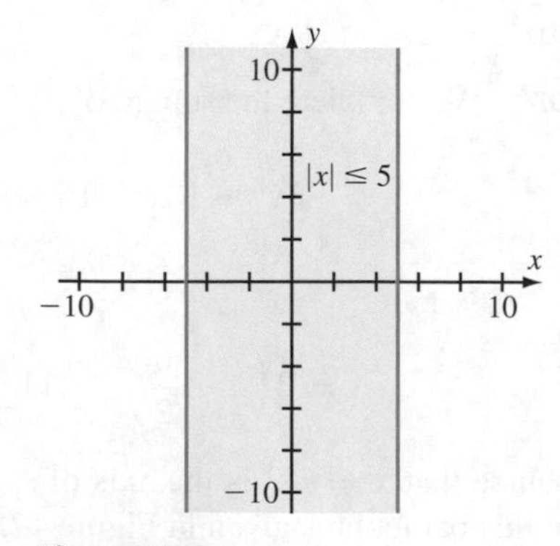

Figure 15b
$\{(x, y) : |x| \leq 5\}$

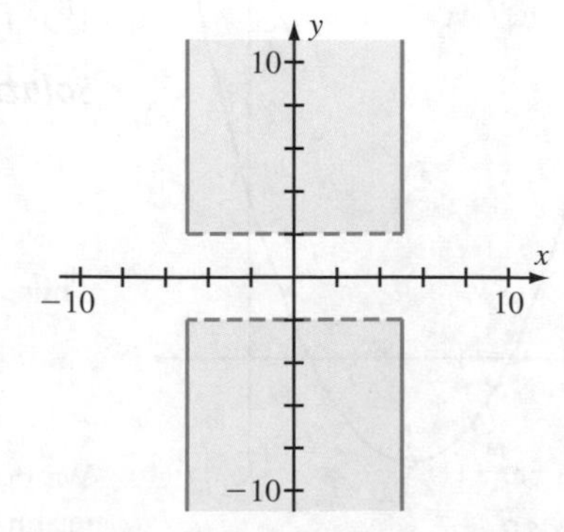

Figure 15c
$|y| > 2$

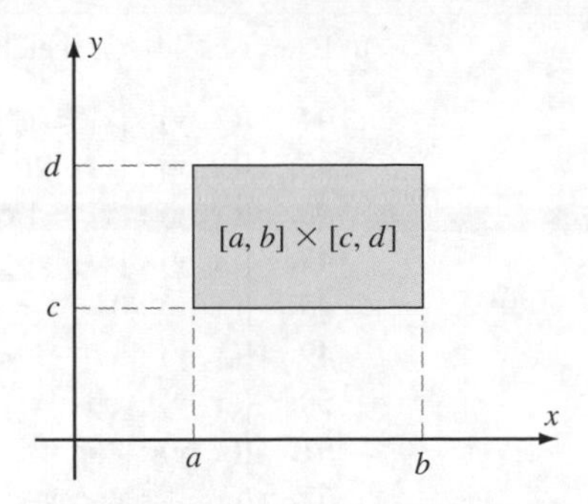

Figure 16
The rectangle $[a, b] \times [c, d] = \{(x, y) : a \leq x \leq b \text{ and } c \leq y \leq d\}$.

Every time we create a plot (whether by hand, graphing calculator, or computer software), we select a rectangle $[a, b] \times [c, d]$ and graph only those portions of the curve that lie within it. The rectangle we choose is called the *viewing window* or the *viewing rectangle*. Sometimes, as in the case of circles and ellipses, we can find a viewing window that captures the entire curve. For curves such as parabolas and hyperbolas, however, the viewing window we choose will show only part of the curve. Finding appropriate viewing windows for complicated equations can be tricky. Chapter 4 discusses how to use the ideas of calculus to help in this task.

quickquiz

1. Write the equation of a circle passing through the origin and with center 2 units above the origin.
2. Sketch the set $\{(x, y) : |x| < 1, |y| < 1\}$.
3. How can the Cartesian equations of a parabola, a hyperbola, and an ellipse be distinguished from one another?
4. Use the Method of Completing the Square to express $x^2 - 12x$ as the difference of squares.

EXERCISES

Problems for Practice

1. Graph the points $(2, -4)$, $(7, 3)$, $(\sqrt{3}, \sqrt{6})$, $(\pi + 3, \pi - 3)$, $(\pi^2, \sqrt{\pi})$, $(-8/3, -2/\pi)$ on a set of coordinate axes.
2. Graph the set of points that satisfy $y + x = 3$.
3. Graph the set of points that satisfy $y - x = 2$.
4. Which of the points $(2, 1)$, $(3, 0)$, $(4, -1)$, or $(1/2, 9/2)$ is farthest from the origin? Which is nearest to the origin? Which is farthest from $(-5, 6)$? Which is nearest to $(10, 7)$?
5. Let $A = (2, 3)$, $B = (-4, 7)$, and $C = (-5, -6)$. Calculate the distance of each of these points to each of the others.

In Exercises 6–15, the plot of the equation is a circle. Sketch it. Specify the center and radius.

6. $(x + 8)^2 + (y - 1)^2 = 16$
7. $(x - 1)^2 + (y - 3)^2 = 9$
8. $x^2 + y^2 - 6x + 8y - 4 = 0$
9. $4x^2 + 4y^2 - 8y + 16x + 2 = 0$
10. $(x - 3)^2 + y^2 + y = 1$
11. $x^2 + (y + 5)^2 = 2$

12. $x^2 + (y+7)^2 = 1$
13. $3x^2 - 6y + 12x + 3y^2 = 2$
14. $-x^2 - 6y + 6 = y^2 + x$
15. $x^2 + y^2 - y = 0$

In Exercises 16–20, give the equation of the circle that is described in words.

16. The circle with center at the origin and radius 2
17. The circle with center $(-3, 5)$ and radius 6
18. The circle with center $(3, 0)$ and diameter 8
19. The circle with radius 5 and center $(-4, \pi)$
20. The circle with center $(0, -1/4)$ and radius $1/4$

In Exercises 21–24, determine the vertex and axis of symmetry of the parabola.

21. $y = x^2 - 3$
22. $y = 2(3-x)^2 + 4$
23. $y = 2x - x^2$
24. $y = 3x^2 - 6x + 1$

In Exercises 25–35, sketch the set.

25. $\{(x, y) : |x| < 3\}$
26. $\{(x, y) : x^2 < y\}$
27. $\{(x, y) : |x| < 7, |y+4| > 1\}$
28. $\{(x, y) : |x| \leq 5, |y| > 2\}$
29. $\{(x, y) : x^2 + y^2 > 16\}$
30. $\{(x, y) : x < 2y, x \geq y - 3\}$
31. $\{(x, y) : (x-2)^2 + y^2 \geq 4\}$
32. $\{(x, y) : x + 5y \geq 4, x \leq 2, y \geq -8\}$
33. $\{(x, y) : |x - y| < 1, x \geq 4\}$
34. $\{(x, y) : x^2 + y^2 = 4, y > 0, x > 1\}$
35. $\{(x, y) : x^2 + y^2 < 9, x < 0, y > -1\}$

Further Theory and Practice

36. Find a point that is equidistant from the two points $(2, 3)$ and $(8, 10)$.
37. Find a point that is equidistant from the three points $(2, 3)$, $(8, 2)$, and $(7, 9)$.
38. Show that there can be no point equidistant from $(1, 2)$, $(3, 4)$, $(8, 15)$, and $(6, -3)$.
39. Describe all points that are equidistant from $(1, 0)$, $(0, 1)$, and $(0, -1)$.

Which of the equations in Exercises 40–43 are circles? Which are not? Give precise reasons for your answers.

40. $5x^2 - 5y^2 + x + 2y = 5$
41. $5x^2 + 5y^2 = 2x - 3y + 6$
42. $x^2 - y^2 + 6x = -2y^2 - 7$
43. $x^2 + 2y^2 + x - 5y = 7 + 3x^2$

In Exercises 44–53, sketch the set.

44. $\{(x, y) : |x| < 1, |y| < 1\}$
45. $\{(x, y) : |x| \cdot |y| < 1\}$
46. $\{(x, y) : |x| < 1 \text{ or } |y| < 1\}$
47. $\{(x, y) : |x + y| \leq 1\}$
48. $\{(x, y) : 0 < x \leq y < 2x \leq 1\}$
49. $\{(x, y) : 3x - 1 < 0 \leq y + 2x + 5\}$
50. $\{(x, y) : x^2 < y^2\}$
51. $\{(x, y) : x < y^2 < x^2\}$
52. $\{(x, y) : x < y < \sqrt{x}\}$
53. $\{(x, y) : y > 2x, x - 3y = 6\}$
54. A circle of radius $\sqrt{65}$ passes through the points $(0, -6)$ and $(3, -5)$. What is its center?
55. From Example 1 we know that the graph of $y = 2x$ is a line. This line intersects each planar circle in zero, one, or two points. Give an algebraic reason that explains this geometric fact.
56. Find the equation of the circle that passes through the points $(-3, 4)$, $(1, 6)$, and $(9, 0)$.
57. Two different circles can intersect in zero, one, or two points. Give an algebraic reason that explains this geometric fact.
58. Plot the four curves $y = x$, $y = x^2$, $y = x^3$, and $y = x^4$ in the viewing window $[0, 1] \times [0, 1]$. Sketch the graph of $y = x^N$ $(0 \leq x \leq 1)$ for some value of $N > 4$. What inequality is illustrated by these plots?
59. Plot the loçus of points P with an ordinate that is equal to distance of the point P to the point $(0, 1)$.

Calculator/Computer Exercises

60. Graph $y = x^4 - 32x^3 + 224x^2 + 512x - 4096$ in the viewing rectangle $[-6, 22] \times [-5000, 12000]$. You will see four points at which the graph crosses the x-axis. Zoom in to approximate the coordinates of these points to three decimal places.
61. In the viewing rectangle $[0, 15] \times [0, 2.6]$, graph

$$y = (ax^2 + 7x + 2)/(Ax^2 - 3x + 12)$$

for $a = A = 2$, for $a = A = 3$, for $a = 2$, $A = 3$, and for $a = 3$, $A = 2$.

a. Zoom out in the x direction by a factor of ten. Repeat. Identify the horizontal lines that these four curves approach as x tends to infinity.

b. Divide each term in the numerator and denominator of $(ax^2 + 7x + 2)/(Ax^2 - 3x + 12)$ by x^2. Use the resulting expression to explain algebraically why each curve approaches a horizontal line as x tends to infinity.

62. The graph of $y = (ax^3 + bx^2 + cx + d)/x$ $(a, d \neq 0)$ is called the *trident of Newton*.

a. Graph the trident $y = (10x^3 + 1)/x$, the parabola $y = 10x^2$, and the hyperbola $y = 1/x$ in the viewing window $[-2.5, 2.5] \times [-30, 65]$.

b. Repeat the three graphs in the window $[-0.5, 0.5] \times [-50, 50]$.

c. Repeat the three graphs in the window $[-100, 100] \times [0, 100000]$.

d. Explain the reasons for what you observe in parts b and c.

63. For any two positive constants a and b, the graph of $y = a \cdot x/(b + x^2)$ is called the *serpentine of Newton*. The graphs of the equations $y = (a/b)x$ and $y = a/x$ closely approximate the graph of the serpentine in *appropriately chosen* viewing rectangles. Choose specific values of a and b and graph the serpentine in a small window centered at $(0, 0)$ and in a large window centered at $(0, 0)$. Use these windows to illustrate the stated approximations.

64. Choose a, b, and c with $a < b < c$. Let $\mathcal{C}_{A,B,C}$ be the curve with a Cartesian equation of

$$y = A \cdot |x - a| + B \cdot |x - b| + C \cdot |x - c|$$

for $-a - 1 \leq x \leq c + 1$. Plot $\mathcal{C}_{A,B,C}$ for $A = B = C = 1$. Repeat for several other positive values of A, B, and C. Based on your graphs, devise a strategy for determining the minimum y-coordinate that a point on $\mathcal{C}_{A,B,C}$ might have when the coefficients A, B, and C are positive.

1.3 Lines and Their Slopes

Slopes

When a highway sign reads "Caution: 6% Grade," it tells us that for every mile of horizontal motion, the road drops 0.06 mile. This information is important for truckers who are carrying heavy loads. In mathematical language, the road has *slope* -0.06. Now look at the line in Figure 1. We want to compute its slope, or the ratio of vertical motion to horizontal motion (*rise over run*). To do this, we choose two points on the line. The *rise* is the difference of y-coordinates, or $y_2 - y_1$. The *run* is the difference of x-coordinates, or $x_2 - x_1$. The slope is the ratio of these two quantities:

$$\text{Slope} = \frac{\text{rise}}{\text{run}} = \frac{\Delta y}{\Delta x} = \frac{y_2 - y_1}{x_2 - x_1}.$$

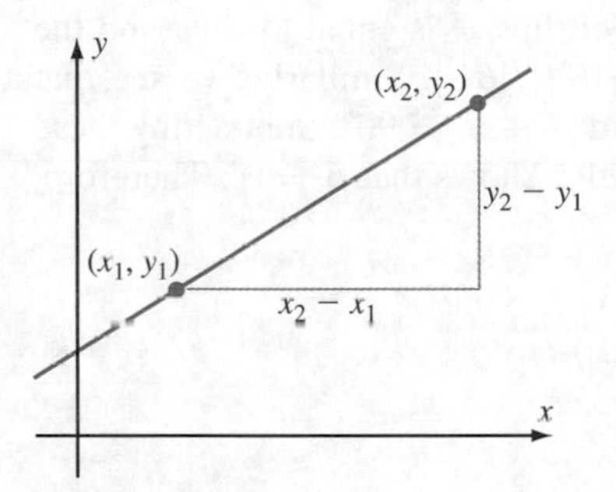

Figure 1
$\text{Slope} = \dfrac{\text{rise}}{\text{run}} = \dfrac{\Delta y}{\Delta x} = \dfrac{y_2 - y_1}{x_2 - x_1}$

inSIGHT

We subtract the x-coordinates in the same order as the y-coordinates: $x_2 - x_1$ and $y_2 - y_1$. However, we *can* reverse the order in *both* the numerator and the denominator, because this introduces two minus signs that cancel each other.

If we compute the slope of the line in Figure 1 using two other points $(\tilde{x}_1, \tilde{y}_1)$ and $(\tilde{x}_2, \tilde{y}_2)$, do we obtain the same slope? The two triangles ΔABC and $\Delta \tilde{A}\tilde{B}\tilde{C}$ are similar

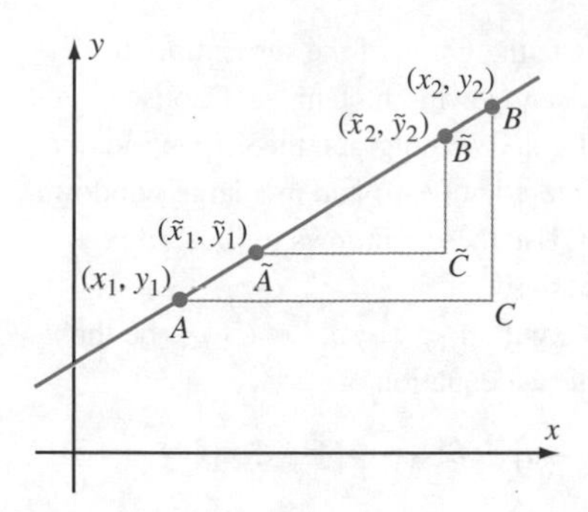

Figure 2
ΔABC and $\Delta\tilde{A}\tilde{B}\tilde{C}$ are similar.

(see Figure 2), so the ratios of the corresponding sides are the same. In other words, rise over run is the same, and we *do* obtain the same slope if we compute using two other points.

Example 1 Find the slope of the line in Figure 3.

Solution If we use the two points (3, 1) and (−3, 4), then

$$\text{Slope} = \frac{\text{rise}}{\text{run}} = \frac{4-1}{-3-3} = -\frac{1}{2}.$$

Had we used another pair of points, such as (5, 0) and (−1, 3), we would have obtained the same slope. Check that even if you reverse the order of the points, you still obtain slope −1/2. ■

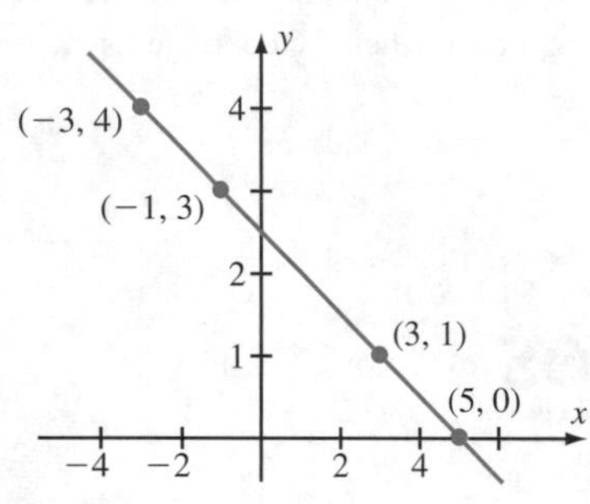

Figure 3

A positive slope indicates that the line *rises* relative to left-to-right motion. A negative slope, as in Example 1, indicates that the line *falls* relative to left-to-right motion. The equation of a horizontal line is of the form $y = b$ for some constant b; the slope of such a line is zero because $\Delta y = 0$. A vertical line has an equation of the form $x = a$ for some constant a; the slope of such a line is *undefined* because $\Delta x = 0$.

Slopes of Perpendicular Lines

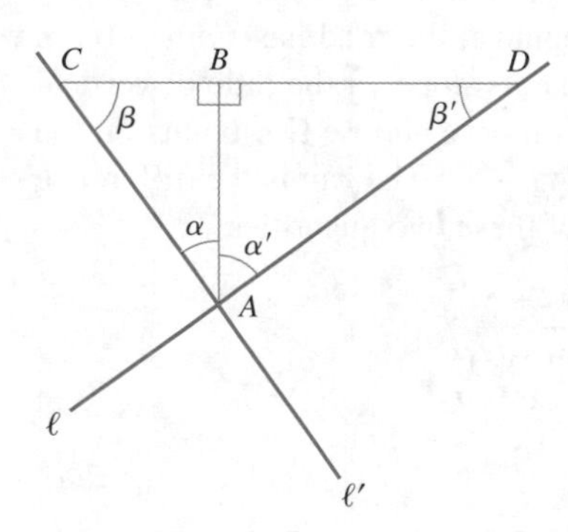

Figure 4

How are the slopes of two perpendicular lines related? Look at the perpendicular lines ℓ and ℓ', shown in Figure 4. Neither line is vertical or horizontal, so they both have nonzero slope. From point C to point A, the run on line ℓ' is equal to $|\overline{BC}|$ and the rise is the negative of $|\overline{BA}|$. Thus, $-\text{slope } \ell' = |\overline{BA}|/|\overline{BC}|$. Similarly, we see that $\text{slope } \ell = |\overline{BA}|/|\overline{BD}|$. Notice that $\alpha + \alpha' = 90°$ and $\beta' + \alpha' = 90°$. Subtracting these two equations gives $\alpha = \beta'$. The same reasoning also shows that $\beta = \alpha'$. Therefore, ΔABC and ΔDBA are similar. Thus,

$$\frac{|\overline{BA}|}{|\overline{BC}|} = \frac{|\overline{BD}|}{|\overline{BA}|} = \frac{1}{(|\overline{BA}|/|\overline{BD}|)},$$

or

$$-\text{slope } \ell' = \frac{1}{\text{slope } \ell}. \tag{1.2}$$

It is also true that if the slopes of two lines ℓ and ℓ' satisfy equation (1.2), then ℓ and ℓ' are mutually perpendicular. In conclusion:

Theorem 1 Two lines, neither of which is horizontal or vertical, are mutually perpendicular if and only if their slopes are negative reciprocals.

Example 2 Find the slope of any line perpendicular to the one shown in Figure 5.

Solution We start by finding the slope of the line in the figure: $(6-2)/(3-(-4)) = 4/7$. The slope of any perpendicular line is the negative reciprocal of this number, or $-7/4$. ■

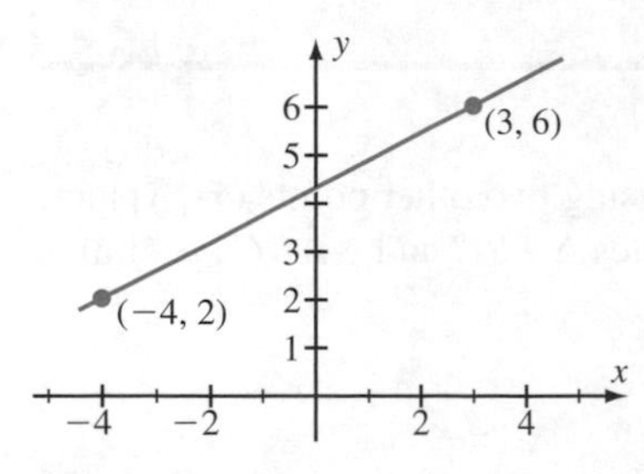

Figure 5

Slopes of Parallel Lines

If lines ℓ and ℓ' are parallel, then the two right triangles shown in Figure 6 are congruent. It follows that slope $\ell =$ slope ℓ'. The converse is also easily deduced:

Theorem 2 Two lines are parallel if and only if they have the same slope.

Example 3 Verify that the line through $(3, -7)$ and $(4, 6)$ is parallel to the line through $(2, 5)$ and $(4, 31)$.

Solution We compute that the slope of the first line is $(6-(-7))/(4-3) = 13$, and the slope of the second line is $(31-5)/(4-2) = 13$. Because the slopes are equal, the lines are parallel. ■

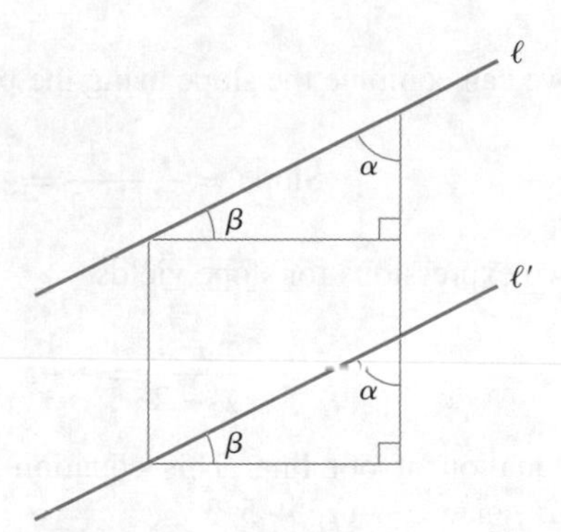

Figure 6

Equations of Lines

We need to understand how to write the equations that represent lines. We also want to determine whether a given equation represents a line. There are many geometrical ways to describe a line. For instance, a line is determined by two points through which it passes (see Figure 7a). A line is also determined by a point and a slope (see Figure 7b).

Both of these descriptions of a line require two pieces of information: two points or a point and a slope. This observation motivates the mathematical description of a line.

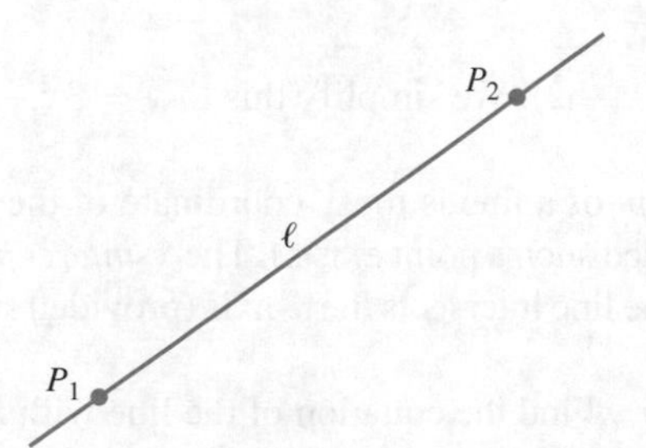

Figure 7a
Line ℓ is determined by P_1 and P_2.

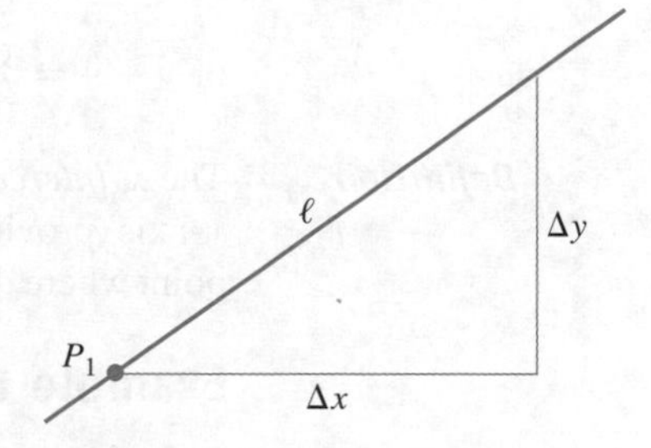

Figure 7b
Line ℓ is determined by P_1 and $m = \Delta y/\Delta x$.

Basic Rule for Obtaining the Equation of a Line

To write the equation of a line, write its slope in two different ways and equate.

Example 4 Write the equation of the line passing through points (2, 1) and (−4 , 3).

Solution If we let (x, y) be a variable point on the line, then we can compute the slope using (2, 1) and (x, y):

$$\text{Slope} = \frac{y-1}{x-2}.$$

Alternatively, we can compute the slope using the points (2, 1) and (−4, 3):

$$\text{Slope} = \frac{3-1}{-4-2} = -\frac{1}{3}.$$

Equating the two expressions for slope yields

$$\frac{y-1}{x-2} = -\frac{1}{3},$$

which is the equation of our line. This equation can also be written as $y - 1 = -(1/3) \cdot (x - 2)$, or $y = -x/3 + 5/3$. ■

Example 5 Find the equation of the line passing through the point (2, −4) and having slope 3.

Solution Let (x, y) be a variable point on the line. We can compute the slope using (2, −4) and (x, y):

$$\text{Slope} = \frac{y-(-4)}{x-2} = \frac{y+4}{x-2}.$$

We also know that the slope is 3. Equating the two expressions for slope, we find that

$$\frac{y+4}{x-2} = 3,$$

or $y + 4 = 3(x - 2)$. We simplify this to $y = 3x - 10$. ■

Definition The *x-intercept* of a line is the x-coordinate of the point where the line intersects the x-axis (provided such a point exists). The *y-intercept* of a line is the y-coordinate of the point where the line intersects the y-axis (provided such a point exists).

Example 6 Find the equation of the line with x-intercept 3 and y-intercept −5.

Solution The line with x-intercept 3 crosses the x-axis at point (3, 0). The line also crosses the y-axis at point (0, −5). If we let (x, y) be a variable point on the line, then we can obtain one expression for the slope from (0, −5) and (x, y):

$$\text{Slope} = \frac{y-(-5)}{x-0} = \frac{y+5}{x}.$$

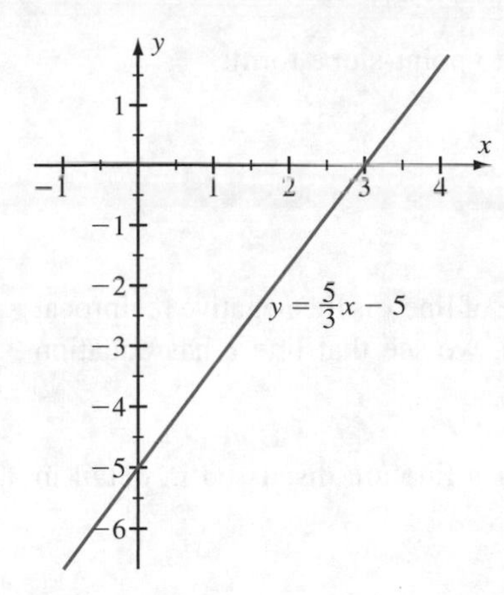

Figure 8

We can obtain another expression for the slope from (3, 0) and (0, −5):

$$\text{Slope} = \frac{0-(-5)}{3-0} = \frac{5}{3}.$$

Equating the two expressions for slope gives

$$\frac{y+5}{x} = \frac{5}{3},$$

or $y = (5/3)x - 5$ (see Figure 8). ■

There are many special forms for equations of lines. In the following discussion, we consider several of these forms. The discussion excludes vertical lines, which have equations of the form $x = a$.

The Point-Slope Form of a Line

Suppose line ℓ has slope m and passes through point (x_0, y_0). Let (x, y) be a variable point on line ℓ. One expression for slope is

$$\text{Slope} = \frac{y - y_0}{x - x_0}.$$

The slope is also m, so equating these two expressions for slope gives

$$\frac{y - y_0}{x - x_0} = m,$$

or $y - y_0 = m(x - x_0)$. By isolating y, we obtain the *point-slope form*

$$y = m(x - x_0) + y_0$$

for the line with slope m and passing through point (x_0, y_0).

Example 7 What is the slope of the line $4x - 2y = 7$?

Solution We write the line in point-slope form:

$$-2y = -4x + 7, \text{ or}$$
$$y = 2x - \frac{7}{2}.$$

There are now many ways to write the line in point-slope form. The two most natural are $y = 2(x - 0) + (-7/2)$ and $y = 2(x - 7/4) + (0)$. Either way, we see that the line has slope 2. As additional information, we also see that the line passes through points $(0, -7/2)$ and $(7/4, 0)$. ■

Example 8 Write the equation of the line ℓ that is perpendicular to the line $2x + 3y + 5 = 0$ and passes through point $(6, -8)$.

Solution The line $2x + 3y + 5 = 0$ can be put into point-slope form:

$$3y = -2x - 5, \text{ or}$$
$$y = -\frac{2}{3}x - \frac{5}{3}.$$

So the given line has slope $-2/3$. Therefore, the slope of line ℓ is the negative reciprocal of $-2/3$, or $3/2$. Using the point-slope form again, we see that line ℓ has equation $y = (3/2) \cdot (x - 6) - 8$. ■

The following special forms for the equation of a line are discussed in detail in Exercises 33 and 34.

The Slope-Intercept Form of a Line

The line with slope m and y-intercept $(0, b)$ has equation

$$y = mx + b.$$

This is simply the point-slope form using point $(0, b)$.

The Intercept Form of a Line

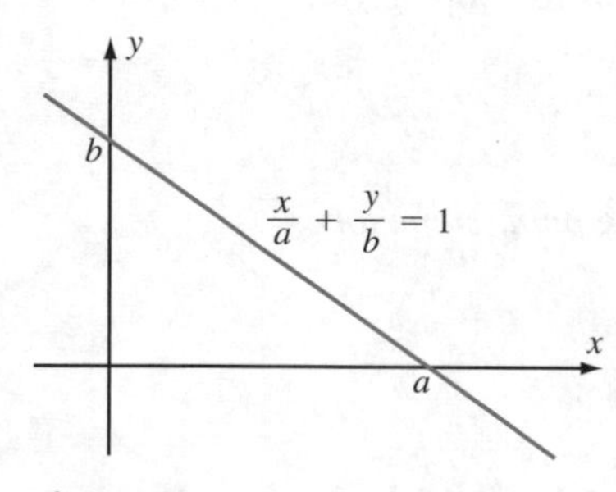

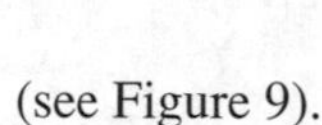

Figure 9

The line with x-intercept a and y-intercept b $(a, b \neq 0)$ has equation

$$\frac{x}{a} + \frac{y}{b} = 1$$

(see Figure 9).

Example 9 What is the intercept form of the line defined by the equation $2x - 3y + 6 = 0$?

Solution First, we rewrite the equation as $2x - 3y = -6$. Dividing both sides of the equation by -6 yields the intercept form equation:

$$\frac{x}{-3} + \frac{y}{2} = 1.$$

From this equation, we see at a glance that the line intersects the x-axis at $(-3, 0)$ and the y-axis at $(0, 2)$.

An alternative method is to substitute $y = 0$ into the equation to obtain the x-intercept $a = -3$. Substituting $x = 0$ into the equation leads to the y-intercept $b = 2$. The intercept form of the equation is an immediate consequence of these calculations. ■

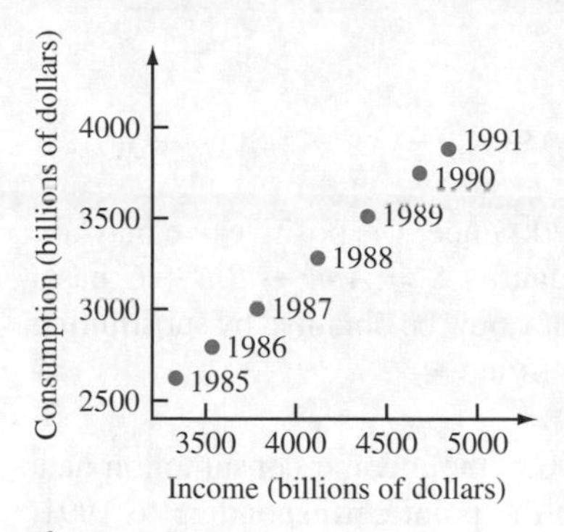

Figure 10
Total U.S. Income and Consumption, 1985–1991

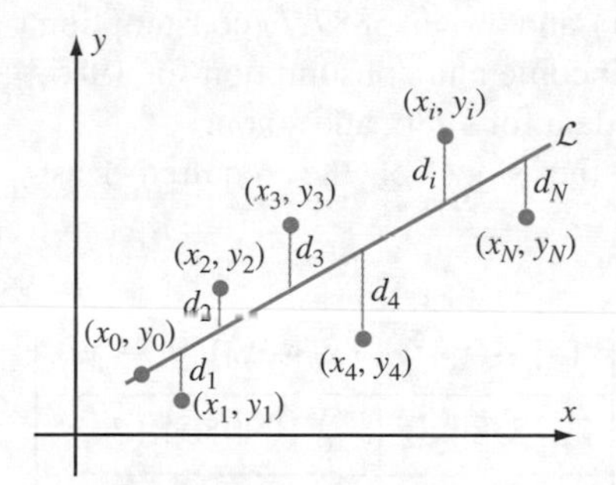

Figure 11

Least Squares Lines

Table 1 displays total annual American income and consumption in billions of dollars for the years 1985–1991.

	1985	1986	1987	1988	1989	1990	1991
Income (x)	3325.3	3526.2	3776.6	4070.8	4384.3	4679.8	4834.4
Consumption (y)	2629.0	2797.4	3009.4	3296.1	3523.1	3748.4	3887.7

Table 1

Figure 10 shows a plot of the seven data points (x, y). (Economists refer to the data of Table 1 as *time-series data*. Figure 10 is called a *time-series graph.*) It is apparent that the plotted points lie close to a straight line. The question arises: What equation $y = mx + b$ should we use to describe the approximate relationship between the observed values of x and y? Certainly we may pass a line through each pair of plotted points. There are, however, 21 pairs of plotted points, and no two slopes of the lines that pass through these point pairs are exactly the same.

Finding a *line of best fit* is a common problem in nearly every field in which numerical data is obtained. The method that is used most often is the *Method of Least Squares*. Suppose we have $N + 1$ data points $(x_0, y_0), (x_1, y_1), \ldots, (x_N, y_N)$ that have a plot that is approximately linear. For any line $\mathcal{L}$ through the point (x_0, y_0), let d_i denote the vertical distance of the data point (x_i, y_i) to $\mathcal{L}$ (as in Figure 11). The *least squares line* (or the *regression line*) *through* (x_0, y_0) is the line for which the sum $d_1^2 + d_2^2 + \cdots + d_N^2$ is minimized.

Theorem 3 Given the $N + 1$ data points $(x_0, y_0), (x_1, y_1), \ldots, (x_N, y_N)$, the *least squares line through* (x_0, y_0) is given by $y = m(x - x_0) + y_0$ where

$$m = \frac{(x_1 - x_0) \cdot (y_1 - y_0) + (x_2 - x_0) \cdot (y_2 - y_0) + \cdots + (x_N - x_0) \cdot (y_N - y_0)}{(x_1 - x_0)^2 + (x_2 - x_0)^2 + \cdots + (x_N - x_0)^2}. \quad \textbf{(1.3)}$$

Proof For any real number m, the equation of line $\mathcal{L}$ through point (x_0, y_0) with slope m is $y = m(x - x_0) + y_0$. The vertical distance of the observation (x_i, y_i) to $\mathcal{L}$ is

$$d_i = |y_i - (m(x_i - x_0) + y_0)| = |(y_i - y_0) - m(x_i - x_0)|.$$

We therefore seek the value m for which the sum

$$S = ((y_1 - y_0) - m(x_1 - x_0))^2 + \cdots + ((y_N - y_0) - m(x_N - x_0))^2$$

is minimized. By expanding each square and collecting like powers of m, we find that

$$S = Am^2 + Bm + C$$

with

$$A = (x_1 - x_0)^2 + (x_2 - x_0)^2 + \cdots + (x_N - x_0)^2$$

and

$$B = -2((x_1 - x_0)(y_1 - y_0) + (x_2 - x_0)(y_2 - y_0) + \cdots + (x_N - x_0)(y_N - y_0)).$$

We do not calculate C because its value is not needed. Since A is positive, we may use Theorem 1 from Section 1.2 to conclude that the quantity $S = Am^2 + Bm + C$ has a minimum value when $m = -B/2A$. Formula (1.3) may now be obtained by substituting the expressions for A and B into $-B/2A$ and simplifying. ■

Because d_i measures the amount by which the approximating line misses the ith observation, the expression $d_1 + \cdots + d_n$ represents the total absolute error. This may seem to be a simpler measure of error than the sum of the squares of the errors, but, by using $d_1^2 + \cdots + d_n^2$, we can reduce the problem of minimizing the error to a problem that we have already solved—finding the vertex of a parabola.

Example 10 Refer to Table 1 and Figure 10. Use the income-consumption data for 1985–1991 to pass a least squares line through the point corresponding to 1991. Use this line to predict the level of consumption in 1995, a year in which total personal income was \$6115.1 billion. In fact, total personal consumption for 1995 was actually \$4924.9 billion. What is the relative error of the regression line prediction?

Solution We set $x_0 = 4834.4$ (income for 1991) and $y_0 = 3887.7$ (consumption for 1991). Let $(x_1, y_1) = (3325.3, 2629.0)$, or the income and consumption for 1985. Continue with $(x_2, y_2) = (3526.2, 2797.4)$, or the data for 1986, and so on.

Table 2 lists all computations needed for the slope of the required least squares line.

i	x_i	y_i	$x_i - x_0$	$y_i - y_0$	$(x_i - x_0)^2$	$(x_i - x_0)(y_i - y_0)$
1	3325.3	2629.0	−1509.1	−1258.7	2277382.81	1899504.17
2	3526.2	2797.4	−1308.2	−1090.3	1711387.24	1426330.46
3	3776.6	3009.4	−1057.8	−878.3	1118940.84	929065.74
4	4070.8	3296.1	−763.6	−591.6	583084.96	451745.76
5	4384.3	3523.1	−450.1	−364.6	202590.01	164106.46
6	4679.8	3748.4	−154.6	−139.3	23901.16	21535.78
Total					5917287.02	4892288.37

Table 2

If we set $m = 4892288.37/5917287.02 \approx 0.82678$, then the equation for the least squares line is

$$y = 0.82678(x - 4834.4) + 3887.7.$$

The predicted value of y corresponding to $x = 6115.1$ is

$$0.82678(6115.1 - 4834.4) + 3887.7 = 4946.6$$

(see Figure 12). The relative error is $100 \times |4924.9 - 4946.6|/4946.6\%$, or about 0.44%.

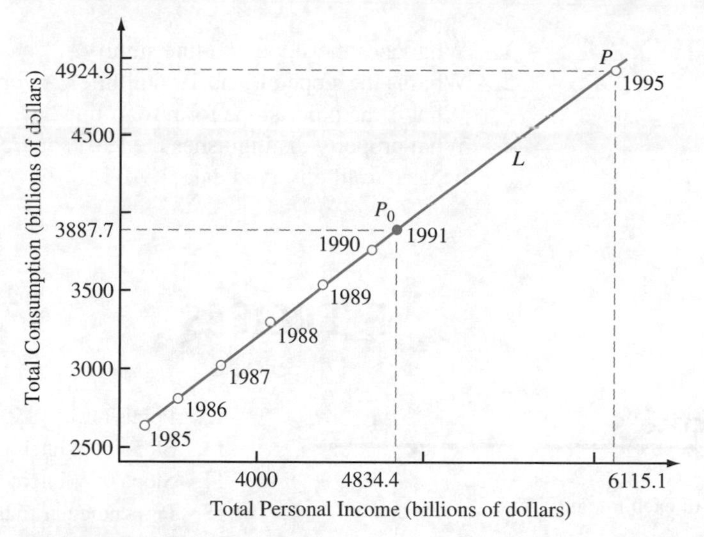

Figure 12
The regression line L through P_0 based on the observations obtained from 1985–1991 may be used to predict consumption in 1995 (plotted as point P).

inSIGHT

We can calculate the sum of the absolute errors $SAE = d_1 + d_2 + \cdots + d_6$ for each m. The graph of the points (m, SAE) appears in Figure 13a. Observe that when we use the sum of the absolute errors, we obtain a jagged curve. By contrast, the sum of the square errors, $SSE = d_1^2 + d_2^2 + \cdots + d_6^2$, results in a smooth graph, which is a parabola (as we learned in the proof of Theorem 3 and as is shown in Figure 13b). Chapter 4 returns to this application and shows how calculus provides us with a powerful tool for finding the lowest point on the graph of a smooth curve. Chapter 13 uses calculus to solve an even more general problem about lines of best fit.

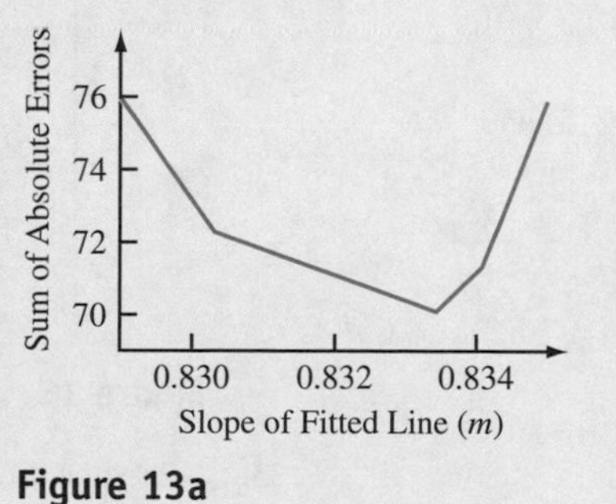

Figure 13a

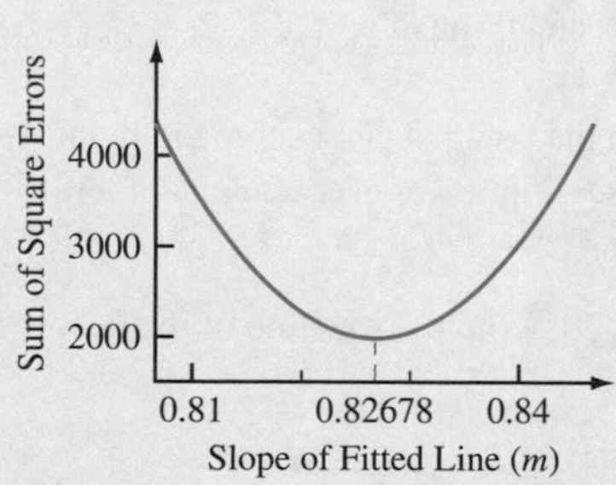

Figure 13b

quickquiz

1. What does the slope of a line signify?
2. What is the slope of a horizontal line? A vertical line?
3. What is the point-slope form for a line?
4. What property distinguishes the least squares line from other lines that might be used to fit observed data?

EXERCISES

Problems for Practice

1. Find the slopes of each line in Figure 14.

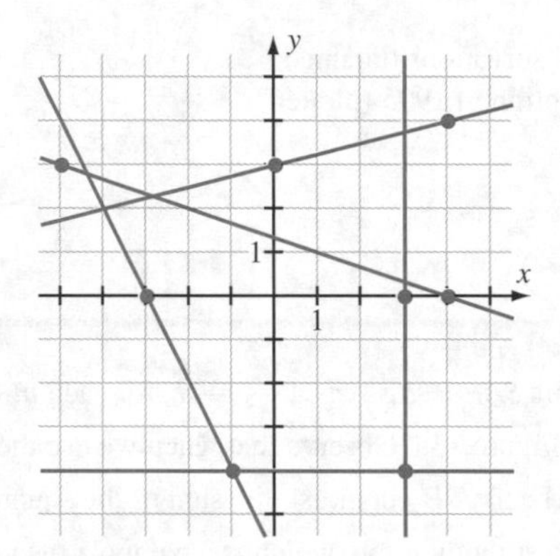

Figure 14

2. Sketch, on the same set of axes, lines passing through point (1, 3) and having slopes −3, −2, −1, 0, 1, 2, and 3.
3. Sketch, on the same set of axes, lines having slope −3 and passing through points (−2, −7), (−1, 0), (3, 0), and (5, 1).
4. Sketch the line that passes through point (−2, 5) and that rises 7 units for every 2 units of left-to-right motion.

In Exercises 5–20, write the equation of the line.

5. Slope 5, point (−3, 7)
6. Slope −2, point (4.1, 8.3)
7. Slope −4, y-intercept (0, 9)
8. Slope π, y-intercept $(0, \pi^2)$
9. x-intercept (9/7, 0), y-intercept (0, 18/5)
10. x-intercept (−5, 0), y-intercept $(0, 2\pi)$
11. Points (−2, 7/11), (18, 14)
12. Points (1.3, −8.7), (−5.5, 2.1)
13. Slope 3, x-intercept −4
14. Slope 0, y-intercept 2
15. Perpendicular to the line $y - 3x = 5$, y-intercept −5
16. Parallel to the line $3y - 6x + 7 = 0$, passing through point (2, 1)
17. Parallel to the line $-2x + 4y - 8 = 0$, x-intercept 4
18. Perpendicular to the line $4x + 7y = 5$, passing through point (1, 1)
19. Slope −1/3, y-intercept −1/8
20. Slope −4, x-intercept −4
21. Is the line through points (1, 2) and (4, 7) parallel to the line through points (−4, −3) and (−7, 6)? Is it perpendicular to the line through points (7, 9) and (5, −7)?
22. Write the equation of the line perpendicular to the line in Figure 15 and passing through point (7, −8).

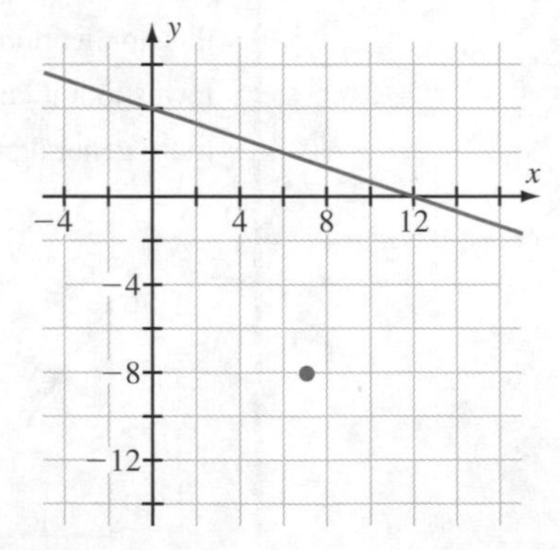

Figure 15

23. Is the line through points (−3, 8) and (2, 23) parallel to the line through points (−4, 6) and (3, 28)? Is it perpendicular to the line through points (−2, 4) and (7, 1)?

In Exercises 24–31, sketch the line on a set of coordinate axes.

24. $y - 2x = 4$
25. $2x + 8y = 6$
26. $x = 2$
27. $y = -7$
28. $x/3 + y/9 = 1$
29. $-x/5 - y/13 = 1$
30. $3x - 9y + 14 = 0$
31. $7x + 8y - 15 = 0$
32. What are the slopes of the lines in Figure 16?

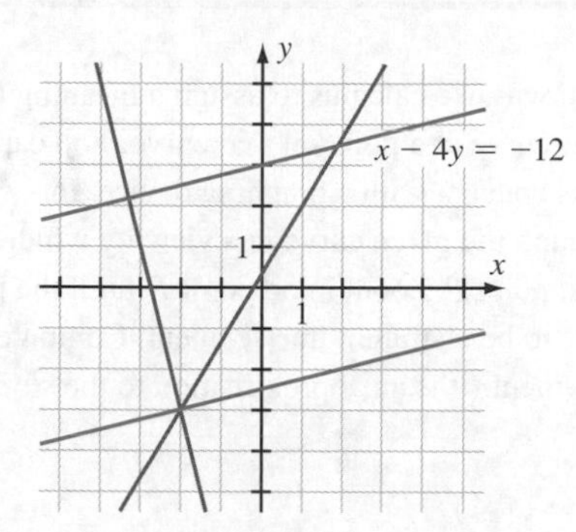

Figure 16

Further Theory and Practice

33. Imitate the verification of the point-slope form to show that the slope-intercept form for a line is correct.

34. Imitate the verification of the point-slope form to show that the intercept form for a line is correct.

35. Find a point (a, b) so that the line through (a, b) and $(5, -7)$ has slope 2.

36. Find a point (a, b) so that the line through (a, b) and $(4, 2)$ is parallel to the line through $(3, 6)$ and $(2, 9)$.

37. Find a point (a, b) so that the line through (a, b) and $(-2, 7)$ is perpendicular to the line through $(-2, 7)$ and $(4, 9)$.

38. Find a point (a, b) with distance 1 unit from $(4, 6)$ so that the line through (a, b) and $(4, 6)$ is perpendicular to the line through $(4, 6)$ and $(-8, 4)$.

39. Find a point (a, b) with distance 1 unit from the origin and a point (c, d) with distance 5 units from the origin so that the line through (a, b) and (c, d) has slope 2.

40. Find a point (a, b) on the line through $(-2, 7)$ and $(9, -4)$ so that the line through (a, b) and $(2, 1)$ has slope 8.

41. Find the point on the line $3x - 8y = 4$ that is nearest to point $(2, 8)$.

42. Where does the line $x - 7y = -15$ intersect the circle $(x - 3)^2 + (y + 1)^2 = 25$?

43. Find the point of intersection of the lines $x + 2y = 4$ and $2x + y = 5$. Find a line through the point of intersection that makes equal angles with each of the two given lines.

44. Physical therapists recommend that ramps for people who use wheelchairs rise not more than 1 in. for each foot of forward motion. Formulate this recommendation in terms of slope. If the entrance to a certain public building is 30 ft from the sidewalk and the front door is 8 ft off the ground (up a steep flight of stairs), how can a suitable wheelchair ramp be built?

45. Let (x_1, y_1) and (x_2, y_2) be distinct points of the plane. Demonstrate that the equation of the line passing through these two points can be written in the form

$$\frac{y - y_1}{x - x_1} = \frac{y_2 - y_1}{x_2 - x_1}.$$

This is sometimes called the *two-point form* of a line.

46. Refer to Exercise 45 for the two-point form of the equation of a line. Use that form to write the equations of the lines passing through each pair of points.

a. $(2, 7), (6, -4)$
b. $(12, 1), (-4, -4)$
c. $(-7, -4), (9, 0)$
d. $(1, -5), (-5, 1)$

47. For what values of y_0 is the distance between the points of intersection of $y = y_0$ with $y = 2x + 1$ and $y = x + 2$ equal to 1,000,000?

48. Determine the regression line through $P_0 = (x_0, y_0)$ that is based on the observations $P_- = (x_0 - h, y_-)$ and $P_+ = (x_0 + h, y_+)$. How is the slope of the regression line related to the slopes of the segments $\overline{P_-P_0}$ and $\overline{P_0P_+}$?

49. The following table records several paired values of automobile mileage (x), measured in thousands of miles, and hydrocarbon emissions per mile (y), measured in grams.

x	5.013	10.124	15.060	24.899	44.862
y	0.270	0.277	0.282	0.310	0.345

Plot these points. Determine the regression line through (44.862, 0.345). If this pattern continues for higher-mileage cars, about how many grams of hydrocarbons per mile would a car with 100,000 miles emit?

50. Let x denote the perimeter of a rectangle, and let y denote the area. Calculate x and y for the five rectangles with (width, height) of (7, 42), (8, 23), (11, 13), (18, 9), and (26, 8). Plot the five ordered pairs (x, y) you have calculated. The plot is nearly linear. Find the least squares line through the origin

($x = y = 0$ when the width and height are both zero). For a square of side length 2, the perimeter x is 8. What is the regression line approximation of the area y? What is the relative error? What went wrong?

51. Suppose X and Y are items or goods that can be purchased at prices p_X and p_Y, respectively. Suppose x represents the units of good X and y the units of good Y that a consumer might purchase. In economics, the first quadrant of the xy-plane is called *commodity space*. If the consumer has a fixed amount (C) that he may allot to the purchase of goods X and Y, then the locus of all points (x, y) that represent purchasable combinations of these two goods is called the consumer's *budget line*. (It is actually a line segment.) Determine a Cartesian equation for the budget line. What are its intercepts? What is its slope? If the consumer's circumstances change so that he has a different amount (C') that he can use toward the purchase of goods X and Y, what is the relationship of the new budget line to the old one?

52. The graph of $y = x^2$ is said to be *concave up*. One way to define this property is: If $P = (a, a^2)$ and $Q = (b, b^2)$ are any two points on the graph of $y = x^2$, then the line segment joining P and Q lies above the graph of $y = x^2$.

a. What is the equation of the line that passes through P and Q?

b. Express the upward concavity of the parabola $y = x^2$ as an inequality involving a, b, and x for $a < x < b$. What does this inequality become when x is $(a + b)/2$?

53. Prove that three distinct points in the plane are collinear if and only if the slope determined by any pair of the points is the same as the slope determined by any other pair.

54. If $Ax + By + C = 0$ is the equation of line ℓ (not both A and B are zero) and $P = (p, q)$ is a point in the plane, then prove that the distance from the point to the line is

$$\frac{|Ap + Bq + C|}{\sqrt{A^2 + B^2}}.$$

(The distance of P to ℓ is defined as the distance of P to Q where Q is the point on ℓ such that the line segment $\overline{PQ}$ is perpendicular to ℓ.)

55. Let $a_1, a_2, a_3, \ldots, a_N$ be a collection of a finite number of points in the plane that are not all collinear. Assume that N is at least 3. Prove that there is a line that passes through only two points, and no others.

56. A *lattice point* in the plane is a point with integer coordinates. Suppose P and Q are lattice points. Under what circumstances is the midpoint of the line segment $\overline{PQ}$ a lattice point? Suppose $P_1, \ldots, P_5$ are five distinct lattice points, with no three collinear. Show that at least one pair of the given points determines a line that passes through a sixth lattice point.

Calculator/Computer Exercises

In later chapters, we will use calculus to assign a meaning to the slope of a planar curve at a point on that curve. You can begin to explore this concept with a graphing device. In Exercises 57–62, graph the given curve in a viewing window containing the given point P. Zoom in on point P until the graph of the curve appears to be a straight line segment. Compute the slope of the line segment (it is an approximation to the slope of the curve at P).

57. $y = x^2$, $P = (1, 1)$

58. $y = x^2$, $P = (2, 4)$

59. $y = 2x/(x^2 + 1)$, $P = (0, 0)$

60. $y = 2x/(x^2 + 1)$, $P = (1/2, 4/5)$

61. $y = 2x/(x^2 + 1)$, $P = (1, 1)$

62. $y = \sqrt{x}$, $P = (1, 1)$

63. Graph the parabola $y = \sqrt{x}$. Add the upper half of the circle centered at $P = (5/4, 0)$ with radius 1 to your graph. Notice that the parabola and the circle appear to be tangent at the point of intersection Q. What are the coordinates of the point Q? What is the equation of the line through P and Q? What is the line ℓ through Q that is perpendicular to $\overline{PQ}$? Add the plot of ℓ to your graph.

64. Graph $y = 1/(1 + x^2)$. Add the lower half of the circle centered at $P = (9/4, 3)$ with radius $5\sqrt{5}/4$ to your graph. Notice that the curve and the circle appear to be tangent at the point of intersection Q. What is the equation of the line through P and Q? What is the line ℓ through Q that is perpendicular to $\overline{PQ}$? Add the plot of ℓ to your graph.

65. Let $P_0 = (x_0, y_0) = (4, 16)$, $(x_1, y_1) = (1, 2)$, and $(x_2, y_2) = (2, 6)$. Obtain an expression for the sum of the squares of the errors, $d_1^2 + d_2^2$, associated with a line through P_0 that has slope m. With m measured along the horizontal axis and $d_1^2 + d_2^2$ measured along the vertical axis, graph $d_1^2 + d_2^2$ in the viewing window $[4.5, 5] \times [0, 1.2]$. Use your graph to indicate the slope of the regression line $\mathcal{L}$ through P_0. Now plot the sum of the absolute errors $d_1 + d_2$ in the viewing window $[3.5, 6] \times [0, 7]$. Use your plot to find the value of m that minimizes $d_1 + d_2$. Finally, on the same coordinate axes, plot the three data points, the line through P_0 that minimizes $d_1 + d_2$, and $\mathcal{L}$.

1.4 Functions and Their Graphs

Suppose S and T are sets (collections of objects). A *function* f on the set S with values in the set T is a *rule* that assigns to each element of S a *unique* element of T. We write $f : S \to T$ and say "f maps S into T." The set S is called the *domain* of f, and the set T is called the *range* of f. Notice that there are three distinct parts to a function: a domain, a range, and a rule that assigns one, and only one, value in the range to each value in the domain. For an element x in S, we write $f(x)$ (read "f at x") for the element in T that is assigned to x by the function f. We say that x is the *argument* of $f(x)$. The *image* of f is the set $\{f(x) : x \in S\}$ of all values in T that the function f assumes. It is useful to think of a function as an "input-output machine": The value x is the input, the value $f(x)$ is the output. The domain of the function is the set of input values, the image is the set of output values.

Examples of Functions of a Real Variable

The domains and ranges of the functions encountered in calculus are usually sets of real numbers; in fact, these sets will often consist of one or more intervals in $\mathbb{R}$. A function with a domain that is a set of real numbers is a *function of a real variable*. A function with a range that is a set of real numbers is a *real valued function*. Often, we use *arrow notation*, $x \mapsto f(x)$ (the symbol $\mapsto$ is read as "is mapped to"). If we refer to a function f by giving a formula for $f(x)$ without specifying the domain or range of f, then we imply that the domain of f is the largest set of real values x for which $f(x)$ makes sense *as a real number*. If it is not explicitly specified, then the range of a real-valued function f is taken to be the entire real line $\mathbb{R}$. The theorems of calculus often enable us to determine the image of f.

Example 1 Let $F(x) = \sqrt{9 - x^2}$. What is the domain of F?

Solution For $F(x)$ to make sense as a real number, the inequality

$$9 - x^2 \geq 0$$

must be satisfied. Therefore, $x^2 \leq 9$, or $-3 \leq x \leq 3$. In summary, the domain of F is the interval $[-3, 3]$. ■

Example 1 illustrates one of the most common considerations in determining the domain of a function: The argument of an even root such as $\sqrt{\ }$ or $\sqrt[4]{\ }$ must be non-negative. The example that follows illustrates another situation that must often be considered: The denominator of a fraction may not be zero.

Example 2 Consider the two functions G and H given by $G(x) = x^2 + 2x + 4$ and $H(x) = (x^3 - 8)/(x - 2)$. Are they the same function?

Solution The function G has the entire real line $\mathbb{R}$ as its domain. The function H is not defined at $x = 2$ because we may not replace x with 2 in the denominator of $(x^3 - 8)/(x - 2)$. The domain of H is $\{x \in \mathbb{R} : x \neq 2\}$. For every value x in the domain of H, the factor $(x - 2)$ is *nonzero* and may be cancelled from the numerator and denominator of $(x^3 - 8)/(x - 2)$. Thus, for x in the domain of H,

$$H(x) = \frac{x^3 - 8}{x - 2} = \frac{(x^2 + 2x + 4)(x + 2)}{(x - 2)} = x^2 + 2x + 4 = G(x).$$

In summary, the two functions G and H take the same value at each point in the domain of H, but they are not the same function because they have different domains. ■

inSIGHT

Suppose S' is a set that consists of some of but not all the points of a set S. Given a function G defined on S, we can create a function on the subset S' by assigning the output value $G(x)$ to the input value $x \in S'$. This function is called the *restriction* of G to S'. The notation $G|_{S'}$ is sometimes used to denote the restriction of G to S'. In Example 2, the domain S of G equals $\mathbb{R}$, the subset S' of S is $\{x \in \mathbb{R} : x \neq 2\}$, and H is the restriction of G to S'.

Piecewise-Defined Functions

A function is often given by different rules over different subintervals of its domain. Such a function is said to be *piecewise-defined*, or a *multicase function*.

Example 3 Schedule X from the 1995 U.S. income tax Form 1040 is reproduced here. This form helps determine the income tax $T(x)$ of a single filer with taxable income x.

Over	But not over	The tax is	Of the amount over
$0	$23350	15%	$0
23350	56550	$3502.50 + 28%	23350
56550	117950	12798.50 + 31%	56550
117950	256500	31832.50 + 36%	117950
256500		81710.50 + 39.6%	256500

Write T using mathematical notation.

Solution The function T is defined for all positive values of x, but there are five different cases. We use the following notation:

$$T(x) = \begin{cases} 0.15x & \text{if } 0 < x \leq 23350 \\ 3502.50 + 0.28(x - 23350) & \text{if } 23350 < x \leq 56550 \\ 12798.50 + 0.31(x - 56550) & \text{if } 56550 < x \leq 117950 \\ 31832.50 + 0.36(x - 117950) & \text{if } 117950 < x \leq 256500 \\ 81710.50 + 0.396(x - 256500) & \text{if } 256500 < x. \end{cases}$$

For a given value of x, there are two steps used to determine $T(x)$. First, we must determine the interval in which the value x lies. Second, we must apply the formula that corresponds to that interval. For example, a single filer with taxable income of $30,000 would observe that $x = 30000$ lies in $(23350, 56550]$ and would compute his tax to be

$$T(30000) = 3502.50 + 0.28(30000 - 23350) = \$5364.50.$$

The *absolute value function* $x \mapsto |x|$, which is a multicase function that will often be used, can be written as

$$x \mapsto \begin{cases} x & \text{if } x \geq 0 \\ -x & \text{if } x < 0 \end{cases}.$$

Another example of a piecewise-defined function that occurs frequently is the *signum,* or *sign,* function:

$$\text{signum}(x) = \begin{cases} 1 & \text{if } x > 0 \\ 0 & \text{if } x = 0 \\ -1 & \text{if } x < 0 \end{cases}.$$

Graphs of Functions

In calculus, it is often useful to represent functions visually. Pictures, or *graphs,* can help us think about functions. We graph functions in the xy-plane. For now, we will only graph functions with domains and ranges that are subsets of the real numbers. The elements of a function's domain are thought of as points of the x-axis. A function's values are measured on the y-axis.

Definition

The *graph* of f is the set of all points (x, y) in the xy-plane for which x is in the domain of f and $y = f(x)$. Thus, if f is a real-valued function with domain $S \subset \mathbb{R}$, then the graph of f is the curve $\{(x, y) : x \in S, y = f(x)\}$.

The graphs of some basic functions are shown in Figure 1. If a number x_0 is not in the domain of a function f, then there is no point of the form (x_0, y) that lies on the graph of f. In other words, the vertical line $y = x_0$ does not intersect the graph of f when x_0 is outside the domain of f. If x_0 *is* in the domain of f, then there is one and only one point of the form (x_0, y) that lies on the graph of f—namely, the unique point (x_0, y_0) with $y_0 = f(x_0)$. These considerations lead to the following geometric test, which indicates whether a given curve is the graph of a function.

Vertical Line Test

If every vertical line drawn through a curve intersects that curve only once, then the curve is the graph of a function.

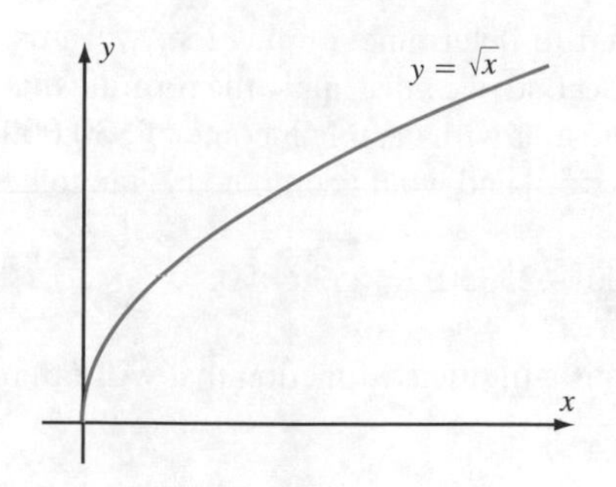

Figure 1a

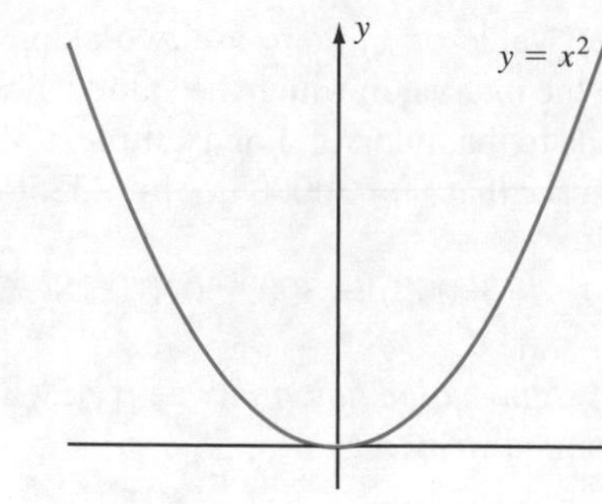

Figure 1b

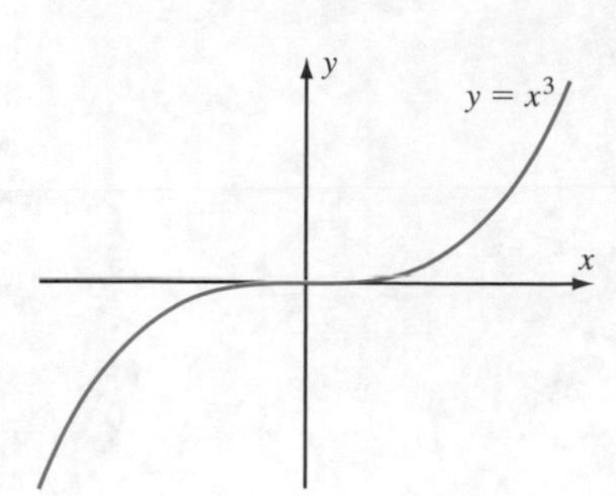

Figure 1c

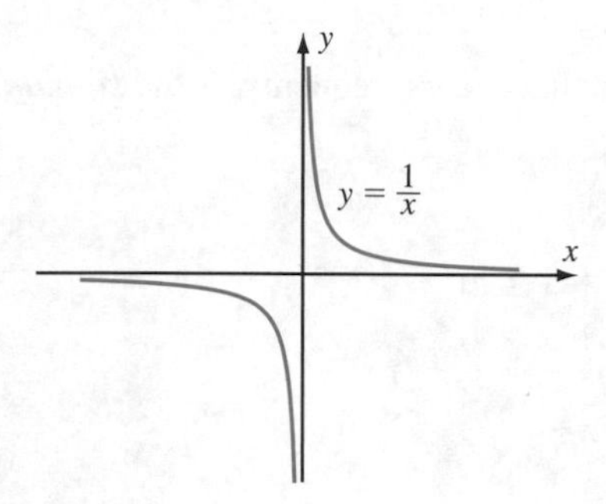

Figure 1d

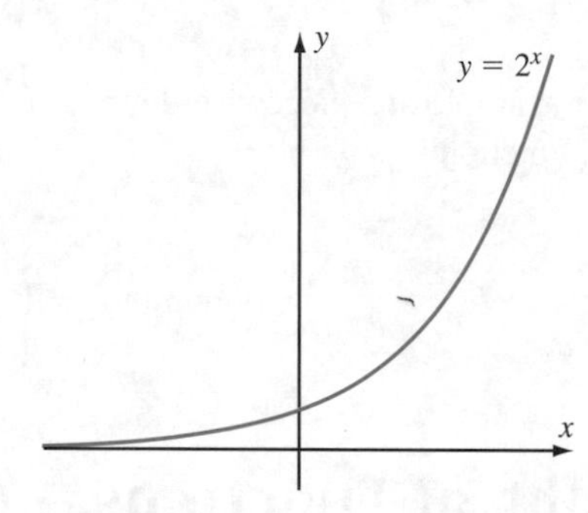

Figure 1e

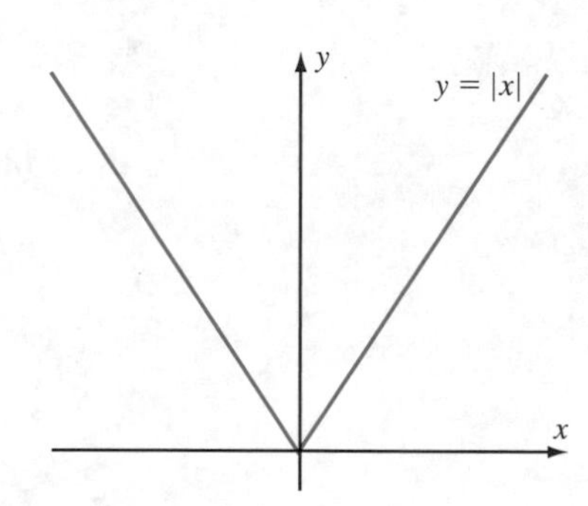

Figure 1f

Example 4 Is the circle defined by $x^2 + y^2 = 10$ the graph of a function?

Solution Observe that for $x = 3$, there are two y values, $y = 1$ and $y = -1$, for which (x, y) is on the given circle. Therefore, the curve cannot be the graph of a function. ■

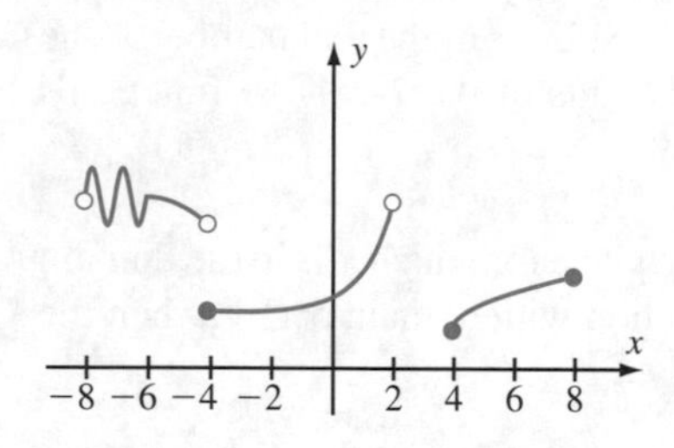

Figure 2

Example 5 Is the curve in Figure 2 the graph of a function?

Solution For each x that lies in either the open interval $(-8, 2)$ or the closed interval $[4, 8]$, there is only one corresponding y value. Thus, the curve passes the Vertical Line Test and is the graph of a function $x \mapsto f(x)$ (even though we do not know a formula for $f(x)$). The domain of f is $(-8, 2) \cup [4, 8]$. ■

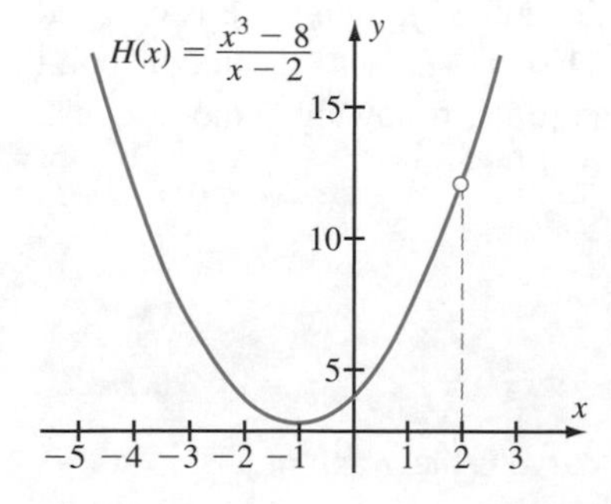

Figure 3

Chapter 4 explains some techniques for graphing functions. For now, it is a good idea to learn to recognize and sketch the graphs of basic functions (such as those plotted in Figure 1) and to use a plotting device for more complicated functions. Sometimes it is useful to add detail to a curve drawn by a plotting device. For instance, the function H from Example 2 is graphed in Figure 3. An open dot indicates that the number 2 is not in the domain of H.

Example 6 Graph the signum function, which was defined earlier in this section.

Solution The part of the graph of signum to the left of $x = 0$ lies on the horizontal line $y = -1$. The part to the right of $x = 0$ lies on the horizontal line $y = 1$. A closed

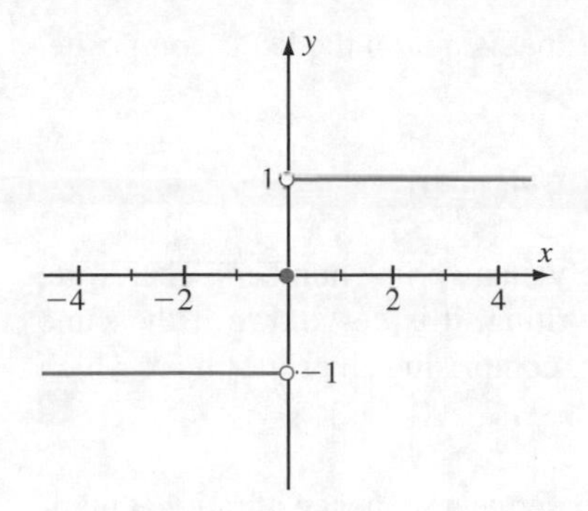

Figure 4
$y = \text{signum}(x)$

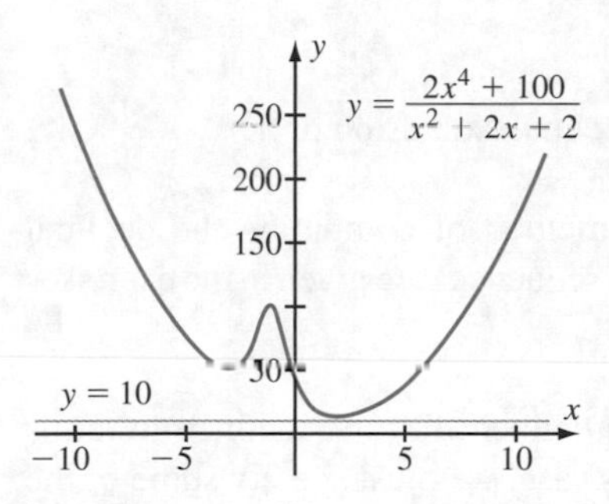

Figure 5

dot indicates that $(0, 0)$ is on the graph of signum. Because $(0, -1)$ and $(0, 1)$ are not on the graph of signum, we use open dots (see Figure 4). ■

Example 7 One popular graphing calculator uses the rectangle $[-10, 10] \times [-10, 10]$ as its default viewing window. Explain why no parts of the graphs of $f(x) = \sqrt{x^2 - x - 132}$ and $g(x) = (2x^4 + 100)/(x^2 + 2x + 2)$ appear in this viewing window.

Solution The function f may be written as

$$f(x) = \sqrt{x^2 - x - 132} = \sqrt{(x + 11)(x - 12)}.$$

For x to be in the domain of f, $x + 11$ and $x - 12$ must have the same sign. This happens only for $x \leq -11$ (both factors nonpositive) and for $x \geq 12$ (both factors nonnegative). No part of the graph of f will appear in any viewing window of the form $[-10, 10] \times [y_{\min}, y_{\max}]$ because $[-10, 10]$ is not in the domain of f.

The problem with the graph of g is different. By completing the square in the denominator, we see that $g(x) = (2x^4 + 100)/((x + 1)^2 + 1)$. The denominator is greater than or equal to 1, hence it is never 0. The domain of g is, therefore, the set of all real numbers. In particular, $[-10, 10]$ is in the domain of g, and we can be sure that some viewing rectangle of the form $[-10, 10] \times [y_{\min}, y_{\max}]$ will contain part of the graph of g. Indeed, Figure 5 shows the graph of g in the viewing window $[-10, 10] \times [0, 250]$. We see that $g(x) > 10$ for every value of x in $[-10, 10]$. In other words, the rectangle $[-10, 10] \times [-10, 10]$ lies entirely below the graph of g. ■

Sequences

A function f with a domain that is the set $\mathbb{Z}^+$ of positive integers is called an *infinite sequence*. In general, a function f is a *sequence* if its domain is a finite or infinite set of consecutive integers. We can write a sequence f by *listing* its values:

$$f(1),\ f(2),\ f3), \ldots.$$

Thus, a sequence f can be thought of as a list, with $f(1)$ being the first element of the list, $f(2)$ the second element, and so on. It is common to abandon the function notation altogether and write the sequence as

$$f_1,\ f_2,\ f_3, \ldots,$$

or sometimes as $\{f_n\}_{n=1}^{\infty}$, or even as $\{f_n\}$. We refer to f_n as the nth *term* of the sequence f. Occasionally it is useful to begin a sequence with an index different from 1. An example is $\{3n - 5\}_{n=4}^{\infty}$, which denotes the sequence 7, 10, 13, 16, ...

Example 8 A positive integer that is greater than 1 is *composite* if it is the product of two integers greater than 1. The integers greater than 1 that are not composite are *prime*. Let $\{p_n\}$ be the sequence of prime numbers, ordered by increasing size. List the first 18 terms of this sequence.

Solution We can verify that each integer in the list

$$2, 3, 5, 7, 11, 13, 17, 19, 23, 29, 31, 37, 41, 43, 47, 53, 59, 61, \ldots$$

is prime and that each positive integer less than 62 that is not on the list is composite. Thus,

$$p_1 = 2, p_2 = 3, p_3 = 5, \ldots, p_{18} = 61.$$

According to a theorem of Euclid, there are infinitely many prime numbers. Therefore, $\{p_n\}$ is an infinite sequence. At the time of this writing, it is not known if the same holds true for the sequence 3, 5, 11, 17, 29, 41, ..., comprising all primes p for which $p + 2$ is also prime. ■

Example 9 Use the decimal expansion of π to define a sequence of rational numbers $\{q_n\}$ that become progressively closer to π.

Solution We can define such a sequence $\{q_n\}$ by $q_1 = 3.1$, $q_2 = 3.14$, $q_3 = 3.141$, $q_4 = 3.1415$, $q_5 = 3.14159$, and so on. Thus,

$$q_n = \text{ the first } (n+1) \text{ digits of the decimal expansion of } \pi.$$

This is not a constructive definition because no method of computing the decimal expansion of π has been given. Several constructive sequences are given in the exercises for this section. ■

A sequence $\{s_n\}$ is *inductively* (or *recursively*) defined if s_{n+1} is defined in terms of some or all of its predecessors $s_1, \ldots, s_n$. In this case, we must specify some of the first few values of the sequence $\{s_n\}$. This is known as an *initialization*.

Example 10 The Fibonacci sequence f_n is defined by $f_{n+2} = f_{n+1} + f_n$ for $n \geq 1$. The values f_1 and f_2 are both initialized to be 1. What is f_7?

Solution We calculate:

$$\begin{aligned} f_3 &= f_2 + f_1 = 1 + 1 = 2 \\ f_4 &= f_3 + f_2 = 2 + 1 = 3 \\ f_5 &= f_4 + f_3 = 3 + 2 = 5 \\ f_6 &= f_5 + f_4 = 5 + 3 = 8 \\ f_7 &= f_6 + f_5 = 8 + 5 = 13. \end{aligned}$$

■

Functions from Data

In practice, many functional relationships are not initially discovered in the form of mathematical expressions. For example, suppose that several observations or measurements of two variables x and y are made and that y depends on x by means of an undetermined function f. One way to find an expression for $f(x)$ is to plot a sequence $\{(x_n, y_n)\}$ of the observed values. The result is known as a *scatter plot,* or a *scatter diagram.* Comparison of the scatter plot with known graphs can suggest a suitable formula for f. This formula is then said to be a *mathematical model* of the observations.

Example 11 In 1929, the American astronomer Edwin Hubble made the startling announcement that the universe is expanding. Hubble measured the distances and recession velocities of several galaxies relative to a fixed galaxy. (*Recession velocity* is the speed at which two galaxies move away from each other.) Some of Hubble's data is tabulated here.

	Virgo	Pegasus	Perseus	Ursa Major 1	Leo	Gemini	Bootes	Hydra	Coma Berenices
d	22	68	108	255	315	405	685	1100	137
v	0.75	2.40	3.20	9.30	12.00	14.40	24.50	38.00	

In the table, d represents distance in millions of light years and v represents velocity in thousands of miles per second. Find a model for v as a function of d. What is the recession velocity of Coma Berenices?

Solution Figure 6 shows a plot of the tabulated data. Since d is the *independent* variable and v the *dependent* variable, we refer to the horizontal axis as the d-axis, the vertical axis as the v-axis, and the plane they determine as the dv-plane. Notice that the plotted points lie near a line that passes through the origin of the dv-plane. Let H denote the slope of the least squares line through the origin that is fitted to the tabulated observations. According to equation (1.3) from Section 1.3,

$$\begin{aligned} H &= \frac{d_1 v_1 + d_2 v_2 + \cdots + d_8 v_8}{d_1^2 + d_2^2 + \cdots + d_8^2} \\ &= \frac{(22)(0.75) + (68)(2.4) + \cdots + (1100)(38.0)}{22^2 + 68^2 + \cdots + 1100^2} = 3.512 \times 10^{-2}. \end{aligned}$$

Thus, $v = f(d)$ where $f(d) = H \cdot d = 3.512 \times 10^{-2} \cdot d$. The graph of f is given in Figure 7, with Hubble's observations superimposed.

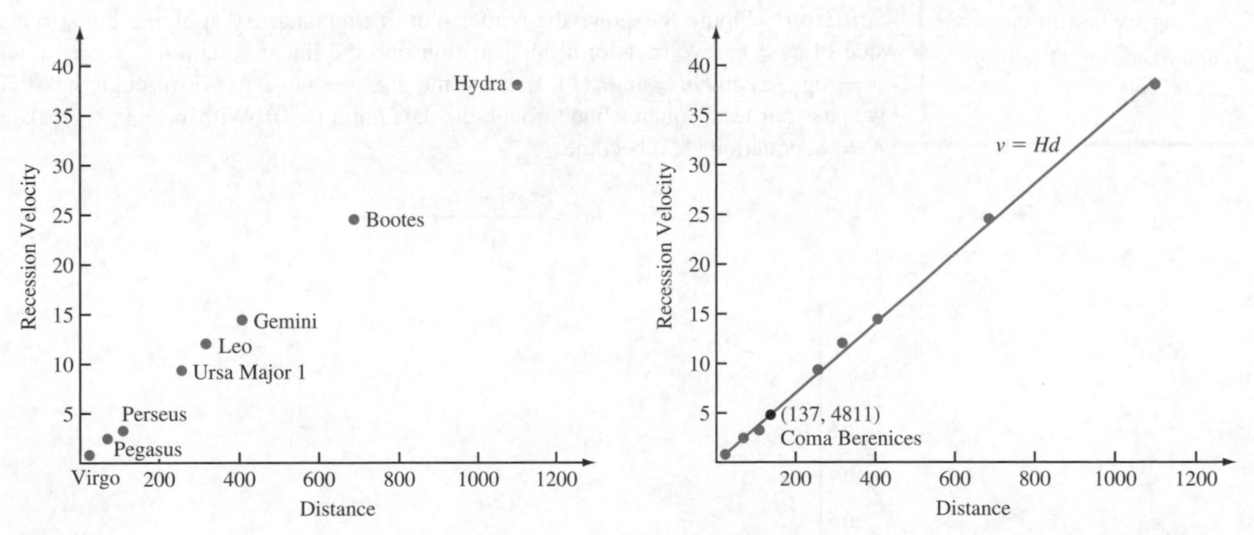

Figure 6

Figure 7

Because d is 137 for Coma Berenices, we calculate that the recession velocity is $f(137) = 3.512 \times 10^{-2} \cdot 137 = 4.811$, or, remembering the units of v, 4811 mi/s. In fact, Hubble measured the recession velocity of Coma Berenices to be 4700 mi/s. ■

The constant H is known as *Hubble's constant*. A dimension analysis shows that $1/H$ represents a time because

$$v = \frac{\text{distance}}{\text{time}},$$
$$v = H \cdot d, \quad \text{and}$$
$$d = \text{distance}.$$

In the Big Bang theory of the universe, $1/H$ represents the age of the universe. Refinements in astronomical measurements have led to corrections to Hubble's data and to revised estimates of H. Nevertheless, the age of the universe remains a matter of debate among astronomers. Data obtained from the Hipparcos Project in May 1997 suggest that the universe is approximately 11×10^9 years old.

The Method of Least Squares can often be applied in instances when a scatter plot is not linear. For example, suppose $z = A \cdot k^x$, where A is a known constant and k is a fixed base that is to be determined. We apply $\log_{10}$ to each side of $z = A \cdot k^x$:

$$\begin{aligned} \log_{10}(z) &= \log_{10}(A \cdot k^x) \\ &= \log_{10}(A) + \log_{10}(k^x) && \text{Using the logarithm law, } \log_{10}(u \cdot v) = \log_{10}(u) + \log_{10}(v) \\ &= \log_{10}(A) + x \cdot \log_{10}(k). && \text{Using the logarithm law, } \log_{10}(u^v) = v \cdot \log_{10}(u) \end{aligned}$$

If we set $y = \log_{10}(z)$, $m = \log_{10}(k)$, and $b = \log_{10}(A)$, then we have the linear relationship

$$y = m \cdot x + b$$

in the variables x and y.

Example 12 A radioactive isotope of gold is used to diagnose arthritis. Let $f(x)$ denote the blood serum gold at time x in days as a fraction of the initial dose at time $x = 0$. Measured values of $f(x)$ for $x = 1, 2, \ldots, 8$ are listed in the table.

Days	0	1	2	3	4	5	6	7	8
Blood Serum Gold	1.0	0.91	0.77	0.66	0.56	0.49	0.43	0.38	0.34

Find a positive constant k such that k^x is a model for $f(x)$. Use the model to predict the fraction of the initial dose that remains in the blood serum after ten days.

Solution Figure 8a shows the scatter plot of this data. By applying $\log_{10}$ to each side of $z = k^x$, we transform this equation into the linear equation $y = m \cdot x$ with $y = \log_{10}(z)$ and $m = \log_{10}(k)$. Because the line $y = m \cdot x$ passes through the origin, we pass our least squares line through the data point (0, 0). With $x_0 = 0$, $y_0 = 0$, and $N = 8$, equation (1.2) becomes

$$m = \frac{x_1 y_1 + x_2 y_2 + \cdots + x_8 y_8}{x_1^2 + x_2^2 + \cdots + x_8^2}.$$

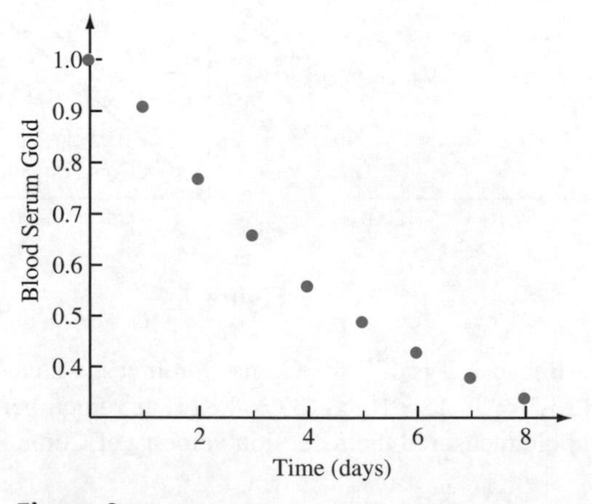

Figure 8a

The values that we need for this formula can be found in the following table.

	x	z	y	xy	x^2
	1	0.91	−0.0409	−0.0409	1
	2	0.77	−0.1135	−0.2270	4
	3	0.66	−0.1804	−0.5413	9
	4	0.56	−0.2518	−1.0072	16
	5	0.49	−0.3098	−1.5490	25
	6	0.43	−0.3665	−2.1991	36
	7	0.38	−0.4202	−2.9415	49
	8	0.34	−0.4685	−3.7481	64
Sum				−12.2545	204

Our estimate of m is $-12.2545/204$, or $m = -0.06$.

Figure 8b shows the transformed data points, along with the least squares line $y = -0.06x$. Our estimate for m provides an estimate for the base k, namely $k = 10^m = 10^{-0.06} \approx 0.87$. In Figure 8c, the graph of $y = 0.87^x$ is superimposed on the original scatter plot. The predicted value of the serum gold concentration ten days after the initial dose is $f(10) = 0.87^{10} \approx 0.25$.

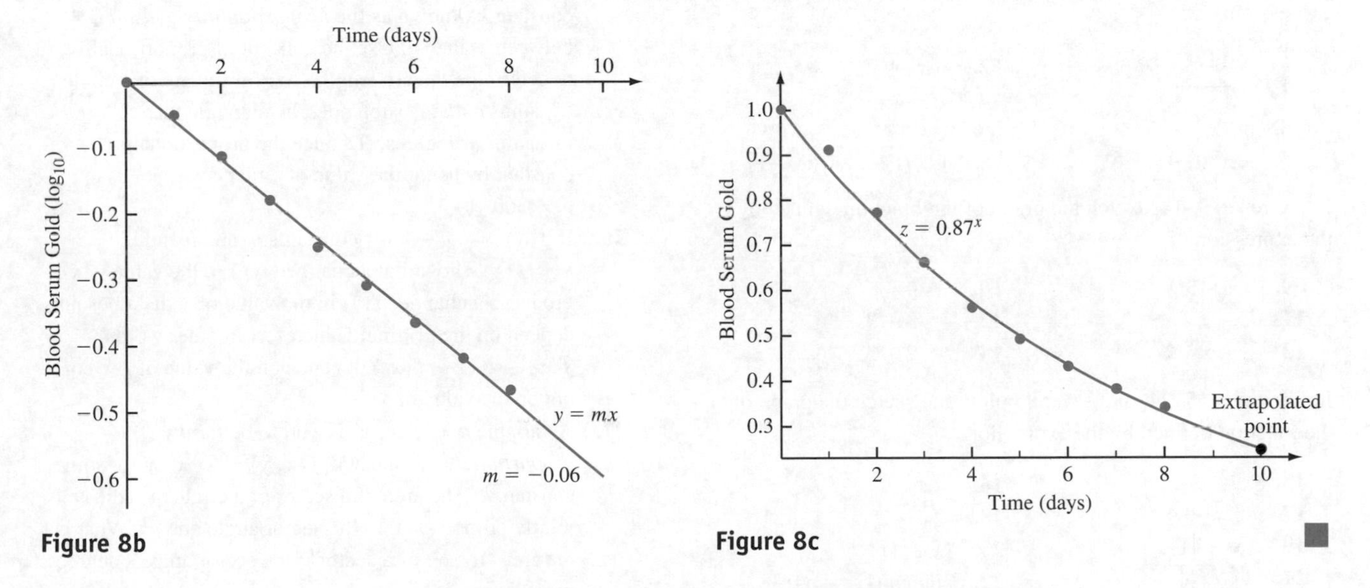

Figure 8b

Figure 8c

Functions of Several Variables

Many variables depend on several other variables. For example, the volume V of a right circular cone of height h and base radius r depends on both h and r. We say that V is a function of h and r, and we write $V(h, r)$. In this case, we can describe V by the rule

$$V(h, r) = \frac{1}{3}\pi r^2 h.$$

Similar notation is used for functions of three or more variables. Thus, the volume V of a rectangular box is a function of its height h, width w, and depth d:

$$V(h, w, d) = hwd.$$

Although we will occasionally encounter multivariable functions in the next several chapters, we will not study the calculus of multivariable functions until Chapter 13.

quickquiz

1. How does the domain of a function appear in its graph?
2. How does the image of a function appear in its graph?
3. How can we tell when a curve in the plane is the graph of a function?
4. Define the sequence of positive powers of 2 inductively.

EXERCISES

Problems for Practice

In Exercises 1–8, state the domain of the function defined by the expression.

1. $x/(1+x)$
2. $\sqrt{x^2+2}$
3. $\sqrt{x^2-2}$
4. $\sqrt{2-x^2}$
5. $1/(x^2-1)$
6. $\sqrt{x}/(x^2+x-6)$
7. $\sqrt{x^2-4x+5}$
8. $1/\sqrt{(x^2-4)(x-1)}$

In Exercises 9–14, sketch the graph of the function defined by the expression.

9. x^2+1
10. x^2-1
11. $1-x^2$
12. $x^2/3+3$
13. $3-x^2/2$
14. $x^2/2-3$

In Exercises 15–24, plot several points and sketch the graph of the function defined by the expression.

15. x^{-2}
16. $\sqrt{x^2}$
17. $\sqrt{2x+4}$
18. $\sqrt{x-2}$
19. $(x-4)^{-1/2}$
20. $(x+1)^{-3}$
21. $1/\sqrt{x+1}$
22. $\text{signum}(|x^2-x|)$
23. $\begin{cases} x^2 & \text{if } x < 1 \\ 2-x^2 & \text{if } x \geq 1 \end{cases}$
24. $\begin{cases} x^2-4 & \text{if } x < -2 \\ x+2 & \text{if } -2 \leq x < 2 \\ x^2 & \text{if } 2 \leq x \end{cases}$

Further Theory and Practice

25. We say that y *is proportional to* x if $y = kx$ for some constant k (known as the *proportionality constant* between x and y). Use the concept of proportionality to determine the arc length $s(\alpha)$ for the arc of a circle of radius r that is subtended by an angle that measures α degrees. (Deduce the proportionality constant by using the value of s that corresponds to $\alpha = 360°$.)

26. If $f(x) = mx + b$ for constants m and b and if $y = f(x)$, show that a change (Δx) in the value of x produces a change (Δy) in the value of y that does not depend on the original value of x. In other words, $f(x_0 + \Delta x) - f(x_0)$ depends on the value of Δx but not on the value of x_0.

27. A variable $u = f(x, y)$ is said to be *jointly proportional* in x and y if $f(x, y) = k \cdot x \cdot y$ for some constant k. The area of a sector of a circle of radius r is jointly proportional to the sector angle and r^2. What is the area $A(r, \alpha)$ of a sector if the sector angle's degree measure is α?

28. Calculate the area $A(\ell, \alpha)$ of an isosceles triangle having two sides of length ℓ enclosing an angle α.

29. Let r and s be the roots of $x^2 + Ax + B$. Express the coefficients A and B as functions of r and s.

30. Sketch the graph of the tax function T for $0 < x \leq 100000$ (see Example 3).

31. Figure 9, next page, shows the graph of a function f. Give a formula for f.

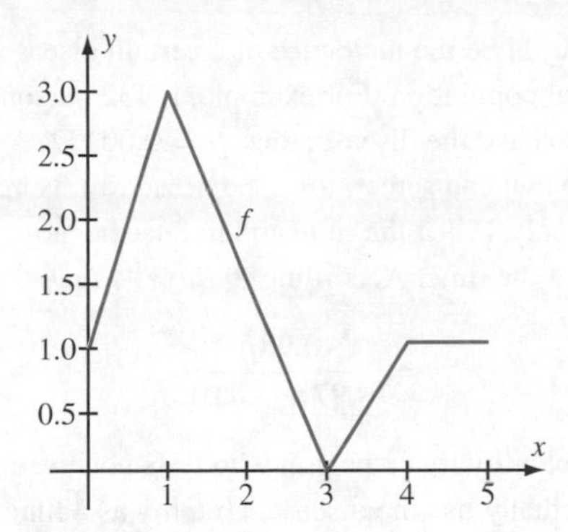

Figure 9

32. The domain of the function f that is graphed in Figure 9 is $[0, 5]$. For every $x \in [0, 5]$, let $m(x)$ be the slope of the graph of f at $(x, f(x))$, provided the slope exists. Otherwise $m(x)$ is not defined. What is the domain of m? Sketch the graph of the slope function m. For $x \in [0, 5]$, let $A(x)$ be the area under the graph of f and over the interval $[0, x]$. Sketch the graph of the area function A.

33. Let T denote the tax function described in Example 3. In this exercise, we will restrict the domain of T to $0 < x \leq 117950$. For each positive $x \notin \{23350, 56550, 117950\}$, let $m(x)$ denote the slope of the graph of T at the point $(x, T(x))$. Let $A(x)$ be the area under the graph of m (not T) and over the interval $(0, x]$. Complete the template.

$$m(x) = \begin{cases} ? & \text{if } ? \\ ? & \text{if } ?. \\ ? & \text{if } ? \end{cases}$$

Describe the function m for $0 < x < 117950$. Repeat for the function A. What is the relationship of A to T?

In Exercises 34–40, give a recursive definition of the sequence.

34. $f_n = 2n, n = 1, 2, 3, \ldots$

35. $f_n = 2^n, n = 1, 2, 3, \ldots$

36. $f_n = 1 + 2 + \cdots + n, n = 1, 2, 3, \ldots$

37. $f_n = n!, n = 1, 2, 3, \ldots$ (The product of the consecutive integers $1, 2, 3, \ldots, n$ is denoted $n!$)

38. $f_n = 2^{((-1)^n)}, n = 1, 2, 3, \ldots$

39. $f_n = n(n + 1)/2, n = 1, 2, 3, \ldots$

40. $\{1, 5, 17, 53, 161, \ldots\}$

41. For any real number x, the *greatest integer in* x is denoted by $\lfloor x \rfloor$ and defined to be the unique integer satisfying $\lfloor x \rfloor \leq x < \lfloor x \rfloor + 1$. For example, $\lfloor 3.2 \rfloor = 3$ and $\lfloor -3.2 \rfloor = -4$. Notice that $\lfloor x \rfloor = x$ if and only if x is an integer. The function $x \mapsto \lfloor x \rfloor$ is called the *greatest integer function*. (The expression $\lfloor x \rfloor$ is sometimes read as "the floor of x.") The *integer part* of x, denoted by $\text{Int}(x)$, is that part of the decimal expansion of x to the left of the decimal point. Express $\lfloor x \rfloor$ in terms of $\text{Int}(x)$. Graph $x \mapsto \lfloor x \rfloor$ and $x \mapsto \text{Int}(x)$ for $-3 \leq x \leq 3$.

42. A kilowatt-hour is the amount of energy consumed in 1 hour at the constant rate of 1000 watts. At time $t = 0$ hour, a three-way lamp is turned on at the 50-watt setting. An hour later, the lamp is turned up to 100 watts. Forty minutes after that, the lamp is turned up to 150 watts. Ninety minutes later, it is turned off. Let $E(t)$ be the (cumulative) energy consumption in kilowatt-hours expressed as a function of time measured in hours. Graph $E(t)$ for $0 \leq t \leq 4$.

43. A loan of P dollars is paid back by means of monthly installments of m dollars for n full years. Find a formula for the function $I(P, m, n)$ that gives the total amount of interest paid over the life of the loan.

44. Initially, a solution of salt water has total mass 500 kg and is 99% water. Over time, the water evaporates. Just for fun, estimate to within 25 kg the mass of the solution at the time when water makes up 98% of the solution. More precisely, find a formula for the total mass $m(p)$ of the solution as a function of p, the percentage contribution of water to the total mass. Exactly what is $m(98)$? Graph $m(p)$.

45. Recursively define two sequences $\{c_n\}_{n=1}^{\infty}$ and $\{C_n\}_{n=1}^{\infty}$ by initializing $c_0 = C_0 = 1$ and, for each nonnegative integer n, setting

$$c_{n+1} = c_0 c_n + c_1 c_{n-1} + \cdots + c_{n-1} c_1 + c_n c_0$$

and

$$C_{n+1} = \frac{4n+2}{n+2} C_n.$$

Compute c_n and C_n for $0 \leq n \leq 8$. As you might guess from the computed values, the two sequences are actually the same. However, the easier formula is not always the easiest to use. Prove that each *Catalan number* (as these numbers are called) is an integer.

A *Boolean function* is a function with a range that is a two-element set. Typically we use $\{0,1\}$ for the range. A statement $P(x)$ that can be either true or false can be regarded as a Boolean function. We make such a function numerical by using 1 for true and 0 for false. Thus, the function $x \mapsto (x^2 < x)$ has the value 1 for $-1 < x < 1$ and 0 for all other values of x. Boolean functions can be convenient tools for expressing (and sometimes graphing) multicase functions. For instance, we can use Boolean functions to describe the function

$$F(x) = \begin{cases} -x+3 & \text{if } x < 0 \\ 3 & \text{if } 0 < x < 2 \\ x+1 & \text{if } x \geq 2 \end{cases}$$

by one formula:

$$F(x) = -x \cdot (x < 0) + 3(0 < x)(x < 2) + (x+1) \cdot (x \geq 2).$$

In Exercises 46–49, express the multicase function by means of Boolean functions. Plot the function.

46. $x \mapsto |x|$

47. $H(x) = \begin{cases} 1 & \text{if } x \geq 0 \\ 0 & \text{if } x < 0 \end{cases}$ The *Heaviside function.*

48. $R(x) = \begin{cases} 1 & \text{if } x > 1 \\ x & \text{if } 0 < x \leq 1 \\ 0 & \text{if } x \leq 0 \end{cases}$ The ramp function.

49. $b(x) = \begin{cases} 0 & \text{if } x \leq 0 \\ x & \text{if } 0 < x \leq 1 \\ 2-x & \text{if } 1 < x \leq 2 \\ 0 & \text{if } 2 < x \end{cases}$ The triangular bump function.

Calculator/Computer Exercises

50. In the table, x represents cigarette consumption per adult (in hundreds) for 1962 for eight countries. The variable y represents mortality per 100,000 due to heart disease in 1962.

	x	y
Australia	32.20	238.1
Belgium	17.00	118.1
Canada	33.50	211.6
Great Britain	27.90	194.1
Ireland	27.70	187.3
Netherlands	18.14	124.7
United States	39.00	256.9
West Germany	18.90	150.3

Plot the eight points (x, y). Find a function $f(x) = mx + 12.5$ that could be used to model the relationship between cigarette consumption and mortality due to heart disease.

51. Let $p \in [0, 1]$ be the incidence of a certain disease in the general population. For example, if 132 persons out of 100,000 have the disease, then $p = 0.00132$. A procedure that can screen for the disease results in false positives only 1% of the time and in false negatives only 2% of the time. According to *Bayes's law,*

$$y = \frac{0.98p}{0.97p + 0.01}$$

is the probability that a person who tests positive for the disease actually has the disease. Graph y as a function of p in an appropriate window. What does this graph tell you about the value of *routine* screenings of the general population?

52. The *astronomical unit* (AU) is the mean distance of Earth to the sun. The table gives the mean distances (D) in astronomical units between the sun and the six planets that were known at the time of Johannes Kepler. The table also provides the planets' periods of orbit about the sun (T) in years. Plot the six points (D, T). In the window $[0, 10] \times [0, 30]$, graph the function $f_q(x) = x^q$ for each $q = m/n$ with $1 \leq m$ and $n \leq 3$. For what value of q does the graph of f_q fit the data well? Notice that $T < D$ for $D < 1$ and $T > D$ for $D > 1$. Explain why this indicates that $q > 1$. Which points show that $q < 2$?

	T	D
Mercury	0.240	0.387
Venus	0.615	0.723
Earth	1.000	1.000
Mars	1.880	1.524
Jupiter	11.860	5.203
Saturn	29.547	9.539

53. In 1615, Kepler became interested in the shapes of wine casks and, in particular, which dimensions resulted in maximum volume. He noticed that as a variable approached a maximum, its *increment* became negligible. In this exercise, we use the function $f(x) = 5 - 24x^2 + 20x^3 - 3x^4$ to investigate Kepler's observation. Graph this function in the window $[2, 5] \times [20, 135]$. Using the graph, determine the value x_0 of x for which $f(x)$ is maximized.
Graphical methods had not been introduced in Kepler's time, so he relied on numerical computation. In doing so, he noticed that the increment of a variable near its

maximum was quite different from its increment elsewhere. Let h be a very small positive number. Set

$$F(h, x) = f(x) - f(x - h).$$

This *backward difference* of f records the increment at x by which f changes as its argument decreases by h. Choose a value x_1 in the interval $(2, x_0)$. For $h = 10^{-4}$, 10^{-5}, 10^{-6}, and 10^{-8}, tabulate and compare the values $F(h, x_0)$ and $F(h, x_1)$. How do the increments compare in size to h? Describe the difference between the increments at x_0 and x_1.

54. First Algorithm for Square Root Extraction In this exercise, we compute the first several terms of a sequence $\{m_n\}$ of rational numbers that has $\sqrt{2}$ as its limit. Let I_1 be the interval $[1, 2]$. It contains $\sqrt{2}$. Let $m_1 = 3/2$ be the midpoint of I_1 so that m_1 divides I_1 into two subintervals. Let I_2 be the subinterval that contains $\sqrt{2}$. Let m_2 be the midpoint of I_2 so that m_2 divides I_2 into two subintervals. Let I_3 be the subinterval that contains $\sqrt{2}$. Continue this process. Calculate m_n for $1 < n < 8$. For each n, what is the length of the interval I_n? Without calculating $\sqrt{2}$, what is the maximum error that can result if m_n is used to approximate $\sqrt{2}$?

55. Second Algorithm for Square Root Extraction In this exercise, we investigate a 4000-year-old method for approximating the square root of a positive number η. For simplicity, assume that η lies in the interval $(1, 4)$. Set $x_0 = 3/2$. For $n \geq 0$, define

$$x_{n+1} = \frac{1}{2}\left(x_n + \frac{\eta}{x_n}\right).$$

The sequence $\{x_n\}$ has $\sqrt{\eta}$ as its limit.

a. Use the inductive definition of x_{n+1} to obtain the identity $|x_{n+1} - \sqrt{\eta}| = (x_n - \sqrt{\eta})^2/2x_n$.

b. If x_n gives $\sqrt{\eta}$ to k decimal places, use part a to deduce that x_{n+1} gives $\sqrt{\eta}$ to $2k$ decimal places.

c. Use the algorithm from part b to compute $\sqrt{3.75}$ to six decimal places.

56. Third Algorithm for Square Root Extraction

a. If $d^2 = 2s^2$, verify that $2(s + d)^2 = (2s + d)^2$.

b. Let a first square have side length s_1 and diagonal length d_1. Let a second square have side length $s_2 = s_1 + d_1$. Show that the diagonal length d_2 of the second square is $d_2 = 2s_1 + d_1$.

c. Let this process be continued indefinitely. Thus, if an nth square has side and diagonal lengths s_n and d_n, respectively, construct an $(n + 1)$th square with side and diagonal lengths $s_{n+1} = s_n + d_n$ and $d_{n+1} = 2s_n + d_n$. Show that $d_n^2 - 2s_n^s = (-1)^n$.

d. Let $s_1 = 1$. Note that $d_1 = \sqrt{2}$. What would happen if we did not know how to approximate $\sqrt{2}$? We could use an ancient Greek algorithm. Let $S_1 = D_1 = 1$. Let $S_{n+1} = S_n + D_n$, and $D_{n+1} = 2S_n + D_n$ for $n \geq 1$. These are the recursion formulas from part c. Let $r_n = D_n/S_n$. Tabulate r_1 through r_8. Compare r_8 with $\sqrt{2}$. Use the equation from part c to explain why r_n is a good rational approximation to $\sqrt{2}$ for large n.

57. Viète's Algorithm for π Let $q_1 = 1/\sqrt{2}$. For $n \geq 2$, let $q_n = q_1(1 + q_{n-1})^{1/2}$. Let $Q_n = q_1 q_2 \cdot \ldots \cdot q_n$ and $p_n = 2/Q_n$. Calculate p_n for $1 \leq n \leq 8$.

58. Arithmetic-Geometric Mean Method for π This method, which is a very fast procedure for calculating the digits of π, presupposes that a fast method is already available for calculating the digits of $\sqrt{2}$. Suppose that $a_0 = \sqrt{2}$, $b_0 = 0$, and $p_0 = 2 + \sqrt{2}$ have been initialized. For $n \geq 0$, let

$$a_{n+1} = \frac{1}{2}\left(\sqrt{a_n} + \frac{1}{\sqrt{a_n}}\right)$$

$$b_{n+1} = \sqrt{a_n} \cdot \frac{(1 + b_n)}{(a_n + b_n)}$$

$$p_{n+1} = p_n \cdot b_{n+1} \cdot \frac{(a_{n+1} + 1)}{(1 + b_{n+1})}.$$

Calculate p_n for $n = 1, 2,$ and 3.

1.5 Combining Functions

Most of the functions that we encounter are built up from simple functions. This building is accomplished using arithmetic and other operations. Although these two statements may seem perfectly obvious, it is worth noting explicitly how some of these operations work. The functions that we will be considering from now until Chapter 12 will have both domain and range that are subsets of the real number system. We will not always state this fact explicitly.

Arithmetic Operations

Let c be a constant. Suppose that f and g are functions with the same domain S. For $s \in S$,

$$\begin{array}{ll} (f+g)(s) & \text{means } f(s)+g(s); \\ (f-g)(s) & \text{means } f(s)-g(s); \\ (f\cdot g)(s) & \text{means } f(s)\cdot g(s); \\ (cf)(s) & \text{means } c\cdot f(s); \text{ and} \\ (f/g)(s) & \text{means } f(s)/g(s), \text{ as long as } g(s)\neq 0. \end{array}$$

inSIGHT

If f and g are functions, then $f+g$, $f-g$, $f\cdot g$, and f/g are new functions. The rules defining the new functions are expressed in terms of the original functions f and g. The next two examples show how these simple ideas are put into practice.

Example 1 Let $f(x) = 2x$ and $g(x) = x^3$. Calculate $f+g$, $f-g$, $f\cdot g$, and f/g.

Solution We calculate that

$$\begin{aligned} (f+g)(x) &= f(x)+g(x) = 2x+x^3; \\ (f-g)(x) &= f(x)-g(x) = 2x-x^3; \\ (f\cdot g)(x) &= f(x)\cdot g(x) = 2x\cdot x^3 = 2x^4; \text{ and} \\ \left(\frac{f}{g}\right)(x) &= \frac{f(x)}{g(x)} = \frac{2x}{x^3} = \frac{2}{x^2}, \text{ as long as } x\neq 0. \end{aligned}$$ ■

Given a function f, we let f^2 denote the function $f\cdot f$. That is, $f^2(x) = f(x)^2$. We similarly define f^k for other positive integers k. For instance, if $f(x) = 2x-5$, then $f^4(x) = (2x-5)^4$.

Example 2 Let $f(x) = x^2$ and $g(x) = x+2$. Compute

$$\left(\frac{f+(2g)\cdot f}{g^2}\right)(3).$$

Solution To perform this calculation, we apply the rules to write

$$\left(\frac{f+(2g)\cdot f}{g^2}\right)(3) = \frac{f(3)+2\cdot g(3)\cdot f(3)}{g(3)^2} = \frac{9+2\cdot 5\cdot 9}{5^2} = \frac{99}{25}.$$ ■

Polynomial Functions

A polynomial function p is a function of the form

$$x \mapsto p(x) = a_N x^N + a_{N-1}x^{N-1} + \cdots + a_1 x + a_0$$

where N is a nonnegative integer and a_N does not equal zero. We say that N is the *degree* of the polynomial $p(x)$ and that a_N is the *leading coefficient*. According to this definition, if c is a nonzero constant, then the constant function $q(x) = c$ is a degree zero polynomial. A degree one polynomial p has the form $p(x) = a_1 x + a_0$ and is sometimes called a *linear polynomial* because its graph is a line. Notice that all polynomials are built from the constant functions and the identity function $x \mapsto x$ by means of the arithmetic operations.

If $p(r) = 0$, then we say that p *vanishes* at r and that r is a *root*, or a *zero*, of p. If r is a root of a polynomial p, then there is a polynomial q of degree one less than that of p such that $p(x) = (x - r) \cdot q(x)$. The polynomial q may be found by division. The Fundamental Theorem of Algebra asserts that every polynomial with real-valued coefficients can be factored into a product of polynomials with real-valued coefficients that have degree one or two. A degree two polynomial $ax^2 + bx + c$ that cannot be factored as a product of two degree one polynomials with real coefficients is said to be *irreducible*. These are the quadratic polynomials with roots

$$\frac{-b + \sqrt{b^2 - 4ac}}{2a} \quad \text{and} \quad \frac{-b - \sqrt{b^2 - 4ac}}{2a},$$

which have nonzero imaginary parts. Such nonreal roots occur if and only if $b^2 < 4ac$.

Example 3 Factor the polynomial $x^3 + 3x^2 + 7x + 10$ as a product of polynomials with real coefficients.

Solution Because the degree of the given polynomial is greater than two, it can be factored. A plot of the equation $y = x^3 + 3x^2 + 7x + 10$ reveals that -2 is a root, as shown in Figure 1. It follows that $x - (-2) = x + 2$ is a factor. Division of $x + 2$ into $x^3 + 3x^2 + 7x + 10$ yields another factor, $x^2 + x + 5$:

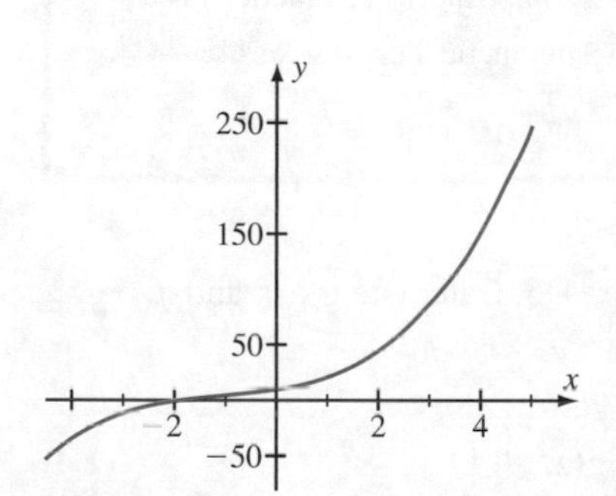

Figure 1
$y = x^3 + 3x^2 + 7x + 10$

$$\begin{array}{r}
x^2 + x + 5 \\
x + 2 \,\big)\, \overline{x^3 + 3x^2 + 7x + 10} \\
\underline{x^3 + 2x^2 } \\
x^2 + 7x \\
\underline{x^2 + 2x } \\
5x + 10 \\
\underline{5x + 10} \\
0
\end{array}$$

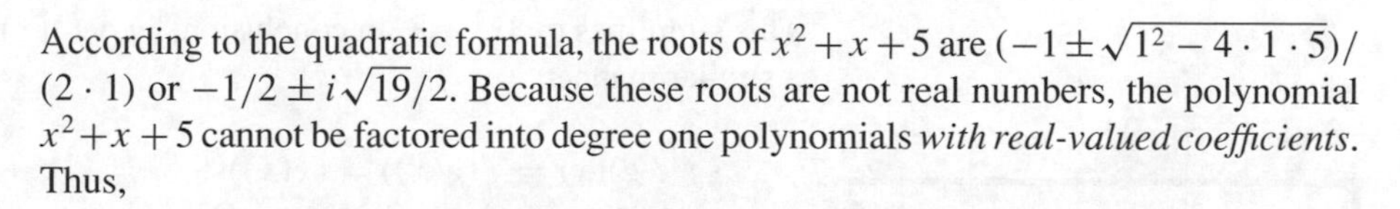

According to the quadratic formula, the roots of $x^2 + x + 5$ are $(-1 \pm \sqrt{1^2 - 4 \cdot 1 \cdot 5})/(2 \cdot 1)$ or $-1/2 \pm i\sqrt{19}/2$. Because these roots are not real numbers, the polynomial $x^2 + x + 5$ cannot be factored into degree one polynomials *with real-valued coefficients*. Thus,

$$x^3 + 3x^2 + 7x + 10 = (x + 2)(x^2 + x + 5)$$

is the complete factorization. ■

Composition of Functions

Another way of combining functions is by *functional composition*. Suppose that f and g are functions and that the domain of g contains the range of f. This means that if x is in the domain of f, then g may be applied to $f(x)$ (see Figure 2). The result of these two operations, one following the other, is called g *composed* with f, or the *composition* of g with f. We write $g \circ f$ for the composition and define

$$(g \circ f)(x) = g(f(x)).$$

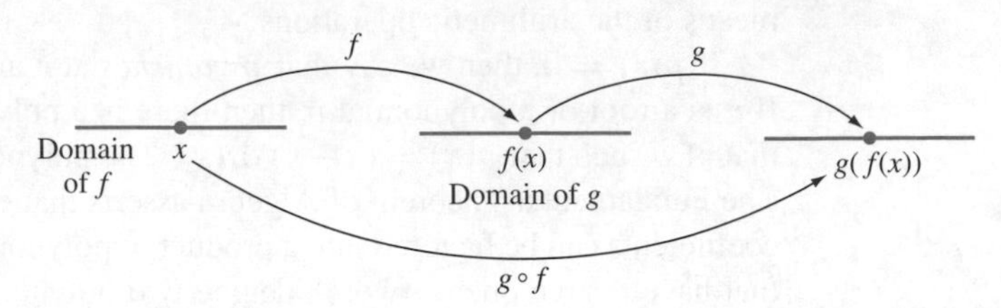

Figure 2

inSIGHT

> Working with composition of functions can be confusing if the concept of composition is not clearly understood. The expression $(g \circ f)(x)$ may *look* like multiplication of functions, but it is not. A good rule when dealing with compositions of functions is to use parentheses to organize formulas. Also, simplify parenthetical expressions by starting from the inside and working out.

Example 4 Let $f(x) = x^2 + 1$ and $g(x) = 3x + 5$. Calculate $g \circ f$ and $f \circ g$.

Solution We calculate that

$$(g \circ f)(x) = g(f(x)) = g(x^2 + 1). \tag{1.4}$$

We have started to work *inside* the parentheses: The first step is to substitute the definition of f, namely $x^2 + 1$, into the equation. The definition of g now says that the evaluation of g at *any* argument is obtained by multiplying that argument by 3 and then adding 5. This is the rule no matter what the argument of g is. In the present case, we are applying g to $x^2 + 1$. Therefore, the right side of equation (1.4) equals $3(x^2 + 1) + 5$. This simplifies to $3x^2 + 8$. In conclusion, $(g \circ f)(x) = 3x^2 + 8$. We compute $f \circ g$ in a similar manner:

$$(f \circ g)(x) = f(g(x)) = (g(x))^2 + 1 = (3x + 5)^2 + 1 = 9x^2 + 30x + 26. \quad \blacksquare$$

> Example 4 shows that $f \circ g$ can be different from $g \circ f$. In general, *composition of functions is not commutative.*

Example 5 How can we write the function $r(x) = (2x + 7)^3$ as the composition of two functions? If $u(t) = 3/(t^2 + 4)$, how can we find two functions v and w for which $u = v \circ w$?

Solution We can think of the function r as two operations applied in sequence. Read the function aloud: First we double and add seven, then we cube. Thus, define $f(x) = 2x + 7$ and $g(x) = x^3$ to get $r(x) = (g \circ f)(x)$, or $r = g \circ f$.

Similarly, read aloud the way we evaluate u at t: First we square t and add four, then we divide three by the quantity just obtained. As a result, we define $w(t) = t^2 + 4$ and $v(t) = 3/t$. It follows that $u(t) = (v \circ w)(t)$, or $u = v \circ w$. ■

Of course, there is nothing to prevent us from composing three or more functions. We define

$$(H \circ G \circ F)(x) = H(G(F(x))).$$

Example 6 Write the function r from Example 5 as the composition of three functions instead of two.

Solution Read the function aloud: *First* we double, *then* we add seven, *then* we cube. We set $F(x) = 2x$, $G(x) = x + 7$, and $H(x) = x^3$. We get $r(x) = (H \circ G \circ F)(x)$. ■

Vertical and Horizontal Translations

It is useful to obtain the graphs of the functions $x \mapsto f(x + h)$ and $x \mapsto f(x) + k$ from the graph of the function f. A point on the graph of $x \mapsto f(x + h)$ is of the form $(x, f(x + h))$. A horizontal shift, or *translation,* by h of this point results in $(x + h, f(x + h))$, which is a point on the graph of f. Vertical translations are treated similarly. We record these observations as a theorem.

Theorem 1 The graph of $x \mapsto f(x + h)$ is obtained by shifting the graph of f horizontally by an amount h. The shift is to the left if $h > 0$ and to the right if $h < 0$ (see Figure 3). The graph of $x \mapsto f(x) + k$ is obtained by shifting the graph of f vertically by an amount k. The shift is up if $k > 0$ and down if $k < 0$ (see Figure 4).

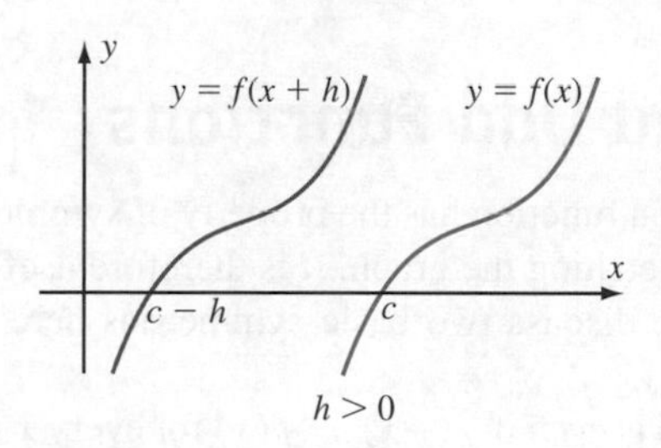

Figure 3a

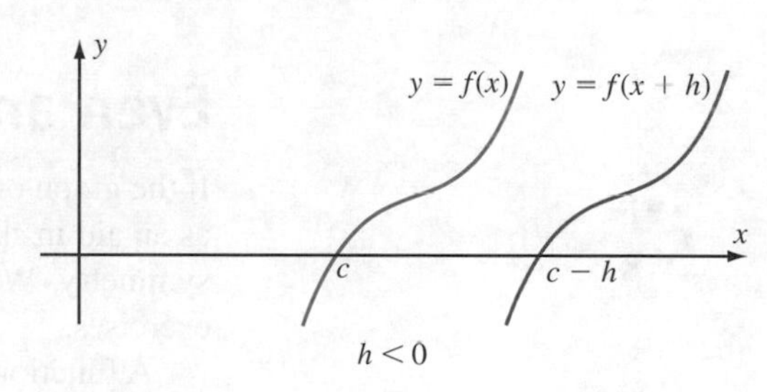

Figure 3b

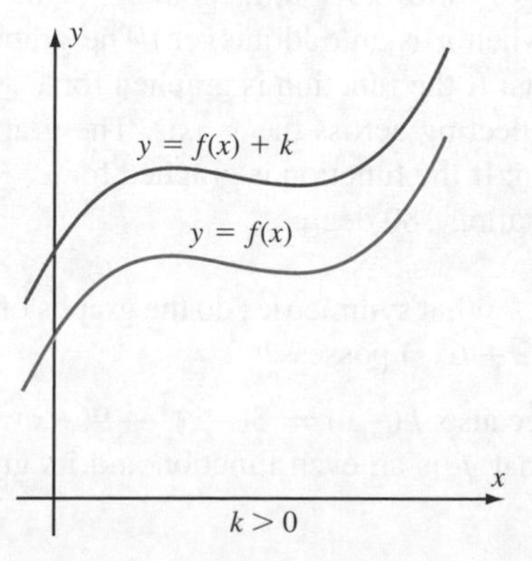

Figure 4a

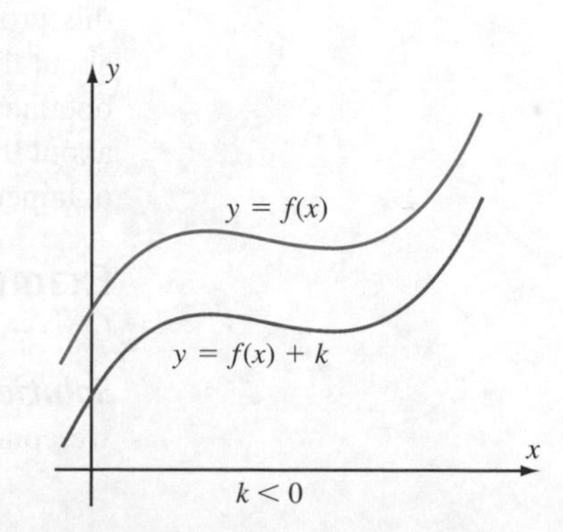

Figure 4b

Example 7 Describe the relationship between the graph of $f(x) = x^2 + 6x + 13$ and the parabola $y = x^2$.

Solution We use the Method of Completing the Square to write

$$\begin{aligned} f(x) &= x^2 + 6x + 13 \\ &= \left(x^2 + 6x + \left(\frac{6}{2}\right)^2 \right) + 13 - \left(\frac{6}{2}\right)^2 \\ &= (x+3)^2 + 4. \end{aligned}$$

Thus, we obtain the graph of f by shifting the parabola $y = x^2$ to the left 3 units and up 4 units. The graph of f is therefore a parabola with vertex at $(-3, 4)$ (see Figure 5).

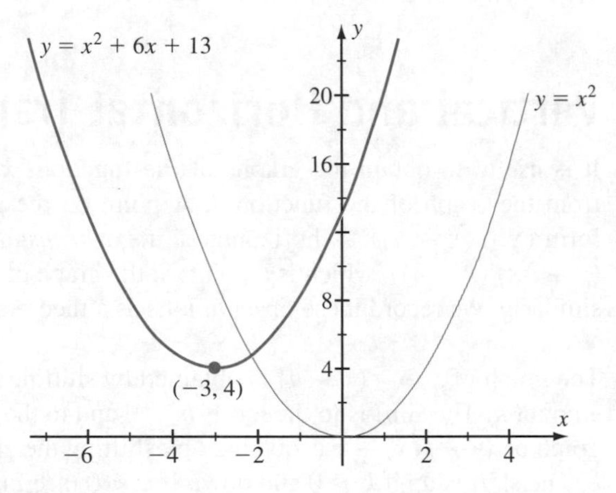

Figure 5

Even and Odd Functions

If the graph of a function has the property of symmetry, then that property can be used as an aid in sketching the graph. It is therefore useful to identify algebraic criteria for symmetry. We discuss two basic symmetries here and explore several others in the exercises.

A function is *even* if $f(-x) = f(x)$ for every x in its domain. (This is true because $f(x) = x^n$ satisfies this property when n is an even integer.) A function is *odd* if $f(-x) = -f(x)$ for every x in its domain. (This is true because $f(x) = x^n$ satisfies this property when n is an odd integer.) The graph of an even function is symmetric about the y-axis: If the function is graphed for $x \geq 0$, then its complete graph may be obtained by reflecting across the y-axis. The graph of an odd function is symmetric about the origin: If the function is graphed for $x \geq 0$, then the complete graph may be obtained by rotating 180 degrees.

Example 8 What symmetries do the graphs of $f(x) = 3x^4 - 9x^2 - 17$ and $g(x) = (4x^3 - 10x)/(2 + 6x^2)$ possess?

Solution Because $f(-x) = 3(-x)^4 - 9(-x)^2 - 17 = 3x^4 - 9x^2 - 17 = f(x)$, we conclude that f is an even function and its graph is symmetric about the y-axis.

Similarly we calculate that

$$\begin{aligned} g(-x) &= \frac{4(-x)^3 - 10(-x)}{2 + 6(-x)^2} \\ &= \frac{-4x^3 + 10x}{2 + 6x^2} \\ &= -\left(\frac{4x^3 - 10x}{2 + 6x^2}\right) = -g(x). \end{aligned}$$

Thus, g is an odd function. Its graph is symmetric about the origin (see Figures 6 and 7).

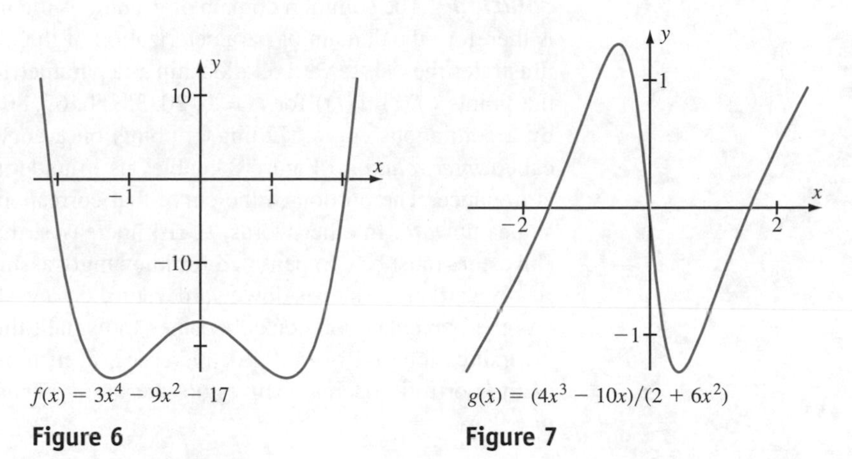

Figure 6

Figure 7

Pairing Functions—Parametric Curves

Given two functions f and g with a common domain I, we can plot the point $(f(t), g(t))$ as t varies through I. The resulting graph is called a *parametric curve*. This notion is made precise by the following definition.

Definition Suppose $\mathcal{C}$ is a curve in the plane and I is an interval of real numbers. If f and g are functions with domain I, and if the plot of the points $(f(t), g(t))$ for t in I coincides with $\mathcal{C}$, then $\mathcal{C}$ is said to be *parameterized* by the equations $x = f(t)$ and $y = g(t)$. These equations are called *parametric equations* for $\mathcal{C}$. The variable t is a *parameter* and I is the *domain* of the parameterization of $\mathcal{C}$. We say that $\mathcal{C}$ is a *parametric curve*.

Notice that y is not given as a function of x. Instead, each of the variables x and y is given as a function of the parameter t.

Example 9 Purchasing a stock amounts to purchasing a share of a company. The value of such an investment rises or falls with the value of the company. Purchasing a bond amounts to making a loan; interest is paid to the bondholder until the loan is repaid. In general, bonds are not as rewarding as stocks, but they are less risky. The risk $f(t)$ and reward (i.e., investment return) $g(t)$ of a combined investment portfolio of stocks and bonds depend on the percentage t of the portfolio that is made up of stocks. The following two tables present risk and reward data for several values of t.

t	0	20	35	50	65	80	100
$f(t)$	0.1179	0.1121	0.1186	0.1325	0.1513	0.1735	0.2085

t	0	20	35	50	65	80	100
$g(t)$	8.333	9.583	10.920	12.100	13.125	14.250	15.630

Sketch the risk-reward curve.

Solution The common domain of f and g is the interval $I = [0, 100]$. This interval is therefore the domain of parameterization of the risk-reward curve. Figure 8, which illustrates the risk-reward relationship as a parametric curve, was produced by plotting the points $(f(t), g(t))$ for $t = 0, 20, 35, 50, 65, 80, 100$ and then joining the points by a continuous curve. (Filling in points on a curve between plotted data points is called *interpolation*.) Figure 8 enables us to understand the risk-reward relationship at a glance. The portion of the curve that corresponds to values of t greater than 20 slopes upward. In other words, as risk increases, the investor's return also increases. (Investors must be compensated for knowingly assuming greater risk.) For $t < 20$, the risk-reward curve slopes downward; return on investment decreases as risk increases. (As the percentage of stocks becomes too small, the portfolio suffers from a lack of diversification and overall risk increases, even though the risks of the components of the portfolio decrease. Investors are often insensitive to, or unaware of, this sort of risk.)

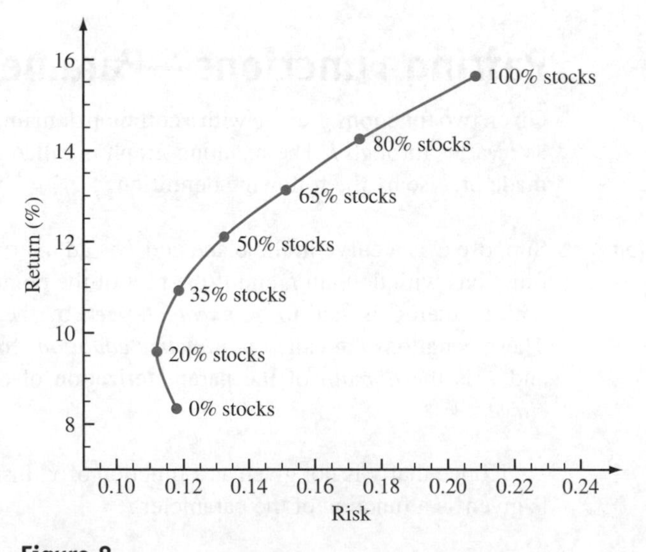

Figure 8
In a mixed portfolio of stocks and bonds, the expected return of the portfolio and the risk of the portfolio are functions of the percentage of stocks.

■

If a curve C is defined by parametric equations $x = f(t)$ and $y = g(t)$, then it is often useful to imagine that the parameter t denotes time. We may then think of a particle tracing out the curve $\mathcal{C}$ and occupying the point $(f(t), g(t))$ at time t. In this context, we may refer to $\mathcal{C}$ as a *path* or a *trajectory*.

Example 10 A particle moves in the xy-plane with coordinates given by

$$x = 2 - t^2 \quad \text{and} \quad y = 1 + 3t^2, \qquad -1 \le t \le 2.$$

Describe the particle's path $\mathcal{C}$.

In Example 10, we eliminated the parameter t to deduce that the parametric curve $\mathcal{C}$ is the graph of a Cartesian equation. Although this method is often useful, it is not always possible to eliminate the parameter. When we do eliminate the parameter as an aid in sketching the curve, we should not neglect the information that the parameterization carries.

Solution We can rewrite the equation for x as $t^2 = 2 - x$. Substituting this equation into the parametric equation for y results in $y = 1 + 3(2 - x)$, or $y = 7 - 3x$. However, the curve $\mathcal{C}$ is not the entire straight line with Cartesian equation $y = 7 - 3x$. To see this, we analyze the motion of the particle. It starts at the point $(1, 4)$ when $t = -1$. As t increases from -1 to 0, the x-coordinate increases to 2 and the y-coordinate decreases to 1. Therefore, the line segment from $(1, 4)$ to $(2, 1)$ ($\overline{AB}$ in Figure 9) is traced for $-1 \le t \le 0$. Then, as t increases from 0 to 2, x decreases to -2 and y increases to 13 ($\overline{BC}$ of Figure 9). We conclude that $\mathcal{C}$ is the line segment $\overline{ABC}$ with endpoints $(-2, 13)$ and $(2, 1)$. Segment $\overline{AB}$ of $\mathcal{C}$ is traced once in each direction.

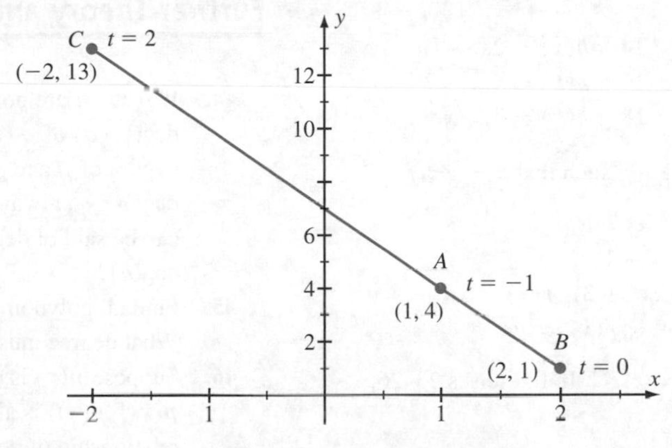

Figure 9

quickquiz

1. What features of the domain and range must we check when combining functions using arithmetic operations?
2. What features of the domain and range must we check when combining functions using composition?
3. How does the graph of $x \mapsto x^3 + 2$ compare with the graph of $x \mapsto (x - 1)^3 + 4$?
4. Do either of the two functions $x \mapsto (x^2 + 1)/(3x^4 + 5)$ or $x \mapsto (x^3 + 1)/(3x^4 + 5)$ have a graph that is symmetric with respect to the y-axis? With respect to the origin?
5. Describe the parametric curve $x = |t|$, $y = t$.

EXERCISES

Problems for Practice

Let $F(x) = x^2 + 5$, $G(x) = (x+1)/(x-1)$, and $H(x) = 2x - 5$. In Exercises 1–12, calculate the function.

1. $F + G$ **2.** $F - 3H$
3. $G \circ H$ **4.** $H \circ G$
5. $H \cdot G$ **6.** G/F
7. $F \circ G \circ H$ **8.** $H \circ F - H \cdot F$
9. $F \cdot (G + H)$ **10.** $G \circ (F + K)$
11. $(F + G)/H^2$ **12.** $H \circ H \circ H - H \circ H$

In Exercises 13–18, write the function h as the composition of two functions. (There is more than one correct way to do this.)

13. $h(x) = (x-2)^2$ **14.** $h(x) = 2x + 7$
15. $h(x) = (x^3 + 3x)^4$ **16.** $h(x) = \sqrt{x - 7}$
17. $h(x) = x^2 + 4x + 4$ **18.** $h(x) = 3/\sqrt{x}$

In Exercises 19–22, find a function g such that $h = g \circ f$.

19. $h(x) = 3x^2 + 6x + 4$, $f(x) = x + 1$
20. $h(x) = x^2 + 4$, $f(x) = x - 1$
21. $h(x) = (x^2+1)/(x^4 + 2x^2 + 3)$, $f(x) = x^2 + 1$
22. $h(x) = 2x^2 + x - \sqrt[6]{x} + 1$, $f(x) = \sqrt{x}$

Let $f(x) = \sqrt{2x + 5}$ and $g(x) = x^{-1/3}$. In Exercises 23–26, calculate the given expression.

23. $(f \circ g)(1/8)$ **24.** $(g \circ f)(2)$
25. $f^2(11) \cdot g^3(54)$ **26.** $(g \circ g)(512)$

In Exercises 27–30, write the polynomial as a product of irreducible polynomials of degree one or two.

27. $x^2 + 4x - 5$ **28.** $x^3 + x^2 - 4x - 4$
29. $x^4 + 2x^3 - 2x^2 - 8x - 8$ **30.** $x^4 + 3x^2 + 2$

In Exercises 31–34, two functions f and g are given. Write $g(x)$ as $f(x + h)$ for some constant h. Describe the relationship between the plots of f and g.

31. $f(x) = 2x + 1$, $g(x) = 2x + 5$
32. $f(x) = 1 - 3x$, $g(x) = 7 - 3x$
33. $f(x) = x^2 + 4$, $g(x) = x^2 - 6x + 13$
34. $f(x) = \sqrt{1 - x^2}$, $g(x) = \sqrt{2x - x^2}$

In Exercises 35–38, two functions f and g are given. Write $g(x)$ as $f(x + h) + k$ for some constants h and k. Describe the relationship between the plots of f and g.

35. $f(x) = x^2$, $g(x) = x^2 + 2x + 5$
36. $f(x) = 3x^2$, $g(x) = 3x^2 + 12x$
37. $f(x) = \sqrt{1 - x^2}$, $g(x) = 1 + \sqrt{2x - x^2}$
38. $f(x) = x/(1 - x)$, $g(x) = (2x + 1)/x$

In Exercises 39–43, describe the curve that has the parametric equation. Sketch the curve.

39. $x = 2t + 1$, $y = 6t - 4$
40. $x = 7$, $y = t^2 + 1$, $-1 \le t \le 2$
41. $x = 1/t$, $y = 3$, $0 < t \le 1$
42. $x = 12t^2 + 1$, $y = 2t$
43. $x = 1/(1 + t^2)$, $y = 1 + t^2$

Further Theory and Practice

44. If p and q are polynomials, express the degree $\deg(p \cdot q)$ of $p \cdot q$ in terms of the degrees $\deg(p)$ and $\deg(q)$ of p and q. Do the same for $\deg(p \circ q)$. Is $\deg(p \circ q)$ always equal to $\deg(q \circ p)$? What can be said of $\deg(p \pm q)$ in terms of $\deg(p)$ and $\deg(q)$?

45. Find all polynomials p such that $(p \circ p)(x) = x$. *Hint:* What degree must p have?

46. Suppose $p(x)$ is a polynomial. Show that $f(x) = p(x + p(x))$ is also a polynomial. What is the relationship of $\deg(f)$ to $\deg(p)$? Is there any relationship between the roots of p and f?

47. A function that is a quotient of polynomials is called a *rational function*. Explain why every rational function can be written in the form $f(x) + g(x)/h(x)$, where f, g, and h are polynomial functions and where the degree of $g(x)$ is *strictly less than* the degree of $h(x)$. Write $(3x^5 + 2x^4 - x^2 + 6)/(x^3 - x + 3)$ in this form.

48. An *affine* function is one of the form $f(x) = ax + b$, where a and b are constants. Prove that the composition of two affine functions is affine.

In Exercises 49–52, find a function g such that $g \circ f = h$.

49. $f(x) = x + 1$, $h(x) = x^2 + 2x + 3$
50. $f(x) = 2x + 3$, $h(x) = (x + 5)/(x - 5)$
51. $f(x) = x^2 - 9$, $h(x) = 2x^2$
52. $f(x) = (x - 1)/x^2$, $h(x) = x^2/(x - 1)$

In Exercises 53–56, find a function f such that $g \circ f = h$.

53. $g(x) = x^2 + 2$, $h(x) = (x - 4)^2 + 2$
54. $g(x) = (x - 1)/(x + 1)$, $h(x) = x^2/(x^2 + 2)$
55. $g(x) = x^3 + 1$, $h(x) = x^2$
56. $g(x) = 4x + 5$, $h(x) = 8x$

In Exercises 57–60, find a function with a graph that is the curve $\mathcal{C}$.

57. $\mathcal{C}$ is obtained by translating the graph of $y = x^2$ 3 units right.
58. $\mathcal{C}$ is obtained by translating the graph of $y = x^3 + 2x$ 3 units down and 2 units right.
59. $\mathcal{C}$ is obtained by reflecting the graph of $y = (x^3 + 1)/(x^2 + 1)$ about the y-axis.
60. $\mathcal{C}$ is obtained by reflecting the graph of $y = (x + 1)/(x^4 + 1)$ about the origin.
61. Explain the difference between $f \circ f$ and $f \cdot f$. Let $f(x) = x^p$ for some fixed power p. Is it ever true that $f \circ f = f \cdot f$?
62. What composition property of the function $f(x) = (1 - x)/(1 + x)$ is illustrated by Figure 10?

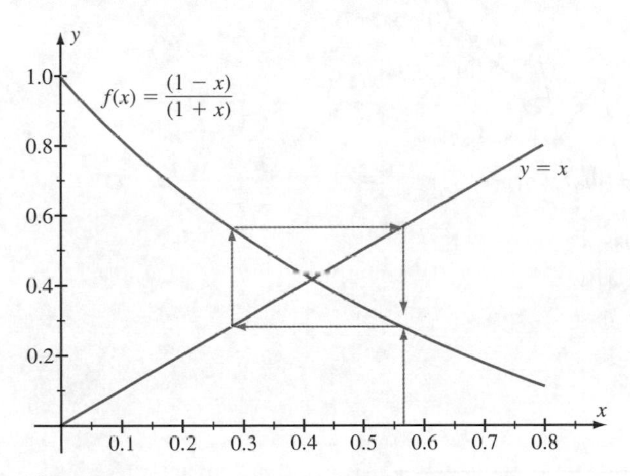

Figure 10

63. Let $f(x) = x^5$. Find a function g such that $g(f(x)) = \pi$. Find a function h such that $h(f(x)) = x$. Find a function k such that $k(f(x)) = x + \pi$.
64. Let $f(x) = x^3 + 1$. Find a function g such that $g(f(x)) = 1/x$. Find a function h such that $h(f(x)) = 1/(1 + x)$. Find a function k such that $k(f(x)) = (f \circ f \circ f)(x)$.
65. Let $f(x) = \begin{cases} 2x & \text{if} \quad x \in [0, 3] \\ 12 - 2x & \text{if} \quad x \in (3, 6] \\ 0 & \text{if} \quad x \notin [0, 6] \end{cases}$. Graph f.

Notice that f is a one-tooth function. Show that $f \circ f$ is a two-tooth function. Graph it.

66. To study how a function f changes at a point x, we can measure its values at the endpoints of a small interval $[x - h/2, x + h/2]$, centered at x. In this case, h is a fixed small positive number: The smaller h is, the more *localized* about x is the information that is gathered. However, h cannot be zero because no change is measured. The function r defined by

$$r(x) = \frac{f(x + h/2) - f(x - h/2)}{h}$$

measures the *average rate of change* of f over the interval $[x - h/2, x - h/2]$. If $f(x) = mx + b$, then what is $r(x)$? If $f(x) = ax^2 + mx + b$, then what is $r(x)$? If f is a polynomial of degree n, then show that r is a polynomial of degree $n - 1$.

67. Let F, G, and H be functions. Verify that

$$H \circ G \circ F = H \circ (G \circ F) = (H \circ G) \circ F.$$

68. Let p be a positive constant. Find two functions F and G that allow you to express the power law for logarithms,

$$\log_{10}(x^p) = p \cdot \log_{10}(x),$$

in the form $\log_{10} \circ G = F \circ \log_{10}$. Can you do the same for the addition law for logarithms:

$$\log_{10}(p \cdot x) = \log_{10}(p) + \log_{10}(x)?$$

In Exercises 69–72, describe the curve that has the given parametric equations.

69. $x = t^2$, $y = t^3$
70. $x = 2t^2 + t - 1$, $y = t + 1$
71. $x = \log_{10}(t)/\log_{10}(2)$, $y = 3t$
72. $x = 3\log_{10}(t)$, $y = 5\log_{10}(t)$
73. Annual California strawberry production $f(t)$ and annual American per capita strawberry consumption $g(t)$ are plotted for $t \in \{1976, 1977, 1978, \ldots, 1996\}$ in Figures 11a and 11b, next page. (As is usually done in such plots, consecutive data points are connected by line segments.) Use the 20 given data points (10 on each curve) to plot the production-consumption relationship described by the parametric equations $x = f(t)$ and $y = g(t)$. What is the equation of the least squares line that approximates the plotted data $\{(f(t), g(t))\}$ and that passes through the point $(1.29, 5.94)$ corresponding to $t = 1996$?

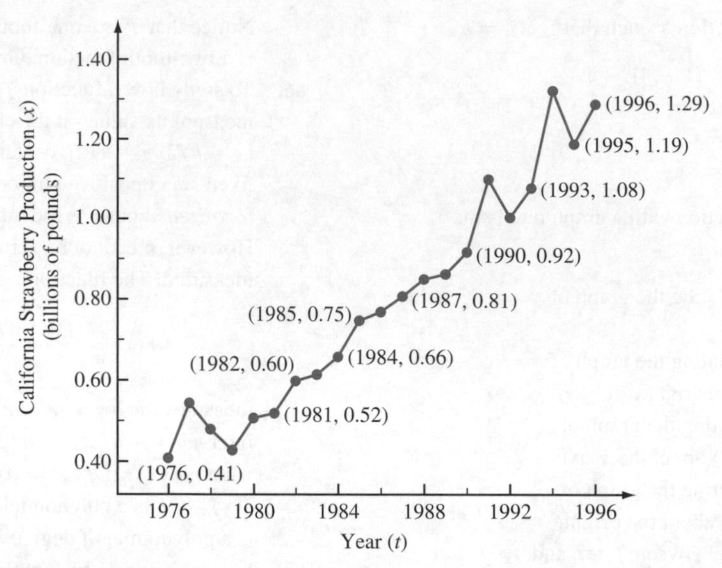

Figure 11a

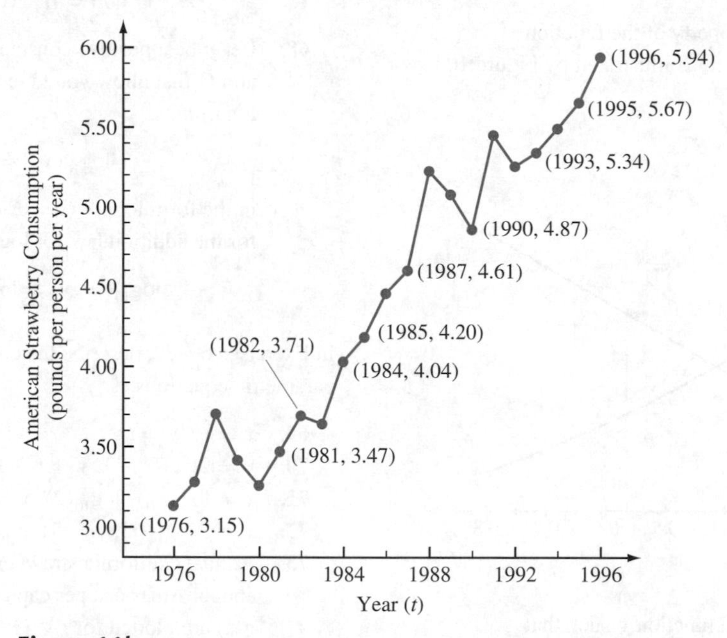

Figure 11b

74. Suppose $A = (x_1, y_1)$ and $B = (x_2, y_2)$ are any two distinct points in the plane. Let

$$f(t) = (1 - t) \cdot x_1 + t \cdot x_2$$
$$g(t) = (1 - t) \cdot y_1 + t \cdot y_2.$$

Show that $x = f(t)$, $y = g(t)$, $0 \leq t \leq 1$, is a parameterization of line segment $\overline{AB}$.

75. Every graph $\mathcal{G}$ of a Cartesian equation $y = \varphi(x)$ may be parameterized. To show this, define $f(t) = t$ and $g(t) = \varphi(t)$ and let I denote the domain of the function φ. Let $\mathcal{C}$ be the curve defined by $x = f(t)$, $y = g(t)$ for t in I. Prove that a point (x_0, y_0) is on the graph $\mathcal{G}$ if and only if it is on the curve $\mathcal{C}$.

76. Let $S(x) = -x$. Express the property that f is an even function by means of the composition of S and f. Express the property that g is an odd function by means of the composition of S and g.

77. Show that the function $R : \mathbb{R}^2 \to \mathbb{R}^2$ defined by $R(x, y) = (2x_0 - x, y)$ is a reflection about the line $x = x_0$. *Hint:* Every point in the plane can be written as $(x_0 \pm d, y)$ where $d \geq 0$. Calculate the image of this point under R. Suppose the function f satisfies the identity $f(2x_0 - x) = f(x)$. Show that the graph of f is symmetric about the line $x = x_0$.

78. Let p be any polynomial.
   a. Show that
      i. $q(x) = (p(x) + p(-x))/2$ is an even polynomial, and
      ii. $r(x) = (p(x) - p(-x))/2$ is an odd polynomial.

   Since $p = q + r$, this shows that every polynomial can be written as the sum of an even polynomial and an odd polynomial.
   b. Show that q contains only even powers of x and r contains only odd powers.
   c. If p is even, deduce that the coefficient of each odd power of x in $p(x)$ is zero. If p is odd, deduce that the coefficient of each even power of x in $p(x)$ is zero.
   d. If p is even, show that there is a polynomial $s(x)$ such that $p(x) = s(x^2)$.
   e. If p is odd, show that $p(0) = 0$. Deduce that there is an even polynomial $t(x)$ such that $p(x) = x \cdot t(x)$.

Calculator/Computer Exercises

In Exercises 79–82, graph the curves $\mathcal{C}$ and $\mathcal{C}'$ in the same viewing window.

79. $\mathcal{C} = \{(x, y) : y = x^2\}$; $\mathcal{C}'$ is obtained by translating $\mathcal{C}$ 3 units right.

80. $\mathcal{C} = \{(x, y) : y = x^3 + 2x\}$; $\mathcal{C}'$ is obtained by translating $\mathcal{C}$ 3 units down and 2 units right.

81. $\mathcal{C} = \{(x, y) : y = (x^3 + 1)/(x^2 + 1)\}$; $\mathcal{C}'$ is obtained by reflecting $\mathcal{C}$ about the y-axis.

82. $\mathcal{C} = \{(x, y) : y = (x + 1)/(x^4 + 1)\}$; $\mathcal{C}'$ is obtained by reflecting $\mathcal{C}$ about the origin.

83. Graph $f(x) = x^4 - 32x^3 + 224x^2 + 512x - 4096$ in the viewing rectangle $[-6, 22] \times [-5000, 12000]$. You will see four points at which the graph crosses the x-axis.
   a. The graph appears to be symmetric about a vertical line $x = x_0$. Find the value of x_0. (If the graph is symmetric about this vertical line, the part to the left of $x = x_0$ will be superimposed on the part to the right when it is reflected through the line $x = x_0$. Imagine that $x = x_0$ is the spine of an opened book and that the page on the left is turned over.)
   b. Let $\rho(x) = 2x_0 - x$. Graph both f and $f \circ \rho$ on $[x_0, 22] \times [-5000, 12000]$. What conclusions can be drawn from the resulting graph? Refer to Exercise 77.
   c. Let $p(x) = f(x_0 - x)$. Show that p is an even function. There is a degree two polynomial q such that $p(x) = q(x^2)$. Refer to Exercise 78, part d. Use the quadratic formula twice to obtain exact expressions for the four roots of p. Find exact expressions for the four roots of f.

Let $h = 0.005$. The average rate of change $r(x)$ of the function f over the interval $[x - h, x + h]$ of length $2h$ is given in Exercise 66. In Exercises 84–88, plot the given function f. Calculate the average rate of change $r(x_0)$ at the given point. Then plot the line through $(x_0, f(x_0))$ with slope $r(x_0)$. The two plots should appear to be tangent to each other. The reasons for this are discussed in Chapter 3.

84. $f(x) = x^2$, $x_0 = 1$
85. $f(x) = \sqrt{x}$, $x_0 = 3$
86. $f(x) = 1/x$, $x_0 = 1$
87. $f(x) = 2x - \sqrt{x}$, $x_0 = 2$
88. $f(x) = (2x + 1)/(1 + x^2)$, $x_0 = 2$

In Exercises 89–90, plot the parametric equations $x = f(t)$, $y = g(t)$, $t \in I$, for the given f, g, and I. Follow the accompanying directions.

89. $f(t) = t^4 + t + 1$, $g(t) = t^3 - t$, $I = [-1, 0]$. Find the points for which $y = 18x/25$ and the values of the parameter that correspond to these points.

90. $f(t) = t^4 + t + 1$, $g(t) = t^3 - 2t$, $I = [-7/4, 3/2]$. A point P is a double point of a parametric curve if there are two values of t in I such that $P = (f(t), g(t))$. Find the double point of the given curve and the two values that parameterize that point.

1.6 Trigonometry

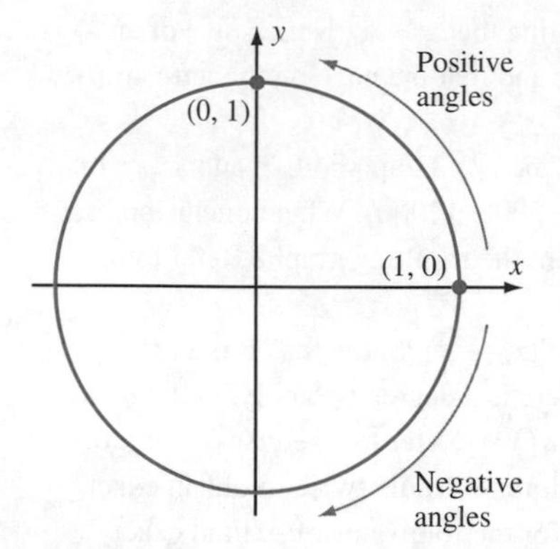

Figure 1
The unit circle

When first learning trigonometry, we study triangles and measure angles in degrees. In calculus, however, it is convenient to study trigonometry in a more general setting and to measure angles differently.

In calculus, we measure angles by rotation along the unit circle in the plane, beginning at the positive x-axis. Counterclockwise rotation corresponds to positive angles, and clockwise rotation corresponds to negative angles (see Figure 1). The *radian measure* of an angle is defined as the length of the arc of the unit circle that it subtends with the positive x-axis (with an appropriate $+$ or $-$ sign).

In degree measure, one full rotation about the unit circle is $360°$; in radian measure, one full rotation about the circle is the circumference of the unit circle, namely 2π. Let us use the symbol θ to denote an angle. The principle of proportionality tells us that

$$\frac{\text{degree measure of } \theta}{360°} = \frac{\text{radian measure of } \theta}{2\pi}.$$

In other words,

$$\text{radian measure of } \theta = \frac{\pi}{180} \mathrm{R} \cdot (\text{degree measure of } \theta)$$

and

$$\text{degree measure of } \theta = \frac{180}{\pi} \mathrm{D} \cdot (\text{radian measure of } \theta).$$

Figure 2 shows several angles that include both the radian and degree measures. This book always refers to radian measure for angles unless degrees are explicitly specified. Doing so makes the calculus formulas simpler. Thus, for example, if we refer to "the angle $2\pi/3$," then it should be understood that this is in radian measure. Likewise, if we refer to "the angle 3," it is also understood to be radian measure. We sketch this last angle (see Figure 3) by noting that a full rotation is $2\pi = 6.28\ldots$, and 3 is a little less than half of that.

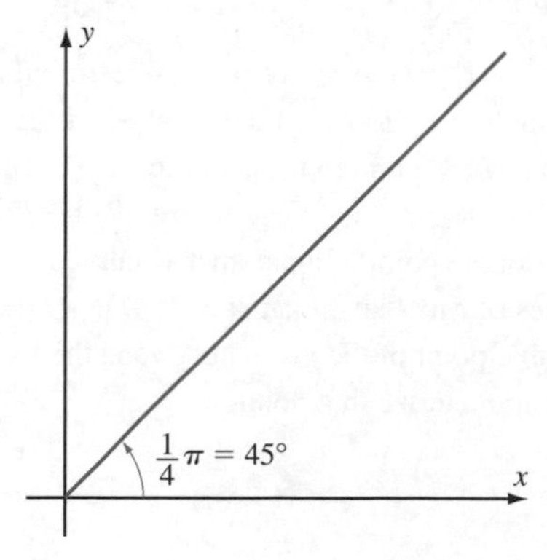

Figure 2a

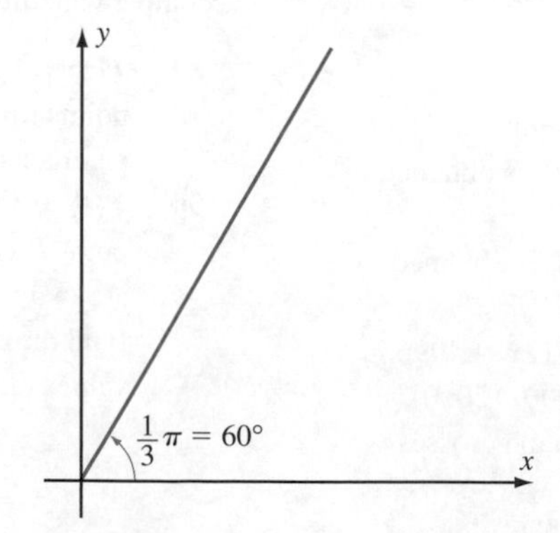

Figure 2b

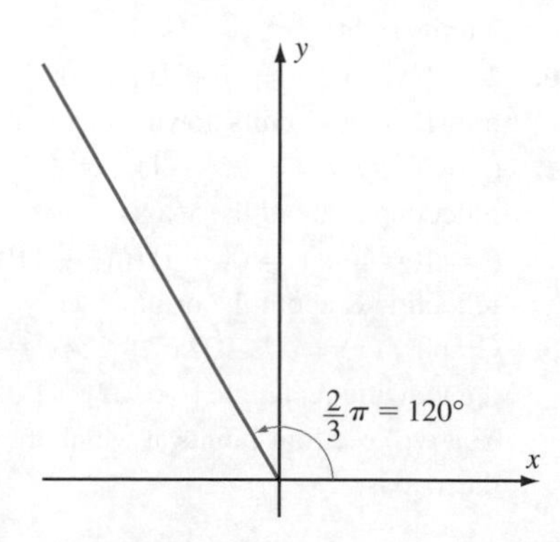

Figure 2c

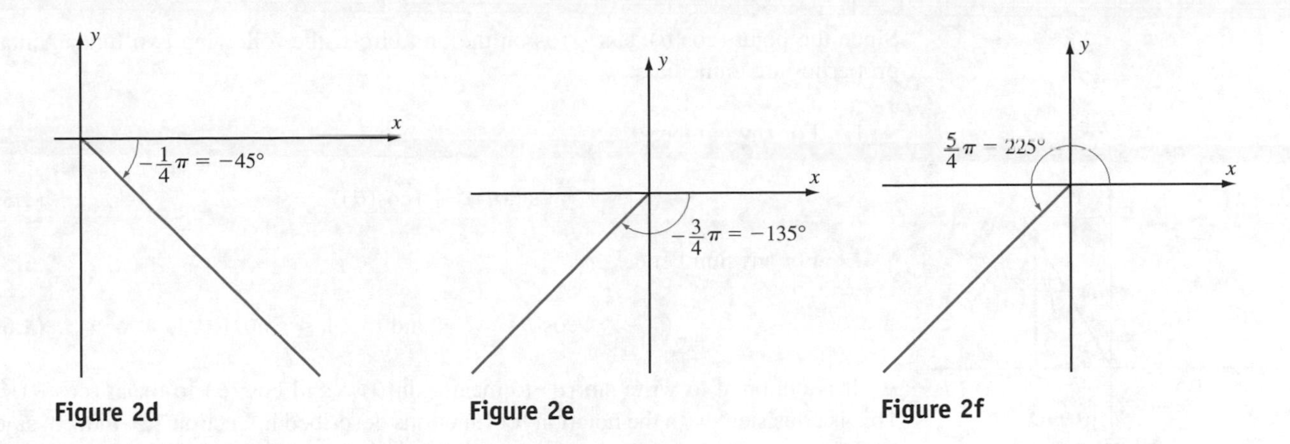

Figure 2d **Figure 2e** **Figure 2f**

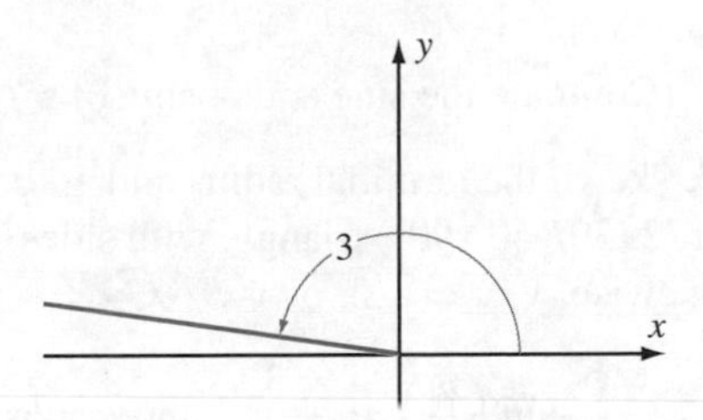

Figure 3

Sine and Cosine Functions

Definition Let $A = (1, 0)$ denote the point of intersection of the unit circle with the positive x-axis. Let θ be *any* real number. A unique point $P = (x, y)$ on the unit circle is associated with θ by rotating $\overline{OA}$ by θ radians. (Remember that the rotation is counterclockwise for $\theta > 0$ and clockwise for $\theta < 0$.) (See Figure 4.) The radius $\overline{OP}$ is called the *terminal radius* of θ, and P is called the *terminal point* corresponding to θ. The number y is called the *sine* of θ and is written $\sin(\theta)$. The number x is called the *cosine* of θ and is written $\cos(\theta)$.

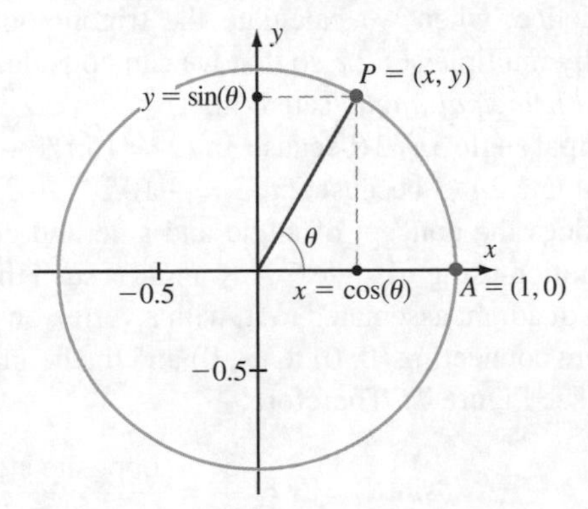

Figure 4

Since the point $(\cos(\theta), \sin(\theta))$ is on the unit circle, the following two fundamental properties are immediate:

1. For any number θ,
$$(\sin(\theta))^2 + (\cos(\theta))^2 = 1. \tag{1.5}$$

2. For any number θ,
$$-1 \le \cos(\theta) \le 1 \quad \text{and} \quad -1 \le \sin(\theta) \le 1. \tag{1.6}$$

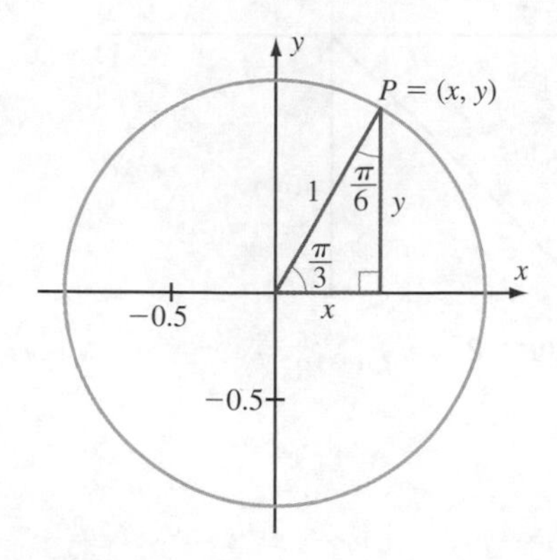

Figure 5

It is common to write $\sin^2(\theta)$ to mean $(\sin(\theta))^2$ and $\cos^2(\theta)$ to mean $(\cos(\theta))^2$. This is consistent with the notation for functions described in Section 1.5. Indeed, sine and cosine are both functions with domain equal to $\mathbb{R}$ and image equal to $[-1, 1]$.

Example 1 Compute the sine and cosine of $\pi/3$.

Solution We sketch the terminal radius and associated triangle (see Figure 5). This is a $\pi/6$-$\pi/3$-$\pi/2$ (30°-60°-90°) triangle with sides of ratio $1 : \sqrt{3} : 2$. Thus, $1/x = 2$, or $x = 1/2$. Likewise, $y/x = \sqrt{3}$, or $y = \sqrt{3}x = \sqrt{3}/2$. It follows that

$$\sin\left(\frac{\pi}{3}\right) = \frac{\sqrt{3}}{2} \quad \text{and} \quad \cos\left(\frac{\pi}{3}\right) = \frac{1}{2}. \quad ■$$

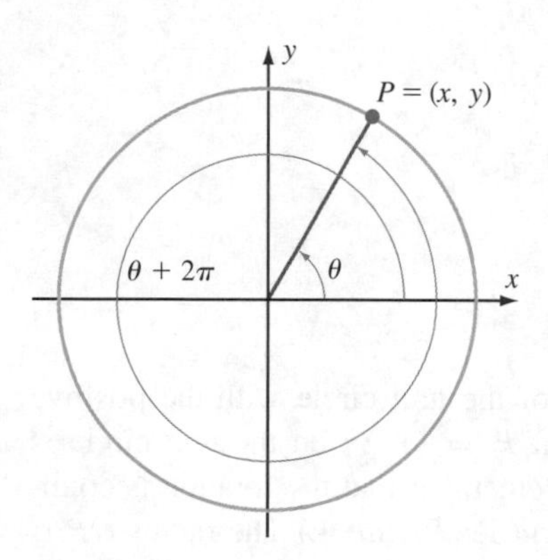

Figure 6

If θ is any number, then θ and $\theta + 2\pi$ have the same terminal radius and the same terminal point (adding 2π simply adds one more trip around the circle—see Figure 6). As a result,

$$\sin(\theta) = \sin(\theta + 2\pi)$$

and

$$\cos(\theta) = \cos(\theta + 2\pi).$$

We say that the sine and cosine functions have *period* 2π. In other words, the functions repeat themselves every 2π units.

In practice, when we calculate the trigonometric functions of an angle θ, we reduce it by multiples of 2π so that we can consider an equivalent angle θ', called the *associated principal angle,* satisfying $0 \le \theta' < 2\pi$. For instance, $15\pi/2$ has associated principal angle $3\pi/2$ (because $3\pi/2 = 15\pi/2 - 3 \cdot 2\pi$) and $-10\pi/3$ has associated principal angle $2\pi/3$ because $2\pi/3 = -10\pi/3 + 2 \cdot 2\pi$.

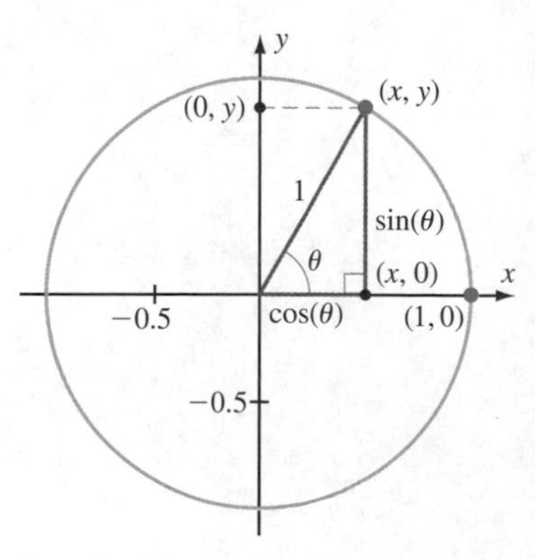

Figure 7

How does the concept of angle and sine and cosine presented here relate to the classical notion using triangles? Any angle θ such that $0 \le \theta < \pi/2$ has a right triangle in the first quadrant associated to it, with a vertex on the unit circle such that the base is the segment connecting $(0, 0)$ to $(x, 0)$ and the height is the segment connecting $(x, 0)$ to (x, y) (see Figure 7). Therefore,

$$\sin(\theta) = y = \frac{y}{1} = \frac{\text{opposite side of triangle}}{\text{hypotenuse}}$$

and

$$\cos(\theta) = x = \frac{x}{1} = \frac{\text{adjacent side of triangle}}{\text{hypotenuse}}.$$

Thus, for angles θ between 0 and $\pi/2$, the new definition of sine and cosine using the unit circle is equivalent to the classical definition using adjacent and opposite sides and the hypotenuse. For other angles θ, the classical approach is to reduce to this special case by subtracting multiples of $\pi/2$. The approach using the unit circle is considerably clearer because it makes the signatures of sine and cosine obvious.

Other Trigonometric Functions

In addition to sine and cosine, there are four other trigonometric functions—tangent, cotangent, secant, and cosecant—defined as follows:

$$\tan(\theta) = \frac{y}{x} = \frac{\sin(\theta)}{\cos(\theta)}$$
$$\cot(\theta) = \frac{x}{y} = \frac{\cos(\theta)}{\sin(\theta)}$$
$$\sec(\theta) = \frac{1}{x} = \frac{1}{\cos(\theta)}$$
$$\csc(\theta) = \frac{1}{y} = \frac{1}{\sin(\theta)}.$$

Whereas sine and cosine have domain the entire real line, $\tan(\theta)$ and $\sec(\theta)$ are undefined at odd multiples of $\pi/2$ (because cosine is zero there). Also, the functions $\cot(\theta)$ and $\csc(\theta)$ are undefined at even multiples of $\pi/2$ (because sine is zero there).

Figure 8 shows the graphs of the six trigonometric functions. Compare the graph of the cosine and sine functions. We obtain the graph of $y = \cos(\theta)$ by shifting the

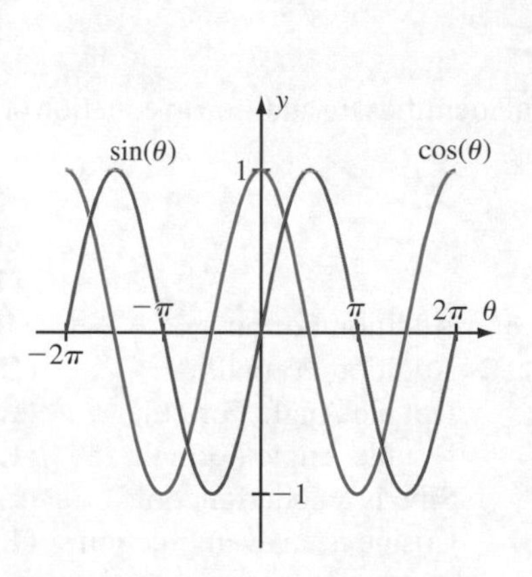

Figure 8a

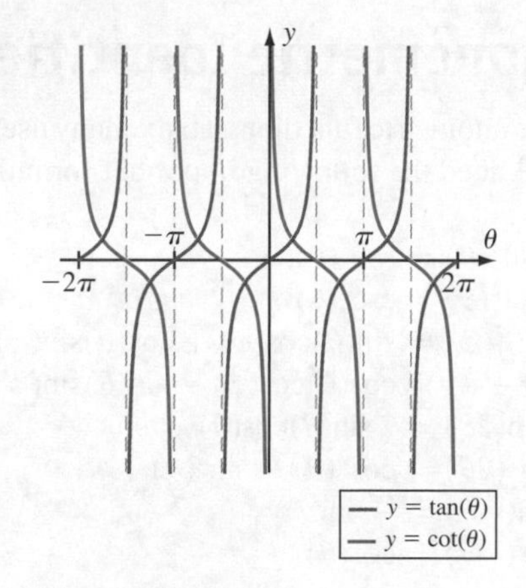

Figure 8b

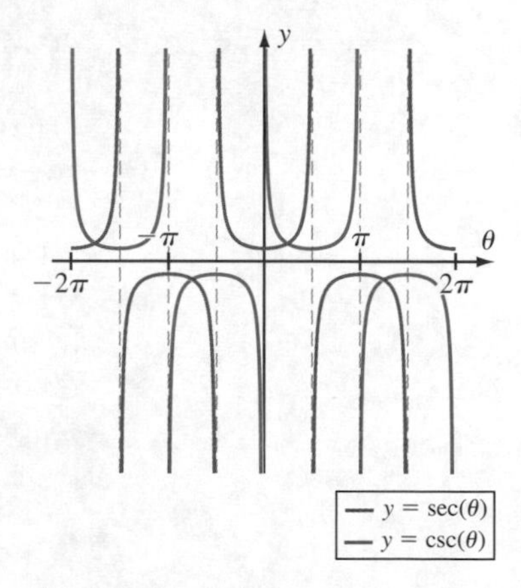

Figure 8c

graph of $y = \sin(\theta)$ by $\pi/2$ to the left. According to Theorem 1 from Section 1.5, this means that

$$\sin\left(x + \frac{\pi}{2}\right) = \cos(x).$$

This identity is a special case of the Addition Formula for the sine, which is given in equation (1.9) below.

Example 2 Compute all trigonometric functions for the angle $\theta = 11\pi/4$.

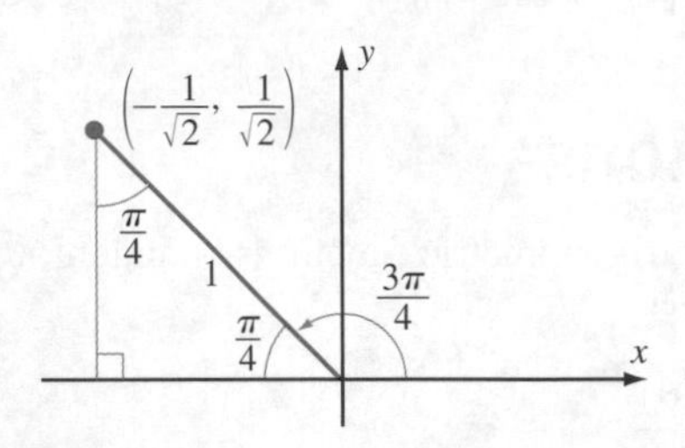

Figure 9

Solution Because the principal associated angle is $3\pi/4$, we deal with that angle. Figure 9 shows that the triangle associated to this angle is an isosceles right triangle with hypotenuse 1. Therefore, $x = -1/\sqrt{2}$ and $y = 1/\sqrt{2}$. It follows that

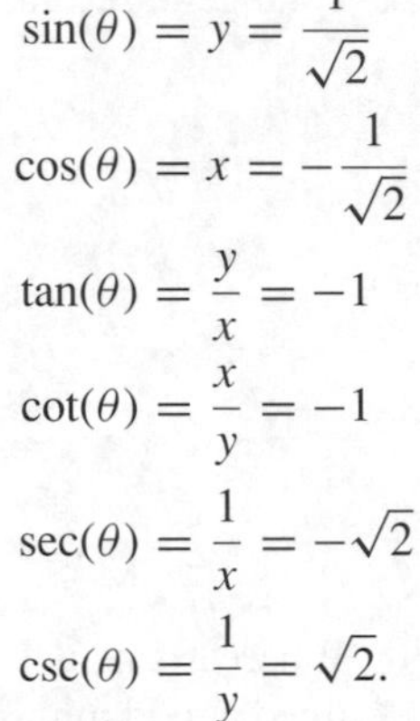

$$\sin(\theta) = y = \frac{1}{\sqrt{2}}$$

$$\cos(\theta) = x = -\frac{1}{\sqrt{2}}$$

$$\tan(\theta) = \frac{y}{x} = -1$$

$$\cot(\theta) = \frac{x}{y} = -1$$

$$\sec(\theta) = \frac{1}{x} = -\sqrt{2}$$

$$\csc(\theta) = \frac{1}{y} = \sqrt{2}.$$

■

Similar calculations allow us to complete Table 1 for the values of the trigonometric functions at the principal angles that are multiples of $\pi/6$ or $\pi/4$.

Trigonometric Identities

The trigonometric functions satisfy many useful identities. In addition to equation (1.5), we will need the following important formulas.

Identity	Name	No.
$1 + \tan^2(\theta) = \sec^2(\theta)$		**(1.7)**
$1 + \cot^2(\theta) = \csc^2(\theta)$		**(1.8)**
$\sin(\theta + \phi) = \sin(\theta)\cos(\phi) + \cos(\theta)\sin(\phi)$	Addition Formula	**(1.9)**
$\cos(\theta + \phi) = \cos(\theta)\cos(\phi) - \sin(\theta)\sin(\phi)$	Addition Formula	**(1.10)**
$\sin(2\theta) = 2\sin(\theta)\cos(\theta)$	Double Angle Formula	**(1.11)**
$\cos(2\theta) = \cos^2(\theta) - \sin^2(\theta)$	Double Angle Formula	**(1.12)**
$\sin(-\theta) = -\sin(\theta)$	Sine is an odd function	**(1.13)**
$\cos(-\theta) = \cos(\theta)$	Cosine is an even function	**(1.14)**
$\sin^2\left(\frac{\theta}{2}\right) = \frac{1 - \cos(\theta)}{2}$	Half-Angle Formula	**(1.15)**
$\cos^2\left(\frac{\theta}{2}\right) = \frac{1 + \cos(\theta)}{2}$	Half-Angle Formula	**(1.16)**

Angle	Sine	Cosine	Tangent	Cotangent	Secant	Cosecant
0	0	1	0	undef	1	undef
$\pi/6$	$1/2$	$\sqrt{3}/2$	$1/\sqrt{3}$	$\sqrt{3}$	$2/\sqrt{3}$	2
$\pi/4$	$\sqrt{2}/2$	$\sqrt{2}/2$	1	1	$\sqrt{2}$	$\sqrt{2}$
$\pi/3$	$\sqrt{3}/2$	$1/2$	$\sqrt{3}$	$1/\sqrt{3}$	2	$2/\sqrt{3}$
$\pi/2$	1	0	undef	0	undef	1
$2\pi/3$	$\sqrt{3}/2$	$-1/2$	$-\sqrt{3}$	$-1/\sqrt{3}$	-2	$2/\sqrt{3}$
$3\pi/4$	$\sqrt{2}/2$	$-\sqrt{2}/2$	-1	-1	$-\sqrt{2}$	$\sqrt{2}$
$5\pi/6$	$1/2$	$-\sqrt{3}/2$	$-1/\sqrt{3}$	$-\sqrt{3}$	$-2/\sqrt{3}$	2
π	0	-1	0	undef	-1	undef
$7\pi/6$	$-1/2$	$-\sqrt{3}/2$	$1/\sqrt{3}$	$\sqrt{3}$	$-2/\sqrt{3}$	-2
$5\pi/4$	$-\sqrt{2}/2$	$-\sqrt{2}/2$	1	1	$-\sqrt{2}$	$-\sqrt{2}$
$4\pi/3$	$-\sqrt{3}/2$	$-1/2$	$\sqrt{3}$	$1/\sqrt{3}$	-2	$-2/\sqrt{3}$
$3\pi/2$	-1	0	undef	0	undef	-1
$5\pi/3$	$-\sqrt{3}/2$	$1/2$	$-\sqrt{3}$	$-1/\sqrt{3}$	2	$-2/\sqrt{3}$
$7\pi/4$	$-\sqrt{2}/2$	$\sqrt{2}/2$	-1	-1	$\sqrt{2}$	$-\sqrt{2}$
$11\pi/6$	$-1/2$	$\sqrt{3}/2$	$-1/\sqrt{3}$	$-\sqrt{3}$	$2/\sqrt{3}$	-2

Table 1

Example 3 Prove identity (1.7).

Solution Using equation (1.5), we have

$$\begin{aligned}
\tan^2(\theta) + 1 &= \frac{\sin^2(\theta)}{\cos^2(\theta)} + 1 \\
&= \frac{\sin^2(\theta)}{\cos^2(\theta)} + \frac{\cos^2(\theta)}{\cos^2(\theta)} \\
&= \frac{\sin^2(\theta) + \cos^2(\theta)}{\cos^2(\theta)} \\
&= \frac{1}{\cos^2(\theta)} \\
&= \sec^2(\theta).
\end{aligned}$$

■

Identities (1.5)–(1.16) were recorded by the astronomer Ptolemy (ca. 85–165 CE) of Alexandria but were discovered at a much earlier date by Hipparchus (ca. 180–125 BCE). There are many other trigonometric identities of a more specialized nature. For instance, if α and β are the numbers in $(0, \pi/2)$ such that $\tan(\alpha) = 1/5$ and $\tan(\beta) = 1/239$, then

$$4\alpha - \beta = \frac{\pi}{4}.$$

This identity was discovered in 1706 by John Machin (1680–1751 CE), who used it to calculate more than 100 digits of the number π.

Modeling with Trigonometric Functions

Although the subject of trigonometry was invented by the ancient Greek astronomers for measurement, the trigonometric functions have proved to be useful in many contexts. For example, because the sine function is cyclic with period 2π (that is, the sine function repeats itself after 2π), it is often used to model cyclic phenomena.

Example 4 The table shows the normal monthly mean temperatures of St. Louis, Missouri, in degrees Fahrenheit.

Jan.	Feb.	March	April	May	June	July	Aug.	Sept.	Oct.	Nov.	Dec.
29	34	43	56	66	75	79	77	70	58	45	34

Model temperature (T) as a function of time (t) by using $T(t) = b + A\sin(\omega \cdot t + \phi)$ for suitable constants b, A, ω, and ϕ.

Solution If the unit of time is chosen to be one month, then the data in the table correspond to $0 \leq t \leq 12$. The mean temperature of January corresponds to $t = 1/2$, the mean temperature of February to $t = 3/2$, and so on. Because

$$\sin(\omega(t_0 + 2\pi/\omega)) = \sin(\omega t_0 + 2\pi) = \sin(\omega t_0),$$

we see that whatever the value of t_0, $T(t_0) = T(t_0 + 2\pi/\omega)$. In other words, the function T repeats itself after $2\pi/\omega$. This is called the *period* of T. Since the period of weather is 12 when measured in months, we set $2\pi/\omega = 12$, or $\omega = \pi/6$. The range of mean temperatures is $79 - 29 = 50$. The midpoint of this range is 54. Because the middle value of $b + A\sin(\pi \cdot t/6 + \phi)$ is b, we set $b = 54$. Inequality (1.6) implies that

$$54 - A \leq 54 + A\sin\left(\frac{\pi}{6}t + \phi\right) \leq 54 + A.$$

To obtain the extreme values 29 and 79, we set $A = 25$. Thus,

$$T(t) = 54 + 25\sin\left(\frac{\pi}{6}t + \phi\right).$$

It only remains to find ϕ. The maximum value of the function $54 + 25\sin(\pi \cdot t/6 + \phi)$ occurs when the sine takes on its maximum value 1, which occurs when its argument is $\pi/2$. Assuming that the high mean temperature of 79 occurs in the middle of July when $t = 13/2$, we set $\pi \cdot (13/2)/6 + \phi = \pi/2$. This gives $\phi = -7\pi/12$. Our model is

$$T(t) = 54 + 25\sin\left(\frac{\pi}{6}t - \frac{7\pi}{12}\right).$$

Figure 10 shows how well this corresponds to the measured data.

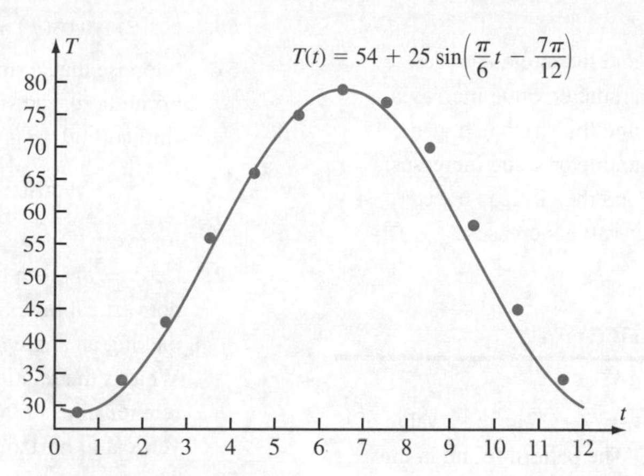

Figure 10

quickquiz

1. What is the domain of the sine function?
2. What is the image of cosine?
3. Which trigonometric functions are even? Odd?
4. Simplify $\sin(2\theta)/\sin(\theta)$.

EXERCISES

Problems for Practice

In Exercises 1–4, calculate each of the six trigonometric functions at angle θ without using a calculator. Use Table 1, page 63, to verify your calculations.

1. $\theta = \pi/6$
2. $\theta = \pi/4$
3. $\theta = 2\pi/3$
4. $\theta = 4\pi/3$

In Exercises 5–14, calculate the given expression without using a calculator.

5. $\sin(\pi/3)\sin(\pi/6)$
6. $\cos(0) - \cos(\pi)$
7. $\cos(\pi/6) + \cos(\pi/3)$
8. $\sin(\pi/4)\cos(\pi/4)$
9. $\tan(\pi/3)/\tan(\pi/6)$
10. $\cos(2\pi/3)\csc(2\pi/3)$
11. $\sin(\pi \cdot \sin(\pi/6))$
12. $\sec(-\pi/3)^{\csc(-\pi/2)}$
13. $\sin(19\pi/2)^{\cos(33\pi)}$
14. $4\tan(\pi/4) - \sin(17\pi/2)$

In Exercises 15–20, θ is a number between 0 and $\pi/2$. Calculate the trigonometric function from the information.

15. $\cos(\theta)$, $\sin(\theta) = 1/3$
16. $\tan(\theta)$, $\cos(\theta) = 3/5$
17. $\sin(2\theta)$, $\cos(\theta) = 4/5$
18. $\sin(\theta/2)$, $\cos(\theta) = 3/7$
19. $\cos(\theta/2)$, $\sin(\theta) = 5/12$
20. $\cos(\theta + \pi)$, $\cos(\theta) = 0.1$

In Exercises 21–24, state which of the six trigonometric functions are positive when evaluated at θ in the indicated interval.

21. $\theta \in (0, \pi/2)$
22. $\theta \in (\pi/2, \pi)$
23. $\theta \in (\pi, 3\pi/2)$
24. $\theta \in (3\pi/2, 2\pi)$

In Exercises 25–28, graph the function.

25. $f(t) = \sin(2t)$, $-2\pi \le t \le 2\pi$
26. $f(t) = \cos(t/2)$, $-2\pi \le t \le 2\pi$

27. $f(t) = \sin(t - \pi/6), 0 \leq t \leq 2\pi$
28. $f(t) = \cos(2t + \pi/3), 0 \leq t \leq 2\pi$

In Exercises 29–31, use the cosine and sine functions to give a parameterization of the unit circle such that the domain of parameterization is $[0, 2\pi]$ and the additional requirements are satisfied.

29. The initial point is $(1, 0)$, and the circle is traced counterclockwise as the parameter value increases.
30. The initial point is $(0, 1)$, and the circle is traced counterclockwise as the parameter value increases.
31. The initial point is $(0, 1)$, and the circle is traced clockwise as the parameter value increases.

Further Theory and Practice

32. For each $\theta \in \{0, \pi/6, \pi/4, \pi/3, \pi/2\}$, find a value of n such that $\sin(\theta) = \sqrt{n}/2$. (The pattern found in this exercise is sometimes used as a memory aid.)
33. The length $s(r, \theta)$ of an arc of a circle of radius r subtended by an angle of radian measure θ is proportional to r and to θ. What is $s(r, \theta)$?
34. The area $A(r, \theta)$ of a sector of a circle of radius r subtended by an angle of radian measure θ is jointly proportional to θ and to r^2. What is $A(r, \theta)$?
35. The equations $\sin(2\theta) = \sin(\theta)\cos(\theta)$ and $\sin(2\theta) = 2\sin(\theta)\cos(\theta)$ are fundamentally different. Explain.
36. Find constants A and B such that $\sin(3\theta) = A\sin(\theta) - B\sin^3(\theta)$ for every θ.

In Exercises 37–44, use one or more of the basic trigonometric identities to derive the given identity.

37. $\tan(\theta + \phi) = \dfrac{\tan(\theta) + \tan(\phi)}{1 - \tan(\theta)\tan(\phi)}$
38. $\sin(\theta)\sin(\phi) = \dfrac{\cos(\theta - \phi) - \cos(\theta + \phi)}{2}$
39. $\cos(\theta)\cos(\phi) = \dfrac{\cos(\theta - \phi) + \cos(\theta + \phi)}{2}$
40. $\sin(\theta)\cos(\phi) = \dfrac{\sin(\theta + \phi) + \sin(\theta - \phi)}{2}$
41. $\sin(\theta) = \cos\left(\dfrac{\pi}{2} - \theta\right)$ 42. $\sin(\theta + \pi) = -\sin(\theta)$
43. $\cos(\theta + \pi) = -\cos(\theta)$ 44. $\tan(\theta + \pi) = \tan(\theta)$
45. Use the identity $\pi/3 + \pi/4 = 7\pi/12$ to calculate the six trigonometric functions at $7\pi/12$.
46. Use the Half-Angle Formulas to evaluate the six trigonometric functions at $\pi/8$.

In Exercises 47–50, derive the identity by using the identities in Exercises 38–40.

47. $\sin(\theta) + \sin(\phi) = 2\sin\left(\dfrac{\theta + \phi}{2}\right)\cos\left(\dfrac{\theta - \phi}{2}\right)$
48. $\sin(\theta) - \sin(\phi) = 2\cos\left(\dfrac{\theta + \phi}{2}\right)\sin\left(\dfrac{\theta - \phi}{2}\right)$
49. $\cos(\theta) + \cos(\phi) = 2\cos\left(\dfrac{\theta + \phi}{2}\right)\cos\left(\dfrac{\theta - \phi}{2}\right)$
50. $\cos(\theta) - \cos(\phi) = -2\sin\left(\dfrac{\theta + \phi}{2}\right)\sin\left(\dfrac{\theta - \phi}{2}\right)$
51. Suppose that A and B are constants. Use the Addition Formula for the sine to find an amplitude C and a phase shift ϕ (both in terms of A and B) such that

$$A\cos(\theta) + B\sin(\theta) = C\sin(\theta + \phi)$$

for every θ.
52. Let $y = mx + b$ be the equation of a nonhorizontal, nonvertical line. Such a line intersects the x-axis making an angle $\phi \in [0, \pi)$ with the positive x-axis. What is the relationship between m and ϕ? Can this relationship, suitably interpreted, be extended to vertical or horizontal lines? Why is ϕ independent of b?
53. The *angle of elevation* of a point above the earth is the acute angle between the horizontal and the line of sight. Suppose two points on the earth's surface have distance ℓ between them. Line ℓ is small enough for us to assume that the earth is flat between the two points. Suppose that θ and ϕ are the angles of elevation of a mountain peak relative to these two points (see Figure 11). Show that the height h of the mountain is given by

$$h = \frac{\ell}{|\cot(\theta) - \cot(\phi)|}.$$

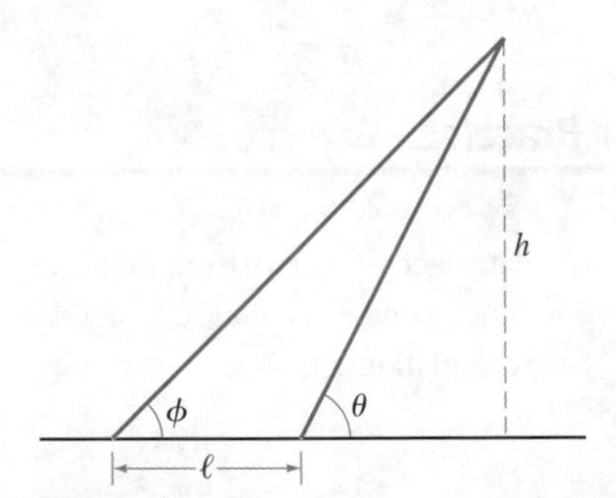

Figure 11

Suppose that a and b are positive constants. In Exercises 54–57, identify the curve that is parameterized by the equations. In each case, use a trigonometric identity to eliminate the parameter. (It may help to consider the special case $a = b = 1$ first.)

54. $x = a \cdot \cos(\theta), y = b \cdot \sin(\theta), \theta \in [0, 2\pi]$
55. $x = a \cdot \cos^2(\theta), y = b \cdot \sin^2(\theta), \theta \in [0, 2\pi]$
56. $x = a \cdot \sec(\theta), y = b \cdot \tan(\theta), \theta \in [0, \pi/2)$
57. $x = a \cdot \tan(\theta), y = b \cdot \cot(\theta), \theta \in (0, \pi/2)$

58. Suppose that h and k are constants and that a and b are positive constants. Parameterize the ellipse with center (h, k) and semi-axes a and b parallel to the x- and y-axes.

A function f is said to have *period* p if there is a smallest positive number p such that $f(x+p) = f(x)$ for all x in the domain of f. In Exercises 59–63, find the period of the function.

59. $x \mapsto \sin(x + \sqrt{3})$ **60.** $x \mapsto \cos(2\pi x)$
61. $x \mapsto \tan(x)$ **62.** $x \mapsto \sin(x) + \tan(x)$
63. $x \mapsto \tan(2x) + \sin(3x)$

64. The lowest monthly normal temperature of Philadelphia is 31°F and occurs in January. The highest monthly normal temperature of Philadelphia is 77°F and occurs in July. Find a model of temperature T as a function of time t that has the form $T(t) = b + A\sin(\omega t + \phi)$.

65. The lowest monthly normal temperature of Nome is 4°F and occurs in December. The highest monthly normal temperature of Nome is 51°F and occurs in July. Find a model of temperature T as a function of time t that has the form $T(t) = b + A\sin(\omega t + \phi)$.

66. The number of visitors to a tourist destination is seasonal. However, the long-term trend at this particular destination is down. A model for the number of tourists y as a function of time t measured in years is

$$y(t) = 5 \cdot 10^4 \frac{2 + \cos(2\pi t)}{1 + t^{1/4}}.$$

Here $t = 0$ corresponds to a time at which there were 150,000 tourists. Is y a periodic function of t? (Refer to the instructions for Exercise 59 for the definition of "period.") Sketch the graph of y as a function of t for $0 \leq t \leq 4$. From the sketch, y has *local maxima* at $t =$ 1, 2, 3, and so on. Define in your own words what the statement "a local maximum of y occurs at t_0" means.

67. A polygon is *regular* if all sides have equal length. For example, an equilateral triangle is a regular 3-gon (triangle) and a square is a regular 4-gon (quadrilateral). A polygon is said to be inscribed in a circle if all of its vertices lie on the circle.

a. Show that the perimeter $p(n, r)$ of a regular n-gon inscribed in a circle of radius r is

$$p(n, r) = 2rn\sin\left(\frac{\pi}{n}\right).$$

b. Show that the area $A(n, r)$ of a regular n-gon inscribed in a circle of radius r is

$$A(n, r) = r^2 n \sin\left(\frac{\pi}{n}\right)\cos\left(\frac{\pi}{n}\right) = \frac{1}{2}r^2 n \sin\left(\frac{2\pi}{n}\right).$$

Calculator/Computer Exercises

68. Graph $y = 1 - \cos(2x)$ and $y = \sin(x)$ in $[0.2, 1] \times [0, 1.5]$. Successively zoom in to the point of intersection to find a value $x_0 \in (0.2, 1)$ that satisfies $1 - \cos(2x_0) = \sin(x_0)$. Give x_0 to three decimal places. What is $\sin(x_0)$ to three decimal places? Use a trigonometric identity to find the exact value of $\sin(x_0)$.

69. Tabulate the values of the sequence $a_n = n \cdot \sin(1/n)$ for $n = 10^k$, $1 \leq k \leq 6$. What do you suppose the limit of a_n is as n tends to infinity? Let $f(x) = \sin(x)/x$ and notice that $a_n = f(1/n)$. Graph f in a viewing window that is appropriate for illustrating the limiting behavior of a_n for n tending to infinity. Explain how the limiting behavior of a_n can be determined from the graph.

70. What is the domain of $\tilde{f}(x) = \sin(x)/x$? Graph this function in the rectangle $[-2\pi, 2\pi] \times [-1/2, 1]$. Use the graph of $\tilde{f}$ to find a multicase function f that is defined on the entire real line, that has the same value as $\tilde{f}$ when evaluated at a point in the domain of $\tilde{f}$, and that has a graph with no jumps or gaps.

71. Graph $f(x) = 90x + \cos(x)$ in the viewing windows $[-10, 10] \times [-1000, 1000]$ and $[-0.0001, 0.0001] \times [0.5, 1.5]$. Explain why the graph of f appears to be a straight line in each of these windows. Which straight lines do these graphs appear to coincide with? Sketch the graph of the function f in a way that better displays its behavior.

72. In later chapters, we will investigate the approximation of complicated functions by simpler ones. For instance, the function $x \mapsto x$ is a good approximation to the function $x \mapsto \sin(x)\cos(x^2)$ for $|x|$ sufficiently small. Use a graphing utility to find an interval on which x approximates $\sin(x)\cos(x^2)$ with an absolute error of at most 0.01.

73. The monthly normal temperatures in degrees Fahrenheit for Raleigh, North Carolina, are 40, 42, 49, 59, 67, 74, 78, 77, 71, 60, 50, and 42 (starting with January). Plot these temperatures. Find and plot a continuous model.

Cosine waves are functions of the form $x \mapsto A\cos(\nu \cdot x + \phi)$ where A is the *amplitude* of the wave, ν the *frequency*, and ϕ the *phase shift*. The physical superposition of two waves is obtained mathematically by adding the two wave functions. Exercises 74–76 concern the superposition of cosine waves. The identities from Exercises 49 and 50 are particularly useful in this context.

74. For $\phi = 0, \pi/3, 2\pi/3$, and $3\pi/2$, graph the superposition $x \mapsto \cos(2x) + \cos(2x + \phi)$, $0 \leq x \leq 2\pi$. Each of these four curves is the graph of a wave of the form $x \mapsto A \cdot \cos(2x + \theta)$ for some amplitude A and phase shift θ, each depending on ϕ. Use your graph to determine A and θ when $\phi = 2\pi/3$. Use the identity from Exercise 49 to obtain a formula for A and θ. Calculate the magnitude of the superposed wave at $x = 2$ for the four given values of ϕ. Each of these signals is less than 1.5. For what value of ϕ is the magnitude of the superposed wave at $x = 2$ as large as it can be?

75. Suppose that ν_1 and ν_2 are both large compared with their absolute difference $|\nu_1 - \nu_2|$. The superposition $t \mapsto \cos(\nu_1 \cdot t) + \cos(\nu_2 \cdot t)$ might be described as a high-frequency cosine wave in which the beats have low-frequency *amplitude modulation*. Use Exercise 49 to express the superposition in the form $A(t) \cdot \cos(\omega \cdot t)$ where the frequency ω is about the same size as the individual frequencies ν_1 and ν_2 and where the amplitude $A(t)$ of the beats varies in time according to a low-frequency cosine wave. Illustrate with the graph of the superposed wave for $\nu_1 = 8$ and $\nu_2 = 6$. What is ω in this case? Add the graphs of $\pm 2\cos(t)$ to your figure. What is the frequency of the modulated amplitudes?

76. An *amplitude modulation* (AM) radio transmitter broadcasts an audio tone (or *baseband signal*) $\cos(\nu \cdot t)$ by modulating the amplitude of a very high-frequency *carrier signal* $A \cdot \cos(\omega \cdot t)$ by a positive factor $(1 + m \cdot \cos(\nu \cdot t))$, where the *modulation index* m is less than 1. Thus, the emitted signal has the form

$$S(t) = A \cdot (1 + m \cdot \cos(\nu \cdot t)) \cos(\omega \cdot t).$$

a. Use the identity from Exercise 49 to express $S(t)$ as a superposition of three cosine waves. The largest and smallest frequencies of these three waves are called *sidebands*.

b. Set $A = 1$, $m = 1/2$, $\nu = 2$, and $\omega = 8$. Graph the signal $S(t)$ and envelopes $\pm(1 + m \cdot \cos(\nu \cdot t))$ for $0 \le t \le 2\pi$. Explain how the sidebands can be determined visually from your graph.

c. Suppose that $\nu_{max} < \omega$ is the highest baseband frequency. The largest and smallest frequencies broadcast are called the *upper* and *lower sidebands*. What are they in terms of ω and ν_{max}? Their difference is called the *bandwidth*.

d. Carrier frequencies for AM radio range from 550 kHz (kilohertz), or 550×10^3 cycles per second, to 1610 kHz. Typically, ν_{max} can be as high as 15 kHz. An AM receiver works by recovering the audio source $\cos(\nu \cdot t)$ from the received signal $S(t)$, a process known as *demodulation*. Demodulation requires that the bandwidths of different radio stations that broadcast to the same location not overlap. What must the minimum frequency separation between the carrier signals of two such stations be?

77. Snell's inequality, discovered in 1621, states that

$$t \le \tan\left(\frac{t}{3}\right) + 2\sin\left(\frac{t}{3}\right),$$

for $0 < t < 3\pi/2$. An even older inequality,

$$\frac{3\sin(t)}{2 + \cos(t)} < t,$$

for all $t > 0$, was obtained in 1458 by Nicholas Cusa (1401–1464). Graphically illustrate each inequality for $0 < t < 0.01$.

Summary of Key Topics

Number Systems (Section 1.1)

In calculus, we deal with the three types of real numbers:

1. Terminating decimal expansions
2. Nonterminating decimal expansions that, after a finite number of terms, repeat a single block of terms endlessly
3. Nonterminating, nonrepeating decimal expansions

The first two types are rational numbers, and the third type consists of irrational numbers.

Sets of Real Numbers (Section 1.1)

Sets of numbers are usually described with the notation $\{x : P(x)\}$ and are graphed on a number line. Often a set is described by inequalities that must first be solved before the set can be easily sketched. The most frequently encountered sets are intervals.

$$
\begin{aligned}
(a, b) &= \{x : a < x < b\} \\
[a, b] &= \{x : a \leq x \leq b\} \\
[a, b) &= \{x : a \leq x < b\} \\
(a, b] &= \{x : a < x \leq b\} \\
(a, \infty) &= \{x : a < x\} \\
[a, \infty) &= \{x : a \leq x\} \\
(-\infty, a) &= \{x : x < a\} \\
(-\infty, a] &= \{x : x \leq a\} \\
(-\infty, \infty) &= \{x : -\infty < x < \infty\}
\end{aligned}
$$

Points and Loci in the Plane (Section 1.2)

In the plane, points are located using ordered pairs of real numbers, (x, y). The distance between two points (a_1, b_1) and (a_2, b_2) is given by the formula

$$d = \sqrt{(a_2 - a_1)^2 + (b_2 - b_1)^2}.$$

In general, we are interested in graphing sets of points that are described by equations. Such a graph is called the locus of the equation. For a given equation, a point (x, y) is on the graph if and only if the numbers x and y satisfy the equation simultaneously. An important instance is given by equations of the form

$$(x - h)^2 + (y - k)^2 = r^2$$

with $r > 0$. Such an equation describes a circle with center (h, k) and radius r.

Lines and Their Slopes (Section 1.3)

If ℓ is a line in the plane and if (x_1, y_1) and (x_2, y_2) are points on ℓ, then the slope of ℓ is given by the expression

$$\frac{y_2 - y_1}{x_2 - x_1}.$$

Parallel lines have the same slopes; perpendicular lines have slopes that are negative reciprocals. It takes two pieces of data to describe a line in the plane:

- a point through which the line passes and the slope, or
- two points through which the line passes, or
- two intercepts, or
- the slope and an intercept.

The following are three particularly useful forms of equations for lines.

1. *The Point-Slope Form* A line with slope m and passing through the point (x_0, y_0) has equation

$$y = m(x - x_0) + y.$$

2. *The Slope-Intercept Form* A line with slope m and y-intercept $(0, b)$ has equation

$$y = mx + b.$$

3. *The Intercept Form* A line with intercepts $(a, 0)$ and $(0, b)$ where a and b are both nonzero has equation

$$\frac{x}{a} + \frac{y}{b} = 1.$$

The equation for a line (except a vertical one) can always be obtained by finding two expressions for the slope and equating them.

Functions and Their Graphs (Section 1.4)

If S and T are sets, then a function is a rule that assigns to each element of S a unique element of T. We call S the domain of the function and T the range of the function. In calculus, S and T are usually sets of real numbers, and functions are usually described by formulas. Frequently, the formula is given and it is up to the reader to infer the domain and range of the function. The graph of a function consists of all points (x, y) in the plane such that x is in the domain of f and $y = f(x)$.

Combining Functions (Section 1.5)

Most functions that we encounter are built up from simple functions by using operations, such as addition, subtraction, multiplication, division, and composition, on functions. Composition is the most subtle. If f and g are functions and if the domain of g contains the range of f, then we define $(g \circ f)(x) = g(f(x))$. Care should be taken not to confuse composition with multiplication.

Trigonometric Functions (Section 1.6)

Radian measure of an angle is related to the degree measure of an angle by the formula

$$x \text{ radians} = \frac{180x}{\pi} \text{ degrees}$$

If θ is a real number, let $P(\theta)$ be the point of the unit circle cut off by rotating the positive x-axis by θ radians. The rotation is counterclockwise for $\theta > 0$ and clockwise for $\theta < 0$. We define $\cos(x)$ and $\sin(x)$ to be the abscissa and ordinate, respectively, of $P(\theta)$. The other trigonometric functions are defined as quotients: $\tan(x) = \sin(x)/\cos(x)$, $\cot(x) = \cos(x)/\sin(x)$, $\csc(x) = 1/\sin(x)$, and $\sec(x) = 1/\cos(x)$. There are many identities among the trigonometric functions, and most can be derived from

$$\cos^2(\theta) + \sin^2(\theta) = 1,$$
$$\sin(\theta + \phi) = \sin(\theta)\cos(\phi) + \cos(\theta)\sin(\phi),$$

and

$$\cos(\theta + \phi) = \cos(\theta)\cos(\phi) - \sin(\theta)\sin(\phi).$$

genesis & DEVELOPMENT

In this first chapter, we reviewed three fundamental developments that underlie calculus: the real number line, the mathematical notion of a function, and the principles of analytic geometry.

Irrational Numbers and Successive Approximation

The history of the real number line begins in the Golden Age of ancient Greece. The first important step came with the realization that not every number is rational. Some time before 410 BCE, the followers of Pythagoras (ca. 580–500 BCE) discovered that $\sqrt{2}$ is irrational. The elegant argument outlined in Exercise 57, Section 1.1, is the earliest proof that has been preserved. It was recorded by Aristotle (384–322 BCE) but is certainly older.

The mathematicians of ancient Greece devised a number of methods for successively approximating $\sqrt{2}$ and other irrational numbers. One of these algorithms is based on finding integer solutions to the equation $x^2 - 2y^2 = 1$ (see Exercise 56, Section 1.4). Archimedes (ca. 287–212 BCE) used the same idea to approximate $\sqrt{3}$ to within 2.74×10^{-7}. He obtained his estimate, $1351/780$, from the solution $x = 1351$, $y =$, 780 of the equation $x^2 - 3y^2 = 1$.

Equations of the form $x^2 - D \cdot y^2 = 1$, with D a nonsquare integer, are called *Pell equations,* and Archimedes seems to have known a thing or two about their solution. In 1773, a copy of a previously unknown message from Archimedes to the mathematicians of Alexandria was discovered in a German library. In his letter, Archimedes posed a riddle concerning the numbers of eight types of cattle. The key to the cattle problem, as it has come to be known, is the solution in integers of the Pell equation $x^2 - 4729494 \cdot y^2 = 1$. In 1880, A. Amthor obtained the least positive integer solution

$$x_0 = 109931986732829734979866232821433543901088049,$$
$$y_0 = 50549485234315033074477819735540408986340,$$

which enabled him to show that the solution to Archimedes' problem has 206,545 decimal digits. In 1981, a Cray I supercomputer at the Lawrence Livermore National Laboratory calculated this large number, using about 10 minutes of computing time. Ironically, the life span of the computer that solved this ancient problem was only about a dozen years. Having been purchased in the late 1970s for $19,000,000, it was auctioned off in 1993 for $10,000.

The Number π

The irrational number π also featured prominently in the mathematics of the ancient Greeks. Archimedes proved that if A and C are the area and circumference of a circle of radius r, then the ratios A/r^2 and $C/2r$ are the *same* constant π. (It is not known who discovered this fact or even when the discovery was made.) Archimedes also approximated π to two decimal places by deriving the inequalities $3 + 10/71 < \pi < 3 + 1/7$. The computation of the digits of π has attracted the interest of mathematicians ever since. Sir Isaac Newton (1642–1727), for example, calculated π to 15 decimal places. By 1761, 113 digits of π were known. In that year, the Swiss mathematician Johann Heinrich Lambert (1728–1777) published a proof that π is irrational. In 1806, Adrien-Marie Legendre (1752–1833) proved the stronger result that π^2 is irrational, and he conjectured even more.

We say that a real number is *algebraic* if it is a root of some polynomial with rational coefficients. Otherwise, we say that the number is *transcendental.* Every rational number is algebraic, but an irrational number may be algebraic or transcendental. For example, the irrational number $\sqrt{2}$ is algebraic because it is a root of $x^2 - 2$. At the time Legendre ventured his conjecture, mathematicians did not even know if transcendental numbers existed. It was not until 1844 that Joseph Liouville (1809–1882) established the existence of transcendental numbers. Thirty years later, Georg Cantor's (1845–1918) novel ideas about infinite sets put the matter in a totally new light.

Using simple but original ideas, Cantor proved that algebraic numbers can be put into one-to-one correspondence with natural numbers and that transcendental numbers cannot. We say that algebraic numbers are *countable,* or *denumerable,* because they have this property and that transcendental numbers are *uncountable,* or *nondenumerable*. In a very real sense, Cantor proved that there are more transcendental numbers than algebraic numbers. He did not, however, show that any one particular number is transcendental.

Indeed, it is usually very difficult to show that a given number is transcendental. In 1882, Carl Louis Ferdinand von Lindemann (1852–1939) finally proved Legendre's conjecture that π is transcendental. Meanwhile, the digit hunters continued their work. In September 1999, a supercomputer at the University of Tokyo was used to calculate more than 200 billion digits of π.

Zeno's Paradoxes

The fundamental idea of calculus, that which distinguishes it from algebra and classical geometry, is the infinite process. We have mentioned the algorithms by which Greek mathematicians created *infinite* sequences, which tend to $\sqrt{2}$ and $\sqrt{3}$ in the limit. We have discussed the manner in which Archimedes inscribed an *infinite* sequence of regular polygons in a circle in order to *successively approximate* its area and circumference. A number of other infinite processes have been outlined in the exercises of Chapter 1. The idea is not to actually perform an infinite number of operations but to see what results in the limit. That is the essence of calculus. Although the mathematicians of ancient Greece who originated these ideas were well aware of the power of the infinite process, they were also concerned with the logical difficulties that can ensue.

Zeno of Elea (ca. 496–435 BCE) lived in the age of Pericles. Little is known of Zeno's life. As to his death, it is written that Zeno was tortured and executed for his part in a conspiracy against the state. Although the original writings of Zeno have not survived, numerous references to them have. Zeno is best known today for four paradoxes—the *Dichotomy,* the *Achilles,* the *Arrow,* and the *Stade*—that Aristotle retold.

According to the *Dichotomy,* "There is no motion because that which is moved must arrive at the middle before it arrives at the end, and it must traverse the half of the half before it reaches the middle, and so on ad infinitum." Thus, if a line is infinitely divisible, motion cannot begin. The *Achilles* asserts that, given a head start, a tortoise will not be overtaken by the runner Achilles because Achilles must first reach the point P_1 at which the tortoise started. In the time required for Achilles to arrive at P_1, the tortoise will have advanced to a point P_2, which Achilles must reach before overtaking the tortoise. By the time Achilles has run from P_1 to P_2, the tortoise will have moved ahead to a point P_3, and so on, ad infinitum. The four paradoxes are "immeasurably subtle and profound," in the words of Bertrand Russell. Aristotle called them fallacies but was not able to refute them. Indeed, it would be a long time before mathematicians created a framework in which they could undertake the infinite processes of calculus free from such logical difficulties.

Analytic Geometry

After the decline of Greek civilization, mathematical progress came slowly. The next significant contribution to the foundation of calculus came in the 14th century when philosophers in England and France attempted to analyze accelerated motion. In doing so, the Norman scholar Nicole Oresme (1323–1382) identified the real numbers with a geometric continuum. After another long hiatus, François Viète (1540–1603) further demonstrated the power of geometrically interpreting the real numbers. His work set the stage for the development of analytic geometry by Pierre de Fermat (1601–1665) and René Descartes (1596–1650).

Analytic geometry, as developed in the 17th century by Fermat and Descartes, was the final breakthrough that permitted the discoveries of calculus. The crucial idea that the solution set of an equation such as $F(x, y) = 0$ can be identified with a curve in the plane appeared in the *Géométrie,* which Descartes published in 1637. In the words of Descartes, "I believed that I could borrow all that was best both in geometrical analysis and in algebra, and correct all the defects of the one by help of the other."

Descartes led something of a fugitive existence. He left France for Holland, living there the better part of his adult life but changing residence 24 times in 21 years. His motto was "He who hides well lives well." In 1649, Descartes accepted an invitation to tutor Sweden's Queen Christina in philosophy, but he died of pneumonia soon after his arrival. The mortal remains of Descartes had a macabre afterlife that mirrored his transient lifestyle. Issues of religious affiliation led to an initial burial in a Stockholm cemetery ordinarily reserved for unbaptized children. Sixteen years later, a delegation was dispatched to repatriate the remains of France's most distinguished philosopher. As the journey was long, the coffin was short. The body of Descartes, less its skull, returned home to be interred first in the Church of St. Paul, Paris, and then, in order, the (former) Church of Ste. Genevieve-du-Mont, the courtyard of the Louvre, the Pantheon, the garden of the Musée des Monuments Français, and, finally, the Church of St. Germain-des-Pres, Paris. In a chapel to the south of the choir, a marble slab bearing an 1819 inscription marks the spot. The skull of Descartes, returned to France three years too late for a reunion, is now in the collection of Paris's Musée de l'Homme.

The Scientific Revolution

Prior to the 16th century, most scientific teaching in Europe consisted of scholars passing a fixed collection of theories and ideas to their students. These theories were not to be questioned or analyzed—merely learned. The method of experimentation and proof had not yet become a universally accepted part of scientific activity. Ancient authority ruled the sciences.

During the Age of Reason, a revolution took place. The so-called rationalist movement in philosophy demanded that new ideas be demonstrated or proved. Descartes became one of the most influential exponents of the movement. Today, the rationalist methodology is universally accepted. When a calculus instructor insists on *proving* assertions in class, it is because of a deeply held conviction that students should not simply learn facts by rote but should truly understand the concepts being taught.

If the 16th century had brought a revolution in scientific thinking, then the 17th century ushered in the era of *quantitative* relationships. In *The Assayer,* Galileo (1564–1642) published the manifesto of the new science:

> *Philosophy is written in this grand book, the universe which stands continually open to our gaze. But the book cannot be understood unless one first learns to comprehend the language and read the letters in which it is composed. It is written in the language of mathematics and its characters are triangles, circles, and other geometric figures.*

Scientists now looked for *functional* relationships between the variables of nature. Galileo determined the distance an object falls under gravity as a function of time. Johannes Kepler (1571–1630) discovered that the time in which a planet orbits the sun is a function of its mean distance to the sun (Kepler's Third Law). Christiaan Huygens (1629–1695) determined centrifugal force as a function of radius and velocity, providing a link between the laws of Kepler and Galileo. Before the end of the 17th century, Newton formulated the law of universal gravitation, demonstrating that Kepler's laws and Galileo's law are but different aspects of the same general phenomenon.

The realization that natural phenomena could be expressed through functional relationships was not at all limited to motion. The Dutch scientist Willebrord Snell (1580–1626) found that the angle of refraction of a ray of light is a function of the angle of incidence (Snell's law). Evangelista Torricelli (1608–1647) demonstrated that the height of mercury in a vertical tube is a function of atmospheric pressure. Robert Boyle (1627–1691) postulated the functional relationship among the pressure, volume, and temperature of a gas. Robert Hooke (1635–1703) expressed the force exerted by a spring as a function of the amount of its compression or extension. The age of functions had arrived. Before the end of the 17th century, calculus would be invented as a tool to study these functions.

The Completeness Property of the Real Numbers

Although scientific observation validated the phenomena predicted by calculus, the new discipline quickly came under serious criticism concerning the foundation of the entire subject. Just as the ancient Greeks had opened the door to Zeno's paradoxes when they admitted the infinite process, so had the mathematicians of the 17th century set themselves up for a new set of paradoxes by their lack of rigor in working with the real number line.

The real number system, $\mathbb{R}$, has familiar algebraic properties. For example, in calculating the sum $a + b$, we can add a to b or we can add b to a. By virtue of its algebraic properties, we say that $\mathbb{R}$ is a *field*. Because $\mathbb{R}$ also has the binary relation $>$ with its familiar properties, we say that $\mathbb{R}$ is an *ordered field*. It is not the only one, however; the rational number system $\mathbb{Q}$ is another ordered field. The property that distinguishes the real field from the rational field is the *completeness property* of the real numbers, which guarantees that there are no holes or gaps in the real number line. Here is one way to describe the completeness of the real numbers: If $\{I_n\}$ is a sequence of closed bounded nonempty *nested* real intervals, then there is at least one point in $\mathbb{R}$ that is in every I_n. Nested means that $I_{n+1} \subset I_n$, or each interval is a subinterval of its predecessor.

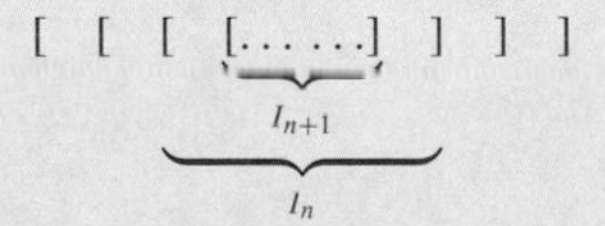

The real numbers satisfy this property and are complete; the rationals are not. In 1872, several mathematicians, including Georg Cantor, Julius Richard Dedekind (1831–1916), and Karl Weierstrass (1815–1897), enlarged $\mathbb{Q}$, preserving its algebra and order relations, to obtain a complete ordered field $\mathbb{R}$. Although many further discoveries would be made, by 1872 everything was in place for a rigorous development of calculus, a subject that was then already 200 years old.

2

Limits

PREVIEW

The concept of limit is the basis for a solid understanding of calculus, which is why we will now spend an entire chapter studying limits before proceeding to calculus. The notion of limit originated in Greece approximately 2500 years ago. It arose then for many of the same reasons that we study limits in a modern calculus course: to understand continuous motion and to measure lengths, areas, volumes, and other physical quantities.

Consider a circle $\mathcal{C}$ of radius r. We all learn in geometry that its area A is πr^2 and its circumference C is $2\pi r$, but what exactly do we mean by the area and circumference of a circle? Precise answers can be understood only in terms of limits.

One approach is to approximate the area and circumference of $\mathcal{C}$ by a region with an area and perimeter that we do know. Figure 1 shows an equilateral triangle (a regular 3-gon) $\mathcal{P}_3$ inscribed in $\mathcal{C}$. We know the area and perimeter of the triangle $\mathcal{P}_3$, but it is not a very good approximation to the circle. We add sides, one at a time, to the inscribed figure to create inscribed polygons with more and more sides of equal length. It seems clear that the inscribed polygon is becoming closer and closer to the circle. Figure 2 shows that the area inside the inscribed 6-gon is fairly close to the area of the circle, and the area inside the inscribed 12-gon is *extremely* close to that of the circle. When the polygon has one billion sides, we could not possibly draw a figure illustrating the difference between the inscribed polygon and the circle, *but there is a difference.* No

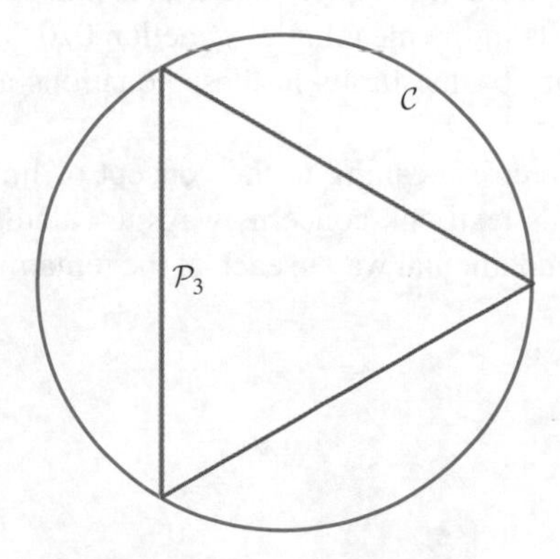

Figure 1
An equilateral triangle inscribed in circle $\mathcal{C}$

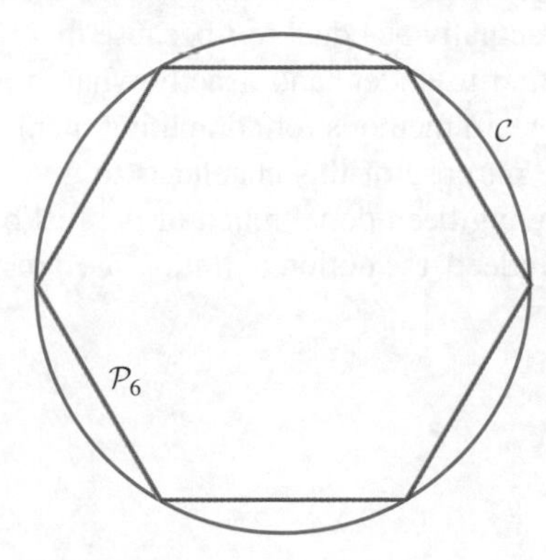

Figure 2a
A 6-gon inscribed in circle $\mathcal{C}$

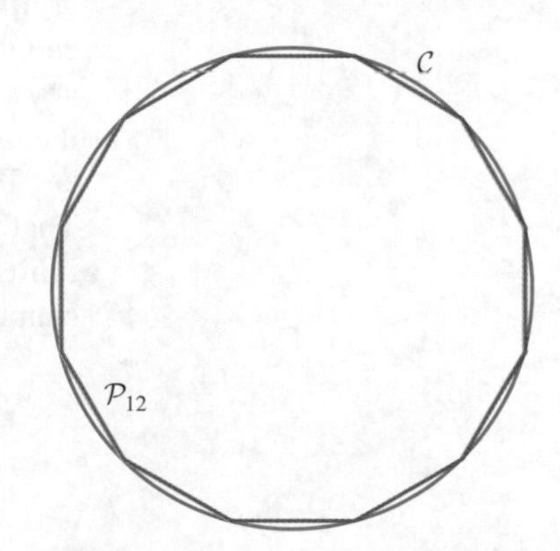

Figure 2b
A 12-gon inscribed in circle $\mathcal{C}$

matter how small the sides of the polygon become, the polygon will have many small flat sides of equal length, *even though* the circle is round. It is *correct* to say that when the side length of the polygon is close to zero (that is, the number of sides of the polygon is large), the area inside the polygon *is close to* the area inside the circle. It is *incorrect* to say that the area of the polygon is ever *equal* to that of the circle.

The correct mathematical terminology is that as the number of sides n tends to infinity, the *limit* of the area A_n inside the polygon $\mathcal{P}_n$ equals the area A inside the circle and the *limit* of the perimeter C_n of $\mathcal{P}_n$ equals the circumference C of the circle. We write this as

$$A = \lim_{n \to \infty} A_n \quad \text{and} \quad C = \lim_{n \to \infty} C_n. \qquad \textbf{(2.1)}$$

For another application of the *limit* concept, suppose that a training film is taken of a runner. The film shows elapsed time and distance markers that allow us to measure the distance $s(t)$ the athlete has run in any given time t. It is easy to compute the *average* speed of the runner between any two points: Simply divide the total distance run by the total elapsed time. However, the speed of the runner will vary from one instant of time to another. How can the runner's speed $v(t)$ at a given instant of time t be calculated? For any choice of time $u \neq t$, the average speed of the runner in the small time interval $[t, u]$ is

$$\frac{s(u) - s(t)}{u - t}.$$

For u close to t, this will give a good approximation to $v(t)$, because the runner's speed does not change much in the small time interval $[t, u]$. If we press this point further, then we intuitively arrive at the concept of *instantaneous velocity at* t:

$$v(t) = \lim_{u \to t} \frac{s(u) - s(t)}{u - t}. \qquad \textbf{(2.2)}$$

As we will see, equations (2.1) and (2.2) actually serve to *define* the concepts of area, circumference, and instantaneous velocity. These examples convey an important idea: The concept of *limit* involves a variable that approaches arbitrarily close to a number but does not actually reach it. In equation (2.1), the natural number n can get arbitrarily large, but it is never actually equal to infinity. In equation (2.2), the number u cannot actually be equal to t because that results in the meaningless fraction $0/0$. The key is first to understand exactly what is meant by the limits in these equations and then to learn methods for computing them.

The purpose of this chapter is to give a precise meaning to the concept of limit. That having been done, much of the rest of this textbook concerns ways to calculate limits. Indeed, the notion of limit is used in a fundamental way in each of the remaining chapters.

2.1 The Concept of Limit

This section includes a number of examples that informally explore the idea of limit. The understanding of limits gained by reading this section is intuitive, but it is adequate for most purposes in calculus. Section 2.2 presents a precise formulation of the limit concept.

Informal Definition of Limit

Let f be a real-valued function that is defined in an open interval to the left of a real number c and in an open interval to the right of c. We say, "$f(x)$ has the limit ℓ as x tends to c," and we write

$$\lim_{x \to c} f(x) = \ell$$

if the values $f(x)$ are close to ℓ when x is close to, but not equal to, c. Sometimes we write

$$f(x) \to \ell \quad \text{as} \quad x \to c$$

and say "$f(x)$ tends to ℓ as x tends to c" (see Figure 1).

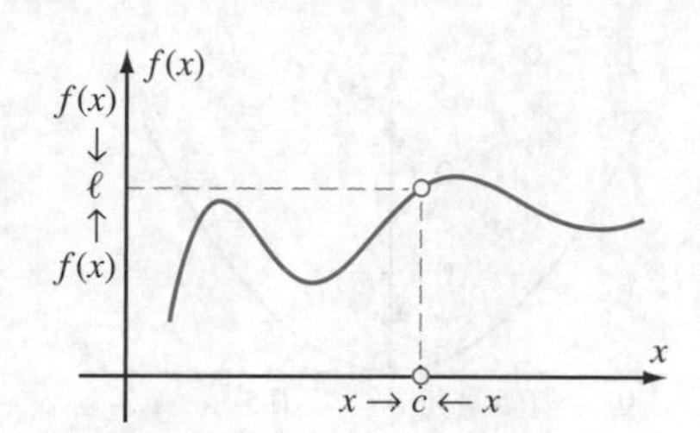

Figure 1
$f(x)$ has the limit ℓ as x tends to c.

inSIGHT

The value $f(c)$ plays no role in the definition of the limit of $f(x)$ as x tends to c. *We do not even assume that* f *is defined at* c. A definition of "limit" that requires f to be defined at c would not apply to many important applications. Even when $f(c)$ is defined, the value $f(c)$ does not have to be related to $\lim_{x\to c} f(x)$ in any way. In a sense, $\lim_{x\to c} f(x)$ is what we *anticipate* that $f(x)$ will equal at $x = c$, not necessarily what $f(x)$ actually equals when $x = c$.

Example 1 Discuss the limit of $f(t) = t^2$ as t approaches 0. Suppose that $g(t)$ is not defined when $t = 0$ but $g(t) = t^2$ for $t \neq 0$. Does $g(t)$ have a limit as t approaches 0? Does the function

$$h(t) = \begin{cases} t^2 & \text{if } t \neq 0 \\ 1 & \text{if } t = 0 \end{cases}$$

have a limit as t approaches 0?

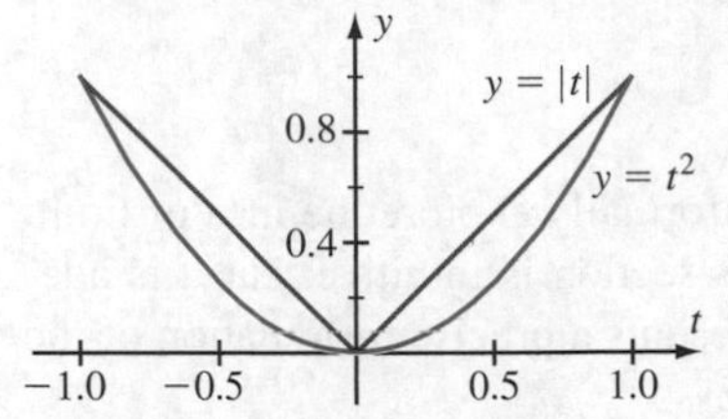

Figure 2

Solution When $|t| < 1$, we have $t^2 = |t| \cdot |t| < 1 \cdot |t| = |t|$. Thus, when $|t|$ is less than 1 (as is ultimately the case when t approaches 0), the number $f(t) = t^2$ is even closer to 0 than is $|t|$ (see Figure 2). We conclude that as t approaches 0, so does $f(t)$. We write this as $\lim_{t\to 0} f(t) = 0$.

Next, we consider g. The fact that 0 is not in the domain of g does not affect whether $g(t)$ has a limit at $t = 0$. For $t \neq 0$, we have $g(t) = t^2$, and our calculation of the limit of $f(t)$ at $t = 0$ allows us to conclude that $\lim_{t\to 0} g(t) = \lim_{t\to 0} t^2 = 0$.

Finally, we turn to the function h. The definition of "limit" tells us that the equation $h(0) = 1$ has *nothing* to do with the existence or value of $\lim_{t\to 0} h(t)$. When t is near 0 but unequal to 0, we have $h(t) = t^2$. Therefore, $\lim_{t\to 0} h(t) = \lim_{t\to 0} t^2 = 0$. Figure 3 illustrates the three limits that we have calculated.

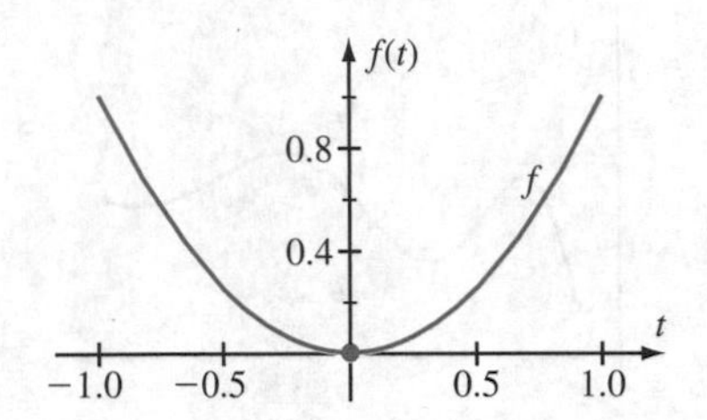

Figure 3a
$\lim_{t\to 0} f(t) = 0$

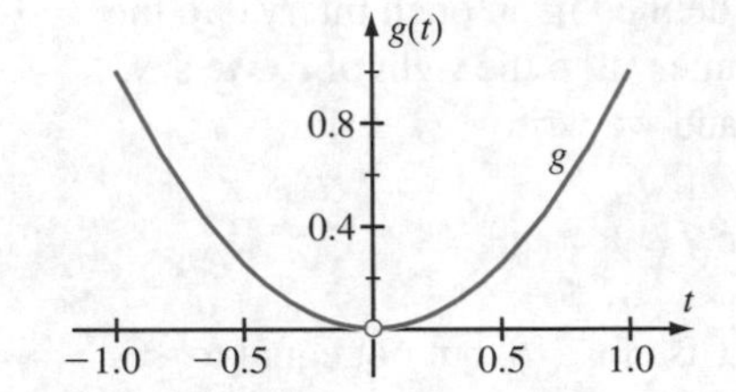

Figure 3b
$\lim_{t\to 0} g(t) = 0$

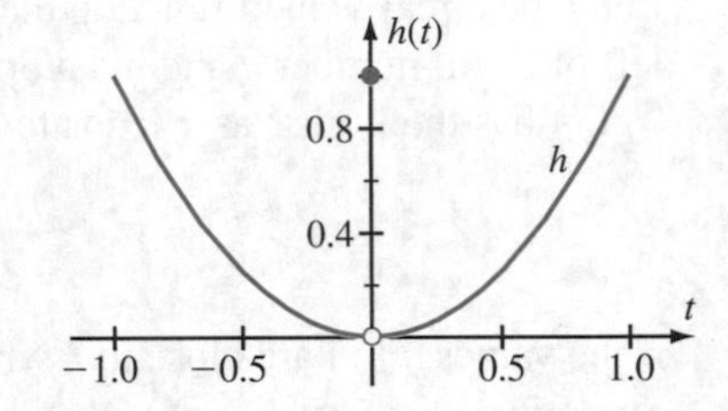

Figure 3c
$\lim_{t\to 0} h(t) = 0$

■

Example 2 Let $f(x) = (x^2 - 6x + 6)/x$. Evaluate $\lim_{x\to 5} f(x)$.

Solution We look at the components of f. As x tends to 5, the expression x^2 tends to 25 and the expression $6x$ tends to 30. We conclude that

$$\lim_{x\to 5} f(x) = \frac{25 - 30 + 6}{5} = \frac{1}{5}.$$

A numerical investigation, as shown in the table, supports this conclusion.

Approach from the left side of 5

$x \to 5^-$	4.99	4.999	4.9999	4.99999
$f(x)$	0.19240	0.19924	0.19992	0.19999

Approach from the right side of 5

5.00001	5.0001	5.001	5.01	$5^+ \leftarrow x$
0.20000	0.20008	0.20076	0.20760	$f(x)$

■

inSIGHT

In Example 2, we used the plausible idea that the limit of a sum or difference is the sum or difference of the limits, the limit of a product is the product of the limits, and the limit of a quotient is the quotient of the limits (as long as we do not divide by zero). These limit calculations are treated in more detail in Section 2.2.

One-Sided Limits

inSIGHT The notation for one-sided limits arises from the $c = 0$ case. As x approaches 0 from the left, x assumes negative values; hence, $x \to 0^-$. As x approaches 0 from the right, x assumes positive values; therefore, $x \to 0^+$. Keep this in mind to help remember the notation.

Implicit in the limit definition is the fact that a function should have just one limit at a point c. Suppose that $f(x)$ is close to a number ℓ_L when x is close to c and to the left of c and that $f(x)$ is close to a different number ℓ_R when x is close to c and to the right of c. We say that f has a *left limit* ℓ_L at c and a *right limit* ℓ_R at c, but we do *not* say that f has a limit at c (see Figure 4). These *one-sided limits* are written as $\lim_{x \to c^-} f(x) = \ell_L$ (or $f(x) \to \ell_L$ as $x \to c^-$) for the left limit and $\lim_{x \to c^+} f(x) = \ell_R$ (or $f(x) \to \ell_R$ as $x \to c^+$) for the right limit. From these informal definitions, it is clear that

$$\lim_{x \to c} f(x) = \ell \quad \text{is equivalent to} \quad \lim_{x \to c^-} f(x) = \ell_L, \ \lim_{x \to c^+} f(x) = \ell_R, \quad \text{and} \quad \ell_L = \ell_R.$$

In other words, if both left and right limits exist and are the same, then the limit *does* exist.

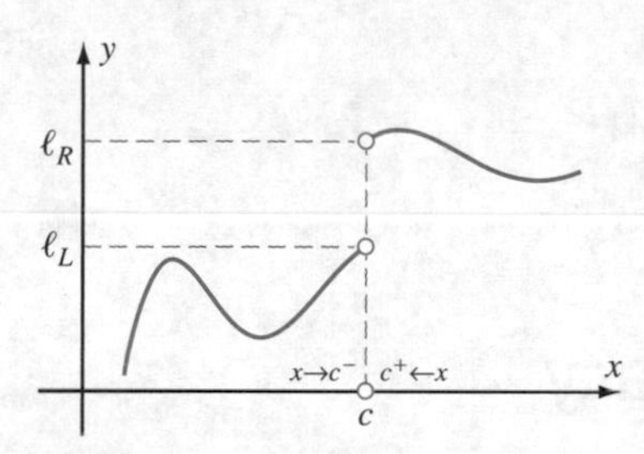

Figure 4
$\lim_{x \to c^+} f(x) = \ell_R$,
$\lim_{x \to c^-} f(x) = \ell_L$

Example 3 Let H and G be defined by

$$H(x) = \begin{cases} 1 & \text{if } x < 3 \\ 2 & \text{if } x \geq 3 \end{cases} \quad \text{and} \quad G(s) = \begin{cases} s^2 + 1 & \text{if } s < 4 \\ 5s - 3 & \text{if } s > 4 \end{cases}.$$

Does $H(x)$ have a limit as x tends to 3? Does $G(s)$ have a limit as s tends to 4?

Solution Refer to Figure 5 for the graph of H. As x approaches (but does not equal) 3 from the left, $H(x)$ tends to 1. On the other hand, as x approaches (but does not equal) 3 from the right, $H(x)$ tends to 2. Thus, $\lim_{x \to 3^-} H(x) = 1$ and $\lim_{x \to 3^+} H(x) = 2$. However, $H(x)$ does not have a limit as x tends to 3 because there is *no single value* that $H(x)$ approaches as x tends to 3.

When s is near 4 but to the left of 4, s^2 is near 16 and $s^2 + 1$ is near 17. Thus, $\lim_{s \to 4^-} G(s) = 17$. On the other hand, when s is near 4 but to the right of 4, $5s$ is near 20 and $5s - 3$ is near 17. Thus, $\lim_{s \to 4^+} G(s) = 17$. Because $G(s)$ tends to 17 from the left *and* the right, we can say that $\lim_{s \to 4} G(s) = 17$. This limit can also be seen from the graph of G (see Figure 6).

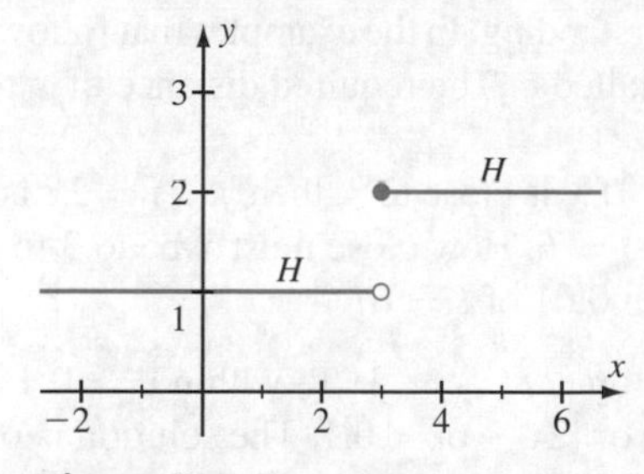

Figure 5
$H(x) = \begin{cases} 1 & \text{if } x < 3 \\ 2 & \text{if } x \geq 3 \end{cases}$

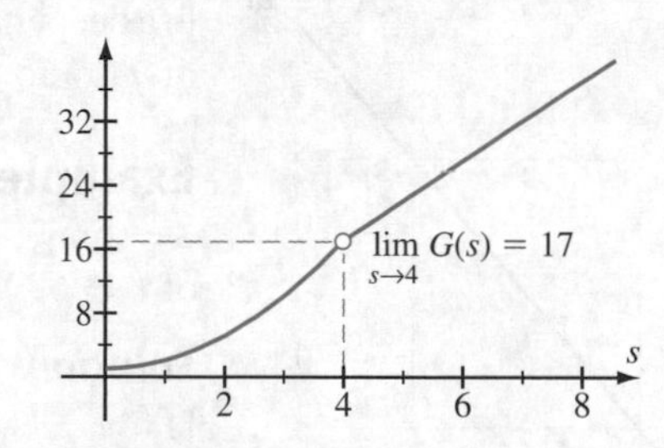

Figure 6
$G(s) = \begin{cases} s^2 + 1 & \text{if } s < 4 \\ 5s - 3 & \text{if } s > 4 \end{cases}$

■

Limits Using Algebraic Manipulations

In many situations, some algebraic manipulations are required in order to evaluate a limit.

Example 4 Does

$$f(x) = \frac{x^2 - 4}{x + 2}$$

have a limit as x approaches -2?

Solution This type of limit often occurs in calculus. The function f is *not* defined at the point where the limit is being calculated. We therefore cannot attempt a shortcut by simply substituting $x = -2$. Instead, we attack this problem by noticing that

$$f(x) = \frac{(x+2)(x-2)}{(x+2)}.$$

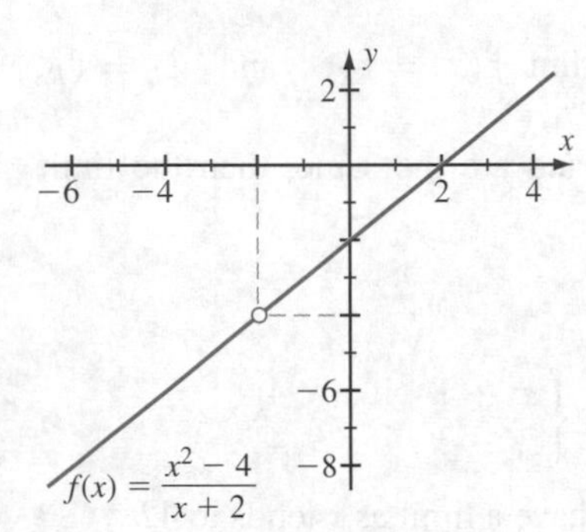

Figure 7

As long as $x \neq -2$, then $x + 2 \neq 0$. We can divide this common nonzero factor from the numerator and denominator. As a result, $f(x) = x - 2$ when $x \neq -2$. Remember, in calculating the limit of f at -2, we consider the values of $f(x)$ for x *close to, but not equal to,* -2. Therefore, as $x \to -2$, $f(x)$ tends to the same limit as $x - 2$ does. In summary,

$$\lim_{h \to -2} \frac{x^2 - 4}{x + 2} = \lim_{x \to -2} (x - 2) = -4.$$

Refer to Figure 7. ■

Specified Degrees of Accuracy

We turn now to a practical, computation-oriented approach to the limit concept. By $\lim_{x \to c} f(x) = \ell$, we mean that $f(x)$ tends to ℓ when x tends to c. Now we ask:

> How close does x need to be to c to force $f(x)$ to stay within 0.1 of ℓ? How close does x need to be to c to force $f(x)$ to stay within 0.01 of ℓ? and so on.

This method of thinking about limits is important for applications while also leading to a deeper understanding. In the examples that follow, the bound on the desired distance of $f(x)$ to ℓ is called ϵ. The required distance of x to c is called δ.

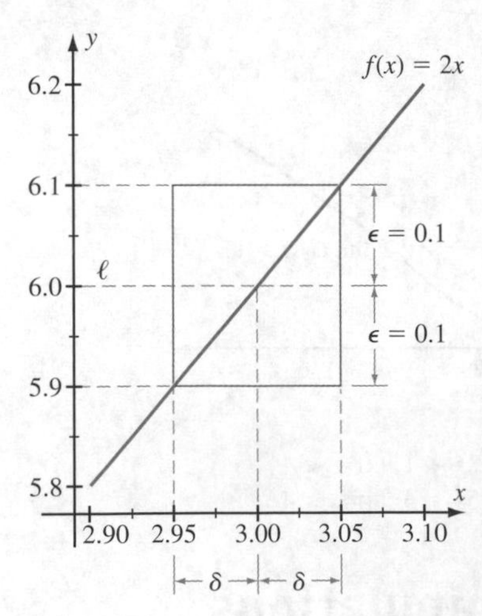

Figure 8
$\delta = 0.05, \epsilon = 0.1$

Example 5 If x is close to 3, then $f(x) = 2x$ is close to 6. Therefore, we say that $\ell = \lim_{x \to 3} f(x) = 6$. How close must x be to 3 to force $f(x) = 2x$ to be within 0.1 of $\ell = 6$? Within 0.01 of $\ell = 6$?

Solution To force $f(x) = 2x$ to within $\epsilon = 0.1$ of $\ell = 6$, we solve the inequality $|f(x) - \ell| < \epsilon$, or $|2x - 6| < 0.1$. The solution is obtained by multiplying both sides by $1/2$, which gives us $|x - 3| < 0.05$. Therefore, provided x stays within $\delta = 0.05$ of $c = 3$, $f(x)$ will stay within $\epsilon = 0.1$ of $\ell = 6$ (see Figure 8). Similarly, to make $f(x)$ stay within $\epsilon = 0.01$ of $\ell = 6$, we solve the inequality $|2x - 6| < 0.01$. The solution is $|x - 3| < 0.005$. Therefore, $f(x) = 2x$ stays within $\epsilon = 0.01$ of $\ell = 6$, provided that x stays within $\delta = 0.005$ of $c = 3$. There is clearly a pattern here. If we want to make $f(x) = 2x$ stay within a certain distance ϵ of $\ell = 6$, then we restrict x to lie within $\delta = \epsilon/2$ of $c = 3$. We can achieve any desired *degree of accuracy* by taking ϵ as small as we like. Our calculation offers compelling evidence of a quantitative nature that justifies the intuitive feeling that $\lim_{x \to 3} 2x = 6$. ■

Example 6 Let $f(x) = x^2 + 4$. How close to 0 does x have to be to force $f(x)$ to be within $\epsilon = 0.1$ of $\ell = 4$?

Solution Although it is intuitively clear that $\lim_{x \to 0} f(x) = 4$, let us examine the matter from the point of view of approximation, as in Example 5. To force $f(x) = x^2 + 4$ to be within $\epsilon = 0.1$ of $\ell = 4$, we solve the inequality

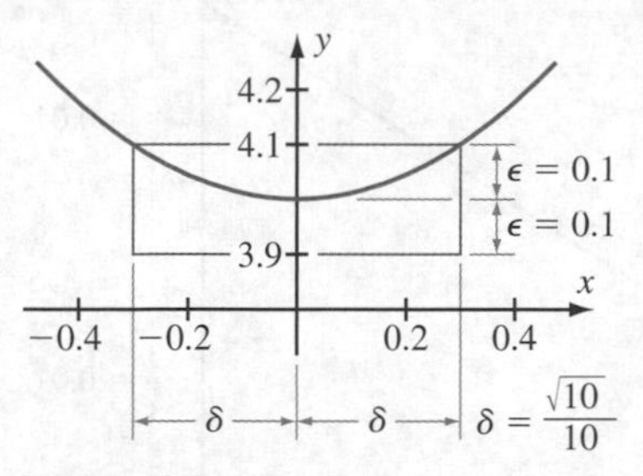

Figure 9

$$|f(x) - 4| < 0.1.$$

The solution is $|(x^2 + 4) - 4| < 0.1$, or $|x^2| < 0.1 = 10/100$, or $|x| < \sqrt{10}/10$. Since we are interested in the distance of x from 0, we write

$$|x - 0| < \frac{\sqrt{10}}{10}.$$

Thus, $f(x) = x^2 + 4$ is within $\epsilon = 0.1$ of $\ell = 4$ when x is within $\delta = \sqrt{\epsilon} = \sqrt{10}/10$ of $c = 0$ (see Figure 9). We arrive at our conclusion by working backward from the desired estimate on the function x^2. Let us now check our work: If $|x - 0| < \sqrt{10}/10$, then $|x^2| < 10/100$, or $|(x^2 + 4) - 4| < 0.1$. This means that $|f(x) - 4| < 0.1$, which is the desired result. Once we become accustomed to this process, we can work backward with confidence and not bother to check the work by repeating the calculation in the forward direction. ■

The degree of closeness for x to c in Example 6 (namely, $\delta = \sqrt{\epsilon}$) is different from that in Example 5 (namely, $\delta = \epsilon/2$). Different functions approach their limits at different rates. This will be one of the main points in the limit problems that we study.

Graphical Methods

Examples 5 and 6 provide algebraic solutions to the question, "For a given small number ϵ, how can we take x close enough to c to force $f(x)$ to be within ϵ of ℓ?" We may also answer this question by graphical methods. Indeed, the condition that $f(x)$ stays within ϵ of ℓ when x stays within δ of c is easily visible when we plot f in the viewing rectangle $[c - \delta, c + \delta] \times [\ell - \epsilon, \ell + \epsilon]$: The graph exits the window only through the lateral sides—not through the top or bottom of the viewing window.

Example 7 Let $f(x) = (x^2 - 6x + 6)/x$. In Example 2, we observed that $\lim_{x \to 5} f(x) = 1/5$. Find a positive number δ such that $f(x)$ stays within 0.01 of $1/5$ when x stays within δ of 5.

Solution In this example, we set $c = 5$, $\ell = 1/5$, and $\epsilon = 0.01$. We begin by making an arbitrary choice of δ: $\delta = 0.05$. Figure 10, next page, shows the graph of f in the viewing window $[5 - 0.05, 5 + 0.05] \times [1/5 - 0.01, 1/5 + 0.01]$. Since the graph of f exits this window through its top and bottom, we must use a smaller value for δ. To determine such a value, we observe that on the horizontal axis in Figure 10, the distance between tick marks is 0.004. From the plot, we can see that $f(x)$ stays within $\epsilon = 0.01$ of $\ell = 1/5$, provided that x stays within 3×0.004 of $c = 5$. We therefore set $\delta = 0.012$. The resulting graph of f in the viewing rectangle $[5 - 0.012, 5 + 0.012] \times [1/5 - 0.01, 1/5 + 0.01]$ shows that our choice $\delta = 0.012$ suffices (see Figure 11).

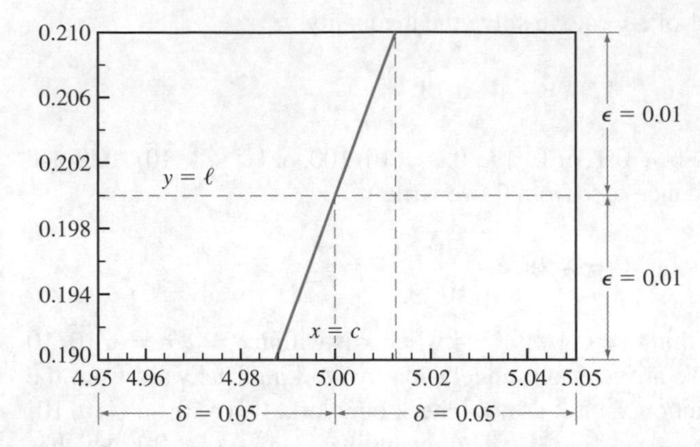

Figure 10
Viewing rectangle $[5 - \delta, 5 + \delta] \times [1/5 - \epsilon, 1/5 + \epsilon]$

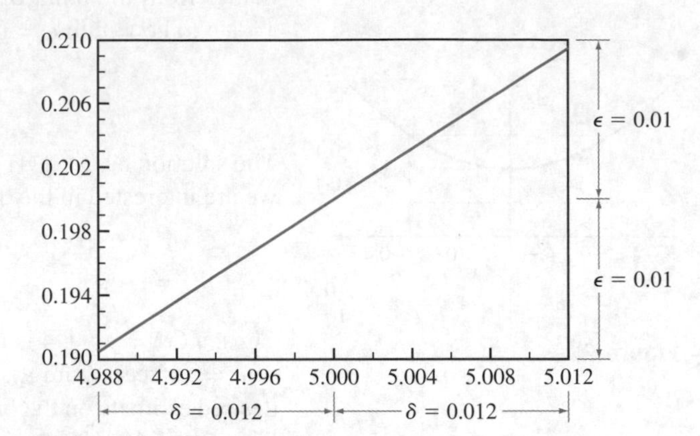

Figure 11
Viewing rectangle $[5 - \delta, 5 + \delta] \times [1/5 - \epsilon, 1/5 + \epsilon]$

Some Applications of Limits

Numerical calculations of the type we have been examining come up frequently in applications. Following are some examples.

Example 8 An unmanned spacecraft is to be sent to a distant planet. The ground control personnel compute that an error of size d degrees in the spacecraft's initial angle of trajectory will result in the spacecraft missing its target by $E(d) = 30 \cdot (10^d - 1)$ miles. The spacecraft must land within 0.1 mile of its target for the mission to be considered successful. What error in the measurement d of the trajectory angle is allowable?

Solution To answer this question, we require $30 \cdot (10^d - 1) < 0.1$. This inequality is equivalent to $10^d - 1 < 1/300$, or $10^d < 1 + 1/300$. The solution set is $d < \log_{10}(1 + 1/300) \approx 0.00144524073$. Thus, $d = 0.00144$ will suffice. Trajectory angle measurements on this flight can be measured to within an accuracy of 0.00144 degrees without affecting the success of this mission (see Figure 12).

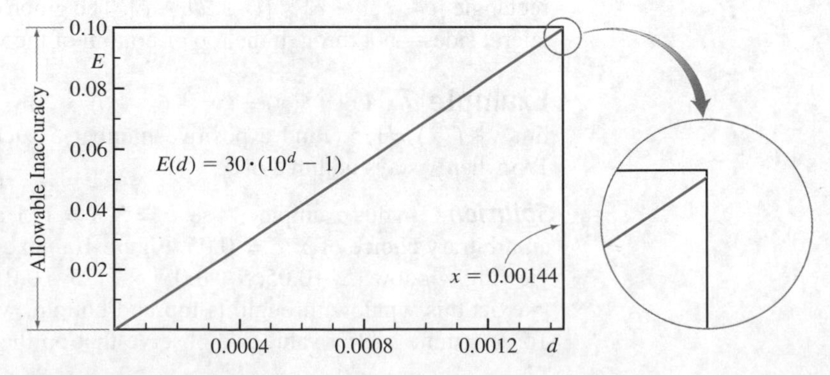

Figure 12

Example 9 In a controlled reaction of hydrogen with bromine, the rate R at which hydrogen bromide is produced (in moles per liter per second) is given by $R(c) = 0.08\sqrt{c}$ where c is the concentration of bromine (in moles per liter). How close to 0.16 must the bromine concentration be maintained to ensure that the rate of production of hydrogen bromide is within 0.001 of 0.032? Answer in the form of an interval $(0.16 - \delta, 0.16 + \delta)$.

Solution We require that $|0.08\sqrt{c} - 0.032| < 0.001$. As we learned in Example 3 from Section 1.1, this inequality is equivalent to $-0.001 < 0.08\sqrt{c} - 0.032 < 0.001$, or, after adding 0.032 to each side, $0.031 < 0.08\sqrt{c} < 0.033$. Dividing by the positive number 0.08 results in $0.3875 < \sqrt{c} < 0.4125$. Finally, by squaring each side, we obtain $0.15015625 < c < 0.17015625$. Let $\delta_1 = 0.16 - 0.15015625 = 0.00984375$ and $\delta_2 = 0.17015625 - 0.16 = 0.01015625$. We have found that $R(c)$ is within 0.001 of 0.032 for c in the interval $I = (0.16 - \delta_1, 0.16 + \delta_2)$. The interval $(0.16 - \delta, 0.16 + \delta)$ we seek must be contained in I and centered at 0.16. We therefore take δ to be the *smaller* of δ_1 and δ_2: $\delta = 0.00984375$. In summary, the rate of production of hydrogen bromide is within 0.001 of 0.032 when the concentration of bromine is within 0.00984375 of 0.16. A plot of $R(c)$ in the viewing window $[0.16 - 0.00984375, 0.16 + 0.00984375] \times [0.032 - 0.001, 0.032 + 0.001]$ confirms our analysis (see Figure 13).

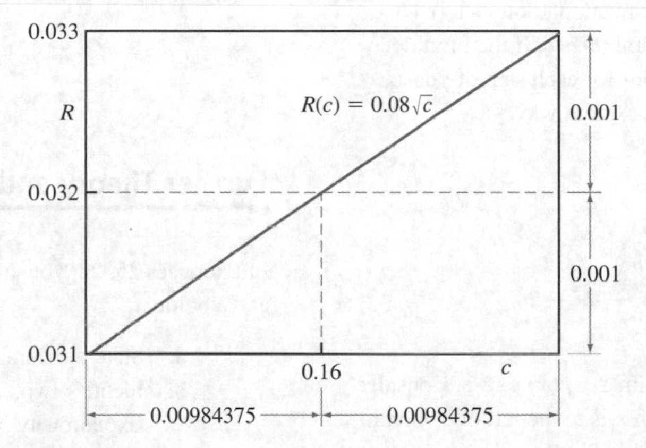

Figure 13 ■

quickquiz

1. Can $\lim_{x\to 0} f(x)$ exist if f does not have the point 0 in its domain?
2. Calculate $\lim_{t\to 5}(t^2 - 25)/(t - 5)$.
3. What role do ϵ and δ play in the numerical idea of limit?
4. If the square of a certain number is to be within 0.0001 of 4, then how close must the number be to 2?

EXERCISES

Problems for Practice

In Exercises 1–8, decide whether the indicated limit exists. If the limit does exist, compute it.

1. $\lim_{x\to 2}(x+3)$ **2.** $\lim_{s\to 9}(s^2-6s+10)$
3. $\lim_{h\to 4}(3h^2+2h+1)$ **4.** $\lim_{s\to 0}(2+s)/s$
5. $\lim_{h\to 1}(h-3)/(h+1)$ **6.** $\lim_{x\to -3}(x-3)/(x^2-9)$
7. $\lim_{x\to 2} g(x)$ for

$$g(x)=\begin{cases}4 & \text{if } x<0\\ -4 & \text{if } x\geq 0\end{cases}$$

8. $\lim_{x\to -1} f(x)$ for

$$f(x)=\begin{cases}6 & \text{if } x\leq -1\\ 10 & \text{if } x>-1\end{cases}$$

In Exercises 9–14, some algebraic manipulation is necessary to determine whether the indicated limit exists. If the limit does exist, compute it and supply reasons for each step of your answer. If the limit does not exist, explain why.

9. $\lim_{x\to 5}(x^2-25)/(x-5)$
10. $\lim_{t\to 7}(t+7)/(t^2-49)$
11. $\lim_{t\to -7}(t+7)/(t^2-49)$
12. $\lim_{h\to 0}(h^6+h^8)/(h^8-3h^3)$
13. $\lim_{x\to -4}(x^2+6x-8)/(x^2-2x-24)$
14. $\lim_{x\to -3}(x^2-9)^2/(x+3)^2$

In Exercises 15–18, determine if $\lim_{x\to c} f(x)$ exists. Consider separately the values f takes when x is to the left of c and the values f takes when x is to the right of c. If the limit exists, compute it. Justify your assertions.

15. $f(x)=\begin{cases}x^2-3x & \text{if } x<2\\ \dfrac{-x}{x-1} & \text{if } x\geq 2\end{cases}, c=2$

16. $f(x)=\begin{cases}\dfrac{x^2+6x-7}{x^2-1} & \text{if } x<1\\ (x+2)^2 & \text{if } x\geq 1\end{cases}, c=1$

17. $f(x)=\begin{cases}\dfrac{x^2-4}{3} & \text{if } x\leq 5\\ \dfrac{x^2-3x-10}{x^2-9x+20} & \text{if } x>5\end{cases}, c=5$

18. $f(x)=\begin{cases}\dfrac{x^2-9}{x^2+9} & \text{if } x\leq 3\\ \dfrac{(x-3)^2}{x^2-9} & \text{if } x>3\end{cases}, c=3$

In Exercises 19–22, calculate how close x needs to be to c to force $f(x)$ to be within 0.01 of ℓ.

19. $f(x)=x+1, c=2, \ell=3$
20. $f(x)=2x-3, c=-2, \ell=-7$
21. $f(x)=5x+2, c=3, \ell=17$
22. $f(x)=\begin{cases}4x-3 & \text{if } x\leq 1\\ x & \text{if } x>1\end{cases}, c=1, \ell=1$

In Exercises 23 and 24, demonstrate that the limit does not exist by showing that the degree of closeness 0.1 cannot be achieved for any ℓ.

23. $\lim_{x\to 2} f(x)$ where

$$f(x)=\begin{cases}5 & \text{if } x\leq 2\\ 3 & \text{if } x>2\end{cases}$$

24. $\lim_{x\to -5} f(x)$ where

$$f(x)=\begin{cases}4x & \text{if } x\leq -5\\ 3x-4 & \text{if } x>-5\end{cases}$$

Further Theory and Practice

In Exercises 25–28, you are given four functions f, g, h, and k and a point c.

a. State the domain of each function.
b. Identify two of these functions that are the same. Explain why. In what way do the other functions differ from these two functions?
c. For each function, decide if it has a limit at c. Identify the limit if it exists.

25. $g(x)=\dfrac{x^2-1}{x-1}, h(x)=\begin{cases}\dfrac{x^2-1}{x-1} & \text{if } x\neq 1\\ 2 & \text{if } x=1\end{cases},$

$f(x)=x+1, k(x)=\begin{cases}\dfrac{x^2-1}{x-1} & \text{if } x\neq 1\\ 1 & \text{if } x=1\end{cases}, c=1$

26. $f(x)=\dfrac{x^2+x-6}{x-2}, h(x)=\begin{cases}\dfrac{x^2+x-6}{x-2} & \text{if } x\neq 2\\ 0 & \text{if } x=2\end{cases},$

$g(x)=x+3, k(x)=\begin{cases}\dfrac{x^2+x-6}{x-2} & \text{if } x\neq 2\\ 5 & \text{if } x=2\end{cases}, c=2$

27. $f(x) = \begin{cases} \dfrac{|x|}{\text{signum}(x)} & \text{if } x \neq 0 \\ 0 & \text{if } x = 0 \end{cases}$, $h(x) = \dfrac{|x|}{\text{signum}(x)}$,

$g(x) = \begin{cases} \dfrac{|x|}{\text{signum}(x)} & \text{if } x \neq 0 \\ 1 & \text{if } x = 0 \end{cases}$, $k(x) = x$, $c = 0$

28. $f(x) = \begin{cases} \dfrac{x^3 - x^2}{x^2} & \text{if } x \neq 0 \\ -1 & \text{if } x = 0 \end{cases}$, $h(x) = \dfrac{x^3 - x^2}{x^2}$,

$g(x) = \begin{cases} \dfrac{x^3 - x^2}{x^2} & \text{if } x \neq 0 \\ 0 & \text{if } x = 0 \end{cases}$, $k(x) = x - 1$, $c = 0$

29. The Heaviside function $H(x)$ is defined by

$$H(x) = \begin{cases} 1 & \text{if } x > 0 \\ 0 & \text{if } x < 0 \end{cases}$$

Compute $\lim_{x \to 0^+} H(x)$ and $\lim_{x \to 0^-} H(x)$. Does $\lim_{x \to 0} H(x)$ exist?

30. Suppose a sample of gas is held at constant pressure. Let V_0 denote the volume when the temperature is 0 degrees centigrade. The *Law of Charles and Gay-Lussac* relates the volume V and temperature T (measured in degrees centigrade) of the given gas sample by the equation $V = V_0 + V_0 \cdot T/273$. Use this law to show that T belongs to an interval of the form $T > \tau$. The number τ is known as *absolute zero*. Discuss the one-sided limit of V as T tends to absolute zero.

31. One foot is 0.3048 meter. How precisely should a length be measured in meters to guarantee accuracy of 0.001 foot?

32. One ounce is 28.350 grams. How precisely should weight be measured in ounces to guarantee accuracy of 0.001 gram?

33. Let $\lfloor x \rfloor$ be the greatest integer that is less than or equal to x. For instance, $\lfloor 3/2 \rfloor = 1$, $\lfloor 0 \rfloor = 0$, $\lfloor -1/2 \rfloor = -1$. Discuss each limit.

a. $\lim_{x \to 0} \lfloor x \rfloor$

b. $\lim_{x \to 1/2} \lfloor x \rfloor$

c. $\lim_{x \to 1} 1/\lfloor x \rfloor$

d. $\lim_{x \to -1/2} 1/\lfloor x \rfloor$

34. Let n be an integer. For the greatest integer function $\lfloor x \rfloor$ defined in Exercise 33, discuss each limit.

a. $\lim_{x \to n^+} \lfloor x \rfloor$

b. $\lim_{x \to n^-} \lfloor x \rfloor$

c. $\lim_{x \to n} \lfloor x \rfloor$

35. Graph the greatest integer function $\lfloor x \rfloor$, defined in Exercise 33. What is the graphical indication that $\lim_{x \to n} \lfloor x \rfloor$ does not exist for an integer n?

36. The graph of a function f that is defined on $(1, 5]$ appears in Figure 14. Determine all values of c at which the limit $\lim_{x \to c} f(x)$ exists. Determine all values of c at which the left limit $\lim_{x \to c^-} f(x)$ exists. Determine all values of c at which the right limit $\lim_{x \to c^+} f(x)$ exists.

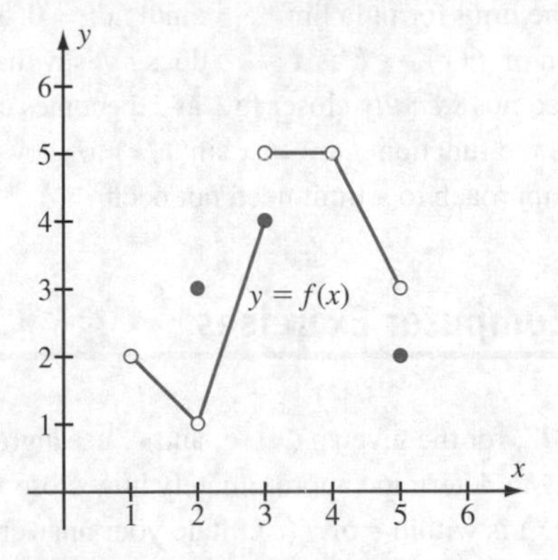

Figure 14

37. Suppose a particle moving in the xy-plane is at the point $(t^2, 0)$ at time t.

a. Compute the average velocity of the particle from $t = 1$ to $t = 2$.

b. Compute the average velocity from $t = 1$ to $t = 1.5$.

c. Compute the average velocity from $t = 1$ to $t = 1.1$.

d. Compute the average velocity from $t = 1$ to $t = 1 + h$ where h is a small nonzero number.

e. What is the limit of these average velocities as $h \to 0$?

38. Repeat Exercise 37 for a particle in the xy-plane that is at point $(3t^2, 4t^2)$ at time t.

39. Graph the function $f(x) = x^4$ for $1 \le x \le 3$.

a. What is the slope of the line passing through $(2, f(2))$ and $(3, f(3))$?

b. What is the slope of the line passing through $(2, f(2))$ and $(2.5, f(2.5))$?

c. What is the slope of the line passing through $(2, f(2))$ and $(2.1, f(2.1))$?

d. What is the slope of the line passing through $(2, f(2))$ and $(2 + h, f(2 + h))$ where h is a small nonzero number?

e. What is the limit of these slopes as $h \to 0$?

40. In Example 8, we determined that $E(d)$ would be in an acceptable range for d satisfying

$$d < \log_{10}\left(1 + \frac{0.1}{30}\right) = 0.0014452\ldots.$$

Ordinarily, the number 0.0014452... would be rounded to three significant digits as 0.00145. Explain why the value 0.00144 was chosen instead.

41. Discuss each limit.

a. $\lim_{x\to 0} x^{5/3}/|x|$

b. $\lim_{x\to 1} \sqrt{|x-1|}\,/\,\sqrt{|x^2-1|}$

42. Justify the limit formula $\lim_{x\to 0} x\sin(1/x) = 0$. The definition of $f(x) \to \ell$ as $x \to c$ does *not* say that $f(x)$ becomes *steadily* closer to ℓ as x becomes closer to c. Use the function $f(x) = x\sin(1/x)$ to show that such an approach to a limit need not occur.

Calculator/Computer Exercises

In Exercises 43–48, for the given $f(x)$, c, and ϵ, use a graphing or zooming utility to determine approximately how close x must be to c so that $f(x)$ is within ϵ of $f(c)$. State your answer in the form $|x-c| < \delta$ for a δ that you determine.

43. $f(x) = 4/x^2,\ c = 2,\ \epsilon = 0.01$

44. $f(x) = x^2,\ c = 1,\ \epsilon = 0.1$

45. $f(x) = x^2,\ c = 10,\ \epsilon = 0.1$

46. $f(x) = 1/x,\ c = 0.2,\ \epsilon = 0.01$

47. $f(x) = 1/x,\ c = 5,\ \epsilon = 0.01$

48. $f(x) = \sqrt{x^2+16},\ c = 3,\ \epsilon = 0.01$

49. A machinist must do all of his work to within a tolerance of 10^{-3} mm. The calibrations on his machine are such that if the machinist's settings are accurate to within m mm, then the dimensions of the product have a tolerance within $1.2|m| + 100m^2$ mm. What accuracy of the settings is required (that is, how small must the machinist make m) to produce work of the desired tolerance?

In Exercises 50–54, a function f is given, as is a point c that is *not* in the domain of f. Analyze $\lim_{x\to c} f(x)$ as follows.

a. Evaluate $f(x)$ at values of x that are very close to, but unequal to, c. Make a table of your results.

b. Formulate a guess for the value of the limit.

c. Find a value δ such that the values of $f(x)$ are within 0.01 of the limit when x is within δ of c.

d. Graph f near $x = c$. Use a zooming utility to verify visually that the limit of f at c exists.

50. $f(x) = \sin(x)/x,\ c = 0$

51. $f(x) = \cos(x)/(x-\pi/2),\ c = \pi/2$

52. $f(x) = \sin(x)/(x-\pi),\ c = \pi$

53. $f(x) = (x - \sin(x))/x^3,\ c = 0$

54. $f(x) = (\cos(x) - 1)/x^2,\ c = 0$

55. The position of a moving body is given by $p(t) = t^2 + 2$ inches.

a. How small does h need to be for the average velocity (distance traveled divided by time elapsed) between $t = 2$ and $t = 2 + h$ to be between 3.9 and 4.1?

b. How small does h need to be for the average velocity between $t = 2$ and $t = 2 + h$ to be between 3.99 and 4.01?

c. Let $\epsilon > 0$ be a small positive number. How small does h need to be to guarantee that the average velocity between $t = 2$ and $t = 2 + h$ is between $4 - \epsilon$ and $4 + \epsilon$?

d. What do parts a–c tell you about the instantaneous velocity of the body at time $t = 2$?

56. Repeat Exercise 55 with $p(t) = t^3 - 4t$ and average velocities calculated with left endpoint $t = 3$.

57. A spring moves so that the amount of extension x is given by $x(t) = \sin(t)$. (A negative value of x corresponds to compression.) In this exercise, you will approximate the instantaneous velocity $v(t)$ of the spring at time t by its average velocity $\bar{v}(t)$ over the time interval $[t - 0.0001, t + 0.0001]$.

a. Write $\bar{v}(t)$ explicitly and graph $\bar{v}(t)$ for $0 \le t \le 2\pi$.

b. When, approximately, is $\bar{v}(t)$ the greatest? The most negative?

c. When, approximately, is $\bar{v}(t)$ zero?

d. Graph x and $\bar{v}$ in the viewing window $[0, 2\pi] \times [-1.1, 1.1]$. For what subinterval I of $[0, 2\pi]$ is $\bar{v} < 0$? What is the behavior of x as t increases through this subinterval I? What does $\bar{v}(t) < 0$ mean?

e. Based on the graph of $\bar{v}$, what function does $\bar{v}$ appear to approximate?

2.2 Limit Theorems

In Section 2.1, we discussed limits intuitively and numerically. Now we learn a precise definition of the statement

$$\lim_{x\to c} f(x) = \ell.$$

The definition is written with mathematical symbols, but it means, "$f(x)$ can be forced arbitrarily close to ℓ by making x be sufficiently close to c." We measure the distance of $f(x)$ to ℓ by the absolute difference $|y - \ell|$. When we say that $f(x)$ can be forced arbitrarily close to ℓ, we mean that we can make $|f(x) - \ell|$ smaller than ϵ, no matter how small the positive number ϵ is. Thus, "$f(x)$ can be forced arbitrarily close to ℓ" translates as "for any given positive ϵ, we can make $|f(x) - \ell| < \epsilon$."

"By making x be sufficiently close to c" means that we are considering only those x that lie within a specified distance δ of c. In other words, we refer to those x for which $0 < |x - c| < \delta$. Now we are ready for the complete definition.

Definition

Let f be a real-valued function that is defined in an open interval just to the left of a real number c and in an open interval just to the right of c. We say, "The limit of $f(x)$ is equal to ℓ as x approaches c," and we write $\lim_{x\to c} f(x) = \ell$ if, for any $\epsilon > 0$, there is a $\delta > 0$ such that

$$|f(x) - \ell| < \epsilon$$

for all values of x such that

$$0 < |x - c| < \delta.$$

The limit can also be written as $f(x) \to \ell$ as $x \to c$. Refer to Figure 1.

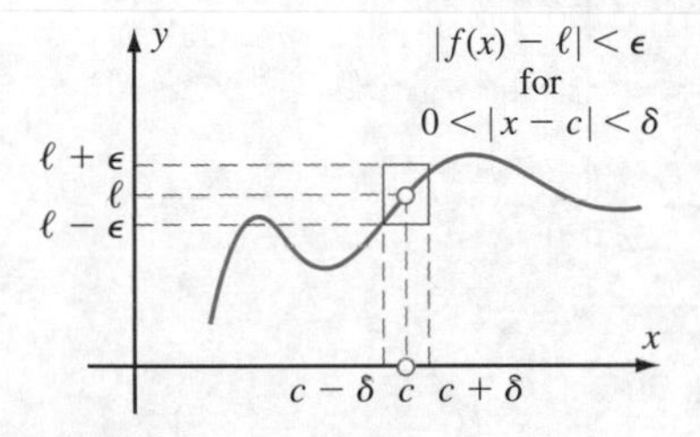

Figure 1
$f(x)$ has the limit ℓ as x tends to c.

This definition is often referred to as the ϵ-δ *definition of limit*. Notice that the inequality $0 < |x - c| < \delta$ means we are considering values of x that are *near to* c but *not equal to* c. An ϵ-δ verification of a limit starts with the assumption that an arbitrary positive ϵ is given. Then a δ corresponding to that ϵ must be found. This is exactly what we did in the quantitative examples from Section 2.1 in which specific values of ϵ were given. As you read the next example, notice that we find δ in terms of ϵ for a general value of ϵ.

Example 1 Let $f(x) = 3x + 2$. Using the rigorous definition of limit, verify that $\lim_{x\to 4} f(x) = 14$.

Solution Let $\epsilon > 0$. We must find a $\delta > 0$ such that $0 < |x - 4| < \delta$ implies $|f(x) - 14| < \epsilon$. As is usually the case, we work backward to find a δ that corresponds to the given ϵ. First, notice that there is a direct relationship between $|x - 4|$ and $|f(x) - 14|$:

$$|f(x) - 14| = |3x + 2 - 14| = |3x - 12| = 3|x - 4|.$$

From this, we see that $|f(x) - 14| < \epsilon$ if $|x - 4| < \epsilon/3$. Therefore, $\epsilon/3$ is the value for δ that we want. That is, if $0 < |x - 4| < \epsilon/3$, then $|f(x) - 12| < \epsilon$. ■

Here is one way to use the information from Example 1. Suppose we want to force $f(x) = 3x + 2$ to be within 0.06 of 14. We set $\epsilon = 0.06$. The calculation in the example tells us that $\delta = \epsilon/3 = 0.02$. In short, if x is within $\delta = 0.02$ of $c = 4$, then $f(x) = 3x + 2$ is forced to within $\epsilon = 0.06$ of $\ell = 14$ (see Figure 2). Now suppose instead that we want to force $f(x)$ to be within 0.003 of 14. To do this, we simply set $\epsilon = 0.003$. Our computation tells us that when x is within $\delta = \epsilon/3 = 0.001$ of $c = 4$, $f(x) = 3x + 2$ is forced to within $\epsilon = 0.003$ of $\ell = 14$. By refining our degree of accuracy, we do not have to keep repeating the calculation. We have done the calculation once and for all, but with symbols ϵ and δ instead of specific numbers. We can substitute numbers later, depending on the degree of accuracy desired.

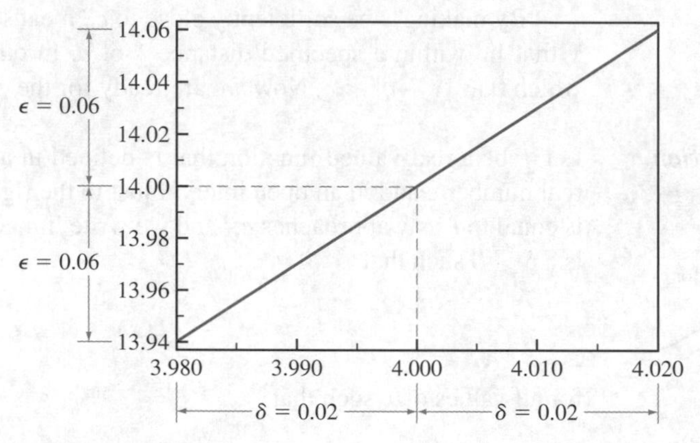

Figure 2

In Example 1, our intuition allows us to easily determine the required limits. Why, then, has the ϵ-δ definition been introduced? The answer is that there are limits that are *not* obvious. (See, for example, the trigonometric limits at the end of this section.) Having a rigorous definition allows us to verify limits in such situations.

One-Sided Limits

Many useful functions that do not have limits at certain points *do* have one-sided limits at those points. Furthermore, many functions are defined on an interval of the form (a, b), $[a, b]$, $(a, b]$, $[a, b)$ or on a half line of the form (a, ∞), $[a, \infty)$, $(-\infty, b)$, $(-\infty, b]$. The concept of one-sided limit is necessary to investigate the behavior of such a function at an endpoint of its domain of definition.

Definition Suppose f is a function defined in an interval just to the right of c. If for every $\epsilon > 0$ there is a $\delta > 0$ such that $|f(x) - \ell| < \epsilon$ for each x that satisfies $c < x < c + \delta$, then we say that f has ℓ as a *limit from the right,* or f has *right limit* ℓ at c, and we write

$$\lim_{x \to c^+} f(x) = \ell.$$

Figure 3 shows a function f with a right limit ℓ at c.

Definition Let f be a function defined in an interval just to the left of c. If for every $\epsilon > 0$ there is a $\delta > 0$ such that $|f(x) - L| < \epsilon$ for each x that satisfies $c - \delta < x < c$, then we say that f has L as a *limit from the left,* or f has *left limit* L at c, and we write

$$\lim_{x \to c^-} f(x) = L.$$

Figure 4 illustrates a function f with left limit L at c.

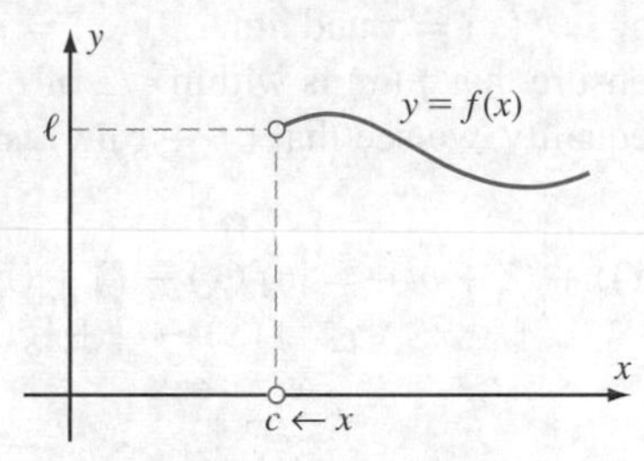

Figure 3
$\lim_{x \to c^+} f(x) = \ell$

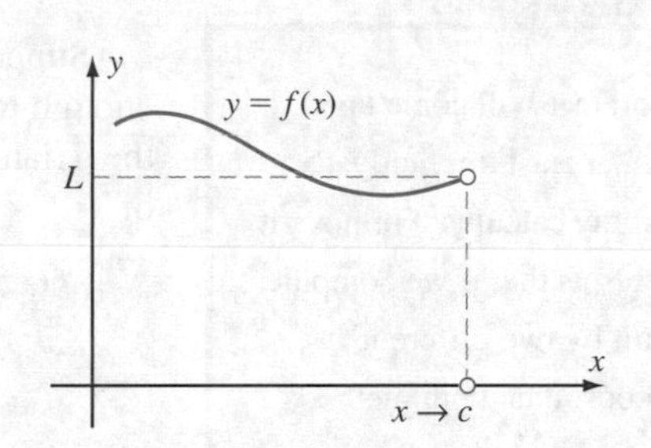

Figure 4
$\lim_{x \to c^-} f(x) = L$

By comparing the definitions of one-sided and two-sided limits, we see that

$$\lim_{x \to c} f(x) = \ell \quad \text{is equivalent to} \quad \lim_{x \to c^-} f(x) = \ell \quad \text{and} \quad \lim_{x \to c^+} f(x) = \ell.$$

Example 2 What are the one-sided limits of f at 7 if f is defined by

$$f(x) = \begin{cases} 4x & \text{if } x < 7 \\ 0 & \text{if } x = 7 \\ -6 & \text{if } x > 7 \end{cases}?$$

Solution As with all limit problems, we first look at this one intuitively. To the right of $c = 7$, f is constantly equal to -6, so it appears that f has right limit -6 at $c = 7$. To the left of $c = 7$, f is equal to $4x$, so it appears that f has left limit $4 \cdot 7 = 28$ at $c = 7$. Now let us carefully verify these assertions by using the definitions. Refer to Figure 5.

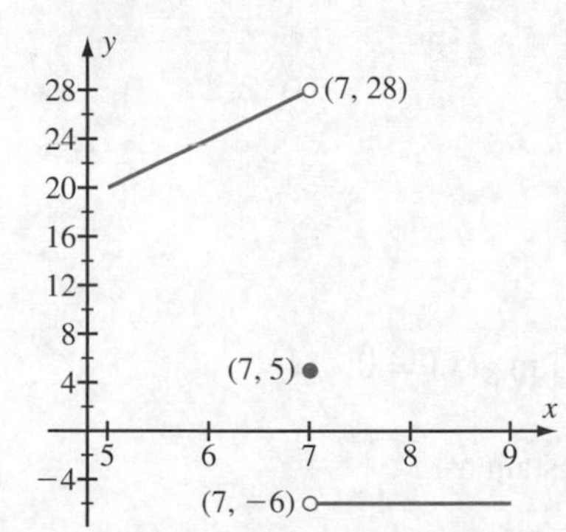

Figure 5

$$f(x) = \begin{cases} 4x & \text{if } x < 7 \\ 5 & \text{if } x = 7 \\ -6 & \text{if } x > 7 \end{cases}$$

First we consider the right limit. Let $\epsilon > 0$. Choose $\delta = \epsilon$. If $7 < x < 7 + \delta$, then

$$|f(x) - (-6)| = |-6 - (-6)| = 0 < \epsilon.$$

Therefore, $\lim_{x \to 7^+} f(x) = -6$.

For the left limit, let $\epsilon > 0$, and define $\delta = \epsilon/4$. If $7 - \delta < x < 7$, then

$$|f(x) - 28| = |4x - 28| = 4|x - 7| < 4\delta = \epsilon.$$

By definition, then, $\lim_{x\to 7^-} f(x) = 28$. Notice in this example that the ordinary (two-sided) limit, $\lim_{x\to 7} f(x)$, does not exist. ■

Basic Limit Theorems

There are a few simple theorems that save a great deal of work in calculating limits. Each theorem is intuitively plausible. The first tells us that a function can have *at most* one limit at a point c.

Theorem 1 Let $a < c < b$. A function f with a domain that contains a set of the form $(a, c) \cup (c, b)$ cannot have two distinct limits at c.

inSIGHT

Theorem 1 will come up often, at least implicitly, as we study calculus. For now, it assures us that if we compute a limit by two different methods, then both methods will give the same answer.

Suppose $\lim_{x\to c} f(x) = \ell$ and $\lim_{x\to c} g(x) = m$. Given $\epsilon > 0$, we can take x close enough to c to ensure that $f(x)$ is within $\epsilon/2$ of ℓ and $g(x)$ is within $\epsilon/2$ of m. Using the Triangle Inequality, we see that $(f+g)(x)$ is within ϵ of $\ell + m$ for these values of x:

$$\begin{aligned} |(f(x)+g(x)) - (\ell + m)| &= |(f(x) - \ell) + (g(x) - m)| \\ &\le |f(x) - \ell| + |g(x) - m| \qquad \text{Triangle Inequality} \\ &< \frac{\epsilon}{2} + \frac{\epsilon}{2} \\ &= \epsilon. \end{aligned}$$

This tells us that the limit of a sum $f + g$ at c is the sum of the limits of f and g at c. We state this result, as well as some similar assertions, in the following theorem, which tells us that limits of algebraic combinations of functions behave in the expected way.

Theorem 2 If f and g are two functions, c is a real number, and $\lim_{x\to c} f(x)$ and $\lim_{x\to c} g(x)$ exist, then the following equations hold

a. $\lim\limits_{x\to c}(f+g)(x) = \lim\limits_{x\to c} f(x) + \lim\limits_{x\to c} g(x)$ and

$\lim\limits_{x\to c}(f-g)(x) = \lim\limits_{x\to c} f(x) - \lim\limits_{x\to c} g(x)$;

b. $\lim\limits_{x\to c}(f\cdot g)(x) = \left(\lim\limits_{x\to c} f(x)\right)\cdot\left(\lim\limits_{x\to c} g(x)\right)$;

c. $\lim\limits_{x\to c}\left(\dfrac{f}{g}\right)(x) = \dfrac{\lim_{x\to c} f(x)}{\lim_{x\to c} g(x)}$, provided that $\lim\limits_{x\to c} g(x) \neq 0$;

d. $\lim\limits_{x\to c}(\alpha\cdot f(x)) = \alpha\cdot\left(\lim\limits_{x\to c} f(x)\right)$ for any constant α.

By successively applying Theorem 2b as many times as necessary, we can handle the product of any finite number of functions. For example, because we know that $\lim_{x\to c} x = c$, we can deduce that $\lim_{x\to c} x^n = c^n$ for any $n \in \mathbb{Z}^+$. Similar remarks apply to any finite sum of functions. From these observations and the other parts of Theorem 2, we obtain the next theorem.

Theorem 3 If $p(x)$ is a polynomial, then

$$\lim_{x \to c} p(x) = p(c).$$

If $q(x)$ is also a polynomial and if $q(c) \neq 0$, then

$$\lim_{x \to c} \frac{p(x)}{q(x)} = \frac{p(c)}{q(c)}.$$

Functions of the form $x \mapsto p(x)/q(x)$ where $p(x)$ and $q(x)$ are polynomials are called *rational functions*. Theorem 3 gives a large class of functions with limits that can be obtained by *evaluation*. Later in this section, we will see other important functions that have this property.

Example 3 Use Theorem 3 to compute $\lim_{x \to -3}(2x^2 + 1)/(6x - 5)$

Solution Let $p(x) = 2x^2 + 1$ and $q(x) = 6x - 5$. Notice that $q(-3) = 6(-3) - 5 = -23$. Since this quantity is nonzero, it follows from Theorem 3 that

$$\lim_{x \to -3} \frac{2x^2 + 1}{6x - 5} = \lim_{x \to -3} \frac{p(x)}{q(x)} = \frac{p(-3)}{q(-3)} = -\frac{19}{23}.$$

■

Notice that Theorems 2 and 3 allow us to calculate limits without reference to the ϵ-δ definition. Our next theorem can be used in the same way.

Theorem 4 Let n be a positive integer and c a real number. We assume further that c is positive if n is even. Then,

$$\lim_{x \to c} \sqrt[n]{x} = \sqrt[n]{c}.$$

Example 4 Calculate $\lim_{x \to 9}(\sqrt{x} - 3)/(x - 9)$.

Solution Because the limit of the denominator is 0, Theorem 2c cannot be applied to calculate the limit of the quotient. We must first prepare the quotient by means of an algebraic manipulation that is frequently helpful:

$$\frac{\sqrt{x} - 3}{x - 9} = \frac{\sqrt{x} + 3}{\sqrt{x} + 3} \cdot \frac{\sqrt{x} - 3}{x - 9} = \frac{x - 9}{(\sqrt{x} + 3)(x - 9)} = \frac{1}{\sqrt{x} + 3}.$$

Therefore,

$$\lim_{x \to 9} \frac{\sqrt{x} - 3}{x - 9} = \lim_{x \to 9} \frac{1}{\sqrt{x} + 3} \overset{\text{Theorem 2c}}{=} \frac{1}{\lim_{x \to 9}(\sqrt{x} + 3)} \overset{\text{Theorem 2a}}{=} \frac{1}{(\lim_{x \to 9} \sqrt{x}) + 3} \overset{\text{Theorem 4}}{=} \frac{1}{6}.$$

■

A Rule That Tells When a Limit Does Not Exist

Sometimes it is useful to know a rule that shows that a certain limit does not exist.

Theorem 5

If $\lim_{x\to c} g(x) = 0$ and if $f(x)$ does not have 0 as its limit as x approaches c, then $\lim_{x\to c} f(x)/g(x)$ does not exist.

inSIGHT

Here is an equivalent formulation of Theorem 5 that is frequently useful: If $\lim_{x\to c} f(x)/g(x)$ exists and if $\lim_{x\to c} g(x) = 0$, then $\lim_{x\to c} f(x) = 0$.

Proof Let $h(x) = f(x)/g(x)$. To show that $h(x)$ does not have a limit as x approaches c, we write $f(x) = g(x) \cdot h(x)$ and argue by contradiction. If h did have a limit ℓ as x approached c, then Theorem 2b would give

$$\lim_{x\to c} f(x) = \lim_{x\to c} h(x) \cdot \lim_{x\to c} g(x) = \ell \cdot 0 = 0,$$

contradicting the hypothesis about f. ■

Example 5 Does $\lim_{x\to -1}(2+x)/(1+x)^2$ exist?

Solution We notice that

$$\lim_{x\to -1} (2+x) = 2 + (-1) = 1 \neq 0 \quad \text{and} \quad \lim_{x\to -1} (x+1)^2 = ((-1)+1)^2 = 0^2 = 0.$$

According to Theorem 5, $\lim_{x\to -1}(2+x)/(1+x)^2$ does not exist. ■

The Pinching Theorem

Theorem 6

The Pinching Theorem Suppose f, g, and h are functions with domains that each contain $S = (a, c) \cup (c, b)$. Assume further that $g(x) \leq f(x) \leq h(x)$ for all $x \in S$ (see Figure 6). If $\lim_{x\to c} g(x) = \ell$ and $\lim_{x\to c} h(x) = \ell$, then $\lim_{x\to c} f(x) = \ell$.

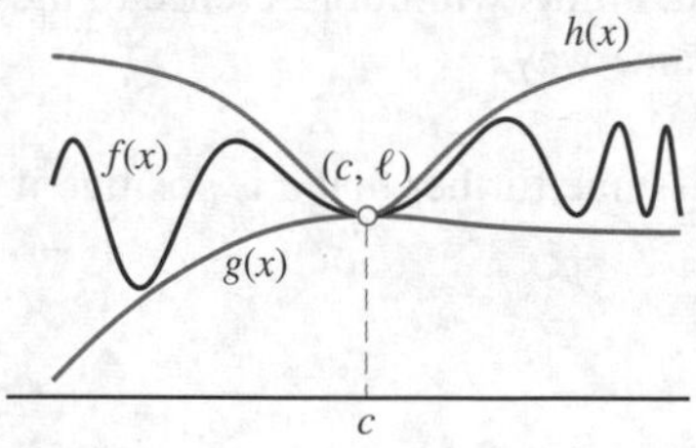

Figure 6
$\lim_{x\to c} f(x) = \ell$

The truth of the Pinching Theorem is geometrically clear (refer again to Figure 6). We may apply the Pinching Theorem in one-sided situations as well. For example, if $g(x) \leq f(x) \leq h(x)$ for all $x \in (c, b)$, if $\lim_{x\to c^+} g(x) = \ell$, and if $\lim_{x\to c^+} h(x) = \ell$, then we may conclude that $\lim_{x\to c^+} f(x) = \ell$.

Example 6 Let $f(x) = x \cdot \sin(1/x)$ when $x \neq 0$. Use the Pinching Theorem to compute $\lim_{x\to 0} f(x)$.

Solution Notice that

$$f(x) \leq \left| x \cdot \sin\left(\frac{1}{x}\right) \right| = |x| \cdot \left| \sin\left(\frac{1}{x}\right) \right| \leq |x|$$

and

$$-|x| \leq -|x| \cdot \left| \sin\left(\frac{1}{x}\right) \right| = -\left| x \cdot \sin\left(\frac{1}{x}\right) \right| \leq f(x).$$

In short,

$$-|x| \leq f(x) \leq |x|. \tag{2.3}$$

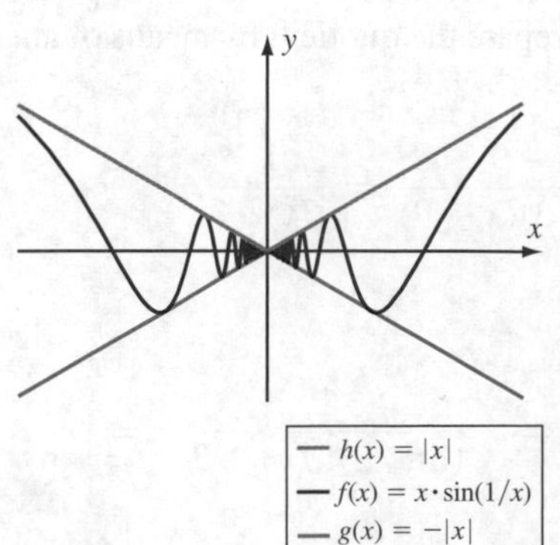

Figure 7

We have "pinched" f between the functions $h(x) = |x|$ and $g(x) = -|x|$ (see Figure 7).

Now it is easy to see that

$$\lim_{x\to 0} |x| = 0 \quad \text{and} \quad \lim_{x\to 0}(-|x|) = 0. \tag{2.4}$$

Equations (2.3) and (2.4), together with the Pinching Theorem, allow us to see that $\lim_{x\to 0} x \cdot \sin(1/x) = 0$. ■

Some Important Trigonometric Limits

Theorem 7 For every c, $\lim_{t\to c} \sin(t) = \sin(c)$ and $\lim_{t\to c} \cos(t) = \cos(c)$.

Proof Let t be an angle, measured in radians, between 0 and $\pi/2$. Figure 8 exhibits the unit circle. Because t represents a small *positive* angle measured in radians, the length of the arc that the angle subtends is t. We see that $0 \le \sin(t) = |\overline{AB}| < |\overline{AC}| < t$. A similar picture shows that $t < \sin(t) \le 0$ when t is negative. Therefore, $-|t| < \sin(t) < |t|$ for all t in $(-\pi/2, \pi/2)$. By the Pinching Theorem,

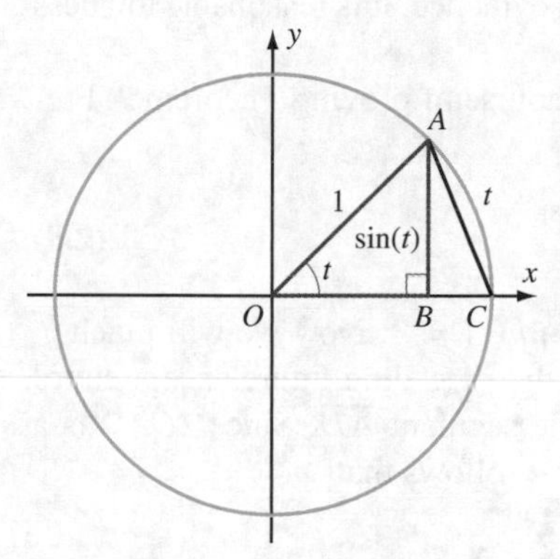

Figure 8
$0 \le \sin(t) = |\overline{AB}| < |\overline{AC}| < t$

$$\lim_{t\to 0} \sin(t) = 0. \tag{2.5}$$

Because

$$\cos(t) = \cos\left(\frac{t}{2} + \frac{t}{2}\right) \overset{(1.9)}{=} \cos^2\left(\frac{t}{2}\right) - \sin^2\left(\frac{t}{2}\right) \overset{(1.4)}{=} 1 - 2\sin^2\left(\frac{t}{2}\right),$$

we may conclude that

$$\lim_{t\to 0} \cos(t) = \lim_{t\to 0}\left(1 - 2\sin^2\left(\frac{t}{2}\right)\right) = 1 - 2\cdot 0^2 = 1. \tag{2.6}$$

Now let c be arbitrary. For every t, let $h = t - c$. Notice that $h \to 0$ as $t \to c$. We have

$$\begin{aligned}
\lim_{t\to c} \sin(t) &= \lim_{t\to c} \sin(c + t - c) \\
&= \lim_{h\to 0} \sin(c + h) && \text{Definition of } h \\
&= \lim_{h\to 0}(\sin(c)\cos(h) + \cos(c)\sin(h)) && \text{Addition Formula (1.8)} \\
&= \lim_{h\to 0} \sin(c)\cos(h) + \lim_{h\to 0} \cos(c)\sin(h) && \text{Theorem 2a} \\
&= \sin(c) \cdot \lim_{h\to 0} \cos(h) + \cos(c) \cdot \lim_{h\to 0} \sin(h) && \text{Theorem 2d} \\
&= \sin(c) \cdot 1 + \cos(c) \cdot 0 && \text{Equations (2.5) and (2.6)} \\
&= \sin(c).
\end{aligned}$$

A similar calculation using Addition Formula (1.9) shows that $\lim_{t\to c} \cos(t) = \cos(c)$ for every real c. ■

inSIGHT

Because $\csc(x) = 1/\sin(x)$, $\sec(x) = 1/\cos(x)$, $\tan(x) = \sin(x)/\cos(x)$, and $\cot(x) = \cos(x)/\sin(x)$, we may use Theorem 2c to conclude that if f is any trigonometric function, then

$$\lim_{x\to c} f(x) = f(c)$$

at each point c in the domain of f.

We now have a large class of functions with limits that can be computed simply by evaluation: polynomials, rational functions, nth roots, trigonometric functions, and all the combinations thereof that can be formed with the arithmetic operations described in Theorem 2. However, many limits in calculus are not obtained so easily. The next theorem gives two important examples.

Theorem 8 If t is measured in *radians,* then

$$\lim_{t\to 0}\frac{\sin(t)}{t}=1 \quad\text{and}\quad \lim_{t\to 0}\frac{1-\cos(t)}{t^2}=\frac{1}{2}. \tag{2.7}$$

Proof Let $f(t)=\sin(t)/t,\ t\neq 0$. On a calculator set to radian mode, we compute that

x	0.1	0.01	0.001	0.0001
$f(x)$	0.9983341664	0.9999833334	0.9999998333	0.9999999983

to ten decimal places. On the basis of this numerical evidence, it is reasonable to guess that $\lim_{t\to 0} f(t)=1$. Let us investigate why.

Refer again to Figure 8. As we observed in the course of proving Theorem 7, Figure 8 shows that $\sin(t)<t$ or, equivalently,

$$f(t)<1. \tag{2.8}$$

This pinches f from above by the constant function $h(t)=1$. Now we will pinch f from below. Refer to Figure 9, which is similar to the preceding figure except chord AC of Figure 8 has been replaced with tangent line segment $\overline{AD}$. Since $\angle OAD$ is a right angle, we see that ΔABD is similar to ΔOBA. It follows that

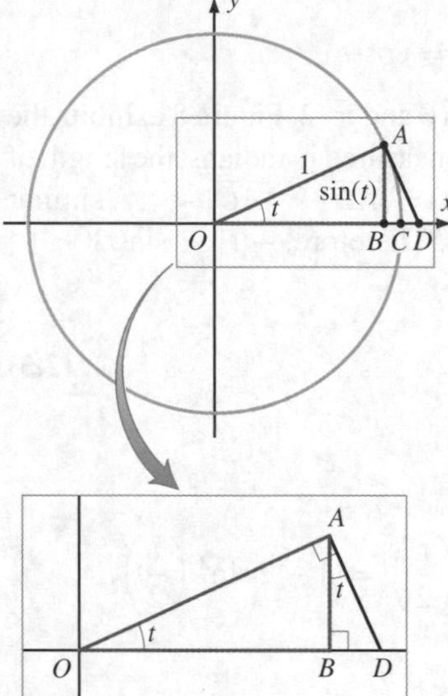

Figure 9
$t=\angle AOD=\angle BAD=\frac{\pi}{2}-\angle ADO$

$$\frac{|\overline{BD}|}{|\overline{AD}|}=\frac{\sin(t)}{1}.$$

Therefore, $|\overline{BC}|<|\overline{BD}|=|\overline{AD}|\cdot\sin(t)$. However, ΔOAC shows that $|\overline{AD}|=\tan(t)$. It follows that

$$|\overline{BC}|<\tan(t)\cdot\sin(t).$$

Therefore,

$$|\overline{BC}|+\sin(t)<\tan(t)\cdot\sin(t)+\sin(t)=\sin(t)(1+\tan(t)). \tag{2.9}$$

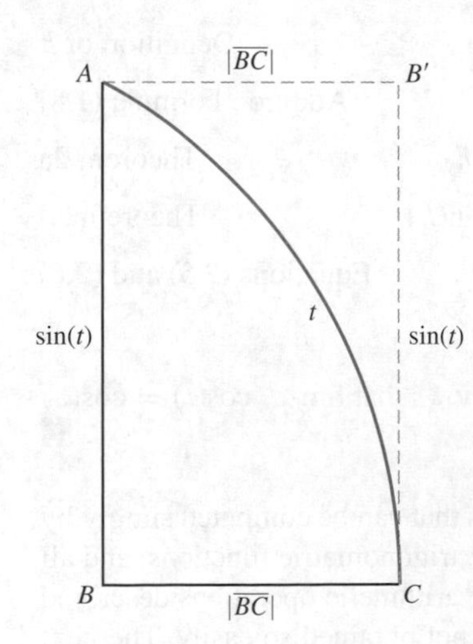

Figure 10

Now observe that $t<|\overline{BC}|+\sin(t)$ (see Figure 10). This inequality, together with inequality (2.9), shows that

$$t<\sin(t)(1+\tan(t)).$$

Divide each side of this inequality by the positive quantity $t(1+\tan(t))$ to obtain

$$\frac{1}{1+\tan(t)}<\frac{\sin(t)}{t}=f(t).$$

Thus, for small *positive* t, we have pinched f from below by $g(t)=1/(1+\tan(t))$. Since $\lim_{t\to 0}1/(1+\tan(t))=1/(1+\tan(0))=1$, the Pinching Theorem tells us that we have $\lim_{t\to 0^+}f(t)=1$. But $f(-t)=f(t)$. It follows that $\lim_{t\to 0^-}f(t)=\lim_{t\to 0^+}f(t)=1$. We conclude that the two-sided limit of $f(t)$ at $t=0$ exists and equals 1, which is the first limit formula of Theorem 8.

For the second limit formula, we use the Half-Angle Formula $1 - \cos(t) = 2\sin^2(t/2)$. We have

$$\frac{1-\cos(t)}{t^2} = \frac{2\sin^2(t/2)}{t^2} = \frac{1}{2}\left(\frac{\sin(t/2)}{t/2}\right)^2.$$

Let $x = t/2$ and observe that $x \to 0$ as $t \to 0$. By the first part of this proof, we conclude that

$$\lim_{t\to 0}\frac{1-\cos(t)}{t^2} = \lim_{x\to 0}\frac{1}{2}\left(\frac{\sin(x)}{x}\right)^2 = \frac{1}{2}\cdot 1^2 = \frac{1}{2}.$$ ■

Example 7 Show that

$$\lim_{t\to 0}\frac{1-\cos(t)}{t} = 0.$$

Solution By Theorem 8 and Theorem 2b, we have

$$\lim_{t\to 0}\frac{1-\cos(t)}{t} = \lim_{t\to 0}\frac{1-\cos(t)}{t^2}\cdot t = \lim_{t\to 0}\frac{1-\cos(t)}{t^2}\cdot\lim_{t\to 0} t = \frac{1}{2}\cdot 0 = 0.$$ ■

quickquiz

1. In the rigorous definition of limit, why do we specify that $0 < |x - c|$?
2. If $\lim_{x\to c} f(x) = \ell$, then what is $\lim_{x\to c}(3f(x) + 1)$? State precise reasons.
3. Discuss one-sided limits for the function $f(x) = \sqrt{x+3}$ at the point $c = -3$.
4. If f has both a left limit and a right limit at c, does $\lim_{x\to c} f(x)$ necessarily exist? Explain.
5. What does the Pinching Theorem state?

EXERCISES

Problems for Practice

In Exercises 1–4, first use your intuition to guess the limit. Then use the rigorous definition of limit to verify your guess.

1. $\lim_{x\to 3} 2x$
2. $\lim_{x\to 1}(6x - 1)$
3. $\lim_{x\to -2}(x^2 - 4)/(x - 2)$
4. $\lim_{x\to 2}(x^2 - 4)/(x - 2)$

In Exercises 5–8, use the rigorous definition of limit to check that the indicated limit does not exist.

5. $\lim_{x\to 5} f(x)$ where

$$f(x) = \begin{cases} 3x & \text{if } x < 5 \\ 4 & \text{if } x > 5 \end{cases}$$

6. $\lim_{x\to -3} g(x)$ where

$$g(x) = \begin{cases} x^2 & \text{if } x < -3 \\ 4x & \text{if } x > -3 \end{cases}$$

7. $\lim_{x\to 1} H(x)$ where

$$H(x) = \begin{cases} 0 & \text{if } x < 1 \\ -4 & \text{if } x > 1 \end{cases}$$

8. $\lim_{x\to 0} f(x)$ where

$$f(x) = \begin{cases} x - 1 & \text{if } x < 0 \\ x^3 & \text{if } x \geq 0 \end{cases}$$

In Exercises 9–12, use your intuition to guess $\lim_{x\to 4^-} f(x)$ and $\lim_{x\to 4^+} f(x)$. Then use the precise definition of limit to verify your answer.

9. $f(x) = \begin{cases} x^2 + 1 & \text{if } x < 4 \\ -x - 3 & \text{if } x > 4 \end{cases}$

10. $f(x) = \begin{cases} 2x+3 & \text{if } x < 4 \\ x+7 & \text{if } x > 4 \end{cases}$

11. $f(x) = \begin{cases} x^3 & \text{if } x < 4 \\ 64 & \text{if } x = 4 \\ 4x^2 & \text{if } x > 4 \end{cases}$

12. $f(x) = \begin{cases} 8 & \text{if } x < 4 \\ 5 & \text{if } x = 4 \\ 2x & \text{if } x > 4 \end{cases}$

In Exercises 13–20, use Theorem 2 to evaluate the limit. Explicitly record each fact you use.

13. $\lim_{x\to 4}(2x+6)$
14. $\lim_{x\to -3}(-4x+5)$
15. $\lim_{x\to 1}(x^2-6)$
16. $\lim_{x\to 2}(x+1)/x$
17. $\lim_{x\to 0}(x^2+2)/(x+1)$
18. $\lim_{x\to 7}(x+3)(x-7)$
19. $\lim_{x\to 4}(x-5)x/(x+1)$
20. $\lim_{x\to -2}(x^3+1)^3$

In Exercises 21–28, state whether the limit exists. Then justify your answer using Theorems 3, 4, and 5.

21. $\lim_{x\to 1}\dfrac{(x-2)^2}{x+1}$
22. $\lim_{x\to 4}\dfrac{x+4}{x-4}$
23. $\lim_{x\to 3}\dfrac{x^2-9}{x-3}$
24. $\lim_{x\to -2}\dfrac{x^2-4}{x-2}$
25. $\lim_{x\to -1}\dfrac{x^2-1}{x+1}$
26. $\lim_{x\to 16}\dfrac{1/\sqrt{x}}{1/(x^{1/4}-10)}$
27. $\lim_{x\to \pi}\dfrac{x-\sqrt[3]{x}}{\sqrt{x}-3}$
28. $\lim_{x\to \pi}\dfrac{x^2-9}{(x-\pi)^2}$

Use the Pinching Theorem to evaluate the limits in Exercises 29–34.

29. $\lim_{x\to 0} x^3\cos(1/x)$
30. $\lim_{x\to 1}(x-1)\cdot\sin(1/(x-1))$
31. $\lim_{x\to 5}(1+(x-5)^2\sin(\csc(\pi x)))$
32. $\lim_{x\to -9}(4+(x+9)\cos(1/x))$
33. $\lim_{x\to 2}((x+1)+(x-2)^2\sin(\sec(\pi/x)))$
34. $\lim_{x\to -1}(x+1)(x-6)^2$

In Exercises 35–38, decide whether the limit can be determined from the given information. If the answer is yes, then find the limit.

35. $2-|x-2|^3 \le f(x) \le 2+|x-2|^2$, $\lim_{x\to 2} f(x)$
36. $2-|x-2|^3 \le f(x) \le 2+|x-2|^2$, $\lim_{x\to 4} f(x)$
37. $-3|x|-|x-5|^2 \le f(x) \le 6|x|+|x-5|^2$, $\lim_{x\to 0} f(x)$
38. $-|x|-2|x|^2 \le f(x) \le |x|^2+|x|^4$, $\lim_{x\to 2} f(x)$

In Exercises 39–42, investigate the one-sided limits of the function at the endpoints of its domain of definition.

39. $f(x)=\sqrt{(x-1)(2-x)}$
40. $f(x)=(x-1)/(x^2-1)$, $-1<x<1$
41. $f(x)=(x^2-4)/(x^2+2x)$, $-2<x<2$
42. $f(x)=\sqrt{x}+1/\sqrt{1-x}$, $0<x<1$

Further Theory and Practice

In Exercises 43–52, use your intuition to decide whether the limit exists. Justify your answer by using the rigorous definition of limit.

43. $\lim_{x\to 4}|x|/x$
44. $\lim_{x\to 0}|x|/x$
45. $\lim_{x\to 7}(|x|-3x)$
46. $\lim_{x\to 1}(x^2-1)/|x-1|$
47. $\lim_{x\to -3}(x-|x|)^2$
48. $\lim_{x\to 1} g(x)$ where

$$g(x) = \begin{cases} 6 & \text{if } x \le -3 \\ -2x & \text{if } x > -3 \end{cases}$$

49. $\lim_{x\to 0} H(x)$ where

$$H(x) = \begin{cases} x^2 & \text{if } x \le 0 \\ x & \text{if } x > 0 \end{cases}$$

50. $\lim_{x\to 4} f(x)$ where

$$f(x) = \begin{cases} x^2-1 & \text{if } x < 4 \\ 3x+3 & \text{if } x \ge 4 \end{cases}$$

51. $\lim_{x\to 5} g(x)$ where

$$g(x) = \begin{cases} x+1 & \text{if } x < 5 \\ x-1 & \text{if } x \ge 5 \end{cases}$$

52. $\lim_{x\to 2} g(x)$ where

$$g(x) = \begin{cases} \dfrac{x^2+x-6}{x-2} & \text{if } x < 2 \\ \dfrac{x^3-2x^2+x-2}{x-2} & \text{if } x > 2 \end{cases}$$

In Exercises 53–62, use the basic limits of Theorem 8 to evaluate the limit. Note: x° means "x degrees."

53. $\lim_{x\to 0}\dfrac{\sin(3x)}{x}$
54. $\lim_{x\to 0}\dfrac{\tan(2x)}{x}$
55. $\lim_{x\to 0}\dfrac{\sin(2x)}{\sin(x)}$
56. $\lim_{x\to 0}\dfrac{\sin^2(x)}{1-\cos(x)}$
57. $\lim_{x\to 0}\dfrac{1-\cos(x)}{\tan(x)}$
58. $\lim_{x\to 0}\dfrac{\sin(x)-\tan(x)}{x^3}$
59. $\lim_{x\to 0}\dfrac{\sin(x^\circ)}{x}$
60. $\lim_{x\to 0}\dfrac{1-\cos(x^\circ)}{x^2}$

61. $\lim_{x\to 0} \dfrac{\sin(2x)\tan(3x)}{x^2}$ **62.** $\lim_{x\to 0} \dfrac{\sin^2(3x^2)}{x^4}$

In Exercises 63–69, use algebraic manipulation (as in Example 5) to evaluate the limit.

63. $\lim_{x\to 1} \left(\dfrac{x^2-1}{\sqrt{x}-1}\right)$ **64.** $\lim_{x\to 4} \left(\dfrac{x-4}{\sqrt{x}-2}\right)$

65. $\lim_{h\to 0} \dfrac{h}{\sqrt{1+2h}-1}$ **66.** $\lim_{t\to 0} \left(\dfrac{\sqrt{3+t}-\sqrt{3}}{t}\right)$

67. $\lim_{h\to 0} \left(\dfrac{2}{h\sqrt{4+h}} - \dfrac{1}{h}\right)$

68. $\lim_{x\to 4} \left(\dfrac{\sqrt{8-x}-2}{\sqrt{5-x}-1}\right)$

69. $\lim_{x\to 4} \dfrac{\sqrt{x}-2}{\sqrt{x+5}-3}$

In Exercises 70–76, evaluate the one-sided limits.

70. $\lim_{x\to 3^+} \dfrac{x-3}{|x-3|}$

71. $\lim_{x\to 3^-} \dfrac{x-3}{|x-3|}$

72. $\lim_{x\to 2^-} \left(\dfrac{6+x^2}{5} + \sqrt{13-2x}\right)$

73. $\lim_{x\to 4^-} \dfrac{x+\sqrt{16-3x}}{x+\sqrt{16-x^2}}$

74. $\lim_{x\to 0^+} \dfrac{\sqrt{\sin(x)}}{\sqrt{x}}$

75. $\lim_{x\to 0^+} \dfrac{\sin\left(\sqrt{x}\right)}{\sqrt{x}}$

76. $\lim_{x\to 1^-} \dfrac{\left(\sqrt{x}-1\right)\sqrt{x^2-3x+2}}{(1-x)^{3/2}}$

77. Let f and g be functions with domain $(-1, 1)$. Prove that if $|f(x)| \le 10^6$ for all x and if $\lim_{x\to 0} g(x) = 0$, then $\lim_{x\to 0}(f(x)\cdot g(x)) = 0$. Explain how you use the hypothesis on f. Does the value 10^6 play any role? Can this hypothesis be dropped altogether?

78. Prove the following variant of Theorem 5: If $\lim_{x\to c} f(x)/g(x)$ exists and $\lim_{x\to c} g(x) = 0$, then $\lim_{x\to c} f(x) = 0$.

In Exercises 79–82, use the Pinching Theorem to establish the required limit.

79. If $0 \le f(x) \le x^2$ for all x, then $\lim_{x\to 0} f(x)/x = 0$.

80. If $|g(x) - f(x)| \le \sqrt{|x|}$ and if $\lim_{x\to 0} f(x) = \ell$, then $\lim_{x\to 0} g(x) = \ell$.

81. If $\lim_{x\to c} |f(x)| = 0$, then $\lim_{x\to c} f(x) = 0$.

82. If $\lim_{x\to c} f(x) = 0$, then $\lim_{x\to c} f^3(x) = 0$.

83. Suppose f is defined on $(0, \infty)$. Suggest a definition of $\lim_{x\to\infty} f(x)$ in terms of a one-sided limit as $1/x \to 0^+$.

84. Let $f : (-1, 1) \to (0, \infty)$ be a function and suppose that $\lim_{x\to 0} f(x) = \ell$. Prove that $\ell \ge 0$. Give an example of such an f for which $\ell = 0$.

Computer/Calculator Exercises

85. Using Theorem 3, we have $\lim_{x\to 2} x^3/(x^2 - 2.9x + 2)$ equals 40. As $x \to 2^+$, about how close to 2 must x be to ensure that

$$\left|\frac{x^3}{x^2-2.9x+2} - 40\right| < 0.1?$$

Illustrate this by a graph of $f(x) = x^3/(x^2 - 2.9x + 2)$ in an appropriate viewing rectangle.

86. It can be shown that

$$\lim_{x\to\pi^+} \frac{\sqrt{x-\pi}}{\sqrt{\sin(2x)}} = \frac{1}{\sqrt{2}}.$$

About how close to π must x be to ensure that

$$\left|\frac{\sqrt{x-\pi}}{\sqrt{\sin(2x)}} - \frac{1}{\sqrt{2}}\right| < 0.01?$$

Illustrate this by a graph of $f(x) = \sqrt{x-\pi}/\sqrt{\sin(2x)}$ in an appropriate viewing rectangle.

87. Calculate $\ell = \lim_{x\to 2}(x^3 - 3x^2 + 2x + 1)$. About how close to 2 must x be to ensure that

$$|(x^3 - 3x^2 + 2x + 1) - \ell| < 0.1?$$

Illustrate this by a graph of $f(x) = x^3 - 3x^2 + 2x + 1$ in an appropriate viewing rectangle.

88. Verify that $\lim_{x\to 1} 2x^3/(x+1) = 1$. For $\epsilon = 1/100$, use a computer algebra system to determine exactly what δ (in the definition of limit) should be.

89. For Exercises 29–34, use a graphing utility to visually verify the pinching that is taking place.

90. Graph $f(x) = \sin(x)/x$ in the viewing window $[-\pi/2, \pi/2] \times [0, 1]$. Set your graphing device to degree mode to graph $g(x) = \sin(x^\circ)/x$ where x° means "x degrees." What is the relationship between f and g? Use this relationship to obtain $\lim_{x\to 0} g(x)$. Explain the appearance of the graph of g. Remember to reset your graphing device to radian mode.

91. Let $f(x) = (1 - \cos(x))/x^2$ and $g(x) = (1 - \cos(x^\circ))/x^2$ where x° means "x degrees." Compare the limits of f and g at 0. Explain why there is a constant α such that $f(x) = \alpha \cdot g(180x/\pi)$. What is α? Use the equality $f = \alpha \cdot g$ to explain the different limits of f

and g. Remember to reset your graphing device to radian mode.

In Exercises 92–95, use a graphing utility to illustrate the Pinching Theorem for the specified functions $G(x)$, $F(x)$, and $H(x)$ and for point c. For each exercise, complete the following.

a. Specify a suitable viewing rectangle R.
b. Print or sketch the graphs of G, F, and H in R.
c. Identify which are the pinching functions.
d. Ascertain that the functions have a common limit as $x \to c$.
e. Determine $\lim_{x\to c} F(x)$.

92. $G(x) = 2$, $F(x) = 2 + (x-1)^2 \cos^2(3x)$, $H(x) = 2 + |x-1|$, $c = 1$

93. $G(x) = \begin{cases} 3 - (x-2)^2 & \text{if } -\infty < x < 2 \\ 3 & \text{if } 2 \le x < \infty \end{cases}$, $F(x) = 3 + (x-2)^3/|2x-4|$, $H(x) = \begin{cases} 3 & \text{if } -\infty < x < 2 \\ 3 + (x-2)^2 & \text{if } 2 \le x < \infty \end{cases}$, $c = 2$

94. $G(x) = -|x-1|^3$, $F(x) = \begin{cases} (x-1)^3 & \text{if } -\infty < x < 3 \\ 7 + \sqrt{x-2} & \text{if } 3 \le x < \infty \end{cases}$, $H(x) = 7 + \sqrt{|x-2|}$, $c = 3$

95. $G(x) = 8 - 20|x-3|$, $F(x) = \begin{cases} (x-1)^3 & \text{if } -\infty < x < 3 \\ 7 + \sqrt{x-2} & \text{if } 3 \le x < \infty \end{cases}$, $H(x) = 8 + \frac{3}{2}|x-3|$, $c = 3$.

2.3 Continuity

When we drive along a high-speed road, we frequently cannot see very far ahead. We learn to drive as though the portion of the road immediately beyond our vision is the obvious extension of the portion of the road that we *can* see. What we are doing is computing the limit of what we see. Usually, the actual state of the road coincides with what we have anticipated it will be (that is, the road is continuous). If the road has been damaged, or if a bridge is out, then the actual state of the road does *not* coincide with what we have anticipated (the road is discontinuous). In this section, we develop mathematical analogues of these ideas.

The Definition of Continuity at a Point

Here is a precise definition of continuity of a function at a point.

Definition Suppose a function f is defined on an open interval that contains the point c. We say that f is *continuous* at c, provided that

$$\lim_{x\to c} f(x) = f(c).$$

Refer to Figure 1.

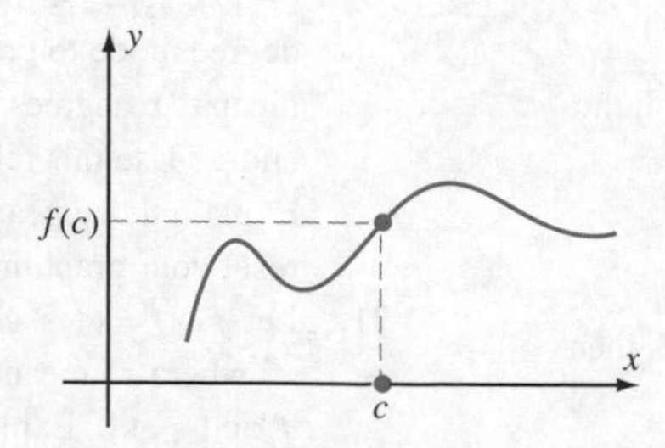

Figure 1
$f(x)$ has the limit $f(c)$ as x tends to c.

> **inSIGHT**
>
> When we say that f is continuous at c, we are asserting *three* things:
>
> 1. f is defined at c;
> 2. there is a certain value $\lim_{x \to c} f(x)$ that we *expect* f to take at c (that is, $\lim_{x \to c} f(x)$ exists); and
> 3. the expected value $\lim_{x \to c} f(x)$ agrees with the value $f(c)$, which f *actually* takes at c (that is, $\lim_{x \to c} f(x) = f(c)$).

We only consider continuity of a function at points of its domain. For instance, we would not say that the function $f(x) = \sin(x)/x$ is continuous at 0, because 0 is not in the domain of f. If a point c is in the domain of f and f is not continuous at c, then we say that f is *discontinuous* at c. Such a point is called a *point of discontinuity* of f. If c is a point of discontinuity of f, then either f does not have a limit at c or the limit exists but is not equal to $f(c)$.

When a function f is continuous at *every point* of its domain, then we say that f is a *continuous function*. The theorems about limits that were developed in Section 2.2 show that polynomials, rational functions, nth roots, and trigonometric functions are continuous functions.

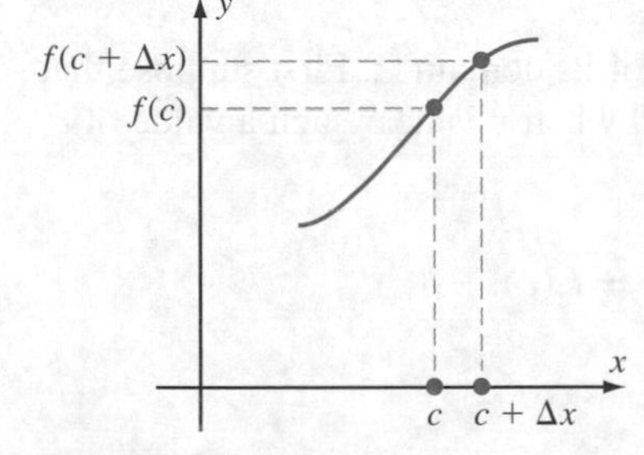

Figure 2
A small input error near a point of continuity results in a small output error.

An Equivalent Formulation of Continuity

The continuity of f at c can be expressed in another form. Let Δx denote an increment in x. Write $x = c + \Delta x$ and observe that $x \to c$ is equivalent to $\Delta x \to 0$. With this notation, the continuity of f at c is equivalent to

$$\lim_{\Delta x \to 0} f(c + \Delta x) = f(c).$$

Imagine an experiment to determine the mass $y = f(x)$ of a compound that is produced by a chemical reaction with a substance that has mass x. Suppose the value of x in the experiment is taken to be c. Inevitably there will be a small measurement error Δx. With care, the *actual* value of x, $c + \Delta x$, will be close to the desired value c, but it will usually not be *exactly* c. Although the result of the experiment is intended to be $f(c)$, it is really $f(c + \Delta x)$. However, the result of the experiment is not necessarily unreliable. If c is a value at which f is continuous, then the difference between $f(c + \Delta x)$ and $f(c)$ can be made small by controlling the size of the measurement error Δx. In other words, sufficiently small errors in the independent variable x result in small errors in the dependent variable y (see Figure 2). By contrast, if c is a point at which f is discontinuous, then a very small nonzero value of Δx can result in a substantial error, $f(c + \Delta x) - f(c)$, in the outcome of the experiment (see Figure 3).

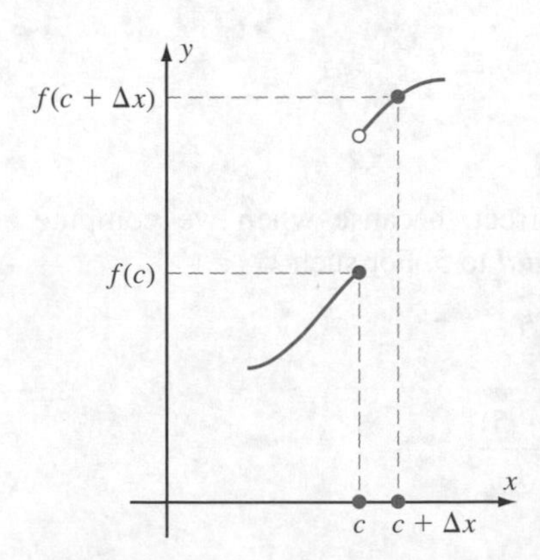

Figure 3
A small input error near a point of discontinuity can result in a large output error.

Example 1 Let

$$F(x) = \begin{cases} -4 & \text{if } x < 1 \\ -x - 3 & \text{if } x \geq 1 \end{cases}.$$

Verify that F is continuous at every $c \in \mathbb{R}$.

Solution If $c < 1$, then

$$\lim_{x\to c} F(x) = \lim_{x\to c}(-4) = -4 = F(c).$$

If $c > 1$, then

$$\lim_{x\to c} F(x) = \lim_{x\to c}(-x-3) = -c-3 = F(c).$$

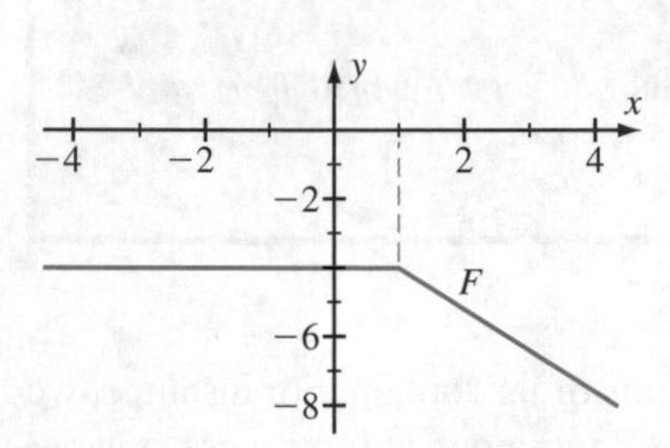

Figure 4

If $c = 1$, then we see, by checking both on the right and on the left, that

$$\lim_{x\to 1^-} F(x) = \lim_{x\to 1^-}(-4) = -4 = F(1) \quad \text{and}$$
$$\lim_{x\to 1^+} F(x) = \lim_{x\to 1^+}(-x-3) = -4 = F(1).$$

Therefore, F is continuous at every $c \in \mathbb{R}$ (see Figure 4). ■

Example 2 Let

$$F(x) = \begin{cases} \dfrac{x^2-6x+5}{x-5} & \text{if } x \neq 5 \\ 4 & \text{if } x = 5 \end{cases}.$$

Is F a continuous function?

Solution We must investigate F at each point c of its domain $\mathbb{R}$. First suppose that $c \neq 5$. Because the denominator of $F(x)$ is never 0 when x is near such a value of c, we may apply Theorem 2c from Section 2.2:

$$\lim_{x\to c} F(x) = \frac{c^2-6c+5}{c-5} = F(c).$$

For $c = 5$, we calculate

$$\begin{aligned}\lim_{x\to 5} F(x) &= \lim_{x\to 5} \frac{x^2-6x+5}{x-5} \\ &= \lim_{x\to 5} \frac{(x-5)\cdot(x-1)}{x-5} \\ &= \lim_{x\to 5}(x-1) \\ &= 4 \\ &= F(5).\end{aligned}$$

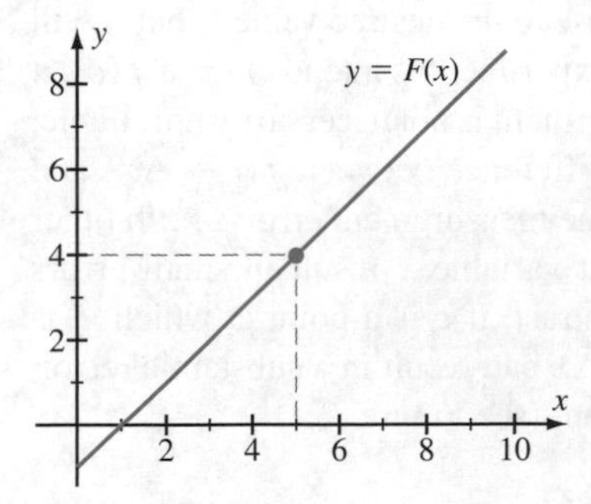

Figure 5

$$F(x) = \begin{cases} \frac{x^2-6x+5}{x-5} & \text{if } x \neq 5 \\ 4 & \text{if } x = 5 \end{cases}$$

Notice that the first two equalities here are correct, because when we compute $\lim_{x\to 5} F(x)$, we use values of x *near* 5 but *not equal* to 5. For such x,

$$\begin{aligned}F(x) &= \frac{x^2-6x+5}{x-5} \\ &= \frac{(x-1)(x-5)}{x-5} \\ &= x-1.\end{aligned}$$

Thus, F is continuous at $c = 5$. Since F is continuous at each point of its domain, it is a continuous function. Figure 5 shows the graph of F in the viewing window $[0, 10] \times [-1, 9]$. ■

Example 3 Let

$$G(x) = \begin{cases} x^2 & \text{if } x < 3 \\ 9 & \text{if } x = 3 \\ 0 & \text{if } x > 3 \end{cases} \quad \text{and} \quad H(x) = \begin{cases} x^2 & \text{if } x < 3 \\ 0 & \text{if } x = 3 \\ 9 & \text{if } x > 3 \end{cases}.$$

Refer to Figures 6 and 7. Discuss the continuity of G and H at $c = 3$.

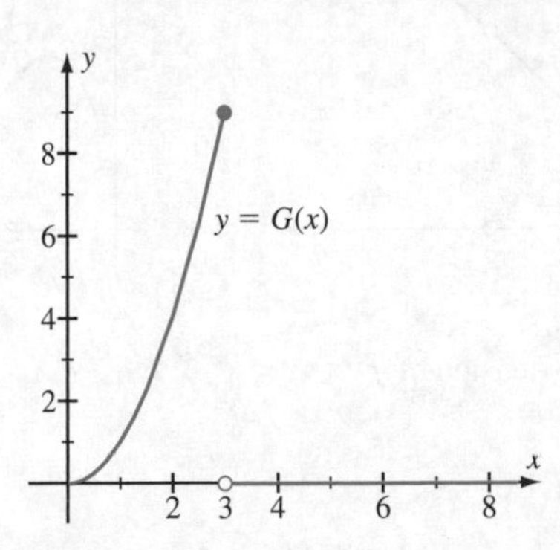

Figure 6

$$G(x) = \begin{cases} x^2 & \text{if } x < 3 \\ 9 & \text{if } x = 3 \\ 0 & \text{if } x > 3 \end{cases}$$

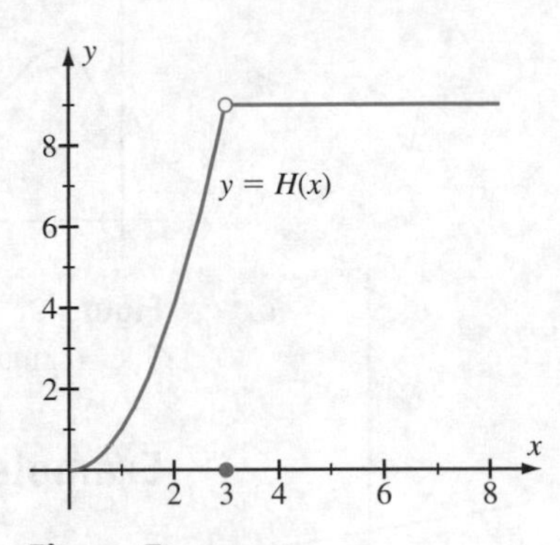

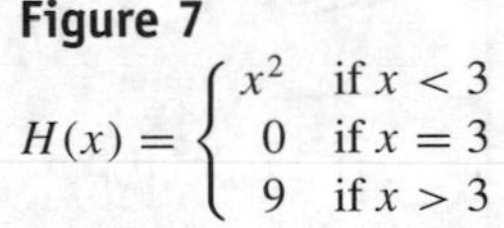

Figure 7

$$H(x) = \begin{cases} x^2 & \text{if } x < 3 \\ 0 & \text{if } x = 3 \\ 9 & \text{if } x > 3 \end{cases}$$

Solution Observe that $\lim_{x\to 3} G(x)$ does not exist because $\lim_{x\to 3^+} G(x) = 0$ and $\lim_{x\to 3^-} G(x) = 9$. So it is certainly not the case that $\lim_{x\to 3} G(x) = G(3)$. Thus, G is discontinuous at $c = 3$. Although $\lim_{x\to 3} H(x)$ exists, $\lim_{x\to 3} H(x) = 9 \neq H(3)$. Therefore, H is also discontinuous at $c = 3$. ■

Example 4 Show that the function

$$F(x) = \begin{cases} \dfrac{\sin(x)}{x} & \text{if } x \neq 0 \\ 1 & \text{if } x = 0 \end{cases}$$

is continuous at $c = 0$.

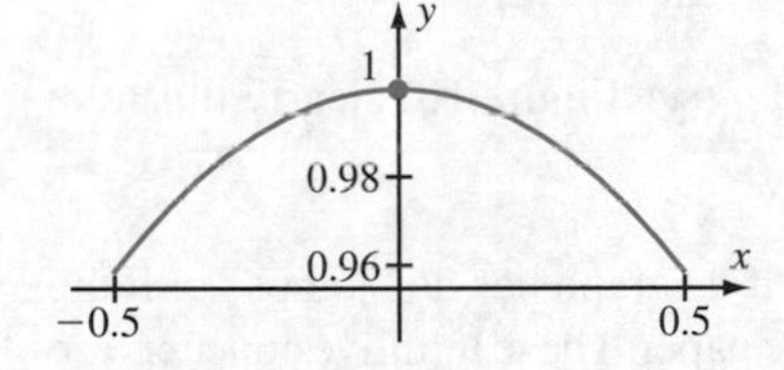

Figure 8

$$F(x) = \begin{cases} \sin(x)/x & \text{if } x \neq 0 \\ 1 & \text{if } x = 0 \end{cases}$$

Solution Theorem 8 from Section 2.2 tells us that $\lim_{x\to 0} F(x) = 1 = F(0)$. Thus, F is continuous at 0 (see Figure 8). ■

Continuous Extensions

Consider a continuous function f defined on $(a, c) \cup (c, b)$. Can we assign a value to f at c to create a continuous function on (a, b)? More precisely, is there a continuous function F that is defined on the entire interval (a, b) such that $F(x) = f(x)$ for $x \neq c$? The answer is that F exists precisely when $\lim_{x\to c} f(x)$ exists. In this case,

$$F(x) = \begin{cases} f(x) & \text{if } x \neq c \\ \lim_{x\to c} f(x) & \text{if } x = c \end{cases}.$$

We refer to F as the *continuous extension* of f. This situation is illustrated in Figure 9. Because the domain of F is larger than the domain of f, F is not the same function as f.

Refer back to Example 2 and you will see that the given function F is a continuous extension of $x \mapsto (x^2 - 6x + 5)/(x - 5)$. Similarly, in Example 4, the given function F is a continuous extension of $x \mapsto \sin(x)/x$.

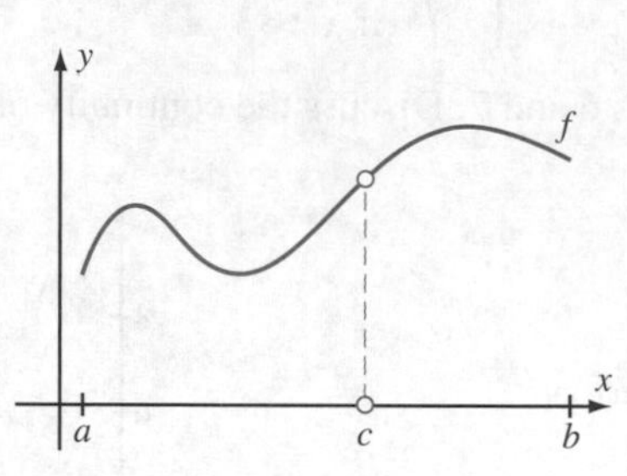

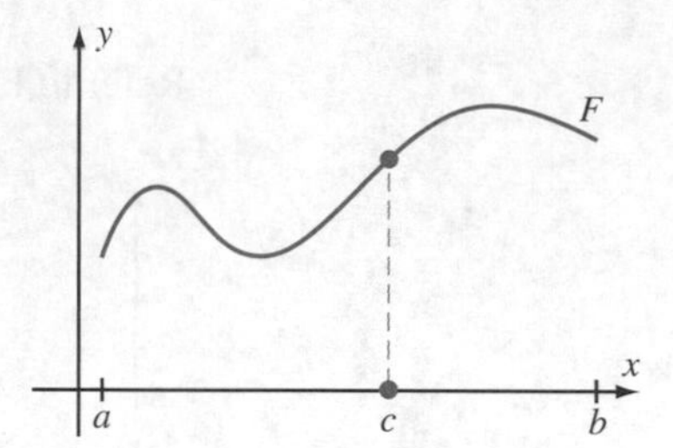

Figure 9
The continuous extension of f

Example 5 Let

$$f(x) = \begin{cases} x^2 - 14 & \text{if } x < -4 \\ x + 7 & \text{if } x > -4 \end{cases} \quad \text{and} \quad g(x) = \begin{cases} x^2 - 3 & \text{if } x < -4 \\ 9 - x & \text{if } x > -4 \end{cases}.$$

Is it possible to define a continuous extension of f? Of g?

Solution Notice that $\lim_{x \to -4} f(x)$ does not exist. Therefore, no continuous extension exists. Observe that as x approaches -4 from the left, the value $g(x) = x^2 - 3$ approaches 13. In addition, as x approaches -4 from the right, the value $g(x) = 9 - x$ also approaches 13. Thus, $\lim_{x \to -4} g(x) = 13$, and the function

$$G(x) = \begin{cases} g(x) & \text{if } x \neq -4 \\ 13 & \text{if } x = -4 \end{cases}$$

is a continuous extension of g. The graphs of f and g (see Figure 10) clearly illustrate the different behaviors of the functions at $x = -4$. ■

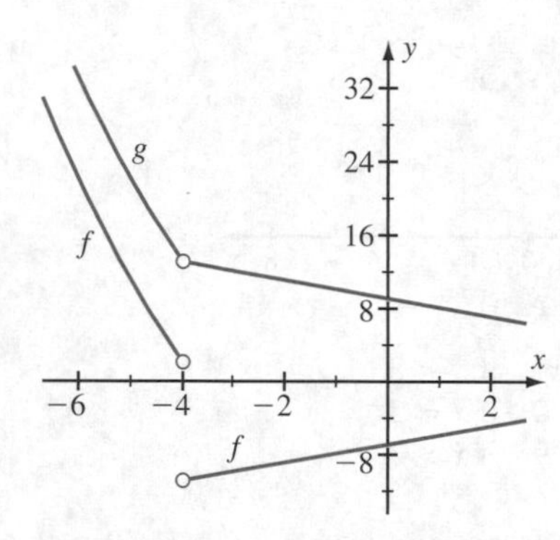

Figure 10
$f(x) = \begin{cases} x^2 - 14 & \text{if } x < -4 \\ x + 7 & \text{if } x > -4 \end{cases}$

$g(x) = \begin{cases} x^2 - 3 & \text{if } x < -4 \\ 9 - x & \text{if } x > -4 \end{cases}$

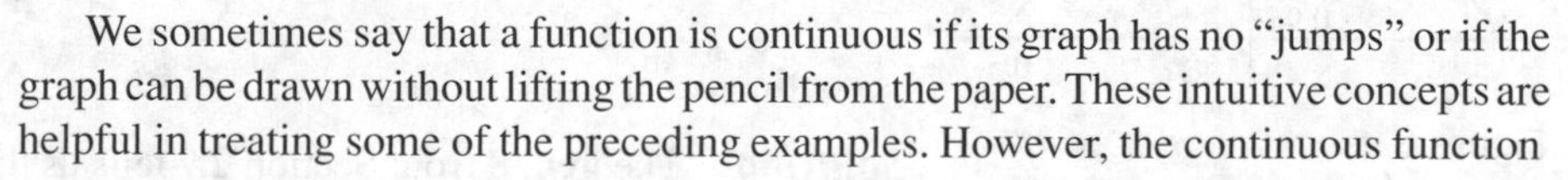

We sometimes say that a function is continuous if its graph has no "jumps" or if the graph can be drawn without lifting the pencil from the paper. These intuitive concepts are helpful in treating some of the preceding examples. However, the continuous function

$$F(x) = \begin{cases} x \cdot \sin\left(\dfrac{1}{x}\right) & \text{if } x \neq 0 \\ 0 & \text{if } x = 0 \end{cases}$$

that we considered in Example 6 from Section 2.2 is much more difficult to understand in this informal manner. The graph oscillates rapidly near the origin, as shown in Figure 11, and the truth is that we cannot really draw the graph *with or without* lifting the pencil from the paper. It is precisely because of functions like these that we have gone to the trouble of learning careful definitions of "limit" and "continuous." These careful definitions allow us to see in Example 4 that the function f is continuous at $c = 0$. Furthermore, our precise notion of continuity enables us to say exactly at which points the function is continuous and at which points it is not. The "lifting the pencil from the paper" description of continuity treats the graph as a whole and does not allow us to isolate the points of continuity from the points of discontinuity.

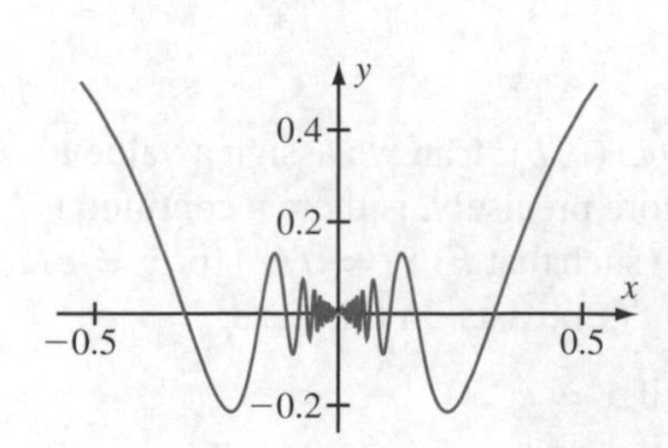

Figure 11
$F(x) = \begin{cases} x \cdot \sin(1/x) & \text{if } x \neq 0 \\ 0 & \text{if } x = 0 \end{cases}$

One-Sided Continuity

Definition Suppose f is a function with a domain that contains $[c, b)$. If

$$\lim_{x \to c^+} f(x) = f(c),$$

then we say that f is *continuous from the right* at c, or f is *right-continuous* at c. Figure 12 exhibits a function that is right-continuous at c.

Definition Suppose f is a function with a domain that contains $(a, c]$. If

$$\lim_{x \to c^-} f(x) = f(c),$$

then we say that f is *continuous from the left* at c, or f is *left-continuous* at c. Figure 13 shows a function that is left-continuous at c.

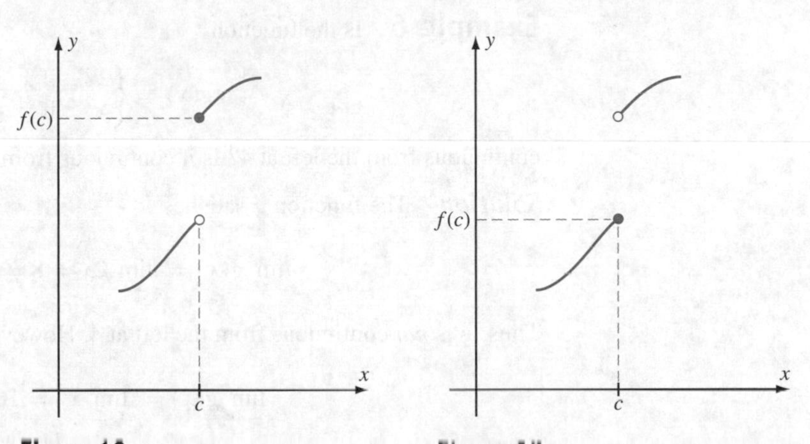

Figure 12
f is right-continuous at c.

Figure 13
f is left-continuous at c.

Notice that, just as in the definition of continuity, the notions of left and right continuity at c entail three requirements:

1. The point c is in the domain of f.
2. The limit exists.
3. The one-sided limit must equal $f(c)$.

Notice also that "f is continuous at c" is equivalent to "f is left-continuous at c and right-continuous at c." If $\lim_{x \to c^+} f(x)$ and $\lim_{x \to c^-} f(x)$ both exist but are different numbers, f is said to have a *jump discontinuity* at c (see Figure 14, next page).

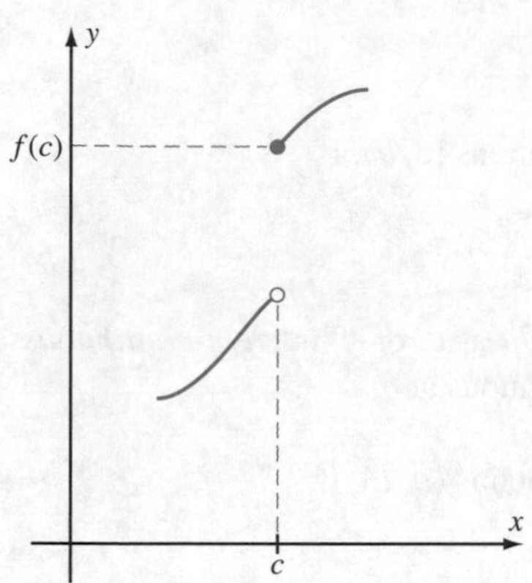

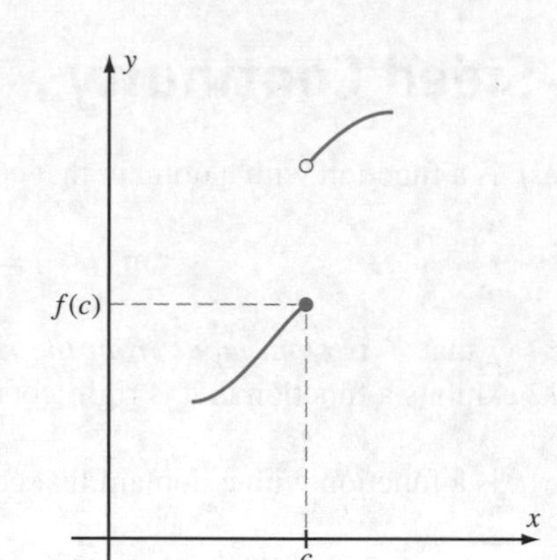

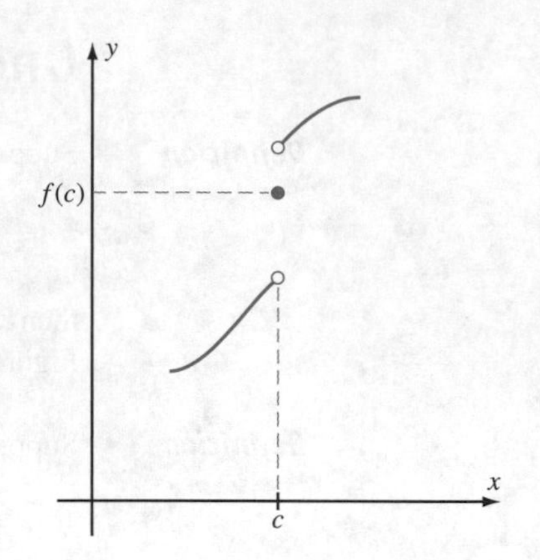

Figure 14
Examples of functions with jump discontinuities

Example 6 Is the function

$$g(x) = \begin{cases} 2x & \text{if } x < 4 \\ x^2 & \text{if } x \geq 4 \end{cases}$$

continuous from the left at 4? Is it continuous from the right at 4?

Solution The function g satisfies

$$\lim_{x \to 4^-} g(x) = \lim_{x \to 4^-} 2x = 8 \neq 16 = g(4).$$

Thus, g is *not* continuous from the left at 4. However,

$$\lim_{x \to 4^+} g(x) = \lim_{x \to 4^+} x^2 = 16 = g(4).$$

Therefore, g *is* continuous from the right at 4. ■

Some Theorems about Continuity—Arithmetic Operations and Composition

Our work with the concept of continuity would be tedious if we had to check, from the definition, whether each function that we encounter is continuous. Fortunately, the limit theorems from Section 2.2 give rise to analogous theorems for continuity.

Theorem 1 Let α be a constant.

a. If f and g are functions that are continuous at $x = c$, then so are $f + g$, $f - g$, $\alpha \cdot f$, and $f \cdot g$. If $g(c) \neq 0$, then f/g is also continuous at $x = c$.
b. If f and g are both right continuous at $x = c$, then so are $f + g$, $f - g$, $\alpha \cdot f$, and $f \cdot g$. If $g(c) \neq 0$, then f/g is also right continuous at $x = c$. The same is true if right continuity is replaced with left continuity.

Theorem 1 follows from the corresponding results about limits in Theorem 2 from Section 2.2.

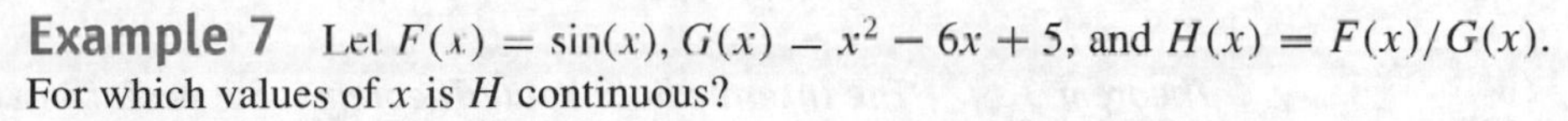

Example 7 Let $F(x) = \sin(x)$, $G(x) = x^2 - 6x + 5$, and $H(x) = F(x)/G(x)$. For which values of x is H continuous?

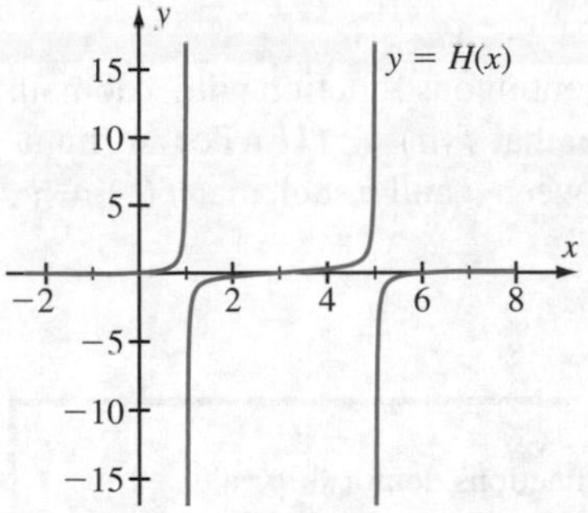

Figure 15
$H(x) = \sin(x)/(x^2 - 6x + 5)$

Solution As was mentioned earlier in this section, the limit theorems from Section 2.2 tell us that trigonometric functions and polynomials are continuous. In particular, the functions F and G are continuous at every point of the real line. Therefore, the function H is continuous at every value of x for which $G(x) = x^2 - 6x + 5 = (x-1)(x-5)$ is nonzero. Thus, H is continuous at each point different from 1 and 5. The points 1 and 5 are not in the domain of H, so we do not speak about the continuity of H at these points. However, the plot of H in the viewing window $[-2, 8] \times [-15, 15]$ (see Figure 15) indicates that H does not have a continuous extension at the exceptional points. ■

If the image of a function f is contained in the domain of a function g, then we may consider the composition $(g \circ f)(x) = g(f(x))$. This equation defines an operation that consists of first applying f and then applying g. In general, using the limits of f and g to calculate the limit of the composite function $g \circ f$ can be tricky. Exercises 55–57 at the end of this section explore the possibilities. The following theorem assures us that nothing unexpected happens when both f and g are continuous. Composition preserves continuity.

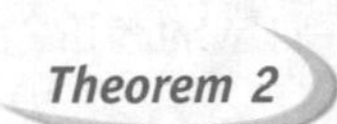

Theorem 2 Suppose the image of a function f is contained in the domain of a function g. If f is continuous at c and g is continuous at $f(c)$, then the composed function $g \circ f$ is continuous at c.

Example 8 Is $x \mapsto \cos(x^2 + 1)$ continuous at every x?

Solution The functions $f(x) = x^2 + 1$ and $g(x) = \cos(x)$ have the entire real line as their domain, and they are continuous at each point. Therefore, $(g \circ f)(x) = \cos(x^2 + 1)$ is defined for every value of x, and, by Theorem 2, it is continuous at every x. ■

Advanced Properties of Continuous Functions

The next two theorems about continuity depend on a deeper property of the real numbers—the *completeness property*. When we imagine the real numbers, we are accustomed to picturing a line (the real number line) that has no gaps in it. The analytic formulation of this idea is the completeness property of the real numbers. There are a number of equivalent ways to express the property precisely. One formulation is the *nested interval property*, which is discussed in Genesis & Development, Chapter 1. Another is the *least upper bound property* of the real numbers—see Genesis & Development, Chapter 2. A third formulation of the completeness property is discussed in Section 2.6.

Like many theorems in calculus, the two theorems that follow require that we use the notion of a function f being continuous on a closed interval $[a, b]$. This means that f is continuous at every point in the open interval (a, b), $f(x)$ is right continuous at

a, and $f(x)$ is left continuous at b. In particular, if f is continuous on $[a, b]$, then the functional values $f(a)$ and $f(b)$ at the endpoints are the same as the one-sided limits of f at the endpoints.

Theorem 3

The Intermediate Value Theorem Suppose f is a continuous function with a domain that contains the closed interval $[a, b]$. Suppose further that $f(a) \neq f(b)$. For any number γ between $f(a)$ and $f(b)$, there is a number c between a and b such that $f(c) = \gamma$.

In simple language, Theorem 3 states that continuous functions do not skip values. If f is continuous and if f takes the values α and β, then f takes all the values *between* α and β (see Figure 16). In even plainer language, the graph of f does not have jumps—a jump would correspond to skipped values. This explains why we sometimes say that a continuous function can be graphed without lifting the pencil from the paper. Although we now understand this statement to be purely informal, we can use it to guide our intuition.

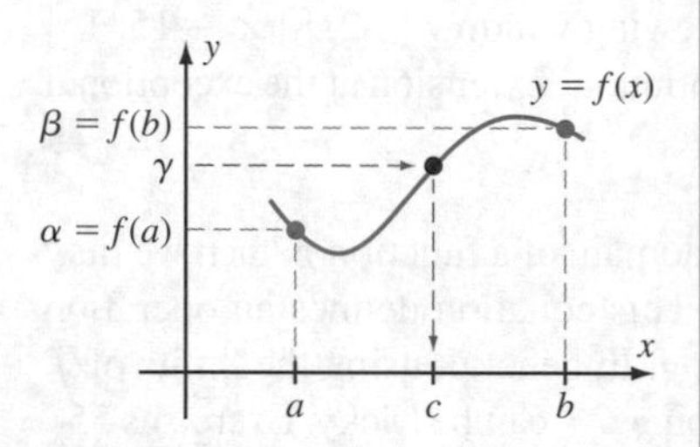

Figure 16

Example 9 Is there a point in the interval $(0, 2)$ at which the function $f(x) = (x^3 + 1)^2$ takes the value 10?

Solution Because it is a polynomial, the function $f(x) = (x^3 + 1)^2$ is continuous. We can easily find values a and b such that $\gamma = 10$ is between $f(a)$ and $f(b)$. For example, if $a = 0$ and $b = 2$, then $f(a) = 1 < 10 < 81 = f(b)$. According to the Intermediate Value Theorem, we can be sure that there is a number c between 0 and 2 such that $f(c) = 10$. ■

inSIGHT

The Intermediate Value Theorem gives us conditions under which we can be certain that the equation $f(c) = \gamma$ has a solution c. In Example 9, we can actually find a formula for the value c for which $f(c) = \gamma$. With the aid of a little algebra, we obtain $c = \sqrt[3]{\sqrt{10} - 1} \simeq 1.2931\ldots$. When we use the Intermediate Value Theorem, we usually cannot find an explicit formula for the value of a solution c of the equation $f(c) = \gamma$. In Chapter 4, we will learn how to find good approximations to the solutions of such equations.

Example 10 Mr. Woodman weighs 150 pounds at noon on January 1 ($t = 1$), 130 pounds at noon on January 15 ($t = 15$), and 140 pounds at noon on January 30 ($t = 30$). His weight is a continuous function of time t. How often during the month of January can we be sure that he will weigh 132 pounds?

Solution Let $w(t)$ be Mr. Woodman's weight in pounds at time t where t is measured in days. Our hypotheses may be written as

$$w(1) = 150, \qquad w(15) = 130, \qquad w(30) = 140.$$

Because 132 is between 150 and 130, there is a time t_1, $1 < t_1 < 15$, such that $w(t_1) = 132$. Likewise, 132 is between 130 and 140. Thus, there is a time t_2, $15 < t_2 < 30$, such that $w(t_2) = 132$. In other words, we have found that between noon on January 1 and noon on January 30, Mr. Woodman will weigh 132 pounds *at least* two times. As Figure 17 illustrates, Mr. Woodman *may* weigh 132 at other times between January 1 and January 30. Without further information, however, we can only say for certain that there are at least two times.

inSIGHT

As Example 10 illustrates, the Intermediate Value Theorem says nothing about uniqueness. It gives us conditions under which there is sure to be a solution c of the equation $f(c) = \gamma$, but it does *not* tell whether there is only one solution or several solutions.

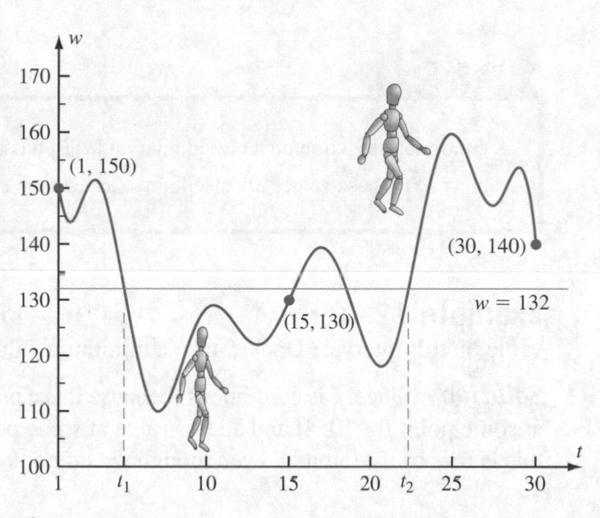

Figure 17

■

We have already mentioned that if n is a positive integer, then $x \mapsto \sqrt[n]{x}$ is a continuous function (Theorem 4 from Section 2.2). Until now we have implicitly assumed the existence of nth roots. How can we be sure, however, that there is a number c such that $c^{17} = \pi$? The Intermediate Value Theorem can be used to prove the existence of such roots, as is shown in the next example.

Example 11 Let γ be a positive number and n a positive integer. Show that there is a number c such that $c^n = \gamma$.

Solution Let $f(x)$ be defined on $[0, \infty)$ by $f(x) = x^n$. We know that f is a continuous function by Theorem 3 from Section 2.2. Let $a = 0$, $b = \gamma + 1$, $\alpha = f(a)$, and $\beta = f(b)$. Then,

$$\alpha = f(0) = 0 < \gamma < \gamma + 1 < (\gamma + 1)^n = f(b) = \beta.$$

The Intermediate Value Theorem tells us that there is a $c \in (a, b)$ such that $f(c) = \gamma$. Thus, $c^n = \gamma$. ■

The Intermediate Value Theorem is said to be an *existence theorem* because it asserts the existence of a value without pinpointing it. We will encounter several theorems of this type in our study of calculus. Indeed, the theorem that follows asserts the existence of a highest and a lowest point on the graph of a function that is continuous on a closed bounded interval.

Definition Suppose f is a function with domain S. If there is a point $\alpha \in S$ such that $f(\alpha) \le f(x)$ for all $x \in S$, then the point α is called a *minimum* for the function f and $m = f(\alpha)$ is called the *minimum value* of f. If there is a point $\beta \in S$ such that $f(\beta) \ge f(x)$ for all $x \in S$, then the point β is called a *maximum* for the function f and $M = f(\beta)$ is called its *maximum value*.

Theorem 4 ***The Extreme Value Theorem*** If f is a continuous function that is defined on the closed interval $[a, b]$, then f has a minimum $\alpha \in [a, b]$ and a maximum $\beta \in [a, b]$.

A continuous function on a closed interval $[a, b]$ has a greatest value $f(\beta)$ and a least value $f(\alpha)$. These values are also known as *extreme values,* or *extrema,* of f.

Example 12 Let $f(x) = x^3 - 7x^2 + 3x - 1$ on $[0, 4]$. Does f take a maximum value on this interval? Does f take a minimum value?

Solution Since f is continuous (because it is a polynomial), it takes a greatest value at some point $\beta \in [0, 4]$ and a least value at some point $\alpha \in [0, 4]$. We can be sure that this is true by Theorem 4, even though we do not know how to find α and β yet. ■

inSIGHT

One important application of calculus is to actually *find* α and β. This, in turn, will enable us to solve many interesting problems. (Chapter 4 covers this idea in more detail.)

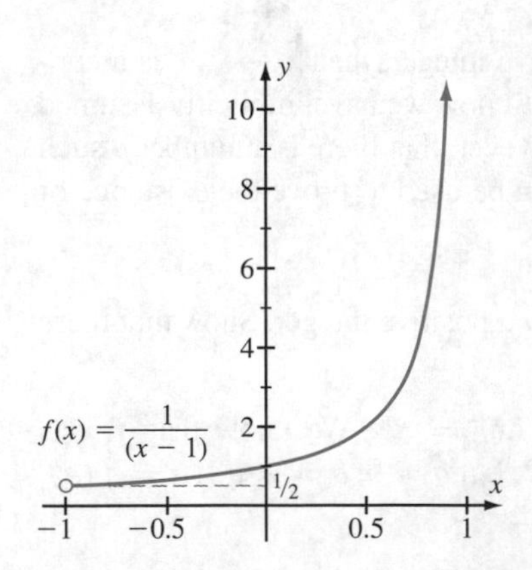

Figure 18

Example 13 Let $f(x) = 1/(1 - x)$ (see Figure 18). Are there maximum and minimum values for f on the interval $(-1, 1)$?

Solution The function f is continuous on the interval $(-1, 1)$. However, f has no greatest value on the interval $(-1, 1)$ because $f(x)$ becomes large without bound as x approaches 1 from the left (for instance, $f(0.9) = 10$, $f(0.9999) = 10000$, etc.). Also, f has no least value on $(-1, 1)$, although this is so for a more subtle reason—if x is near -1 and to the right of -1, then $f(x)$ is nearly $1/2$. Indeed, the values of f become arbitrarily close to $1/2$ when x is sufficiently close to -1 on the right, but f never actually assumes the value $1/2$ at any point of the interval $(-1, 1)$. So, f does not assume a minimum value on $(-1, 1)$. These facts do not contradict the Extreme Value Theorem, since the interval $(-1, 1)$ is not closed (that is, it does not include its endpoints). ■

quickquiz

1. What does it mean for a function f to be continuous at a point of its domain? What does it mean for f to be continuous?
2. At what points is the function $x \mapsto (\sin(x^2) + 3)/(2 - x^4)$ continuous?
3. What operations on functions preserve continuity?
4. Let $f : [0, 3] \to (0, 1]$. Does f necessarily have a minimum value? What if f were continuous?
5. Suppose that $f : [1, 2] \to (0, 1]$ is continuous and that $f(1) = 1/2$ and $f(2) = 1/4$. Does $1/f$ assume the value π?

EXERCISES

Problems for Practice

In Exercises 1–14, determine at which values of its domain the function is continuous. Justify your answer.

1. $f(x) = x^2 + 4$
2. $H(x) = 2x - 9$
3. $g(x) = 1/(x + 1)$
4. $H(x) = x/(x^2 - x - 5)$
5. $f(x) = 1/(x^2 + 1)$
6. $f(x) = 1/(1 - x^2)$
7. $H(x) = (x^2 - 5x + 6)/(x^2 + x - 12)$
8. $f(x) = \begin{cases} x^2 + 1 & \text{if } x < 0 \\ 1/(2x + 1) & \text{if } x \geq 0 \end{cases}$
9. $g(x) = \begin{cases} x + 1 & \text{if } x \leq 3 \\ 6 & \text{if } x > 3 \end{cases}$
10. $k(x) = \begin{cases} x^2 - 2 & \text{if } x \leq 3 \\ 8 & \text{if } x > 3 \end{cases}$
11. $H(x) = \begin{cases} (x + 1)^4 & \text{if } x < -4 \\ -20x + 1 & \text{if } x \geq -4 \end{cases}$
12. $k(x) = \begin{cases} (x^2/4) - 7 & \text{if } x < 6 \\ 2 & \text{if } x = 6 \\ 9 - x & \text{if } x > 6 \end{cases}$
13. $g(x) = \begin{cases} 2x - 19 & \text{if } x < 7 \\ -5 & \text{if } x = 7 \\ 2 - x & \text{if } x > 7 \end{cases}$
14. $k(x) = \begin{cases} 2x + 1 & \text{if } x < 0 \\ 4 & \text{if } x = 0 \\ x^2 - 2 & \text{if } x > 0 \end{cases}$

In Exercises 15–18, define $F(2)$ to obtain a continuous extension F of f.

15. $f(x) = \begin{cases} x^2 + 7 & \text{if } x < 2 \\ x^3 + 3 & \text{if } x > 2 \end{cases}$
16. $f(x) = \begin{cases} 5/(x^2 + 6) & \text{if } x < 2 \\ 1/x & \text{if } x > 2 \end{cases}$
17. $f(x) = \begin{cases} x^6/(18 - x) & \text{if } x < 2 \\ x^4/(x + 2) & \text{if } x > 2 \end{cases}$
18. $f(x) = \begin{cases} x^3/(x^3 + 1) & \text{if } x < 2 \\ x^4/(x^4 + 2) & \text{if } x > 2 \end{cases}$

For Exercises 19–22, determine the points at which the function is left continuous, the points at which the function is right continuous, and the points at which the function is continuous. Give reasons for your answers.

19. $f(x) = \begin{cases} 2 & \text{if } x \leq 5 \\ 3 & \text{if } x > 5 \end{cases}$
20. $g(x) = \begin{cases} x^2 - 15 & \text{if } x < -4 \\ 1 & \text{if } x \geq -4 \end{cases}$
21. $f(x) = \begin{cases} x + 1 & \text{if } x < 3 \\ 2 & \text{if } x \geq 3 \end{cases}$
22. $g(x) = \begin{cases} x^2 & \text{if } x < 0 \\ 2x & \text{if } x \geq 0 \end{cases}$

In Exercises 23–28, a continuous function f is defined on a closed bounded interval I. Determine the extreme values of the function f. Sketch the graph of f and label the points at which f assumes its extreme values.

23. $f(x) = x^2 + 2x + 3$, $I = [-3, -2]$
24. $f(x) = x^2 + 2x + 3$, $I = [-2, 1]$
25. $f(x) = x^2 + 2x + 3$, $I = [1, 2]$
26. $f(x) = 11 + 6x - x^2$, $I = [1, 4]$

27. $f(x) = 2 - \cos(x)$, $I = [0, 2\pi]$
28. $f(x) = x + \sin(x)$, $I = [-\pi, \pi]$

Further Theory and Practice

29. For each function, define $f(x)$ for $1 \le x \le 3$ so that f is continuous at every $x \in \mathbb{R}$.

a. $f(x) = \begin{cases} x^2 + 16 & \text{if } x < 1 \\ x - 4 & \text{if } x > 3 \end{cases}$

b. $f(x) = \begin{cases} (x+1)/(x^2+2) & \text{if } x < 1 \\ x^2 & \text{if } x > 3 \end{cases}$

c. $f(x) = \begin{cases} 1 & \text{if } x < 1 \\ -6 & \text{if } x > 3 \end{cases}$

30. At which points are the following functions continuous? At which points are they discontinuous? Give reasons for your answers.

a. $f(x) = \begin{cases} \sin(\pi \cdot x/2) & \text{if } x \notin \mathbb{Z} \\ 0 & \text{if } x \in \mathbb{Z} \end{cases}$

b. $f(x) = \begin{cases} \cos(\pi \cdot x/2) & \text{if } x \notin \mathbb{Z} \\ 1 & \text{if } x \in \mathbb{Z} \end{cases}$

31. Graph the functions

$$f(x) = \frac{x^2 - 1}{x - 1},\ g(x) = \begin{cases} x + 1 & \text{if } x \neq 1 \\ 2 & \text{if } x = 1 \end{cases},$$

$$h(x) = x + 1, \text{ and } k(x) = \begin{cases} x + 1 & \text{if } x \neq 1 \\ 1 & \text{if } x = 1 \end{cases}$$

over the interval [0, 2] centered at $c = 1$. For each function, compute its left and right limits at $c = 1$, determine if it has a limit at $c = 1$, and determine if it is continuous at $c = 1$. (Pay particular attention to the domains of the functions.) Are any of these functions identical?

32. Consider the single filer tax function T that is given in Example 3 from Section 1.4. Show that T is a continuous function.

33. Consider the single filer tax function T that is given in Example 3 from Section 1.4. There are five tax brackets that correspond to five marginal tax rates: 15%, 28%, 31%, 36%, and 39.6%. Suppose the tax brackets are not altered but the marginal tax rates are reduced to 10%, 15%, 25%, 30%, and 35%. What is the new tax function (assuming it is continuous)?

34. Prove that a function f is continuous at c if and only if f is both left continuous at c and right continuous at c.

35. If $x \in \mathbb{R}$, then let $\lfloor x \rfloor$ denote the greatest integer that does not exceed x. (See Exercise 33 from Section 2.1 for a discussion of this function.)

a. Sketch the graph of $\lfloor x \rfloor$.
b. What is $\lim_{x \to 3^-} \lfloor x \rfloor$? Prove it.
c. What is $\lim_{x \to 2^+} \lfloor 4 - 2x \rfloor$? Prove it.
d. What is $\lim_{x \to 0^-} x \cdot \lfloor 1/x \rfloor$? Prove it.
e. What is $\lim_{x \to 0^-} \lfloor -x \rfloor$? Prove it.

36. The function $g(x) = -(x-2)^2 + 4$ assumes a maximum value on the interval (0, 4). What is that value? However, g assumes no minimum value on this interval. Why does this not contradict the Extreme Value Theorem?

37. Suppose a train pulls out of a station and comes to a stop at the next station 80 km away exactly 1 h later. Assuming the velocity of the train is a continuous function, prove that at some moment the train must have been traveling at a velocity of precisely 60 km/h.

38. Let p be a polynomial of odd degree and leading coefficient 1. It can be shown that $\lim_{x \to \infty} p(x) = \infty$ and $\lim_{x \to -\infty} p(x) = -\infty$. Use these facts and the Intermediate Value Theorem to prove that p has at least one real root. Give an example of such a p with exactly one real root.

39. Show that $10x^3 - 7x^2 + 20x - 14 = 0$ has a root in (0, 1). Deduce that $10\sin^3(x) - 7\sin^2(x) + 20\sin(x) - 14 = 0$ has a solution.

40. Locate, within an accuracy of two decimal places, a root of $p(x) = 3x^5 - 7x^2 + 1$ that lies between -1 and 2. Use the Intermediate Value Theorem and your calculator.

41. Because $p(x) = x^3 + x + 1$ has odd degree, we can be sure (by Exercise 38) that it has at least one real root. Determine a finite interval that contains a root of p. Explain why.

42. Use the Intermediate Value Theorem to show that $x/2 = \sin(x)$ has a positive solution. (Consider the function $f(x) = x/2 - \sin(x)$ and the points $a = \pi/6$ and $b = 2.000001$.)

43. Show that the polynomial p defined by $p(x) = 10x^4 + 46x^2 - 39x^3 - 39x + 36$ has at least two roots between 1 and 3. (Evaluate p at suitable points.)

The function p is continuous on the closed interval $I = [-1, 3]$ and $p(-1) = 4$ and $p(3) = -2$. In Exercises 44–47, answer true or false. If the statement is true, explain why. Otherwise, sketch a function p for which the statement is false.

44. p has at least one root in I.
45. p has at most one root in I.
46. p has exactly one root in I.
47. p has an odd number of roots in I.

In Exercises 48–53, answer true or false. If the statement is true, explain why. Otherwise, sketch a function that demonstrates that it is false. In each exercise, the function f is defined for $a \le x \le b$.

48. If f is continuous, then $|f|$ is continuous on (a, b). (Note: $|f|$ is defined by $|f|(x) = |f(x)|$.)
49. If $|f|$ is continuous, then f is continuous.
50. If f is continuous, then f^2 is continuous.
51. If f^2 is continuous, then f is continuous.
52. If f is continuous *and* bounded on $(a, b]$, then f has a maximum value.
53. If f has no maximum on $[a, b]$, then f is discontinuous at some point of interval $[a, b]$.
54. For what positive values of α does the equation $\alpha x = \sin(x)$ have a positive solution in x? (Apply the Intermediate Value Theorem to $f(x) = \alpha x - \sin(x)$. To determine if $f(a) < 0$ for some positive value of a, use $\lim_{x\to 0}(\sin(x))/x = 1$.)

Exercises 55–57 illustrate the following theorem: If the range of f is contained in the domain of g, if $\lim_{x\to c} f(x) = \ell$, and if $\lim_{y\to\ell} g(y) = L$, then $\lim_{x\to c} g(f(x)) = g(\ell)$ *or* $\lim_{x\to c} g(f(x)) = L$ *or* $\lim_{x\to c} g(f(x))$ does not exist.

55. Let

$$g(x) = \begin{cases} 0 & \text{if } x \neq 0 \\ 1 & \text{if } x = 0 \end{cases}$$

and $f(x) \equiv 0$. Let $c = 0$. Evaluate ℓ and L. Show that $\lim_{x\to 0} g(f(x)) = g(\ell)$, and compare this result with Theorem 2.

56. Let g be defined as in Exercise 55. Let $f(x) = x$. Let $c = 0$. Evaluate ℓ and L. Show that $\lim_{x\to 0} g(f(x)) = L$, and compare this result with Theorem 2.

57. Let g be defined as in Exercise 55. Let

$$f(x) = \begin{cases} x & \text{if } x \text{ is irrational} \\ 0 & \text{if } x \text{ is rational} \end{cases}.$$

Show that $\lim_{x\to 0} g(f(x))$ does not exist, and compare this result with Theorem 2.

58. Let $f : [0, 1] \to [0, 1]$ be a continuous function. Prove that there is a number c, $0 \le c \le 1$, such that $f(c) = c$. Such a value is said to be a *fixed point* of f. (*Hint:* Think about the function $g(x) = f(x) - x$.)

59. A hiker walks up a mountain path. He starts at the bottom of the path at 8:00 AM and reaches the top at 6:00 PM. The next morning, he starts down the same path at 8:00 AM and reaches the bottom at 6:00 PM. Show that there is at least one time of the day such that on each day the hiker's elevation was the same. Illustrate this by giving a representative sketch of the ascent and descent elevation functions in the rectangle $[0\,\text{h}, 10\,\text{h}] \times [0, L]$ where L is the elevation of the peak.

60. Prove that if f is a continuous function on $[0, 1]$ and $f(0) = f(1)$, then there is a value c in $(0, 1)$ such that $f(c) = f(c + 1/2)$. This is a special case of the *Horizontal Chord Theorem*. (*Hint:* Apply the Intermediate Value Theorem to the function g defined on $[0, 1/2]$ by $g(x) = f(x + 1/2) - f(x)$.)

61. Let

$$f(x) = x^2 - \frac{\left(\cos\left(\frac{\pi}{2}\sqrt[1001]{x}\right)\right)^3}{1000000}.$$

The left column in the table shows how to decompose f, step by step, into simpler components. Complete the table by citing a reason for the continuity of each function. As an example, we have already filled in the first line.

Function	Reason for Continuity
$F(x) = \sqrt[1001]{x}$	Theorem 4, Section 2.2
$G(x) = (\pi/2) \cdot F(x)$	
$H(x) = \cos(x)$	
$J(x) = H(G(x))$	
$K(x) = x^3$	
$L(x) = K(J(x))$	
$M(x) = L(x)/1000000$	
$N(x) = x^2$	
$f(x) = N(x) - M(x)$	

Evaluate $\lim_{x\to 0} f(x)$.

Calculator/Computer Exercises

62. Locate, to four decimal places of accuracy, the maximum and minimum values of the function $h(x) = x^4 - 5x^3 + 7x + 9$ on the interval $[0, 4]$.
63. Explain why $p(\tan(x)) = c$ has a real solution x for any real value of c and any cubic real polynomial p. Approximate a solution to $\tan^3(x) + 3\tan(x) = 19$.
64. Graph $f(x) = (x^3 - 3x + 1)/(x^3 - 3x - 1)$ in the viewing rectangle $[-3, 3] \times [-15, 15]$. Locate the points at which f does not have a continuous extension. (Give your approximations to four decimal places.)

In Exercises 65–68, graph the given function $f(x)$ over an interval centered about the given point c and determine if f has a continuous extension at c.

65. $f(x) = x/|x|$, $c = 0$
66. $f(x) = (x^4 - 6x^3 + 7x^2 + 4x - 4)/(x - 2)$, $c = 2$
67. $f(x) = (x^4 - 6x^3 + 7x^2 + 4x - 4)/(x - 2)^2$, $c = 2$
68. $f(x) = x^2 \sin(x)/(\sqrt{1 + x^2} - 1)$, $c = 0$
69. Use a computer algebra system to factor

$$x^4 - 10x^3 + 38x^2 - 64x + 40.$$

Use the factorization to show that

$$p(x) = x^4 - 10x^3 + 38x^2 - 64x + 50 \geq 10.$$

For what values of γ does $p(x) = \gamma$ have a real solution? To four decimal places, find two solutions to $p(x) = 20$.

70. Let $f(x) = (3x - \sin(2x))/x$ for $-2 \leq x \leq 4$, $x \neq 0$. Why is the Extreme Value Theorem not applicable to f? Locate the extreme values of f to four decimal places of accuracy.

2.4 Infinite Limits and Asymptotes

Consider the graph of the curve $y = 1/x$ in Figure 1. As $x \to 0^+$, y becomes arbitrarily large and positive. Pictorially, the graph approaches the positive y-axis. Also, as $x \to 0^-$, y becomes negative without bound. Pictorially, the graph approaches the negative y-axis. In this section, we develop a precise language for describing this type of behavior of graphs.

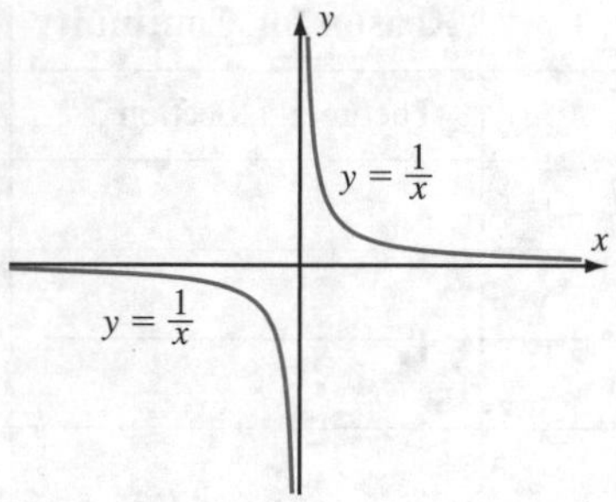

Figure 1

Infinite-Valued Limits

It is convenient to *extend* the notion of limit. Up to now, we have agreed that the limits $\lim_{x\to 0^+} 1/x$ and $\lim_{x\to 0^-} 1/x$ do not exist. For the limit $\lim_{x\to 0^+} 1/x$ to exist, there would have to be a real number ℓ to which the values of $1/x$ would tend as x approached 0 from the right. As the graph of $y = 1/x$ in Figure 1 shows, no such real number can exist. For similar reasons, $\lim_{x\to 0^-} 1/x$ cannot equal a real number. The definition of "limit" that we now give extends our existing definition so that ℓ is allowed to be infinite.

Definition Let f be a function that is defined on an interval just to the left of c and also on an interval just to the right of c. We write

$$\lim_{x\to c} f(x) = +\infty$$

if $f(x)$ becomes arbitrarily large and positive as $x \to c$. We write

$$\lim_{x\to c} f(x) = -\infty$$

if $f(x)$ becomes arbitrarily large and negative without bound as $x \to c$.

More rigorously, $\lim_{x\to c} f(x) = +\infty$ if for any $N > 0$ there is a $\delta > 0$ such that $f(x) > N$ whenever $0 < |x - c| < \delta$. Also, $\lim_{x\to c} f(x) = -\infty$ if for any $M > 0$ there is a $\delta > 0$ such that $f(x) < -M$ whenever $0 < |x - c| < \delta$.

The one-sided limits $\lim_{x\to c^+} f(x) = \pm\infty$ and $\lim_{x\to c^-} f(x) = \pm\infty$ are defined similarly.

> The symbol ∞ is used many times in this section. *It does not represent a real number.* It is merely a convenient piece of notation.

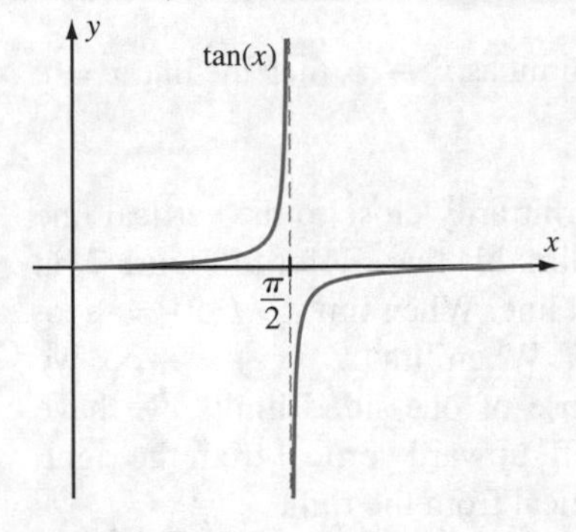

Figure 2
$\lim_{x\to\pi/2^+}\tan(x) = -\infty$

The definitions of $\lim_{x\to c} f(x) = \pm\infty$ provide an analytic basis with which to investigate infinite limits. In practice, the graph of a function with an infinite limit will clearly indicate this behavior. Often we determine where a denominator of $f(x)$ vanishes in order to locate possible points c at which $\lim_{x\to c} f(x)$ might be $\pm\infty$. For example, the denominator $\cos(x)$ in $\sin(x)/\cos(x)$ vanishes at $x = \pi/2$ and

$$\lim_{x\to\pi/2^+} \tan(x) = \lim_{x\to\pi/2^+} \frac{\sin(x)}{\cos(x)} = -\infty,$$

as shown in Figure 2.

Example 1 Analyze the following limits:

a. $\displaystyle\lim_{x\to -1} \frac{1}{(x+1)^2}$ b. $\displaystyle\lim_{x\to\pi/2^-} \sec(x)$

c. $\displaystyle\lim_{x\to\pi/2^+} \sec(x)$ d. $\displaystyle\lim_{x\to 2} -\frac{1}{(x-2)^4}$

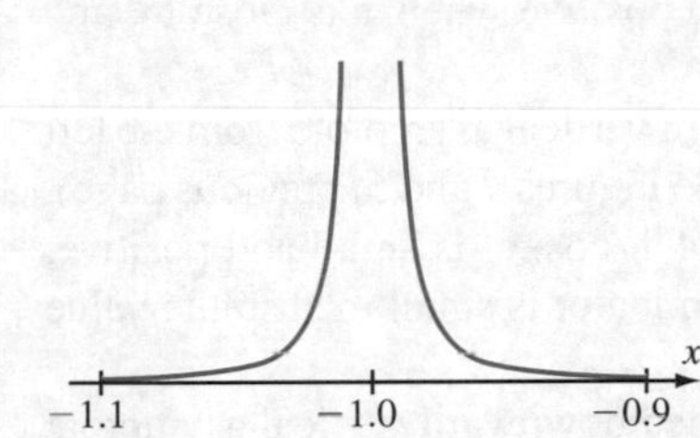

Figure 3
The plot of $1/(x+1)^2$ in the viewing rectangle $[-1.1, -0.9] \times [0, 10000]$

Solution By inspection, we find:

a. $\lim_{x\to -1} 1/(x+1)^2 = +\infty$ (see Figure 3),
b. $\lim_{x\to\pi/2^-} \sec(x) = +\infty$ (see Figure 4),
c. $\lim_{x\to\pi/2^+} \sec(x) = -\infty$ (see Figure 5), and
d. $\lim_{x\to 2} -1/(x-2)^4 = -\infty$ (see Figure 6).

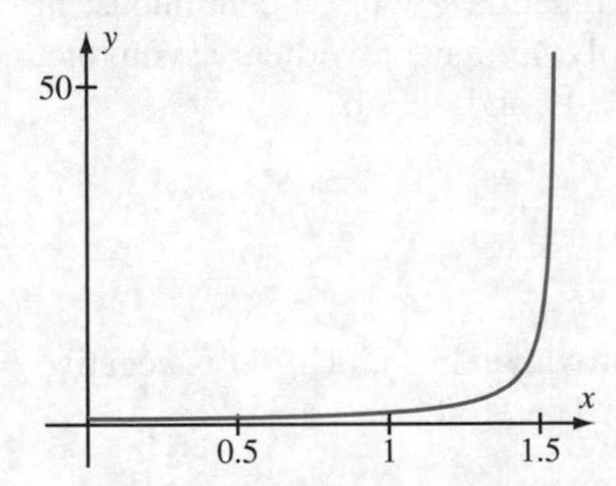

Figure 4
The plot of $\sec(x)$ in the viewing rectangle $[0, \pi/2] \times [0, 50]$

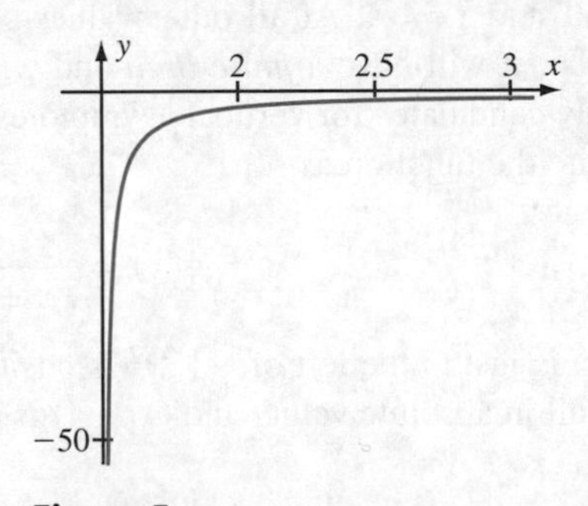

Figure 5
The plot of $\sec(x)$ in the viewing rectangle $[\pi/2, \pi] \times [-50, 0]$

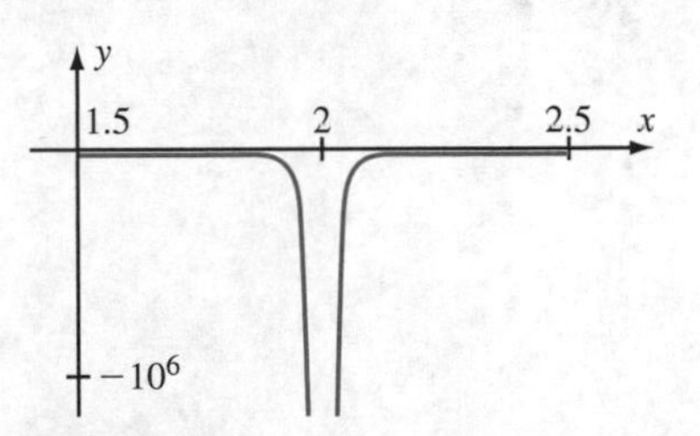

Figure 6
The plot of $-1/(x-2)^4$ in the viewing rectangle $[1.5, 2.5] \times [-10^6, 0]$

■

Vertical Asymptotes

The close connection between the figures and our perception of these limits suggests the following definition.

Definition If a function f has a one-sided or two-sided *infinite* limit as $x \to c$, then the line $x = c$ is a *vertical asymptote* of f.

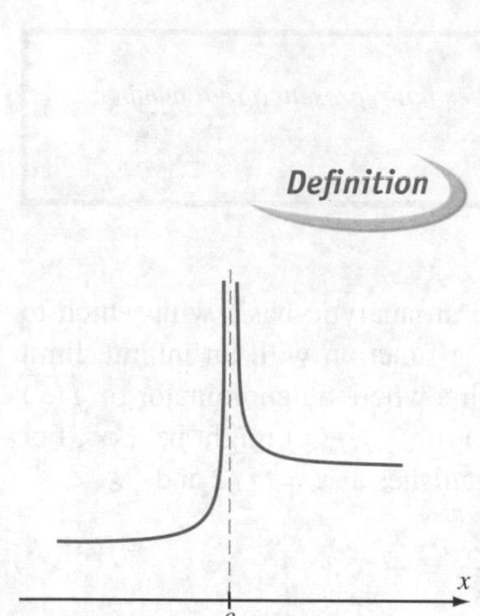

Figure 7
$x = c$ is a vertical asymptote.

In this circumstance, the graph of f becomes arbitrarily close to the vertical line $x = c$ and the sketch of the graph of f should exhibit this line. Refer to Figure 7, in which the vertical asymptote is rendered as a dotted line. When $\lim_{x\to c} f(x) = +\infty$, we sometimes say the asymptote is *upward vertical*. When $\lim_{x\to c} f(x) = -\infty$, we say the asymptote is *downward vertical*. In the case of one-sided limits, we have four possible behaviors: upward vertical from the left, upward vertical from the right, downward vertical from the left, and downward vertical from the right.

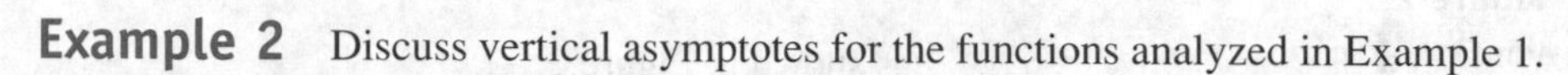

Example 2 Discuss vertical asymptotes for the functions analyzed in Example 1.

Solution The graph of $y = 1/(x+1)^2$ has $x = -1$ as an upward vertical asymptote, since the denominator becomes arbitrarily small and positive when x is close to, and on either side of, -1 (see Figure 3).

The function $y = \sec(x)$ has $x = \pi/2$ as an upward vertical asymptote from the left *and* a downward vertical asymptote from the right (see Figures 4 and 5, previous page). When x is just to the left of $\pi/2$, the denominator of $1/\cos(x)$ is small and positive; when x is just to the right of $\pi/2$, however, the denominator is small in absolute value and negative.

Finally, $y = -1/(x-2)^4$ has $x = 2$ as a two-sided downward vertical asymptote (see Figure 6). The denominator is small and positive when x is close to 2. ■

Example 3 Examine the function $g(x) = x^2/(x^2-3x-4)$ for vertical asymptotes.

Solution The denominator factors as $(x+1)\cdot(x-4)$; hence, it vanishes only at $x = -1$ and $x = 4$. At all other values of x, the function g will be continuous; in particular, it will have a *finite limit* and will certainly *not* have a vertical asymptote. The only candidates for vertical asymptotes are $x = -1$ and $x = 4$.

Write the function as

$$g(x) = \frac{x^2}{(x+1)(x-4)}.$$

When x is just to the left of -1, x^2 is positive (approximately 1), $(x+1)$ is negative and small in absolute value, and $(x-4)$ is negative. Therefore,

$$\lim_{x\to 1^-} g(x) = \lim \frac{+}{(\text{small } -)\cdot(-)} = +\infty.$$

Similar reasoning shows that

$$\lim_{x\to 1^+} g(x) = \lim \frac{+}{(\text{small } +)\cdot(-)} = -\infty$$

(the only difference this time is that $(x+1) > 0$).

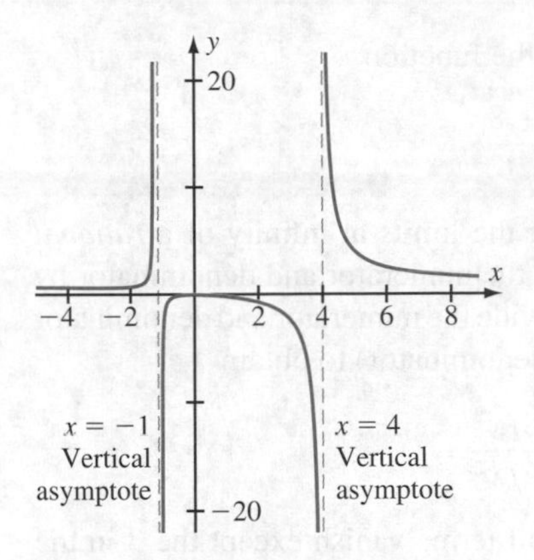

Figure 8
$g(x) = x^2/(x^2 - 3x - 4)$

When x is just to the left of 4, x^2 is positive, $(x + 1)$ is positive, and $(x - 4)$ is negative and small in absolute value. Therefore,

$$\lim_{x \to 4^-} g(x) = \lim \frac{+}{(+) \cdot (\text{small } -)} = -\infty.$$

Similar reasoning shows that

$$\lim_{x \to 4^+} g(x) = \lim \frac{+}{(+) \cdot (\text{small } +)} = +\infty.$$

We conclude that g has $x = -1$ and $x = 4$ as vertical asymptotes. It should be noted that the asymptotes are one-sided. We obtain Figure 8 by using the preceding information and plotting some points. ■

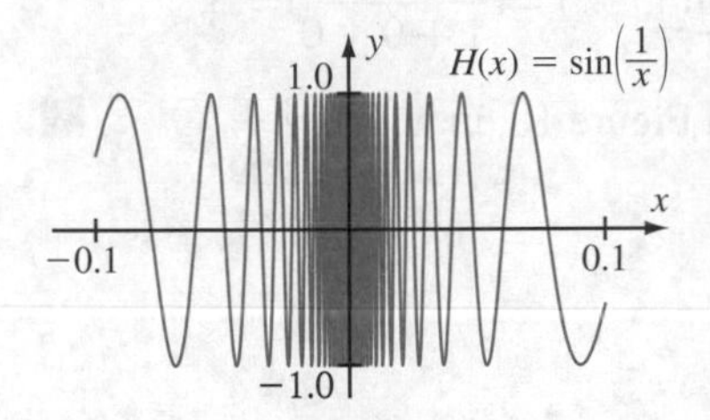

Figure 9
Infinitely many oscillations and no asymptotes

inSIGHT

The function g in Example 3 has a vertical asymptote at each point at which it is undefined. The same is true of the functions in Example 1. However, it is incorrect to think that a function must have a vertical asymptote at a point at which it is not defined. Consider the function $H(x) = \sin(1/x)$, which is undefined at $x = 0$. Since $|H(x)|$ is never greater than 1, H certainly cannot have $+\infty$ or $-\infty$ as a limit. In fact, H has no limit of any kind at 0, as shown in Figure 9.

Limits at Infinity

Now we turn to the problem of identifying horizontal asymptotes. This requires another notion of limit.

Definition Assume that the domain of f contains an interval of the form (a, ∞), and let α be a finite real number. We say that

$$\lim_{x \to +\infty} f(x) = \alpha$$

if $f(x)$ approaches α when x becomes arbitrarily large and positive.

Now assume that the domain of g contains an interval of the form $(-\infty, b)$. Let β be a finite real number. We say that

$$\lim_{x \to -\infty} g(x) = \beta$$

if $g(x)$ approaches β when x is negative and when x becomes arbitrarily large in absolute value.

More rigorously, $\lim_{x \to +\infty} f(x) = \alpha$ if, for any $\epsilon > 0$, there is an N such that $|f(x) - \alpha| < \epsilon$ whenever $N < x < \infty$. Also, $\lim_{x \to -\infty} g(x) = \beta$ if, for any $\epsilon > 0$, there is an M such that $|g(x) - \beta| < \epsilon$ whenever $-\infty < x < M$.

Example 4 Examine the limits at infinity for the function

$$f(x) = \frac{3x^2 - 6x + 8}{x^2 + 4x + 6}.$$

Solution A good rule of thumb when studying the limits at infinity of a *rational function* (the quotient of polynomials) is to divide the numerator and denominator by the highest power of x that appears. We therefore divide the numerator and denominator of f by x^2 (the highest power that appears in the denominator) to obtain

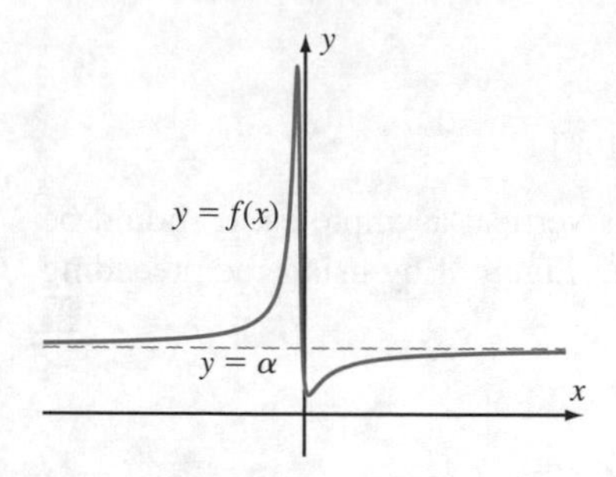

Figure 10
$\lim_{x\to-\infty} f(x) = \alpha$ and $\lim_{x\to+\infty} f(x) = \alpha$

$$f(x) = \frac{3 - 6/x + 8/x^2}{1 + 4/x + 6/x^2}.$$

Now it is clear that as $x \to +\infty$ or $x \to -\infty$, all terms vanish except the 3 in the numerator and the 1 in the denominator. Thus, we see that

$$\lim_{x\to+\infty} f(x) = \frac{3+0+0}{1+0+0} = 3 \quad \text{and} \quad \lim_{x\to-\infty} f(x) = \frac{3+0+0}{1+0+0} = 3.$$

This limiting behavior is visible in the plot of f in Figure 10, in which $\alpha = 3$. ■

Horizontal Asymptotes

Definition If either

$$\lim_{x\to+\infty} f(x) = \alpha \quad \text{or} \quad \lim_{x\to-\infty} f(x) = \alpha,$$

then we say that the line $y = \alpha$ is a *horizontal asymptote* of f.

If the line L defined by $y = \alpha$ is a horizontal asymptote of f, then the graph of f becomes arbitrarily close to L. When we sketch the graph of f, we include the horizontal asymptote L (see Figure 10). It may happen that both $\lim_{x\to+\infty} f(x) = \alpha$ and $\lim_{x\to-\infty} f(x) = \alpha$. In this case, the line $y = \alpha$ is certainly a horizontal asymptote. However, the line would still be called an asymptote if only one of these limits existed. If $\lim_{x\to+\infty} f(x) = \alpha$ and $\lim_{x\to-\infty} f(x) = \alpha'$ with $\alpha \neq \alpha'$, then f has two horizontal asymptotes. In Figure 10, we would say that $y = \alpha$ is a right upper asymptote and a left lower asymptote. Of course, another graph could have a right lower asymptote or a left upper asymptote. The next example shows that a horizontal asymptote need not be either upper or lower.

Example 5 Analyze the graph of $g(x) = \sin(2x)/(1 + x^2)$ for horizontal asymptotes.

Solution We divide the numerator and denominator by the highest power of x that appears in the denominator (namely, x^2):

$$g(x) = \frac{\sin(2x)/x^2}{(1/x^2) + 1}.$$

Because $|\sin(2x)| \leq 1$, it follows that $\sin(2x)/x^2 \to 0$ as $x \to \pm\infty$. Similarly, $1/x^2 \to 0$ as $x \to \pm\infty$. Using limit rules that are parallel to those from Section 2.2 for finite

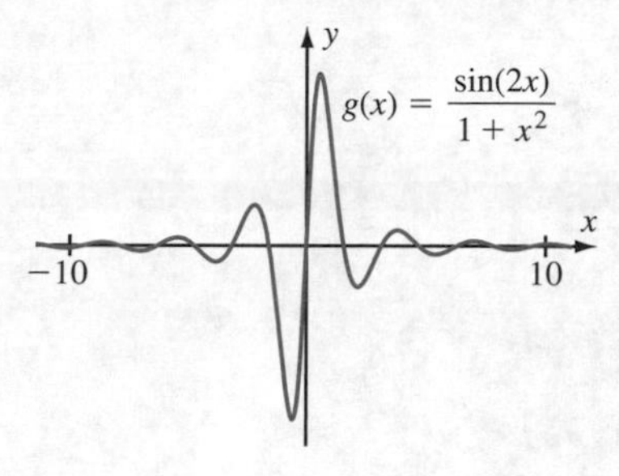

Figure 11

limits, we obtain

$$\lim_{x\to+\infty} g(x) = \frac{0}{0+1} = 0 \quad \text{and} \quad \lim_{x\to-\infty} g(x) = \frac{0}{0+1} = 0.$$

These limits show that the graph of g has the line $y = 0$ as a left and a right horizontal asymptote. Because the graph of $g(x)$ oscillates about its asymptote, we do not speak of $y = 0$ as either an upper or a lower asymptote (see Figure 11). ■

Example 6 Determine all horizontal and vertical asymptotes for the function

$$f(x) = \frac{7x^3 - 1}{2x^3 + 12x^2 + 18x}.$$

Solution First, if we divide the numerator and denominator by x^3, we obtain

$$f(x) = \frac{7 - 1/x^3}{2 + 12/x + 18/x^2}.$$

It is then clear that

$$\lim_{x\to+\infty} f(x) = \frac{7}{2} \quad \text{and} \quad \lim_{x\to-\infty} f(x) = \frac{7}{2}.$$

Therefore, f has the line $y = 7/2$ as a horizontal asymptote.

For vertical asymptotes, we factor the denominator of $f(x)$:

$$f(x) = \frac{7x^3 - 1}{2x(x+3)^2}.$$

We notice that f is continuous except at 0 and -3. When x is close to and *on either side of* -3, $7x^3 - 1$ is negative, $2x$ is negative, and $(x+3)^2$ is positive and arbitrarily small. It follows that

$$\lim_{x\to-3} f(x) = \lim \frac{(-)}{(-)\cdot(\text{small }+)} = +\infty.$$

Therefore, the line $x = -3$ is an upward vertical asymptote.

Also, when x is just to the left of 0, $7x^3 - 1$ is negative, $2x$ is negative and arbitrarily small in absolute value, and $(x+3)^2$ is positive. Therefore,

$$\lim_{x\to0^-} f(x) = \lim \frac{(-)}{(\text{small }-)\cdot(+)} = +\infty.$$

Similarly, when x is just to the right of 0, $7x^3 - 1$ is negative, $2x$ is positive and arbitrarily small, and $(x+3)^2$ is positive. Thus,

$$\lim_{x\to0^+} f(x) = \lim \frac{(-)}{(\text{small }+)\cdot(+)} = -\infty.$$

We conclude that the line $x = 0$ is a vertical asymptote for f. Notice that the asymptotic behavior on the left of 0 is different from that on the right of 0: On the left, the graph goes up; on the right, it goes down. The graph of f in Figure 12, next page, exhibits the vertical and horizontal asymptotes that we have found.

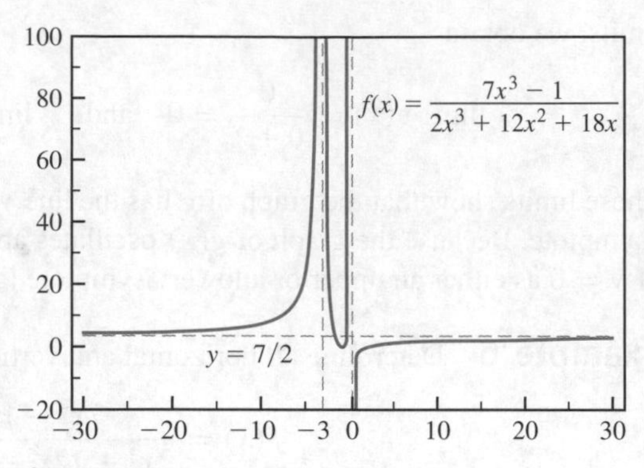

Figure 12

When we seek horizontal asymptotes of a rational function

$$\frac{a_n x^n + a_{n-1}x^{n-1} + \cdots + a_1 x + a_0}{b_m x^m + b_{m-1}x^{m-1} + \cdots + b_1 x + b_0} \qquad (a_n, b_m \neq 0),$$

it is useful to divide each term by x^m. The resulting expression can be easily analyzed for asymptotes. The result is shown in Table 1.

Comparison of Degrees	Horizontal Asymptotes
$n > m$	No asymptotes
$n = m$	$y = a_n/b_m$ as $x \to \pm\infty$
$n < m$	$y = 0$ as $x \to \pm\infty$

Table 1

quickquiz

1. What does the presence of a vertical asymptote for a function f tell us about f?
2. What does the presence of a horizontal asymptote for a function g tell us about g?
3. Does the function $f(x) = \sin(x)$ have either horizontal or vertical asymptotes? Why or why not?
4. Discuss asymptotes for the function $g(x) = (x^2 - 5)/(2x^2 - 8)$.

EXERCISES

Problems for Practice

In Exercises 1–20, determine whether the limit exists. If it does exist, then compute it. In each case, give reasons for your answer.

1. $\lim_{x\to 6} 3/(x-6)^2$
2. $\lim_{x\to -\infty} 4x/(4x-7)$
3. $\lim_{x\to +\infty}(x^4+3x-92)/(x^5-7x^2+44)$
4. $\lim_{x\to -1^+}(x+1)^{-1}$
5. $\lim_{x\to +\infty}(x+\sqrt{x})/(x-\sqrt{x})$
6. $\lim_{x\to \pi} 3/(\sin(x))$
7. $\lim_{x\to \infty}(x+\cos(x))/(x-\sin(x))$
8. $\lim_{x\to +\infty} \sqrt{x}/(x+1)$
9. $\lim_{x\to 1^+} 1/\sqrt{x-1}$
10. $\lim_{x\to 0^+} \csc(x)$
11. $\lim_{x\to +\infty}(x^2-4x+9)/(3x^2-8x+18)$
12. $\lim_{x\to 3^-} \cos(x)/(x-3)$
13. $\lim_{x\to 0^+} \csc^2(x)$
14. $\lim_{x\to +\infty}(x^2+5x-7)/(3x^2+4)$
15. $\lim_{x\to 0} 3/\sqrt{|x|}$
16. $\lim_{x\to +\infty}(\sqrt{x}-5)/(\sqrt{x}+4)$
17. $\lim_{x\to 0^-} \cot(x)$
18. $\lim_{x\to -\infty} x^{4/3}/(x^2+\sin(x))$
19. $\lim_{x\to 2^-} \tan(\pi/x)$
20. $\lim_{x\to +\infty} \sin(x)/x$

In Exercises 21–34, find all horizontal and vertical asymptotes for the function. Then plot several points and obtain a sketch of the graph.

21. $f(x)=\dfrac{x}{x-7}$
22. $g(x)=\dfrac{x^2-1}{x^2-4}$
23. $m(x)=\dfrac{\sqrt{|x|}}{x}$
24. $p(x)=\dfrac{\sqrt{|x|}}{x+3}$
25. $k(x)=\dfrac{(x-1)^{2/3}}{(x^2+8)^{1/3}}$
26. $r(x)=\dfrac{x}{x^2+1}$
27. $k(x)=\dfrac{1}{x^2+1}$
28. $h(x)=\dfrac{x^2}{x^2-7x+12}$
29. $k(x)=\dfrac{x^{1/3}}{|x|}$
30. $f(x)=\dfrac{x}{(x+1)(x+4)}$
31. $h(x)=\dfrac{(x+4)^{-1/3}}{x\cdot(x+1)}$
32. $h(x)=\dfrac{x^3+1}{x^3-1}$
33. $k(x)=\cot(x),\ -\dfrac{\pi}{2}<x<\dfrac{\pi}{2}$
34. $q(x)=\sec(x),\ -\dfrac{\pi}{2}<x<\dfrac{\pi}{2}$

Further Theory and Practice

35. Formulate and prove a version of Theorem 1 (the Limit Uniqueness Theorem) from Section 2.2 for $\lim_{x\to +\infty} f(x)$ and $\lim_{x\to -\infty} f(x)$.
36. Formulate and prove a version of Theorem 2 (the sum, difference, product, quotient, and scalar multiplication rules for limits) from Section 2.2 for the limits $\lim_{x\to +\infty} f(x)$ and $\lim_{x\to -\infty} f(x)$.
37. Formulate and prove a version of the Pinching Theorem from Section 2.2 for the limits $\lim_{x\to +\infty} f(x)$ and $\lim_{x\to -\infty} f(x)$.
38. We do not attempt to formulate sum, difference, product, or quotient rules for infinite valued limits (as in the first definition of "limit" in this section) because there are serious ambiguities in doing arithmetic with the symbols $+\infty$ and $-\infty$. Give examples to illustrate these ambiguities.
39. Prove that if
$$\lim_{x\to c} F(x)=\ell\in\mathbb{R}$$
and if
$$\lim_{x\to c} G(x)=+\infty,$$
then
$$\lim_{x\to c}(F(x)+G(x))=+\infty.$$
Similarly, if
$$\lim_{x\to c} H(x)=-\infty,$$
then prove that
$$\lim_{x\to c}(F(x)+H(x))=-\infty.$$

In Exercises 40–45, find all horizontal and vertical asymptotes for each function and sketch the graph for each.

40. $f(x)=\sin(x)/x$
41. $g(x)=x/\sin(x)$
42. $m(x)=|x|^{1/2}/\sin(x)$
43. $f(x)=\cos^2(x)/x$
44. $k(x)=|x|/\sin(x)$
45. $g(x)=1/|\sin(x)|$

In Exercises 46–48, graph the function f. Refer to the definition of "vertical asymptote" to decide if $x=0$ is a vertical asymptote of the graph of f.

46. $f(x)=\sin(1/x)$
47. $f(x)=|\sin(1/x)/x|$
48. $f(x)=(1.1+\sin(1/x))/x$

49. Let $f(x) = x/10^{100}$ and $g(x) = 10^{100}/x$. What are $f(10^{80})$ and $g(10^{80})$? Now compute $\lim_{x\to\infty} f(x)$ and $\lim_{x\to\infty} g(x)$. Is it significant that $f(x)$ is much smaller than $g(x)$ for all $0 < x < 10^{80}$? (The number 10^{80} was chosen simply because it is large. It is on the order of the number of protons in the universe, as computed by Sir Arthur Eddington.)

Exercises 50–52 concern physical laws that entail natural one-sided limits and vertical asymptotes. In each exercise, sketch the graph and indicate the vertical asymptote.

50. $m = m_0/\sqrt{1-(v/c)^2}$ (Einstein's Relativistic Mass Law, which relates the mass m of an object at velocity v to its rest mass m_0 and to the speed of light c)

51. $P = RT/(V-b) - a/V^2$ (Van der Waals's equation relating the pressure P and volume V of 1 mole of a gas at constant temperature T. The constant R is the universal gas constant, and the constant a depends on the particular gas under consideration. The constant b is the volume occupied by the gas molecules themselves.)

52. $H = 2gR^2/(2gR - v_0^2) - R$ (The equation for the height H a missile rises above Earth's surface as a function of its initial velocity v_0; g is the acceleration due to Earth's gravity and R is Earth's radius.)

53. If f is a continuous function on a closed interval $[a, b]$, can the graph of f have a vertical asymptote $x = c$ where c is a value in $[a, b]$? Explain your reasoning.

54. In an enzyme reaction, the reaction rate V is given in terms of substrate amount S by

$$V = \frac{V_* S}{K + S}.$$

Identify V_* as a limit as $S \to \infty$. Although V does not have a maximum value, V_* is called the *maximum velocity* (where *velocity* refers to reaction rate). Explain why.

In Exercises 55–58, you are asked to investigate the notion of a *skew asymptote,* which is also called a *slant asymptote* and an *oblique asymptote*.

55. We say that the graph of a function f has the line $y = mx + b$ as its skew asymptote if either

$$\lim_{x\to+\infty} (f(x) - (mx + b)) = 0$$

or

$$\lim_{x\to-\infty} (f(x) - (mx + b)) = 0.$$

a. If $m = 0$, prove that this notion of skew asymptote coincides with the notion of horizontal asymptote.

b. If $f(x) = p(x)/q(x)$ is a quotient of polynomials (a rational function), then prove that a *necessary condition* for the graph of f to have $mx + b$ as a skew asymptote, with $m \neq 0$, is that the degree of p is precisely one greater than the degree of q.

c. If the degree of p exceeds the degree of q by precisely two, then we could say that the graph of f is asymptotic to a parabola. Explain this idea in more detail, and give a precise definition.

56. Find all horizontal, vertical, and skew asymptotes of the graphs of the following functions. Then plot some points and give a sketch.

a. $f(x) = \dfrac{x^2}{x-5}$

b. $g(x) = \dfrac{x^{3/2}}{x^{1/2}+7}$

c. $k(x) = \dfrac{(x^2+4x+3)^{4/3}}{(x-9)^{5/3}}$

d. $m(x) = (x+5)\cdot\dfrac{\sin(1/x)}{(1/x)}$

57. Find all parabolic asymptotes for the functions $F(x) = x^3/(x+3)$, $G(x) = x^4/(x^2+5x)$, and $H(x) = (x^{7/3} + x^2 + x^{1/3})/(x^{1/3}+1)$. (Refer to Exercise 55, part c, for terminology.) Sketch the graphs of H and its parabolic asymptote.

58. Consider the following three attempts at determining the skew asymptote of the function

$$f(x) = \frac{x^2+3x+2}{x-1}.$$

Determine which argument is correct and explain what is wrong with the others.

a. Since

$$\begin{aligned} f(x) &= \frac{x^2+3x+2}{x-1} \\ &= \frac{x^2}{x}\,\frac{(1+3/x+2/x^2)}{(1-1/x)} \\ &= x\,\frac{(1+3/x+2/x^2)}{(1-1/x)} \end{aligned}$$

and since the fraction tends to 1 as $x \to \infty$, the skew asymptote must be $x \cdot 1$.

b. Since

$$\begin{aligned} f(x) &= \frac{x^2+3x+2}{x-1} = \frac{x}{x}\,\frac{(x+3+2/x)}{(1-1/x)} \\ &= \frac{(x+3+2/x)}{(1-1/x)} \end{aligned}$$

and since the numerator tends to $x+3$ and the denominator to 1, the skew asymptote must be $x+3$.

c. Since

$$f(x) = \frac{x^2+3x+2}{x-1} = x+4+\frac{6}{x-1},$$

the skew asymptote must be $x+4$.

Calculator/Computer Exercises

Sometimes the asymptotic behavior of a graph takes place very slowly. In Exercises 59 and 60, you are asked to perform a numerical calculation to determine how slowly the limit is being achieved.

59. We know that

$$\lim_{x\to+\infty} \frac{\sqrt{x}}{x+1} = 0.$$

How large must x be to guarantee that the function is smaller than 0.001?

60. We know that

$$\lim_{x\to 0^-} \frac{|x|^{1/4}+6}{x^{1/3}} = -\infty.$$

How close must x be to 0 to guarantee that $(|x|^{1/4}+6)/x^{1/3} < -10^8$?

In Exercises 61–63, the function f has one or more horizontal asymptotes. Plot f and its horizontal asymptote(s). Specify another window in which f and its right horizontal asymptote appear to nearly coincide. Repeat for the left horizontal asymptote if it is different.

61. $f(x) = (3x^2 + x\cos(x))/(x^2+1)$
62. $f(x) = (2x^3 + \sin(x))/(|x|^3+1)$
63. $f(x) = \sqrt{x^4 + x\sin(x)}/(x^2+2x+2)$

In Exercises 64–66, the function f has one or more vertical asymptotes. Plot f and its vertical asymptote(s).

64. $f(x) = (x^2+x+3)/(x^3-x-3)$
65. $f(x) = (4+\cos(x))/(\sin(x)+\cos(2x)),\ 0 \le x \le 3$
66. $f(x) = (x^4+1)/(x^3-3x+1)$

2.5 Limits of Sequences

This section addresses limits of infinite sequences of real numbers. Recall that an infinite sequence f of real numbers is a function with domain $\mathbb{Z}^+$ and range $\mathbb{R}$ (as discussed in Section 1.4). Following the common convention for sequences, we will write f_n instead of $f(n)$ and refer to the sequence as $\{f_n\}$. Our interest here is whether the numbers f_n tend to some limit as n tends to infinity. Before giving a precise definition of "limit," let us apply our intuition to some examples.

Example 1 We suppose that a quantity of radioactive material is decaying. At the beginning of each week, there is half as much material as there was at the beginning of the previous week. The initial quantity is 5 g. Use sequence notation to express the amount of material at the beginning of the jth week. Does this sequence have a limit?

Solution The amount a_1 of material at the start of the first week is 5 g. The amount a_2 of material at the start of the second week is 5/2 g. The amount a_3 of material at the start of the third week is 5/4 g. Similarly, the amounts at the beginnings of the fourth and fifth weeks are $a_4 = 5/8$ g and $a_5 = 5/16$ g. In general, the amount a_j of material at the start of the jth week is given by $a_j = 5 \cdot (1/2)^{j-1}$. The magnitude of each term is half of its predecessor. Our intuitive conclusion is that the amount of radioactive material tends to 0 as the number of weeks tends to infinity. ■

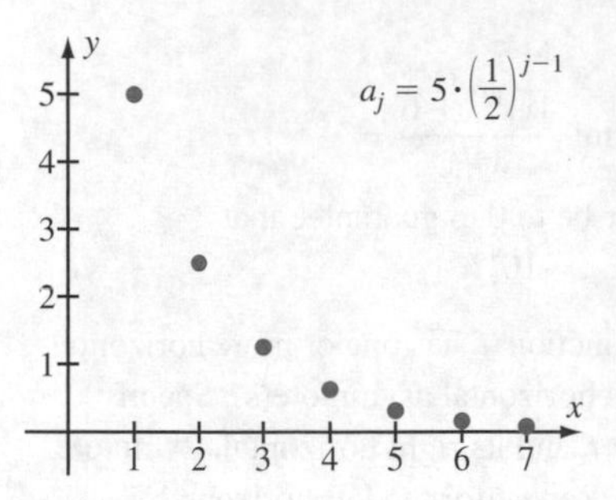

Figure 1

InSIGHT

A sequence $\{a_j\}_{j=1}^{\infty}$ can be graphed in the same way as any other real-valued function. We simply plot the points $(1, a_1)$, $(2, a_2)$, $(3, a_3)$, and so on. Figure 1 shows the points $(1, 5)$, $(2, 5/2)$, $(3, 5/4)$, $(4, 5/8)$, $(5, 5/16)$, $(6, 5/32)$, and $(7, 5/64)$ of the sequence $\{a_j\}_{j=1}^{\infty}$ from Example 1. Each point is half the distance to the x-axis of the preceding point. The graph confirms our intuitive conclusion that a_j tends to 0 as j tends to infinity.

Example 2 Does the sequence $a_n = (-1)^n$ tend to a limit?

Solution We may write out this sequence as $-1, 1, -1, 1, \ldots$. The sequence does not seem to tend to any limit: Half of the time the value is 1, and the other half the value is -1. The sequence does not *become* and *remain* close to a single value. Therefore, we say that it has no limit (see Figure 2). ■

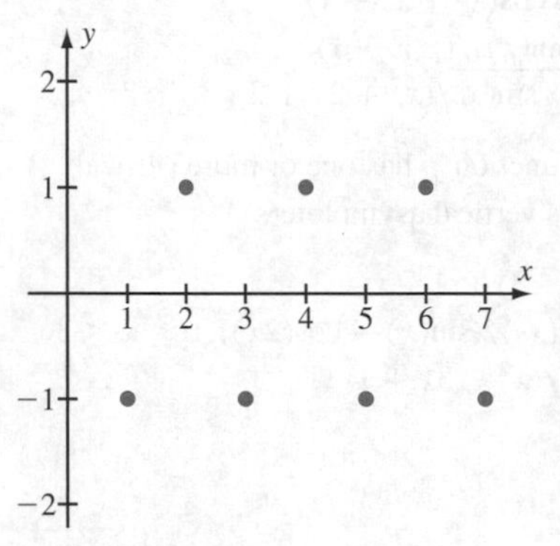
Figure 2

InSIGHT

In Example 1, we used j as the index variable of a sequence, whereas in Example 2 we used n. There is no significance to the letter used because the index variable merely serves as a placeholder for the integers $1, 2, 3, \ldots$. The letters i, j, k, m, and n are all commonly used as index variables of sequences.

A Precise Discussion of Convergence and Divergence

Definition Suppose $\{a_n\}_{n=1}^{\infty}$ is a sequence and ℓ is a real number. We say the sequence has *limit* ℓ (or *converges* to ℓ) if for each $\epsilon > 0$ there is an integer N such that if $n \geq N$, then $|a_n - \ell| < \epsilon$ (see Figure 3).

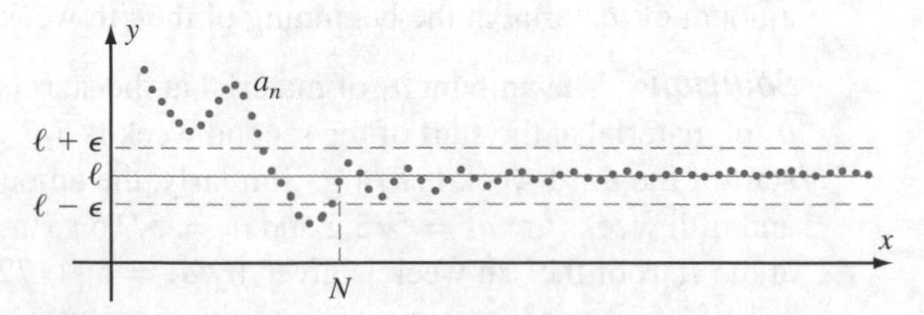

Figure 3
$|a_n - \ell| < \epsilon$ for $n \geq N$

When the sequence $\{a_n\}_{n=1}^{\infty}$ has limit ℓ, we write

$$\lim_{n\to\infty} a_n = \ell,$$

or

$$a_n \to \ell \quad \text{as} \quad n \to \infty.$$

A sequence that has a limit is *convergent*. If a sequence does not converge, then we say that it *diverges,* and we call it a *divergent* sequence.

> In Section 2.4, we discussed the *limit at infinity,* $\lim_{x\to\infty} f(x)$, for a function f of a continuous real variable x. The definition of $\lim_{n\to\infty} a_n$ is very similar. The difference is that a_n is defined for only positive integer values of n.

Example 3 Discuss convergence for the sequence $1, 1/2, 1/3, 1/4, \ldots$.

Solution Let $a_n = 1/n$ for all n. The terms a_n get progressively closer to 0. Let us now prove that $\lim_{n\to\infty} a_n = 0$. We set $\ell = 0$ and suppose that ϵ is a given positive number. We reason backward to select N. To guarantee that $|a_n - \ell| < \epsilon$, we need $|1/n - 0| < \epsilon$. This simply means that $1/n < \epsilon$. This last inequality is satisfied if $n > 1/\epsilon$. Therefore, we select N to be the first integer that is greater than $1/\epsilon$. If $n \geq N$, then $n > 1/\epsilon$ is true. Therefore, $1/n < \epsilon$ and $|a_n - \ell| < \epsilon$. In conclusion, $\lim_{n\to\infty} a_n = 0$. ■

The Tail of a Sequence

Intuitively, we say that $\{a_n\}_{n=1}^{\infty}$ converges to ℓ if a_n is as close as we please to ℓ when n is large enough. A crucial feature of this idea is that convergence depends only on what $\{a_n\}_{n=1}^{\infty}$ does when n is large. The values of the first ten thousand (or first million, or first billion, etc.) terms are irrelevant. If we fix any N, then we refer to the terms

$$a_{N+1}, a_{N+2}, a_{N+3}, \ldots$$

as the *tail end* of the sequence $\{a_n\}$. Of course, since N is an arbitrary positive integer, there are infinitely many tail ends of a sequence.

Example 4 Find the limit of the sequence defined by

$$a_n = \begin{cases} 0 & \text{if } 1 \leq n \leq 100000 \\ 10 & \text{if } n \geq 100001 \end{cases}.$$

Solution The sequence $\{a_n\}_{n=1}^{\infty}$ converges to 10. To see this, pick $\epsilon > 0$. Let $N = 100001$. If $n \geq N$, then $a_n = 10$ and $|a_n - 10| = |10 - 10| = 0 < \epsilon$, as desired. ■

Some Special Sequences

A number of special sequences occur repeatedly in our work. We now collect several of these sequences for easy reference. Some of these sequences diverge because their terms grow without bound. It is convenient to extend the limit notation to include these types of divergent sequences.

Definition If for any $M > 0$ there is an N such that $a_n > M$ for all indices $n \geq N$, then we write

$$\lim_{n\to\infty} a_n = \infty$$

and say that the sequence $\{a_n\}$ *tends to infinity*. If for any $M < 0$ there is an N such that $a_n < M$ for all indices $n \geq N$, then we write

$$\lim_{n\to\infty} a_n = -\infty$$

and say that the sequence $\{a_n\}$ *tends to* $-\infty$.

Theorem 1 Let r be any real number. Assume that $p > 0$.

a. $\lim_{n\to\infty} n^p = \infty$.
b. $\lim_{n\to\infty} 1/n^p = 0$.
c. If $|r| > 1$, then the sequence $\{r^n\}_{n=1}^{\infty}$ diverges and $\lim_{n\to\infty} |r|^n = \infty$.
d. If $|r| < 1$, then $\lim_{n\to\infty} r^n = 0$.
e. $\lim_{n\to\infty} p^{1/n} = 1$.

Although we will not give a formal proof of this theorem, let us at least note that each conclusion of Theorem 1 is plausible. For example, part a states that as n grows without bound, so does any positive power of n. Part c states that when a number greater than 1 is raised to successively higher exponents, then the resulting powers grow without bound. Parts b and d follow from parts a and c, using the following intuitive rule.

If $\{a_n\}$ is a sequence of nonzero numbers such that $\lim_{n\to\infty} a_n = \infty$, then $\lim_{n\to\infty} \dfrac{1}{a_n} = 0$.

The next example illustrates the use of Theorem 1.

Example 5 Discuss the convergence of $\{1/\sqrt{n}\}$, $\{\sqrt[100]{n}\}$, $\{(-99/100)^n\}$, and $\{(101/100)^n\}$.

Solution Theorem 1 can be applied to each of these sequences. The sequence $\{1/\sqrt{n}\}$ converges by Theorem 1b with $p = 1/2$. We write $\sqrt[100]{n} = n^{1/100}$ and conclude that $\{\sqrt[100]{n}\}$ diverges to infinity by using Theorem 1a with $p = 1/100$. The sequence $\{(-99/100)^n\}$ can be written as $\{r^n\}$ with $r = -99/100$. Since $|r| = 99/100 < 1$, we conclude that the sequence $\{(-99/100)^n\}$ converges to 0 by Theorem 1d. Finally, we write $\{(101/100)^n\}$ as $\{r^n\}$ with $r = 101/100 > 1$. This sequence tends to infinity by Theorem 1c. ■

inSIGHT

A numerical investigation of the convergence of a sequence requires some caution. The evaluations $(101/100)^{50} = 1.6446\ldots$ and $(101/100)^{150} = 4.4484\ldots$ hardly suggest that the sequence $\{(101/100)^n\}$ tends to infinity. As the table indicates, it is necessary to calculate $(101/100)^n$ for rather large values of n to obtain convincing numerical evidence.

n	200	500	1000	10000
$(101/100)^n$	7.3160	1000	20959	1.6358×10^{43}

Limit Theorems

There are a few rules that enable us to easily evaluate the limits of many common sequences. The limit rules for the sequences that follow are similar to the limit rules for the functions in Section 2.2.

Theorem 2 Suppose $\{a_n\}_{n=1}^{\infty}$ and $\{b_n\}_{n=1}^{\infty}$ are convergent sequences.

a. $\lim_{n\to\infty}(a_n \pm b_n) = \lim_{n\to\infty} a_n \pm \lim_{n\to\infty} b_n$.
b. $\lim_{n\to\infty}(a_n \cdot b_n) = (\lim_{n\to\infty} a_n) \cdot (\lim_{n\to\infty} b_n)$.
c. $\lim_{n\to\infty}(a_n/b_n) = \lim_{n\to\infty} a_n / \lim_{n\to\infty} b_n$, provided that $\lim_{n\to\infty} b_n \neq 0$.
d. $\lim_{n\to\infty} \alpha \cdot a_n = \alpha \cdot \lim_{n\to\infty} a_n$ for any real number α.
e. $\lim_{n\to\infty} \alpha = \alpha$ for any real number α.
f. $\lim_{n\to\infty} a_n$ is unique.
g. If $\lim_{n\to\infty} a_n = \lim_{n\to\infty} b_n$ and $a_n \leq c_n \leq b_n$ for all n, then $\{c_n\}$ converges to the same limit as $\{a_n\}_{n=1}^{\infty}$ and $\{b_n\}_{n=1}^{\infty}$.

From now on, we will usually calculate the limit of a sequence by using Theorem 2, together with the known limits of several basic sequences, such as those given in Theorem 1. In general, although we will not explicitly mention Theorem 2f, we will use it frequently, because it shows that if we calculate a limit by any particular method, the answer will be the same as the answer obtained by a different method. Theorem 2g is the Pinching Theorem for sequences.

Example 6 Compute $\lim_{j\to\infty}(3/j + 1)$.

Solution We apply Theorem 2 as follows:

$$\lim_{j\to\infty}\left(\frac{3}{j}+1\right) = \lim_{j\to\infty}\frac{3}{j} + \lim_{j\to\infty} 1 \qquad \text{Theorem 2a}$$

$$= 3 \cdot \lim_{j\to\infty}\frac{1}{j} + \lim_{j\to\infty} 1 \qquad \text{Theorem 2d}$$

$$= 3 \cdot 0 + \lim_{j\to\infty} 1 \qquad \text{Example 3}$$

$$= 1. \qquad \text{Theorem 2e}$$

■

Example 7 Use Theorem 2 to analyze the limit

$$\lim_{j\to\infty} \frac{j^3+4j-6}{3j^3+2j}.$$

Solution The numerator and the denominator of this limit become ever larger with j. Therefore, $\{j^3+4j-6\}$ and $\{3j^3+2j\}$ are not convergent sequences, and Theorem 2c cannot be applied directly. Before calling on Theorem 2, we divide the numerator and denominator by the largest power of j that appears (namely, j^3):

$$\begin{aligned}
\lim_{j\to\infty} \frac{j^3+4j-6}{3j^3+2j} &= \lim_{j\to\infty} \frac{1+4j^{-2}-6j^{-3}}{3+2j^{-2}} \\
&= \frac{\lim_{j\to\infty}(1+4j^{-2}-6j^{-3})}{\lim_{j\to\infty}(3+2j^{-2})} && \text{Theorem 2c} \\
&= \frac{\lim_{j\to\infty} 1 + 4\cdot \lim_{j\to\infty} j^{-2} - 6\cdot \lim_{j\to\infty} j^{-3}}{\lim_{j\to\infty} 3 + 2\lim_{j\to\infty} j^{-2}} && \text{Theorem 2a and 2d} \\
&= \frac{1-4\cdot 0+6\cdot 0}{3+2\cdot 0} && \text{Theorem 1a} \\
&= \frac{1}{3}. && \text{Simplify}
\end{aligned}$$

■

Example 8 Calculate $\lim_{j\to\infty}(2^j+3^j)/4^j$.

Solution

$$\begin{aligned}
\lim_{j\to\infty} \frac{2^j+3^j}{4^j} &= \lim_{j\to\infty} \frac{(2/4)^j+(3/4)^j}{(4/4)^j} && \text{Multiply the numerator and denominator by } 4^{-j} \\
&= \lim_{j\to\infty} \frac{(1/2)^j+(3/4)^j}{1} \\
&= \frac{0+0}{1} = 0. && \text{Theorem 1d}
\end{aligned}$$

■

Geometric Series

If an arrow is shot at a target 1 unit away, it must first travel half the distance, then half the remaining distance (or $1/4$ unit), then half the remaining distance (or $1/8$ unit), and so on. By this reasoning, we conclude that the distance the arrow travels is $1/2+1/4+1/8+1/16+\cdots$, without end. On the other hand, the distance the arrow travels is clearly 1 unit. In other words, it appears that

$$1/2+1/4+1/8+1/16+\cdots = 1.$$

First, we must give a meaning to the left side of this equation. We certainly do *not* mean that infinitely many numbers are added. After all, how could such an infinite sum be accomplished? We can, however, add the N numbers

$$1/2+1/4+1/16+\cdots+1/2^N$$

for any positive integer N, no matter how large. We then define $1/2 + 1/4 + 1/8 + 1/16 + \cdots$ to be the limit

$$1/2 + 1/4 + 1/8 + 1/16 + \cdots = \lim_{N\to\infty} (1/2 + 1/4 + 1/16 + \cdots + 1/2^N)$$

as the number of terms N tends to infinity. Our discussion suggests that this limit exists and is equal to 1. Theorem 3, which follows, allows us to verify this and similar limits. Chapter 10 is devoted to the study of other types of infinite sums.

Definition A limit of the form

$$\lim_{N\to\infty} (1 + r + r^2 + \cdots + r^{N-1})$$

is called a *geometric series*. Notice that a geometric series arises by adding the terms of a geometric progression $1, r, r^2, r^3 \ldots$.

Theorem 3 Suppose that $r \neq 1$.

a. For each positive integer N,

$$1 + r + r^2 + \cdots + r^{N-1} = \frac{r^N - 1}{r - 1}. \qquad (2.10)$$

b. If $|r| < 1$, then

$$\lim_{N\to\infty} (1 + r + r^2 + \cdots + r^{N-1}) = \frac{1}{1 - r}. \qquad (2.11)$$

Proof For $r \neq 1$, equation (2.10) is equivalent to

$$(r - 1)(r^{N-1} + \cdots + r^2 + r + 1) = r^N - 1.$$

We verify this by expanding the product on the left side:

$$\begin{aligned} r \cdot r^{N-1} &+ r \cdot r^{N-2} + \cdots + r \cdot r + r \cdot 1 \\ &- 1 \cdot r^{N-1} - \cdots - 1 \cdot r^2 - 1 \cdot r - 1 \cdot 1. \end{aligned}$$

After cancellation of the terms in the aligned columns, we are left with $r^N - 1$. This proves Theorem 3a.

For the second assertion, we have

$$\begin{aligned} \lim_{N\to\infty} (1 + r + r^2 + \cdots + r^{N-1}) &= \lim_{N\to\infty} \frac{r^N - 1}{r - 1} && \text{Theorem 3a} \\ &= \frac{1}{r - 1} \cdot \lim_{N\to\infty} r^N - \frac{1}{r - 1} && \text{Theorem 2a, 2d, 2e} \\ &= \frac{1}{r - 1} \cdot 0 - \frac{1}{r - 1} && \text{Theorem 1d} \\ &= \frac{1}{r - 1}. && \text{Simplify} \end{aligned}$$

■

Example 9 Evaluate $1/2 + 1/4 + 1/8 + 1/16 + \cdots$.

Solution We compute successive sums: $1/2$, $1/2 + 1/4$, $1/2 + 1/4 + 1/8$, and so on. When we have added N terms, we obtain

$$\frac{1}{2} + \frac{1}{4} + \frac{1}{8} + \cdots + \frac{1}{2^N} = \left(1 + \left(\frac{1}{2}\right) + \left(\frac{1}{2}\right)^2 + \left(\frac{1}{2}\right)^3 + \cdots + \left(\frac{1}{2}\right)^N\right) - 1.$$

Letting $N \to \infty$, we obtain

$$\begin{aligned}\frac{1}{2} + \frac{1}{4} + \frac{1}{8} + \frac{1}{16} + \cdots &= \lim_{N\to\infty}\left(\frac{1}{2} + \frac{1}{4} + \frac{1}{8} + \cdots + \frac{1}{2^N}\right)\\ &= \lim_{N\to\infty}\left(1 + \left(\frac{1}{2}\right) + \left(\frac{1}{2}\right)^2 + \left(\frac{1}{2}\right)^3 + \cdots + \left(\frac{1}{2}\right)^N\right) - 1\\ &= \frac{1}{1 - 1/2} - 1 \qquad \text{Theorem 3b}\\ &= 1.\end{aligned}$$

Zeno's paradoxes, to which this example is related, are discussed in the Genesis & Development for Chapter 1. ■

Example 10 What rational number does the repeating decimal expansion $0.999\ldots$ represent?

Solution The number $0.999\ldots$ represents

$$\begin{aligned}\lim_{N\to\infty}\left(\frac{9}{10} + \frac{9}{100} + \frac{9}{1000} + \cdots + \frac{9}{10^N}\right) &= \frac{9}{10}\lim_{N\to\infty}\left(1 + \frac{1}{10} + \frac{1}{100} + \cdots + \frac{1}{10^{N-1}}\right)\\ &= \frac{9}{10} \cdot \frac{1}{1 - 1/10} \qquad \text{Theorem 2b}\\ &= 1.\end{aligned}$$

■

A LOOK BACK

When we refer to an infinite sum $a_1 + a_2 + a_3 + \cdots$, we actually mean the limit $\lim_{N\to\infty}(a_1 + a_2 + \cdots + a_N)$ of a *finite* sum with an increasing number of terms N. Chapters 9 and 10 are devoted to a detailed study of such sums.

A LOOK FORWARD

Composition with Continuous Functions

It is often useful to transform one limit into another by using a continuous function, as our next theorem explains.

Theorem 4 Let $\{b_j\}_{j=1}^{\infty}$ be a sequence that converges to a limit ℓ. Suppose $\{b_n\}$ is contained in the domain of a function f that is continuous at ℓ. Thus, $\lim_{j\to\infty} f(b_j) = f(\lim_{j\to\infty} b_j) = f(\ell)$.

Example 11 Evaluate $\lim_{j\to\infty}\sqrt{1+2/j}$.

Solution We set $b_j = 1+2/j$. Therefore, $\lim_{j\to\infty} b_j = 1 = \ell$. Let $f(t) = \sqrt{t}$. Since f is continuous on the interval $(0, \infty)$ and this interval contains ℓ, Theorem 4 applies:

$$\lim_{j\to\infty}\sqrt{1+\frac{2}{j}} = \lim_{j\to\infty} f(b_j) = f\left(\lim_{j\to\infty} b_j\right) = f(1) = 1. \quad \blacksquare$$

quickquiz

1. What does it mean for a sequence to converge to 3?
2. What does it mean for a sequence to diverge?
3. Discuss convergence for the sequence $\{\sin(j\pi/2)\}_{j=1}^{\infty}$.
4. For what values of p does $\{n^p/(n^3+1)\}$ tend to infinity?

EXERCISES

Problems for Practice

In Exercises 1–15, determine whether the sequence $\{a_n\}$ converges. If it does, state the limit. Explain your reasoning, but do not attempt to supply proofs using the rigorous definition.

1. $a_n = n/(n^2+1)$
2. $a_n = n+5$
3. $a_n = n/(n+1)$
4. $a_n = 3-(-1)^n$
5. $a_n = 1/(n+2)$
6. $a_n = 1 \quad 1/n$
7. $a_n = \cos(n\pi)$
8. $a_n = \sin(n\pi)$
9. $a_n = 3^{-n}+2^{-n}$
10. $a_n = 1/k - 4^{-n}$
11. $a_n = (3n^3-5)/(4n^2+5)$
12. $a_n = (n^2-n)/(n^2+n)$
13. $a_n = 4+(-1)^n/(n^2+1)$
14. $a_n = n^{1/100}$
15. $a_n = (1/100)^{1/n}$

In Exercises 16–21, compute the limit using the rules from Theorems 1, 3, and 4. Explicitly state each theorem you use.

16. $\lim_{n\to\infty}(1+1/n)$
17. $\lim_{j\to\infty}(2^{-j}+3)$
18. $\lim_{k\to\infty}(k^2-5k+2)/(4k^2+6k+3)$
19. $\lim_{m\to\infty}(1+1/m)\cdot(2-6/m^2)$
20. $\lim_{n\to\infty}(10^{-n}+10^n)/10^n$
21. $\lim_{n\to\infty}(2^n-3^n)/(3^n+4^n)$

Use the Pinching Theorem to evaluate the limits in Exercises 22–25.

22. $\lim_{n\to\infty}\sin(1/n)/n$
23. $\lim_{k\to\infty}\cos(2k)/2^k$
24. $\lim_{j\to\infty}\sqrt{j+1/j^2}$
25. $\lim_{m\to\infty}(2+\cos(m))/m^2$

Evaluate the limits in Exercises 26–35. Explicitly state each theorem you use.

26. $\lim_{n\to\infty}(n-2n^2)/(n+n^2)$
27. $\lim_{j\to\infty}\cos(1/j)$
28. $\lim_{k\to\infty}(2-(1/2)^k)/(4+(1/3)^k)$
29. $\lim_{n\to\infty}(2^{-n})n/(n+4)$
30. $\lim_{n\to\infty}\tan(n^2\pi/(4n^2-n))$
31. $\lim_{j\to\infty}\sqrt{4j+1/\sqrt{j}}$
32. $\lim_{k\to\infty}(\sin(2^k)+2^k)/2^k$
33. $\lim_{m\to\infty}\cos(\pi\sin(\pi m/(2m+2)))$
34. $\lim_{n\to\infty}n^8/(n^2+2n^4)^2$
35. $\lim_{j\to\infty}4j/\sqrt{j^2+5j+2}$

In Exercises 36–38, evaluate the limit.

36. $1+1/3+1/9+1/27+\cdots$
37. $1-1/3+1/9-1/27+\cdots$
38. $1+1/\sqrt{3}+1/3+1/(3\sqrt{3})+\cdots$
39. Express the limit 1.11111111... as a ratio of integers.
40. Express the limit 1.01010101... as a ratio of integers.

Further Theory and Practice

In Exercises 41–44, determine whether the limit exists and what the limit is. Then verify your answer using the precise definition of limit.

41. $\lim_{n\to\infty}1/(n+7)$
42. $\lim_{k\to\infty}k/(k^2+1)$
43. $\lim_{m\to\infty}(2m+3)/(m+5)$
44. $\lim_{j\to\infty}\sin(1/j)$

In Exercises 45–50, evaluate the limit.

45. $1/4 + 1/8 + 1/16 + 1/32 + \cdots$

46. $1/\sqrt{2} + 1/2 + 1/(2\sqrt{2}) + 1/4 + \cdots$

47. $16/3 + 32/9 + 64/27 + 128/81 + \cdots$

48. $3.512121212\ldots$

49. $123.01232323\ldots$

50. $\alpha^2/(1+\alpha^2)^3 + \alpha^3/(1+\alpha^2)^5 + \alpha^4/(1+\alpha^2)^7 + \cdots$

51. Find a divergent sequence $\{a_n\}$ such that $\{a_n^2\}$ is convergent.

52. Prove that if $\{a_j\}_{j=1}^\infty$ converges to a real number ℓ, then $\lim_{j\to\infty}(a_j - \ell) = 0$.

53. Prove that if $\{a_j\}_{j=1}^\infty$ converges to a real number ℓ, then $\lim_{j\to\infty}(a_j - a_{j+1}) = 0$.

54. If $\{a_j\}_{j=1}^\infty$ and $\{b_j\}_{j=1}^\infty$ are sequences satisfying $|a_j - b_j| < 1/j$ for every j, prove that either both sequences converge to the same limit or both sequences diverge.

55. Suppose that $a_n \to \ell \neq 0$. Show that $\lim_{n\to\infty}(a_n/a_{n+1})$ converges.

56. Let α be a positive real number. Set $a_1 = 1$ and $a_n = (\alpha + a_{n-1})/2$ for $n > 1$. Find a formula for $|\alpha - a_n|$. Prove that $\{a_n\}$ converges.

57. If $p(x)$ is a polynomial of degree m and $q(x)$ is a polynomial of degree n, prove the following assertions.

a. If $m < n$, then $\lim_{j\to\infty} p(j)/q(j) = 0$.

b. If $m = n$, then $\lim_{j\to\infty} p(j)/q(j) = p_m/q_n$ where p_m is the leading coefficient of p and q_n is the leading coefficient of q.

c. If $m > n$, then $\lim_{j\to\infty} p(j)/q(j) = \infty$.

58. If p is a polynomial, show that $\lim_{k\to\infty} p(k+1)/p(k) = 1$.

59. Many doubly indexed sequences $\{a_{n,m}\}$ arise in mathematics. Often we need to know if

$$\lim_{n\to\infty}\lim_{m\to\infty} a_{n,m} = \lim_{m\to\infty}\lim_{n\to\infty} a_{n,m}.$$

Show that this equality can be false by considering $a_{n,m} = n/(n+m)$.

60. Suppose that ξ is a fixed positive number. For each integer $n \geq 1$, define

$$a_n = \frac{\lfloor\xi\rfloor + \lfloor 2\xi\rfloor + \lfloor 3\xi\rfloor + \cdots + \lfloor n\xi\rfloor}{n^2}$$

where, for any real η, the symbol $\lfloor\eta\rfloor$ denotes the greatest integer less than or equal to η. Investigate $\lim_{n\to\infty} a_n$.

Calculator/Computer Exercises

There are many sequences with convergence properties that are *not* obvious. The sequences in Exercises 61–65 converge, but it is quite tricky to determine what the limit is and to prove the answer. Approximate the limit to four decimal places. State what value of j you are using to approximate the limit and why you believe your answer is correct to the stated accuracy.

61. $\lim_{j\to\infty} j\cdot\sin(1/j)$

62. $\lim_{j\to\infty} j^2\cdot(1-\cos(1/j))$

63. $\lim_{j\to\infty} j^{1/j}$

64. $\lim_{j\to\infty}(1-1/j)^j$

65. $\lim_{j\to\infty}(\sqrt{j}/\sqrt{j+1})^{\sqrt{j}}$

66. Graph the function $f(x) = x^{1/x}$ for $x > 0$. (The value of $x^{1/x}$ for irrational x will be discussed in Section 2.6. In any event, a graphing utility will only plot f for rational values of x because all calculations are carried to a finite number of decimal places.) What does the limiting behavior of $f(x)$ as $x \to +\infty$ appear to be? What does this behavior tell you about the limit of the sequence $\{j^{1/j}\}_{j=1}^\infty$?

67. Graph the function $f(x) = x^x$ for $x > 0$. What does the limiting behavior of $f(x)$ as $x \to 0^+$ appear to be? What does this behavior tell you about the limit of the sequence $\{j^{1/j}\}_{j=1}^\infty$?

68. Here is an unusual way to use a sequence to solve a quadratic equation: Consider the equation $x^2 - 6x + 5 = 0$. Write $x^2 = 6x - 5$, or $x = 6 - 5/x$. For the x on the right, substitute $6 - 5/x$. The result is

$$x = 6 - \frac{5}{6 - 5/x}.$$

Again, for the x on the right, substitute $x = 6 - 5/x$. The result is

$$x = 6 - \cfrac{5}{6 - \cfrac{5}{6 - 5/x}}.$$

This process, which we may continue indefinitely, suggests that x is the limit of the sequence

$$6\,,\ 6 - \frac{5}{6}\,,\ 6 - \frac{5}{6 - 5/6}\,,\ 6 - \cfrac{5}{6 - \cfrac{5}{6 - 5/6}}\,, \ldots.$$

This is called a *continued fraction expansion*. Compute the first 20 terms of this sequence. Do they seem to be converging to something? Is the limit a root of the original quadratic equation? Can you discover the other root in the same way?

2.6 Exponential Functions

Many sequences have the property that each term is greater than or equal to the terms that have come before it. Similarly, it is often the case that each term of a sequence is less than or equal to the terms that have come before it. This section introduces some terminology to describe these properties.

Definition A sequence $\{a_n\}$ is said to be *increasing* if $a_n \leq a_{n+1}$ for every n. A sequence $\{b_n\}$ is said to be *decreasing* if $b_n \geq b_{n+1}$ for every n. Either type of sequence is said to be *monotone* or *monotonic* (see Figure 1).

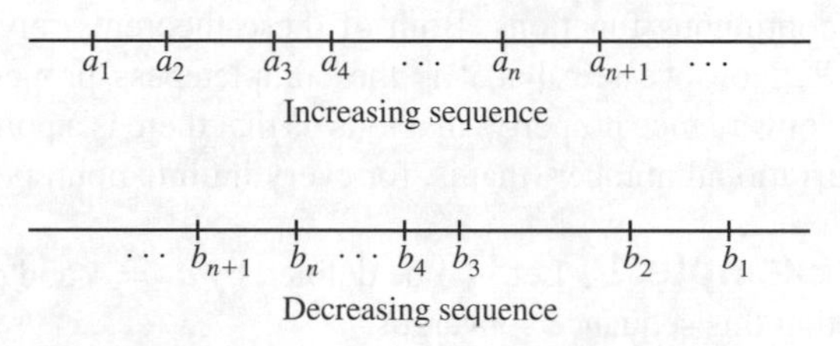

Figure 1

The Monotone Convergence Property of the Real Numbers

Recall from Chapter 1 that every irrational number has an infinite nonrepeating decimal expansion. What exactly is meant by this? Consider the irrational number π. When we write $\pi = 3.141592653\ldots$, we mean that π is the limit of the sequence

$$3,\ 3.1,\ 3.14,\ 3.141,\ 3.1415,\ 3.14159,\ 3.141592,\ 3.1415926,\ \ldots.$$

The terms of this sequence are increasing. They do not, however, grow arbitrarily large. In fact, every term in the sequence is less than 3.2. Here is a definition for a sequence that does not have arbitrarily large terms.

Definition A sequence $\{a_n\}$ is *bounded above* if there is a real number U such that $a_n \leq U$ for all n. We refer to U as an *upper bound* for the sequence $\{a_n\}$ and say that $\{a_n\}$ is *bounded above* by U. Similarly, a sequence $\{a_n\}$ is *bounded below* if there is a real number L such that $L \leq a_n$ for all n. We refer to L as a *lower bound* for the sequence $\{a_n\}$ and say that $\{a_n\}$ is *bounded below* by L. If $\{a_n\}$ is bounded both above and below, then we simply say that $\{a_n\}$ is *bounded*. In this final case, there is a number $M > 0$ such that $|a_n| \leq M$ for all n. We say that $\{a_n\}$ is *bounded* by M.

Every bounded monotonic sequence $\{a_n\}_{n=1}^{\infty}$ is contained in a closed interval: If $\{a_n\}$ is *increasing* and bounded above by M, then $\{a_n\}$ is contained in $[a_1, M]$. If $\{a_n\}$ is *decreasing* and bounded below by M, then $\{a_n\}$ is contained in $[M, a_1]$. Our next theorem guarantees that a bounded increasing sequence actually converges to a real number, even though we may not be able to specify in advance what that number is.

Theorem 1 ***Monotone Convergence Property*** If $\{a_n\}_{n=1}^{\infty}$ is monotonic and bounded, then $\{a_n\}_{n=1}^{\infty}$ converges to some real number ℓ. If $\{a_n\}_{n=1}^{\infty}$ lies in a closed interval I, then ℓ belongs to I.

The monotone convergence property is one way of expressing the completeness of the real number system. Equivalent versions of the completeness property may be found in the Genesis & Developments for Chapters 1 and 2. It is important to understand this property's many consequences that are crucial for calculus. In Section 2.3, we learned the Intermediate Value Theorem and the Extreme Value Theorem for continuous functions: Both of these theorems rely on the completeness property of $\mathbb{R}$. Now we see that it is the completeness property (in the form of the monotone convergence property) that tells us that there is a point on the real number line for every irrational number (that is, for every infinite nonrepeating decimal expansion).

Example 1 Let $\{a_n\}$ be defined by $a_1 = 1$ and $a_n = a_{n-1} + 1/n^n$ for $n \geq 2$. Show that this sequence converges.

Solution According to the recursive definition of the sequence $\{a_n\}$, we have $a_n > a_{n-1}$ for each n. In other words, the sequence $\{a_n\}$ is increasing. The table provides the first several terms to five decimal places.

n	1	2	3	4	5	6	7
a_n	1.00000	1.25000	1.28704	1.29094	1.29126	1.29129	1.29129

The tabulated values of $\{a_n\}$ indicate that $\{a_n\}$ is increasing and bounded. To *prove* that $\{a_n\}$ is bounded, notice that $1/n^n \leq 1/2^n$ for $n \geq 2$. It follows that

$$a_2 = 1 + \frac{1}{2^2} < 1 + \frac{1}{2} + \frac{1}{4}$$
$$a_3 = a_2 + \frac{1}{3^3} < a_2 + \frac{1}{2^3} < 1 + \frac{1}{2} + \frac{1}{4} + \frac{1}{8}$$
$$a_4 = a_3 + \frac{1}{4^4} < a_3 + \frac{1}{2^4} < 1 + \frac{1}{2} + \frac{1}{4} + \frac{1}{8} + \frac{1}{16}$$

and, in general,

$$a_n < 1 + \frac{1}{2} + \frac{1}{4} + \frac{1}{8} + \cdots + \frac{1}{2^n}.$$

Therefore, by Example 9 from Section 2.5,

$$a_n < 1 + \frac{1}{2} + \frac{1}{4} + \frac{1}{8} + \cdots = 2$$

for every n. The sequence $\{a_n\}$ is bounded by 2 and, therefore, converges to a number $\ell \leq 2$. ■

INSIGHT

> The monotone convergence property allows us to deduce that $\{a_n\}$ converges to some real number ℓ, but it does not tell us what the number ℓ is. We can, however, compute ℓ to any required accuracy. A practical rule of thumb in numerical work is to terminate computation when several successive calculations agree to the required accuracy. For example, if we require three decimal places of ℓ, then we can be reasonably sure that $\ell = 1.291\ldots$, because the computed values a_5, a_6, and a_7 agree on those digits.

Limits of Recursive Sequences

If $b_n = f(b_{n-1})$ for some continuous function f and if $\{b_n\}$ converges to a point b that is in the domain of f, then the limit must be a root of the equation $f(x) = x$:

$$f(b) = f\left(\lim_{n\to\infty} b_n\right) = f\left(\lim_{n\to\infty} b_{n-1}\right) \overset{\text{Continuity}}{=} \lim_{n\to\infty} f(b_{n-1}) = \lim_{n\to\infty} b_n = b.$$

A root of the equation $f(x) = x$ is called a *fixed point* of f. The following example illustrates how we can use this idea to evaluate a limit.

Example 2 Let $b_1 = 1$. For $n \geq 2$, let $b_n = \sqrt{2 + b_{n-1}}$. The sequence $\{b_n\}$ is bounded and increasing. Evaluate its limit.

Solution The monotone convergence property asserts that $b_n \to \ell$ for some number ℓ. Let $f(x) = \sqrt{2+x}$. Since $b_n = f(b_{n-1})$, we know that ℓ is a fixed point of f: $\ell = \sqrt{2+\ell}$. Thus, $\ell^2 = 2 + \ell$, or $\ell^2 - \ell - 2 = 0$. The two roots of this quadratic equation are -1 and 2. Because $\{b_n\}$ is a sequence of positive terms, we may rule out -1 as a limit. We conclude that $b_n \to 2$. Figure 2 illustrates how we may use the plots of the equations $y = f(x)$ and $y = x$ to understand the limiting behavior of the sequence $\{b_n\}$.

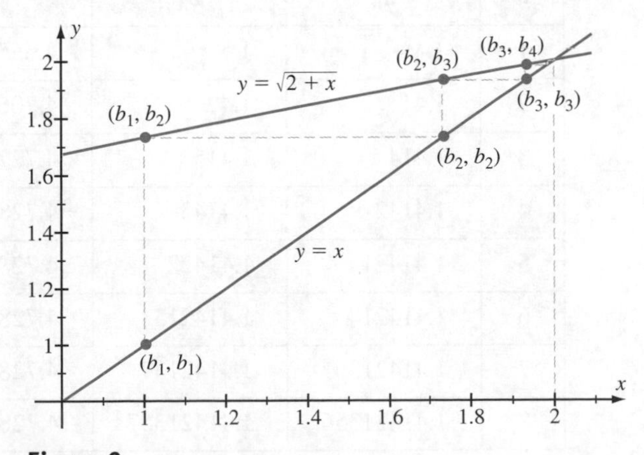

Figure 2

■

Irrational Exponents

Suppose a is a positive number. What does it mean to write a^x if x is an irrational number? We do know what a^x means when x is equal to an integer m:

$$a^m = \underbrace{a \cdot a \cdots a}_{m \text{ factors}} \quad \text{if } m > 0, \quad a^m = 1 \text{ if } m = 0, \quad \text{and} \quad a^m = \frac{1}{\underbrace{a \cdot a \cdots a}_{|m| \text{ factors}}} \quad \text{if } m < 0.$$

We also know what a^x means for *all* rational numbers x. For a positive integer q, the (positive) qth root $a^{1/q}$ of a was shown to exist by the Intermediate Value Theorem (see Section 2.3). The theory of exponents can therefore be extended to rational numbers: If p is any integer and if q is a positive integer, then $a^{p/q}$ is defined by

$$a^{p/q} = (a^p)^{1/q} = \sqrt[q]{a^p}.$$

However, this still leaves us with the problem of defining a^x for x irrational. What, for example, do we mean by the expression $3^{\sqrt{2}}$?

We answer this question by using what we already know about rational exponents. To be specific, we approximate $3^{\sqrt{2}}$ by numbers of the form 3^x where x is a rational approximation of $\sqrt{2}$. The better x approximates $\sqrt{2}$, the better 3^x approximates $3^{\sqrt{2}}$. Let us see how this idea can be put into practice.

The number $\sqrt{2} = 1.41421356\ldots$ is between the rational numbers $1.4 = 14/10$ and $1.5 = 15/10$. Therefore, $3^{\sqrt{2}}$, however it is defined, should be between $3^{14/10} = 4.655536\ldots$ and $3^{15/10} = 5.196152\ldots$. We can narrow the range in which $3^{\sqrt{2}}$ must lie by using a better rational approximation of $\sqrt{2}$. Thus, $\sqrt{2}$ is between $1.41 = 141/100$ and $1.42 = 142/100$. Therefore, $3^{\sqrt{2}}$ should be between $3^{141/100} = 4.706965\ldots$ and $3^{142/100} = 4.758961\ldots$. We now know what the first decimal place of $3^{\sqrt{2}}$ must be. Continuing this process, we let a_n be the rational number obtained by terminating the decimal expansion of $\sqrt{2}$ after the nth decimal place. Let b_n be the rational number obtained by augmenting the digit in the last decimal place of a_n by 1: $b_n = a_n + 1/10^n$. The following table lists the results.

n	a_n	b_n	3^{a_n}	3^{b_n}
1	1.4	1.5	4.655536722	5.196152423
2	1.41	1.42	4.706965002	4.758961394
3	1.414	1.415	4.727695035	4.732891793
4	1.4142	1.4143	4.728733930	4.729253463
5	1.41421	1.41422	4.728785881	4.728837832
6	1.414214	1.414215	4.728806661	4.728811856
7	1.4142136	1.4142137	4.728804583	4.728805103
8	1.41421356	1.41421357	4.728804376	4.728804427

From these numerical calculations, we see that $3^{\sqrt{2}}$ must be 4.728804 to six decimal places. Furthermore, we see that we can actually define $3^{\sqrt{2}}$ precisely by *passing to the limit.*

Definition

If a is a positive number and x is an irrational number, we define a^x by

$$a^x = \lim_{n\to\infty} a^{x_n} \tag{2.12}$$

where $\{x_n\}$ is a sequence of rationals such that $x_n \to x$.

By taking an increasing sequence $\{x_n\}$ of rational numbers that approaches x, we can be sure that the sequence $\{a^{x_n}\}$ is convergent (by the monotone convergence property). It is important to know, however, that the limit $\lim_{n\to\infty} a^{x_n}$ does not depend on the particular sequence $\{x_n\}$ that is chosen to approach x. We will not prove this last fact now. For the time being, we will accept that all the usual laws of exponentials follow from equation (2.12):

$$a^{x+y} = a^x a^y, \quad a^{-x} = \frac{1}{a^x}, \quad \text{and} \quad (a^x)^y = a^{xy} \quad (a > 0, x, y \in \mathbb{R}).$$

In Chapter 6, we use calculus to provide an alternative view of irrational exponents and prove these exponent laws. In our alternative approach, equation (2.12) will result from the continuity of the exponential function $x \mapsto a^x$.

Graphs of Exponential Functions

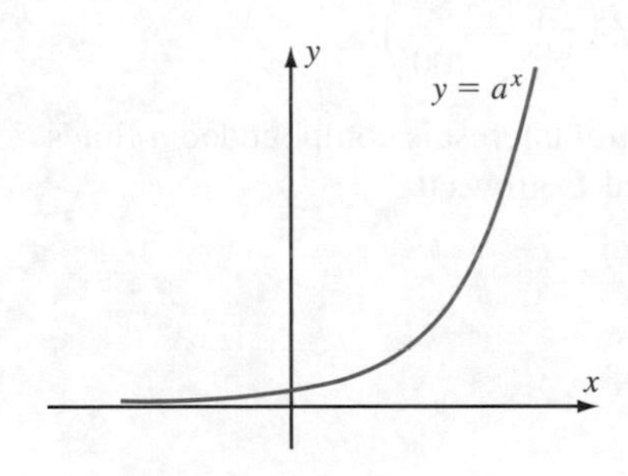

Figure 3
The graph of $y = a^x$ for $a > 1$

Figure 3 shows the graph of $x \mapsto a^x$ $(a > 1)$, and Figure 4 shows the graph of $x \mapsto a^x$ $(0 < a < 1)$. Notice that

$$\lim_{x\to-\infty} a^x = 0 \quad \text{and} \quad \lim_{x\to\infty} a^x = \infty$$

for $a > 1$ and

$$\lim_{x\to-\infty} a^x = \infty \quad \text{for } a > 1 \quad \text{and} \quad \lim_{x\to\infty} a^x = 0$$

for $0 < a < 1$. As the figures indicate, the exponential functions $x \mapsto a^x$ are continuous: For each c, we have

$$\lim_{x\to c} a^x = a^c.$$

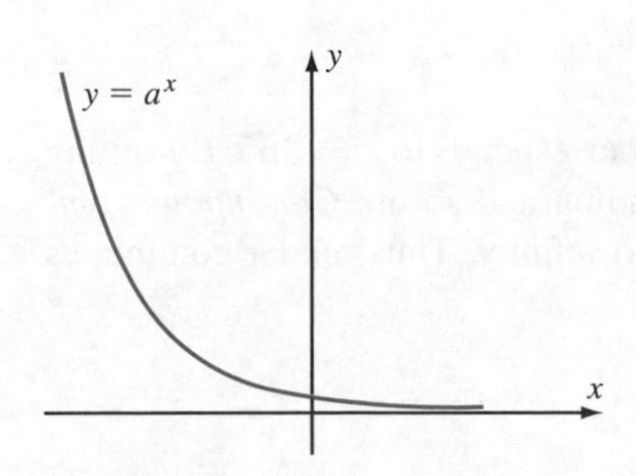

Figure 4
The graph of $y = a^x$ for $0 < a < 1$

Earlier in this chapter, we observed that the power functions $x \mapsto x^p$ are continuous functions for p rational. If p is irrational, then the power function $x \mapsto x^p$ is also continuous on $(0, \infty)$; that is,

$$\lim_{x\to c} x^p = c^p \quad \text{for } c > 0.$$

Compound Interest

If an amount of P dollars is put in the bank at r percent *simple interest* per year, then, after one year, the account has

$$P + \frac{r}{100}P = P \cdot \left(1 + \frac{r}{100}\right) \text{ dollars.}$$

This represents the return of original amount P (known as the *principal*) plus the interest $(r/100) \cdot P$. If the interest is *compounded* n times during the year, however, then the year is divided into n equal pieces, and at the end of each time interval of length $1/n$, an interest payment of r/n percent is added to the account. Each time this fraction of the interest is added to the account, the money in the account is multiplied by

$$1 + \frac{r/n}{100}.$$

Since this is done n times during the year, the account holds

$$P \cdot \left(1 + \frac{r}{100n}\right)^n \text{ dollars}$$

at the end of the year.

For example, under semiannual (twice yearly) compounding, the account is worth $P \cdot (1 + r/100)$ after six months. This becomes the principal on which interest is paid for the second half of the year. Therefore, at the end of the year, the account is worth

$$P \cdot \left(1 + \frac{r}{200}\right)\left(1 + \frac{r}{200}\right) = P \cdot \left(1 + \frac{r}{200}\right)^2.$$

The same reasoning shows that if r percent *annual* interest is compounded n times a year, then at the end of t years, the initial principal P grows to

$$P \cdot \left(1 + \frac{r}{100n}\right)^{nt}.$$

The next theorem summarizes our findings.

Theorem 2 If the interest on principal P accrues at r percent compounded n times per year, then the value of the deposit after t years is

$$P \cdot \left(1 + \frac{r}{100n}\right)^{nt}.$$

Compounding is often performed daily, which corresponds to $n = 365$. By setting $n = 365 \times 24$, we could obtain hourly compounding and so on. *Continuous compounding* is the limiting result of letting n tend to infinity. Thus, under continuous compounding, the account is worth

$$\lim_{n\to\infty} P\left(1 + \frac{r}{100n}\right)^n$$

at the end of one year.

To evaluate this limit, we set $u = r/100$, obtaining $\lim_{n\to\infty}(1 + u/n)^n$. We start with the case $u = 1$, tabulating the sequences $\{a_n\}$ and $\{c_n\}$ defined by

$$a_n = \left(1 + \frac{1}{n}\right)^n \quad \text{and} \quad c_n = \left(1 - \frac{1}{n}\right)^{-n}$$

for several values of n (see the following table).

n	a_n	c_n
7	2.546499 . . .	2.941897 . . .
8	2.565784 . . .	2.910285 . . .
9	2.581174 . . .	2.886507 . . .
10	2.593742 . . .	2.867971 . . .
10^6	2.718280 . . .	2.718283 . . .
10^7	2.7182816 . . .	2.7182819 . . .

The sequence $\{a_n\}$ appears to be increasing, which agrees with our intuition about compounding: The more frequent the compounding, the better the return. Another glance at the table suggests that a_n does not diverge to infinity. In fact, it appears that $a_n < c_n$ for each n and that $\{c_n\}$ is a decreasing sequence. Assuming these observations are true, we can conclude that the bounded monotone sequences $\{a_n\}$ and $\{c_n\}$ converge. Indeed, the last few rows of the table suggest that $\{a_n\}$ and $\{c_n\}$ converge to the *same* limit. We record these observations in the following theorem, which we prove in Chapter 6.

Theorem 3 There is a real number e such that

$$e = \lim_{n\to\infty}\left(1 + \frac{1}{n}\right)^n \tag{2.13}$$

and

$$e = \lim_{n\to\infty}\left(1 - \frac{1}{n}\right)^{-n}. \tag{2.14}$$

The number e is one of the basic constants of calculus. It is an irrational number with a decimal expansion of

$$e = 2.71828182845904523536\ldots$$

to 20 places. The function $u \mapsto e^u$ is called the *exponential function* and is denoted by exp. Thus, $\exp(u)$ is an alternative notation for e^u.

It is often convenient to express e as the limit of a continuous variable. To this end, we have

$$e = \lim_{x\to\infty}\left(1+\frac{1}{x}\right)^{x} \tag{2.15}$$

and

$$e = \lim_{x\to\infty}\left(1-\frac{1}{x}\right)^{-x}. \tag{2.16}$$

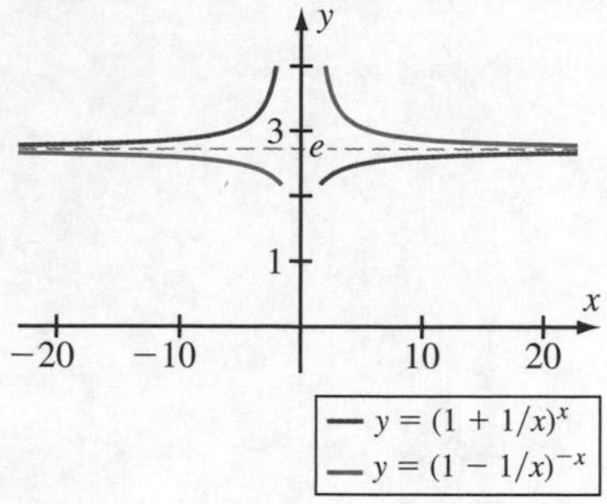

Formulas (2.15) and (2.16) indicate that the graphs of

$$x \mapsto \left(1+\frac{1}{x}\right)^{x},\ x>0, \quad \text{and} \quad x \mapsto \left(1-\frac{1}{x}\right)^{-x},\ x>0,$$

Figure 5

have a common horizontal asymptote $y = e$ (see Figure 5). In our next theorem, we use formulas (2.15) and (2.16) to evaluate the limit $\lim_{x\to\infty}(1+u/x)^x$ for every value of u.

Theorem 4 For every real number u, we have

$$e^u = \lim_{x\to\infty}\left(1+\frac{u}{x}\right)^{x}. \tag{2.17}$$

Proof Because equation (2.17) is clear for $u = 0$, we may suppose that $u \neq 0$. Since power functions are continuous, we have

$$e^u \overset{(2.15)}{=} \left(\lim_{t\to\infty}\left(1+\frac{1}{t}\right)^{t}\right)^{u} = \lim_{t\to\infty}\left(1+\frac{1}{t}\right)^{ut}.$$

Set $x = ut$. Observe that $1/t = u/x$. If $u > 0$, then $x \to \infty$ as $t \to \infty$ and

$$e^u = \lim_{t\to\infty}\left(1+\frac{1}{t}\right)^{ut} = \lim_{x\to\infty}\left(1+\frac{u}{x}\right)^{x},$$

as required. The $u < 0$ case is proved analogously, starting with formula (2.16) instead of (2.15).

We can now state our main result about continuous compounding. ■

Theorem 5 If the interest on principal P accrues at an annual rate of r percent, compounded continuously, then the money accumulated after t years is $P \cdot e^{rt/100}$.

Example 3 If \$5000 is placed in a savings account with 6% annual interest, compounded continuously, then how large is the account after 4.5 years?

Solution If $M(t)$ is the amount of money in the account at time t, then the preceding discussion tells us that $M(t) = 5000 \cdot e^{6t/100}$. After 4.5 years, the size of the account is

$$M\left(\frac{9}{2}\right) = 5000 \cdot e^{6\cdot(9/2)/100} \approx \$6549.82.$$

■

It is interesting to compare the solution to Example 3 with the behavior of the account when simple interest is paid. Under these circumstances, after 4.5 years, the size of the account is

$$5000 \cdot \left(1 + \frac{6}{100}\right)^{9/2} \approx \$6499.00.$$

Thus, although compound interest is advantageous, we see that compounding a great many times (even continuously) does not make a large difference in the profits earned over a relatively short time. Financial advisors often speak, however, of the "magic" of compounding, because the benefits of compounding over long periods of time can be impressive, as our next example illustrates.

Example 4 A woman wishes to set up an endowment to pay her nephew \$50,000 in cash on the day of his 21st birthday. The endowment is set up on the day of his birth and is locked in at 9% annual interest, compounded continuously. How much principal should be put into the account to yield the desired sum?

Solution Let P be the initial principal deposited in the account on the day of the nephew's birth. After 21 years at 9% annual interest, compounded continuously, the principal is to have grown to \$50,000:

$$50000 = P \cdot e^{(0.09) \cdot 21}.$$

Solving for P gives

$$P = 50000 \cdot e^{-0.09 \cdot 21} = 50000 \cdot e^{-1.89} \approx 7553.59.$$

If the account were drawing only simple annual interest of 9%, then the compounding would be done only at the end of every year. The initial principal would be determined by the equation $50000 = P \cdot (1 + .09)^{21}$. Solving this equation for P, we find that a deposit of \$8184.90 would be required, which is 8.35% more than the sum needed under continuous compounding. ■

Exponential Functions in Modeling

The exponential function is particularly useful for modeling certain types of growth and decay. We say that the function

$$x \mapsto Ae^{kx}$$

grows exponentially if $k > 0$ and *decays exponentially* if $k < 0$. The constant k is called the *growth* (or *decay*) *constant*. Various population sizes show exponential growth. Exponential decay is used to model the decrease in the amount of radiation and the decline of the purchasing power of the dollar.

A characteristic of exponential growth $y(t) = Ae^{kt}$ is that there is a *doubling time* τ during which $y(t)$ doubles in magnitude: $y(t + \tau) = 2y(t)$ for all values of t. The doubling time τ satisfies the relationship $e^{k\tau} = 2$. We can solve for τ (as in Exercise 55) to obtain

$$\tau = \frac{0.6931471806\ldots}{k} \quad \text{or} \quad \text{growth constant} \times \text{doubling time} = 0.6931471806\ldots. \qquad \textbf{(2.18)}$$

Example 5 The first eight American census figures are listed in the table.

Year ($t + 1790$)	1790	1800	1810	1820	1830	1840	1850	1860
Population (in millions)	3.929	5.308	7.240	9.638	12.860	17.063	23.191	31.443

Source: U.S. Bureau of the Census

Use $\tau = 23.5$ as the doubling time of the population to find an exponential growth model.

Solution In exponential growth (or decay), given by $y(t) = Ae^{kt}$, the value of A is the initial value $y(0)$. From the given value $\tau = 23.5$, we solve the equation $23.5 = 0.6931471806/k$ to obtain $k = 0.0294956$. Our model is therefore $y(t) = 3.929e^{0.0294956 \cdot t}$, which is plotted in Figure 6. The horizontal axis represents time t measured in years with the origin $t = 0$ corresponding to 1790. The census data points are also plotted in Figure 6. Notice how closely our model fits the data.

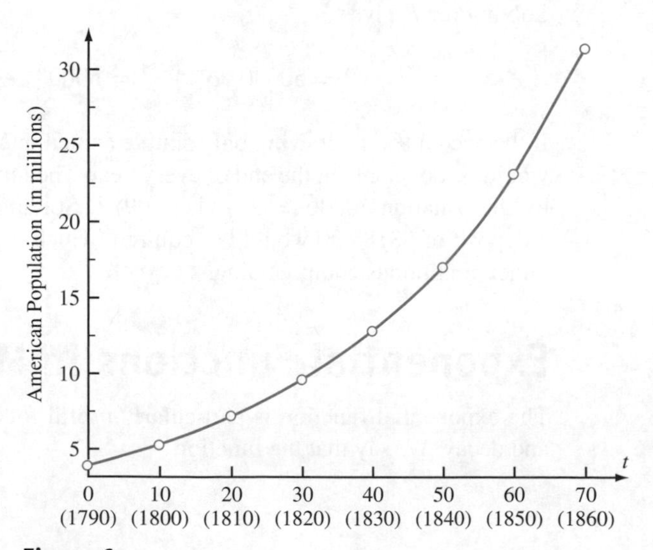

Figure 6

■

quickquiz

1. What is a monotone sequence? Do all monotone sequences converge?
2. What do we mean by 4^π?
3. What is the value of $\lim_{n\to\infty}(1+1/n)^n$?
4. Sketch the graphs of $x \mapsto e^x$ and $x \mapsto (e/\pi)^x$.

EXERCISES

Problems for Practice

In Exercises 1–6, simplify the expression.

1. $\sqrt{2}^{\sqrt{3}} \cdot \sqrt{2}^{\sqrt{3}}$ **2.** $4^\pi \cdot 4^e$

3. $(1/8)^{-\pi/3}$ **4.** $(8^{\sqrt{3}} \cdot 4^{\sqrt{7}})/2^\pi$

5. $\left(\sqrt{11}^{\sqrt{2}}\right)^{\sqrt{2}}$ **6.** $\left(\sqrt{e}^{\sqrt{2}}\right)^2$

In Exercises 7–12, make a rough sketch of the function. Label salient points.

7. $f(x) = 3^x + 1$ **8.** $x \mapsto 2^{-x}$

9. $g(x) = 3 - e^x$ **10.** $x \mapsto |e^x - 1|$

11. $x \mapsto e^{|x|}$ **12.** $f(x) = 1 - 1/2^{x-1}$

In Exercises 13–16, find the limit.

13. $\lim_{x\to\infty}(e^{2x} - e^{-2x})/(e^{2x} + e^{-2x})$

14. $\lim_{x\to e^-} \pi^{1/(x-e)}$

15. $\lim_{x\to e^+} \pi^{1/(x-e)}$

16. $\lim_{x\to(\pi/2)^-}(1/\sqrt{2})^{\tan(x)}$

17. Find all asymptotes of each equation.

a. $y = (e^{7x} + e^{-7x})/(e^{7x} - e^{-7x})$

b. $y = e^{-x^2}$

c. $y = 3x/(x - e)$

d. $y = (3^x + 2^x)/(3^x - 3)$

18. An annual interest rate of 6% is paid on an initial investment of \$1000. How much is the investment worth in 1 year under the following conditions?

a. Simple interest

b. Semiannual compounding

c. Quarterly compounding

d. Daily compounding

e. Continuous compounding

19. An annual interest rate of 7% is paid on an initial investment of \$1000. How much is the investment worth in 5 years under the following conditions?

a. Simple interest

b. Semiannual compounding

c. Quarterly compounding

d. Daily compounding

e. Continuous compounding

20. If an amount of income A is to be received at a future date, the *present value* of that payment is the amount P, which will grow to A under continuous compounding at the current interest rate by the time the payment is received. Mr. Woodman wants to give \$100,000 to his son Chip when Chip turns 25 years old. Given that current interest rates are 5% and Chip has just turned 18, what is the present value of the gift?

Use Theorem 4 to calculate the limits in Exercises 21–24.

21. $\lim_{n\to\infty}\left(1+\frac{2}{n}\right)^n$ **22.** $\lim_{n\to\infty}\left(1-\frac{e}{n}\right)^n$

23. $\lim_{n\to\infty}\left(1-\frac{1}{n}\right)^n$ **24.** $\lim_{n\to\infty}\left(\frac{n+1}{n}\right)^n$

In Exercises 25–28, a convergent sequence $\{a_n\}$ is defined recursively. Calculate its limit by using the method from Example 2.

25. $a_1 = 1, a_n = \sqrt{2 + 3a_{n-1}}$

26. $a_1 = 3/5, a_n = \sin(\pi a_{n-1}/3)$

27. $a_1 = 1, a_n = (a_{n-1} + 6)/(a_{n-1} + 2)$

28. $a_1 = 1, a_n = (a_{n-1}/2) + (1/a_{n-1}^2)$

Further Theory and Practice

29. Suppose $P(t) = Ae^{kt}$ where k is a positive constant. Show that there is a number τ (known as the doubling time) in the open interval $(0, 1/k)$ such that

$$P(t+\tau) = 2P(t)$$

for every t. Find a formula involving k and τ.

30. Suppose $P(t) = mt + b$. Can P have a doubling time (in the sense of Exercise 29)? Explain.

31. The volume of harvestable timber in a young forest grows exponentially with a yearly rate of increase equal to 3.5%. What percentage increase is expected in 10 years? What is the doubling time?

32. If at time $t = 0$ an object is placed in an environment that is maintained at a constant temperature T_∞, then according to Newton's Law of Heating and Cooling, the temperature $T(t)$ of the object at time t is

$$T(t) = T_\infty + (T(0) - T_\infty)e^{-kt}$$

for some constant $k > 0$. What is the limiting temperature of the object as $t \to \infty$? What asymptote does the graph of T have? Sketch the graph of T assuming that $T(0) > T_\infty$. Sketch the graph of T assuming that $T(0) < T_\infty$.

33. Evaluate $\lim_{k\to\infty}(1 + 3/k)^{2k}$.

34. If $a_n \to 0$ and $\{b_n\}$ is a bounded sequence, show that $a_n \cdot b_n \to 0$.

35. Let $\{a_j\}_{j=1}^\infty$ be a convergent sequence. Prove that there is a number $M > 0$ such that $|a_j| \le M$ for all j. (In other words, if $\{a_j\}_{j=1}^\infty$ is a convergent sequence, then it is bounded.) *Hint:* If $a_j \to \ell$, can infinitely many terms of $\{a_j\}$ be greater than $\ell + 1$?

36. Suppose that $a_n \to 1$. Since any power of 1 is also 1, it might seem as if $a_n^n \to 1$. In fact, for any $v > 0$, it is possible to find a sequence $\{a_n\}$ such that $a_n \to 1$ but $a_n^n \to v$. Explain by using one of the sequences studied in this section.

37. Suppose that $\alpha > 1$. Let $x_0 = 1$. For $n > 1$, let $x_n = (\alpha + x_{n-1})/2$. Show that $\{x_n\}$ is a bounded increasing sequence. To what does $\{x_n\}$ converge?

38. For a given positive integer initialization h_1, the *Hail Stone Sequence* is defined for $n > 1$ by $h_n = h_{n-1}/2$ if h_{n-1} is even and by $h_n = 3h_{n-1} + 1$ if h_{n-1} is odd. Is it possible to initialize this sequence so that it is monotone? Explain.

39. Suppose $\{r_k\}$ is a sequence of positive numbers such that $r_{k+1} = 1 + 1/r_k$. Suppose $r_k \to \ell$. What is ℓ?

40. By considering the numbers $a = (\sqrt{2})^{\sqrt{2}}$ and $b = a^{\sqrt{2}}$, deduce that an irrational number to an irrational power can be rational.

41. Let

$$s_N = \frac{\sin^2(1)}{2} + \frac{\sin^2(2)}{4} + \frac{\sin^2(3)}{8} + \cdots + \frac{\sin^2(N)}{2^N}.$$

Prove that the sequence $\{s_N\}$ is convergent.

42. Let $s_n = 1 + 1/2! + 1/3! + \cdots + 1/n!$

a. Show that $k! \ge 2^{k-1}$ for $k \in \mathbb{Z}^+$.

b. Use part a to show that

$$s_n < 1 + 1/2! + 1/3! + \cdots < 1 + 1/2 + 1/4 + \cdots.$$

What real number does $1 + 1/2 + 1/4 + \cdots$ represent?

c. Prove that $\{s_n\}$ converges. Cite the theorems that you use.

d. Compute $\ell = \lim_{k\to\infty} s_k$ to several decimal places. Identify the irrational number ℓ exactly. No proof is required—by computing enough decimal places, you can discover the relationship between ℓ and a known constant.

43. Let $x_0 = 0.4$, and let $x_n = \pi \sin(x_{n-1})/2$ for $n \ge 1$. Sketch a graph (similar to the plot in Figure 2) that illustrates convergence of x_n. To what does the sequence converge?

44. Suppose M and P_0 are constants such that $M > P_0 > 0$. The size of a population P is given as a function of time t by

$$P(t) = \frac{P_0 M}{P_0 + (M - P_0)e^{-kt}}$$

for some positive constant k. (This is known as the *logistic growth formula*.) What is the initial population size? Find $\lim_{t\to\infty} P(t)$. What horizontal asymptote does the graph of P have?

45. If the air resistance to a falling object is proportional to the velocity of that object, then the velocity of that object when dropped from a height is

$$v(t) = kg(1 - e^{-t/k})$$

in the downward direction. What is the limiting velocity v_∞ as $t \to \infty$? (This is known as the *terminal velocity*.) Sketch the graph of v. What horizontal asymptote does it have?

46. When a drug is intravenously introduced into a patient's bloodstream at a constant rate, the concentration C of the drug in the patient's body is typically given by

$$C(t) = \frac{\alpha}{\beta}(1 - e^{-\beta t})$$

where α and β are positive constants. In the long run, what is the concentration of the drug in the patient's bloodstream? Sketch the graph of C. What horizontal asymptote does it have?

47. The diffusion of a solute through a cell membrane is described by Fick's law:

$$c(t) = (c(0) - C)\exp\left(\frac{-kAt}{V}\right) + C.$$

Here $c(t)$ is the concentration of the solute, C is the concentration outside the cell, A is the area of the cell membrane, V is the volume of the cell, and k is a positive constant. In the long run, what is the concentration $c(t)$ of the solute? Sketch the graph of c when $c(0) > C$ and when $c(0) < C$. What horizontal asymptote does each graph have?

48. Investment advisors often refer to the following rule of thumb: If an investment is left to compound at a constant annual rate of $r\%$, then the time required for the investment to double in value is $72/r$ years. Explain where this rule of 72 comes from. What is the exact rule that it approximates? What integer is more exact than 72? (The number 72 is used because $72/r$ is an integer for the common rates $r = 4, 6,$ and 8.)

49. Evaluate the following.

a. $\lim_{n\to\infty}(1 + 1/n^2)^{n^2}$

b. $\lim_{n\to\infty}(n/(n-2))^{2n}$

50. Let a and b be positive numbers with $a > b$. Define two sequences as follows: $a_1 = a$, $b_1 = b$. For $n > 1$, let

$$a_n = \frac{a_{n-1} + b_{n-1}}{2} \quad \text{and} \quad b_n = \sqrt{a_{n-1}b_{n-1}}.$$

a. Prove the *Arithmetic-Geometric Mean Inequality*

$$\sqrt{xy} \le \frac{x+y}{2} \quad (\text{for all } x, y > 0)$$

starting from $(\sqrt{x} - \sqrt{y})^2 \ge 0$.

b. Use the Arithmetic-Geometric Mean Inequality to prove, by induction, that $a_{n+1} \ge a_n \ge b_n \ge b_{n+1}$.

c. Deduce that $\{a_n\}$ and $\{b_n\}$ converge to a common limit. (This limit is denoted $AGM(a, b)$ and is called the Arithmetic-Geometric Mean of a and b.)

Calculator/Computer Exercises

51. Suppose that due to inflation the purchasing power of \$1.00 decreases at the constant annual rate of 4.5%. This means the value of the dollar decreases by the factor $e^{-0.045}$ every year.

a. Find the value $v(t)$ of \$1.00 in t years.

b. Graph v.

c. In how many years will the value of \$1.00 be \$0.50? \$0.10?

52. The *learning curve* is sometimes exponential in form. Suppose that after t weeks in a classroom, the class average (out of 100) on a standard aptitude test is

$$A(t) = 100(1 - e^{Kt}).$$

A particular class has an average score of 75 after 6 weeks. Graph the learning curve. What average score can we predict for the class after 12 weeks? When will the average score exceed 90?

53. Moore's law (after Gordon Moore, cofounder of Intel) asserts that the number of transistors that can be pressed onto a fixed area doubles every 18 months. Assuming that this observation remains valid, in how many months will the number of transistors that can be pressed onto a fixed area increase by a factor of 100?

54. Graph $y = \exp(x) - 2$. Zoom in to approximate, to five decimal places, the value of x at which the curve crosses the x-axis. Use your approximation to obtain equation (2.18) to five decimal places.

55. Formula (2.18) is obtained by finding the value γ such that $e^\gamma = 2$. Let $f(x) = e^x - 2$. Since $f(0) = -1 < 0$ and $f(1) = e - 2 > 0$, the Intermediate Value Theorem tells us that γ belongs to the interval $I_1 = [0, 1]$. If $m_1 = 1/2$ is the midpoint of I_1, then m_1 divides I_1 into two subintervals, one of which contains γ. Let I_2 be the subinterval that contains γ: It is the one for which $f(x)$ has opposite signs at the endpoints. Let m_2 be the midpoint of I_2 (m_2 divides I_2 into two subintervals). Let I_3 be the subinterval that contains γ. Continue this process. Calculate m_n for $1 \le n \le 10$. For each n, what is the length of the interval I_n? What is the maximum error that can result if m_n is used to approximate γ?

56. A life insurance company uses Figure 7 in their promotional literature. Explain how the company obtained the numbers \$78,350, \$61,390, \$48,100, and \$37,690 (which were rounded to the nearest \$5). Explain what an annual inflation rate of $r\%$ means in terms of the amount of money necessary at the end of the year to purchase what \$1 would have purchased at the beginning of the year.

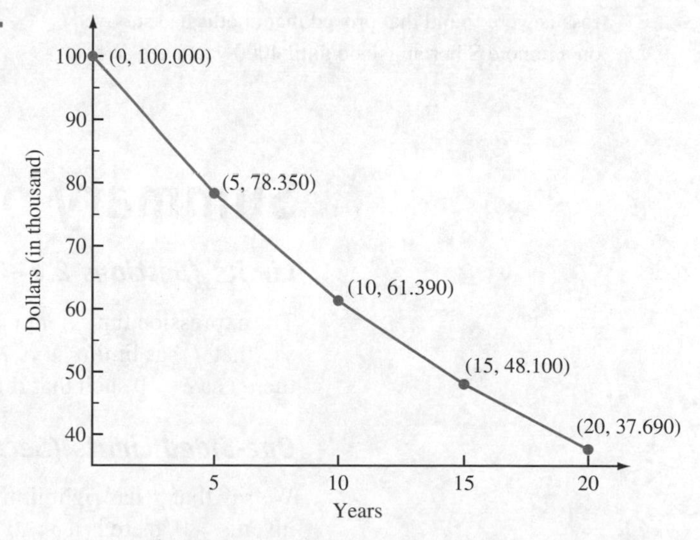

Figure 7

57. The amount of ^{226}Ra decays exponentially. The half-life of ^{226}Ra is 1620 years: In other words, the amount of

^{226}Ra is halved every 1620 years. An area is contaminated with a level that is five times greater than the maximum safe level. Without a cleanup, how long will this area remain unsafe?

58. On August 24, 1995, the treasury department reported that, at some instant of time that morning, the U.S. national debt stood at \$4,959,228,737,168.46. According to the same report, the national debt was increasing by \$9600 per second *at the time of the report*. In an article the next day, *USA Today* used this information to project that the national debt would be $\$6.71 \times 10^{12}$ five years later. Suppose the national debt is increasing at an annual rate of $r\%$ compounded continuously. Calculate r using the data supplied by the treasury. Using an exponential growth law, what will the debt be five years after the treasury report?

Radiocarbon Dating Two isotopes of carbon, ^{12}C and ^{14}C, are found in a known ratio in organic matter. After an organism dies, carbon is no longer metabolized. Due to radioactive decay, the amount $r(t)$ of ^{14}C decreases according to the law

$$r(t) = r(0) \cdot e^{-t1.212856\times10^{-4}}.$$

In radiocarbon dating, the amount of (stable) ^{12}C in a sample is measured and used to determine $r(0)$. The amount $r(t)$ of ^{14}C is also measured. The formula allows us to solve for t and to determine the time that has elapsed since death. Use this information for Exercises 59–64.

59. Until relatively recently, mammoths were thought to have become extinct 10,000 years ago. In the late 1980s, fossils were found that proved mammoths had survived on a remote Siberian island until 4000 years ago. What percentage of $r(0)$ is the amount of ^{14}C in these fossils?

60. In 1994, a parka-clad mummified body of a girl was found in a subterranean meat cellar near Barrow, Alaska. Radiocarbon analysis showed that the girl died around CE 1200. What percentage of $r(0)$ is the amount of ^{14}C in the mummy?

61. Prehistoric cave art has recently been found in a number of new locations. At Chauvet, France, the amount of ^{14}C in charcoal samples is $0.0879 \cdot r(0)$. About how old are the cave drawings?

62. In 1995, the Dead Sea Scroll text known as 4Q258 was radiocarbon dated. The result suggested a date of about CE 180. Paleographic evidence places the date at about 100 BCE. What fraction of r did the daters find? What fraction of $r(0)$ should it have been to validate the paleographic evidence?

63. The oldest North American mummy, the Spirit Cave Man, was discovered near Fallon, Nevada, in 1940. In the days before radiocarbon dating, Nevada State Museum anthropologists estimated the mummy to be about 2000 years old. In 1996, the mummy was removed from a sealed box to test a new method of counting carbon atoms in hair. To the surprise of experts, the radiocarbon dating placed the mummy's age at 9400 years. In what fraction of $r(0)$ did the test result? (The dating has been validated by applying standard radiocarbon procedures to textile samples that were buried with the mummy.)

64. The skeletal remains of a human ancestor, Lucy, are reported to be 3,180,000 years old. Explain why the radiocarbon dating of matter of this age would be futile.

Summary of Key Topics

Limits (Sections 2.1–2.2)

The expression $\lim_{x\to c} f(x) = \ell$ means that $f(x)$ approaches ℓ as x approaches c. We say that f has limit ℓ at c. More precisely, $\lim_{x\to c} f(x) = \ell$ means that given $\epsilon > 0$, there is a $\delta > 0$ such that if $0 < |x - c| < \delta$, then $|f(x) - \ell| < \epsilon$.

One-Sided Limits (Section 2.2)

We say that f has right limit ℓ at c if the domain of f contains the interval (c, b), and given $\epsilon > 0$, there is a $\delta > 0$ such that if $c < x < c + \delta$, then $|f(x) - \ell| < \epsilon$. The definition of left limit is similar. These limits are written $\lim_{x\to c^+} f(x)$ and $\lim_{x\to c^-} f(x)$, respectively. Left and right limits have properties similar to those of two-sided limits. A function has limit ℓ at c if and only if it has left limit ℓ at c and right limit ℓ at c.

Uniqueness of Limits (Section 2.2)

A function has at most one limit at a point: If $\lim_{x\to c} f(x) = \ell$ and $\lim_{x\to c} f(x) = m$, then $\ell = m$.

Algebraic Properties of Limits (Section 2.2)

a. $\lim_{x\to c}(f(x) + g(x)) = \lim_{x\to c} f(x) + \lim_{x\to c} g(x)$
b. $\lim_{x\to c}(f(x) \cdot g(x)) = (\lim_{x\to c} f(x)) \cdot (\lim_{x\to c} g(x))$
c. $\lim_{x\to c} f(x)/g(x) = \lim_{x\to c} f(x)/\lim_{x\to c} g(x)$, provided that $\lim_{x\to c} g(x) \neq 0$
d. $\lim_{x\to c}(\alpha \cdot f(x)) = \alpha \cdot \lim_{x\to c} f(x)$ for any constant $\alpha \in \mathbb{R}$

Nonexistence of Limits (Section 2.2)

If $\lim_{x\to c} f(x) \neq 0$ and $\lim_{x\to c} g(x) = 0$, then $\lim_{x\to c} f(x)/g(x)$ does not exist.

The Pinching Theorem (Section 2.2)

If $g(x) \leq f(x) \leq h(x)$ and $\lim_{x\to c} g(x) = \lim_{x\to c} h(x) = \ell$, then $\lim_{x\to c} f(x) = \ell$.

Two Important Trigonometric Limits (Section 2.2)

$$\lim_{x\to 0} \frac{\sin(x)}{x} = 1 \quad \text{and} \quad \lim_{x\to 0} \frac{1 - \cos(x)}{x^2} = \frac{1}{2}$$

Continuity (Section 2.3)

A function f is continuous at c if $\lim_{x\to c} f(x) = f(c)$.

One-Sided Continuity (Section 2.3)

A function f with a domain that contains $[c, b)$ is said to be right continuous at c if $\lim_{x\to c^+} f(x) = f(c)$. The definition of left continuous is similar. Left and right continuous functions have properties similar to those for continuous functions. A function is continuous at c if and only if it is both left and right continuous at c.

Properties of Continuous Functions (Section 2.3)

If the functions f and g are continuous at c, then so are $f + g$, $f - g$, and $f \cdot g$. Furthermore, f/g is continuous at c provided that $g(c) \neq 0$.

The Intermediate Value Theorem (Section 2.3)

If f is a continuous function with a domain that contains $[a, b]$, if $f(a) = \alpha$ and $f(b) = \beta$, and if γ is a real number between α and β, then there is a number c between α and β such that $f(c) = \gamma$.

The Extreme Value Theorem (Section 2.3)

If f is a continuous function with a domain that contains $[a, b]$, then there are numbers α and β in $[a, b]$ such that $f(\alpha) \leq f(x) \leq f(\beta)$ for all $x \in [a, b]$.

Infinite Limits and Limits at Infinity (Section 2.4)

The expression $\lim_{x\to c} f(x) = \infty$ means that $f(x)$ gets arbitrarily large as x approaches c. More precisely, given $K > 0$, there is a $\delta > 0$ such that if $0 < |x - c| < \delta$, then $f(x) > K$. The definitions of the one-sided infinite limits $\lim_{x\to c^+} f(x) = \infty$ and $\lim_{x\to c^-} f(x) = \infty$ are similar. Analogous definitions exist for $\lim_{x\to c} f(x) = -\infty$, $\lim_{x\to c^+} f(x) = -\infty$, and $\lim_{x\to c^-} f(x) = -\infty$. In any of these cases, we say that $x = c$ is a vertical asymptote of the graph of f.

The expression $\lim_{x\to\infty} f(x) = \ell$ means that $f(x)$ gets arbitrarily close to ℓ as x gets arbitrarily large through positive values. More precisely, given $\epsilon > 0$, there is a $K > 0$ such that if $x > K$, then $|f(x) - \ell| < \epsilon$. The definition of $\lim_{x\to-\infty} f(x) = \ell$ is analogous. In either case, we say that $y = \ell$ is a horizontal asymptote of the graph of f.

Limits of Sequences (Section 2.5)

The expression

$$\lim_{j\to\infty} a_j = \ell$$

means that a_j gets arbitrarily close to the real number ℓ as the positive integer variable j gets arbitrarily large. More precisely, given $\epsilon > 0$, there is a $J > 0$ such that if $j > J$, then $|a_j - \ell| < \epsilon$. We say that the sequence $\{a_j\}$ is convergent.

The algebraic properties of limits discussed in Section 2.2 are also valid for limits of sequences, as is the Pinching Theorem.

The Monotone Convergence Property (Section 2.6)

A sequence $\{a_j\}$ is said to be monotone if it is decreasing ($a_{j+1} \le a_j$ for every j) or if it is increasing ($a_{j+1} \ge a_j$ for every j). A sequence $\{a_j\}$ is said to be bounded if there is a K such that $|a_j| < K$ for every j. The monotone convergence property of the real number system states that every monotone bounded sequence is convergent.

Exponential Functions (Section 2.6)

If $a > 0$ and x is irrational, then we may define a^x by

$$a^x = \lim_{j\to\infty} a^{x_j}$$

where $\{x_j\}$ is a sequence of rational numbers that converges to x. All the usual rules of exponents remain valid:

$$a^{x+y} = a^x a^y, \quad a^{-x} = \frac{1}{a^x}, \quad \text{and} \quad (a^x)^y = a^{xy}.$$

The Number e (Section 2.6)

The number e is defined by

$$e = \lim_{n\to\infty} \left(1 + \frac{1}{n}\right)^n.$$

It is an irrational number with a decimal representation that begins with 2.7182818284590452. In general, for any real number u,

$$e^u = \lim_{x\to\infty} \left(1 + \frac{u}{x}\right)^x.$$

genesis & DEVELOPMENT

It has not always been the case that mathematical standards require concepts to be clearly defined and results to be correctly demonstrated. Even though the *limit* is basic to calculus, a rigorous treatment of limits was not developed until the 19th century. By that time, most of the topics that are studied in a modern calculus course had long been known. For 200 years, mathematicians relied on an intuitive understanding of limits, but it eventually became clear that a more precise viewpoint was necessary.

The Controversy over Infinitesimals

The fundamental concept of differential calculus is the limit

$$\ell = \lim_{\Delta x \to 0} \frac{f(c + \Delta x) - f(c)}{\Delta x}.$$

We have already used a limit of this form in our discussion of instantaneous velocity (see Section 2.1). We will consider such limits in much greater detail in Chapter 3. For a long time, the most common approach to such a limit was through the mechanism of *infinitesimals* (quantities that are infinitely small). The limit ℓ was envisioned as a quotient in which the numerator was the infinitesimal increment dy in the variable $y = f(x)$ caused by incrementing the value c of the variable x by the infinitesimal amount dx. For example, if $y = x^2$, then

$$dy = (c + dx)^2 - c^2 = 2c \cdot dx + (dx)^2$$

and

$$\ell = \frac{dy}{dx} = \frac{2c \cdot dx + (dx)^2}{dx} = 2c + dx.$$

Under the rules of infinitesimal algebra, a noninfinitesimal quantity remains unchanged when an infinitesimal is added to it. Thus, the limit ℓ is $2c$.

Although the infinitesimal method yields the correct answer, a logical dilemma is present: If $2c + dx = 2c$, then dx is really 0, in which case the quotient $\frac{dy}{dx}$ (called a *fluxion* in the terminology of Newton) is not algebraically permitted. On the other hand, if the quotient $\frac{dy}{dx}$ is permitted because $dx \neq 0$, then $2c + dx \neq 2c$. It is evident that Newton envisioned some sort of limiting process, although he never stated his method of reasoning clearly enough to resolve the logical dilemma.

In 1734, George Berkeley (1685–1753), Bishop of Cloyne, Ireland, published *The Analyst,* a tract addressed to an "infidel mathematician," generally presumed to be the astronomer Edmond Halley. In it, Berkeley attacked the hypocrisy of those who scorned religious belief but who fell back on mathematical dogma:

> *The Imagination, which faculty derives from sense, is very much strained and puzzled to frame clear ideas of the least particles of time. . . . The further the mind analyseth and pursueth these fugitive ideas the more it is lost and bewildered; the objects at first fleeting and minute, soon vanishing out of sight. . . . All these points, I say, are supposed and believed by certain rigorous exactors of evidence in religion, men who pretend to believe no further than they can see. That men who have been conversant only about clear points should with difficulty admit obscure ones might not seem altogether unaccountable. But he who can digest a second or third fluxion, a second or third difference, need not, methinks, be squeamish about any point in divinity. . . . Nothing is easier than to devise expressions or notations for fluxions and infinitesimals, . . . but if we . . . set ourselves attentively to consider the things themselves which are supposed to be expressed or marked thereby, we shall discover much emptiness, darkness, and confusion; nay if I mistake not, direct impossibilities and contradictions. . . . I have no controversy about your conclusions, but only about your logic and method: how you demonstrate? what objects you are conversant with, and whether you conceive them clearly? what principles you proceed upon; how sound they may be; and how you apply them?*

The criticisms of Berkeley initiated a flurry of interest in the foundations of calculus. There was also a second stimulus that pointed to the need for more precision. In 1747, Jean le Rond d'Alembert (1717–1783) published both the law governing the motion of a vibrating string and its general solution.

Although d'Alembert's equation did not come into question, his solution to it did. In 1748, Euler and, in 1753, Daniel Bernoulli (1700–1782) proposed different solutions. Mathematics was in the embarrassing situation of having three of its greatest proponents debate mathematical truths as they would points of metaphysics.

Bolzano and the Intermediate Value Theorem

By the end of the 18th century, mathematicians had started to look for ways to formulate the concepts and theorems of calculus without reference to *infinitesimals*. The approaches taken by Bernard Bolzano (1781–1848) and Augustin-Louis Cauchy (1789–1857) stand out. Bolzano gave the first "modern" definition of "continuity" in his 1817 paper on the Intermediate Value Theorem. According to Bolzano, f is *continuous at* x "if the difference $f(x+\omega) - f(x)$ can be made smaller than any given magnitude by taking ω as small as is wished." In the title of his paper, Bolzano stressed that he was giving an *analytic* proof of the Intermediate Value Theorem. The theorem was known previously, but it was always considered self-evident on *geometric* grounds. Bolzano realized that the theorem must depend on both the properties of the functions under consideration and on the completeness property of the real numbers. Here, for the first time, we find explicit awareness that a completeness property of the real number system requires proof. In his paper, Bolzano stated two equivalent completeness properties. The second of these is now known as the *least upper bound property* of the real number system.

Let S be a bounded set of real numbers. A number U is said to be the *least upper bound* of S if U is the smallest number such that $x \leq U$ for each x in S. For example, if S is the sequence $\{1 - 1/n : n \in \mathbb{Z}^+\}$, then 1 (and therefore any number greater than 1) is an upper bound for S. If r is a number smaller than 1, then an n can be found such that $r < 1 - 1/n < 1$. The existence of a member of S larger than r rules out r as an upper bound of S. It follows that 1 is the least upper bound of S.

The least upper bound property states that every nonempty set of real numbers that is bounded above has a least upper bound. The rational number system $\mathbb{Q}$ does not have this property. For example, if S is the set consisting of each rational number x with $x^2 < 2$, then every rational number larger than $\sqrt{2}$ is an upper bound but not a least upper bound. In the real number system, the same set S would have $\sqrt{2}$ as a least upper bound. The least upper bound property, the monotone convergence property, and the nested interval property (the latter was discussed in the Genesis & Development for Chapter 1) are equivalent properties of the real number system.

If f is continuous on $[a, b]$ and γ is between $f(a)$ and $f(b)$, then the Intermediate Value Theorem asserts that there is a value c in (a, b) such that $f(c) = \gamma$. Here is how Bolzano used the least upper bound property to prove the Intermediate Value Theorem: Assume without loss of generality that $f(a) < f(b)$. Let $S = \{x \in [a, b] : f(x) < \gamma\}$. Since a is in S, S is a nonempty set. Also, $S \subset [a, b]$, so b is an upper bound of S. Let U be the least upper bound of S. By continuity, $f(U) = \lim_{x \to U^-} f(x) \leq \gamma$. However, if $f(U)$ were strictly less than γ, then the right limit $\lim_{x \to U^+} f(x) = f(U) < \gamma$ shows that there are points x just to the right of U for which $f(x) < \gamma$. This cannot be, since U is an upper bound of S. Therefore, $f(U) = \gamma$.

Bolzano worked and published in relative obscurity; his ideas only received their due long after his death. Bolzano was ordained in 1804, the year he became professor of theology at the University of Prague, where he lectured until his dismissal for heresy in 1819. Thereafter, Bolzano was forced to submit to police supervision and his writing was curtailed.

Cauchy and Weierstrass

Credit, if not priority, for the movement toward rigor in calculus must be given to Augustin-Louis Cauchy, the most prominent mathematician in the greatest center of mathematics of his day. His *Cours d'Analyse* (1821) and subsequent Ecole Polytechnique lecture notes alerted a generation of mathematicians to the possibility of attaining a rigor in calculus that was comparable to that of classical geometry. Although the verbal definition of "limit" that Cauchy gave in his *Cours* was scarcely better than that given by his predecessors, and certainly worse than that given by Bolzano, it is clear from his translation of his verbal definition into mathematical inequalities that he meant exactly the definition of limit as given in Section 2.2. Indeed, it was Cauchy who introduced the symbols ϵ and δ (with ϵ standing for "*erreur*" and δ for "*différence*"). Many of the theorems of calculus received their first modern formulations in Cauchy's work.

A completely correct presentation of the foundations of calculus was finally given by Karl Weierstrass. We have already mentioned Weierstrass in the Genesis & Development for Chapter 1 in connection with his construction of the real number system. Weierstrass had an unusual career as a research mathematician. Born in the village of Ostenfelde, Weierstrass entered the University of Bonn in 1834 as a law student. He did not begin a serious study of mathematics, however, until 1839, when he was 24. From 1840 until 1856, Weierstrass taught science, mathematics, and physical fitness at a high school. He became a professor at the University of Berlin in 1864, at the relatively advanced age of 49. It was in his lectures at the University of Berlin that Weierstrass clarified the fundamental structures of mathematical analysis. Of primary importance in this regard was Weierstrass's careful delineation of different types of convergence. In the words of David Hilbert (1862–1943),

> *It is essentially a merit of the scientific activity of Weierstrass that mathematicians are now in complete agreement and certainty concerning types of analytic reasoning based on the concepts of irrational number and, more generally, limit.*

At one time, mathematical theory required successive revision as mathematicians delved deeper into existing theory. The formulation of a precise definition of limit has given a permanent character to the foundation of calculus.

3

The Derivative

If f is a function of a variable x, an increment Δx in x will usually cause the value of f to change. The numerator of the quotient

$$\frac{f(x + \Delta x) - f(x)}{\Delta x}$$

represents the change of f as x varies from x to $x + \Delta x$. The denominator Δx represents the change in x. Therefore, the quotient itself represents the *average* rate of change of f as its argument varies from x to $x + \Delta x$. We use the limit

$$\lim_{\Delta x \to 0} \frac{f(x + \Delta x) - f(x)}{\Delta x} \tag{3.1}$$

to define the *instantaneous rate of change* of f at x. This chapter shows that this limit also appears as the slope of the tangent line to the graph of f at the point $(x, f(x))$.

The tangent line problem and the notion of instantaneous rate of change led mathematicians to create and develop a branch of mathematics called *differential calculus*. In this chapter, we learn about differential calculus: We name the limit in line (3.1) a *derivative,* study it in its own right, and learn many of its useful applications. After considering the tangent line problem and the notion of instantaneous rate of change in further detail, we then learn methods of computing derivatives for many common functions. These methods allow us to evaluate the derivatives of these functions by following rules that avoid reference to the definition of "limit."

3.1 Rates of Change and Tangent Lines

Imagine automobile A driving down a road, as shown in Figure 1. We are standing at the side of the road, and we wish to determine the velocity of car A at any given moment. Here is how we proceed: We record the position $p(t)$ of car A as time elapses. Time $t = 0$ corresponds to the time that our observations begin (see Table 1).

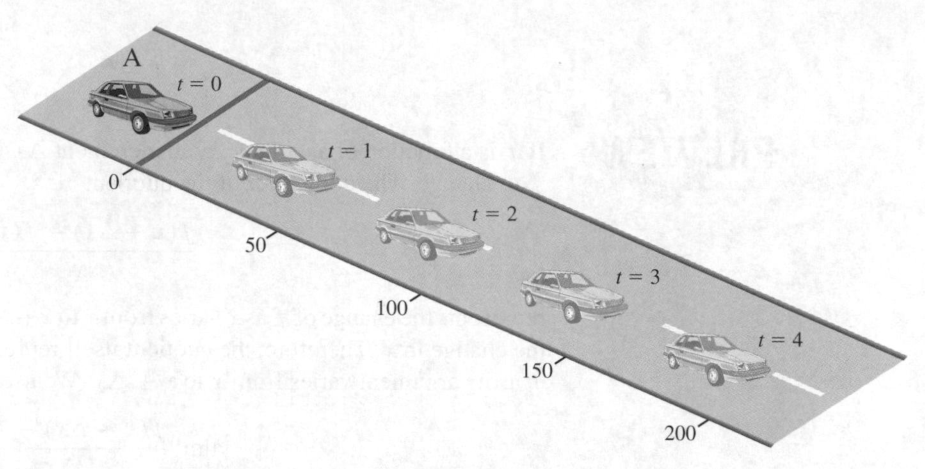

Figure 1

Time (seconds)	0	1	2	3	4
Distance (feet)	0	50	100	150	200

Table 1

Analyzing the speed of car A is simple, because it is traveling at a constant rate: 50 feet in 1 second, 100 feet in 2 seconds, 150 feet in 3 seconds, 200 feet in 4 seconds, and so on. In all cases,

$$\text{rate} = \frac{\text{distance traveled}}{\text{time elapsed}} = 50 \text{ ft/s}.$$

Suppose that a second car B travels down the road, as shown in Figure 2, with position $p(t) = 8t^2$ feet where t is measured in seconds. Some values of $p(t)$ are recorded in Table 2.

With each passing second, car B covers more ground than in the previous second. For example, car B moves $8 - 0 = 8$ feet in the first second, $32 - 8 = 24$ feet in the second second, $72 - 32 = 40$ feet in the third second, and $128 - 72 = 56$ feet in the fourth second. In other words, car B is *accelerating*. Since the velocity of car B is ever changing, how can we calculate the velocity of car B at a given instant of time, say $t = 1$?

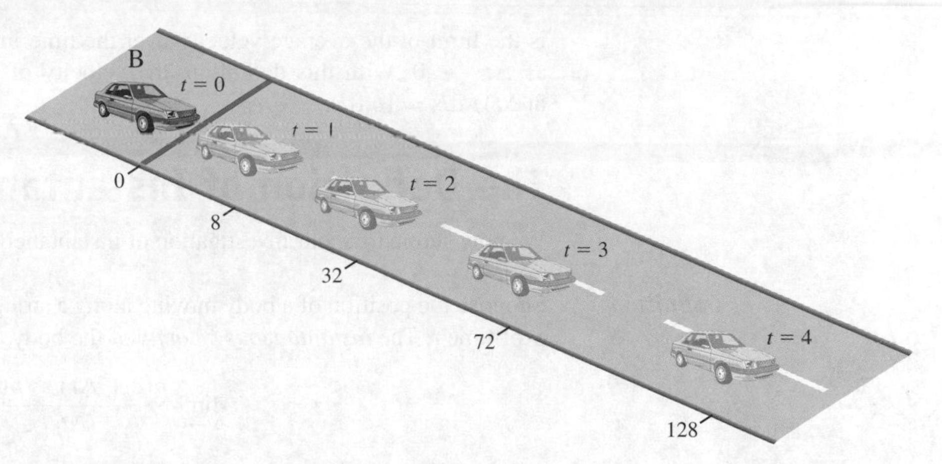

Figure 2

Time (seconds)	0	1	2	3	4
Distance (feet)	0	8	32	72	128

Table 2

To answer this question, we begin by calculating the *average velocity* of car B between times $t = 1$ and $t = 2$:

$$\text{rate} = \frac{\text{distance traveled}}{\text{time elapsed}} = \frac{(32 - 8)\text{ ft}}{(2 - 1)\text{ s}} = 24\text{ ft/s}.$$

Although this calculation gives a rough approximation to the velocity at time $t = 1$, it does not tell us the *exact* velocity at the instant $t = 1$. However, it does suggest an idea: Instead of calculating average velocity over a long time interval (of length 1 second), let us calculate average velocity over a short time interval. This should give a better approximation to the velocity at the instant $t = 1$. We use the symbol Δt to denote the elapsed time. Over the time interval from $t = 1$ to $t = 1 + \Delta t$, car B travels from $p(1)$ to $p(1 + \Delta t)$. The average velocity, in feet per second, is

$$\begin{aligned}\text{rate} &= \frac{\text{distance traveled}}{\text{elapsed time}} \\ &= \frac{p(1 + \Delta t) - p(1)}{\Delta t} \\ &= \frac{8(1 + \Delta t)^2 - 8 \cdot 1^2}{\Delta t} \\ &= \frac{8 + 16 \cdot \Delta t + 8(\Delta t)^2 - 8}{\Delta t} \\ &= \frac{16 \cdot \Delta t + 8(\Delta t)^2}{\Delta t} \\ &= 16 + 8(\Delta t).\end{aligned}$$

The average velocity over the time interval from $t = 1$ to $t = 1 + \Delta t$ is equal to $16 + 8(\Delta t)$ ft/s. If we let Δt tend to 0, then the average velocity, $16 + 8(\Delta t)$ ft/s, has limit 16 ft/s. It is therefore reasonable to say that the velocity of car B *at the instant* $t = 1$

is the limit of the average velocity over the time interval from $t = 1$ to $t = 1 + \Delta t$ as $\Delta t \to 0$. With this definition, the velocity of car B at $t = 1$ is $\lim_{\Delta t \to 0}(16 + 8(\Delta t))$ ft/s $= 16$ ft/s.

The Definition of Instantaneous Velocity

We now summarize our investigation of instantaneous velocity with a definition.

Definition Suppose the position of a body moving along a straight path is described by a function p of time t. The *instantaneous velocity* of the body at a time c is given by

$$\lim_{\Delta t \to 0} \frac{p(c + \Delta t) - p(c)}{\Delta t}, \tag{3.2}$$

provided that this limit exists and is finite.

Our first two examples show that limit formula (3.2) agrees with our intuitive understanding of velocity in some simple situations.

Example 1 Suppose the position of a third car C is given by $p(t) = \alpha$ ft for all values of t. Since $p(t)$ is constant, car C is not moving. Common sense tells us that the car's velocity is 0 for every value of time. Verify that formula (3.2) also tells us that the velocity is 0.

Solution Let c denote any instant of time. We calculate:

$$\lim_{\Delta t \to 0} \frac{p(c + \Delta t) - p(t)}{\Delta t} = \lim_{\Delta t \to 0} \frac{\alpha - \alpha}{\Delta t} = \lim_{\Delta t \to 0} \frac{0}{\Delta t} = \lim_{\Delta t \to 0} 0 = 0. \quad \blacksquare$$

Example 2 Suppose the position of car A is given in feet by $50t$ where t is measured in seconds (as shown in Table 1). Verify that the instantaneous velocity of car A is 50 ft/s for every value of t.

Solution Since $p(t) = 50t$, the instantaneous velocity in feet per second at time $t = c$ is

$$\lim_{\Delta t \to 0} \frac{p(c + \Delta t) - p(c)}{\Delta t} = \lim_{\Delta t \to 0} \frac{50(c + \Delta t) - 50c}{\Delta t} = \lim_{\Delta t \to 0} \frac{50\Delta t}{\Delta t} = \lim_{\Delta t \to 0} 50 = 50. \quad \blacksquare$$

inSIGHT

In Example 2, we can replace $p(t) = 50t$ with $p(t) = \alpha t$ where α is an arbitrary constant. By doing so, we see that when position is given by $p(t) = \alpha t$, velocity is equal to α for every value of t.

Our next two examples call for velocity calculations that require the precise definition of instantaneous velocity, as given by formula (3.2).

Example 3 If the position of a car B is given by $p(t) = 8t^2$, then what is its velocity at time $t = c$? At $t = 2$? At $t = 3$?

Solution The instantaneous velocity of car B at time $t = c$ is, by formula (3.2),

$$\begin{aligned} \lim_{\Delta t \to 0} \frac{p(c + \Delta t) - p(c)}{\Delta t} &= \lim_{\Delta t \to 0} \frac{8(c + \Delta t)^2 - 8c^2}{\Delta t} \\ &= \lim_{\Delta t \to 0} \frac{(8c^2 + 16c\Delta t + 8(\Delta t)^2) - 8c^2}{\Delta t} \\ &= \lim_{\Delta t \to 0} (16c + 8\Delta t) = 16c. \end{aligned}$$

This computation repeats, for a general instant c of time, the calculation that we performed at the beginning of this section for $c = 1$. Of course, if we set $c = 1$ in the velocity formula $16c$, we get 16, which is the same value we obtained in the earlier discussion. By setting $c = 2$, we see that the velocity of car B is $16 \cdot 2$, or 32, when $t = 2$. Without any further limit computation, we also see that when $t = 3$, the velocity of car B is $16 \cdot 3$, or 48. ■

In Example 3, we can replace $p(t) = 8t^2$ with $p(t) = \alpha t^2$ where α is an arbitrary constant. By doing so, we see that when position is given by $p(t) = \alpha t^2$, velocity is equal to $2\alpha c$ for every value $t = c$.

Example 4 Relative to a specified horizontal coordinate system, the position of a point Q on a piston is given by

$$p(t) = 18 - 2t^2 + 4t,$$

in inches, for $0 \le t \le 4$ seconds (see Figure 3). What is the instantaneous velocity of the piston at time $t = 3$?

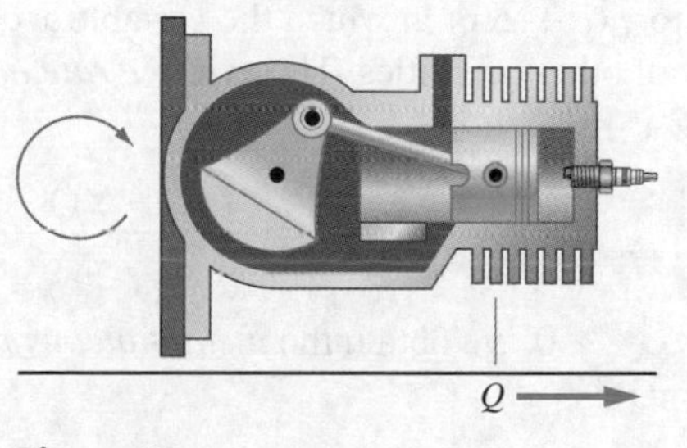

Figure 3a
Positive velocity

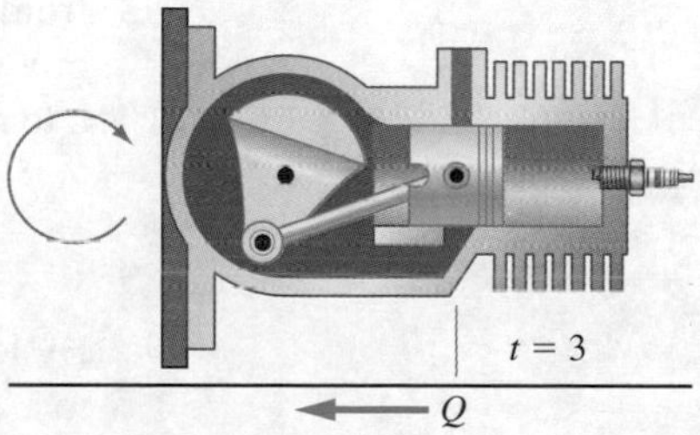

$t = 3$
Q

Figure 3b
Negative velocity

Solution We calculate $p(3) = 18 - 2 \cdot 3^2 + 4 \cdot 3 = 12$ and

$$\begin{aligned} p(3 + \Delta t) &= 18 - 2(3 + \Delta t)^2 + 4(3 + \Delta t) \\ &= 18 - 2(9 + 6\Delta t + (\Delta t)^2) + 12 + 4\Delta t = 12 - 8\Delta t - 2(\Delta t)^2. \end{aligned}$$

It follows that $p(3 + \Delta t) - p(3) = -8\Delta t - 2(\Delta t)^2$. According to the definition, the instantaneous velocity at time $t = 3$ is

$$\lim_{\Delta t \to 0} \frac{p(3 + \Delta t) - p(3)}{\Delta t} = \lim_{\Delta t \to 0} \frac{-8\Delta t - 2(\Delta t)^2}{\Delta t} = \lim_{\Delta t \to 0} (-8 - 2\Delta t) = -8 \text{ in./s.}$$

■

We can understand the meaning of an object's negative velocity by looking at the *definition* of velocity. If $\Delta t > 0$, then the average velocity

$$\frac{p(c + \Delta t) - p(c)}{\Delta t}$$

at time $t = c$ will be positive if $(p(c + \Delta t) - p(c)) > 0$ (the object is moving forward) and negative if $(p(c + \Delta t) - p(c)) < 0$ (the object is moving backward). Passing to the limit as $\Delta t \to 0$, we see that positive instantaneous velocity corresponds to forward motion, whereas negative instantaneous velocity corresponds to backward motion. In Example 4, the piston is moving to the left (in the negative direction in the coordinate system) at the instant $t = 3$ (see Figure 3b). In everyday speech, we do not distinguish between speed and velocity. In mathematics and physics, however, *speed* refers to the absolute value of *velocity*. Thus, speed is always nonnegative.

Instantaneous Rate of Change

We have just defined instantaneous velocity as a limit of average velocities. The mathematical construction behind this definition arises in many different contexts. Let f be a function. A change in x from c to $c + \Delta x$ will produce a change in the value of f from $f(c)$ to $f(c + \Delta x)$ in which the variable x could represent a time measurement or a variety of other quantities. The *average rate of change* of $f(x)$ as x changes from c to $c + \Delta x$ is

$$\frac{\Delta f}{\Delta x} = \frac{f(c + \Delta x) - f(c)}{\Delta x}.$$

By letting $\Delta x \to 0$, we obtain the *instantaneous rate of change* of $f(x)$ with respect to x at $x = c$.

Definition The *instantaneous rate of change* of $f(x)$ at $x = c$ is

$$\lim_{\Delta x \to 0} \frac{f(c + \Delta x) - f(c)}{\Delta x}, \qquad \textbf{(3.3)}$$

provided this limit exists and is finite.

Notice that the limits in formulas (3.2) and (3.3) are of the same type. This is because instantaneous velocity is a particular example of the instantaneous rate of change of a function. In formula (3.2), the function p represents position, and the variable t represents time. By contrast, in formula (3.3), f is a general function of a general variable x. In economics, $f(x)$ might be the cost of producing x units of a commodity. The instantaneous rate of change of f would then be called the *marginal cost* of producing that item. In pharmacology, f might refer to the measurable effect of a drug on a patient. The instantaneous rate of change of f with respect to time would be called the *sensitivity* of the patient to the drug.

Example 5 The volume V of a sphere of radius r is given by

$$V = \frac{4}{3}\pi r^3 \quad (r > 0).$$

What is the instantaneous rate of change of V with respect to r?

Solution Let $\alpha = 4\pi/3$. We are required to calculate

$$\lim_{\Delta r \to 0} \frac{\Delta V}{\Delta r} = \lim_{\Delta r \to 0} \frac{\alpha(r + \Delta r)^3 - \alpha r^3}{\Delta r} = \alpha \lim_{\Delta r \to 0} \frac{(r^3 + 3r^2 \cdot \Delta r + 3r(\Delta r)^2 + (\Delta r)^3) - r^3}{\Delta r}.$$

Since the numerator simplifies to $3r^2 \cdot \Delta r + 3r(\Delta r)^2 + (\Delta r)^3$, it follows that

$$\lim_{\Delta r \to 0} \frac{\Delta V}{\Delta r} = \lim_{\Delta r \to 0} \alpha \frac{(\Delta r)(3r^2 + 3r(\Delta r) + (\Delta r)^2)}{\Delta r} = \alpha \lim_{\Delta r \to 0} (3r^2 + 3r(\Delta r) + (\Delta r)^2) = \alpha \cdot 3r^2.$$

Because $\alpha = 4\pi/3$, the instantaneous rate of change is $4\pi r^2$. ■

A LOOK BACK

A careful examination of Examples 1, 2, 3, and 5 shows that there is a pattern to the instantaneous rates of change of the functions $f(x) = \alpha x^n$ for $n = 0, 1, 2,$ and 3. Example 1 shows that the rate of change of $f(x) = \alpha x^0$ is 0, which can be written as $0 \cdot \alpha x^{0-1}$. Example 2 shows that the rate of change of $f(x) = \alpha x^1$ is α, which can be written as $1 \cdot \alpha x^{1-1}$. Example 3 shows that the rate of change of $f(x) = \alpha x^2$ is $2\alpha x$, which can be written as $2 \cdot \alpha x^{2-1}$. Example 5 shows that the rate of change of $f(x) = \alpha x^3$ is $3\alpha x^2$, which can be written as $3 \cdot \alpha x^{3-1}$. In summary:

$$\text{If } f(x) = \alpha x^n \quad (n = 0, 1, 2, 3), \quad \text{then} \quad \lim_{\Delta x \to 0} \frac{f(c + \Delta x) - f(c)}{\Delta x} = n\alpha c^{n-1}. \qquad \textbf{(3.4)}$$

Sums of Functions

The following theorem tells us that we can calculate the instantaneous rate of change of a sum by adding the instantaneous rates of change of its summands. It also tells us that if α is a constant, then the instantaneous rate of change of $\alpha f(x)$ is α times the instantaneous rate of change of $f(x)$.

Theorem 1

The Sum Rule Suppose α and β are constants. If r and s are the instantaneous rates of change of $f(x)$ and $g(x)$ at a point $x = c$, then $\alpha \cdot r + \beta \cdot s$ is the instantaneous rate of change of $\alpha \cdot f(x) + \beta \cdot g(x)$ at $x = c$.

Theorem 1 is an immediate consequence of Theorem 2 from Section 2.2.

Example 6 What is the instantaneous rate of change of $F(x) = 7x^3 - 5x^2$ at $x = -1$?

Solution If we let $f(x) = x^3$, $g(x) = x^2$, $\alpha = 7$, and $\beta = -5$, then $F = \alpha \cdot f + \beta \cdot g$. By the limit formula in line (3.4), the instantaneous rate of change of $f(x)$ at $x = -1$ is $3 \cdot (-1)^2$, and the instantaneous rate of change of $g(x)$ at $x = -1$ is $2 \cdot (-1)$.

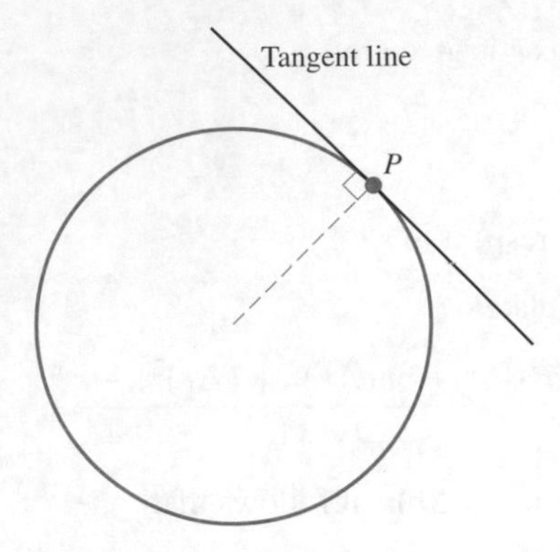

Figure 4

According to Theorem 1, the instantaneous rate of change of $F(x)$ at $x = -1$ is $7 \cdot 3 \cdot (-1)^2 + (-5) \cdot 2 \cdot (-1) = 31$. ■

The Concept of Tangent Line

It is easy to understand the idea of the tangent line to a circle. Here are three good ways to think about the tangent line to the circle C at the point P (illustrated in Figure 4).

1. The tangent line is perpendicular to the radius at P and passes through P.
2. The tangent line passes through P and intersects C at only that one point.
3. The tangent line passes through P, and the circle lies on one side of the tangent line.

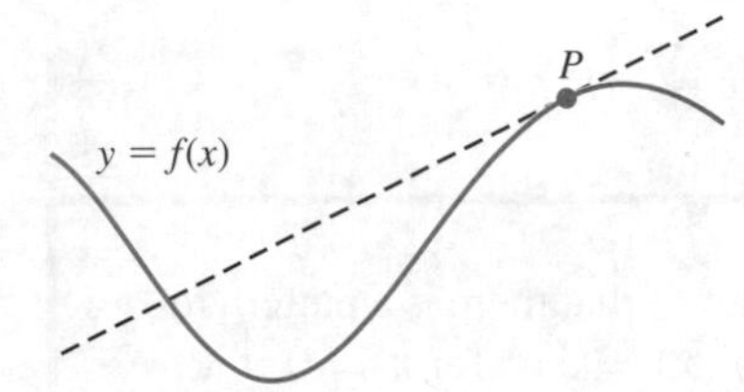

Figure 5

All of these descriptions are correct, and all of them uniquely determine the tangent line to the circle C at the point P. However, each is useless in determining the tangent line to the curve in Figure 5 at the point P. Intuition tells us that the dotted line in the figure is the tangent line, even though this line *in no sense* satisfies statements 1, 2, or 3. Therefore, we need a new mathematical description of tangent line.

We use a limiting process to determine the tangent line to the graph of $y = f(x)$ at a point $P = (c, f(c))$. Consider Figure 5. The dotted line indicates what our intuition tells us the tangent line is—the dotted line through P that most nearly approximates the curve. The trouble is that it takes *two* pieces of information to determine a line, and we only know one—the point P through which the line passes. We get a second piece of information—namely, the slope—by considering nearby *secant lines,* such as the line containing segment $\overline{P\tilde{P}}$ (see Figure 6).

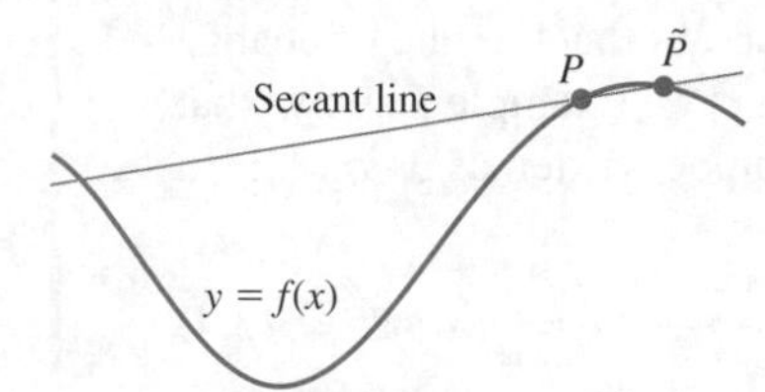

Figure 6

A secant line is a line passing through two points of the curve—in this case $P = (c, f(c))$ and $\tilde{P} = (c + \Delta x, f(c + \Delta x))$. The slope of this secant line is

$$\frac{f(c + \Delta x) - f(c)}{(c + \Delta x) - c} = \frac{f(c + \Delta x) - f(c)}{\Delta x}.$$

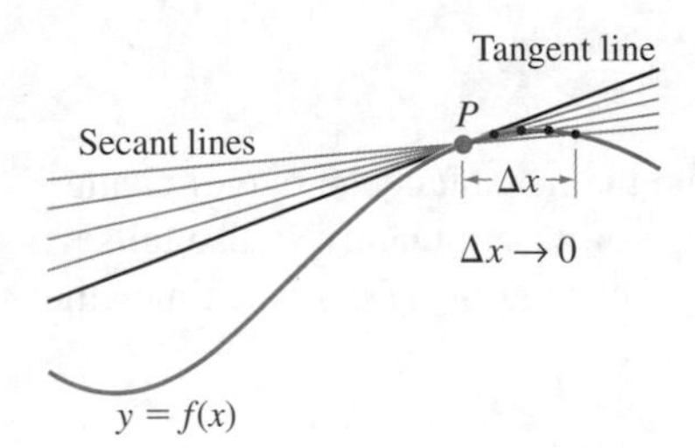

Figure 7

What we see geometrically is that as $\Delta x \to 0$ (that is, as $\tilde{P} \to P$), the slope of the secant line tends to the slope of the tangent line at P. In other words,

$$\begin{pmatrix} \text{slope of} \\ \text{tangent line} \\ \text{at } P = (c, f(c)) \end{pmatrix} = \lim_{\Delta x \to 0} \frac{f(c + \Delta x) - f(c)}{\Delta x}$$

(see Figure 7). These considerations lead to the definition of the tangent line to the graph of a function at a point P.

Definition

Let f be defined on an open interval containing the point c. Suppose that the limit

$$m = \lim_{\Delta x \to 0} \frac{f(c + \Delta x) - f(c)}{\Delta x} \tag{3.5}$$

exists and is finite. The *tangent line* to the graph of f at $P = (c, f(c))$ is the line with the equation

$$y = m(x - c) + f(c) \tag{3.6}$$

where m is given by limit formula (3.5).

inSIGHT

> Compare limit (3.3) with the right side of equation (3.5). The expressions are identical. In other words, the *slope of the tangent line* to the graph of f at $(c, f(c))$ is exactly the same as the *instantaneous rate of change* of $f(x)$ with respect to x at $x = c$. This may seem surprising at first—especially when $f(x)$ has some other physical meaning. In Example 5, for instance, we calculated the instantaneous rate of change of a function that represents volume. Whatever the physical interpretation of $f(x)$ may be, when we graph the equation $y = f(x)$, the value $f(x)$ takes on a second meaning: It is the height of the point $(x, f(x))$ above or below the x-axis. The instantaneous rate of change of height with respect to length is the slope. Therefore,
>
> $$\lim_{\Delta x \to 0} \frac{f(c + \Delta x) - f(c)}{\Delta x}$$
>
> represents both (1) the instantaneous rate of change of f with respect to x when $x = c$ (a rate of change of volume) *and* (2) the slope of the tangent line to the graph of f at the point $(c, f(c))$ (a rate of change of height).

Finding the Tangent Line to a Curve

The following two examples require the point-slope equation of a line. Review Section 1.3, if needed.

Example 7 Find the equation of the tangent line to the graph of $y = x^2$ at the point $P = (3, 9)$.

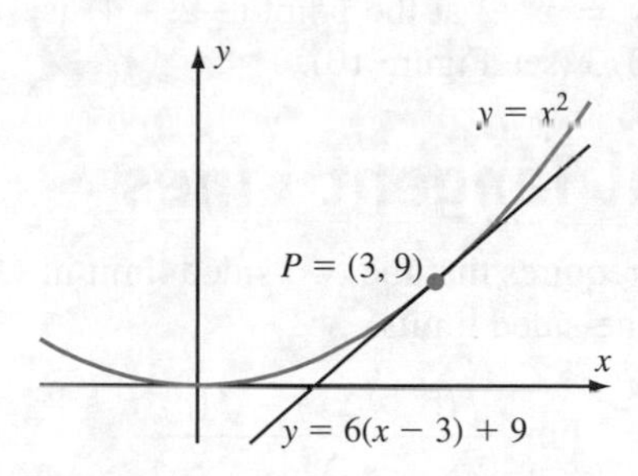

Figure 8

Solution Let $f(x) = x^2$. The tangent line will have slope

$$m = \lim_{\Delta x \to 0} \frac{f(3 + \Delta x) - f(3)}{\Delta x}.$$

To evaluate this limit, we can use line (3.4) with $\alpha = 1$, $n = 2$, and $c = 3$. We obtain $m = 2 \cdot 1 \cdot 3^1 = 6$. This limit calculation tells us that the slope of the required tangent line is 6. According to formula (3.6), the equation of this tangent line is $y = 6(x - 3) + 9$. Figure 8 shows the graph of f and its tangent line. ■

Example 8 Find the tangent line to the curve $y = 1/x$ at the point $P = (-3, -1/3)$.

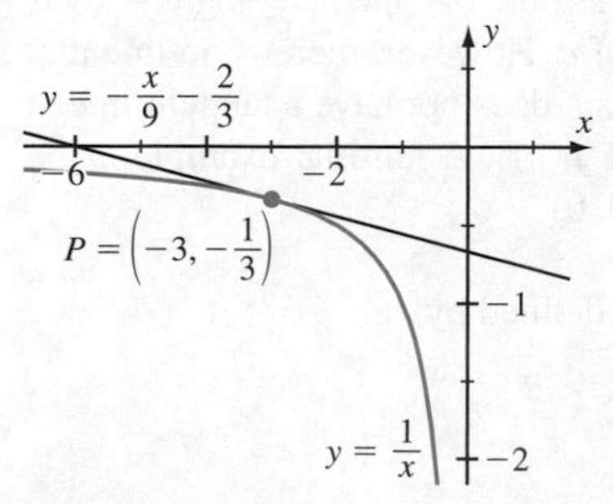

Figure 9

Solution Let $f(x) = 1/x$. Let c be any nonzero number. We calculate that

$$\lim_{\Delta x \to 0} \frac{f(c + \Delta x) - f(c)}{\Delta x} = \lim_{\Delta x \to 0} \frac{1}{\Delta x}\left(\frac{1}{c + \Delta x} - \frac{1}{c}\right)$$

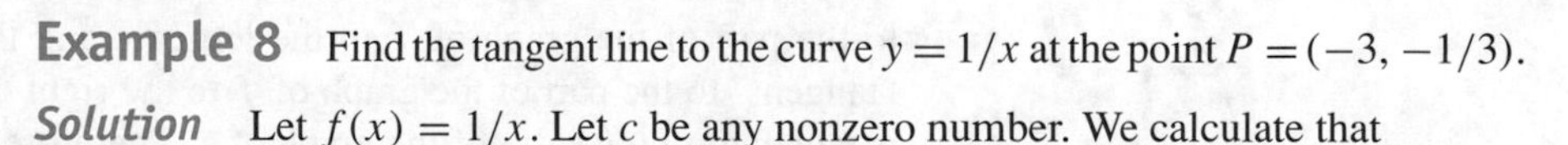

$$= \lim_{\Delta x \to 0} \frac{1}{\Delta x}\left(\frac{-\Delta x}{c(c + \Delta x)}\right) = -\lim_{\Delta x \to 0} \frac{1}{c(c + \Delta x)} = -\frac{1}{c^2}.$$

By setting $c = -3$, we see that the slope of the tangent line at P is $-1/(-3)^2$, or $-1/9$. Since the tangent line passes through P and has slope $-1/9$, the line has equation $y = (-1/9)(x - (-3)) + (-1/3)$, or $y = -x/9 - 2/3$. Figure 9 illustrates the graph of $y = 1/x$ and its tangent line at P. ■

inSIGHT

> Example 8 tells us that the instantaneous rate of change of $1/x$ is $-1/x^2$. Using this fact, together with Theorem 1, we deduce that for any constant α, the instantaneous rate of change of α/x is $-\alpha/x^2$. We may state this result as follows: For $n = -1$, the instantaneous rate of change of αx^n is $n\alpha c^{n-1}$. In other words, formula (3.4) is valid for $n = -1$ (in addition to the values $n = 0, 1, 2, 3$ that have already been investigated).

Normal Lines to Curves

A line L is said to be *perpendicular,* or *normal,* to a curve $y = f(x)$ at a point $P = (c, f(c))$ if L is perpendicular to the tangent line to the curve at the point P. If the tangent line is the horizontal line $y = f(c)$, then the normal line is the vertical line $x = c$. Otherwise the slope of the tangent line is nonzero. Its negative reciprocal is therefore the slope of the normal line (by Theorem 1 from Section 1.3).

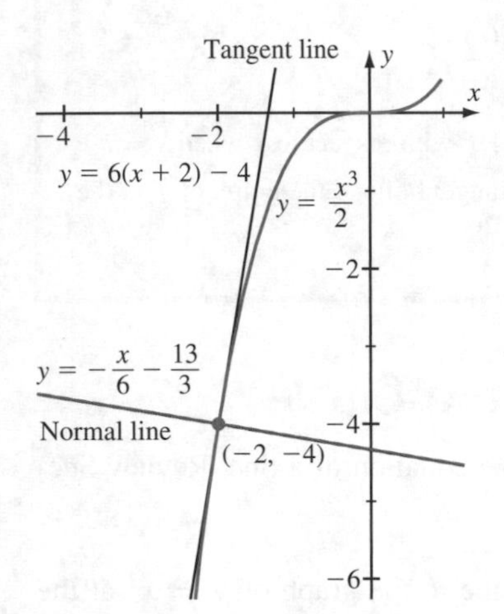

Figure 10

Example 9 Find the line perpendicular to the curve $y = x^3/2$ at the point $(-2, -4)$.

Solution Let $f(x) = x^3/2$ and $c = -2$. Using formula (3.4) with $\alpha = 1/2$ and $n = 3$, we have

$$\lim_{\Delta x \to 0} \frac{f(c + \Delta x) - f(c)}{\Delta x} = \frac{3}{2}c^2.$$

Therefore, the slope of the tangent line to the graph at point $(-2, -4)$ is $3 \cdot (-2)^2/2$, or 6. The slope of the normal line will be the negative reciprocal of this number, or $-1/6$. Therefore, the normal line to the graph of $y = x^3/2$ at the point $(-2, -4)$ is $y = (-1/6)(x - (-2)) + (-4)$, or $y = -x/6 - 13/3$ (see Figure 10). ■

Corners, Cusps, and Vertical Tangent Lines

The definition of the tangent line of f at $(c, f(c))$ requires that the two-sided limit in formula (3.5) exists. This limit fails to exist if the one-sided limits

$$\ell_L = \lim_{\Delta x \to 0^-} \frac{f(c + \Delta x) - f(c)}{\Delta x} \quad \text{and} \quad \ell_R = \lim_{\Delta x \to 0^+} \frac{f(c + \Delta x) - f(c)}{\Delta x}$$

exist and are finite but unequal. In this case, the line $y - f(c) = \ell_L(x - c)$ is "tangent" to the part of the graph of f to the left of c and the line $y - f(c) = \ell_R(x - c)$ is "tangent" to the part of the graph of f to the right of c. However, there is no line that is tangent to *both* sides of the graph of f. Therefore, f does not have a tangent line at $(c, f(c))$. Instead, we say that f has a *corner* at $(c, f(c))$. A familiar example is the V-shaped graph of $y = |x|$, which has a corner at $(0, 0)$.

Example 10 Does the graph of the function f defined by

$$f(x) = \begin{cases} x^2 & \text{if } x < 2 \\ 5 - \dfrac{x^2}{4} & \text{if } x \geq 2 \end{cases}$$

have a tangent line at the point $(2, 4)$?

Solution The left and right limits of $f(x)$ at $x = 2$ exist and are equal to 4. Because these limits exist and agree, f is continuous at 2. For $c = 2$, we have

$$\begin{aligned}\lim_{\Delta x \to 0^-} \frac{f(c+\Delta x) - f(c)}{\Delta x} &= \lim_{\Delta x \to 0^-} \frac{((2+\Delta x)^2 - 4)}{\Delta x} \\ &= \lim_{\Delta x \to 0^-} \frac{(\Delta x)^2 + 4\Delta x}{\Delta x} \\ &= \lim_{\Delta x \to 0^-} (4 + \Delta x) \\ &= 4.\end{aligned}$$

On the other hand,

$$\begin{aligned}\lim_{\Delta x \to 0^+} \frac{f(c+\Delta x) - f(c)}{\Delta x} &= \lim_{\Delta x \to 0^+} \frac{(5 - (1/4)(2+\Delta x)^2) - (5 - (1/4)\cdot 2^2)}{\Delta x} \\ &= \lim_{\Delta x \to 0^+} \frac{-\Delta x - (1/4)(\Delta x)^2}{\Delta x} \\ &= \lim_{\Delta x \to 0^+} \left(-1 - \frac{1}{4}\Delta x\right) \\ &= -1.\end{aligned}$$

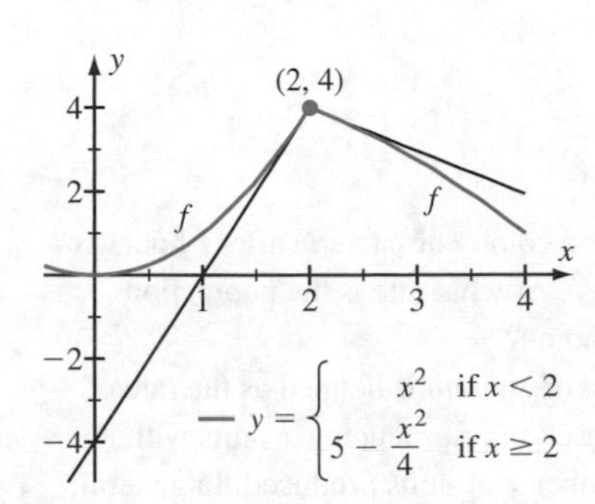

Figure 11

Since the one-sided limits exist but are unequal, the graph of f has a corner at point (2, 4) (see Figure 11). The half lines that are "one-sided" tangents are included in Figure 11. The graph does *not* have a tangent line at point (2, 4). ■

The final example in this section concerns two functions for which formula (3.5) does not evaluate to a finite limit.

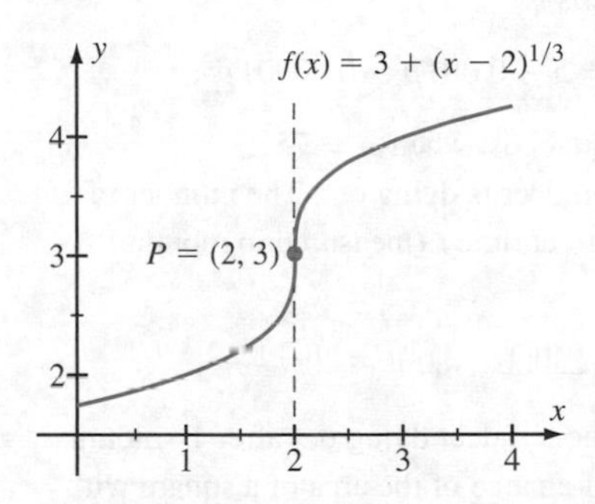

Figure 12

Example 11

Discuss the tangent line of the graph of $f(x) = 3 + (x-2)^{1/3}$ at the point $P = (2, 3)$. Does the graph of $g(x) = 3 + |x-2|^{1/3}$ have a tangent line at P?

Solution Since

$$\frac{f(2+\Delta x) - f(2)}{\Delta x} = \frac{(3 + (\Delta x)^{1/3}) - 3}{\Delta x} = \frac{1}{(\Delta x)^{2/3}},$$

we may use Theorem 5 from Section 2.2 to conclude that

$$\lim_{\Delta x \to 0} \frac{f(2+\Delta x) - f(2)}{\Delta x} = +\infty.$$

Thus, our definition of tangent line does not apply. Nevertheless, the graph of f indicates the presence of a vertical tangent line, $x = 2$, at P (see Figure 12). It is possible to develop the concept of the tangent line to include vertical tangent lines, but we will not do so because we do not need the greater generality.

Figure 13 shows the graph of $g(x)$. The presence of a tangent line at a point of a graph should be thought of as a *measure of smoothness* of the graph. The graph of g is certainly not smooth at P. In fact, the quotient

$$\frac{g(2+\Delta x) - g(2)}{\Delta x} = \frac{(3 + |\Delta x|^{1/3}) - 3}{\Delta x} = \frac{|\Delta x|^{1/3}}{\Delta x}$$

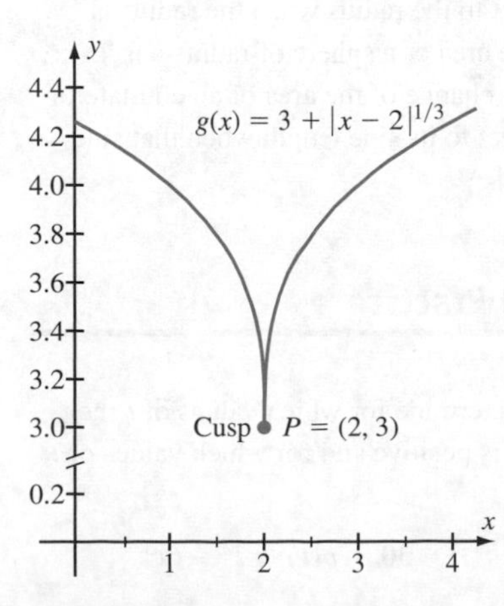

Figure 13

does not even have an infinite limit as $\Delta x \to 0$. (Verify that the limit is $+\infty$ as $x \to 0^+$ and $-\infty$ as $x \to 0^-$.) The graph of g does not have a tangent line at P. We say that the point P is a *cusp* of the graph of g. ■

quickquiz

1. How does instantaneous velocity differ from average velocity?
2. If a moving body has instantaneous velocity that is always negative, what can you conclude about the motion?
3. The position of a particle at time t is $p(t)$. If $p(1) = 3$ and if the velocity of the particle is -2 when $t = 1$, what is the tangent line to the graph of p at the point $(1, 3)$?
4. What is the tangent line of $y = x^2$ at the point $(2, 4)$?

EXERCISES

Problems for Practice

In Exercises 1–6, the function $p(t)$ describes the position of an object at time t. Calculate the instantaneous velocity at time c.

1. $p(t) = 5, c = 2$ **2.** $p(t) = 6t + 8, c = 3$
3. $p(t) = -5t^2 + 2, c = 1$ **4.** $p(t) = 4t^3, c = 4$
5. $p(t) = 1/t, c = 1/2$ **6.** $p(t) = t(t + 1), c = 0$

In Exercises 7–10, $p(t)$ describes the position of a moving body at time t. Determine whether, at time $t = 4$, the body is moving forward, backward, or neither.

7. $p(t) = 6t + 3$ **8.** $p(t) = t^2 - 8t$
9. $p(t) = -t^3 + 5t^2$ **10.** $p(t) = 1/t$

In Exercises 11–14, find the slope of the tangent line to the graph of the function at point P.

11. $f(x) = x^2, P = (3, 9)$
12. $g(x) = x^3 - 6x, P = (1, -1)$
13. $H(x) = 3x^2 + 6, P = (-1, 9)$
14. $f(x) = -4x^2 + x + 1, P = (2, -13)$

In Exercises 15–18, find the equation of the tangent line to the graph of the function at point P.

15. $f(x) = 2x^2, P = (5, 50)$
16. $f(x) = x^3/6, P = (2, 4/3)$
17. $f(x) = -3x^2 + 5, P = (-2, -7)$
18. $f(x) = 1/x, P = (-1, -1)$

In Exercises 19–22, find the equation of the normal line to the graph of the function at the point P.

19. $f(x) = 2x^2, P = (5, 50)$
20. $f(x) = x^3/6, P = (2, 4/3)$
21. $f(x) = -3x^2 + 5, P = (-2, -7)$
22. $f(x) = 1/x, P = (2, 1/2)$

23. The population of a colony of bacteria after t hours is $B(t) = 5000 + 6t^3$. At what rate is the population changing after 2 hours?

24. The *marginal cost* of an item is defined as the rate of change of the cost $C(x)$ of producing x units with respect to the number x of units produced. In general, the larger the quantity produced, the cheaper each unit is. Suppose, for example, that the cost, in cents, of producing p pencils is

$$C(p) = 5 - 0.001p - 0.00001p^2.$$

What is the marginal cost when $p = 75$?

25. A large herd of reindeer is dying out. The number of reindeer in the herd at time t (measured in months), $0 \leq t \leq 15$, is

$$r(t) = 25000 - 800t - 40t^2 - t^3.$$

At what rate are the reindeer dying out after 11 months?

26. What is the rate of change of the area of a square with respect to its side length when the side length is 8 cm?

27. What is the rate of growth of the surface area of a sphere with respect to the radius when the radius is 8 in.? (The surface area of a sphere of radius r is $4\pi r^2$.)

28. What is the rate of change of the area of an equilateral triangle with respect to its side length when that side length is 8 in.?

Further Theory and Practice

For Exercises 29 and 30, determine for which values of t the velocity of a moving body is positive and for which values of t the velocity is negative.

29. $p(t) = t^2 + t$ **30.** $p(t) = t^2 - 6t^3$

31. Six points are labeled on the graph of the function f (see Figure 14). The instantaneous rates of change of f with respect to x at the six points are $-3, -1, 0, 3, 10$, and 20. Match each point to the corresponding rate of change.

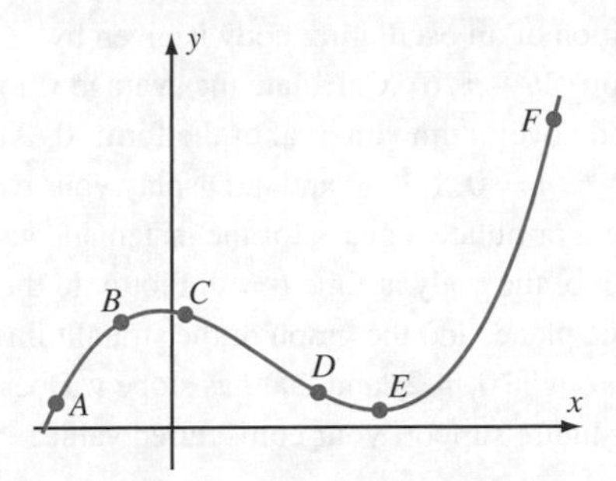

Figure 14

32. The simultaneous effect of money supply and interest rate policy on unemployment U and inflation I is illustrated by a *Phillip's curve*. For the Phillip's curve in Figure 15, describe the rate of change of inflation with respect to unemployment. Where does an incremental increase in unemployment decrease inflation the most? Where does an incremental decrease in unemployment increase inflation the most? Where does an incremental increase in unemployment decrease inflation the least? Where does an incremental decrease in unemployment increase inflation the least?

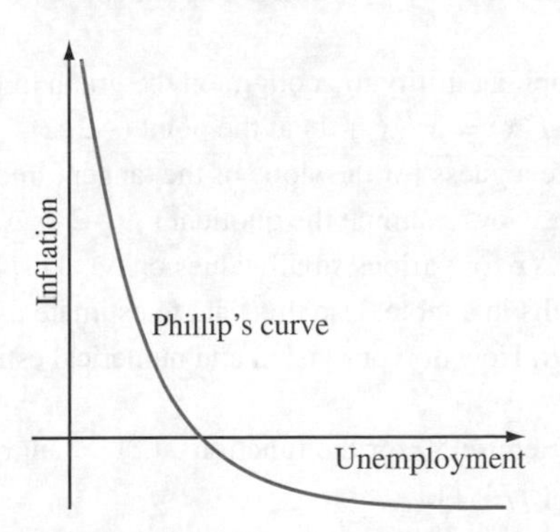

Figure 15

33. When a visiting team hits a home run at Wrigley Field in Chicago, it is traditional to throw the ball back onto the field. Suppose that the height H above field level of a ball thrown back by a Cubs fan is given, in feet, by

$$H(t) = 18 + 13.8t - 16t^2$$

where t is measured in seconds.

a. How high above field level is the fan sitting?

b. What is the rate at which the ball rises as a function of time? Explain why this is *not* the velocity of the ball.

c. At what time does the ball reach its maximum height? What is this maximum height?

d. What is the average rate of change of H during the ball's upward trajectory?

e. Over what time interval is the average rate of change of H equal to 0?

f. How long is the ball in the air?

g. What is the rate of change of H at the moment the ball hits the ground?

h. What is the average rate of change of H over the entire trajectory of the ball?

i. Is there a time when the rate of change of H is equal to the overall average that you determined in part h?

34. Let $f(x) = x^2$ and suppose a and b are different constants. Find a formula for the point of intersection of the tangent line to the graph of f at $x = a$ and the tangent line to the graph of f at $x = b$.

35. Let $f(x)$ be a quadratic polynomial. Show that the tangent line to the graph of f at any point P can only intersect the graph of f at P.

36. Find the equation of the line with positive slope that is tangent to the graph of $f(x) = 3x^3 + 12$ and that passes through the origin.

37. Find the equations of the tangent lines to the graph of $f(x) = 3x^2$ that pass through the point $(1, 9)$.

38. Under what conditions on a and b does the graph of $f(x) = ax^2 + b$ have a tangent that passes through the origin?

39. The instantaneous rate of change of velocity is acceleration. For the position function $p(t) = t^3$, what is the acceleration at time $t = 1$?

40. Suppose f is a function with a graph that has a tangent line at each point. If $g(x) = f(x) + \alpha$ for some constant α, show that the graph of g has a tangent line at each point and that the slope of the tangent line to the graph of g at $(c, g(c))$ is the same as the slope of the tangent line to the graph of f at $(c, f(c))$. Explain this geometrically.

41. Finding a Function from the Tangent Lines to Its Graph If f is a continuous function with $f(2) = 5$ and if the slope of the tangent line to the graph of f at $(c, f(c))$ is -2 for $-\infty < c < 1$, 1 for $1 < c < 3$, and -1 for $3 < c < \infty$, find f.

42. Let P be any point of the form $(c, 1/c)$ in the first quadrant. Let L be the line tangent to $f(x) = 1/x$ at P. Show that the area of the triangle formed by L and the positive axes is independent of P. Compute that area.

43. Suppose that f is defined on an open interval centered at c. Suppose also that

$$\ell_R = \lim_{h \to 0^+} \frac{f(c+h) - f(c)}{h}$$

and

$$\ell_L = \lim_{h \to 0^-} \frac{f(c+h) - f(c)}{h}$$

exist. Let

$$T_R(x) = f(c) + \ell_R(x - c) \text{ for } x \geq c, \text{ and}$$
$$T_L(x) = f(c) + \ell_L(x - c) \text{ for } x \leq c.$$

Define $\alpha_f(c)$ to be the radian measure of the angle through which T_R must be rotated counterclockwise about $(c, f(c))$ to coincide with T_L. You may think of $\alpha_f(c)$ as the angle of the corner at $P = (c, f(c))$.

a. For what values of $\alpha_f(c)$ is there actually a corner at P? Explain.

b. For what value of $\alpha_f(c)$ is there a tangent line at P? Explain.

c. For what value of $\alpha_f(c)$ is there a cusp at P? Explain.

d. If the graph of f has a vertical tangent at P, is $\alpha_f(c)$ defined? Explain.

44. Suppose the graph of f has a nonvertical tangent T_P at each point P on it. Thus, for each P, there are two numbers, $m(P)$ and $b(P)$, such that $T_P(x) = m(P)x + b(P)$.

a. Define the tangents to the graph of f at infinity and minus infinity—$\lim_{P \to \infty} T_P$ and $\lim_{P \to -\infty} T_P$, respectively—in an appropriate way.

b. Show that if $f(x) = 1/x$, then $\lim_{P \to \infty} T_P$ and $\lim_{P \to -\infty} T_P$ are horizontal asymptotes of the graph of f.

c. Is the converse to part b true? Think of $f(x) = \sin(x)/x$. Explain your answer.

Calculator/Computer Exercises

45. The trajectory of a fly ball is such that the height, in feet, above ground is $H(t) = 4 + 72t - 16t^2$ when time t is measured in seconds.

a. Compute the average velocity in the following time intervals.

i. [2, 3] **ii.** [2, 2.1]
iii. [2, 2.01] **iv.** [2, 2.001]

b. Compute the instantaneous velocity at $t = 2$.

46. Use the trajectory described in Exercise 45.

a. Compute the average velocity in the following time intervals.

i. [3, 3.1] **ii.** [3, 3.01]
iii. [2.9, 3] **iv.** [2.99, 3]

b. Compute the instantaneous velocity at $t = 3$.

47. The position of an oscillating body is given by $p(t) = \sin(2t + \pi/6)$. Calculate the average velocity of the body over a time interval of the form $[0, \Delta t]$ for $\Delta t = 10^{-n}$, $n = 0, 1, 2, 3$, and 4. Display your results in a table. Formulate a guess for the instantaneous velocity v of the body at time $t = 0$. Plot p. In the same coordinate plane, add the graph of the straight line that passes through $(0, 1/2)$ and that has slope v. Does the resulting figure support your conjectured value? Explain.

48. The position of a moving body is given by $p(t) = (2.718281828)^t$. Calculate the average velocity of the body over a time interval of the form $[1, 1 + \Delta t]$ for a sequence of small values of Δt. Display your results in a table. Formulate a guess for the instantaneous velocity m of the body at time $t = 1$. Plot p. In the same coordinate plane, add the graph of the straight line that passes through $(1, 2.718281828)$ and that has slope m. Does the resulting figure support your conjectured value? Explain.

49. Repeat Exercise 48 with $t = 2$ and $t = 3$. For a general value of t, what is the apparent relationship between the position of the body $p(t)$ and the instantaneous velocity at t?

50. Use a graphing utility to zoom in on the graph of the function $f(x) = x/(x + 1)$ at the point $(-2, 2)$. Formulate a guess for the slope of the tangent line L at that point. Now examine the quotient $(f(-2 + \Delta x) - f(-2))/\Delta x$ for various small values of Δx. Display your results in a table. Use this data to estimate the slope of L. How do your visual and numerical estimates compare?

51. Repeat Exercise 50 for the function $f(x) = \tan(x)$ at the point $(\pi/4, 1)$.

52. Suppose $a > 0$. This exercise investigates tangent lines to the exponential functions $x \mapsto a^x$.

a. Let $f(x) = a^x$. Use the formula for the slope of a tangent line to show that the slope of the tangent line to f at point $(c, f(c))$ is

$$f(c) = \lim_{h \to 0} \frac{a^h - 1}{h}.$$

b. Show that the slope of the tangent line to f at point $(0, 1)$ is

$$\lim_{h \to 0} \frac{a^h - 1}{h}.$$

c. The limit in part b cannot be computed by substituting $h = 0$ in the expression because that results in the meaningless expression $0/0$. We will learn how to compute this limit in Section 3.6. For now, we investigate it numerically. To be specific, let $a = 2$. To identify the limit, graph the function $h \mapsto (2^h - 1)/h$ in a window with a horizontal range that is a small interval centered at 0. From the graph (zooming in if necessary), identify

$$\lim_{h \to 0} \frac{2^h - 1}{h}$$

to four decimal places.

d. Use a graphing utility (and the result of part c) to graph $f(x) = 2^x$ and the tangent lines to the graph of f at points $(0, 1)$ and $(2, 4)$ in the same viewing window.

e. Repeat parts c and d with $a = 3$. Use $(-1, 1/3)$ and $(1, 3)$ as the points of tangency.

3.2 The Definition of the Derivative

In Section 3.1, we saw that the limit

$$\lim_{\Delta x \to 0} \frac{f(c + \Delta x) - f(c)}{\Delta x}$$

represents the instantaneous rate of change of the function f at the point c. It is also the slope of the tangent line to the graph of f at $(c, f(c))$. Later in this chapter, and especially in Chapter 4 and beyond, we will see many other applications of this limit. Because of its importance, we will now name this limit and study ways to evaluate it.

Definition Let f be a function that is defined in an open interval that contains a point c. If the limit

$$\lim_{\Delta x \to 0} \frac{f(c + \Delta x) - f(c)}{\Delta x}$$

exists and is finite, then we say that f is *differentiable* at c. We call this limit the *derivative* of the function f at the point c, and we denote it by the symbol $f'(c)$. The process of calculating $f'(c)$ is called *differentiation*.

inSIGHT

The property of differentiability can also be stated in geometric terms: The function f is differentiable at c if and only if the graph of f has a *nonvertical* tangent line at the point $(c, f(c))$.

Although we have defined the derivative $f'(c)$ by

$$f'(c) = \lim_{\Delta x \to 0} \frac{f(c + \Delta x) - f(c)}{\Delta x}, \tag{3.7}$$

it is often convenient to express $f'(c)$ in other ways. Often the letter h is used instead of Δx to indicate an increment:

$$f'(c) = \lim_{h \to 0} \frac{f(c + h) - f(c)}{h}. \tag{3.8}$$

Also, if we set $x = c + \Delta x$, then $\Delta x = x - c$ and

$$\frac{f(c + \Delta x) - f(c)}{\Delta x} = \frac{f(x) - f(c)}{x - c}.$$

The statement $\Delta x \to 0$ is equivalent to $x \to c$. Thus, the definition of the derivative of f at c can also be stated as

$$f'(c) = \lim_{x \to c} \frac{f(x) - f(c)}{x - c}. \tag{3.9}$$

Example 1 Calculate $f'(5)$ for $f(x) = 4x - 7$.

Solution Here $c = 5$. By definition, we have

$$f'(5) = \lim_{\Delta x \to 0} \frac{f(5 + \Delta x) - f(5)}{\Delta x} = \lim_{\Delta x \to 0} \frac{(4(5 + \Delta x) - 7) - (4 \cdot 5 - 7)}{\Delta x} = \lim_{\Delta x \to 0} \frac{4 \cdot \Delta x}{\Delta x} = \lim_{\Delta x \to 0} 4 = 4.$$

■

inSIGHT

One of this chapter's goals is to develop formulas that allow us to calculate derivatives without referring to the limit definition. In Example 1, we can write $f(x) = f_1(x) + f_2(x)$ where $f_1(x) = 4x$ and $f_2(x) = -7$. Formula (3.4) tells us that the instantaneous rate of change of $f_1(x)$ at $c = 5$ is $1 \cdot 4 \cdot 5^0$, or 4. The instantaneous rate of change of $f_2(x)$ is clearly 0, since $f_2(x)$ is constant. Theorem 1 from Section 3.1 tells us that $f'(5) = 4 + 0 = 4$.

Other Notations for the Derivative

Several notations are used for the derivative $f'(c)$ of f at c. For example, $f'(c)$ is often denoted by $D(f)(c)$. *Leibniz notation* for the derivative of f at c is also widely used and can take the following forms:

$$\frac{df}{dx}(c) \quad \text{or} \quad \frac{d}{dx} f(c) \quad \text{or} \quad \left. \frac{df}{dx} \right|_{x=c}.$$

Leibniz notation is named after Gottfried Wilhelm von Leibniz (1646–1716), one of the "inventors" of calculus. Leibniz notation tells us explicitly what the variable is (in this case, the variable is x). This information can be a helpful reminder when several variables are under simultaneous consideration. It is also a suggestive notation. After all,

$$\frac{df}{dx}(c) = \lim_{\Delta x \to 0} \frac{f(c + \Delta x) - f(c)}{\Delta x} = \lim_{\Delta x \to 0} \frac{\Delta f}{\Delta x}. \tag{3.10}$$

To Leibniz, the symbol $\frac{df}{dx}$ in formula (3.10) actually represented a ratio of the quantities df and dx. Although these quantities are known as *differentials,* we will not give any meaning to them as separate entities. The only *formal* meaning that we will give to them is when they appear together as a quotient, as in formula (3.10). Later we will find that there are circumstances in which manipulation of these differentials will be useful.

A curve C in the xy-plane is not necessarily the graph of a function. However, such a curve can still have a tangent line. When we think of the variables x and y as being related by virtue of the ordered pair (x, y) being on the curve $\mathcal{C}$, we write

$$\frac{dy}{dx}(x_0) \quad \text{or} \quad \frac{dy}{dx}(x_0, y_0) \tag{3.11}$$

for the slope of a tangent line to the curve at the point $(x_0, y_0) \in \mathcal{C}$. If there is more than one value of y such that (x_0, y) is on $\mathcal{C}$, then the second expression in line (3.11) should be used (see Figure 1).

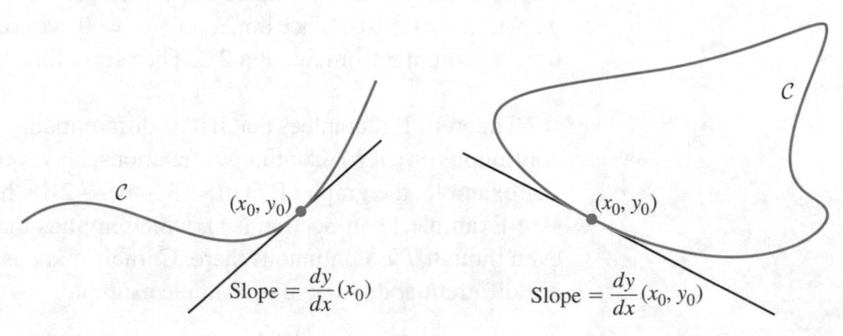

Figure 1

If g were a function of the time variable t, then we might adopt *Newton notation* and write the derivative as $\dot{g}(t)$. This notational device is named after Sir Isaac Newton (1642–1727), the other principal "inventor" of calculus.

The Derived Function

Definition If f has a derivative at every point of S, then we say that f is *differentiable on* S. (In the most common situation encountered in this text, S is an open interval.) In this case, we can form a function f' on S that is defined by

$$f'(x) = \lim_{\Delta x \to 0} \frac{f(x + \Delta x) - f(x)}{\Delta x} \qquad \text{for } x \in S.$$

This function f' is said to be the *derived function* of f, or the *derivative* of f. Other notations for this function are

$$\frac{df}{dx}, \quad \frac{d}{dx}f, \quad \text{and} \quad D(f).$$

Continuity and Differentiability

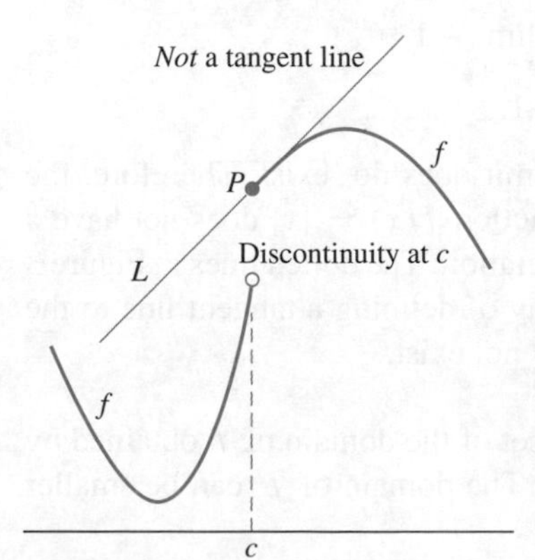

Figure 2
The graph of f has no tangent line at $P = (c, f(c))$.

Continuity and differentiability are properties of a function that are reflected in its graph. If a function f is continuous at c, then we know that if we follow the points $(x, f(x))$ as $x \to c$, those points approach $P = (c, f(c))$. Continuity may be interpreted as *predictability*.

We know that when a function f is differentiable at c, then its graph has a well-defined tangent line at $(c, f(c))$. Differentiability may be interpreted as a degree of *smoothness*, which is a stronger property than predictability. If f is discontinuous at c, then there is no line that is tangent to the graph of f at $P = (c, f(c))$, as Figure 2 suggests. In Figure 2, line L through P is the only candidate to be a tangent line.

However, line L cannot be regarded as a tangent line because it does not have one of the properties we expect: For x just to the left of c, the point $(x, f(x))$ on the graph of f is far from L. We state our conclusions as a theorem.

Theorem 1 If f is not continuous at a point c in its domain, then f is not differentiable at c. If f is differentiable at a point c in its domain, then f is continuous at c.

Proof The two statements of Theorem 1 are logically equivalent. We prove the second: Suppose that f is differentiable at a point c in its domain. By definition, $\lim_{x\to c}(f(x)-f(c))/(x-c)$ exists. Since $\lim_{x\to c}(x-c) = 0$, we conclude that $\lim_{x\to c}(f(x)-f(c)) = 0$ by Theorem 5 from Section 2.2. Therefore, $\lim_{x\to c} f(x) = f(c)$. ■

Theorem 1 guarantees that if f is differentiable on an interval (a, b), then f is also continuous on (a, b). Continuous functions, however, are not necessarily differentiable. For example, the graph of $f(x) = 3 + (x - 2)^{1/3}$ has a *vertical* tangent at $P = (2, 3)$ (see Example 11 in Section 3.1), which implies that f is *not* differentiable at $x = 2$, even though f *is* continuous there. Corners and cusps are also points of continuity but not differentiability. Here is a simple example.

Example 2 Let $f(x) = |x|$. Show that $f'(0)$ does not exist.

Solution The relevant limit as $\Delta x \to 0$ through *positive* values is

$$\begin{aligned}\lim_{\Delta x\to 0^+} \frac{f(0+\Delta x) - f(0)}{\Delta x} &= \lim_{\Delta x\to 0^+} \frac{|\Delta x| - 0}{\Delta x} \\ &= \lim_{\Delta x\to 0^+} \frac{\Delta x}{\Delta x} \\ &= \lim_{\Delta x\to 0^+} 1 \\ &= 1.\end{aligned}$$

However, as $\Delta x \to 0$ through *negative* values, the limit is

$$\begin{aligned}\lim_{\Delta x\to 0^-} \frac{f(0+\Delta x) - f(0)}{\Delta x} &= \lim_{\Delta x\to 0^-} \frac{|\Delta x| - 0}{\Delta x} \\ &= \lim_{\Delta x\to 0^-} \frac{-\Delta x}{\Delta x} \\ &= \lim_{\Delta x\to 0^-} -1 \\ &= -1.\end{aligned}$$

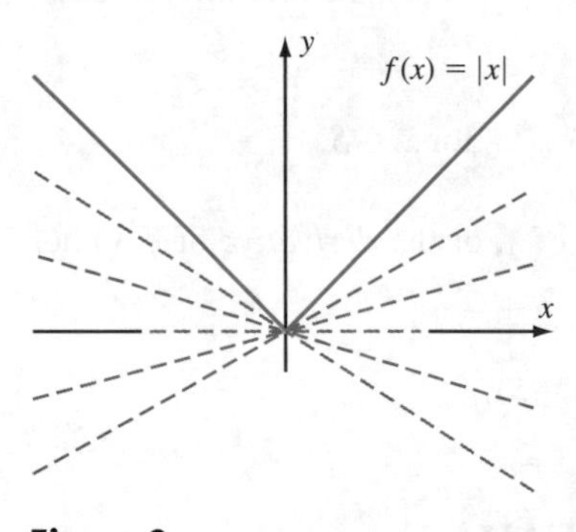

Figure 3

Since the left and right limits do not agree, the limit does not exist. Therefore, the derivative does not exist at 0. The fact that the function $f(x) = |x|$ does not have a derivative at $x = 0$ has a simple geometrical interpretation: The dotted lines in Figure 3 show that there is no geometrically satisfactory way of defining a tangent line to the graph of f at $(0, 0)$, thus signifying that $f'(0)$ does not exist. ■

The domain of a derived function f' is the subset of the domain of f obtained by removing the points where f is not differentiable. The domain of f' can be smaller than the domain of *f*.

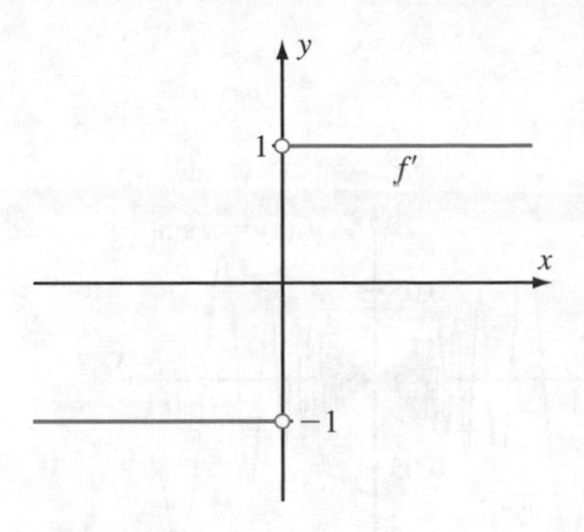

Figure 4
The graph of f' where $f(x) = |x|$

Example 3 Let $f(x) = |x|$. Completely describe the function f'.

Solution Example 2 shows that $f'(0)$ does not exist. Therefore, 0 is not in the domain of f'. If $x > 0$, then the graph of f near the point $(x, f(x))$ is a straight line segment of slope 1. If $x < 0$, then the graph of f near the point $(x, f(x))$ is a straight line segment of slope -1. These considerations suggest that $f'(x) = 1$ for $x > 0$ and $f'(x) = -1$ for $x < 0$, which we verify using the limit definition of $f'(x)$. We conclude that the domain of the derived function f' is the union, $(-\infty, 0) \cup (0, \infty)$, of the two open half lines to the left and right of 0 and

$$f'(x) = \begin{cases} -1 & \text{if } x < 0 \\ 1 & \text{if } x > 0 \end{cases}.$$

Figure 4 shows the graph of f'. ■

A LOOK BACK

Differentiability is a stronger property than continuity (see Theorem 1). Continuous functions are appealing for their predictability. Differentiable functions have this predictability *and* a powerful geometric property—a well-defined tangent.

Using Graphing Utilities to Investigate Differentiability

Suppose that a function f is continuous and that the differentiability of f at a point c is in question. Define a new function ϕ by

$$\phi(x) = \frac{f(x) - f(c)}{x - c} \quad \text{for } x \neq c.$$

We know that ϕ is a continuous function but the point c is *not* in its domain. According to the derivative definition and our knowledge from Section 2.3, the function ϕ can be continuously extended at c if and only if

$$\lim_{x \to c} \phi(x) = \lim_{x \to c} \frac{f(x) - f(c)}{x - c}$$

exists. That is, ϕ can be continuously extended at c if and only if f is differentiable at c. The next example shows how we can use this equivalence to determine the differentiability of f by looking at the graph of ϕ.

Example 4 Determine whether the functions

$$f_1(x) = \sqrt{1 - \cos(x)} \quad \text{and} \quad f_2(x) = \begin{cases} x \sin\left(\dfrac{1}{x}\right) & \text{if } x \neq 0 \\ 0 & \text{if } x = 0 \end{cases}$$

are differentiable at $x = 0$.

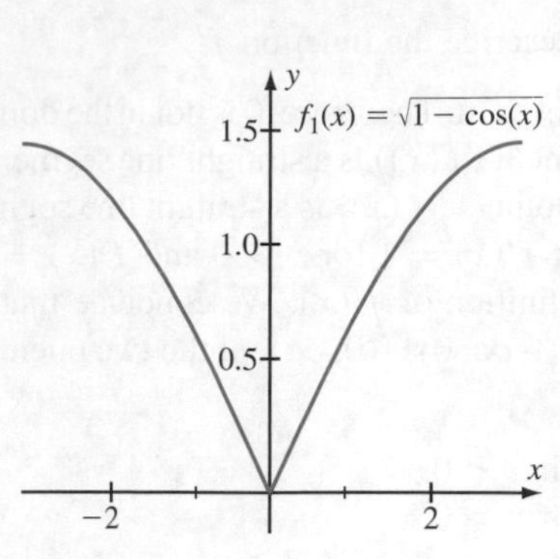

Figure 5

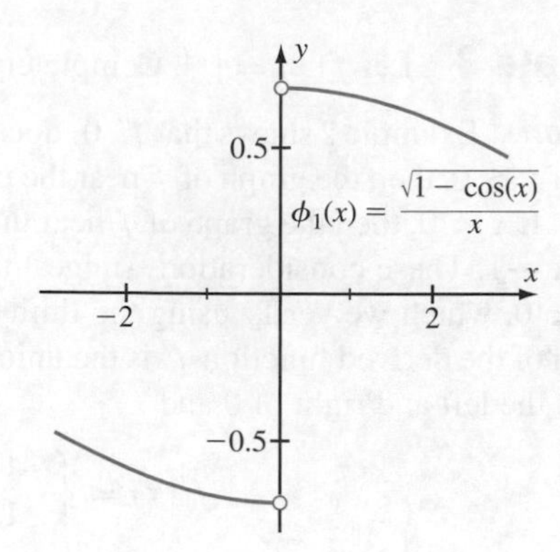

Figure 6

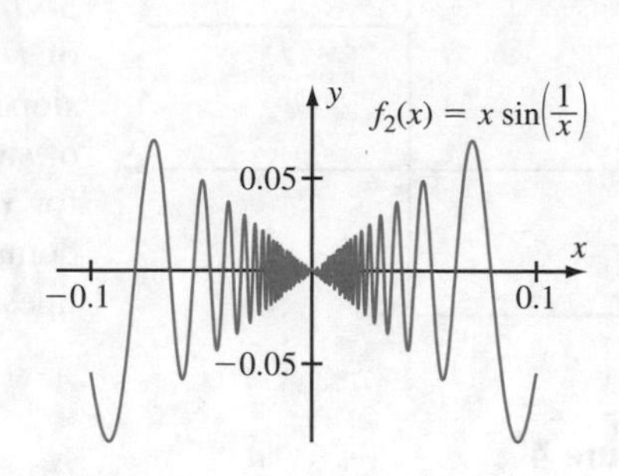

Figure 7

Solution The graph of f_1 does not seem to be smooth at $x = 0$ (see Figure 5). This suggests that f_1 is not differentiable at $x = 0$. Let

$$\phi_1(x) = \frac{f_1(x) - f_1(0)}{x - 0} = \frac{\sqrt{1-\cos(x)} - 0}{x - 0} = \frac{\sqrt{1-\cos(x)}}{x} \quad \text{for } x \neq 0.$$

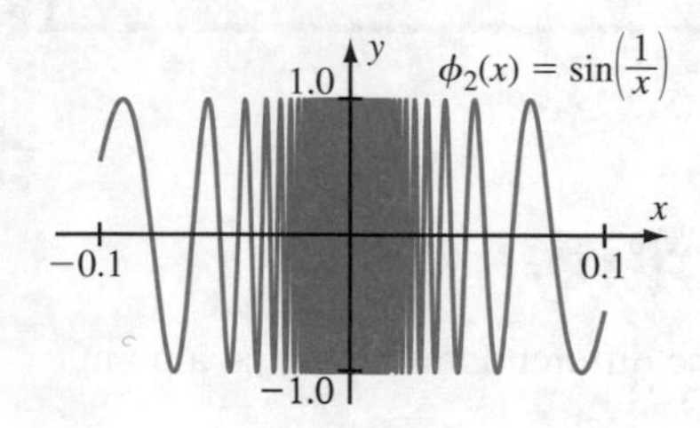

Figure 8

The graph of ϕ_1 shows that $\phi_1(x)$ cannot be continuously extended to $x = 0$ (see Figure 6). This confirms that f_1 is not differentiable at $x = 0$.

The graph of f_2 indicates that f is continuous at $x = 0$, but differentiability is far from transparent (see Figure 7). Let

$$\phi_2(x) = \frac{x\sin(1/x) - 0}{x - 0} = \sin\left(\frac{1}{x}\right) \quad \text{for } x \neq 0.$$

It is evident from the graph of ϕ_2 that ϕ_2 cannot be continuously extended at $x = 0$ (see Figure 8). This shows that f_2 is not differentiable at $x = 0$. ■

Derivatives of Sine and Cosine

In Section 2.2 (Theorem 8 and Example 7), we went to some trouble to obtain the limit formulas

$$\lim_{\Delta x \to 0} \frac{\sin(\Delta x)}{\Delta x} = 1 \quad \text{and} \quad \lim_{\Delta x \to 0} \left(\frac{1-\cos(\Delta x)}{\Delta x}\right) = 0. \tag{3.12}$$

We can now use these limit formulas to evaluate the derivatives of sine and cosine.

Theorem 2 For each x,

$$\frac{d}{dx}\sin(x) = \cos(x) \tag{3.13}$$

and

$$\frac{d}{dx}\cos(x) = -\sin(x). \tag{3.14}$$

Proof To prove equation (3.13), we start with the definition of the derivative and calculate as follows:

$$\begin{aligned}\frac{d}{dx}\sin(x) &= \lim_{\Delta x\to 0}\frac{\sin(x+\Delta x)-\sin(x)}{\Delta x}\\ &= \lim_{\Delta x\to 0}\frac{(\sin(x)\cos(\Delta x)+\cos(x)\sin(\Delta x))-\sin(x)}{\Delta x}\\ &= \lim_{\Delta x\to 0}\frac{\sin(x)(\cos(\Delta x)-1)}{\Delta x}+\lim_{\Delta x\to 0}\frac{\cos(x)\sin(\Delta x)}{\Delta x}\\ &= \sin(x)\lim_{\Delta x\to 0}\frac{(\cos(\Delta x)-1)}{\Delta x}+\cos(x)\lim_{\Delta x\to 0}\frac{\sin(\Delta x)}{\Delta x}\\ &= \sin(x)\cdot 0+\cos(x)\cdot 1\\ &= \cos(x).\end{aligned}$$

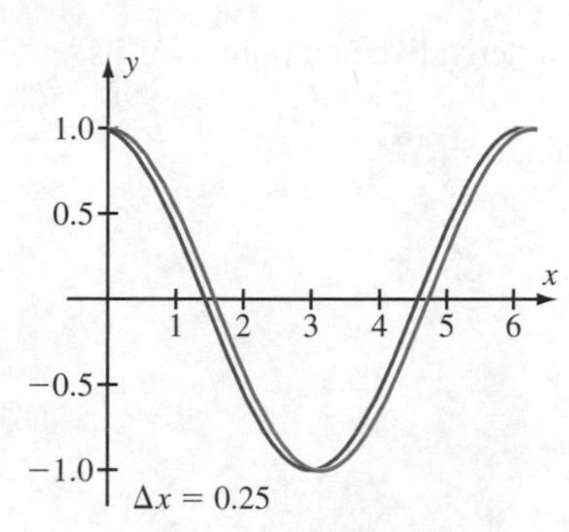

Figure 9a

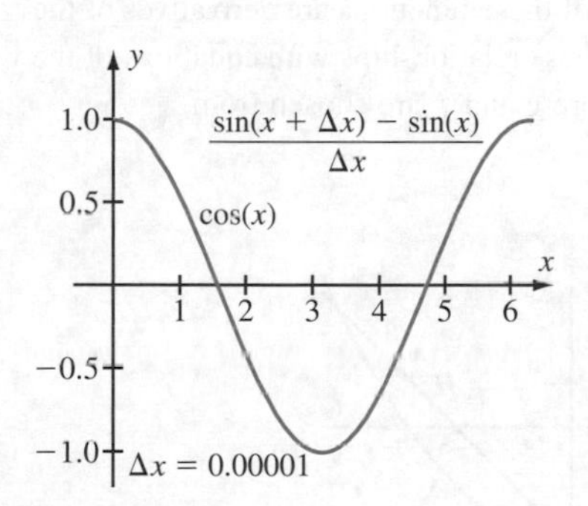

Figure 9b

In a similar manner, we can use the trigonometric identity

$$\cos(x+\Delta x)=\cos(x)\cos(\Delta x)-\sin(x)\sin(\Delta x)$$

to derive formula (3.14). ■

Example 5 Graphically illustrate the differentiation formula

$$\frac{d}{dx}\sin(x)=\cos(x).$$

Solution By definition,

$$\frac{d}{dx}\sin(x)=\lim_{\Delta x\to 0}\frac{\sin(x+\Delta x)-\sin(x)}{\Delta x}.$$

We can *approximate* this limit by substituting a small value for Δx. Figure 9a illustrates the approximation when $\Delta x = 0.25$. Although the graph is not extremely accurate, it does give the idea. In Figure 9b, we set $\Delta x = 10^{-5}$ and graphed *both* the approximate derivative $x \mapsto (\sin(x+10^{-5})-\sin(x))/10^{-5}$ and the exact derivative $x \mapsto \cos(x)$. Only one curve appears because the two graphs *coincide* within the accuracy of the plotter that we have used. ■

Derivatives of Expressions

In calculus, we often use expressions as a shorthand for functions. Thus, the expression $x^3+\cos(\sqrt{x+1})$ is frequently interpreted to be the function f defined by $f(x) = x^3+\cos(\sqrt{x+1})$ for all values of x for which the expression makes sense. With such an interpretation, we write

$$\frac{d}{dx}(x^3+\cos(\sqrt{x+1}))$$

as an equivalent form of $\frac{df}{dx}(x)$. This shorthand has the advantage of making clearer the relationship between the expressions defining f and f'. With this notation, some of the limits that we learned in Section 3.1 can be expressed as derivative formulas:

$$\frac{d}{dx}x^n = nx^{n-1} \quad \text{for } n = -1, 0, 1, 2, \text{ and } 3.$$

quickquiz

1. How does the concept of derivative differ from the concept of instantaneous velocity?
2. What does it mean for a function *not* to have a derivative at a point c of its domain?
3. What is the derivative of $f(x) = x^2$ at the point $c = 5$?
4. What is the derivative of $f(x) = \cos(x)$ at the point $c = \pi/3$?

EXERCISES

Problems for Practice

In Exercises 1–6, compute the indicated derivative for the given function.

1. $f'(-3)$, $f(x) = x^2$
2. $g'(2)$, $g(t) = t^3/3$
3. $\dfrac{dF}{dt}(8)$, $F(t) = 5t + 6$
4. $D(f)(0)$, $f(x) = \sin(x)$
5. $g'(5)$, $g(t) = 8t^3 - 8t$
6. $D(H)(1)$, $H(x) = 10/x$

In Exercises 7–10, calculate $g'(x)$.

7. $g(x) = -\pi \cdot x$
8. $g(x) = x^2 - 4x$
9. $g(x) = 4x^3 + 6x^2 + 1$
10. $g(x) = 2x + \sin(x)$

In Exercises 11–14, find a point x where $f'(x) = 6$.

11. $f(x) = x^2 + 5$
12. $f(x) = 12\sin(x)$
13. $f(x) = x^3 + 1$
14. $f(x) = 3x^2 - 6x + 50$

In Exercises 15–18, sketch the graph of $f'(x)$.

15. $f(x) = 3 - 2x^3$
16. $f(x) = x^2 + 3x$
17. $f(x) = x^2 + |x|$
18. $f(x) = 3x - |x|$

Further Theory and Practice

In Exercises 19–22, calculate the derivative of the function at point c.

19. $f(x) = |x|^{3/2}$, $c = 0$
20. $F(x) = x \cdot |x|$, $c = 0$
21. $f(x) = \sqrt{x}$, $c = 9$
22. $g(x) = 1/\sqrt{x}$, $c = 1/16$
23. Figure 10 illustrates the graphs of four functions F, G, H, and K. Three of these functions are derivatives of the others. Identify these relationships with equations of the form $g = f'$ where g and f are chosen from $\{F, G, H, K\}$.

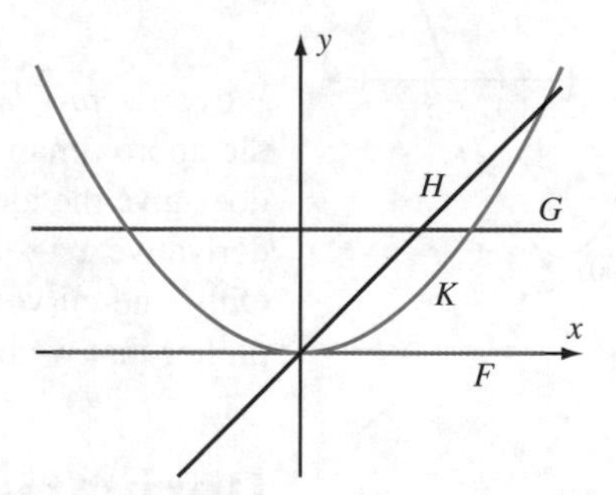

Figure 10

24. If $f(x) = |x - 2| + |x - 3|$, what is the domain of f'? Sketch the graph of f'.
25. Find a degree one polynomial function $p(x)$ such that $p(2) = 6$ and $p'(4) = -5$.
26. Find a degree two polynomial function such that $p(3) = 4$, $p(5) = -10$, and $p'(-3) = 7$.

In Exercises 27–30, decide whether the function is differentiable at $x = 0$. Give a reason for your answer.

27. $f(x) = \begin{cases} x^2 & \text{if } x \le 0 \\ x & \text{if } x > 0 \end{cases}$

28. $g(x) = \begin{cases} x^3 & \text{if } x \le 0 \\ x^2 & \text{if } x > 0 \end{cases}$

29. $f(x) = \begin{cases} |x| & \text{if } x \le 0 \\ x^2 + x & \text{if } x > 0 \end{cases}$

30. $g(x) = \sin(|x|)$

In Exercises 31–34, identify the limit as a derivative $f'(c)$ by identifying f and c. (You need not evaluate the limit.)

31. $\lim_{h \to 0} \left(\frac{\sqrt{4+h} - 2}{h} \right)$

32. $\lim_{h \to 0} \left(\frac{1/(5+h) - 1/5}{h} \right)$

33. $\lim_{\Delta x \to 0} \frac{1}{\sqrt{3 + \Delta x} + \sqrt{3}}$

34. $\lim_{h \to 0} \left(\frac{1 - \sqrt{(1+h)}}{h\sqrt{(1+h)}} \right)$

35. Describe the difference in the tangent lines to the graph of f at points $(a, f(a))$ and $(b, f(b))$ if $f'(a) > 0$ and $f'(b) < 0$.

36. Graph $f(x) = \sqrt{8x}$ and $g(x) = x^2$ for $0 \le x \le 2$. Both graphs begin at the same point and end at the same point. Describe the behavior of $f'(x)$ and $g'(x)$ as x increases.

37. Suppose a positive differentiable function f satisfies $f'(x) = f(x)$ for every x. Suppose also that $f(0) = 1$. Draw the tangent line to the graph of f at the point $(0, 1)$. Do you think $f(2)$ is greater than or less than $f(1)$? Explain your reasoning. Roughly locate $(2, f(2))$ and draw the tangent line to the graph of f at that point. Continue with $x = 3$. Sketch the graph of f for $x > 0$. Continue your analysis for $x < 0$.

38. If $f(-x) = f(x)$ for every x, then f is called an *even* function. If $f(-x) = -f(x)$ for every x, then f is called an *odd* function. Assume that f is differentiable. Show that if f is even, then f' is odd. Show that if f is odd, then f' is even.

39. If f is an even function that is differentiable at 0, what value must $f'(0)$ assume? Explain.

40. Evaluate $f'(0)$ if $f(x) = x^\alpha$ where $\alpha \ge 1$. Show that $f(x) = x^\alpha$ is not differentiable at 0 if $\alpha < 1$.

41. Let $f(x) = x/(1 + x^2)$. Use the identity

$$\frac{\frac{x}{1+x^2} - \frac{c}{1+c^2}}{x - c} = -\frac{xc - 1}{(1+x^2)(1+c^2)}$$

to compute $f'(c)$.

42. Let $f(x) = x/(1 + x^2)^2$. Use the identity

$$\frac{\frac{x}{(1+x^2)^2} - \frac{c}{(1+c^2)^2}}{x - c} = -\frac{cx^3 + c^2x^2 + 2xc + xc^3 - 1}{(1+x^2)^2(1+c^2)^2}$$

to compute $f'(c)$.

43. Suppose f is differentiable at c. Let $g(x) = f(x + k)$ where k is a constant. Show that g is differentiable at $c - k$ and evaluate $g'(c - k)$ in terms of $f'(c)$.

44. Find a continuous function f that is differentiable on $(-\infty, 0) \cup (0, \infty)$ such that $f'(x) = H(x)$ for $x \ne 0$. Here, H is the Heaviside function:

$$H(x) = \begin{cases} 0 & \text{if } x < 0 \\ 1 & \text{if } x > 0 \end{cases}.$$

45. In economics, if $q(p)$ is the demand for a product at price p—that is, the number of units of the product that are sold at price p—then

$$E(p) = -\lim_{\Delta p \to 0} \frac{(q(p + \Delta p) - q(p))/q(p)}{\Delta p/p}$$

is called the *elasticity of demand*. Compute $E(p)$ in terms of the derivative of the demand function q. The elasticity of demand for mass-produced cigars has been estimated to be 1.89 across their price range. If the price of a cigar is increased from \$1.00 to \$1.20, then by approximately what percent will demand change? How does the sign of 1.89 (namely, positive) tell you whether demand is increased or decreased by the price hike?

46. Let $C(x)$ denote the cost of producing x units of a commodity. Although x usually refers to a nonnegative integer, it is common in economics to treat x as a continuous variable. The marginal cost $M(x)$ refers to the cost of producing an $(x + 1)$th unit after x units are produced.

a. Describe M in terms of C.

b. Write the limit definition of $C'(x)$. What is the relationship between this definition and $M(x)$?

c. Under what circumstances can we plausibly use $C'(x)$ as an expression for the marginal cost? (This is commonly done in economics.)

47. **Continuation** The cost $C(x)$ of producing x units of a commodity is illustrated in Figure 11.

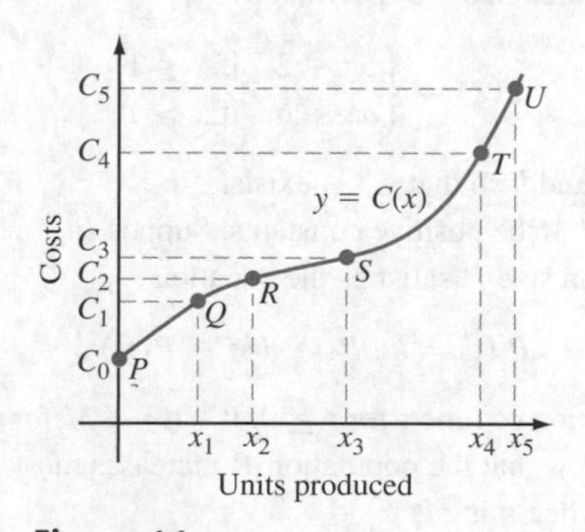

Figure 11

a. Before even 1 unit of the commodity is produced, various fixed costs (such as setting up a manufacturing plant) are incurred. How do these fixed costs show up in Figure 11 (page 173)?

b. The arc of $y = C(x)$ between P and Q is nearly a straight line. What is the equation of that line? What, approximately, is $C'(x)$ between P and Q?

c. The arc of $y = C(x)$ between R and S is nearly a straight line. What is the equation of that line? What, approximately, is $C'(x)$ between R and S?

d. The arc of $y = C(x)$ between T and U is nearly a straight line. What is the equation of that line? What, approximately, is $C'(x)$ between T and U?

e. The *unit* costs of many production expenses are reduced when the number x of units increases. Economists refer to such diminished unit costs as *economies of scale*. For example, if a manufacturer receives components in a cargo container, then it takes a certain number of worker hours to transfer the container from the shipper's truck to the warehouse receiving dock. This number is the same whether the container holds 1000 components or 5000 components. What part of the graph corresponds to the producer's economies of scale? Referring to values of $C'(x)$, what is the mathematical relationship that describes this phenomenon?

f. A producer of commodities is a consumer of raw materials or components. When demand for such materials increases, the producer's suppliers can raise prices. How is this phenomenon reflected in Figure 11 (page 173)? Referring to values of $C'(x)$, what is the mathematical relationship that describes this phenomenon?

48. Let the function f be specified by

$$f(x) = \begin{cases} x^2 + 2 & \text{if } x \leq 1 \\ ax + b & \text{if } x > 1 \end{cases}.$$

Select a and b so that $f'(1)$ exists.

49. Let k and M be positive constants. Suppose a population size P satisfies the equation

$$P'(t) = k \cdot P(t) \cdot (M - P(t))$$

as a function of time t for $t \geq 0$. If $P(t) < M$ for all $t \geq 0$, show that the population P increases most rapidly when it is $M/2$.

50. Let $F(t)$ denote the probability that a randomly selected individual does not live beyond the age of t. It can then be shown that the probability of a randomly selected individual not living beyond age $u > t$, given that he has already lived to age t, is

$$\frac{F(u) - F(t)}{1 - F(t)}.$$

Let $M(t)$ denote the instantaneous rate of change of the probability of an individual dying at age t, given that he has lived to age t. Actuaries refer to $M(t)$ as the *force of mortality* at age t. What is M in terms of F?

51. Prove that the function

$$f(x) = \begin{cases} x \cdot \sin\left(\dfrac{1}{x}\right) & \text{if } x \neq 0 \\ 0 & \text{if } x = 0 \end{cases}$$

is continuous at $x = 0$ but not differentiable at $x = 0$.

52. Prove that the function

$$f(x) = \begin{cases} x^2 \cdot \sin\left(\dfrac{1}{x}\right) & \text{if } x \neq 0 \\ 0 & \text{if } x = 0 \end{cases}$$

is differentiable at $x = 0$. Compute $f'(0)$.

53. Let f be differentiable at c. Let $y = ax + b$ be the equation of the tangent line to the graph of f at $(c, f(c))$. Prove that

$$\lim_{x \to c} \frac{f(x) - (ax + b)}{x - c} = 0.$$

Calculator/Computer Exercises

In Exercises 54–57, you are given a function f, a viewing rectangle R, and a point c. Use a graphing utility to graph both f and the tangent to the graph of f at $(c, f(c))$ in R. (Both graphs should appear together in R.)

54. $f(x) = \sqrt{x}$, $R = [0, 4] \times [0, 2]$, $c = 2$

55. $f(x) = x^2 - 2x$, $R = [0, 1] \times [-1, 0]$, $c = 1/2$

56. $f(x) = \sin(x)$, $R = [0, \pi/2] \times [0, 1]$, $c = \pi/4$

57. $f(x) = 1/x$, $R = [1/2, 2] \times [0, 2]$, $c = 1$

For Exercises 58–61, a function f and four values c, ξ, η, and ς are provided. Find an appropriate viewing rectangle centered about the point $P = (c, f(c))$. Graph f and the three secant lines passing through P that are determined by $(\xi, f(\xi))$, $(\eta, f(\eta))$, and $(\varsigma, f(\varsigma))$. Estimate $f'(c)$.

58. $f(x) = \sqrt{x + 1}$, $c = 0$, $\xi = 0.5$, $\eta = 0.3$, $\varsigma = 0.1$

59. $f(x) = x/(x + 1)$, $c = 1$, $\xi = 1.8$, $\eta = 1.5$, $\varsigma = 1.2$

60. $f(x) = 1 + \tan(x)$, $c = 0$, $\xi = -0.5$, $\eta = -0.3$, $\varsigma = -0.1$

61. $f(x) = \sin(x)$, $c = \pi/4$, $\xi = 1.2$, $\eta = 1$, $\varsigma = 0.8$

In Exercises 62–65, approximate $f'(c)$ for f and c in the following way: Find a small viewing window with $P = (c, f(c))$ near the center. The window should be small enough so that the graph of f appears to be a straight line. Let Q and R be the endpoints of the graph of f as it exits this window. Use the slope of $\overline{QR}$ as the approximation to $f'(c)$.

62. $f(x) = \tan\left(\dfrac{\pi}{4}\sin\left(\dfrac{\pi}{2}(x)\right)\right)$, $c = 1$

63. $f(x) = \dfrac{x^2+x+1}{x^2+2}$, $c = 1$

64. $f(x) = \dfrac{x-1}{x}$, $c = 1$

65. $f(x) = \csc(\pi x)$, $c = 0.999$

In Exercises 66–71, a function f and a point c are given. Graph the function

$$\phi(x) = \frac{f(x) - f(c)}{x - c}$$

in an appropriate viewing window centered about the line $x = c$. Use the graph of ϕ to decide whether $f'(c)$ exists. Explain the reason for your answer. If you answer that $f'(c)$ exists, use the graph of ϕ to approximate the value of $f'(c)$.

66. $f(x) = |x|$, $c = 0$

67. $f(x) = \sqrt{x^3 - 3x + 2}$, $c = 1$

68. $f(x) = \sqrt{x^4 - 2x^3 + 2x - 1}$, $c = 1$

69. $f(x) = \sqrt[3]{\cos(x)}$, $c = \pi/2$

70. $f(x) = \begin{cases} \sin^2(x)/x & \text{if } x \neq 0 \\ 1 & \text{if } x = 0 \end{cases}$, $c = 0$

71. $f(x) = \begin{cases} \sin(x)/x & \text{if } x \neq 0 \\ 1 & \text{if } x = 0 \end{cases}$, $c = 0$

In Exercises 72–75, graph function f in the indicated viewing rectangle R. From this graph, you will be able to detect at least one point at which f may not be differentiable.

a. By zooming in, if necessary, identify each point c for which $f'(c)$ does not exist.

b. Sketch or print your final graph and explain what feature of the graph indicates that f is not differentiable at c.

72. $f(x) = x\sqrt{|\cos(x)|}$, $R = [-3, 3] \times [-3, 3]$

73. $f(x) = (\cos(x))^{4/5}$, $R = [0, 3] \times [0, 1]$

74. $f(x) = x(\sin(x))^{1/3}$, $R = [1, 5] \times [-5, 2.5]$

75. $f(x) = \sqrt{|x|} \cdot (\cos(x^2))^{1/3}$, $R = [-1, 2] \times [-1.4, 1]$

76. Graph $f(x) = \cos(x)$ and $g(x) = 10^4(\cos(x + .0001) - \cos(x))$ in the viewing rectangle $[0, 2\pi] \times [-1, 1]$. What is the approximate relationship of g to f'? The function g is a good approximation to what well-known function? What formula do these graphs illustrate?

77. Let $f(x) = \sin(2x)$. Graph

$$g(x) = 10^4(f(x + 10^{-4}) - f(x))$$

for $0 \leq x \leq \pi$. Use your graph to identify $f'(x)$.

3.3 Rules for Differentiation

In Sections 3.1 and 3.2, we learned how to differentiate some specific functions. For example, the rate of change of a constant function is 0. Therefore:

The Derivative of a Constant Is Zero

If α is a real number and if $f(x) = \alpha$ for all x, then $f'(x) = 0$ for all x.

We have learned some other differentiation rules as well:

$$\frac{d}{dx}\sin(x) = \cos(x) \quad \text{and} \quad \frac{d}{dx}\cos(x) = -\sin(x).$$

For any constant α,

$$\frac{d}{dx}(\alpha x) = \alpha, \quad \frac{d}{dx}(\alpha x^2) = 2\alpha x, \quad \frac{d}{dx}(\alpha x^3) = 3\alpha x^2, \quad \text{and} \quad \frac{d}{dx}\left(\frac{\alpha}{x}\right) = -\frac{\alpha}{x^2}.$$

If we want to know the derivative of any other function f, we must calculate

$$f'(x) = \lim_{\Delta x \to 0} \frac{f(x + \Delta x) - f(x)}{\Delta x}.$$

This is the definition of the derivative, and all information about the derivative follows from it. Nevertheless, working directly with this limit can be difficult or tedious. Therefore, it is important to know some shortcuts, or rules, for calculating derivatives. In this section, we learn two types of rules. First, we learn formulas for the derivatives of basic functions such as those given above. We then learn rules that tell us how to differentiate algebraic combinations of basic functions.

Addition, Subtraction, and Scalar Multiplication

Let f and g be functions with known derivatives. Suppose α is a constant. Theorem 1 from Section 3.1 tells how to compute the derivatives of $f + g$ and $\alpha \cdot f$.

The Derivative of a Sum (or Difference) Is the Sum (or Difference) of the Derivatives

If $f'(c)$ and $g'(c)$ exist, then so do $(f + g)'(c)$ and $(f - g)'(c)$. Moreover,

$$(f + g)'(c) = f'(c) + g'(c) \quad \text{and} \quad (f - g)'(c) = f'(c) - g'(c).$$

The Derivative of a Constant Times a Function Is the Constant Times the Derivative of the Function

If $f'(c)$ exists and if α is any constant, then $(\alpha \cdot f)'(c)$ exists. Moreover,

$$(\alpha \cdot f)'(c) = \alpha \cdot f'(c).$$

Of course, these three equations may also be expressed using Leibniz notation.

$$\frac{d}{dx}(f + g) = \frac{d}{dx}f + \frac{d}{dx}g$$

$$\frac{d}{dx}(f - g) = \frac{d}{dx}f - \frac{d}{dx}g$$

$$\frac{d}{dx}(\alpha \cdot f) = \alpha \cdot \frac{d}{dx}f$$

Definition If f and g are functions and if α and β are constants, then we refer to $\alpha \cdot f + \beta \cdot g$ as a *linear combination* of f and g.

Theorem 1 from Section 3.1 gives us the rule for differentiating linear combinations.

The Derivative of a Linear Combination of Functions Is the Linear Combination of the Derivatives of the Functions

If $f'(c)$ and $g'(c)$ exist and if α and β are any constants, then $(\alpha \cdot f + \beta \cdot g)'(c)$ exists. Moreover,

$$(\alpha \cdot f + \beta \cdot g)'(c) = \alpha \cdot f'(c) + \beta \cdot g'(c).$$

Example 1 Calculate $\frac{d}{dx}(x^3 + 6\cos(x))$.

Solution We recall that $\frac{d}{dx}\cos(x) = -\sin(x)$ and $\frac{d}{dx}x^3 = 3x^2$. Therefore,

$$\begin{aligned}\frac{d}{dx}(x^3 + 6\cos(x)) &= \frac{d}{dx}(x^3) + \frac{d}{dx}(6\cos(x)) \\ &= 3x^2 + 6\frac{d}{dx}\cos(x) \\ &= 3x^2 - 6\sin(x).\end{aligned}$$ ■

Example 2 shows that the rule for differentiating a sum or difference can be applied repeatedly to differentiate sums or differences of any number of terms.

Example 2 Calculate the derivative of $6s^3 - 7s^2 + 2s - 5$ with respect to s.

Solution We reduce our differentiation problem to simpler differentiations that we already know. Thus,

$$\begin{aligned}\frac{d}{ds}((6s^3 - 7s^2) + (2s - 5)) &= \frac{d}{ds}(6s^3 - 7s^2) + \frac{d}{ds}(2s - 5) \\ &= (6 \cdot 3s^2 - 7 \cdot 2s) + (2 - 0) = 18s^2 - 14s + 2.\end{aligned}$$ ■

Products and Quotients

Now we turn to the differentiation of products and quotients. Since the derivative of a sum and a difference both turned out to be easy and rather obvious, one might assume that the derivative of a product or a quotient is equally simple. This assumption would be incorrect. For example, let $f(x) = x^2$ and $g(x) = x$. We have

Derivative of product		Product of derivatives
$\frac{d}{dx}(x^2 \cdot x) = \frac{d}{dx}(x^3) = 3x^2$	and	$\frac{d}{dx}(x^2) \cdot \frac{d}{dx}(x) = 2x \cdot 1 = 2x.$

Plainly, the derivative of a product is *not* the product of the derivatives.

Also,

Derivative of quotient		Quotient of derivatives
$\frac{d}{dx}\left(\frac{x^2}{x}\right) = \frac{d}{dx}(x) = 1$	and	$\frac{\frac{d}{dx}(x^2)}{\frac{d}{dx}(x)} = \frac{2x}{1} = 2x.$

So, the derivative of a quotient is *not* the quotient of the derivatives.

The correct formulas are a bit surprising but not difficult to use.

Theorem 1 ***The Product Rule*** If $f'(c)$ and $g'(c)$ exist, then $(f \cdot g)'(c)$ also exists and

$$(f \cdot g)'(c) = f'(c) \cdot g(c) + f(c) \cdot g'(c).$$

In words: *We differentiate a product by adding the second function times the derivative of the first to the first function times the derivative of the second.*

A proof of the Product Rule is provided in this chapter's Genesis & Development. In the meantime, let us see how this rule is used.

Example 3 Differentiate $H(x) = x \cdot \cos(x)$.

Solution We think of H as the product $f(x) \cdot g(x)$, where $f(x) = x$ and $g(x) = \cos(x)$. Therefore,

$$\begin{aligned} H'(x) &= f'(x) \cdot g(x) + f(x) \cdot g'(x) \\ &= \left(\frac{d}{dx}x\right) \cdot \cos(x) + x \cdot \frac{d}{dx}\cos(x) \\ &= 1 \cdot \cos(x) + x \cdot (-\sin(x)) \\ &= \cos(x) - x \cdot \sin(x). \end{aligned}$$

■

Sometimes it is necessary to differentiate a product of three or more functions. To treat such a problem, we use the associative property of multiplication and apply the Product Rule repeatedly, as in the next example.

Example 4 Differentiate $r \cdot \sin(r) \cdot \cos(r)$ with respect to r.

Solution We write

$$\begin{aligned} \frac{d}{dr}(r \cdot \sin(r) \cdot \cos(r)) &= \frac{d}{dr}((r \cdot \sin(r)) \cdot \cos(r)) \\ &= \left(\frac{d}{dr}(r \cdot \sin(r))\right) \cdot \cos(r) + (r \cdot \sin(r)) \cdot \frac{d}{dr}\cos(r) \\ &= \left(\left(\frac{d}{dr}r\right) \cdot \sin(r) + r \cdot \left(\frac{d}{dr}\sin(r)\right)\right) \cdot \cos(r) \\ &\qquad + (r \cdot \sin(r)) \cdot \frac{d}{dr}\cos(r) \\ &= (1 \cdot \sin(r) + r \cdot \cos(r)) \cdot \cos(r) + r \cdot \sin(r) \cdot (-\sin(r)) \\ &= r\cos^2(r) - r\sin^2(r) + \sin(r)\cos(r). \end{aligned}$$

■

We now learn to differentiate the quotient f/g of two functions. First we investigate the special case obtained by taking $f(x) = 1$. Because this case is useful, we record it separately.

Theorem 2 ***The Reciprocal Rule*** If $g'(c)$ exists and if $g(c) \neq 0$, then $(1/g)'(c)$ exists. Moreover,

$$\left(\frac{1}{g}\right)'(c) = \frac{-g'(c)}{g^2(c)}.$$

A proof of the Reciprocal Rule is provided in Section 3.5. For now, let us explore some of its consequences.

Example 5 Show that

$$\frac{d}{dx}\sec(x) = \sec(x)\tan(x).$$

Solution Since $\sec(x) = 1/\cos(x)$, the Reciprocal Rule applies.

$$\begin{aligned}
\frac{d}{dx}\sec(x) &= -\frac{\frac{d}{dx}\cos(x)}{\cos^2(x)} \\
&= -\frac{-\sin(x)}{\cos^2(x)} \\
&= \frac{\sin(x)}{\cos^2(x)} \\
&= \frac{1}{\cos(x)} \cdot \frac{\sin(x)}{\cos(x)} \\
&= \sec(x) \cdot \tan(x)
\end{aligned}$$

■

Theorem 3 ***The Quotient Rule*** If $f'(c)$ and $g'(c)$ exist and if $g(c) \neq 0$, then $(f/g)'(c)$ exists. Moreover,

$$\left(\frac{f}{g}\right)'(c) = \frac{g(c) \cdot f'(c) - f(c) \cdot g'(c)}{g^2(c)}.$$

Proof Notice that the quotient f/g may be written as the product $f \cdot (1/g)$. The Reciprocal Rule tells us that $(1/g)'(c)$ exists. The Product Rule then implies that $(f \cdot (1/g))'(c)$ exists and

$$\begin{aligned}
\left(\frac{f}{g}\right)'(c) &= \left(f \cdot \left(\frac{1}{g}\right)\right)'(c) \\
&= f'(c) \cdot \left(\frac{1}{g}\right)(c) + f(c) \cdot \left(\frac{1}{g}\right)'(c) \\
&= f'(c) \cdot \left(\frac{1}{g(c)}\right) + f(c) \cdot \left(\frac{-g'(c)}{g^2(c)}\right) \\
&= f'(c) \cdot \left(\frac{g(c)}{g^2(c)}\right) + f(c) \cdot \left(\frac{-g'(c)}{g^2(c)}\right) \\
&= \frac{g(c) \cdot f'(c) - f(c) \cdot g'(c)}{g^2(c)}.
\end{aligned}$$

Combine terms, using common denominator $g^2(c)$ ■

Example 6 Calculate the derivative of

$$H(x) = \frac{x^2 + 1}{x^3}.$$

Solution Observe that $H(x) = f(x)/g(x)$ where $f(x) = x^2+1$ and $g(x) = x^3$. Thus,

$$\begin{aligned} H'(x) &= \frac{g(x) \cdot f'(x) - f(x) \cdot g'(x)}{(g(x))^2} \\ &= \frac{(x^3) \cdot (2x) - (x^2+1) \cdot (3x^2)}{(x^3)^2} \\ &= \frac{-x^4 - 3x^2}{x^6} = -\frac{(x^2+3)}{x^4}. \end{aligned}$$

Notice that this calculation makes sense only if $g(x) \neq 0$—that is, when $x \neq 0$. ■

Numeric Differentiation

The rules in this section enable us to differentiate extremely complicated functions. The application of these rules, however, can be laborious. Moreover, even when we use a differentiation rule to obtain an exact evaluation of a derivative, we may have to approximate constants such as $\sqrt{2}$ and π that appear in our answer. It is, therefore, convenient to have a method for approximating the numerical value $f'(c)$ right from the outset. Such a procedure is known as *numeric differentiation*.

Suppose that f is a function defined on the interval (a, b) and differentiable at $c \in (a, b)$. Since

$$f'(c) = \lim_{h \to 0} \frac{f(c+h) - f(c)}{h},$$

we can approximate $f'(c)$ by the difference quotient $(f(c+h) - f(c))/h$ for a small value of h. When $h > 0$ and c is fixed, the quotient

$$D_+ f(c, h) = \frac{f(c+h) - f(c)}{h}$$

is known as a *forward difference quotient*. The *backward difference quotient* is defined by

$$D_- f(c, h) = \frac{f(c) - f(c-h)}{h}.$$

The *central difference quotient* is defined by

$$D_0 f(c, h) = \frac{f(c + h/2) - f(c - h/2)}{h}.$$

Each of the three difference quotients can be used to approximate $f'(c)$. However, the situation illustrated by Figure 1 is typical: For a given value of h, the central difference quotient usually gives the best approximation to the derivative.

When we use the approximation

$$D_0 f(c, h) \approx f'(c),$$

the question of *how* to choose h in an optimal way is not so easy to answer. Consider $f(x) = x^4$ and $c = 1/2$. The exact value of $f'(c)$ is $1/2$. Now consider Table 1, which shows difference quotients that were computed using ten significant digits, of which six are exhibited.

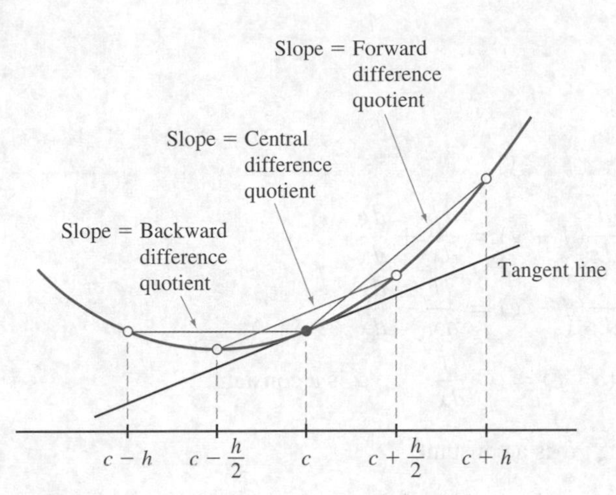

Figure 1

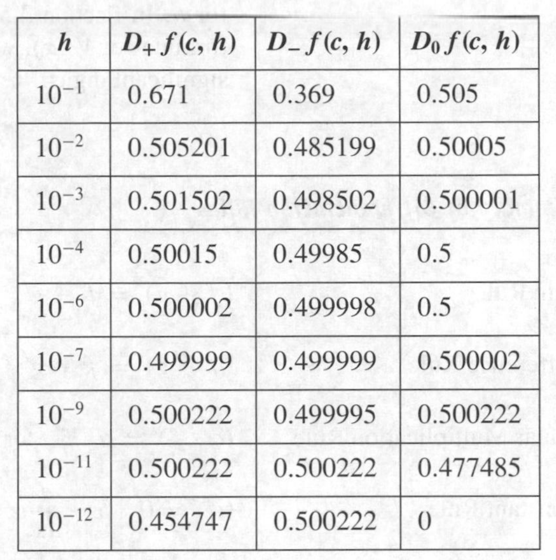

h	$D_+ f(c, h)$	$D_- f(c, h)$	$D_0 f(c, h)$
10^{-1}	0.671	0.369	0.505
10^{-2}	0.505201	0.485199	0.50005
10^{-3}	0.501502	0.498502	0.500001
10^{-4}	0.50015	0.49985	0.5
10^{-6}	0.500002	0.499998	0.5
10^{-7}	0.499999	0.499999	0.500002
10^{-9}	0.500222	0.499995	0.500222
10^{-11}	0.500222	0.500222	0.477485
10^{-12}	0.454747	0.500222	0

Table 1

As you read down each column, you will notice that with decreasing h, each difference quotient seems to approach the limit 0.5 at first. Observe that the central difference quotient converges quickest. In each column, however, accuracy is degraded when h becomes *too small*. This is the result of *loss of significance,* a problem discussed in Section 1.1.

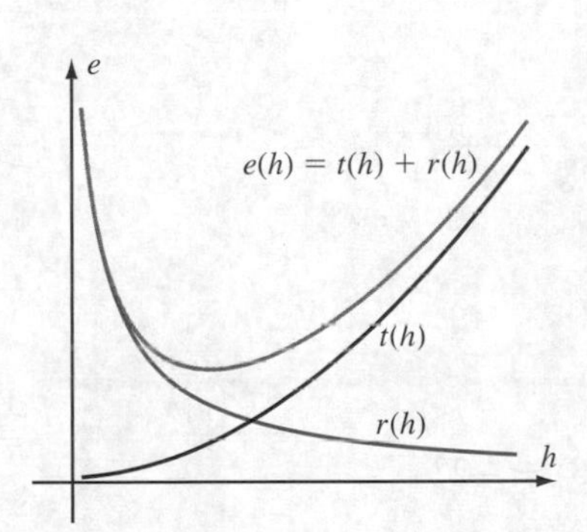

Figure 2

It is difficult to predict the value of h at which accuracy begins to deteriorate. There are two types of error that result when we use $D_0 f(c, h)$ to approximate $f'(c)$. As we learned in Chapter 1, there is a *roundoff error* $r(h)$. Terminating the limit process

$$f'(c) = \lim_{h \to 0} D_0 f(c, h)$$

by stopping at a particular value of h also introduces a *truncation error* $t(h)$. The *total error* $e(h)$ is given by $e(h) = t(h) + r(h)$. Figure 2 shows typical graphs of these error functions. Unfortunately, there is no way to ensure that $t(h)$ and $r(h)$ are *both* very small. One strategy is to let h become smaller and smaller until either the desired accuracy is attained or the total error begins to increase.

Example 7 Let $V(x) = \sqrt{2 + \sin(x)}$. Approximate $V'(0.4174)$ to four significant digits.

Solution Table 2 presents the results of our central difference quotient calculations. (Ten digits have been carried in the computations, six digits are tabulated, but no more than four are significant.)

h	0.1	0.01	0.001	0.0001	0.00001	0.000001
$D_0 V(0.4174, h)$	0.294630	0.294708	0.294709	0.294710	0.294700	0.295000

Table 2

Since the leftmost four digits of $D_0V(0.4174, h)$ agree when $h = 10^{-2}$ and 10^{-3}, we can be reasonably certain that $V'(0.4174) = 0.2947$ to four significant digits and that the last column represents a decline in accuracy caused by using too small a value for h. In Section 3.5, we will show that $V'(x) = \cos(x)/(2\sqrt{2+\sin(x)})$. With this formula for $V'(x)$, we can verify that our numerical differentiation is correct to four significant digits. ■

Summary of Differentiation Rules

Sum Rule	$(f+g)' = f' + g'$ or $\dfrac{d}{dx}(f+g) = \dfrac{df}{dx} + \dfrac{dg}{dx}$
Difference Rule	$(f-g)' = f' - g'$ or $\dfrac{d}{dx}(f-g) = \dfrac{df}{dx} - \dfrac{dg}{dx}$
Scalar Multiplication Rule	$(\alpha \cdot f)' = \alpha \cdot f'$ or $\dfrac{d}{dx}(\alpha \cdot f) = \alpha \cdot \dfrac{df}{dx}$ α is a constant.
Constant Rule	$(\alpha)' = 0$ or $\dfrac{d}{dx}\alpha = 0$ α is a constant.
Product Rule	$(f \cdot g)' = f' \cdot g + f \cdot g'$ or $\dfrac{d}{dx}(f \cdot g) = \dfrac{df}{dx} \cdot g + f \cdot \dfrac{dg}{dx}$
Reciprocal Rule	$\left(\dfrac{1}{g}\right)' = -\dfrac{g'}{g^2}$ or $\dfrac{d}{dx}\left(\dfrac{1}{g}\right) = -\dfrac{\frac{dg}{dx}}{g^2}$
Quotient Rule	$\left(\dfrac{f}{g}\right)' = \dfrac{g \cdot f' - f \cdot g'}{g^2}$ or $\dfrac{d}{dx}\left(\dfrac{f}{g}\right) = \dfrac{g \cdot \frac{df}{dx} - f \cdot \frac{dg}{dx}}{g^2}$
Approximate Derivative Rule	$f'(c) \approx \dfrac{f(c+h/2) - f(c-h/2)}{h}$

quickquiz

1. What is $f'(3)$ if $f(x) = 4g(x) + 7$ and $g'(3) = -2$?
2. What is the derivative of $x^2 \sin(x)$?
3. What is the derivative of $\sin(x)/x$?
4. Approximate the derivative of $\tan(x)$ at $x = 0$. Use the central difference quotient approximation with $h = 0.01$.

EXERCISES

Problems for Practice

In Exercises 1–16, compute the derivative of the given function.

1. $4x^3 + 3x^2$

2. $x^3 \cdot \cos(x)$

3. $(x^2 - 5x) \cdot (4x^3 + x^2)$

4. $7(x + x^2)\sin(x)$

5. $5\sin(x) - 6x\cos(x)$

6. $\cos^2(x)$

7. $\dfrac{x}{\sin(x)}$

8. $\dfrac{\sin(x)}{x}$

9. $\dfrac{1}{x^3 + x^2 + 1}$

10. $\dfrac{\sin(x)}{x - 5}$

11. $\dfrac{x^2 + 1}{x^2 + 2}$

12. $(x^3 + 5) \cdot \sec(x)$

13. $\dfrac{3x}{x^2+1}$

14. $\dfrac{9}{3\cos(x)+x^3}$

15. $\dfrac{x\sin(x)}{x+\sin(x)}$

16. $\dfrac{x^2+\cos(x)}{3x-\sin(x)}$

To calculate the derivative of each function in Exercises 17–22, you must apply the Product or Quotient Rule several times. Use parentheses to organize your calculations. (Do not expand products prior to differentiation. For example, after expansion, the product in Exercise 19 becomes $x^6+3x^4+2x^2$, which can be routinely differentiated but which does not provide practice in successively applying the rules of differentiation.)

17. $\dfrac{(x^2+4)(x-5)}{x-1}$

18. $\dfrac{\sin(x)}{(x+1)(x+3)}$

19. $x(x^2+1)(x^3+2x)$

20. $\sin^3(x)$

21. $\sin^2(x)\cdot\sec(x)$

22. $(x^2+1)^3$

In Exercises 23–26, find the tangent line to the graph of the function at point *P*.

23. $f(x)=(3x^2-5)/(x+4),\ P=(1,-2/5)$
24. $f(x)=(x-\pi/2)\cdot\cos(x),\ P=(\pi,-\pi/2)$
25. $f(x)=\sin(x)/(x+1),\ P=(0,0)$
26. $f(x)=x^2\cdot\cos(x),\ P=(2\pi/3,-2\pi^2/9)$

In Exercises 27–30, use the information to estimate $f'(c)$ at point c.

27. $f(4)=5.7$ and $f(4.1)=6.2,\ c=4$
28. $f(4)=5.7$ and $f(4.1)=6.2,\ c=4.1$
29. $f(\pi+0.01)=f(\pi)+0.2,\ c=\pi$
30. $f(3.47)=2.61$ and $f(3.49)=2.67,\ c=3.48$

Further Theory and Practice

31. Let $f_n(x)=x^n$ for each nonnegative integer n. Notice that $f_{n+1}(x)=xf_n(x)$. Use the Product Rule to show that

$$f'_{n+1}(x)=f_n(x)+xf'_n(x). \tag{3.15}$$

We know that the equation

$$\frac{d}{dx}x^n=nx^{n-1} \tag{3.16}$$

is valid for $n=0$, 1, 2, and 3. Use equation (3.15) to show that whenever equation (3.16) is true for a particular nonnegative integer value of n, it is also true for $n+1$. Deduce that equation (3.16) is true for all nonnegative integers. What is $\frac{d}{dx}x^{10}$?

In Exercises 32–35, use any method to calculate the derivative of the expression.

32. $\sec^2(x)$
33. $\tan(x)$
34. x^{-10}
35. $\sin(2x)$

In Exercises 36–39, use the information to find the requested derivative.

36. Given $f(2)=3$, $f'(2)=5$, $g(2)=8$, and $g'(2)=-6$, find $(f\cdot g)'(2)$ and $(f/g)'(2)$.
37. Given $f(3)=-4$ and $f'(3)=6$, find $(1/f)'(3)$ and $(6f)'(3)$.
38. Given $f(-1)=5$, $g(-1)=4$, $f'(-1)=0$, and $g'(-1)=-8$, find $(f+g)'(-1)$, $(f\cdot g)'(-1)$, $(f^2)'(-1)$, and $(g^3)'(-1)$.
39. Given $f(4)=2$, $g(4)=1$, $f'(4)=-5$, and $g'(4)=-9$, find $(f\cdot g^2)'(4)$, $(g\cdot f^2)'(4)$, $(1/f^2)'(4)$, and $(g/f)'(4)$.
40. Find the points on the graph of the function $f(x)=x^3-2x^2-8x+3$ at which the tangent is parallel to $y=4-9x$.
41. Find all values of c for which the tangent lines to the graphs of $f(x)=x^2-7x+9$ and $g(x)=9/x$ at $(c,f(c))$ and $(c,g(c))$ are parallel.
42. Find all values of c for which the tangent lines to the graphs of $f(x)=x^3-8x+3$ and $g(x)=4/x$ at $(c,f(c))$ and $(c,g(c))$ are parallel.
43. Figure 3 shows the graph of $f(x)=2x/(x+1)$. In this exercise, we need to find every value of x such that the tangent line to the graph of f at $(x,f(x))$ passes through the point $(1,3)$.

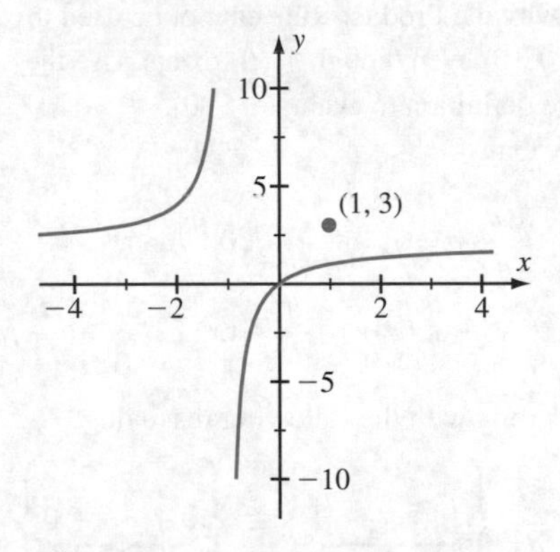

Figure 3
The graph of $f(x)=\frac{2x}{x+1}$

a. What is the domain of f? Sketch the graph of f. How many solutions do there appear to be? Sketch the tangents that satisfy the requirement. Estimate the values of x at the points of tangency. What are their relationships with respect to $x = -1$?

b. What is the slope of the tangent line to the graph of f at $(c, f(c))$? By inspection, the tangents to the graph of f all slope upward. Is this property consistent with the algebraic expression that gives the slopes of the tangents? Explain.

c. Write the equation of the tangent line to the graph of f at $(c, f(c))$.

d. The property of the tangent line passing through point $(1, 3)$ imposes an equation that c must satisfy. What is this equation? How many solutions does it have? (This number should equal the number of solutions that you predicted in part a.) Find the explicit solutions.

44. If f is a differentiable function, find a formula for $(f^2)'(x)$. Next find a formula for $(f^3)'(x)$. What do you expect the formula for $(f^n)'(x)$ to be? (Predict by analogy but do not prove this last formula. These matters will be taken up later in the chapter.)

45. Show that if α is a root of multiplicity 3 of the polynomial p, then α is a root of p' of multiplicity 2. (The multiplicity of α is k if $(x-\alpha)^k$ is a factor of p, but $(x-\alpha)^{k+1}$ is not.)

46. Show that if $f(x) = |x|$, then $(f^2)'(0)$ exists even though $f'(0)$ does not.

47. The functions $f(x) = x^\alpha$ are not differentiable at 0 for values of $\alpha < 1$. For which of these values of α is f^2 differentiable at 0?

48. Let $f(x) = xg(x)$ where g is the continuous function defined by $g(x) = x\sin(1/x)$ for $x \neq 0$ and $g(0) = 0$. Explain why the Product Rule cannot be used to evaluate $f'(0)$ even though $f'(0)$ exists. Use the derivative definition to evaluate $f'(0)$.

49. We know that

$$\frac{d}{dx}x = 1, \quad \frac{d}{dx}1 = 0, \quad \text{and}$$

$$\frac{d}{dx}\{f_1(x) + \cdots + f_n(x)\} = \frac{d}{dx}f_1(x) + \cdots + \frac{d}{dx}f_n(x).$$

What is wrong with the following reasoning?

$$\frac{d}{dx}x = \frac{d}{dx}\left\{\underbrace{1 + \cdots + 1}_{x \text{ summands}}\right\} = \left\{\underbrace{0 + \cdots + 0}_{x}\right\} = 0.$$

50. Show that if $f(x) = x^2$, then the central difference quotient $D_0 f(c, h)$ for approximating $f'(c)$ is exactly equal to $f'(c)$.

51. Show that the central difference quotient $D_0 f(c, h)$ is an average of a forward and a backward difference quotient.

52. Evaluate the limit

$$L = \lim_{h\to 0} \frac{\sqrt{1+h} - 1}{h}$$

by identifying it as a derivative. Obtain another limit formula by expressing L as a limit of central difference quotients.

53. Assume that g is differentiable and nonvanishing and that $1/g$ is differentiable. Derive the Reciprocal Rule for g from the Product Rule applied to g and $1/g$.

54. Suppose that a consumer's satisfaction with purchasing x units of good X and y units of good Y can be quantified. The locus of points (x, y) for which the consumer's satisfaction is a fixed value is called an *indifference curve:* The consumer is equally satisfied at all points on the curve. Figure 4 illustrates a typical indifference curve. Most consumers have a fixed budget for the purchase of goods X and Y. The locus of points that represent the combination of goods X and Y that the consumer can actually purchase on this budget is a straight line called the *budget line*. The point of tangency between the budget line and an indifference curve is called the *consumer equilibrium*. If $(2, 1/5)$ is the equilibrium point for a consumer whose indifference curve is

$$3y(x^2+1) + 5x(y+1) = 15,$$

find the consumer's budget line in the form

$$Ax + By = C.$$

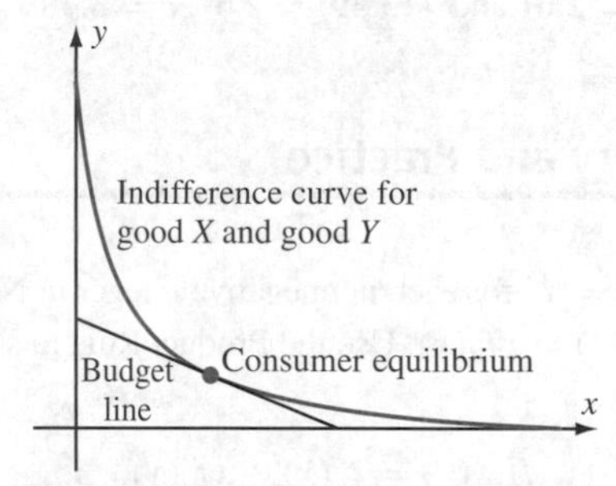

Figure 4

55. In economics, if $q(p)$ is the demand for a product at price p—that is, the number of units of the product that are sold at price p—then

$$E(p) = -q'(p) \cdot \frac{p}{q(p)}$$

is called the *elasticity of demand*.

a. Let Δp denote a small change in p, and $\Delta q = q(p + \Delta p) - q(p)$ the corresponding change in q. Show that for Δp sufficiently small we have the approximation

$$E(p) \approx -\frac{\Delta q / q(p)}{\Delta p / p}.$$

b. Elasticity of demand for potatoes at standard supermarket prices has been estimated to be 0.31. Approximately what percentage change in potato consumption will result from a 2% increase in the price of potatoes?

c. Over a fixed period, a television repair service recorded the number of repairs that it made at three different prices.

Price	Contracts Obtained
\$90	96
\$80	200
\$70	300

Estimate the elasticity of demand at $p = \$70$ and at $p = \$90$. Use a central difference quotient to estimate elasticity of demand at $p = \$80$. (Elasticity of demand for television repair was estimated to be 3.84 in the 1970s.)

Calculator/Computer Exercises

In Exercises 56–59, a function f and a point c are given. Prepare a table of the forward, backward, and central difference quotients, $D_+ f(c, h)$, $D_- f(c, h)$, and $D_0 f(c, h)$, respectively, for $h = 10^{-n}$, $1 \leq n \leq 5$.

56. $f(x) = \sin(\pi x)$, $c = 1/4$

57. $f(x) = \cos(\pi x)$, $c = 1$

58. $f(x) = (x^3 + 3x^2 + 2)/(x^2 - 2)$, $c = 2$

59. $f(x) = \sqrt{x + 1/x}$, $c = 3$

In Exercises 60–63, find $f'(c)$ (exactly) for function f and value c of x. Then approximate $f'(c)$ to five decimal places by (a) finding a floating point evaluation of the exact answer and (b) using a central difference quotient $D_0 f(c, h)$. Record the value of h used.

60. $f(x) = \sqrt{2x}$, $c = 1$

61. $f(x) = \sin(x)$, $c = \pi/6$

62. $f(x) = \cos(x)$, $c = \pi/4$

63. $f(x) = 1/x$, $c = 1/\sqrt[4]{2}$

64. Let $f(x) = x^4/4$. Graph $x \mapsto D_0 f(x, 10^{-5})$ for $-1.5 \leq x \leq 1.5$. What is $f'(x)$? Add the graph of $f'(x)$ to your viewing window. (The graphs should nearly coincide.)

65. Let $f(x) = \cos(x)$. Graph $x \mapsto D_0 f(x, 10^{-5})$ for $0 \leq x \leq 2\pi$. What is $f'(x)$? Add the graph of $f'(x)$ to your viewing window. (The graphs should nearly coincide.)

66. Let $f(x) = \sin^2(x)$. Graph $x \mapsto D_0 f(x, 10^{-5})$ for $0 \leq x \leq 2\pi$. Now compute f' and add the graph to the viewing window. (The graphs should nearly coincide.)

67. Let $f_a(x) = a^x$. Let $L(a) = D_0 f_a(0, 10^{-5})$. Compute $L(2.70)$ and $L(2.73)$. Graph L for $2.70 \leq a \leq 2.73$ to see that there is a value a_0 in the interval $[2.70, 2.73]$ such that $L(a_0) = 1$. Narrow down the interval in which a_0 must lie until you have found a_0 to three significant digits. What is

$$\left.\frac{d}{dx} a_0^x\right|_{x=0} ?$$

3.4 Differentiation of Some Basic Functions

In Section 3.3, we learned how to differentiate functions that are formed arithmetically from simpler functions, provided we know how to differentiate the simpler ones. In this section, we build the library of basic differentiable functions. Your work in calculus will be easier if you commit the derivatives in this section to memory.

Powers of *x*

We have learned to differentiate x^p for the integers $p = -1, 0, 1, 2$, and 3. For these values of p, we know that

$$\frac{d}{dx}x^p = p \cdot x^{p-1}. \tag{3.17}$$

The next example shows that formula (3.17) is also valid for $p = 1/2$.

Example 1 Show that

$$\frac{d}{dx}\sqrt{x} = \frac{1}{2\sqrt{x}}$$

for all $x > 0$.

Solution We verify this formula by referring to the derivative definition. Because

$$\begin{aligned}\frac{\sqrt{x+\Delta x}-\sqrt{x}}{\Delta x} &= \frac{\left(\sqrt{x+\Delta x}-\sqrt{x}\right)\left(\sqrt{x+\Delta x}+\sqrt{x}\right)}{\Delta x\left(\sqrt{x+\Delta x}+\sqrt{x}\right)}\\ &= \frac{(x+\Delta x)-x}{\Delta x\left(\sqrt{x+\Delta x}+\sqrt{x}\right)}\\ &= \frac{1}{\sqrt{x+\Delta x}+\sqrt{x}},\end{aligned}$$

we conclude that

$$\frac{d}{dx}\sqrt{x} = \lim_{\Delta x\to 0}\frac{\sqrt{x+\Delta x}-\sqrt{x}}{\Delta x} = \lim_{\Delta x\to 0}\frac{1}{\sqrt{x+\Delta x}+\sqrt{x}} = \frac{1}{2\sqrt{x}}.$$ ■

Now that we have verified formula (3.17) for a noninteger value of p, it is reasonable to wonder if the formula is valid for all values of p. Indeed it is. We record this fact as a theorem for now, but we defer the proof until Chapter 6. In the meantime, you should use this differentiation formula wherever it is needed.

Theorem 1 ***The Power Rule*** If p is any real number, then

$$\frac{d}{dx}x^p = px^{p-1}.$$

Example 2 In an adiabatic (no heat in or out) argon system, the third power of pressure P is inversely proportional to the fifth power of volume V. At a pressure of 1000 atm (atmospheres), the volume of the argon sample is 34 cm^3. What is the rate of change of pressure with respect to volume when the sample is 50 cm^3?

Solution There is a proportionality constant k such that $P^3 = k/V^5$, or $P = cV^{-5/3}$, where $c = k^{1/3}$. We solve for c:

$$1000 \text{ atm} = c \cdot (34 \text{ cm}^3)^{-5/3},$$

Example 2 shows that differentiating the polynomial $3x^5 - 7x^9$ results in another polynomial, $15x^4 - 63x^8$, of one degree less. A similar calculation establishes the same result for any polynomial.

or

$$c = 1000 \cdot 34^{5/3} \text{ atm cm}^5 = 3.6 \times 10^5 \text{ atm cm}^5.$$

Therefore,

$$\begin{aligned} \frac{dP}{dV} &= 3.6 \times 10^5 \left(-\frac{5}{3}\right) V^{-5/3-1} \text{ atm cm}^5 \\ &= -6.0 \times 10^5 V^{-8/3} \text{ atm cm}^5 \end{aligned}$$

and

$$\left.\frac{dP}{dV}\right|_{V=50\,\text{cm}^3} = -6.0 \times 10^5 \times (50\,\text{cm}^3)^{-8/3} \text{ atm cm}^5 \approx -17 \text{ atm cm}^{-3}. \quad \blacksquare$$

Example 3 Differentiate $3/x^{12}$ and $3x^5 - 7x^9$ with respect to x.

Solution We have

$$\frac{d}{dx}\left(\frac{3}{x^{12}}\right) = 3\frac{d}{dx}(x^{-12}) = 3 \cdot (-12)x^{-13} = -\frac{36}{x^{13}}$$

and

$$\begin{aligned} \frac{d}{dx}(3x^5 - 7x^9) &= \frac{d}{dx}(3x^5) - \frac{d}{dx}(7x^9) \\ &= 3\frac{d}{dx}(x^5) - 7\frac{d}{dx}(x^9) \\ &= 3 \cdot 5x^4 - 7 \cdot 9x^8 && \text{Power Rule} \\ &= 15x^4 - 63x^8. \end{aligned} \quad \blacksquare$$

Theorem 2 The derivative p' of a polynomial p is a polynomial. If p is not a constant, then the degree of p' is one less than the degree of p. If p is a constant, then p' is identically 0.

Trigonometric Functions

We continue to build our library of differentiation formulas by finding the derivatives of the remaining trigonometric functions. We have already seen how to differentiate sine and cosine (Section 3.2) and secant (Example 5 from Section 3.3). We can now easily handle the other trigonometric functions. Before doing so, recall that we often write $\sin^2(x)$ or $\tan^3(x)$ rather than $(\sin(x))^2$ or $(\tan(x))^3$.

Example 4 Use the Quotient Rule to show that

$$\frac{d}{dx}\tan(x) = \sec^2(x), \quad \frac{d}{dx}\cot(x) = -\csc^2(x), \quad \text{and}$$

$$\frac{d}{dx}\csc(x) = -\csc(x)\cot(x).$$

Solution We will only write the proof of the first of these equations. The other two can be obtained in a similar manner and are left as Exercises 63 and 64.

$$\begin{aligned}\frac{d}{dx}\tan(x) &= \frac{d}{dx}\left(\frac{\sin(x)}{\cos(x)}\right)\\ &= \frac{\cos(x)\cdot\frac{d}{dx}(\sin(x)) - \sin(x)\cdot\frac{d}{dx}(\cos(x))}{\cos^2(x)}\\ &= \frac{\cos^2(x)+\sin^2(x)}{\cos^2(x)}\\ &= \frac{1}{\cos^2(x)}\\ &= \sec^2(x).\end{aligned}$$

■

Example 5 Find the tangent line to the graph of the curve $y = x \cdot \cot(x)$ at the point $(\pi/6, \pi\sqrt{3}/6)$.

Solution We first find the slope:

$$\frac{dy}{dx} = \frac{dy}{dx}(x\cdot\cot(x)) = 1\cdot\cot(x) + x\cdot(-\csc^2(x)) = \cot(x) - x\cdot\csc^2(x).$$

Evaluating at $x = \pi/6$, we find the slope to be $\sqrt{3} - (\pi/6)\cdot 2^2 = \sqrt{3} - 2\pi/3$. Since the line also passes through $(\pi/6, \pi\sqrt{3}/6)$, the point-slope form of its equation is

$$y = (\sqrt{3} - 2\pi/3)\cdot\left(x - \frac{\pi}{6}\right) + \frac{\pi\sqrt{3}}{6}.$$

■

The Derivative of the Natural Exponential Function

In Section 2.6, we introduced the number e and derived the basic formula

$$e^x = \lim_{n\to\infty}\left(1 + \frac{x}{n}\right)^n.$$

The function $x \mapsto e^x$ is sometimes called the *natural exponential function* (or the *exponential function,* for short). It is often abbreviated as exp (in the same way that cos is used as an abbreviation for cosine). Thus, $\exp(x)$ and e^x denote the same value. In many applications, it is necessary to know the derivative of the natural exponential function.

Theorem 3 For every x,

$$\frac{d}{dx}e^x = e^x.$$

Proof A complete proof is given in Chapter 6. At this time, we simply present convincing evidence of the validity of this theorem. Let $f(x) = e^x$. We first examine

$$f'(0) = \lim_{h\to 0}\frac{e^h - e^0}{h} = \lim_{h\to 0}\frac{e^h - 1}{h}.$$

Table 1 shows values of the difference quotient for small *positive* h. Table 2 is similar, except that small *negative* values of h have been used.

h	$\lim_{h\to 0}(e^h - 1)/h$
10^{-3}	1.00050016
10^{-4}	1.00005000
10^{-5}	1.00000500
10^{-6}	1.00000050
10^{-7}	1.00000005
10^{-8}	1.00000000

Table 1

h	$\lim_{h\to 0}(e^h - 1)/h$
-10^{-3}	0.99950016
-10^{-4}	0.99995000
-10^{-5}	0.99999500
-10^{-6}	0.99999950
-10^{-7}	0.99999995
-10^{-8}	0.99999999

Table 2

The numerical evidence for the equation $f'(0) = 1 = f(0)$ is certainly compelling. For now, let us accept that $f'(0) = 1$. The derivative of f at a general value x can be reduced to the $x = 0$ case as follows:

$$\begin{aligned} f'(x) &= \lim_{h\to 0} \frac{f(x+h) - f(x)}{h} \\ &= \lim_{h\to 0} \frac{e^{x+h} - e^x}{h} \\ &= \lim_{h\to 0} \frac{e^x e^h - e^x}{h} \\ &= e^x \lim_{h\to 0} \frac{e^h - 1}{h} \\ &= e^x \lim_{h\to 0} \frac{f(0+h) - f(0)}{h} \\ &= e^x f'(0) \\ &= e^x. \end{aligned}$$

■

Example 6 Suppose an object vibrates about its equilibrium position according to the formula $x(t) = e^{-t}\sin(t)$ mm. (Resistance to movement gives rise to the "damping" factor e^{-t}.) At what speed is the object moving when $t = 1$?

Solution We compute $x'(t)$ by means of the Quotient Rule:

$$\begin{aligned} x'(t) &= \frac{d}{dt}\left(\frac{\sin(t)}{e^t}\right) \\ &= \frac{e^t \frac{d}{dt}\sin(t) - \sin(t)\frac{d}{dt}e^t}{(e^t)^2} \\ &= \frac{e^t\cos(t) - e^t\sin(t)}{e^{2t}} \\ &= \frac{\cos(t) - \sin(t)}{e^t}. \end{aligned}$$

Therefore,

$$x'(1) = \frac{\cos(1) - \sin(1)}{e} \approx -0.1108.$$

■

Summary of Derivatives of Special Functions

Power Rule

If p is any real number, then $\frac{d}{dx}x^p = px^{p-1}$.

Trigonometric Functions

$$\frac{d}{dx}\sin(x) = \cos(x) \quad \frac{d}{dx}\tan(x) = \sec^2(x) \quad \frac{d}{dx}\sec(x) = \sec(x)\tan(x)$$

$$\frac{d}{dx}\cos(x) = -\sin(x) \quad \frac{d}{dx}\cot(x) = -\csc^2(x) \quad \frac{d}{dx}\csc(x) = -\csc(x)\cot(x)$$

Natural Exponential Function $\frac{d}{dx}e^x = e^x$

quickquiz

1. What is the rule for differentiating x^p with respect to x?
2. What are the derivatives of each of the six trigonometric functions?
3. What is $\lim_{x\to 0}(e^x - 1)/x$?
4. What is the derivative of xe^x?

EXERCISES

Problems for Practice

In Exercises 1–20, compute the derivative of the function.

1. $f(x) = 8x^{10} - 6x^{-5}$
2. $g(x) = x^{-9}\cot(x)$
3. $H(x) = \csc(x)\cot(x)$
4. $f(x) = x/(\tan(x) + \sec(x))$
5. $g(x) = \sec(x)/(\sqrt{x} + 1)$
6. $H(x) = \csc^2(x)$
7. $f(x) = \sec(x) - \tan(x)$
8. $g(x) = \csc(x) + \cot(x)$
9. $H(x) = x\cot(x) - \csc(x)$
10. $f(x) = x^2\tan(x) + \cot(x)$
11. $g(x) = \tan(x)\sec(x)$
12. $H(x) = (x^{-5} + 7x^6)\csc(x)$
13. $f(x) = x^{-5}e^x$
14. $g(x) = \cot(x)/(5x - \csc(x))$
15. $H(x) = \tan^2(x)$
16. $f(x) = \sqrt{x^5} - 3x^{-7}$
17. $g(x) = x/(2 + \tan(x))$
18. $H(x) = \cos(x)/(3 - \sec(x))$
19. $f(x) = x^2/(e^x + \sin(x))$
20. $g(x) = (\sqrt[4]{x} - 7x^{-2})(\sin(x) + \tan(x))$

In Exercises 21–26, find the tangent line to the graph of the function at the indicated point.

21. $f(x) = x^{-3}\cdot\sin(x)$, $x = \pi$
22. $f(x) = x^7 + \tan(x)$, $P = (0, 0)$
23. $f(x) = 3\tan(x)\cdot\sec(x)$, $x = \pi/3$

24. $f(x) = e^x, x = 0$
25. $f(x) = x\tan(x), x = \pi$
26. $f(x) = \csc(x)/x, x = \pi/2$

In Exercises 27–30, compute the derivative of the function.

27. $f(x) = 2x + e^x$
28. $f(x) = e^{2x}$
29. $f(x) = e^{-x}\cos(x)$
30. $f(x) = (e^x + 1)/(e^x - 1)$

Further Theory and Practice

In Exercises 31–34, find a polynomial with a derivative that is the given polynomial.

31. $7x^6 - 4x + 6$
32. $x^9 - 2x^3 - 1$
33. $x^8 + 6x^5 - 3x$
34. $10x^6 + x^2 + 4x - 3$

In Exercises 35–38, find a function with a derivative that is the given function.

35. $\cos(x)$
36. $3\sec^2(x)$
37. $-8\csc^2(x)$
38. $\sin(x)(5\sec^2(x) - 1)$
39. Find a linear combination $f(x)$ of $\cos(x)$ and $\sin(x)$ such that

$$f'(x) + f(x) = 3\cos(x) - 6\sin(x).$$

40. If α is a root of a polynomial p of multiplicity m, show that α is a root of p' of multiplicity $m - 1$.
41. What is the equation of the line that is tangent to the graph of $y = e^x$ at the point (t, e^t)? What are the x- and y-intercepts of this line?
42. Find a function with a derivative that is $\tan^2(x)$.
43. Let p be a fixed number. Let C be any number. Show that $y(x) = Cx^p$ is a solution of the differential equation

$$\frac{dy}{dx} = p\frac{y}{x}.$$

In Exercises 44–47, compute $f'(c)$ for the given f and c.

44. $f(x) = (x^2 + 1)/(2x + 1)^2, c = -1$
45. $f(x) = (\cos(x) - \sin(x))/(\cos(x) + \sin(x)), c = \pi/4$
46. $f(x) = (x^5 + 2x^4 - 5x^3 + x^2 + x - 3)\cdot(2x^3 - 4x^2 + 7x - 2), c = 1$
47. $f(x) = (x^5 + 2x^4 - 5x^3 + x^2 + x - 3)/(2x^3 - 4x^2 + 7x - 2), c = 1$
48. Let $f(x) = \sec^2(x) - \tan^2(x)$. Use the Product Rule to differentiate each summand of f. Simplify to show that $f'(x) = 0$ for each x in the domain of f. How can you arrive at this simple result more quickly?
49. Let $f_n(x) = x^n e^x$ for every positive integer n. Find f_n' in terms of f_n and f_{n-1}.
50. Let

$$\cosh(x) = \frac{e^x + e^{-x}}{2}$$

and

$$\sinh(x) = \frac{e^x - e^{-x}}{2}.$$

Express $\cosh'(x)$ in terms of $\sinh(x)$ and $\sinh'(x)$ in terms of $\cosh(x)$.
51. Let $f(x) = \tan(x)$. For every $t \in (-\pi/2, \pi/2)$, write the equation of the tangent line to the graph of f at $(t, f(t))$ in the form

$$x = M(t)\,(y - \tan(t)) + t.$$

What function is $M(t)$? What line results when you compute

$$x = \lim_{t\to(\pi/2)^-} (M(t)(y - \tan(t)) + t)?$$

Is this line an asymptote of f?
52. Suppose $y = x^p$. Explain how an awareness of units can be used to predict the correct value of α in the equation $\frac{dy}{dx} = px^\alpha$.
53. Evaluate

$$\lim_{h\to 0} \frac{e^{h/2} - e^{-h/2}}{h}$$

by relating the expression to a derivative.
54. Find all values of c such that the intercepts of the tangent line to $y = e^x$ at (c, e^c) are equidistant from the origin.
55. Show that

$$\frac{d}{dx}|x|^p = p\frac{|x|^p}{x}.$$

56. Suppose $2p \geq p^2$. Let

$$f(x) = \frac{x^p}{x^2 + 1}.$$

At what points does the graph of f have a horizontal tangent line?
57. Suppose p is a degree n polynomial and q is a degree m polynomial. What is the degree of $(p \cdot q)'$? Of $(p \circ q)'$? Of the numerator and denominator of $(p/q)'$ (before performing any cancellations)?

58. Use the Method of Mathematical Induction to prove the formula

$$\frac{d}{dx}x^k = k \cdot x^{k-1} \qquad \text{for } k = 1, 2, 3, \ldots.$$

59. Let $g(x) = e^{kx}$ where k is a constant. Use the derivative definition to show that

$$g'(x) = g'(0)g(x).$$

60. If $q(p)$ is the demand for a product at price p—that is, the number of units of the product that are sold at price p—then $E(p) = -q'(p) \cdot p/q(p)$ is called the elasticity of demand for the product at price p.

a. Suppose a and b are positive constants and $q(p) = a - b \cdot p$. What is $E(p)$?

b. Suppose a and b are positive constants and $q(p) = a/p^b$. What is $E(p)$?

c. For the two demand functions considered in parts a and b, explain why you would not expect $E(p)$ to depend on the value of a.

61. Suppose f, g, and t carry units of mass, distance, and time, respectively. In the Product Rule for $(f \cdot g)'(t)$, there is a term $f'(t) \cdot g(t)$ and a term $f(t) \cdot g'(t)$ but no term involving $f'(t) \cdot g'(t)$. Explain why you would not expect one by considering units. (Use the derivative definition to understand units of derivatives.)

62. Suppose f, g, and t carry units. Show that the Quotient Rule for $(f/g)'(t)$ is dimensionally correct—that is, the Quotient Rule leads to the correct units for the derivative of a quotient.

For Exercises 63 and 64, follow the calculation of Example 4 and verify the formula.

63. $\dfrac{d}{dx}\cot(x) = -\csc^2(x)$

64. $\dfrac{d}{dx}\csc(x) = -\csc(x)\cot(x)$

Calculator/Computer Exercises

65. Numerically verify

$$\left.\frac{d}{dx}e^{-x}\right|_{x=0} = -1$$

by computing appropriate central difference quotients.

66. Graph $f(x) = 10^4(e^{x+10^{-4}} - e^{x-10^{-4}})/2$ and $x \mapsto e^x$ in the viewing rectangle $[-1, 2] \times [0, 7.5]$. Explain what differentiation law the two graphs illustrate.

For the functions in Exercises 67–70, graph f and f' in the given viewing rectangle R. Complete the table.

Interval where f increases	Interval where f decreases	Point(s) at which f has a horizontal tangent
Interval where $f' > 0$	Interval where $f' < 0$	Point(s) at which $f' = 0$

Use your table to draw inferences that relate the sign of f' to the behavior of f. (These relationships will be studied in Chapter 4.)

67. $f(x) = 2x - 5/(x^2 + 1)$, $R = [-2, 2] \times [-5.5, 5.5]$

68. $f(x) = \sin(x)$, $R = [0, 2\pi] \times [-1, 1]$

69. $f(x) = \sin(x)/(2 + \sin(x))$, $R = [0, 2\pi] \times [-1, 1]$

70. $f(x) = (x^2 - 2)/(x^2 + 1)$, $R = [-1, 3] \times [-2, 2]$

71. Let $g_k(x) = e^{kx}$. Plot

$$k \mapsto D_0 g_k(0, 10^{-4}).$$

Use your plot to determine $g_k'(0)$.

72. Choose three values of k and three values of x. For each of the three values of k, numerically differentiate $x \mapsto e^{kx}$ at each of the three values of x. Based on your data, what do you think the function

$$k \mapsto e^{-kx}\frac{d}{dx}e^{kx}$$

is? What formula do you conjecture for

$$\frac{d}{dx}e^{kx}?$$

73. For what values of $x > 1$, if any, does $x \mapsto e^x$ grow faster than $x \mapsto x^{20}$? When, if ever, does e^x catch up in size to x^{20}?

74. Graph $f(x) = (x + 1)^{4/3}(x - 1)^{2/3}$ for $-1.5 \le x \le 1.5$. Does the graph of f have a tangent line at the point $(-1, 0)$? At the point $(1, 0)$? Use the Power Rule to explain the behavior of the graph of f at these two points.

75. For fixed constants p_0, τ_0, μ, and Δ, the *Hugoniot curve* $\mathcal{H}$ is the curve in the $p\tau$-plane defined by the equation

$$p(\tau - \mu^2\tau_0) - p_0(\tau_0 - \mu^2\tau) + 2\mu^2\Delta = 0.$$

This curve arises in the study of gas dynamics produced by combustion.

a. Find $\dfrac{d\tau}{dp}$.

b. Let $Q = (p_1, \tau_1)$ be a point on the Hugoniot curve $\mathcal{H}$. What is the equation of the line that is tangent to $\mathcal{H}$ at Q?

c. Assume that $\Delta < 0$. (This is the mathematical condition for an *exothermic* reaction.) Show that the point (p_0, τ_0) lies below the Hugoniot curve.

d. Continue to suppose that $\Delta < 0$. Solve for those points $Q = (p_1, \tau_1)$ on the Hugoniot curve $\mathcal{H}$ such that the line tangent to $\mathcal{H}$ at Q passes through the point $P = (p_0, \tau_0)$. The points that you find are called *Chapman-Jouguet points*. The tangent lines that you find are called *Rayleigh lines*.

e. Illustrate the Hugoniot curve and its Rayleigh lines for the specifications $p_0 = 2.1$, $\tau_0 = 1.7$, $\mu = 1/2$, and $\Delta = -2$. Label the Chapman-Jouguet points.

3.5 The Chain Rule

In principle, we know how to differentiate $H(x) = (x^3 + x)^{100}$. All we have to do is multiply out the expression, then differentiate. However, this would require a great deal of work that can be avoided.

Instead, we think of H as the composition of two functions: $H = g \circ f$ where $g(u) = u^{100}$ and $f(x) = x^3 + x$. (Recall that, in Section 1.5, we learned how to recognize compositions of functions.) We know how to differentiate both f and g. Let us now see how to use this information to differentiate their composition H.

A Rule for Differentiating the Composition of Two Functions

Imagine car A driving down the road with position given by $a(t)$. Imagine a second car B driving down the road with position given by $b(t)$. At a given instant, if B is traveling at a rate of 8 mi/h ($\frac{db}{dt} = 8$) and A is traveling twice as fast as B, then it is clear that A is traveling at the rate of 16 mi/h. In other words, if we know the rate of change of a with respect to b (in this case, 2) and we know the rate of change of b with respect to t (in this case, 8), then we know the rate of change of a with respect to t (namely, $2 \cdot 8 = 16$). Notice that the third quantity is the *product* of the first two. Schematically, we have

$$\begin{pmatrix} \text{rate of change of} \\ a \\ \text{with respect to} \\ t \end{pmatrix} = \begin{pmatrix} \text{rate of change of} \\ a \\ \text{with respect to} \\ b \end{pmatrix} \times \begin{pmatrix} \text{rate of change of} \\ b \\ \text{with respect to} \\ t \end{pmatrix},$$

or

$$\frac{da}{dt} = \frac{da}{db} \cdot \frac{db}{dt}.$$

This example suggests the following formula:

> The derivative of a composition is the product of the derivatives of the component functions.

This formula, known as the *Chain Rule,* gives us a way to differentiate the composition of two differentiable functions. The formulation of the Chain Rule in Leibniz notation is particularly suggestive because it looks as though we are cancelling fractions. As a matter of fact we are not—derivatives expressed in Leibniz notation are *not* fractions. However, the difference quotients that converge to the derivative *are* fractions, and we *can* cancel those. This observation, if executed with some care, can be used to prove the Chain Rule. A somewhat different proof is provided in this chapter's Genesis & Development.

Theorem 1 ***Chain Rule*** Suppose the range of the function f is contained in the domain of the function g. If f is differentiable at c and if g is differentiable at $f(c)$, then $g \circ f$ is differentiable at c. Moreover,

$$\frac{d}{dx}(g \circ f)\bigg|_{x=c} = \left(\frac{dg}{du}\bigg|_{u=f(c)}\right) \cdot \left(\frac{df}{dx}\bigg|_{x=c}\right). \tag{3.18}$$

Recall from Section 3.2 that the vertical bar is notation that indicates where the derivative is being evaluated. It is important to notice that the two derivatives on the right side of formula (3.18) are evaluated at different points. When we calculate $(g \circ f)(x) = g(f(x))$, we see that g is evaluated at $f(x)$ and that f is evaluated at x. This evaluation scheme is preserved in the application of the Chain Rule: In computing the derivative of $g \circ f$ at c, the derivative of g is evaluated at $f(c)$ and the derivative of f is evaluated at c.

With primes denoting derivatives, the Chain Rule becomes

$$(g \circ f)'(c) = g'(f(c)) \cdot f'(c).$$

inSIGHT

It is sometimes useful to think of a composition as having an "outside" function and an "inside" function. The outside function is the *last* function applied: In the composition $g \circ f$, it would be g. In applying the Chain Rule to such a composition, we differentiate the outside function first, holding the inside function fixed. *Then* we differentiate the inside function, which would be f in the composition $g \circ f$. Keep this idea in mind when reading the examples that follow.

Example 1 Differentiate $H(x) = \sin(3x^2 + 1)$.

Solution We write $H = g \circ f$ where $g(u) = \sin(u)$ is the outside function and $f(x) = 3x^2 + 1$ is the inside function:

$$\frac{dg}{du} = \cos(u) \quad \text{and} \quad \frac{df}{dx} = 6x.$$

As a result,

$$\begin{aligned} \frac{dH}{dx} &= \left.\frac{dg}{du}\right|_{u=f(x)} \cdot \frac{df}{dx} \\ &= \cos(f(x)) \cdot (6x) \\ &= (\cos(3x^2 + 1)) \cdot 6x. \end{aligned}$$

Using the idea from the Insight, we have the following solution scheme:

$$\underbrace{H'(x)}_{\substack{\text{Derivative}\\\text{of}\\\text{composed}\\\text{function}}} = \underbrace{\cos}_{\substack{\text{Derivative}\\\text{of}\\\text{outside}\\\text{function}}}\underbrace{(3x^2 + 1)}_{\substack{\text{Inside}\\\text{function}}} \cdot \underbrace{6x}_{\substack{\text{Derivative}\\\text{of}\\\text{inside}\\\text{function}}}.$$

■

Example 2 Differentiate $H(x) = (x^3 + x)^{100}$.

Solution We write $H = g \circ f$ where $g(u) = u^{100}$ and $f(x) = x^3 + x$. According to the Chain Rule,

$$\frac{dH}{dx} = \left(\left.\frac{dg}{du}\right|_{u=f(x)}\right) \cdot \left(\frac{df}{dx}\right).$$

Since

$$\frac{dg}{du} = 100u^{99} \quad \text{and} \quad \frac{df}{dx} = 3x^2 + 1,$$

we know that

$$\begin{aligned} \frac{dH}{dx} &= (100 \cdot u^{99})|_{u=f(x)} \cdot (3x^2 + 1) \\ &= (100 \cdot (f(x))^{99}) \cdot (3x^2 + 1) \\ &= (100(x^3 + x)^{99}) \cdot (3x^2 + 1). \end{aligned}$$

■

inSIGHT

The basic idea of the Chain Rule is that the derivative of a composition is the product of the derivatives. The statement of the Chain Rule is a bit complicated, however, because we have to be careful about *where* we evaluate the derivatives. Notice where $\frac{df}{dx}$ is evaluated: It must be evaluated at the same point where f itself is evaluated—that is, at x. Notice where $\frac{dg}{du}$ is evaluated: It must be evaluated at the same point where g is evaluated—that is, at $f(x)$.

Example 3 Find the tangent line to the graph of $p(x) = \tan^7(x)$ at the point $x = \pi/4$.

Solution Writing $\tan^7(x)$ as $(\tan(x))^7$ makes it clearer that $u \mapsto u^7$ is the outside function. Thus, $p = g \circ f$ where $g(u) = u^7$ and $f(x) = \tan(x)$. Now $g'(u) = 7u^6$ and $f'(x) = \sec^2(x)$. Notice that $g'(f(x)) = 7f(x)^6 = 7f^6(x)$. We can now compute $(g \circ f)'(x)$:

$$\begin{aligned} p'(x) &= g'(f(x)) \cdot f'(x) \\ &= 7 \cdot f^6(x) \cdot \sec^2(x) \\ &= 7 \cdot \tan^6(x) \cdot \sec^2(x). \end{aligned}$$

The slope of the tangent line to the graph of p at $x = \pi/4$ will be

$$p'\left(\frac{\pi}{4}\right) = 7\tan^6\left(\frac{\pi}{4}\right)\sec^2\left(\frac{\pi}{4}\right) = 7 \cdot 1^6 \cdot (\sqrt{2})^2 = 14.$$

Since $p(\pi/4) = 1$, the tangent line passes through $(\pi/4, 1)$. In conclusion, the tangent line has equation $y = 14(x - \pi/4) + 1$. ■

inSIGHT

Make sure you understand the difference between $g'(f(x))$ and $(g \circ f)'(x)$. Because f plays a role in the formation of the function $g \circ f$, its rate of change $f'(x)$ is a component of the rate of change $(g \circ f)'(x)$. On the other hand, $g'(f(x))$ stands for the rate of change of g at the point $f(x)$—the rate of change of f plays no role here.

Examples 2 and 3 involve an application of the Chain Rule that arises frequently. In each, we considered a function f that has been raised to a fixed power p. Because this type of composition occurs so often, it is worthwhile stating this special case of the Chain Rule.

The Chain Rule Applied to Powers

$$\frac{d}{dx} f(x)^p = p \cdot f(x)^{p-1} \cdot f'(x)$$

Example 4 Use the Chain Rule applied to powers to derive the Reciprocal Rule from Section 3.3.

Solution If we take $p = -1$ in the formula for the Chain Rule applied to powers, then we have

$$\frac{d}{dx}\left(\frac{1}{f(x)}\right) = \frac{d}{dx} f(x)^{-1} = (-1) \cdot f(x)^{-1-1} \cdot f'(x) = -\frac{f'(x)}{f(x)^2},$$

as required. ■

Multiple Compositions

Of course some functions are the composition of more than two simpler functions. To differentiate such a function, we must apply the Chain Rule several times

Example 5 Differentiate $Q(x) = \sin^3(5x)$.

Solution We notice that we can write Q as $Q = H \circ (G \circ F)$ where $H(s) = s^3$, $G(u) = \sin(u)$, and $F(x) = 5x$. With this grouping, Q is the composition of two functions. By the Chain Rule,

$$\frac{dQ}{dx} = \left(\frac{dH}{ds}\bigg|_{s=G\circ F(x)}\right) \cdot \left(\frac{d(G\circ F)}{dx}\right). \qquad \textbf{(3.19)}$$

Now we apply the Chain Rule a second time to the composition $G \circ F$ to obtain

$$\frac{d(G\circ F)}{dx} = \left(\frac{dG}{du}\bigg|_{u=F(x)}\right) \cdot \left(\frac{dF}{dx}\right).$$

Substituting this last equation into equation (3.19) yields

$$\frac{dQ}{dx} = \left(\frac{dH}{ds}\bigg|_{s=G\circ F(x)}\right) \cdot \left(\frac{dG}{du}\bigg|_{u=F(x)}\right) \cdot \left(\frac{dF}{dx}\right).$$

The philosophy of differentiating the outside function first and working toward the inside applies even in the case of multiple compositions. Review Example 5 with this idea in mind.

Finally, we calculate that

$$\frac{dH}{ds}(s) = 3s^2, \quad \frac{dG}{du}(u) = \cos(u), \quad \text{and} \quad \frac{dF}{dx}(x) = 5.$$

Therefore,

$$\begin{aligned}\frac{dQ}{dx} &= 3[G(F(x))]^2 \cdot \cos(f(x)) \cdot 5 \\ &= 15\sin^2(5x) \cdot \cos(5x).\end{aligned}$$

■

The Chain Rule gives insight into new kinds of rate of change problems, as in the next example.

Example 6 An oil slick spreads so that it forms a disk centered about the point of contamination. As the fixed volume of spilled oil disperses and the thickness of the slick decreases, the radius of the slick increases at the rate of 2 ft/min. How fast is the area of polluted surface water growing when the radius is 500 ft?

In Example 6, we differentiated πr^2 with respect to the variable t—not the variable r. Therefore, we had to use the Chain Rule. Leibniz notation aided us in keeping track of the terms.

Solution The area of polluted surface water is given by

$$A = \pi r^2.$$

The radius r of the slick is a function of time t (with $\frac{dr}{dt} = 2$ ft/min). It follows that A depends on t as well. We use Leibniz notation and differentiate both sides of the equation $A = \pi r^2$ with respect to t to obtain

$$\frac{dA}{dt} = \frac{dA}{dr} \cdot \frac{dr}{dt} = (\pi \cdot 2r) \cdot \frac{dr}{dt}.$$

Substituting $r = 500$ ft and $\frac{dr}{dt} = 2$ ft/min into this equation gives

$$\frac{dA}{dt} = \pi \cdot 1000\,\text{ft} \cdot 2\,\text{ft/min} = 2000\pi\ \text{ft}^2/\text{min}.$$

Therefore, the surface area of the oil slick is increasing at the rate of 2000π ft²/min. ■

quickquiz

1. State the Chain Rule in Leibniz notation.
2. State the Chain Rule in prime notation.
3. How does the Chain Rule differ from the Product Rule?
4. What is the derivative of $f(x) = \cos(x^2 + 3x)$?

EXERCISES

Problems for Practice

In Exercises 1–18, calculate the derivative of the given function.

1. $f(x) = (x^2 + 3x)^{10}$
2. $g(x) = \sin(9x)$
3. $H(x) = \tan(\sin(x))$
4. $f(x) = \left(\dfrac{x}{x+1}\right)^4$
5. $g(x) = \sec(\cos(x))$
6. $H(x) = \cot(5 - x^5)$
7. $f(x) = \sin(x^8 + x^2)$
8. $g(x) = e^{5x}$
9. $H(x) = (\sin(x) + x^2)^6$
10. $f(x) = \sin^8(x)$
11. $g(x) = \tan(x^3)$
12. $H(x) = \sin(x^2 - 3x)$
13. $f(x) = \exp(x^2 + 1)$
14. $g(x) = \cos\left(\dfrac{x}{x^2+1}\right)$
15. $H(x) = \cos(\sqrt{x})$
16. $f(x) = \dfrac{1}{\tan^5(x)}$
17. $g(x) = \tan(1 - 7x)$
18. $H(x) = \sqrt{e^x + x^2}$

In Exercises 19–22, compute $(f \circ g)'$ and $(g \circ f)'$.

19. $f(x) = x^4 + 7x^2$, $g(x) = \sqrt{x}$
20. $f(x) = 3x^5 - 2x^2$, $g(x) = \sin(x)$
21. $f(x) = x/(x + 1)$, $g(x) = x^3$
22. $f(x) = \sin(x)$, $g(x) = \tan(x)$

In Exercises 23–26, use the information about f and g to compute $(g \circ f)'(c)$.

23. $g'(3) = 2$, $f(2) = 3$, $f'(2) = 8$, $c = 2$
24. $g'(9) = -3$, $f(-2) = 9$, $f'(-2) = 4$, $c = -2$
25. $g'(1/2) = 1/3$, $f(6) = 1/2$, $f'(6) = 0$, $c = 6$
26. $g'(-5) = -7$, $f(-6) = -5$, $f'(-6) = -9$, $c = -6$

Further Theory and Practice

In Exercises 27–30, let $F(x) = x + x^2$, $G(x) = \sin(x)$, and $H(x) = x/(x^2 + 3)$. Calculate the derivative.

27. $(F \circ G \circ H)'(2)$
28. $\frac{d}{dx}\left(\frac{F \circ G}{H}\right)(3)$
29. $(F \circ F \circ F)'(6)$
30. $\frac{d}{dx}\left(G \circ \left(\frac{1}{F \circ H}\right)\right)(4)$

Dimensional analysis refers to the attention we give to the units of mathematical formulas. The basic dimensions of mechanics are length, mass, and time. Trigonometric functions are unitless because they represent the ratio of two lengths—in other words, the units cancel. Arguments of trigonometric functions are angles and are, thus, unitless. Likewise, exponential functions are unitless and take unitless arguments. Exercises 31–34 concern dimensional analysis.

31. An oscillation about equilibrium is given by $x(t) = A\sin(\omega t)$ where t is measured in seconds and x in meters. What units do the constants A and ω bear? Is $x'(t)$ dimensionally correct? Explain. What is the role of the Chain Rule?
32. Suppose t represents time and y represents distance. If we change the units of either t or $y = f(t)$ or both, then we change the numerical value of $\frac{dy}{dt}$. For example, 60 mi/h is the same speed as 88 ft/s. What rule of calculus accounts for the numerical change in $\frac{dy}{dt}$ when we change the units of y? Explain. What about t? Explain.
33. An object oscillates about its equilibrium position according to the formula

$$x(t) = Ae^{-kt}\sin(\omega t)$$

where t is measured in seconds and x in centimeters. Describe the units of the positive constants k, A, and ω. Is $x'(t)$ dimensionally consistent? Explain.

34. According to the theory of relativity, the mass m of an object at speed v is given by

$$m(v) = \frac{m_0}{\sqrt{1 - v^2/c^2}}$$

where c is the speed of light and m_0 is its rest mass. What is $m'(v)$? Identify each application of the Chain Rule in this computation. Suppose that an incorrect computation without the Chain Rule results in

$$m'(v) = -\frac{1}{2}m_0\left(1 - \frac{v^2}{c^2}\right)^{-3/2}.$$

Use dimensional analysis to show that this formula must be wrong.

35. A 100 g sample of the isotope ^{14}C will have mass

$$m(t) = 100 \cdot e^{-0.00121286 \cdot t}\text{ g}$$

at time $t > 0$ where t is measured in years. At what rate is the mass of the ^{14}C sample changing? What is the relationship between $m(t)$ and $m'(t)$? What are

$$\lim_{t\to\infty} m(t) \quad\text{and}\quad \lim_{t\to\infty} m'(t)?$$

What does $\lim_{t\to\infty} m'(t)$ tell you about the decay process?

36. The diameter of a certain spherical balloon is decreasing at the rate of 2 in./s. How fast is the volume of the balloon decreasing when the diameter is 8 ft?

37. The leg of an isosceles right triangle increases at the rate of 2 in./min. At the moment when the hypotenuse is 8 in., how fast is the area changing?

38. If $f : \mathbb{R} \to \mathbb{R}$ is a differentiable function, define $g(x) = 1/f(1/x)$ for $x \neq 0$. Assume that f never takes the value 0. Compute $g'(x)$ for $x \neq 0$.

39. Suppose f is a differentiable function. Let $g(t) = f(t^2)$. Explain the difference, if any, between $f'(t^2)$ and $g'(t)$. Between $f'(t)$ and $g'(\sqrt{t})$.

40. The formulas for $(f \circ g)'$ and $(g \circ f)'$ both involve f' and g'. In what way do these formulas differ?

In Exercises 41–50, find $f'(x)$ for the function f.

41. $f(x) = \sec^5(7x)$

42. $f(x) = \tan(\sqrt{x}) + \sqrt{\tan(x)}$

43. $f(x) = e^{\cos(2x)}$

44. $f(x) = (5x^3 + 1)^2/\sqrt{x^2 + 1}$

45. $f(x) = \cos(\sqrt{2x^2 + 3})$

46. $f(x) = \sin^3(7x^2)$

47. $f(x) = \sqrt{\sin^2(3x) + \cos^2(3x)}$

48. $f(x) = \tan(\sin(e^{3x}))$

49. $f(x) = \sqrt{\tan(5x) + \sqrt{2x + 1}}$

50. $f(x) = (x^2 + \sin(x/3))^2(x^3 + 1)^{4/3}$

51. If

$$f(x) = \begin{cases} (cx + 1)^3 & \text{if } x \leq 0 \\ x + 1 & \text{if } x > 0 \end{cases},$$

can we choose c so that f is a differentiable function?

52. If

$$f(x) = \begin{cases} e^{ax} & \text{if } x \leq 0 \\ 1 + \sin(bx) & \text{if } x > 0 \end{cases},$$

how must we choose a and b so that f is a differentiable function?

53. Suppose M and P_0 are constants such that $M > P_0 > 0$. The size of a population P is given as a function of time t by

$$P(t) = \frac{P_0 M}{P_0 + (M - P_0)e^{-kMt}}$$

for some positive constant k. (This is known as *logistic growth*.)

a. What is the initial population size? What is the limiting size as $t \to \infty$?

b. Find $P'(t)$. Verify the differential equation

$$P'(t) = k \cdot P(t) \cdot (M - P(t)).$$

c. Does the population growth rate have a limit as t tends to infinity?

54. If the air resistance to a falling object is proportional to the velocity of that object, then the velocity of that object when dropped from a height is

$$v(t) = kg(1 - e^{-t/k})$$

in the downward direction. What is the limiting (or terminal) velocity v_∞ as $t \to \infty$? Calculate the acceleration $v'(t)$ of the object. What is the limit of the acceleration (rate of change of velocity)?

55. If the air resistance to a falling object is proportional to the square of the velocity of that object, then the velocity of that object when dropped from a height is

$$v(t) = \sqrt{\frac{g}{\kappa}}\left(\frac{e^{2t\sqrt{g\kappa}} - 1}{e^{2t\sqrt{g\kappa}} + 1}\right)$$

in the downward direction. Here κ is a positive constant that depends on the aerodynamic properties of the object as well as on the density of the air. What is the limiting (or terminal) velocity v_∞ as $t \to \infty$? Calculate the acceleration $v'(t)$ of the object. What is the limit of the acceleration? Verify that v satisfies the differential equation

$$v' = g - \kappa \cdot v^2.$$

56. A drug is intravenously introduced into a patient's bloodstream at a constant rate, beginning at $t = 0$. The concentration C of the drug in the patient's body is given by the formula

$$C(t) = \frac{\alpha}{\beta}(1 - e^{-\beta t})$$

where α and β are positive constants.

a. Show that $C(t)$ is a solution of the differential equation $C'(t) = \alpha - \beta \cdot C(t)$.

b. In the long run, what is the concentration of the drug in the patient's bloodstream?

c. What is the rate at which the drug is administered? What is the rate at which the body eliminates the drug from the bloodstream? (These rates can be determined from the differential equation in part a.)

Computer/Calculator Exercises

57. Let $f(x) = \tan(x^3 + x + 1)$. To five decimal places, find the unique value of x_0 in $(0, \pi/2)$ for which $f'(x_0) = 20$.

58. Graph $f(x) = \sin(x^2 + 1)$ and f' in the same viewing window for $-2 \le x \le 2$. For x in this closed interval, where are $x \mapsto \sin(x)$ and $x \mapsto x^2 + 1$ increasing most rapidly? Where is f increasing most rapidly? Explain why this value of x is different from the previous two.

59. Where are $g(x) = \sin(x)$ and $f(x) = \pi \cos^2(x)$ increasing most rapidly for x in the interval $I = [0, \pi]$? Where is $g \circ f$ increasing most rapidly? Explain why this value of x is different from the previous two.

60. The average temperature, in degrees Fahrenheit, of the town of Rainy Lake is modeled by

$$F(t) = 47 - 43\cos\left(\frac{2\pi}{365}(t - 33)\right).$$

In this expression, t is measured in days with $t = 0$ corresponding to the start of the new year. On what day is the average temperature the greatest? On what days is the average temperature increasing the fastest? Decreasing the fastest? Stable? What are the net temperature changes on the days during which the temperature is stable?

61. One gram of the isotope ^{238}U decreases according to the exponential decay law $m(t) = e^{-kt}$ g where t is measured in years. After 4.51 billion years, 0.5 g remains. Approximately what is the rate of change of this ^{238}U sample when its mass is $1/8$ g?

62. Refer to Exercise 35 for the mass of a 100 g sample of the isotope ^{14}C.

a. Is there ever a time at which the mass decreases at the rate of 1 g per millennium? (A millennium is 1000 years.) If so, when? If not, why not?

b. Is there ever a millennium during which the mass is reduced by exactly 1 g? If so, which millennium (expressed as a time interval $[t_0, t_0 + 1000]$)? If not, why not?

63. A $T_0 = 350$°F casserole dish is plunged into a large basin of hot water with temperature $T_\infty = 115$°F. The temperature of the dish satisfies Newton's law of cooling:

$$T(t) = T_\infty + (T_0 - T_\infty)e^{-kt}.$$

In 25 seconds, the temperature of the dish is 175°F.

a. Graph the temperature T.

b. Graph the rate T' at which the temperature decreases.

c. What is the rate of change of T when T is equal to 175°F? 155°F? 135°F? 125°F? 115.5°F?

64. Two ceramic objects are removed from a $T_0 = 450$°F oven into a room at $T_\infty = 72$°F. They are to be glued together, but the glue is not reliable when applied to objects with temperatures greater than 85°F. Five minutes after being taken from the oven, the temperature T of the ceramic objects is 300°F. Refer to Exercise 63 for Newton's law of cooling. Graph the temperature T. Graph the rate T' at which the temperature decreases. What is the rate of change of T when the objects are cool enough to be glued? If the thermometer used is accurate to three significant digits, when does the temperature of the objects measure 75°F?

65. After jumping from an airplane, a sky diver has velocity given by

$$v(t) = -256(1 - e^{-t/8}) \text{ ft/s}.$$

Eight seconds into her descent, she opens her parachute. For $t \ge 8$, her velocity is given by

$$v(t) = -16 - (240e - 256)e^{-2t+15} \text{ ft/s}.$$

Graph $v(t)$. Is it everywhere continuous? Differentiable? Graph acceleration (the rate of change of velocity). Is it everywhere continuous? Differentiable?

3.6 Inverse Functions and the Natural Logarithm

Figures 1 and 2 show the graphs of two polynomials p and q. Although both functions have the same domain $S = [0, 1]$, and both take values in $T = [0, 1]$, the graphs of p and q reveal a fundamental difference between these two functions. If you are given a number $t \in T$, you can determine the value of $s \in S$ for which $q(s) = t$, because there is one and only one value of s such that $q(s) = t$. Stated geometrically, the horizontal line $y = t$ intersects the graph of q exactly once. In short, if you know $q(s)$, then you can determine what s is.

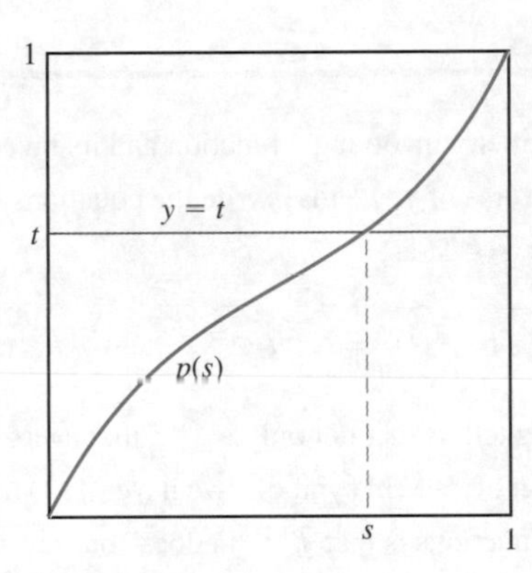

Figure 1

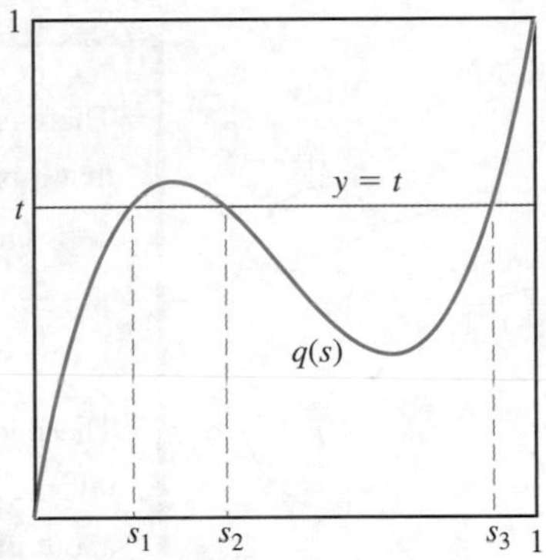

Figure 2
$q(s_1) = q(s_2) = q(s_3) = t$

If you are given a number t between 0 and 1, however, there could be as many as three values of s for which $p(s) = t$. Stated somewhat differently, you *cannot* identify the value of s by knowing the value of $p(s)$. These considerations lead to the main definition of this section.

The Inverse of a Function

Definition Let $f : S \to T$ be a function. We say that f has an *inverse* (is *invertible*) if there is a function $g : T \to S$ such that

$$(f \circ g)(t) = t \text{ for all } t \in T \quad \text{and} \quad (g \circ f)(s) = s \quad \text{for all } s \in S.$$

There can be, at most, one function g with these properties. If this function exists, then we call it the inverse of f and denote it by f^{-1} (read "f inverse"). Notice that the domain of f^{-1} is the range of f and the range of f^{-1} is the domain of f.

Example 1 Let $f : [0, \infty) \to [1, \infty)$ be defined by $f(s) = \sqrt{1 + s^2}$. Let $g : [1, \infty) \to [0, \infty)$ be defined by $g(t) = \sqrt{t^2 - 1}$. Verify that $g = f^{-1}$.

Solution Let S denote the domain $[0, \infty)$ of f, and T the domain $[1, \infty)$ of g. To show that $g = f^{-1}$, we must verify two identities: $(f \circ g)(t) = t$ for all $t \in T$ and

$(g \circ f)(s) = s$ for all $s \in S$. Starting with $t \in T = [1, \infty)$, we have

$$(f \circ g)(t) = f(g(t)) = \sqrt{1 + g(t)^2} = \sqrt{1 + \left(\sqrt{t^2 - 1}\right)^2} = \sqrt{t^2} = |t| \overset{\text{Since } t \geq 1.}{=} t.$$

Similarly, for $s \in S = [0, \infty)$, we have

$$(g \circ f)(s) = \sqrt{f(s)^2 - 1} = \sqrt{\left(\sqrt{1 + s^2}\right)^2 - 1} = \sqrt{s^2} = |s| \overset{\text{Since } s \geq 0.}{=} s.$$

Having verified the the two required identities, we conclude that $g = f^{-1}$. ■

inSIGHT

There is a symmetry involving a function and its inverse. Once we have determined that g is the inverse of f, we may write the equations $(f \circ g)(t) = t$, $t \in T$, and $(g \circ f)(s) = s$, $s \in S$, as

$$f(f^{-1}(t)) = t, \quad t \in T, \quad \text{and} \quad f^{-1}(f(s)) = s, \quad s \in S.$$

These identities tell us that not only is f^{-1} the inverse function of f, but also f is the inverse function of f^{-1}. In symbols, we have $(f^{-1})^{-1} = f$. A convenient way to think about inverse functions is that f^{-1} "undoes" or "reverses" what f does. Symmetrically, f undoes what f^{-1} does (see Figure 3).

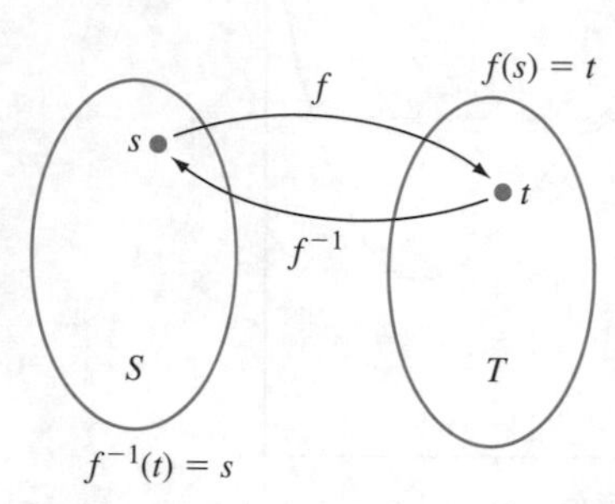

Figure 3

To understand inverses, we must have a good understanding of functional composition. This will help avoid confusing reciprocals with inverses. Take this opportunity to review composition of functions in Section 1.5.

Basic Rule for Finding Inverses

Suppose that f is an invertible function and that an explicit formula is given for $f(s)$. If t is in the image of f, then the equation $f(s) = t$ has a *unique* solution s in the domain of f. Since $f(f^{-1}(t)) = t$, we conclude that $s = f^{-1}(t)$. Thus, to find the inverse function f^{-1}, let $f(s) = t$ and solve this equation for s in terms of t. After s is isolated on the left side of the equation, the right side is $f^{-1}(t)$.

The next two examples illustrate this technique.

Example 2 Find the inverse of the invertible function $f(s) = 2s$, $s \in \mathbb{R}$.

Solution We rewrite $f(s) = t$ as $2s = t$, or $s = t/2$. Since $s = f^{-1}(t)$, we conclude that $f^{-1}(t) = t/2$. We may check this solution as follows:

$$(f \circ f^{-1})(t) = 2 \cdot f^{-1}(t) = 2 \cdot \left(\frac{t}{2}\right) = t$$

and

$$(f^{-1} \circ f)(s) = \frac{1}{2} \cdot f(s) = \frac{1}{2} \cdot 2s = s.$$

■

INSIGHT

It is a common error to suppose that f^{-1} means the *reciprocal* of f—that is, the function $1/f$. This is *not* so. The function $1/f$ is the *multiplicative inverse* of f (that is, $f \cdot 1/f = 1$), but it is not the inverse with respect to composition. The function $f(s) = 2s$, $s \in \mathbb{R}$, from Example 2 has reciprocal

$$\left(\frac{1}{f}\right)(t) = \frac{1}{2t} \quad \text{for} \quad t \neq 0.$$

Call this function g. Notice that

$$(f \circ g)(t) = f(g(t)) = 2 \cdot g(t) = 2 \cdot \left(\frac{1}{2t}\right) = \frac{1}{t} \neq t.$$

Therefore, g, the reciprocal of f, is certainly not the inverse of f.

To summarize:

1. $f(t)^{-1}$ means $1/f(t)$, the reciprocal of the number $f(t)$.
2. $f^{-1}(t)$ means the value of the inverse function f^{-1} at t.

Example 3 Let $f : \mathbb{R} \to \mathbb{R}$ be defined by $f(x) = (x^3/8) + 1$. Find f^{-1}.

Solution Let $f(x) = y$—that is, $x^3/8 + 1 = y$. We must solve the equation $x^3/8 + 1 = y$ for x in terms of y. We find that $x^3 = 8(y - 1)$, or $x = 2(y - 1)^{1/3}$. Since $x = f^{-1}(y)$, we conclude that $f^{-1}(y) = 2(y - 1)^{1/3}$. Finally, we check our work by calculating

$$(f \circ f^{-1})(y) = f(f^{-1}(y)) = \frac{1}{8}(f^{-1}(y))^3 + 1 = \frac{1}{8}(2(y-1)^{1/3})^3 + 1$$
$$= \frac{1}{8} \cdot 8 \cdot (y - 1) + 1 = y$$

and

$$(f^{-1} \circ f)(x) = 2(f(x) - 1)^{1/3} = 2\left(\left(\frac{1}{8}x^3 + 1\right) - 1\right)^{1/3} = x.$$

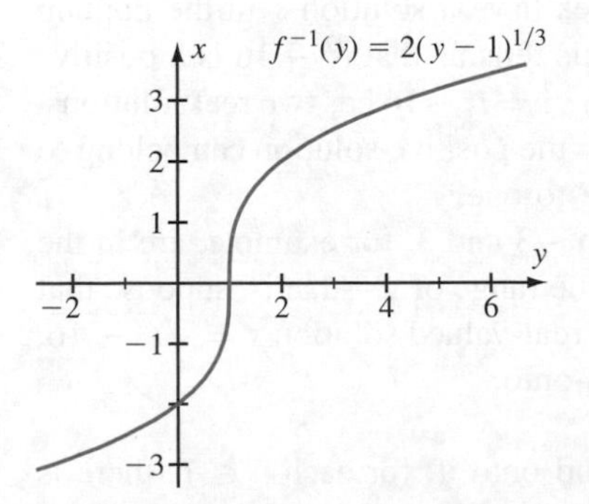

Figure 4

In plotting f^{-1}, we follow the convention of measuring the argument of the function along the horizontal axis (see Figure 4). The variable y is, therefore, the label for the *horizontal* axis. ■

Examples 2 and 3 show how to find a formula for an inverse function when it is *possible* to do so. You should be aware, however, that even if an equation $f(s) = t$ has a unique solution s for a given t, it may be impossible to find an explicit formula that expresses s in terms of t. This can happen for degree five and greater polynomials f. Even when we cannot find an explicit formula for f^{-1}, we can still turn to its graph for useful information. Fortunately, the graph of an inverse function f^{-1} can always be obtained from the graph of f, as we shall see later in this section.

Recognizing an Invertible Function

It is important to understand why some functions do not have inverses.

Definition If a function $f : S \to T$ has the property that $f(s_1) \neq f(s_2)$ whenever s_1 and s_2 belong to S and $s_1 \neq s_2$, then we say that f is *one-to-one*. In plain language, a function is one-to-one if it takes different elements of its domain to different elements of its image.

The one-to-one property concerns the solution of the equation $f(s) = t$. A function $f : S \to T$ is one-to-one if, for each $t \in T$, there is *no more than one* $s \in S$ such that $f(s) = t$. The one-to-one property does not address the question of whether we can find a value s that solves the equation $f(s) = t$. It simply says that *if* the equation $f(s) = t$ can be solved for s, then the solution is unique.

Definition If a function $f : S \to T$ has the property that for every t in T there is an s in S for which $f(s) = t$, then we say that f is *onto*. In other words, a function is onto if its image is the same as its range.

The *onto* property of f concerns the solvability of the equation $f(s) = t$. A function $f : S \to T$ is onto if, for each $t \in T$, there is *at least one* $s \in S$ such that $f(s) = t$. The onto property does *not* address the question of whether the equation $f(s) = t$ has a unique solution s.

Example 4 Let $f : [3, \infty) \to (0, \infty)$ be defined by $f(s) = \sqrt{s^2 + 16}$. Let $g : \mathbb{R} \to [4, \infty)$ be defined by $g(s) = \sqrt{s^2 + 16}$. Is f onto? One-to-one? What about g?

Solution If s belongs to the domain of f, then $s \geq 3$ and $f(s) = \sqrt{s^2 + 16} \geq \sqrt{9 + 16} = 5$. Thus, if $0 < t < 5$, then t belongs to the range of f, but the equation $f(s) = t$ does not have a solution s in the domain of f. Thus, f is *not* onto. If t belongs to the range of f and the equation $f(s) = t$ does have a solution s in the domain of f, then $s^2 = t^2 - 16$ holds for some $s \geq 3$. This tells us that $t^2 - 16$ is a positive number that is at least 9. It follows that the equation $s^2 = t^2 - 16$ has two real solutions, $s = \sqrt{t^2 - 16}$ and $s = -\sqrt{t^2 - 16}$. However, only the positive solution can belong to the domain $[3, \infty)$ of f. We conclude that f *is* one-to-one.

The function g is *not* one-to-one because both -3 and 3, for example, are in the domain of g, and $g(-3) = g(3)$. Next, let t be in the range of g—that is, suppose that $t \geq 4$. The equation $\sqrt{s^2 + 16} = g(s) = t$ has the real-valued solution $s = \sqrt{t^2 - 16}$, which is in the domain of g. We conclude that g *is* onto. ■

inSIGHT

Example 4 shows that the properties one-to-one and onto are independent. A function can be one-to-one but not onto, or it can be onto but not one-to-one.

A function $f : S \to T$ is both one-to-one and onto if, for each $t \in T$, there is exactly one $s \in S$ such that $f(s) = t$. Given $t \in T$, denote this unique solution s

by $g(t)$. We have $f(g(t)) = t$ for $t \in T$ and $g(f(s)) = g(t) = s$ for $s \in S$. These two identities show that f is invertible and $g = f^{-1}$. Conversely, an invertible function is both one-to-one and onto: The unique solution to the equation $f(s) = t$ for $t \in T$ is $s = f^{-1}(t)$. We state this as a theorem.

Theorem 1 A function f is invertible if and only if it is both one-to-one and onto.

Example 5 Let $f : (-\pi, \pi) \to [-1, 1]$ be defined by $f(x) = \sin(x)$. Explain why f is not invertible.

Solution The function f is not one-to-one. For example, $f(3\pi/4) = f(\pi/4) = 1/\sqrt{2}$ (see Figure 5). According to Theorem 1, the function f cannot be invertible. ■

The following geometric formulation of Theorem 1 is known as the *Horizontal Line Test*.

Horizontal Line Test

If $f : S \to T$, then f is invertible if and only if for each t in T the horizontal line $y = t$ intersects the graph of f exactly once.

inSIGHT

If U is a subset of the domain of f, then we call the function $u \mapsto f(u)$, $u \in U$, the *restriction* of f to U and denote it by $f|_U$. The function g from Example 6 is the restriction to $U = [-\pi/2, \pi/2]$ of the function f from Example 5. Example 6 shows us that we can obtain an invertible function by restricting the domain of a noninvertible function. This technique will be important when we define the inverse trigonometric functions in Chapter 6.

Example 6 Let $g : [-\pi/2, \pi/2] \to [-1, 1]$ be defined by $g(x) = \sin(x)$. Use the Horizontal Line Test to discuss the invertibility of this function.

Solution Figure 6 shows the graph of g. For each y_0 in the range of g—that is, for each y_0 that satisfies $-1 \le y_0 \le 1$—the horizontal line $y = y_0$ intersects the graph of g at exactly one point (x_0, y_0) with x_0 in the domain of g. Therefore, the function g *is* invertible. ■

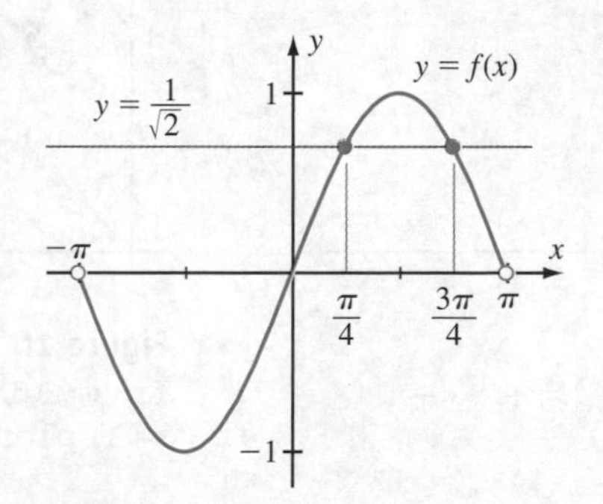

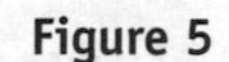

Figure 5

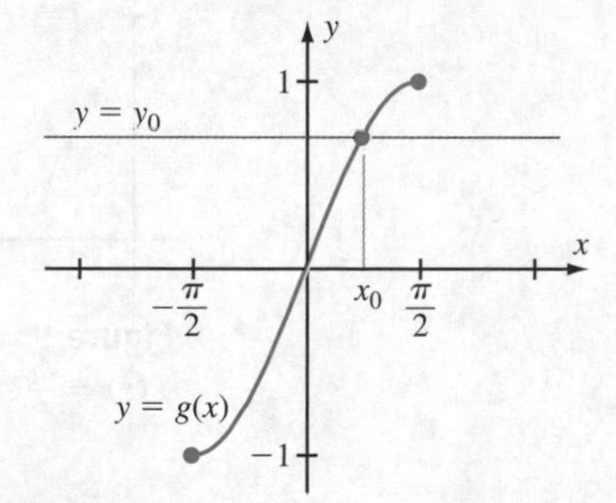

Figure 6

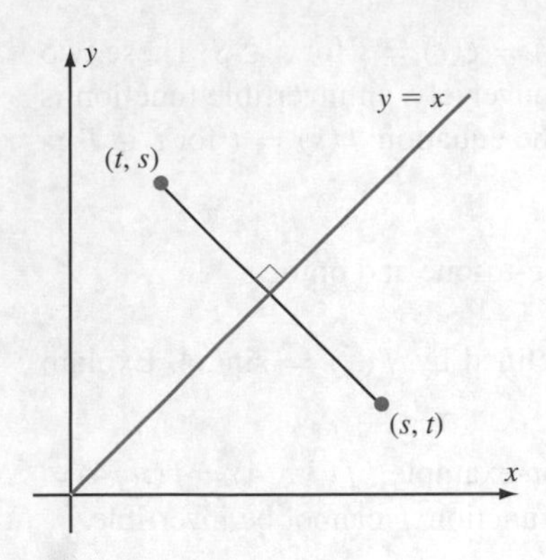

Figure 7

The Graph of the Inverse Function

Suppose that $f : S \to T$ is invertible and that (s, t) is a point on the graph of f. This means $t = f(s)$ and, therefore, $s = f^{-1}(t)$. In other words, the point (t, s) is on the graph of f^{-1}. Figure 7 shows the geometrical connection between the points (s, t) and (t, s): The points are reflections of each other through the line $y = x$. (To verify this assertion, notice that the line segment between the points (s, t) and (t, s) has slope $(s - t)/(t - s) = -1$, which is the negative reciprocal of the slope of the line $y = x$. Also, (s, t) and (t, s) are equidistant from the line $y = x$ by symmetry.) This illustrates the following important principle (see also Figure 8):

> The graph of f^{-1} is the reflection in the line $y = x$ of the graph of f.

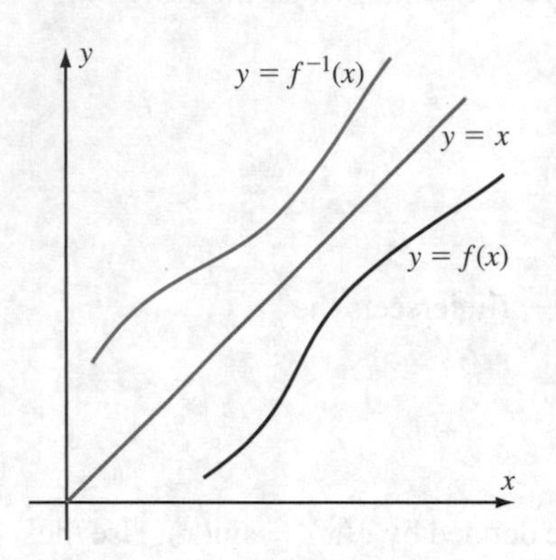

Figure 8

It is easy to use the parametric plot feature of a graphing utility to graph the inverse of a function. (Review the discussion of parametric equations in Section 1.5, if necessary.) Simply plot the parametric equations $x = f(t)$ and $y = t$. For any point (x, y) on the graph, we have $y = t = f^{-1}(x)$. It is *not* necessary to find an explicit formula for f^{-1}.

Example 7 Figure 9 shows the invertible function $f : [0, 1] \to [1, 2]$, defined by $f(s) = (16/9)s^3 - (8/3)s^2 + (17/9)s + 1, 0 \le s \le 1$. Sketch the graph of the inverse function f^{-1}.

Solution The inverse function f^{-1} maps the interval [1, 2] to [0, 1] (see Figure 10). The command in the caption for Figure 10 shows how to plot an inverse function using the computer algebra system Maple.

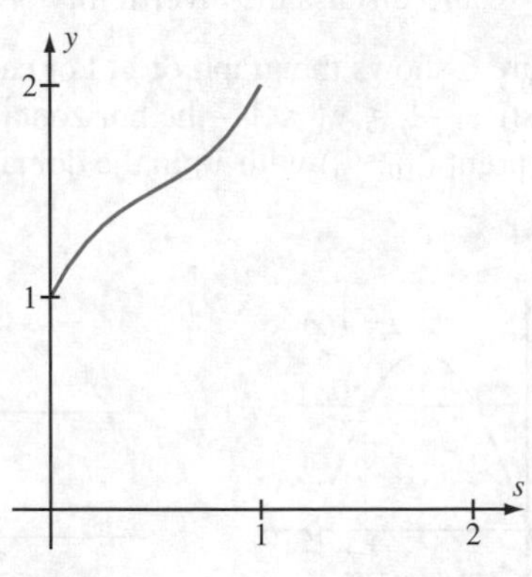

Figure 9

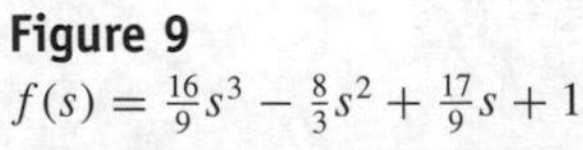

$f(s) = \frac{16}{9}s^3 - \frac{8}{3}s^2 + \frac{17}{9}s + 1$

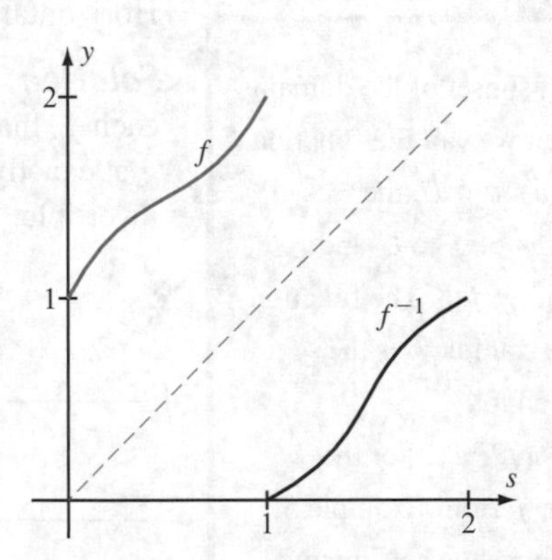

Figure 10

```
f:=s->16/9*s^3-8/3*s^2+17/9*
s+1: plot([f(s),s,s=0 . .1]);
```
■

An Application—The Natural Logarithm

A function f is *increasing* if $f(s) < f(u)$ for $s < u$. Given two different numbers, let s denote the smaller; u, the larger. The inequality $f(s) < f(u)$ shows that $f(s) \neq f(u)$. Thus, an *increasing function* is one-to-one. Similarly, a *decreasing function,* that is, a function f that satisfies $f(s) > f(u)$ for $s < u$, is also one-to-one.

As an example, consider the exponential function $x \mapsto \exp(x) = e^x$ from $\mathbb{R}$ to $\mathbb{R}^+$. The number e is defined by $e = \lim_{n\to\infty}(1 + 1/n)^n = 2.718281828\ldots$ (as in Sections 2.6 and 3.4). Since e is greater than 1, it follows that $1 < e^x$ for $x > 0$. In particular, if $s < u$, then $1 < e^{u-s}$. By multiplying each side of this inequality by the *positive* quantity e^s, we obtain the inequality $e^s < e^u$ for $s < u$. In other words, the exponential function is increasing and, therefore, invertible.

The inverse of the exponential function is called the *natural logarithm* and is denoted by $x \mapsto \ln(x)$. The domain of the natural logarithm is the image of the exponential function, namely, $\mathbb{R}^+$. The range of the natural logarithm is the domain of the exponential function, namely, $\mathbb{R}$. For every real s and positive t, we have

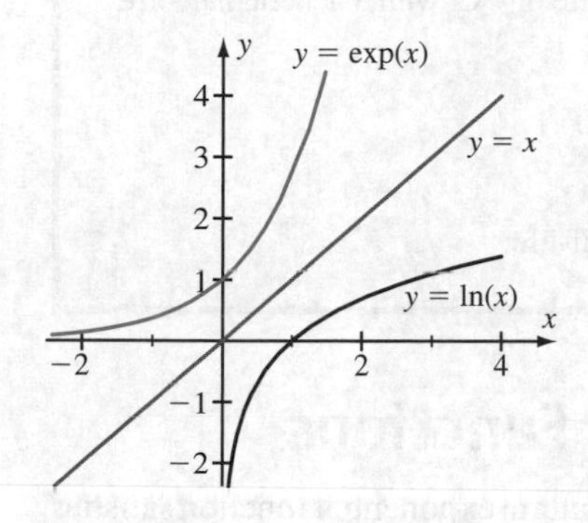

Figure 11

$$\ln(e^s) = s \quad \text{and} \quad e^{\ln(t)} = t.$$

By taking $s = 0$ and $s = 1$, we obtain

$$\ln(1) = 0 \quad \text{and} \quad \ln(e) = 1.$$

The graph of the natural logarithm is the reflection of the graph of the natural exponential function (see Figure 11). Notice that

$$\lim_{x\to 0^+} \ln(x) = -\infty \quad \text{and} \quad \lim_{x\to\infty} \ln(x) = \infty.$$

The basic algebraic properties of the natural logarithm are

$$\ln(xy) = \ln(x) + \ln(y), \quad x, y > 0,$$

and

$$\ln(x^p) = p\ln(x).$$

These properties can be deduced from the corresponding exponent laws for the natural exponential function. We investigate these important properties in some detail in Chapter 6.

Example 8 The mass of radioactive ^{14}C is given by $m(t) = Ae^{-kt}$ where t is measured in years. If 1000 g of ^{14}C is present at time $t = 0$ and if 500 g remains after 5715 years, what mass will remain at a time $t > 0$?

Solution By evaluating at $t = 0$, we see that $m(0) = Ae^0 = A$. Thus, $A = m(0) =$ 1000 g and $m(t) = 1000e^{-kt}$ g. Into this equation, substitute $t = 5715$ and $m(5715) =$ 500 to obtain $500 = 1000e^{-k\cdot 5715}$, or $1/2 = e^{-k\cdot 5715}$. Therefore,

$$\ln(1/2) = \ln(e^{-k\cdot 5715}) = -k \cdot 5715,$$

or $k = \ln(2)/5715 \approx 1.213 \times 10^{-4}$. Thus,

$$m(t) = 1000e^{-t \cdot 1.213 \times 10^{-4}} \text{ g.}$$ ■

inSIGHT

The half-life τ of a radioactive substance is the length of time required for the mass of the substance to be reduced to half the initial value: $m(\tau)/m(0) = 1/2$. In the computation of Example 8, replacing 5715, the half-life of ^{14}C, with a general half-life shows that the mass $m(t)$ of *any* radioactive substance is

$$m(t) = m(0)e^{-\ln(2)t/\tau}$$

where $m(0)$ is the initial mass at $t = 0$ and τ is its half-life.

Derivatives of Exponential Functions

We know from Section 3.4 that the derivative of the natural exponential function satisfies

$$\frac{d}{dx}e^x = e^x.$$

We are now able to obtain a formula for the derivative of the general exponential function $x \mapsto a^x$ for any positive a.

Theorem 2 If a is a positive constant, then

$$\frac{d}{dx}a^x = a^x \ln(a).$$

Proof We reduce the computation to one we already know. Notice that $a^x = (e^{\ln(a)})^x = e^{x \cdot \ln(a)}$. Thus, $a^x = e^{x \ln(a)}$. We calculate

$$\frac{d}{dx}a^x = \frac{d}{dx}e^{x \ln(a)} \overset{\text{Chain Rule}}{=} e^{x \ln(a)} \frac{d}{dx}(x \ln(a)) = e^{x \ln(a)} \ln(a) = a^x \ln(a).$$ ■

inSIGHT

The proof of Theorem 2 employs a formula that is important in its own right. The equation

$$a^x = e^{x \ln(a)}$$

shows that we can always rewrite an exponential function so that e becomes the base.

Continuity and Differentiability of Inverse Functions

The next theorem guarantees that we can differentiate the inverse of a differentiable invertible function.

Theorem 3 Suppose S and T are open intervals in $\mathbb{R}$ and $f : S \to T$ is one-to-one and onto, hence invertible. Therefore,

a. If f is continuous on S, then f^{-1} is continuous on T.
b. If f is differentiable on S and f' is never 0 on S, then f^{-1} is differentiable on T.

Although we will not prove this theorem, it is not too difficult to understand why it must be true. If $f'(c)$ exists and $f'(c) \neq 0$, then the graph of f has a nonhorizontal tangent line ℓ at $(c, f(c))$. The reflection through $y = x$ of the line ℓ is, therefore, a nonvertical tangent line L to the graph of f^{-1} at $(f(c), c)$. This is precisely the geometric criterion for the *existence* of the derivative of f^{-1} at $f(c)$.

Our geometric considerations also allow us to deduce the value of $(f^{-1})'$. Reflection through the line $y = x$ has the effect of interchanging x and y coordinates. Therefore, if $\Delta y/\Delta x$ is the slope of line ℓ, then $\Delta x/\Delta y$ is the slope of the reflected line L (see Figure 12). In other words, the slope of L is the reciprocal of the slope of ℓ. Since $f'(c)$ is the slope of ℓ, it follows that L must have slope $1/f'(c)$.

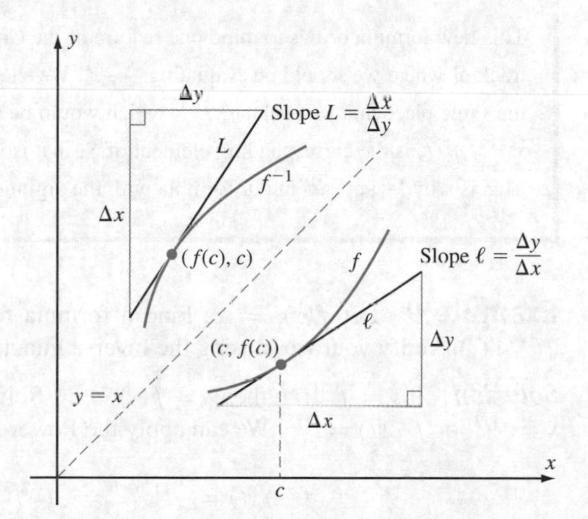

Figure 12

Let us verify our deduction. To begin, differentiate both sides of the basic equation

$$(f \circ f^{-1})(t) = t$$

with respect to t. We use the Chain Rule on the left side to get

$$f'(f^{-1}(t)) \cdot (f^{-1})'(t) = 1,$$

or

$$(f^{-1})'(t) = \frac{1}{f'(f^{-1}(t))}.$$

If we let $s = f^{-1}(t)$, then we can write this formula in Leibniz notation as

$$\left.\frac{d(f^{-1})}{dt}\right|_t = \frac{1}{\left.\frac{df}{ds}\right|_{s=f^{-1}(t)}}.$$

We now have an equation for the derivative of the inverse of a function, which we summarize in the following theorem.

Theorem 4 ***Inverse Function Derivative Rule*** Let f be an invertible function that is defined on an open interval containing the point s. Let $t = f(s)$. If f is differentiable at $s = f^{-1}(t)$ and if $\frac{df}{ds}$ is nonzero at this point, then the derivative of f^{-1} at t is given by

$$\left.\frac{d(f^{-1})}{dt}\right|_t = \frac{1}{\left.\frac{df}{ds}\right|_{s=f^{-1}(t)}}.$$

inSIGHT

This new formula brings to mind one feature of the Chain Rule: We must keep careful track of where we should be evaluating $\frac{d(f^{-1})}{dt}$. We should evaluate this derivative at the same place that we evaluate f^{-1}, which would be at an element $t \in T$. Since $s = f^{-1}(t)$ is the corresponding element of S, that is where we would evaluate f, or $\frac{df}{ds}$. This is why $\frac{df}{ds}$ appears in our formula with the argument $f^{-1}(t)$.

Example 9 Let $f(s) = s^3$. Find a formula for $f^{-1}(t)$ and use it to calculate $(f^{-1})'(t)$. Verify your work using the Inverse Function Derivative Rule.

Solution If $s = f^{-1}(t)$, then $t = f(s) = s^3$. Solving this equation for s, we obtain $s = t^{1/3}$, or $f^{-1}(t) = t^{1/3}$. We can apply the Power Rule directly to f^{-1}:

$$(f^{-1})'(t) = \frac{1}{3}t^{1/3-1} = \frac{1}{3}t^{-2/3} \quad \text{for } t \neq 0.$$

Next we use the Inverse Function Derivative Rule to verify this formula. The rule tells us that if

$$\left.\frac{df}{ds}\right|_{s=f^{-1}(t)} \neq 0,$$

then

$$\frac{d(f^{-1})}{dt} = \frac{1}{\left.\frac{df}{ds}\right|_{s=f^{-1}(t)}} = \frac{1}{\left.3s^2\right|_{s=f^{-1}(t)}} = \frac{1}{3(t^{1/3})^2} = \frac{1}{3}t^{-2/3},$$

which is in agreement with our direct calculation. ■

Example 10 Consider the invertible function $f(x) = (x^5 + x + 2)^{5/2}$. Calculate $(f^{-1})'(32)$.

Solution In this example, we cannot compute an explicit formula for the function f^{-1}. However, the Inverse Function Derivative Rule provides us with a formula for

$(f^{-1})'(32)$ in terms of $f'(x)$ where x is the unique solution of $f(x) = 32$. In this case, we can spot that $x = 1$, since $f(1) = 4^{5/2} = 32$. We calculate

$$\left.\frac{df}{dx}\right|_x = \frac{5}{2}\cdot(x^5 + x + 2)^{3/2}\cdot(5x^4 + 1)$$

and conclude that

$$\left.\frac{d(f^{-1})}{dy}\right|_{y=32} = \frac{1}{\left.\frac{df}{dx}\right|_{x=f^{-1}(32)}} = \frac{1}{(5/2)\cdot(x^5+x+2)^{3/2}\cdot(5x^4+1)|_{x=1}}$$
$$= \frac{1}{(5/2)\cdot 8\cdot 6} = \frac{1}{120}.$$

■

inSIGHT

In Example 10, we can evaluate $x = f^{-1}(y)$ for the particular value $y = 32$ even though we cannot evaluate $f^{-1}(y)$ for all values of y. If the equation $f(x) = y$ cannot be solved exactly, then we turn to the "solve" command of a calculator or computer to find an approximate solution. For example, suppose that $(f^{-1})'(35)$ is required. The command `fsolve( (x^5+x+2)^(5/2)=35, x)` in the computer algebra system Maple tells us that $x = 1.0234$ is the solution of the equation $f(x) = 35$ (to four decimal places). Therefore,

$$\left.\frac{d(f^{-1})}{dy}\right|_{y=35} = \frac{1}{\left.\frac{df}{dx}\right|_{x=f^{-1}(35)}} \approx \frac{1}{(5/2)\cdot(x^5+x+2)^{3/2}\cdot(5x^4+1)|_{x=1.0234}} \approx \frac{1}{136.86}.$$

The Derivative of the Natural Logarithm

The formula for the derivative of an inverse function can be used to calculate the derivative of the natural logarithm function.

Theorem 5 Show that

$$\frac{d}{dt}\ln(t) = \frac{1}{t}$$

for every $t > 0$.

inSIGHT

Because the natural logarithm is one of the important functions of calculus, you should memorize the formula for its derivative.

Proof If $f(s) = e^s$, then $f^{-1}(t) = \ln(t)$. Applying the Inverse Function Derivative Rule results in

$$\frac{d(f^{-1})}{dt}(t) = \frac{1}{\left.\frac{df}{ds}\right|_{s=f^{-1}(t)}} = \frac{1}{e^s|_{s=f^{-1}(t)}} = \frac{1}{e^{f^{-1}(t)}} = \frac{1}{e^{\ln(t)}} = \frac{1}{t}.$$

■

Example 11 Find the equation of the tangent line to $y = \ln(x)$ at its x-intercept.

Solution Earlier in this section, we observed that $\ln(1) = 0$. Since the natural logarithm is an increasing function, the point $(1, 0)$ is the only x-intercept of the curve $y = \ln(x)$. Since

$$\frac{d}{dx}\ln(x)\bigg|_{x=1} = \frac{1}{x}\bigg|_{x=1} = 1,$$

the point-slope equation for the required tangent line is $y = 1\cdot(x-1)+0$, or $y = x-1$. (Use your graphing calculator to illustrate this example.) ■

quickquiz

1. When does the inverse of a function exist?
2. What is the formula for the derivative of the inverse?
3. How does $\ln(3e^4)$ simplify?
4. What is the derived function of $x \mapsto 2^x$?
5. Express π^7 as a power of e.

EXERCISES

Problems for Practice

In Exercises 1–12, determine whether the function is invertible. If the function *is* invertible, find the inverse. If it is *not* invertible, explain why not.

1. $f : [0, \infty) \to [1, \infty),\ f(s) = s^2 + 1$
2. $g : [0, 2] \to [-1, 1],\ g(s) = (s-1)^2$
3. $f : [0, 1] \to [0, 2],\ f(s) = s^2 + s$
4. $g : (0, \infty) \to (1, \infty),\ g(s) = s^4 + 1$
5. $f : [1, 10) \to [1, 1000),\ f(s) = s^3$
6. $h : \mathbb{R} \to \mathbb{R},\ h(s) = s^4 - 2s$
7. $h : (4, \infty) \to (1, 16/15),\ h(s) = s^2/(s^2 - 1)$
8. $g : [0, 1] \to [0, 1/2],\ g(s) = s/(s+1)$
9. $h : (1, 6) \to (2, 3),\ h(s) = \sqrt{s+3}$
10. $f : (1, 4) \to (4, 5),\ f(s) = \sqrt{s} + 3$
11. $g : [1, \infty) \to (0, 1],\ g(s) = 1/(s^2 + 1)$
12. $f : (0, \infty) \to (0, 1],\ h(s) = 1/(1 + s^4)$
13. Examine the graphs in Figure 13, and determine which represent invertible functions. If the graph represents an invertible function, draw a graph of the inverse.

In Exercises 14–17, assume that $f : \mathbb{R} \to \mathbb{R}$ is invertible and differentiable, with a differentiable inverse. Compute $(f^{-1})'(4)$ from the given information.

14. $f^{-1}(4) = 3,\ f'(3) = 6$
15. $f^{-1}(4) = 1,\ f'(1) = 2$
16. $f^{-1}(4) = 7,\ f'(7) = -8$
17. $f^{-1}(4) = -1,\ f'(-1) = 3$

In Exercises 18–21, calculate the derivative of the inverse of the function by the method given in Theorem 3.

18. $f : [0, \infty) \to [4, \infty),\ f(s) = s^2 + 4$
19. $f : (0, 3) \to (2, 245),\ f(s) = s^5 + 2$
20. $f : (1, 3] \to (2, 12],\ f(s) = s^2 + s$
21. $f : [1, 8] \to [1/64, 1],\ f(s) = 1/s^2$

In Exercises 22–25, simplify the expression.

22. $\ln(e^2)$
23. $e^{\ln(3)}$
24. $\ln(7\sqrt{e})$
25. $\exp(-3\ln(2))$

In Exercises 26–37, calculate $f'(x)$ for the given $f(x)$.

26. $\ln(2x)$
27. $\ln(1 + \exp(x))$
28. 10^x
29. $2^{\ln(x)}$
30. $x/\ln(x)$
31. $\sqrt{\ln(x)}$
32. $\ln^3(x)$
33. $\cos(\ln(x))$
34. $\ln(\ln(x))$
35. $\sin(2 \cdot 3^x)$
36. $2^{(x^\pi)}$
37. $2^{(\pi^x)}$

In Exercises 38–40, evaluate the derivative f' of the function f in two ways. First, apply the Chain Rule to $f(x)$ without simplifying $f(x)$ in advance. Second, simplify $f(x)$, and then differentiate the simplified expression. Verify that the two expressions are equal.

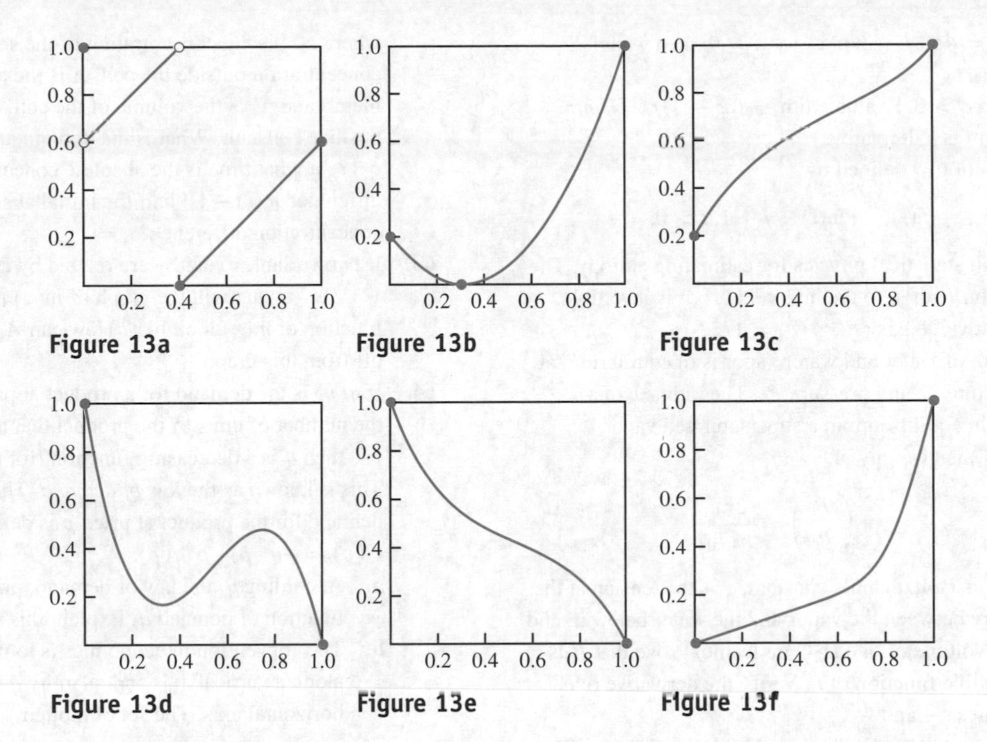

Figure 13a Figure 13b Figure 13c

Figure 13d Figure 13e Figure 13f

38. $f(x) = \ln(3x)$
39. $f(x) = \ln(x^2)$
40. $f(x) = (2^x)^3$

Further Theory and Practice

41. Suppose that $f : S \to T$ is invertible. Let c denote a constant, and define $f + c$ to the function for which $(f + c)(x) = f(x) + c$. What is the domain and range of $f + c$? Explain why $f + c$ is invertible. What is the relationship between $(f + c)^{-1}$ and f^{-1}?
42. Suppose that $f : (a, b) \to (c, d)$ is invertible. Does it follow that f^2 is invertible? If $c > 0$, does your answer change?
43. Prove that if $f : (a, b) \to (c, d)$ and $g : (c, d) \to (\alpha, \beta)$ are invertible, then $g \circ f$ is invertible. If all functions involved are differentiable, derive a formula for $((g \circ f)^{-1})'$ in terms of f' and g'.
44. For each f in Exercises 18–21, write the equation of the tangent line to the graph of f^{-1} at the point $(f(2), 2)$.

In Exercises 45–48, find $f^{-1}(c)$ for f and c, but do not try to calculate $f^{-1}(t)$ for a general value of t. Calculate $(f^{-1})'(c)$.

45. $f(s) = s^5 + 2s^3 + 2s + 3,\ c = 3$
46. $f(s) = s^3 + 2s - 7,\ c = 5$
47. $f(s) = \pi s - \cos(\pi s/2),\ c = \pi$
48. $f(s) = (s^3 + s^2 + 1)/(s^2 + 1),\ c = 3/2$
49. Suppose A and k are positive constants. If a variable y grows exponentially according to $y = Ae^{kt}$, then the doubling time τ is defined by the property that the value of y doubles when t is increased by τ. Express the rate constant k in terms of the doubling time τ of y.
50. A population is growing exponentially according to the law $P(t) = Ae^{kt}$ where A and k are positive constants. What is the rate of change of P when P equals $2A$?
51. A population grows exponentially according to the law $P(t) = P(0) \cdot 2^{kt}$ where t is measured in years. What is the doubling time of the population (as defined in Exercise 49)? Find $P'(t)$ when $P(t) = 2 \cdot P(0)$.
52. Suppose the function f is invertible and $P = (2, 3)$ is a point on its graph. Suppose the slope of the tangent to the graph of f at P is $1/7$. Write the equation of a tangent line to the graph of $y = f^{-1}(x)$.
53. Suppose $u(x)$ is always positive and v is an arbitrary function. Let $f(x) = u(x)^{v(x)}$. In this expression for f, both the base and the exponent are variable. Express f in the form $f(x) = e^{g(x)}$. Use this form to find $f'(x)$.
54. Is there a value c such that the tangent lines to the graphs of $y = e^x$ and $y = \ln(x)$ at (c, e^c) and $(c, \ln(c))$ are parallel? Why or why not?
55. Is there a value c such that the tangent lines to the graphs of $y = e^x$ and $y = \ln(x)$ at (c, e^c) and $(c, \ln(c))$ are perpendicular? Why or why not?

56. If $f(x) = 4e^{x-1} + 3\ln(x)$ for $x > 0$, then what is $(f^{-1})'(4)$?

57. Suppose $a > 0$. Evaluate $\lim_{x\to 0}(a^x - 1)/x$. (*Hint:* This limit is a derivative.)

58. The function f defined by

$$f(t) = t\ln(t) - t + 1,\; t > 0,$$

is used in statistical physics for estimating entropy. The restriction F of f to the interval $[1, \infty)$ is invertible. For what value c is $(F^{-1})'(c) = 1/2$?

59. A system of water and water vapor is in equilibrium at temperature T_0 and pressure P_0. The radius r of a droplet in equilibrium in a supersaturated vapor $(P > P_0)$ is given by

$$\ln\left(\frac{P}{P_0}\right) = \frac{2\tau v}{rkT_0}$$

where k is Boltzmann's constant, τ is the tension at the boundary between the vapor and the water below it, and v is the volume of one H_2O molecule. Prove that P is an invertible function of r. Verify the derivative rule that relates $\frac{dr}{dP}$ and $\frac{dP}{dr}$.

60. The concentration C of a drug in a patient's bloodstream is

$$C(t) = \frac{\alpha}{\beta}(1 - e^{-\beta t})$$

when it is administered intravenously. The constants α and β are positive.

a. What are the dimensions of the constants β and α? (*Hint:* The dimension of C is mass/volume. An exponential is unitless.)

b. Compare $C(s)$ with $C(u)$ for $s < u$. Is the concentration of the drug increasing? Why or why not?

c. What is the horizontal asymptote of the graph of C?

d. Is C invertible? If so, what is the domain and formula for the inverse function? If not, why not?

e. At what time is the concentration equal to $0.9 \cdot \alpha/\beta$?

61. Suppose the air resistance to an object dropped from height h is proportional to the square of its velocity (with proportionality constant κ). Its height at time t is given by

$$y(t) = h + t\sqrt{\frac{g}{\kappa}} - \frac{1}{\kappa}\ln\left(\frac{1 + e^{2t\sqrt{g\kappa}}}{2}\right).$$

What is the velocity of the object at time t?

62. According to Fick's law, the diffusion of a solute through a cell membrane is described by

$$c(t) = (c(0) - C)\exp\left(-\frac{kAt}{V}\right) + C$$

where $c(t)$ is the concentration of the solute, C is the concentration outside the cell, A is the area of the cell membrane, V is the volume of the cell, and k is a positive constant. What is the domain and formula for c^{-1}? At what time is the absolute concentration difference $|c(t) - C|$ half the initial absolute concentration difference?

63. If two variables x and y are related by an equation $y = Ax^m$, what will the graph of $\ln(y)$ plotted as a function of $\ln(x)$ look like? How can A and m be read off from this graph?

64. If $q(p)$ is the demand for a product at price p—that is, the number of units of the product that are sold at price p—then q is a decreasing function (for most products). This is known as the *law of demand*. The elasticity of demand for the product at price p is defined as $E(p) = -q'(p) \cdot p/q(p)$.

a. According to the law of demand, price p is a function of demand q. Explain this mathematically.

b. The convention in economics is to plot price p along a vertical axis and quantity q along a horizontal axis. The set of plotted points $(q(p), p)$ is called the *demand curve*. At a point $(q(p), p)$ on the demand curve, express the slope of the tangent line in terms of the demand function q.

c. Let m be the slope of the line that is perpendicular to the demand curve at the point $(q(p), p)$. Express m in terms of the elasticity of demand.

Calculator/Computer Exercises

65. Graph $f(x) = x^2$ and $f^{-1}(x) = \sqrt{x}$ in $[0, 1] \times [0, 1]$. Use equal scaling on each axis. Add the tangent lines to f at $(2/3, 4/9)$ and to f^{-1} at $(4/9, 2/3)$ to this viewing window. Explain how the Inverse Function Derivative Rule is illustrated by this window.

66. Graph $x \overset{f}{\mapsto} x^3 + 1$ and $x \overset{f^{-1}}{\mapsto} (x - 1)^{1/3}$ in $[0, 9] \times [0, 9]$. Add to the viewing window the line ℓ that is tangent to the graph of f at $(3/2, f(3/2))$. Add the line L that is tangent to the graph of f^{-1} at $(f(3/2), 3/2)$.

67. The half-life of ^{60}C (cobalt 60) is 5.3 yr. What is the rate at which a sample of cobalt is decaying when its mass is 10 g? At what time thereafter will its instantaneous rate of change be 1 g/yr?

68. Strontium 90 has a half-life of 28 yr. Cobalt 60 has a half life of 5.3 yr. A certain sample of cobalt 60 is 200 g at the same instant a certain sample of strontium 90 is

100 g. Are the masses of the two samples ever equal? If so, when? If not, why not? Are the instantaneous rates of change of the masses of the two samples ever equal? If so, when? If not, why not?

In Exercises 69–72, for function f and point c, use a solve or zoom utility to approximate the solution of $f(s) = c$. Approximate $(f^{-1})'(c)$.

69. $f(s) = s^3 + 2s + \sin(s)$, $c = 1$

70. $f(s) = 2s + \sqrt{s^3 + 1}$, $c = 4$

71. $f(s) = s^5 + 4s - 2$, $c = 15$

72. $f(s) = s^7 + s^3 - 2$, $c = 1$

73. Approximate the value c such that the tangent lines to the graphs of $y = e^x$ and $y = \ln(x)$ at (c, e^c) and $(c, \ln(c))$ are parallel. Illustrate this with a graph.

74. Use a central difference quotient to approximate the derivative of 10^x at $x = 0$. Use your computation to evaluate $\ln(10)$ to five decimal places.

75. Graph $f(x) = 10^5(\ln(x + 10^{-5}) - \ln(x - 10^{-5}))/2$ for $0.1 \le x \le 10$. Add the graph of $x \mapsto 1/x$ to the same viewing window. Which differentiation rule do the graphs illustrate?

76. A topic of current interest in population biology is the effect of decreasing habitat on the variation of species. In a recent experiment, formica plates were placed on the ocean floor. The number of species y of marine organisms that attached themselves to the plates was tabulated as a function of the plate size in cubic centimeters. When the points $(\ln(x), \ln(y))$ were plotted, a straight line passing through $(2.303, 1.792)$ and $(9.210, 3.466)$ resulted. Determine y as a function of x and graph this function.

77. Biologists have wondered if the brain weight–to–body weight ratio distinguishes humans from other animals. In one study, the average relationships between prenatal brain weight y (in grams) and prenatal body weight x (in grams) were determined. The results are listed in the table.

	Brain Weight–to–Body Weight Equation
Macaques	$\ln(y) = -2.056 + 1.0164 \ln(x)$
Humans	$\ln(y) = -2.211 + 1.0160 \ln(x)$

In the same viewing window, plot y as a function of x for both humans and macaques. What is $\frac{dy}{dx}$ for each species when $x = 500$ g?

78. Fresh from the pot, the temperature of a cup of coffee at a chain of restaurants is 80°C. As the coffee cools, its temperature is given by

$$T(t) = 18 + 62 \cdot e^{-kt}$$

where k is a constant. Plot t. If the coffee cools from 70°C to 60°C in 10 minutes, in how many more minutes will the temperature be 44°C?

In Exercises 79–81, use a parametric plotter to graph the inverse function of function f. Do not attempt to find a formula for f^{-1}.

79. $f(x) = x^5 - 3x^3 + x^2 + 6x - 8$, $0 \le x \le 2$

80. $f(x) = x^3 - 3x^2 + 3x$, $0 \le x \le 1$

81. $f(x) = 3 + \cos(x) + \cos^2(x)$

82. Let

$$f(s) = \sqrt{s^3 + 2s + 4\sin^4\left(\frac{\sqrt{s}}{2}\right)}$$

for $1 \le s \le 8$. Plot f^{-1}. To your plot, add the graph of the tangent line at the point $(6, f^{-1}(6))$.

3.7 Higher Derivatives

The derivative of a function is itself a function. For example, the derivative of the function

$$p(t) = t^3 - 5t^2 + 7$$

is

$$p'(t) = 3t^2 - 10t.$$

Because the function p' is differentiable, we may consider its derivative $(p')'(t) = 6t - 10$. We simplify the notation by writing p'' for $(p')'$. Of course, the second derivative p'' is also a function, and we may differentiate it to obtain the *third derivative*

$$p'''(t) = 6.$$

This procedure may be carried on indefinitely: The derivative of the $(k-1)$th derivative is called the kth derivative. Writing more than two or three primes becomes tedious, however, and it is even more tedious to read. So we have other ways to write the higher derivatives.

Notation for Higher Derivatives

Definition The derivative of f' is called the *second derivative* of f. In general, the derivative of the $(k-1)$th derivative of f, if it exists, is called the kth derivative. We write the kth derivative of f as

$$f''^{\cdots\prime}(k \text{ primes}) \quad \text{or} \quad f^{(k)} \quad \text{or} \quad \frac{d^k f}{dx^k} \quad \text{or} \quad \frac{d^k}{dx^k} f \quad \text{or} \quad D^k(f).$$

Sometimes roman numerals are used. For example, $f^{(\text{iv})}$ might be used instead of $f^{(4)}$. The number k is called the *order* of $f^{(k)}$. To distinguish f' from higher order derivatives, we often call f' the *first derivative* of f. The function f itself is sometimes called the *zeroth derivative* of f.

Example 1 Define $p(x) = 5x^4 - 8x^2 - 7x - 11$. Find $p^{(3)}(4)$.

When we calculate a higher derivative, we do not substitute a value for x until we are finished differentiating. If you substitute the value of x too soon, you end up with a constant function and the next derivative will be 0.

Solution We have, in order,

$$p^{(1)}(x) = 20x^3 - 16x - 7, \qquad p^{(2)}(x) = 60x^2 - 16, \quad \text{and} \quad p^{(3)}(x) = 120x.$$

Therefore, $p^{(3)}(4) = 480$. ■

When f is a function and $f^{(1)}(c), f^{(2)}(c), \ldots, f^{(k)}(c)$ exist, we say that f is k *times differentiable* at c. Suppose that f is two times differentiable at c. This means $f'(c)$ and $f''(c)$ exist. In particular, the function f' is differentiable at c. According to the derivative definition, the function f' is defined on an open interval centered at c. Thus, hidden in this definition is the requirement that f is differentiable not only at c but also on an open interval centered at c. Similarly, if f is k times differentiable at c, then $f, f', \ldots, f^{(k-1)}$ are differentiable in an open interval centered at c.

Velocity and Acceleration

If p is position and t is time, then $v(t) = p'(t)$ is the rate of change of position with respect to time, or *velocity*. The rate of change of velocity with respect to time is *acceleration* $a(t)$. Thus, $a(t) = v'(t) = p''(t)$: Acceleration is the first derivative of velocity with respect to time and the second derivative of position with respect to time.

Example 2 If position p (measured in feet) at time t (measured in seconds) is given by $p(t) = \tan(t)$ for small values of t, then find the acceleration at time $t = 0.01$ s.

Solution We compute

$$\frac{d^2p}{dt^2} = \frac{d}{dt}\left(\frac{dp}{dt}\right) = \frac{d}{dt}(\sec^2(t)) \overset{\text{Chain Rule}}{=} 2\sec(t)\cdot\left(\frac{d}{dt}\sec(t)\right) = 2\sec^2(t)\tan(t).$$

At time $t = 0.01$, we have

$$a(0.01) = \left.\frac{d^2p}{dt^2}\right|_{0.01} = 2\sec^2(0.01)\tan(0.01) \approx 0.02 \text{ ft/s}^2.$$

■

Newton's Notation

Yet another notation for higher derivatives is Newton's. If $p(t)$ is a function of the time variable t, then the first and second derivatives are represented by

$$\dot{p}(t) \quad\text{and}\quad \ddot{p}(t).$$

This notation is often used in physics and engineering.

Example 3 An arrow is shot straight up. Its height (in feet) at time t (in seconds) is given by the formula $H(t) = -16t^2 + 160t + 5$. Compute its velocity and its acceleration. Discuss the physical significance of these quantities.

Solution The velocity is $v(t) = \dot{H}(t) = -32t + 160$ ft/s. The acceleration is $a(t) = \ddot{H}(t) = -32$ ft/s^2. Some interesting physical data may be read from this information. For instance, when the arrow reaches its maximum height, at that instant, it has neither upward nor downward motion: Its velocity is 0. We solve $0 = \dot{H}(t) = -32t + 160$ to find that the maximum height occurs when $t = 5$. At this time, the height is $H(5) = 405$ ft. At $t = 5$, the *acceleration* is -32 ft/s^2—the acceleration is constantly equal to this value—but the velocity is 0. ■

inSIGHT

Although one can compute second, third, fourth, and higher derivatives, the physical significance of these quantities is increasingly difficult to understand. In some areas of engineering, the third and fourth derivatives—called *jerk* and *surge*—are useful. When you are in an elevator that is not working properly and your stomach feels as if it were left on the previous floor, then you are experiencing jerk. In the past, engineers have built jerk meters to subject space vehicles to vibration tests.

Approximation of Second Derivatives

Higher derivatives can also be approximated by difference quotients. We illustrate the procedure by approximating the second derivative. We start by writing $f''(c)$ as $(f')'(c)$. We can approximate $(f')'(c)$ by the central difference quotient $D_0(f')(c, h)$

of f' (as discussed in Section 3.3). Thus,

$$f''(c) \approx \frac{f'(c+h/2) - f'(c-h/2)}{h}.$$

Next, we approximate $f'(c+h/2)$ and $f'(c-h/2)$ by the central difference quotients $D_0 f(c+h/2, h)$ and $D_0 f(c-h/2, h)$:

$$f''(c) \approx \frac{\frac{f(c+h)-f(c)}{h} - \frac{f(c)-f(c-h)}{h}}{h},$$

or

$$f''(c) \approx \frac{\big(f(c+h) - f(c)\big) - \big(f(c) - f(c-h)\big)}{h^2}.$$

This is the *second central difference quotient* approximation to $f''(c)$. We may also use the algebraic simplification

$$f''(c) \approx \frac{f(c+h) - 2f(c) + f(c-h)}{h^2}.$$

Because the number $f(c+h) + f(c-h)$ is very close to $2f(c)$ when h is small, the subtraction of these nearly equal numbers will usually result in a loss of significant digits. Therefore, the second central difference quotient approximation must be implemented with care.

Example 4 Using a certain approximation of air resistance, we can show that the height of an object propelled upward from ground level is

$$f(t) = 33.33 \cdot \ln(54.25 \cos(0.5424t - 1.552)) \text{ m}, \quad 0 \le t \le 2.862 \text{ s}.$$

What is the approximate acceleration of the object at time $t = 2$?

Solution We are asked to approximate $f''(c)$ for $c = 2$. (Because the formula for $f(t)$ is based on an approximation, using the differentiation rules of calculus to obtain an "exact" value of $f''(2)$ would suggest an accuracy that cannot be obtained.) Figure 1 is a screen capture of a Maple computer algebra system session. In the first line, the second central difference quotient has been defined and named $D2$. The next line sets the number of digits used in the computations to 15 to help guard against loss of significance (refer to Section 1.1). In the third line, f has been defined. The fourth line evaluates the second central difference quotients that correspond to $h = 0.01$, $0.001, \ldots, 0.00000001$. The first four values of h indicate that $f''(2) \approx -12.3$. The two smallest values of h, 10^{-7} and 10^{-8}, result in wildly inaccurate approximations due to loss of significance. We conclude that $f''(2) \approx -12.3$. (The negative sign indicates that the object is decelerating, as we would expect.) The last line of the Maple session evaluates $f''(2)$ using the derivative rules of calculus. The result confirms our numerical approximation.

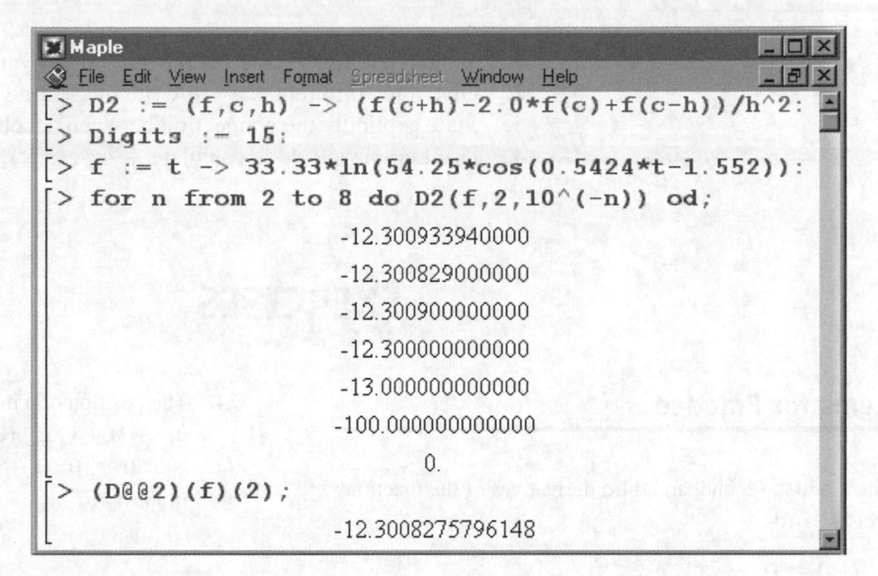

Figure 1 ■

Leibniz's Rule

The Product Rule tells us how to compute the first derivative of a product $f \cdot g$. In 1695, Leibniz derived a rule for computing *higher* derivatives of a product. Leibniz's Rule tells us how to calculate $(f \cdot g)^{(n)}(c)$ in terms of $f'(c)$, $f''(c)$, ..., $f^{(n)}(c)$ and $g'(c)$, $g''(c)$, ..., $g^{(n)}(c)$. For now, we will state Leibniz's Rule only for second derivatives. Exercise 33 contains the general rule and asks you to verify it for $n = 2$ and 3.

Theorem 1 ***Leibniz's Rule for Second Derivatives*** If f and g are two times differentiable at c, then so is $f \cdot g$ and

$$(f \cdot g)''(c) = f''(c)g(c) + 2f'(c)g'(c) + f(c)g''(c).$$

Example 5 A mass on a spring oscillates about its equilibrium position according to the formula

$$x(t) = e^{-t/2} \cos\left(t\sqrt{2}\right) \text{ cm}$$

where t is measured in seconds. What is the acceleration of the mass at time t?

Solution If $f(t) = e^{-t/2}$ and $g(t) = \cos(t\sqrt{2})$, then $f^{(1)}(t) = -e^{-t/2}/2$, $f^{(2)}(t) = e^{-t/2}/4$, $g^{(1)}(t) = -\sqrt{2}\sin(t\sqrt{2})$, and $g^{(2)}(t) = -2\cos(t\sqrt{2})$. Thus,

$$\begin{aligned}
\frac{d^2}{dt^2} e^{-t/2} \cos\left(t\sqrt{2}\right) &= g(t) f^{(2)}(t) + 2 f^{(1)}(t) g^{(1)}(t) + f(t) g^{(2)} \\
&= \frac{1}{4} e^{-t/2} \cos\left(t\sqrt{2}\right) + 2\left(-\frac{1}{2} e^{-t/2}\right)\left(-\sqrt{2}\sin\left(t\sqrt{2}\right)\right) - 2e^{-t/2} \cos\left(t\sqrt{2}\right) \\
&= \sqrt{2} e^{-t/2} \sin\left(t\sqrt{2}\right) - \frac{7}{4} e^{-t/2} \cos\left(t\sqrt{2}\right) \text{ cm/s}^2.
\end{aligned}$$

■

quickquiz

1. What is the significance of the second derivative?
2. Give three different ways to write the second derivative of a function f.
3. State explicitly the connection between acceleration and velocity.
4. Use Leibniz's Rule to calculate $\frac{d^2}{dx^2}x^2\cos(x)$.

EXERCISES

Problems for Practice

Compute the first, second, and third derivatives of the functions in Exercises 1–10.

1. $f(x) = 4x^3 - 7x^{-5} + 2x^{5/2}$
2. $g(t) = t/(t+5)$
3. $g(t) = \tan(t)$
4. $f(x) = (2x-5)\sin(x)$
5. $k(x) = (x+1)/(x-1)$
6. $H(x) = (x^2 - x)(x+5)$
7. $k(x) = \sin(x^3 - 3x)$
8. $H(t) = (t^2+1)^{5/3}$
9. $H(t) = t\tan(t)$
10. $f(s) = 5s^2 - 6s + 7$

In Exercises 11–20, calculate the requested derivative.

11. $f^{(5)}$ where $f(x) = \sin(x)$
12. $\dfrac{d^2 f}{dx^2}$ where $f(x) = x^2/(x-1)$
13. $\ddot{g}(t)$ where $g(t) = (3t^2 - 5t)^{-6}$
14. $f'''(t)$ where $f(t) = \sin(t)\cos(2t)$
15. $H^{(4)}(x)$ where $H(x) = -8x^6 + 7x^5 - 9x^2 + 11$
16. $\dfrac{d^3 f}{dx^3}$ where $f(x) = \sec(5x)$
17. $f''(t)$ where $f(t) = \cos(4x+3)$
18. $\dfrac{d^2}{dt^2}(t^{-5/3} + 4t^4 - 8t^{7/2})$
19. $g''(x)$ where $g(x) = \cos(\sin(x))$
20. $H^{(2)}(x)$ where $H(x) = x^3\tan(x)$
21. The velocity of a car at time t (in seconds) is $v(t) = t^2 - 5t$ ft/s. What is the acceleration at $t = 3$? What is the acceleration at $t = 6$?
22. What is acceleration when velocity is constant?
23. The position of a bicycle at time t (measured in minutes) is $p(t) = 2t^3 + t^2 + 6t$ m for $0 < t < 10$. What is the acceleration of the bicycle at time $t = 3$? At time $t = 6$?
24. The position of a moving body is described by $p(t) = at^2 + bt + c$. If $p(0) = 3$, what is c? If $p'(0) = 6$, what is b? If $p''(0) = -5$, what is a? Where is the body at time $t = 6$?

Further Theory and Practice

25. Let
$$f(x) = \begin{cases} x^3 & \text{if } x > 0 \\ -x^3 & \text{if } x \le 0 \end{cases}.$$
    a. Prove that $f'(0)$ exists and equals 0.
    b. Prove that $f''(0)$ exists and equals 0.
    c. Prove that $f'''(0)$ does not exist.
26. If f and g are twice differentiable, then calculate formulas for
$$\left(\frac{f}{g}\right)'' \quad \text{and} \quad (f \cdot g)'''.$$
27. Suppose $p(x)$ is a polynomial. Prove that $p(x)$ has degree precisely k if and only if $p^{(k+1)}(x) = 0$ for all x and $p^{(k)}(x) \neq 0$ for some x.
28. If $p(x)$ is a degree two polynomial such that $p(0) = 2$, $p'(0) = 4$, and $p''(0) = -7$, then find p explicitly.
29. Let $p(x)$ be a degree three polynomial such that $p(1) = 1$, $p'(2) = -2$, $p''(-1) = -14$, and $p'''(5) = 6$. What polynomial is p?
30. On the evening news, we sometimes hear that the rate of inflation is increasing more slowly than it was last year. Let $V(t)$ be the value of the dollar at time t, measured against 1960 dollars. What derivative of the function V is the newscaster describing? Is the derivative positive or negative?
31. If $p(x)$ is a polynomial and if $|p''(x)| \le 1$ for all x, prove that p is of degree at most two and its leading coefficient is less than or equal to $1/2$ in absolute value.
32. A source emits waves of speed s and frequency υ_0. If the velocity at which the source moves toward a fixed

receptor is u, then the frequency ν of the received wave is given by

$$\nu = \nu_0 \cdot \left(1 + \frac{u}{s}\right),$$

according to the *Doppler effect*. (The same formula applies when the source moves away from the listener; in this case $u < 0$.) A train accelerates at the constant rate a as it passes by a listener. What is the rate of change of the frequency of the train's whistle (as heard by the listener) as the train approaches the listener? As it recedes from the listener?

33. Let n be a positive integer. If k is a positive integer no greater than n, the expression

$$\binom{n}{k} = \frac{n(n-1)(n-2)\cdots(n-k+1)}{k\cdot(k-1)\cdots 3\cdot 2\cdot 1}$$

is called a *binomial coefficient*. For example, if $n = 7$ and $k = 3$, then $n - k + 1 = 5$ and

$$\binom{n}{k} = \frac{7\cdot 6\cdot 5}{3\cdot 2\cdot 1} = 35.$$

By definition, $\binom{n}{0} = 1$. Leibniz's Rule states that $(f \cdot g)^{(n)}(c)$ is the sum of the $n + 1$ expressions that are obtained by substituting $k = 0, 1, 2, \ldots, n$ in the formula $\binom{n}{k} \cdot f^{(n-k)}(c) \cdot g^{(k)}(c)$. Explicitly write Leibniz's Rule for $n = 1$, 2, and 3. Observe that the $n = 1$ case is the Product Rule and verify Leibniz's Rule for $n = 2$ and 3.

In Exercises 34–38, use Leibniz's Rule to compute the given derivative.

34. $D^2(f \cdot g)(x)$ where $f(x) = x^3$ and $g(x) = \cos(x)$

35. $\dfrac{d^2}{dx^2}(x^2\sqrt{x^2+1})$

36. $(f \cdot g)''(x)$ where $f(x) = x + \tan(x)$ and $g(x) = x^2 + \sin(x)$

37. $D^2(f \cdot g)(x)$ where $f(x) = x^3$ and $g(x) = \cos(x)$

38. $\dfrac{d^2}{dx^2}((x+3)^2(2x+1)^3)$

39. In an important investigation of the gravitational field in 1782, A. M. Legendre introduced the Legendre polynomials P_n. The following formula, discovered by Benjamin Olinde Rodrigues (1794–1850 or 1851), may be used to define the Legendre polynomials:

$$P_n(x) = \frac{1}{2^n n!}\frac{d^n}{dx^n}(x^2-1)^n.$$

Calculate $P_1(x)$, $P_2(x)$, $P_3(x)$, and $P_4(x)$.

40. Suppose that f is invertible with inverse g, that f is twice differentiable, that $f(a) = b$, and that $f'(a) \neq 0$. Show that

$$g''(b) = -\frac{f''(a)}{(f'(a))^3}.$$

41. Prove that any polynomial $p(x)$ of degree k satisfies

$$p(x) = p(0) + p'(0)\cdot x + \frac{p''(0)}{2!}\cdot x^2 + \cdots + \frac{p^{(k)}(0)}{k!}\cdot x^k$$

for all k.

42. Let $p(x)$ be a polynomial. Prove that p has a double root at $x = a$ if and only if $p(a) = 0$ and $p'(a) = 0$. Prove that p has a root of order k at a if and only if $p(a) = p'(a) = \cdots = p^{(k-1)}(a) = 0$.

43. Define the function

$$f(x) = ((x^{100} + x^{150})^{14} + (x^{93} + 16) + 8x^{90})^{42}.$$

Compute $f^{(83)}(0)$. (*Hint:* No calculations are necessary.)

44. Suppose that

$$P(t) = \frac{P_0 M}{P_0 + (M - P_0)e^{-kMt}}.$$

Verify that $P(t_*) = M/2$ if t_* is the solution of the equation $P''(t) = 0$.

45. When the air resistance to an object that is dropped from height h is proportional to the square of its velocity, the height of the object is given by

$$y(t) = h + t\sqrt{\frac{g}{\kappa}} - \frac{1}{\kappa}\ln\left(\frac{1 + e^{2t\sqrt{g\kappa}}}{2}\right).$$

Verify that y satisfies the differential equation

$$y''(t) = -g + \kappa(y'(t))^2.$$

Calculator/Computer Exercises

In Exercises 46–49, you are given a function $f(x)$, a value c, and an initial viewing rectangle. Graph the four functions

$$f(x)$$
$$g(x) = f(c) + f'(c)(x - c)$$
$$h(x) = g(x) + \frac{1}{2}D^2(f)(c)(x-c)^2$$
$$k(x) = h(x) + \frac{1}{6}D^3(f)(c)(x-c)^3$$

in viewing rectangle R. List the functions $\{g, h, k\}$ in the order of best approximation to $f(x)$ near c. (You may have to investigate this numerically if you are using a calculator because the graphical resolution may be inadequate. If so, tabulate evaluations of the four functions for several values of x near c to support your conclusion.)

46. $f(x) = \sqrt{x}, c = 0.5, R = [0, 1] \times [0, 1]$

47. $f(x) = \sin(x), c = \pi/4, R = [0, \pi] \times [0, 1]$

48. $f(x) = x^3, c = 1, R = [0, 2] \times [0, 8]$

49. $f(x) = 8x^4 - 8x + 1, c = 1/2, R = [0, 1] \times [-1, 1]$

In Exercises 50–53, use second central difference quotients to approximate $f''(c)$ to four decimal places.

50. $f(x) = \sqrt{2x}, c = 1$

51. $f(x) = \sin(x), c = \pi/6$

52. $f(x) = \sin(\pi x), c = 1/6$

53. $f(x) = \sqrt{x + 1/x}, c = 1$

54. Use a computer algebra system to evaluate

$$F_n(x) = \frac{d^n}{dx^n}(x^n \cos(x))$$

for $1 \leq n \leq 5$. If $F_n(x) = p(x)\cos(x) + q(x)\sin(x)$, what can you say about p and q? Consider which of the terms involved in these formulas are even or odd.

55. Let f and g each be k times differentiable functions and suppose that $f \circ g$ makes sense. If c is in the domain of g, can you find a formula for $(f \circ g)^{(k)}(c)$? Use a computer algebra system to calculate such a formula for several values of k and see whether you can determine a pattern. This matter has been studied in detail by Faà di Bruno, and the correct formula for $(f \circ g)^{(k)}(c)$ is named in his honor.

3.8 Implicit Differentiation and Related Rates

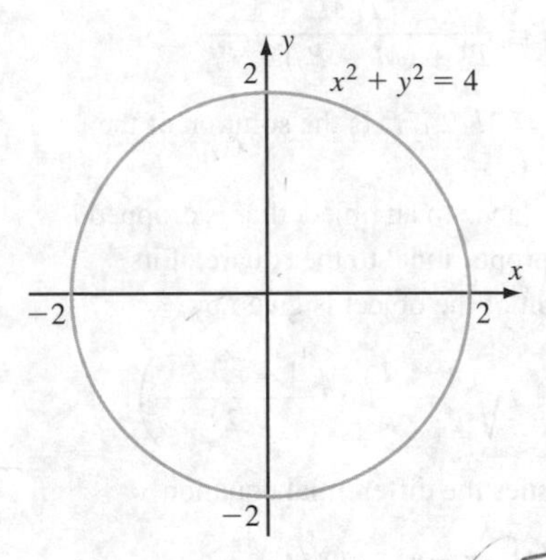

Figure 1

Examine the graph of the circle $x^2 + y^2 = 4$, shown in Figure 1. We know that at each point of the graph there is a tangent line to the curve. However, we *cannot* calculate the equation of the tangent line using the techniques of calculus that we have learned so far. This is because the circle is not the graph of a function. One way to remedy this problem is to solve for y in terms of x. By doing so, we obtain two curves:

$$y = +\sqrt{4 - x^2} \quad \text{and} \quad y = -\sqrt{4 - x^2}.$$

Thus, we have realized the circle as the union of the graphs of two functions (see Figure 2). Finding the slope of the tangent line to the circle at the point $(1, \sqrt{3})$ is the same as finding the slope of the tangent line to the graph of $f(x) = +\sqrt{4 - x^2}$ at that point. Similarly, finding the slope of the tangent line to the circle at the point $(\sqrt{2}, -\sqrt{2})$ is the same as finding the slope of the tangent line to $g(x) = -\sqrt{4 - x^2}$ at that point. So, for example,

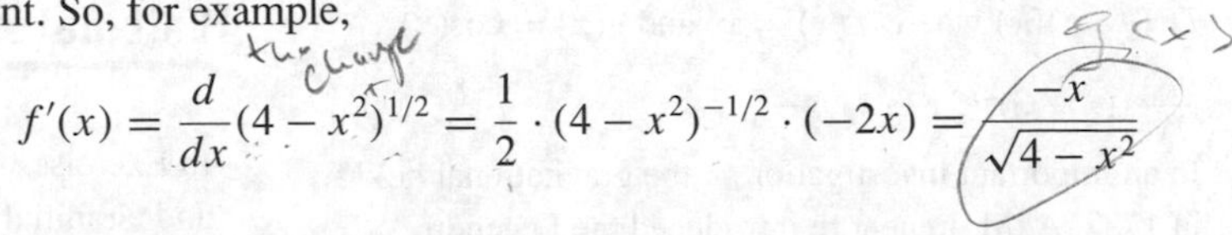

$$f'(x) = \frac{d}{dx}(4 - x^2)^{1/2} = \frac{1}{2} \cdot (4 - x^2)^{-1/2} \cdot (-2x) = \frac{-x}{\sqrt{4 - x^2}}$$

and

$$g'(x) = -f'(x) = \frac{x}{\sqrt{4 - x^2}}.$$

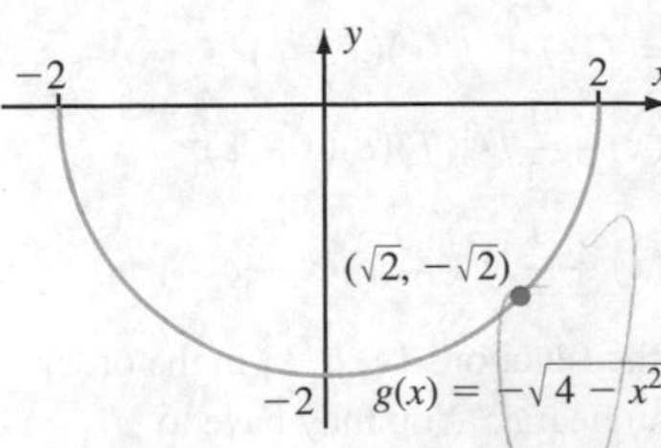

Figure 2

The slopes of the tangent line to the circle at $(1, \sqrt{3})$ and $(\sqrt{2}, -\sqrt{2})$ are, therefore, $f'(1) = -1/\sqrt{3}$ and $g'(\sqrt{2}) = 1$.

There is, however, a much simpler way to handle the situation in which the equation does not express y explicitly as a function of x. Called the *Method of Implicit Differentiation,* this technique eliminates the need to solve for y in terms of x.

The Method of Implicit Differentiation

If $F(x, y) = C$ is a given equation and if $P = (x_0, y_0)$ satisfies this equation, then we may find

$$\left.\frac{dy}{dx}\right|_P,$$

if it exists, by differentiating the equation *without* first solving for y in terms of x. When we do this, we treat y as though it were a differentiable function of x on an open interval centered at x_0. We say that we have *implicitly differentiated* F with respect to x.

When we discuss the calculus of parametric curves in Chapter 12, we will have theoretical ways for justifying the Method of Implicit Differentiation. In the present section, we only concentrate on learning how to use the method.

Example 1 Use the Method of Implicit Differentiation to find the slopes of the tangent lines to the curve $x^2 + y^2 = 4$ at the points $(1, \sqrt{3})$ and $(\sqrt{2}, -\sqrt{2})$.

Solution We apply $\frac{d}{dx}$ to both sides of the equation (it is best to use the Leibniz notation with this technique). Remembering to treat y as though it were a function of x, we have

$$\frac{d}{dx}(x^2 + y^2) = \frac{d}{dx}(1)$$

or

$$2x \cdot \frac{dx}{dx} + 2y \cdot \frac{dy}{dx} = 0.$$

Take particular notice that we have applied the Chain Rule to the first term, even though this was not strictly necessary. There was no need to write $\frac{dx}{dx}$, but in doing so we made our treatment of the term x^2 the same as our treatment of the term y^2. As we work our first examples, this notation will lead to less confusion. Since $\frac{dx}{dx} = 1$, our equation becomes

$$2x + 2y \cdot \frac{dy}{dx} = 0,$$

or

$$(2y) \cdot \frac{dy}{dx} = -2x.$$

We may solve for $\frac{dy}{dx}$ by dividing both sides by $2y$ if $y \neq 0$. The result is

$$\frac{dy}{dx} = -\frac{x}{y} \quad \text{if } y \neq 0.$$

inSIGHT

Compare Example 1 with the discussion at the beginning of this section. Notice that the Method of Implicit Differentiation does not require separate consideration of the two functions $f(x) = \sqrt{4 - x^2}$ and $g(x) = -\sqrt{4 - x^2}$, which have graphs that make up the curve $x^2 + y^2 = 4$.

It is now simply a matter to find the slope of the tangent line to the curve at point $(1, \sqrt{3})$. At this point, we have

$$\frac{dy}{dx} = -\frac{x}{y} = -\frac{1}{\sqrt{3}}.$$

This is the same answer we obtained in the introduction to this section, but the technique is easier.

Similarly, we find the slope of the tangent line to the curve at $(\sqrt{2}, -\sqrt{2})$ to be

$$\frac{dy}{dx} = -\frac{x}{y} = -\frac{\sqrt{2}}{-\sqrt{2}} = 1. \quad ■$$

Knowing how to use the Method of Implicit Differentiation requires knowing when the method cannot be applied. Suppose that $P = (x_0, y_0)$ is a point on the curve $F(x, y) = C$. There are two reasons that

$$\left.\frac{dy}{dx}\right|_P$$

may fail to exist. In using the Method of Implicit Differentiation, we assume that y is a function of x on an open interval centered at x_0. However, this may not actually be the case. In Example 1, the Method of Implicit Differentiation provided us with the slope of the tangent line at all points on the circle $x^2 + y^2 = 4$, *except* for those with ordinate 0. There are two such points: $(-2, 0)$ and $(2, 0)$. Neither $x_0 = -2$ nor $x_0 = 2$ is the center of an x-interval on which y is defined.

Even if y is a function of x on an interval centered at x_0, it may be that y is not differentiable at P. There may be no tangent line at P or the tangent line may be vertical. This is the other reason that the Method of Implicit Differentiation may not provide a value for $\frac{dy}{dx}$.

Implicit differentiation always expresses the derivative $\frac{dy}{dx}$ as a fraction involving x or y or both. The points on the curve where the numerator of this fraction is nonzero but where the denominator is zero correspond to places where the tangent line is vertical or where it fails to exist. (No simple rule is available for points at which both the numerator and denominator are zero.)

Example 2 At what points on the curve $(y - 1)^5 + y(x - 2) = 0$ does $\frac{dy}{dx}$ not exist?

Solution If we differentiate the equation $(y - 1)^5 + y(x - 2) = 0$ implicitly, we obtain

$$5(y - 1)^4 \cdot \frac{dy}{dx} + \frac{dy}{dx} \cdot (x - 2) + y(1 - 0) = 0,$$

or

$$(5(y - 1)^4 + (x - 2)) \cdot \frac{dy}{dx} = -y.$$

Thus,

$$\frac{dy}{dx} = -\frac{y}{5(y - 1)^4 + (x - 2)} \quad \text{if } 5(y - 1)^4 + (x - 2) \neq 0.$$

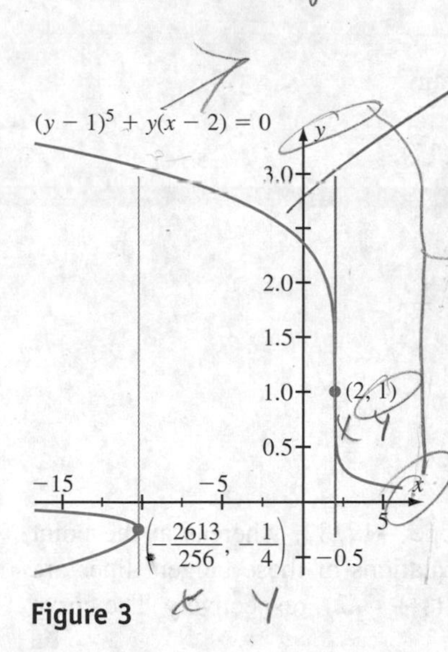

Figure 3

We must investigate the points (x, y) that satisfy the equations $5(y-1)^4 + (x-2) = 0$ (because we cannot solve for $\frac{dy}{dx}$ at such points) and $(y-1)^5 + y(x-2) = 0$ (because we consider only points on the given curve). The first of these equations gives us

$$x - 2 = -5(y-1)^4. \tag{3.20}$$

Substituting this formula for $x - 2$ into the second equation results in $(y-1)^5 - 5y(y-1)^4 = 0$. Factoring, we obtain $(y-1)^4(y-1-5y) = 0$. The solutions to this last equation are $y = 1$ and $y = -1/4$. Substituting these values back into equation (3.20) allows us to determine that $x = 2$ when $y = 1$ and $x = -2613/256$ when $y = -1/4$. We have graphed portions of the curve $(y-1)^5 + y(x-2) = 0$ in Figure 3. We see that, although y is a function of x near the point $(2, 1)$, the derivative $\frac{dy}{dx}$ fails to exist because the tangent line is vertical. On the other hand, we cannot even define y as a function of x at the point $(-2613/256, -1/4)$. Therefore, $\frac{dy}{dx}$ has no meaning at this point. ■

Example 3 Find the slope of the tangent line to the curve $8xy^{9/2} = 9 + x^{-7/5}$ at the point $(1, 1)$.

Solution Using implicit differentiation we have

$$\frac{d}{dx}8xy^{9/2} = \frac{d}{dx}(9 - x^{-7/5}),$$

or

$$8y^{9/2} + 8x \cdot \frac{9}{2} \cdot y^{7/2} \cdot \frac{dy}{dx} = 0 + \frac{7}{5} \cdot x^{-12/5}.$$

To make the algebra simpler, we substitute the point $(1, 1)$ *before* we solve for $\frac{dy}{dx}$ (but notice that we *have* finished differentiating). We obtain $8 + 36 \cdot \frac{dy}{dx} = 7/5$, or $\frac{dy}{dx} = -11/60$. ■

INSIGHT

You might have noticed that the equation considered in Example 3 is not of the form $F(x, y) = C$, as specified in the Method of Implicit Differentiation. However, it is easy to rearrange the equation to be in this form: $8xy^{9/2} + x^{-7/5} = 9$. As Example 3 shows, however, it is not necessary to bring all variables to one side of the equation to use the Method of Implicit Differentiation. We may differentiate the equation as it is presented.

Example 4 Find the equation of the tangent line to the curve

$$xy^4 + x = 17 \tag{3.21}$$

at each point where $x = 1$.

Solution If we substitute $x = 1$ into the equation, we obtain $y^4 + 1 = 17$. This simplifies to $y^4 = 16$, or $y = \pm 2$. Thus, there are two points, $(1, 2)$ and $(1, -2)$, with

$x = 1$. Now we differentiate equation (3.21) to obtain

$$\frac{d}{dx}(xy^4 + x) = \frac{d}{dx}(17).$$

Performing the differentiations, we have

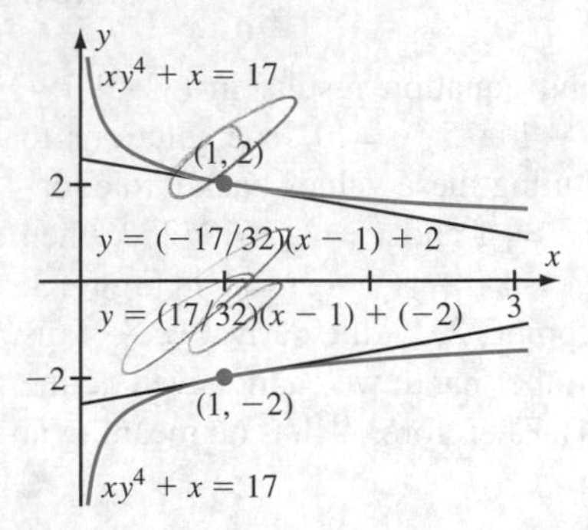

Figure 4

$$y^4 + x \cdot 4y^3 \cdot \frac{dy}{dx} + 1 = 0,$$

or

$$\frac{dy}{dx} = \frac{-1 - y^4}{4xy^3}.$$

Thus, at the point $(1, 2)$, the tangent line has slope $-17/32$; whereas at the point $(1, -2)$, the tangent line has slope $17/32$. The equations of these tangent lines are $y = (-17/32)(x - 1) + 2$ and $y = (17/32)(x - 1) + (-2)$, respectively. The curve and its tangents are illustrated in Figure 4. ■

inSIGHT

Notice that when we apply the Method of Implicit Differentiation to find the slope of the tangent line to $F(x, y) = C$ at (x_0, y_0), the expression for

$$\left.\frac{dy}{dx}\right|_{(x_0, y_0)}$$

will, in general, involve y_0 as well as x_0. As we see in Example 4, there may be more than one value of y for a given x_0 such that (x_0, y) is on the curve. The value of $\frac{dy}{dx}$ will depend on which value of y is used. Geometrically, this corresponds to the vertical line $x = x_0$ intersecting the curve $F(x, y) = 0$ in more than one point. The tangents to the curve at these points need not be parallel, which is why the value of y must usually be specified in the expression for $\frac{dy}{dx}$ that arises from the Method of Implicit Differentiation. This is unlike the situation for the explicitly defined functions that we considered prior to this section where the vertical line $x = x_0$ intersected the curve in only one point and $\frac{dy}{dx}$ was determined by x_0 alone.

Related Rates

The Method of Implicit Differentiation is often used to solve related rates problems wherein two variables are related by means of an equation. When one variable changes, so must the other to satisfy the equation. The question becomes: How can the rate of change of one variable be used to calculate the rate of change of the other variable? The next example illustrates the idea.

Example 5 The relationship between pressure P (measured in atmospheres) and volume V (measured in liters) of 1 mole of carbon dioxide at 0°C is approximated by van der Waal's equation

$$\left(P + \frac{a}{V^2}\right)(V - b) = 22.40$$

where $a = 3.592$ and $b = 0.04267$. At a certain time t_0, the volume V of a carbon dioxide sample is 4 L and the pressure is changing at the rate of 0.1 atm/s. What is the rate of change of V at t_0?

Solution

1. The quantities involved are the pressure P and the volume V, which are functions of time t.
2. The relationship between the variables is expressed by van der Waal's equation. We can use the form that has been given without explicitly solving for one variable in terms of the other.
3. We implicitly differentiate with respect to t. Using the Product Rule, we obtain

$$\left(\frac{dP}{dt} - 2\frac{a}{V^3}\frac{dV}{dt}\right)(V - b) + \left(P + \frac{a}{V^2}\right)\frac{dV}{dt} = 0.$$

4. We know that at t_0, $\frac{dP}{dt} = 0.1$ and $V = 4$. The constants a and b are given. To use the equation obtained in step 3, we also need to know the value of P at t_0. We use van der Waal's equation to determine the value of P:

$$\left(P + \frac{3.592}{4^2}\right)(4 - 0.04267) = 22.40,$$

or $P = 5.436$. We now substitute all known values into the equation of step 3:

$$\left(0.1 - 2\frac{3.592}{4^3}\frac{dV}{dt}\right)(4 - 0.04267) + \left(5.436 + \frac{3.592}{4^2}\right)\frac{dV}{dt} = 0.$$

We solve for $\frac{dV}{dt}$, obtaining $\frac{dV}{dt} = -7.586 \times 10^{-2}$ L/s. We see that the volume of the carbon dioxide sample is *decreasing* (indicated by the minus sign, but also consistent with common sense) at the rate of 7.586×10^{-2} L/s. ■

Basic Steps for Solving a Related Rates Problem

1. Identify the quantities (or functions) that are varying, and identify the variable (this is often time) with respect to which the change in these quantities is taking place.
2. Establish a relationship (an equation) between the quantities isolated in step 1.
3. Differentiate the equation from step 2 with respect to the variable identified in step 1. Be sure to apply the Chain Rule carefully.
4. Substitute the numerical data, and solve for the unknown rate of change.

Example 6 A sample of a new polymer is in the shape of a cube. It is subjected to heat so that its expansion properties may be studied. The surface area of the cube is increasing at the rate of 6 in.^{2}/min. How fast is the volume increasing at the moment when the surface area is 150 in.^{2}?

Solution The primary interest of this example is that surface area and volume are indirectly related. We use the Chain Rule, but we do not have to differentiate implicitly.

1. The functions are surface area A and volume V. It is useful to introduce the function s for side length because A and V are related by means of s. The variable is time t.
2. We have $V = s^3$ and $A = 6s^2$. (The latter equation arises because a cube has six faces, each with area s^2.)
3. We differentiate both equations from step 2. Thus,

$$\frac{dV}{dt} = 3s^2 \cdot \frac{ds}{dt} \quad \text{and} \quad \frac{dA}{dt} = 6 \cdot 2s \cdot \frac{ds}{dt}.$$

   We use the second of these equations to solve for $\frac{ds}{dt}$:

$$\frac{ds}{dt} = \frac{1}{12s} \cdot \left(\frac{dA}{dt}\right),$$

   which we substitute into the first:

$$\frac{dV}{dt} = 3s^2 \cdot \left(\frac{1}{12s} \cdot \left(\frac{dA}{dt}\right)\right) = \frac{s}{4} \cdot \frac{dA}{dt}.$$

4. We notice that, at the instant the problem is posed, $A = 150$, hence $s = 5$. Also, $\frac{dA}{dt} = 6$. Substituting this information into the last equation in step 3 gives

$$\frac{dV}{dt} = \frac{5}{4} \cdot 6 = \frac{15}{2} \text{ in.}^3/\text{min}.$$

   The volume of the cube of polymer is expanding at the rate of $15/2$ in.^{3}/min. ■

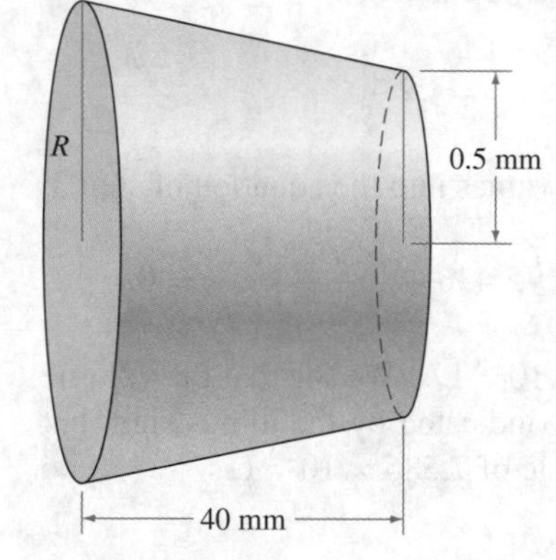

Figure 5

Example 7 A blood vessel is slightly tapered in shape, like a truncated cone (see Figure 5). The effect of a certain pain-relieving drug is to increase the larger radius R at the rate of 0.01 mm/min. During this expansion process, the length of the blood vessel remains fixed at 40 mm, and the smaller radius remains fixed at 0.5 mm. At the moment when the larger radius is 2 mm, how is the volume of the blood vessel changing?

Solution

1. The functions involved in this problem are the larger radius R and the volume of the blood vessel V. The variable is t.
2. To calculate the volume of the blood vessel for a given R, we use the cross-sectional diagram, shown in Figure 6. The edge of the blood vessel is given by a line of slope $(0.5 - R)/40$ and y-intercept R. The line has equation

$$y = \left(\frac{R - 0.5}{-40}\right) \cdot x + R.$$

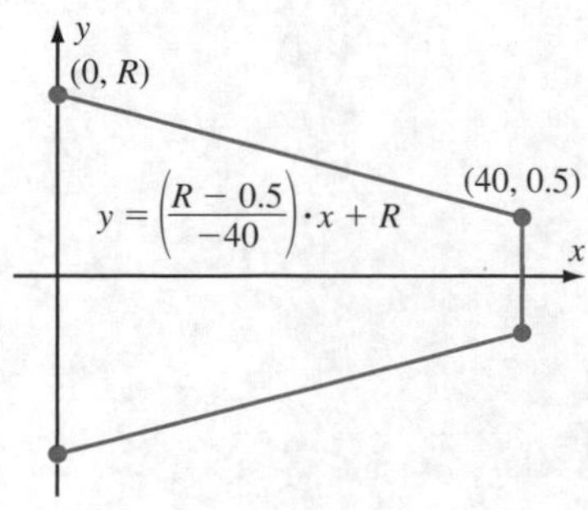

Figure 6

The line has x-intercept $40R/(R-0.5)$. Thus,

$$\begin{aligned}
V &= \begin{pmatrix} \text{volume of cone with} \\ \text{radius} = R, \\ \text{height} = \frac{40R}{R-0.5} \end{pmatrix} - \begin{pmatrix} \text{volume of cone with} \\ \text{radius} = 0.5, \\ \text{height} = \frac{40R}{R-0.5} - 40 \end{pmatrix} \\
&= \left(\frac{1}{3}\pi R^2\left(\frac{40R}{R-0.5}\right)\right) - \left(\frac{1}{3}\pi(0.5)^2\left(\frac{40R}{R-0.5} - 40\right)\right) \\
&= \frac{1}{3}\pi\left(R^2 \cdot \frac{40R}{R-0.5} - 0.25 \cdot \frac{20}{R-0.5}\right) \\
&= \frac{1}{3}\pi\,\frac{40R^3-5}{R-0.5} \\
&= \frac{40\pi}{3}\,\frac{(R-0.5)(R^2+0.5R+0.25)}{R-0.5} \\
&= \frac{40\pi}{3} \cdot (R^2+0.5R+0.25).
\end{aligned}$$

To arrive at the second-to-last line, we divided $R-0.5$ into $40R^3-5 = 40(R^3-0.125)$, obtaining the factor $R^2+0.5R+0.25$.

3. We may now differentiate:

$$\frac{dV}{dt} = \frac{40\pi}{3}(2R+0.5) \cdot \frac{dR}{dt}.$$

4. We substitute the values $R = 2$ and $\frac{dR}{dt} = 0.01$:

$$\begin{aligned}
\frac{dV}{dt} &= \frac{40\pi}{3}(2 \cdot 2 + 0.5) \cdot 0.01 \\
&= \frac{180\pi}{3} \cdot 0.01 \\
&= 0.6\pi.
\end{aligned}$$

Thus, the volume of the blood vessel is increasing at the rate of 0.6π mm/min. ■

Calculating Higher Derivatives

We can use the Method of Implicit Differentiation to calculate higher derivatives.

Example 8 Consider the equation $\sin(y) + x = 1/\sqrt{2}$. Calculate $\frac{d^2y}{dx^2}$ at the point $(0, \pi/4)$.

Solution Differentiating both sides with respect to x yields

$$\cos(y) \cdot \frac{dy}{dx} + 1 = 0. \qquad \textbf{(3.22)}$$

We do not yet substitute the coordinates of the point we wish to study. We first must differentiate again:

$$\left(-\sin(y)\cdot\frac{dy}{dx}\right)\cdot\left(\frac{dy}{dx}\right)+\cos(y)\cdot\left(\frac{d^2y}{dx^2}\right)+0=0. \tag{3.23}$$

Now that we have finished differentiating, it is correct to put in our numerical values. Setting $y=\pi/4$ in equation (3.22) yields

$$\frac{1}{\sqrt{2}}\cdot\left.\frac{dy}{dx}\right|_{(0,\pi/4)}+1=0,$$

or

$$\left.\frac{dy}{dx}\right|_{(0,\pi/4)}=-\sqrt{2}.$$

Now we substitute this information about $\frac{dy}{dx}$, together with $x=0$ and $y=\pi/4$, into equation (3.23) to obtain

$$\left(-\frac{1}{\sqrt{2}}\right)\cdot(-\sqrt{2})^2+\left(\frac{1}{\sqrt{2}}\right)\cdot\left(\left.\frac{d^2y}{dx^2}\right|_{(0,\pi/4)}\right)=0,$$

or

$$\left.\frac{d^2y}{dx^2}\right|_{(0,\pi/4)}=2.$$

■

inSIGHT

There is a subtlety in going from equation (3.22) to equation (3.23). When we differentiated the term $\cos(y)\cdot\frac{dy}{dx}$, we had to apply the Product Rule. The differentiation of the factor $\cos(y)$ led to another factor of $\frac{dy}{dx}$, while the differentiation of the factor $\frac{dy}{dx}$ led to a factor of $\frac{d^2y}{dx^2}$. In this step, we must be careful to keep track of both types of derivative that arise.

quickquiz

1. What is the purpose of implicit differentiation?
2. At what stage in the implicit differentiation process do you substitute the values of x and y?
3. Calculate $\frac{dy}{dx}$ for the equation $y^4+y=x^2$.
4. Suppose x and y are positive variables related by the equation $2x+y^2=10$. If x is a function of t and $\frac{dx}{dt}=4$ for all t, what is $\frac{dy}{dt}$ when $x=3$?

EXERCISES

Problems for Practice

In Exercises 1–6, use implicit differentiation to calculate $\frac{dy}{dx}$ at the point P.

1. $xy^2 + yx^2 = 6,\ P = (1, 2)$
2. $\sin(\pi xy) - xy^2 + 2y = 1,\ P = (1, 1)$
3. $x^{3/5} + 4y^{3/5} = 12,\ P = (32, 1)$
4. $xy^3 + 5yx^4 - 3xy^5 = -16,\ P = (2, 2)$
5. $x^4 - y^4 = -15,\ P = (1, 2)$
6. $-5x^3 = (x + 2y)/(x - y),\ P = (1, 2)$

In Exercises 7–10, use implicit differentiation to find the tangent line to the curve at the point P.

7. $xy/4 = \sqrt{x} + \sqrt{y},\ P = (4, 4)$
8. $x^3 - y^3 = 9,\ P = (2, -1)$
9. $\sin^2(x) + \cos^2(y) = 1,\ P = (\pi/4, \pi/4)$
10. $x^{1/2} - y^{1/2} = 1,\ P = (9, 4)$

In Exercises 11–14, use implicit differentiation to find the normal line to the curve at the point P.

11. $xy^4 - 2x^3y^2 + 4xy = -16,\ P = (2, 2)$
12. $x^4 + 4y^6 = 20,\ P = (2, 1)$
13. $x\sin(y/2) - y = 0,\ P = (\pi, \pi)$
14. $x^3 - 3xy^2 = 2x^2,\ P = (3, 1)$

In Exercises 15–20, find $\frac{dy}{dx}$ and $\frac{d^2y}{dx^2}$ at the point P by implicit differentiation.

15. $y^{1/3} - x^{1/3} = 1,\ P = (1, 8)$
16. $2xy - xy^3 = 12,\ P = (-3, 2)$
17. $x^4 + 2y^4 = 18,\ P = (-2, 1)$
18. $xy^2 = 9,\ P = (1, 3)$
19. $xy - y^2x = x - 1,\ P = (1, 1)$
20. $\cos(x^2 + y^2) = 0,\ P = (\sqrt{\pi}/2, \sqrt{\pi}/2)$

In Exercises 21–24, find a tangent line to the curve at the point P.

21. $e^{xy} = 2y^2 - 1,\ P = (0, -1)$
22. $2^{x-y} = xy^3,\ P = (2, 1)$
23. $\ln(xy - 1) + y^2 - 2x = 2,\ P = (1, 2)$
24. $xe^{x-1} - \ln(xy) = 1,\ P = (1, 1)$
25. The diagonal of a square is increasing at a rate of 2 in./min. At the moment when the diagonal measures 5 in., how fast is the area of the square increasing?
26. The side length of an equilateral triangle is decreasing at a rate of 3 cm/min. How fast is the area decreasing at the moment when the area is 10 cm^2?
27. Particle A moves down the x-axis in the positive direction at a rate of 5 units per second. Particle B walks up the y-axis in the positive direction at a rate of 8 units per second. At the moment when A is at $(4, 0)$ and B is at $(0, 9)$, how rapidly is the distance between A and B changing?
28. A police officer stands 50 ft from the edge of a straight highway while a car speeds down the highway. Using a radar gun, the officer ascertains that at a particular instant, the car is 500 ft from him and that the distance between himself and the car is changing at a rate of 120 ft/s. At that moment, how fast is the car traveling down the highway?
29. A 5 ft, 10 in. woman is walking away from a wall at the rate of 4 ft/s. A light is attached to the wall at a height of 10 ft. How fast is the length of the woman's shadow changing at the moment when she is 12 ft from the wall?
30. The radius of a certain cone is increasing at a rate of 6 cm/min, while the height is decreasing at a rate of 4 cm/min. At the instant when the radius is 9 cm and the height is 12 cm, how is the volume changing?
31. A lighthouse is 100 ft tall. It keeps its beam focused on a boat that is sailing away from the lighthouse at the rate of 300 ft/min. If θ denotes the acute angle between the beam of light and the surface of the water, how fast is θ changing at the moment the boat is 1000 ft from the lighthouse?
32. A jogger runs around a track that has the elliptical shape $x^2 + 2y^2 = 40000$. The unit of measurement is meters. As part of wind sprint training, he runs in such a way that $\frac{dy}{dt} = y$. What is the rate of change of his x-coordinate at the moment when $y = 100$?

Further Theory and Practice

33. Show that the tangent lines to the curve $x^2 - 4xy + y^2 = 9$ at the points where the curve crosses the x-axis are parallel.
34. Find all points on the curve $xy - 2x^2 + y^2 = 8$ where the tangent line has slope $-2, -1, 0, 1,$ or 2. Show that no tangent line to the curve is vertical.

35. The curve $x^2 - xy + y^2 = 4$ is an ellipse. Notice that $P = (a, b)$ is on the ellipse if and only if $P' = (-a, -b)$ is on the ellipse. Assuming that P and P' are both points on the ellipse, show that these tangent lines at P and P' are parallel.

36. Suppose (x_0, y_0) is a point on the ellipse $x^2/a^2 + y^2/b^2 = 1$. Show that the tangent line to the ellipse at (x_0, y_0) is given by

$$\frac{x_0 x}{a^2} + \frac{y_0 y}{b^2} = 1.$$

37. For each positive value of n, the equation

$$x^3 - 3nxy + y^3 = 0$$

defines a curve known as a *Folium of Descartes* (see Figure 4 in this chapter's Genesis & Development for a plot). Show that $P = (a, b)$ is on the Folium of Descartes if and only if $P' = (b, a)$ is. In this case, what is the product of the slopes of the tangent lines at P and P'?

38. Let S and T be open intervals in $\mathbb{R}$. Let $f : S \to T$ be a function that is differentiable and invertible. Show how to derive the formula for $\frac{d}{dt} f^{-1}$ (which we learned in Theorem 2 from Section 3.6) by applying implicit differentiation to the equation $t = f(s)$.

39. The graph of the curve $x^y = y^x$ $(x, y > 0)$ consists of the ray $x = y > 0$ and a curve C. Find $\frac{dy}{dx}$ at the point P on C that has abscissa 2. (*Hint:* Take the natural logarithm of each side first.)

40. Use the Method of Implicit Differentiation to find the slope of the curve $x^{2/3} + y^{2/3} = 25$ at the points $(27, 64)$ and $(27, -64)$. Verify your answers by explicitly solving for y and differentiating the explicit expression. Why must you use *two* explicit formulas for y?

41. Let a be a positive constant. Let T be *any* tangent line to

$$\sqrt{x} + \sqrt{y} = a.$$

Compute the sum s of the intercepts of T on the coordinate axes. (In what remarkable way does s depend on T?)

42. Suppose that $a \neq 0$. The locus of

$$x^{2/3} + y^{2/3} = a^{2/3}$$

is called an astroid $\mathcal{A}_a$. Let T be a tangent line to the astroid $\mathcal{A}_a$. Find a formula for the distance between the two coordinate axis intercepts of T. (How does this distance depend on T?)

43. Saha's equation

$$\frac{1-y}{y^2} = \frac{A\exp(b/x)}{x^{3/2}}$$

describes the degree of ionization within stellar interiors. In this equation, A and b are constants, y represents the fraction of ionized atoms in the star, and x represents stellar temperature in degrees Kelvin. Find $\frac{dy}{dx}$.

44. Let p and q be nonzero integers. Apply implicit differentiation to the equation $x^p - y^q = 0$ to obtain a proof of the formula

$$\frac{d}{dx}(x^{p/q}) = \frac{p}{q} \cdot x^{(p/q)-1}.$$

45. By the *demand curve* for a given commodity, we mean the set of all points (p, q) in the pq-plane where q is the number of units of the product that can be sold at price p. The elasticity of demand for the product at price p is defined to be $E(p) = -q'(p) \cdot p/q(p)$.

a. Suppose that a demand curve for a commodity is given by

$$p + q + 2p^2q + 3pq^3 = 1000$$

where p is measured in dollars and the quantity q of items sold is measured by the 1000. For example, the point $(p, q) = (6, 3.454)$ is on the curve, which means that 3454 items are sold at \$6. What is the slope of the demand curve at the point $(6, 3.454)$?

b. What is elasticity of demand for the product of part a at $p = \$6$?

46. The sum of the lengths of two pieces of wire remains constantly equal to 100 in., no matter what the temperature. One piece of wire is used to form a square, and the other to form a circle. Under the influence of heat, the area of the circle is increasing at the rate of 10 in.^{2}/min. At the moment when the circle has area 20 in.^{2}, how is the area of the square changing?

47. A rectangle is such that its length increases at the rate of 5 in./min and its width decreases at the rate of 7 in./min. At the moment when the length is 6 and the width is 8, is the area increasing or decreasing?

48. A 6 ft tall archer shoots an arrow into the air, at an angle of 60 deg with the horizontal and with an initial velocity of 100 ft/s. At what rate is the distance between the archer's feet and the arrow increasing at the moment that the arrow is at height 100 ft and going up? Is the answer different when the arrow is coming down?

49. In a hemispherical tank of radius 20 ft, the volume of water is $\pi h^2(60-h)/3$ ft^3 when the depth of water is h ft at the deepest point. If the water is draining at a rate of 5 ft^3/min, how fast is the area of the water on the surface decreasing when the water is 10 ft deep?

50. Two cells, one cubical in shape and the other spherical, live symbiotically in a solution. Their relationship is such that the sum of their volumes always remains the same. Due to a change of environment, the radius of the spherical cell begins to increase at a rate of 0.0001 mm/h. At the moment when the spherical cell has diameter 0.002 mm, how fast is the side length of the cubical cell increasing?

Computer/Calculator Exercises

In Exercises 51–54, find the slope of the curve at point P. In each case, it is necessary to determine the ordinate of P. State your answer with five significant digits.

51. $x^3 - 3xy + y^3 = 0$, $P = (10.000, *)$
52. $x^y - y^x = 0$, $x \neq y$, $P = (3.0000, *)$
53. $x^3 - x^2y^2 + y^3 = 0$, $P = (-1.4649, *)$
54. $\ln(xy) + y = 5$, $P = (2.124, *)$

In Exercises 55–58, find the tangent line of the curve at point P. In each case, it is necessary to determine the ordinate of P. State your answer with five significant digits.

55. $x^3 - 2xy + y^3 = 0$, $P = (10.000, *)$
56. $x^y - y^x = 0$, $x \neq y$, $P = (3.1416, *)$
57. $x^3 - x^2y^2 + y^3 = 0$, $P = (-2.0125, *)$
58. $e^{xy} + y = 10$, $P = (1.5727, *)$

In Exercises 59–62, use an implicit plotting utility to plot the curve in a viewing rectangle that contains point P. Add a plot of the tangent line to the curve at P. (If you are using Maple, load the plotting package by typing `with(plots);`. This provides easy access to the "`implicitplot`" command.)

59. $x^3 - 2xy + y^3 = 0$, $P = (0.5, 0.9304\ldots)$
60. $x^y - y^x = 0$, $x \neq y$, $P = (3.7500, 2.0860)$
61. $x^3 - x^2y^2 + y^3 = 0$, $P = (2.1125, 1.9289)$
62. $xe^{xy} + y = 2$, $P = (1.0000, 0.44285)$
63. Use a graphing utility's zoom feature to locate the points on the curve $x^4 + x + 5xy^3 = 1$ where the tangent line is vertical. Now use the Method of Implicit Differentiation to find those points.
64. Use a graphing utility's zoom feature to locate the points on the curve $x^3 - 6xy^2 = 4$ where the tangent line is vertical. Now use the Method of Implicit Differentiation to find those points.
65. Refer to Exercise 45. Let $q(p)$ be the demand function—that is, the number of units of the commodity that can be sold at price p. Find the point on the demand curve that corresponds to $p = \$4$. Differentiate implicitly to find $q'(4)$. Find $q(4.2)$ and $q(3.8)$. Use these values to approximate $q'(4)$.
66. Saha's equation describes the degree of ionization within stellar interiors. For our own sun,

$$\frac{1-y}{y^2} = \frac{0.787 \times 10^9 \exp(158000/x)}{x^{3/2}}$$

where y represents the fraction of ionized atoms in the sun and x represents solar temperature in degrees Kelvin. Find $\frac{dy}{dx}$ when $x = 6 \times 10^6$.

3.9 Differentials and Approximation of Functions

Loosely interpreted, the equation

$$f'(c) = \lim_{\Delta x \to 0} \frac{f(c+\Delta x) - f(c)}{\Delta x}$$

says that

$$f'(c) \approx \frac{f(c+\Delta x) - f(c)}{\Delta x}$$

when Δx is small and nonzero. A choice of a fairly small value of Δx often gives a good approximation. With a little algebraic manipulation, this approximation of $f'(c)$ can be turned into an approximation of $f(c + \Delta x)$:

$$f(c + \Delta x) \approx f(c) + f'(c) \cdot \Delta x. \tag{3.24}$$

We may interpret equation (3.24) as follows: If we know the values of $f(c)$ and $f'(c)$, then we can estimate the value of $f(x)$ at a nearby point $x = c + \Delta x$. Sometimes we abbreviate $f(c + \Delta x) - f(c)$ as $\Delta f(c)$. With this notation, approximation (3.24) becomes

$$\Delta f(c) \approx f'(c)\Delta x, \tag{3.25}$$

which is illustrated in Figure 1. We call this approximation scheme the *Method of Increments*.

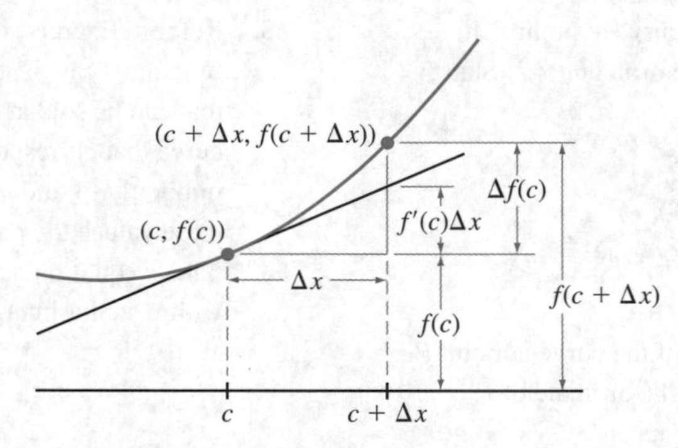

Figure 1
$f(c + \Delta x) - f(c) = \Delta f(c) \approx f'(c)\Delta x$

Example 1 Use approximation (3.24) to estimate the value of $\sqrt{4.1}$.

Solution If $f(x) = x^{1/2}$, then $f'(x) = (1/2)x^{-1/2}$. Choose $c = 4$ and $\Delta x = 4.1 - c = 0.1$. The selection of c here is no accident. It is easy for us to calculate f and f' at this particular c. Moreover, $c = 4$ is near 4.1, and we shall see that this increases our accuracy. According to approximation (3.24),

$$f(c + \Delta x) \approx f(c) + f'(c)\Delta x = c^{1/2} + \frac{1}{2}c^{-1/2}\Delta x.$$

As a result, $(4.1)^{1/2} \approx 4^{1/2} + (1/2) \cdot 4^{-1/2} \cdot 0.1 = 2.025$. A calculator shows that our simple calculation is accurate to within 0.001. ■

inSIGHT

In Example 1, if we had estimated $10^{1/2}$ using the Method of Increments with $c = 4$ and $\Delta x = 6$, then we would have obtained $10^{1/2} \approx 4^{1/2} + (1/2)4^{-1/2}(6) = 3.5$. The true value of $10^{1/2}$ is 3.1622777, to seven decimal places. Thus, our approximation is off by more than 0.33. A much better approximation is obtained by taking $c = 9$ and $\Delta x = 1$. The resulting approximation is

$$10^{1/2} \approx 9^{1/2} + (1/2)9^{-1/2}(1) = 3.166\ldots,$$

which is correct to two decimal places. The accuracy of the Method of Increments strongly depends on the size of the increment Δx: *In general, the smaller Δx, the more effective the method is.*

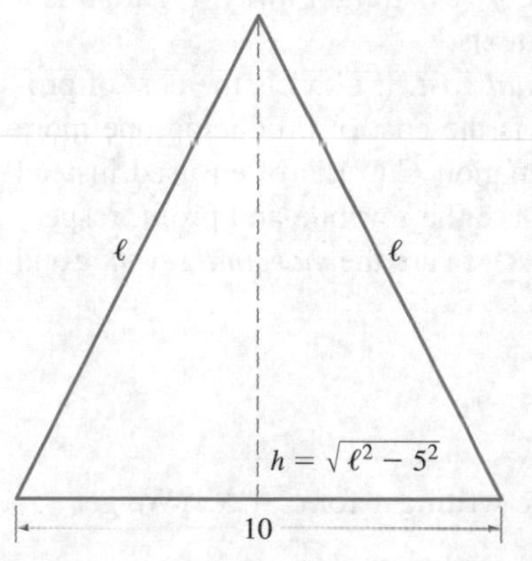

Figure 2

Example 2 A machine shop builds triangular jigs of isosceles shape. The bases are precut and have length exactly 10 in. However, during a rush period, the measurement of the other sides gets sloppy, and they are off by as much as 0.1 in. By about how much will the area of the resulting jig deviate from the desired area?

Solution The ideal length of the two equal sides is denoted by ℓ in Figure 2. The height of the triangle is $h = \sqrt{\ell^2 - 5^2}$. Therefore, the area is

$$A(\ell) = \frac{1}{2} \cdot 10 \cdot h = 5 \cdot \sqrt{\ell^2 - 25}.$$

Notice that $A'(\ell) = 5\ell(\ell^2 - 25)^{-1/2}$. If the length measurement is off by $\Delta\ell$, then the area is

$$A(\ell + \Delta\ell) \approx A(\ell) + A'(\ell) \cdot \Delta\ell = A(\ell) + 5\ell \cdot (\ell^2 - 25)^{-1/2} \cdot \Delta\ell.$$

Thus, the increment in area is

$$\Delta A \equiv A(\ell + \Delta\ell) - A(\ell) \approx 5\ell \cdot (\ell^2 - 25)^{-1/2} \cdot \Delta\ell.$$

Since $|\Delta\ell| \leq 0.1$, the approximate deviation ΔA of the area can be as much as

$$5\ell \cdot (\ell^2 - 25)^{-1/2} \cdot 0.1.$$

For example, if the desired length ℓ is 20 in. and the error measurement is 0.1 in., then the error in area is approximately $5 \cdot 20 \cdot (400 - 25)^{-1/2} \cdot 0.1 \approx 0.5164$ in.^{2}

In this example, the important feature of the Method of Increments is that it gives us a formula that will work for *all* triangles under consideration, and it involves a minimum of calculation. ■

Increments in Economics

If we take $\Delta x = 1$ from equation 3.24, we obtain $f(c+1) \approx f(c) + f'(c)$, or

$$f(c+1) - f(c) \approx f'(c). \tag{3.26}$$

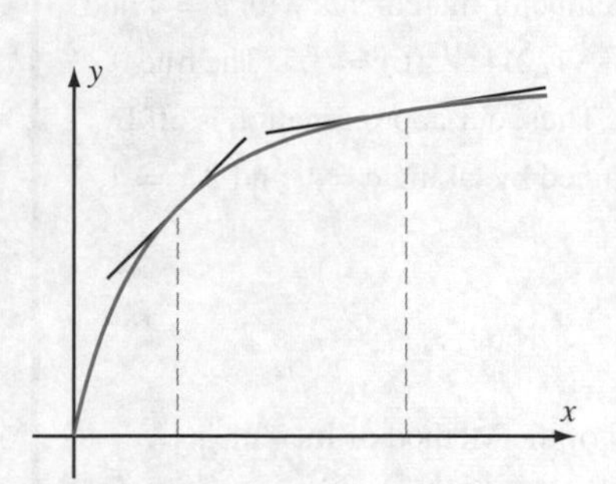

Figure 3
Level of output as a function of the quantity of labor input. The slope of the tangent line equals the marginal product of labor.

This approximation is often used in economics. For example, if $f(x)$ represents the quantity of goods produced as a function of the amount of labor x, then f is called the *product of labor*. The increment in production that is achieved by adding one more unit of labor is known as the *marginal product of labor* (*MPL*). Thus,

$$MPL(c) = f(c+1) - f(c).$$

If we compare this with approximation (3.26), we obtain

$$MPL(c) \approx f'(c).$$

Indeed, most economics texts define $MPL(c)$ to be $f'(c)$ straight away. Figure 3 is a graphic that is typically found in macroeconomics texts.

Another example of this construction is *marginal cost*. If $C(x)$ is the cost of producing x units of an item, then the marginal cost is the cost of producing one more item: $C(x+1) - C(x)$. In practice, the approximation $C'(x)$ is often used instead of $C(x+1) - C(x)$. Similarly, if $R(x)$ and $P(x)$ are the revenue and profit, respectively, produced by selling x items, then $R'(x)$ and $P'(x)$ are the *marginal revenue* and *marginal profit,* respectively.

Linearization

Let us look at approximation (3.24) again, this time writing x for $c + \Delta x$. We get

$$f(x) = f(c + \Delta x) \approx f(c) + f'(c) \cdot \Delta x = f(c) + f'(c) \cdot (x - c),$$

or

$$f(x) \approx f(c) + f'(c) \cdot (x - c)$$

for x close to c. The right side of this approximation is a linear function of x, which we denote by L:

$$L(x) = f(c) + f'(c) \cdot (x - c). \tag{3.27}$$

The graph of L is the tangent line to the graph of f at $(c, f(c))$.

Definition

The function L defined by equation (3.27) is called the *linearization* of f at c. The approximation $L(x) \approx f(x)$ for x close to c is called the *tangent line approximation* to f at c. Sometimes the term *best linear approximation* is used instead.

In the left viewing window in Figure 4, we see the graph of a function f together with its linearization at c. In the middle view, we have zoomed in around the point $(c, f(c))$. Notice how good an approximation L is of f in this viewing window. In the right view, we have zoomed in still further. Distinguish between L and f. There is, to be sure, a difference between these two functions, but it is too minute to show up on screen.

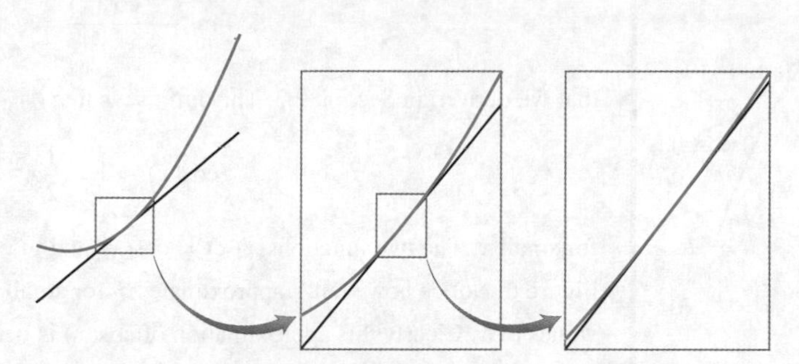

Figure 4

Important Linearizations

It is useful to know certain linearizations because of the frequency with which they recur throughout calculus.

Example 3 Show that

$$(1+u)^p \approx 1 + pu$$

for values of u near 0.

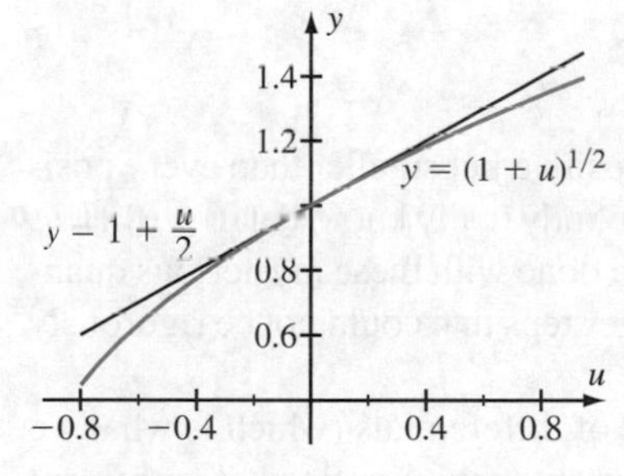

Figure 5

Solution Let $f(x) = (1+x)^p$ and $c = 0$. Since

$$\frac{d}{dx}(1+x)^p = p(1+x)^{p-1}\frac{d}{dx}(1+x) = p(1+x)^{p-1},$$

we have $f'(c) = p$. Since $f(c+\Delta x) = (1+(c+\Delta x))^p = (1+\Delta x)^p$, equation (3.24) takes the form $(1+\Delta x)^p \approx 1 + p \cdot \Delta x$. Replacing Δx with u gives us the desired result. Figure 5 illustrates the approximation when $p = 1/2$. ■

The differential approximations for $\sin(x)$, $\cos(x)$, and $\tan(x)$ at $x = 0$ are

$$\sin(x) \approx x, \quad \cos(x) \approx 1, \quad \text{and} \quad \tan(x) \approx x,$$

because

$$\begin{aligned}
\sin(\Delta x) &\approx \sin(0) + \sin'(0) \cdot \Delta x = \Delta x \\
\cos(\Delta x) &= \cos(0) + \cos'(0) \cdot \Delta x = 1 \\
\tan(\Delta x) &= \tan(0) + \tan'(0) \cdot \Delta x = \sec^2(0) \cdot \Delta x = \Delta x.
\end{aligned}$$

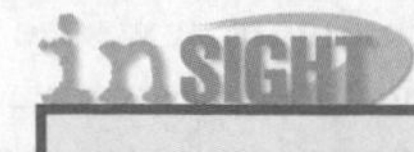

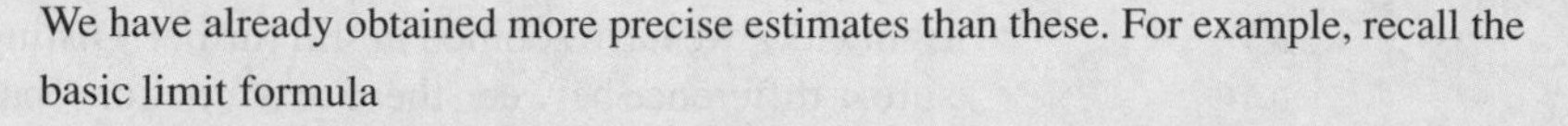

We have already obtained more precise estimates than these. For example, recall the basic limit formula

$$\lim_{x\to 0}\frac{1-\cos(x)}{x^2}=\frac{1}{2}$$

that we derived in Section 2.3. This limit says that

$$\cos(x)\approx 1-\frac{1}{2}x^2$$

for small x. The two functions $f(x)=\cos(x)$ and $g(x)=1-\frac{1}{2}x^2$ are shown in Figure 6. Notice how well g approximates f for small values of x—but *only* for small values of x. Clearly this approximation of $\cos(x)$ is better than the linearization $L(x)=1$. In Chapter 10, we will learn a general approach, called the Taylor expansion, for obtaining estimates that are more precise than the tangent line approximation.

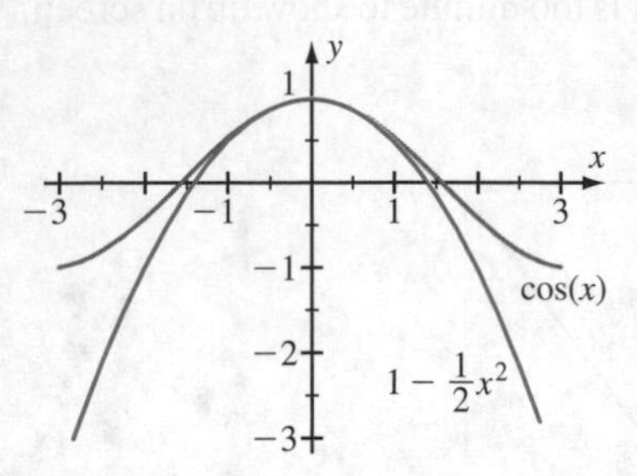

Figure 6

Differentials

In the early days of calculus, the subject was not rigorous. There was no concept of "limit;" instead the derivative was obtained as a quotient of infinitesimally small quantities:

$$\frac{f(x+dx)-f(x)}{dx}.$$

In this expression, dx represents a quantity that is positive but smaller than every positive real number. What does this mean? In the past, nobody really knew, but nevertheless calculations of derivatives of specific functions were done with these mysterious quantities. Many times these calculations involved strange steps that could not be rigorously justified.

It took nearly 200 years for the puzzling theory of differentials (which is what we call expressions such as dx) to be replaced by the more precise notions of increment Δx and of limit. More recently, in 1960, it was determined that the notion of differential could be made precise, too. However, the modern theory of infinitesimals is a subject for an advanced course in mathematical logic, and we cannot discuss it here. What you do need to know is that the *notation* of differentials is still used today, purely as an intuitive device. It is reassuring to know that differentials can be treated rigorously, but for us they will simply be an economical way for expressing the method of increments.

Equation (3.25) tells us that a small change in x from c, by an amount Δx, causes f to change by an amount that can be estimated by $f'(c)\Delta x$. As Δx becomes smaller, the estimate becomes more and more accurate. Thus, when Δx becomes "infinitesimal," the estimate (3.25) becomes an equality. We represent the infinitesimal increment in x as dx and the resulting infinitesimal change in f as df. Thus, approximation (3.25) becomes

$$df(x)=f'(x)\,dx. \qquad \textbf{(3.28)}$$

It is tempting to divide this equation by dx to yield the identity

$$\frac{df}{dx} = f'(x).$$

This explains the origin of the Leibniz notation for the derivative. For the purposes of calculus, however, we think of equation (3.28) as nothing more than another way to write approximation (3.25). Indeed, approximation (3.25) is sometimes referred to as a *differential approximation.*

The following list shows the rules of differentiation in differential form (c represents a constant).

$$\begin{aligned} d(c) &= 0 \\ d(cu) &= c\,d(u) \\ d(u+v) &= du + dv \\ d(uv) &= u\,dv + v\,du \\ d(u/v) &= (v\,du - u\,dv)/v^2 \\ d(1/v) &= -dv/v^2 \\ d(u^p) &= pu^{p-1}\,du \\ d(\ln(u)) &= (1/u)\,du \end{aligned} \qquad \begin{aligned} d(\sin(u)) &= \cos(u)\,du \\ d(\cos(u)) &= -\sin(u)\,du \\ d(\tan(u)) &= \sec^2(u)\,du \\ d(\sec(u)) &= \sec(u)\tan(u)\,du \\ d(\cot(u)) &= -\csc^2(u)\,du \\ d(\csc(u)) &= -\csc(u)\cot(u)\,du \\ d(\exp(u)) &= \exp(u)\,du \\ d(a^u) &= a^u\ln(a)\,du \end{aligned}$$

A LOOK BACK

It may help to remember that there is only one key idea behind all the terminology that has been introduced. *Tangent line approximation, best linear approximation,* and *differential approximation* all refer to the same thing: If f is differentiable at c, if (x, y) is on the tangent line to the graph of f at $(c, f(c))$, and if x is close to c, then $f(x)$ is close to y.

quickquiz

1. What is the basic formula for approximating by increments?
2. What is the differential approximation?
3. Use differentials to approximate $7.9^{1/3}$.

EXERCISES

Problems for Practice

In Exercises 1–14, use the Method of Increments to estimate the value of the function f at the point x using the known value at the initial point c. Use a calculator to check the accuracy of your calculation.

1. $f(x) = x^{1/2}$, $c = 4$, $x = 3.9$
2. $f(x) = \sin(x)$, $c = 0$, $x = 0.02$
3. $f(x) = \sin(x) - \cos(x)$, $c = \pi/4$, $x = \pi/3$
4. $f(x) = x^{-1/3}$, $c = 8$, $x = 8.07$
5. $f(x) = (x^2 + 1)^{1/3}$, $c = 0$, $x = 1$
6. $f(x) = (1 + x)^{-1/4}$, $c = 15$, $x = 16$
7. $f(x) = \tan(x)$, $c = \pi/4$, $x = 0.8$
8. $f(x) = \sqrt{x}$, $c = 9$, $x = 8.95$
9. $f(x) = \cot(x)$, $c = \pi/3$, $x = \pi/4$
10. $f(x) = x^{2/3}$, $c = 8$, $x = 8.15$
11. $f(x) = x^{-3/2}$, $c = 4$, $x = 4.21$

12. $f(x) = \ln(x),\ c = e^3,\ x = 20$
13. $f(x) = \sin(\sqrt{\pi x}), c = \pi/16, x = 0.2$
14. $f(x) = \sqrt{\sin(x)},\ c = \pi/4,\ x = 0.75$

In Exercises 15–20, choose an appropriate function f and point c, and use the differential approximation of f to estimate the given number. Compute the error.

15. $\sqrt{24}$
16. $\sqrt[3]{-7.5}$
17. $\sqrt{1+\sqrt{9.1}}$
18. $\sin(12\pi/25)$
19. $\cos(59°)$
20. $\tan(0.7)$

Further Theory and Practice

21. If $f(x) = (g \circ h)(x)$, derive an expression for df in terms of g and h.

22. Suppose that f, g, and h are differentiable functions and that $f = g \cdot h$. Suppose that at $x = 4$ the following information is known:

$$g(x) = 6 \quad \text{and} \quad dg = 7\,dx$$
$$h(x) = 5 \quad \text{and} \quad dh = -2\,dx.$$

What is df at $x = 4$?

23. Let f be a differentiable even function. What is the linearization $L(x)$ of f at 0?

24. In general, the function $p_f(x) = f(0) + f'(0)x + (1/2)f''(0)x^2$ is a more accurate approximation of a twice differentiable function f near 0 than is the linearization $L_f(x)$. Let f be a twice differentiable odd function. Show that $L_f = p_f$.

25. Explain how the linearizations of the differentiable functions f and g at c may be used to discover the product rule for $(f \cdot g)'(c)$.

26. Explain how the linearizations of the differentiable functions f and g at c may be used to discover the quotient rule for $(f/g)'(c)$ at a point c for which $g(c) \neq 0$. Suppose $(u, v) \mapsto F(u, v)$ is a given function of two variables. Suppose y is an unknown function of x that satisfies both the differential equation

$$y'(x) = F(x, y(x))$$

and the initial condition $y(x_0) = y_0$. (Together, these two equations constitute an *initial value problem.* We will study such problems in greater detail in Chapter 6.)

The Method of Increments can be used to approximate $y(x_1)$ where $x_1 = x_0 + \Delta x$:

$$y(x_1) \approx y(x_0) + y'(x_0)\Delta x = \underbrace{y_0 + F(x_0, y_0)\Delta x}_{y_1}.$$

We call this *Euler's Method* of approximating the unknown function y. In Exercises 27–30, an initial value problem is given. Calculate the Euler's Method approximation y_1 of $y(x_1)$.

27. $\dfrac{dy}{dx} = x + y,\ y(1) = 2,\ x_1 = 1.2$

28. $\dfrac{dy}{dx} = x - y,\ y(-2) = -1,\ x_1 = -2.15$

29. $\dfrac{dy}{dx} = x^2 - 2y,\ y(0) = 3,\ x_1 = 1/4$

30. $\dfrac{dy}{dx} = 1 + y/x,\ y(2) = 1/2,\ x_1 = 3/2$

By the *demand curve* for a given commodity, we mean the set of all points (p, q) in the pq-plane where q is the number of units of the product that can be sold at price p. In Exercises 31–34, use the differential approximation to estimate the demand $q(p)$ for a commodity at a given price p.

31. Suppose that a demand curve for a commodity is given by

$$p + q + 2p^2q + 3pq^3 = 1000$$

where p is measured in dollars and the quantity q of items sold is measured by the 1000. For example, the point $(p, q) = (6.75, 3.248)$ is on the curve, which means 3248 items are sold at \$6.75. What is the slope of the demand curve at the point $(6.75, 3.248)$? Approximately how many units will be sold if the price is increased to \$6.80? Decreased to \$6.60?

32. Suppose a demand curve for a commodity is given by

$$2 \cdot p^2 \cdot q + p \cdot \frac{\sqrt{q}}{100} = 500005$$

where p is measured in dollars. Approximately how many units of the commodity can be sold at \$5.10?

33. The demand curve for a commodity is given by

$$2 \cdot p \cdot q + p \cdot \sqrt{q} = 802000$$

where p is measured in dollars. Approximately how many units of the commodity can be sold at \$9.75?

34. The demand curve for a commodity is given by

$$\frac{p^2 \cdot q}{10} + 5 \cdot p \cdot \sqrt{q} = 39000$$

where p is measured in dollars. Approximately how many units of the commodity can be sold at \$1.80?

Calculator/Computer Exercises

In Exercises 35–38, a function f, a point c, an increment Δx, and a positive integer N are given. Use the Method of Increments to estimate $f(c + \Delta x)$. Let $h = \Delta x/N$, and use the Method of Increments to obtain an estimate y_1 of $f(c+h)$. With $c+h$ as the base point and y_1 as the value of $f(c+h)$, use the Method of Increments to obtain an estimate y_2 of $f(c+2h)$. Continue this process until you obtain an estimate y_N of $f(c+N\cdot h) = f(c+\Delta x)$. We say that we have taken N *steps* to obtain the approximation. The number h is said to be the *step size*. Using a calculator or computer, evaluate $f(c+\Delta x)$ directly and compare the accuracy of the one-step and N-step approximations.

35. $f(x) = x^{1/3}, c = 27, \Delta x = 0.9, N = 3$
36. $f(x) = \sqrt{x}, c = 4, \Delta x = 0.5, N = 5$
37. $f(x) = \ln(x), c = e, \Delta x = 3 - e, N = 2$
38. $f(x) = 1/\sqrt[3]{x}, c = -8, \Delta x = 1, N = 4$

Refer to the instructions for Exercises 27–30 for terminology. In Exercises 39–42, an initial value problem is given, along with its exact solution. Verify that the given solution is correct by substituting it into the given differential equation and into the initial value condition. Calculate the Euler's Method approximation $y_1 = y_0 + F(x_0, y_0)\Delta x$ of $y(x_1)$. Let $m_1 = (F(x_0, y_0) + F(x_1, y_1))/2$ and $z_1 = y_0 + m_1\Delta x$. This is the *Improved Euler's Method* approximation of $y(x_1)$. Calculate z_1. Determine which of the two approximations to $y(x_1)$, y_1, or z_1 is more accurate.

39. $\dfrac{dy}{dx} = x + y, y(1) = 2, x_1 = 1.2$; Exact solution: $y(x) = 4\exp(x-1) - x - 1$
40. $\dfrac{dy}{dx} = x - y, y(-2) = -1, x_1 = -2.15$; Exact solution: $y(x) = 2\exp(-2-x) + x - 1$
41. $\dfrac{dy}{dx} = x^2 - 2y, y(0) = 3, x_1 = 1/4$; Exact solution: $y(x) = x^2/2 - x/2 + 1/4 + 11/4\exp(-2x)$
42. $\dfrac{dy}{dx} = 1 + y/x, y(2) = 1/2, x_1 = 3/2$; Exact solution: $y(x) = x\ln(x) + x(1/4 - \ln(2))$

In Exercises 43–45, a demand curve is given. For the given price p_0, solve the demand equation for q. Then use the differential approximation to estimate the demand at price p_1. Find the exact demand at price p_1. What is the relative error that the differential approximation causes?

43. $p^2\cdot q + 2p\cdot q^{1/4} = 250100, p_0 = 5.10, p_1 = 5$
44. $2\cdot p\cdot q + p\cdot q^{1/3} = 160200, p_0 = 9.75, p_1 = 10$
45. $p^2\cdot q/10 + 5\cdot p\cdot q^{1/5} = 1280200, p_0 = 1.80, p_1 = 2$

Summary of Key Topics

Instantaneous Velocity (Section 3.1)

If the position of a moving body at time t is given by a function $p(t)$, then the quantity

$$\lim_{\Delta t\to 0}\frac{p(t+\Delta t) - p(t)}{\Delta t},$$

if it exists, is called the instantaneous velocity of the moving body.

Tangent and Normal Lines (Section 3.1)

If f is a function with a domain that contains point (a, b) and if $c \in (a, b)$, then the slope of the tangent line to the graph of f at c is equal to $f'(c)$, provided that the derivative exists. Thus, the tangent line has equation

$$y = f'(c)\cdot(x - c) + f(c).$$

The normal line to the curve at the point c is the line through $(c, f(c))$, which is perpendicular to the tangent line at that point.

Yet another interpretation of the derivative involves rates of change. If f is a function of a parameter x, then $f'(c)$ represents the instantaneous rate of change of f with respect to x at the point c, provided the derivative exists.

The Derivative (Section 3.2)

In general, if f is a function with a domain that contains (a, b), if $c \in (a, b)$, and if

$$\lim_{\Delta x \to 0} \frac{f(c + \Delta x) - f(c)}{\Delta x}$$

exists, then this quantity is called the derivative of f at the point c. It is denoted by any of the notations

$$f'(c) \quad \text{or} \quad \left.\frac{df}{dx}\right|_c \quad \text{or} \quad \dot{f}(c) \quad \text{or} \quad D(f)(c).$$

The process of calculating the derivative is called differentiation. A function that possesses the derivative is called differentiable.

Derivatives of Special Functions (Sections 3.1, 3.4, 3.6)

By direct calculation, we can determine the derivatives of a number of particular functions. The derivative for any real number p is

$$\frac{d}{dx} x^p = p \cdot x^{p-1}.$$

Also,

$$\frac{d}{dx} \sin(x) = \cos(x)$$

$$\frac{d}{dx} \cos(x) = -\sin(x)$$

$$\frac{d}{dx} \tan(x) = \sec^2(x)$$

$$\frac{d}{dx} \cot(x) = -\csc^2(x)$$

$$\frac{d}{dx} \sec(x) = \sec(x) \cdot \tan(x)$$

$$\frac{d}{dx} \csc(x) = -\csc(x) \cdot \cot(x)$$

$$\frac{d}{dx} e^x = e^x$$

$$\frac{d}{dx} a^x = \ln(a) a^x$$

$$\frac{d}{dx} \ln(x) = \frac{1}{x}.$$

Rules for Taking Derivatives (Section 3.3)

We have

$$\begin{aligned} (f+g)' &= f' + g' \\ (f-g)' &= f' - g' \\ (\alpha \cdot f)' &= \alpha \cdot (f') \qquad \alpha \text{ is a constant} \\ (f \cdot g)' &= f' \cdot g + f \cdot g' \\ \left(\frac{f}{g}\right)' &= \frac{g \cdot f' - f \cdot g'}{g^2} \end{aligned}$$

The Chain Rule (Section 3.5)

If $g(u)$ and $f(x)$ are differentiable expressions, if the range of f is in the domain of g, and if c is in the domain of f, then $g \circ f$ is differentiable at c and

$$\frac{d}{dx}(g \circ f)\bigg|_{x=c} = \frac{dg}{du}\bigg|_{u=f(c)} \cdot \frac{df}{dx}\bigg|_{x=c}.$$

In prime notation, we can write the chain rule as

$$(g \circ f)'(x) = g'(f(x)) \cdot f'(x).$$

The Inverse of a Function and Its Derivative (Section 3.6)

A function $f : S \to T$ is called invertible if there is a function $g : T \to S$ such that $(g \circ f)(s) = s$ for all $s \in S$ and $(f \circ g)(t) = t$ for all $t \in T$. We call g the inverse of f, and we denote it by the symbol f^{-1}. If S and T are open intervals, if $f : S \to T$ is invertible and differentiable, and if f' never vanishes on S, then f^{-1} is differentiable on T. The derivative is given by the formula

$$\frac{d}{dt} f^{-1}\bigg|_{(t)} = \frac{1}{\frac{df}{ds}\big|_{s=f^{-1}(t)}}.$$

The exponential function $\exp(x) = e^x$ has an inverse function. It is called the natural logarithm and is denoted by $x \mapsto \ln(x)$.

Higher Derivatives (Section 3.7)

The derivative of a given function f is a new function f', the derived function. This new function can in turn be differentiated, provided the derivative exists. The resulting function is called the second derivative. It is denoted by

$$f'' \quad \text{or} \quad f^{(2)} \quad \text{or} \quad \frac{d^2 f}{dx^2} \quad \text{or} \quad \ddot{f} \quad \text{or} \quad D^2 f \quad \text{or} \quad D^2(f).$$

The second derivative function can, in turn, be differentiated, and so on. The derivative of the $(k-1)$th derivative function is called the kth derivative function. It is denoted by the symbols

$$f^{(k)} \quad \text{or} \quad \frac{d^k f}{dx^k} \quad \text{or} \quad D^k f \quad \text{or} \quad D^k(f) \quad \text{or} \quad f^{\prime\prime\cdots\prime} \ (k \text{ primes}).$$

Leibniz's Rule for the second derivative of a product asserts that

$$(f \cdot g)^{(2)} = f^{(2)} g^{(0)} + 2 f^{(1)} g^{(1)} + f^{(0)} g^{(2)}.$$

Implicit Differentiation (Section 3.8)

If $f(x, y) = C$ is an equation expressing y implicitly as a function of x, then we can differentiate it without first solving for y as a function of x. This is the Method of Implicit Differentiation.

Related Rates (Section 3.8)

The Chain Rule (together with the Method of Implicit Differentiation, when needed) can be used to solve problems related to different rates of change.

Differential Approximation (Section 3.9)

If f is differentiable at c, then f may be approximated at a nearby point $c + \Delta x$ by

$$f(c + \Delta x) \approx f(c) + f'(c) \cdot \Delta x.$$

Writing Δf for $f(c + h) - f(c)$, we have

$$\Delta f \approx f'(c) \cdot \Delta x,$$

which is written in terms of differentials as

$$df = f'(c)\, dx.$$

genesis & DEVELOPMENT

The Tangent Problem in Antiquity

Apollonius (ca. 260 BCE–?) was the greatest geometer of antiquity. His most important work was *Conics,* written in eight books. The first four of these books have been preserved in Greek, the next three in ninth-century Arabic translations. Book VIII has been lost. Before Apollonius, the geometers of ancient Greece did not find an acceptable definition of the tangent line to a curve. Apollonius, however, found a sophisticated way to approach the difficulty. His insight was to start with the *normal line* $\mathcal{N}$. Given a curve $\mathcal{C}$ and a point Q not on $\mathcal{C}$, Apollonius constructed $\mathcal{N}$ by solving a *minimization* problem. He found the point P on $\mathcal{C}$ with minimum distance to Q (as shown in Figure 1). Once Apollonius obtained the normal line $\mathcal{N} = \overline{PQ}$, he took the the tangent to be the line through P perpendicular to $\mathcal{N}$.

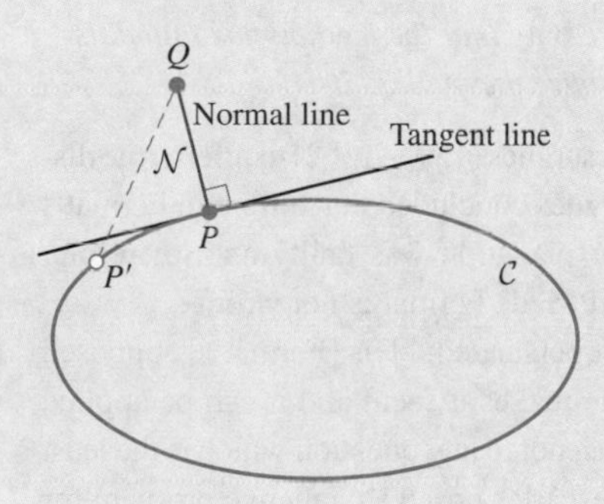

Figure 1
P is the point on $\mathcal{C}$ for which $|\overline{QP'}| > |\overline{QP}|$ for all $P' \neq P$ on $\mathcal{C}$.

The Solutions of Fermat and Descartes to the Tangent Problem

The introduction of analytic geometry revived the tangent problem in the 1630s. Descartes described the problem of finding tangents to curves as "the most useful and general that I have ever wanted to know in geometry." To find the tangent at a point P of a curve C, Descartes would find a circle that just *touched* C at P *without crossing* C. He let O be the center of this circle. Descartes took the line through P perpendicular to $\overline{OP}$ to be the tangent of C. In effect, Descartes had "reduced" the problem of finding a tangent line to the seemingly more difficult problem of finding a tangent circle. This method appeared in *La Géométrie* (1637), accompanied by illustrative examples that did not reveal how cumbersome a method it really was.

We will illustrate the method of Descartes by finding the tangent line to $y = \sqrt{x}$ at the point $P = (c, \sqrt{c})$. We seek the circle with center $(a, 0)$ and radius r that just touches the given parabola at P. After writing a as $c + h$, the equation of the circle becomes

$$(x - (c + h))^2 + y^2 = r^2.$$

The parameters h and r are to be determined in terms of the given value c. At the point P of intersection of the circle and the parabola $y = \sqrt{x}$, the ordinate satisfies $y^2 = x$. Therefore, the abscissa c of P satisfies the second degree equation

$$Q(x) = (x - (c + h))^2 + x - r^2 = 0.$$

In other words, $x - c$ is a factor of $Q(x)$. Descartes reasoned that since the circle *touches,* but does not *cross,* the parabola, the root c must have *even* multiplicity m. Because the degree of Q is 2, the only possibility is that $m = 2$ and $Q(x) = (x - c)^2$. Thus, $(x - c)^2 = x^2 + (1 - 2(c + h))x + (c + h)^2 - r^2$. Expanding the left side, we obtain

$$x^2 - 2cx + c^2 = x^2 + (1 - 2(c + h))x + ((c + h)^2 - r^2).$$

Each power of x must appear with the same coefficient on both sides of this identity, which leads to the simultaneous equations $-2c = 1 - 2(c + h)$ and $c^2 = (c + h)^2 - r^2$. Solving these equations, we obtain $h = 1/2$. The line $\mathcal{N}$ determined by the two points $P = (c, \sqrt{c})$ and $(c + h, 0)$, therefore, has slope $(\sqrt{c} - 0)/(c - (c + h))$, which equals $-2\sqrt{c}$. The slope of the line perpendicular to $\mathcal{N}$ (namely, the tangent line) is therefore $1/(2\sqrt{c})$ (see Figure 2).

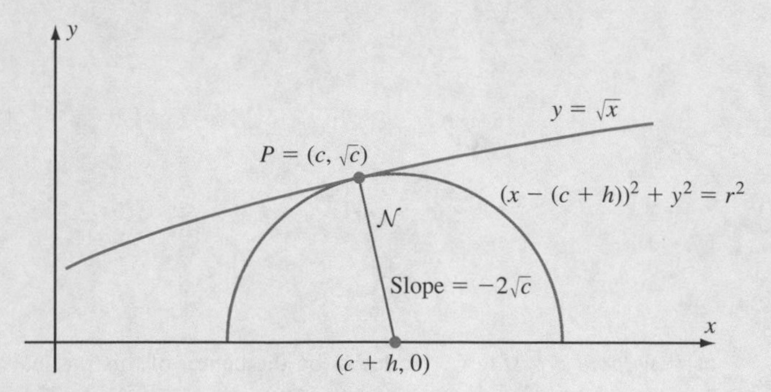

Figure 2

The method of Descartes becomes unwieldy when tried on even slightly more complicated examples. One contemporary protested that the method gives rise to computations that are a "labyrinth from which it is extraordinarily difficult to emerge." When Fermat learned of Descartes's work on tangents in 1637, he published his own method for finding the tangent line to the graph of a function f at c, a method that he had discovered some time earlier. Fermat did not use the slope m of the tangent line as a parameter. Instead, he used the length s of the *subtangent* (see Figure 3). Since m and s are related by the equation $m = f(c)/s$, the difference is not significant. Using the notation of Figure 3, we have $s/h = f(c)/(H - f(c))$ by similar triangles. This allows us to solve

$$m = \frac{f(c)}{s} = \frac{H - f(c)}{h}.$$

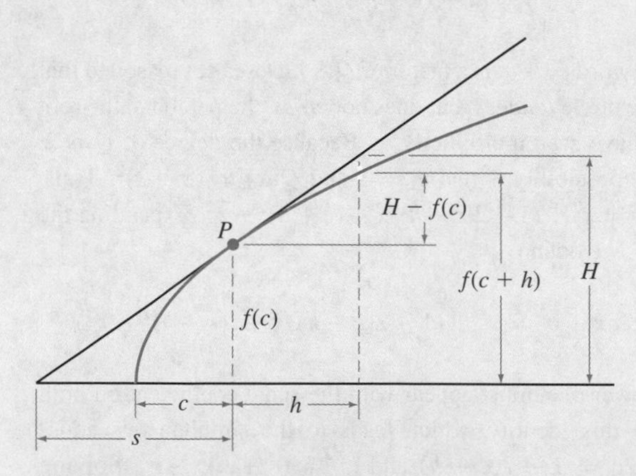

Figure 3

Fermat argued that when h is very small, $H \approx f(c+h)$. He called this approximation an *adequation*. Fermat made the substitution $H \approx f(c+h)$, obtaining

$$m \approx \frac{f(c+h) - f(c)}{h}.$$

For a polynomial f, Fermat expanded $f(c+h)$ to obtain a polynomial in h with constant term $f(c)$:

$$f(c+h) = f(c) + f_1(c)h + f_2(c)h^2 + \cdots.$$

Thus,

$$m \approx \frac{f_1(c)h + f_2(c)h^2 + \cdots}{h}.$$

He then eliminated h from the denominator by cancellation:

$$m \approx f_1(c) + f_2(c)h + \cdots.$$

At this stage, Fermat either set $h = 0$ or simply discarded the terms with h as a factor—it is not clear from his writing. In any event, Fermat changed the adequation to the equality $m = f_1(c)$. When f is a polynomial, it is easy to calculate each of the terms $f_1(c), f_2(c), \ldots$. Of course, it turns out that $f_1(c) = f'(c)$.

Fermat and Descartes became embroiled in a dispute over the tangent problem. It began with a letter of January 18, 1638, in which Descartes attributed the success of Fermat's method to luck. When other French mathematicians sided with Fermat, Descartes wrote:

> *I admire that [Fermat's method] . . . has found defenders; when M. Descartes will have understood the method, then he will cease to admire* only *that the method has found its defenders and also admire the method itself.*

Descartes chose Gérard Desargues (1593–1662) to referee the dispute. In April 1638, Desargues concluded that although Fermat's method was correct, his explanation was faulty and, therefore, "M. des Cartes is right and M. de Fermat is not wrong."

Descartes was still not persuaded: "His [Fermat's] supposed rule is not so general as he makes it seem and it can be applied only to the easiest problems, not to any question which is the least bit difficult." With that, Descartes posed a challenge problem for Fermat: to find the tangent lines of $x^3 + y^3 = nxy$, a curve that is now known as the *Folium of Descartes* (see Figure 4). Descartes must have expected that Fermat's method would be as futile as his own when applied to this curve. In fact, Fermat had not the slightest difficulty. He quickly sent Descartes a solution, together with a revised explanation of his method. Descartes conceded, not too gracefully: "[H]ad you so explained [your method] in the first place, I would not have objected at all." In fact, Descartes continued to object in his correspondence with others. "Is it not a great marvel," he wrote, "that in six months Fermat has found a new slant to justify his method." In a subsequent letter, Descartes described Fermat's tangent construction as "the most ridiculous gibberish." By 1641, Descartes finally realized

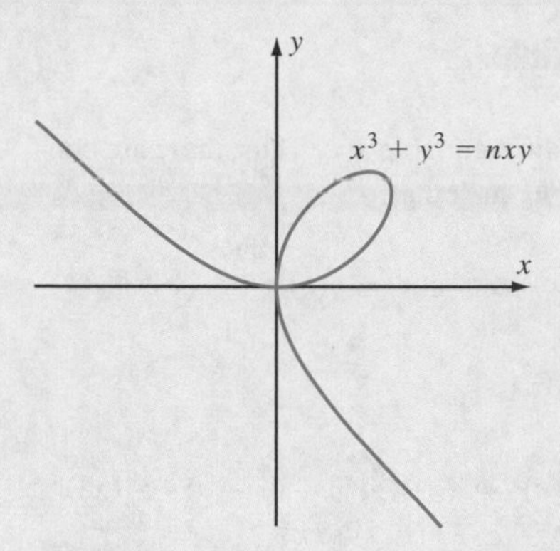

Figure 4
Folium of Descartes

the futility of attacking the mathematics of Fermat and switched tactics: "I believe that he [Fermat] does know mathematics but I still maintain that in philosophy he reasons badly."

Newton's Method of Differentiation

So far, we have described the tangent problem only from a geometric point of view. The kinematic viewpoint is also important. Among the early contributors to the differential calculus, Roberval and Torricelli are noteworthy for being the first to link the tangent problem with the notion of instantaneous velocity. To Sir Isaac Newton, that connection would be fundamental.

The mathematical language of Newton has not, for the most part, survived. Newton called the variables x and y *fluents*, imagining them to trace a curve by their movement. The velocities of the *fluents* were called *fluxions* and denoted by $\dot{x}$ and $\dot{y}$. Newton's infinitesimals were called *moments of fluxions* and represented by $\dot{x}o$ and $\dot{y}o$ where o is not 0 but rather an "infinitely small quantity." To illustrate, let us use

$$F(x, y) = x^3 - ax^2 + axy - y^3 = 0,$$

the example given by Newton in his *Methodus Fluxionum* (1670). Newton substituted $x + \dot{x}o$ for x and $y + \dot{y}o$ for y in the equation $F(x, y) = 0$. On expanding the left side of the equation $F(x + \dot{x}o, y + \dot{y}o) = 0$, remembering that $F(x, y) = 0$, we arrive at

$$(3x^2 - 2ax + ay)\dot{x}o + (ax - 3y^2)\dot{y}o + a \cdot \dot{x}o \cdot \dot{y}o \\ + (\dot{x}o)^2(3x - a + \dot{x}o) - (\dot{y}o)^2(3y + \dot{y}o) = 0.$$

Newton divided by o, obtaining

$$(3x^2 - 2ax + ay)\dot{x} + (ax - 3y^2)\dot{y} + o(a\dot{x}\dot{y} \\ + \dot{x}^2(3x - a + \dot{x}o) - \dot{y}^2(3y + \dot{y}o)) = 0.$$

He then *discarded* all terms with o as a factor. In his words,

> *But whereas* o *is supposed to be infinitely little . . . , the terms which are multiplied by it will be nothing in respect to the rest. Therefore I reject them and there remains:* $(3x^2 - 2ax + ay)\dot{x} + (ax - 3y^2)\dot{y} = 0.$

The result is that

$$\frac{\dot{y}}{\dot{x}} = -\frac{3x^2 - 2ax + ay}{ax - 3y^2},$$

which is the slope of the tangent line.

Newton, like Fermat before him and Leibniz after him, employed the suspect procedure of dividing by a quantity that he would later take to be zero. Newton's attempt to explain this undesirable situation is not very convincing. From his *Principia* of 1686, we find:

> *Those ultimate ratios with which quantities vanish are not truly the ratios of ultimate quantities, but limits toward which the ratios of quantities, decreasing without limit, do always converge, and to which they approach nearer than by any given difference, but never go beyond, nor in effect attain to, until the quantities have diminished in infinitum. Quantities, and the ratio of quantities, which in any finite time converge continually to equality, and before the end of that time approach nearer the one to the other than by any given difference, become ultimately equal.*

Bishop Berkeley, the critic who is quoted at length in the Genesis & Development for Chapter 2, had this to say in response:

> *The great author of the method of fluxions felt this difficulty and therefore he gave in to those nice abstractions and geometrical metaphysics without which he saw nothing could be done on the received principles . . . . It must, indeed, be acknowledged that he used fluxions like the scaffold of a building, as things to be laid aside or got rid of . . . . And what are these fluxions? . . . They are neither finite quantities nor quantities infinitely small, nor yet nothing. May we not call them the ghosts of departed quantities?*

As we have described in the Genesis & Development for Chapter 2, this controversy eventually led to the formalization of the limit concept.

The Product Rule and the Chain Rule

The rules of differential calculus that have been presented in this chapter were developed by Newton and, independently, at a later date, by Leibniz. The Product Rule, for example, appeared in Leibniz's work as $d(xv) = xdv + vdx$. To Leibniz, it was a matter of discarding a *second order infinitesimal*: $d(xv)$ is the infinitesimal increment in xv that arises by incrementing x by dx and v by dv. Thus,

$$\begin{aligned} d(xv) &= (x+dx)(v+dv) - xv = xdv + vdx + (dxdv) \\ &= xdv + vdx. \end{aligned}$$

Leibniz introduced the notation dx in a manuscript of November 11, 1675. In that manuscript, Leibniz struggled to obtain the rules of differential calculus such as the Product Rule. It is not easy to determine Leibniz's progress on that particular day. His initial guess was that $d(xv) = dx \cdot dv$. By the end of the manuscript, he seems to have been aware that $d(xv)$ is not $dx \cdot dv$. His initial guess for the Quotient Rule was also wrong. By the end of November 1675, however, Leibniz had discovered a correct calculus of differentials.

Newton evidently felt constrained to account for the $dxdv$ term that Leibniz simply discarded in the derivation of the product rule. He offered the following demonstration (which we have converted into Leibniz's infinitesimal notation):

$$\begin{aligned} d(xv) &= \left(x + \frac{1}{2}dx\right)\left(v + \frac{1}{2}dv\right) - \left(x - \frac{1}{2}dx\right)\left(v - \frac{1}{2}dv\right) \\ &= xdv + vdx. \end{aligned}$$

Berkeley specifically cited this derivation as an example of Newton's mathematical sleight of hand. In 1862, Sir William Rowan Hamilton (1805–1865), himself a major figure in the development of calculus and physics, was also troubled by this computation:

> *It is very difficult to understand the* logic *by which Newton proposes to prove [the Product Rule]. His mode of getting rid of [dxdv] appeared to me ... to involve so much artifice, as to deserve to be called sophistical.... But by what right or by what reason other than to give an unreal air of simplicity to the calculation does he prepare the products thus?... [I]t quite masks the notion of a limit.... Newton does not seem to have cared for being very consistent in his philosophy, if he could anyway get hold of truth—or what he considered to be such.*

In the early 1900s, Constantine Caratheodory (1873–1950) developed differential calculus in such a way that the standard rules became particularly easy to prove. Following Caratheodory's approach, f is said to be differentiable at c if and only if there is a function Φ_f that is continuous at c such that

$$f(x) = f(c) + (x-c)\cdot\Phi_f(x).$$

Thus, the derivative of f at c is defined by $f'(c) = \Phi_f(c)$. Let us use this interpretation of the derivative to derive two theorems whose proofs we deferred earlier in this chapter.

Proof of the Product Rule

Suppose that f and g are differentiable at c. Thus, there are continuous functions Φ_f and Φ_g such that

$$f(x) = f(c) + (x-c)\cdot\Phi_f(x) \quad\text{and}\quad g(x) = g(c) + (x-c)\cdot\Phi_g(x).$$

Therefore,

$$\begin{aligned} (f\cdot g)(x) &= f(x)g(x) \\ &= (f(c) + (x-c)\cdot\Phi_f(x))(g(c) + (x-c)\cdot\Phi_g(x)) \\ &= f(c)g(c) + (x-c)\cdot f(c)\cdot\Phi_g(x) \\ &\quad + (x-c)\cdot g(c)\cdot\Phi_f(x) + (x-c)^2\cdot\Phi_f(x)\cdot\Phi_g(x) \\ &= (f\cdot g)(c) + (x-c)\cdot(f(c)\cdot\Phi_g(x) \\ &\quad + g(c)\cdot\Phi_f(x) + (x-c)\cdot\Phi_f(x)\cdot\Phi_g(x)). \end{aligned}$$

Since the function

$$\Phi_{f\cdot g}(x) = f(c)\cdot\Phi_g(x) + g(c)\cdot\Phi_f(x) + (x-c)\cdot\Phi_f(x)\cdot\Phi_g(x)$$

is continuous, it follows that $f\cdot g$ is differentiable and

$$\begin{aligned} (f\cdot g)'(c) &= \Phi_{f\cdot g}(c) \\ &= f(c)\cdot\Phi_g(c) + g(c)\cdot\Phi_f(c) \\ &\quad + (c-c)\cdot\Phi_f(c)\cdot\Phi_g(c) \\ &= f(c)\cdot g'(c) + g(c)\cdot f'(c). \end{aligned}$$

Proof of the Chain Rule

Recall that the Chain Rule asserts that if f is differentiable at c and if g is differentiable at $f(c)$, then $g\circ f$ is differentiable at c and $(g\circ f)'(c) = g'(f(c))\cdot f'(c)$. Assuming the stated differentiability hypotheses for f and g, there are continuous functions Φ_f and Φ_g such that

$$f(x) = f(c) + (x-c)\cdot\Phi_f(x), \qquad f'(c) = \Phi_f(c),$$

and

$$g(x) = g(f(c)) + (x - f(c))\cdot\Phi_g(x), \qquad g'(f(c)) = \Phi_g(f(c)).$$

Therefore,

$$\begin{aligned} g(f(x)) &= g(f(c)) + (f(x) - f(c))\cdot\Phi_g(f(x)) \\ &= g(f(c)) + ((x-c)\cdot\Phi_f(x))\cdot\Phi_g(f(x)) \\ &= g(f(c)) + (x-c)\cdot(\Phi_f(x)\cdot(\Phi_g\circ f)(x)). \end{aligned}$$

Since $x \mapsto \Phi_f(x)\cdot(\Phi_g\circ f)(x)$ is continuous at c, it follows that $g\circ f$ is differentiable at c and that

$$\begin{aligned} (g\circ f)'(c) &= \Phi_f(c)\cdot(\Phi_g\circ f)(c) \\ &= \Phi_g(f(c))\cdot\Phi_f(c) \\ &= g'(f(c))\cdot f'(c). \end{aligned}$$

4

Applications of the Derivative

PREVIEW

The need to maximize and minimize functions often arises in science, engineering, and commerce. A company developing a new product might want to maximize its profit. An ecologist might want the company to minimize its consumption of resources. In short, procedures for extremizing functions are among the most vital applications of mathematics. In this chapter, we learn that the derivative provides a powerful tool for analyzing the extreme values of functions.

We learn that to maximize or minimize a differentiable function f, we must first find the roots of the equation $f'(x) = 0$. Unfortunately, solving such an equation can present formidable difficulties. Here again, the derivative is often an essential aid. This chapter shows us how to use differential calculus to locate roots of equations.

One of the main considerations of this chapter is the use of the derivative to aid in sketching a function's graph. We will learn to put together several uses of the derivative to sketch the graphs of functions. The availability of a graphing device, though useful, is no substitute for this knowledge.

4.1 The Derivative and Graphing

In this section, we begin to investigate the properties of the graph of a function f that can be deduced from the derivative f'. Our first example is simple enough to be analyzed without calculus. It does, however, suggest ways in which powerful methods of calculus can be used to examine more complicated functions.

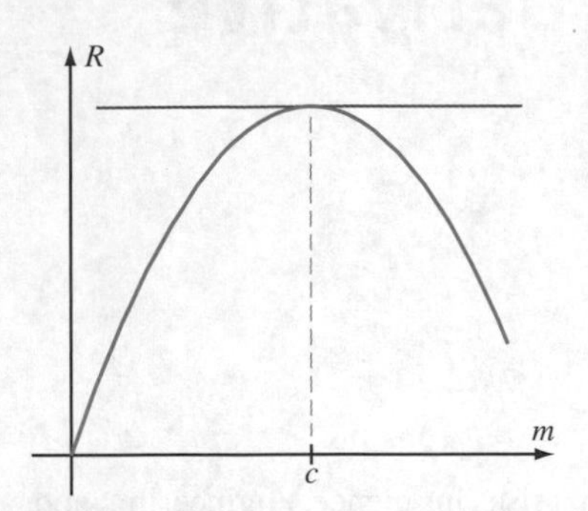

Figure 1
$R(m) = k \cdot m \cdot (2c - m)$

Example 1 Population biologist Raymond Pearl observed that the mass m of a yeast culture grows at the rate $R(m) = k \cdot m \cdot (2c - m)$ where c and k are positive constants. For what value of the mass m is its rate of change R greatest?

Solution Figure 1 shows the graph of R as a function of m. Since the rate R is a quadratic expression in m, we may complete the square as follows:

$$\begin{aligned} R(m) &= k \cdot m \cdot (2c - m) \\ &= kc^2 - k(c^2 - 2cm + m^2) \\ &= kc^2 - k(m - c)^2. \end{aligned}$$

Because $k \cdot (m - c)^2 > 0$ for $m \neq c$, we see that

$$R(m) = kc^2 - k(m - c)^2 < kc^2 - 0 = kc^2 = R(c)$$

for $m \neq c$. Thus, the maximum value of R is attained when $m = c$. ■

inSIGHT

In Example 1, the tangent line to the graph of R at $(c, R(c))$ appears to be horizontal or, equivalently, has zero slope. Indeed, we calculate that

$$\frac{dR}{dm} = \frac{d}{dm}(kc^2 - k(m - c)^2) = -2k(m - c),$$

which is zero for $m = c$. We see that the maximum value of R occurs at *precisely* the value at which $\frac{dR}{dm}$ is zero. As we will learn in this section, it is no coincidence that the maximum value of R is located at a point where the derivative vanishes.

Maxima and Minima

To use the derivative for identifying maximum and minimum values of a function, we must first develop an appropriate notion of maximum and minimum. Remember that $f'(c)$ is defined as the limit of $(f(x) - f(c))/(x - c)$ *as x approaches c*. Therefore, we cannot expect $f'(c)$ to contain information about f far away from c. The information that $f'(c)$ contains must be *local* in nature.

Definition Let f be a function with domain S. We say that f has a *local maximum* at the point $c \in S$ if there is a $\delta > 0$ such that $f(x) \leq f(c)$ for all $x \in S$ such that $|x - c| < \delta$. We call $f(c)$ a *local maximum value* for f. The term *relative maximum,* which has the same meaning as *local maximum,* is also used. If $f(x) \leq f(c)$ for *all* $x \in S$, then we say that f has an *absolute maximum* at c and that $f(c)$ is the *absolute maximum value* for f. Another term for *absolute maximum* is *global maximum*. Refer to Figure 2.

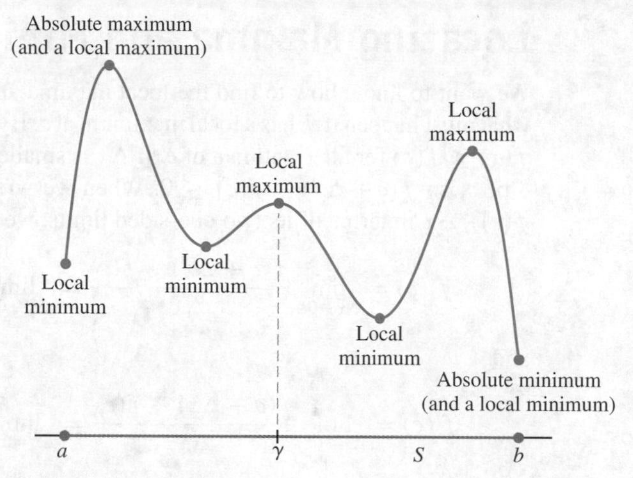

Figure 2

Definition Let f be a function with domain S. We say that f has a *local minimum* (or *relative minimum*) at the point $c \in S$ if there is a $\delta > 0$ such that $f(x) \geq f(c)$ for all $x \in S$ such that $|x - c| < \delta$. We call $f(c)$ a *local minimum value* for f. If $f(x) \geq f(c)$ for *all* $x \in S$, then we say that f has an *absolute minimum* (or *global minimum*) at c and that $f(c)$ is the *absolute minimum value* for f. Refer to Figure 2.

The term *extremum* refers to either a local maximum or a local minimum. The plural forms of "extremum," "maximum," and "minimum" are "extrema," "maxima," and "minima," respectively.

If f has a local maximum at c, then f takes its greatest value at c *only when compared with nearby points*. If f has a local minimum at c, then f takes its least value at c when compared with values of $f(x)$ for x *near* c. By contrast, an *absolute* maximum or minimum at c is determined by comparing $f(c)$ with the values of f at *all* points of the function's domain. In Figure 2, $f(a)$ is *not* the least value that the function f takes, but it is the least when compared with nearby values in the domain of f. Therefore, f has a local minimum at a but not an absolute minimum. Similarly, $f(\gamma)$ is not the greatest value that the function f takes, so f does not have an absolute maximum at γ. However, $f(\gamma)$ is the greatest value when compared with nearby values. Therefore, f has a local maximum at γ.

If $f(c)$ is the greatest value of f when compared with *all* points of the domain, then it is certainly the greatest value when compared with all nearby points. In other words, an absolute maximum is a local maximum. The converse is not necessarily true, as Figure 2 shows. Similarly, if an absolute minimum occurs at a point c, then a local minimum also occurs at c, although the converse is not true.

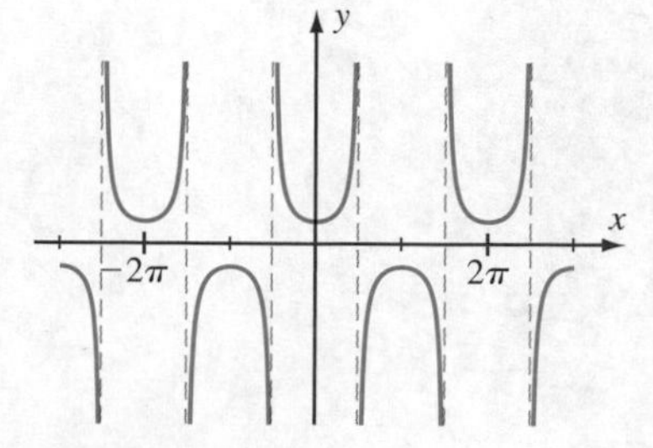

Figure 3
$f(x) = \sec(x)$

Example 2
Discuss local and absolute maxima and minima for the function $f(x) = \sec(x)$, $-3\pi \leq x \leq 3\pi$.

Solution Examine the graph of f in Figure 3. The function f has many local maxima, for instance at -3π, $-\pi$, π, and 3π. It also has many local minima, for instance at

-2π, 0, and 2π. None of these local maxima and minima are absolute maxima and minima because $\sec(x)$ takes both positive and negative values that are large in absolute value, without bound. ■

Locating Maxima and Minima

We want to know how to find the local maxima and minima of a function. Let us see what must happen if f has a local maximum at c. By definition, there is a $\delta > 0$ such that $f(x) \leq f(c)$ for all x within δ of c. If Δx is smaller than δ, then $x = c + \Delta x$ is within δ of c; so $f(c + \Delta x) - f(c) \leq 0$. When we write $f'(c) = \lim_{\Delta x \to 0}(f(c + \Delta x) - f(c))/\Delta x$ in terms of its two one-sided limits, we obtain

$$f'(c) = \lim_{\Delta x \to 0^+} \frac{f(c + \Delta x) - f(c)}{\Delta x} = \lim_{\Delta x \to 0^+} \frac{\text{negative numerator}}{\text{positive denominator}} \leq 0$$

and

$$f'(c) = \lim_{\Delta x \to 0^-} \frac{f(c + \Delta x) - f(c)}{\Delta x} = \lim_{\Delta x \to 0^-} \frac{\text{negative numerator}}{\text{negative denominator}} \geq 0.$$

Taken together, these two inequalities imply that $f'(c) = 0$. Similar reasoning leads to the same conclusion when f has a local minimum at c. These observations constitute Fermat's Theorem.

Theorem 1 ***Fermat's Theorem*** Let f be defined on an open interval that contains the point c. Suppose f is differentiable at c. If f has a local extremum at c, then $f'(c) = 0$.

inSIGHT

Fermat's Theorem has a geometric interpretation: The graph of a differentiable function has a horizontal tangent at a local extremum (as we observed in Figure 1). Fermat's Theorem does not apply to endpoints, such as a and b in Figure 2.

We must be careful when applying Fermat's Theorem. Notice the direction of the implication: *If a local extremum occurs at* c, *then* $f'(c) = 0$. Suppose we differentiate a function f and find the places where $f'(x) = 0$. Fermat's Theorem says that these are the points we should be considering if we wish to identify the local extrema of f. However, Fermat's Theorem does *not* tell us that these candidates must actually be extrema. Figure 4 shows some of the possibilities that can occur. Notice in particular that $f'(\gamma) = 0$, but f does not have a local extremum at γ. In summary, Fermat's Theorem *cannot* be used for concluding that a local extremum occurs at a point; it can only be used to locate *candidates* for a local extremum.

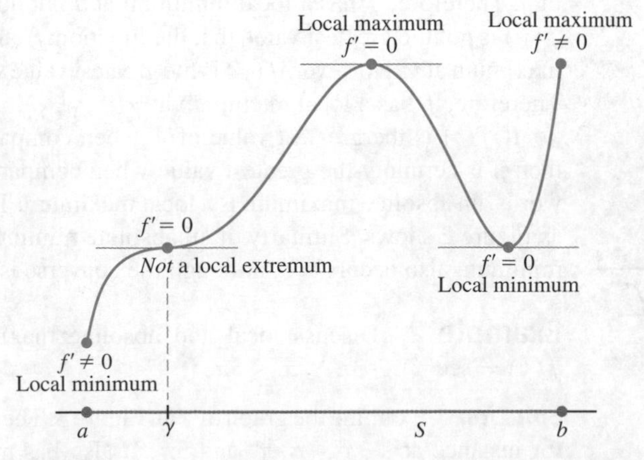

Figure 4

Example 3 Use the first derivative to locate local and absolute maxima and minima for the function $f(x) = \cos(x)$.

Solution First note that f is differentiable on the entire real line; so Fermat's Theorem applies. We calculate that $f'(x) = -\sin(x)$; therefore, $f'(x) = 0$ when $x = \ldots, -3\pi, -2\pi, -\pi, 0, \pi, 2\pi, \ldots$. Are there local maxima or local minima or neither at these points? First, there are local minima (indeed absolute minima) at the points $\ldots, -3\pi, -\pi, \pi, \ldots$. This is true because the value of f is -1 at these points, and that is certainly the absolute minimum value for cosine ($-1 \le \cos(x) \le 1$ for all x). Second, there are local maxima (indeed absolute maxima) at the points $\ldots, -2\pi, 0, 2\pi, \ldots$. This is true because the value of f is $+1$ at these points, and that is certainly an absolute maximum value for cosine. Are there any other local maxima and minima for f? If there were, then Fermat's Theorem guarantees that f' would vanish there. However, we have already checked all the points where f' vanishes. So we have found all local maxima and minima. ■

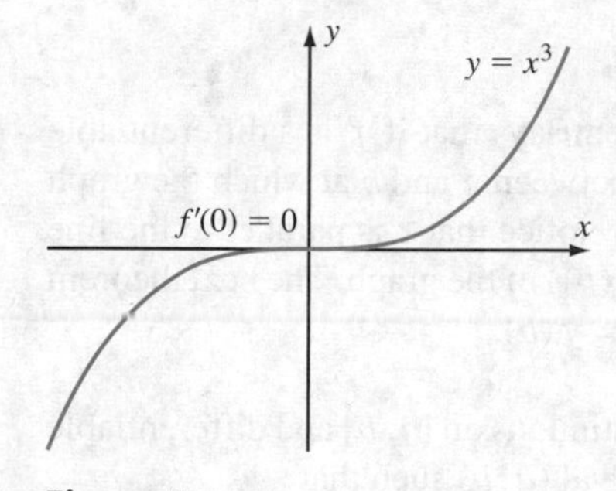

Figure 5

Example 4 Use Fermat's Theorem to determine whether the function $f(x) = x^3$ has any local or absolute extrema.

Solution Observe that f is differentiable at every point. Therefore, we may use Fermat's Theorem. We calculate that $f'(x) = 3x^2$. Thus, the only zero of f' is the point $x = 0$. If f has a local extremum, then it must occur at $x = 0$. A glance at the graph of f in Figure 5 shows that there is neither a local maximum nor a local minimum at 0. Does this contradict Fermat's Theorem? The theorem says that *if* f has a local maximum or minimum at $x = c$, then $f'(c) = 0$. It does *not* say that if $f'(c) = 0$, there must be a local maximum or a local minimum at c. Thus, there is no contradiction. ■

inSIGHT

It is important to understand that we really did use Fermat's Theorem in Example 4. We used it to conclude that no value of x, except *possibly* $x = 0$, could be a local extremum of f. Once we have used Fermat's Theorem to rule out most points, we must then investigate the points that remain. Fermat's Theorem says *nothing* about these remaining points.

Rolle's Theorem and the Mean Value Theorem

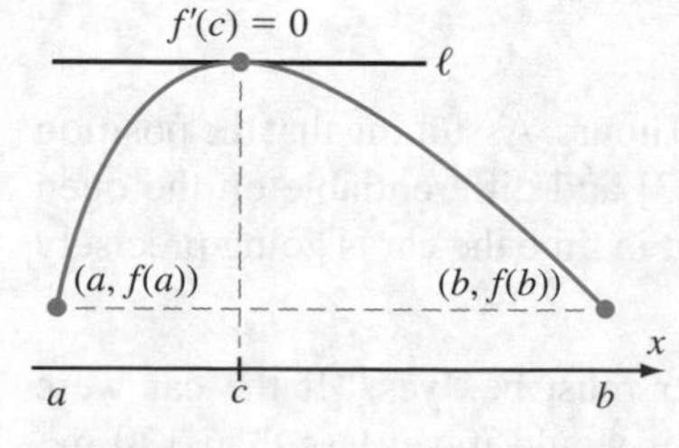

Figure 6
Rolle's Theorem

In Section 4.2, we will resume our investigation of extreme values. For now, we continue with our study of what the derivative says about the graph of a function f. Suppose that f is continuous on an interval $[a, b]$ and that $f(a) = f(b)$. The Extreme Value Theorem (see Section 2.3) guarantees that f has an absolute minimum and an absolute maximum on the interval $[a, b]$. If one of these extreme values is assumed at a point c inside the open interval (a, b) and if f is differentiable at c, then Fermat's Theorem tells us that $f'(c) = 0$ (see Figure 6). If not, one of the extreme values occurs at a and the other at b, and the maximum and minimum values of f are the same because $f(a) = f(b)$. We deduce that f is constant and $f'(c) = 0$ for any choice of c in the interval (a, b). Either way, there is a point in the interior of the interval at which f' is zero. This theorem is named after Michel Rolle, the mathematician who first proved it for polynomials.

Theorem 2 **Rolle's Theorem** Let f be a function that is continuous on $[a, b]$ and differentiable on (a, b). If $f(a) = f(b)$, then there is a number $c \in (a, b)$ such that $f'(c) = 0$.

Example 5 A ball is thrown straight up from ground level. It returns under the force of gravity to ground level at some time t_0. Assume only that the motion is described by a differentiable function. Is the velocity of the ball ever zero?

Solution If $h(t)$ is height, then $h(0) = h(t_0) = 0$. By Rolle's Theorem, there is a time c between 0 and t_0 when $h'(c) = 0$. In other words, the velocity of the ball is zero at time c. Our physical intuition reassures us on this point: The time c is the time when the ball reaches its highest point. At that instant, the ball is traveling neither up nor down. ■

When interpreted geometrically, Rolle's Theorem says that if f is a differentiable function with $f(a) = f(b)$, then there is a point c between a and b at which the graph of f has a horizontal tangent line ℓ (see Figure 6). Notice that ℓ is parallel to the line segment that joins the endpoints $(a, f(a))$ and $(b, f(b))$ of the graph. The next theorem states that there is an analogous result when $f(a) \neq f(b)$.

Theorem 3 **Mean Value Theorem** If f is a function that is continuous on $[a, b]$ and differentiable on (a, b), then there is a number c in the open interval (a, b) such that

$$f'(c) = \frac{f(b) - f(a)}{b - a}.$$

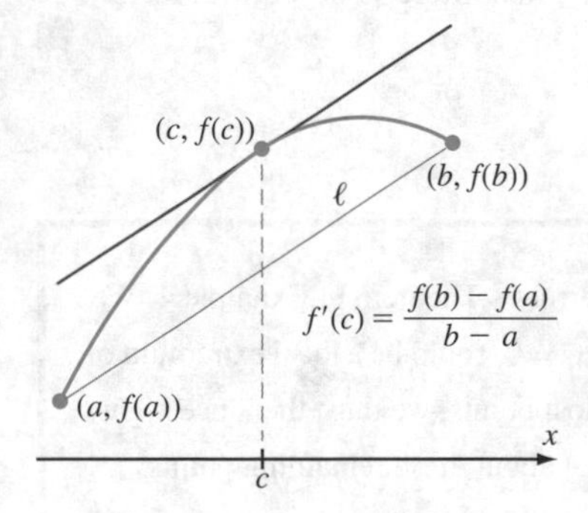

Figure 7

inSIGHT

Look at Figure 7. The Mean Value Theorem says that there is a number c between a and b for which the tangent line at $(c, f(c))$ is parallel to the line ℓ passing through the points $(a, f(a))$ and $(b, f(b))$. This assertion is really Rolle's Theorem in disguise. To see this, simply rotate Figure 7 so that the line ℓ is horizontal (see Figure 8)—the picture is essentially the same as that for Rolle's Theorem. Informally, the Mean Value Theorem is a "rotated" version of Rolle's Theorem. See Exercise 29 for an analytical proof.

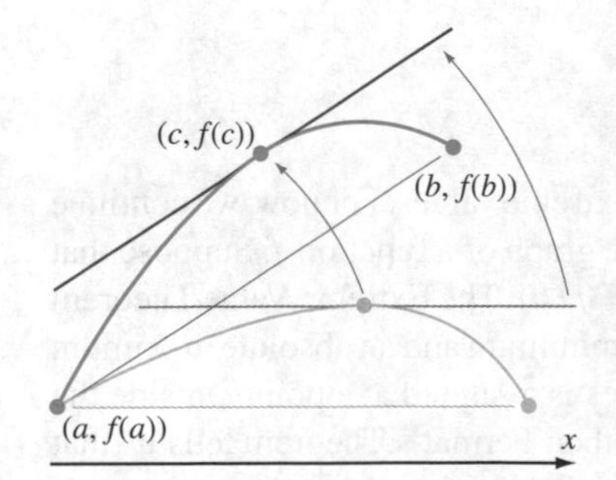

Figure 8

Example 6 An automobile travels 120 miles in 3 hours. Assuming that the position function p is continuous on the closed interval $[0, 3]$ and differentiable on the open interval $(0, 3)$, can we conclude that at some moment in time the car is going precisely 40 miles per hour?

Solution Common sense tells us that the answer must be "yes." If the car were always traveling at less than 40 mi/h, then in 3 h, it would travel less than 120 mi. Similarly, if the car were always traveling faster than 40 mi/h, then in 3 h it would go more than 120 mi. Therefore, the car either travels at a constant speed of 40 mi/h, in which case the answer to the question is obviously "yes," or there is a time when the car travels at a speed less than 40 mi/h and another time when it travels at more than 40 mi/h. Since we expect the speed to be a continuous function, we expect it to assume the intermediate value 40.

Let us use the Mean Value Theorem to give a short, more direct answer. We are told that $p(3) = p(0) + 120$. By the Mean Value Theorem, there is a point $c \in (0, 3)$ such that

$$p'(c) = \frac{p(3) - p(0)}{3 - 0} = \frac{(120 + p(0)) - p(0)}{3} = \frac{120}{3} = 40,$$

as required. ■

An Application of the Mean Value Theorem

We know that the derivative of a constant function is identically zero. The Mean Value Theorem enables us to see that the converse is also true.

Theorem 4 Let f be a differentiable function on an interval (α, β). If $f'(x) = 0$ for each x in (α, β), then f is a constant function.

Proof Let a and b be any two points in (α, β). By the Mean Value Theorem, there is a c between a and b such that

$$\frac{f(b) - f(a)}{b - a} = f'(c);$$

but by hypothesis, $f'(c) = 0$. Thus, $f(b) - f(a) = 0$, or $f(b) = f(a)$. Therefore, f takes the same value at any two points a and b in the interval (α, β). This is another way of saying that f is constant. ■

Theorem 5 If F and G are differentiable functions such that $F'(x) = G'(x)$ for every x in (α, β), then there is a constant C such that $F(x) = G(x) + C$ for every x in the interval (α, β).

Proof Theorem 4, applied to the function $f(x) = F(x) - G(x)$, tells us that there is a constant C such that $f(x) = C$ for each x in (α, β). ■

Example 7 Suppose that $F'(x) = 2x$ and $F(1) = 10$. Find $F(4)$.

Solution The function $G(x) = x^2$ certainly has derivative $2x$. According to Theorem 5, the function F that we seek must have the form $F(x) = x^2 + C$ where C is a constant. Now we use the information $F(1) = 10$ to determine the specific value of C: $F(1) = 1^2 + C = 10$, or $C = 9$. Therefore, $F(x) = x^2 + 9$ and $F(4) = 16 + 9 = 25$. ■

quickquiz

1. What is a local minimum? An absolute maximum?
2. True or false: If $f'(c) = 0$, then a local extremum of $f(x)$ occurs at $x = c$. Explain.
3. Describe the geometric interpretation of the Mean Value Theorem.
4. Find the local maxima and minima of the function $f(x) = x^2 - x - 1$.

EXERCISES

Problems for Practice

In Exercises 1–6, locate all local maxima and minima for the function. (Calculus is not needed for these exercises.)

1. $f(x) = (x-2)^2 + 3$
2. $f(x) = 2 - 4(x-6)^2$
3. $f(x) = 1/(x^2+1)$
4. $f(x) = 1/((x-2)^2+4)$
5. $f(x) = 3\sin(4x)$
6. $f(x) = \exp(-|x|)$

For Exercises 7–16, use Fermat's Theorem to locate all *candidates* for local maxima and minima. Test *values* of the function to the left and right of each candidate to determine if it is a local maximum, a local minimum, or neither.

7. $f(x) = 2x^2 - 24x + 36$
8. $f(x) = 12x^2 + 48x$
9. $f(x) = x^4 - 2x^2 + 1$
10. $f(x) = \cot(x)$
11. $f(x) = (x-3)(x+5)^3$
12. $f(x) = x/(x^2+1)$
13. $f(x) = x - \ln(x)$
14. $f(x) = x + 1/x$
15. $f(x) = e^x - x$
16. $f(x) = xe^x$

In Exercises 17–22, find all functions with the given function as derivative.

17. 5
18. $3x$
19. $x^2 + \pi$
20. $4x^{1/2} + 3$
21. $\cos(x)$
22. $3\sin(x) - x$

In Exercises 23–28, verify that the hypotheses of the Mean Value Theorem hold for the function f and interval I. By applying the theorem, what assertion about existence can you make?

23. $f(x) = x^4 + 7x^3 - 9x^2 + x,\ I = [0, 1]$
24. $f(x) = x^2 \cdot \sin(x),\ I = [0, \pi/2]$
25. $f(x) = x^{1/5},\ I = [1, 32]$
26. $f(x) = -(x-2)^2 + 4,\ I = [-2, 4]$
27. $f(x) = -(x-1)^4 + 1,\ I = [0, 2]$
28. $f(x) = x^2 + 2x + 3,\ I = [1, 4]$

Further Theory and Practice

29. Analytic Proof of the Mean Value Theorem Suppose that f satisfies the hypotheses of the Mean Value Theorem. Define g by

$$g(x) = f(x) - \left(f(a) + \frac{f(b)-f(a)}{b-a} \cdot (x-a) \right).$$

Apply Rolle's Theorem to g. Deduce the assertion of the Mean Value Theorem.

In Exercises 30–33, a continuous function f is given on an interval $I = [a, b]$. Sketch the graph of $y = f(x)$ for $x \in I$. In each case, explain why there can be no c in (a, b) such that $f'(c) = (f(b) - f(a))/(b-a)$. Explain why this does not contradict the Mean Value Theorem.

30. $f(x) = |x-5|,\ I = [4, 7]$
31. $f(x) = (x^2 - 2x + 1)^{1/4},\ I = [0, 2]$
32. $f(x) = \begin{cases} x-5 & \text{if } -3 \le x \le -1 \\ -x-7 & \text{if } -1 < x \le 2 \end{cases},\ I = [-3, 2]$
33. $f(x) = \exp(-|x|),\ I = [-1, 1]$

In Exercises 34–37, find a value c, the existence of which is guaranteed by Rolle's Theorem applied to the function f on the interval $I = [a, b]$.

34. $f(x) = e^x \sin(x),\ a = 0,\ b = \pi$
35. $f(x) = x^3 - x,\ a = 0,\ b = 1$
36. $f(x) = \sin(x) + \cos(x),\ a = -\pi/4,\ b = 3\pi/4$
37. $f(x) = (x^2 + x)/(x^2 + 1),\ a = -1,\ b = 0$

In Exercises 38–41, find a value c, the existence of which is guaranteed by the Mean Value Theorem applied to the function f on the interval $I = [a, b]$.

38. $f(x) = x/(x-1),\ a = 2,\ b = 4$
39. $f(x) = Ax^2 + B,\ A \neq 0,\ a, b$ arbitrary
40. $f(x) = x^3 + 3x - 1,\ a = 1,\ b = 5$
41. $f(x) = x + 1/x,\ a = 1,\ b = 2$
42. Consider the function $f(x) = \sin(1/x)$ defined for x in the interval $(0, 1)$. Use the Mean Value Theorem to explain why the function f' takes on all real values.
43. Differentiate $f(x) = \cos^2(x)$ and $g(x) = \sin^2(x)$. Use the results of these differentiations to prove the identity $\cos^2(x) + \sin^2(x) = 1$.
44. Differentiate $f(x) = \sec^2(x)$ and $g(x) = \tan^2(x)$. Use the results of these differentiations to prove the identity $1 + \tan^2(x) = \sec^2(x)$.
45. Suppose that f satisfies the conditions of the Mean Value Theorem on $[a, b]$. Show that if $0 < h < b - a$, then there exists a $\theta \in (0, 1)$ such that $f(a+h) = f(a) + h \cdot f'(a + \theta h)$.
46. Show that $3x^4 - 4x^3 + 6x^2 - 12x + 5 = 0$ has at most two real-valued solutions.
47. Show that $x^3 - 3x^2 + 4x - 1 = 0$ has exactly one real root.
48. Show that $x^9 + 3x^3 + 2x + 1 = 0$ has exactly one real root.

49. Consider the polynomial $p(x) = x^3 + ax^2 + b$. Use Rolle's Theorem to show that p cannot have three negative roots.
50. Use Rolle's Theorem to prove that $p(x) = x^3 + ax^2 + b$ can have no negative root if $a < 0$.
51. Consider the polynomial $p(x) = x^3 + ax + b$. Show that p cannot have three negative roots.
52. Use Rolle's Theorem to prove that $p(x) = x^3 + ax + b$ can have only one real root if $a > 0$.
53. Use the Mean Value Theorem to show that if $f(a) \geq g(a)$ and if $f'(x) > g'(x)$ for $x > a$, then $f(x) > g(x)$ for $x > a$.
54. Bernoulli's inequality states that $(1 + x)^p > 1 + px$ for $x > 0$ and $p > 1$. Use Exercise 53 to prove this.
55. Use the Mean Value Theorem to show that $(1 + x)^p < 1 + px$ for $x > 0$ and $p < 1$.
56. Use the Mean Value Theorem to show that $\sin(x) < x$ for $x > 0$.

Discrete Dynamical Systems Suppose that Φ is a continuous function. The collection of sequences $\{x_n\}_{n \geq 0}$ that satisfy $x_{n+1} = \Phi(x_n)$ for $n \geq 0$ is called the *discrete dynamical system associated with* Φ. Notice that each element $\{x_n\}$ of a discrete dynamical system is determined by its first element x_0. A number x_* such that $\Phi(x_*) = x_*$ is called an *equilibrium point* of the dynamical system. We say that an equilibrium point x_* of Φ is *stable* if there is a $\delta > 0$ such that each element $\{x_n\}$ of the dynamical system associated with Φ converges to x_* provided that $|x_* - x_0| < \delta$. Exercises 57–62 concern these ideas.

57. Suppose that x_* is an equilibrium point of the dynamical system. Show that if $x_N = x_*$, then $x_n = x_*$ for all $n \geq N$.
58. Suppose that $\{x_n\}_{n \geq 0}$ is an element of the discrete dynamical system associated with Φ. Suppose that

$$\xi = \lim_{n \to \infty} x_n.$$

Show that ξ is an equilibrium point of the discrete dynamical system associated with Φ.
59. Suppose that x_* is an equilibrium point of the dynamical system Φ. Suppose that Φ' exists and is continuous on an open interval centered at x_*. Suppose also that $|\Phi'(x_*)| < 1$. Show that there are positive numbers δ and K with $K < 1$ such that

$$|\Phi'(x)| < K \text{ for every } x \text{ in } (x_* - \delta, x_* + \delta).$$

Illustrate this with a sketch.
60. **Continuation** Suppose that x_0 is in the interval $(x_* - \delta, x_* + \delta)$. Deduce that x_1 is also in this interval by observing that

$$|x_1 - x_*| = |\Phi(x_0) - \Phi(x_*)| \leq K|x_0 - x_*|.$$

Explain the reasoning behind this observation. Illustrate the position of x_*, x_0, and x_1 with a sketch.
61. **Continuation** Follow the steps in Exercise 60 to deduce that

$$|x_n - x_*| \leq K^n|x_0 - x_*|.$$

62. **Continuation** Prove the following theorem: If x_* is an equilibrium point of the dynamical system Φ such that (1) Φ' exists and is continuous on an open interval centered at x_* and (2) $|\Phi'(x_*)| < 1$, then x_* is a stable equilibrium of Φ.
63. Suppose that p is a differentiable function such that $p(a) = p(b) = 0$. Let k be *any* real number. Use the function $x \mapsto e^{-kx}p(x)$ to show that there is a c between a and b such that $p'(c) = kp(c)$.
64. Suppose that $p(x)$ is a polynomial of degree d with k distinct real roots. Apply Rolle's Theorem to these roots in sequential pairs to draw the conclusion that p' has at least $k - 1$ distinct real roots. Apply Rolle's Theorem again to draw a conclusion about the number of distinct real roots of p''. Continue this process until you obtain an assertion about the number of distinct real roots of the dth derivative of p. Since the dth derivative is a constant function, what can you conclude about the relationship between k and d?
65. Suppose that $a < b$, that $f(a) = g(b) = 0$, that f and g are continuous functions on $[a, b]$, and that f and g are differentiable functions on (a, b). Show that the equation

$$f'(x)g(x) + f(x)g'(x) = 0$$

has a solution $x = c$ in the interval (a, b).
66. Use the Mean Value Theorem to prove that

$$\lim_{x \to +\infty} \left(\sqrt{x+1} - \sqrt{x}\right) = 0.$$

67. Consider $f(x) = \sqrt{1 + x}$ on the interval $[h/2, h]$. Use the Mean Value Theorem to prove that

$$\lim_{h \to 0} \left(\frac{\sqrt{1+h} - \sqrt{1+h/2}}{h} \right) = \frac{1}{4}.$$

68. Suppose that $f : \mathbb{R} \to \mathbb{R}$ is a differentiable function and that $|f'(x)| \leq C_1$ for all x, where C_1 is a numerical constant. Prove that

$$|f(s) - f(t)| \leq C_1 \cdot |s - t|$$

for all $s, t \in \mathbb{R}$.
69. Suppose that f satisfies the hypotheses of Rolle's Theorem on $[a, b]$. Let $g(x) = f(x + (b - a)/2) - f(x)$. Use the Intermediate Value Theorem to find an $a_1 \in [a, (a + b)/2)$ such that $g(a_1) = 0$.

Let $b_1 = a_1 + (b - a)/2$. Show that $f(a_1) = f(b_1)$. Notice that $[a_1, b_1]$ has half the length of $[a, b]$. Explain how this method can be repeated indefinitely so that from a pair (a_n, b_n) with $f(a_n) = f(b_n)$, we obtain a pair (a_{n+1}, b_{n+1}) with $f(a_{n+1}) = f(b_{n+1})$, $[a_{n+1}, b_{n+1}] \subset [a_n, b_n]$, and $b_{n+1} - a_{n+1} = (b_n - a_n)/2$. Deduce that there exists a point $c \in (a, b)$ for which

$$f'(c) = \lim_{n\to\infty} \frac{f(b_n) - f(a_n)}{b_n - a_n} = 0.$$

Computer/Calculator Exercises

In Exercises 70–73, calculate and plot the derivative f' of the function f. Use this plot to locate all candidates for local extrema of f. Add the plot of f to the window containing the graph of f'. From this second plot, determine the behavior of f at each candidate for a local extremum.

70. $f(x) = x^2 - 2x\ln(1 + x^2) + x - 4$
71. $f(x) = x^3 - 2x + \cos(x)$
72. $f(x) = \sin^2(x) - x^3 + 5x + 20$
73. $f(x) = x - 2\exp(-x^2)$

In Exercises 74–77, approximate the value c guaranteed by the application of the Mean Value Theorem to the given function f on the given interval $[a, b]$. Graph the function, the tangent line at $(c, f(c))$, and the line segment between $(a, f(a))$ and $(b, f(b))$.

74. $f(x) = \sin(x), a = 0, b = \pi/2$
75. $f(x) = x^5 + x^3 + 1, a = 0, b = 2$
76. $f(x) = x\sin(1/x), a = 1/(4\pi), b = 1/(2\pi)$
77. $f(x) = x^4 - 2x^3 + x^2 - 2x + 13, a = -1, b = 3$

4.2 Maxima and Minima of Functions

In Section 4.1, we learned that the derivative is a useful tool for *finding* candidates for local extrema. However, we did not learn much about *identifying* these extrema—the subject of this section. We begin with a consideration of what the derivative tells us about the graph of a function. Recall the following definition from Section 3.6.

Definition We say that a function f is *increasing* on an interval I if $f(\alpha) < f(\beta)$ whenever α and β are points in I with $\alpha < \beta$ (see Figure 1). We say that f is *decreasing* if $f(\alpha) > f(\beta)$ whenever $\alpha < \beta$ (see Figure 2).

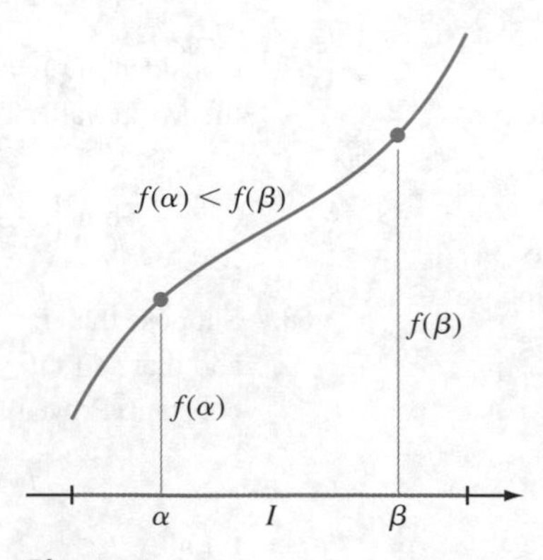

Figure 1
f is increasing on I.

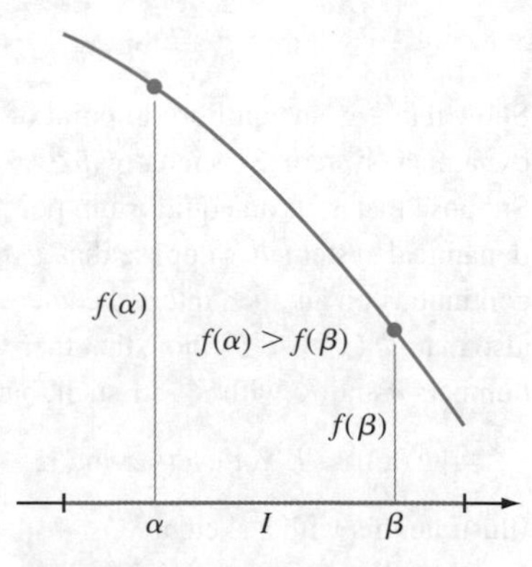

Figure 2
f is decreasing on I.

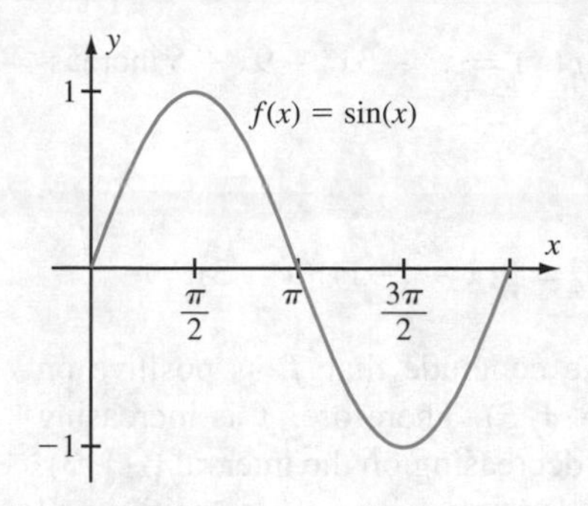

Figure 3

Example 1 Where is the function $f(x) = \sin(x)$ increasing or decreasing on the interval $(0, 2\pi)$?

Solution As Figure 3 shows, the function f is increasing on the intervals $(0, \pi/2)$ and $(3\pi/2, 2\pi)$. It is decreasing on the interval $(\pi/2, 3\pi/2)$. ■

The statements about the sine function in Example 1 are obvious because it is a familiar function. What we need is a technique for telling where *any* function is increasing or decreasing.

Using the Derivative to Tell When f Is Increasing or Decreasing

Theorem 1 If $f'(x) > 0$ for each x in an interval I, then f is increasing on I. If $f'(x) < 0$ for each x in I, then f is decreasing on I.

Proof For the first assertion, let $\alpha < \beta$ be points in I with $\alpha < \beta$. By the Mean Value Theorem, we know there is a number c between α and β such that

$$\frac{f(\beta) - f(\alpha)}{\beta - \alpha} = f'(c);$$

but by hypothesis, we know that $f'(c) > 0$ and $\beta - \alpha > 0$. It follows that

$$f(\beta) - f(\alpha) = \underbrace{f'(c)}_{\text{Positive}} \cdot \underbrace{(\beta - \alpha)}_{\text{Positive}} > 0,$$

or $f(\beta) > f(\alpha)$. Therefore, f is increasing on I. We may similarly prove the second assertion of Theorem 1. ■

Example 2 Examine the function $f(x) = x^3 - 6x^2 + 13x - 7$ to determine intervals on which it is increasing or decreasing.

Solution The graph of f in Figure 4 suggests that f is increasing on the interval $[-1, 5]$ over which it is plotted. However, the plot of f over a finite interval cannot exhibit the behavior of f over the entire real line. Instead, we appeal to Theorem 1. The function $f(x)$ satisfies

$$\begin{aligned} f'(x) &= 3x^2 - 12x + 13 \\ &= 3(x^2 - 4x) + 13 \\ &= 3(x^2 - 4x + 4) + 1 = 3(x-2)^2 + 1 \end{aligned}$$

for every x. Thus, $f'(x) > 0$ for all $x \in \mathbb{R}$. By Theorem 1, f is increasing on all $\mathbb{R}$. ■

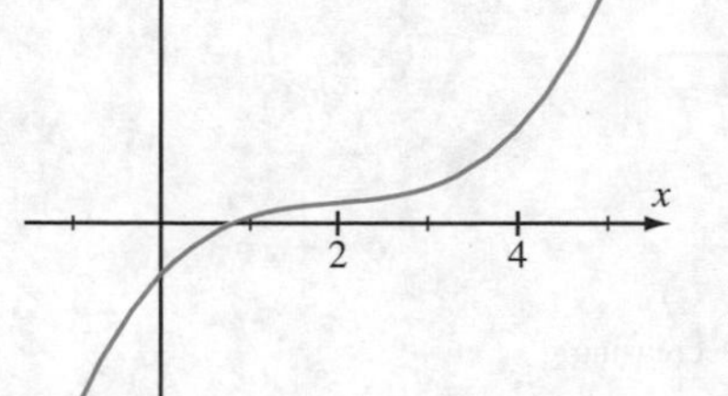

Figure 4

It is sometimes helpful to notice that if $a < b$, then the two factors of $(x-a)(x-b)$ have the same sign if and only if $x < a$ (both negative) or $x > b$ (both positive). We therefore see that

$$(x-a)(x-b) \begin{cases} > 0 & \text{if } x < a \\ < 0 & \text{if } a < x < b. \\ > 0 & \text{if } x > b \end{cases}$$

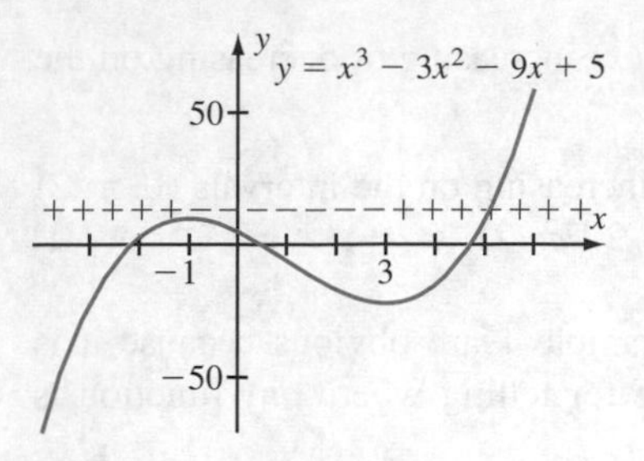

Figure 5

Example 3 On which intervals is the function $f(x) = x^3 - 3x^2 - 9x + 5$ increasing? On which intervals is it decreasing?

Solution Begin by calculating

$$f'(x) = 3x^2 - 6x - 9 = 3(x + 1)(x - 3) = 3(x - (-1))(x - 3).$$

Using the information preceding this example, we conclude that f' is positive on $(-\infty, -1)$ and on $(3, \infty)$ and f' is negative on $(-1, 3)$. Therefore, f is increasing on each of the intervals $(-\infty, -1)$ and $(3, \infty)$ and decreasing on the interval $(-1, 3)$. Figure 5 exhibits the graph; it includes $+$ signs to indicate where f' is positive and $-$ signs to indicate where f' is negative. ■

The First Derivative Test for Extrema

Now that we have learned to identify where a differentiable function is increasing or decreasing, we can apply our knowledge to determine where a differentiable function has a local maximum or minimum.

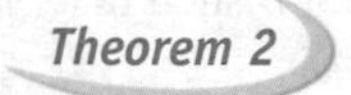

First Derivative Test Let f be a differentiable function on an open interval and suppose that $f'(c) = 0$ at some point c inside this interval.

a. If $f'(x) < 0$ for $x < c$ and $f'(x) > 0$ for $x > c$, then f has a local minimum at c.
b. If $f'(x) > 0$ for $x < c$ and $f'(x) < 0$ for $x > c$, then f has a local maximum at c.
c. If f' does not change sign at c (even though $f'(c) = 0$), then f has *neither* a local minimum *nor* a local maximum at c.

The rationale for the test is obvious. Look at Figure 6: The hypotheses in statement a imply that f is decreasing just to the left of c and increasing just to the right of c. The hypotheses of statement b imply that f is increasing just to the left of c and decreasing just to the right of c. Finally, the hypothesis of statement c says that either f is increasing both to the left and to the right of c or f is decreasing both to the left and to the right of c.

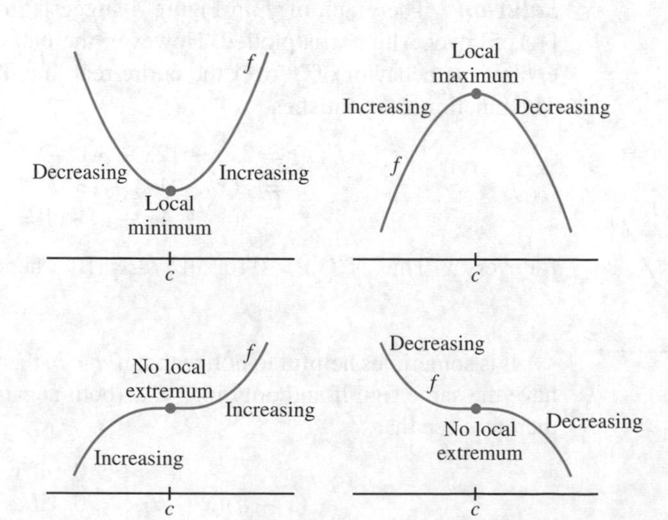

Figure 6

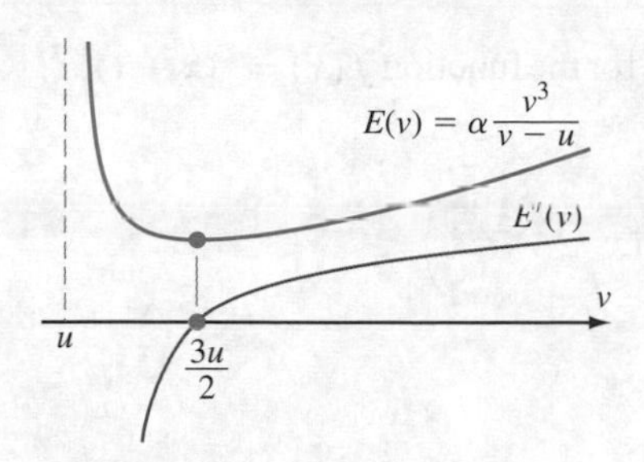

Figure 7
$E'(v) < 0$ for $v < 3u/2$ and $E'(v) > 0$ for $v > 3u/2$

Example 4 Suppose a river's current has speed u and a fish is swimming upstream with speed v relative to the water. The energy E expended in such a migration is

$$E(v) = \alpha \frac{v^3}{v-u}$$

where α is a positive constant. For what value of v is E minimized?

Solution If v were equal to u, then the fish would be at a standstill relative to land. If v were less than u, then the fish would be carried downstream by the current. Since the fish is swimming upstream, we infer that $v > u$. In other words, the domain of E is the open interval (u, ∞). Also, E is differentiable at all points. By Fermat's Theorem, any local extremum will occur at a zero of E'. Using the Quotient Rule, we calculate

$$E'(v) = \alpha \frac{(v-u) \cdot 3v^2 - v^3(1-0)}{(v-u)^2} = \alpha \frac{2v^3 - 3uv^2}{(v-u)^2} = 2\alpha v^2 \frac{v - 3u/2}{(v-u)^2}.$$

Setting $E'(v) = 0$ yields $v = 3u/2$. The sign of $E'(v)$ is the same as the sign of $v - 3u/2$. Thus, $E'(v) < 0$ for $v < 3u/2$ and $E'(v) > 0$ for $v > 3u/2$. According to the First Derivative Test, E has a local minimum at $v = 3u/2$. The graphs of E and E' in Figure 7 bear this out. ■

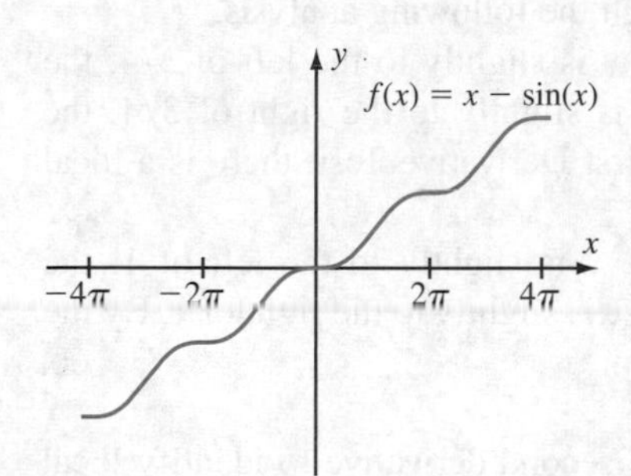

Figure 8

Example 5 Find the local extrema of the function $f(x) = x - \sin(x)$.

Solution Because f has domain the entire real line and f is differentiable at all x, the local extrema will all be zeros of f'. We calculate the derivative: $f'(x) = 1 - \cos(x)$. The zeros of f' are $\ldots, -4\pi, -2\pi, 0, 2\pi, 4\pi, \ldots$. These points are the only candidates for local extrema. Let us first examine 0. If x is just to the left of 0, then $\cos(x) < 1$ and $f'(x) > 0$. If x is just to the right of 0, then $\cos(x) < 1$ and $f'(x) > 0$. Thus, statement c of the First Derivative Test tells us that $c = 0$ is *not* a local extremum. Since f' is 2π-periodic, the same analysis will apply to the other critical points: None of the points is a local extremum. In fact, we see that $f'(x) > 0$ except at the integer multiples of 2π. The graph will go continually uphill from left to right, as shown in Figure 8. ■

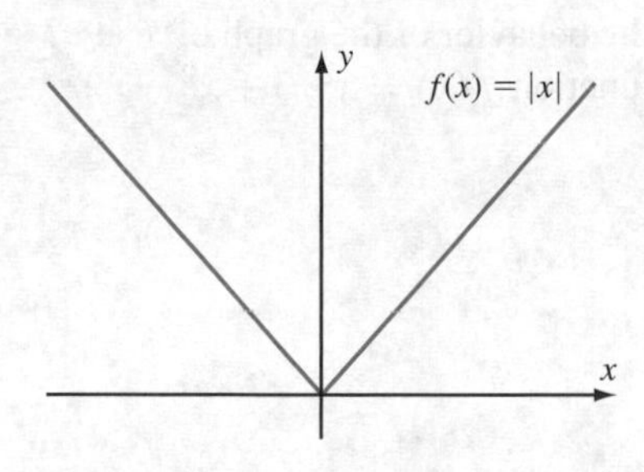

Figure 9

Critical Points

It is important to notice that many of the functions we encounter in practice are not differentiable at every point of their domains. For example, $f(x) = |x|$ is not differentiable at $x = 0$, but it *does* have an absolute minimum there (see Figure 9). Therefore, our tests for local extrema should take into account points of nondifferentiability. We begin with a definition.

Definition Let c be a point in an open interval on which f is continuous. We call c a *critical point* for f if one of the following two conditions holds:

1. f is not differentiable at c, or
2. f is differentiable at c and $f'(c) = 0$.

Notice the following consequence of the First Derivative Test: If the domain of f is an open interval, then the critical points are the only places where we will have to search for local extrema of f.

Example 6 Find and analyze the critical points for the function $f(x) = x(x-1)^{1/3}$.

Solution As long as $x \neq 1$,

$$f'(x) = 1(x-1)^{1/3} + x\left(\frac{1}{3}(x-1)^{-2/3}\right)$$

$$= \frac{(x-1) + x/3}{(x-1)^{2/3}}$$

$$= \frac{(4/3)x - 1}{(x-1)^{2/3}}.$$

Figure 10

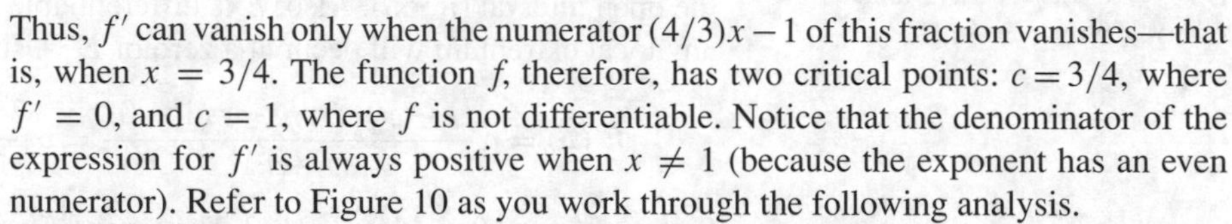

Thus, f' can vanish only when the numerator $(4/3)x - 1$ of this fraction vanishes—that is, when $x = 3/4$. The function f, therefore, has two critical points: $c = 3/4$, where $f' = 0$, and $c = 1$, where f is not differentiable. Notice that the denominator of the expression for f' is always positive when $x \neq 1$ (because the exponent has an even numerator). Refer to Figure 10 as you work through the following analysis.

Begin with the critical point $c = 3/4$. When x is slightly to the left of $3/4$, the numerator of f' is negative; so $f' < 0$. When x is slightly to the right of $3/4$, the numerator of f' is positive; so $f' > 0$. By the First Derivative Test, there is a local minimum at $c = 3/4$.

Next, consider the critical point $c = 1$. When x is slightly to the left of 1, the numerator of f' is positive. The same is true when x is slightly to the right of 1. By the First Derivative Test, $c = 1$ is not a local extremum. ■

In later sections, we will learn how to use the second derivative to identify local maxima and minima. We will also learn how to locate *absolute* maxima and minima.

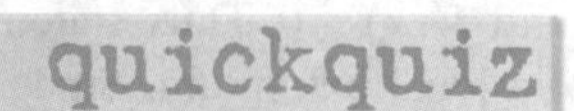

1. What does it mean to say that a function f is increasing on an interval I?
2. What condition might we check to see if a differentiable function f is decreasing on an interval I?
3. If $f'(c) = 0$, what may we conclude about the behavior of the graph of f at c?
4. Find and analyze the local extrema for the function $f(x) = x^3/3 + x^2 - 15x + 5$.

EXERCISES

Problems for Practice

In Exercises 1–12, use the first derivative to determine the intervals on which the function is increasing and on which the function is decreasing.

1. $f(x) = x^3 + 3x^2 - 45x + 2$
2. $f(x) = 4x^3 - 6x^2 + 8$
3. $f(x) = -x/(x^2 + 5)$
4. $f(x) = x/(x + 1)$
5. $f(x) = (x + 1)/(x - 1)$
6. $f(x) = x(x + 2)^2$
7. $f(x) = (x+1)^2(x+2)^2$
8. $f(x) = \cos(x) + \sin(x)$
9. $f(x) = 2\cos^2(x) + 3x$
10. $f(x) = xe^{-x}$
11. $f(x) = e^x - x$
12. $f(x) = x\ln(x)$

In Exercises 13–20, find all critical points of the function. Then use the First Derivative Test to determine whether each critical point is a local maximum, a local minimum, or neither.

13. $f(x) = x^2 + x$
14. $f(x) = 3x^3 + x^2 + 4$
15. $f(x) = x^5 - 5x^4$
16. $f(x) = (x - 2)/(x + 7)$

17. $f(x) = 1/(x^4 + 6)$
18. $f(x) = (x - 3)^2(x + 6)^2$
19. $f(x) = x^2 e^{-x}$
20. $f(x) = x^2 \ln(x)$

Find and test the critical points of the functions in Exercises 21–28.

21. $f(x) = |x - 7| + x^2$ **22.** $f(x) = 5x - |8 - 3x|$
23. $f(x) = |\sin(x)|$ **24.** $f(x) = x^{1/3} - x^{1/5}$
25. $f(x) = (x - 4)^{1/5} \cdot (3x + 6)^{2/3}$
26. $f(x) = \sqrt{|2x + 9|}$
27. $f(x) = x^{1/3} - |x|$
28. $f(x) = x^{1/5} \cdot (x - 5)$
29. Figure 11 shows the graph of function f'. Determine on what intervals f is increasing and on what intervals f is decreasing.

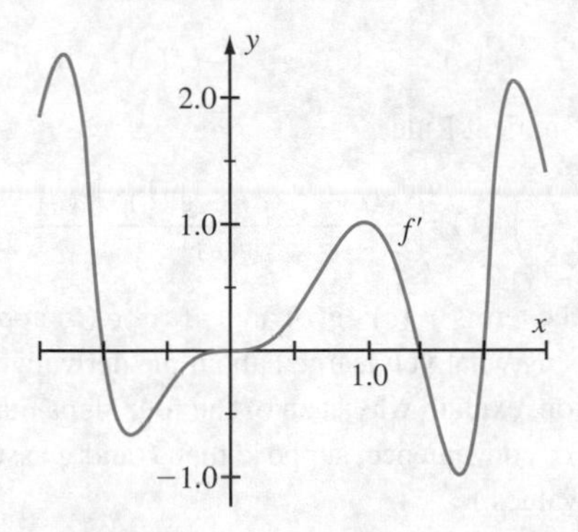

Figure 11

30. Figure 12 shows the graph of function f'. Determine the points where $f(x)$ changes (as x moves to the right) from increasing to decreasing. Determine the points where f changes from decreasing to increasing.

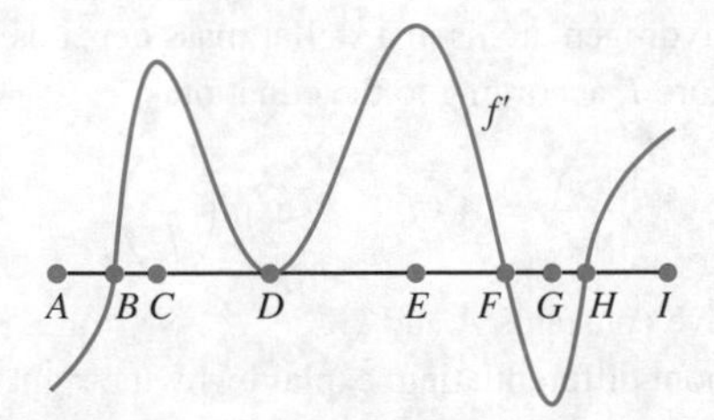

Figure 12

Further Theory and Practice

31. Let f be differentiable on $\mathbb{R}$. Suppose that $f'(2) > 0$. Is $f(2.000001) > f(2)$? Explain your answer.

32. True or false: If $f : \mathbb{R} \to \mathbb{R}$ is a polynomial and if f has two local minima, then f has a local maximum. Explain your answer.

33. If $f : \mathbb{R} \to \mathbb{R}$ is differentiable and $f' > 0$ everywhere, then prove that f is one-to-one. If $f' < 0$ everywhere, then prove that f is one-to-one. What if $f' > 0$, except at finitely many points?

34. If f is increasing on an interval I, does it follow that f^2 is increasing? What if the range of f is $(0, \infty)$?

35. If a differentiable function f is increasing on an interval I, prove that f^3 is increasing.

36. Suppose that a real-valued function f is defined and increasing on an interval (a, b). Let J be the image of f. Without using calculus, prove that $f^{-1} : J \to (a, b)$ exists and is increasing. Now assume that f' exists and is positive on the interval (a, b). Give a calculus proof that f^{-1} is increasing.

37. Without using calculus, prove that the composition of two increasing functions is increasing. Now assume that f and g are differentiable functions with positive derivatives and that $g \circ f$ is defined. Use calculus to show that $g \circ f$ is increasing.

38. Let $f(x) = ax^2 + bx + c$ with $a \neq 0$. By completing the square, determine the interval on which f is increasing and the interval on which f is decreasing. Use the derivative of f to verify your conclusions.

39. The position of an automobile at time t is given by

$$p(t) = 4t^3 - 7t^2 + 4t - 9 \text{ meters.}$$

For what times will the car be traveling forward, and for what times will it be going in reverse?

40. The reaction rate V of a common enzyme reaction is given in terms of substrate level S by

$$V = \frac{V_* S}{K + S} \quad (S \geq 0).$$

Show that V is an increasing function of S. It follows that V has no absolute maximum value. Is V bounded? What is

$$\lim_{S \to \infty} V?$$

Explain how you can graphically determine the Michaelis constant K by using the value $\lim_{S \to \infty} V$.

41. A polynomial f has consecutive roots of multiplicity 1 at a and b. What can you say about the sign of $f'(a)f'(b)$? Explain your answer.

42. Figure 13 shows the graph of function f''. It is known that $f'(A) = 0$, $f'(B) = 0$, $f'(C) = 0$, and $f'(D) = 0$. Determine whether f has any local extreme values at the points A, B, C, and D. Explain your reasoning.

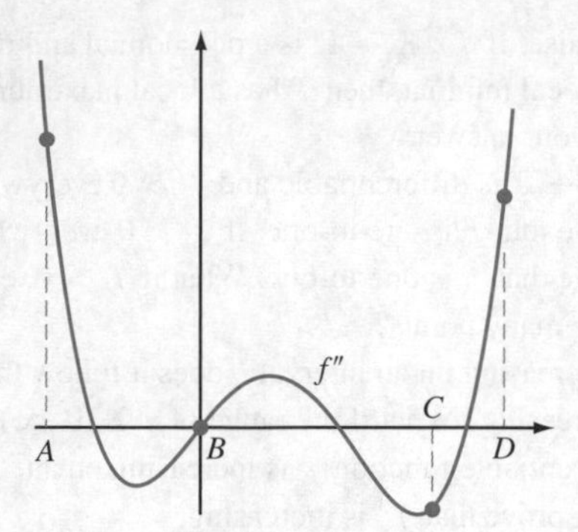

Figure 13

43. In economic theory, the *product of labor* $f(n)$ measures commodity output as a function of the level n of labor input. The *marginal product of labor* (MPL) represents the extra output that results from one additional unit of work. Economists usually assume that the function MPL has a certain property. The three graphs in Figure 14 illustrate this property in different ways. What is the property, and how do the graphs illustrate it?

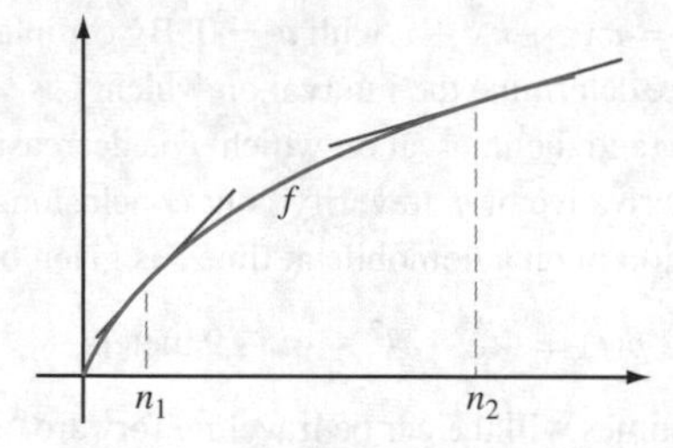

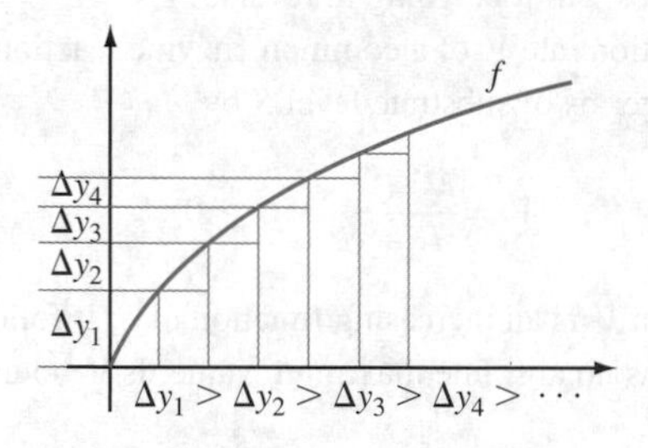

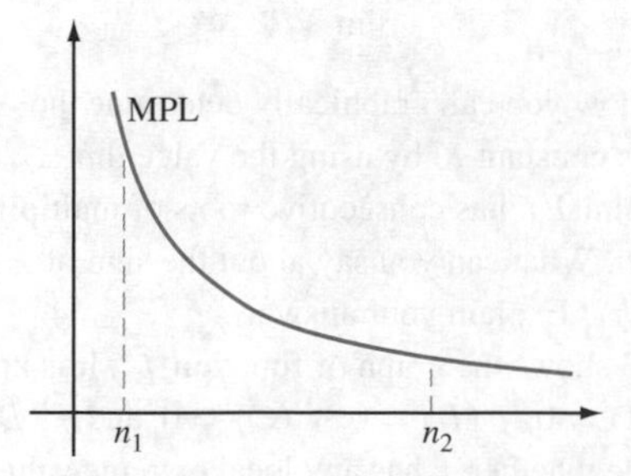

Figure 14

44. Give necessary and sufficient algebraic conditions on the coefficients of the polynomial

$$p(x) = ax^3 + bx^2 + cx + d$$

that will guarantee that p is increasing on the entire real line.

45. Let $f : \mathbb{R} \to (0, \infty)$ be a differentiable function. Explain why the critical points of f will be the same as the critical points of f^k for any positive integer k. Explain why the local maxima for one function will be the same as those for the other. Do the same for local minima.

46. Let $f : \mathbb{R} \to (0, \infty)$ be a differentiable function. Compare the local extrema of f with those for $1/f$. The local maxima for f become *what* for $1/f$? The local minima for f become *what* for $1/f$?

47. In Section 3.3, we learned the Product Rule

$$(f \cdot g)'(c) = f'(c) \cdot g(c) + f(c) \cdot g'(c)$$

and the Quotient Rule

$$\left(\frac{f}{g}\right)'(c) = \frac{g(c) \cdot f'(c) - g'(c) \cdot f(c)}{g(c)^2}.$$

Each of the terms $f'(c) \cdot g(c)$ and $f(c) \cdot g'(c)$ appears twice. Using what you learned about the derivative in this section, explain why each of the four signs makes sense. (For convenience, suppose that f and g assume positive values.)

48. If T is the temperature, in degrees Kelvin, of a white dwarf, then the rate at which T changes as a function of time t is given by $\frac{dT}{dt} = -\alpha T^{7/2}$ where α is a positive constant. Do white dwarfs cool down or heat up? Is the rate at which a white dwarf's temperature changes increasing or decreasing? Explain your answers.

49. According to Saha's equation, the fraction $x \in (0, 1)$ of ionized hydrogen atoms in a stellar mass depends on temperature T, according to the equation

$$\frac{1 - x}{x^2} = A \cdot T^{-3/2} \cdot \exp\left(\frac{k}{T}\right)$$

for positive constants A and k.

a. Without differentiating, explain why the right side of this equation is a decreasing function of T.

b. Without differentiating, explain why the left side of this equation is a decreasing function of x.

c. Without differentiating, explain why x is an *increasing* function of T.

d. Use part c to explain why Saha's equation determines x as a function of T.

e. By differentiating implicitly, verify that x is an increasing function of T.

50. Show that the function $y \mapsto y\exp(y)$ is increasing for $y > 0$. Deduce that for every positive x there is a unique y such that $y\exp(y) = x$. This relationship implicitly determines a function that is often denoted by W and that is called Lambert's W function:

$$W(x)\exp(W(x)) = x$$

for $x > 0$. Use implicit differentiation to show that W is an increasing function. Show that

$$W'(x) = \frac{W(x)}{x(1 + W(x))}.$$

51. In this exercise, you will show that

$$\lim_{n\to\infty} n^{1/n} = 1,$$

following a proof of Alan Beardon. Let $f(x) = px - x^p$. Show that $f(1) \le f(x)$. Deduce that $x^p \le px + 1 - p$. Set $x = \sqrt{n}$ and $p = 1/n$, and use the Pinching Theorem to show that $\lim_{n\to\infty} n^{1/n} = 1$.

52. *Darboux's Theorem* (after Jean Gaston Darboux, 1842–1917) states that if f' exists at every point of an open interval containing $[a, b]$ and if γ is between $f'(a)$ and $f'(b)$, then there is a c in the interval (a, b) such that $f'(c) = \gamma$. (The Intermediate Value Theorem tells us this if f' is continuous, but we do not need that assumption.) In this exercise, you will prove Darboux's Theorem. Suppose, for definiteness, that $f'(a) < \gamma < f'(b)$. Define g on $[a, b]$ by $g(x) = f(x) - \gamma x$.

a. Explain why g has a minimum value that occurs at some point c in (a, b).

b. Show that $f'(c) = \gamma$.

c. Use Darboux's Theorem to prove that if f increases on the interval (l, c) and decreases on the interval (c, r) (or vice versa), then $f'(c) = 0$.

Calculator/Computer Exercises

In Exercises 53–60, use the first derivative to determine the intervals on which the function is increasing and on which the function is decreasing. Use a graphing or solving utility as needed.

53. $f(x) = x^4 - x^2 - 7.1x + 3.2$

54. $f(x) = x^4 - 4x^3 + 5.94x^2 - 3.872x$

55. $f(x) = 2x^5 - 1.2x^4 - 1.7x^3 + 24.5x^2 - 53.7x + 1.2$

56. $f(x) = (x^3 - 5x^2 + 1)\tan(x),\ -\pi/2 < x < \pi/2$

57. $f(x) = (x^2 + x + 1)/(x^4 + 1)$

58. $f(x) = x\ln(x)$

59. $f(x) = e^{-x}\ln(x)$

60. $f(x) = x^2 + e^{-x}$

In Exercises 61–64, plot f and f' over the indicated interval. Identify the intervals on which the sign of f' remains constant. Identify the points where the sign of f changes. Use these observations to identify the intervals on which f is increasing, the intervals on which f is decreasing, the critical points of f, and any local extrema that f may have.

61. $f(x) = \sqrt{|\sin^2(x) - \cos(x)|},\ -\pi/2 < x < \pi/2$

62. $f(x) = \ln(2 + x - \sin^2(x)),\ 0 < x < 2\pi$

63. $f(x) = \exp(x^2 - 2x)\ln(x)\ln(1 - x),\ 0 < x < 1$

64. $f(x) = (x^2 - x)/(1 + \tan^2(x)),\ -\pi/2 < x < \pi/2$

65. Suppose that f is a differentiable function. Let S be the function $S(x) = \text{signum}(f'(x))$. Explain how to use the graph of S to determine the local extrema of f. Explain how to use the graph of S to distinguish between relative minima and relative maxima of f. For each function f, graph S, and use the graph of S to enumerate the local extrema of f and to classify each local extremum.

a. $f(x) = x^5 - 12x^4 + 55x^3 - 120x^2 + 124x - 48$

b. $f(x) = 4x^4 - 24x^3 + 51x^2 - 44x + 12$

c. $f(x) = x^4 - x^3 - 7x^2 + x + 6$

4.3 Applied Maximum-Minimum Problems

In this section, we learn how to apply the techniques of calculus to study a variety of practical problems, all of which involve finding the maximum or minimum of some quantity. We will certainly make use of the ideas introduced in Section 4.2. In this section, however, we concentrate on finding absolute extrema. Our first example illustrates the important role that endpoints play in such a search.

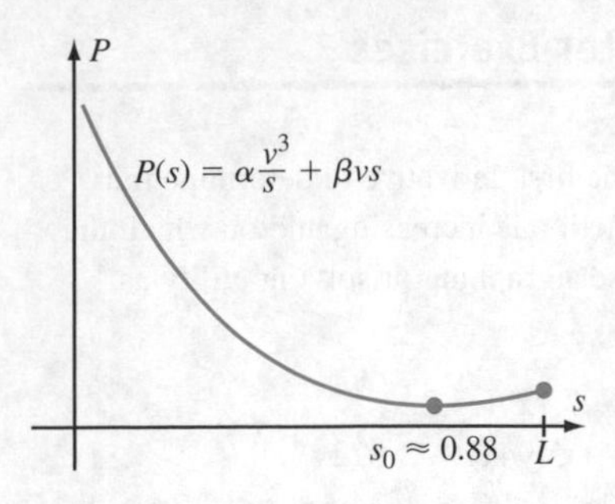

Figure 1

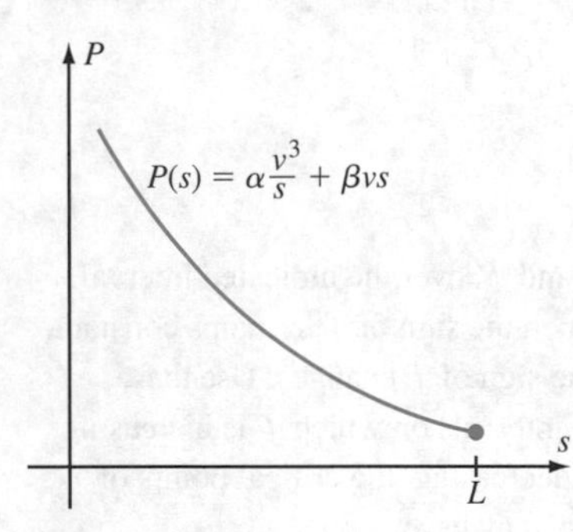

Figure 2

Example 1 The total mechanical power P (in watts) that a human requires for walking a fixed distance at constant speed v (in kilometers per hour) with step length s (in meters) is

$$P(s) = \alpha\frac{v^3}{s} + \beta vs \qquad (0 < s \le L)$$

where L is the maximum step size. In this example, we use $L = 1$ m. The constants α and β vary from person to person; let us use $\alpha = 1.4$ and $\beta = 45$—values that are typical for the chosen units and distance. For what s is $P(s)$ minimized if $v = 5$? If $v = 6$?

Solution We calculate that

$$P'(s) = \frac{d}{ds}\left(\alpha\frac{v^3}{s} + \beta vs\right) = -\alpha\frac{v^3}{s^2} + \beta v = \frac{\beta v}{s^2}\left(s^2 - \frac{\alpha}{\beta}v^2\right) = \frac{\beta v}{s^2}(s+s_0)(s-s_0)$$

where $s_0 = v\sqrt{\alpha/\beta}$. Thus, s_0 is the unique positive solution of the equation $P'(s) = 0$. For $v = 5$, we have $s_0 = v\sqrt{\alpha/\beta} = 5\sqrt{1.4/45} \approx 0.88$, which is less than L and, therefore, in the domain of P. Since $P'(s) < 0$ for $0 < s < s_0$ and $P'(s) > 0$ for $s > s_0$, we conclude that P decreases on the interval $(0, s_0)$ and increases on the interval $(s_0, L]$. Therefore, P has absolute minimum at $s = s_0$. The graph of P confirms our analysis (see Figure 1).

When $v = 6$, the value of $s_0 = v\sqrt{\alpha/\beta}$ is about 1.06, which is greater than L and, therefore, outside the domain of P. Our previous observation of the sign of P' tells us that P is decreasing on its domain $(0, L]$ and, therefore, has a minimum when $s = L$ (see Figure 2). ■

Example 1 shows that when we determine the absolute extrema of a function, we must consider its domain. If a function is defined at an endpoint of its domain, then an extreme value of the function may occur there.

Closed Intervals

Let us consider the absolute extrema of a continuous function f on a *closed* interval $[a, b]$. The Extreme Value Theorem (see Section 2.3) guarantees that f will have both an absolute maximum and an absolute minimum in $[a, b]$. Also, the techniques of Section 4.2 guarantee that if these extrema are inside the *open* interval (a, b), then they will be found among the critical points. It is possible, however, that the absolute maximum or absolute minimum of f could be at a or at b. We therefore rely on the following procedure.

Basic Rule for Finding Extrema of a Continuous Function on a Closed Bounded Interval

To find the extrema of a continuous function f on a closed interval $[a, b]$, we should test the points in (a, b) where f is not differentiable, the points in (a, b) where f' exists and equals zero, and the endpoints a and b. In brief, we should test the *critical points* and the *endpoints*.

Example 2 A man builds a rectangular garden along the side of his house. He will put fencing on three sides (see Figure 3). If he has 200 ft of fencing available, then what dimensions will yield the garden of greatest area?

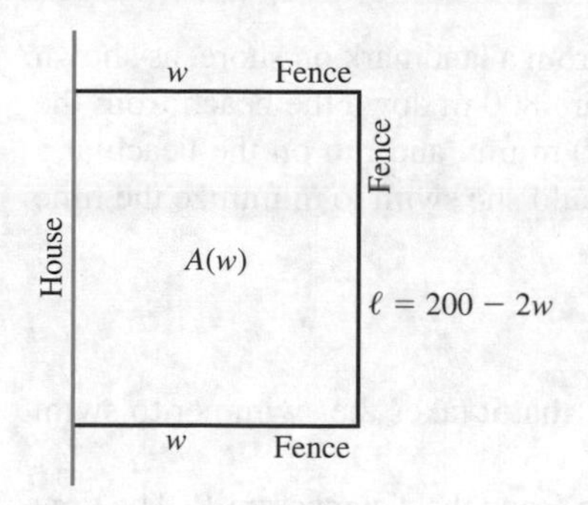

Figure 3

Solution The function that we must maximize is A, the area inside the fence. If ℓ denotes length and w denotes width, as in Figure 3, then $A = \ell \cdot w$. The trouble is that A seems to be a function of two variables, and we have no techniques for maximizing such a function. Therefore, we must use the additional piece of information that $\ell + 2w = 200$. Since $\ell = 200 - 2w$, we may substitute this expression into the formula for A. Thus, our job is to maximize the function

$$A(w) = (200 - 2w) \cdot w = 200w - 2w^2, \quad 0 \leq w \leq 100.$$

Notice that we are careful to write the range of values of w: This step is necessary to apply the basic rule. In this case, we have determined the possible values of w by observing that the relations $\ell + 2w = 200$ and $\ell \geq 0$ imply the inequality $w \leq 100$.

Because $A(w)$ is a polynomial, it is differentiable at all values of w. Also, $\frac{dA}{dw} = 200 - 4w$. Clearly $\frac{dA}{dw}$ vanishes at $w = 50$. Thus, $w = 50$ is a critical point.

By the basic rule, the maximum will be either the critical point $w = 50$ or at one of the endpoints $w = 0$ or $w = 100$. Finally, we calculate that $A(0) = 0$, $A(50) = 5000$, and $A(100) = 0$. Clearly the maximum occurs when $w = 50$. For this value of w, $\ell = 200 - 2 \cdot 50 = 100$. ■

inSIGHT

Even Example 2, which is straightforward, contains all of the following major ingredients of the examples that we will see in this section.

1. Determine the function f to be maximized or minimized.
2. Identify the relevant variable(s).
3. If f depends on more than one variable, find relationships among the variables that will allow substitution into the expression for f. The methods developed in this section require you to obtain a function of one variable to be maximized or minimized.
4. Determine the set of allowable values for the variable (this is usually an interval).
5. Find all places where the function is not differentiable, all places where the derivative is zero, and the endpoints. These are the points to test.
6. Use the First Derivative Test, or simply substitute in values as in Example 2, to determine which of the points found in step 5 solves the maximum-minimum problem being studied.

We should follow the six steps listed in the Insight in every extremization problem that concerns a continuous function on a closed bounded interval. The next example is a calculus classic. An interesting variant of this problem will be given in Example 5. Still other versions may be found in the exercises. All of the examples that follow have been inspired by Fermat's derivation of the Law of Refraction in optics, an application that is discussed in this chapter's Genesis & Development.

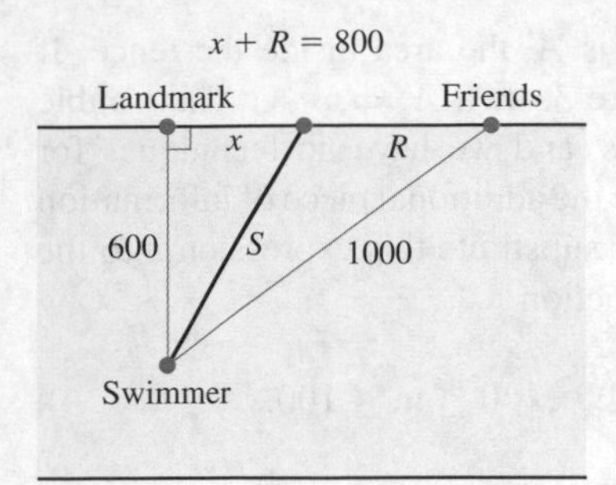

Figure 4

Example 3 A swimmer is 600 m straight out from a landmark on shore, as shown in Figure 4. She wants to meet some friends who are 800 m down the beach from the landmark. The swimmer can swim at a rate of 100 m/min and run on the beach at a rate of 200 m/min. Toward what point on shore should she swim to minimize the time it takes to join her friends?

Solution

1. The function to be minimized is the time T that it takes the swimmer to swim and run to her friends.
2. The relevant variables are the distance swum S and the distance run R. The time spent swimming is $S/100$, and the time spent running is $R/200$. Therefore,
$$T = \frac{S}{100} + \frac{R}{200}. \tag{4.1}$$
3. Let x be the distance between the point where the swimmer hits the beach and the landmark. By the Pythagorean Theorem,
$$x = (S^2 - 600^2)^{1/2}.$$
Also, $x + R = 800$, or
$$R = 800 - x = 800 - (S^2 - 600^2)^{1/2}.$$
Substituting this expression for R into equation (4.1) gives
$$T(S) = \frac{S}{100} + \frac{800 - (S^2 - 600^2)^{1/2}}{200}. \tag{4.2}$$
4. From Figure 4, we know that the swimmer cannot swim any less than her 600 m distance from the shore, nor can she swim more than the hypotenuse $\sqrt{600^2 + 800^2} = 1000$ of the larger right triangle in Figure 4. Therefore, the range of values for S is $600 \le S \le 1000$.
5. The function T is differentiable at all points of $(600, 1000)$, and we compute
$$\begin{aligned}\frac{dT}{dS} &= \frac{1}{100} - \frac{(1/2)\cdot(S^2 - 600^2)^{-1/2}\cdot 2S}{200}\\ &= \frac{2}{200} - \frac{S}{200\cdot(S^2 - 600^2)^{1/2}}\\ &= \frac{1}{200\cdot(S^2 - 600^2)^{1/2}}\cdot(2\cdot(S^2 - 600^2)^{1/2} - S).\end{aligned}$$
This expression can vanish only if
$$2\cdot(S^2 - 600^2)^{1/2} = S,$$
or
$$4\cdot(S^2 - 600^2) = S^2,$$

or

$$3S^2 = 4 \cdot 600^2.$$

Thus, the only root of $\frac{dT}{dS}$ in the interval (600, 1000) is $S = 400\sqrt{3}$. This critical point and the endpoints $S = 600$ and $S = 1000$ are the candidates for the minimum we seek.

6. Using equation (4.2), we calculate that $T(600) = 10$, $T(1000) = 10$, and

$$T\left(400\sqrt{3}\right) = 4 + 3\sqrt{3} \approx 9.196 < 10.$$

Thus, $S = 400\sqrt{3}$ gives the minimum. The value of x that we seek is therefore $x = \sqrt{(400\sqrt{3})^2 - 600^2} = 200\sqrt{3}$. ■

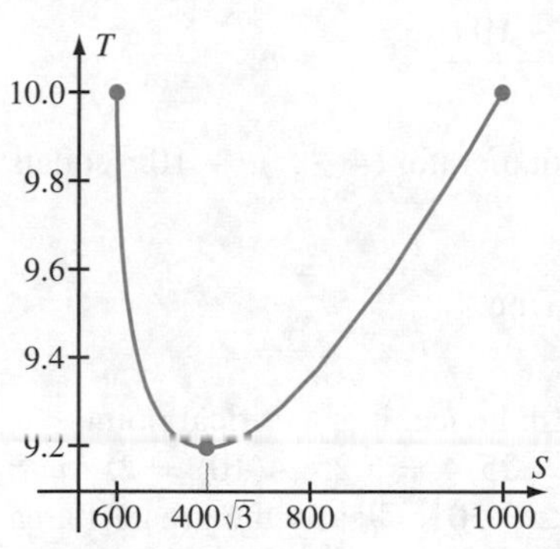

Figure 5
$T(S) = \frac{S}{100} + \frac{800 - (S^2 - 600^2)^{1/2}}{200}$

inSIGHT

The graph of T from Example 3 appears in Figure 5. By repeatedly zooming in on the low point of the graph, we can determine the coordinates (692.8, 9.196). Such a graphical procedure provides a relatively efficient way of *approximating* extreme values. More efficient algorithms for approximating locations of extreme values are discussed in textbooks on numerical analysis.

Examples with the Solution at an Endpoint

Example 4 A 10 in. piece of wire can be bent into a circle or a square, or it can be cut into two pieces. In this case, a circle will be formed from the first piece and a square from the second. What is the maximal area that can result?

Solution

1. The function to be maximized is the sum of the two areas (one of which will be zero if the wire is not cut). Call this function A.
2. The relevant variables are the lengths c (for the part of the wire that will form the circle) and s (for the part of the wire that will form the square). If the circle is made from the piece of length c, then the circumference of the circle is c, and the radius of the circle is $c/(2\pi)$. Therefore,

$$\text{area of the circle} = \pi \cdot \text{radius}^2 = \pi \cdot \left(\frac{c}{2\pi}\right)^2 = \frac{c^2}{4\pi}.$$

The square is manufactured from the piece of wire of length s. The perimeter of the square is then s, and the side length is $s/4$. The area of the square is $(s/4)^2$, or $s^2/16$. In conclusion, the function that we need to maximize is

$$A = \frac{c^2}{4\pi} + \frac{s^2}{16}.$$

3. The variables c and s are related by the equation $c + s = 10$, or $s = c - 10$. Substituting this into the formula for A gives

$$A = \frac{c^2}{4\pi} + \frac{(c-10)^2}{16}. \tag{4.3}$$

4. Clearly $0 \le c \le 10$, since the wire is 10 in. long.
5. The function A is a polynomial; hence, it is differentiable everywhere. The only candidates for the maximum we seek are the endpoints 0 and 10 and the zeros of A'. We calculate

$$A'(c) = \frac{2c}{4\pi} + \frac{2(c-10)}{16}$$
$$= \frac{(4+\pi)c - 10\pi}{8\pi}.$$

The only root of $A'(c) = 0$ occurs when the numerator $(4+\pi)c - 10\pi$ equals zero, or

$$c = \frac{10\pi}{4+\pi} \approx 4.399.$$

This number is certainly in the interval [0, 10]; hence, it is a critical point.
6. Referring to equation (4.3), we see that $A(0) = 25/4 = 6.25$, $A(10) = 25/\pi \approx 7.958$, and $A(10\pi/(4+\pi)) = 25/(4+\pi) \simeq 3.501$. Clearly the greatest area is achieved when $c = 10$, or when we use the entire wire to make a circle but no square. The plot of equation (4.3), in Figure 6, confirms our analysis. ■

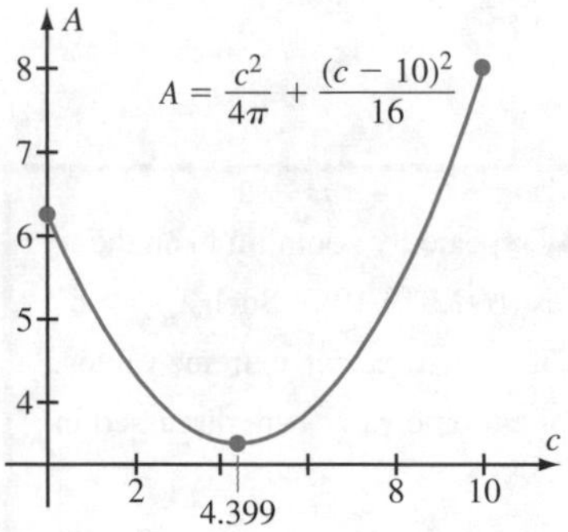

Figure 6

Our next example appears to be quite similar to Example 3. However, there is a surprising and instructive difference.

Example 5 A jogger is at point P, 4 mi due south of point O. His destination, point R, is 6 mi due east of O. He can run along roads from P to O and from O to R at the rate of 5 mi/h. He can also hike across the field from P to a point Q at the rate of 3 mi/h and then follow the road to point R at the rate of 5 mi/h (see Figure 7). How quickly can the jogger get from P to R?

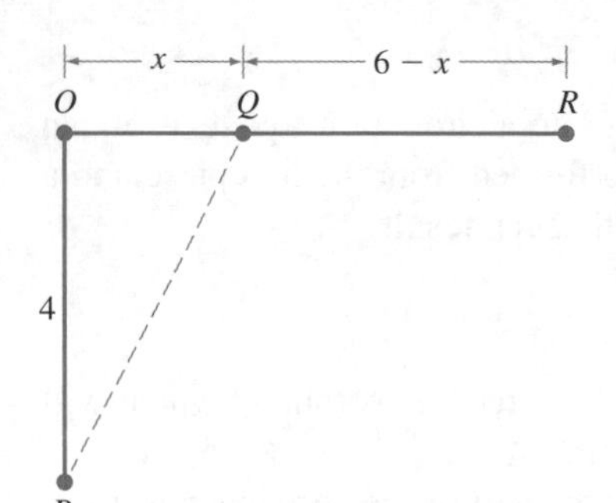

Figure 7

Solution

1. The function to be minimized is the sum of the times required to run from P to Q and from Q to R. Let us call this total time T.
2. Let x be the distance of Q from O.
3. When $x = 0$, the runner jogs 4 mi north and then 6 mi east, all at 5 mi/h. His total time is 2 h along this route. Thus, $T(0) = 2$. Now, suppose that $x > 0$. The distance of Q to R is $6 - x$ and the time required to jog this distance is $(6-x)/5$. By the Pythagorean Theorem, the distance of P to Q is $\sqrt{4^2 + x^2}$, and the time required to hike this distance is $(\sqrt{4^2 + x^2})/3$. Thus, the time T required for the journey is

$$T(x) = \frac{\sqrt{4^2 + x^2}}{3} + \frac{6-x}{5}$$

where $0 < x \le 6$. (All values of $T(x)$ are stated in hours.)

4. Clearly $0 \le x \le 6$.
5. We calculate

$$T'(x) = \frac{x}{3\sqrt{16+x^2}} - \frac{1}{5} = \frac{5x - 3\sqrt{16+x^2}}{15\sqrt{16+x^2}}$$

On multiplying both numerator and denominator by $5x + 3\sqrt{16+x^2}$, we obtain

$$T'(x) = \frac{25x^2 - 9(16+x^2)}{15\sqrt{16+x^2}\left(5x+3\sqrt{16+x^2}\right)} = \frac{16(x^2-9)}{15\sqrt{16+x^2}\left(5x+3\sqrt{16+x^2}\right)}.$$

The only point x in the domain of T for which $T'(x) = 0$ is $x = 3$. The candidates for the location of an extremum are, therefore, 3 and the endpoints 0 and 6.

6. Since $T'(x) < 0$ for x in the interval (0, 3) and since $T(x) > 0$ in the interval (3, 6), we deduce that T decreases on (0, 3) and increases on (3, 6). If T has a minimum, then it must occur at $x = 3$ or $x = 0$, because T is discontinuous at $x = 0$. In fact, $T(0) = 2$ and

$$T(3) = \frac{\sqrt{4^2+3^2}}{3} + \frac{6-3}{5} = \frac{34}{15} \approx 2.2667.$$

We conclude that $T(0) = 2$ is the minimum time in which to get from P to R. ■

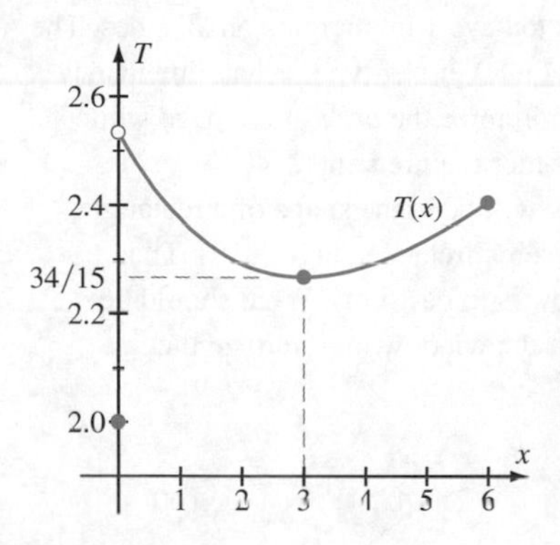

Figure 8

inSIGHT

The graph of T from Example 5 appears in Figure 8. Notice that T is discontinuous. The Extreme Value Theorem, therefore, does not apply to T, and we cannot be certain, prior to our analysis, that T has an extreme value on [0, 6]. In fact, T does not assume a maximum value, but, as we have found, T does assume a minimum value on [0, 6].

Examples 4 and 5 demonstrate that checking the endpoints is not simply an academic exercise. As we have seen, the solution to a perfectly reasonable problem can occur at an endpoint.

quickquiz

1. Where do we seek maxima and minima of a continuous function on a closed interval?
2. Must maxima and minima occur at critical points of the function?
3. Why is the derivative relevant when finding the extrema of a function?
4. When is the derivative irrelevant when finding the extrema of a function?

EXERCISES

Problems for Practice

Solve the applied maximum-minimum problems in Exercises 1–20. Some may not have a solution; others may have their solution at the endpoint of the interval of definition.

1. A rectangle is to have perimeter 100 ft. What dimensions will give it the greatest area?
2. What positive number plus its reciprocal gives the least sum?
3. What first quadrant point on the curve $xy^2 = 1$ is closest to the origin?
4. Find the longest vertical chord between the curves $y = x^3/3 - 3x^2/2 + 2x - 3/4$ and $y = 3x^3/4 - 15x^2 + 5x, 0 \le x \le 3$.
5. A box (with no lid) is to be constructed from a sheet of cardboard by cutting squares from the corners and folding up the sides (see Figure 9). If the original sheet of cardboard measures 20 in. by 20 in., then what should be the size of the squares removed to maximize the volume of the resulting box?

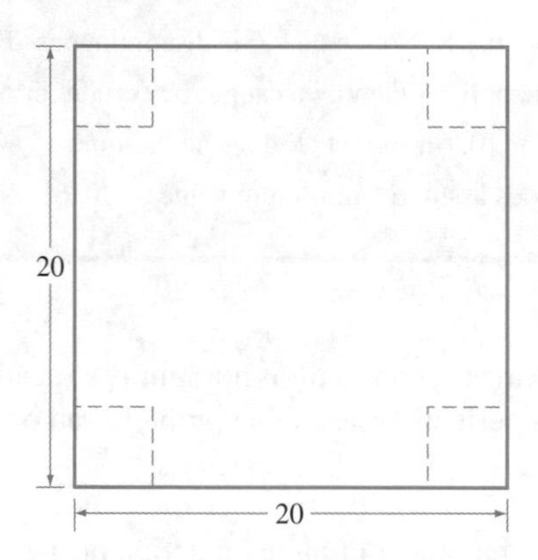

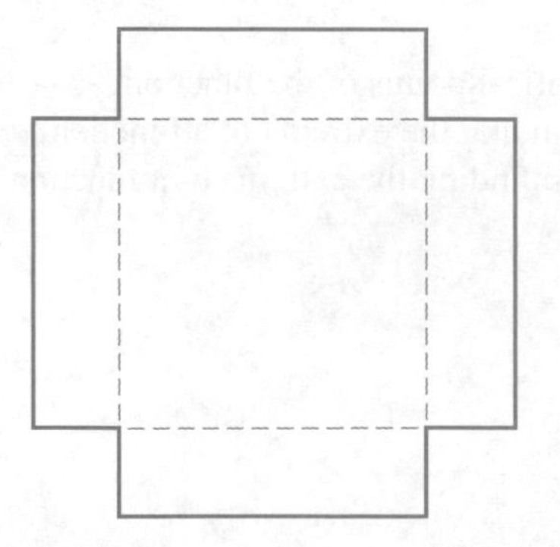

Figure 9

6. Of all right triangles with hypotenuse 100 in., which has the greatest area?
7. A line with negative slope passes through the point $(1, 2)$ and has x-intercept a and y-intercept b. For what slope is the product ab minimized?
8. A cylindrical can is constructed with tin sides and an aluminum top and bottom. If aluminum costs $0.002/in.^{2}, if tin costs $0.001/in.^{2}, and if the can is to hold 20 in.^{3}, what dimensions will minimize the cost of the can?
9. Which number in the set $\{x : x \ge 0\}$ exceeds its cube by the greatest amount?
10. A right triangle is to contain an area of 50 cm^2. What dimensions will minimize the sum of the lengths of its two legs?
11. A printed page is to have 1 in. margins on all sides. The page should contain 80 in.^{2} of type. What dimensions of the page will minimize the *area* of the page while still meeting the other requirements?
12. A church window will be in the shape of a rectangle surmounted by a semicircle (see Figure 10). The area of the window is to be 15 m^2. What should be the dimensions of the window to minimize the perimeter?

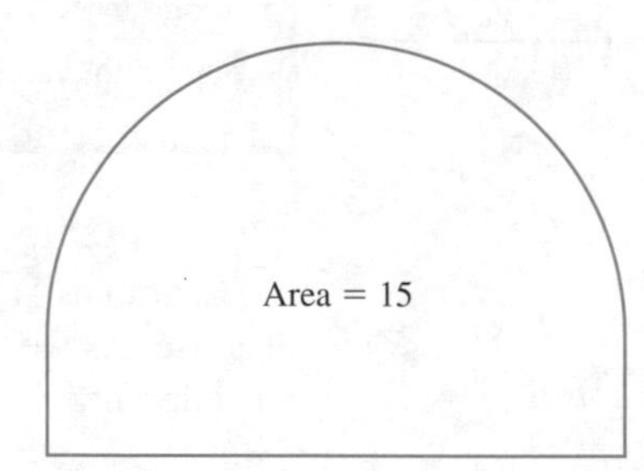

Figure 10

13. A rectangle has its base on the x-axis and its upper corners are on the graph of $y = 4 - x^2$. What dimensions for the rectangle will give it maximal area?
14. A box with square base and rectangular sides is being designed. The material for the sides costs $0.10/in.^{2}. The material for the top and bottom costs $0.04/in.^{2}. If the box is to hold 100 in.^{3}, what dimensions will minimize the cost of materials for the box?

15. An open-topped rectangular planter is to hold 3m^3. Its concrete base is square. Each brick side is rectangular. If the unit cost of brick is 12 times that of concrete, what dimensions result in the cheapest material cost?
16. What is the right triangle of least area that can be inscribed in a circle of radius 3 m?
17. A garden is to be in a pentagonal shape consisting of a rectangle adjoining an equilateral triangle, as in Figure 11. If the garden is to enclose $100\ \text{m}^2$, what dimensions will minimize the amount of fence needed to enclose the garden?

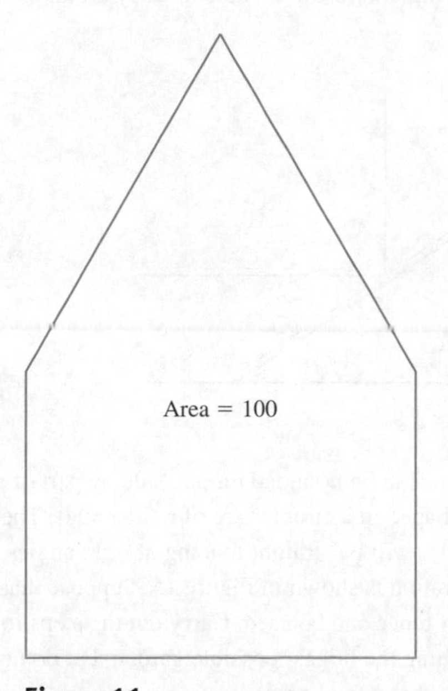

Figure 11

18. What is the rectangle (with sides parallel to the axes) of greatest area that can be inscribed in the ellipse $x^2 + 4y^2 = 16$?
19. A cosmetic container is to be in the shape of a right circular cylinder surmounted by a hemisphere (see Figure 12). If the surface area is 5π in.^{2}, what is the greatest volume that this container could possibly hold?
20. Assume that if the price of a certain book is p dollars, then it will sell x copies where $x = 7000 \cdot (1 - p/35)$. Suppose the dollar cost of producing those x copies is $15000 + 2.5x$. Finally, assume that the company will not sell this book for more than \$35. Determine the price for the book that will maximize profit.

Figure 12
Cylinder surmounted by hemisphere

Further Theory and Practice

21. A right circular cone of height h and base radius r has volume $\pi r^2 h/3$ and lateral surface area $\pi r\sqrt{r^2 + h^2}$. What is the greatest volume that such a cone can have if its surface area, including the base, is 2π?
22. A rectangle is to have one corner on the positive x-axis, one corner on the positive y-axis, one corner at the origin, and one corner on the line $y = -5x + 4$. Which such rectangle has greatest area?
23. In a model for optimizing the angle of release θ of a basketball shot, suppose that a and b are positive constants. Let θ_0 be the value of θ in the range $\{\theta \in (0, \pi/2) : \tan(\theta) > b/a\}$ for which
$$f(\theta) = (a\sin(\theta)\cos(\theta) - b\cos^2(\theta))^{-1}$$
is minimized. What is $\tan(\theta_0)$?
24. A piece of paper is circular in shape, with radius 6 in. A sector is cut from the circle and the two straight edges taped together to form a cone (refer to Figure 13, next page). What angle for the sector will maximize the volume inside the cone?
25. A wire of length L can be shaped into a circle or a square, or it can be cut into two pieces, one of which is formed into a circle and the other a square. How is the minimum total area obtained?

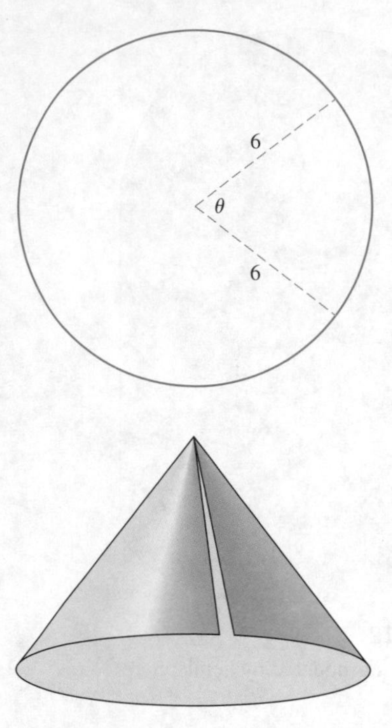

Figure 13

26. A wire of length L can be shaped into a (closed) semicircle or a square. The wire can also be divided into two pieces, with one piece forming a (closed) semicircle, and the other, a square. Find the largest and smallest areas that are possible.

27. A wire of length L can be shaped into a (closed) semicircle or a circle. The wire can also be divided into two pieces, with one piece forming a (closed) semicircle, and the other, a circle. Find the largest and smallest areas that are possible.

28. A battery that produces V volts with internal resistance r ohms is connected, in series, to an external device with resistance R ohms. The power output of the battery is then $V^2R/(r+R)^2$. What value of R results in maximum power?

29. The drag on an airplane at a given altitude is given by $(a + b/v^4)v^2$ where a and b are positive constants and v is the velocity of the plane. At what speed is drag minimized?

30. In a certain process, the optimal time t for removing an object from a heat source x units away is obtained by maximizing

$$H = \frac{A}{\sqrt{t}} \exp\left(-\frac{cx^2}{t}\right)$$

where A and c are positive constants. What value of t maximizes H?

31. The potential energy of a diatomic molecule is $U = A(b/r^{12} - 1/r^6)$ where A and b are positive constants and r is the interatomic distance. What value of r minimizes U?

32. Carpenters will need to carry rods of various lengths around a hall corner, as shown in Figure 14. One hall is 6 ft wide, and the other is 8 ft wide. Ignoring the thickness of the rod, and assuming that the rod is carried parallel to the floor, determine the longest rod that the carpenters will be able to carry around the corner.

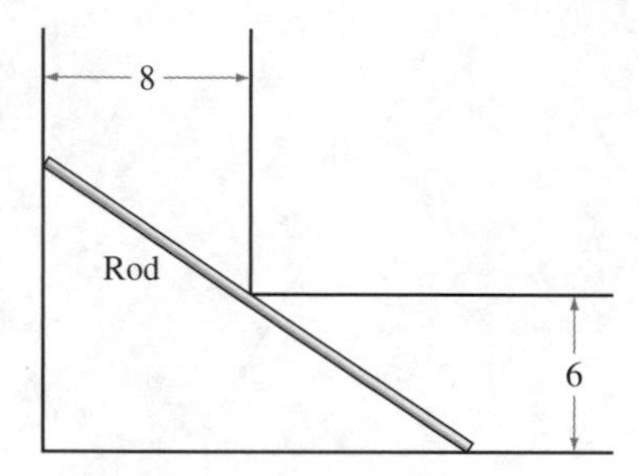

Figure 14

33. A garden is to be bounded on one side by part of a river that is shaped in a circular arc of radius 80 ft. The other three sides will be straight fencing at right angles. The configuration is shown in Figure 15. Suppose that 100 ft of fence can be used. Carry out the steps for determining the largest possible garden, but do not solve for the critical points.

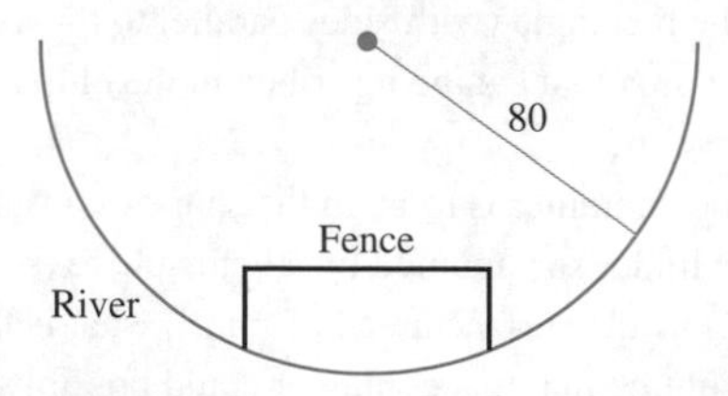

Figure 15

34. When a person coughs, air pressure in the lungs increases and the radius r of the windpipe decreases from the normal value ρ. Units can be chosen so that the flow F through the windpipe resulting from a cough is given by

$$F(r) = (\rho - r)r^4$$

for $0 < r \le \rho$. What value of r results in the most productive cough?

35. In Figure 16, B is a vascular branch point on a main artery from the heart H. The branch angle $\theta \in (0, \pi/2)$ determines the resistance Ω to blood flow to the organ O along the path HBO, according to

$$\Omega(\theta) = \frac{a - b\cot(\theta)}{R^4} + \frac{b}{r^4 \sin(\theta)}.$$

For what value of $\cos(\theta)$ is resistance minimized?

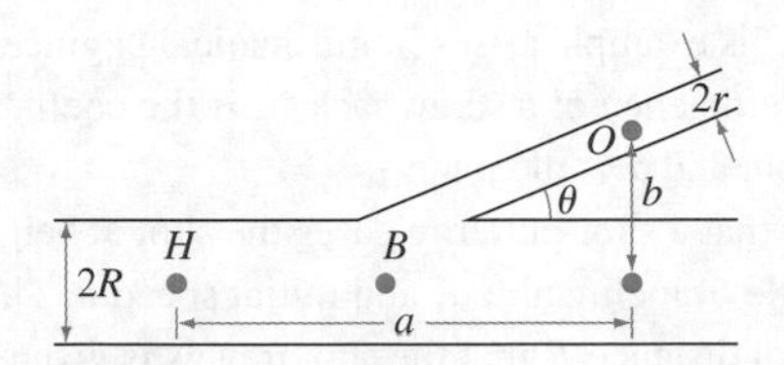

Figure 16

36. This problem was first proposed in the early 1900s by E. V. Huntington, a professor at Harvard University: A circular pool has radius r and center at point O. Suppose that Earl is at point P at the edge of the pool. He wishes to get to the diametrically opposite point R in the least amount of time. He can swim from P to Q and then run to R. Suppose that Earl's maximum swimming speed is u and his maximum running speed is v.

a. Find an expression $f(\theta)$ for the time required for Earl to swim from P to Q in a straight line and then to run from Q to R along the edge of the pool. (θ is the angle $\angle POR$.)

b. What is the maximum number of critical points that f can have in the interval $(0, \pi)$?

c. Compute $f'(\pi)$. Based on the sign of $f'(\pi)$, can f have an absolute minimum at a critical point in $(0, \pi)$? Explain your reasoning. Sketch possible graphs of f.

d. How quickly can Earl get from P to R?

37. V. L. Klee's Problem: If 100 m of straight fencing is already in place, how large a rectangular region can be enclosed if an additional 200 m of fencing is purchased?

Exercises 38–40 concern the most economical dimensions of a cylindrical tin can that holds a predetermined volume V. Exercise 38 is a calculus classic. Exercises 39 and 40 are based on an article by Professor P. L. Roe of the University of Michigan.

38. What are the dimensions if the amount of metal used in the can is to be minimized? Express your answer as a height h to radius r ratio.

39. When the circular top and bottom is cut from a large sheet of metal, there will be wastage. Thus, the amount of metal necessary to make a can is actually larger than the amount of metal used in the can itself. Suppose that the top and bottom circles are each cut from squares with side lengths that are the diameter of the circle. What is h/r when the amount of metal (including wastage) is minimized?

40. Suppose the metal necessary to make the cylindrical can is $4\sqrt{3}r^2 + 2\pi r h$, including wastage. In forming the can, a side join of length h and two circular joins of length $2\pi r$ each are necessary. If a length k join costs the same as a unit area of the metal, then the total cost C of production is

$$C = 4\sqrt{3}r^2 + 2\pi r h + (4\pi r + h)/k.$$

Write C as a function of r and find an equation for the critical point of C. Show that there are constants a, b, c, and d such that

$$\frac{h}{r} = \frac{ark + b}{crk + d}$$

when r is the critical point. In the case of very inexpensive side joins ($k \to \infty$), what is the approximate ratio h/r? In the case of very expensive side joins ($k \to 0$), what is the approximate ratio h/r?

41. Suppose that a uniform rod of length ℓ and mass m can rotate freely about one end. If a point mass $6m$ is attached to the rod a distance x from the pivot, then the period of small oscillations is equal to

$$2\pi\sqrt{\frac{2}{3g}} \cdot \sqrt{\frac{\ell^2 + 18x^2}{12x + \ell}}.$$

For what value of x is the period least?

42. A volume v_0 of gas is held at pressure p_0 in a reservoir. The gas is discharged through a nozzle of opening area A into a region at lower pressure p. The rate of discharge (in units of weight/time) is given by

$$A\sqrt{\frac{2g\gamma}{\gamma - 1}\frac{p_0}{v_0}\left(\left(\frac{p}{p_0}\right)^{2/\gamma} - \left(\frac{p}{p_0}\right)^{(\gamma+1)/\gamma}\right)}$$

where γ is the adiabatic constant of the gas. What is p/p_0 when the rate of discharge is greatest?

43. When a boat travels through open water, it creates a trailing wedge of interior angle $2\beta_0$ in which there is constructive interference between waves. This is known as the Kelvin wedge. The value of $\tan(\beta_0)$ is the maximum value of

$$\frac{\tan(\alpha)}{2+\tan^2(\alpha)}.$$

Show that $\sin(\beta_0) = 1/3$. (The Kelvin wedge, therefore, encloses an angle of about 39°, independent of the speed of the boat.)

44. A road is to be built from $P = (0, b)$ to $Q = (a, 0)$. The cost of new road construction is c_1 per unit distance. There is an existing straight road from the origin $(0, 0)$ to Q. To take advantage of this existing route, highway engineers are considering building a road from P to some point $R = (c, 0)$ on the existing road. In that case, the existing road between R and Q must be upgraded to handle the increased traffic flow. The cost of this upgrade is c_2 per unit distance where $c_2 < c_1$. Determine the minimum cost route.

Calculator/Computer Exercises

45. Centrifugal acceleration at the θth parallel is proportional to

$$\frac{\sin(2\theta)}{\sqrt{0.997+0.00328\cos(2\theta)}} \quad (0 \le \theta \le \pi/2).$$

At what parallel is this expression maximized?

46. Suppose that for a beam of length 11 m, the deflection from horizontal of a point a distance x from one end is proportional to

$$2x^4 - 33x^3 + 1331x.$$

Determine the point at which the deflection is greatest.

47. Two heat sources separated by a distance 10 are located on the x-axis. The heat received by any point on the x-axis from each of these sources is inversely proportional to the square of its distance from the source. Suppose that the heat received 1 unit away from one source is twice that received 1 unit away from the other source. What is the relative location of the coolest point between the two sources?

48. Let α_0 denote the solution of $\tan(\alpha_0) = 1/3$ in $(0, \pi/2)$. Plot

$$E(\alpha) = \frac{\tan(\alpha)(1-3\tan(\alpha))}{3+\tan(\alpha)}$$

for $0 < \alpha \le \alpha_0$. Use the plot of E to find the maximum value of $E(\alpha)$ to three decimal places. Calculate $E'(\alpha)$. Show that $E(\alpha)$ is maximized when α satisfies

$$\tan(\alpha)^2 + 6\tan(\alpha) - 1 = 0.$$

Solve for $\tan(\alpha)$ and find the *exact* maximum value of $E(\alpha)$. (This example arises in mechanical engineering. E is the efficiency of a screw jack, μ is the coefficient of friction, and α is the pitch.)

49. Suppose that a shot-putter releases the shot at height h, with angle of inclination α, and initial speed v. The horizontal distance R that the shot travels is given by

$$R = \frac{v^2\sin(2\alpha) + v\sqrt{v^2\sin^2(2\alpha) + 8gh\cos^2(\alpha)}}{2g}.$$

Use a computer algebra system to find the value of α that maximizes R.

50. An underground pipeline is to be built between two points $P = (1, 4)$ and $Q = (2, 0)$. The subsurface rock formation under P is separated from that rock that lies under Q by a curve $\mathcal{C}$ of the form $y = x^2$. The cost per unit distance of laying pipeline from P to the graph of $\mathcal{C}$ is 5000. The unit cost from the graph of $\mathcal{C}$ to Q is 3000. Assuming that the pipeline will consist of two straight line segments, analyze the minimum cost route.

51. A tetramer is a protein with four subunits. In the study of tetramer binding, the equation

$$Y = \frac{x + 3x^2 + 3x^3 + 10x^4}{1 + 4x + x^2 + 4x^3 + 10x^4}$$

expresses a typical relationship between saturation Y and ligand concentration $x (x \ge 0)$. Ordinarily, the variable Y/x is plotted as a function of x. Explain why Y/x has an absolute maximum value. Find this value to three decimal places.

4.4 Concavity

We have learned that we can use the sign of the first derivative f' of a differentiable function f to determine where the graph of f rises and where it falls. However, much more can be said about the shape of a graph. In this section, we study the properties of a graph that are revealed by the second derivative of f.

Definition Let the domain of a differentiable function f contain an open interval I. If f' (the slope of the tangent line to the graph) increases as x moves from left to right on I, then the graph of f is said to be *concave up* on I (see Figure 1). If f' (the slope of the tangent line to the graph) decreases as x moves from left to right on I, then the graph of f is said to be *concave down* on I (see Figure 2).

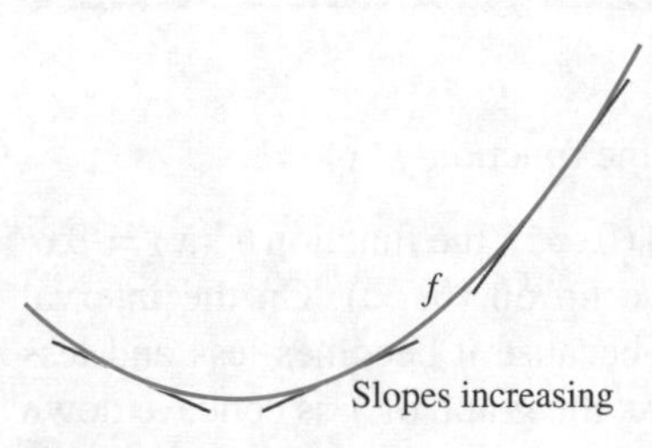

Figure 1
The graph of f is concave up.

There are many ways to think about the concept of concavity. Another way to define "concave up" for a point $P = (x, f(x))$ on the graph of f is to require that the tangent line to the graph at the point P lie below the graph just to the left and to the right of P. (Refer to Figure 1.) Similarly, "concave down" means that if P is a point of the graph of f, then the tangent line to the graph at P should lie above the graph just to the left and to the right of P (as shown in Figure 2). Alternatively, we could define the graph of f to be concave up if the graph of f lies below every chord joining two points of the graph of f. With this definition, we would define the graph of f to be concave down if its graph lies above every chord joining two of its points (see Figure 3). These other definitions represent alternative points of view and will not be discussed further.

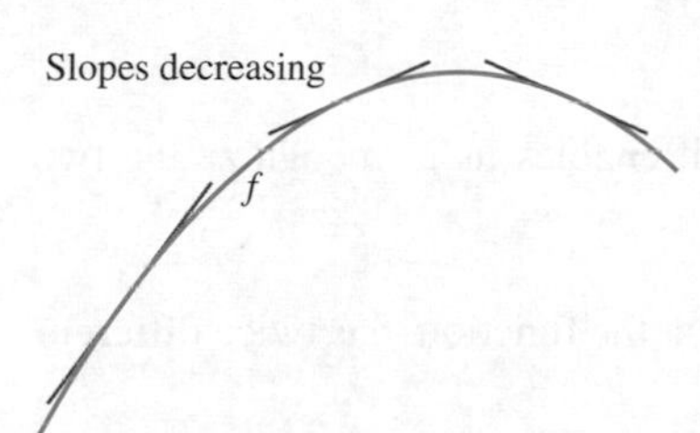

Figure 2
The graph of f is concave down.

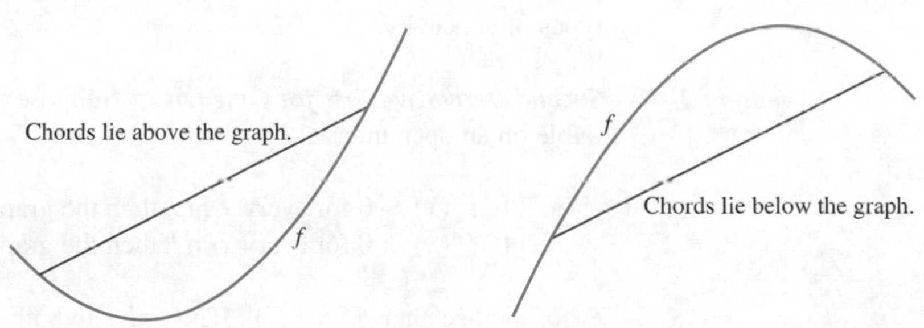

Figure 3a
The graph of f is concave up.

Figure 3b
The graph of f is concave down.

Example 1 Discuss concavity for the graphs of $f(x) = 1 + x^2$ and $g(x) = -x^4$.

Solution We calculate that $f'(x) = 2x$, which is an increasing function on the entire real line. Therefore, the graph of f is concave up on the interval $(-\infty, \infty)$ (see Figure 4). Similarly, the function $g(x) = -x^4$ satisfies $g'(x) = -4x^3$, which is a decreasing function on $(-\infty, \infty)$. Therefore, the graph of g is concave down on $(-\infty, \infty)$, as can also be seen in Figure 4. ■

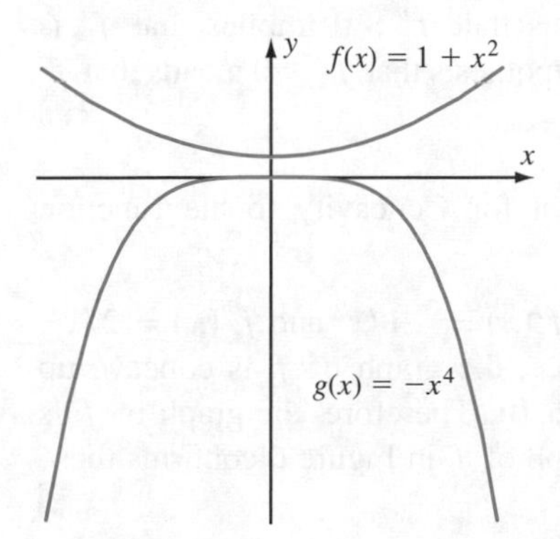

Figure 4

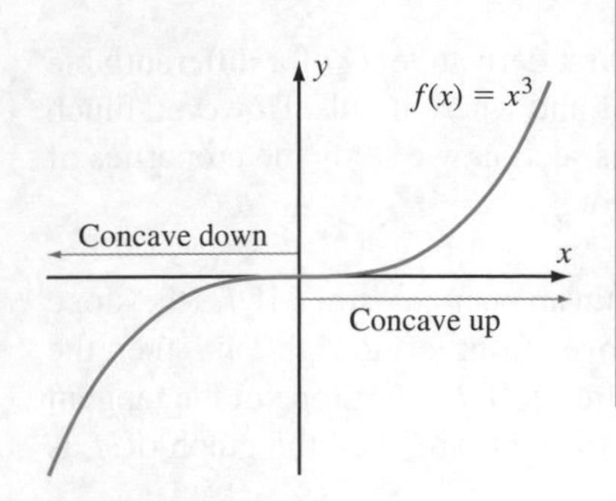

Figure 5

inSIGHT

At first, it is easy to confuse "increase" and "decrease" of a given function with the "concave up" and "concave down" properties of that function. The first two properties are independent of the second two. This means that a function can be increasing and concave down (g on the interval $(-\infty, 0)$) or decreasing and concave down (g on the interval $(0, \infty)$) or increasing and concave up (f on the interval $(0, \infty)$) or decreasing and concave up (f on the interval $(-\infty, 0)$). Refer to Figure 4 on the previous page.

Example 2 Discuss concavity for the graph of the function $f(x) = x^3$.

Solution First of all, $f'(x) = 3x^2$. On the interval $(0, \infty)$, the function $f'(x) = 3x^2$ is increasing. Therefore, the graph of f is concave up on $(0, \infty)$. On the interval $(-\infty, 0)$, the function $f'(x) = 3x^2$ is decreasing, because it becomes less and less positive when x is negative and increasing. Therefore, the graph of f is concave down on $(-\infty, 0)$. Refer to Figure 5. ■

Example 2 shows that the graph of a function can be concave up on one interval in its domain and concave down on another.

Using the Second Derivative to Test for Concavity

An important use of the second derivative is that it enables us to recognize the two types of concavity.

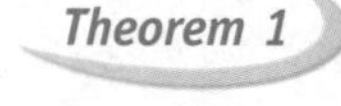

Second Derivative Test for Concavity Suppose that the function f is twice differentiable on an open interval I.

a. If $f''(x) > 0$ for every x in I, then the graph of f is concave up on I.
b. If $f''(x) < 0$ for every x in I, then the graph of f is concave down on I.

Proof Notice that $f'' = (f')'$. Thus, the hypothesis that $f'' > 0$ implies that f' is increasing or that f is concave up. Similarly, the hypothesis that $f'' < 0$ means that f' is decreasing or that f is concave down. ■

Example 3 Apply the Second Derivative Test for Concavity to the function $f(x) = 1/x$.

Solution Provided that $x \neq 0$, we calculate that $f'(x) = -1/x^2$ and $f''(x) = 2/x^3$. Notice that $f'' > 0$ on the interval $(0, \infty)$. Hence, the graph of f is concave up on that interval. Also, $f'' < 0$ on the interval $(-\infty, 0)$. Therefore, the graph of f is concave down on that interval. A glance at the graph of f in Figure 6 confirms these calculations. ■

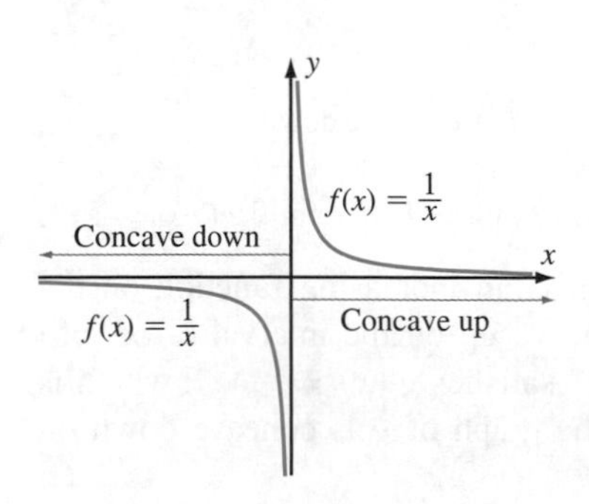

Figure 6

Points of Inflection

Definition Let f be a continuous function on an open interval I. If the graph of f changes concavity (from positive to negative or from negative to positive) at a point c in I, then c is called a *point of inflection* (or an *inflection point*). Refer to Figure 7.

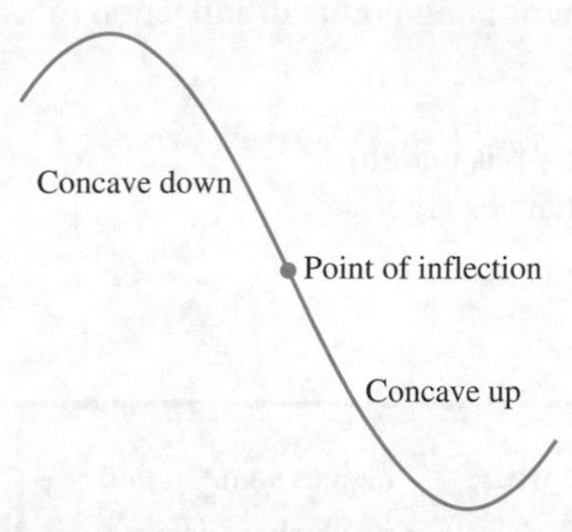

Figure 7

Notice that the function $f(x) = 1 + x^2$ from Example 1 (graphed in Figure 4, page 277) is concave up everywhere; hence, it has no points of inflection. Similarly, the function $g(x) = -x^4$ from Example 1 (also in Figure 4) has no points of inflection because it is concave down everywhere. On the other hand, the function $f(x) = x^3$ from Example 2 (see Figure 5) changes concavity at $c = 0$, so 0 is a point of inflection. Finally, the graph of $f(x) = 1/x$ from Example 3 (see Figure 6) is concave down to the left of 0 and concave up to the right of 0; however, 0 is not in the domain of f, so 0 is not an inflection point of f.

Example 4 Examine the graph of $f(x) = x - (x - 1)^3$ for concavity and points of inflection.

Solution We calculate that $f'(x) = 1 - 3(x-1)^2$ and $f''(x) = -6(x-1)$. Therefore, $f'' > 0$ on the interval $(-\infty, 1)$ and $f'' < 0$ on the interval $(1, \infty)$. We conclude that the graph of f is concave up when $x < 1$ and concave down when $x > 1$. It follows that 1 is a point of inflection for the graph of f, as shown in Figure 8. ■

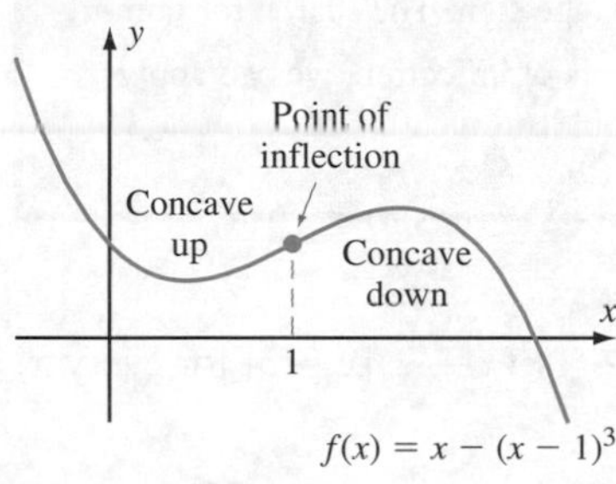

Figure 8

inSIGHT

The critical points of the function f from Example 4 are the solutions of the equation $f'(x) = 1 - 3(x - 1)^2 = 0$. We see that the critical points of f are located at $x = 1 \pm 1/\sqrt{3}$. However, the inflection point is at $x = 1$. *Inflection points* are *not* the same as *critical points*.

Example 5 Examine the function $f(x) = x^{1/3}$ for concavity and points of inflection.

Solution For $x \neq 0$, we have $f'(x) = (1/3) \cdot x^{-2/3}$ and $f''(x) = (-2/9) \cdot x^{-5/3}$. Therefore, $f'' > 0$ on the interval $(-\infty, 0)$, and $f'' < 0$ on the interval $(0, \infty)$. Thus, the graph of f is concave up on $(-\infty, 0)$ and concave down on $(0, \infty)$. Therefore, $c = 0$ is a point of inflection for the graph of f. Notice that f is not differentiable at 0, but 0 is still an inflection point, because f is continuous and the concavity changes there (see Figure 9). ■

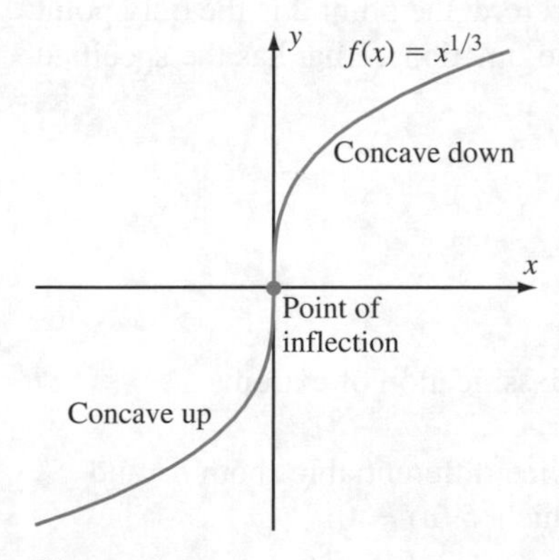

Figure 9

Strategy for Determining Points of Inflection

Suppose that f has a point of inflection at c and that f'' exists on both sides of c. Since the graph of f is concave up on one side of c and concave down on the other, we conclude that as x moves from one side of c to the other, $f''(x)$ changes its sign. If

$f''(c)$ exists, then we expect its value to be 0 and that is true (by Darboux's Theorem, Section 4.2, Exercise 52). For instance, in Example 4, the inflection point of the function $f(x) = x - (x - 1)^3$ occurs at $x = 1$, which is a root of the equation $f''(x) = 0$. We have also seen (in Example 5) that an inflection point can occur at a point where $f''(x)$ fails to exist. *One of these two possibilities must occur at a point of inflection.* These considerations lead to the following strategy for determining points of inflection of a continuous function f on an open interval I.

1. Locate all points of I at which $f''(x) = 0$ or f'' is undefined.
2. At each of these points, check to see if f'' changes sign.

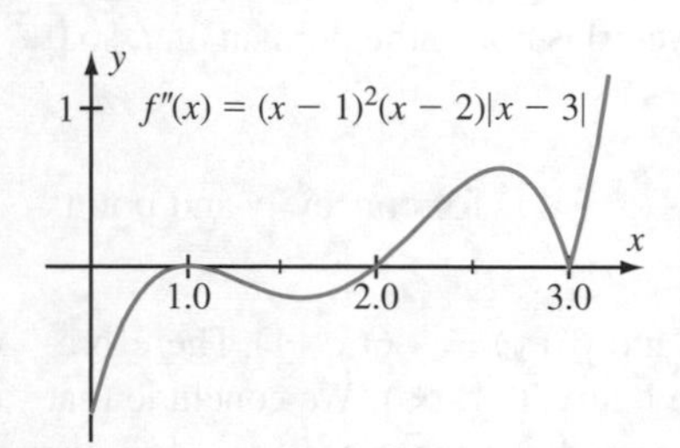

Figure 10

inSIGHT

Recall that for a local extremum, we search for a point where f' changes sign. To find such points, we must first locate the points at which $f'(x) = 0$ or at which f' does not exist. These are the points at which f' *might* change sign. Having located these points, we must then check each one to see if f' really does change sign. The search for points of inflection is completely analogous. To determine points of inflection, we can apply the very same procedure, using f'' instead of f'.

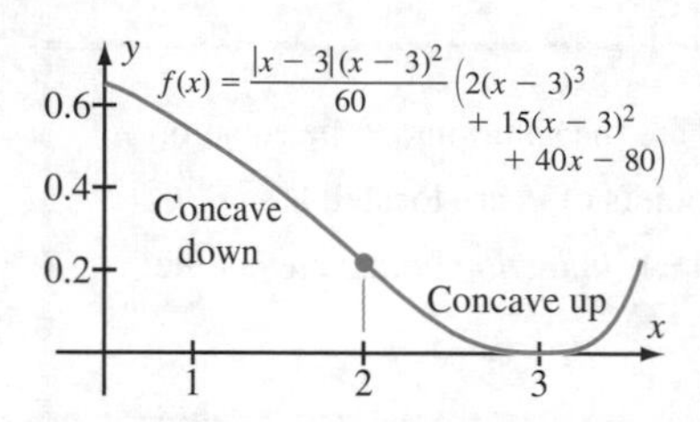

Figure 11
Point of inflection

Example 6 A function f satisfies $f''(x) = (x - 1)^2(x - 2)|x - 3|$ for every x. Determine all the points of inflection of f.

Solution By hypothesis, $f''(x)$ exists for every x. Therefore, a point of inflection c must satisfy $f''(c) = 0$. Clearly the points 1, 2, and 3 are the only candidates. Since the factors $(x - 2)^2$ and $|x - 3|$ are nonnegative, we see that the sign of $f''(x)$ is the same as the sign of $(x - 2)$. This tells us that $f''(x)$ changes sign only at $x = 2$ (as can be seen from the plot of f'' in Figure 10). Therefore, the point 2 is the only point of inflection of f. Figure 11 contains the plot of one function f that has the specified second derivative. It supports our conclusions. ■

The Second Derivative Test at a Critical Point

We now apply our knowledge of concavity to the classification of extrema.

Theorem 2 ***Second Derivative Test for Extrema*** Let f be twice differentiable (both f' and f'' exist) on an open interval containing a point c at which $f'(c) = 0$.

a. If $f''(c) > 0$, then c is a local minimum.
b. If $f''(c) < 0$, then c is a local maximum.
c. If $f''(c) = 0$, then no conclusion is possible *from this test.*

Proof The inequality $f''(c) > 0$ means $(f')'(c) > 0$, so f' is *increasing* at c. Since $f'(c) = 0$, it follows that f' must be going from negative to positive as x passes from left to right through c. By the First Derivative Test, there is a local minimum at c. Similarly, $f''(c) < 0$ means $(f')'(c) < 0$, so f' is *decreasing* at c. Since $f'(c) = 0$, f' must be going from positive to negative as x passes from left to right through c. By the First Derivative Test, there is a local maximum at c.

To understand the last statement of the Second Derivative Test for Extrema, notice that $f'(0) = 0$ and $f''(0) = 0$ for each of the three functions $f(x) = x^4$, $f(x) = -x^4$, and $f(x) = x^3$. The first function obviously has a minimum at 0, the second has a maximum, and the third has neither (see Figure 12). By themselves, the two equations $f'(c) = 0$ and $f''(c) = 0$ do not distinguish among any of the three possible behaviors. ■

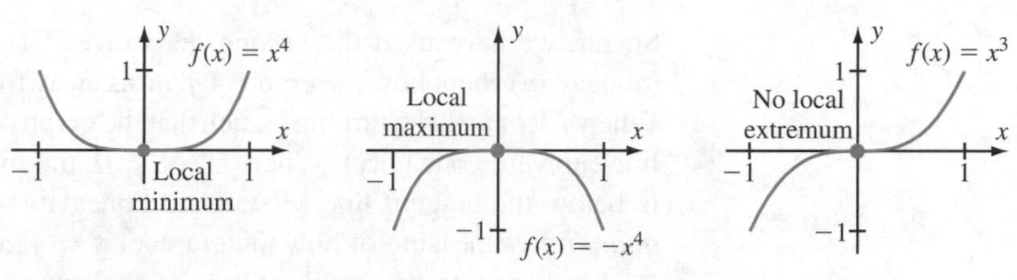

Figure 12
$f'(0) = 0$ and $f''(0) = 0$ for all three functions.

inSIGHT

It is at first surprising that $f''(c) > 0$ corresponds to a *minimum* while $f''(c) < 0$ corresponds to a *maximum*. Psychologically, this seems backward. However, the inequality $f''(c) > 0$ means that the graph is concave up at c, which indicates a minimum rather than a maximum. Similarly, $f''(c) < 0$ means that the graph is concave down at c, which indicates a maximum rather than a minimum.

Do not interpret the third part of the Second Derivative Test for Extrema to mean that no conclusion is possible *at all*. We are able to analyze the behavior of each of the three functions $f(x) = x^4$, $f(x) = -x^4$, and $f(x) = x^3$ at its critical point even though the Second Derivative Test for Extrema is inconclusive.

Example 7 Use the Second Derivative Test for Extrema to examine the critical points of $f(x) = 2x^3 + 15x^2 + 24x + 23$.

Solution This function is a polynomial; it is defined on the entire real line and is differentiable everywhere. Therefore, the only critical points are the zeros of f'. Now

$$f'(x) = 6x^2 + 30x + 24 = 6(x + 1) \cdot (x + 4).$$

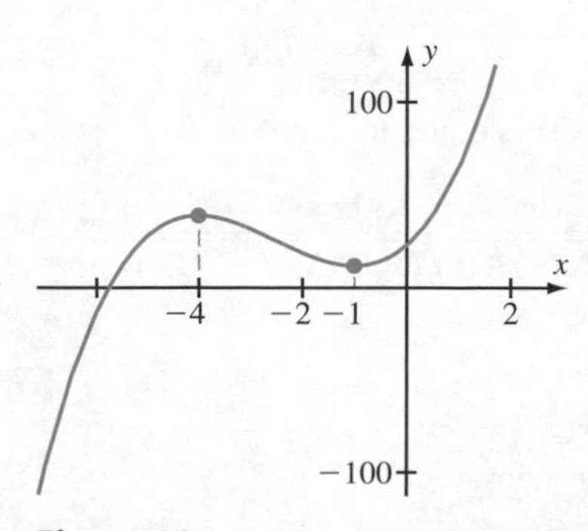

Figure 13

Thus, the zeros of f' are -1 and -4. These are the only critical points of f. To use the Second Derivative Test, we calculate that $f''(x) = 12x + 30$. Since $f''(-1) = 18 > 0$, the Second Derivative Test says that there is a local minimum at $c = -1$. Because $f''(-4) = -18 < 0$, the Second Derivative Test says that there is a local maximum at $c = -4$. Refer to Figure 13. ■

A LOOK BACK

Although the Second Derivative Test for Extrema is convenient, it does not supersede the First Derivative Test (see Section 4.2). This is because the First Derivative Test (1) requires less computation than the Second Derivative Test, (2) can be applied when the second derivative does not exist, and (3) can be applied when the second derivative exists but is zero.

Enrichment: Curvature*

So far, we have used the second derivative $f''(c)$ to tell us *qualitative* information (concavity) about how the graph of f turns away from its tangent line at $P = (c, f(c))$. When $f''(c) > 0$, the turning is such that the graph near the point P lies above the tangent line (upward concavity). When $f''(c) < 0$, the turning is such that the graph near P is below the tangent line (downward concavity). Now we will define *curvature*—a *quantitative* measure of how the graph of $y = f(x)$ turns.

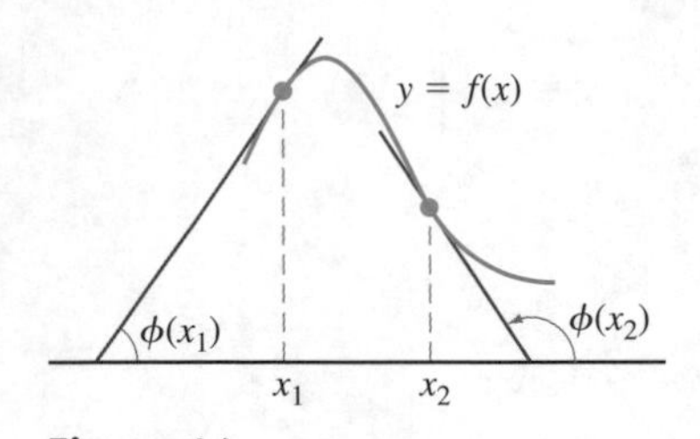

Figure 14

Let $\mathcal{C}$ denote the graph of $y = f(x)$. For each x, let $\phi(x)$ be the angle that the tangent line to $\mathcal{C}$ at $(x, f(x))$ makes with the positive x-axis (see Figure 14). Now look at Figure 15. In following the road from R to S, car B must undergo the same change in angle $\Delta\phi$ as car A does when it travels from P to Q. All other things being equal, it is easier for car B to negotiate its turn than for car A, because car B can effect the required change of angle over a greater distance. Curvature captures this idea. Informally, curvature is the instantaneous rate of change of angle with respect to distance along the curve.

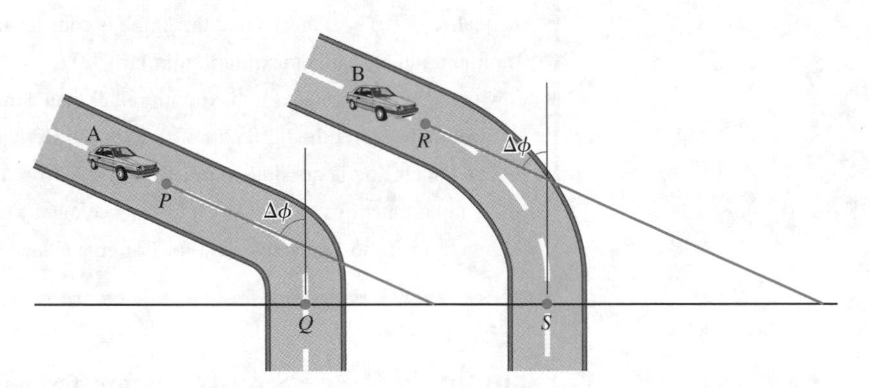

Figure 15

Definition The *curvature* $\kappa(x)$ of $y = f(x)$ at the point $(x, f(x))$ is equal to

$$\kappa(x) = \lim_{\Delta x \to 0} \frac{|\phi(x + \Delta x) - \phi(x)|}{\sqrt{(\Delta x)^2 + (\Delta y)^2}} = \lim_{\Delta x \to 0} \frac{|\phi(x + \Delta x) - \phi(x)|}{\sqrt{(\Delta x)^2 + (f(x + \Delta x) - f(x))^2}}.$$

*Knowledge of this topic is not required for subsequent sections.

Now we must do two things: (1) find a convenient way to compute $\kappa(x)$, and (2) make sure that this quantity corresponds to our intuitive understanding of curvature. Our next theorem accomplishes the first task.

Theorem 3 Suppose that f is twice differentiable. The curvature $\kappa(x)$ of the graph of f at the point $(x, f(x))$ is given by

$$\kappa(x) = \frac{|f''(x)|}{(1+(f'(x))^2)^{3/2}}. \tag{4.4}$$

Proof Since $\tan(\phi(x)) = f'(x)$, implicit differentiation of this equation with respect to x results in $\sec^2(\phi(x)) \cdot \phi'(x) = f''(x)$. It follows that

$$\phi'(x) = \frac{f''(x)}{\sec^2(\phi(x))}. \tag{4.5}$$

Now we use the identity $\sec^2(\phi(x)) = 1+\tan^2(\phi(x)) = 1+(f'(x))^2$ to rewrite formula (4.5) as

$$\phi'(x) = \frac{f''(x)}{1+(f'(x))^2}. \tag{4.6}$$

We are now ready to calculate $\kappa(x)$:

$$\kappa(x) = \lim_{\Delta x \to 0} \frac{|\phi(x+\Delta x) - \phi(x)|}{\sqrt{(\Delta x)^2 + (f(x+\Delta x) - f(x))^2}} = \lim_{\Delta x \to 0} \frac{\left|\frac{\phi(x+\Delta x)-\phi(x)}{\Delta x}\right|}{\sqrt{1+\left(\frac{f(x+\Delta x)-f(x)}{\Delta x}\right)^2}} = \frac{|\phi'(x)|}{\sqrt{1+(f'(x))^2}}.$$

Into the numerator of the last expression, we substitute the formula for $\phi'(x)$ that is given by equation (4.6). The required formula for $\kappa(x)$ results. ■

When we learn a new formula, it is always a good idea to see how it applies to familiar cases. The first thing to notice is that if $f(x) = mx + b$, then $f''(x) = 0$ for all x. Therefore, $\kappa(x) = 0$ for each x, which is just what curvature should be at each point of a straight line. In Exercise 55, you are asked to use formula (4.4) to show that the curvature of the circle $x^2 + y^2 = r^2$ is $1/r$. As expected, the curvature of a circle is the same at all points and it is inversely proportional to the radius: the larger the circle, the smaller the curvature.

Example 8 At which of the points $(0, 1)$ and $(1, e)$ is the curvature of $y = \exp(x)$ greater?

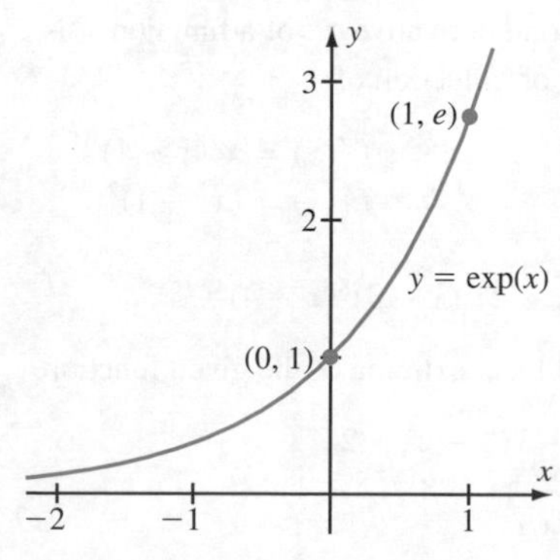

Figure 16

Solution Figure 16 shows the graph of $y = \exp(x)$ with equal scaling along the axes. It appears, from a careful inspection of the figure, that curvature is greater at the point $(0, 1)$. Let us verify our estimate numerically. If $f(x) = \exp(x)$, then $f'(x) = f''(x) = \exp(x)$ and

$$\kappa(x) = \frac{|f''(x)|}{(1+f'(x)^2)^{3/2}} = \frac{e^x}{(1+e^{2x})^{3/2}}.$$

Thus,

$$\kappa(0) = \frac{e^0}{(1+e^0)^{3/2}} = \frac{1}{2\sqrt{2}} \approx 0.35 \quad \text{and} \quad \kappa(1) = \frac{e^1}{(1+e^2)^{3/2}} = 0.11.$$

Our calculation confirms what our intuition suggests: Curvature is greater at the point $(0, 1)$ than at the point $(1, e)$. ■

Example 9 Show that the maximum curvature of the parabola $y = ax^2 + bx + c$ is $2|a|$ and that this maximum occurs at $x = -b/(2a)$.

Solution The curvature of $f(x) = ax^2 + bx + c$ is

$$\kappa(x) = \frac{|f''(x)|}{(1 + (f'(x))^2)^{3/2}} = \frac{|2a|}{(1 + (2ax + b)^2)^{3/2}}.$$

Although we can apply the methods of calculus to maximize the function κ, it is easiest to notice that the smallest the denominator can be is 1, which happens when $2ax + b = 0$. In other words, the largest curvature is $|2a|/(1 + 0)^{3/2} = |2a|$, which occurs for $x = -b/(2a)$. This calculation confirms what our intuition suggests: The maximum curvature of a parabola occurs at its vertex. ■

quickquiz

1. What does it mean to say that a function f is concave up on an interval I?
2. What does it mean to say that a function f is concave down on an interval I?
3. State the Second Derivative Test for a local maximum or a local minimum.
4. Suppose that $f''(x) = (x + 2)^4(x - 3)^5$. Does $f(x)$ have a point of inflection at $x = -2$? At $x = 3$?

EXERCISES

Problems for Practice

In Exercises 1–20, determine on which intervals the function is concave up or concave down. Find all critical points and points of inflection. Identify all local maxima and minima.

1. $f(x) = x^3 + 9x^2 - 21x + 15$
2. $f(x) = -2x^3 + 12x^2 + 7$
3. $f(x) = 2x^3 - 3x^2 - 12x + 1$
4. $f(x) = x^4 - x^3 - 6x^2$
5. $f(x) = x^4 - 7x^3 + 9$
6. $f(x) = 3x^5 - 10x^3 + 15x - 8$
7. $f(x) = 2x^5 + 15x^4 + 30x^3 - 42$
8. $f(x) = x/(x - 4)$
9. $f(x) = x/\sqrt{x^2 + 1}$
10. $f(x) = (x^2 - 1)/(x^2 + 1)$
11. $f(x) = x^{1/3}(x - 5)$
12. $f(x) = (x - 4)^{1/3} \cdot x^{2/3}$
13. $f(x) = (x + 1)^2/x$
14. $f(x) = 1/(x^2 + 1)$
15. $f(x) = \cos(x) + x$
16. $f(x) = e^x$
17. $f(x) = \ln(x)$
18. $f(x) = xe^x$
19. $f(x) = x \ln(x)$
20. $f(x) = x^2 \ln(x)$

Further Theory and Practice

In Exercises 21–26, the derivative f' of a function f is given. Determine and classify all local extrema of f.

21. $f'(x) = x(x - 1)$
22. $f'(x) = x^2(x - 1)$
23. $f'(x) = x^2 - 1$
24. $f'(x) = (x^2 - 1)^2$
25. $f'(x) = (x^2 - 1)^5$
26. $f'(x) = (x - 1)(x - 2)^2(x - 3)^3(x - 4)^4$

In Exercises 27–32, the second derivative f'' of a function f is given. Determine all points of inflection of f.

27. $f''(x) = x(x - 1)$
28. $f''(x) = x^2(x - 1)$
29. $f''(x) = x^2 - 1$
30. $f''(x) = (x^2 - 1)^2$
31. $f''(x) = (x^2 - 1)^5$
32. $f''(x) = (x - 1)(x - 2)^2(x - 3)^3(x - 4)^4$

In Exercises 33–36, find all local extrema of the given function.

33. $f(x) = x^6 - 3x^5 + 3x^4 - x^3 + 2$
34. $f(x) = (x^3 - 3x^2 - 4x - 8)/(x + 1)$
35. $f(x) = (6x^2 - x^3)^{1/3}$
36. $f(x) = (x^4 + 4x^3 - 18x^2 - 21x - 48)/(x + 1)$
37. For a positive constant a, the Witch of Agnesi is the curve with the equation $y = 8a^3/(4a^2 + x^2)$. Find the

points of inflection of this curve. On what intervals is this curve concave up? Concave down?

38. Explain why a point of inflection of a function f cannot be a local extremum even if it is a critical point of f.

39. Suppose that f is a positive function such that $f'' > 0$. What is the concavity of f^2?

40. Suppose that f and g are twice differentiable functions that are concave up. Is $f + g$ concave up? Is $f \cdot g$ concave up? Assuming that the range of f is contained in the domain of g, is $g \circ f$ concave up?

41. Let k and c be positive constants. Suppose that a yeast population has mass $m(t)$ that, for all values of t, satisfies the inequalities $0 < m(t) < 2c$ and the differential equation

$$m'(t) = k \cdot m(t) \cdot (2c - m(t)).$$

By completing the square on the right side of this equation, determine the ordinate of the point of inflection of the graph of $y = m(t)$. Discuss the concavity of this graph.

42. For every positive x, the number $W(x)$ is defined to be the unique positive number such that

$$W(x)\exp(W(x)) = x.$$

The function $x \mapsto W(x)$ is known as Lambert's W function. Show that

$$W'(x) = \frac{1}{x + \exp(W(x))}.$$

Deduce that W is an increasing function, and show that the graph of W is concave down.

43. The luminosity L of a cooling white dwarf (star) has the equation

$$L(t) = L(0) \cdot \left(\frac{T(t)}{T(0)}\right)^{7/2}$$

where the temperature T satisfies

$$\frac{dT}{dt} = -\alpha T^{7/2}$$

for some positive constant α.

a. Show that the graph of T is concave up. *Hint:* Do not solve for T. Investigate T'' by differentiating T' implicitly.

b. Show that the graph of L is concave up.

c. Figure 17 illustrates the graph of L as it would commonly be given in an astrophysics text. In this plot, units of luminosity have been chosen so that $L(0) = 1$. The plot does not appear to be concave up. Explain why there is no inconsistency with part b.

44. The product of labor $f(n)$ and the marginal product of labor (MPL) were defined in Section 4.2, Exercise 43.

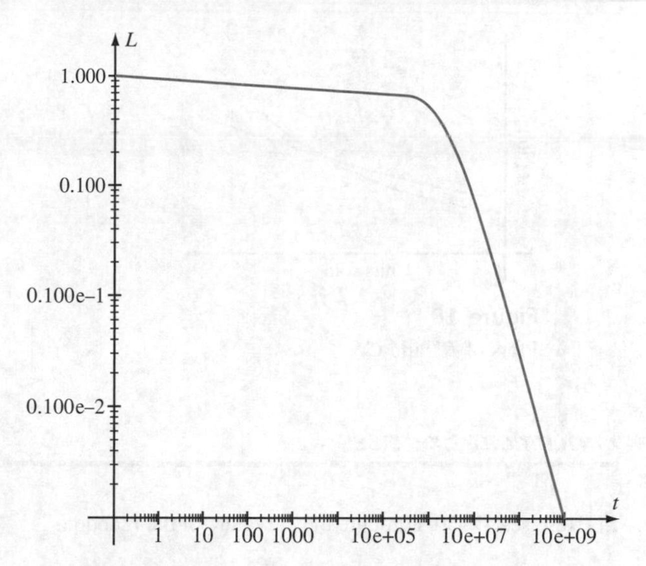

Figure 17

Economists usually assume that the graph of the function f has a certain property. The three graphs that appear in Figure 14 from Section 4.2 illustrate this property in different ways. What is the property? Explain how the graphs illustrate this property.

45. Suppose two goods, X and Y, are available to a consumer. Let $U(x, y)$ denote the satisfaction (or *utility*) that a consumer derives from purchasing x units of good X and y units of good Y. For each fixed value of x or y, $U(x, y)$ is an increasing function of the other variable. The locus of points $\{(x, y) : U(x, y) = s\}$ at which a consumer's satisfaction $U(x, y)$ remains constant at level s is called an *indifference curve*. The *marginal rate of substitution* $MRS_{XY}(x, y)$ of good X for good Y is the slope of the indifference curve at (x, y).

a. Is MRS_{XY} positive or negative? Explain. What does this say about the geometry of indifference curves?

b. As an individual "moves down" an indifference curve and has fewer and fewer units of Y, each remaining unit of Y becomes more and more valuable to him. He, therefore, requires more and more units of good X to replace each additional unit of Y. What does this say about MRS_{XY}? What does this say about the geometry of indifference curves? Sketch a typical indifference curve.

46. Let $R(q)$ denote the revenue realized by selling q units of an item. Let $C(q)$ denote the total cost incurred in selling those q units. Suppose that the profit $R - C$ is maximized at q_0. The second derivatives R'' and C'' are plotted in Figure 18, next page. Identify which curve is R'' and which curve is C''. Explain your reasoning.

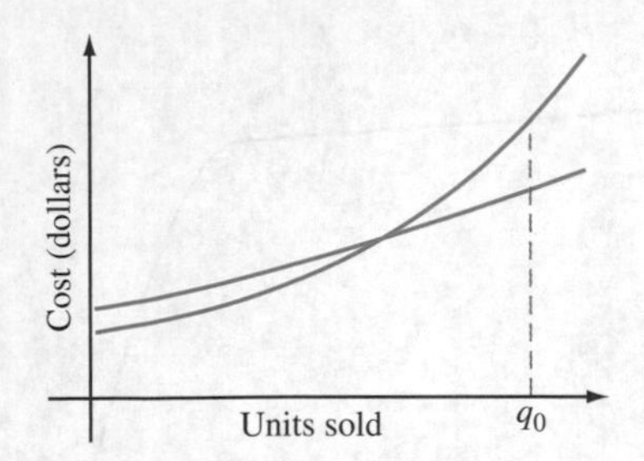

Figure 18
Plots of R'' and C''

Enrichment Exercises

In Exercises 47–50, compute the curvature of the function.

47. $f(x) = x^3 + 2x + 1$ **48.** $f(x) = \sin(x)$
49. $f(x) = \sqrt{x}$ **50.** $f(x) = x/(x^2 + 1)$

In Exercises 51–54, find the maximum curvature of the function.

51. $f(x) = \ln(x)$ **52.** $f(x) = x^3 + 2x + 1$
53. $f(x) = \sin(x)$ **54.** $f(x) = e^x$
55. Show that the curvature of the circle $x^2 + y^2 = r^2$ is $1/r$ at all points.
56. If f is twice differentiable at c and if c is an inflection point of f, what can you say about the curvature of f at c?
57. Let $g(x) = f(x + a) - b$. Show that the curvature of the graph of g at $(x, g(x))$ is the same as that of the graph of f at $(x + a, f(x + a))$. This says that curvature is *not changed* by horizontal or vertical translations of a graph, which agrees with our intuitive concept of curvature.
58. Let f be a twice differentiable function and let k be a constant. Calculate the curvature at c of $y = f(c) + f'(x) \cdot (x - c) + k \cdot (x - c)^2$. What is the parabola that passes through the point $P_c = (c, f(c))$, has the same tangent line at P_c as the graph of f, and has the same curvature at P_c?
59. Suppose that p is a polynomial of degree two or more. Let κ be the curvature function of its graph. Show that $\lim_{x \to \pm\infty} \kappa(x) = 0$. Use this to show that the curvature function of a polynomial is bounded.

Calculator/Computer Exercises

In Exercises 60–65, approximate the critical points and inflection points of the function f to three decimal places. Determine the behavior of f at each critical point.

60. $f(x) = x\sin(x)$
61. $f(x) = x/(1 + \exp(x))$
62. $f(x) = x^4 + 3x^2 - 2x + 4$
63. $f(x) = x^4 - 2\exp(x) + 3x$
64. $f(x) = 6x^5 - 45x^4 + 80x^3 + 30x^2 - 30x + 2$
65. $f(x) = (x^2 + 1)/(x^2 + x + 1)$
66. Analyze the critical points and the points of inflection of

$$f(x) = 2x^6 + \frac{204}{5}x^5 + 333x^4 + 1388x^3 + 3120x^2 + 3600x + 1.$$

67. Let f be a twice-differentiable function. Let C be the function $C(x) = \text{signum}(f''(x))$. Explain how the graph of C may be used to determine the concavity of the graph of f. Explain how the graph of C may be used to determine the points of inflection of f. Illustrate by plotting $f(x) = 6x^6 + x^5 - 60x^4 - 35x^3 + 120x^2 + 52x + 160$ and $300\text{signum}(f''(x))$ in the viewing window $[-2.3, 2.8] \times [-310, 310]$. (The factor 300 provides a clearer view.)

Enrichment Exercise

68. Graph both $f(x) = x^2\exp(-x)$ and the curvature function of f for $-1 \le x \le 5$. Indicate the point at which the graph of f has maximum curvature. Label points of inflection of f.

4.5 Graphing Functions

Curve sketching is one of the most important techniques of analytical thinking. The skills of creating and understanding graphs are used in all of the physical sciences and in many of the social sciences as well. In previous sections, we learned that certain aspects of the graph of a function f can be determined from the first and second derivatives of f. We also learned that graphs of functions can have vertical asymptotes or horizontal asymptotes or both. In this section, we combine all these concepts and use them to sketch the graphs of functions. Even if you have a graphing calculator or software with graphing capabilities, the knowledge that you will gain in this section will be useful because the *best way* to learn how to interpret a graph is to learn how it is created.

Basic Strategy of Curve Sketching

The following steps can be used to sketch the graphs of an extensive range of functions.

Basic Steps in Sketching a Graph

1. Determine the domain and (if possible) the range of the function.
2. Find all horizontal and vertical asymptotes.
3. Calculate the first derivative and find the critical points for the function.
4. Find the intervals on which the function is increasing or decreasing.
5. Calculate the second derivative and find the intervals on which the function is concave up or concave down.
6. Identify all local maxima, local minima, and points of inflection.
7. Plot these points, as well as the y-intercept (if applicable) and any x-intercepts (if they can be computed). Sketch the asymptotes.
8. Connect the points plotted in step 7, keeping in mind concavity, local extrema, and asymptotes.

If you follow this systematic procedure in sketching your graphs, you will find that you can handle a variety of interesting examples.

Example 1 Graph the function $f(x) = 5x/(x-2)^2$.

Solution We follow the steps for sketching a graph.

1. The domain of f consists of all real numbers except 2. When x is near the point 2, x takes arbitrarily large positive values.
2. We notice that $\lim_{x\to+\infty} f(x) = \lim_{x\to-\infty} f(x) = 0$. Therefore, the line $y = 0$ is a horizontal asymptote for the graph. Also, $\lim_{x\to 2} f(x) = +\infty$. Therefore, the line $x = 2$ is an upward vertical asymptote for f.
3. We calculate that

$$f'(x) = \frac{(x-2)^2 \cdot 5 - 5x \cdot 2(x-2)}{(x-2)^4} = \frac{-5(x+2)}{(x-2)^3}.$$

   The first derivative is defined at all points *of the domain* of f. Since f' vanishes at -2, this is the only critical point. In step 6, we will determine whether f has a local extremum at $x = -2$.
4. The first derivative can change sign only at -2 (the critical point) and $+2$ (the point where f is undefined). Since $f'(-3) = -1/25$, we conclude that $f' < 0$ on $(-\infty, -2)$; hence, f is decreasing there. Since $f'(0) = 5/4 > 0$, we conclude that $f' > 0$ on $(-2, 2)$; hence, f is increasing there. Finally, since $f'(3) = -25 < 0$, we see that $f' < 0$ on $(2, \infty)$; hence, f is decreasing there.

5. The second derivative is

$$f''(x) = \frac{(x-2)^3 \cdot (-5) - (-5(x+2)) \cdot (3(x-2)^2)}{(x-2)^6}$$
$$= \frac{10 \cdot (x+4)}{(x-2)^4}.$$

Notice that the denominator of f'' is always positive on the domain of f. Clearly $f'' < 0$ on the interval $(-\infty, -4)$ because the numerator is negative. Hence, f is concave down on that interval. Also, $f''(x) > 0$ when $x > -4$ (except at $x = 2$, the point where f, f', and f'' are undefined). Hence, f is concave up on each of the intervals $(-4, 2)$ and $(2, \infty)$.

6. Because $f''(-2) = 5/64 > 0$, there is a local minimum at the critical point $x = -2$. We will label the point $(-2, f(-2)) = (-2, -5/8)$ on our graph. From step 5, we know the concavity changes at $x = -4$. Therefore, f has a point of inflection at $x = -4$. We will label the point $(-4, f(-4)) = (-4, -5/9)$ on our graph. The concavity does not change at $x = 2$.

7. The y-intercept is $(0, f(0)) = (0, 0)$. Since $x = 0$ is the only solution to $f(x) = 0$, the point $(0, 0)$ is also the only x-intercept. The points $(0, 0)$, $(-2, -5/8)$, and $(-4, -5/9)$ are plotted in Figure 1.

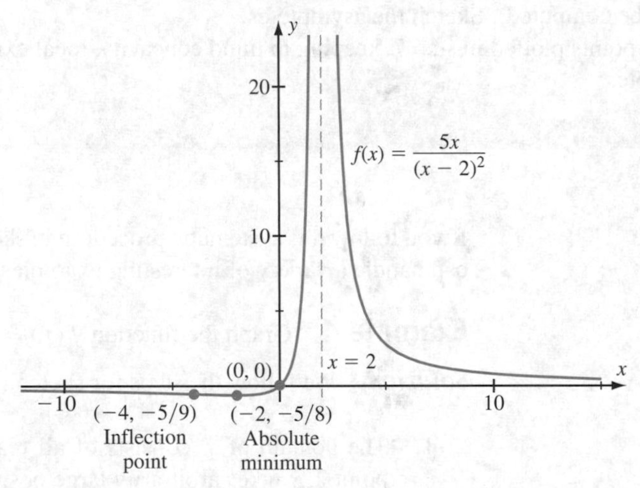

Figure 1

8. Figure 1 is the final sketch of the graph of f. All the pertinent data are labeled. We can deduce from all the information gathered that f has an absolute minimum at $x = -2$. There is no absolute maximum. ■

Example 2 Sketch the graph of $g(x) = 4x^3 + x^4$.

Solution

1. Since g is a polynomial, the domain of g consists of all real numbers. When x is large, the term x^4 forces g to take arbitrarily large positive values.

2. By the comment in step 1, g has no finite limit as $x \to \pm\infty$. Thus, there are no horizontal asymptotes. Also, since g is a polynomial, it is continuous at every point. Consequently, there are no vertical asymptotes.

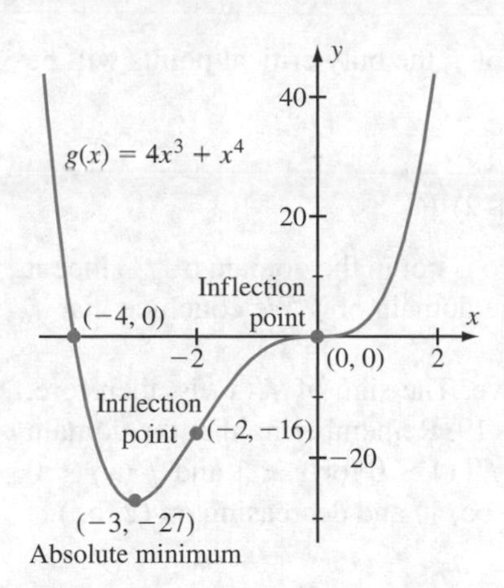

Figure 2

inSIGHT

In Examples 1 and 2, we stated that the information gathered allowed us to determine that a particular point is an absolute minimum of the function under consideration. The graphs of each function certainly support our conclusion about these points. Note, however, that since a graph can only be given in a finite viewing rectangle, the graphs cannot, by themselves, prove assertions concerning absolute extrema. If you have not already done so, it would be instructive for you to explain why $(-2, -5/8)$ must be an absolute minimum of the function f from Example 1 and why $(-3, -27)$ must be an absolute minimum of the function g from Example 2.

3. We have $g'(x) = 12x^2 + 4x^3 = 4x^2(3 + x)$. Since g' is defined for all x, the only critical points will be the zeros of g': 0 and -3. In step 6, we will determine whether g has a local extremum at either of these points.
4. The first derivative can change sign only at -3 and at 0. Since $g'(-4) = -64 < 0$, we conclude that $g' < 0$ on $(-\infty, -3)$; hence, g is decreasing there. Since $g'(-1) = 8 > 0$, we conclude that $g' > 0$ on $(-3, 0)$; hence, g is increasing there. Since $g'(1) = 16 > 0$, we conclude that $g' > 0$ on $(0, \infty)$; hence, g is also increasing there.
5. The second derivative is $g''(x) = 24x + 12x^2 = 12x(2 + x)$. The function g'' can only change sign at the points where it vanishes, namely, 0 and -2. Since $g''(-3) = 36 > 0$, we conclude that $g'' > 0$ on $(-\infty, -2)$; hence, g is concave up there. Since $g''(-1) = -12 < 0$, we conclude that $g'' < 0$ on $(-2, 0)$; hence, g is concave down there. Since $g''(1) = 36 > 0$, we conclude that $g'' > 0$ on $(0, \infty)$; hence, g is concave up there.
6. By step 5, we know the concavity changes at -2 and at 0, which are inflection points. We will label the points $(-2, g(-2)) = (-2, -16)$ and $(0, g(0)) = (0, 0)$ on our graph. In step 4, we saw that $g' < 0$ to the left of 3 and $g' > 0$ to the right of 3. Therefore, a local minimum occurs at the critical point -3. We will label the point $(-3, g(-3)) = (-3, -27)$ on our graph. We have already noted that the other critical point 0 is an inflection point.
7. The intercepts $(0, 0)$ and $(-4, 0)$ will also be labeled in our graph.
8. Figure 2 shows the sketch of the graph of g. All pertinent data have been indicated. We can deduce that there is an absolute minimum at -3. There is no absolute maximum. ■

Example 3 Sketch the graph of $f(x) = (4x + 1)/\sqrt{x^2 - 3x + 2}$.

Solution

1. Since $x^2 - 3x + 2 = (x - 1)(x - 2)$ is negative between its two roots, the interval $(1, 2)$ is not in the domain of f. The endpoints of this interval are also excluded because the denominator is 0 at these values. The domain of f consists of the union of the open intervals $(-\infty, 1)$ and $(2, \infty)$. The function f is continuous.
2. Since $\lim_{x\to 1^-} f(x) = \infty$ and $\lim_{x\to 2^+} f(x) = \infty$, we know that $x = 1$ is an upward vertical asymptote from the left and $x = 2$ is an upward vertical asymptote from the right. (It makes no sense to consider $\lim_{x\to 1^+} f(x)$ and $\lim_{x\to 2^-} f(x)$, since such limits involve x in the interval $(1, 2)$, which is not in the domain of f.) By writing
$$f(x) = \frac{x(4 + 1/x)}{\sqrt{x^2(1 - 3/x + 2/x^2)}} = \frac{x}{|x|}\frac{(4 + 1/x)}{\sqrt{1 - 3/x + 2/x^2}}$$
and noting that $x/|x| = 1$ when $x > 0$ and -1 when $x < 0$, we see that $\lim_{x\to\infty} f(x) = 4$ and $\lim_{x\to -\infty} f(x) = -4$. Therefore, the lines $y = 4$ and $y = -4$ are horizontal asymptotes.
3. After some simplification, we find that
$$f'(x) = -\frac{1}{2}\frac{14x - 19}{(x^2 - 3x + 2)^{3/2}}.$$

Since f' is defined for all x in the domain of f, the only critical points will be the zeros of f'. The expression

$$-\frac{1}{2}\frac{14x - 19}{(x^2 - 3x + 2)^{3/2}}$$

has one zero, $x = 19/14$. However, this zero is not in the domain of f' since it is between 1 and 2 and, therefore, not in the domain of f. We conclude that f has no critical points.

4. The denominator of $f'(x)$ is always positive. The sign of $f'(x)$ is, therefore, opposite to the sign of its numerator $14x - 19$. Remembering that the domain of f is $(-\infty, 1) \cup (2, \infty)$, we deduce that $f'(x) > 0$ for $x < 1$ and $f'(x) < 0$ for $x > 2$. Therefore, f is increasing on $(-\infty, 1)$ and decreasing on $(2, \infty)$.
5. After simplification, we find that

$$f''(x) = \frac{1}{4}\frac{56x^2 - 156x + 115}{(x^2 - 3x + 2)^{5/2}}.$$

Note that f'' exists at every point in the domain of f. Since $156^2 - 4 \cdot 56 \cdot 115 = -1424 < 0$, the quadratic formula tells us that f'' has no real zeros.

6. By step 5, we know that there are no points of inflection. Since $f''(0) > 0$, we deduce that the graph of f is concave up everywhere.
7. The intercepts $(-1/4, 0)$ and $(0, \sqrt{2}/2)$ and the vertical and horizontal asymptotes are plotted in Figure 3.

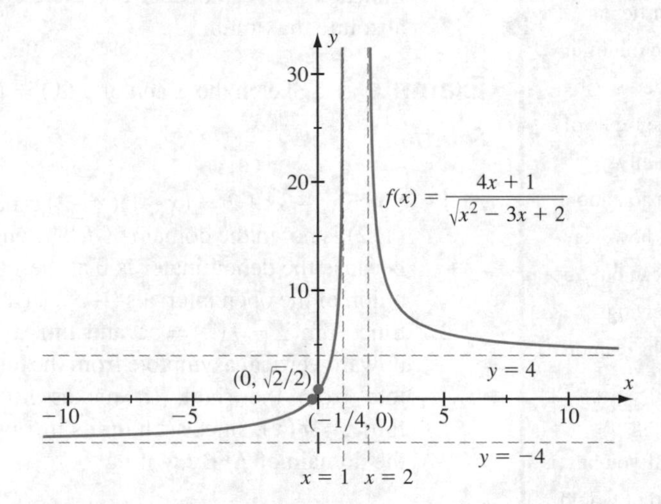

Figure 3

8. The final sketch of the graph of f appears in Figure 3. There are no local extrema. Notice that although $f(x) > -4$ for all x in the domain of f, f never takes on this value. Therefore, -4 is not the minimum value of f. ■

Periodic Functions

A function f is said to be *periodic* if there is a positive number p such that

$$f(x + p) = f(x), \quad x \in \mathbb{R}. \tag{4.7}$$

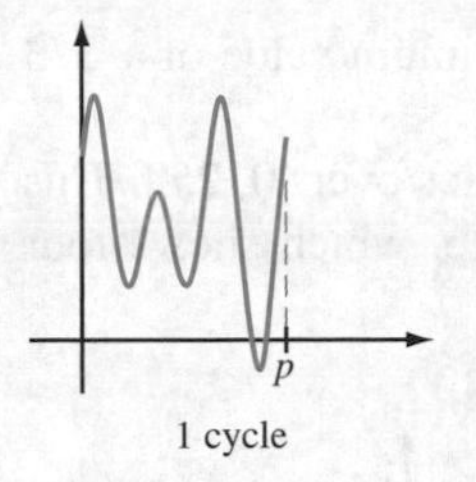

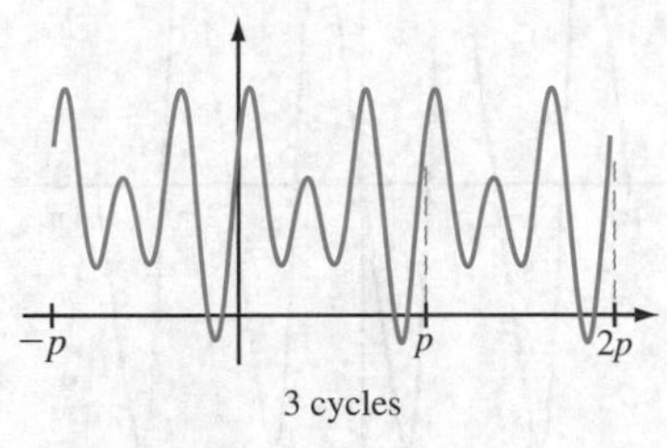

Figure 4

The smallest positive number p for which equation (4.7) is true is called the *period* of *f*. The sine, cosine, secant, and cosecant functions all have period 2π. However, we observe that

$$\tan(x+\pi) = \frac{\sin(x+\pi)}{\cos(x+\pi)} = \frac{\sin(x)\cos(\pi)+\cos(x)\sin(\pi)}{\cos(x)\cos(\pi)-\sin(x)\sin(\pi)}$$
$$= \frac{\sin(x)(-1)+0}{\cos(x)(-1)-0} = \frac{\sin(x)}{\cos(x)} = \tan(x),$$

so the tangent (and cotangent) have period π.

To graph a function f with a period p, we plot f over the interval $[0, p]$ and then translate the plot of this restriction (see Figure 4). If f is continuous on $[0, p]$, then it attains a maximum value and a minimum value according to the Extreme Value Theorem. The next example illustrates these points.

Example 4 Sketch the graph of the function $f(x) = \sin(x)/(2+\cos(x))$.

Solution

1. The function $f(x)$ is defined for all real x since its denominator is never 0. Observe that $f(x+2\pi) = f(x)$ for all x, which tells us that f is periodic and the entire graph of f is determined by the part that lies over $[0, 2\pi]$.
2. Since f is continuous at every x, there are no vertical asymptotes. Since f is periodic, we conclude that f has no horizontal asymptotes.
3. We calculate that
$$f'(x) = \frac{1+2\cos(x)}{(2+\cos(x))^2}.$$
The zeros of f' occur when $\cos(x) = -1/2$. The solutions of this equation in the interval $[0, 2\pi]$ are $x = 2\pi/3$ and $x = 4\pi/3$.
4. Observe that the sign of $f'(x)$ is the same as the sign of $1 + 2\cos(x)$. This expression is positive for x in the intervals $[0, 2\pi/3)$ and $(4\pi/3, 2\pi]$; it is negative for x in the interval $(2\pi/3, 4\pi/3)$. Therefore, f is increasing on $[0, 2\pi/3)$ and $(4\pi/3, 2\pi]$, and f is decreasing on $(2\pi/3, 4\pi/3)$.
5. After differentiation and algebraic simplification, we find that
$$f''(x) = -2\sin(x)\frac{1-\cos(x)}{(2+\cos(x))^3}.$$
Since the expressions $1-\cos(x)$ and $2+\cos(x)$ are nonnegative, we see that the signs of $f''(x)$ and $\sin(x)$ are opposite. Therefore, $f''(x) < 0$ for $0 < x < \pi$ and $f''(x) > 0$ for $\pi < x < 2\pi$. We conclude that the graph of f is concave down on the interval $(0, \pi)$ and concave up on the interval $(\pi, 2\pi)$.
6. Since $2\pi/3$ is in the interval on which $f'' < 0$, the Second Derivative Test tells us that $f(x)$ has a local maximum at $x = 2\pi/3$. Similarly, we see that f has a local minimum at $4\pi/3$. From the formula for $f''(x)$ that we obtained in step 5, we notice that the sign of $f''(x)$ changes at each point where $\sin(x)$ vanishes. Therefore, the points 0, π, and 2π are the points of inflection that f has in the interval $[0, 2\pi]$.
7. The y-intercept is the origin. The x-intercepts are the points of the form $(k\pi, 0)$ where k is an integer. Since $f(0) = 0$, $f(2\pi/3) = \sqrt{3}/3$, $f(4\pi/3) = -\sqrt{3}/3$,

and $f(2\pi) = 0$, we deduce that f has an absolute maximum value of $\sqrt{3}/3$ and an absolute minimum value of $-\sqrt{3}/3$.

8. Figure 5 contains a sketch of that part of the graph that lies over $[0, 2\pi]$. This portion of the graph is repeated periodically in Figure 6, which shows four cycles of f.

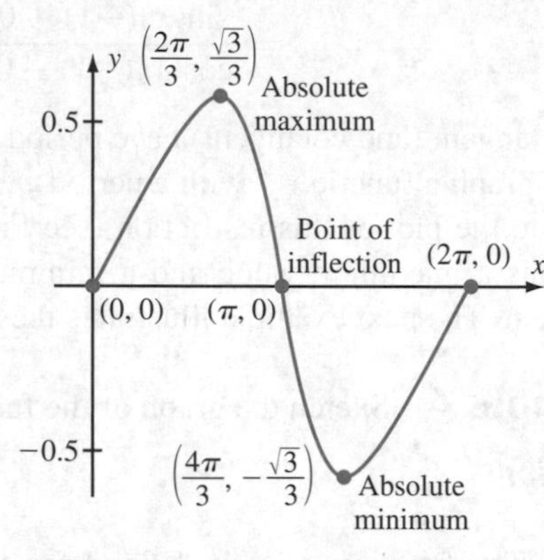

Figure 5

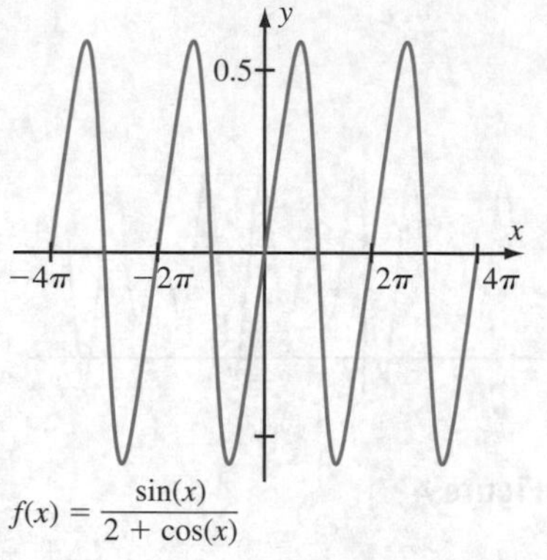

$f(x) = \dfrac{\sin(x)}{2 + \cos(x)}$

Figure 6

Skew Asymptotes

Figure 7 shows the line $y = 2x + 3$ and the graph of $f(x) = 2x + 3 + 1/x$. Notice that the graph of f becomes closer to the straight line $y = 2x + 3$ as $x \to \infty$. This leads to a definition of skew asymptotes.

Definition

The line $y = mx + b$ is a *skew asymptote* of f if

$$\lim_{x\to\infty} (mx + b - f(x)) = 0$$

or if

$$\lim_{x\to-\infty} (mx + b - f(x)) = 0.$$

Other terms for this line are *slant asymptote* and *oblique asymptote*.

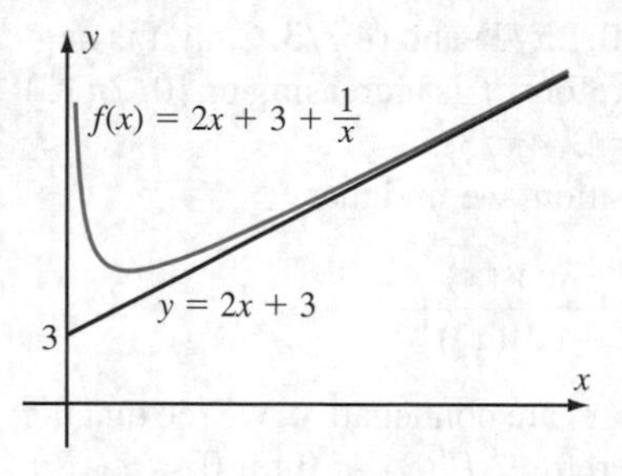

Figure 7

The most common situation in which a skew asymptote occurs is in graphing a rational function with a numerator of degree one more than its denominator. In this case, division allows us to determine the skew asymptote. Step 2 of the eight basic steps in sketching a graph should thus be expanded to include *all* asymptotes.

Example 5 Sketch the graph of $f(x) = x^3/(x^2 - 4)$.

Solution We follow the steps for sketching a graph.

1. The domain of f consists of all real numbers, except $x = +2$ and $x = -2$.
2. The lines $x = 2$ and $x = -2$ are vertical asymptotes. Since

$$\frac{x^3}{x^2 - 4} = x + \frac{4x}{x^2 - 4},$$

we see that $y = x$ is a skew asymptote:

$$\lim_{x\to\pm\infty}(f(x)-x) = \lim_{x\to\pm\infty}\left(x+\frac{4x}{x^2-4}-x\right) = \lim_{x\to\pm\infty}\frac{4x}{x^2-4} = \lim_{x\to\pm\infty}\frac{4x/x^2}{1-4/x^2} = 0.$$

3. We calculate that

$$\frac{d}{dx}\left(\frac{x^3}{x^2-4}\right) = \frac{x^2(x^2-12)}{(x^2-4)^2}.$$

The first derivative is defined at *all* points of the domain of f. Since f' vanishes at 0 and $\pm\sqrt{12}$, these are the only critical points.

4. The first derivative changes sign at $-\sqrt{12}$ and $\sqrt{12}$ but not at 0. We see that $f' > 0$ on $(-\infty, -\sqrt{12})$ and on $(\sqrt{12}, \infty)$, and $f' < 0$ on $(-\sqrt{12}, -2)$, $(-2, 2)$, and $(2, \sqrt{12})$.

5. Differentiating the expression in step 3, we find that the second derivative of f is

$$f''(x) = 8x\frac{x^2+12}{(x^2-4)^3}.$$

The factor x^2+12 is always positive and doesn't affect the sign of f''. The sign of f'' can change only at a zero of f'' ($x = 0$) or at a point where f'' doesn't exist ($x = \pm 2$). To determine the sign of f'' on the intervals $(-\infty, -2)$, $(-2, 0)$, $(0, 2)$, and $(2, \infty)$, we need only evaluate f'' at one point in each interval. Since $f''(-3) = -504/125$, $f''(-1) = 104/27$, $f''(1) = -104/27$, and $f''(3) = 504/125$, we deduce that f is concave up over $(-2, 0)$ and $(2, \infty)$ and concave down over $(-\infty, -2)$ and $(0, 2)$.

6. From step 5, we know the concavity changes at -2, 0, and 2. Since -2 and 2 are not in the domain of f, $(0, 0)$ is the only point of inflection. From step 4, we see that a local maximum occurs at $-\sqrt{12}$ and a local minimum at $\sqrt{12}$.

7. The graph of f crosses the axes only at $(0, 0)$. In Figure 8, we have plotted this point together with the point of inflection and the local extrema. We have also drawn in the vertical asymptotes and the skew asymptote.

8. The final sketch of the graph of f is in Figure 8. All of the pertinent data have been indicated. ■

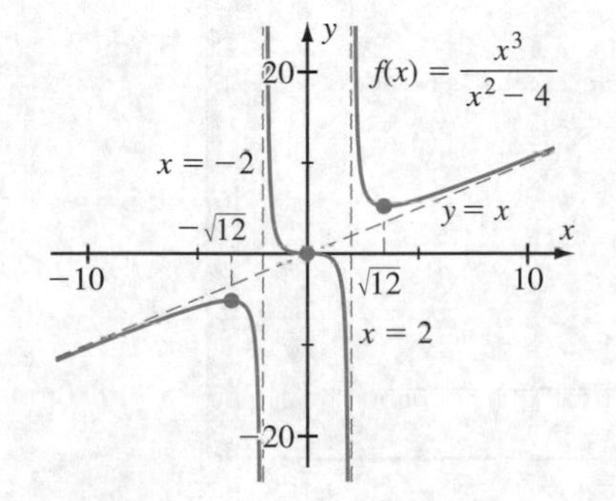

Figure 8

Graphing Calculators and Software

Graphing calculators and plotting software are convenient tools. However, we must understand their limitations and, when needed, be able to correct the plots that they render. Figure 9 on the next page is a screen capture in which a computer algebra system has been used to plot

$$f(x) = \frac{x(\tan(x)-\sin(x))^2}{\tan(\sin(x))-\sin(\tan(x))}$$

over the interval $[-0.01, 0.01]$. Even though 20 digits have been carried in the computation, a loss of significance (see Section 1.1) has resulted in a wildly inaccurate plot. Although a computer algebra system does allow us to increase the number of digits until a reliable plot is attained, this resource may not be an option with a graphing calculator.

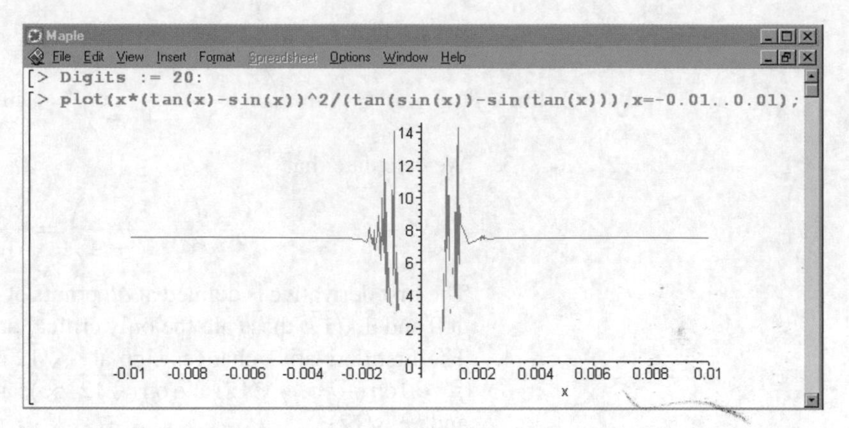

Figure 9

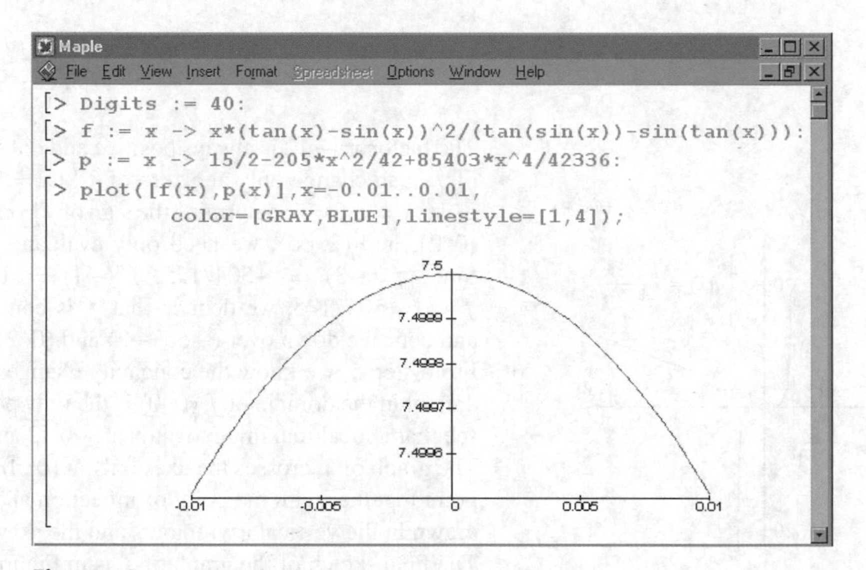

Figure 10

In any case, an important approximation technique of calculus (to be learned in Chapter 10) shows that

$$f(x) = \frac{15}{2} - \frac{205}{42}x^2 + \frac{85403}{42336}x^4 + E(x)$$

where the summand $E(x)$ is negligible over the interval $[-0.1, 0.1]$. By dropping the term $E(x)$, we obtain a good polynomial approximation that is easily and accurately plotted. Although Figure 10 shows only one curve, we are actually viewing both the plot of f, rendered using 40 digits, and the plot of $y = 15/2 - 205x^2/42 + 85403x^4/42336$. The differences between the values of $f(x)$ and the approximating polynomial would not have been visible had we not plotted the second curve with a dashed line style.

There are many situations in which calculator or computer plots can be deceptive. Often the solution is to find a better viewing window. In Figure 11, we plotted the function $f(x) = 100x$ and the function $g(x) = 100x + \sin(x^4)$ over the interval $[0, 2]$. The graphs of f and g over this interval appear to be the same line segment. Yet when we zoom in to the subinterval $[1.1, 1.2]$, as in Figure 11, the graphs of f and g appear to be two different line segments. Of course, no part of the graph of g is actually a line segment. Over the interval [10, 12], the plots of f and g once again appear virtually indistinguishable (Figure 12). When we zoom in to the subinterval [10.1, 10.11], as in Figure 12, we see the expected sinusoidal oscillations that the plot of g has about the line $y = 100x$.

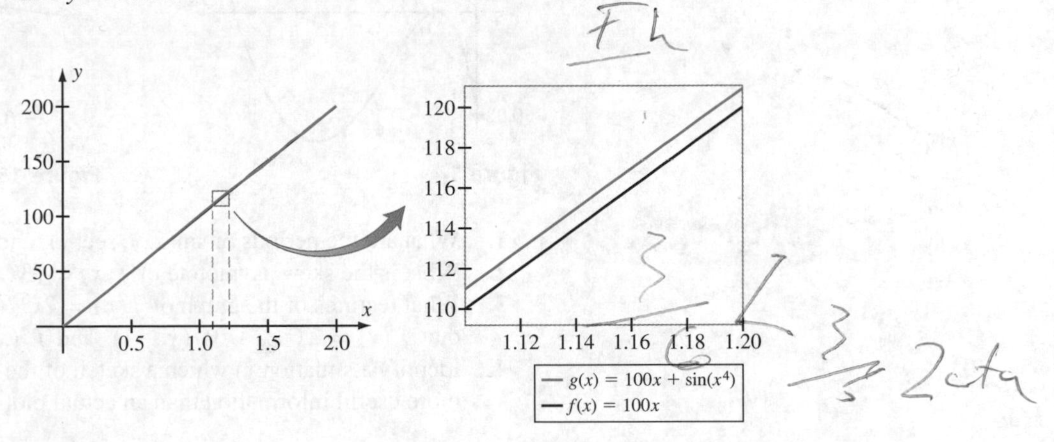

Figure 11

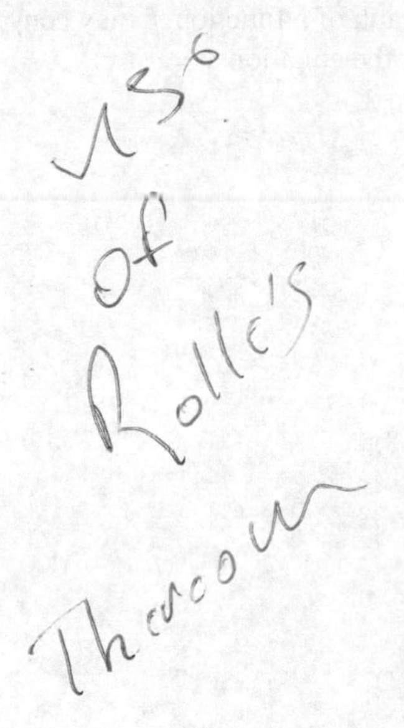

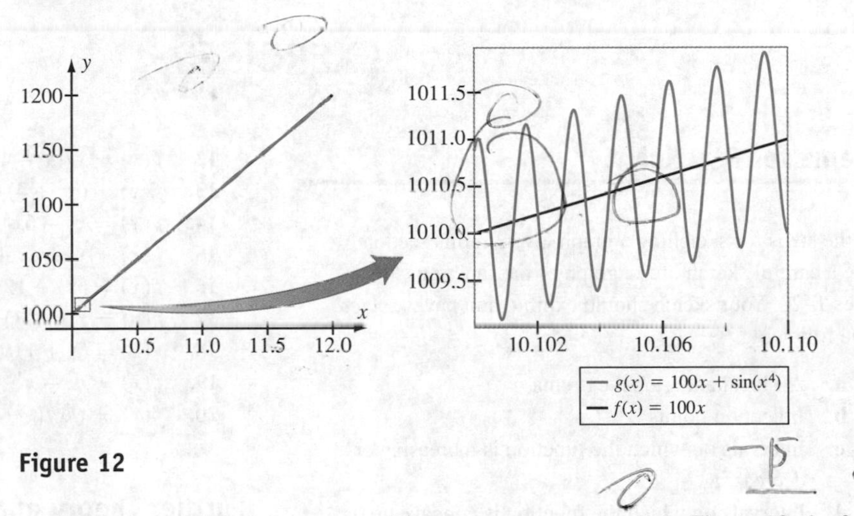

Figure 12

Sometimes a sketch is more useful than an accurate graph. In Figure 13, the function

$$f(x) = \frac{(x-2)^2(x-3)}{x^2-1}$$

has been plotted in the viewing rectangle $[-15, 15] \times [-50, 50]$. This plot reveals many features of f: a local maximum near $x = -5$, a local minimum near $x = 1/2$, vertical asymptotes $x = -1$ and $x = 1$, and a skew asymptote $y = x - 7$. However, this plot hides important information and, as a result, makes $f(x)$ appear to be increasing for $x > 1$. When we plot f over the interval [1.8, 3.2] (see Figure 14, next page), we discover two local extrema, a point of inflection, and a second x-intercept, none of which is visible in Figure 13. We may incorporate all the information about f into a *sketch,* as in Figure 15, next page. However, Figure 15 was obtained by distorting the distance between the two x-intercepts. There is simply no scale available that allows *all* features of the graph of f to be clearly visible in *one* viewing window.

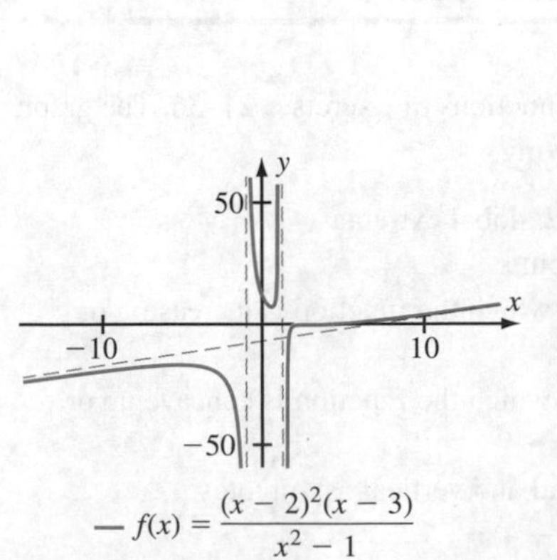

Figure 13

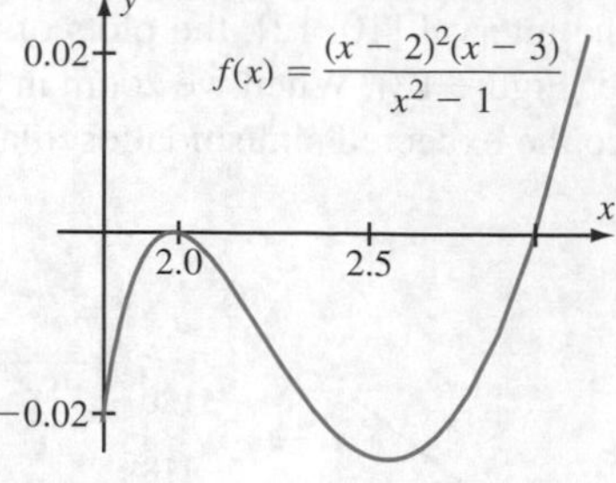

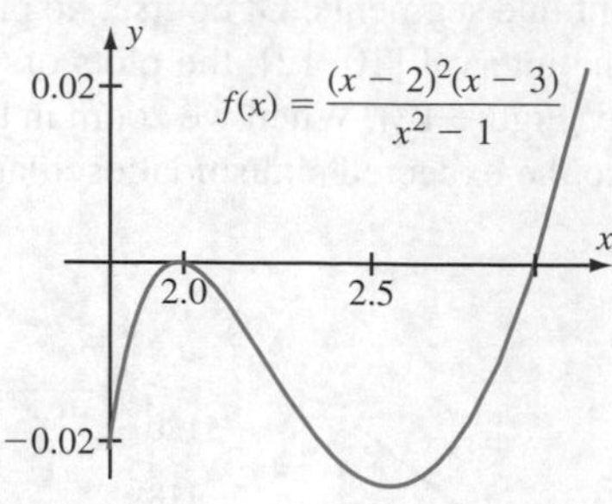

Figure 14

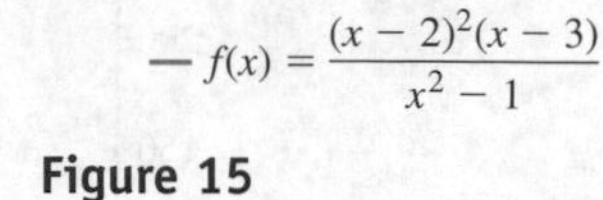

Figure 15

quickquiz

1. What are the periods of $\sin(2x)$, $\sec(x)$, and $\tan(x)$?
2. What is the skew asymptote of $f(x) = (3x^2 - 2x + 5\cos(x))/x$?
3. What features of the graph of $f(x) = 2x^3/(x-1)^2$ can be identified, given that $f'(x) = 2x^2(x-3)/(x-1)^3$ and $f''(x) = 12x/(x-1)^4$?
4. Identify a situation in which a sketch of the graph of a function f may convey more useful information than an actual plot of the equation $y = f(x)$.

EXERCISES

Problems for Practice

Follow the steps for sketching a graph given in this section to create a careful sketch of the graph of the functions in Exercises 1–20. Your sketch should exhibit, and have *labeled*, all of the following.

a. All local and global extrema
b. Inflection points
c. Intervals on which the function is increasing or decreasing
d. Intervals on which the function is concave up or concave down
e. All horizontal and vertical asymptotes

1. $f(x) = x^3 - 3x^2 - 9x + 7$
2. $f(x) = x^3(x-8)$
3. $f(x) = (x^{1/3} + 1)/x$
4. $f(x) = x^{-4/3}(x+2)$
5. $f(x) = x^{1/3}(x-2)$
6. $f(x) = \cos(x) - \sin(x)$
7. $f(x) = (x+1)(x-2)^2$
8. $f(x) = |x|/(x+1)$
9. $f(x) = \sin(x) - x$
10. $f(x) = (x+1)/(x+2)^2$
11. $f(x) = x/(x^2+4)$
12. $f(x) = x/(x-4)^2$
13. $f(x) = (x^2+4)/(x^2-4)$
14. $f(x) = x^5 - 6x^4 + 8x^3$
15. $f(x) = x^{1/3}/(x+3)$
16. $f(x) = (x^3 - 12x)/(x^2+4)$
17. $f(x) = x(x+1)^{1/5}$
18. $f(x) = (x+1)/(x-1)$
19. $f(x) = x^2 - x^{1/2}$
20. $f(x) = \sqrt{x}/(\sqrt{x} - 1)$

Further Theory and Practice

Sketch the graph of the functions in Exercises 21–36. The graph should include the following.

a. All local and global extrema
b. Inflection points
c. Intervals on which the function is increasing or decreasing
d. Intervals on which the function is concave up or concave down
e. All horizontal and vertical asymptotes.

21. $f(x) = x^{1/3}/(x-4)$
22. $f(x) = (1/2) \cdot x^{2/3} - x^{1/3}$

23. $f(x) = x^2(1 - x^2)^{1/2}$
24. $f(x) = x^2/(x^{1/3} + 8)$
25. $f(x) = \sqrt{(x^2 + x - 20)/(x^2 - 2x - 3)}$
26. $f(x) = |x^2 - 9|$
27. $f(x) = (x - 5)\sqrt{|x + 2|}$
28. $f(x) = \cos^4(x)$
29. $f(x) = (x^3 + x + 1)/x$
30. $f(x) = \sin(x) + \cos(x)$
31. $f(x) = \tan(x) + \sec(x)$
32. $f(x) = |x^{1/3} - 4|$
33. $f(x) = |x| \cdot (x + 1)^2$
34. $f(x) = (x + 1)/\sqrt{(x - 1)^2}$
35. $f(x) = (13x + 14)/(x^2 - 1)$
36. $f(x) = (3x + 2)/|x^2 - 1|$

In Exercises 37–38, repeat the instructions for Exercises 21–36. Also determine the skew asymptote of the given function.

37. $f(x) = (x^2 - 4)/(2x)$
38. $f(x) = (x^3 + x^2 + 2)/x^2$
39. If $f : \mathbb{R} \to \mathbb{R}$ is twice differentiable and if $f''(x) \geq 1$ for all x, then prove that the graph of f cannot have any horizontal asymptotes.
40. A graph is called *symmetric with respect to the y-axis* if whenever (x, y) lies on the graph, $(-x, y)$ lies on the graph. We test for symmetry with respect to the y-axis by replacing x with $-x$ in the equation for the graph. If the equation remains unchanged, the symmetry property holds. Explain the reasoning behind this test.

    A graph is called *symmetric with respect to the origin* if whenever (x, y) lies on the graph, $(-x, -y)$ lies on the graph. We test for symmetry with respect to the origin by replacing x with $-x$ and y with $-y$ in the equation for the graph. If the equation remains unchanged, the symmetry property holds. Explain the reasoning behind this test.

    Test the following equations for symmetry in the y-axis and symmetry in the origin. Give a sketch of each graph.

    a. $9y^2 - x^2 = 36$ **b.** $x^3 + y^2 = 4$
    c. $x^2 - 4y^2 = 16$ **d.** $x + \sin(y) = 3y$

Calculator/Computer Exercises

Follow the steps for sketching a graph given in this section to create a careful sketch of the graph of the functions in Exercises 41–46. Your sketch should exhibit, and have *labeled,* all of the following.

a. All local and global extrema
b. Inflection points
c. Intervals on which the function is increasing or decreasing
d. Intervals on which the function is concave up or concave down
e. All horizontal, vertical, and skew asymptotes

41. $f(x) = x^4 + x^3 + x^2 + x - 4$
42. $f(x) = x^5 + x^3 + 2x^2$
43. $f(x) = x^5 + x^4 - 5x^3 - 5x^2 + 6x + 6$
44. $f(x) = \dfrac{(x - 1)(x - 4)}{(x - 3)(x - 2)}$
45. $f(x) = x^2 \sin(\pi x) + \sqrt{1 - x^2}$
46. $f(x) = \dfrac{(x - 3)^2(x - 4)}{(x - 1)(x - 2)}$

In Exercises 47–50, use a computer algebra system to create a careful sketch of the graph of the functions. Your sketch should exhibit, and have *labeled,* all of the following.

a. All local and global extrema
b. Inflection points
c. Intervals on which the function is increasing or decreasing
d. Intervals on which the function is concave up or concave down
e. All horizontal, vertical, and skew asymptotes

47. $f(x) = \dfrac{\sqrt{4 - x^2}}{x^2 + 1}$
48. $f(x) = \dfrac{x^3 + x + 1}{x^2 + 4x + 5}$
49. $f(x) = \dfrac{x^3 + 1.1x + 2.3}{\sqrt{16 - x^2}}$
50. $f(x) = \dfrac{x^2 - x + 2}{\sqrt{x^4 + 2x^2 + 2}}$
51. Plot
$$f(x) = \frac{x^2 - 1}{x^2 + 1} \quad \text{and} \quad g(x) = \frac{x^4 - 102x^2 + 100}{x^4 - 99x^2 - 100}$$
over the intervals [0, 8] and [8, 10]. Explain why the graphs over the first interval are so similar and why the graphs over the second interval become very different.
52. Plot the function
$$f(x) = \frac{3x^5 + 10000x^4}{x^5 + 10000x^4 + x + 10000}$$
over the interval $[-15, 15]$. What appears to be the asymptote for f? Analyze f for asymptotes, and plot f in a window that better illustrates its asymptote.
53. Plot $f(x) = 12x^5 - 2565x^4 + 146200x^3 + 1$ for $x \in I = [0, 300]$. Plot f' and f'' for $x \in I$. Is f increasing on I? Is $f' > 0$ on I? Is the graph of f concave up on I? Is $f'' > 0$ on I?

4.6 L'Hôpital's Rule

Consider the limit

$$\lim_{x\to c} \frac{f(x)}{g(x)}. \tag{4.8}$$

If $\lim_{x\to c} f(x)$ exists and if $\lim_{x\to c} g(x)$ exists and is not zero, then the limit (4.8) is straightforward to evaluate (see Section 2.2). However, when $\lim_{x\to c} g(x) = 0$, the situation is more complicated. For example, if $f(x) = \sin(x)$ and $g(x) = x$, then we know that the limit of the quotient as $x \to 0$ exists and equals 1 (see Section 2.2). However, if $f(x) = x$ and $g(x) = x^2$, then the limit of the quotient as $x \to 0$ does not exist.

In this section, we learn a rule for evaluating limits of the type in (4.8) when either $\lim_{x\to c} f(x) = \lim_{x\to c} g(x) = 0$ or $\lim_{x\to c} f(x) = \lim_{x\to c} g(x) = \infty$. In these cases, when the limits of the numerator and denominator of (4.8) are evaluated independently, the quotient takes the form $\frac{0}{0}$ or $\frac{\infty}{\infty}$. Such forms of the limit are called *indeterminate forms* because the symbols $\frac{0}{0}$ and $\frac{\infty}{\infty}$ have no meaning. The limit might actually exist and be finite, or it might not exist. One cannot analyze such a form simply by evaluating the limits of the numerator and denominator and forming their quotient.

L'Hôpital's Rule for the Indeterminate Form $\frac{0}{0}$

Suppose that f' and g' are continuous at c and $f(c) = g(c) = 0$. Provided there are no divisions by zero, we have

$$\lim_{x\to c} \frac{f(x)}{g(x)} = \lim_{x\to c} \frac{f(x)-f(c)}{g(x)-g(c)} = \lim_{x\to c} \frac{\frac{f(x)-f(c)}{x-c}}{\frac{g(x)-g(c)}{x-c}} = \frac{\lim_{x\to c} \frac{f(x)-f(c)}{x-c}}{\lim_{x\to c} \frac{g(x)-g(c)}{x-c}} = \frac{f'(c)}{g'(c)}$$

and

$$\lim_{x\to c} \frac{f'(x)}{g'(x)} = \frac{\lim_{x\to c} f'(x)}{\lim_{x\to c} g'(x)} = \frac{f'(c)}{g'(c)}.$$

Therefore,

$$\lim_{x\to c} \frac{f(x)}{g(x)} = \lim_{x\to c} \frac{f'(x)}{g'(x)}.$$

This observation suggests the following procedure for calculating limit (4.8).

Theorem 1 ***L'Hôpital's Rule*** Let f and g be differentiable functions on $(a, c) \cup (c, b)$. If

$$\lim_{x\to c} f(x) = \lim_{x\to c} g(x) = 0,$$

then

$$\lim_{x\to c} \frac{f(x)}{g(x)} = \lim_{x\to c} \frac{f'(x)}{g'(x)},$$

provided that the limit on the right exists as a finite or an infinite limit.

A proof of l'Hôpital's Rule is outlined in the exercises. Meanwhile, the following are some examples in which the rule is used.

insight

When we apply l'Hôpital's Rule to a limit of type (4.8), we calculate the derivative of the numerator f and the derivative of the denominator g. We do *not* calculate the derivative of the quotient f/g.

Example 1 Evaluate

$$\lim_{x\to 1} \frac{\ln(x)}{x^2-1}.$$

Solution We first notice that both the numerator and the denominator have limit 0 as $x \to 1$. Thus, the quotient is indeterminate at 1 and is of the form $\frac{0}{0}$. L'Hôpital's Rule applies, so the limit equals

$$\lim_{x\to 1} \frac{\ln(x)}{x^2-1} = \lim_{x\to 1} \frac{\frac{d}{dx}\ln(x)}{\frac{d}{dx}(x^2-1)},$$

provided this last limit exists. We calculate

$$\lim_{x\to 1} \frac{\frac{d}{dx}\ln(x)}{\frac{d}{dx}(x^2-1)} = \lim_{x\to 1} \frac{1/x}{2x} = \frac{1}{2}.$$

The requested limit has value 1/2. ■

Example 2 Evaluate the limit

$$\lim_{x\to 0} \frac{x}{x-\sin(x)}.$$

Solution As $x \to 0$, both numerator and denominator tend to 0, so the quotient is indeterminate of the form $\frac{0}{0}$. Thus, l'Hôpital's Rule applies. Our limit equals

$$\lim_{x\to 0} \frac{\frac{d}{dx}x}{\frac{d}{dx}(x-\sin(x))} = \lim_{x\to 0} \frac{1}{1-\cos(x)} = +\infty.$$

We conclude that the original limit equals $+\infty$. ■

Example 3 Calculate

$$\lim_{x\to 0} \frac{\sin(3x)}{\sin(2x)} \quad \text{and} \quad \lim_{x\to \pi/6} \frac{\sin(3x)}{\sin(2x)}.$$

Solution With the first limit, we may proceed as in Examples 1 and 2. We summarize the calculations:

$$\lim_{x\to 0} \frac{\sin(3x)}{\sin(2x)} = \lim_{x\to 0} \frac{\frac{d}{dx}\sin(3x)}{\frac{d}{dx}\sin(2x)} = \lim_{x\to 0} \frac{3\cos(3x)}{2\cos(2x)} = \frac{3}{2}.$$

For the second limit, we calculate $\lim_{x\to\pi/6} \sin(3x) = \sin(\pi/2) = 1$ and $\lim_{x\to\pi/6} \sin(2x) = \sin(\pi/3) = \sqrt{3}/2$. Therefore, l'Hôpital's Rule does not apply. However, we do not need it because

$$\lim_{x\to \pi/6} \frac{\sin(3x)}{\sin(2x)} = \frac{\lim_{x\to\pi/6} \sin(3x)}{\lim_{x\to\pi/6} \sin(2x)} = \frac{1}{\sqrt{3}/2} = \frac{2}{\sqrt{3}}.$$

■

It is tempting to apply l'Hôpital's Rule to any limit that is written as a quotient. However, l'Hôpital's Rule is *only* valid when its hypotheses are satisfied. If we had applied l'Hôpital's Rule to the second limit in Example 3, then we would have obtained

$$\lim_{x\to\pi/6}\frac{\sin(3x)}{\sin(2x)} = \lim_{x\to\pi/6}\frac{\frac{d}{dx}\sin(3x)}{\frac{d}{dx}\sin(2x)} = \lim_{x\to\pi/6}\frac{3\cos(3x)}{2\cos(2x)} = \frac{0}{1} = 0,$$

which is wrong. The error here is that we did not check that the numerator and denominator both have limit zero; in other words, we did not check that the quotient is indeterminate at 0.

When we apply l'Hôpital's Rule, we may end up with another indeterminate form. Sometimes we need to apply l'Hôpital's Rule two or more times to evaluate a limit.

Example 4 Evaluate the limit

$$\lim_{x\to\pi}\frac{1+\cos(x)}{(x-\pi)^2}.$$

Solution Notice that both the numerator and denominator tend to 0 as $x \to \pi$. Therefore, the quotient is indeterminate at π and of the form $\frac{0}{0}$. So, we may apply l'Hôpital's Rule to evaluate the limit:

$$\lim_{x\to\pi}\frac{-\sin(x)}{2(x-\pi)},$$

provided this limit exists. However, both the numerator and denominator of this last expression tend to 0 as $x \to \pi$, so we apply l'Hôpital's Rule again. The last limit equals

$$\lim_{x\to\pi}\frac{-\cos(x)}{2},$$

provided this limit exists. The limit does exist, and it equals $1/2$. We conclude that

$$\lim_{x\to\pi}\frac{1+\cos(x)}{(x-\pi)^2} = \frac{1}{2}.$$

■

L'Hôpital's Rule for the Indeterminate Form $\frac{\infty}{\infty}$

There is another version of l'Hôpital's Rule that we can use for the indeterminate form $\frac{\infty}{\infty}$.

Theorem 2 Let $f(x)$ and $g(x)$ be differentiable functions on $(a, c) \cup (c, b)$. If $\lim_{x\to c} f(x)$ and $\lim_{x\to c} g(x)$ both exist and equal $+\infty$ or $-\infty$ (they may have the same sign or different

insight

In Example 5, a formal limit computation results in $\lim_{x\to 0} x^2 \cdot \ln(|x|) = 0\cdot(-\infty)$. Although the expressions $0\cdot\infty$ and $0\cdot(-\infty)$ are also called indeterminate forms, l'Hôpital's Rule does not directly address them. An algebraic manipulation is required to rewrite the limit in a way that is appropriate for l'Hôpital's Rule.

signs), then

$$\lim_{x\to c}\frac{f(x)}{g(x)} = \lim_{x\to c}\frac{f'(x)}{g'(x)},$$

provided this last limit exists either as a finite or an infinite limit.

Example 5 Evaluate the limit $\lim_{x\to 0} x^2 \cdot \ln(|x|)$.

Solution This may be rewritten as $\lim_{x\to 0} \ln(|x|)/(1/x^2)$. Notice that the numerator tends to $-\infty$ and the denominator tends to $+\infty$ as $x \to 0$. Thus, the quotient is indeterminate at 0 and of the form $\frac{-\infty}{+\infty}$. Thus, we may apply Theorem 2 to see that the limit equals

$$\lim_{x\to 0}\frac{1/x}{-2x^{-3}} = \lim_{x\to 0} -\frac{x^2}{2} = 0.$$

■

L'Hôpital's Rule for Limits at ∞

Theorem 3

Let f and g be differentiable functions. If $\lim_{x\to+\infty} f(x) = \lim_{x\to+\infty} g(x) = 0$ or if $\lim_{x\to+\infty} f(x) = \pm\infty$ and $\lim_{x\to+\infty} g(x) = \pm\infty$, then

$$\lim_{x\to+\infty}\frac{f(x)}{g(x)} = \lim_{x\to+\infty}\frac{f'(x)}{g'(x)},$$

provided this last limit exists either as a finite or an infinite limit. The same result holds for the limit as $x \to -\infty$.

Example 6 Evaluate $\lim_{x\to+\infty} x^2/e^x$.

Solution We first notice that both the numerator and the denominator tend to $+\infty$ as $x \to +\infty$. Thus, the quotient is indeterminate at $+\infty$ and of the form $\frac{\pm\infty}{+\infty}$. Therefore, Theorem 3 applies. Our limit equals $\lim_{x\to+\infty} 2x/e^x$. Again, the numerator and denominator tend to $+\infty$ as $x \to +\infty$, so we again apply Theorem 3. The limit equals $\lim_{x\to+\infty} 2/e^x = 0$. We conclude that $\lim_{x\to+\infty} x^2/e^x = 0$. ■

The Use of the Logarithm

We can use the natural logarithm to reduce an expression involving exponentials to one involving a product or a quotient.

The Indeterminate Form 0^0

Example 7 Evaluate $\lim_{x\to 0^+} x^x$.

Solution We study the limit of $f(x) = x^x$ by considering $\ln(f(x)) = x \cdot \ln(x)$. We rewrite this as

$$\lim_{x\to 0^+} \ln(f(x)) = \lim_{x\to 0^+}\frac{\ln(x)}{1/x}.$$

Both numerator and denominator tend to $\pm\infty$, so the quotient is indeterminate and of the form $\frac{-\infty}{\infty}$. Thus, l'Hôpital's Rule applies. The limit equals

$$\lim_{x\to 0^+} \frac{1/x}{-1/x^2} = \lim_{x\to 0^+} (-x) = 0.$$

The only way that $\ln(f(x))$ can tend to 0 is if $f(x) = x^x$ tends to 1. We conclude that $\lim_{x\to 0^+} x^x = 1$. ■

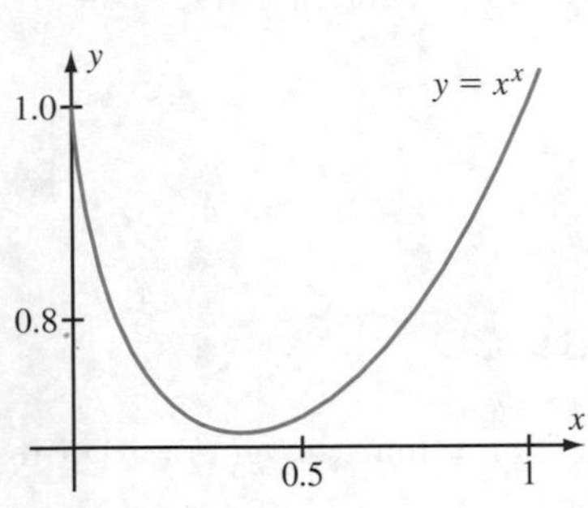

Figure 1

inSIGHT

The limit in Example 7 is not obvious. For example, if $p > 0$ is a *fixed* power, then $\lim_{x\to 0^+} x^p = 0$. A plot of x^x for $0 < x < 1$ (see Figure 1) provides us with visual evidence that our calculation is correct. Moreover, numerical evidence, shown in the table, also supports our conclusion.

x	0.1	0.01	0.001	0.0001	0.00001
x^x	0.7943	0.9550	0.9931	0.9991	0.9999

The Indeterminate Form 1^∞

Example 8 Evaluate $\lim_{x\to\infty}(1 + 3/x)^x$.

Solution Let $f(x) = (1 + 3/x)^x$, and consider $\ln(f(x)) = \ln((1 + 3/x)^x) = x \cdot \ln(1 + 3/x)$. This expression is indeterminate of the form $\infty \cdot 0$.

We rewrite it as

$$\lim_{x\to\infty} \frac{\ln(1 + 3/x)}{1/x},$$

so that both the numerator and denominator tend to 0. L'Hôpital's Rule applies, and we have

$$\lim_{x\to\infty} \ln(f(x)) = \lim_{x\to\infty} \frac{(1/(1 + 3/x))(-3/x^2)}{-1/x^2} = 3 \lim_{x\to\infty} \frac{1}{(1 + 3/x)} = 3.$$

But the only way that $\ln(f(x))$ can tend to 3 is if $f(x)$ tends to e^3. We conclude that $\lim_{x\to\infty}(1 + 3/x)^x = e^3$, which agrees with formula (2.17) from Section 2.6. ■

The Indeterminate Form ∞^0

Example 9 Calculate $\lim_{x\to\infty} x^{1/x}$.

Solution We study the limit of $f(x) = x^{1/x}$ by considering $\ln(f(x)) = \ln(x)/x$, which gives us an indeterminate form that is suitable for l'Hôpital's Rule. We have

$$\lim_{x\to\infty} \ln(f(x)) = \lim_{x\to\infty} \frac{\ln(x)}{x} = \lim_{x\to\infty} \frac{\frac{d}{dx}\ln(x)}{\frac{d}{dx}x} = \lim_{x\to\infty} \frac{1/x}{1} = 0.$$

The only way that $\ln(f(x))$ can tend to 0 as x tends to infinity is if $f(x) = x^{1/x}$ tends to 1. We conclude that $\lim_{x\to\infty} x^{1/x} = 1$. ■

Putting Terms over a Common Denominator

Many times a simple algebraic manipulation will put a limit into a form that we can study using l'Hôpital's Rule.

Example 10 Evaluate the limit

$$\lim_{x\to 0}\left(\frac{1}{\sin(x)} - \frac{1}{x}\right).$$

insight

Notice that the original expression in Example 10 leads to the indeterminate form $\infty - \infty$. L'Hôpital's Rule shows us that even though the fractions have individual infinite limits, their difference involves a subtle cancellation and yields a finite limit.

Solution We put the fractions over a common denominator to rewrite our limit as

$$\lim_{x\to 0}\left(\frac{x - \sin(x)}{x \cdot \sin(x)}\right).$$

Both numerator and denominator vanish as $x \to 0$. Thus, the quotient has indeterminate form $\frac{0}{0}$. By l'Hôpital's Rule, the limit is equal to

$$\lim_{x\to 0}\frac{1 - \cos(x)}{\sin(x) + x\cos(x)}.$$

This quotient is still indeterminate; we apply l'Hôpital's Rule again to obtain

$$\lim_{x\to 0}\frac{\sin(x)}{2\cos(x) - x\sin(x)} = 0.$$

■

A LOOK BACK

We have encountered seven types of indeterminate forms. The two basic indeterminate forms $\frac{0}{0}$ and $\frac{\infty}{\infty}$ can be handled directly with l'Hôpital's Rule. The indeterminate form $0\cdot\infty$ is usually converted to one of the basic forms by placing the reciprocal of one of the factors in the denominator. Thus, $0\cdot\infty$ is symbolically presented as

$$\frac{\infty}{1/0} \quad \text{or} \quad \frac{0}{1/\infty}.$$

Some algebraic manipulation is necessary before l'Hôpital's Rule can be applied to the indeterminate form $\infty - \infty$. The three indeterminate forms that involve exponents, namely, 0^0, 1^∞, and ∞^0, are converted to the indeterminate form $0\cdot\infty$ by applying the logarithm.

quickquiz

1. What important hypothesis must be verified before l'Hôpital's Rule can be applied to the limit of a quotient?
2. How can we apply l'Hôpital's Rule to evaluate $\lim_{x\to 0^+} x \cdot \ln(x)$?
3. For which indeterminate forms is it helpful to use the logarithm?
4. Evaluate $\lim_{x\to+\infty}(\sqrt{x+1} - \sqrt{x})$.

EXERCISES

Problems for Practice

In Exercises 1–16, use a form of l'Hôpital's Rule to find the limit, if it exists.

1. $\lim_{x\to 0} \frac{1-e^x}{x}$ **2.** $\lim_{x\to +\infty} \frac{\ln(x)}{\sqrt{x}}$

3. $\lim_{x\to 5} \frac{\ln(x/5)}{x-5}$ **4.** $\lim_{x\to 0} \frac{x+\sin(5x)}{x-3\sin(4x)}$

5. $\lim_{x\to \pi/2} \frac{\ln(\sin(x))}{(\pi-2x)^2}$ **6.** $\lim_{x\to 0} \frac{\cos(x)-1}{e^x-1}$

7. $\lim_{x\to -1} \frac{\cos(x+1)-1}{x^3+x^2-x-1}$

8. $\lim_{x\to 0} \frac{e^x-e^{-x}}{x}$

9. $\lim_{x\to 0} \frac{e^x-e^{-x}}{x^2}$ **10.** $\lim_{x\to -1} \frac{x+x^2}{\ln(2+x)}$

11. $\lim_{x\to 1} \frac{\ln(x)}{x-\sqrt{x}}$ **12.** $\lim_{x\to 0} \frac{\sin^2(3x)}{1-\cos(4x)}$

13. $\lim_{x\to +\infty} \frac{\sin(3/x)}{\sin(9/x)}$ **14.** $\lim_{x\to +\infty} \frac{\sin^2(2/x)}{3/x}$

15. $\lim_{x\to -\infty} \frac{\ln(1+1/x)}{\sin(1/x)}$ **16.** $\lim_{x\to \pi/2} \frac{\tan(2x)}{\cot(x)}$

In Exercises 17–24, apply l'Hôpital's Rule repeatedly to evaluate the limit, if it exists.

17. $\lim_{x\to\infty} x^2/e^{3x}$
18. $\lim_{x\to\infty} x^3/e^{2x}$
19. $\lim_{x\to 0} \sin^2(x)/x^2$
20. $\lim_{x\to 1}(x-1)^3/(\ln(x))^2$
21. $\lim_{x\to 1} (x-1)^2/\cos^2(\pi x/2)$
22. $\lim_{x\to 0}(1-\cos(4x))/\sin^2(x)$
23. $\lim_{x\to 0} (e^{2x}-e^{3x})/x^2$
24. $\lim_{x\to 0} (e^x-e^{-x}-2x)/\sin(x^2)$

In Exercises 25–32, use an algebraic manipulation to put the limit in a form that can be treated using l'Hôpital's Rule. Evaluate the limit.

25. $\lim_{x\to+\infty} x\cdot e^{-2x}$ **26.** $\lim_{x\to 0}(e^x-e^{-x})\cdot x^{-1}$
27. $\lim_{x\to\infty} e^{-x}\ln(x)$ **28.** $\lim_{x\to+\infty} x^{-2}\ln(x)$
29. $\lim_{x\to 1^+}(x-1)^{-1}\ln(x)$
30. $\lim_{x\to 0}\sin(x)\cot(3x)$
31. $\lim_{x\to -2}(x+2)^2\tan(\pi x/4)$
32. $\lim_{x\to\pi}\tan(x/2)\sin(3x)$

In Exercises 33–40, use the logarithm to reduce the given limit to one that can be handled with l'Hôpital's Rule.

33. $\lim_{x\to 0^+} x^{\sqrt{x}}$ **34.** $\lim_{x\to 0^+}(\ln(x))^x$
35. $\lim_{x\to 0^+}(\sqrt{x})^{\sqrt{x}}$ **36.** $\lim_{x\to+\infty}(x\cdot\ln(x))^{-1/x}$
37. $\lim_{x\to+\infty} x^{1/x}$ **38.** $\lim_{x\to -1^+}(2+x)^{\ln(x+1)}$
39. $\lim_{x\to+\infty} x^{1/\sqrt{x}}$ **40.** $\lim_{x\to-\infty}(x^2)^{(e^x)}$

In Exercises 41–46, put the fraction over a common denominator and use l'Hôpital's Rule to evaluate the limit, if it exists.

41. $\lim_{x\to\pi/2}\left(\frac{1}{\cos(x)}-\frac{1}{\cos(3x)}\right)$

42. $\lim_{x\to 0}\left(\frac{1}{x}-\frac{1}{\ln(x+1)}\right)$

43. $\lim_{x\to 0}\left(\frac{x}{1-\cos(x)}-\frac{2}{x}\right)$

44. $\lim_{x\to 0}\left(\frac{1}{3x}-\frac{1}{3\sin(x)}\right)$

45. $\lim_{x\to 0}\left(\frac{1}{\sin x}-\frac{1}{\sin(2x)}\right)$

46. $\lim_{x\to 0}\left(\frac{1}{e^x+e^{-x}-2}-\frac{1}{x^2}\right)$

In Exercises 47–52, use an algebraic manipulation to reduce the limit to one that can be treated with l'Hôpital's Rule.

47. $\lim_{x\to+\infty}(\sqrt{4x-5}-2\sqrt{x})$
48. $\lim_{x\to+\infty}(\sqrt{x^2+6x}-x)$
49. $\lim_{x\to-\infty}(x+1)\ln(x/(x+1))$
50. $\lim_{x\to+\infty}((x^2+x)^{1/2}-(x^2+1)^{1/2})$
51. $\lim_{x\to+\infty}(e^{2x}-e^x)^{1/2}-e^x$
52. $\lim_{x\to+\infty}((x^6+4x)^{1/2}-(x^3+1))$

Further Theory and Practice

53. Use l'Hôpital's Rule to check that $\lim_{h\to 0^+}(1+h)^{1/h}=e$. Now calculate $\lim_{h\to 0^+}(1+hx)^{1/h}$ for any fixed x.

54. Use l'Hôpital's Rule to calculate

$$\lim_{h\to 0}\frac{f(x+h)-2f(x)+f(x-h)}{h^2}$$

for a function f that is twice continuously differentiable at x.

55. Let a and b be constants with $b\neq 1$. Use l'Hôpital's Rule to calculate

$$\lim_{x\to 0}\frac{a\sin(x)-\sin(ax)}{\tan(bx)-b\tan(x)}.$$

56. For any real number k and any positive number a, prove that

$$\lim_{x\to+\infty} \frac{x^k}{e^{ax}} = 0.$$

57. Use l'Hôpital's Rule to evaluate each one-sided limit.

a. $\lim_{x\to\pi/2^-} \frac{\cot(x)}{x-\pi/2}$ **b.** $\lim_{x\to\pi^+} \frac{\tan(x/2)}{(x-\pi)^{-1}}$

c. $\lim_{x\to0^+} \frac{\ln(1+x)}{\ln(1+3x)}$ **d.** $\lim_{x\to0^+} \frac{\sin(x)}{\ln(1+x)}$

e. $\lim_{x\to0^+} \frac{\ln(x)}{1/\sqrt{x}}$ **f.** $\lim_{x\to0^+} \frac{\ln(x)}{1/\sin(x)}$

58. For any constants α and β with $\beta > 0$, show that

$$\lim_{x\to\infty} \frac{\ln(x)^\alpha}{x^\beta} = 0.$$

59. Evaluate $\lim_{x\to\infty} \frac{x^{(x+1)/x}}{\sqrt{1+x^2}}$.

60. Cauchy's Mean Value Theorem states: If $f(x)$ and $g(x)$ are functions that are continuous on the closed interval $[a, b]$ and differentiable on the open interval (a, b) and if $g(a) \neq g(b)$, then there is a point $\xi \in (a, b)$ such that

$$\frac{f(b)-f(a)}{g(b)-g(a)} = \frac{f'(\xi)}{g'(\xi)}.$$

Complete the following outline to obtain a proof of Cauchy's Mean Value Theorem.

a. Define the function

$$r(t) = (f(b)-f(a))g(t) - (g(b)-g(a))f(t) + (f(a)(g(b)-g(a)) - g(a)(f(b)-f(a))).$$

b. Check that $r(a) = r(b) = 0$.

c. Apply Rolle's Theorem to the function r on the interval $[a, b]$ to conclude that there is a point $\xi \in (a, b)$ such that $r'(\xi) = 0$.

d. Rewrite the conclusion of part c to obtain the conclusion of Cauchy's Mean Value Theorem.

61. Write the equation of the line through the points $(g(a), f(a))$ and $(g(b), f(b))$. If $P = (x, y)$ is a point in the plane, calculate the vertical distance from P to this line. Now give a geometric interpretation of the function r in Exercise 60, part a. Explain the motivation of the proof of Cauchy's Mean Value Theorem.

62. Use Cauchy's Mean Value Theorem to prove the first form of l'Hôpital's Rule (Theorem 1).

Calculator/Computer Exercises

In Exercises 63–66, investigate the given limit numerically and graphically.

63. $\lim_{x\to0}(1-\cos(x))/\sin(x)^2$

64. $\lim_{x\to0} \tan(2\sin(x))/\sin(2\tan(x))$

65. $\lim_{x\to0^+} x^{(1-\sin(x)/x)}$

66. $\lim_{x\to0}(1+\tan(2x))^{1/x}$

In Exercises 67–69, demonstrate the assertion of l'Hôpital's Rule by plotting f/g and f'/g' in a neighborhood of c.

67. $f(x) = 1-\cos(x)$, $g(x) = x\sin(x)$, $c = 0$

68. $f(x) = 1+x-e^x$, $g(x) = x\sin(x)$, $c = 0$

69. $f(x) = 1+\cos(x)$, $g(x) = (x-\pi)\sin(x)$, $c = \pi$

70. Example 7 shows that $\lim_{x\to0^+} x^x = 1$. Plot $y = x^x$ for values of x that have the form $-p/q$ where p and q are positive integers with q odd. Let $\{x_n\}$ be a sequence of such negative rational numbers. If $\lim_{n\to\infty} x_n = 0$, then what can be said of $\lim_{n\to\infty} x_n^{x_n}$?

4.7 The Newton-Raphson Method

All quadratic equations $ax^2 + bx + c = 0$ can be solved explicitly by means of the quadratic formula:

$$x = \frac{-b \pm \sqrt{b^2-4ac}}{2a}.$$

There are formulas for solving cubic and fourth-degree equations as well. However, there are no elementary formulas for solving general polynomial equations of degree five and higher. If we turn to *transcendental* equations, such as

$$\frac{\pi}{3} = x - \frac{1}{2}\sin(x),$$

then matters are even more obscure. Yet there is often a need to solve such equations, as we will see in several applied problems.

If we settle for an approximate solution of an equation (with, say, a specified number of decimal places of accuracy), then calculus enables us to find sufficiently accurate solutions for many equations that come up in practice. The technique that we will study in this section is called the *Newton-Raphson Method* (or sometimes just *Newton's Method*).

The Geometry of the Newton-Raphson Method

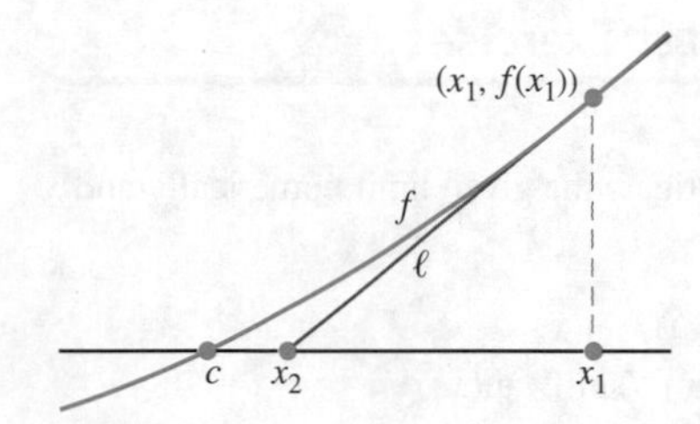

Figure 1

Refer to Figure 1 to see the fundamental idea behind the Newton-Raphson Method: The graph of a differentiable function f is approximated well, near a point $(x_1, f(x_1))$, by the tangent line ℓ at the point $(x_1, f(x_1))$. The figure suggests that if x_1 is near a point c where the graph of f crosses the x-axis, then the point x_2 at which ℓ intersects the axis will be even closer. This simple observation is the basic idea of the Newton-Raphson Method.

Example 1 Let P denote the point at which the graphs of $y = \cos(x)$ and $y = x$ cross. A very rough estimate of the x-coordinate of P is $\pi/3$. Apply the fundamental idea behind the Newton-Raphson Method to find a better approximation.

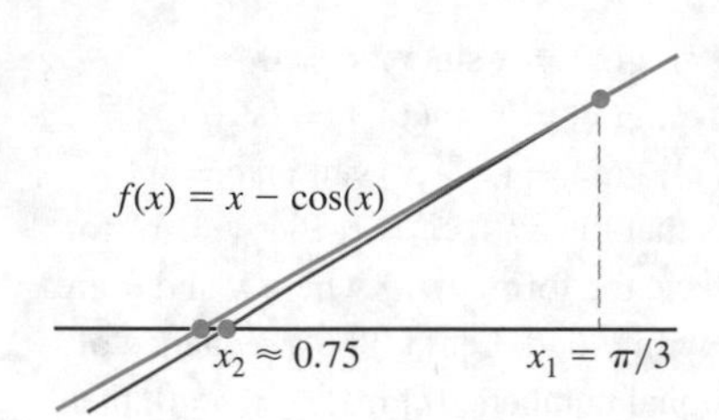

Figure 2

Solution The idea of the Newton-Raphson Method is to use a tangent line approximation to improve an initial estimate of a root of an equation $f(x) = 0$. We start by casting the current problem into this form. The two curves intersect at the value of x for which $x = \cos(x)$ or $x - \cos(x) = 0$. If we define $f(x) = x - \cos(x)$, then the value of x that we are seeking is the solution of $f(x) = 0$. Let us begin with the given estimate $x_1 = \pi/3$ of the point at which the graph of f crosses the x-axis (see Figure 2). The tangent line at $(x_1, f(x_1)) = (\pi/3, \pi/3 - 1/2)$ has slope

$$f'\left(\frac{\pi}{3}\right) = 1 + \sin(x)\big|_{x=\pi/3} = 1 + \frac{\sqrt{3}}{2}.$$

The equation of the tangent line is, therefore,

$$y - \left(\frac{\pi}{3} - \frac{1}{2}\right) = \left(1 + \frac{\sqrt{3}}{2}\right)\left(x - \frac{\pi}{3}\right).$$

The line intersects the x-axis when $y = 0$—that is, at

$$x_2 = \frac{\pi}{3} + \frac{1/2 - \pi/3}{1 + \sqrt{3}/2} \approx 0.75395527.$$

Examine the table:

	x	$\cos(x)$
x_1	1.0471976	0.5
x_2	0.75395527	0.72898709

We see that x_1 was a rather poor estimate: The values of x and $\cos(x)$ are quite far apart for that value of x. However, applying the idea of the Newton-Raphson Method results in a remarkable improvement: The values of x and $\cos(x)$ differ by less than 0.025 at x_2. ■

If even greater accuracy is required, we can replace x_1 by x_2 and repeat the procedure to get a better approximation x_3. We keep repeating until the desired degree of accuracy is obtained. That procedure is the Newton-Raphson Method.

Calculating with the Newton-Raphson Method

To put the Newton-Raphson Method to use, we need to develop an *algorithm*. If we are given a differentiable function f and an initial estimate x_1 of a root c of f, then the tangent line at $(x_1, f(x_1))$ has equation

$$y = f'(x_1)(x - x_1) + f(x_1).$$

Let x_2 be the x-intercept of this tangent line. By substituting $y = 0$ and $x = x_2$ in the equation of the tangent line, we find that

$$x_2 = x_1 - \frac{f(x_1)}{f'(x_1)},$$

provided that $f'(x_1) \neq 0$. Let us summarize what we have learned.

The Newton-Raphson Method

If f is a differentiable function, then the $(j+1)$th estimate x_{j+1} for a zero of f is obtained from the jth estimate x_j by the formula

$$x_{j+1} = x_j - \frac{f(x_j)}{f'(x_j)},$$

provided that $f'(x_j) \neq 0$.

We can recast the Newton-Raphson Method in a form that highlights its iterative nature. Let

$$\Phi(x) = x - \frac{f(x)}{f'(x)}. \quad \textbf{(4.9)}$$

Starting from a first estimate x_1 of a root c of $f(x) = 0$, we generate subsequent approximations, $x_2, x_3, \ldots$, by the formula

$$x_{j+1} = \Phi(x_j). \quad \textbf{(4.10)}$$

Example 2 Use the Newton-Raphson Method to determine $\sqrt{3}$ to within an accuracy of 10^{-7}.

Solution This problem is equivalent to finding the positive value where the function $f(x) = x^2 - 3$ vanishes. For a first estimate, we take $x_1 = 1.7$. Notice that $f'(x) = 2x$. According to the Newton-Raphson Method, we have

$$\begin{aligned} x_2 &= x_1 - \frac{f(x_1)}{f'(x_1)} \\ &= x_1 - \frac{x_1^2 - 3}{2x_1} \\ &= 1.7 - \frac{(1.7)^2 - 3}{2 \cdot (1.7)} \\ &\approx 1.7323529. \end{aligned}$$

A calculator check reveals that x_2 already agrees with $\sqrt{3}$ to three decimal places of accuracy. We apply the Newton-Raphson Method a second time:

$$\begin{aligned} x_3 &= x_2 - \frac{f(x_2)}{f'(x_2)} \\ &\approx 1.7323529 - \frac{(1.7323529)^2 - 3}{2 \cdot (1.7323529)} \\ &\approx 1.73205083. \end{aligned}$$

The value of x_3 agrees with $\sqrt{3}$ to seven decimal places of accuracy, which we can verify with a calculator. ■

Accuracy of the Newton-Raphson Method

If we do not already know the answer in advance, or if we do not have a calculator handy, how do we determine the accuracy of the approximation? If the Newton-Raphson Method is to be useful in practice, then this is clearly a question that must be answered.

Suppose that our sequence $\{x_j\}$ of Newton-Raphson estimates lies in an interval on which the values of $|f'(x)|$ are bounded by a positive constant C_1. Suppose also that x_j and x_{j+1} agree to k decimal places of accuracy. We have

$$x_{j+1} = x_j + \epsilon$$

where $|\epsilon| < 5 \times 10^{-(k+1)}$. Thus,

$$x_j - \frac{f(x_j)}{f'(x_j)} = x_j + \epsilon,$$

or

$$f(x_j) = -f'(x_j) \cdot \epsilon.$$

It follows that

$$\begin{aligned} |f(x_j)| &= |f'(x_j) \cdot \epsilon| \\ &= |f'(x_j)| \cdot |\epsilon| \\ &\leq C_1 \cdot \left(5 \times 10^{-(k+1)}\right) \\ &= 5 \cdot C_1 \cdot 10^{-(k+1)}. \end{aligned}$$

This inequality allows us to estimate how close $f(x_j)$ is to zero. In practice, we may simply continue the iterations until two or three successive approximations are within $5 \times 10^{-(k+1)}$ of each other. This is usually the signal that any further refinement will take place beyond the kth decimal place.

If f has a continuous second derivative, then we can say more about the rate at which the Newton-Raphson iterates $\{x_j\}$ approach a root c of f. Let us denote the absolute difference $|c - x_j|$ by ε_j. If $f'(c)$ and $f''(c)$ are nonzero, and if the Newton-Raphson iterates $\{x_j\}$ converge to c, then it may be shown that

$$\lim_{j \to \infty} \frac{\varepsilon_{j+1}}{\varepsilon_j^2} = \left| \frac{f''(c)}{2f'(c)} \right|. \tag{4.11}$$

We use this result only to explain why the Newton-Raphson Method often yields a very accurate approximation after a few iterations. Notice that the right side of equation (4.11) is a positive number—let us call it C. Equation (4.11) says that $\varepsilon_{j+1} \approx C \cdot \varepsilon_j^2$ for large j. Thus, when the Newton-Raphson Method is successful, the $(j+1)$th error ε_{j+1} is approximately proportional to the square ε_j^2 of the jth error. For example, if ε_j is on the order of 10^{-3}, then ε_{j+1} will be on the order of 10^{-6} and ε_{j+2} will be on the order of 10^{-12}. A loose rule of thumb is that in a successful application of the Newton-Raphson Method, the number of decimal places of accuracy doubles with each iteration. The technical description is that the Newton-Raphson Method is a *second-order process.*

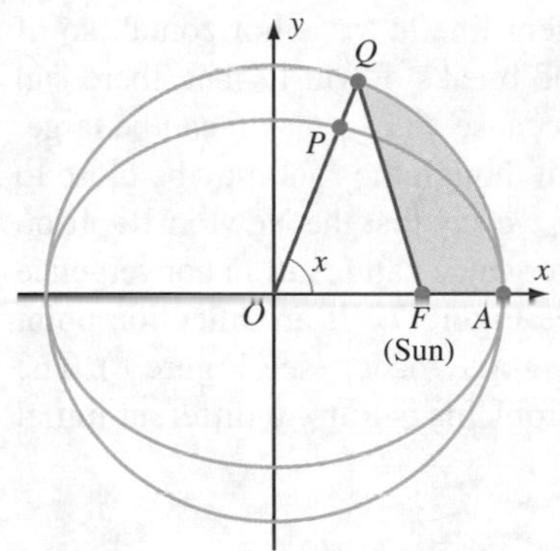

Figure 3
When a planet is at position P, its mean anomaly M is defined by $M = 2(\text{Area}(AFQ)/\overline{OA}^2)$.

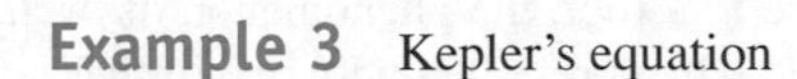

Example 3 Kepler's equation

$$M = x - e \cdot \sin(x)$$

relates the mean anomaly M and the eccentric anomaly x of a planet traveling in an elliptic path of eccentricity e (see Figure 3). It is important in predictive astronomy to be able to solve for x if M is given. Kepler expressed his opinion that an exact solution for x is impossible: "[I]f anyone would point out my mistake and show me the way, then he would be a great Apollonius in my opinion." Instead of finding an exact solution, find x to four decimal places when $M = 1.0472$ and $e = 0.2056$ (the eccentricity of Mercury's orbit).

Solution Let $f(x) = M - x + e \cdot \sin(x) = 1.0472 - x + 0.2056 \cdot \sin(x)$. Since $f(0) = 1.0472 > 0$ and $f(\pi/2) = -0.3180 < 0$, the Intermediate Value Theorem tells us that f has a root c in the interval $(0, \pi/2)$. We calculate $f'(x) = -1 + 0.2056 \cdot \cos(x)$. Because $f'(x) < -1 + 0.2056 \cdot 1 < 0$, we can deduce that f is decreasing and that the equation $f(x) = 0$ has *only* one solution. The graph of f in Figure 4 confirms our analysis. From the graph, it appears that $x_1 = 1.2$ is a good first estimate. We let

$$\Phi(x) = x - \frac{1.0472 - x + 0.2056 \cdot \sin(x)}{-1 + 0.2056 \cdot \cos(x)}$$

and calculate

$$x_2 = \Phi(x_1) = 1.2419527$$
$$x_3 = \Phi(x_2) = 1.2417712$$
$$x_4 = \Phi(x_3) = 1.2417712.$$

Since $|x_4 - x_3| < 5 \times 10^{-5}$, we can be reasonably sure that $c = 1.24177$ is correct to four decimal places. ■

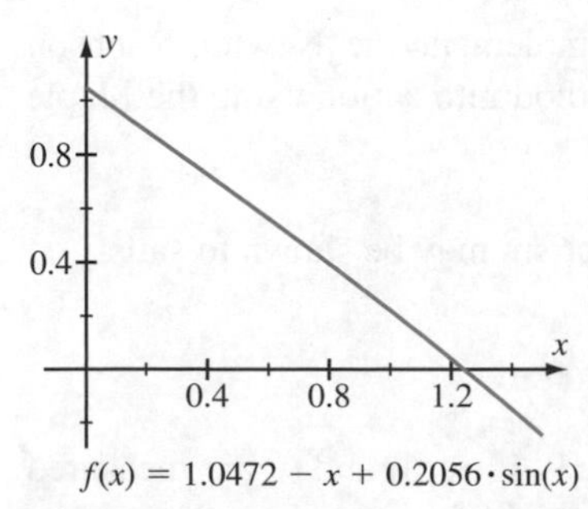

Figure 4

inSIGHT

> With the Newton-Raphson Method and a scientific calculator or computer, solving Kepler's equation is quite routine. That wasn't always so, however. Writing in 1914, the American astronomer Forrest Moulton commented, "Astronomers have devoted much attention to this equation, and several hundred methods of [approximating its solution] have been discovered."

Pitfalls of the Newton-Raphson Method

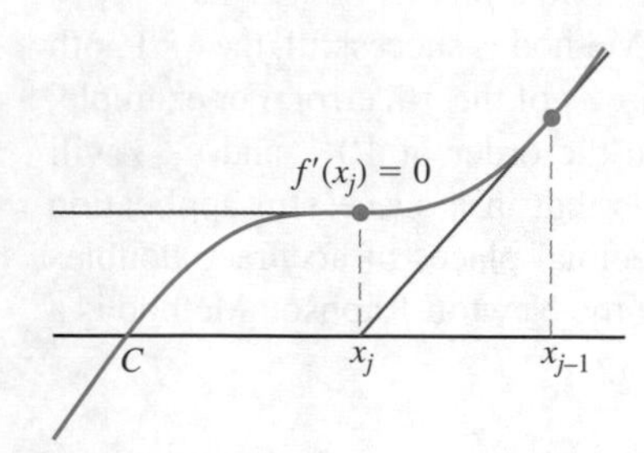

Figure 5

Obviously the Newton-Raphson Method will go badly wrong if at some stage $f'(x_j) = 0$. Look at Figure 5. We see that the tangent line at x_j is horizontal, so of course it never intersects the x-axis, and the method breaks down. In fact, there can be trouble if $f'(x_j)$ is small, even if it is not zero, because $f(x_j)/f'(x_j)$ can be large. If this is the case, then x_{j+1} will be far from x_j, even though the root may be close to x_j, as illustrated in Figure 6. In these circumstances, we say that the Newton-Raphson Method *diverges*. The *overshoot* that causes this divergence can result in convergence to a root $\tilde{C}$ other than the root C that is sought (see Figure 7). If an inflection point is located close to a root, the phenomenon of *cycling* may occur (see Figure 8). The best thing to do when presented with one of these problems is to try a different initial estimate.

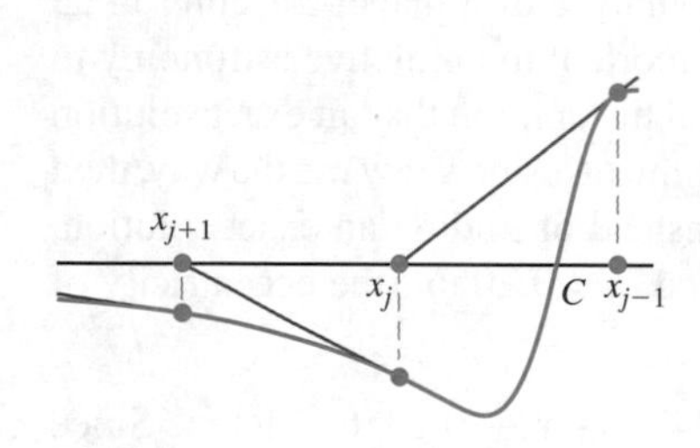

Figure 6

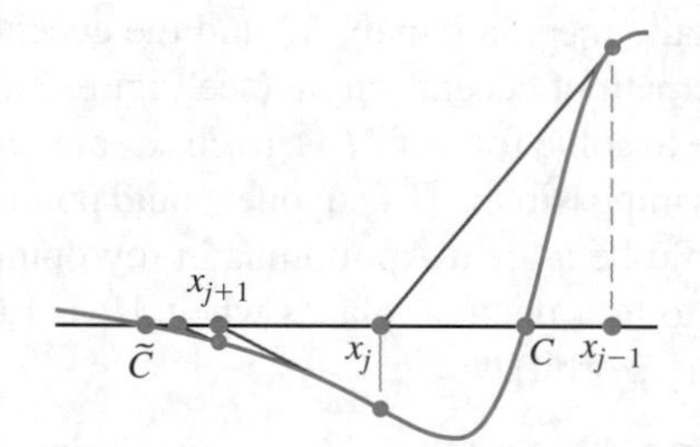

Figure 7

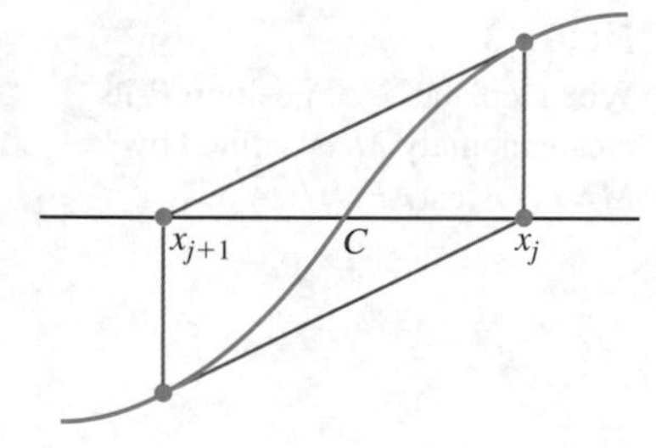

Figure 8

A Computer Implementation

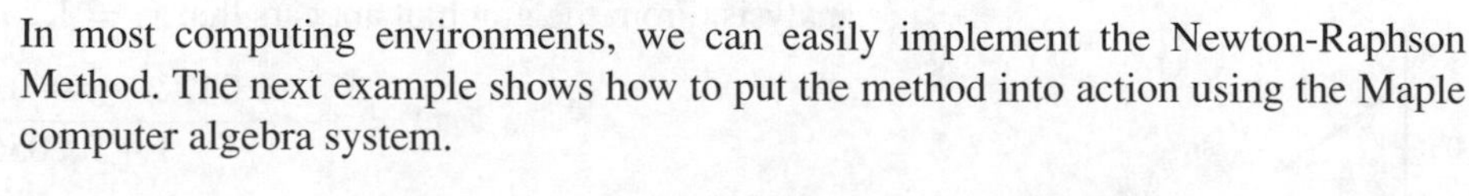

In most computing environments, we can easily implement the Newton-Raphson Method. The next example shows how to put the method into action using the Maple computer algebra system.

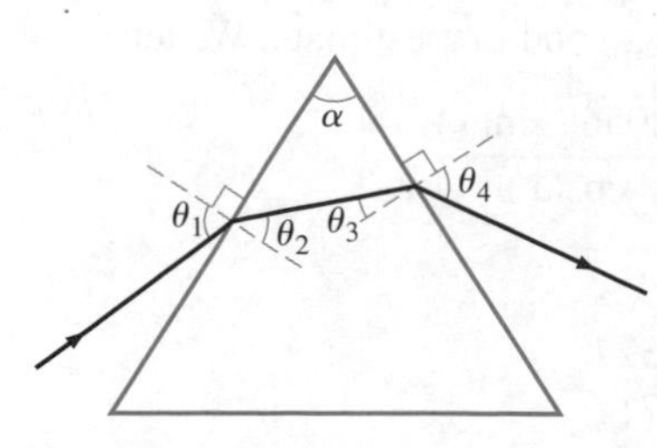

Figure 9
Refraction of light by a triangular prism

Example 4 The index of refraction r of a glass prism may be shown to satisfy

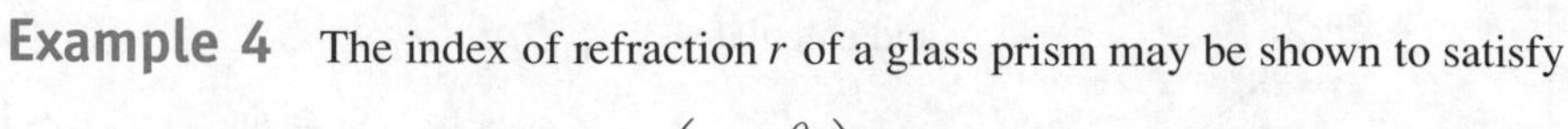

$$r \sin\left(\alpha - \frac{\theta_1}{r}\right) = \sin(\theta_4)$$

where the angles α, θ_1, and θ_4 are as shown in Figure 9. If α, θ_1, and θ_4 are measured to be 45°, 46°, and 54°, respectively, then what is r to eight decimal places?

Solution Let

$$f(r) = r \sin\left(\frac{\pi}{4} - \frac{46\pi}{180r}\right) - \sin\left(\frac{54\pi}{180}\right).$$

The solution to $f(r) = 0$ will give the desired angle of refraction. As in most computations that involve the derivatives of trigonometric functions, we convert to radian measurement of angles. Figure 10 shows a solution using Maple. Since $|x_5 - x_4| < 5 \times 10^{-9}$, we can be reasonably certain that $r = 2.08020645$ is correct to eight decimal places. (It should be noted that if α, θ_1, and θ_4 have not been measured to more than two significant digits, then we should be satisfied with $r = 2.1$ because the extra digits in our answer have no reliability.)

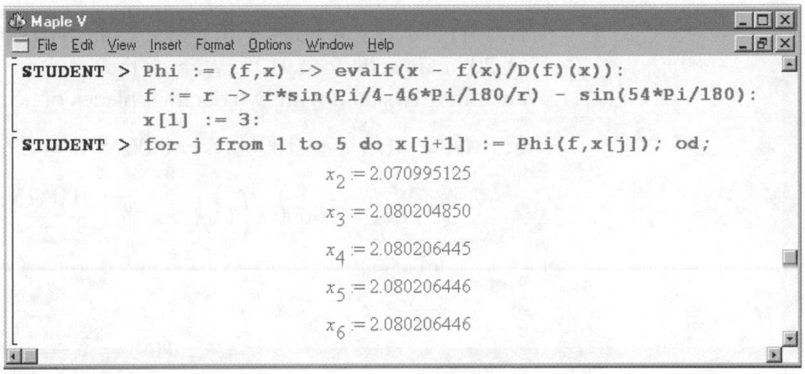

```
Maple V
File Edit View Insert Format Options Window Help
STUDENT > Phi := (f,x) -> evalf(x - f(x)/D(f)(x)):
          f := r -> r*sin(Pi/4-46*Pi/180/r) - sin(54*Pi/180):
          x[1] := 3:
STUDENT > for j from 1 to 5 do x[j+1] := Phi(f,x[j]); od;
                    x2 := 2.070995125
                    x3 := 2.080204850
                    x4 := 2.080206445
                    x5 := 2.080206446
                    x6 := 2.080206446
```

Figure 10

An Application in Economics: Bond Valuation

Revenue-producing entities (such as corporations and taxing authorities) raise capital by issuing bonds. When the bond is sold to an investor, the issuer agrees to pay back (or *redeem*) the *face value* P to the investor at some specified date (at which time the bond is said to *mature*). If the *coupon rate* of the bond is $100r\%$, then until the bond matures, the issuer makes semiannual interest payments amounting to $rP/2$ each. For example, a bond with a \$5000 face value that pays \$125 in interest every six months is said to have a 5% coupon rate because $r \cdot 5000/2 = 125$ implies that $r = 0.05$. In the following discussion, it is important to remember that once a bond has been issued, the redemption value P and the coupon rate r do not change.

Because bondholders often want to sell bonds before their maturity, determining the market value of a bond is an important problem. The market price V of a bond is rarely its face value P. One reason for this is that changes in interest rates cause bonds to fluctuate in price. As interest rates vary, the market price V of a bond must adjust so that it remains competitive with other bonds. Suppose, for example, that interest rates have gone up since bond B was issued. New bonds that are otherwise similar to B will have a higher coupon rate. The only way to sell bond B is to offer it at a market price V that is *below* its face value P. Analogously, if interest rates have gone down since bond B was issued, new bonds will have a lower coupon rate than B has. Bond B can, therefore, command a selling price V that is higher than its face value P.

A bondholder will enjoy a *capital gain* (or suffer a *capital loss*) if the price V at which he buys a bond is less than (or greater than) its redemption price P. This capital gain or loss must be considered in the determination of the *effective yield* x of the bond, which is expressed by the formula

$$V = P \cdot \left((1+x)^{-n} + \frac{r}{2} \cdot \frac{1-(1+x)^{-n}}{\sqrt{1+x}-1} \right).$$

(The derivation of this formula does not concern us here.) In practice, the investor will know the price V at which a bond is offered as well as its coupon rate r, its face value P, and its maturity n. To make an informed investment decision, he must determine the effective yield x. Because it is not feasible to solve exactly for x, a root approximation method, such as the Newton-Raphson Method, is necessary.

Example 5 A bond with a face value \$10,000, coupon rate 6.75%, and maturity of 20 years sells for \$9125. What is the effective yield of the bond? (Bond yields are customarily stated with two decimal places of accuracy.)

Solution Let $P = 10000$, $r = 0.0675$, $n = 20$, and $V = 9125$. Let

$$f(x) = 10000 \cdot \left((1+x)^{-20} + \frac{0.0675}{2} \cdot \frac{1-(1+x)^{-20}}{\sqrt{1+x}-1} \right) - 9125$$

and

$$\Phi(x) = x - \frac{f(x)}{f'(x)}.$$

Since there is a capital gain, the effective yield is greater than r. Therefore, it makes sense to start our approximation process with a number x_1 greater than r. If we choose $x_1 = 0.07$, then we calculate

$$x_2 = \Phi(0.07) = 0.07711$$
$$x_3 = \Phi(0.07711) = 0.07753$$
$$x_4 = \Phi(0.07753) = 0.07753.$$

Clearly f is quite complicated and Φ is even more so. Carrying out the Newton-Raphson Method would be rather laborious even with a scientific calculator. The values of $\Phi(x_j)$ recorded above were obtained with a computer algebra system. The agreement between x_3 and x_4 to two decimal places indicates that it is time to stop the algorithm. The effective yield x is 7.75%. ■

quickquiz

1. What is the purpose of the Newton-Raphson Method?
2. State the algorithm for the Newton-Raphson Method.
3. What is the accuracy of the Newton-Raphson Method?
4. If we wish to apply the Newton-Raphson Method to calculate $4^{1/5}$ by setting $x_1 = 1$ and letting $x_{n+1} = \Phi(x_n)$ for $n \geq 1$, what function Φ can we use?

EXERCISES

Problems for Practice

In Exercises 1–6, use the Newton-Raphson Method and the given initial estimate to find a root of the function f to an accuracy of 10^{-3}. Write each iteration of the Newton-Raphson Method explicitly, and record how many iterations were required.

1. $f(x) = x^2 - 6, x_1 = 2$
2. $f(x) = x^3 + 2, x_1 = -1$
3. $f(x) = x^2 - 5x + 2, x_1 = 4$
4. $f(x) = x^3 + x + 1, x_1 = -1$
5. $f(x) = x^5 - 31, x_1 = 2$
6. $f(x) = x^4 - 80, x_1 = 3$

In Exercises 7–12, calculate the indicated root, using the Newton-Raphson Method, to an accuracy of 10^{-2}.

7. $4^{1/3}$
8. $(-5)^{1/3}$
9. $6^{1/2}$
10. $(0.5)^{1.5}$
11. $\sqrt{10}$
12. $(-5)^{1/5}$

Further Theory and Practice

13. If f is a nonconstant linear function, how many iterations of the Newton-Raphson Method are required before the root of f is found? Does your answer depend on the initial guess?
14. Suppose you are applying the Newton-Raphson Method to a function f and that after five iterations you land precisely on a zero of f. What value will the sixth and subsequent iterations have?
15. Suppose that a given function f satisfies $f(c) = 0$ and that in addition there is an interval I about r such that

$$\left| \frac{f(x) f''(x)}{f'(x)^2} \right| < 1.$$

It is known that if the initial estimate is a point chosen from the interval I, then the Newton-Raphson Method will certainly converge to c. Test each function for this condition on the given interval.

- **a.** $f(x) = \sin x, c = 0, I = (-\pi/4, \pi/4)$
- **b.** $f(x) = x^2 - x, c = 1, I = (1/2, 2)$
- **c.** $f(x) = x^3 + 8, c = -2, I = (-3, -1)$
- **d.** $f(x) = 3x^2 - 12, c = 2, I = (1.5, 2.5)$

16. It is interesting to observe that division can be accomplished by an implementation of the Newton-Raphson Method that does not involve division at all. Suppose β is a nonzero number. Let $f(x) = \beta - 1/x$. What is the sequence of Newton-Raphson iterates?
17. A Babylonian approximation to $\sqrt{c}$ from the second millennium BCE consists of starting with a first estimate x_1 and then computing the subsequent approximations that are defined iteratively by

$$x_{j+1} = \frac{1}{2}\left(x_j + \frac{c}{x_j} \right).$$

Show that this ancient algorithm is the Newton-Raphson Method applied to $f(x) = x^2 - c$. Of course, there was no notion of derivative when this algorithm was discovered. Considering that x_{j+1} is the average of x_j and c/x_j, what might Babylonian mathematicians have had in mind when they devised this algorithm?

18. Define

$$g(x) = \begin{cases} (x-4)^{1/2} & \text{if } x \geq 4 \\ -(4-x)^{1/2} & \text{if } x < 4 \end{cases}.$$

Prove that, for any initial guess other than $x = 4$, the Newton-Raphson Method will not converge.

19. If f is a function that is always concave up or always concave down and if f has a root, then the Newton-Raphson Method always converges to a root no matter what the first estimate. Explain why.

In Exercises 20–23, refer to Exercises 57–62 from Section 4.1 for background on discrete dynamical systems. Assume that f'' exists and is continuous.

20. Let $\Phi(x) = x - f(x)/f'(x)$. Suppose that $f'(x_*) \neq 0$. Show that x_* is an equilibrium point of the dynamical system associated to Φ if and only if x_* is a root of f.
21. Suppose that x_* is a root of f, that $|\Phi'(x_*)| < 1$, and that Φ' is continuous on an open interval centered at x_*. Show that the sequence $\{x_n\}$ of Newton-Raphson iterates will converge to x_*, provided that x_1 is sufficiently close to x_*.
22. Show that $|\Phi'(x_*)| < 1$ is equivalent to

$$\left| \frac{f(x_*) f''(x_*)}{f'(x_*)^2} \right| < 1.$$

23. Deduce that according to the hypotheses in the instructions, the Newton-Raphson iterates will converge to the root x_* of f, provided that x_1 is chosen sufficiently close to x_*.

Calculator/Computer Exercises

24. Apply the Newton-Raphson Method to the function

$$f(x) = x^5 + 4x^4 + 6x^3 + 4x^2 + 1,$$

using the initial estimate $x = 0$. What happens? Now try the initial estimate $x = 1$. What happens this time? Sketch the graph of f. Use this graph to explain what went wrong. Now use the initial guess -3 to see how things improve. Calculate the root of f to an accuracy of 10^{-3}.

In Exercises 25–28, use the Newton-Raphson Method to approximate a root of the function $f(x)$ to within 5×10^{-8}. State the initialization chosen and the number of iterations required for the stated accuracy.

25. $f(x) = x^5 + x^3 + 2x - 5$

26. $f(x) = 2x + \cos(x)$

27. $f(x) = 5x^5 - 9x^4 + 15x^3 - 27x^2 + 10x - 18$

28. $f(x) = x^7 + 3x^4 - x^2 + 3x - 2$

In Exercises 29 and 30, find the largest real root of the function $f(x)$.

29. $f(x) = x + \sin(4.6x)/x$

30. $f(x) = x + \tan(x)$, $x \in (-\pi, \pi)$

31. Use the Newton-Raphson Method to approximate a nonzero solution of $\sqrt{x^3+1} = \sqrt[3]{x^2+1}$ to within 5×10^{-5}.

32. The polynomial $f(x) = x^{11} + x^7 - 1$ has one root γ in the interval $[0, 1]$. Find this root to five decimal places using the Newton-Raphson Method.

In Exercises 33–36, graph the polynomial $p(x)$ in the indicated viewing rectangle. Identify the root c that has multiplicity 2. Use the Newton-Raphson Method with initial estimate $x_1 = c + 1/2$ to obtain iterates $x_2, x_3, \ldots, x_N$. Terminate the process when x_N approximates c to three decimal places. (Notice that the convergence is slow.) Record the value of N so that it can be used for comparison in Exercise 37.

33. $p(x) = x^4 - 2x^3 + 3x^2 - 4x + 2$, $[-2, 3] \times [-5, 50]$

34. $p(x) = x^4 + x^3 - 6x^2 - 4x + 8$, $[-3, 0] \times [-5, 25]$

35. $p(x) = 4x^4 - 4x^3 - 35x^2 + 36x - 9$, $[-3.2, 3.2] \times [-125, 70]$

36. $p(x) = x^4 - 8x^3 + 23x^2 - 28x + 12$, $[0, 4] \times [-5, 15]$

37. A Variant of Newton-Raphson for Roots of Higher Multiplicity Each polynomial in Exercises 33–36 has one root c of multiplicity $m = 2$. Approximate each such root using the initial estimate $x_1 = c + 1/2$ and the iteration formula

$$x_{j+1} = x_j - m\frac{p(x_j)}{p'(x_j)}.$$

In each case, terminate the process when the iterate x_M approximates c to three decimal places. Compare the number M of iterates with the corresponding number N that results when the unmodified Newton-Raphson Method is used.

38. The *atomic packing factor* (PF) of a crystal is the volume of a unit cube that is occupied by atoms. The ion size ratio of Na^+ and Cl^- is r. From the geometry of the NaCl crystal, the PF of NaCl, which is known to be $2/3$, is given by

$$\frac{2\pi(1+r^3)}{3(1+r)^3}.$$

Use the Newton-Raphson Method to approximate r to within 5×10^{-6}.

39. The period T of a pendulum of length ℓ and maximum deviation φ (measured in radians) from the vertical is given approximately by

$$T = 2\pi\sqrt{\frac{\ell}{g}}\left(1 + \frac{1}{16}\varphi^2 + \frac{11}{3072}\varphi^4\right).$$

If $\ell = 0.8555$ m and $T = 2.000$ s, what is φ? Use $g = 9.807$ m/s^2 to compute φ to four significant digits.

40. A boat starts for the opposite shore of a 2 km wide river so that (1) the boat always heads toward the point that is directly across from the original launch point, and (2) the speed of the boat is three times faster than the current of the river. If we set up on an xy-coordinate system so that the boat's initial position is (2,0), the landing site is (0,0), and the current is in the positive y-direction, then the path of the boat is given by

$$y = \left(\frac{x}{2}\right)^{2/3} - \left(\frac{x}{2}\right)^{4/3}.$$

What are the x-coordinates at the two points at which the boat is 0.137 km downstream?

41. Use the Newton-Raphson Method to determine x to within 5×10^{-3}.

a. A bond with a face value \$10,000, coupon rate 6.75%, and maturity of 20 years sells for \$9125. What is the effective yield x of the bond?

b. A bond with a face value \$10,000, coupon rate 6.5%, and maturity of 10 years sells for \$10,575. What is the effective yield of the bond?

In Exercises 45–48, two \$10,000 bonds with the same maturity are offered. Determine which is the better investment by calculating the effective yield of each.

42. Price $=$ \$10531, coupon rate $= 7\%$ or Price $=$ \$91467, coupon rate $= 5\%$, $n = 10$

43. Price $=$ \$8681, coupon rate $= 5\%$ or Price $=$ \$10674, coupon rate $= 7\%$, $n = 20$

44. Price $=$ \$9518, coupon rate $= 6.5\%$ or Price $=$ \$8705, coupon rate $= 6\%$, $n = 30$

45. Price $=$ \$10679, coupon rate $= 8\%$ or Price $=$ \$11052, coupon rate $= 9\%$, $n = 8$

46. The polynomial $x^4 + 3.6995x^3 - 4.04035x^2 - 16.13693x + 6.83886$ has two negative roots that are quite close to each other and two positive roots. Use the Newton-Raphson Method to find all four roots to five significant floating point digits. For each root, specify the initial approximations and the number of iterations required to achieve the stated accuracy. Use a graphing utility to obtain good initial estimates.

47. Figure 8 (page 310) illustrates the phenomenon of cycling with the function $f(x) = x + \sin(x)$ and an initial estimate $x_1 \in (0, 2)$ that leads, after two iterations, to $x_3 = x_1$. What equation does x_1 satisfy? Use the Newton-Raphson Method to approximate x_1. (Ironically, the authors successfully used the Newton-Raphson Method to produce an illustration of how the method might fail.)

48. For every positive x, the number $W(x)$ is defined to be the unique positive number such that

$$W(x)\exp(W(x)) = x.$$

The function $x \mapsto W(x)$ is Lambert's W function. Compute $W(1)$ and $W(e)$.

49. In biological kinetics, the equation

$$Y = \frac{x + 3x^2 + 3x^3 + 10x^4}{1 + 4x + x^2 + 4x^3 + 10x^4}$$

represents a typical relationship between saturation Y and dimensionless ligand concentration x ($x \geq 0$). Explain why Y/x has an absolute maximum value. Use the Newton-Raphson Method to find the absolute maximum value of Y/x.

4.8 Antidifferentiation and Applications

It is useful in calculus to be able to reverse the differentiation process. That is, given a function f, we wish to find a function F such that $F' = f$. In this section, we discuss this technique and find some applications for it.

Antidifferentiation

Let f be defined on an open interval I. If F is a differentiable function such that $F'(x) = f(x)$ for all $x \in I$, then F is called an *antiderivative* for f on I. It is possible for f to have more than one antiderivative. Indeed, if F has derivative f and if C is any constant, then $(F + C)'$ also equals f because the derivative of a constant is 0. According to Theorem 5 from Section 4.1, *all* the antiderivatives of f will differ from F by a constant. We now introduce special notation to describe the collection of antiderivatives of f.

Definition Let f be a continuous function defined on an open interval I. If F is a function that is differentiable on I and satisfies $F' = f$ on I, then we call F an *antiderivative* for f. We denote the collection of all antiderivatives of f by

$$\int f(x)\,dx.$$

This expression is called the *indefinite integral* of *f*. It is convenient to write

$$\int f(x)\,dx = F(x) + C \qquad \textbf{(4.12)}$$

where C is an arbitrary constant. We call C the *constant of integration*. The right side of equation (4.12) is interpreted as the collection of all functions that have the form $F(x) + C$ for some constant C.

Example 1 Calculate $\int x^3\,dx$.

Solution Since differentiation of a power of x involves reducing the exponent by one, we reverse our reasoning and guess that x^4 will be an antiderivative for x^3 (one power higher). However,

$$\frac{d}{dx}x^4 = 4x^3,$$

which is off by a factor of 4. We therefore adjust our guess to

$$F(x) = \frac{1}{4}x^4.$$

Now we have

$$F'(x) = \frac{1}{4} \cdot 4x^3 = x^3,$$

as desired. Since all other antiderivatives for x^3 will differ from this one by a constant, we write our solution as

$$\int x^3\,dx = \frac{1}{4}x^4 + C.$$

■

You may be surprised that the antidifferentiation process looks a bit indirect. In fact it is, but we will learn to antidifferentiate in an organized way. Chapter 7 is devoted to a variety of techniques for obtaining the antiderivatives of functions.

A LOOK FORWARD

Antidifferentiating Powers of x

Example 1 is an instance of a general principle that is worth isolating.

Indefinite Integrals of Powers of x

For any number $m \neq -1$, we have

$$\int x^m \, dx = \frac{1}{m+1} x^{m+1} + C. \qquad \textbf{(4.13)}$$

Indeed, for $m \neq -1$,

$$\frac{d}{dx}\left(\frac{1}{m+1} \cdot x^{m+1}\right) = \frac{1}{m+1} \cdot (m+1)x^m = x^m.$$

Observe that we must exclude $m = -1$ because the factor $m + 1$ in the denominator is 0 for $m = -1$. However, in Section 3.6, we learned that

$$\frac{d}{dx} \ln(x) = \frac{1}{x}.$$

Therefore,

$$\int x^{-1} \, dx = \ln(x) + C.$$

We can restate some of the basic rules of differentiation in a form that is helpful for antidifferentiaton.

Theorem 1 Let f and g be functions defined on an open interval I. Let F be an antiderivative for f and G an antiderivative for g. Then

a. $\int (f(x) + g(x)) \, dx = F(x) + G(x) + C$; and
b. $\int \alpha \cdot f(x) \, dx = \alpha \cdot F(x) + C$ for any constant α.

inSIGHT

You may wonder why Theorem 1a does not have two constants of integration (since there are two indefinite integrals). The answer is that we can represent the sum of two arbitrary constants by another (single) arbitrary constant. Similarly, since a multiple of an arbitrary constant is another arbitrary constant, there is no need to write the constant in Theorem 1b as $\alpha \cdot C$.

Example 2 Calculate $\int (3x^5 - 7x^{3/2} - 4) \, dx$.

Solution According to Theorem 1a, we can solve this problem by antidifferentiating in pieces. We can use formula (4.13) for each.

1. An antiderivative of x^5 is $x^6/6$. Therefore, an antiderivative of $3x^5$ is $3x^6/6 = x^6/2$.
2. An antiderivative for $7x^{3/2}$ is $7 \cdot (2/5)x^{5/2} = (14/5)x^{5/2}$.
3. An antiderivative for 4 is $4x$. (Explain this step using Theorem 1b and formula (4.13) with $\alpha = 4$, $f(x) = x^0$, and $m = 0$.)

Putting these calculations together gives the antiderivative

$$F(x) = \frac{1}{2}x^6 - \frac{14}{5}x^{5/2} - 4x.$$

We check our work by differentiating:

$$\frac{d}{dx}\left(\frac{1}{2}x^6 - \frac{14}{5}x^{5/2} - 4x\right) = \frac{1}{2} \cdot 6x^5 - \frac{14}{5} \cdot \frac{5}{2} \cdot x^{3/2} - 4 = 3x^5 - 7x^{3/2} - 4.$$

Since all other antiderivatives differ by a constant from the one we have found, we write the solution as

$$\int (3x^5 - 7x^{3/2} - 4)\, dx = \frac{1}{2}x^6 - \frac{14}{5}x^{5/2} - 4x + C.$$ ■

Example 3 Calculate the indefinite integral

$$\int \frac{\sqrt{t} - t^2}{t^4}\, dt.$$

Solution Algebraic manipulations can often simplify an integration problem. We rewrite the indefinite integral as

$$\int \frac{t^{1/2}}{t^4}\, dt - \int \frac{t^2}{t^4}\, dt = \int t^{-7/2}\, dt - \int t^{-2}\, dt.$$

It is a simple matter of applying the rule for indefinite integrals of rational powers to obtain

$$-\frac{2}{5}t^{-5/2} + t^{-1} + C.$$ ■

Antidifferentiation of Other Functions

We may also antidifferentiate trigonometric functions, provided we remember the *differentiation formulas* (now is a good time to review Chapter 3's Summary of Key Topics).

Example 4 Calculate $\int \sin(5x)\, dx$.

Solution We recall that the derivative of $\cos(5x)$ is $-5\sin(5x)$. We need to adjust our guess to eliminate the factor of -5. Therefore, the antiderivative of $\sin(5x)$ should be $-(1/5)\cos(5x)$. We check that

$$\frac{d}{dx}\left(-\frac{1}{5}\cos(5x)\right) = -\frac{1}{5}(-5\sin(5x)) = \sin(5x),$$

as desired. We write our solution as

$$\int \sin(5x)\,dx = -\frac{1}{5}\cos(5x) + C.$$

■

Velocity and Acceleration

If a body moves in a linear path with position $p(t)$ at time t, then the instantaneous velocity of the body at time t is the instantaneous rate of change of position, or the derivative of p:

$$v(t) = \frac{d}{dt}p(t).$$

Similarly, the acceleration of the body at time t is the instantaneous rate of change of velocity, or the derivative of v:

$$a(t) = \frac{d}{dt}v(t) = \frac{d}{dt}\left(\frac{d}{dt}p(t)\right) = \frac{d^2}{dt^2}p(t).$$

In the language of antiderivatives, we have

> Velocity is an antiderivative of acceleration.
> Position is an antiderivative of velocity.

inSIGHT

If we measure distance p in feet and time t in seconds, then the average rate of change of p with respect to t is expressed as

$$\frac{\Delta p \text{ ft}}{\Delta t \text{ s}}.$$

This explains why the unit of measure for *velocity* is feet per second (ft/s).

Likewise, if velocity v is measured in ft/s, then acceleration is measured by considering

$$\frac{\Delta v \text{ ft/s}}{\Delta t \text{ s}}.$$

This explains why the unit of measure for *acceleration* is feet per second per second (ft/s^2).

Example 5 A police car accelerates from rest at a rate of 3 mi/min^2. How far will it have traveled at the moment it reaches a velocity of 65 mi/h?

Solution To solve this problem, we let $p(t)$ be the position of the police car at time t (in minutes). We begin with the information

$$\frac{d^2}{dt^2}p(t) = a(t) = 3.$$

By antidifferentiating, we find that

$$v(t) = \frac{d}{dt}p(t) = 3t + C$$

where C is the constant of integration. We recall that the car begins at rest, which means the initial velocity is 0. Thus,

$$0 = v(0) = \frac{d}{dt}p(0) = 3 \cdot 0 + C, \quad \text{or} \quad C = 0.$$

We have learned that

$$\frac{dp}{dt} = 3t.$$

Antidifferentiating again, we find that

$$p(t) = 3 \cdot \frac{1}{2}t^2 + D$$

where D is a constant of integration. If we call the initial position of the car $p(0) = 0$, then we have

$$0 = p(0) = \frac{3}{2}0^2 + D, \quad \text{or} \quad D = 0.$$

We have learned that

$$p(t) = \frac{3}{2}t^2.$$

To answer the original question, we convert the question to common units of minutes: How far has the police car traveled when it is going $65/60$ mi/min? The time at which the car has reached this velocity is obtained by solving

$$v(t) = 3t = \frac{65}{60}.$$

We find that

$$t = \frac{65}{180} = \frac{13}{36} \text{ min.}$$

The car reaches the desired speed in $13/36$ min. The distance the car has traveled in this time is

$$p\left(\frac{13}{36}\right) = \frac{3}{2}\left(\frac{13}{36}\right)^2 = \frac{169}{864} \approx 0.1956 \text{ mi.}$$

■

Example 6 An object is dropped from a window on a calm day. It strikes the ground precisely 4 s later. From what height was the object dropped?

Solution To answer this question, we need the fact (obtained from experimentation) that the acceleration due to gravity of a falling body near Earth's surface is about 32.16 ft/s^2. If $h(t)$ is the height of the body at time t seconds, then we have

$$\frac{d^2h}{dt^2} = a(t) = -32.16.$$

The minus sign indicates that the body is falling in the direction of *decreasing height*. We antidifferentiate to find that

$$\frac{dh}{dt} = v(t) = -32.16t + C.$$

However, the initial velocity of the falling body is 0, so we find that

$$0 = v(0) = (-32.16) \cdot 0 + C;$$

hence, $C = 0$ and

$$v(t) = \frac{dh}{dt} = -32.16t$$

Antidifferentiating again yields $h(t) = -16.08t^2 + D$ where D is the constant of integration. We do not know the initial height (this is what we *seek*), but we do know that after 4 s the height is 0. Therefore, we may solve

$$0 = h(4) = -16.08 \cdot 4^2 + D,$$

yielding $D = 257.28$. Thus, the height of the falling object at any time t is given by

$$h(t) = -16.08t^2 + 257.28.$$

The answer to the original question is: The initial height is $h(0) = 257.28$ ft. ■

Example 7 It is known that the acceleration due to gravity near the surface of Mars is about 3.72 m/s^2. If an object is propelled downward from a height of 12 m and at a speed of 72 km/h, in how many seconds will it hit the surface of Mars? What is the velocity at impact? (The data given describe the last 12 m of the journey of Mars *Pathfinder,* which blasted into space on December 4, 1996, and entered the Martian atmosphere on July 4, 1997.)

Solution The solution is parallel to that of Example 6, but the acceleration will be different because the events are taking place on Mars.

Let $h(t)$ denote the height of the body at time t. We begin with the equation

$$h''(t) = -3.72 \text{ m/s}^2.$$

We antidifferentiate to find that

$$\int h(t)\, dt = h'(t) = v(t) = (-3.72t + C) \text{ m/s}.$$

We are told that the initial velocity of the falling body is -72 km/h. We convert this to meters per second:

$$v(0) = -72 \text{ km/h} = -72 \cdot \frac{1000 \text{ m}}{3600 \text{ s}} = -20 \text{ m/s}.$$

On the other hand,

$$v(0) = (-3.72 \cdot 0 + C) \text{ m/s} = C \text{ m/s};$$

hence, $-20 = C$. We have learned that

$$v(t) = h'(t) = (-3.72 \cdot t - 20) \text{ m/s}.$$

Antidifferentiating again yields

$$\int v(t)\, dt = h(t) = (-1.86 \cdot t^2 - 20t + D) \text{ m}$$

where D is the constant of integration. We know that the initial height is 12 m. Therefore, we may solve

$$12 \text{ m} = h(0) = (-1.86 \cdot 0^2 - 20 \cdot 0 + D) \text{ m},$$

yielding $D = 12$ m. Thus, the height of the falling object at any time t is given by $h(t) = (-1.86 \cdot t^2 - 20t + 12)$ m.

To answer the original question of when the body strikes the surface of Mars, we solve

$$0 = h(t) = (-1.86 \cdot t^2 - 20t + 12) \text{ m}.$$

Using the quadratic formula, we find that $t = -11.3$ or $t = 0.570$ (to three significant digits). Only the second of these solutions is physically meaningful (since the problem takes place in forward time). Therefore, we conclude that it takes 0.570 s for the body to strike the surface of Mars.

Finally, in the course of our calculation, we determined that

$$v(t) = h'(t) = (-3.72 \cdot t - 20) \text{ m/s}.$$

Thus, the velocity at impact is $v(0.570) = (-3.72 \cdot 0.570 - 20) \text{ m/s} = -22.1$ m/s. ■

quickquiz

1. What is an antiderivative?
2. What is an indefinite integral, and how does it differ from an antiderivative?
3. What arithmetic operations does antidifferentiation respect?
4. How is velocity calculated from acceleration?
5. How is position calculated from acceleration?

EXERCISES

Problems for Practice

In Exercises 1–20, calculate the indefinite integral.

1. $\int (x^2 - 5x)\,dx$
2. $\int (3\sin(x) - 5\cos(x) + 1)\,dx$
3. $\int e^x\,dx$
4. $\int \sec(x)\tan(x)\,dx$
5. $\int \sqrt{x+2}\,dx$
6. $\int x^2 - x^{-2} + x^{1/2} - x^{-1/2}\,dx$
7. $\int (x^2 + x^{-3})/x^4\,dx$
8. $\int x^{3/2}(x^{-3} - 4x^{-2} + 2x^{-1})\,dx$
9. $\int (x+1)^2\,dx$
10. $\int x(x+1)(x+2)\,dx$
11. $\int \exp(e \cdot x)\,dx$
12. $\int x^{1/2}(x+1)\,dx$
13. $\int (x^{-7/3} - 4x^{-2/3})\,dx$
14. $\int \csc^2(x)\,dx$
15. $\int (3\cos(4x) + 2x)\,dx$
16. $\int (3\sin(7x) + 7\sin(3x))\,dx$
17. $\int \sec^2(8x)\,dx$
18. $\int \csc(2x)\cot(2x)\,dx$
19. $\int (3x-2)^3\,dx$
20. $\int x^2(\sqrt{x}+1)^2\,dx$
21. Compute $F(c)$ from the given information.
    a. $F'(x) = 6x^2,\ F(1) = 3,\ c = 0$
    b. $F'(x) = \cos(x),\ F(\pi/2) = -1,\ c = \pi/6$
    c. $F'(x) = 2x + 3,\ F(1) = 2,\ c = -1$
    d. $F'(x) = 6e^{2x},\ F(0) = -1,\ c = 1/2$
22. A car accelerates from rest at the constant rate of 2 mi/min^2. After the car has traveled 10 mi, how fast will it be traveling?
23. A sprinter accelerates during the first 20 m of a race at the rate of 4 m/s^2. Of course, she begins at rest. How fast is she running at the moment she hits the 20 m mark?
24. An object is dropped from a window 100 ft above the ground. At what speed is the object traveling at the moment of impact with the ground?
25. A baseball is thrown straight up with initial velocity 100 ft/s. How high will the ball go before it begins its fall back to Earth? What will be the velocity of the ball, on the way down, when it is at height 25 ft?
26. An automobile is cruising at a constant speed of 55 mi/h. To pass another vehicle, the car accelerates at a constant rate. In the course of a minute, the car covers 1.3 mi. What is the rate at which the car is accelerating? What is the speed of the car at the end of this minute?
27. A car traveling at a speed of 50 mi/h must come to a halt in 1200 ft. If the vehicle will decelerate at a constant rate, what should that rate be?
28. A baseball is dropped from the top of a building. When it strikes the ground, its instantaneous velocity is -75 ft/s. How tall is the building?
29. A baseball is dropped from the top of a building that is 375 ft tall. How long will it take the ball to strike the ground?
30. The acceleration due to gravity near the surface of the moon is about -5.3064 ft/s^2. If an object is dropped from the top of a cliff on the moon and takes 5 seconds to strike the surface of the moon, how high is the cliff?

Further Theory and Practice

Use trigonometric identities to compute the indefinite integrals in Exercises 31–35.

31. $\int \sin(x)\cos(x)\,dx$
32. $\int (\cos^2(x) - \sin^2(x))\,dx$
33. $\int \tan^2(x)\,dx$
34. $\int \sqrt{\dfrac{1+\cos(2x)}{2}}\,dx$
35. $\int \cos^2(x)\,dx$
36. Suppose that f is continuous on $\mathbb{R}$, that f is positive on $(-\infty, -3)$, $(2, 4)$, and $(4, \infty)$, and that f is negative on $(-3, 2)$. If F is an antiderivative of f on $\mathbb{R}$, classify the local extrema of F.
37. If G is the antiderivative of g, is $G \circ f$ the antiderivative of $g \circ f$? Why or why not?
38. If $\int g(x)\,dx = G(x) + C$, what is $\int g(f(x))f'(x)\,dx$?
39. Keep the Chain Rule in mind as you calculate the following indefinite integrals.
    a. $\int x(x^2+1)^{100}\,dx$
    b. $\int \cos(x)\sin(x)\,dx$
    c. $\int \sin^2(x)\cos(x)\,dx$
    d. $\int x\exp(x^2)\,dx$
    e. $\int \cos(\sin(x))\cos(x)\,dx$

40. Suppose that F and G are antiderivatives of f on an open interval I. Show that if $x_0 \in I$ and $F(x_0) = G(x_0)$, then $F(x) = G(x)$ for every $x \in I$. Show that this conclusion is false if I is a punctured interval.

41. Find the constant A such that

$$\int e^{kt}(1 + e^{kt})^n \, dt = A(1 + e^{kt})^{n+1} + C$$

for $n \neq -1$ and $k \neq 0$.

42. The error function, denoted $\mathrm{erf}(x)$, is defined to be the antiderivative of $2\exp(-x^2)/\sqrt{\pi}$ with value at 0 of 0. What is the derivative of $x \mapsto \mathrm{erf}(\sqrt{x})$?

43. Refer to Exercise 42 for the definition of the error function erf. Where is erf increasing? On what interval(s) is the graph of erf concave up? Concave down?

44. A subway train can only accelerate or decelerate at a rate of 20 mi/min^2. The distance between stations is 10 mi. The subway authority wants the trip to be made in 20 min. If the engineer will spend the same amount of time accelerating and decelerating, what should be his cruising speed?

45. An arrow is shot at a 60 deg angle to the horizontal with initial velocity 190 ft/s. How high will the arrow travel? What will be the horizontal component of its velocity at height 50 ft (going up)?

46. The engineer of a freight train needs to stop in 1700 ft to avoid striking a barrier. The train is traveling at a speed of 60 mi/min. The engineer applies one set of brakes, which causes him to decelerate at rate of 0.02 mi/min^2 for 15 s. He realizes he is not going to make it, so he applies a second set of brakes, which causes the train to decelerate at a rate of 0.03 mi/min^2. Will he strike the barrier?

47. A woman decides to determine the depth of an empty well by dropping a stone into the well and measuring how long it takes the stone to hit bottom. She drops the stone and hears it hit bottom 6 s later. However, sound travels at the approximate speed of 660 mi/h, which means there is a delay from the time the stone hits bottom to the time that she hears it. Taking this delay into account, determine the depth of the well.

48. A speeding car passes a police officer who is equipped with a radar gun. The officer determines that the car is going 85 mi/h. By the time the officer is ready to give chase, the car has a 15 s lead. The officer has been trained to catch vehicles within 2 min of the beginning of pursuit. At what constant rate does the police car need to accelerate in order to catch the speeding car?

49. At a given moment in time, two race cars are abreast and traveling at the speed of 90 mi/h. Car A begins to accelerate at a constant rate of 6 mi/min^2, and car B begins to accelerate at a constant rate of 9 mi/min^2. The cars are driving on a track with a 2 mi circumference. How long will it take car B to lap car A three times?

50. A train travels 10 mi between stops. The train will accelerate from rest at a constant rate and then decelerate at a constant rate until it comes to a halt. It spends no time at constant velocity between acceleration and deceleration. Describe all possible constant rates of acceleration and deceleration that will take the train from one stop to the next. If the train must travel from one stop to the next in 10 min, and if its maximum possible speed is 120 mi/h, at what rate should it accelerate and decelerate?

51. A model rocket is launched straight up with an initial velocity of 100 ft/s. The fuel on board the rocket lasts for 8 s and maintains the rocket at this upward velocity (countering the negative acceleration due to gravity). After the fuel is spent, gravity takes over. What is the greatest height that the rocket reaches? After how many seconds does it hit the ground? With what velocity does the rocket strike the ground?

52. Visitors to the top of the Empire State Building in New York City are cautioned not to throw objects off the roof. How much damage could a 1 oz weight thrown from the roof do? The height of the building is 1250 ft. Using the quantity mass $\times$ velocity as a measure of the impact of the weight when it strikes, compare your answer with the impact of a 10 oz hammer being swung at a speed of 10 ft/s.

53. Suppose there are two positive constants A and k such that the mass m of a radioactive isotope decays according to the equation

$$m'(t) = -Ae^{-kt}.$$

If $m(0) = A/k$, then at what value of t is $m(t)$ equal to half of $m(0)$?

54. The initial temperature $T(0)$ of an object is 50°C. If the object is cooling at the rate

$$T'(t) = -0.2e^{-0.02t}$$

when measured in degrees centigrade per second, what is the limit of its temperature as t tends to infinity?

55. If the air resistance to a falling object is proportional to the velocity of that object, then the velocity of that object when dropped from a height H is

$$v(t) = -kg(1 - e^{-t/k})$$

where k is a positive constant and g is the constant acceleration due to gravity. Calculate the height y of the object as a function of t.

56. If the air resistance to a falling object is proportional to the square of the velocity of that object, then the downward acceleration of that object is given by

$$a(t) = -4ge^{2\sqrt{g\kappa}t}\left(e^{2\sqrt{g\kappa}t}+1\right)^{-2}$$

where g is the acceleration due to gravity and κ is a positive constant. Find the velocity v of the object as a function of t assuming that $v(0) = 0$. *Hint:* First do Exercise 41.

Calculator/Computer Exercises

57. Suppose that the rate of change of the mass m of a sample of the isotope ^{14}C satisfies

$$m'(t) = -0.1213 \cdot e^{-0.000123t} \text{ g/yr}$$

where t is measured in years. If $m(0) = 1000$ g, for what value of t is $m(t)$ equal to 800 g?

58. The amount y of an isotope satisfies

$$\frac{dy}{dt} = -k \cdot y(0) \cdot e^{-kt} \text{ g/yr}$$

where t is measured in years and $y(0)$ in grams. If, after 1 year, the amount y is one-third the initial amount $y(0)$, what is the value of k?

59. The initial temperature $T(0)$ of an object is 50°C. If the object is cooling at the rate

$$T'(t) = -0.2e^{-0.02t}$$

when measured in degrees centigrade per second, what is the limit of its temperature as t tends to infinity? How much time does it take for the object to cool 4°C (from 50°C to 46°C)? How much additional time elapses before the object cools another 2°C? How much additional time elapses before the object cools another 1°C (from 44°C to 43°C)?

60. Suppose the velocity of an object when dropped from a 100 m height is

$$v(t) = -19.6(1 - e^{-t/2}) \text{ m/s}.$$

(This equation can arise when air resistance is proportional to velocity.) Calculate the height y of the object as a function of t. Use the Newton-Raphson Method to compute the time T of descent. Plot $y(t)$ for $0 \le t \le T$.

61. Suppose the acceleration of an object when dropped from a 100 m height is

$$a(t) = -39.2 \cdot e^{4.427t}(1 + e^{4.427t})^{-2}.$$

(This equation can arise when air resistance is proportional to the square of the velocity.) Calculate the downward velocity v if $v(0) = 0$. (Refer to Exercise 41.) Plot $v(t)$ as a function of t. What is the terminal velocity

$$v_\infty = \lim_{t\to\infty} v(t)?$$

Use the Newton-Raphson Method to determine the value of t for which $v(t) = 0.999v_\infty$.

62. After jumping from an airplane, a skydiver has velocity given by

$$v(t) = -192(1 - e^{-t/6}) \text{ ft/s}.$$

The skydiver opens his parachute 12 s into his descent. For $t > 12$, his velocity is given by

$$v(t) = (16 - 192(1 - e^{-2}))e^{24-2t} - 16 \text{ ft/s}.$$

a. Graph the skydiver's velocity. Is it everywhere continuous? Differentiable?

b. Graph the skydiver's acceleration. Is it everywhere continuous? Differentiable?

c. Determine the skydiver's terminal velocity

$$v_\infty = \lim_{t\to\infty} v(t).$$

d. If the skydiver jumped from 3000 ft, calculate his height function.

e. How long does the first half of his jump—that is, the first 1500 ft—take? How long does the second half take?

63. The sine integral Si is defined to be the antiderivative of $\sin(x)/x$ such that $\text{Si}(0) = 0$. Analyze the graph of $\text{Si}(x)$ over $-4\pi \le x \le 4\pi$ for intervals of increase and decrease and for upward and downward concavity. Explain your analysis. Use a computer algebra system to graph $\text{Si}(x)$ over this interval.

4.9 Enrichment: Applications to Economics*

Many basic problems in economics and finance require calculus for their solution. In this section, we examine a number of these topics.

Profit Maximization

Let p be the price of a product and let x denote the number of units of that product that are sold. For an ordinary item, if the price p increases, then the sales level x decreases. In other words, x is a *decreasing* function of p:

$$x = \mathcal{D}(p). \tag{4.14}$$

This equation is called the *demand equation,* and the function $\mathcal{D}$ is known as the *demand function*. Since $\mathcal{D}$ is a decreasing function, it has an inverse:

$$p = \mathcal{D}^{-1}(x) \tag{4.15}$$

(see Figure 1). (Some economics texts call $\mathcal{D}^{-1}$ the demand function. In a few texts, *both* $\mathcal{D}$ and $\mathcal{D}^{-1}$ are called the demand function.)

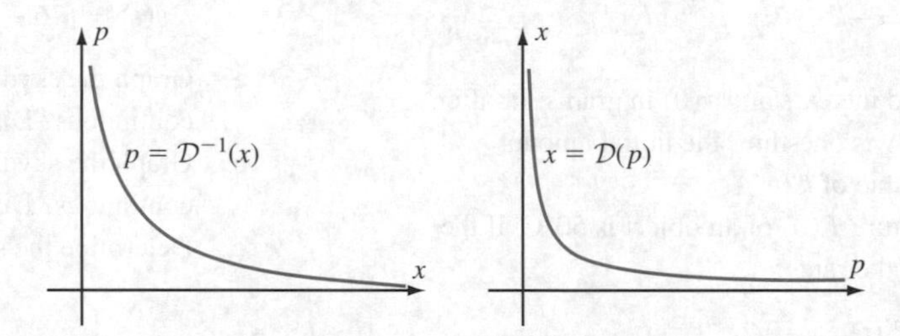

Figure 1
Equivalent representations of the demand equation

The cost $C(x)$ of producing x units of a product includes (1) *fixed costs* C_0 that are independent of the number of units produced and (2) *incremental costs* that rise with each additional unit produced. For example, if a new automobile motor is produced, then there will be design and testing costs that are the same, however many engines are manufactured. There are also incremental costs. Each new motor requires a certain amount of raw materials, labor, and plant costs. The derivative $C'(x)$ is called the *marginal cost* (as discussed in Section 3.9). In a simple model, each additional item produced will account for a new cost equal to m. In this model,

$$C(x) = \underbrace{C_0}_{\text{Fixed costs}} + \underbrace{m \cdot x}_{\text{(Marginal cost)} \cdot \text{(Quantity)}}$$

*Knowledge of this section is not required for subsequent sections.

Notice that $m = C'(x)$ is the marginal cost in *this* model. Often marginal cost is nonconstant. For example, the cost of labor in producing a new aircraft type decreases as workers become better acquainted with the assembly tasks.

The total revenue R is the number of units sold times the price per unit: in symbols, $R = x \cdot p$. Using equations (4.14) and (4.15), we may express revenue as

$$R = x \cdot \mathcal{D}^{-1}(x), \quad \text{or} \quad R = \mathcal{D}(p) \cdot p. \tag{4.16}$$

The profit P is equal to revenue less cost: $P = R - C$. We therefore have

$$P = x \cdot \mathcal{D}^{-1}(x) - C(x), \quad \text{or} \quad P = \mathcal{D}(p) \cdot p - C(\mathcal{D}(p)). \tag{4.17}$$

Figure 2 illustrates typical cost, revenue, and profit plots, graphed as functions of p. Observe that cost is a decreasing function of p, because the number of units produced decreases as p increases. There are two intervals, $(0, p_1)$ and (p_4, ∞), over which $C > R$. At these prices, $P = R - C$ is negative, which means a loss ensues. On the interval $(0, p_1)$, the price per item is too low to generate profits, and on the interval (p_4, ∞), the high price suppresses sales. The price p_2 at which revenue is maximized is usually not the same price p_3 that maximizes profit. Determining this value p_3 is clearly an important business problem.

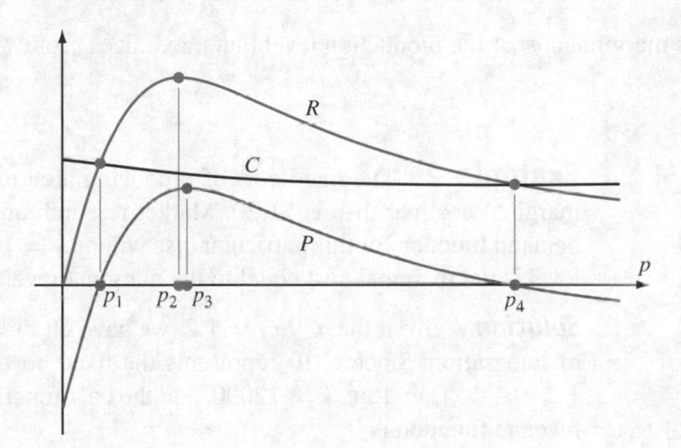

Figure 2

Example 1 The fixed costs of publishing a book include copyediting, typesetting, artwork, and page composition. Each book costs an incremental marginal cost that includes paper, printing, binding, material for the cover, and production. Suppose the total dollar cost for producing x copies of a particular book is $C(x) = 12000 + 3x$. To determine a retail price, the publisher must make a projection of sales. Past experience indicates that the number x of books sold is related to price p by the demand equation $x = 6000 \cdot (1 - p/30)$. What price for the book will optimize profit?

Solution The demand equation is $x = \mathcal{D}(p) = 6000 \cdot (1 - p/30)$. Equation (4.17) gives us the profit P as a function of p:

$$\begin{aligned} P &= \mathcal{D}(p) \cdot p - C(\mathcal{D}(p)) = 6000 \cdot \left(1 - \frac{p}{30}\right) \cdot p - \left(12000 + 3 \cdot 6000 \cdot \left(1 - \frac{p}{30}\right)\right) \\ &= 6600p - 200p^2 - 30000. \end{aligned}$$

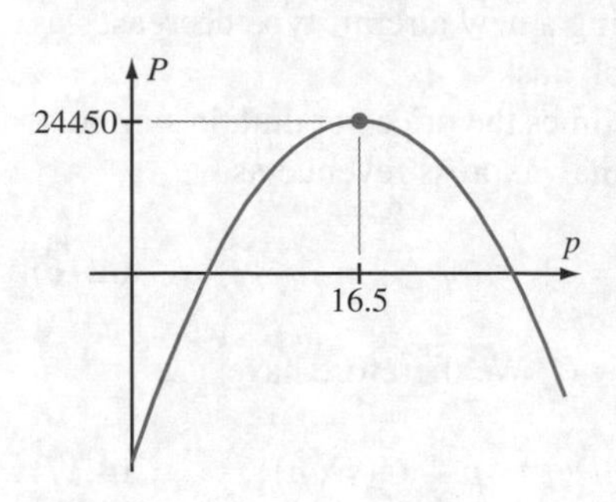

Figure 3
$P = 6600p - 200p^2 - 30000$

The domain for the function P is $p > 0$. Since P is differentiable at all points in its domain, the only critical points are the zeros of $P'(p) = 6600 - 400p$. Thus, $p = 6600/400 = 33/2$ is the unique critical point of P. Note that $P(p)$ is a quadratic in p with negative leading coefficient. Its graph is a parabola that opens downward (see Figure 3). The critical point $p = 16.5$, therefore, gives an absolute maximum. Thus, selling the book at \$16.50 will maximize the profit (and this maximum profit is $P(16.5) = 24450$). Notice that the graph of P in Figure 3 indicates that the publisher will *lose* money if the book is priced too low or too high. ■

If the revenue and cost functions are differentiable functions of x, then so is the profit function $P = R - C$. At the production level x_0 at which profit is maximized, we have $P'(x_0) = R'(x_0) - C'(x_0) = 0$, so

$$R'(x_0) = C'(x_0)$$

at the value x_0 that maximizes profit. In business, this equation is often stated as the maximum profit principle.

Maximum Profit Principle

Marginal revenue equals marginal cost at the production level that maximizes profit.

Example 2 The fixed cost of producing a certain compact disc is \$12,000. The marginal cost per disc is \$1.20. Market research and previous sales suggest that the demand function for this particular disc will be $x = 19000 - 600p$. At what production level is the marginal cost equal to the marginal revenue?

Solution Given that $C'(x) = 1.2$, we have $C(x) = 1.2x + k$ where k is a constant of integration. Since $C(0)$ represents the fixed costs, we see that $12000 = C(0) = 1.2 \cdot 0 + k$. Therefore, $k = 12000$, and the cost function is $C(x) = 12000 + 1.2x$. The revenue function is

$$R = x \cdot p = \frac{x(19000 - x)}{600},$$

and the profit function is

$$P = R - C = \frac{x(19000 - x)}{600} - (12000 + 1.2x).$$

We compute

$$P'(x) = \frac{190}{6} - \frac{x}{300} - 1.2$$

and solve $P'(x) = 0$ to find the critical point $x = 300 \cdot (190/6 - 1.2) = 9140$. Since $P''(x) = -1/300 < 0$, we conclude that the profit curve is maximized at its critical point $x = 9140$. The maximum profit principle tells us that marginal revenue equals marginal cost at this production level. Indeed,

$$R'(9140) = \frac{d}{dx}\left(\frac{x(19000-x)}{600}\right)\bigg|_{x=9140} = \frac{190}{6} - \frac{x}{300}\bigg|_{x=9140} = 1.2 = C'(x).$$

Economic Lot Size

The following dilemma arises in manufacturing. If the anticipated demand for a product during a specified time period (such as a year) is met with one production run, then the manufacturer faces large storage costs. On the other hand, since each production run entails start-up costs, multiple production runs will reduce the storage costs while increasing the costs of production. What, then, is the number of production runs that will minimize the total costs to the manufacturer? This is an optimization problem that calculus can solve.

Let N denote the number of units produced each year in x production runs of equal size N/x. Let c be the start-up cost of each production run, and let μ be the marginal cost of each unit. The cost of each production run is then $c + \mu \cdot (N/x)$, and the total production cost PC over the year is

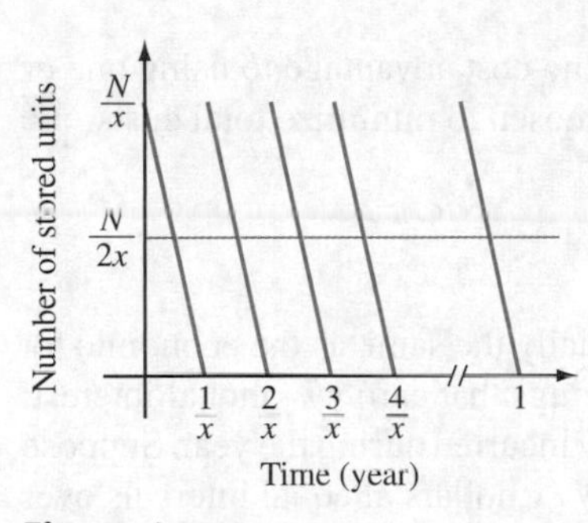

Figure 4

$$PC(x) = x \cdot \left(c + \mu \cdot \frac{N}{x}\right) = cx + \mu N.$$

We assume that the rate of sales of the item is constant throughout the year. Under this assumption, we plot the number of stored units as a function of time, as in Figure 4. As can be seen, the average number of units in storage is $N/(2x)$. If σ is the annual cost of storing one unit, then the total storage cost SC is

$$SC(x) = \sigma \cdot \frac{N}{2x}.$$

The total cost TC of production and storage is $PC + SC$, or

$$TC(x) = cx + \mu N + \sigma \cdot \frac{N}{2x}. \tag{4.18}$$

If x_0 is the value of x that minimizes $TC(x)$, then the quantity N/x_0 of each production run is the *economic lot size*.

For the moment, let us consider $(0, \infty)$ to be the domain of the function TC. We calculate

$$TC'(x) = c - \sigma \cdot \frac{N}{2x^2}$$

and find a critical point at

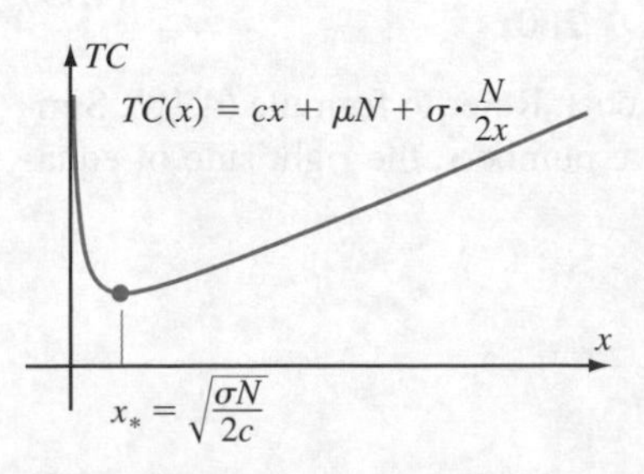

Figure 5

$$x_* = \sqrt{\frac{\sigma N}{2c}}.$$

We see that $TC'(x) < 0$ for x less than x_* and $TC'(x) > 0$ for x greater than x_*. In other words, TC decreases on $(0, x_*)$ and increases on (x_*, ∞). Figure 5 shows a sketch of TC.

In fact, the domain of TC is actually the set of integers $\{1, 2, 3, \ldots\}$, and it is not very likely that the critical point x_* will turn out to be an integer. Nevertheless, the graph tells us how to find the economic lot size x_0. If $x_* < 1$, then $x_0 = 1$ minimizes

$TC(x)$. If $x_* > 1$ is not an integer, then x_0 will either be the greatest integer $\lfloor x_* \rfloor$ smaller than x_* or x_0 will be $\lfloor x_* \rfloor + 1$. We must calculate TC at both values to see which results in the minimum.

Example 3 A software manufacturer anticipates selling 30,000 copies of a certain program per year. The annual storage cost per package is \$1.20 per year. The production start-up costs are \$600, and the marginal cost is \$3.40. How many production runs will minimize total costs?

Solution In this problem, $N = 30000$, $c = 600$, $\mu = 3.40$, and $\sigma = 1.20$. The total cost function TC is given by

$$TC(x) = 600x + 3.40 \cdot 30000 + 1.20 \cdot \frac{30000}{2x},$$

and the critical number x_* is

$$x_* = \sqrt{\frac{\sigma N}{2c}} = \sqrt{\frac{1.20 \cdot 30000}{2 \cdot 600}} = 5.4772.$$

Since $TC(5)$ and $TC(6)$ both equal 108600, there is no cost advantage to using one or the other. Either five or six production runs can be chosen to minimize total costs. ■

The Square Root Formula

There is a problem in personal finance that is essentially the same as the economic lot size problem. Imagine a retired person living on savings that earn $r\%$ annual interest. Suppose that living expenses are N dollars, uniformly incurred during the year. Suppose also that our retiree makes x equal withdrawals of N/x dollars at equal intervals over the year. Let τ denote the transaction cost of each withdrawal. (This cost includes actual monetary charges as well as the value of time and trouble imposed by the transaction.) How frequently should the retiree make his withdrawals?

The situation is very similar to the manufacturer's problem we just analyzed. If we interpret the vertical axis as "Cash holdings" instead of "Number of stored units," then Figure 4 depicts the retiree's cash holdings as a function of time. The average cash on hand is $N/(2x)$, which results in an annual loss of

$$\frac{r}{100} \cdot \frac{N}{2x}$$

in interest. The total cost of the transactions is, therefore,

$$\text{transaction costs} = x \cdot \tau + \frac{rN}{200x}. \tag{4.19}$$

Compare this expression with the formula for total cost. Refer to formula (4.18). Similar analysis shows that as x varies over the positive numbers, the right side of equation (4.19) is minimized at

$$x_* = \sqrt{\frac{rN}{200\tau}}.$$

This equation is known as the *square root formula* in economics. If x_* is not an integer, as is likely, then the number of withdrawals should be determined as in the economic lot size problem, as the next example illustrates.

Example 4 A retiree withdraws \$10,000 a year from his savings account. If his transaction cost is \$30 for each withdrawal and if his money earns 5.5% interest annually, then how many times a year should he withdraw his money?

Solution We set $N = 10000$, $\tau = 30$, and $r = 5.5$. The total transaction cost is

$$C(x) = 30x + \frac{5.5 \cdot 10000}{200x} = 30x + \frac{275}{x}.$$

The square root formula gives $\sqrt{5.5 \cdot 10000/(200 \cdot 30)} \approx 3.028$ for the critical value of x. Since $C(3) = 30 \cdot 3 + 275/3 \approx 181.67$ and $C(4) = 30 \cdot 4 + 275/4 = 188.75$, we conclude that $x = 3$ minimizes the cost function. ■

Average Cost

If the cost of producing x units of a commodity is $C(x)$, then the average cost $\overline{C}(x)$ of producing those x units is the cost divided by the number of units, or

$$\overline{C}(x) = \frac{C(x)}{x}.$$

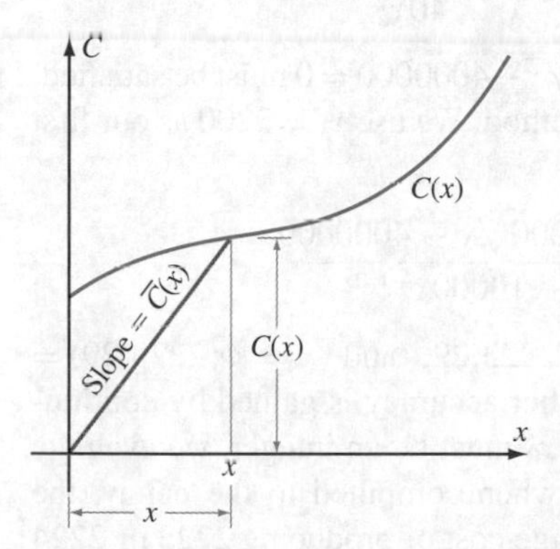

Figure 6

In a simple model in which the fixed costs are c and in which the marginal cost is a constant μ, the cost function is given by $C(x) = c + \mu x$ and the average cost is given by $\overline{C}(x) = (c + \mu x)/x = \mu + c/x$, which decreases as x increases.

More often, a cost function has a shape like the example illustrated in Figure 6. Although the marginal cost $C'(x)$ (the slope of the cost function) initially decreases, it ultimately begins to increase. There can be a number of causes: The initial decrease might be due to learning curve benefits. As x becomes large, however, a manufacturer might have to pay overtime wages or start new shifts as production passes a certain level. It could also be that the increased production of the plant may drive up the prices of raw materials. Any of these factors can lead to greater marginal costs.

Average cost has a simple geometric interpretation that can be understood from Figure 6. If we draw a line segment between the origin and the point $(x, C(x))$, then the slope of that line segment is $C(x)/x$, which is, by definition, the average cost $\overline{C}(x)$. If you imagine the point $(x, C(x))$ moving to the right along the curve, then you can see the slope $\overline{C}(x)$ initially *decrease*. However, it is clear that $\overline{C}(x)$ eventually *increases*. These geometric considerations suggest that for a typical cost function, the average cost $\overline{C}(x)$ will have a local minimum. If $\overline{C}(x)$ has a local minimum at $x = x_0$, then $\overline{C}'(x_0) = 0$, or

$$\left.\frac{d}{dx}\frac{C(x)}{x}\right|_{x=x_0} = \left.\frac{xC'(x) - C(x)}{x^2}\right|_{x=x_0} = 0.$$

Therefore,

$$C'(x_0) = \frac{C(x_0)}{x_0} = \overline{C}(x_0).$$

This leads us to our second economic principle:

Minimum Average Cost Principle

When the average cost is minimized, it equals the marginal cost.

Example 5 Suppose that the cost function for producing a certain washing machine is $C(x) = 100000 + 1000\sqrt{x} + x^2/40$. Find the production level x_0 at which the average cost is minimized. What is $\overline{C}(x_0)$? What is the marginal price at this production level? Interpret your answer graphically.

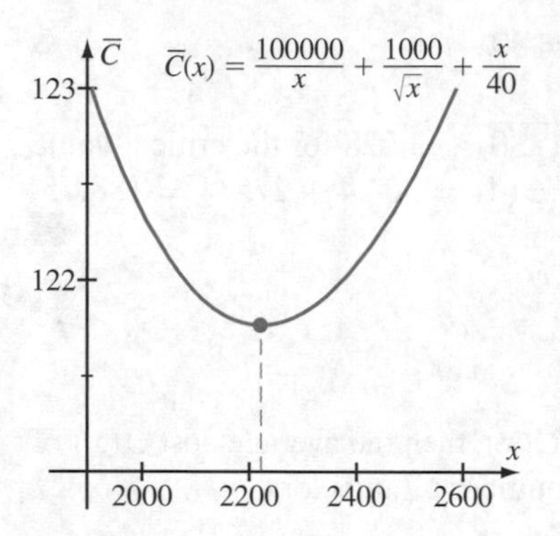

Figure 7

Solution The plot of $\overline{C}$ in Figure 7 indicates that the minimum value of $\overline{C}(x)$ will occur for x between 2200 and 2300. To be more precise, we solve for the critical point of $\overline{C}$. We have

$$\overline{C}(x) = \frac{100000 + 1000\sqrt{x} + x^2/40}{x} = \frac{100000}{x} + \frac{1000}{\sqrt{x}} + \frac{x}{40}$$

and

$$\overline{C}'(x) = -\frac{100000}{x^2} - \frac{500}{x^{3/2}} + \frac{1}{40} = \frac{x^2 - 20000\sqrt{x} - 4000000}{40x^2}.$$

At a critical point, the equation $f(x) = x^2 - 20000\sqrt{x} - 4000000 = 0$ must be satisfied. We find the solution using the Newton-Raphson Method. We use $x_1 = 2200$ as our first approximation. If we set

$$\Phi(x) = x - \frac{f(x)}{f'(x)} = x - \frac{x^2 - 20000\sqrt{x} - 4000000}{2x - 10000x^{-1/2}},$$

then $x_2 = \Phi(2200) = 2223.43$, $x_3 = \Phi(2223.43) = 2223.29$, and $x_4 = \Phi(2223.29) = 2223.29$. We stop the process here because no further accuracy is gained by continuing it. We have found that $x_0 = 2223.29$. Because x_0 must be an integer, we evaluate $\overline{C}(2223) \approx 121.77$ and $\overline{C}(2224) = 121.77$. Thus, when computed to the penny, the minimal average cost is \$121.77, which is the average cost of producing 2223 or 2224 machines. Since $C(x) = 100000 + 1000\sqrt{x} + x^2/40$, we have

$$C'(2223.29) = \left. \frac{500}{\sqrt{x}} + \frac{1}{20}x \right|_{x=2223.29} = 121.77.$$

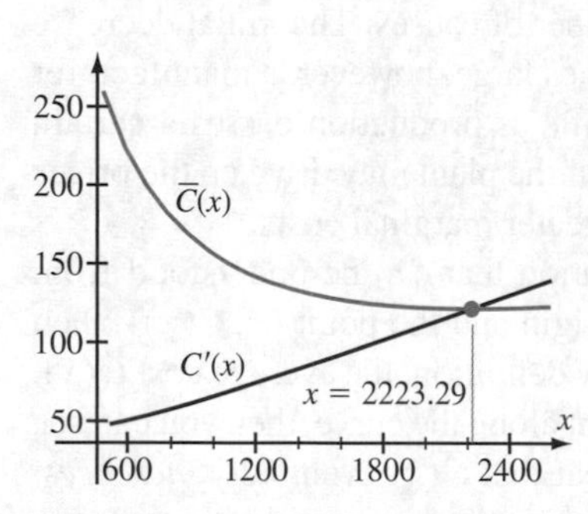

Figure 8

According to the minimum average cost principle, this number would equal the lowest average cost. The graphical interpretation is that the plots of C' and $\overline{C}$ intersect at $x = x_0$, as is shown in Figure 8. ■

Example 5 uses two different applications of calculus. The first application tells us that the minimum average cost can be found by solving the equation $\overline{C}'(x) = 0$. The second application, the Newton-Raphson Method, is what permits us to solve the equation.

Elasticity of Demand

We have already considered the role that the demand equation $x = \mathcal{D}(p)$ plays in the determination of profits. We now examine the relationship between price p and demand x in greater detail. In particular, we are interested in the sensitivity of demand to price.

Suppose that x_0 is the demand of a commodity sold at price p_0. If we change the price by an amount Δp, then the change Δx in demand will be

$$\Delta x = \mathcal{D}(p_0 + \Delta p) - \mathcal{D}(p_0).$$

The percentage change in demand is, therefore,

$$100\frac{\Delta x}{x_0} = \frac{100(\mathcal{D}(p_0 + \Delta p) - \mathcal{D}(p_0))}{x_0}.$$

Economists measure how sensitive demand is to price by considering the ratio of the percentage change in demand to the percentage change in price:

$$\frac{100\frac{\Delta x}{x_0}}{100\frac{\Delta p}{p_0}} = \frac{p_0}{x_0}\frac{\mathcal{D}(p_0 + \Delta p) - \mathcal{D}(p_0)}{\Delta p}. \qquad \textbf{(4.20)}$$

The *elasticity of demand* $E(p_0)$ of the commodity at price p_0 is defined as the *negative* of the limit of this ratio as Δp tends to 0:

$$E(p_0) = -\lim_{\Delta p \to 0} \frac{100\frac{\Delta x}{x_0}}{100\frac{\Delta p}{p_0}}.$$

In view of equation (4.20), we have

$$E(p_0) = -\lim_{\Delta p \to 0} \frac{p_0}{x_0}\frac{\mathcal{D}(p_0 + \Delta p) - \mathcal{D}(p_0)}{\Delta p} = -\frac{p_0}{x_0}\mathcal{D}'(p_0).$$

Finally, because $x_0 = \mathcal{D}(p_0)$, we obtain

$$E(p_0) = -p_0\frac{\mathcal{D}'(p_0)}{\mathcal{D}(p_0)}. \qquad \textbf{(4.21)}$$

Since $\mathcal{D}$ is almost always a decreasing function, the derivative $\mathcal{D}'(p_0)$ is usually negative. Therefore, $E(p_0)$ is generally positive. If $E(p_0) < 1$, then demand is said to be *inelastic* at price p_0. If $E(p_0) > 1$, then demand is said to be *elastic* at price p_0. If $E(p_0) = 1$, then demand is said to have *unit elasticity*.

Example 6 A product is selling at a price at which elasticity of demand is 2. Approximately what effect on sales will a 3% price increase have?

Solution From the definition of elasticity of demand, we have

$$2 = E(p_0) = -\lim_{\Delta p \to 0} \frac{100\frac{\Delta x}{x_0}}{100\frac{\Delta p}{p_0}} \approx -\frac{100\frac{\Delta x}{x_0}}{100(0.03)},$$

or

$$\frac{\Delta x}{x_0} \approx -0.06.$$

In words, there will be a *decrease* in demand of approximately 6%. ■

There is an important relationship between marginal revenue and elasticity of demand. The formula for revenue as a function of price is $R = \mathcal{D}(p) \cdot p$, as given by the second equation in line (4.16). If we differentiate both sides with respect to p and evaluate at p_0, we obtain

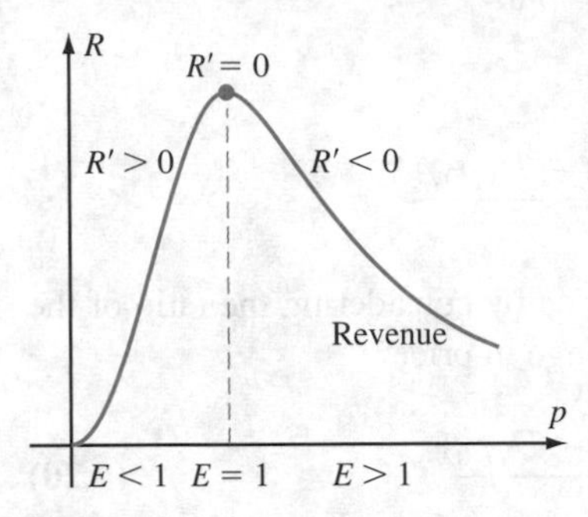

Figure 9

$$\left.\frac{dR}{dp}\right|_{p=p_0} = \mathcal{D}'(p) \cdot p + \mathcal{D}(p)|_{p=p_0} = p_0\mathcal{D}'(p_0) + \mathcal{D}(p_0)$$

$$= \mathcal{D}(p_0)\left(\frac{p_0\mathcal{D}'(p_0)}{\mathcal{D}(p_0)} + 1\right) = x_0(1 - E(p_0)).$$

We see that $R'(p_0) > 0$ when $E(p_0) < 1$ and $R'(p_0) < 0$ when $E(p_0) > 1$. In ordinary language, this means that revenue is an *increasing* function of price when the demand is *inelastic* and a *decreasing* function of price when the demand is *elastic*. Since marginal revenue is 0 when elasticity of demand is equal to 1, we obtain a third business principle:

Maximum Revenue Principle

Revenue is maximized when the demand has unit elasticity.

A typical revenue function is plotted in Figure 9.

quickquiz

1. What is the relationship between marginal revenue and marginal cost at a production level that maximizes profits?
2. If the start-up cost of a manufacturing process is \$1000, if 9000 units are to be produced during the year, and if the annual storage cost is \$2 per unit, how many production runs minimize total costs?
3. Suppose the cost of making 1000 floppy drives is \$30,000 and these production numbers represent the minimum average cost. About how much does it cost to produce the 1001st floppy drive? Justify your estimate.
4. If elasticity of demand is 2, will a price increase of 5% result in increased or decreased revenues?

EXERCISES

Problems for Practice

In Exercises 1–4, calculate the marginal cost for the cost function at the given production level x.

1. $C(x) = 1000 + 2x$, $x = 1000$
2. $C(x) = 6000 + 3\sqrt{x}$, $x = 10000$
3. $C(x) = 8500 + 20(x - \sqrt{x})$, $x = 100$
4. $C(x) = 3000 + \ln(x + 1)$, $x = 99$

In Exercises 5–8, for a production process with the given cost function, calculate the following.

a. The fixed cost
b. The cost of producing the 100th unit
c. The marginal cost when $x = 100$
d. The average cost of producing 100 units

5. $C(x) = 2000 + 3x$
6. $C(x) = 6000(1 + \sqrt{x}/200)$
7. $C(x) = 12000((x + 1)/(\sqrt{x} + 10))$
8. $C(x) = 8500 \ln(x + 100)$

In Exercises 9–12, calculate the production level that maximizes profits for the cost and demand functions.

9. $C(x) = 1000 + 2x$, $\mathcal{D}^{-1}(x) = 1602 - 8x$
10. $C(x) = 800 + 3x$, $\mathcal{D}^{-1}(x) = 26403 - 12x$
11. $C(x) = 1200 + 8x$, $\mathcal{D}(p) = 51 - p/8$
12. $C(x) = 7000 + x$, $\mathcal{D}(p) = 2402 - 2p$

In Exercises 13–16, an annual production cost PC is expressed as a function of the number x of production runs. For the total annual production N and annual unit storage cost σ, find the most economical number of production runs.

13. $PC(x) = 3x + 200$
14. $PC(x) = 2.5x + 64000$
15. $PC(x) = 8x + 9000$
16. $PC(x) = 145x + 80000$
17. Alfred withdraws \$12,000 a year from his savings. If his transaction cost is \$90 for each withdrawal and if his savings earn 6% interest annually, then how often should Alfred withdraw his money?
18. Cecilia withdraws \$10,000 a year from her savings. If her transaction cost is \$18 for each withdrawal and if her savings earn 9% interest annually, then how often should Cecilia withdraw her money?
19. Orville withdraws \$10,000 a year from his savings. If his transaction cost is \$45 for each withdrawal and if his savings earn 10% interest annually, then how often should Orville withdraw his money?
20. Donna withdraws \$8000 a year from her savings. If her transaction cost is \$72 for each withdrawal and if her savings earn 5% interest annually, then how often should Donna withdraw her money?
21. Suppose the demand equation of a particular commodity is $p = Ae^{-kx}$ for positive constants A and k. What production level maximizes *revenue*?

In Exercises 22–25, calculate the production level that minimizes average cost for the given cost function $C(x)$ (if one exists).

22. $C(x) = 12000 + 2\sqrt{x} + 1.2x$
23. $C(x) = 16000 + x^2/2000$
24. $C(x) = 900 + x/100 + x^2/400$
25. $C(x) = 16000 + 0.02x + x^3/8000$

In Exercises 26–28, use the demand equation to calculate the elasticity of demand as a function of p. For what prices is demand inelastic? Elastic?

26. $x = 19000 - 600p$
27. $x = 200000 - 50000\sqrt{p}$
28. $x = 96000 - 20p^2$

In Exercises 29–32, use the elasticity of demand function to determine the price p at which revenue is maximized (assuming a typical concave down revenue curve).

29. $E(p) = \sqrt{p}/(12 - 2\sqrt{p})$
30. $E(p) = 9p^3/(12000 - 3p^3)$
31. $E(p) = 2p^2/(4800 - p^2)$
32. $E(p) = 20000/(60000 - 20000 \ln(p))$

In Exercises 33–36, use the information to approximate the percentage demand increase or decrease in which the indicated price change results.

33. A price increase from \$2.50 to \$3.00; $E(2.50) = 1/2$
34. A price increase from \$3.00 to \$3.30; $E(3.00) = 1.5$
35. A price decrease from \$6.00 to \$5.40; $E(6.00) = 1$
36. A price decrease from \$3.00 to \$2.40; $E(3.00) = 1/10$
37. Blackberry Hill sells 2000 hamburgers a week at \$3.75. Past sales figures indicate that the demand function is linear. After raising the price of a hamburger to \$4.50, the restaurant records a sales drop to 1700 hamburgers per week.
    a. Determine the demand equation.

b. What is the change in revenue that resulted from the price increase?

c. Compute the elasticity of demand at \$3.75 per hamburger and at \$4.50 per hamburger.

d. Using only the value of $E(4.50)$ calculated in part c, determine whether raising the price beyond \$4.50 will increase the revenue.

e. Using only the value $E(4.50)$, approximate the actual change in demand that will result in a further increase to \$4.95.

38. Roadside Records sells 20,000 compact discs a week at \$15.00. Past sales figures indicate that the demand function is linear. After raising the price of a disc to \$16.50, the retailing chain registers a sales drop to 16,000 discs per week.

a. Determine the demand equation.

b. What is the change in revenue that resulted from the price increase?

c. Compute the elasticity of demand $E(p)$. What is $E(15.00)$? $E(16.50)$?

d. Will revenue increase or decrease if prices are reduced from the current price of \$16.50? Use only your computation from part c to explain your answer.

e. Use the value $E(16.50)$ to approximate the change in demand that will result from an increase in price from \$16.50 to \$17.00.

f. Sometimes retailers need to (temporarily) maximize revenue, even if profits suffer. At what price should Roadside sell its discs to maximize revenue? Use your determination of $E(p)$ in part c and explain.

Further Theory and Practice

39. A cost function is sketched in Figure 10. What is the significance of $|\overline{OA}|$? What significance does the slope of $\overline{OB}$ have? How fast is the marginal cost changing at point P?

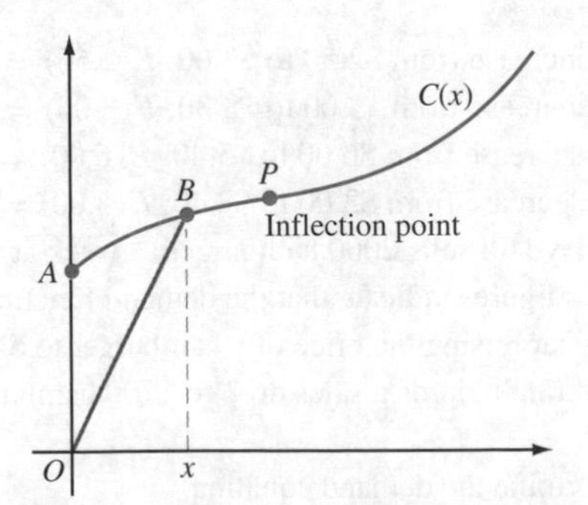

Figure 10

40. The elasticity of demand of a commodity is graphed in Figure 11. At what price is revenue maximized? Explain your answer.

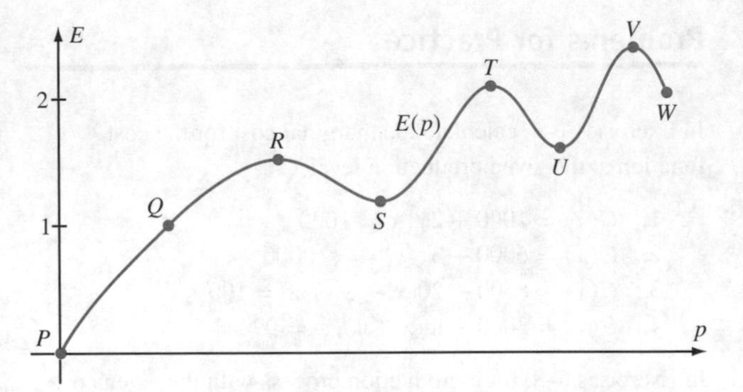

Figure 11

41. Suppose the profit function P is a differentiable function of production level x on $(0, \infty)$ with P negative for sufficiently large values of x. What theorem guarantees that the marginal profit is somewhere 0? What can be said about marginal cost and marginal revenue at such a point?

42. If the marginal cost of a product is decreasing on the interval I, what can you say about the graph of the cost function over I? Explain your answer.

43. Suppose the marginal cost function of a production process crosses the x-axis at $x = \xi$. What significance does the point $(\xi, C(\xi))$ have for the graph of the cost function?

44. Suppose that a cost function is twice differentiable and that it has an inflection point at c. What can you say about the marginal cost in a sufficiently small interval centered about c?

45. Suppose the fixed costs of a manufacturing process are nonzero. Explain why the average cost function will have a minimum on any interval of the form $(0, N]$.

46. Relate the signs of R' and $E - 1$. Describe the function

$$p \mapsto \text{signum}(R'(p) \cdot (E(p) - 1)).$$

47. Several economics texts imply that it is necessary for the graph of a cost function C to have an inflection point in order for the average cost function $\overline{C}$ to have a local minimum. Show that such a statement is false by considering the function $C(x) = a + bx^2$.

48. Several economics texts imply that if the graph of a cost function C has an inflection point at c, then the average cost function $\overline{C}$ will have a local minimum at c. Show that such a statement is false by considering the function $C(x) = c + x\exp(-x)$.

In Exercises 49–52, calculate the production level that minimizes average cost for the given cost function C.

49. $C(x) = 10 - \sqrt{x} + 2x$

50. $C(x) = 800 + 1.5x + 2\sqrt{x}$

51. $C(x) = 1000 + x^{3/2}$

52. $C(x) = 12000 + 10x\ln(x)$ $(x \geq 1)$

53. Suppose $C(x) = a + bx + cx^2$ with a and c positive. Show that the value of b does not affect the production level that minimizes average cost.

Exercises 54–56 concern a manufacturing process for which the cost function $C(x) = c + \mu x$ is linear. Suppose also that the demand equation, $x = a - bp$ is linear.

54. Find a formula for the price at which profits are maximized.

55. Suppose the marginal cost increases from μ to $\mu + \delta$. How much should the manufacturer increase the price of his product to maximize his profit?

56. Suppose the fixed costs increase from c to $c + \Delta$. How much should the manufacturer increase the price of his product to maximize his profit?

57. Suppose that the minimum average cost of a manufacturing process with differentiable cost function C occurs at level ξ. If the government imposes a new tax of $\$\mu$ per unit on the manufacturer, then the average cost is still minimized at production level ξ. Explain why. (Relate the new cost function to the old one and the new average cost function to the old one.)

58. Suppose that the cost of each production run of u units is $5000 + 25\sqrt{u}$. Suppose that the annual storage is $6 per unit. If the annual demand is 10,000 units at a steady rate, then in how many equally spaced batches should this commodity be produced to minimize total costs?

Exercises 59 and 60 concern the economic order quantity.

59. A company annually uses N units of a required item at a steady rate. The company incurs fixed expenses of c for each order and pays μ for each unit. Its annual storage cost is s per unit. Determine the number of orders x_* that minimizes costs. The size of each order, N/x_*, is called the *economic order quantity*. *Hint:* Follow the discussion of the economic lot size.

60. A company annually uses 100,000 units of a required item at a steady rate. The company incurs fixed expenses of $100 for each order and pays $0.27 for each unit. Its annual storage cost is $0.072 per unit. Determine the number x_* of orders that minimizes total costs.

61. The following problem once frequently appeared in calculus books: "The speed of waves of length L in deep water is proportional to

$$\sqrt{\frac{L}{a} + \frac{a}{L}}$$

where a is a certain constant. Show that the speed is the lowest when $L = a$." (Quoted from *The Calculus for the Practical Man,* J.E. Thompson, 1931.) Use the square root formula to deduce this. Explain your reasoning.

62. If the demand equation is the linear $x = a - bp$, show that demand is inelastic if $p < a/(2b)$ and elastic if $p > a/(2b)$.

63. Use the revenue function to explain why demand is generally inelastic on an interval of the form $(0, N)$ and elastic on (N, ∞).

64. Suppose that the demand function $\mathcal{D}$ is twice differentiable. Show that if $\mathcal{D}$ is decreasing and if the graph of $\mathcal{D}$ is concave down, then the elasticity of demand is an increasing function of price.

65. If $\mathcal{D}$ is a decreasing function and E is an increasing function, show that the graph of revenue is concave down.

66. Express elasticity of demand as a function of x.

67. If revenue R is a differentiable function, then elasticity of demand can be defined by

$$E(p) = 1 - \frac{R'(p)}{\mathcal{D}(p)}.$$

Does formula (4.21) follow from this definition?

68. Suppose that $E(p)$ is the elasticity of demand associated with a demand equation $x = \mathcal{D}(p)$. Use the demand function $\mathcal{D}$ to find an antiderivative of $E(p)/p$.

69. Suppose that $x = \mathcal{D}(p)$ is a demand equation with elasticity of demand $E(p)$.

a. Let A be a positive constant and set

$$\mathbf{D}(p) = A \cdot \mathcal{D}(p).$$

Show that the elasticity of demand of $\mathbf{D}$ is also E.

b. If $x = \mathbf{D}(p)$ is a demand equation with the same elasticity of demand as $x = \mathcal{D}(p)$, then show that

$$\mathbf{D}(p) = A \cdot \mathcal{D}(p)$$

for some constant A. (Suppose that the demand equation of a particular commodity is $p = Ae^{-kx}$ for positive constants A and k.)

70. Suppose that the cost function is linear:

$$C(x) = c + \mu x.$$

Express the production level that maximizes profit in terms of Lambert's W function, which is defined impicitly by the equation

$$W(x)e^{W(x)} = x.$$

Hint: Make the change of variable $t = 1 - xk$ in the equation for the critical point of the profit function. The resulting equation has the form $t \exp(t) = G(\mu/A)$ where G is a linear function.

Calculator/Computer Exercises

71. A garden equipment manufacturer determines that the demand equation for its premium garden hose is

$$x = 124721 - 21180\sqrt{p}.$$

The fixed cost of production is \$30,000 and the marginal cost is \$1.50 for each unit. Graph the cost and revenue as functions of demand over an interval that exhibits both breakeven points. To the nearest unit, what is the demand that maximizes profit?

72. The Fiberall Grain Company has determined through market surveys that the weekly demand function of their new cereal Harvest Bits will be

$$x = 96658 - 622p^3.$$

The cost of producing x boxes of Harvest Bits is $\$60000 + 0.40x$. Graph the cost and revenue as functions of demand over an interval that exhibits both breakeven points. To the nearest unit, what is the demand that maximizes profit? What is the corresponding price per box?

73. Suppose that the demand equation of a particular commodity is $p = 10e^{-x/40000}$. Suppose that the cost function is linear:

$$C(x) = 30000 + 2x.$$

Plot the profit function over an interval that exhibits its maximum. Find the demand x that maximizes profit.

In Exercises 74–76, determine the demand x that minimizes the average cost of production for cost function C. Graph $\overline{C}(x)$ and $C'(x)$ over an interval that shows their intersection. Explain the significance of the point of intersection.

74. $C(x) = 100000 + 0.01x^2 + 0.0005x^3$

75. $C(x) = 60000 + 0.02\sqrt{x} + 0.0001x^2$

76. $C(x) = \begin{cases} 360 & \text{if } 0 \le x < 1 \\ 500 + 100(x - \ln(x)) & \text{if } 1 \le x \end{cases}$

77. Plot the elasticity of demand for the demand function of Exercise 71. Determine the price range for which demand is inelastic and the price range for which demand is elastic. What demand maximizes revenue?

78. Plot the elasticity of demand for the demand function of Exercise 72. Determine the price range for which demand is inelastic and the price range for which demand is elastic. What demand maximizes revenue?

Summary of Key Topics

Local and Absolute Maxima and Minima (Section 4.1)

Let f be a function with domain $S \subseteq \mathbb{R}$. We say that f has a local maximum at the point $c \in S$ if there is a $\delta > 0$ such that $f(c) \ge f(x)$ for all $x \in S$ such that $|x - c| < \delta$. We say that f has a local minimum at the point $c \in S$ if there is a $\delta > 0$ such that $f(c) \le f(x)$ for all $x \in S$ such that $|x - c| < \delta$.

We call c an absolute minimum or maximum if the preceding inequalities hold for all x in S. The locations of local or global maxima and minima are called local or global extrema. If f is differentiable on an open interval I and if $c \in I$ is an extremum, then $f'(c) = 0$.

Rolle's Theorem and the Mean Value Theorem (Section 4.1)

Let f be continuous on $[a, b]$ and differentiable on (a, b) and suppose that $f(a) = f(b) = 0$. There is a number $c \in (a, b)$ such that $f'(c) = 0$.

More generally, let f be continuous on $[a, b]$ and differentiable on (a, b). There is a $c \in (a, b)$ such that

$$f'(c) = \frac{f(b) - f(a)}{b - a}.$$

A consequence of the Mean Value Theorem is that if f is differentiable on an open interval I and $f' = 0$ on I, then f is a constant function. As a consequence, if two functions f and g have the same derivative function, then they differ by a constant.

Critical Points (Section 4.2)

A point where either $f' = 0$ or f' is undefined is called a *critical point*.

Increasing and Decreasing Functions (Section 4.2)

If $f(a) < f(b)$ whenever a and b are elements of an interval I and $a < b$, then f is increasing on I. If $f(a) > f(b)$ whenever a and b are elements of an interval I and $a < b$, then f is decreasing on I.

If $f' > 0$ on an open interval I, then f is increasing on I. If $f' < 0$ on an open interval I, then f is decreasing on I.

The First Derivative Test for Extrema (Section 4.2)

Let c be a critical point for a differentiable function f.

1. If there is a $\delta > 0$ such that $f' < 0$ on $(c - \delta, c)$ and $f' > 0$ on $(c, c + \delta)$, then c is a local minimum.
2. If there is a $\delta > 0$ such that $f' > 0$ on $(c - \delta, c)$ and $f' < 0$ on $(c, c + \delta)$, then c is a local maximum.

Applied Maximum-Minimum Problems (Section 4.3)

The ability to locate extrema of functions enables us to solve a variety of applied problems. In these applied problems, we often must find the extrema of a continuous function on a closed and bounded interval. We seek the extrema at the endpoints and the critical points.

Concavity, Points of Inflection, Second Derivative Tests (Section 4.4)

If f' increases on an open interval I, then the graph of f is said to be concave up on I. If f' decreases on an open interval I, then the graph of f is said to be concave down on I.

If $f'' > 0$ on I, then f is concave up on I. If $f'' < 0$ on I, then f is concave down on I. A point of inflection for a function is a point where the concavity changes from down to up or up to down.

If f is twice differentiable on an open interval containing c and if c is a critical point, then the Second Derivative Test for Extrema states:

1. If $f''(c)$ exists and is positive, then c is a local minimum.
2. If $f''(c)$ exists and is negative, then c is a local maximum.
3. If $f''(c)$ exists and equals zero, then no conclusion is possible.

Curvature (Section 4.4)

If f is twice differentiable, we can measure the rate at which the graph of f turns at c by the curvature

$$\kappa(c) = \frac{|f''(c)|}{(1+(f'(c))^2)^{3/2}}.$$

Graphing (Sections 4.1, 4.2, 4.4, 4.5)

We note that differentiable functions are continuous, thus making graphing simpler. The concepts of increasing, decreasing, critical point, point of inflection, concave up, and concave down enable us to draw very sophisticated sketches of curves.

L'Hôpital's Rule (Section 4.6)

If c is either a real number or $\pm\infty$ and if $\lim_{x\to c} f(x) = \lim_{x\to c} g(x) = 0$ or $\lim_{x\to c} f(x) = \lim_{x\to c} g(x) = \infty$, then

$$\lim_{x\to c} \frac{f(x)}{g(x)} = \lim_{x\to c} \frac{f'(x)}{g'(x)}.$$

The Newton-Raphson Method (Section 4.7)

If x_0 is a good initial estimate to a root c of a differentiable function f, then the sequence of iterates $\{x_j\}$ obtained by

$$x_{j+1} = x_j - \frac{f(x_j)}{f'(x_j)}$$

is often used to approximate c.

Antidifferentiation (Section 4.8)

Given a function f, it is useful to be able to determine functions F such that $F' = f$. We call F an antiderivative for f and write $\int f(x)\,dx = F(x) + C$. The expression $\int f(x)\,dx$, which is called an indefinite integral, can be used to study problems about velocity and acceleration of moving bodies.

Applications to Economics (Section 4.9)

The equation $x = \mathcal{D}(p)$, which relates sales x to price p, is called the demand equation. If the cost of producing x items is $C(x)$, then calculus can be applied to maximize the profit function

$$P = x \cdot \mathcal{D}(x) - C(x) \quad \text{or} \quad P = \mathcal{D}(p) \cdot p - C(\mathcal{D}(p)).$$

In general, marginal revenue will equal marginal cost at a production level that maximizes profit.

The minimum average cost is the ordinate of the point of intersection of the graph of the average cost function and the graph of the marginal cost function.

When storage costs are considered alongside production costs, the methods of calculus can be used to minimize the total cost of annual production processes. Under standard assumptions, the most economical number of production runs is 1, $\lfloor\sqrt{(\sigma N)/(2c)}\rfloor$, or $\lfloor \sqrt{(\sigma N)/(2c)} \rfloor + 1$ where N is the number of items that are produced annually, σ is the annual storage cost per item, and c is the fixed cost of each production run.

Elasticity of demand E relates the percentage demand change that results from a percentage change in price. It is defined by

$$E(p) = -p\frac{\mathcal{D}'(p)}{\mathcal{D}(p)}.$$

Marginal revenue is related to demand by

$$R'(p) = x(1 - E(p)).$$

genesis & DEVELOPMENT

Fermat's Life

In the center of Toulouse, France, lies a beautiful civic building called the Capitole. In this building is a large statue of Pierre de Fermat; he is seated, and a sign declares him to be the father of differential calculus. A muse stands nearby, showing her abundant appreciation for his mental powers. Whereas mathematicians today either work for universities, government, or industry, in Fermat's day many mathematicians were amateurs. Fermat was a judge by profession and a rather wealthy man. He was a cultured man and a supporter of the arts. He wrote elegantly in his native French and was fluent in Italian, Spanish, Latin, and classical Greek. Although he developed a reputation as a classicist and a poet, his special passion was for mathematics.

Fermat was born in the small town of Beaumont-de-Lomagne in the year 1601. Sometime in the 1620s, Fermat studied at the University of Bordeaux with followers of François Viète, the leading French mathematician of an earlier generation. Fermat went on to study law at the university at Orléans, attaining his degree in 1631. He served as magistrate in a special court in Castres that had been created to settle disputes among Protestants and Catholics. It was in Castres, in 1665, that Fermat died. Ten years later, his remains were transferred to the Church of the Augustines in Toulouse.

Fermat's Investigation of Extrema

In 1629, Fermat began his work on the reconstruction of the lost books of Apollonius, a project that Viète, working from the surviving commentaries of Pappus, had launched. It was while working on this project that Fermat conceived the idea of analytic geometry, sometime before 1636. He wrote of his discovery that year and circulated it in Paris in 1637, prior to the publication of Descartes's *La Géometrie*. That same year, Fermat also published his method of maxima and minima. In 1615, Kepler had stated, based on numerical evidence, that there is an "imperceptible" variation near an extremum. Today we would say that this is true of a *differentiable* function f at a local extremum c, and we would express it more precisely by the formula $f'(c) = 0$. Indeed, it was Fermat who, in 1629, first discovered this necessary condition for a local extremum.

After demonstrating his method on a simple classical problem, Fermat proposed and answered the following problem: Divide a line segment of length b into two pieces such that the product of the length of one by the square of the length of the other is a maximum. In other words, maximize

$$f(x) = (b - x) \cdot x^2 = bx^2 - x^3 \quad (0 \le x \le b).$$

Suppose that this maximum occurs at $x = a$. Fermat set $f(a) = f(a + h)$ and worked with the equation

$$\frac{f(a+h) - f(a)}{h} = 0,$$

or

$$\frac{b(a+h)^2 - (a+h)^3 - (ba^2 - a^3)}{h} = 0.$$

This simplifies to $(2ab - 3a^2) + (b - 3a)h - h^2 = 0$. Either by discarding the remaining terms that involve h or by setting $h = 0$, Fermat concluded that $2ab - 3a^2 = 0$ and $a = 2b/3$. Fermat did not use language suggesting that he considered h to be infinitesimal or even small. Rather his approach was an algebraic one that he developed before his discovery of analytic geometry. The derivative had not yet been invented, and it was this very investigation that led Fermat to introduce it, thus staking his claim as "father of differential calculus."

Fermat's Investigation of Points of Inflection

The notion of point of inflection arose very early in the study of the tangent problem. In letters written to Fermat in 1636, Roberval suggested that the curve known as the *conchoid* has two points at which there are no tangent lines. What Roberval had discovered and was referring to is that the conchoid has two points of inflection. The graph of the conchoid (together with one of its

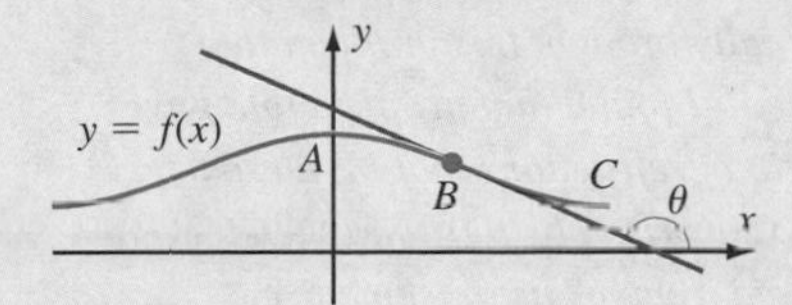

Figure 1
A conchoid and a tangent at a point of inflection

tangent lines at an inflection point) appears in Figure 1. Roberval called attention to the change in concavity at B by describing the conchoid as "convex on the outside" from A to B and "convex on the inside" from B through C to infinity.

Fermat addressed this issue in a letter to Mersenne dated April 20, 1638. Fermat's discussion indicates that he continued to think of his Tangent Method as a by-product of his Method of Maxima/Minima. He wrote to Mersenne: "[I]n order to properly graph the [conchoid], it is appropriate that we investigate the points of inflection, where the curvature becomes concave or convex, or vice versa. This question is elegantly resolved by the Method of Maxima/Minima by virtue of the following general lemma." Fermat continued by asserting that among all tangents to the conchoid, the tangent at the inflection point is distinguished by minimizing the angle θ made with the positive x-axis, which is the same as minimizing $\tan(\theta) = f'(x)$. Since Fermat's Theorem gives $(f')'(x) = 0$ as a necessary condition for this minimum, Fermat had, in effect, discovered a necessary condition for $(x, f(x))$ to be a point of inflection—namely, $f''(x) = 0$.

Fermat's Principle of Least Time

Although Fermat had little interest in physics, he did turn his consideration to one particular question in optics. Therein lies another bitter dispute with Descartes. This particular quarrel began with an early correspondent of Fermat, Jean de Beaugrand. Beaugrand's major work, *Géostatique* (1636), had received severe criticism from Descartes, who wrote:

> *Although I have seen many squarings of the circle, perpetual motions, [etc.], . . . I can say nonetheless that I have never seen so many errors combined in one postulate. I can therefore conclude that the contents of* Géostatique *are so irrelevant, ridiculous, and mistaken, that I wonder if any honest man has ever troubled himself to read it.*

Beaugrand retaliated by circulating three pamphlets in which he charged Descartes with plagiarism. Also, using his position as *Secrétaire du Roi,* Beaugrand obtained a printer's copy of Descartes's *La Dioptrique* prior to its publication. Beaugrand distributed copies in the hope that embarrassing prepublication criticism would ensue. Fermat was one of the recipients.

In a 1637 commentary on *La Dioptrique,* Fermat drew attention to errors that Descartes had made in his derivations of the Laws of Reflection and Refraction. Descartes was not pleased. The situation between the two grew more tense when, after reading *La Géometrie,* Fermat sent Descartes documents that established his priority in the matters of analytic geometry and the tangent problem.

In 1658, eight years after the death of Descartes, the editor of the *Letters of Descartes,* Claude Clerselier, found two letters from Fermat among the papers of Descartes. Clerselier asked Fermat for copies of any other letters that he might have sent to Descartes. Since Fermat had only written the two that had been found, Fermat interpreted the request to be a reopening of the optics controversy.

Fermat argued that there were three fundamental flaws that invalidated Descartes's treatment of refraction: Descartes did not understand what a valid demonstration should be, Descartes had not argued consistently even within his own incorrect logical framework, and Descartes had based his derivation on the patently false notion that the denser the medium, the easier the passage of light through it. Experts warned Fermat to be more cautious, and experimentation later confirmed the Law of Refraction that Descartes had published.

At about this time, Fermat received the *Traité de la Lumière,* by Marin Cureau de la Chambre (1594–1669). In it was a derivation of the Law of Reflection based on the method of minimization. Suppose that a ray of light emitted from A is reflected by the straight line $\overline{EF}$ and arrives at B (see Figure 2). The Law of Reflection states that the angle of incidence $\angle ACG$ equals the angle of reflection $\angle GCB$. Cureau derived this from the postulate that light travels from A to $\overline{EF}$ to B in the *least distance*. That is, for any other point D on $\overline{EF}$, $|\overline{AC}| + |\overline{CB}| < |\overline{AD}| + |\overline{DB}|$. This inequality follows easily from classical geometry.

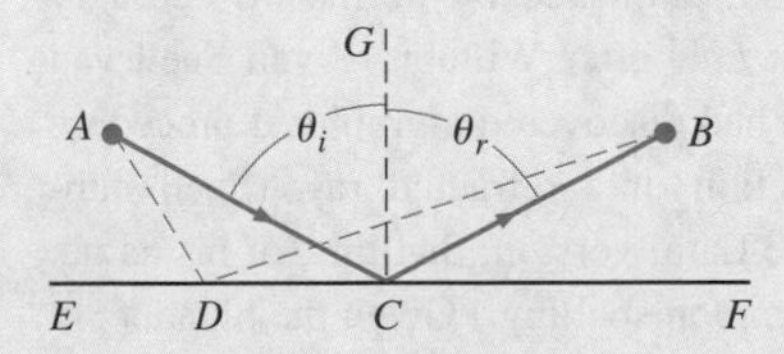

Figure 2

Fermat saw that if he reformulated Cureau's principle as a *least time* postulate, then he would be able to apply calculus to the more complicated phenomenon of refraction. With refraction, we suppose that the points A and B are in different media,

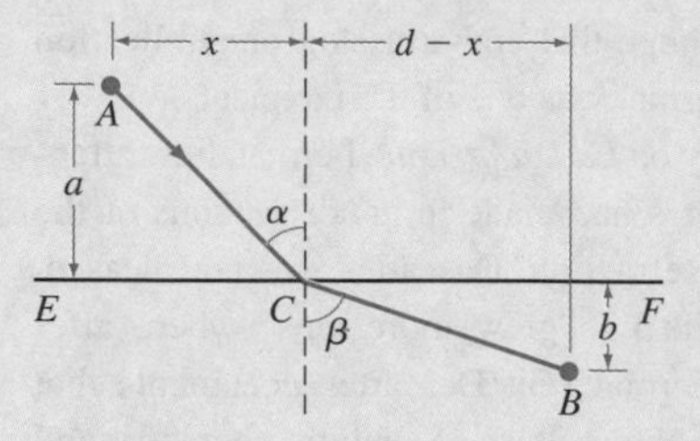

Figure 3

separated by boundary $\overline{EF}$ (as in Figure 3). Light travels with speed v_A in the medium containing point A and with speed v_B in the medium containing B. The Law of Refraction asserts that

$$\frac{\sin(\alpha)}{\sin(\beta)} = \frac{v_A}{v_B},$$

where α is the angle of incidence and β is the angle of refraction (as in Figure 3). Fermat supposed that the point C at which the ray of light impinged on $\overline{EF}$ was such that the light ray traveled from A to C to B in *least time*. This general principle is now known as *Fermat's Principle of Least Time*. From this principle, Fermat derived the Law of Refraction. We see from Figure 3 that the total time required for the light ray to reach B from A by impinging on $\overline{EF}$ at a point with abscissa x beyond A is

$$f(x) = \frac{\sqrt{a^2 + x^2}}{v_A} + \frac{\sqrt{b^2 + (d - x)^2}}{v_B}.$$

We therefore set

$$f'(x) = \frac{x}{v_A\sqrt{a^2 + x^2}} - \frac{d - x}{v_B\sqrt{b^2 + (d - x)^2}} = 0,$$

or

$$\frac{\sin(\alpha)}{v_A} - \frac{\sin(\beta)}{v_B} = 0,$$

which is the Law of Refraction.

The Law of Refraction, published by Fermat in February 1662, is known as *Snell's Law* after Willebrord van Snell van Roijen (1580–1626), who had discovered a graphical procedure for determining the direction of a refracted ray. (Snell published his papers using the Latin version, Snellius, of his name, which accounts for the common spelling.) Given the blunders in Descartes's "derivation" of Snell's Law, Fermat was astounded to find himself reaching the same conclusion. As he wrote to Clerselier,

> *The reward of my work has been the most extraordinary, unexpected, and fortunate that I have ever obtained. Because, after running through all the equations, multiplications, antitheses, and other operations of my method, and having finally brought this problem to a conclusion [as enclosed], I found that my principle gave precisely the same ratio of refractions that M. Descartes had obtained. I was so surprised by this unexpected event that I scarcely recovered from my astonishment.*

Fermat's Principle of Least Time was not accepted by his contemporaries. Certainly, the physical mechanism is not obvious. How does light do it? In Feynman's rhetorical words, "Does [light] *smell* the nearby paths and check them against each other?" Clerselier wrote to Fermat on May 6, 1662: "The principle which you take as the basis of your demonstration . . . is nothing but a moral principle and not a physical one; it is not, and cannot be, the cause of any effect of Nature." Nor was Clerselier the only one to express such an opinion. Huygens, for example, the greatest expert at the time, thought little of Fermat's reasoning. In Fermat's last scientific letter, it is not known to whom, he wrote, "[The Cartesians] prefer to leave the matter undecided . . . . Let posterity be the judge."

Rolle's Theorem and the Mean Value Theorem

Michel Rolle (1652–1719) published his *Traité d'Algèbre* in 1690. In a follow-up one year later, Rolle demonstrated that if $P(x)$ is a polynomial, then $P'(x)$ has a real root between any two real roots of $P(x)$. He did not, however, state his result, which we now call *Rolle's Theorem,* in the language of derivatives. Instead, he used his Method of Cascades, a formal *algebraic* operation on a polynomial $P(x)$. Indeed, Rolle was a vigorous opponent of infinitesimal methods, which he did not believe to have a solid foundation.

The Mean Value Theorem first appeared (in geometric form) in the *Geometria Indivisibilibus Continuorum* of 1635 of Bonaventura Cavalieri (ca. 1598–1647). Its analytic formulation is sometimes attributed to Lagrange and sometimes to the physicist André-Marie Ampère (1775–1836), who used it in a paper written in 1806.

The Newton-Raphson Method

Newton's method of approximating real roots of polynomials appeared in a paper of 1669, but the first published account appeared in Wallis's *Algebra* of 1685. Wallis demonstrated the technique with Newton's example: $y^3 - 2y - 5 = 0$. Newton chose 2 as the first approximation to the root and substituted $2 + p$ for y, obtaining the equation $p^3 + 6p^2 + 10p - 1 = 0$ for the error p. By neglecting the terms involving p^2 and p^3, Newton solved $p = 0.1 + q$ (where q is an error term). He substituted the right side into the cubic polynomial for p, obtaining

$q^3 + 6.3q^2 + 11.23q + 0.061 = 0$. He carried out the next step in the same way, writing $q = -0.061/11.23 + r$, or $q = -0.0054 + r$. Newton continued with one more iteration, obtaining a term $s = -0.00004854$. The resulting approximation to the root is

$$\begin{aligned} 2 + p &= 2 + 0.1 + q = 2 + 0.1 - 0.0054 + r \\ &= 2 + 0.1 - 0.0054 - 0.00004854 = 2.09455146. \end{aligned}$$

(Check Newton's accuracy with the "solve" feature of your calculator or computer software.)

Notice that in Newton's original algorithm, the function under consideration changes at every step. Moreover, the derivative does not explicitly appear. In a booklet that appeared in 1690, Joseph Raphson (ca. 1648–ca. 1715) introduced the important modifications that transformed Newton's method into the form we now use. Raphson's improvement is that the function $\Phi(x) = x - f(x)/f'(x)$ remains the same at every step of the iteration.

5

The Integral

Up to now, our discussion of calculus has been limited to the derivative and its applications. In this chapter, we learn about one of the most powerful and pervasive tools in all of mathematical science: the definite integral of a function f. The definite integral represents a limit of the form

$$\lim_{\Delta x \to 0} \left(f(x_1)\,\Delta x + f(x_2)\,\Delta x + \cdots + f(x_N)\,\Delta x \right)$$

where Δx is an increment of x, as usual.

Like the derivative, the definite integral involves the key concept of calculus: the *limit*. However, addition and multiplication are the algebraic operations of the definite integral. This contrasts with the definition of the derivative, $f'(x) = \lim_{\Delta x \to 0}(f(x + \Delta x) - f(x))/\Delta x$, which involves subtraction and division.

In Section 5.1, we motivate the notion of integral by considering the problem of calculating area. We learn that if f is a positive function defined on the interval $[a, b]$, then we can think of the definite integral as the area that lies under the graph of $y = f(x)$ for $a \leq x \leq b$. However, keep in mind that the integral is useful in analyzing many different mathematical and physical problems. We begin with area simply because it provides us with an accessible and geometrical place to start, but area is only the first of many interpretations of the definite integral that we will learn.

The contrasting algebraic operations involved in the derivative and the integral suggest that the processes of differentiation and integration are in some way *inverse* to one another. The Fundamental Theorem of Calculus links the two main branches of calculus by making the inverse relationship between derivative and integral precise. In doing so, the theorem provides a rule allowing us to calculate many definite integrals without referring to the limit definition. For the definite integrals that lie beyond the power of the Fundamental Theorem of Calculus, it is important to have efficient *approximation* methods. In this chapter, we study several of these numerical techniques as well.

5.1 Introduction to Integration—The Area Problem

We calculate the area of a rectangle by multiplying length by width. We find other areas, such as the area of a trapezoid or of a triangle, by performing geometric constructions to reduce the problem to the calculation of the area of a rectangle (see Figure 1). If we wish to calculate the area of the region under the graph of a function $y = f(x)$, then there is no hope of dividing the region and reassembling it into *one* rectangle (see Figure 2). Instead we attempt to divide the region into *many* rectangles.

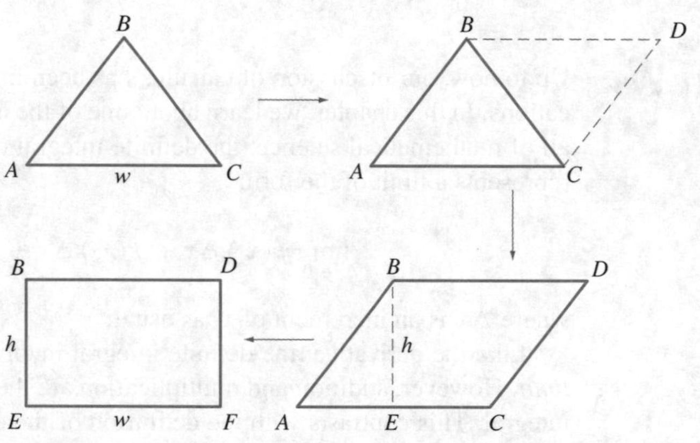

Figure 1
Area $(\Delta ABC) = \frac{1}{2}\text{area}(EBDF) = \frac{1}{2}hw$

Figure 3 suggests the procedure that we will follow: The area of the shaded region is the approximate sum of the areas of the rectangles. Of course, there is some inaccuracy: Notice that the rectangles do not quite fit. Parts of some of the rectangles protrude from the region while other parts fall short. If we take a great many rectangles, as in Figure 4, it appears that the sum total of all these errors is quite small. We might hope that as the number of rectangles becomes large, the total error tends to zero. If so, this procedure would tell us the area of the region in Figure 2.

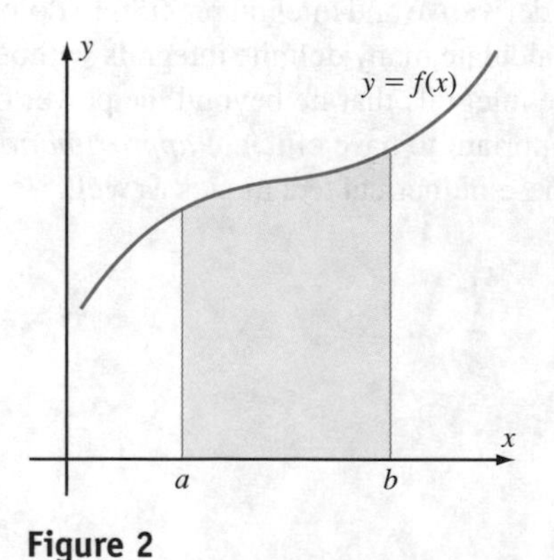

Figure 2

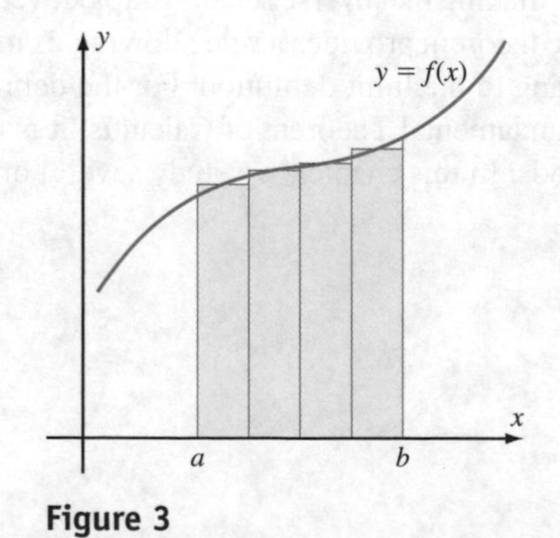

Figure 3

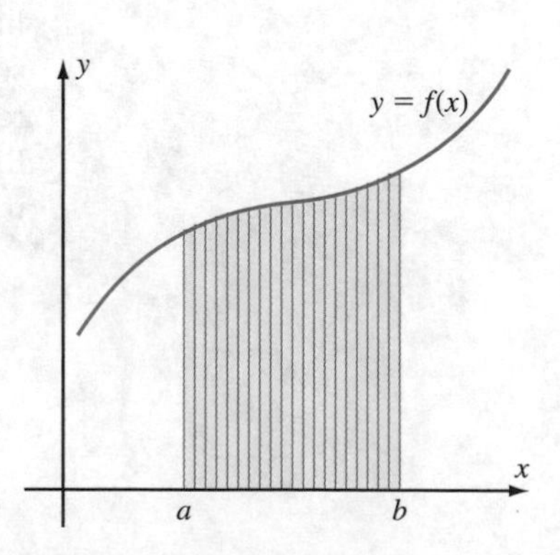

Figure 4

Approximating an area by subdividing into simple regions (rectangles or triangles) is an ancient program. It is the method by which Archimedes obtained the formula for the area inside a circle. This chapter presents the method precisely. Before we begin, however, it will be convenient to learn a notation for the sum of a large or indefinite number of terms.

Summation Notation

We read the notation $\sum_{i=M}^{N} a_i$, where the a_i are real numbers and M and N are integers with $N \geq M$, as "the sum of the numbers a_i for i equal to M, $M+1$, $M+2, \ldots$, ending with i equal to N." Thus,

$$\sum_{i=M}^{N} a_i = a_M + a_{M+1} + a_{M+2} + \cdots + a_N.$$

Example 1 Evaluate the sums $\sum_{i=1}^{5} i^2$ and $\sum_{i=5}^{8} (3i+4)$.

Solution The sum $\sum_{i=1}^{5} i^2$ is a shorthand notation for $1^2 + 2^2 + 3^2 + 4^2 + 5^2 = 55$. The expression $\sum_{i=5}^{8} (3i+4)$ means $(3 \cdot 5+4)+(3 \cdot 6+4)+(3 \cdot 7+4)+(3 \cdot 8+4)$, which sums to 94. ■

The symbol Σ, which is the Greek letter "sigma" (the analogue to the Roman "s"), stands for "summation." There is a certain amount of flexibility inherent in the sigma notation. The sum can begin at any integer M and can end at any integer N with $N \geq M$. The letter i in the sum $\sum_{i=M}^{N} a_i$ is called the *index of summation*. Since the summation index i is simply a placeholder for the integers from M to N, another letter could be used with no change in meaning. Thus,

$$\sum_{i=M}^{N} a_i = \sum_{j=M}^{N} a_j.$$

Also, the distributive law and the associativity of addition can be used to manipulate finite sums. Thus, if $a_M, a_{M+1}, \ldots, a_N$, $b_M, b_{M+1}, \ldots, b_N$, and α are real numbers, then

1. $\sum_{i=M}^{N} (a_i + b_i) = \sum_{i=M}^{N} a_i + \sum_{i=M}^{N} b_i$ and
2. $\sum_{i=M}^{N} \alpha \cdot a_i = \alpha \cdot \sum_{i=M}^{N} a_i$.

Example 2 Evaluate the sum $\sum_{j=1}^{4} (3j^2 - 5j)$.

Solution The expression $\sum_{j=1}^{4}(3j^2 - 5j)$ may be rewritten as

$$3 \cdot \sum_{j=1}^{4} j^2 - 5 \cdot \sum_{j=1}^{4} j = 3(1^2+2^2+3^2+4^2) - 5(1+2+3+4) = 3 \cdot 30 - 5 \cdot 10 = 40.$$

■

Some Special Sums

There are certain sums that we frequently use. For example, we have already encountered the sum $1 + r + r^2 + \cdots + r^{N-1}$, which arises when we add the terms of the geometric sequence $\{r^j\}_{j=0}^{N-1}$ (see Section 2.5). Using sigma notation, we can express formula (2.10) for the sum of a geometric series as

$$\sum_{j=0}^{N-1} r^j = \frac{r^N - 1}{r - 1}.$$

The following are two more sums that will be useful in this section. (Proofs of these formulas are outlined in Exercises 39–41.)

$$\sum_{j=1}^{N} j = \frac{N(N+1)}{2} \tag{5.1}$$

$$\sum_{j=1}^{N} j^2 = \frac{N(N+1)(2N+1)}{6} \tag{5.2}$$

Example 3 Use formulas (5.1) and (5.2) to calculate the sums $\sum_{j=1}^{100} j$ and $\sum_{j=4}^{9} j^2$.

Solution By formula (5.1),

$$\sum_{j=1}^{100} j = \frac{100 \cdot (100+1)}{2} = 5050.$$

In other words, the sum of the first 100 positive integers is 5050.

Formula (5.2) shows how to evaluate the sum of consecutive squares when the first term is 1^2. To apply formula (5.2) to the given sum, we must first rewrite it:

$$\sum_{j=4}^{9} j^2 = \left(\sum_{j=1}^{9} j^2\right) - \left(\sum_{j=1}^{3} j^2\right) = \frac{9 \cdot 10 \cdot 19}{6} - \frac{3 \cdot 4 \cdot 7}{6} = 271.$$

■

Approximation of Area

Let f be a positive, continuous function defined on an interval $[a, b]$. Our goal is to determine the area of the region lying above the x-axis, below the graph of the function $y = f(x)$, and between the vertical lines $x = a$ and $x = b$, as shown in Figure 2, page 348.

Definition Let N be a positive integer. The *uniform partition of order* N of the interval $[a, b]$ is the set $\{x_0, x_1, \ldots, x_N\}$ of equally spaced points:

$$x_j = a + j \cdot \frac{b-a}{N} \quad (0 \le j \le N).$$

Notice that $x_0 = a$, $x_N = b$, and the uniform partition breaks the interval $[a, b]$ into the N subintervals

$$I_1 = [x_0, x_1], \quad I_2 = [x_1, x_2], \quad \ldots, \quad I_N = [x_{N-1}, x_N]$$

of equal length. We let Δx denote the common length $(b-a)/N$ of these subintervals. Refer to Figure 5.

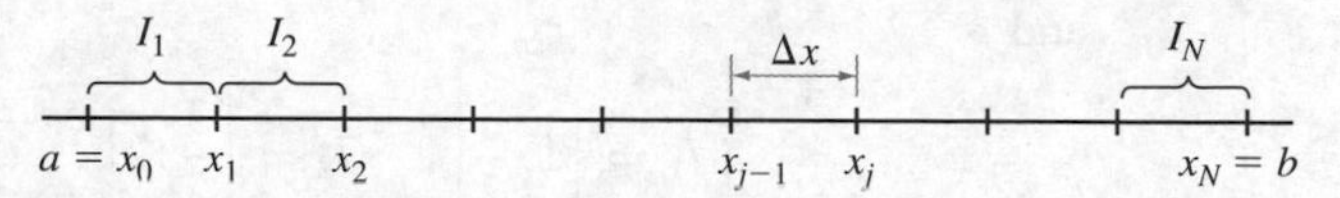

Figure 5
In the uniform partition of order N of the interval $[a, b]$, each subinterval has width $\Delta x = (b - a)/N$.

We will erect a rectangle over each of the subintervals $I_1, I_2, \ldots, I_N$. How tall should each rectangle be? We want each rectangle to be tall enough to touch the graph. One convenient way to do this is to take the height of the jth rectangle (over subinterval $I_j = [x_{j-1}, x_j]$) to be $f(x_j)$, as shown in Figure 6. The upper right corner of the rectangle, therefore, just touches the graph. Thus, the jth rectangle has dimensions base $= \Delta x$ and height $= f(x_j)$. The area of the jth rectangle is therefore $A_j =$ (height) $\cdot$ (base) $= f(x_j) \cdot \Delta x$. The sum of these rectangular areas,

$$\sum_{j=1}^{N} A_j = \sum_{j=1}^{N} f(x_j) \cdot \Delta x, \tag{5.3}$$

should give an approximation to the area bounded below by the x-axis, above by the curve $y = f(x)$, and on the left and right by the vertical lines $x = a$ and $x = b$. Formula (5.3) is known as the *right endpoint approximation*.

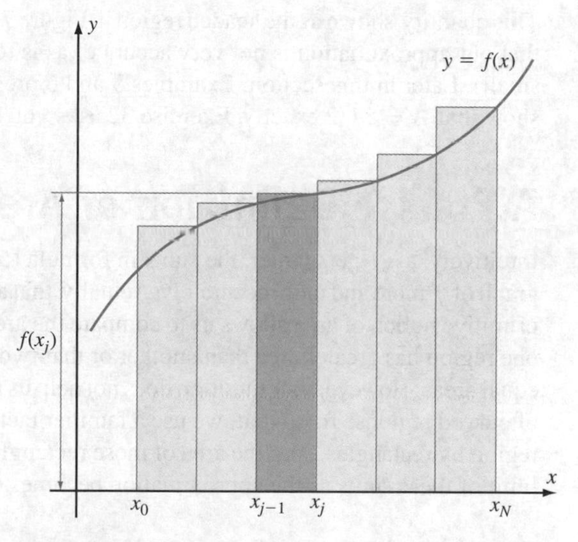

Figure 6
The height of the jth rectangle is $f(x_j)$.

Example 4 Use the uniform partition $\{1, 3/2, 2, 5/2, 3, 7/2, 4\}$ to approximate the area A of the region bounded by the graph of $f(x) = x^2 + x$, the x-axis, and the vertical lines $a = 1$ and $b = 4$.

Solution Observe that

$$x_0 = 1, \quad x_1 = \frac{3}{2}, \quad x_2 = 2, \quad x_3 = \frac{5}{2}, \quad x_4 = 3, \quad x_5 = \frac{7}{2}, \quad x_6 = 4$$

and

$$I_1 = \left[1, \frac{3}{2}\right], \quad I_2 = \left[\frac{3}{2}, 2\right], \quad I_3 = \left[2, \frac{5}{2}\right],$$

$$I_4 = \left[\frac{5}{2}, 3\right], \quad I_5 = \left[3, \frac{7}{2}\right], \quad I_6 = \left[\frac{7}{2}, 4\right].$$

There are six subintervals, each with length $\Delta x = 1/2$. The sum of the corresponding rectangular areas is

$$\begin{aligned} A_1 + A_2 + A_3 + A_4 + A_5 + A_6 &= f(x_1)\cdot\Delta x + f(x_2)\cdot\Delta x + f(x_3)\cdot\Delta x \\ &\quad + f(x_4)\cdot\Delta x + f(x_5)\cdot\Delta x + f(x_6)\cdot\Delta x \\ &= \left(\left(\frac{3}{2}\right)^2 + \frac{3}{2}\right)\cdot\left(\frac{1}{2}\right) + (2^2+2)\cdot\left(\frac{1}{2}\right) \\ &\quad + \left(\left(\frac{5}{2}\right)^2 + \frac{5}{2}\right)\cdot\left(\frac{1}{2}\right) + (3^2+3)\cdot\left(\frac{1}{2}\right) \\ &\quad + \left(\left(\frac{7}{2}\right)^2 + \frac{7}{2}\right)\cdot\left(\frac{1}{2}\right) + (4^2+4)\cdot\left(\frac{1}{2}\right) \\ &= \frac{265}{8} \\ &= 33.125. \end{aligned}$$

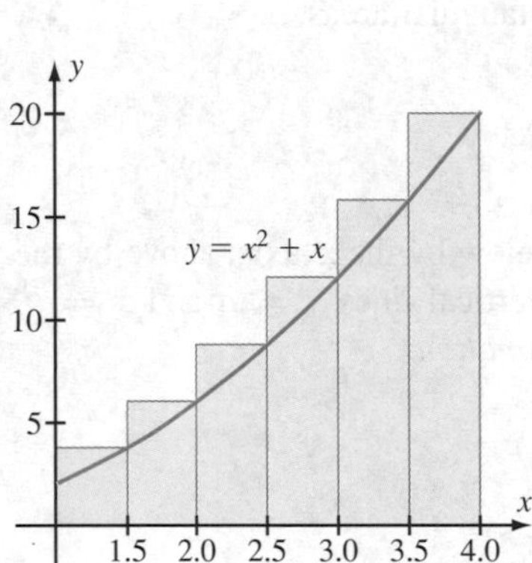

Figure 7

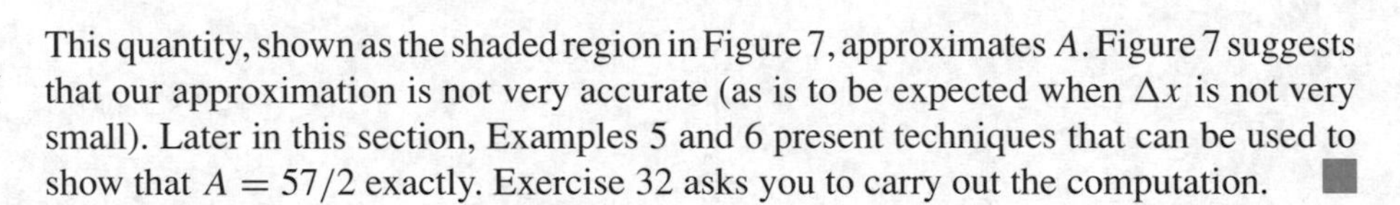

This quantity, shown as the shaded region in Figure 7, approximates A. Figure 7 suggests that our approximation is not very accurate (as is to be expected when Δx is not very small). Later in this section, Examples 5 and 6 present techniques that can be used to show that $A = 57/2$ exactly. Exercise 32 asks you to carry out the computation. ■

A Precise Definition of Area

Intuitively, as N gets larger, the sums in formula (5.3) approximate the area under the graph of f more and more accurately. Actually, that area is not yet a defined concept. Our primitive notion of area allows us to compare the areas of different regions—to say that one region has greater area than another or that two regions have equal areas or nearly equal areas. However, our intuition does not help us assign numerical values to the areas of curved regions. To do that, we use a familiar tactic of calculus: We approximate the region by rectangles, sum the area of those rectangles, as in formula (5.3), and take the limit of these sums as the approximation becomes ever closer.

Definition

The *area* A of the region that is bounded above by the graph of f, below by the x-axis, and laterally by the vertical lines $x = a$ and $x = b$, is defined as the limit

$$A = \lim_{N\to\infty} \sum_{j=1}^{N} f(x_j)\cdot\Delta x. \qquad \textbf{(5.4)}$$

(As defined in Section 2.5, this means that for every $\epsilon > 0$, we can force the area of the rectangular approximation ($\sum_{j=1}^{N} f(x_j) \cdot \Delta x$) to be within ϵ of the quantity A by taking a sufficiently large number N of rectangles.)

inSIGHT

Suppose that we did not define the area to be this exotic-looking limit. What would the area of a region bounded by curves *mean?* The answer is that it would not mean anything—there would be no prior definition for this concept. If we wish to consider area, we must first define the concept. Having done so, we must show that our intuitive ideas about area are consistent with our definition.

Granted that *some* definition is needed, we may still question the particular definition adopted. Does limit (5.4) exist? *Yes,* for continuous functions, it does exist. Would the limit be different if we had determined the height of the rectangle with base $I_j = [x_{j-1}, x_j]$ by using some point in I_j other than the right endpoint x_j? *No,* the limit would be the same. The left endpoint, the midpoint, or any other point of I_j would serve just as well in limit (5.4). We discuss these issues in greater detail in Section 5.2.

We now use this definition of area to calculate some areas.

Example 5 Calculate the area under the graph of $y = 2x$ and above the x-axis between the vertical lines $x = 0$ and $x = 6$.

Solution In this example, we take $a = 0$, $b = 6$, and $f(x) = 2x$. When we divide the interval $[0, 6]$ into N equal subintervals, the common length is $\Delta x = (6-0)/N = 6/N$. The uniform partition of order N is $\{x_0, x_1, \ldots, x_j, \ldots, x_N\}$ where $x_j = x_0 + j \cdot \Delta x = 0 + 6j/N$. The corresponding approximation to area is

$$\sum_{j=1}^{N} f(x_j) \cdot \Delta x = \sum_{j=1}^{N} \left(2 \cdot \frac{6j}{N}\right) \cdot \frac{6}{N} = \frac{72}{N^2} \cdot \sum_{j=1}^{N} j \overset{(5.1)}{=} \frac{72}{N^2} \cdot \frac{N \cdot (N+1)}{2}$$
$$= 36\frac{N+1}{N} = 36\left(1 + \frac{1}{N}\right).$$

Clearly, this last expression has limit 36 as N tends to infinity. Thus, according to our definition, the area of the region under $y = x$, above the x-axis, and between $x = 0$ and $x = 6$ is 36. ■

inSIGHT

The test of any elaborate new concept is twofold: Is it consistent with simpler concepts when those simpler concepts are applicable? Does it succeed in any cases for which the simpler concepts fail? In Example 5, the region for which we calculated area is a right triangle with base 6 and altitude 12. The familiar formula for the area of a triangle,

$$\text{Area} = \left(\frac{1}{2}\right) \cdot (\text{base}) \cdot (\text{height})$$
$$= \left(\frac{1}{2}\right) \cdot 6 \cdot 12 = 36,$$

confirms that our new definition of area agrees with the standard definition. In the next example, we use limit (5.4) to evaluate an area that *requires* the limit concept.

Example 6 Calculate the area of the region (shaded in Figure 8, next page) that is under the graph of $f(x) = x^2$, above the x-axis, and between the vertical lines $x = 2$ and $x = 6$.

Solution We divide the interval $[2, 6]$ into N subintervals of equal length $\Delta x = (6-2)/N = 4/N$ by means of the partition $\{2 = x_0, x_1, \ldots, x_j, \ldots, x_N = 6\}$ where $x_j = x_0 + j \cdot \Delta x = 2 + 4j/N$. The corresponding approximation to area is

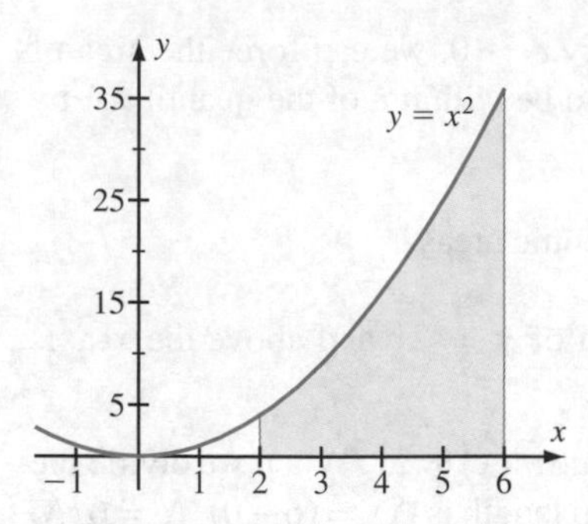

Figure 8

given by

$$\sum_{j=1}^{N} f(x_j)\cdot\Delta x = \sum_{j=1}^{N}\left(2+\frac{4j}{N}\right)^2\cdot\frac{4}{N} \qquad \text{Substitute for } x_j \text{ and } \Delta x$$

$$= \sum_{j=1}^{N}\left(4+16\frac{j}{N}+16\frac{j^2}{N^2}\right)\cdot\frac{4}{N} \qquad \text{Expand the square}$$

$$= \sum_{j=1}^{N}\left(\frac{16}{N}1+\frac{64}{N^2}j+\frac{64}{N^3}j^2\right) \qquad \text{Multiply}$$

$$= \frac{16}{N}\sum_{j=1}^{N}1+\frac{64}{N^2}\sum_{j=1}^{N}j+\frac{64}{N^3}\sum_{j=1}^{N}j^2 \qquad \text{Elementary sum properties}$$

$$= \frac{16}{N}\cdot N+\frac{64}{N^2}\cdot\frac{N(N+1)}{2} + \frac{64}{N^3}\cdot\frac{N(N+1)(2N+1)}{6} \qquad \text{Formulas (5.1) and (5.2)}$$

$$= 16+32\cdot\frac{N+1}{N} + \frac{64}{6}\cdot\frac{N+1}{N}\cdot\frac{2N+1}{N}. \qquad \text{Simplify}$$

We now know that

$$\lim_{N\to\infty}\frac{N+1}{N}=\lim_{N\to\infty}\left(1+\frac{1}{N}\right)=1 \quad\text{and}\quad \lim_{N\to\infty}\frac{2N+1}{N}=\lim_{N\to\infty}\left(2+\frac{1}{N}\right)=2.$$

Thus, the required area of the region is

$$\lim_{N\to\infty}\sum_{j=1}^{N} f(x_j)\cdot\Delta x = 16+32+\frac{64}{6}\cdot 2=\frac{208}{3}.$$

■

Concluding Remarks

The method that we have used to define area can be used to attack many different types of problems in addition to the calculation of area. In the rest of this chapter, we develop a general tool, called the *Riemann integral,* that enables us to study all of these types of problems. We also investigate a method of evaluating Riemann integrals that does not require the elaborate and tedious algebra used to solve Example 6.

quickquiz

1. Write out the sum $\sum_{j=3}^{7}(2j^3-j^4)$.
2. Explicitly calculate $\sum_{j=4}^{150}(3j-j^2)$.
3. The sum $\sum_{j=1}^{4}\exp(-1+3j/4)\cdot(3/4)$ results when the area under a curve $y=f(x)$ and above the interval $[-1,2]$ is approximated by using three rectangles. What is $f(x)$? Sketch the curve and the approximating rectangles.
4. What do we mean when we talk about the area under the graph of a positive function and above an interval of the x-axis?

EXERCISES

Problems for Practice

In Exercises 1–8, write out the sum and perform the addition.

1. $\sum_{j=1}^{6} 3j$
2. $\sum_{j=1}^{7} (-2j^2)$
3. $\sum_{\ell=4}^{6} \ell/(\ell+1)$
4. $\sum_{n=2}^{5} 2n/(n-1)$
5. $\sum_{k=2}^{4} (k^3 - 6k)$
6. $\sum_{m=3}^{6} (2m^2 + 3m)$
7. $\sum_{j=1}^{5} j \cdot \sin(j\pi/2)$
8. $\sum_{j=0}^{5} j^2 \cdot \sin(j\pi/6)$

In Exercises 9–14, use summation notation to express the sum.

9. $2+3+4+5+6$
10. $3+6+9+12+15$
11. $9+13+17+21+25+29$
12. $9+16+25+36$
13. $1/4+1/5+1/6+1/7+1/8$
14. $2/5+3/7+4/9+5/11$

In Exercises 15–22, for the function f, interval I, and positive integer N, use the right endpoint approximation based on a uniform partition of order N to approximate the area of the region that lies under the graph of f and above I.

15. $f(x) = x^2 - 2x$, $I = [3, 6]$, $N = 2$
16. $f(x) = 2x^2 + 1$, $I = [1, 9/2]$, $N = 2$
17. $f(x) = 2 + \sin(2x)$, $I = [\pi/2, 2\pi]$, $N = 3$
18. $f(x) = 4 - 3\cos(x)$, $I = [0, 4\pi]$, $N = 3$
19. $f(x) = x/(x+1)$, $I = [-5, -2]$, $N = 4$
20. $f(x) = -2x/(-x+4)$, $I = [-7, -5]$, $N = 4$
21. $f(x) = x \cdot \sin(x)$, $I = [-\pi, \pi]$, $N = 4$
22. $f(x) = x + \cos(2x)$, $I = [0, 5\pi]$, $N = 10$

Use formulas (5.1) and (5.2) to calculate the sums in Exercises 23–26.

23. $\sum_{j=1}^{16} j^2$
24. $\sum_{j=1}^{20} (j^2 - 4)$
25. $\sum_{j=3}^{40} (4j^2 + j)$
26. $\sum_{j=4}^{25} (6j^2 + 2j + 1)$

Further Theory and Practice

In Exercises 27–30, approximate the area of the region that lies under the graph of f, above the x-axis, and between $x = a$ and $x = b$. First use the right endpoint approximation with two equal subintervals. Then approximate the area by using the two trapezoids that result from connecting $(a, f(a))$ to $(m, f(m))$ and $(m, f(m))$ to $(b, f(b))$ where m is the midpoint of the interval $[a, b]$.

27. $f(x) = 1 + x^3$, $a = -1$, $b = 3$
28. $f(x) = \sqrt{19 - 5x}$, $a = 1$, $b = 3$
29. $f(x) = x + 2\sqrt{(x - x^2)}$, $a = 0$, $b = 1$
30. $f(x) = \sin(x)$, $a = 0$, $b = 2\pi/3$
31. Following the technique from Example 6, calculate the area of the region that lies below the graph of $f(x) = 4 - x^2$ and above the interval $[0, 2]$ of the x-axis.
32. Following the technique from Examples 5 and 6, show that the area of the region that lies below the graph of $f(x) = x^2 + x$ and above the interval $[1, 4]$ of the x-axis is $57/2$.
33. Find p such that $e^p = e \cdot e^2 \cdot e^3 \ldots e^{100}$. Find q so that $\ln(q) = \sum_{n=1}^{100} \ln(n)$.
34. Calculate the sum $S = \sum_{j=1}^{N} (2j-1)$ of the first N odd positive integers.
35. Calculate the sum $S = \sum_{j=1}^{N} (2j)^2$ of the first N even positive square integers. Subtract your value of S from the value of $\sum_{j=1}^{2N} j^2$ to calculate the sum $\sum_{j=1}^{N} (2j-1)^2$ of the first N odd positive square integers.
36. Which is larger, $\sum_{j=1}^{N} j^2$ or $\sum_{j=1}^{N^2} j$? Explain why.
37. Suppose that $\{a_j\}_{j=1}^{\infty}$ is a sequence of real numbers. Suppose that M and N are integers such that $1 < M < N$. The sum

$$S = \sum_{j=M}^{N} (a_j - a_{j-1})$$

is called a *collapsing* or *telescoping sum*. Show that

$$S = a_N - a_{M-1},$$

and explain the reason for the terminology. *Hint:* Look at the cases $N = M + 1$ and $N = M + 2$ to determine the pattern.
38. Evaluate $1/2 + 1/6 + 1/12 + \cdots + 1/(1000 \cdot 1001)$ by writing the jth term as

$$\frac{1}{j \cdot (j+1)} = \frac{1}{j} - \frac{1}{j+1}$$

and using your result from Exercise 37.
39. Write

$$S = 1 + 2 + 3 + \cdots + N$$

forward and backward:

$$\begin{array}{rcccc} S = & 1\ + & 2 + \cdots + & N \\ S = & N\ + & N-1 + \cdots + & 1. \\ \hline 2S = & N+1\ + & N+1 + \cdots + & N+1 \end{array}$$

Add vertically. Derive formula (5.1) by solving for S.

40. Derive formula (5.1) by noticing that

$$\sum_{j=1}^{N}(2j-1) = \sum_{j=1}^{N}(j^2-(j-1)^2).$$

Write the sum on the left in terms of $S = 1+2+3+\cdots+N$. Evaluate the sum on the right using your result from Exercise 37. Solve for S.

41. Derive formula (5.2) by noticing that

$$\sum_{j=1}^{N}(3j^2-3j+1) = \sum_{j=1}^{N}(j^3-(j-1)^3).$$

Write the expression on the left as a combination of three sums, one of which is $3S = 3(1^2+2^2+3^2+\cdots+N^2)$. Use formula (5.1) to evaluate $\sum_{j=1}^{N} 3j$. Evaluate the sum on the right using your result from Exercise 37. Solve for S.

42. Show that

$$\sum_{j=1}^{N}(4j^3-6j^2+4j-1) = \sum_{j=1}^{N}(j^4-(j-1)^4).$$

Evaluate the sum on the right using your result from Exercise 37. Write the expression on the left as a combination of four sums, one of which is $4S = 3(1^3+2^3+3^3+\cdots+N^3)$. Use formula (5.1) to evaluate $\sum_{j=1}^{N} 4j$. Use formula (5.2) to evaluate $\sum_{j=1}^{N} 6j^2$. Solve for S to obtain the identity

$$\sum_{j=1}^{N} j^3 = \left(\frac{N(N+1)}{2}\right)^2.$$

43. Suppose $0 \le a < b$. Use the formula from Exercise 42 to show that the area under the curve $y = x^3$ from $x = 0$ to $x = b$ is $b^4/4$. Deduce that the area under the curve $y = x^3$ from $x = a$ to $x = b$ is $(b^4 - a^4)/4$.

44. Suppose m and k are positive constants. For $f(x) = mx + k$, let $A(b)$ denote the area under the graph of f, above the x-axis, and between $x = 0$ and $x = b$. Calculate $A(b)$ and show that $A'(b) = f(b)$.

45. For $f(x) = x^2$, let $A(b)$ denote the area under the graph of f, above the x-axis, and between $x = 0$ and $x = b$. Following the technique from Example 6, calculate $A(b)$. Show that $A'(b) = f(b)$.

46. Notice that in limit (5.4), $\lim_{N\to\infty} \Delta x = 0$. It may seem that $\lim_{N\to\infty} \sum_{j=1}^{N} f(x_j)\Delta x = 0$ in view of the following calculation:

$$\begin{aligned} \lim_{N\to\infty} \sum_{j=1}^{N} f(x_j)\Delta x &= \lim_{N\to\infty} \sum_{j=1}^{N} f(x_j) \lim_{N\to\infty} \Delta x \\ &= \left(\lim_{N\to\infty} \sum_{j=1}^{N} f(x_j)\right)\cdot 0 \\ &= 0. \end{aligned}$$

Explain what is wrong with this reasoning.

47. For any positive integer k, let

$$S_N(k) = 1^k + 2^k + 3^k + \cdots + N^k.$$

It is known that

$$S_N(k) = \frac{1}{k+1}N^{k+1} + P_k(N)$$

where P_k is a polynomial of degree k. Accept this fact without proof (but notice that for $k = 1$, 2, and 3, this assertion follows from formula (5.1), formula (5.2), and Exercise 42, respectively).

a. Show that $\lim_{N\to\infty} S_N(k)/N^{k+1} = 1/(k+1)$.

b. Suppose $0 \le a < b$. Show that the area under the curve $y = x^k$ from $x = 0$ to $x = b$ is $b^{k+1}/(k+1)$.

c. Deduce that the area under the curve $y = x^k$ from $x = a$ to $x = b$ is $(b^{k+1} - a^{k+1})/(k+1)$.

48. Calculate $\lim_{N\to\infty}(1/N)\sum_{j=1}^{N}\sqrt{1-(j/N)^2}$ by identifying this number as the limit of right endpoint approximations of the area of a familiar region.

49. Let $f : [0, a] \to [0, b]$ be a continuous increasing function ($f(s) < f(t)$ for $0 \le s < t \le b$). Suppose that f is onto. If the area under the graph of f is c, then express the area under the graph of f^{-1} in terms of a, b, and c.

50. Use Exercise 37 and the product formula

$$\sin(A)\cos(B) = \frac{1}{2}(\sin(A+B) + \sin(A-B))$$

with $A = t/2$ and $B = kt$ to show that

$$\sum_{k=1}^{N} \sin\left(\frac{t}{2}\right)\cos(kt) = \frac{1}{2}\left(\sin\left(\left(N+\frac{1}{2}\right)t\right) - \sin\left(\frac{1}{2}t\right)\right).$$

Use this summation formula, together with the product formula, with $A + B = (N + 1/2)t$ and $A - B = -t/2$ to show that

$$\sum_{k=1}^{N} \cos(kt) = \frac{\sin(Nt/2)\cos((N+1)t/2)}{\sin(t/2)}.$$

51. Suppose $0 < b \le \pi/2$. Use the last formula from Exercise 50 with $t = b/N$ to show that the area under the graph of $y = \cos(x)$ over the interval $[0, b]$ is $\sin(b)$.

Calculator/Computer Exercises

A right endpoint approximation can be obtained by a one-line command in computer algebra systems. For example, after having defined the real numbers a and b and the function f, use

```
evalf((b-a)/N*sum(f(a+j*(b-a)/N),j=1..N));
```

in Maple. (Notice that we evaluate $\Delta x \cdot \sum_{j=1}^{N} f(a + j \cdot \Delta x)$, not $\sum_{j=1}^{N}(f(a + j \cdot \Delta x) \cdot \Delta x)$ so that we multiply by Δx one time instead of N times.) In Exercises 52–55, an interval $[a, b]$ and a function f are given. The function is positive on (a, b), and it is 0 at the endpoints. Approximate the area under $y = f(x)$ and over the interval $[a, b]$ by using the right endpoint approximation, starting with $N = 25$ for your first approximation. Increment N by 25 until the first two decimal places of the sum remain the same for three consecutive calculations. Figure 9 shows a Maple implementation for the function $f(x) = 1 - x^x$, $0 \le x \le 1$. (This procedure does not guarantee two decimal places of accuracy. Section 5.7 presents several methods that can be used to achieve a prescribed accuracy.)

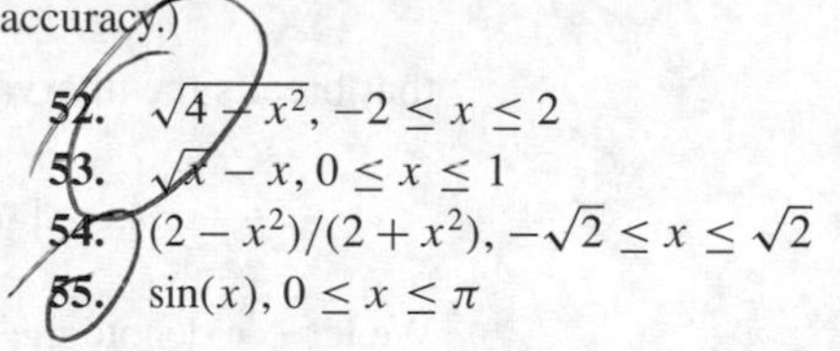

52. $\sqrt{4 - x^2}$, $-2 \le x \le 2$

53. $\sqrt{x} - x$, $0 \le x \le 1$

54. $(2 - x^2)/(2 + x^2)$, $-\sqrt{2} \le x \le \sqrt{2}$

55. $\sin(x)$, $0 \le x \le \pi$

Follow the instructions for Exercises 52–55. However, for Exercises 56–59, you first must locate the left root a and the right root b.

56. $1 + 2x - x^4$ **57.** $2 + x - \exp(x)$

58. $10 - x^2 - 1/x^3$ **59.** $x\exp(-x^2) - 0.1$

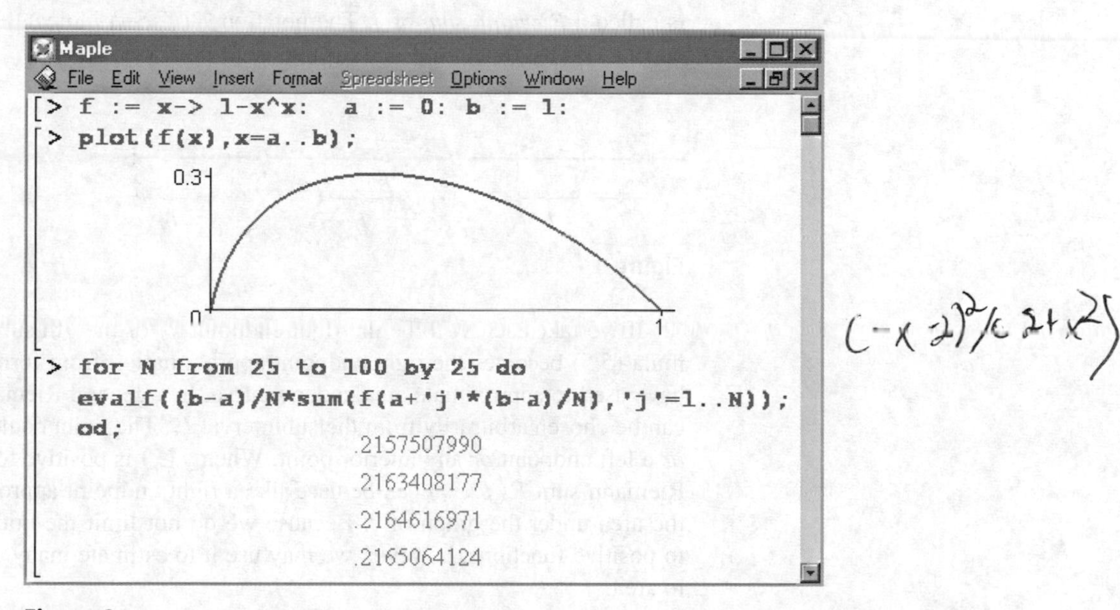

Figure 9

5.2 The Riemann Integral

We now define a mathematical operation on functions that captures the essence of the method from Section 5.1, but that can be applied to several other situations as well.

Riemann Sums

Let f be *any* function defined on the interval $[a, b]$. (We no longer assume that f is a positive function.) Let N be a positive integer. The uniform partition of order N of the interval $[a, b]$ is the set of equally spaced points

$$x_j = a + j \cdot \frac{b-a}{N} \qquad (0 \le j \le N)$$

that breaks the interval $[a, b]$ into the N equal-length subintervals

$$I_1 = [x_0, x_1], \quad I_2 = [x_1, x_2], \quad \ldots, \quad I_N = [x_{N-1}, x_N].$$

We let Δx denote the common length $(b-a)/N$ of these intervals. A *choice of points associated with the uniform partition of order* N is a sequence $\mathcal{S}_N = \{s_1, s_2, \ldots, s_N\}$ of points with s_j in I_j for each $j = 1, \ldots, N$. Figure 1 illustrates one possible choice of points. The expression

$$\mathcal{R}(f, \mathcal{S}_N) = \sum_{j=1}^{N} f(s_j) \cdot \Delta x \tag{5.5}$$

is called a *Riemann sum* of f. The notation $\mathcal{R}(f, \mathcal{S}_N)$ indicates that a Riemann sum depends on the function f and the choice of points $\mathcal{S}_N$.

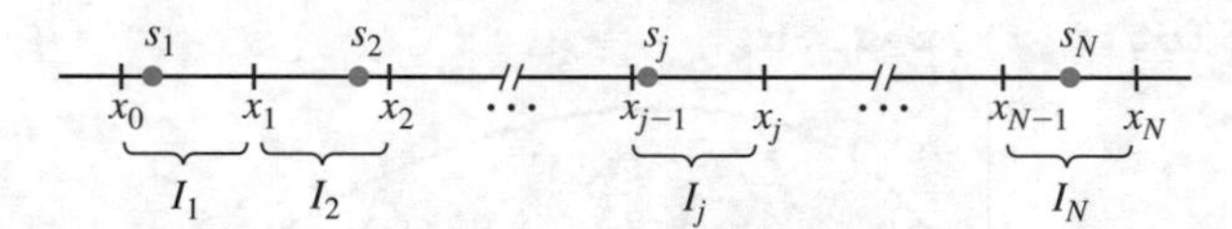

Figure 1

If we take each s_j to be the right endpoint x_j of the jth subinterval I_j, then formula (5.5) becomes the right endpoint approximation from formula (5.3), which we have used to approximate area under a curve. In a general Riemann sum, the point s_j can be chosen arbitrarily from the subinterval I_j. The point could be a right endpoint *or* a left endpoint *or* any interior point. When $f(x)$ is positive for every x in $[a, b]$, a Riemann sum $\mathcal{R}(f, \mathcal{S}_N)$ can be used like a right endpoint approximation to estimate the area under the graph of f. Because we do not limit the notion of Riemann sum to positive functions, however, we may use it to estimate many quantities in addition to area.

Example 1 Write *a* Riemann sum for the function $f(x) = x^2 - 4$ and the interval $[a, b] = [-5, 3]$ using the partition $\{-5, -3, -1, 1, 3\}$.

Solution Observe that

$$I_1 = [-5, -3], \quad I_2 = [-3, -1], \quad I_3 = [-1, 1], \quad I_4 = [1, 3],$$

and $\Delta x = 2$. To form a Riemann sum, we must first make a selection of points s_j $(1 \le j \le 4)$ with s_j in I_j. According to the definition, we may select *any* point s_j in each subinterval. Here is our *arbitrary* choice:

$$s_1 = -3 \in I_1, \quad s_2 = -\frac{3}{2} \in I_2, \quad s_3 = 0 \in I_3, \quad s_4 = 1 \in I_4.$$

The Riemann sum for this collection of points $\mathcal{S}_4$ is

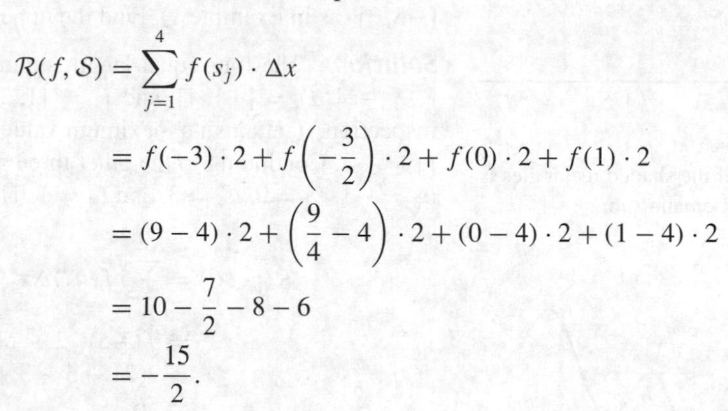

$$\begin{aligned}\mathcal{R}(f, \mathcal{S}) &= \sum_{j=1}^{4} f(s_j) \cdot \Delta x \\ &= f(-3) \cdot 2 + f\left(-\frac{3}{2}\right) \cdot 2 + f(0) \cdot 2 + f(1) \cdot 2 \\ &= (9-4) \cdot 2 + \left(\frac{9}{4} - 4\right) \cdot 2 + (0-4) \cdot 2 + (1-4) \cdot 2 \\ &= 10 - \frac{7}{2} - 8 - 6 \\ &= -\frac{15}{2}.\end{aligned}$$

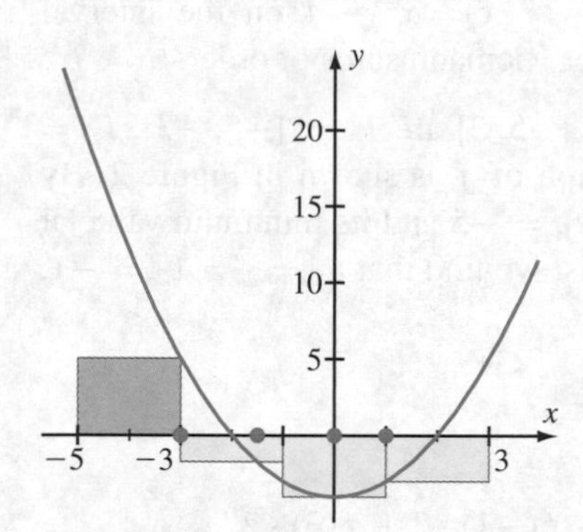

Figure 2

Of course, a different choice of points s_1, s_2, s_3, s_4 would lead to a different Riemann sum. ■

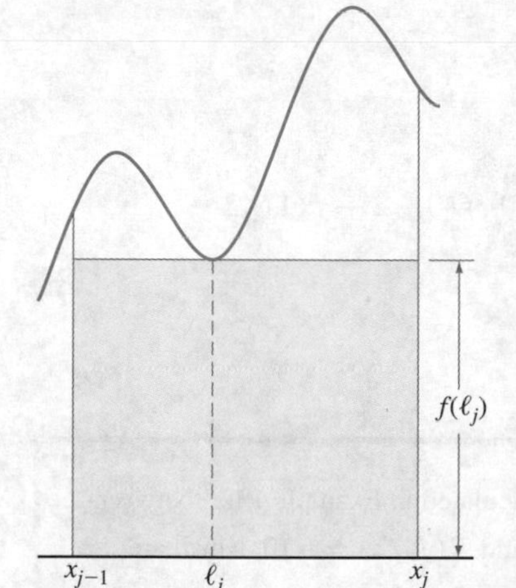

Figure 3a

inSIGHT

> Example 1 shows that a Riemann sum can be negative. If we must give this Riemann sum a geometric meaning, then we can think of it as a *difference of areas,* namely,
>
> $$\mathcal{R}(f, \mathcal{S}_4) = \underbrace{10}_{\text{Area of rectangle above } x\text{-axis}} - \underbrace{(7/2 + 8 + 6)}_{\text{Area of rectangles below } x\text{-axis}}.$$
>
> Refer to Figure 2. Later, however, we will use Riemann sums to represent physical quantities other than area. In those cases, *it is usually not appropriate* to associate Riemann sums with the area concept.

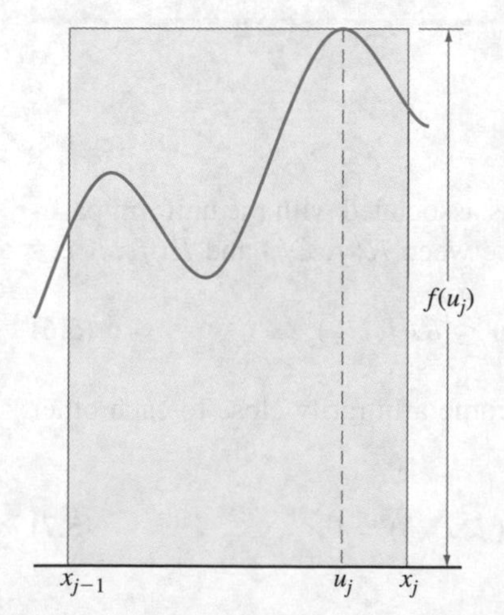

Figure 3b

Suppose f is continuous. Let $\ell_j \in I_j$ be the point at which f attains its minimum value on I_j (see Figure 3a), and let u_j be the point in I_j at which f achieves its maximum value on I_j (see Figure 3b). The Extreme Value Theorem (Section 2.3) guarantees that the points ℓ_j and u_j exist. We denote the resulting choices of points associated to the uniform partition by $\mathcal{L}_N = \{\ell_1, \ell_2, \ldots, \ell_N\}$ and $\mathcal{U}_N = \{u_1, u_2, \ldots, u_N\}$. The Riemann sums

$$\mathcal{R}(f, \mathcal{L}_N) = \sum_{j=1}^{N} f(\ell_j)\Delta x$$

and

$$\mathcal{R}(f, \mathcal{U}_N) = \sum_{j=1}^{N} f(u_j)\Delta x$$

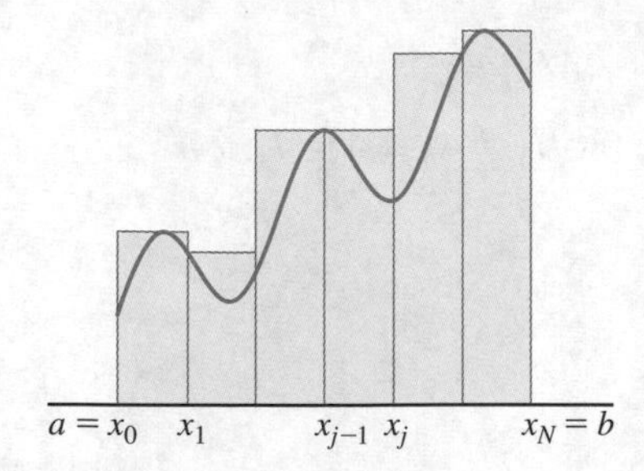

Figure 4
The area of the shaded rectangles is an upper Riemann sum.

are respectively called the *lower Riemann sum* and *upper Riemann sum* of order N. They are the least and greatest Riemann sums of order N, as we will prove in Theorem 1. Figures 4 and 5 illustrate typical lower and upper Riemann sums.

Example 2 Consider the continuous function $f(x) = x^2 - 4$ on the interval $[-5, 3]$ (as in Example 1). Find the upper and lower Riemann sums of order 4.

Solution The four equal-length subintervals of $[-5, 3]$ are $I_1 = [-5, -3]$, $I_2 = [-3, -1]$, $I_3 = [-1, 1]$, and $I_4 = [1, 3]$. The graph of f is shown in Figure 2. By inspection, f attains its maximum value on I_1 at $u_1 = -5$ and its minimum value at $\ell_1 = -3$. Continuing to the other three subintervals, we find that $u_2 = -3$, $\ell_2 = -1$, $u_3 = -1$, $\ell_3 = 0$, $u_4 = 3$, and $\ell_4 = 1$. Therefore,

$$\begin{aligned}\mathcal{R}(f, \mathcal{U}_4) &= \sum_{j=1}^{4} f(u_j)\Delta x \\ &= f(-5)\cdot 2 + f(-3)\cdot 2 + f(-1)\cdot 2 + f(3)\cdot 2 \\ &= 2(21 + 5 - 3 + 5) \\ &= 56\end{aligned}$$

and

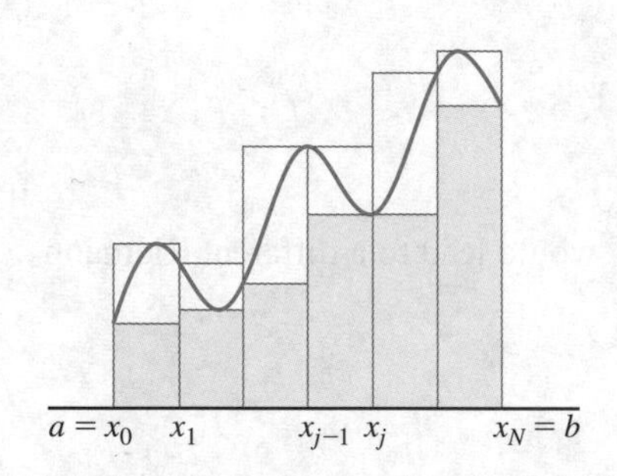

Figure 5
The area of the shaded rectangles is a lower Riemann sum.

$$\begin{aligned}\mathcal{R}(f, \mathcal{L}_4) &= \sum_{j=1}^{4} f(\ell_j)\Delta x \\ &= f(-3)\cdot 2 + f(-1)\cdot 2 + f(0)\cdot 2 + f(1)\cdot 2 \\ &= 2(5 - 3 - 4 - 3) \\ &= -10.\end{aligned}$$

■

inSIGHT

Notice that the Riemann sum $\mathcal{R}(f, \mathcal{S}_4) = -15/2$ calculated in Example 1 lies between the upper and lower Riemann sums $\mathcal{R}(f, \mathcal{U}_4) = 56$ and $\mathcal{R}(f, \mathcal{L}_4) = -10$, which are calculated in Example 2. The following theorem generalizes this observation and states another key fact about upper and lower Riemann sums.

Theorem 1 Suppose that f is continuous on an interval $[a, b]$.

a. If $\mathcal{S}_N = \{s_1, \ldots, s_N\}$ is any choice of points associated with the uniform partition of order N, then $\mathcal{R}(f, \mathcal{S}_N)$ is pinched between $\mathcal{R}(f, \mathcal{L}_N)$ and $\mathcal{R}(f, \mathcal{U}_N)$:

$$\mathcal{R}(f, \mathcal{L}_N) \leq \mathcal{R}(f, \mathcal{S}_N) \leq \mathcal{R}(f, \mathcal{U}_N). \tag{5.6}$$

b. The numbers $\mathcal{R}(f, \mathcal{L}_N)$ and $\mathcal{R}(f, \mathcal{U}_N)$ become arbitrarily close to each other for N sufficiently large. That is,

$$\lim_{N\to\infty} (\mathcal{R}(f, \mathcal{U}_N) - \mathcal{R}(f, \mathcal{L}_N)) = 0. \tag{5.7}$$

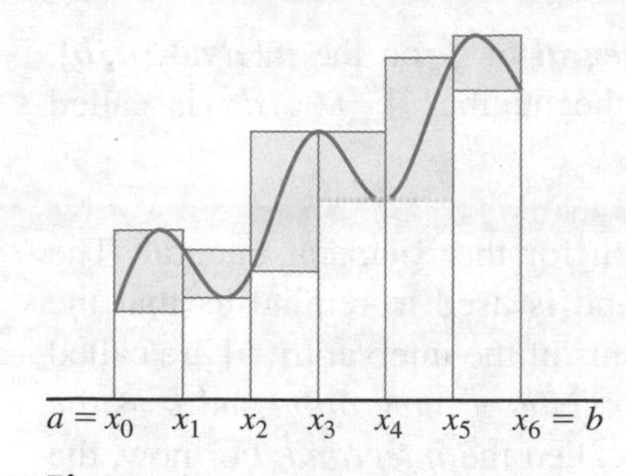

Figure 6
The total area enclosed by the shaded rectangles is $\mathcal{R}(f, \mathcal{U}_6) - \mathcal{R}(f, \mathcal{L}_6)$.

Proof It follows from the definitions of ℓ_j and u_j that

$$f(\ell_j) \le f(s_j) \le f(u_j)$$

for every s_j in I_j. If we multiply each term by the positive quantity Δx and sum over $j = 1, 2, \ldots, N$, then we obtain

$$\sum_{j=1}^{N} f(\ell_j)\Delta x \le \sum_{j=1}^{N} f(s_j)\Delta x \le \sum_{j=1}^{N} f(u_j)\Delta x,$$

which is the assertion of Theorem 1a.

Theorem 1b appears to be plausible when we look at examples. Refer again to Figures 4 and 5, which illustrate typical upper and lower Riemann sums. The area of the shaded rectangles in Figure 6 represents the difference $\mathcal{R}(f, \mathcal{U}_N) - \mathcal{R}(f, \mathcal{L}_N)$ for $N = 6$. In Figure 7, we increase the number of subintervals to 30. Observe that the difference between the upper and lower Riemann sums, again represented by the area of the shaded rectangles, is *significantly smaller*. We can prove Theorem 1b by appealing to an important property of continuous functions on closed, bounded intervals. According to that property (known as *uniform continuity* and proved in texts on advanced calculus), given a positive ϵ, there is a positive δ such that $f(s)$ is within ϵ of $f(t)$, provided that s and t are points in $[a, b]$ within δ of each other. If we take N large enough so that $\Delta x = (b - a)/N < \delta$, then u_j and ℓ_j are within δ of each other for each j. Therefore,

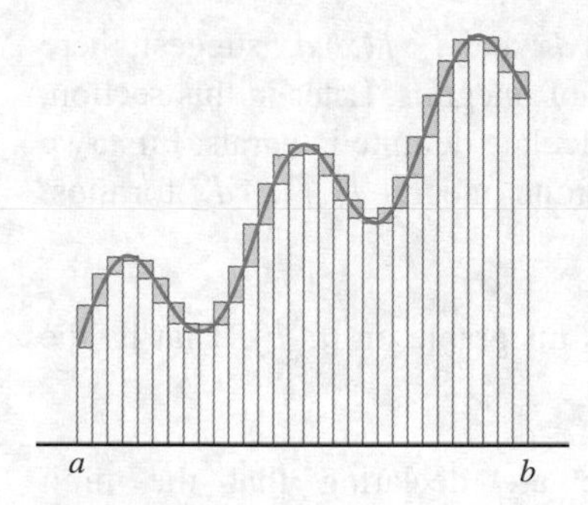

Figure 7
The area of the shaded region decreases as the number N of subdivisions increases.

$$0 \le \mathcal{R}(f, \mathcal{U}_N) - \mathcal{R}(f, \mathcal{L}_N) = \sum_{j=1}^{N} (f(u_j) - f(\ell_j)) \cdot \Delta x < \sum_{j=1}^{N} \epsilon \cdot \Delta x$$

$$= \epsilon \sum_{j=1}^{N} \Delta x = \epsilon \cdot (b - a).$$

Since ϵ is arbitrary, we have shown that $\mathcal{R}(f, \mathcal{U}_N) - \mathcal{R}(f, \mathcal{L}_N)$ can be made arbitrarily small by taking N sufficiently large, which is Theorem 1b. ∎

The Riemann Integral

In Section 5.1, we defined the area under the graph of a positive function to be the *limit* as N tends to infinity of right endpoint approximations. This definition will be the basis of our limit definition for general Riemann sums.

Definition Suppose f is a function defined on the interval $[a, b]$. We say that the Riemann sums $\mathcal{R}(f, \mathcal{S}_N)$ *tend to* the real number ℓ (or that ℓ is the *limit* of the Riemann sums $\mathcal{R}(f, \mathcal{S}_N)$) as N tends to infinity if, for any $\epsilon > 0$, there is a positive integer M such that

$$|\mathcal{R}(f, \mathcal{S}_N) - \ell| < \epsilon$$

for all N greater than or equal to M. If this is the case, we say that f is *integrable* on $[a, b]$, and we denote the limit ℓ by the symbol

$$\int_a^b f(x)\,dx.$$

This numerical quantity is called the *Riemann integral* of f on the interval $[a, b]$. The operation of going from the function f to the number $\int_a^b f(x)\,dx$ is called *integration*.

Let us examine the components of the notation for the Riemann integral. The elongated "s" symbol is called the *integral sign* and is used to remind us that the integral is a limit of *sums*. The left and right endpoints of the interval $[a, b]$ are called the *limits of integration*. We refer to a as the *lower limit of integration* and b as the *upper limit of integration*. The expression $f(x)$ is called the *integrand*. For now, the expression dx serves only to remind us of the variable of integration. We may think of dx as the infinitesimal limit of Δx as Δx tends to zero. In later work, dx will be a useful device for helping us transform integrals into new integrals.

Only the presence of limits of integration serve to distinguish the Riemann integral $\int_a^b f(x)\,dx$ from the *indefinite integral* $\int f(x)\,dx$ from Section 4.8. To emphasize the distinction, we sometimes refer to the Riemann integral $\int_a^b f(x)\,dx$ as a *definite integral*. As the nearly identical notations for $\int f(x)\,dx$ and $\int_a^b f(x)\,dx$ suggest, there is an important relationship between the two types of integrals. Later in this section, we learn that indefinite integrals may be used to calculate definite integrals. First, we state a theorem that assures the existence of the definite integral $\int_a^b f(x)\,dx$ for most of the functions that we encounter in calculus.

Theorem 2 If f is continuous on the interval $[a, b]$, then f is integrable on $[a, b]$. That is, the Riemann integral $\int_a^b f(x)\,dx$ exists.

When we say that $\int_a^b f(x)\,dx$ exists, we are declaring that the limit $\lim_{N\to\infty} \mathcal{R}(f, \mathcal{S}_N)$ exists. The proof of this assertion involves technical details that are best left to a text on advanced calculus. The following simple observations, however, give some idea of what is involved. Suppose f is increasing and positive on $[a, b]$. The areas of the shaded regions in Figures 8a, 8b, and 8c represent $\mathcal{R}(f, \mathcal{L}_N)$ for $N = 1, 2, 4$. The amount by which $\mathcal{R}(f, \mathcal{L}_2)$ exceeds $\mathcal{R}(f, \mathcal{L}_1)$ is represented by the

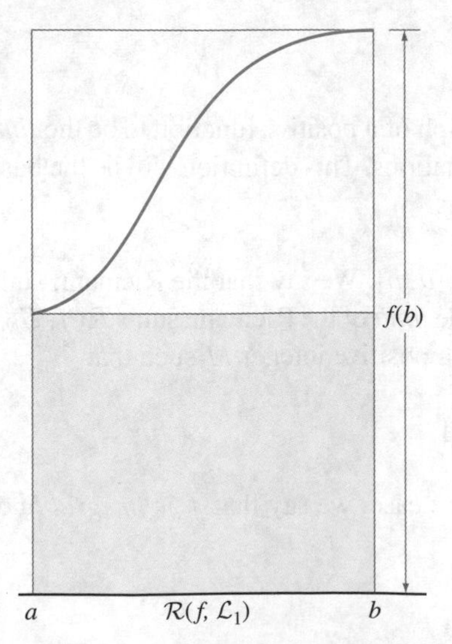

Figure 8a

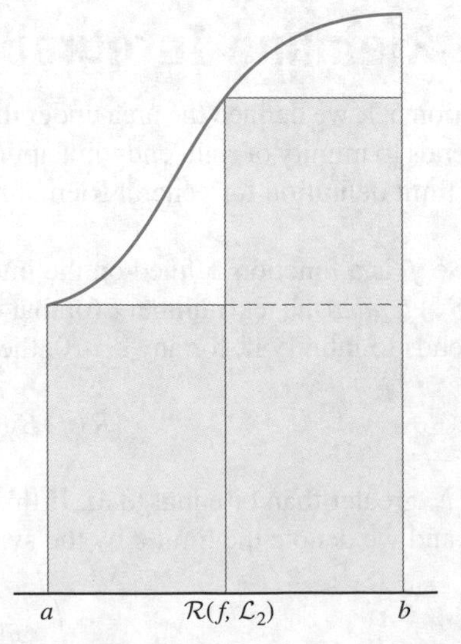

Figure 8b

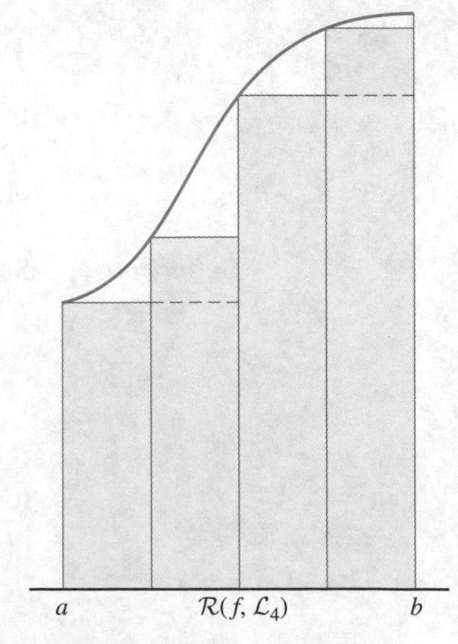

Figure 8c

area of the rectangle that lies above the dashed line in Figure 8b. Similarly, the amount by which $\mathcal{R}(f, \mathcal{L}_4)$ exceeds $\mathcal{R}(f, \mathcal{L}_2)$ is represented by the area of the rectangles that lie above the dashed lines in Figure 8c. If we continue to bisect each subinterval, we obtain an *increasing* sequence of lower Riemann sums that is bounded above by $f(b) \cdot (b-a)$ (the area of the circumscribed rectangle shown in Figure 8a). We also obtain a *decreasing* sequence of upper Riemann sums that is bounded below by $f(a) \cdot (b-a)$, as shown in Figure 9. According to the monotone convergence property of the real numbers (Section 2.6), both sequences of upper and lower Riemann sums converge. We infer from limit (5.7) that the sequences must converge to the same number. Finally, inequality (5.6) tells us that a different choice of points $\mathcal{S}_N$ at each step would lead to a sequence with the same limit, because it would be pinched between the other two.

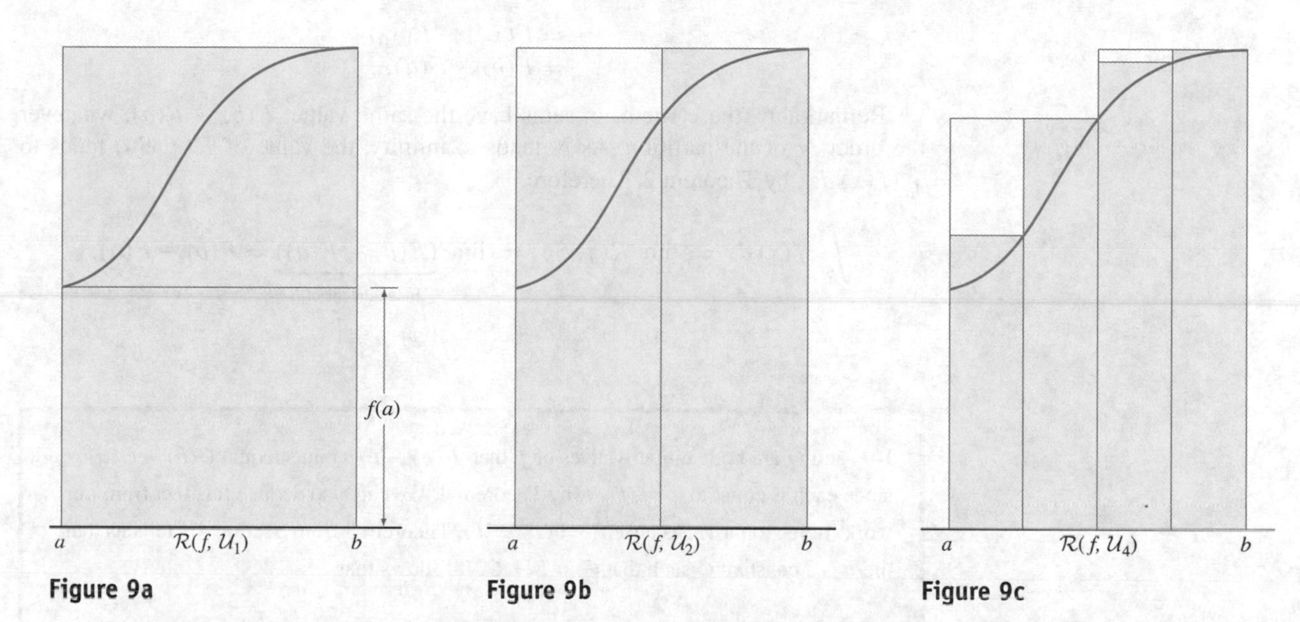

Figure 9a **Figure 9b** **Figure 9c**

Calculating Riemann Integrals

Calculating definite integrals by the method of Riemann sums is tedious and usually difficult. Fortunately, our next theorem provides a way to evaluate many definite integrals without considering Riemann sums. Notice in the proof of the theorem how crucial it is to have flexibility in choosing the points $\{s_j\}$.

Theorem 3 Let f be continuous in $[a, b]$ and suppose that F is an antiderivative of f on (a, b). That is, suppose that F is continuous on $[a, b]$ and $F'(x) = f(x)$ for $a < x < b$. Then,

$$\int_a^b f(x)\,dx = F(b) - F(a). \tag{5.8}$$

Proof Let $\{a = x_0, x_1, \ldots, x_N = b\}$ be the uniform partition of order N. The Mean Value Theorem asserts that there is a point s_j in I_j such that

$$\frac{F(x_j) - F(x_{j-1})}{x_j - x_{j-1}} = F'(s_j).$$

Therefore,

$$F(x_j) - F(x_{j-1}) = F'(s_j) \cdot (x_j - x_{j-1}) = f(s_j)\Delta x.$$

For the Riemann sum corresponding to the choice of points $\mathcal{S}_N = \{s_1, s_2, \ldots, s_N\}$, we have

$$\begin{aligned}\mathcal{R}(f, \mathcal{S}) &= \sum_{j=1}^{N} f(s_j)\Delta x \\ &= \sum_{j=1}^{N} (F(x_j) - F(x_{j-1})) \\ &= F(x_N) - F(x_0) \\ &= F(b) - F(a).\end{aligned}$$

Remarkably, these Riemann sums have the same value, $F(b) - F(a)$, whatever the order N of the partition. As N tends to infinity, the value of $\mathcal{R}(f, \mathcal{S}_N)$ tends to $\int_a^b f(x)\,dx$, by Theorem 2. Therefore,

$$\int_a^b f(x)\,dx = \lim_{N\to\infty} \mathcal{R}(f, \mathcal{S}_N) = \lim_{N\to\infty} \underbrace{(F(b) - F(a))}_{\text{Independent of } N} = F(b) - F(a).$$

■

inSIGHT

If F and G are both antiderivatives of f, then $F(b) - F(a)$ must equal $G(b) - G(a)$ since each is equal to $\int_a^b f(t)\,dt$, by Theorem 3. We can also deduce this fact from our work in Section 4.1: Because $F' = G'$ $(= f)$, Theorem 5 from Section 4.1 tells us that there is a constant C such that $G = F + C$. It follows that

$$G(b) - G(a) = (F(b) + C) - (F(a) + C) = F(b) - F(a).$$

Theorem 3 provides us with a very powerful tool for calculating definite integrals. The right side of equation (5.8) occurs so frequently in calculus that a special notation has been introduced: $F|_a^b$ and $F(x)|_{x=a}^{x=b}$ are both used to symbolize the difference $F(b) - F(a)$. That is,

$$F\Big|_a^b = F(b) - F(a) \quad \text{and} \quad F(x)\Big|_{x=a}^{x=b} = F(b) - F(a).$$

Because Theorem 3 relates the two main branches of calculus (differentiation and integration), it is called the *Fundamental Theorem of Calculus*. Actually, Theorem 3 is only one part of the Fundamental Theorem. Another important part of this theorem is covered in Section 5.4. For now, however, we concentrate on using Theorem 3 to evaluate integrals.

Example 3 Evaluate $\int_1^{27} 1/\sqrt[3]{x}\,dx$.

Solution If $f(x) = x^{-1/3}$, then $F(x) = 3x^{2/3}/2$ is an antiderivative of f and

$$\int_1^{27} \frac{1}{\sqrt[3]{x}}\,dx = F(27) - F(1) = \frac{3}{2}(27^{2/3} - 1^{2/3})$$

$$= \frac{3}{2}(9-1) = 12.$$

■

Example 4 Evaluate $\int_0^{\pi} \sin(x)\,dx$.

Solution Since $F(x) = -\cos(x)$ is an antiderivative of $f(x) = \sin(x)$, it follows that

$$\int_0^{\pi} \sin(x)\,dx = F(\pi) - F(0) = -\cos(\pi) - (-\cos(0)) = -(-1) - (-1) = 2.$$

■

Using the Fundamental Theorem of Calculus to Compute Areas

Let f be a positive continuous function defined on an interval $[a, b]$. In Section 5.1, we defined the area A of the region under the graph of f to be a limit of Riemann sums (as we now call them). Theorem 2 tells us that this limit exists. If we can find an antiderivative F of f, then we can use Theorem 3 to calculate $A = \int_a^b f(x)\,dx = F(b) - F(a)$.

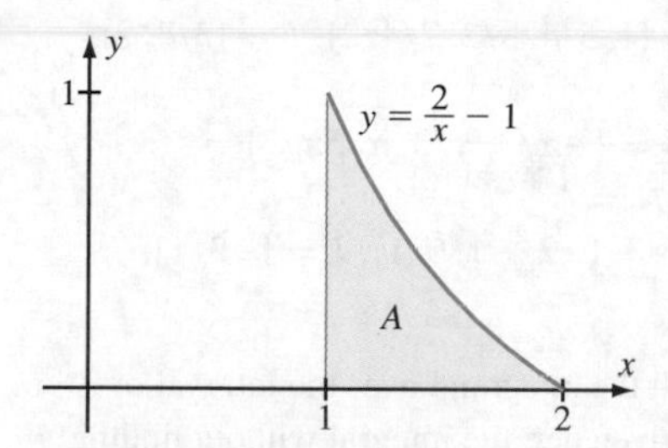

Figure 10

Example 5 Calculate the area A of the region that lies under the graph of $f(x) = x^2$, above the x-axis, and between the vertical lines $x = 2$ and $x = 6$.

Solution In Example 6 from Section 5.1, we found that $A = 208/3$ by means of a somewhat tedious computation. We can now use Theorem 3 to calculate A much more easily. Since $F(x) = x^3/3$ is an antiderivative of $f(x) = x^2$, as we know from formula (4.13), we have

$$A = \int_2^6 x^2\,dx = F(6) - F(2) = \frac{6^3}{3} - \frac{2^3}{3} = \frac{208}{3}.$$

■

inSIGHT

> The minus sign in the expression $F(x)|_{x=a}^{x=b} = F(b) - F(a)$ is the source of many algebraic errors. It is a good idea to enclose $F(a)$ in parentheses when it is negative or when it involves a sum of terms. Doing so will help you avoid distributing the minus sign incorrectly.

Example 6 Calculate the area A of the shaded region in Figure 10.

Solution Since $2\ln(x) - x$ is an antiderivative of $2/x - 1$, we have

$$A = \int_1^2 \left(\frac{2}{x} - 1\right)dx = (2\ln(x) - x)\Big|_{x=1}^{x=2} = (2\ln(2) - 2) - \underbrace{(2\ln(1) - 1)}_{\text{Note parentheses}}.$$

Therefore,

$$A = (2\ln(2) - 2) - (0 - 1) = 2\ln(2) - 2 + 1 = 2\ln(2) - 1.$$

■

quickquiz

1. If $\{s_1, s_2, s_3, s_4\}$ is a choice of points associated with the order 4 uniform partition of the interval $[1, 3]$, what are the possible values of s_3?
2. What is an upper Riemann sum? A lower Riemann sum?
3. Evaluate the definite integral $\int_{-1}^{4} 1\,dx$ by identifying it as an area. Explain the error in the calculation

$$\int_{-1}^{4} 1\,dx = x\Big|_{x=-1}^{x=4} = 4 - 1 = 3.$$

4. Show that $\int_1^4 (x^2 + x)\,dx = 57/2$ (as was asserted in Example 4 from Section 5.1).

EXERCISES

Problems for Practice

In Exercises 1–4, for the given function f, interval I, and uniform partition of order N:

a. Evaluate the Riemann sum $\mathcal{R}(f, \mathcal{S})$ using the choice of points $\mathcal{S}$ that comprises the midpoints of the subintervals of I.

b. Evaluate the definite integral that $\mathcal{R}(f, \mathcal{S})$ approximates.

1. $f(x) = x$, $\mathcal{S} = \{2, 6, 10\}$, $I = [0, 12]$, $N = 4$
2. $f(x) = 2 - x$, $\mathcal{S} = \{-4, -2, 0\}$, $I = [-5, 1]$, $N = 3$
3. $f(x) = x^2$, $\mathcal{S} = \{-1, 1, 3, 5\}$, $I = [-2, 6]$, $N = 4$
4. $f(x) = x^2 + x$, $\mathcal{S} = \{-3, 1, 5, 9\}$, $I = [-5, 11]$, $N = 4$

In Exercises 5–8, for the given function f, interval I, and uniform partition of order N:

a. Evaluate the Riemann sum $\mathcal{R}(f, \mathcal{S})$ using the choice of points $\mathcal{S}$ that comprises the left endpoint of the subintervals of I.

b. Evaluate the definite integral that $\mathcal{R}(f, \mathcal{S})$ approximates.

5. $f(x) = x^3 + 4x$, $\mathcal{S} = \{1, 3, 5\}$, $I = [1, 7]$, $N = 3$
6. $f(x) = \sin(x)$, $\mathcal{S} = \{0, \pi/3, 2\pi/3, \pi\}$, $I = [0, 4\pi/3]$, $N = 4$
7. $f(x) = |x|$, $\mathcal{S} = \{-7, -4, -1, 2\}$, $I = [-7, 5]$, $N = 4$
8. $f(x) = 1/x$, $\mathcal{S} = \{1, 5/4, 3/2, 7/4, 2\}$, $I = [1, 9/4]$, $N = 5$

In Exercises 9–12, for the given function f, interval I, and uniform partition of order N:

a. Evaluate the Riemann sum $\mathcal{R}(f, \mathcal{S})$ using the choice of points $\mathcal{S}$ that comprises the right endpoint of the subintervals of I.

b. Evaluate the definite integral that $\mathcal{R}(f, \mathcal{S})$ approximates.

9. $f(x) = \exp(x)$, $\mathcal{S} = \{-2, 0, 2\}$, $I = [-4, 2]$, $N = 3$
10. $f(x) = 1/x$, $\mathcal{S} = \{1, 5/4, 3/2, 7/4, 2\}$, $I = [3/4, 2]$, $N = 5$
11. $f(x) = \cos(x)$, $\mathcal{S} = \{-\pi/3, \pi/3, \pi, 5\pi/3\}$, $I = [-\pi, 5\pi/3]$, $N = 4$
12. $f(x) = x^2 - x$, $\mathcal{S} = \{-2, -1, 0, 1\}$, $I = [-3, 1]$, $N = 4$

In Exercises 13–19, sketch the integrand over the interval of integration. Evaluate the given definite integral without finding an antiderivative of the integrand.

13. $\int_1^3 \sqrt{2}\,dx$
14. $\int_1^5 (\cos^2(x) + \sin^2(x))\,dx$
15. $\int_{-1}^{3} |x|\,dx$
16. $\int_{-5}^{2} \text{signum}(x)\,dx$
17. $\int_{-2}^{1} (1 + x)\,dx$
18. $\int_{-1}^{1} \sqrt{1 - x^2}\,dx$
19. $\int_{-2}^{3} |x - 1|\,dx$

In Exercises 20–32, evaluate the definite integral by finding an antiderivative of the integrand.

20. $\int_{\pi}^{3\pi} \pi\,dx$
21. $\int_1^2 (6x^2 - 2x)\,dx$
22. $\int_{-1}^{0} x^{99}\,dx$
23. $\int_1^4 \sqrt{x}\,dx$
24. $\int_1^9 (x - 1/\sqrt{x})\,dx$
25. $\int_0^{\pi/4} \sec^2(x)\,dx$
26. $\int_{\pi/3}^{3\pi} \sin(x)\,dx$
27. $\int_0^{\pi/3} \sec(x)\tan(x)\,dx$
28. $\int_1^{16} x^{-3/4}\,dx$
29. $\int_8^{27} x^{1/3}\,dx$
30. $\int_0^1 e^x\,dx$
31. $\int_1^e 1/x\,dx$
32. $\int_0^{\ln(2)} e^{-x}\,dx$

Further Theory and Practice

In Exercises 33–38, use the given function f, interval $[a, b]$, and positive integer N.

- **a.** Compute the upper and lower Riemann sums for f based on the uniform partition of $[a, b]$ of order N.
- **b.** Compute the Riemann sum $\mathcal{R}(f, \mathcal{S})$ using the midpoint of each subinterval for the choice of points.
- **c.** Verify inequalities (5.6).

33. $f(x) = e^x, I = [0, 1], N = 1$
34. $f(x) = \log_{10}(x), I = [10, 100], N = 1$
35. $f(x) = 1/(1 + x), I = [1, 3], N = 2$
36. $f(x) = \sin(x), I = [0, \pi/2], N = 2$
37. $f(x) = \sqrt{1 + x^3}, I = [0, 3], N = 3$
38. $f(x) = 1 + \sin(x), I = [-\pi/2, \pi/2], N = 4$

In Exercises 39–42, calculate the upper and lower Riemann sums for the function f, interval $[a, b]$, and positive integer N. Verify the inequality

$$\mathcal{R}(f, \mathcal{U}_n) - \mathcal{R}(f, \mathcal{L}_n) \le c_1 \cdot \frac{(b_1 - a)^2}{N}$$

where c_1 is the maximum value of $|f'(x)|$ on the interval of integration.

39. $f(x) = \sin(x), I = [0, \pi], N = 3$
40. $f(x) = 18 - x^4, I = [-1, 2], N = 3$
41. $f(x) = e^{|x|}, I = [-2, 1], N = 3$
42. $f(x) = xe^{-x}, I = [0, 4], N = 2$
43. Compute the upper and lower Riemann sums for

$$f(x) = 2x^3 - 9x^2 + 12x + 1$$

and partition $\{0, 1, 2, 3\}$ of interval $[0, 3]$. Verify that $\int_0^3 f(x)\,dx$ is between these two Riemann sums.

44. Let f denote a fixed continuous function on $\mathbb{R}$. Is $\int_a^b f(x)\,dx$ a function of x? Is $\int_a^b f(x)\,dx$ a function of a and b?

45. Let f denote a continuous function on $[a, b]$. Is it true that

$$\int_a^b f(t)\,dt = \int_a^b f(u)\,du?$$

Is it true that

$$\int_a^b f(u)\,dt = \int_a^b f(t)\,du?$$

Explain your answers.

46. Suppose that $a < b$. Obtain a formula for $\int_a^b x^k\,dx$ that is valid for every $k > 0$. Now suppose that $k < 0$. What assumptions on a and b must you make in this case to apply Theorem 3? Evaluate $\int_a^b x^k\,dx$ for $k < 0$, making sure to treat $k = -1$ as a special case.

47. Suppose $b > 0$. A function f is *odd* if $f(-x) = -f(x)$ for all x. It is *even* if $f(-x) = f(x)$. Prove that if f is odd and continuous, then

$$\int_{-b}^{b} f(x)\,dx = 0.$$

Prove that if f is even, then

$$\int_{-b}^{b} f(x)\,dx = 2\int_0^b f(x)\,dx.$$

48. Suppose that f and f'' are both positive functions. Use Figure 11 to show that

$$(b - a)f\left(\frac{a + b}{2}\right) \le \int_a^b f(x)\,dx \le (b - a)\frac{f(a) + f(b)}{2}.$$

(The two triangles with labeled vertices are congruent.)

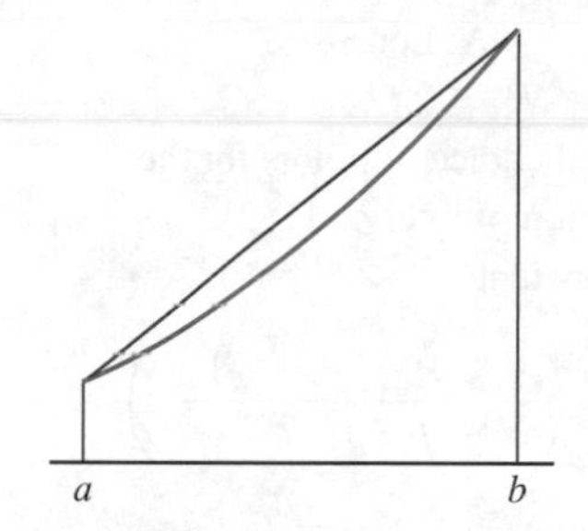

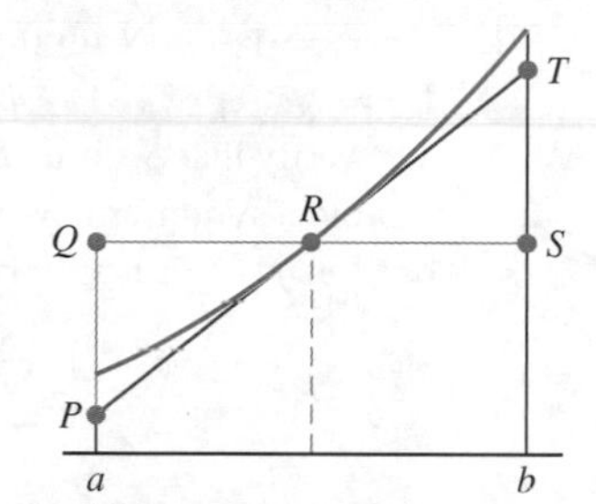

Figure 11
$\int_a^b f(x)\,dx \le (b - a)\frac{f(a)+f(b)}{2}$, $(b - a)f\left(\frac{a+b}{2}\right) \le \int_a^b f(x)\,dx$

Suppose f is continuous on $[a, b]$ and $F'(x) = f(x)$ for $a < x < b$. Let $\{x_0, x_1, x_2\}$ be an order 2 uniform partition of $[a, b]$. In Theorem 3, we used the Mean Value Theorem to obtain points $s_j \in (x_{j-1}, x_j)$ such that

$$f(s_j)\Delta x = (F(x_j) - F(x_{j-1})), \quad 1 \le j \le 2.$$

In Exercises 49 and 50, find the points s_j for the given function f and interval $[a, b]$. Verify that the Riemann sum $\mathcal{R}(f, \{s_1, s_2\})$ equals the integral $\int_a^b f(x)\,dx$.

49. $f(x) = x^2, a = 0, b = 2$
50. $f(x) = 4x^3, a = 1, b = 5$
51. Suppose $a > b > 0$. With respect to the ellipse and circumscribed circle plotted in Figure 12, next page, show that

$$y_E = \frac{b}{a}y_C.$$

By comparing Riemann sums for the ellipse with Riemann sums for the circle, deduce the area enclosed by the ellipse

$$\frac{x^2}{a^2} + \frac{y^2}{b^2} = 1.$$

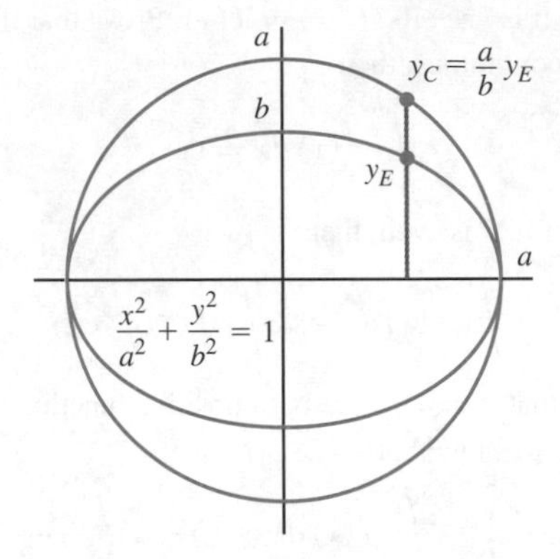

Figure 12

Take as known the area πa^2 of the circumscribed circle.

52. Let $x_j = 1 + j/N$ for $0 \le j \le N$. Let $\mathcal{S}_N = \{s_j = \sqrt{x_{j-1} \cdot x_j} : 1 \le j \le N\}$.

a. Verify that $\mathcal{S}_N$ is a valid choice of points for the order N uniform partition of interval $[1, 2]$.

b. Let $f(x) = 1/x^2$. Show that

$$\mathcal{R}(f, \mathcal{S}_N) = \frac{1}{N} \sum_{j=1}^{N} \left(\frac{1}{N+j-1} - \frac{1}{N+j} \right).$$

c. For the same f, prove that $\mathcal{R}(f, \mathcal{S}_N) = 1/2$ for every N. (Use part b and your result from Exercise 37 from Section 5.1.)

d. Use part c to evaluate $\int_1^2 1/x^2 \, dx$.

53. Let $f(x) = x^2$. For a given positive N, find a number r in $[0, 1]$ for which the Riemann sum $\frac{1}{N} \sum_{j=1}^{N} f(\frac{j+r-1}{N})$ exactly equals the Riemann integral $\int_0^1 f(x)\, dx$.

54. Suppose f is continuous and increasing on $[a, b]$. Sketch a representative graph of f. Illustrate $\mathcal{R}(f, \mathcal{U}_4)$, $\mathcal{R}(f, \mathcal{L}_4)$, and $\mathcal{R}(f, \mathcal{U}_4) - \mathcal{R}(f, \mathcal{L}_4)$. What is the relationship between the choice of points $\{u_1, \ldots, u_4\}$ and the *nodes* $a = x_0, x_1, \ldots, x_4 = b$ of the order 4 uniform partition? What is the relationship between $\{\ell_1, \ldots, \ell_4\}$ and the nodes? Write out $\mathcal{R}(f, \mathcal{U}_4) - \mathcal{R}(f, \mathcal{L}_4)$. What does it simplify to after cancellation? (It is a *collapsing sum*.) In general, show that

$$\mathcal{R}(f, \mathcal{U}_N) - \mathcal{R}(f, \mathcal{L}_N) = (f(b) - f(a)) \cdot \frac{b-a}{N},$$

which tells us that when we use $\mathcal{R}(f, \mathcal{U}_n)$ to approximate $\int_a^b f(x)\, dx$, the error is on the order of $1/N$.

55. Suppose f is continuous on $[a, b]$, differentiable on (a, b), and $|f'(x)| \le C_1$ for all x in (a, b). Use the Mean Value Theorem to prove that

$$f(u_j) - f(\ell_j) \le C_1 \cdot \Delta x$$

(with notation as in the discussion of upper and lower Riemann sums). Deduce that

$$\mathcal{R}(f, \mathcal{U}_N) - \mathcal{R}(f, \mathcal{L}_N) \le \frac{C_1(b-a)^2}{N}.$$

Calculator/Computer Exercises

In Exercises 56–59, approximate the definite integral by using $\mathcal{R}(f, \{s_1, \ldots, s_N\})$ with the uniform partition of order N (where N is given). For each $1 \le j \le N$, let s_j be the midpoint of the jth subinterval. Figure 13 shows how Maple can be used to do this for $\int_0^{\pi/2} \ln(1 + \sin(x))\, dx$, $N = 100$.

56. $\int_0^{\sqrt{3}} \sqrt{1 + x^2}\, dx$, $N = 80$

57. $\int_0^1 \cos^4(\pi x)\, dx$, $N = 100$

58. $\int_0^1 \exp(-x^2)\, dx$, $N = 40$

59. $\int_0^{\pi} x \sin(x)\, dx$, $N = 60$

In Exercises 60–63, plot the function f over the interval $[a, b]$. Calculate the upper and lower Riemann sums $\mathcal{R}(f, \mathcal{U}_{50})$ and $\mathcal{R}(f, \mathcal{L}_{50})$. (The exact value of $\int_a^b f(x)\, dx$ lies between these two estimates.)

60. $f(x) = \sqrt{1 + \sqrt{x}}$, $I = [1, 4]$

61. $f(x) = \sqrt{\sin(x)}$, $I = [1, \pi/2]$

62. $f(x) = 1/(1 + x^{4/3})$, $I = [1, 3]$

63. $f(x) = \cos(x^2)$, $I = [1/2, 1]$

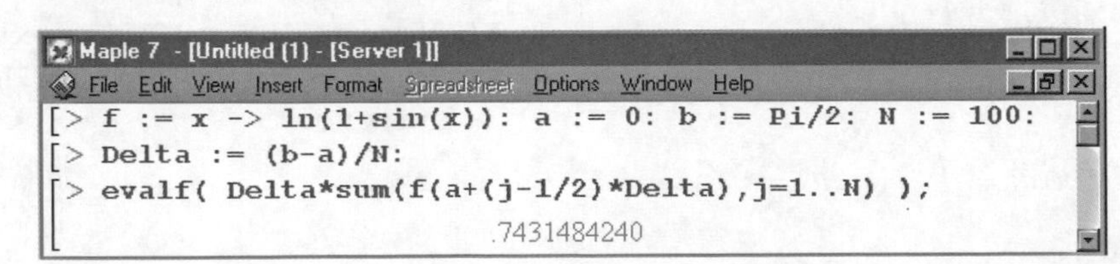

Figure 13

5.3 Rules for Integration

The calculation of integrals can be simplified if we take note of certain rules.

Theorem 1 If f and g are integrable functions on the interval $[a, b]$ and α is a constant, then $f+g$, $f-g$, and $\alpha \cdot f$ are integrable and

a. $\int_a^b (f(x)+g(x))\,dx = \int_a^b f(x)\,dx + \int_a^b g(x)\,dx$ and $\int_a^b (f(x)-g(x))\,dx = \int_a^b f(x)\,dx - \int_a^b g(x)\,dx$;
b. $\int_a^b \alpha \cdot f(x)\,dx = \alpha \cdot \int_a^b f(x)\,dx$;
c. $\int_a^b \alpha\,dx = \alpha \cdot (b-a)$;
d. $\int_a^a f(x)\,dx = 0$; and
e. If $a \le c \le b$, then $\int_a^c f(x)\,dx + \int_c^b f(x)\,dx = \int_a^b f(x)\,dx$.

Notice that Theorem 1a says that $f+g$ and $f-g$ are integrable, provided that f and g can be integrated separately. It also tells how to calculate the integrals of $f+g$ and $f-g$. Likewise, Theorem 1b says that αf is integrable, provided that f is. It tells how to calculate the integral of αf. Rules a and b are consequences of analogous limit rules that we learned in Section 2.2.

Theorem 1c states that integrating a constant α over an interval I results in α times the length of I. In a sense, this rule simply states that the area of a rectangle is the length of the base times the height. This may be proved either by referring to the definition of the integral as a limit or by applying Theorem 3 from Section 5.2 after noting that αx is an antiderivative of α.

Theorem 1d says that the integral over an interval of zero length is zero, whereas rule e says that contiguous intervals of integration may be added. If we think of the integral as representing area, for instance, then rules d and e are geometrically plausible (see Figure 1).

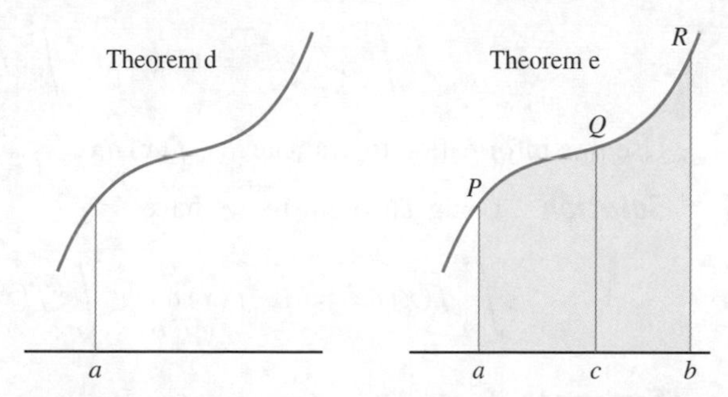

Figure 1
Theorem 1d: Area = base × height = $0 \times f(a) = 0$.
Theorem 1e: Area ($abRP$) = area ($acQP$) + area ($cbRQ$)

Example 1 Calculate the integral

$$\int_1^2 \frac{3+5t^3}{t^2}\,dt.$$

Solution We use Theorem 1a and 1b to rewrite the integral as

$$\int_1^2 \frac{3+5t^3}{t^2}\,dt = \int_1^2 \frac{3}{t^2}\,dt + \int_1^2 5\frac{t^3}{t^2}\,dt = 3\cdot\int_1^2 t^{-2}\,dt + 5\cdot\int_1^2 t\,dt.$$

Since $-t^{-1}$ is an antiderivative of t^{-2} and since $t^2/2$ is an antiderivative of t, we can evaluate each integral using the Fundamental Theorem of Calculus (Theorem 3, Section 5.2):

$$\int_1^2 \frac{3+5t^3}{t^2}\,dt = 3\cdot\int_1^2 t^{-2}\,dt + 5\cdot\int_1^2 t\,dt = 3\cdot\left(-t^{-1}\Big|_{t=1}^{t=2}\right) + 5\left(\frac{t^2}{2}\Big|_{t=1}^{t=2}\right)$$
$$= 3\left(-\frac{1}{2}-(-1)\right) + 5\left(\frac{2^2}{2}-\frac{1}{2}\right) = 9.$$

■

Example 2

Can we use the equations $\int_0^1 x\,dx = 1/2$ and $\int_0^1 x^2\,dx = 1/3$ to evaluate the integral $\int_0^1 x^3\,dx$ by observing that its integrand x^3 is the product of x and x^2?

Solution No. It is tempting, but wrong, to think that we can somehow factor this integral into two simpler integrals that we already know. Using the Fundamental Theorem of Calculus, we obtain

$$\int_0^1 x^3\,dx = \frac{x^4}{4}\Big|_{x=0}^{x=1} = \frac{1}{4} - 0 = \frac{1}{4},$$

which is certainly not $(1/2)\times(1/3)$. Thus,

$$\int_a^b f(x)g(x)\,dx \neq \int_a^b f(x)\,dx \cdot \int_a^b g(x)\,dx,$$

in general. In other words, the integral of a product is *not* the product of the integrals. In Chapter 7, we will learn techniques for handling integrals of products. ■

Example 3

Let f be a continuous function on the interval $[-2, 5]$ such that

$$\int_{-2}^1 f(x)\,dx = 6 \quad\text{and}\quad \int_1^5 f(x)\,dx = -2.$$

Use this information to evaluate $\int_{-2}^5 f(x)\,dx$.

Solution Using Theorem 1e, we have

$$\int_{-2}^5 f(x)\,dx = \int_{-2}^1 f(x)\,dx + \int_1^5 f(x)\,dx = 6 + (-2) = 4.$$

■

Example 4

If g is a continuous function on the interval $[4, 9]$ that satisfies

$$\int_4^9 g(x)\,dx = 8 \quad\text{and}\quad \int_4^7 g(x)\,dx = -7,$$

then calculate $\int_7^9 g(x)\,dx$.

Solution We use Theorem 1e to see that

$$8 = \int_4^9 g(x)\,dx = \int_4^7 g(x)\,dx + \int_7^9 g(x)\,dx = -7 + \int_7^9 g(x)\,dx.$$

Therefore, $\int_7^9 g(x)\,dx = 15$. ■

Example 5 Evaluate $\int_{\pi/2}^{\pi/2} t^{5/2}\sin(t^2)\,dt$.

Solution Theorem 1d tells us that $\int_a^a f(x)\,dx = 0$, no matter what the integrand is. Therefore,

$$\int_{\pi/2}^{\pi/2} t^{5/2}\sin(t^2)\,dt = 0.$$ ■

Reversing the Direction of Integration

inSIGHT

Remember that if f is nonnegative and $a < b$, then the integral of f over $[a, b]$ represents area and is positive. In Example 6, $f(t) = t^2$ is nonnegative, but $\int_1^{-4} t^2\,dt$ is negative. This is because the lower limit of integration is greater than the upper limit.

In the definition of the integral, the upper limit of integration is taken to be greater than or equal to the lower limit. However, it is sometimes convenient to allow the upper limit to be less than the lower limit. If a is *greater* than b, then we *define* $\int_a^b f(x)\,dx$ for a function f integrable on $[b, a]$ by

$$\int_a^b f(x)\,dx = -\int_b^a f(x)\,dx. \qquad \textbf{(5.9)}$$

Example 6 Evaluate $\int_1^{-4} t^2\,dt$.

Solution Notice that $F(t) = t^3/3$ is an antiderivative of t^2. Thus,

$$\int_1^{-4} t^2\,dt = -\int_{-4}^{1} t^2\,dt = -(F(1) - F(-4)) = -\left(\frac{1}{3} - \left(\frac{-64}{3}\right)\right) = -\frac{65}{3}.$$ ■

Let us verify that the Fundamental Theorem of Calculus remains true regardless of the order of the limits of integration.

Theorem 2 If f is continuous on an interval that contains the points a and b and if F is an antiderivative of f, then

$$\int_a^b f(x)\,dx = F(b) - F(a).$$

Proof Suppose $a > b$. (Otherwise this theorem would be the same as the Fundamental Theorem of Calculus.) We have

$$\int_a^b f(x)\,dx \overset{(5.9)}{=} -\int_b^a f(x)\,dx \overset{(5.8)}{=} -(F(a) - F(b)) = F(b) - F(a).$$ ■

Now that we have defined the definite integral even when the lower limit of integration is greater than the upper limit of integration, we may ask if Theorem 1e continues to hold no matter what the order of a, b, and c may be. The answer is "yes," and we state the result as a theorem.

Theorem 3 If f is integrable on an interval that contains the three points a, b, and c, then

$$\int_a^b f(x)\,dx + \int_b^c f(x)\,dx = \int_a^c f(x)\,dx, \tag{5.10}$$

regardless of the order of a, b, and c.

Rather than give a general proof of Theorem 3, we will work out one example of it.

Example 7 Suppose that f is integrable on $[1, 5]$. Show that equation (5.10) is valid for $a = 5$, $b = 3$, and $c = 1$.

Solution We have

$$\begin{aligned}
\int_5^3 f(x)\,dx + \int_3^1 f(x)\,dx &= -\int_3^5 f(x)\,dx - \int_1^3 f(x)\,dx && \text{By formula (5.9)}\\
&= -\left(\int_1^3 f(x)\,dx + \int_3^5 f(x)\,dx\right) && \text{Reorder the terms}\\
&= -\int_1^5 f(x)\,dx && \text{By Theorem 1e}\\
&= \int_5^1 f(x)\,dx. && \text{By formula (5.9)}
\end{aligned}$$

■

Order Properties of Integrals

In this subsection, we collect a number of facts concerning integrals and order relations.

Theorem 4 If f, g, and h are integrable on $[a, b]$ and

$$g(x) \le f(x) \le h(x) \quad \text{for} \quad x \in [a, b],$$

then

$$\int_a^b g(x)\,dx \le \int_a^b f(x)\,dx \le \int_a^b h(x)\,dx. \tag{5.11}$$

In particular, if f is integrable on $[a, b]$ and if m and M are constants such that

$$m \le f(x) \le M$$

for all $x \in [a, b]$, then

$$m \cdot (b - a) \le \int_a^b f(x)\,dx \le M \cdot (b - a). \tag{5.12}$$

Proof If $\mathcal{S}_N = \{s_j\}$ is a choice of points associated with the order N uniform partition of $[a, b]$, then

$$\sum_{j=1}^N g(s_j)\Delta x \le \sum_{j=1}^N f(s_j)\Delta x \le \sum_{j=1}^N h(s_j)\Delta x,$$

or $\mathcal{R}(g, \mathcal{S}_N) \le \mathcal{R}(f, \mathcal{S}_N) \le \mathcal{R}(h, \mathcal{S}_N)$. Line (5.11) results by letting N tend to infinity.

In the particular case when $g(x) = m$ and $h(x) = M$ for all x, line (5.11) simplifies to become line (5.12). ■

The following theorem, which is used often, is obtained by taking $m = 0$ in the first inequality of line (5.12).

Theorem 5 If f is integrable on $[a, b]$ and if $f(x) \geq 0$, then $\int_a^b f(x)\,dx \geq 0$.

The next theorem may be deduced from line (5.11) by taking $g(x) = -|f(x)|$ and $h(x) = |f(x)|$.

Theorem 6 If f and $|f|$ are both integrable on $[a, b]$, then

$$\left| \int_a^b f(x)\,dx \right| \leq \int_a^b |f(x)|\,dx. \tag{5.13}$$

inSIGHT

Thinking of the definite integral as a limit of Riemann sums, we may regard inequality (5.13) as a continuous analogue of the (discrete) triangle inequality $|A_1 + \cdots + A_N| \leq |A_1| + \cdots + |A_N|$. In fact, if we set $A_j = f(s_j) \cdot \Delta x$ for $1 \leq j \leq N$ and let $N \to \infty$, then we obtain an alternative proof of inequality (5.13).

Example 8 Use the inequalities of line (5.12) to estimate $\int_0^2 \sqrt{1 + x^3}\,dx$.

Solution Because $\sqrt{1 + x^3}$ is increasing on [0, 2], we have

$$1 = \sqrt{1 + 0^3} \leq \sqrt{1 + x^3} \leq \sqrt{1 + 2^3} = 3, \quad 0 \leq x \leq 2.$$

We use the estimates of line (5.12) with $a = 0$, $b = 2$, $m = 1$, and $M = 3$ to obtain

$$2 = 1 \cdot (2 - 0) \leq \int_0^2 \sqrt{1 + x^3}\,dx \leq 3 \cdot (2 - 0) = 6.$$

The value of the given integral is between 2 and 6. ■

inSIGHT

The integrand of Example 8 does not have an elementary antiderivative. Therefore, we cannot use the Fundamental Theorem of Calculus to evaluate the integral exactly. In such cases, we must rely on an approximation. As the estimates obtained in Example 8 suggest, Theorem 4 usually gives us only an order of magnitude estimate. In Section 5.7, we study techniques for achieving accurate approximations of definite integrals.

The Mean Value Theorem for Integrals

In Chapter 4, we learned a Mean Value Theorem for derivatives. We will now learn a Mean Value Theorem for integrals.

Theorem 7 ***Mean Value Theorem for Integrals*** Let f be continuous on $[a, b]$. There is a value $c \in (a, b)$ such that

$$f(c) = \frac{1}{b - a} \int_a^b f(x)\,dx. \tag{5.14}$$

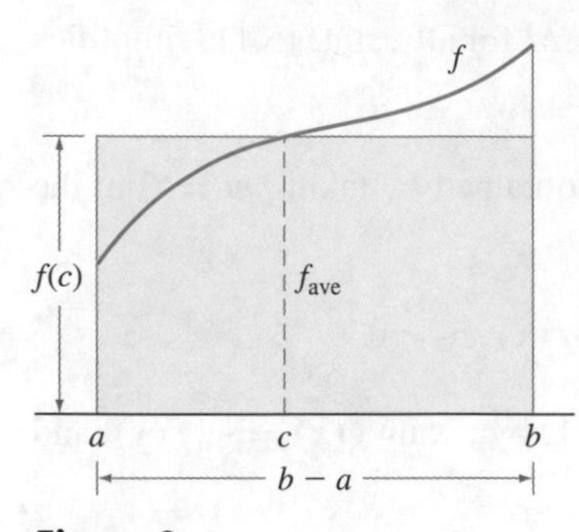

Figure 2
$f_{\text{ave}} \cdot (b-a) =$ area of shaded rectangle $= \int_a^b f(x)\,dx$

Proof Let m and M be the minimum and maximum values of f on $[a, b]$. These values exist by the Extreme Value Theorem (Section 2.3). The function $g(x) = (b-a) \cdot f(x)$ is continuous on $[a, b]$. The minimum and maximum values of g on $[a, b]$ are $\alpha = m \cdot (b-a)$ and $\beta = M \cdot (b-a)$, respectively. According to line (5.12), the number $\int_a^b f(x)\,dx$ lies between α and β. The Intermediate Value Theorem assures us that we can find a point c in the interval (a, b) such that

$$g(c) = \int_a^b f(x)\,dx.$$

Since $g(c) = (b-a)f(c)$, equation (5.14) follows. ■

The *average value* f_{ave} of f on $[a, b]$ is defined as the number

$$f_{\text{ave}} = \frac{1}{b-a} \int_a^b f(x)\,dx.$$

The Mean Value Theorem for Integrals tells us that a *continuous* function assumes its average value. There is no such theorem for discrete functions. For example, if you take two exams and score 80 and 90, then your average is 85. However, you did not score 85 on either exam—the average value 85 is not assumed. In Section 8.3, we motivate the definition of *average value*. For now, we are content with the geometric interpretation of average value: If f is positive on $[a, b]$, then the area under the graph of f over $[a, b]$ equals the area of the rectangle with base $[a, b]$ and height f_{ave}:

$$\int_a^b f(x)\,dx = f_{\text{ave}} \cdot (b-a),$$

which is illustrated in Figure 2.

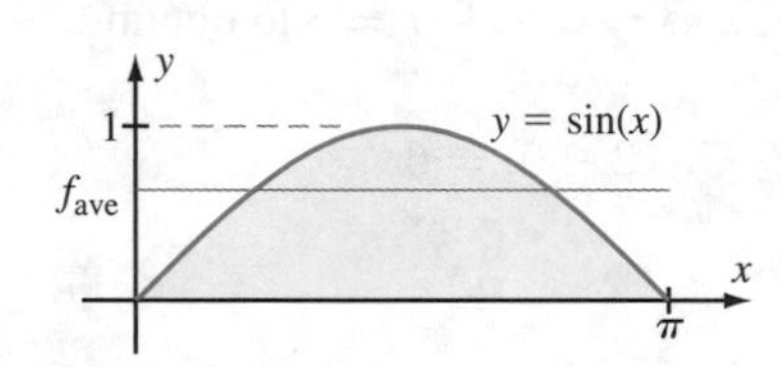

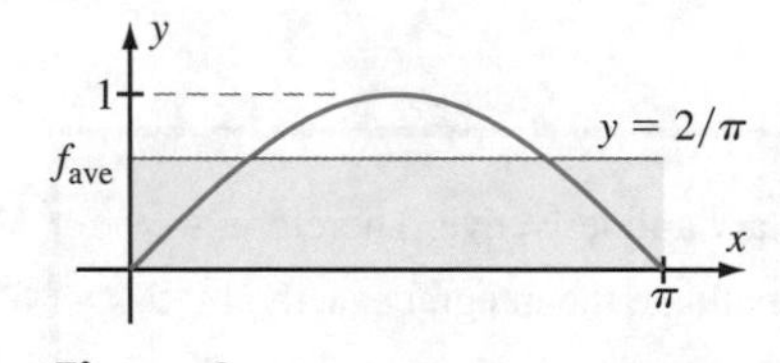

Figure 3
The shaded regions have equal area.

Example 9 Compute the average value of $f(x) = \sin(x)$ for $0 \le x \le \pi$.

Solution The average value of f on $[0, \pi]$ is

$$f_{\text{ave}} = \frac{1}{\pi - 0} \int_0^{\pi} \sin(x)\,dx = \frac{1}{\pi} \cdot \left(-\cos(x)\big|_{x=0}^{x=\pi}\right) = \frac{1}{\pi}(-(-1) - (-1)) = \frac{2}{\pi}.$$

Refer to Figure 3. Notice that the average value is *not* the average of the values of f at the endpoints (which is 0 in this case). It is *not* the average of the maximum and minimum values of f on the interval (which is $1/2$ in this case). Also, it is *not* the value of f at the midpoint of the interval (which is 1 in this case). ■

quickquiz

1. If $\int_0^2 f(x)\,dx = 3$ and $\int_0^7 f(x)\,dx = -5$, then what is $\int_7^2 3f(x)\,dx$?
2. What is $\int_5^5 \tan^3(5x^2 - x)\,dx$?
3. Does the integration process respect multiplication and division in a simple fashion?
4. What is the average value of $f(t) = t^2$ for $0 \le t \le 3$?

EXERCISES

Problems for Practice

1. If $\int_1^3 f(x)\,dx = -8$ and $\int_3^7 f(x)\,dx = 12$, evaluate $\int_1^7 f(x)\,dx$.
2. If $\int_2^{12} g(x)\,dx = -6$ and $\int_2^6 g(x)\,dx = -12$, evaluate $\int_6^{12} g(x)\,dx$.
3. If $\int_{-7}^3 f(x)\,dx = -7$ and $\int_{-7}^3 g(x)\,dx = -4$, evaluate $\int_{-7}^3 (4f(x) - 9g(x))\,dx$.
4. If $\int_4^8 f(x)\,dx = 6$, evaluate $\int_8^4 f(x)\,dx$.
5. If $\int_7^{-2} f(x)\,dx = 6$ and $\int_7^9 f(x)\,dx = -4$, evaluate $\int_9^{-2} f(x)\,dx$.
6. If $\int_{-3}^{-7} g(x)\,dx = 5$ and $\int_{-3}^{-5} g(x)\,dx = 12$, evaluate $\int_{-7}^{-5} g(x)\,dx$.
7. If $\int_9^4 f(x)\,dx = 5$ and $\int_9^4 g(x)\,dx = 15$, evaluate $\int_4^9 (6f(x) - 7g(x))\,dx$.
8. If $\int_3^5 f(x)\,dx = 2$, evaluate $\int_5^3 -4f(x)\,dx$.
9. If $\int_5^{-3} -3f(x)/4\,dx = 7$, evaluate $\int_5^{-3} (6f(x) + 1)\,dx$.
10. If $\int_2^{-9} f(x)\,dx = 5$, evaluate $\int_{-9}^2 (3f(x) - 5x)\,dx$.
11. If $\int_1^4 f(x)/3\,dx = 2$, evaluate $\int_4^1 3f(x)\,dx$.
12. If $\int_0^4 2f(x) - x^2\,dx = 6$, evaluate $\int_0^4 f(x)\,dx$.
13. If $\int_2^1 f(x)\,dx = 0$ and $\int_2^1 g(x)\,dx = 0$, evaluate $\int_2^1 (f(x) - 3g(x) + 5)\,dx$.
14. If $\int_6^8 (3f(x) - x)\,dx = 6$ and $\int_8^6 (2x + 4g(x))\,dx = -8$, evaluate $\int_8^6 (f(x) - 5g(x))\,dx$.

In Exercises 15–20, compute the average value of f over $[a, b]$ and find a point c in (a, b) at which f attains this average value. Illustrate the geometric significance of c with a sketch accompanied by a description.

15. $f(x) = 1 + x$, $a = 1$, $b = 5$
16. $f(x) = x^3$, $a = 0$, $b = 2$
17. $f(x) = \sqrt{x}$, $a = 1$, $b = 4$
18. $f(x) = 1/x^2$, $a = 1/2$, $b = 2$
19. $f(x) = 12x^2 + 5$, $a = 0$, $b = 3$
20. $f(x) = 2x + 2/x$, $a = 1$, $b = e$

Further Theory and Practice

In Exercises 21–24, use the information to determine $\int_a^b f(x)\,dx$ and $\int_a^b g(x)\,dx$.

21. $\int_a^b (f(x) - 3g(x))\,dx = 3$, $\int_a^b (-6g(x) + 9f(x))\,dx = 6$
22. $\int_a^b (f(x) + 4g(x))\,dx = 5$, $\int_a^b 3g(x)\,dx = -2$
23. $\int_a^b (f(x) + 2g(x))\,dx = -9$, $\int_a^b (g(x) - 6f(x))\,dx = 4$
24. $\int_a^b (f(x) - 7g(x))\,dx = -9$, $\int_a^b (g(x) + f(x))\,dx = 0$

In Exercises 25–30, use Theorem 3 from Section 5.2 with one or more of the properties listed in Theorem 1 of this section to calculate the definite integral.

25. $\int_1^3 (3x - 5)/x\,dx$
26. $\int_1^3 x(4x^2 - 2)\,dx$
27. $\int_{-1}^0 (6x^2 + \exp(x))\,dx$
28. $\int_1^2 (t - 2)(t + 3)\,dt$
29. $\int_0^1 3\exp(-x)(2 - \exp(x))\,dx$
30. $\int_{\pi/4}^{\pi/2} (x\sin(x) + 1)/(3x)\,dx$
31. When is it true that
$$\int_a^b f(x)\cdot 1\,dx = \int_a^b f(x)\,dx \cdot \int_a^b 1\,dx?$$
32. Find one example of a function f, a function g, and values of a and b such that
$$\int_a^b \frac{f(x)}{g(x)}\,dx \neq \frac{\int_a^b f(x)\,dx}{\int_a^b g(x)\,dx}.$$

In Exercises 33–36, find the extreme values of the integrand on the given interval of integration. Use these extreme values to obtain numbers A and B such that
$$A \leq \int_a^b f(x)\,dx \leq B.$$

33. $\int_0^1 \sqrt{1 + 8x^5}\,dx$
34. $\int_0^{\pi/2} \sqrt{1 + 3\cos^5(x)}\,dx$
35. $\int_1^4 (x^2 - 4x + 5)^{1/3}\,dx$
36. $\int_0^3 (t + 1)^2/(t^2 + 1)\,dt$
37. Graph $f(x) = \cos^2(x)$ and $g(x) = \sin^2(x)$ in the viewing rectangle $[0, \pi] \times [0, 1]$. It appears that the

areas under the graphs of f and g are the same. Use a trigonometric identity to explain why these areas are the same. Use the identity $\cos^2(x) + \sin^2(x) = 1$ to evaluate the common value of these areas.

38. Suppose f is a continuous function. Define

$$F(x) = \int_a^x f(t)\,dt.$$

Notice that F is a function of the upper limit x of integration, not the variable t of integration. For what values α and β is it true that $F(x+h) - F(x) = \int_\alpha^\beta f(t)\,dt$?

In Exercises 39–42, a function of the form

$$F(x) = \int_1^x f(t)\,dt$$

is given. Calculate $F(x+h) - F(x)$. (If you use Exercise 38, you can do this calculation with one integration.) Use the result to evaluate the difference quotient

$$\frac{F(x+h) - F(x)}{h}.$$

Show that the limit $F'(x)$ of this difference quotient as $h \to 0$ is equal to $f(x)$. (These computations are the focus of Section 5.4.)

39. $F(x) = \int_1^x \pi\,dt$

40. $F(x) = \int_1^x t\,dt$

41. $F(x) = \int_1^x t^2\,dt$

42. $F(x) = \int_1^x 1/\sqrt{t}\,dt$

43. What is wrong with the following calculation:

$$\int_0^1 \sqrt{u}\cdot(u-1)\,du = \frac{u^{3/2}}{3/2}\cdot\left(\frac{u^2}{2} - u\right)\Bigg|_{u=0}^{u=1} ?$$

What is the correct evaluation of the definite integral on the left?

44. Let

$$P_N(x) = \frac{1}{2^N \cdot N!}\cdot\frac{d^N}{dx^N}(x^2-1)^N.$$

This polynomial is called the degree N Legendre polynomial.

a. Calculate P_3 explicitly.

b. If Q is a polynomial of degree 2 or less—that is, if $Q(x) = Ax^2 + Bx + C$ for constants A, B, and C—show that

$$\int_{-1}^1 P_3(x)Q(x)\,dx = 0.$$

c. By setting $A = B = 0$ and $C = 1$ in part b, deduce that P_3 has at least one root in the interval $(-1, 1)$.

d. Let m be the number of roots that P_3 has in $(-1, 1)$. According to part c, m satisfies $1 \le m \le 3$. Let these m roots be $x_1, x_2, \ldots, x_m$. Show that

$$P_3(x)\cdot(x-x_1)\cdot(x-x_2)\cdot\ldots\cdot(x-x_m)$$

does not change sign on $(-1, 1)$.

e. Let $Q(x) = (x-x_1)\cdot(x-x_2)\cdot\ldots\cdot(x-x_m)$. Use the results of parts b and d to show that m cannot be less than 3. This shows that all three roots of P_3 are real and lie in the interval $[-1, 1]$. (A direct proof of this fact is very easy for P_3. However, the method that has been outlined can be used to obtain the same result for every P_N.)

Calculator/Computer Exercises

In Exercises 45–48, for the given integral $\int_a^b f(x)\,dx$, use line (5.12) to find lower and upper estimates ℓ and u such that

$$\ell \le \int_a^b f(x)\,dx \le u.$$

45. $\int_0^2 (1+x)/(1+x^4)\,dx$

46. $\int_0^{3/2} (\sqrt{x} - \sin(x))\,dx$

47. $\int_0^{\pi/2} (3\sin(x^2) + \cos(x))\,dx$

48. $\int_{1/2}^1 \exp(-x)\ln(1+x)\,dx$

In Exercises 49–52, compute the average value of f over $[a, b]$ and find a value of c in (a, b) at which f attains this average value. Illustrate the geometric meaning of the Mean Value Theorem for Integrals with a graph.

49. $f(x) = \sin(x)$, $a = 0$, $b = \pi/2$

50. $f(x) = \sin(x)$, $a = 0$, $b = \pi$

51. $f(x) = x\sqrt{9+x^2}/8$, $a = 0$, $b = 4$

52. $f(x) = (x^3 - 4x + 6)/3$, $a = 0$, $b = 2$

5.4 The Fundamental Theorem of Calculus

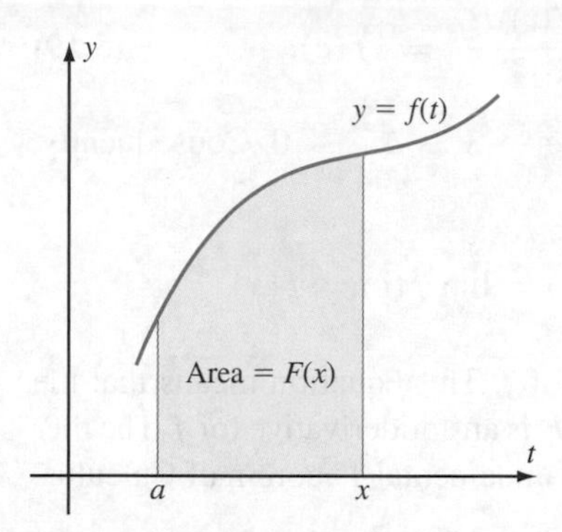

Figure 1

Theorem 3 from Section 5.2 tells us that if f is continuous on $[a, b]$ and *if* f has an antiderivative F on $[a, b]$, then

$$\int_a^b f(x)\,dx = F(b) - F(a). \tag{5.15}$$

This equation is the first part of the Fundamental Theorem of Calculus. In this section, we learn that *every* continuous function f does have an antiderivative. This fact constitutes the second part of the Fundamental Theorem of Calculus.

In this section, we geometrically motivate the second part of the Fundamental Theorem of Calculus by considering areas. Suppose f is a positive continuous function. To f we associate an *area function* F that is defined by

$$F(x) = \int_a^x f(t)\,dt. \tag{5.16}$$

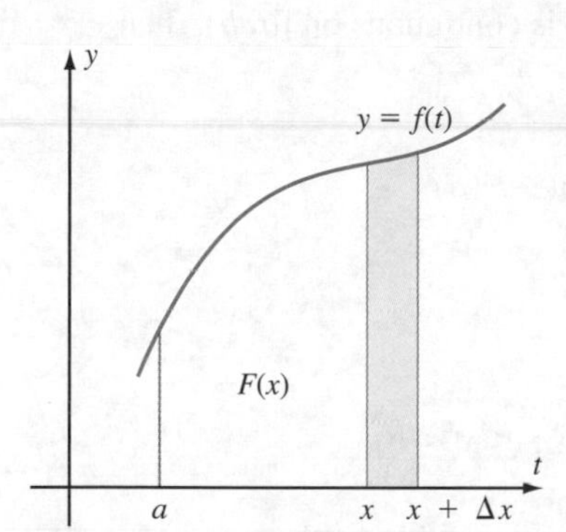

Figure 2
Area = $\int_x^{x+\Delta x} f(t)\,dt$

Figure 1 shows the equation $y = f(t)$ in the ty-plane. As Figure 1 indicates, the value of $F(x)$ is the area of the region that lies below the graph of f, above the t-axis, and between the vertical lines $t = a$ and $t = x$. Notice that F is a function of the upper limit x of integration, *not* of the symbol t that appears within the integral.

The main result of this section is that F is an antiderivative of f. In other words, F is a differentiable function and $F' = f$. To prove these assertions, we refer to the definition of the derivative and look at the difference quotient:

$$\frac{F(x + \Delta x) - F(x)}{\Delta x} = \frac{\int_a^{x+\Delta x} f(t)\,dt - \int_a^x f(t)\,dt}{\Delta x}.$$

We now use Theorem 3 from Section 5.3 to rewrite the numerator:

$$\frac{F(x + \Delta x) - F(x)}{\Delta x} = \frac{\int_x^{x+\Delta x} f(t)\,dt}{\Delta x}. \tag{5.17}$$

The numerator $\int_x^{x+\Delta x} f(t)\,dt$ represents the area of the shaded region in Figure 2. When Δx is small, the integral $\int_x^{x+\Delta x} f(t)\,dt$ is nearly equal to the area of a rectangle of base Δx and height $f(x)$ (see Figure 3). Therefore,

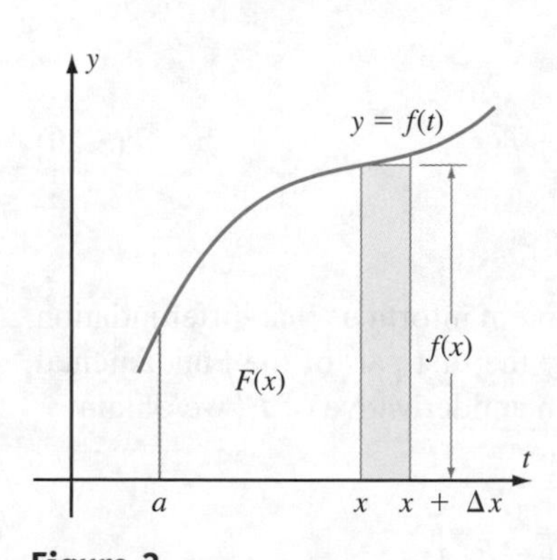

Figure 3
$\int_x^{x+\Delta x} f(t)\,dt$ is approximately the area $f(x)\Delta x$ of the shaded rectangle.

$$\frac{F(x + \Delta x) - F(x)}{\Delta x} \approx \frac{f(x) \cdot \Delta x}{\Delta x}$$

as $\Delta x \to 0$. Thus, we have intuitively deduced that

$$\lim_{\Delta x \to 0} \frac{F(x + \Delta x) - F(x)}{\Delta x} = f(x).$$

We can make this argument rigorous by appealing to the Mean Value Theorem for Integrals, which tells us that there exists a c between x and $x + \Delta x$ such that

$$\frac{\int_x^{x+\Delta x} f(t)\,dt}{\Delta x} = \frac{1}{(x + \Delta x) - x} \int_x^{x+\Delta x} f(t)\,dt = f(c). \tag{5.18}$$

By combining equations (5.17) and (5.18), we find that for any Δx there is a c between x and $x + \Delta x$ such that

$$\frac{F(x+\Delta x) - F(x)}{\Delta x} \overset{(5.17)}{=} \frac{\int_x^{x+\Delta x} f(t)\,dt}{\Delta x} \overset{(5.18)}{=} f(c). \tag{5.19}$$

Since c lies between x and $x + \Delta x$, it follows that $c \to x$ as $\Delta x \to 0$. Consequently, we have the limit equation

$$\lim_{\Delta x \to 0} \frac{F(x+\Delta x) - F(x)}{\Delta x} = \lim_{\Delta x \to 0} f(c) = \lim_{c \to x} f(c) = f(x)$$

where the last equality results from the continuity of f. This equation means that the derivative of F at x exists and equals $f(x)$. In short, F is an antiderivative for f. The theorem that follows summarizes the two parts of the Fundamental Theorem of Calculus.

Theorem 1 ***Fundamental Theorem of Calculus*** Let f be a continuous function on the interval $[a, b]$.

a. If F is *any* antiderivative of f on (a, b) that is continuous on $[a, b]$, then

$$\int_a^b f(t)\,dt = F(b) - F(a).$$

b. The function F defined by

$$F(x) = \int_a^x f(t)\,dt \quad (a \le x \le b)$$

is differentiable on the interval (a, b) and F is an antiderivative of f:

$$F'(x) = f(x) \quad (a < x < b).$$

In Leibniz notation,

$$\frac{d}{dx} \int_a^x f(t)\,dt = f(x). \tag{5.20}$$

Together, the two parts of the Fundamental Theorem inform us that differentiation and integration are *inverse operations*. Let us apply the first part of the Fundamental Theorem to a continuous derivative F'. Since F is an antiderivative of F', we obtain

$$\int_a^x F'(t)\,dt = F(x) - F(a).$$

First differentiate F. (pointing to F')
Then integrate F'. (pointing to the integral)
The result is F (pointing to $F(x)$)
(plus a constant). (pointing to $F(a)$)

Thus, if we differentiate F and integrate the derivative, then we recover the function F with which we started (plus the constant $-F(a)$). The second part of the Fundamental

Theorem, equation (5.20), tells us that if we integrate the function f and then differentiate with respect to the upper limit of integration, then we recover the function f with which we started:

First integrate f.

$$\frac{d}{dx}\int_a^x f(t)\,dt = f(x).$$

The result is f.

Then differentiate.

We therefore think of integration and differentiation as processes that are *inverse* to one another.

Examples Illustrating the First Part of the Fundamental Theorem

Example 1 Compute $\int_0^1 \exp(2t)\,dt$.

Solution According to the first part of the Fundamental Theorem, we need to find an antiderivative for the function $f(t) = \exp(2t)$. Notice that

$$\frac{d}{dt}\exp(2t) = 2\exp(2t).$$

Therefore, $F(t) = (1/2)\exp(2t)$ is a suitable antiderivative and

$$\int_0^1 \exp(2t)\,dt = F(t)\Big|_{t=0}^{t=1} = \frac{1}{2}\exp(2) - \frac{1}{2}\exp(0) = \frac{1}{2}(e^2 - 1).$$ ■

inSIGHT

Remember that a *definite* integral represents a definite number or expression. Including an arbitrary constant of integration in the final answer (as we would for an *indefinite* integral) is wrong. In Example 1, we used the Fundamental Theorem of Calculus to compute the *definite* integral $\int_0^1 f(t)\,dt$. Accordingly, we first computed the *indefinite* integral (or antiderivative) $\int f(t)\,dt$. At this stage, had we added a constant C to the *indefinite* integral $F(t) = \exp(2t)/2$, it would not have mattered:

$$\int_0^1 \exp(2t)\,dt = \frac{1}{2}\exp(2t) + C\Big|_{t=0}^{t=1} = \left(\frac{1}{2}\exp(2) + C\right) - \left(\frac{1}{2}\exp(0) + C\right)$$

$$= \frac{1}{2}e^2 + C - \frac{1}{2} - C = \frac{1}{2}(e^2 - 1).$$

The constant C cancels out and does not appear in the final answer. The definite integral $\int_0^1 \exp(2t)\,dt$ represents a unique number. The *only* correct answer is $(e^2 - 1)/2$. In particular, the answer "$(e^2 - 1)/2 + C$ where C is an arbitrary constant" is incorrect.

Example 2 Calculate $\int_0^{\pi} \sin^2(t)\,dt$.

Solution Since it is not immediately clear what an antiderivative for $\sin^2(t)$ might be, we first rewrite the integrand using the Half-Angle Formula (equation (1.14) from Section 1.6),

$$\sin^2\left(\frac{\theta}{2}\right) = \frac{1-\cos(\theta)}{2}.$$

Setting $\theta = 2t$, we obtain $\sin^2(t) = (1-\cos(2t))/2$. We substitute this identity into our original integral and find the required antiderivative:

$$\int_0^{\pi} \sin^2(t)\,dt = \int_0^{\pi} \frac{1}{2}(1-\cos(2t))\,dt = \frac{1}{2}\left(t - \frac{1}{2}\sin(2t)\right)\Bigg|_{t=0}^{t=\pi} = \frac{\pi}{2}. \quad \blacksquare$$

inSIGHT

The Solution to Example 2 begins with the remark that finding an antiderivative of $\sin^2(t)$ is not obvious. A common error is to think that because $t^3/3$ is an antiderivative of t^2, it must follow that $\sin^3(t)/3$ is an antiderivative of $\sin^2(t)$. *This reasoning is false.* We must use the Chain Rule when we differentiate $\sin^3(t)/3$:

$$\frac{d}{dt}\left(\frac{1}{3}\cdot\sin^3(t)\right) = \frac{1}{3}\cdot 3\,\sin^2(t)\cdot\frac{d}{dt}\sin(t) = \sin^2(t)\cos(t).$$

In conclusion, $\sin^3(t)/3$ is an antiderivative of $\sin^2(t)\cos(t)$, *not* of the given integrand $\sin^2(t)$.

Examples of the Second Part of the Fundamental Theorem

In the second part of the Fundamental Theorem, we define a *function in terms of an integral.* It is important that you understand this part, for it will be the key to many applications of the integral and also the key to our rigorous approach to the logarithm function in Chapter 6.

Example 3 A calculation in elementary mechanics shows that the hydrostatic pressure exerted against the side of a certain swimming pool at depth x is

$$P(x) = \int_0^x t\cdot(\sin(2t)+1)\,dt.$$

What is the rate of change of P with respect to depth when the depth is 4?

Solution We have not yet learned how to calculate an antiderivative of $f(t) = t\cdot(\sin(2t)+1)$. However, the second part of the Fundamental Theorem tells us that we do not need to evaluate the integral that defines $P(x)$ in order to compute $P'(x)$.

Since $P(x) = \int_0^x f(t)\,dt$, we have

$$P'(x) = f(x) = x \cdot (\sin(2x) + 1).$$

In particular, $P'(4) = 4 \cdot (\sin(8) + 1) \approx 7.957$. ■

Example 4 Calculate the derivative of $F(x) = \int_\pi^{x^2} \sin(\sqrt{t})\,dt$ with respect to x.

Solution The Fundamental Theorem tells us how to differentiate an integral with respect to its upper limit of integration. This example involves an upper limit that is a function of the variable of differentiation. We must use the Chain Rule. If we let $u = x^2$, then

$$\frac{d}{dx}F(x) = \frac{d}{dx}\int_\pi^u \sin\left(\sqrt{t}\right)dt = \frac{d}{du}\left(\int_\pi^u \sin\left(\sqrt{t}\right)dt\right) \cdot \frac{du}{dx} = \sin\left(\sqrt{u}\right) \cdot 2x = 2x\sin(|x|),$$

since $\sqrt{x^2} = |x|$. ■

Example 4 illustrates an extension of formula (5.20). If u is a function of x, then

$$\frac{d}{dx}\int_a^{u(x)} f(t)\,dt = f(u(x)) \cdot \frac{d}{dx}u(x). \tag{5.21}$$

Example 5 Calculate the derivative of $F(x) = \int_x^0 \exp(-t^2)\,dt$ with respect to x.

Solution The Fundamental Theorem tells us how to differentiate an integral with respect to its *upper* limit of integration. In this example, the variable of differentiation is the lower limit of integration. There is a simple remedy: We reverse the order of integration. Thus,

$$\frac{d}{dx}\int_x^0 \exp(-t^2)\,dt = -\frac{d}{dx}\int_0^x \exp(-t^2)\,dt = -\exp(-x^2).$$ ■

Example 5 illustrates another extension of formula (5.20). We can combine the ideas of Examples 4 and 5 to state the extension as follows:

$$\frac{d}{dx}\int_{u(x)}^b f(t)\,dt = -f(u(x)) \cdot \frac{d}{dx}u(x). \tag{5.22}$$

Example 6 Let F be defined by $F(x) = \int_{\cos(x)}^{\sin(x)} \sqrt{1-t^2}\,dt$ for $0 < x < \pi/2$. Differentiate $F(x)$ with respect to x.

Solution This example involves *both* upper and lower limits of integration that depend on x. Not one of formulas (5.20), (5.21), and (5.22) is immediately applicable. However, if we choose any fixed number a and write

$$\begin{aligned} F(x) &= \int_{\cos(x)}^{\sin(x)} \sqrt{1-t^2}\,dt = \int_{\cos(x)}^{a} \sqrt{1-t^2}\,dt + \int_a^{\sin(x)} \sqrt{1-t^2}\,dt \\ &= \int_a^{\sin(x)} \sqrt{1-t^2}\,dt - \int_a^{\cos(x)} \sqrt{1-t^2}\,dt, \end{aligned}$$

then we can differentiate each of the two summands using formula (5.21):

$$\begin{aligned}\frac{d}{dx}F(x) &= \sqrt{1-\sin^2(x)}\,\frac{d}{dx}\sin(x) - \sqrt{1-\cos^2(x)}\,\frac{d}{dx}\cos(x)\\ &= \sqrt{\cos^2(x)}\cos(x) - \sqrt{\sin^2(x)}(-\sin(x))\\ &= |\cos(x)|\cos(x) + |\sin(x)|\sin(x)\\ &= \cos^2(x)+\sin^2(x) \qquad \text{because } 0 < x < \pi/2\\ &= 1.\end{aligned}$$

■

INSIGHT

In Chapter 7, we learn how to calculate directly the integral that defines the function F from Example 6. As with Examples 3, 4, and 5, it is not necessary to compute F to obtain F'. It is therefore interesting to observe that our calculation $F'(x) \equiv 1$ actually gives us an indirect way to calculate $F(x)$. Since $G(x) = x$ is also an antiderivative of the constant function 1, Theorem 5 from Section 4.1 tells us that $F(x) = G(x) + C = x + C$ for some constant C. Substituting $x = 0$ into this last equation, we obtain

$$C = F(0) = \int_{\cos(0)}^{\sin(0)} \sqrt{1-t^2}\,dt = -\underbrace{\int_0^1 \sqrt{1-t^2}\,dt}_{\substack{\text{Area of 1/4-unit circle}\\ \text{(see Figure 4)}}} = -\frac{\pi}{4}.$$

Therefore, $F(x) = x - \pi/4$.

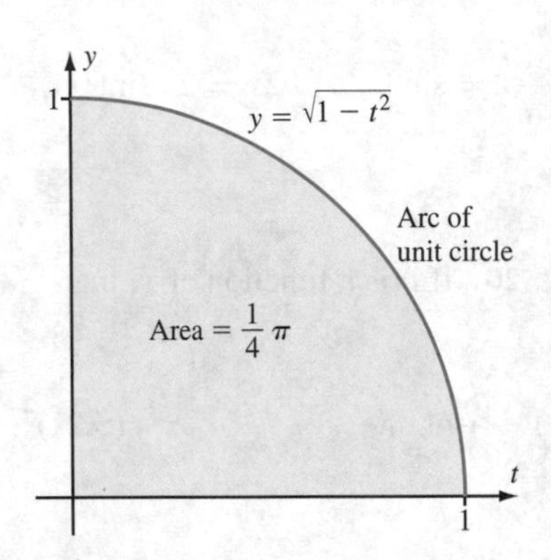

Figure 4

quickquiz

1. State both parts of the Fundamental Theorem of Calculus.
2. Differentiate $f(x) = \int_4^x \sin(t^2)\,dt$.
3. Differentiate $f(x) = \int_x^0 (1+t^2)^{-1}\,dt$.
4. Evaluate $\int_0^1 \frac{d}{dx}\left(xe^{x^2}\right)dx$.

EXERCISES

Problems for Practice

Use the first part of the Fundamental Theorem of Calculus to evaluate the definite integrals in Exercises 1–10.

1. $\int_0^2 (x^4+2)\,dx$
2. $\int_1^4 \sqrt{x}(x+1)\,dx$
3. $\int_{\pi/2}^{\pi} 3\cos(x)\,dx$
4. $\int_{3\pi/4}^{\pi} (\cos(t)-\sin(t))\,dt$
5. $\int_{-2}^{3} (t^2-1)^3\,dt$
6. $\int_{\pi/3}^{\pi/4} \sec^2(s)\,ds$
7. $\int_{-2}^{9} \exp(-x)\,dx$
8. $\int_6^3 (4x^4+x^3)/x\,dx$
9. $\int_{-\pi/3}^{\pi/2} 4\sin(2x)\,dx$
10. $\int_3^{-1} (2s+3)^2 \cdot s\,ds$

In Exercises 11–18, sketch the graph of the function on the interval. Find the area below the graph and above the interval.

11. $f(x) = 3x^2+2,\ I = [3, 7]$
12. $h(t) = 3\cos(3t),\ I = [0, \pi/6]$
13. $f(x) = x^6 - x^2,\ I = [1, 2]$
14. $g(s) = s^{1/2},\ I = [1, 4]$
15. $k(s) = -s^{-1/3},\ I = [-27, -8]$

16. $k(w) = 2w - \sin(w), I = [\pi, 2\pi]$
17. $g(x) = \exp(x), I = [-1, 1]$
18. $k(s) = 1/s, I = [1, e]$

In Exercises 19–26, calculate $F(x) = \int_a^x f(t)\,dt$.

19. $f(t) = 3t^2 + 1, a = -1$
20. $f(t) = \sin(t), a = \pi/4$
21. $f(t) = t^{1/3}, a = 8$
22. $f(t) = \exp(-t), a = -1$
23. $f(t) = (\exp(t) + \exp(-t))/2, a = 0$
24. $f(t) = \sec^2(t), a = 0, 0 \le t < \pi/2$
25. $f(t) = \sec(t)\tan(t), a = 0, 0 \le t < \pi/2$
26. $f(t) = 1/t, a = 1, t > 0$

In Exercises 27–32, a function F of x is given. Calculate its derivative with respect to x.

27. $F(x) = \int_0^x \tan(t^2)\,dt$
28. $F(x) = \int_{\pi/4}^x t\sec(t^2)\,dt$
29. $F(x) = \int_{\pi/4}^{3x} \cot(t)\,dt$
30. $F(x) = \int_x^{-2} \sin(t^3)\,dt$
31. $F(x) = \int_{-1}^{2x} \sqrt{1+t^2}\,dt$
32. $F(x) = \int_x^x t^4 \cdot \tan^2(t)\,dt$

In Exercises 33–36, compute F' and F''. Determine the intervals on which F is increasing, decreasing, concave up, and concave down.

33. $F(x) = \int_0^x t(t-1)\,dt$
34. $F(x) = \int_0^x t\exp(-t)\,dt$
35. $F(x) = \int_1^x t\ln(t)\,dt, x > 0$
36. $F(x) = \int_0^x (\sqrt{1+t^2} - t)\,dt$

Further Theory and Practice

In Exercises 37–40, follow the method from Example 4 to evaluate $F'(x)$.

37. $F(x) = \int_{2x}^{3x} (t^3 + t)^{1/4}\,dt$
38. $F(x) = \int_{-\sqrt{x}}^{\sqrt{x}} (1+t^2)^{-1/2}\,dt$
39. $F(x) = \int_{x-2}^{x-1} 1/t\,dt$
40. $F(x) = \int_{\exp(-x)}^{\exp(x)} \ln(t)\,dt$
41. Let $F(x) = \int_{\ln(1/x)}^{\ln(x)} \sin(t^3)\,dt$ for $x > 0$. Evaluate $F'(x)$ using the Fundamental Theorem of Calculus. What does the value of $F'(x)$ tell you about $F(x)$? How can you deduce this last property without differentiation?
42. For $t > 1$, let $F(t)$ denote the area under the curve $y = 1/x$ and above the interval $[1, t]$. What is $F'(t)$? For what value t is $F(t) = 1$?
43. Suppose a body travels in a line with position $p(t)$ at time t and velocity $v(t)$ at time t. Show that

$$\int_a^b v(t)\,dt = p(b) - p(a).$$

In Exercises 44–47, verify that point P is on the graph of function F and calculate the tangent line to the graph of F at P. Sketch the graph of F and the tangent line at P.

44. $F(x) = \int_1^x 6\sqrt{t}\,dt, P = (4, 28)$
45. $F(x) = \int_1^x 1/t\,dt, P = (e, 1)$
46. $F(x) = \int_4^x t^2\,dt, P = (7, 93)$
47. $F(x) = \int_0^x \cos(t)\,dt, P = (\pi/6, 1/2)$
48. Calculate $F(x) = \int_{-1}^x f(t)\,dt$ where f is the function with the graph that appears in Figure 5.

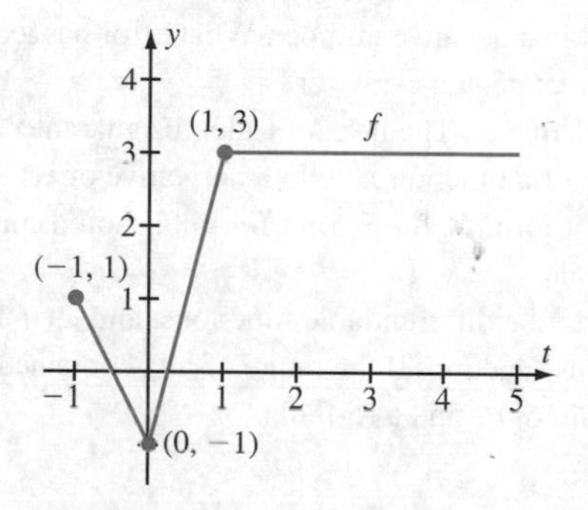

Figure 5

49. Find a function f such that $F(x) = \int_{-1}^x f(t)\,dt$ for the function F with the graph that appears in Figure 6.

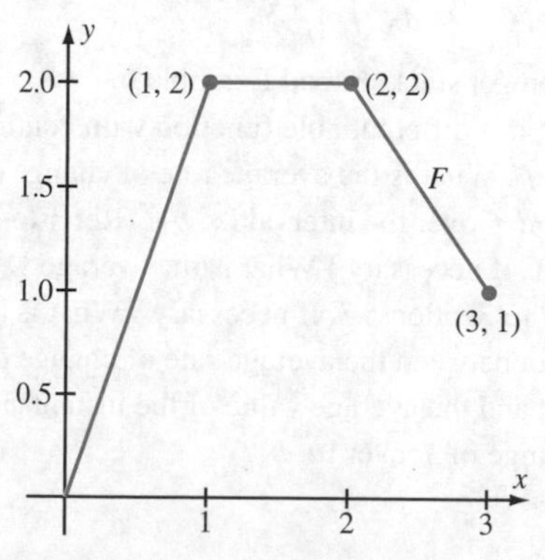

Figure 6

50. For the function f shown in Figure 7, determine on what interval(s) $F(x) = \int_0^x f(t)\,dt$ is increasing, decreasing, concave up, and concave down.

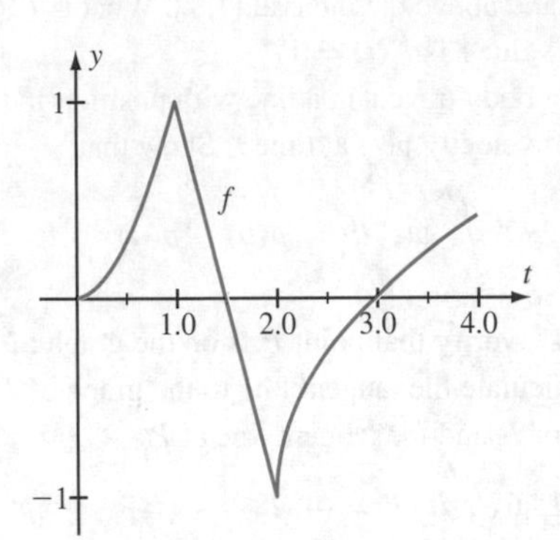

Figure 7

51. Consider the integration

$$\int_{-2}^{2} t^{-4}\,dt = -\frac{1}{3}t^{-3}\Big|_{-2}^{2} = -\frac{1}{12}.$$

Integrating a positive function from left to right should not result in a negative number. What error has led to the incorrect negative answer?

52. The Fundamental Theorem of Calculus guarantees the existence of a function F with a derivative of $|x|$. Write an explicit formula for F. Your formula should contain no integrals.

53. Let g and h be differentiable functions, and let f be a continuous function. If the range of h is contained in the domain of g, find a formula for

$$\frac{d}{dx}\int_a^{g(h(x))} f(t)\,dt.$$

54. Let g and h be differentiable functions and let f be a continuous function. Find a formula for

$$\frac{d}{dx}\int_{h(x)}^{g(x)} f(t)\,dt.$$

Hint: If you get stuck, reread Example 6.

55. Suppose f is a differentiable function with continuous derivative f'. What is the average rate of change of the function f over the interval $[a, b]$? (Refer to Section 3.1, if necessary.) What is the average value of f'? (Refer to Section 5.3, if necessary.) What is the relationship between the average rate of change of f over $[a, b]$ and the average value of the instantaneous rate of change of f over $[a, b]$?

56. Let

$$G(a) = \frac{d}{dx}\int_a^x f(t)\,dt\Big|_{x=0}.$$

What is $G'(a)$?

57. Suppose f and g are functions with continuous derivatives on an interval containing $[a, b]$. Prove that if $f(a) \le g(a)$ and if $f'(x) \le g'(x)$ for all x in $[a, b]$, then $f(x) \le g(x)$ for all x in $[a, b]$.

58. In probability and statistics, the error function (erf) is defined for $x \ge 0$ by

$$\text{erf}(x) = \frac{2}{\sqrt{\pi}}\int_0^x \exp(-t^2)\,dt.$$

Show that the graph of erf is concave down over $[0, \infty)$.

59. *Dawson's integral* is the function defined for $x \ge 0$ by

$$F(x) = \exp(-x^2)\int_0^x \exp(t^2)\,dt.$$

Compute $F'(x)$.

60. The *Fresnel sine integral,* defined by

$$\text{FresnelS}(x) = \int_0^x \sin\left(\frac{\pi}{2}t^2\right)dt,$$

is an important function in the theory of optical diffraction. Determine the intervals on which this function is concave up.

61. The function

$$C(x) = \int_0^x \frac{t^4}{\sqrt{1+t^2}}\,dt \quad (x \ge 0)$$

arises in the computation of pressure within a white dwarf. Show that C is an increasing function with a graph that is concave up.

62. The *sine integral* is the function Si, defined by $\text{Si}(x) = \int_0^x \sin(t)/t\,dt$. Calculate $\lim_{x\to 0}\text{Si}(x)/x$.

63. Let f be continuous. Set $F(x) = \int_a^x f(t)\,dt$. As a consequence of the Fundamental Theorem of Calculus, we know that $F'(x) = f(x)$ does not depend on a. Explain this fact in a more intuitive way by finding the relationship between $F(x)$ and $G(x) = \int_b^x f(t)\,dt$.

Averaging data smooths it out. The Mean Value Theorem for Integrals tells us that for a fixed h, we can think of the integral

$$A_f(x) = \frac{1}{2h}\int_{x-h}^{x+h} f(t)\,dt$$

as an operation that averages the values of f over the interval $[x-h, x+h]$. Figure 8 shows a plot of the Dow-Jones industrial average f, which is an irregularly varying index of the prices of 30 industrial stocks listed on the New York Stock Exchange. Figure 9 shows a plot of the so-called moving

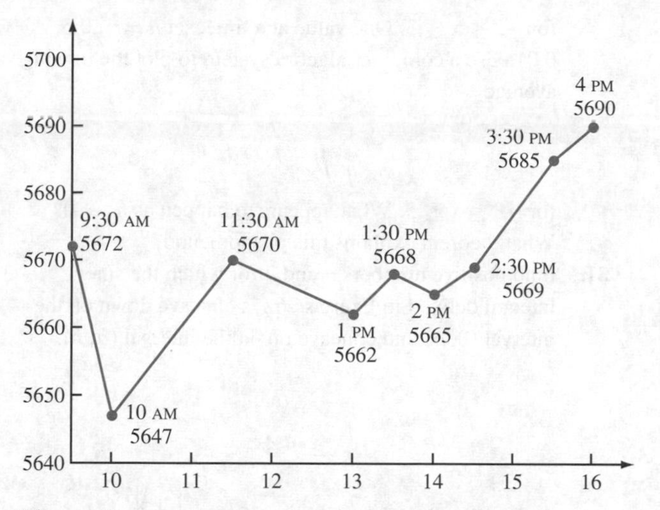

Figure 8
The Dow-Jones industrial average, April 3, 1996

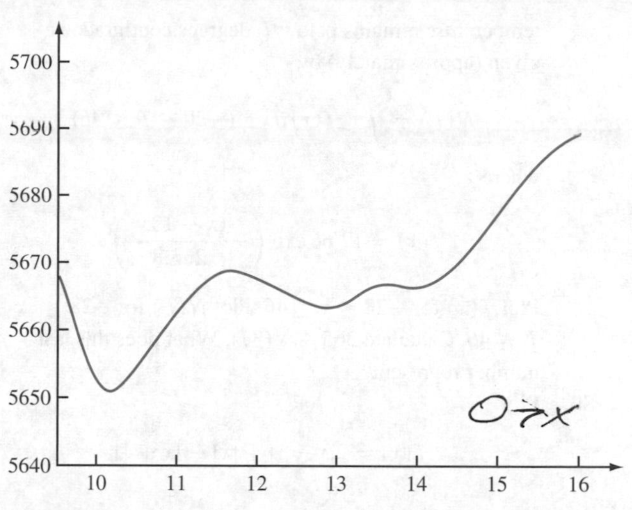

Figure 9
The Dow-Jones industrial average, smoothed out by integrating

average A_f, using $h = 1/3$. The choice of h is arbitrary. With $h = 1/3$, A_f is called a 40-minute moving average. A variety of choices of h have been used in stock analysis. For example, a 150-day moving average of the Standard & Poor's Index appears regularly in *USA Today*. In Exercises 64–66, calculate and plot A_f over $[-1, 1]$ for the function f and the value h.

64. $f(x) = |x|$, $h = 1/2$

65. $f(x) = |x|$, $h = 1/10$

66. $f(x) = \text{signum}(x)$, $h = 1/2$

67. Let $f(x) = \lfloor x \rfloor$ be the greatest integer function. Graph

$$F(x) = \int_0^x f(t)\,dt$$

for $-2 < x < 2$. Where is f continuous? Where is F differentiable?

68. Let

$$f(t) = \begin{cases} t^2 & \text{if } t < 2 \\ t^3 & \text{if } t \geq 2 \end{cases}.$$

What is $F(x) = \int_{-1}^x f(t)\,dt$? What is the function F'? Is it f?

Calculator/Computer Exercises

In Exercises 69–72, plot F' and F''. Determine the intervals on which F is increasing, decreasing, concave up, and concave down.

69. $F(x) = \int_{-3}^x (t^3 - 3t^2 + 3t + 4)\,dt$

70. $F(x) = \int_0^x (t + 1)/(t^4 + 1)\,dt$

71. $F(x) = \int_0^x (t^3 - t)\exp(t)\,dt$

72. $F(x) = \int_1^x (t^2 - 3)\ln(t)\,dt$, $x > 0$

In Exercises 73–76, $F(x) = \int_a^x f(t)\,dt$ is given. For the values of c and h, approximate

$$\int_{c-h/2}^{c+h/2} f(t)\,dt$$

by means of the Riemann sum $\mathcal{R}(f, \mathcal{S}_2)$. Use the midpoints of the subintervals for the choice of points $\mathcal{S}_2$. Use this value in the central difference quotient

$$\frac{F(c + h/2) - F(c - h/2)}{h} = \frac{1}{h}\int_{c-h/2}^{c+h/2} f(t)\,dt$$

to approximate $F'(c)$. Calculate the exact value of $F'(c)$ and compare its decimalization with your approximation.

73. $F(x) = \int_{-1}^x \exp(-t^2)\,dt$, $c = 0$, $h = 0.04$

74. $F(x) = \int_0^x \sin(t^3)\,dt$, $c = \pi/6$, $h = 0.001$

75. $F(x) = \int_1^x t/\sqrt{1+t}\,dt$, $c = 1$, $h = 0.001$

76. $F(x) = \int_0^x \sqrt{1 + \sqrt{t}}\,dt$, $c = 1$, $h = 0.004$

In Exercises 77 and 78, plot $F(x) = \int_0^x f(t)\,dt$ for $0 < x < 2$. Zoom in around the point $(c, F(c))$ until the window is small enough for the graph of F to appear straight. Approximate $F'(c)$ this way. Compute $F'(c)$ exactly.

77. $f(t) = \sqrt{1 + t^2}$, $c = \sqrt{3}$

78. $f(t) = (2t^2 + \cos(\pi t) + 1)/(t^2 + 1)$, $c = 1$

79. In a particular regional climate, the temperature varies between $-28°$C and $46°$C and averages $13°$C. The number of days $N(T)$ in the year on which the temperature remains below T degrees centigrade is given (approximately) by

$$N(T) = \int_{-28}^{T} f(x)\,dx \quad (-28 \le T \le 46)$$

where

$$f(x) = 12.66 \exp\left(-\frac{(x-13)^2}{265.8}\right).$$

Plot $f(x)$ for $-28 \le x \le 46$. Plot $N(T)$ for $-28 \le T \le 46$. Calculate $365 - N(37)$. What does this last number represent?

80. Plot

$$f(x) = 2|x-1| + |x| - |x+1|$$

for $-3 \le x \le 3$. One value at a time, set $h = 1, 0.5, 0.01$. Use a computer algebra system to plot the moving average

$$\frac{1}{h}\int_{x-h}^{x} f(t)\,dt$$

for $-3 \le x \le 3$. What appears to happen as $h \to 0$? What theorem explains this phenomenon?

81. Find positive numbers a and b for which the sine integral defined in Exercise 62 is concave down on the interval $(0, a)$ and concave up on the interval (a, b).

5.5 Integration by Substitution

The Method of Substitution, which is also called "change of variable," provides a way to simplify or transform an integrand. We begin by illustrating the Method of Substitution with an example.

Consider the integral

$$\int \cos(x^2+1)\cdot 2x\,dx.$$

The function $\phi(x) = x^2+1$ and its derivative $\phi'(x) = 2x$ *both* appear in the integrand. Suppose we denote the expression $\phi(x) = x^2+1$ by u. We write

$$u = x^2 + 1. \tag{5.23}$$

We then have $\frac{du}{dx} = 2x$. It is suggestive to write this as

$$\frac{du}{dx}\,dx = 2x\,dx$$

and to abbreviate this last formula as

$$du = 2x\,dx. \tag{5.24}$$

Now we may use equations (5.23) and (5.24) to rewrite $\int \cos(x^2+1)\cdot 2x\,dx$ as $\int \cos(u)\,du$. The transformed integral is considerably simpler. Because an antiderivative for $\cos(u)$ is $\sin(u)$, we have

$$\int \cos(u)\,du = \sin(u) + C.$$

Rewriting this in terms of the original variable x, we have

$$\int \cos(x^2+1)\cdot 2x\,dx = \sin(x^2+1) + C.$$

As with all indefinite integral problems, we can check the answer by differentiation:

$$\frac{d}{dx}(\sin(x^2+1)) = \cos(x^2+1)\cdot\frac{d}{dx}(x^2+1) = \cos(x^2+1)\cdot 2x,$$

as required. To save space, we will not include verifications of our work in later examples. However, you *should* get into the habit of checking the antiderivatives that you obtain.

Let us summarize the Method of Substitution.

Key Steps for the Method of Substitution

1. To apply the Method of Substitution to an integral of the form $\int f(x)\,dx$, find an expression $\phi(x)$ in the integrand that has a derivative $\phi'(x)$ that also appears in the integrand.
2. Substitute u for $\phi(x)$ and du for $\phi'(x)\,dx$.
3. *Do not proceed unless the entire integrand is expressed in terms of the new variable* u. (No xs can remain.)
4. Evaluate the new integral to obtain an answer expressed in terms of u.
5. Resubstitute to obtain an answer in terms of x.

It is an error to omit step 5. Doing so in the example that we worked at the start of this section would result in

$$\int \cos(x^2+1)\cdot 2x\,dx = \int \cos(u)\,du = \sin(u) + C.$$

Because it is common, as a matter of convenience, to write the effect of the change of variable in this manner, we will do so, recognizing that the computation is *incomplete* without step 5. To understand why, carefully examine the equation $\int \cos(x^2+1)\cdot 2x\,dx = \int \cos(u)\,du$. The left side represents a function with the function $x \mapsto \cos(x^2+1)\cdot 2x$ as its derivative. The right side represents a function with the function $u \mapsto \cos(u)$ as its derivative. The two sides are equal *only after* step 5 has been performed.

Some Examples of Indefinite Integration by Substitution

Example 1 Evaluate the indefinite integral $\int \sin^4(x)\cos(x)\,dx$.

Solution We notice that the expression $\sin(x)$, together with its derivative $\cos(x)$, appears in the integrand. This suggests setting

$$u = \sin(x) \quad \text{and} \quad du = \left(\frac{d}{dx}\sin(x)\right)dx = \cos(x)\,dx.$$

The integral then becomes $\int u^4\,du$, which is readily calculated to be $u^5/5 + C$. Resubstituting to obtain an expression in x, we conclude that

$$\int \sin^4(x)\cos(x)\,dx = \frac{1}{5}\sin^5(x) + C.$$

You should verify this answer by differentiating the expression on the right to see that it equals the integrand on the left. ■

Example 2 Compute the integral

$$\int \frac{\sqrt{x}}{(2 + x^{3/2})^3}\,dx.$$

Solution We notice that the derivative of the expression $(2 + x^{3/2})$ is $(3/2)x^{1/2}$ and that the latter is similar (up to a constant multiple) to the numerator. This motivates the substitution

$$u = 2 + x^{3/2} \quad \text{and} \quad du = \frac{d}{dx}(2 + x^{3/2})\,dx = \frac{3}{2}x^{1/2}\,dx.$$

The formula for du may be rewritten as

$$\sqrt{x}\,dx = \frac{2}{3}\,du.$$

Therefore, upon substitution, our integral becomes

$$\int \frac{(2/3)\,du}{u^3} = \frac{2}{3}\int u^{-3}\,du = \frac{2}{3}\left(-\frac{1}{2}u^{-2}\right) + C = -\frac{1}{3}u^{-2} + C.$$

Resubstituting the x-variable gives

$$\int \frac{\sqrt{x}}{(2 + x^{3/2})^3}\,dx = -\frac{1}{3}(2 + x^{3/2})^{-2} + C.$$

You should verify this answer by differentiation. ■

The Method of Substitution for Definite Integrals

The Method of Substitution also applies well to definite integrals; the only new feature is that, when we change variables, we must take the limits of integration into account.

Example 3 Evaluate the integral $\int_3^4 x \cdot \sqrt{25 - x^2}\,dx$.

Solution We observe that the derivative of the expression $25 - x^2$ is $-2x$. This motivates the substitution

$$u = 25 - x^2 \quad \text{and} \quad du = \frac{d}{dx}(25 - x^2)\,dx = -2x\,dx.$$

However, we must not stop here. In the x-integral, the limits of integration are 3 and 4. When $x = 3$, $u = 25 - 3^2 = 16$; when $x = 4$, $u = 25 - 4^2 = 9$. Making our

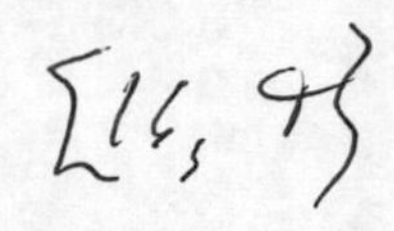

inSIGHT

Example 3 demonstrates that a change of variable can convert an integral of the form $\int_a^b f(x)\,dx$ with $a < b$ into an integral of the form $\int_c^d g(u)\,du$ with $c > d$. In other words, a change of variable can reverse the direction of integration. This is perfectly all right. Calculate in the usual way, and do not omit the minus sign that arises in the calculation of du.

substitutions, we obtain

$$\begin{aligned}\int_3^4 x\cdot\sqrt{25-x^2}\,dx &= -\frac{1}{2}\int_3^4 \sqrt{25-x^2}(-2x)\,dx\\ &= -\frac{1}{2}\int_{16}^{9}\sqrt{u}\,du\\ &= -\left(\frac{1}{2}\cdot\frac{2}{3}u^{3/2}\right)\Bigg|_{u=16}^{u=9}\\ &= \left(-\frac{9^{3/2}}{3}\right)-\left(-\frac{16^{3/2}}{3}\right)\\ &= -\frac{27}{3}+\frac{64}{3}=\frac{37}{3}.\end{aligned}$$

When we apply the Method of Substitution to a definite integral, it is essential that we take into account the effect that the change of variables has on the limits of integration. Example 3 illustrates how to do this *while* the substitution is being performed. An alternative method is to calculate the indefinite integral in the original variable and then use the original limits of integration, as is done in the next example.

Example 4 Calculate $\int_0^2 xe^{x^2}\,dx$.

Solution Let

$$u = x^2 \quad\text{and}\quad du = \frac{d}{dx}(x^2)\,dx = 2x\,dx.$$

When $x = 0$, $u = 0^2 = 0$; when $x = 2$, $u = 2^2 = 4$. Therefore,

$$\int_0^2 xe^{x^2}\,dx = \frac{1}{2}\int_0^4 e^u\,du = \frac{1}{2}(e^4-1).$$

Alternatively, we can use the Method of Substitution to find an antiderivative of the given integrand:

$$\int xe^{x^2}\,dx = \frac{1}{2}\int e^u\,du = \frac{1}{2}e^u + C \overset{\text{Resubstituting } x}{=} \frac{1}{2}e^{x^2}+C.$$

We may omit the additive constant C when calculating the definite integral:

$$\int_0^2 xe^{x^2}\,dx = \frac{1}{2}e^{x^2}\Bigg|_{x=0}^{x=2} = \frac{1}{2}(e^4-1),$$

as we obtained with the first method. Notice that, in this alternative method, we use the original limits of integration after resubstituting the original variable of integration. ■

The Role of the Chain Rule in the Method of Substitution

We have not said much about *why* the Method of Substitution works. All that the method really amounts to is an application of the Chain Rule. For instance, suppose that we want to study the integral

$$\int F(\phi(x)) \cdot \phi'(x)\, dx. \tag{5.25}$$

If G is an antiderivative for F, then

$$(G \circ \phi)'(x) \overset{\text{Chain Rule}}{=} G'(\phi(x))\phi'(x) \overset{G'=F}{=} F(\phi(x)) \cdot \phi'(x).$$

In other words,

$$\int_a^b F(\phi(x)) \cdot \phi'(x)\, dx = G(\phi(x))\Big|_{x=a}^{x=b} = G(t)\Big|_{t=\phi(a)}^{t=\phi(b)}.$$

The Method of Substitution is merely a device for seeing this procedure more clearly: We let

$$u = \phi(x) \quad \text{and} \quad du = \phi'(x)\, dx$$

to transform integral (5.25) to

$$\int F(u)\, du = G(u) + C = G(\phi(x)) + C.$$

When an Integration Problem Seems to Have Two Solutions

It is frustrating when your solution to an exercise is different from the solution that a friend has obtained or from the one in the back of the book. The next example illustrates how this can happen.

Example 5 Compute the indefinite integral

$$\int \sin(x)\cos(x)\, dx.$$

Solution We use the substitution $u = \sin(x)$, $du = \cos(x)\, dx$. The integral becomes

$$\int \sin(x)\cos(x)\, dx = \int u du = \frac{u^2}{2} + C \overset{\text{Resubstituting } x}{=} \frac{\sin^2(x)}{2} + C.$$

This problem can be done in several different ways. Another solution can be derived by setting $u = \cos(x)$, $du = -\sin(x)\, dx$. The integral then becomes

$$\int \sin(x)\cos(x)\, dx = -\int u du = -\frac{u^2}{2} + C = -\frac{\cos^2(x)}{2} + C.$$

Although both of these solution methods are correct, the answer $\sin^2(x)/2 + C$ appears to be quite different from the answer $-\cos^2(x)/2 + C$. How do we reconcile this apparent contradiction?

We observe that

$$\frac{\sin^2(x)}{2} + C = \frac{1 - \cos^2(x)}{2} + C = -\frac{\cos^2(x)}{2} + \left(\frac{1}{2} + C\right).$$

Now the two solutions are starting to look the same. Remember that C is an arbitrary constant. Hence $D = 1/2 + C$ is also an arbitrary constant. Thus, we have discovered that $\sin^2(x)/2 + C$ is precisely the same solution as $-\cos^2(x)/2 + D$.

In summary, although two methods lead to the same solution, it requires a trigonometric identity to see that they are the same. Put another way, any two solutions of our indefinite integral problem will differ by a constant. ■

We conclude with an example in which the Method of Substitution works in a not-so-obvious way.

Example 6 Evaluate the integral

$$\int \frac{x^2}{\sqrt{x+5}}\,dx.$$

Solution We do not immediately see an expression in the integrand whose derivative also appears. We instead set $u = x + 5$, because this substitution *simplifies* the radical. It follows that $du = dx$ and $x = u - 5$. The integral now has the form

$$\int \frac{(u-5)^2}{\sqrt{u}}\,du.$$

If we expand the numerator, this integral becomes

$$\int \frac{u^2 - 10u + 25}{\sqrt{u}}\,du = \int (u^{3/2} - 10u^{1/2} + 25u^{-1/2})\,du = \frac{2}{5}u^{5/2} - \frac{20}{3}u^{3/2} + 50u^{1/2} + C.$$

Resubstituting the x variable gives a final answer of

$$\int \frac{x^2}{\sqrt{x+5}}\,dx = \frac{2}{5}(x+5)^{5/2} - \frac{20}{3}(x+5)^{3/2} + 50(x+5)^{1/2} + C.$$

You should verify this answer by differentiation. ■

quickquiz

1. Explain the main steps of the Method of Substitution.
2. Evaluate $\int \cos^3(x)\sin(x)\,dx$.
3. Evaluate $\int_1^4 x\sqrt{x^2-1}\,dx$.
4. The function that is represented by u in the Method of Substitution is very much like the inside function in the Chain Rule. Explain this statement.

EXERCISES

Problems for Practice

In Exercises 1–10, determine a substitution that will simplify the integral. In each problem, record your choice of u and the resulting expression for du. Then evaluate the integral.

1. $\int \sin(3x)\,dx$
2. $\int \sec^2(4t)\,dt$
3. $\int (x^8+1)^{-5}x^7\,dx$
4. $\int t\sqrt{t^2+4}\,dt$
5. $\int (x^3-5)^{3/2}3x^2\,dx$
6. $\int (t^{1/2}+4)^6 t^{-1/2}\,dt$
7. $\int \sin(s)\cos^4(s)\,ds$
8. $\int (\sqrt{t}+1)/\sqrt{t}\,dt$
9. $\int 8\sin(\pi/t)/t^2\,dt$
10. $\int t^2\sec(t^3)\tan(t^3)\,dt$

Use the Method of Substitution to calculate the indefinite integrals in Exercises 11–22.

11. $\int (x^2+1)^7 \cdot 2x\,dx$ **12.** $\int x^2(5-4x^3)^{-2}\,dx$

13. $\int x \cdot \sqrt{1+x^2}\,dx$ **14.** $\int \cos(3x)\sin(3x)\,dx$

15. $\int \sin(x)/\cos^2(x)\,dx$ **16.** $\int \ln(x)/x\,dx$

17. $\int \sin^5(2x)\cdot\cos(2x)\,dx$ **18.** $\int x\cos(4x^2-5)\,dx$

19. $\int 5x/\sqrt{1+x^2}\,dx$ **20.** $\int \sin(x)\sqrt{\cos(x)}\,dx$

21. $\int \sin(\sqrt{x})/\sqrt{x}\,dx$

22. $\int (\sin(t)-\cos(t))/(\sin(t)+\cos(t))^2\,dt$

Use the Method of Substitution to evaluate the definite integrals in Exercises 23–36.

23. $\int_1^2 (t^2-t)^5(2t-1)\,dt$ **24.** $\int_{-1}^{14}\sqrt{2+x}\,dx$

25. $\int_0^2 x\sqrt{x^2+5}\,dx$ **26.** $\int_{\pi/3}^{\pi}\cos^3(5x)\sin(5x)\,dx$

27. $\int_{\sqrt{\pi}}^{\sqrt{\pi}/2} 2x\cdot\cos(4x^2-\pi)\,dx$

28. $\displaystyle\int_0^8 \frac{\sin\left(\sqrt{t+1}\right)}{\sqrt{t+1}}\,dt$ **29.** $\displaystyle\int_{-1}^0 \frac{x^2}{(x^3-1)^5}\,dx$

30. $\displaystyle\int_0^1 \frac{e^x}{(1+\exp(x))^2}\,dx$ **31.** $\displaystyle\int_{\pi^2/4}^{\pi^2}\frac{\cos\left(\sqrt{x}\right)}{\sqrt{x}}\,dx$

32. $\int_6^4 (2x+5)(x^2+5x)^{-2/3}\,dx$

33. $\int_0^{\pi/4}\sec^3(t)\tan(t)\,dt$

34. $\int_4^9 t^{-3/2}\sin(t^{-1/2})\,dt$

35. $\displaystyle\int_0^{\pi/6}\frac{\sec^2(x)}{(1+\tan(x))^2}\,dx$

36. $\displaystyle\int_1^e \frac{\sqrt{\ln(x)}}{x}\,dx$

Further Theory and Practice

37. Use the substitution $x = a\sin(\theta)$ to calculate

$$\int_{-a}^{a}\sqrt{a^2-x^2}\,dx.$$

What well-known formula results?

38. Use the substitution $x = a\sin(\theta)$ to show that

$$\int_{-a}^{a} b\sqrt{1-(x/a)^2}\,dx = ab\int_{-\pi/2}^{\pi/2}\cos^2(\theta)\,d\theta.$$

Use the trigonometric identity

$$\cos^2(\theta) = \frac{1+\cos(2\theta)}{2}$$

to evaluate the integral on the right. What classical area formula results from these calculations?

Calculate the integrals in Exercises 39–52.

39. $\int x\cdot(2x+3)^{1/2}\,dx$ **40.** $\displaystyle\int \frac{s^2}{(s-1)^3}\,ds$

41. $\displaystyle\int \frac{x}{\sqrt{x+3}}\,dx$ **42.** $\int t^3\cdot(t^2-1)^{1/2}\,dt$

43. $\int s^5\cdot(6+s^2)^{-1/2}\,ds$ **44.** $\int s\sqrt{s+3}\,ds$

45. $\int (s+2)\sqrt{s-5}\,ds$ **46.** $\displaystyle\int \frac{s}{\sqrt{s+4}}\,ds$

47. $\displaystyle\int \frac{2x^3+x}{\sqrt{x^4+x^2+1}}\,dx$ **48.** $\displaystyle\int_0^1 \frac{x^3}{\sqrt{1+x^2}}\,dx$

49. $\displaystyle\int_0^{\pi/3}\frac{\sin(x)}{1-\sin^2(x)}\,dx$ **50.** $\displaystyle\int_1^e \frac{\ln(x^3)}{x}\,dx$

51. $\int_0^1 x\sqrt[3]{1-x}\,dx$ **52.** $\displaystyle\int \frac{\cos(x)-\sin(x)}{\cos(x)+\sin(x)}\,dx$

53. Apply the change of variable $u = x^2+5$ to each integral in the equality

$$\int_{-3}^{2} x\sqrt{x^2+5}\,dx = \int_{-3}^{0} x\sqrt{x^2+5}\,dx + \int_0^2 x\sqrt{x^2+5}\,dx.$$

Verify the resulting equality.

54. Evaluate the integral $\int_{-1}^2 t^2\,dt = 3$. Then perform the substitution $u = t^2$, and evaluate the integral again. Why is the answer different?

55. Let $f:[-a,a]\to\mathbb{R}$ be continuous and odd. Show that

$$\int_{-a}^{a} f(x)\,dx = 0.$$

56. Let $f:[-a,a]\to\mathbb{R}$ be continuous and even. Show that

$$\int_{-a}^{a} f(x)\,dx = 2\int_0^a f(x)\,dx.$$

57. Let $f:[-1,1]\to\mathbb{R}$ be continuous. Evaluate

$$\int_{-\pi/2}^{\pi/2} x\cdot f(\cos(x))\,dx.$$

(See Exercise 55.)

58. Suppose that f is integrable on $[a,b]$. Use a substitution to show that

$$\int_a^b f(x)\,dx = \int_0^{b-a} f(b-x)\,dx.$$

59. Suppose that f is integrable on $[a,b]$. Use a substitution to show that

$$\int_a^b f(x)\,dx = \int_a^b f(a+b-x)\,dx.$$

60. Prove the identity

$$\sin(A)\cos(B) = \frac{\sin(A+B)+\sin(A-B)}{2}.$$

Use this identity to evaluate $\int \sin(\alpha x)\cos(\beta x)\,dx$.

61. Evaluate

$$\int \sec^2(x)\tan(x)\,dx$$

by using the change of variable $u = \tan(x)$. Then evaluate the integral by making the change of variable $v = \sec(x)$. Verify that the two answers that you obtain are equivalent.

62. Prove the following formula:

$$\int_a^b f(u)\,du = \frac{b-a}{2}\int_{-1}^{1} f\left(\frac{(b-a)x+a+b}{2}\right)dx.$$

This formula is an example of the principle of *range reduction,* which is used in numerical analysis.

63. Find constants A and B such that

$$\int \frac{2t^3+3t^2+2t-1}{(t^2+1)(t+1)^2}\,dt = \int\left(\frac{A\cdot t}{(t^2+1)} + \frac{B}{(t+1)^2}\right)dt.$$

Use this simplification to evaluate the integral on the left. (This example shows how beneficial Theorem 1a from Section 5.3 can be.)

Calculator/Computer Exercises

In Exercises 64–67, determine the upper limit of integration b.

64. $\int_0^b 2t\sec^2(t^2)\tan(t^2)dt = \int_0^1 u du$

65. $\int_0^b (4x^3+2x)\sin(x^4+x^2)\,dx = \int_0^\pi \sin(u)\,du$

66. $\int_0^b (1+e^x)\sqrt{x+e^x}\,dx = \int_1^9 \sqrt{u}\,du$

67. $\int_1^b (x+1)\exp(1/x - \ln(x))/x^2\,dx = \int_{1/4}^1 \exp(u)\,du$

5.6 More on the Calculation of Area

If a function f is continuous and nonnegative on an interval $[a, b]$, then $\int_a^b f(x)\,dx \geq 0$. This is clear because each of the summands in the Riemann sum $\sum_{j=0}^N f(s_j)\Delta x$ is nonnegative (since f is nonnegative). Similarly, if f is continuous and $f \leq 0$ on an interval $[c, d]$, then $\int_c^d f(x)\,dx \leq 0$. Again, this is justified by the fact that each summand in the Riemann sum $\sum_{j=0}^N f(s_j)\Delta x$ is nonpositive (since f is nonpositive).

In the first case, the integral represents area (as discussed in Section 5.1). The same ideas show that when $f \leq 0$, the integral represents the *negative* of the area, because the summands $f(s_j)\Delta x$ are negatives of the areas of the corresponding rectangles.

We obtain from these remarks the following guidelines for determining area.

Rules for Calculating Area

1. If $f(x) \geq 0$ for $x \in [a, b]$, then

$$\int_a^b f(x)\,dx$$

equals the area under the graph of f, above the x-axis, and between $x = a$ and $x = b$.

2. If $f(x) \leq 0$ for $x \in [a, b]$, then

$$\int_a^b f(x)\,dx$$

equals the *negative* of the area above the graph of f, below the x-axis, and between $x = a$ and $x = b$.

We may use these observations to solve some fairly sophisticated area problems.

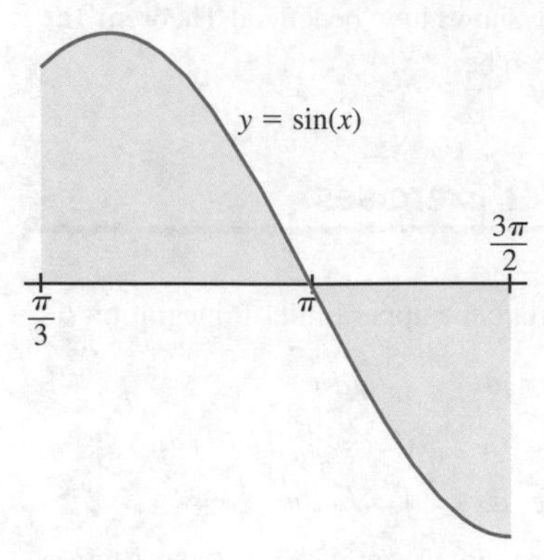

Figure 1

Example 1 What is the area bounded by the graph of $f(x) = \sin(x)$ and the x-axis between the limits $x = \pi/3$ and $x = 3\pi/2$?

Solution Look at Figure 1. Notice that $f(x) \geq 0$ for $x \in [\pi/3, \pi]$ and $f(x) \leq 0$ for $x \in [\pi, 3\pi/2]$. Thus, the (positive) area of the region lying *above* $[\pi/3, \pi]$ is

$$\int_{\pi/3}^{\pi} \sin(x)\,dx = -\cos(x)\Big|_{\pi/3}^{\pi} = -(-1) - \left(-\frac{1}{2}\right) = \frac{3}{2}.$$

On the other hand, the (positive) area of the region lying *below* $[\pi, 3\pi/2]$ is

$$-\int_{\pi}^{3\pi/2} \sin(x)\,dx = -(-\cos(x))\Big|_{\pi}^{3\pi/2} = 0 - (-1) = 1.$$

The total area is the sum of the two component areas, or $3/2 + 1 = 5/2$. ■

Theorem 1e from Section 5.3 suggests a way in which we can calculate areas (and integrals) for piecewise-defined functions.

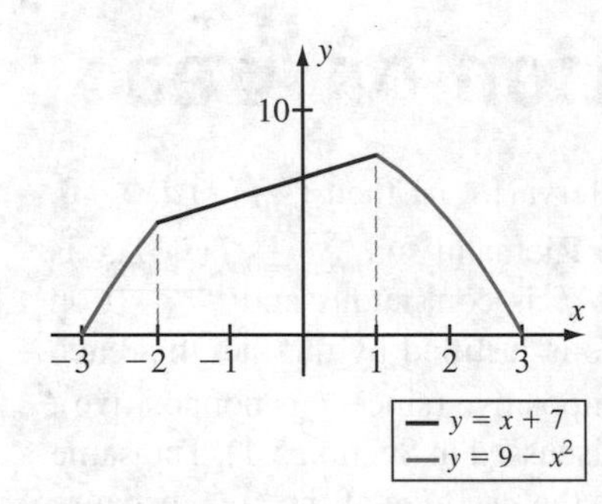

Figure 2

Example 2 Define

$$f(x) = \begin{cases} x + 7 & \text{if } -2 \leq x \leq 1 \\ 9 - x^2 & \text{if } x < -2 \text{ or } x > 1 \end{cases}.$$

What is the area between the graph of f and the x-axis?

Solution From the definition of f, we see that $f \geq 0$ for $-3 \leq x \leq 3$. (The graph of f over this interval appears in Figure 2.) We see that the problem naturally divides into three pieces:

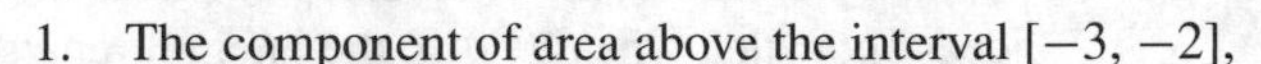

1. The component of area above the interval $[-3, -2]$,
2. The component of area above the interval $[-2, 1]$, and
3. The component of area above the interval $[1, 3]$.

We thus do three separate calculations.

1. When $x \in [-3, -2)$, we have $f(x) = 9 - x^2$ and
$$\int_{-3}^{-2} f(x)\,dx = \left(9x - \frac{x^3}{3}\right)\Bigg|_{x=-3}^{x=-2} = \left(-18 + \frac{8}{3}\right) - \left(-27 + \frac{27}{3}\right) = \frac{8}{3}.$$
2. When $x \in [-2, 1]$, we have $f(x) = x + 7$ and
$$\int_{-2}^{1} f(x)\,dx = \left(\frac{1}{2}x^2 + 7x\right)\Bigg|_{x=-2}^{x=1} = \left(\frac{1}{2} + 7\right) - \left(\frac{4}{2} - 14\right) = \frac{39}{2}.$$
3. When $x \in (1, 3]$, we have $f(x) = 9 - x^2$ and
$$\int_{1}^{3} f(x)\,dx = \left(9x - \frac{x^3}{3}\right)\Bigg|_{x=1}^{x=3} = \left(27 - \frac{27}{3}\right) - \left(9 - \frac{1}{3}\right) = \frac{28}{3}.$$

Thus, the total area between the graph of f and the x-axis is $8/3 + 39/2 + 28/3 = 63/2$. ■

The Area between Two Curves

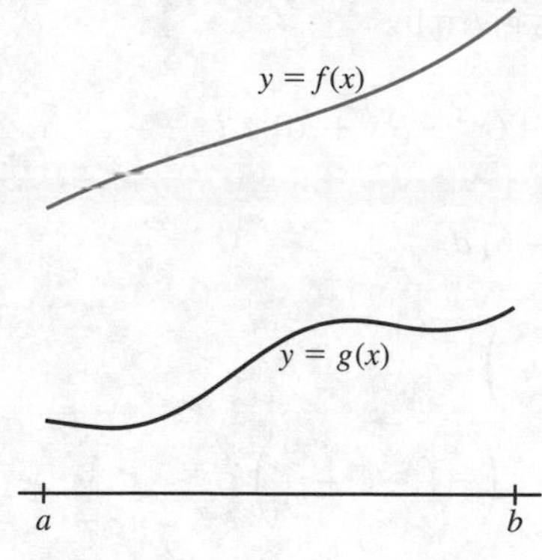

Figure 3

Examples 1 and 2 are special cases of the general problem of finding the area *between two curves*. Suppose f and g are continuous functions with domains that each contain the interval $[a, b]$. Suppose further that $f(x) \geq g(x)$ on this interval (see Figure 3). To estimate the area below the graph of f and above the graph of g on this interval, we let $\mathcal{P} = \{x_0, \ldots, x_N\}$ be a uniform partition of $[a, b]$. We select points $s_j \in I_j = [x_{j-1}, x_j]$ and erect rectangles with base I_j and height $f(s_j) - g(s_j)$, as in Figure 4. The area we wish to estimate is approximately given by the sum of the areas of these rectangles, or

$$\sum_{j=1}^{N} (f(s_j) - g(s_j)) \cdot \Delta x.$$

As Δx tends to 0, we expect that this sum will become a more accurate approximation to the desired area. However, the sum also happens to be a Riemann sum for the integral

$$\int_a^b (f(x) - g(x))\, dx.$$

We have therefore derived the following result, stated as a theorem.

Theorem 1 Let f and g be continuous functions on the interval $[a, b]$ and suppose that $f(x) \geq g(x)$ for all $x \in [a, b]$. The area under the graph of f and above the graph of g on the interval $[a, b]$ is given by

$$\int_a^b (f(x) - g(x))\, dx.$$

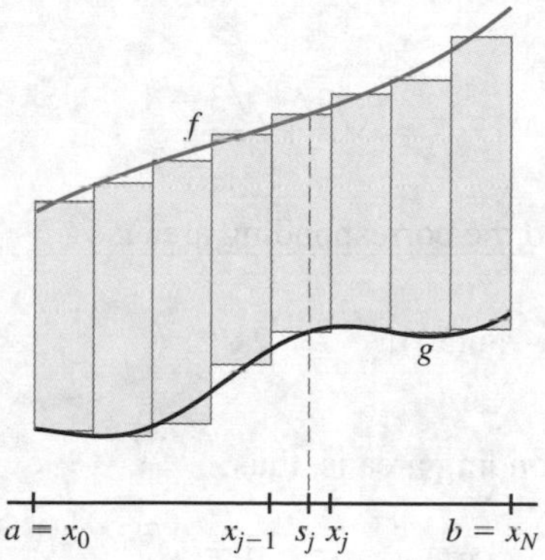

Figure 4

Example 3 Calculate the area A between the curves $f(x) = -x^2 + 6$ and $g(x) = 3x^2 - 8$ on the interval $[-1, 1]$.

Solution We see from the sketch in Figure 5, next page, that $f(x) \geq g(x)$ on this interval. According to Theorem 1, the desired area is given by

$$\int_{-1}^{1} (f(x) - g(x))\, dx = \int_{-1}^{1} ((-x^2 + 6) - (3x^2 - 8))\, dx = \int_{-1}^{1} (-4x^2 + 14)\, dx.$$

Therefore,

$$A = \left(-\frac{4}{3}x^3 + 14x \right) \bigg|_{-1}^{1} = \left(\left(-\frac{4}{3} \right) \cdot 1 + 14 \cdot 1 \right) - \left(\left(-\frac{4}{3} \right) \cdot (-1) + 14 \cdot (-1) \right) = \frac{76}{3}.$$

■

Example 4 Calculate the area between the parabolas $f(x) = -2x^2 + 4$ and $g(x) = x^2 - 9x + 10$.

Solution The region is shaded in Figure 6, next page. We have not specified an interval on which to work, but we can calculate the interval of integration: The parabolas intersect when $-2x^2 + 4 = x^2 - 9x + 10$, or $x^2 - 3x + 2 = 0$. The roots are $x = 1$ and $x = 2$, as can be seen either by factoring or by using the quadratic formula. On the

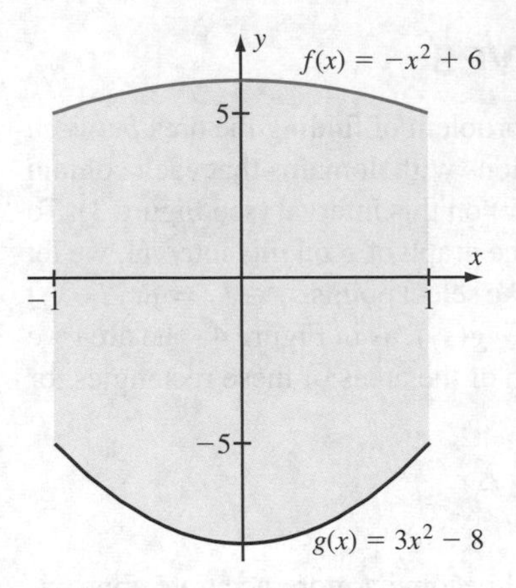

Figure 5

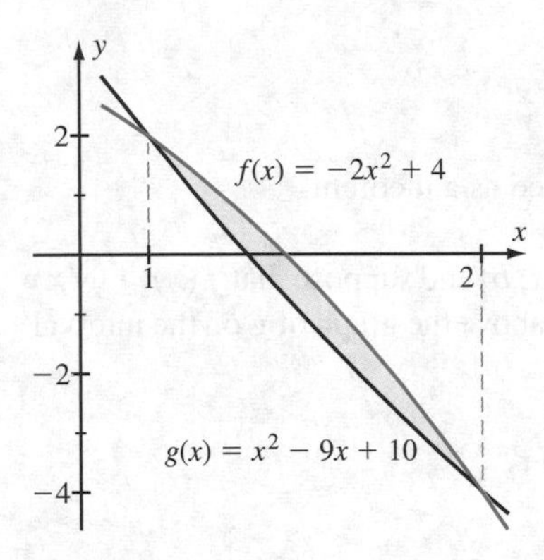

Figure 6

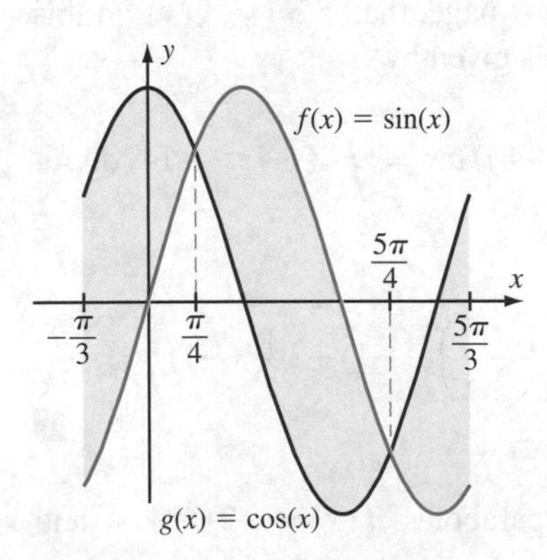

Figure 7

interval $[1, 2]$, $f(x) \geq g(x)$. Thus, the desired area is given by

$$\begin{aligned}
\int_1^2 (f(x) - g(x))\,dx &= \int_1^2 ((-2x^2 + 4) - (x^2 - 9x + 10))\,dx \\
&= \int_1^2 (-3x^2 + 9x - 6)\,dx \\
&= \left(-x^3 - 6x + \frac{9}{2}x^2\right)\Big|_{x=1}^{x=2} \\
&= (-8 - 12 + 18) - \left(-1 - 6 + \frac{9}{2}\right) \\
&= \frac{1}{2}.
\end{aligned}$$

■

Example 5 Calculate the area between the curves $f(x) = \sin(x)$ and $g(x) = \cos(x)$ on the interval $[-\pi/3, 5\pi/3]$.

Solution The graphs of f and g appear in Figure 7. Notice that $f(x) \geq g(x)$ at some points while $g(x) \geq f(x)$ at other points. We need to break up the interval $[-\pi/3, 5\pi/3]$ into subintervals on which either one or the other of these inequalities is true, but not both. Thus, we need to find the points of intersection of the graphs of f and g. Setting $\sin(x) = \cos(x)$, we see that the graphs intersect at the points $x = \pi/4$ and $x = 5\pi/4$ in the interval $[-\pi/3, 5\pi/3]$. We therefore make separate calculations of the area on $[-\pi/3, \pi/4]$, the area on $[\pi/4, 5\pi/4]$, and the area on $[5\pi/4, 5\pi/3]$. On $[-\pi/3, \pi/4]$, we have $\cos(x) \geq \sin(x)$ (as we can see from Figure 7), so the corresponding area is

$$\int_{-\pi/3}^{\pi/4} (\cos(x) - \sin(x))\,dx = (\sin(x) + \cos(x))\Big|_{-\pi/3}^{\pi/4} = \sqrt{2} + \frac{\sqrt{3} - 1}{2}.$$

On $[\pi/4, 5\pi/4]$, we see that $\sin(x) \geq \cos(x)$, and the corresponding area is

$$\int_{\pi/4}^{5\pi/4} (\sin(x) - \cos(x))\,dx = (-\cos(x) - \sin(x))\Big|_{\pi/4}^{5\pi/4} = 2\sqrt{2}.$$

On $[5\pi/4, 5\pi/3]$, $\cos(x) \geq \sin(x)$. The corresponding area is, thus,

$$\int_{5\pi/4}^{5\pi/3} (\cos(x) - \sin(x))\,dx = (\sin(x) + \cos(x))\Big|_{5\pi/4}^{5\pi/3} = \sqrt{2} + \frac{1 - \sqrt{3}}{2}.$$

Adding the three areas that we have computed gives the total area between the curves:

$$\left(\sqrt{2} + \frac{\sqrt{3} - 1}{2}\right) + (2\sqrt{2}) + \left(\sqrt{2} + \frac{1 - \sqrt{3}}{2}\right) = 4\sqrt{2}.$$

■

Reversing the Roles of the Axes

To calculate the area between certain pairs of curves, it is convenient to interchange the roles of the x-axis and the y-axis. Consider the following example.

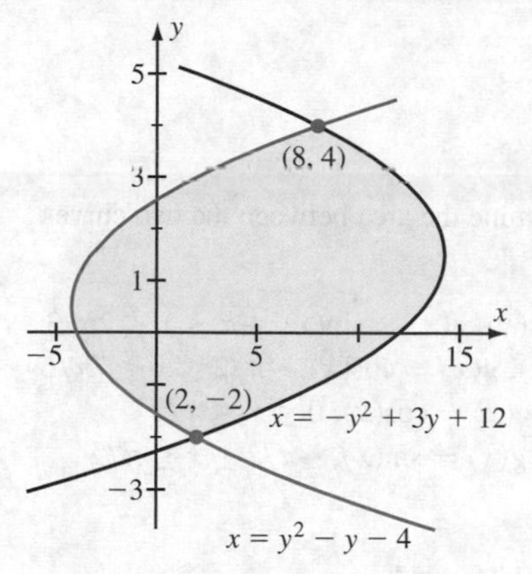

Figure 8

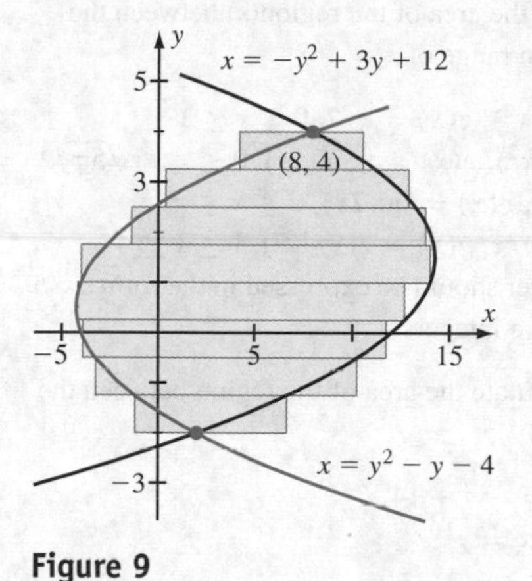

Figure 9

Example 6 Compute the area between the curves $x = y^2 - y - 4$ and $x = -y^2 + 3y + 12$.

Solution The curves in question are parabolas. By setting the expressions for x equal to each other, we obtain

$$y^2 - y - 4 = -y^2 + 3y + 12,$$

or

$$y^2 - 2y - 8 = 0.$$

The roots are $y = -2$ and $y = 4$. The parabolas intersect at $(2, -2)$ and $(8, 4)$. Figure 8 exhibits this information, as well as the area of the region we wish to calculate.

If we were to calculate the area of the region using an x-integral, then we would need to solve for the upper and lower branch of each parabola, which would make the algebra complicated. Instead, we think of x as a function of y and approximate the area by horizontal rectangles (see Figure 9). Reasoning as before, we are led to consider the integral

$$\int_{-2}^{4} ((-y^2 + 3y + 12) - (y^2 - y - 4))\, dy.$$

Notice that we subtract the expression describing the parabola on the *left* from that describing the parabola on the *right*—this corresponds to subtracting lesser y-coordinates from greater y-coordinates, so that we will be computing the *positive* area. Simplifying the integrand of the last integral and then integrating term by term, we obtain

$$\int_{-2}^{4} (-2y^2 + 4y + 16)\, dy = \left(-\frac{2}{3}y^3 + 2y^2 + 16y \right)\Bigg|_{y=-2}^{y=4}$$

$$= \left(\frac{-128}{3} + 32 + 64 \right) - \left(\frac{16}{3} + 8 - 32 \right) = 72$$

for the area between the two parabolas. ■

quickquiz

1. Why does the quantity $\int_a^b f(x)\, dx$ not always equal the area between the graph of f and the x-axis over the interval $[a, b]$?
2. What is the area between the x-axis and the curve $y = \sin(x)$ for $0 \le x \le 2\pi$?
3. What is the area between the y-axis and the curve $x = 1 - y^2$?
4. How do we calculate the area *between* two curves using a single integral (that is, what arithmetic operation is involved and why)?

EXERCISES

Problems for Practice

In Exercises 1–12, calculate the area between the curve and the x-axis on the given interval.

1. $f(x) = \cos(x)$, $I = [\pi/4, 2\pi/3]$
2. $g(x) = 3x^2 - 3x - 6$, $I = [-4, 4]$
3. $h(x) = 2x^2 - 8$, $I = [-5, 7]$
4. $f(x) = \cos(2x + \pi/2)$, $I = [0, 5\pi/4]$
5. $g(x) = \sin(x) + \cos(x)$, $I = [-\pi, 3\pi/2]$
6. $h(x) = x/(x^2+1)^2$, $I = [-1, 3]$
7. $f(x) = x \cdot (1 - x^2)^2$, $I = [-1/2, 1]$
8. $g(x) = \sin(x^{1/2})/x^{1/2}$, $I = [\pi^2/4, 9\pi^2/4]$
9. $h(x) = (\sin(\pi x))^2 \cdot \cos(\pi x)$, $I = [0, 3/2]$
10. $f(x) = (1 - x^2)^3 \cdot x$, $I = [-3, 5]$
11. $g(x) = \sin(3x)\cos(3x)$, $I = [-\pi/2, \pi/3]$
12. $h(x) = 12 - 9x - 3x^2$, $I = [-1, 4]$

In Exercises 13–18, find the area between the graph of the function and the x-axis on the given interval.

13. $f(x) = \begin{cases} -x^2 & \text{if } -3 \le x \le 1 \\ 2x - 3 & \text{if } 1 < x \le 4 \end{cases}$, $I = [-3, 4]$
14. $f(x) = \begin{cases} -x + 3 & \text{if } -2 \le x < 0 \\ -x^2 + 3x + 3 & \text{if } 0 \le x \le 3 \end{cases}$, $I = [-2, 3]$
15. $f(x) = \begin{cases} \sin(x) & \text{if } 0 \le x \le \pi/4 \\ \cos(x) & \text{if } \pi/4 < x \le \pi \end{cases}$, $I = [0, \pi]$
16. $f(x) = \begin{cases} x - 2 & \text{if } -1 \le x \le 2 \\ \sin(\pi x) & \text{if } 2 < x \le 4 \end{cases}$, $I = [-1, 4]$
17. $f(x) = |x - 4|$, $I = [-7, 6]$
18. $f(x) = |x^2 - 4|$, $I = [-2, 3]$

In Exercises 19–26, calculate the points of intersection of the two curves. Sketch both curves on a single set of axes. Find the area of the region(s) bounded by these curves.

19. $f(x) = x^2 + x + 1$, $g(x) = 2x^2 + 3x - 7$
20. $f(x) = x^2$, $g(x) = -2x^2 - 15x - 18$
21. $f(x) = x^2 + 5$, $g(x) = 2x^2 + 1$
22. $f(x) = x^2 + 5x + 9$, $g(x) = -x^2 + 19x - 15$
23. $f(x) = x^3 - 3x^2 - x + 4$, $g(x) = -3x + 4$
24. $f(x) = x^2 + x + 2$, $g(x) = -x^2 + x + 18$
25. $f(x) = -x^4 + x^2 + 16$, $g(x) = 2x^4 - 2x^2 - 20$
26. $f(x) = -x^2 + 4$, $g(x) = 3x$

In Exercises 27–30, determine the area between the two curves over the range of x.

27. $f(x) = -\sqrt{3}\cos(x)$, $g(x) = \sin(x)$, $-\pi \le x \le 2\pi/3$
28. $f(x) = \sqrt{3}\sin(x)$, $g(x) = \cos(x)$, $-\pi/2 \le x \le 7\pi/6$
29. $f(x) = \sin(2x)$, $g(x) = \cos(x)$, $0 \le x \le \pi/2$
30. $f(x) = \cos(2x)$, $g(x) = \sin(x)$, $-\pi/2 \le x \le \pi/2$

Further Theory and Practice

In Exercises 31–34, find the area of the region(s) between the two curves over the given range of x.

31. $f(x) = x/(1 + x^2)$, $g(x) = x/2$, $0 \le x \le 1$
32. $f(x) = x \cdot \cos(x^2)$, $g(x) = x\sin(x^2)$, $0 \le x \le \sqrt{5\pi/2}$
33. $f(x) = 2\sin(x)$, $g(x) = \sin(2x)$, $0 \le x \le \pi$
34. $f(x) = (x^3 - 8)/x$, $g(x) = 7(x - 2)$, $1 \le x \le 4$ (*Hint:* The answer should be expressed in the form $a\sqrt{b}$ where a and b are integers.)

In Exercises 35–38, calculate the area of the region between the pair of curves.

35. $x = y^2 + 6$, $x = -y^2 + 14$
36. $y = (x - 3)/2$, $x = y^2$
37. $x = y^2$, $x = y^3$
38. $x = y$, $x = y^4$

In Exercises 39–42, the integral $\int_a^b f(x)\,dx$ represents the area of a region in the xy-plane. Express the area of the region as an integral of the form $\int_c^d g(y)\,dy$. For example, the integral $\int_0^1 2x\,dx$ represents the area of the triangle with vertices (0, 0), (1, 0), and (1, 2). This area can also be represented as $\int_0^2 (1 - y/2)\,dy$. (See Figure 10, next page.)

39. $\int_0^{\sqrt{2}} \pi\,dx$
40. $\int_0^1 \sqrt{x}\,dx$
41. $\int_0^1 (e - \exp(x))\,dx$
42. $\int_0^2 (x^2 + 2x)\,dx$

In Exercises 43–46, the integral $\int_a^b (f_1(x) - f_2(x))\,dx$ represents the area of a region in the xy-plane that is enclosed by the graphs of f_1 and f_2. Sketch the graphs of f_1 and f_2 for $a \le x \le b$. Express the area of the region as an integral of the form $\int_c^d (g_1(y) - g_2(y))\,dy$. For example, the integral $\int_0^1 (x - x^2)\,dx$ represents the area of the shaded region in Figure 11, next page. This area can also be represented as $\int_0^1 (\sqrt{y} - y)\,dy$.

43. $\int_0^4 \left(\sqrt{x} - x/2\right)dx$
44. $\int_0^1 \left(2\sqrt{x} - 2x^3\right)dx$
45. $\int_{-3}^3 (3 - |x|)\,dx$
46. $\int_{-1}^1 (e - \exp(|x|))\,dx$

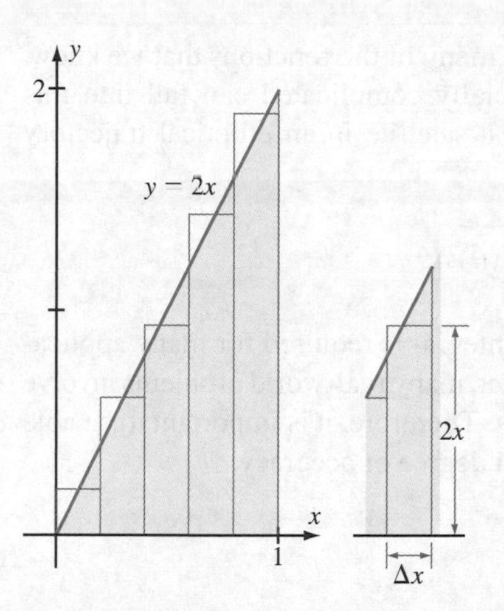

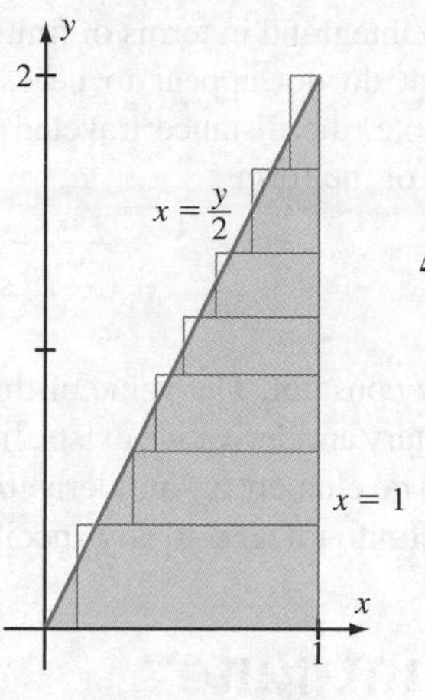

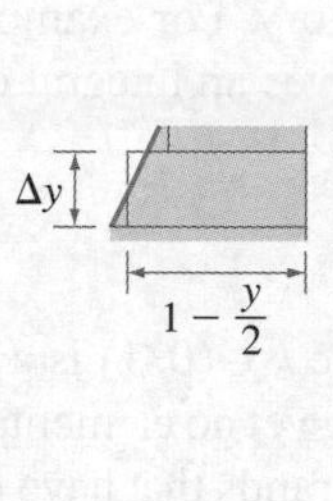

Figure 10

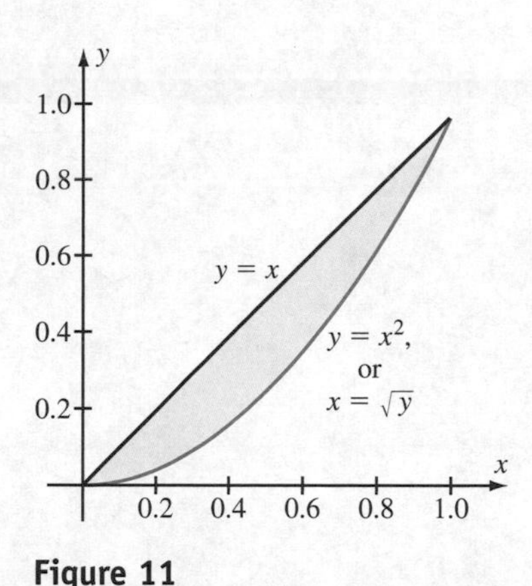

Figure 11

In Exercises 47–50, express the area of the region as a sum of integrals of the form $\int_a^b f(x)\,dx$.

47. The triangle with vertices (1, 0), (3, 0), (2, 1)
48. The triangle with vertices (1, 0), (3, 1), (2, 2)
49. The region enclosed by $y = |x|$ and $y = 2 - x^2$
50. The larger of the two pieces of the disk $x^2 + y^2 \leq 1$ that is formed when the disk is cut by the line $y = x + 1$

In Exercises 51–54, express the area of the region as an integral of the form $\int_c^d g(y)\,dy$ or as a sum of such integrals.

51. The region of Exercise 47
52. The region of Exercise 48
53. The region of Exercise 49
54. The region of Exercise 50
55. Find the area between the curves

$$x - 4y = 1$$

and

$$x^2 + 2xy + y^2 - 2x + 3y = 0.$$

Calculator/Computer Exercises

In Exercises 56–62, plot the graphs as indicated. Then find the abscissas a and b of the two points of intersection. Find the area bounded by the two graphs for $a \leq x \leq b$.

56. $y = x^3 + x,\ y = 6 - x^4,\ -2.2 \leq x \leq 1.4$
57. $y = 1 + 2x,\ y = \exp(x),\ -0.1 \leq x \leq 1.3$
58. $y = \ln(x)/x,\ y = (x - 1)/8,\ 0.9 \leq x \leq 4$
59. $y = \sin(x),\ y = 1 - 3\sin(x),\ 0 \leq x \leq \pi$
60. $y = |x|,\ y = 1 - x^2 - x^3,\ -1 \leq x \leq 0.7$
61. $y = \sqrt{x},\ y = x + x^2,\ 0 \leq x \leq 0.5$
62. $y = \sec^2(x),\ y = x + 2,\ -1 \leq x \leq 1$
63. Approximate the smallest positive solution b to $\sin^2(x) = \cos(x^2)$. Approximate the area between $f(x) = \cos(x^2)$ and $g(x) = \sin^2(x)$ over the interval $[0, b]$.

5.7 Numerical Techniques of Integration

Although the Fundamental Theorem of Calculus provides a powerful rule for evaluating Riemann integrals, many important definite integrals cannot be calculated *exactly*. The inability to calculate an integral exactly occurs when it is impossible to express the antiderivative of the integrand in terms of finitely many of the functions that we know. Even integrands that do not appear to be especially complicated can fall into this category. For example, the distance traveled by a satellite in an elliptical trajectory involves an integral of the form

$$\int_0^{\pi/2} \sqrt{1 - k^2 \sin^2(\theta)}\, d\theta$$

where $k \in (0, 1)$ is a constant. The value of this integral is required for many applications, yet no elementary antiderivative exists. In fact, many real-world problems involve integrands that have no elementary antiderivatives. Therefore, it is important to be able to approximate a definite integral to any specified degree of accuracy.

The Midpoint Rule

Let f be a continuous function on the interval $[a, b]$ and let N denote a positive integer that is to be chosen. To approximate the integral $\int_a^b f(x)\, dx$, we use the uniform partition

$$a = x_0 < x_1 < x_2 < \cdots < x_N = b,$$

which divides the interval $[a, b]$ into N subintervals of equal length

$$\Delta x = \frac{b - a}{N}.$$

Since the definite integral $\int_a^b f(x)\, dx$ is defined as the limit $\lim_{N \to \infty} \sum_{j=1}^N f(s_j)\Delta x$ of Riemann sums, it is natural to choose the points $s_j \in I_j = [x_{j-1}, x_j]$ in a convenient manner and then use the Riemann sum $\sum_{j=1}^N f(s_j)\Delta x$ to approximate the integral. The candidates for s_j that readily come to mind are the endpoints x_{j-1} and x_j, as well as the midpoint

$$\bar{x}_j = \frac{x_{j-1} + x_j}{2} = a + \left(j - \frac{1}{2}\right)\Delta x.$$

Because a continuous function typically either increases over an entire *small* interval or decreases over the interval, the midpoint is the preferred choice (as indicated by the shaded area of Figure 1, next page). The resulting Riemann sum

$$\mathcal{M}_N = \Delta x \cdot (f(\bar{x}_1) + f(\bar{x}_2) + \cdots + f(\bar{x}_N))$$

is called the *midpoint approximation* of order N. In general, the approximation becomes more accurate as N increases. However, we do not want to choose N so large that the calculation of $\mathcal{M}_N$ becomes impractical. We may therefore state the basic approximation problem in this way: How do we determine the smallest value of N that allows us to be sure that $\mathcal{M}_N$ is an acceptable approximation? The following theorem states an error estimate that is the key to solving this problem for the midpoint approximation.

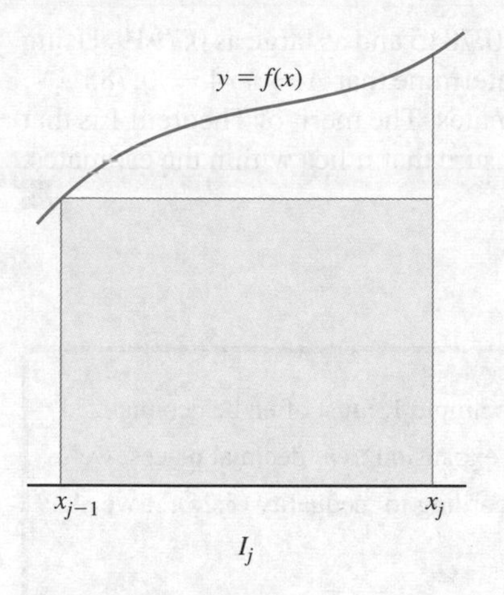

Figure 1a
Left endpoint approximation

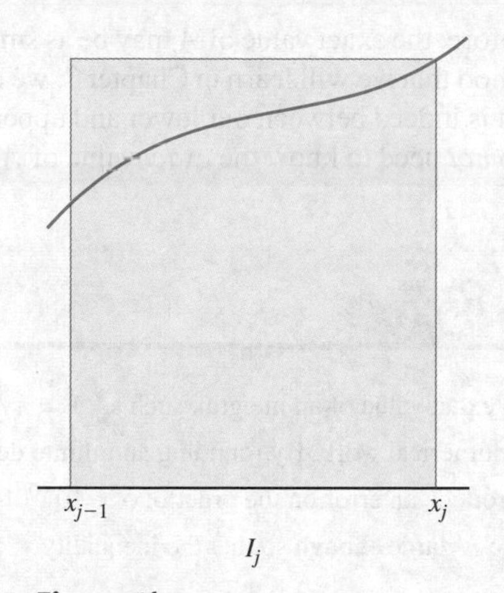

Figure 1b
Right endpoint approximation

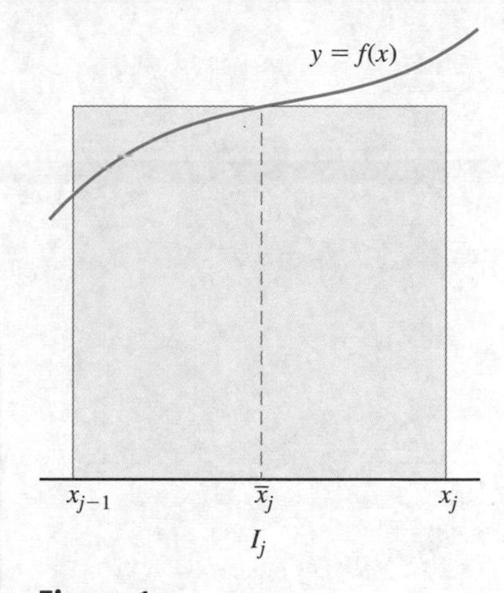

Figure 1c
Midpoint approximation

Theorem 1 ***Midpoint Rule*** Let f be a continuous function on the interval $[a, b]$. If C is a constant such that $|f''(x)| \leq C$ for $a \leq x \leq b$, then

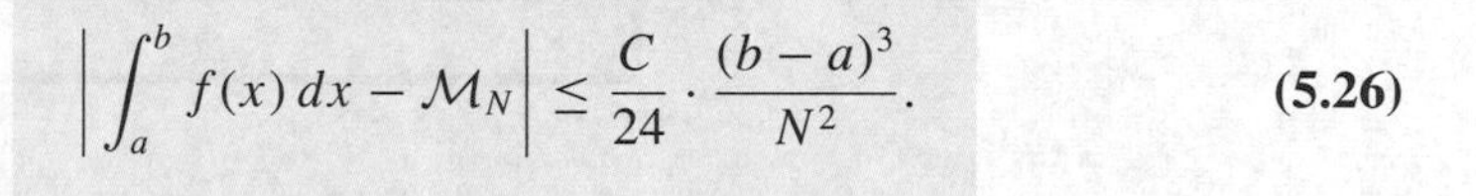

$$\left| \int_a^b f(x)\,dx - \mathcal{M}_N \right| \leq \frac{C}{24} \cdot \frac{(b-a)^3}{N^2}. \tag{5.26}$$

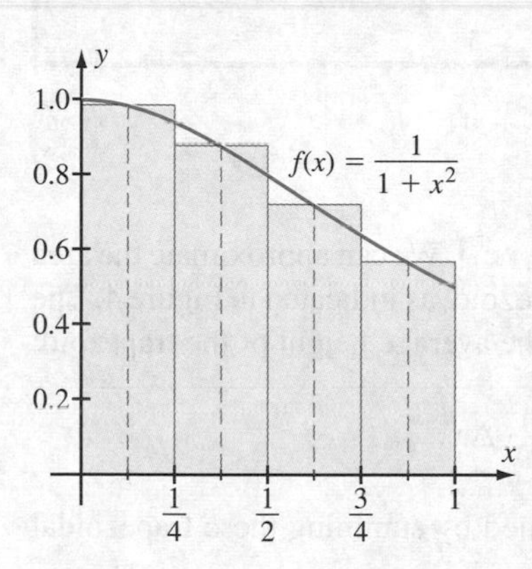

Figure 2

Example 1 Let $f(x) = 1/(1+x^2)$. Estimate $A = \int_0^1 f(x)\,dx$ using the Midpoint Rule with $N = 4$. Based on this approximation, how small might the exact value of A be? How large?

Solution The points $\{0, 1/4, 1/2, 3/4, 1\}$ partition the interval $[0, 1]$ into four subintervals of equal length ($\Delta x = 1/4$). The midpoints of the subintervals are $1/8$, $3/8$, $5/8$, and $7/8$. The Midpoint Rule tells us that A is approximately equal to

$$\mathcal{M}_4 = \frac{1}{4} \cdot \left(\frac{1}{1+(1/8)^2} + \frac{1}{1+(3/8)^2} + \frac{1}{1+(5/8)^2} + \frac{1}{1+(7/8)^2} \right) \approx 0.7867.$$

The number $\mathcal{M}_4$ represents the area of the shaded rectangles in Figure 2. To find out what the exact value of A might be, we calculate that $f'(x) = -2x/(1+x^2)^2$ and

$$f''(x) = \frac{6x^2 - 2}{(1+x^2)^3}.$$

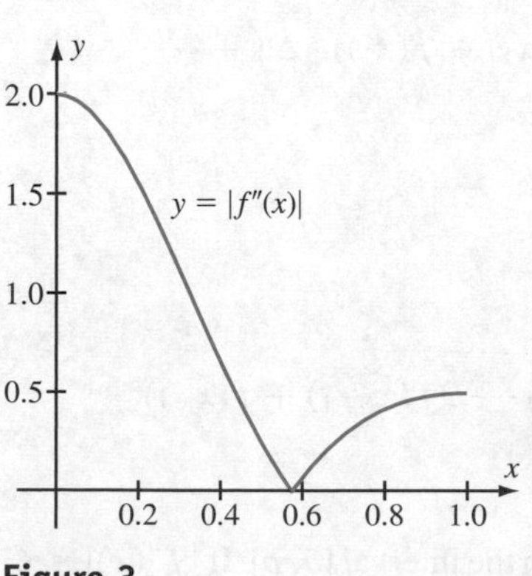

The graph of $|f''|$ in Figure 3 reveals that the maximum value of $|f''(x)|$ for $0 \leq x \leq 1$ occurs at $x = 0$ and is equal to 2. We may therefore use $C = 2$ in inequality (5.26), which tells us that the approximation $\mathcal{M}_4$ differs from A by no more than

$$\frac{C \cdot (b-a)^3}{24N^2} = \frac{2 \cdot 1^3}{24 \cdot 4^2} \approx 0.0052.$$

Figure 3
The plot of $y = |f''(x)|$ where $f(x) = 1/(1+x^2)$

We conclude that

$$\underbrace{0.7867}_{\text{Midpoint approximation}} - \underbrace{0.0052}_{\text{Error estimate}} \leq A \leq \underbrace{0.7867}_{\text{Midpoint approximation}} + \underbrace{0.0052}_{\text{Error estimate}}.$$

Therefore, the exact value of A may be as small as 0.7815 and as large as 0.7919. Using a method that we will learn in Chapter 7, we can determine that $A = \pi/4 = 0.785\ldots$, which is indeed between our lower and upper estimates. The merit of Theorem 1 is that we do *not* need to know the *exact* value of A to be sure that it lies within the estimates. ■

inSIGHT

An exact value of an integral, such as $A = \pi/4$ in Example 1, must often be decimalized in numerical work. By rounding an infinite decimal expansion to m decimal places, we introduce an error on the order of $5 \times 10^{-(m+1)}$. According to inequality (5.26), if we take N large enough so that the inequality

$$\frac{C \cdot (b-a)^3}{24N^2} < 5 \times 10^{-(m+1)}$$

holds, then the midpoint approximation generates an error less than $5 \times 10^{-(m+1)}$. In other words, the midpoint approximation allows us to evaluate an indefinite integral with as much precision as we would be able to obtain from the decimalization of an exact answer.

The Trapezoidal Rule

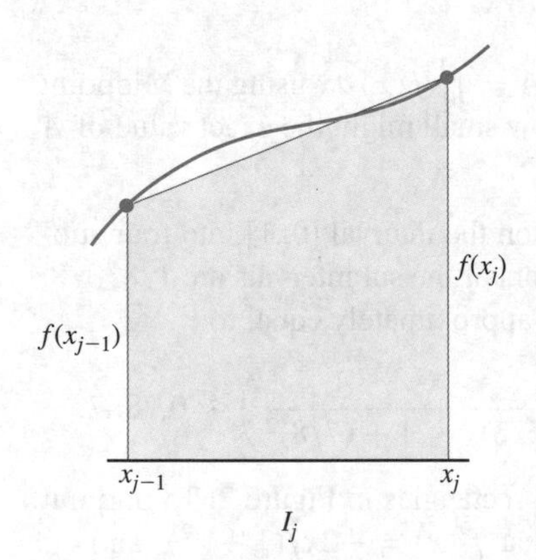

Figure 4

Suppose f is positive over the subinterval $I_j = [x_{j-1}, x_j]$. We can approximate the area under the graph of f and over I_j by the area of a trapezoid, as indicated in Figure 4. The trapezoidal area is the width Δx of the base times the average height of the trapezoid:

$$A_j = \frac{f(x_{j-1}) + f(x_j)}{2} \cdot \Delta x.$$

The order N *trapezoidal approximation* $\mathcal{T}_N$ is defined by summing these trapezoidal areas:

$$\begin{aligned}\mathcal{T}_N &= A_1 + A_2 + \cdots + A_N \\ &= \frac{1}{2} \cdot (f(x_0) + f(x_1)) \cdot \Delta x + \frac{1}{2} \cdot (f(x_1) + f(x_2)) \cdot \Delta x + \cdots \\ &\quad + \frac{1}{2} \cdot (f(x_{N-1}) + f(x_N)) \cdot \Delta x.\end{aligned}$$

After combining terms, we obtain the formula

$$\mathcal{T}_N = \frac{\Delta x}{2} \cdot (f(x_0) + 2f(x_1) + 2f(x_2) + \cdots + 2f(x_{N-1}) + f(x_N)).$$

Theorem 2 ***Trapezoidal Rule*** Let f be a continuous function on the interval $[a, b]$. If $|f''(x)| \le C$ for all x in the interval $[a, b]$, then the order N trapezoidal approximation $\mathcal{T}_N$ is accurate to within $C \cdot (b-a)^3/(12N^2)$. In other words,

$$\left| \int_a^b f(x)\,dx - \mathcal{T}_N \right| \leq \frac{C}{12} \cdot \frac{(b-a)^3}{N^2}. \tag{5.27}$$

It is instructive to compare the error estimates for the midpoint approximation and the trapezoidal approximation:

$$\frac{C \cdot (b-a)^3}{24N^2} \qquad \text{Midpoint Rule}$$

$$\frac{C \cdot (b-a)^3}{12N^2}. \qquad \text{Trapezoidal Rule}$$

As its larger denominator suggests, the Midpoint Rule is usually more accurate than the Trapezoidal Rule. Even so, the Trapezoidal Rule can be more suitable for analyzing discrete data. We illustrate this point with an example from economics.

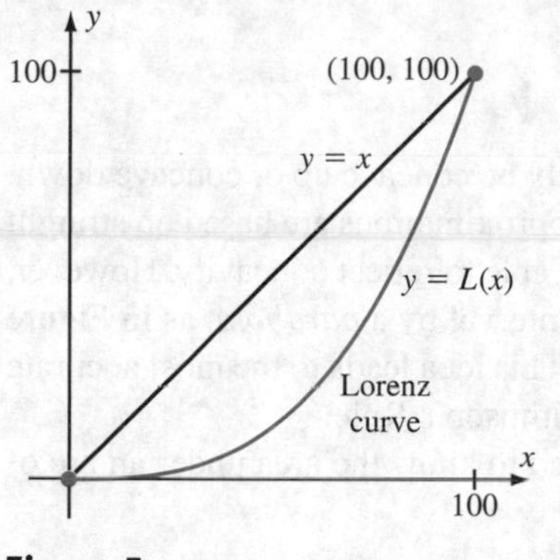

Figure 5

Example 2 For each x between 0 and 100, let $L(x)$ denote the percentage of a nation's total annual income that is earned by the lowest-earning $x\%$ of its inhabitants during a specified year. In 1990, for example, 40% of Americans had income less than or equal to \$29,000. The income of these individuals amounted to 15.4% of all personal income in the United States. Therefore, $L(40) = 15.4$ for the United States in 1990. Economists call the increasing function $L : [0, 100] \to [0, 100]$ a *Lorenz function.* The graph of the equation $y = L(x)$ is called a *Lorenz curve*. Notice that $L(0) = 0$ and $L(100) = 100$. If all personal income were equally distributed, then the Lorenz curve would be $y = x$. In actuality, the Lorenz curve lies below the line $y = x$ for $0 < x < 100$ (see Figure 5). The area A of the region below the Lorenz curve is used to measure the disparity of income distribution. The number $G = 1 - A/5000$ is called the *coefficient of inequality,* or the *Gini coefficient;* it is a number between 0 and 1. If the Lorenz curve is $y = x$ (perfectly equal income distribution), then $A = 5000$ and $G = 0$. As Figure 6 shows, when income distribution becomes increasingly unequal, the Lorenz curve approaches the bottom $(y = 0)$ and right side $(x = 100)$ of the triangle, the area A approaches 0, and the coefficient of inequality G approaches 1. Use the tabulated data to estimate the coefficient of inequality for the United States in 1990.

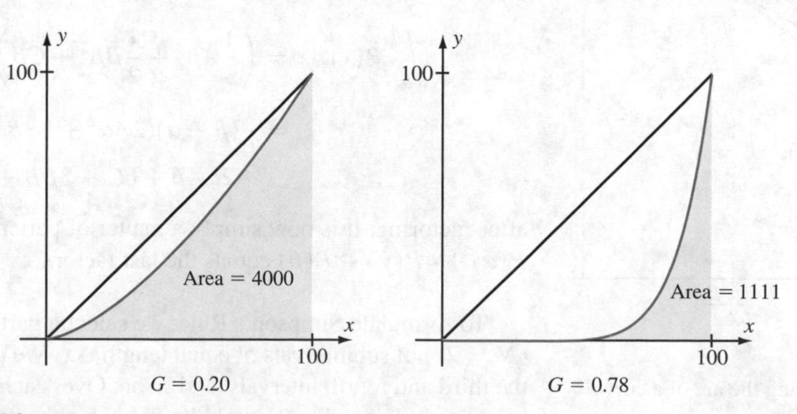

Figure 6

	Lowest Fifth	Second Fifth	Third Fifth	Fourth Fifth	Highest Fifth
Families	4.6	10.8	16.6	23.8	44.2

Percentage Distribution of Total Income 1990
(Source: U.S. Bureau of the Census)

inSIGHT

Example 2 is typical of many discrete data problems in which the Midpoint Rule does not permit effective use of the data. In Example 2, the midpoints of the partitioning subintervals are $\bar{x}_1 = 10$, $\bar{x}_2 = 30, \ldots, \bar{x}_5 = 90$. Notice that the values $L(\bar{x}_j)$ are not available. Yet these are the very values of L that we would need in order to apply the Midpoint Rule.

Solution We know that $L(0) = 0$. The table tells us that $L(20) = 4.6$, $L(40) = 4.6 + 10.8 = 15.4$, $L(60) = 15.4 + 16.6 = 32.0$, $L(80) = 32.0 + 23.8 = 55.8$, and $L(100) = 55.8 + 44.2 = 100$. We therefore use the partition $\{0, 20, 40, 60, 80, 100\}$ of $[0, 100]$ with $N = 5$ and $\Delta x = 20$. According to the Trapezoidal Rule,

$$\int_0^{100} L(x)\,dx \approx \frac{20}{2} \cdot (0 + 2 \cdot 4.6 + 2 \cdot 15.4 + 2 \cdot 32.0 + 2 \cdot 55.8 + 100) = 3156.0.$$

Therefore, the coefficient of inequality is approximately $1 - 3156/5000 \approx 0.3688$. ■

Simpson's Rule

The graph of f over a *small* subinterval will typically be concave up or concave down. Because the Trapezoidal Rule and Midpoint Rule approximations are based on straight line approximations to the graph of f, neither rule is able to reflect concavity. However, if we approximate the graph of f over a small subinterval by a *parabola,* as in Figure 7, then we can take into account the concavity of f. This idea leads to the most accurate of the approximation rules that we will discuss—Simpson's Rule.

To derive a formula for Simpson's Rule, we need to know the area under an arc of a parabola.

Theorem 3 If $P(x) = Ax^2 + Bx + C$ and if $I = [\alpha, \beta]$ is an interval with midpoint γ, then

$$\int_\alpha^\beta P(x)\,dx = \frac{\beta - \alpha}{6} \cdot (P(\alpha) + 4P(\gamma) + P(\beta)).$$

This formula is used only in the derivation of Simpson's Rule and need not be memorized.

Proof We calculate

$$\int_\alpha^\beta P(x)\,dx = \left(\frac{1}{3}A\beta^3 + \frac{1}{2}B\beta^2 + C\beta\right) - \left(\frac{1}{3}A\alpha^3 + \frac{1}{2}B\alpha^2 + C\alpha\right)$$
$$= \frac{1}{6}(\beta - \alpha)(2A\alpha^2 + 3\alpha B + 2\alpha A\beta + 6C + 3\beta B + 2A\beta^2),$$

after factoring. It is now simply a matter of patient algebraic calculation to show that $P(\alpha) + 4P(\gamma) + P(\beta)$ equals the last factor. ■

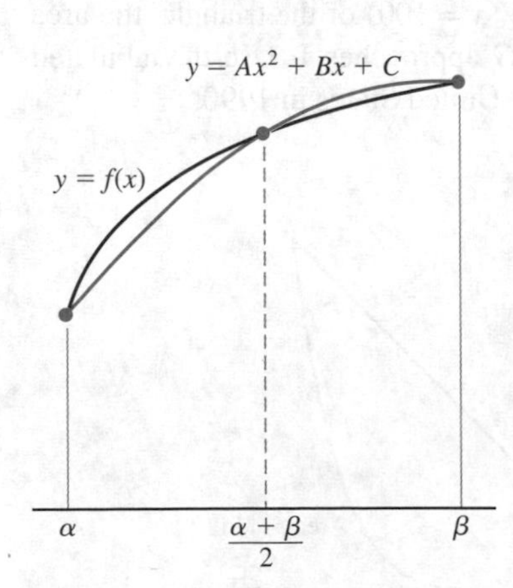

Figure 7
Approximating the arc of a curve with a parabolic arc

To formulate Simpson's Rule, we select a partition of $[a, b]$ with an *even* number ($N = 2\ell$) of subintervals of equal length Δx. We pair up the first and second intervals, the third and fourth intervals, and so on. Over each pair of intervals, we approximate f by a parabola passing through the points on the graph of f corresponding to the three

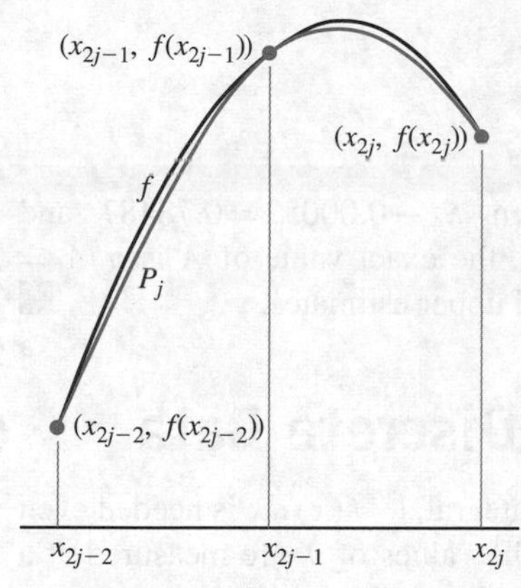

Figure 8

endpoints of the intervals. (Since the equation of a parabola $y = Ax^2 + Bx + C$ has three coefficients, a parabola is uniquely determined by any three points through which it passes.) A typical pair of intervals is

$$[x_{2j-2}, x_{2j-1}] \quad \text{and} \quad [x_{2j-1}, x_{2j}] \quad (j = 1, \ldots, \ell).$$

The approximating parabola P_j passes through the three points $(x_{2j-2}, f(x_{2j-2}))$, $(x_{2j-1}, f(x_{2j-1}))$, and $(x_{2j}, f(x_{2j}))$ (as in Figure 8). By applying Theorem 3, with $\alpha = x_{2j-2}$, $\gamma = x_{2j-1}$, and $\beta = x_{2j}$, we see that the integral of the parabola over the interval $[x_{2j-2}, x_{2j}]$ is

$$\frac{x_{2j} - x_{2j-2}}{6}(P_j(x_{2j-2}) + 4P_j(x_{2j-1}) + P_j(x_{2j}))$$
$$= \frac{2\Delta x}{6}(f(x_{2j-2}) + 4f(x_{2j-1}) + f(x_{2j})).$$

Finally, we add the integrals, from $j = 1$ to $j = \ell$, to obtain Simpson's approximation of $\int_a^b f(x)\,dx$:

$$\mathcal{S}_N = \frac{\Delta x}{3} \cdot (f(x_0) + 4f(x_1) + 2f(x_2) + 4f(x_3) + \cdots + 2f(x_{N-2}) + 4f(x_{N-1}) + f(x_N)).$$

Theorem 4 ***Simpson's Rule*** Let f be continuous on the interval $[a, b]$. Let N be a positive *even* integer. If C is any number such that $|f^{(4)}(x)| \le C$ for $a \le x \le b$, then

$$\left| \int_a^b f(x)\,dx - \mathcal{S}_N \right| \le \frac{C}{180} \cdot \frac{(b-a)^5}{N^4}. \tag{5.28}$$

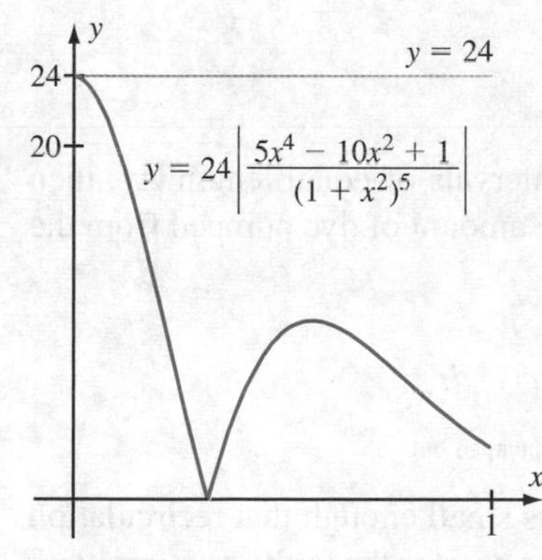

Figure 9

Example 3 Let $f(x) = 1/(1 + x^2)$. Estimate $A = \int_0^1 f(x)\,dx$ using Simpson's Rule with $N = 4$. In what interval of real numbers does A lie? (In Example 1, we answered this question using the Midpoint Rule.)

Solution There are $N = 4$ subintervals for the given partition. Since there is an even number of subintervals, we can apply Simpson's Rule. We have $f(x_0) = 1$, $f(x_1) = 1/(1 + (1/4)^2) = 16/17$, $f(x_2) = 1/(1 + (1/2)^2) = 4/5$, $f(x_3) = 1/(1 + (3/4)^2) = 16/25$, and $f(x_4) = 1/2$. We estimate the integral using Simpson's Rule:

$$\int_0^1 \frac{1}{1 + x^2}\,dx \approx \mathcal{S}_4 = \frac{(1-0)/4}{3}\left(1 + 4 \cdot \frac{16}{17} + 2 \cdot \frac{4}{5} + 4 \cdot \frac{16}{25} + \frac{1}{2}\right) = \frac{0.25}{3} \cdot \frac{8011}{850} \approx 0.78539.$$

With successive applications of the Quotient Rule, we calculate that

$$f^{(4)}(x) = 24 \cdot \frac{5x^4 - 10x^2 + 1}{(1 + x^2)^5}.$$

Using the methods from Section 4.3, we can determine that, for x in $[0, 1]$, $|f^{(4)}(x)|$ attains a maximum value of 24. Plotting $|f^{(4)}|$ also reveals this bound (see Figure 9). Taking $C = 24$ in inequality (5.28) shows that Simpson's Rule with $N = 4$ is accurate

> It is worth noting that the actual error in Example 3 is only $\pi/4 - 0.78539216 = 6.0034\ldots \times 10^{-6}$, a number that is significantly smaller than the upper bound 0.00052 that is furnished by inequality (5.28). In fact, each error estimate given in this section represents a worst-case error. The actual error is often *much* smaller.

to within

$$\frac{24 \cdot 1^5}{180 \cdot 4^4} \approx 0.00052.$$

The exact value of A is therefore between $S_4 - 0.00052 = 0.78487$ and $S_4 + 0.00052 = 0.78591$. As noted in Example 2, the exact value of A is $\pi/4 = 0.78539\ldots$, which indeed is between our lower and upper estimates. ■

Using Simpson's Rule with Discrete Data

In many practical instances, the value of a definite integral $\int_a^b f(x)\,dx$ is needed even though a formula for $f(x)$ is not known. Instead, the values of f are measured at a discrete set of points. Under these circumstances, the integral must be approximated. In Example 2, we considered a discrete data problem that arises in economics. The following example from physiology is also typical.

The rate r at which blood is pumped from the heart is called *cardiac output*. The units of cardiac output are volume/time (such as liters per minute). In the *Impulse Injection Method* of measuring cardiac output, a mass M of dye (or radioactive isotope) is injected into a pulmonary vein. Let $t = 0$ correspond to the time of injection and $t = T$ correspond to the time required to pump all the dye from the heart. The concentration $c(t)$ of dye (in units of mass/volume) that is pumped from the heart into the aorta is monitored. The mass of dye pumped from the heart in the time interval $[t, t + \Delta t]$ is approximately

$$\underbrace{\underset{\frac{\text{mass of dye}}{\text{volume}}}{c(t)} \cdot \underbrace{\underset{\frac{\text{volume}}{\text{time}}}{r} \cdot \underset{\text{time}}{\Delta t}}_{\text{Volume}}}_{\text{Mass of dye}}.$$

If we partition the time interval $[0, T]$ into N subintervals of equal length Δt, then the Riemann sum $\sum_{j=1}^{N} c(t_{j-1}) r\,\Delta t$ approximates the amount of dye pumped from the heart. Letting $\Delta t \to 0$, we obtain

$$\underset{\text{Mass of dye injected}}{M} = \underset{\text{Mass of dye pumped out}}{\int_0^T r \cdot c(t)\,dt}$$

In obtaining this formula, we have assumed that M is small enough that recirculation does not occur before time T. (Blood can travel from the heart to the farthest extremity of the circulatory system and back in as little as 12 seconds.) Cardiac output r is given by

$$r = \frac{M}{\int_0^T c(t)\,dt}.$$

Example 4 A doctor injects 3 mg of dye into a vein near the patient's heart. Based on the aortic concentration readings listed in the table, estimate the patient's cardiac output.

t (in seconds)	0	1	2	3	4	5	6	7	8
$c(t)$ (in milligrams/liter)	0	0.54	2.7	6.6	8.9	6.4	3.1	1.2	0

Solution Since the partition has nine equally spaced points forming an even number ($N = 8$) of equal length subintervals, Simpson's Rule is applicable:

$$\int_0^8 c(t)\,dt \approx \frac{8-0}{3\cdot 8}(1(0)+4(0.54)+2(2.7)+4(6.6)+2(8.9)+4(6.4)+2(3.1)+4(1.2)+1(0))$$
$$\approx 29.453.$$

Therefore, r is approximately $3/29.453$ L/s, or $180/29.453 = 6.11$ L/min. ■

quickquiz

1. Describe the geometric algorithm behind the Midpoint Rule.
2. Describe the geometric algorithm behind the Trapezoidal Rule.
3. Describe the geometric algorithm behind Simpson's Rule.
4. In Simpson's Rule, how does the order of accuracy depend on the number N of subintervals?

EXERCISES

Problems for Practice

In Exercises 1–6, apply the Midpoint Rule, the Trapezoidal Rule, and then Simpson's Rule for approximating the integral of the function f on the interval I with a partition of N equal intervals. Evaluate each approximation.

1. $f(x) = x^2$, $I = [0, 3]$, $N = 2$
2. $f(x) = x^{1/2}$, $I = [1, 4]$, $N = 2$
3. $f(x) = \sin(x)$, $I = [\pi, 2\pi]$, $N = 4$
4. $f(x) = \cos(x)$, $I = [0, \pi]$, $N = 4$
5. $f(x) = \sqrt{1+x}$, $I = [3, 8]$, $N = 4$
6. $1/x$, $I = [1, 2]$, $N = 4$

For Exercises 7–12, calculate the integrals exactly. Compare the exact answer with the approximate answers you obtained in Exercises 1–6. Verify that the errors are within the ranges that are guaranteed by estimates (5.26), (5.27), and (5.28).

7. The integral in Exercise 1
8. The integral in Exercise 2
9. The integral in Exercise 3
10. The integral in Exercise 4
11. The integral in Exercise 5
12. The integral in Exercise 6

Income data for three countries are tabulated below. In each table, the variables x and y have the same meaning as in Example 2. In Exercises 13–16, use the specified approximation method to calculate the coefficient of inequality for the indicated country.

x	20	40	60	80
y	5	20	30	55

Table 1 Income Data, Country A

x	25	50	75
y	15	25	40

Table 2 Income Data, Country B

x	10	20	30	40	50	60	70	80	90
y	4	8	14	22	32	42	56	70	82

Table 3 Income Data, Country C

13. Country A, Trapezoidal Rule
14. Country C, Trapezoidal Rule

15. Country B, Simpson's Rule
16. Country C, Simpson's Rule

Further Theory and Practice

17. How large should N be for the Trapezoidal Rule to approximate $\int_1^9 \exp(-x/2)\,dx$ to within 10^{-4}? How about for Simpson's Rule?
18. For each of the following functions on the interval $[-2, 2]$, give an estimate for how large N must be to guarantee that Simpson's Rule is accurate to within 10^{-3}.
    a. $f(x) = \sin(2x)$
    b. $g(x) = \cos(x^2)$

In Exercises 19 and 20, use Simpson's Rule to estimate cardiac output based on the readings (with t in seconds and $c(t)$ in mg/L) taken after the injection of 5 mg of dye.

19.

t	0	1	2	3	4	5	6	7	8
c	0	1.92	5.74	9.00	10.24	9.02	5.78	2.00	0

20.

t	0	1	2	3	4	5	6	7	8	9	10
c	0	3.8	6.6	8.6	9.8	10	9.4	8.2	6.1	3.1	0

21. Let p be a polynomial of degree three or less. Show that the Simpson's Rule approximation of

$$\int_a^b p(x)\,dx$$

is exact. (*Hint:* Consider the estimate for the error in using Simpson's Rule.)

22. Theorem 3 gives the formula

$$\int_\alpha^\beta P(x)\,dx = \frac{\beta-\alpha}{6}\left(P(\alpha) + 4P\left(\frac{\alpha+\beta}{2}\right) + P(\beta)\right)$$

for $P(x) = Ax^2 + Bx + C$. Explain why the same formula remains true for $P(x) = Ax^2 + Bx + C + Dx^3$. (*Hint:* First consider the case $\alpha = -\beta$.)

23. The value of π is given by

$$\pi = 4\int_0^1 \sqrt{1-x^2}\,dx.$$

In principle, Simpson's Rule can be used to approximate π. In practice, the approximations are not very accurate. For example, with $N = 100$, the approximation is 3.1411..., which is accurate to only three decimal places. Examine the graph of $y = \sqrt{1-x^2}$ for $0 \le x \le 1$. What geometric property does this graph have that graphs of approximating parabolas cannot have? Is the error bound of any use in this exercise?

24. Estimate $\int_{-1/2}^{1/2} \sqrt{|x|}\,dx$ first using Simpson's Rule with $N = 2$ and then using the Midpoint Rule with $N = 2$. Estimate $\int_{-1}^{1} \sqrt{|x|}\,dx$ first using Simpson's Rule with $N = 4$ and then using the Midpoint Rule with $N = 4$. Explain why the Midpoint Rule is the more accurate for this exercise.
25. Suppose f has derivatives of order 4 in an interval containing $[a, b]$. *Chevilliet's form of the error in Simpson's Rule* states that the error in using Simpson's Rule is *exactly*

$$\frac{(b-a)^4}{180N^4}(f'''(\xi) - f'''(\eta))$$

for two points ξ and η in $[a, b]$. Show that the error estimate as stated in the text (sometimes known as *Stirling's form*) follows from Chevilliet's form of the error.

In Exercises 26 and 27, several values of the Lorenz function L have been tabulated. Use trapezoidal approximations to estimate the coefficient of inequality that corresponds to the given data. (Note: The tables represent partitions that are *not* uniform, and the data points (0, 0) and (100, 100) have not been included in the tables.)

26.

x	15	25	50	75	90
$L(x)$	3	19	25	42	70

27.

x	16	28	51	75	88	97
$L(x)$	3	8	24	46	69	88

28. Figure 10 shows American cotton production. This type of graph, in which plotted data points are connected by straight line segments, is a common alternative to bar graphs. Apply the Trapezoidal Rule to find the area

under the graph. How does this area differ from the total cotton production for 1986–1995?

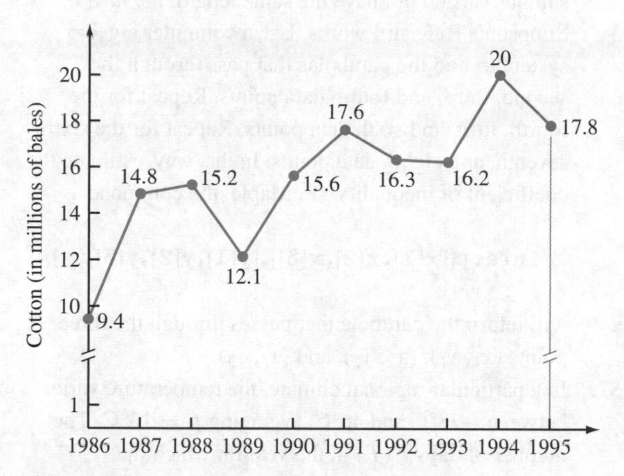

Figure 10
U.S. cotton production

29. Figure 11 shows a map of the province of Manitoba flipped onto its western boundary. Distances are in kilometers. The southern portion of the province, appearing to the right in Figure 11, is essentially a trapezoid. Use Simpson's Rule to estimate the area of the northern portion of the province. Approximately what is the area of Manitoba when estimated in this way? The actual area is 649,953 km^2. What is the percentage error of your approximation? Repeat with the Trapezoidal Rule. Repeat with the Midpoint Rule (using increments of 156 km along the S-axis). What source of error (in addition to the usual) do these area approximations entail?

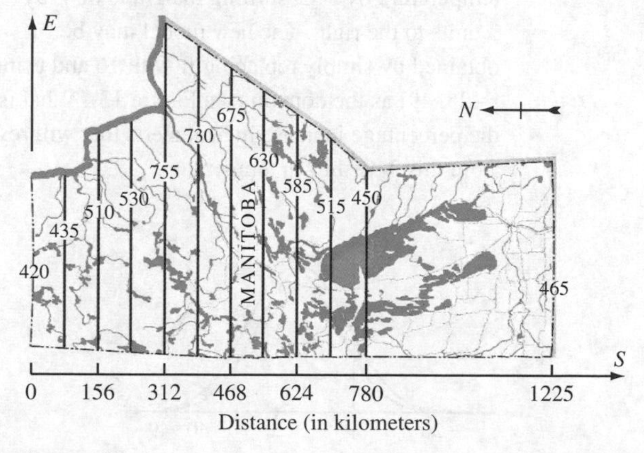

Figure 11

30. A 6 m wide swimming pool is illustrated in Figure 12. Depths are given every 2 m. Estimate the volume of water in the pool when it is filled. Use the Midpoint, Trapezoidal, and Simpson's Rules.

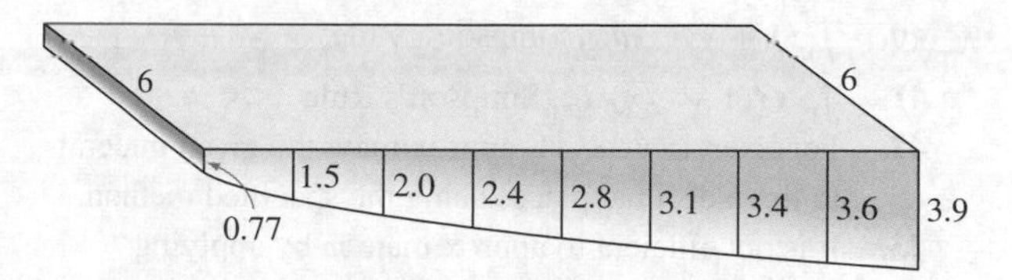

Figure 12

31. Let N be an even integer. The three approximation methods discussed in this section satisfy

$$\mathcal{S}_N = \frac{1}{3}\mathcal{T}_{N/2} + \frac{2}{3}\mathcal{M}_{N/2}. \quad \textbf{(5.29)}$$

Use equation (5.29) to show that

$$(\mathcal{S}_N - \mathcal{T}_{N/2}) = 2(\mathcal{M}_{N/2} - \mathcal{S}_N). \quad \textbf{(5.30)}$$

Equation (5.30) states that (1) one of the estimates $\mathcal{T}_{N/2}$ and $\mathcal{M}_{N/2}$ is greater than $\mathcal{S}_N$ and the other is less than $\mathcal{S}_N$, and (2) the absolute difference $|\mathcal{S}_N - \mathcal{T}_{N/2}|$ is double the absolute difference $|\mathcal{S}_N - \mathcal{M}_{N/2}|$. In Exercises 32–35, calculate $\mathcal{S}_N$, $\mathcal{M}_{N/2}$, and $\mathcal{T}_{N/2}$ for the given even value of N and verify these two statements.

32. $\int_0^4 x^4\,dx$, $N = 4$

33. $\int_0^4 \sqrt{x}\,dx$, $N = 2$

34. $\int_1^7 1/x\,dx$, $N = 6$

35. $\int_0^\pi \sin(x)\,dx$, $N = 4$

36. During a 6-minute span, at intervals of 1 minute, the speedometer of a car read

Time (min)	0	1	2	3	4	5	6
Speed (km/h)	90	80	75	80	80	70	60

Estimate the distance traveled by the car during that 6-minute period.

37. For $p > 0$, let $f_p(x) = x^p/p^2$ $(0 \le x \le 1)$. What is $\lim_{p\to\infty} \int_0^1 f_p(x)\,dx$? What is $M_p = \lim_{x\to 1^-} f_p''(x)$? What is $\lim_{p\to\infty} M_p$? Use this family of functions to explain why the error bound for the Simpson's Rule approximation is very conservative.

Calculator/Computer Exercises

In Exercises 38–41, an integral $\int_a^b f(x)\,dx$ is given. Plot $y = |f^{(k)}(x)|$, $a \le x \le b$, where $k = 2$ if the specified approximation method is the Midpoint or Trapezoidal Rule and $k = 4$ if Simpson's Rule is specified. Use your plot to determine the number of subintervals required to guarantee three decimal places of accuracy using the given method.

38. $\int_1^4 \sqrt{1+\sqrt{x}}\,dx$, Midpoint Rule

39. $\int_{1/4}^4 (1+\sqrt{x})^{-1}\,dx$, Trapezoidal Rule

40. $\int_0^1 (1+x^3)^{-1}\,dx$, Simpson's Rule

41. $\int_1^9 x/(1+\sqrt{x})\,dx$, Simpson's Rule

42. For Exercises 38–41, approximate the given integral to three decimal places using the specified method.

43. It is not efficient to approximate π by applying Simpson's Rule to the area of a semicircle (see Exercise 23). Rewrite the formula

$$\int_0^{1/2} \sqrt{1-x^2}\,dx = \frac{\sqrt{3}}{8} + \frac{\pi}{12}$$

as

$$\frac{\pi}{12} = \int_0^{1/2} \left(\sqrt{1-x^2} - \frac{\sqrt{3}}{4}\right) dx.$$

Apply Simpson's Rule to this formula to approximate π to five decimal places.

In Exercises 44–47, an integral $\int_a^b f(x)\,dx$ and a positive integer N are given. Compute the exact value of the integral, the Simpson's Rule approximation of order N, and the absolute error ϵ. Then find a value c in the interval (a, b) such that $\epsilon = (b-a)^5 |f^{(4)}(c)|/(180 \cdot N^4)$.

44. $\int_1^4 \sqrt{x}\,dx$, $N = 6$ **45.** $\int_1^e 1/x\,dx$, $N = 4$

46. $\int_1^2 \sec^2(\pi x/6)\,dx$, $N = 4$

47. $\int_8^{15} 1/\sqrt{1+x}\,dx$, $N = 4$

48. Data is said to be *normally distributed* with *mean* μ and *standard deviation* $\sigma > 0$ if, for every $a < b$, the probability that a randomly selected data point lies in the interval $[a, b]$ is

$$\frac{1}{\sigma\sqrt{2\pi}} \int_a^b \exp\left(-\frac{1}{2}\left(\frac{x-\mu}{\sigma}\right)^2\right) dx.$$

Suppose the scores on an exam are normally distributed with mean 70 and standard deviation 15. What is the probability that a randomly selected exam has a grade in the range [55, 85]? Use Simpson's Rule with $N = 6$ to approximate this probability.

49. Suppose the scores on an exam are normally distributed with mean 80 and standard deviation 20. (Refer to Exercise 48.) What is the probability that a randomly selected exam has a grade in the range [60, 100]? Use Simpson's Rule with $N = 6$.

50. The partition {0, 16, 28, 51, 75, 88, 97, 100} given in Exercise 27 precludes an application of Simpson's Rule because (1) it is not uniform and (2) it has an odd number of subintervals. To fix the problem of an odd number of intervals, we can use the Trapezoidal Rule on the first subinterval and then *adapt* Simpson's Rule to the remaining six subintervals. Although these subintervals do not have the same length, the *idea* of Simpson's Rule still works. Use a computer algebra system to find the parabolas that pass through the second, third, and fourth data points. Repeat for the fourth, fifth, and sixth data points. Repeat for the sixth, seventh, and eighth data points. In this way, estimate the coefficient of inequality. (In Maple, the command

```
interp([x[1],x[2],x[3]],[y[1],y[2],y[3]],u)
```

will return the parabola that passes through the three points (x_1, y_1), (x_2, y_2), and (x_3, y_3).)

51. In a particular regional climate, the temperature varies between −28°C and 46°C, averaging $\mu = 13$°C. The number of days $F(T)$ in the year on which the temperature remains below T degrees centigrade is given (approximately) by

$$F(T) = \int_{-28}^{T} f(x)\,dx \quad (-28 \le T \le 46)$$

where

$$f(x) = 12.66 \exp\left(-\frac{(x-\mu)^2}{265.8}\right).$$

Notice that $F(T)$ is an *area integral,* which is often used in modeling problems.

a. Use Simpson's Rule to approximate $F(46)$. What should the exact value of $F(46)$ be?

b. Heat alerts are issued when the daily high temperature is 36°C or more. On about how many days a year are heat alerts issued?

c. Suppose that global warming raises the average temperature by 3°C, shifting the graph of f by 3 units to the right. The new model may be obtained by simply replacing μ with 16 and using [−25, 49] as the domain (see Figure 13). What is the percentage increase in heat alerts that will result from this 3°C shift in temperature?

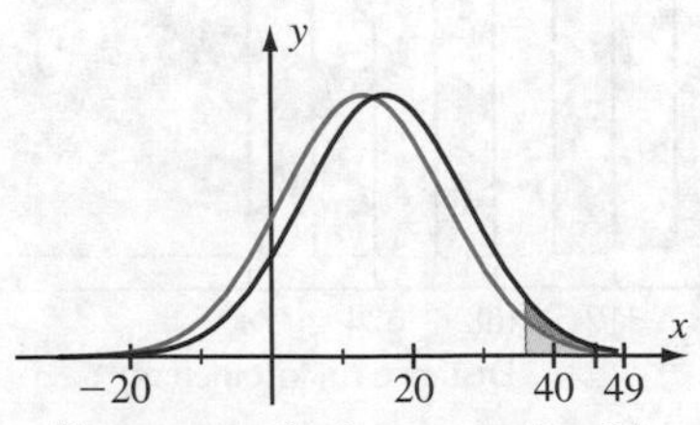

Figure 13
A shift in average temperature by +3°C.

Summary of Key Topics

Sigma Notation and Partitions (Sections 5.1, 5.2)

The notation

$$\sum_{j=m}^{n} a_j$$

means $a_m + a_{m+1} + a_{m+2} + \cdots + a_{n-1} + a_n$. The uniform partition of order N of an interval $[a, b]$ is the set $\{x_0, x_1, \ldots, x_N\}$ of points x_j defined by

$$x_j = a + j \cdot \frac{b-a}{N}.$$

The jth subinterval $[x_{j-1}, x_j]$ is denoted I_j. Each subinterval has length $\Delta x = (b-a)/N$.

Area (Section 5.1)

Let f be a positive continuous function on $[a, b]$. The area of the region under the graph of f, above the x-axis, and between the vertical lines at a and b is the limit of the expression

$$\sum_{j=1}^{N} f(x_j) \cdot \Delta x$$

as N tends to infinity.

The Riemann Integral (Section 5.2)

If f is any function on an interval $[a, b]$ and if for each j a point $s_j \in I_j$ is chosen, then the expression

$$\mathcal{R}(f, \mathcal{S}_N) = \sum_{j=1}^{N} f(s_j) \cdot \Delta x$$

is called a Riemann sum for f. Here we let $\mathcal{S}_N$ denote $\{s_1, \ldots, S_N\}$. If the Riemann sums for f have a limit as the number N of subintervals tends to infinity, then that limit is called the *Riemann integral* for f on $[a, b]$, and f is said to be integrable on $[a, b]$. The Riemann integral of f on $[a, b]$ is denoted by

$$\int_a^b f(x)\, dx$$

and is said to be the *definite integral* of f over the interval $[a, b]$. *A definite integral is always a number*. (By contrast, an antiderivative of f is denoted by $\int f(x)\, dx$ and is said to be an *indefinite integral* of f: It is always a function of x.)

Fundamental Theorem of Calculus (Section 5.2)

Continuous functions are integrable. If f is continuous on $[a, b]$ and has an antiderivative F on (a, b), then the definite integral of f can be calculated by

$$\int_a^b f(x)\,dx = F(b) - F(a).$$

Properties of the Integral (Section 5.3)

The integral satisfies the following five properties. If f and g are integrable on $[a, b]$ and if α is a constant, then

1. $\int_a^b (f(x) + g(x))\,dx = \int_a^b f(x)\,dx + \int_a^b g(x)\,dx$ and $\int_a^b (f(x) - g(x))\,dx = \int_a^b f(x)\,dx - \int_a^b g(x)\,dx$,
2. $\int_a^b \alpha f(x)\,dx = \alpha \cdot \int_a^b f(x)\,dx$,
3. $\int_a^b \alpha\,dx = \alpha(b - a)$,
4. $\int_a^a f(x)\,dx = 0$, and
5. $\int_a^c f(x)\,dx + \int_c^b f(x)\,dx = \int_a^b f(x)\,dx$.

If $b < a$, then we define $\int_a^b f(x)\,dx$ to be equal to $-\int_b^a f(x)\,dx$. If f is continuous on $[a, b]$, then the Mean Value Theorem for Integrals states that there is a c in (a, b) such that

$$f(c) = \frac{1}{b - a}\int_a^b f(x)\,dx.$$

The Fundamental Theorem of Calculus (Section 5.4)

If f is continuous on $[a, b]$ and if the function $F(x)$ is defined by

$$F(x) = \int_a^x f(t)\,dt,$$

then

$$F'(x) = f(x).$$

Substitution (Section 5.5)

If $f(x)$ has the form $F(\phi(x)) \cdot \phi'(x)$, then the calculation of the integral of f can be simplified by a substitution of the form

$$u = \phi(x) \quad \text{and} \quad du = \phi'(x)\,dx.$$

The Calculation of Area (Section 5.6)

If $f(x) \geq 0$ on $[a, b]$, then

$$\int_a^b f(x)\,dx$$

represents the area under the graph of f, above the x-axis, and between $x = a$ and $x = b$. If $f(x) \leq 0$ on $[a, b]$, then the integral represents the negative of the area above the graph of f, below the x-axis, and between $x = a$ and $x = b$.

If $f(x) \geq g(x)$ on $[a, b]$, then the area between the two graphs and between the vertical lines $x = a$ and $x = b$ is given by

$$\int_a^b (f(x) - g(x))\, dx.$$

To calculate the area between two graphs, it is necessary to break up the problem into integrals over intervals on which one of the two functions is greater than or equal to the other.

For some types of area problems, it is useful to integrate in the y-variable, treating x as the dependent variable.

Numerical Integration (Section 5.7)

Methods of approximating definite integrals are particularly useful when an antiderivative of the integrand cannot be found. Let f be a continuous function with a domain that contains the interval $[a, b]$. For a positive integer N, let $\Delta x = (b - a)/N$, $x_j = a + j\Delta x$, and $\bar{x}_j = (x_{j-1} + x_j)/2$. The definite integral $\int_a^b f(x)\, dx$ can be approximated by the Midpoint Rule:

$$\Delta x \cdot (f(\bar{x}_1) + f(\bar{x}_2) + \cdots + f(\bar{x}_N)).$$

The Trapezoidal Rule of approximating $\int_a^b f(x)\, dx$ is

$$\frac{\Delta x}{2}(f(x_0) + 2f(x_1) + 2f(x_2) + \cdots + 2f(x_{N-1}) + f(x_N)).$$

Of these two methods of approximation, the Midpoint Rule is usually more accurate. Still more accurate is Simpson's Rule, which can be applied when N is *even:*

$$\int_a^b f(x)dx \approx \frac{\Delta x}{3}(f(x_0)+4f(x_1)+2f(x_2)+4f(x_3)+\cdots+2f(x_{N-2})+4f(x_{N-1})+f(x_N)).$$

Let C_k be a number such that the kth derivative $f^{(k)}$ satisfies $|f^{(k)}(x)| \leq C_k$ on the interval $[a, b]$ of integration. The errors that result from using the Midpoint Rule, the Trapezoidal Rule, and Simpson's Rule are no larger than

$$\frac{C_2}{24} \cdot \frac{(b-a)^3}{N^2}, \quad \frac{C_2}{12} \cdot \frac{(b-a)^3}{N^2}, \quad \text{and} \quad \frac{C_4}{180} \cdot \frac{(b-a)^5}{N^4},$$

respectively.

genesis & DEVELOPMENT

Archimedes

The Greek geometer Eudoxus (ca. 408–ca. 355 BCE) used the limiting process to rigorously demonstrate area and volume formulas that Democritus (ca. 460–ca. 370 BCE) had discovered earlier. The arguments that Eudoxus introduced came to be known as the Method of Exhaustion. In the hands of Archimedes, the *Method of Exhaustion* became a powerful technique.

In one of his earliest works, Archimedes derived a formula for the area of a parabolic sector. Figure 1 shows an arc AVB of a parabola. The point V is known as the vertex of the segment and is characterized as follows: Among all line segments from the chord AB to the parabolic arc AB that are perpendicular to chord AB, the one terminating at V is longest. Archimedes demonstrated that the area enclosed by line segment $\overline{AB}$ and parabolic segment AVB is $2hb/3$, or $4|\Delta AVB|/3$ (using the notation $|R|$ to denote the area of a planar region R). The introduction of V creates two parabolic subsegments, AV and VB. Each has a vertex, labeled V_1 and V_2 in Figure 2a. Archimedes proved from a few easily derived geometric properties of the parabola that

$$|\Delta AV_1V| + |\Delta VV_2B| = \frac{1}{4}|\Delta AVB|.$$

At the next iteration, illustrated in Figure 2b, Archimedes interposed four new vertices, V_3, V_4, V_5, and V_6, and concluded that

$$|\Delta AV_3V_1| + |\Delta V_1V_4V| = \frac{1}{4}|\Delta AV_1V| \quad \text{and}$$

$$|\Delta VV_5V_2| + |\Delta V_2V_6B| = \frac{1}{4}|\Delta VV_2B|.$$

Combining these two equalities yields

$$\begin{aligned}|\Delta AV_3V_1| &+ |\Delta V_1V_4V| + |\Delta VV_5V_2| + |\Delta V_2V_6B| \\ &= \frac{1}{4}(|\Delta AV_1V| + |\Delta VV_2B|) \\ &= \frac{1}{4^2}|\Delta AVB|.\end{aligned}$$

Continuing, Archimedes exhausted the parabolic segment by a sequence $\{R_n\}$ of regions composed of triangles: $R_0 = \Delta AVB$, $R_1 = \Delta AV_1V \cup \Delta VV_2B$, $R_2 = \Delta AV_3V_1 \cup \Delta V_1V_4V \cup \Delta VV_5V_2 \cup \Delta V_2V_6B$, $R_3 = \ldots$. In general, $|R_n| = 4^{-n}|R_0|$. Using the formula for the sum of a geometric progression, Archimedes obtained

$$\begin{aligned}|R_1 \cup R_2 \cup \ldots \cup R_N| &= \left(1 + \frac{1}{4} + \frac{1}{4^2} + \cdots + \frac{1}{4^N}\right) \times |\Delta AVB| \\ &= \frac{4}{3}\left(1 - \frac{1}{4^{N+1}}\right)|\Delta AVB|.\end{aligned}$$

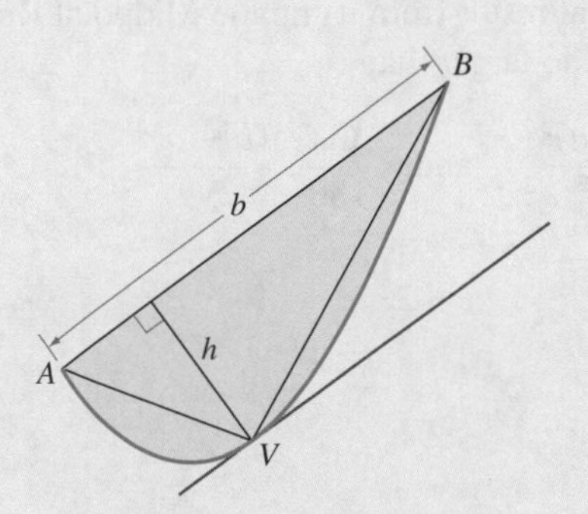

Figure 1
Archimedes proved that the area of the parabolic segment AVB is equal to $(2/3)hb$.

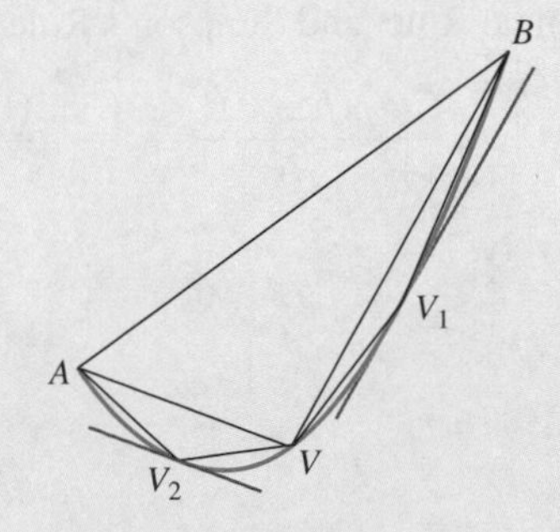

Figure 2a
Vertex V creates two parabolic segments with vertices V_1 and V_2.

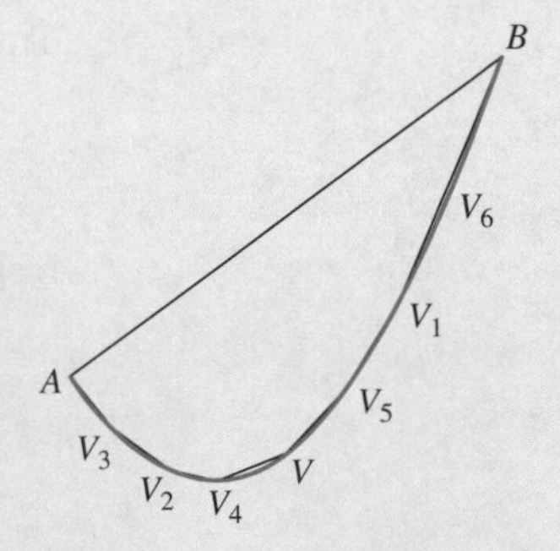

Figure 2b
Repeat the process of adding vertices.

He finished this proof with an airtight argument that $1/4^{N+1} \to 0$ and $|AVB| = 4|\Delta AVB|/3$.

The Method

The translation of Archimedes' works into Latin in the late 1500s introduced European mathematicians to the power of the infinite process. Although these mathematicians found elegant formulas, rigorous proofs, and glimpses of what might be attained, they did not find the methods by which Archimedes *discovered* his results. The 17th century mathematician John Wallis even protested that Archimedes "covered up the traces of his investigations, as if he has grudged posterity the secret of his method of inquiry." So the matter stood for 200 more years.

In 1906, Danish scholar Johan Ludvig Heiberg learned of a manuscript that had been catalogued in a cloister in Constantinople (now Istanbul). The manuscript contained a religious text of the Eastern Orthodox Church and was a so-called *palimpsest*—a parchment of which the original script had been washed and overwritten. Since it was reported that the original mathematical text had been imperfectly washed, Heiberg voyaged from Copenhagen to Constantinople to investigate. What Heiberg uncovered was a mathematical manuscript containing numerous works of Archimedes. It had been copied in Greek in the 10th century only to be washed and written over in the 13th century. As luck would have it, most of the original text could be restored.

Heiberg's chief find was the *Method Concerning Mechanical Theorems, dedicated to Eratosthenes*. Scholars had known from the references of later Greek mathematicians that such a work had existed until at least the fourth century CE, but until Heiberg's discovery, the text was presumed to be lost. In the *Method,* we find, to use the words of Wallis, Archimedes writing for posterity:

> *Archimedes to Eratosthenes, Greeting: ... since I see that you are an excellent scholar... and lover of mathematical research, I have deemed it well to explain to you and put down in this same book, a special method whereby the possibility will be offered to you to investigate any mathematical question by means of mechanics.... [I]f one has previously gotten a conception of the problem by this method, it is easier to produce the proof than to find it without a provisional conception. ... [W]e feel obliged to make the method known partly... in the conviction that there will be instituted thereby a matter of no slight utility in mathematics.*

Theory of Indivisibles

The first half of the 17th century saw a number of attempts to use infinite processes in the computation of areas, volumes, and centers of gravity. Among the most important of the early contributors were Pierre de Fermat and Gilles Roberval in France, Bonaventura Cavalieri and Evangelista Torricelli in Italy, and John Wallis and Isaac Barrow in the United Kingdom. In 1635, Cavalieri introduced his Theory of Indivisibles, a relatively effective albeit cumbersome procedure for calculating areas and volumes. Using his method, Cavalieri became the first to show that $\int_0^1 x^n\,dx = 1/(n+1)$ for natural numbers n. Although Cavalieri's student Evangelista Torricelli became the foremost proponent of the Theory of Indivisibles, Torricelli believed that Cavalieri had only rediscovered a method that Archimedes must have known:

> *I should not dare confirm that this geometry of indivisibles is actually a new discovery. I should rather believe that the ancient geometers availed themselves of this method in order to discover the more difficult theorems, although in their demonstration they may have preferred another way, either to conceal the secret of their art or to afford no occasion for criticism by invidious detractors.*

Fermat and the Integral Calculus

The extension of Cavalieri's formula $\int_0^1 x^n\,dx = 1/(n+1)$ from natural numbers to rational values of n was obtained independently by Fermat and Torricelli. Fermat's clever demonstration runs as follows: Given natural numbers p and $q \neq 0$, let $f(x) = x^{p/q}$. Fix a number $r \in (0, 1)$ and set $x_k = r^{kq}$ for $k = 0, 1, 2, \ldots$. The points $\{x_k\}$ serve to (nonuniformly) partition $[0, 1]$ into infinitely many subintervals. Fermat's idea was to use the sum $\sum_{k=0}^N f(x_k)(x_k - x_{k+1})$ to approximate the area under $y = x^{p/q}$ for $0 \le x \le 1$. Figure 3a, next page, shows the approximation for $r = 0.9$. By taking a larger value of r (with r still less than 1), we obtain a better approximation (see Figure 3b, next page). In the limit, we obtain the exact area:

$$\int_0^1 f(x)\,dx = \lim_{r\to 1^-} \lim_{N\to\infty} \sum_{k=0}^{N} f(x_k)(x_k - x_{k+1}).$$

For $f(x) = x^{p/q}$, we calculate

$$\begin{aligned}\int_0^1 x^{p/q}\,dx &= \lim_{r\to 1^-} \lim_{N\to\infty} \sum_{k=0}^{N} (r^{kq})^{p/q}\left(r^{kq} - r^{(k+1)q}\right)\\ &= \lim_{r\to 1^-} \lim_{N\to\infty} \sum_{k=0}^{N} r^{kp} r^{kq}(1 - r^q)\\ &= \lim_{r\to 1^-} (1 - r^q) \lim_{N\to\infty} \sum_{k=0}^{N} r^{(p+q)k}.\end{aligned}$$

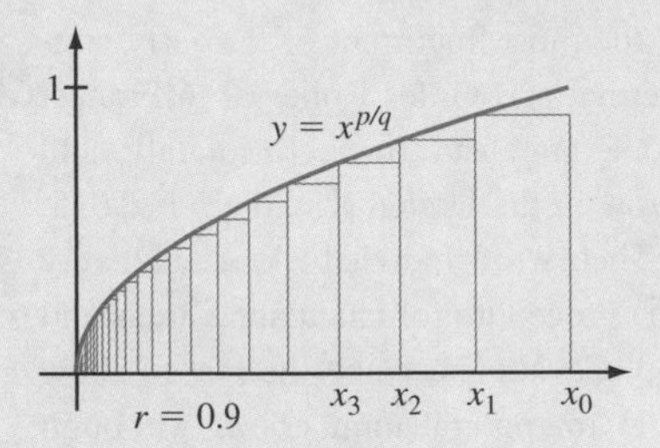

Figure 3a

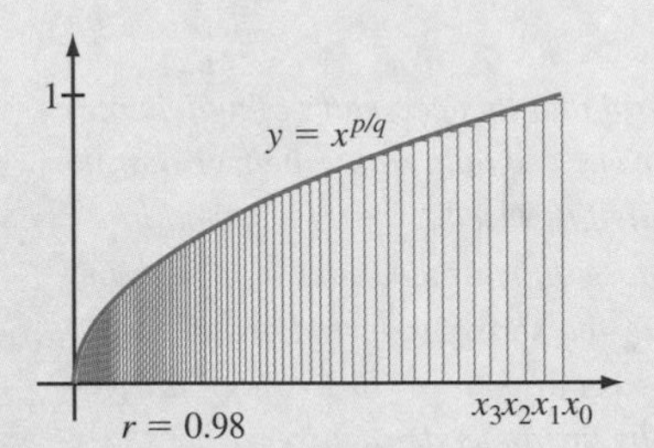

Figure 3b

Since $0 < r^{p+q} < 1$, we may apply formula (2.11) to obtain

$$\int_0^1 x^{p/q}\,dx = \lim_{r\to 1^-} \frac{1-r^q}{1-r^{p+q}}.$$

Although *we* know how to compute the indeterminate form on the right by using l'Hôpital's Rule, Fermat had to resort to the following tricky backward application of formula (2.10):

$$\int_0^1 x^{p/q}\,dx = \lim_{r\to 1^-} \frac{1-r^q}{1-r^{p+q}} = \lim_{r\to 1^-} \frac{(1-r^q)/(1-r)}{(1-r^{p+q})/(1-r)}$$

$$= \frac{\lim_{r\to 1^-}\sum_{k=0}^{q-1} r^k}{\lim_{r\to 1^-}\sum_{k=0}^{p+q-1} r^k} = \frac{q}{p+q} = \frac{1}{p/q+1}.$$

Kepler and Roberval

Before the discovery of the Fundamental Theorem of Calculus, many mathematicians succeeded in evaluating *particular* integrals by ad hoc methods. For example, in his last great astronomical work, the *Epitome Astronomiae Copernicanae* (1618–1621), Kepler calculated what we would now call a limit of Riemann sums for $\int_0^\beta \sin(t)\,dt$. In effect, Kepler discovered the formula

$$\int_0^\beta \sin(t)\,dt = 1 - \cos(\beta).$$

One curve that attracted the attention of prominent mathematicians throughout the 17th century was the *cycloid,* which is the locus of points generated by a fixed point on a circle as the circle rolls along a line at constant speed without spinning or slipping (see Figure 4). Early in the 17th century, Galileo conjectured that the area under one arch of the cycloid is three times the area of the rolling circle. To reach this conjecture, this leading scientist resorted to cut-out paper figures. By 1636, Roberval was able to rigorously verify Galileo's conjecture. Roberval's work is a measure of how much progress had been made in a very short period of time.

Figure 4
A point on a rolling circle generates a cycloid.

Newton and the Fundamental Theorem of Calculus

Sir Isaac Newton's first contact with higher mathematics came as a college senior in 1664 when he was 21 years old. He learned the tangent problem from the *Géométrie* of Descartes, and he learned Cavalieri's Theory of Indivisibles from Wallis's *Arithmetica Infinitorum.* It was Wallis's work that, in the winter of 1664–1665, led Newton to evaluate the area of the region under the graph of $y = (1-t^2)^\alpha$ and over the t-interval $[0, x]$. In doing so, he had already taken the decisive step of letting the endpoint of integration be a variable, a key idea of the Fundamental Theorem of Calculus. Keep in mind that at the time, Newton was still an undergraduate.

When plague closed Cambridge University in 1665 and 1666, Newton returned home to reorganize his discoveries. He wrote his discoveries in "The October 1666 Tract on Fluxions," a manuscript that some contemporaries knew of but that was not published until 1967. It was in this tract that both parts of the Fundamental Theorem of Calculus were first stated. This remarkable manuscript also contains several other theorems, including the Chain Rule and Integration by Substitution.

Notation

The notation that we use today for the indefinite integral was introduced by Gottfried Wilhelm Leibniz (1646–1716) in a manuscript of 1675. At that time, Leibniz did not include the differential. For example, he wrote $\int x^2 = x^3/3$. By the time his work appeared in print in 1686, however, Leibniz included the differential in his notation and stressed its importance. The modern notation $\int_a^b f(x)\,dx$ for the *definite* integral—that is, the positioning of the limits of integration on the integral sign—was introduced in 1822 by Joseph Fourier (1768–1830). It was Fourier's work that indirectly led to the Riemann integral. Fourier's study of heat transfer required the evaluation of definite integrals of the form $\int_0^{2\pi} f(x) \sin(nx)\, dx$ for quite general functions f. However, the definition of integration then in use was far too limited for the desired generality. Georg Friedrich Bernhard Riemann (1826–1866) realized that to further develop Fourier's work, a new concept of the integral was needed. Toward that end, in 1854, Riemann introduced what we now call the Riemann integral.

Bernhard Riemann

Riemann's short life was marked by nervous breakdowns and serious illness—pleurisy, jaundice, and, ultimately, tuberculosis. Although he began his studies in theology, he was able to convince his father to allow him to pursue what was then an "economically valueless" subject—mathematical physics. In quantity, his mathematical output was relatively modest for a mathematician of his stature. Yet his work has assumed fundamental importance in several fields—he is remembered for the Riemann integral in analysis, the Riemann zeta function in number theory, Riemann surfaces and Cauchy-Riemann equations in the theory of functions of a complex variable, and Riemannian manifolds in geometry. The importance that his work has had in physics is also noteworthy. Indeed, Riemannian geometry proved to be essential for Einstein's theory of relativity as well as for much of the physics that has been done since.

6

Differential Equations and Transcendental Functions

PREVIEW

Exponential and logarithm functions occur frequently in such diverse fields as economics, finance, biology, engineering, and the physical and social sciences. By introducing mathematical models in this chapter, we study several applications of these and other transcendental functions.

A common theme in the applications of this chapter is that a function must often be determined from a differential equation that it satisfies. The differential equations that we encounter in Chapter 6 all have the form

$$\frac{dy}{dx} = F(x, y).$$

In this differential equation, the right side, $F(x, y)$, is a given expression. The variable y represents a function of x that is to be determined. Unlike an ordinary algebraic equation in which you determine one or more solution points, a differential equation requires a *function* for its solution. The study of differential equations is an important component of a calculus course. Most students who concentrate on mathematics, physics, or engineering follow their calculus sequence with a more advanced course entirely devoted to the study of differential equations.

The simplest differential equation has the form

$$\frac{dy}{dx} = g(x).$$

Solving this equation amounts to nothing more than finding an antiderivative of $g(x)$. Even when an antiderivative $y(x)$ cannot be constructed from previously known functions, however, the differential equation itself provides a good deal of information about $y(x)$. Using this approach, we take a second, more sophisticated, look at logarithms and exponential functions. Calculus provides us with the tools to better understand several of their properties that, until now, we have accepted on the basis of numerical evidence alone.

6.1 First Order Differential Equations

A *differential equation* is an equation that involves one or more derivatives of an unknown function. For example, if a body of mass m falls under the influence of gravity near the Earth's surface, then its velocity $v(t)$ at time t satisfies the differential equation $\frac{dv}{dt} = -g$ where g is a constant. This differential equation is a *first order differential equation* because the first derivative is the highest order derivative of the unknown function that appears in the equation. In this section, we study first order differential equations that have the form

$$\frac{dy}{dx} = F(x, y) \tag{6.1}$$

where $F(x, y)$ is an expression involving both x and y (in general). For example, $F(x, y)$ could be $\sqrt{2 - xy^3}$. It is also possible for the expression $F(x, y)$ to depend on only one of the variables. Thus, $\frac{dy}{dx} = 1/x$ and $\frac{dy}{dx} = 2y(10 - y)$ are also examples of equation (6.1) because they are obtained by setting $F(x, y) = 1/x$ and $F(x, y) = 2y(10 - y)$, respectively.

Definition We say that a differentiable function y is a *solution* of differential equation (6.1) if $y'(x) = F(x, y(x))$ for every x in some open interval. The graph of a solution is called a *solution curve* of the differential equation.

Our first example shows that there can be infinitely many different functions that satisfy a given differential equation.

Example 1 Let C denote an arbitrary constant. Verify that the function $y(x) = x + Ce^{-x} - 1$ is a solution of the differential equation $\frac{dy}{dx} = x - y$.

Solution We calculate the left side of the differential equation,

$$\frac{dy}{dx} = \frac{d}{dx}(x + Ce^{-x} - 1) = 1 - Ce^{-x},$$

and the right side,

$$x - y = x - (x + Ce^{-x} - 1) = 1 - Ce^{-x}.$$

Since these expressions are equal, it follows that the function $y(x) = x + Ce^{-x} - 1$ satisfies the given differential equation. Notice that this verification does not show how the solution $y(x) = x + Ce^{-x} - 1$ is *found*. Exercise 27 outlines the technique for discovering this solution. ■

inSIGHT

As Example 1 illustrates, the solution of a first order differential equation usually involves an unspecified constant C. Solution curves that correspond to several values of C from the differential equation in Example 1 are graphed in Figure 1. Observe that the plotted solution curves do not intersect. Here is another way to think about this fact: Only one solution curve passes through a given point (x_0, y_0). From an analytic point of view, this means that we can use the equation $y(x_0) = y_0$ to determine the constant C. These considerations lead to the next definition.

y(x) = y0

Definition The pair of equations

$$\frac{dy}{dx} = F(x, y), \quad y(x_0) = y_0 \tag{6.2}$$

is called an *initial value problem* (often abbreviated IVP). We say that a differentiable function y is a *solution* of initial value problem (6.2) if $y(x_0) = y_0$ and $y'(x) = F(x, y(x))$ for all x in some open interval containing x_0. The equation $y(x_0) = y_0$ is called an *initial condition*.

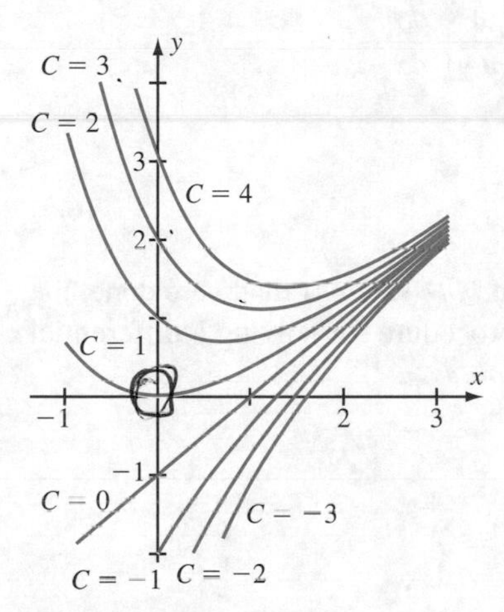

Figure 1

It can be shown that, under mild restrictions on F (satisfied by all examples in this book), initial value problem (6.2) has a *unique* solution.

Example 2 Use the computation from Example 1 to solve the initial value problem $\frac{dy}{dx} = x - y$, $y(0) = 2$.

Solution In Example 1, we verified that $y(x) = x + Ce^{-x} - 1$ satisfies the given differential equation for every constant C. Notice that $y(0) = 0 + Ce^{-0} - 1 = C - 1$. To satisfy the initial condition $y(0) = 2$, we set $C - 1 = 2$ and find that $C = 3$. Thus, $y(x) = x + 3e^{-x} - 1$ solves the given initial value problem. ■

Separable Equations

There is no single technique for solving equation (6.1). Several methods have been developed to handle special cases according to the form of the expression $F(x, y)$. In this section, we study the case in which $F(x, y) = g(x)h(y)$. The expressions

$$F(x, y) = 4\cos(x) = \underbrace{4\cos(x)}_{g(x)} \cdot \underbrace{1}_{h(y)}, \quad F(x, y) = 7y^3 = \underbrace{7}_{g(x)} \cdot \underbrace{y^3}_{h(y)}, \quad \text{and}$$

$$F(x, y) = \frac{2 + x}{1 + y^2} = \underbrace{(2 + x)}_{g(x)} \cdot \underbrace{\left(\frac{1}{1 + y^2}\right)}_{h(y)}$$

are all of this type. When $F(x, y)$ factors as $g(x)h(y)$, the differential equation

$$\frac{dy}{dx} = F(x, y) = g(x)h(y) \tag{6.3}$$

is said to be *separable* because y and x may be separated onto different sides of the equation. Informally, we can rewrite equation (6.3) as

$$\frac{1}{h(y)}\,dy = g(x)\,dx,$$

which, though only suggestive, does hint at the answer $\int \frac{1}{h(y)}\,dy = \int g(x)\,dx$. Let us see how we can obtain this result more carefully. Assuming that $h(y) \neq 0$, we rewrite equation (6.3) as

$$\frac{1}{h(y)}\frac{dy}{dx} = g(x)$$

and integrate with respect to x:

$$\int \frac{1}{h(y)}\frac{dy}{dx}\,dx = \int g(x)\,dx. \tag{6.4}$$

Let $H(u)$ be an antiderivative of $1/h(u)$ and let $G(x)$ be an antiderivative of $g(x)$. Using the Chain Rule, we have

$$\frac{d}{dx}H(y) = H'(y)\frac{dy}{dx} = \frac{1}{h(y)}\frac{dy}{dx}.$$

Therefore, equation (6.4) can be stated as

$$H(y) = G(x) + C. \tag{6.5}$$

(There is no need to add a constant of integration to *both* sides: If that were done, the two constants could be combined into one.) This procedure for solving a differential equation is called the *Method of Separation of Variables*.

Example 3 Solve the initial value problem

$$\frac{dy}{dx} = \frac{x}{1+y^2}, \qquad y(0) = 3.$$

Solution The given differential equation is separable since it may be written as $\frac{dy}{dx} = g(x)h(y)$ with $g(x) = x$ and $h(y) = 1/(1+y^2)$. Following the method of solution just discussed, we separate variables:

$$\int (1+y^2)\underbrace{\frac{dy}{dx}\,dx}_{\text{Treat as } dy} = \int x\,dx,$$

or

$$\frac{1}{3}y^3 + y = \frac{1}{2}x^2 + C.$$

If this equation is to satisfy the given initial condition $y(0) = 3$, then C must satisfy

$$\frac{1}{3}(3)^3 + (3) = \frac{1}{2}(0)^2 + C,$$

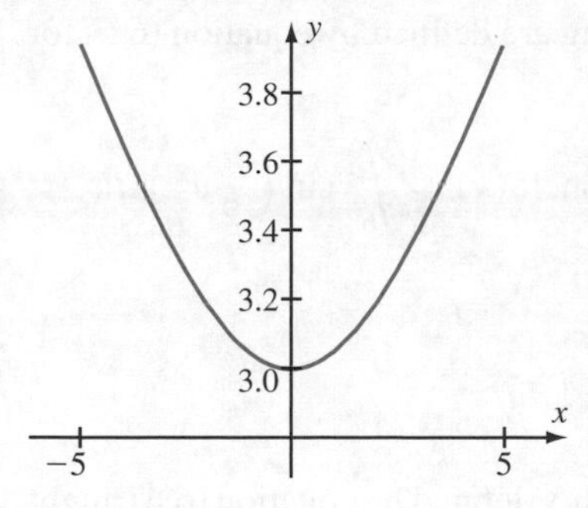

Figure 2
The graph of $\frac{1}{3}y^3 + y = \frac{1}{2}x^2 + 12$

or $C = 12$. Thus, $y^3/3 + y = x^2/2 + 12$ is the solution of the given initial value problem. The solution curve is shown in Figure 2. ■

inSIGHT

Example 3 illustrates a typical feature of the Method of Separation of Variables: In general, the method does not yield the solution $y(x)$ of the differential equation $\frac{dy}{dx} = g(x)h(y)$ in *explicit* form. That is, the Method of Separation of Variables does not ordinarily lead to an equation of the form $y(x) =$ (expression in x). Instead, it usually results in an equation

$$(\text{expression in } y) = (\text{expression in } x)$$

that defines y *implicitly* as a function of x. In Example 3, although we can obtain an explicit solution (see Exercise 50), the formula is so complicated that it is not even obvious that the initial condition is satisfied. It is quite a chore to verify that the explicit solution is correct. By contrast, it is not difficult to implicitly differentiate the equation $y^3/3 + y = x^2/2 + 12$ to verify that the implicitly defined function y does solve the given differential equation.

Equations of the Form $\frac{dy}{dx} = g(x)$

The differential equation $\frac{dy}{dx} = g(x)$ is nothing more than the statement "$y(x)$ is an antiderivative of $g(x)$." Theorem 1 provides the solution of the corresponding initial value problem.

Theorem 1 If g is a continuous function on an open interval containing a, then the initial value problem

$$\frac{dy}{dx} = g(x), \quad y(a) = b$$

has the unique solution

$$y(x) = b + \int_a^x g(u)\,du. \tag{6.6}$$

Proof According to the Fundamental Theorem of Calculus, $\int_a^x g(u)\,du$ is an antiderivative of g. All others have the form $\int_a^x g(u)\,du + C$. Thus, $y(x) = \int_a^x g(u)\,du + C$ for some constant C. We use the initial condition to solve for C,

$$b = y(a) = \int_a^a g(u)\,du + C = 0 + C.$$

We obtain equation (6.6) by substituting $C = b$ in our formula for $y(x)$. ■

Many useful functions in science and engineering are defined by equation (6.6) for particular choices of g: the Fresnel functions

$$\text{FresnelC}(x) = \int_0^x \cos\left(\frac{\pi}{2}u^2\right) du \quad \text{and} \quad \text{FresnelS}(x) = \int_0^x \sin\left(\frac{\pi}{2}u^2\right) du$$

in optics, the Debye function

$$x \mapsto \int_0^x \left(\frac{u}{e^u - 1}\right)^2 e^u \, du$$

in thermodynamics, and so on. Although the function y defined by equation (6.6) might seem complicated or abstract at first, the way it is defined allows us to instantly calculate its derivative. We have accumulated a good deal of experience using the derivative y' to deduce properties of y. It is through those properties that the function y becomes familiar to us. We return to these ideas in Section 6.2.

Examples from the Physical Sciences

In the physical sciences, the most natural way to describe the relationship between two variables is often by means of a differential equation, which is then solved to determine the relationship between the variables. The next three examples illustrate these ideas.

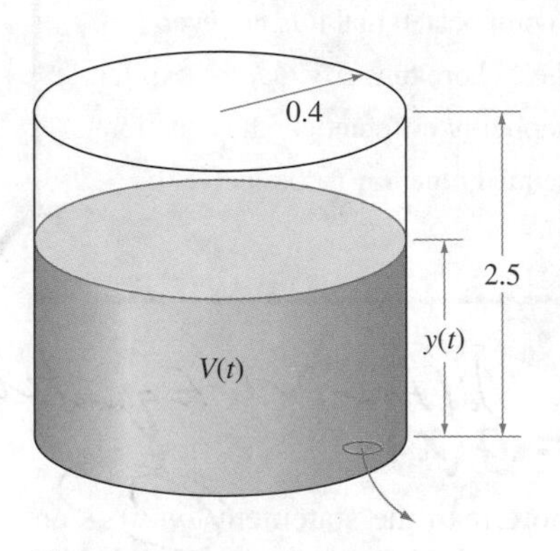

Figure 3
$\frac{dV}{dt} = -2.6 \cdot A \cdot \sqrt{y}$

Example 4 According to Torricelli's law, the rate at which water flows out of a small hole in the bottom of a cylindrical tank is proportional to the area of the hole and to the square root of the height of the water in the tank (see Figure 3). Suppose that the proportionality constant is $-2.6\ \text{m}^{1/2}/\text{s}$. If a tank 2.5 m high has radius 0.4 m, has a hole 2 cm in diameter in its bottom, and is initially full, find the height y of water as a function of time t. How long will it take for the tank to empty?

Solution The area A of the hole in the bottom is $\pi(0.01)^2\ \text{m}^2$. Let $V(t)$ be the volume (in cubic meters) of the water in the tank at time t. The statement concerning the rate at which water flows out of the tank may be written as

$$\frac{dV}{dt} = -2.6 \cdot A \cdot \sqrt{y}.$$

Since $V(t) = \pi(0.4)^2 y(t)$, we have $V'(t) = \pi(0.4)^2 y'(t)$ and, therefore,

$$\pi(0.4)^2 \frac{dy}{dt} = \frac{dV}{dt} = -2.6 \cdot A \cdot \sqrt{y}.$$

We rewrite this as

$$\frac{dy}{dt} = -2.6 \frac{\pi(0.01)^2}{\pi(0.4)^2} \sqrt{y} = -0.001625\sqrt{y}.$$

On separating variables, we have

$$\int y^{-1/2} \frac{dy}{dt}\, dt = \int (-0.001625)\, dt + C,$$

or

$$2y(t)^{1/2} = -0.001625t + C.$$

On substituting $t = 0$ into this equation, we obtain $C = 2\sqrt{y(0)}$. However, $y(0) = 2.5$ because the tank is initially filled to its height of 2.5 m. Therefore, $C = 2\sqrt{2.5}$ and $2y(t)^{1/2} = -0.001625t + 2\sqrt{2.5}$, or

$$y(t) = \left(\sqrt{2.5} - \frac{0.001625}{2}t\right)^2.$$

The tank is empty when $y(t) = 0$, which occurs when $t = 2\sqrt{2.5}/0.001625 \approx 1946$ s. Thus, the tank will be empty in about $1946/60 \approx 32.4$ min. ■

Example 5 In chemical kinetics, the rate at which the concentration $[S]$ of a substance S changes is often described by means of a differential equation. For example, the dimerization of butadiene,

$$C_4H_6 \to \frac{1}{2}C_8H_{12},$$

has reaction rate

$$\frac{d[C_4H_6]}{dt} = -k[C_4H_6]^2$$

where k is a positive constant. If $[C_4H_6] = 1/60$ mol/L when $t = 0$ s and $[C_4H_6] = 1/140$ mol/L when $t = 6000$ s, then determine $[C_4H_6]$ as a function of time. At what time τ is $[C_4H_6] = 1/200$?

Solution Let $y = [C_4H_6]$. We solve as follows:

$$\frac{dy}{dt} = -ky^2 \Longrightarrow \int \frac{1}{y^2}\frac{dy}{dt}\,dt = -\int k\,dt \Longrightarrow -\frac{1}{y(t)} = -kt + C$$

where C is the constant of integration. We can determine the two constants C and k from the two given observations: $y(0) = 1/60$ and $y(6000) = 1/140$. The first data point yields

$$-\frac{1}{1/60} = -k \cdot 0 + C,$$

or $C = -60$. The second data point yields

$$-\frac{1}{1/140} = -k \cdot 6000 - 60,$$

or $k = 1/75$. Thus, $-1/y(t) = -t/75 - 60$. Solving for $y(t)$, we obtain $y(t) = 75/(t + 4500)$. To find τ, we solve the equation $y(\tau) = 1/200$, or $75/(\tau + 4500) = 1/200$. After a bit of algebra, we find that $\tau = 10500$ s. ■

Example 6 Let $y(t)$ and $v(t) = \frac{dy}{dt}$ be the height and velocity, respectively, of a projectile that has been shot straight up from Earth's surface with initial velocity v_0. Newton's Law of Gravitation tells us that

$$\frac{dv}{dt} = -\frac{gR^2}{(R + y)^2}$$

where R is Earth's radius and g is the acceleration due to gravity at Earth's *surface*. Supposing that $v_0 < \sqrt{2gR}$, what is the maximum height attained by the projectile?

Solution At the instant the projectile reaches its maximum height and begins to fall back to Earth, its velocity v is 0. Therefore, our strategy is to find v as a function of y and solve for the value of y for which $v = 0$. As a first step, we use the Chain Rule to express $\frac{dv}{dt}$ in terms of $\frac{dv}{dy}$:

$$\frac{dv}{dt} = \frac{dv}{dy} \cdot \frac{dy}{dt} = \frac{dv}{dy} \cdot v.$$

By equating this expression for $\frac{dv}{dt}$ with the one given by Newton's Law of Gravitation, we obtain

$$v \cdot \frac{dv}{dy} = -\frac{gR^2}{(R+y)^2}.$$

Because this differential equation is separable, we follow the general procedure to obtain

$$\int v \frac{dv}{dy}\, dy = \int \left(-\frac{gR^2}{(R+y)^2} \right) dy,$$

or

$$\frac{1}{2} v(y)^2 = gR^2 (R+y)^{-1} + C.$$

When $y = 0$, we have $v = v_0$. Therefore,

$$\frac{1}{2} v_0^2 = gR^2 (R+0)^{-1} + C,$$

or $C = (1/2)v_0^2 - gR$. It follows that

$$\frac{1}{2} v(y)^2 = gR^2 (R+y)^{-1} + \left(\frac{1}{2} v_0^2 - gR \right). \tag{6.7}$$

We set $v(y) = 0$ in equation (6.7) and solve for y to find that the maximum height is $v_0^2 R/(2gR - v_0^2)$. ■

quickquiz

1. When is the first order differential equation $\frac{dy}{dx} = f(x, y)$ separable?
2. What is an initial value problem?
3. What is the procedure for solving a first order differential equation of the form $\frac{dy}{dx} = g(x)h(y)$?
4. If $\frac{dy}{dx} = \sin(x)/x$ and $y(\pi) = 2$, for what α, β, and C is $y(x) = C + \int_\alpha^\beta \sin(u)/u\, du$?

EXERCISES

Problems for Practice

In Exercises 1–8, verify that the function y satisfies the differential equation. In each expression for y, the letter C denotes a constant.

1. $\frac{dy}{dx} = xy,\ y(x) = Ce^{x^2/2}$
2. $\frac{dy}{dx} = 2xy^2,\ y(x) = 1/(C - x^2)$
3. $\frac{dy}{dx} = x - 3y,\ y(x) = x/3 - 1/9 + Ce^{-3x}$
4. $\frac{dy}{dx} = e^x + y,\ y(x) = e^x(x + C)$
5. $\frac{dy}{dx} = x + y,\ y(x) = Ce^x - x - 1$
6. $\frac{dy}{dx} = x + xy,\ y(x) = Ce^{x^2/2} - 1$
7. $\frac{dy}{dx} = y + x^2,\ y(x) = Ce^x - x^2 - 2x - 2$
8. $\frac{dy}{dx} = (2x - y)/(x + y),\ y(x) = -x + \sqrt{3x^2 + 2C}$

In Exercises 9–24, solve the initial value problem.

9. $y'(x) = 2x,\ y(1) = 3$
10. $y'(x) = 2x + 1,\ y(1) = 5$
11. $y'(x) = \cos(x),\ y(0) = 2$
12. $y'(x) = \sec^2(x),\ y(\pi/4) = 3$
13. $y'(x) = x/y,\ y(0) = 1$
14. $y'(x) = xy^2,\ y(1) = 2$
15. $y'(x) = y^2 \sin(x),\ y(0) = 2$
16. $x^3 \frac{dy}{dx} = \cos^2(y),\ y(0) = 1$
17. $y \frac{dy}{dx} = x/\sqrt{1 + y^2},\ y(0) = 1$
18. $y'(x) = \cos(x)/y,\ y(0) = 2$
19. $\frac{dy}{dx} = y + 1/y,\ y(0) = 3$
20. $(1 + x^2)^4 y'(x) = 6xy,\ y(0) = 1$
21. $x^2 y'(x) = y \sin(1/x),\ y(2/\pi) = 3$
22. $y'(x) = \exp(x + y),\ y(0) = 0$
23. $y'(x) - xy^2 = xy(4x - y),\ y(0) = 2$
24. $\frac{dy}{dx} = x - 1 + yx - y,\ y(2) = 3$

Further Theory and Practice

25. Solve $y'(x) = f'(x)g'(f(x)),\ f(0) = 3,\ g(3) = 2$, and $y(0) = 6$.
26. Show that the equation
$$y^2(x) - 2y(x) = x^3 + 2x$$
determines a solution of the initial value problem
$$\frac{dy}{dx} = \frac{3x^2 + 2}{2y - 2}, \quad y(0) = 0.$$
Observe that the first equation also determines a solution of the initial value problem comprising the same differential equation and the initial condition $y(0) = 2$. The two solutions clearly must differ. Explain how this is possible.
27. A differential equation of the form
$$\frac{dy}{dx} + p(x)y = q(x)$$
is said to be *linear*. Let $P(x)$ be an antiderivative of $p(x)$. Show that y is a solution of the given linear equation if and only if
$$\frac{d}{dx}\left(e^{P(x)}y(x)\right) = e^{P(x)}q(x).$$
Deduce that every solution of the given linear differential equation has the form
$$y(x) = e^{-P(x)}\left(\int e^{P(x)}q(x)\,dx + C\right).$$
Use this technique to derive the solution of
$$\frac{dy}{dx} = e^{-x} - y.$$
28. A differential equation of the form
$$\frac{dy}{dx} + p(x)y = q(x)y^n$$
is called a *Bernoulli equation*. Let $v(x) = y(x)^{1-n}$. Use the Chain Rule to calculate $\frac{dv}{dx}$ in terms of $\frac{dy}{dx}$. Using the original differential equation for y, show that v satisfies a linear equation (see Exercise 27). Illustrate with the equation $\frac{dy}{dx} + y = e^x y^2$. What differential equation does $v(x) = y(x)^{-1}$ satisfy? What is the solution $v(x)$ to that equation? What is $y(x)$?
29. Suppose ϕ is a function of one variable. The equation $\frac{dy}{dx} = \phi(y/x)$ is said to be *homogeneous of degree 0*. Let $w(x) = y(x)/x$. Differentiate both sides of the equation $y(x) = x \cdot w(x)$ with respect to x. By equating the resulting expression for $\frac{dy}{dx}$ with $\phi(y/x)$,

show that $w(x)$ is the solution of a separable differential equation. Illustrate this theory by solving the differential equation $\frac{dy}{dx} = xy/(x^2 + y^2)$. For this example, $\phi(w) = w/(1 + w^2)$.

30. Suppose that ψ is a function of one variable and that a, b, and c are constants. If $\frac{dy}{dx} = \psi(a + bx + cy)$, show that $u(x) = a + bx + cy(x)$ is the solution of a separable equation. Illustrate by solving the equation $\frac{dy}{dx} = 1 + x + y$.

31. The pressure P and temperature T in the outer envelope of a white dwarf (star) are related by the differential equation

$$\frac{dP}{dT} = C\frac{T^{7.5}}{P}, \quad P(0) = 0$$

where C is a constant. Find P as a function of T.

32. The interior temperature T_I (in degrees Kelvin) of a cooling white dwarf (star) satisfies the differential equation

$$\frac{dT_I}{dt} = -k\left(\frac{T_I}{7 \times 10^7 \,{}^\circ\mathrm{K}}\right)^{7/2}$$

where k is a constant with units of degrees Kelvin per year and t is time in years. Solve for T_I. If $k = 6°\mathrm{K/yr}$ and if $T_I(0) = 10^8\,°\mathrm{K}$, in how many years will the star cool to $10^4\,°\mathrm{K}$?

33. A spherical star of radius R is in equilibrium if the pressure gradient $\frac{dP}{dr}$ exactly opposes gravity. In the case of Newtonian gravitation, this amounts to

$$\frac{dP}{dr} = -\frac{Gm(r)\rho(r)}{r^2}, \quad P(R) = 0$$

where G is the gravitational constant, r represents distance to the center, $m(r)$ represents the mass of that part of the star that is within distance r of the center, and $\rho(r)$ represents the mass density at radius r. Compute $P(r)$ under the assumption that the mass density is a constant ρ. Show that the pressure at the center of the star is $2\pi G\rho^2 R^2/3$ (in terms of the stellar radius R) or $(\pi/6)^{1/3} GM^{2/3}\rho^{4/3}$ (in terms of the stellar mass M).

34. Refer to Example 6. The escape velocity is the minimum velocity $v_e = v(R)$ at which the projectile will never return to Earth. Find a formula for v_e in terms of g and R. *Hint:* How is it possible for r to become arbitrarily large? What is the smallest value of $v(0)$ for which this is the case?

35. A bullet is fired into a sand pile with an initial speed of 1250 ft/s. The rate at which its velocity is decreased after entering the sand pile is 12 times the square root of the velocity if units of feet and seconds are used. How long is it before the bullet comes to rest?

36. Refer to Exercise 35. How far does the bullet travel in the sand before coming to rest?

37. When sucrose is dissolved in water, glucose and fructose are formed. The reaction rate satisfies

$$\frac{d[C_{12}H_{22}O_{11}]}{dt} = -5.7 \times 10^{-5} \cdot [C_{12}H_{22}O_{11}].$$

For what value of t is the concentration of sucrose, $[C_{12}H_{22}O_{11}]$, equal to one-third its initial value?

38. The chemical reaction between acetone and iodine,

$$CH_3COCH_3 + I_2 \to CH_3COCH_2I + HI,$$

is governed by the differential equation

$$\frac{d[I_2]}{dt} = -0.41 \times 10^{-3} \cdot [I_2]^2$$

(in mol/L/s). If the initial iodine concentration is c_0, solve for $[I_2]$ as a function of time. Suppose $[I_2]$ is plotted as a function of time. What is the slope of the graph when $[I_2] = 1/10$ mol/L? Suppose $1/[I_2]$ is plotted as a function of time. What is the slope of the graph when $[I_2] = 1/10$ mol/L?

39. In the decomposition of dinitrogen pentoxide,

$$N_2O_5 \to 2NO_2 + \frac{1}{2}O_2,$$

the reaction rate satisfies the equation

$$\frac{d[N_2O_5]}{dt} = -k \cdot [N_2O_5]$$

for some positive constant k. If $[N_2O_5]$ is equal to 2.32 mol/L when $t = 0$ s and if $[N_2O_5]$ is equal to 0.37 mol/L when $t = 3000$ s, then determine $[N_2O_5]$ as a function of time. Illustrate this reaction by means of a suitable *straight line* plot. What is the relationship between the slope of your plot and the rate constant k?

40. The reaction rate of

$$2NO + H_2 \to N_2O + H_2O$$

is given by

$$\frac{d[N_2O]}{dt} = 0.44 \times 10^{-6}(h - [N_2O])^3$$

where h is the amount of H_2 (in moles per liter) that is initially present. Given that the initial concentration of

N_2O is 0, solve for $[N_2O]$ as an explicit function of t. In doing so, you will need to extract a square root. Decide which sign, + or −, to use. Explain your answer. What is $\lim_{t\to\infty} [N_2O]$?

41. The standard normal probability density f satisfies the initial value problem

$$f'(x) = -xf(x), \quad f(0) = \frac{1}{\sqrt{2\pi}}.$$

Find $f(x)$.

42. Suppose that in a certain ecosystem there is one type of predator and one type of prey. Let $x \cdot 100$ denote the number of predators and $y \cdot 1000$ denote the number of prey. The Austrian mathematician A. J. Lotka (1880–1949) and the Italian mathematician Vito Volterra (1860–1940) proposed the following relationship, the *Lotka-Volterra equation,* between the two population sizes:

$$\frac{dy}{dx} = \frac{y \cdot (a - bx)}{x \cdot (cy - d)}.$$

Separate variables and solve this equation. What is the predator-prey relationship if the initial prey population is 1500, the initial predator population is 200, and $a = 6$, $b = 2$, $c = 4$, and $d = 7$? Figure 4 illustrates the typical potato-pancake shape for the solution curve of this Lotka-Volterra equation.

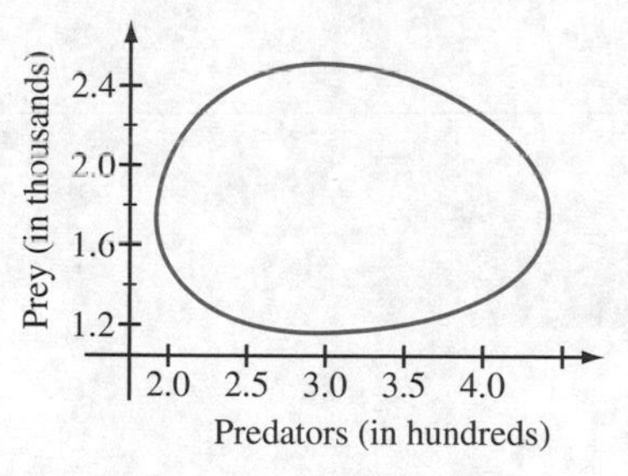

Figure 4
Solution of Lotka-Volterra equation

43. Suppose that λ, A, and p are positive constants. Solve the differential equation

$$\frac{dy}{dt} = \frac{\lambda}{p} \cdot t^{p-1} \cdot (A - y).$$

A solution of this equation is called a *Janoschek growth function*. Such functions are often used to model animal growth. Explain why the parameter A is called "the size of the adult."

44. The Tractrix The tip of a boat is at point $(L, 0)$ on the positive x-axis. Attached to the tip is a chord of length L. A man starting at $(0, 0)$ pulls the boat by the chord as he walks up the y-axis. The chord is always tangent to the path of the tip of the boat. Show that the position $y(x)$ of the tip of the boat is given by

$$\frac{dy}{dx} = -\frac{\sqrt{L^2 - x^2}}{x}, \quad y(L) = 0.$$

The graph of y is called the *tractrix* (see Figure 5).

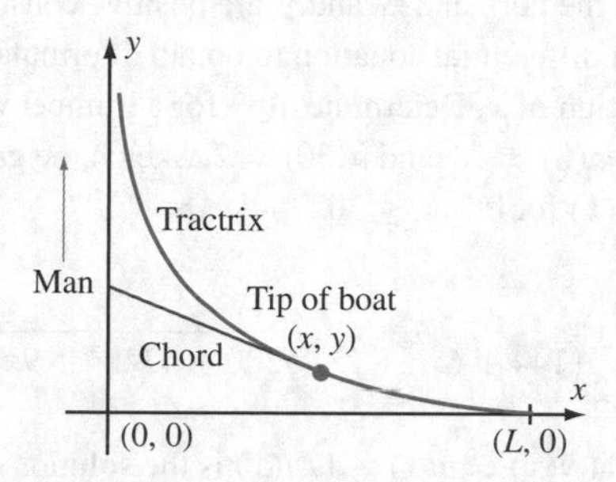

Figure 5

45. Actuaries use mortality tables that show the expected number of survivors of an initial group. Let $L(x)$ denote the number from that group who are living at age x. The instantaneous death rate is $-L'(x)$, and the percentage rate, $-L'(x)/L(x)$, is called the *force of mortality*. The first significant formula was derived by Benjamin Gompertz (1779–1865), who assumed that the force of mortality increases with age according to an expression of the form Bg^x for some constants $B > 0$ and $g > 1$. Solve for $L(x)$ under Gompertz's assumption.

46. William Makeham amended Gompertz's assumption for the force of mortality to take into account accidental deaths (refer to Exercise 45 for background). Under Makeham's assumption, the force of mortality has the form $M + Bg^x$ for some positive constants M, B, and g with $g > 1$. Solve for $L(x)$ under this assumption. The resulting formula is known as *Makeham's Law*.

47. Pierre Lecomte Du Nouy's (1883–1947) equation for the decreasing surface tension T of blood serum is

$$\frac{dT}{dt} = -k\frac{T}{\sqrt{t}}$$

for some constant k. Solve this equation if $T = \tau$ at $t = 0$.

48. Let x and y be measures of two body parts. Typically, x and y have different growth rates. Such a pattern, called *allometric growth,* often occurs because the relative growth rates of x and y are proportional to a common factor $\Phi(t)$: $(1/x)\frac{dx}{dt} = \alpha \cdot \Phi(t)$ and $(1/y)\frac{dy}{dt} = \beta \cdot \Phi(t)$ with $\alpha \neq \beta$. Show that under these conditions, x and y

satisfy the Huxley allometry equation $y = kx^p$ for suitable constants k and p.

49. The shape of a Bessel horn is determined by the equation

$$\frac{da}{dx} = -\gamma \cdot \frac{a}{x + x_0}$$

where $a(x)$ is the bore radius at distance x from the mouth of the horn and x_0 and γ are positive constants. Solve the differential equation to obtain a formula for a as a function of x. Determine $a(x)$ for a trumpet with $\gamma = 0.7$, $a(0) = 10$, and $a(30) = 2$. Sketch the graph of $y = a(x)$ for $0 \le x \le 30$.

50. Let

$$u(x) = \frac{1}{2}\left(144 + 6x^2 + 2\sqrt{5200 + 432x^2 + 9x^4}\right)^{1/3}.$$

Verify that $y(x) = u(x) - 1/u(x)$ is the solution of the initial value problem from Example 3.

51. Verify that $y(x) = \exp(-x^2)\int_0^x \exp(t^2)\,dt$, which is known as *Dawson's integral*, is the solution of the initial value problem $y'(x) = 1 - 2xy$, $y(0) = 0$.

52. For each $x > 0$, there is a unique $y > 0$ such that $ye^y = x$. Denote this value of y by $W(x)$. Show that the function W (called the Lambert W function) is a solution of the differential equation

$$\frac{dy}{dx} = \frac{y}{x(1+y)}.$$

Calculator/Computer Exercises

In Exercises 53–56, solve the initial value problem for $y(x)$. Determine the value of $y(2)$.

53. $\frac{dy}{dx} = (1 + x^2)/(1 + y^4)$, $y(0) = 0$

54. $\frac{dy}{dx} = (x^2)/(1 + \sin(y))$, $y(0) = 0$

55. $y^2 \cdot \frac{dy}{dx} = (1 + y)/(1 + x)$, $y(0) = 0$

56. $\frac{dy}{dx} = x/(1 + e^y)$, $y(0) = 0$

57. The weight y of a goose is determined as a function of time by the initial value problem $\frac{dy}{dt} = 0.09 \cdot \sqrt{t} \cdot (5000 - y)$, $y(0) = 175$ where t is measured in weeks after birth and y is measured in grams. In how many weeks does the goose reach half of its mature weight?

58. Suppose that in a certain ecosystem, $x \cdot 100$ denotes the number of predators and $y \cdot 1000$ denotes the number of prey. Suppose further that the relationship between the two population sizes satisfies the Lotka-Volterra equation

$$\frac{dy}{dx} = \frac{y \cdot (6 - 2x)}{x \cdot (4y - 5)}.$$

Separate variables and solve this equation. If the initial prey population is 1500 and the initial predator population is 200, what are the possible sizes of the population of prey when the predator size is 300?

6.2 A Calculus Approach to the Logarithm

In this section and the next, we take a second look at the exponential and natural logarithm functions. In our new approach, we use calculus to define the exponential and logarithm functions and to develop their properties. In doing so, we fill in some details that so far have been missing. Of course, everything that we learn in this chapter is consistent with the notions that have already been developed.

First, let us summarize the approach we followed in Chapters 2 and 3. In Section 2.6, we introduced a special number that we denoted by e and defined by limit formula (2.13):

$$e = \lim_{n\to\infty}\left(1 + \frac{1}{n}\right)^n.$$

We also used a limit formula, equation (2.12), to extend the definition of e^x to all real values of x. In Section 3.4, we used the limit formula

$$\lim_{h \to 0} \frac{e^h - 1}{h} = 1$$

to obtain the derivative rule for the exponential function: $\frac{d}{dx} e^x = e^x$. In Section 3.6, we defined $x \mapsto \ln(x)$ as the inverse function of $x \mapsto e^x$ and obtained the formula

$$\frac{d}{dx} \ln(x) = \frac{1}{x}$$

by means of the Inverse Function Derivative Rule.

Although the approach that we have just summarized is a valid one, it relies on the existence of the three limits we have mentioned, none of which has been proved. It is instructive to see how a new point of view can finesse these difficulties. For the purposes of this section and the next two, put aside everything you know about the functions $x \mapsto \exp(x)$ and $x \mapsto \ln(x)$. We will now redevelop the logarithm and exponential functions, *starting from scratch*. Our point of departure is a sophisticated definition of the logarithm.

Definition The *natural logarithm function* $x \mapsto \ln(x)$ is the unique solution of the initial value problem

$$\frac{dy}{dx} = \frac{1}{x} \quad (x > 0), \quad y(1) = 0.$$

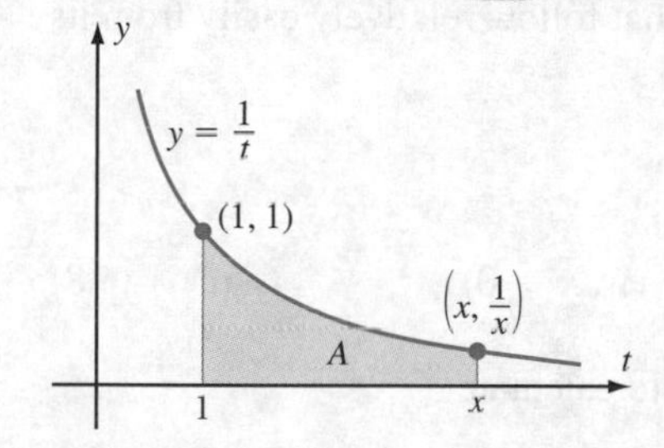

Figure 1
$\ln(x) = A$

As we know from Theorem 1 from Section 6.1, these equations result in the following formula for the natural logarithm:

$$\ln(x) = \int_1^x \frac{1}{t}\, dt, \quad x > 0.$$

Observe that the definite integral $\int_1^x 1/t\, dt$ is a function of the limit of integration x. There is no significance to the particular choice t of the variable of integration.

For $x > 1$, it is perfectly okay to think of $\ln(x)$ as the area of the region that lies under the graph of $y = 1/t$ and above the interval $[1, x]$ (see Figure 1). For $0 < x < 1$, the value $\ln(x)$ is the *negative* of the area between the graph and the x-axis. This is so because the lower limit of integration 1 is greater than the upper limit of integration x in this case (see Figure 2). As you look at Figures 1 and 2, imagine sliding the point labeled x to the left or to the right on the horizontal axis. The area of the shaded region will certainly change. The figures illustrate the functional relationship captured by the formula $\ln(x) = \int_1^x (1/t)\, dt$. Although this definition of the natural logarithm may not be intuitive, especially at first sight, it is a good one to adopt in a calculus course because it leads most quickly to the properties that we need. A number of simple facts about the natural logarithm can be immediately deduced.

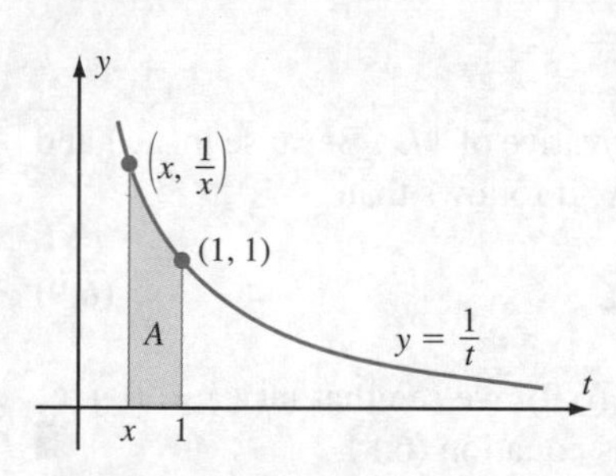

Figure 2
$\ln(x) = -A \; (0 < x < 1)$

Theorem 1 The natural logarithm has the following properties.

a. If $x > 1$, then $\ln(x) > 0$.
b. If $x = 1$, then $\ln(x) = 0$.
c. If $0 < x < 1$, then $\ln(x) < 0$.

d. The natural logarithm has a continuous derivative:

$$\frac{d}{dx}\ln(x) = \frac{1}{x}.$$

e. The natural logarithm is an increasing function—that is, $\ln(x_1) < \ln(x_2)$ if $0 < x_1 < x_2$.

Proof Theorem 1a and 1b may be understood geometrically. If $x > 1$, then $\ln(x)$ represents an area and is therefore positive. If $0 < x < 1$, then

$$\ln(x) = \int_1^x \frac{1}{t}\,dt = -\int_x^1 \frac{1}{t}\,du$$

represents the negative of an area and is therefore negative. Theorem 1c is the initial condition in the definition of the natural logarithm. Theorem 1d is also part of the definition of the natural logarithm plus the observation that $1/x$ is continuous for $x > 0$. Finally, we deduce that $\ln(x)$ is an increasing function of x because its derivative $1/x$ is everywhere positive. ■

Properties of the Natural Logarithm

The natural logarithm has many useful properties that follow relatively easily from its definition as a definite integral.

Theorem 2 The natural logarithm satisfies

$$\ln(w \cdot x) = \ln(w) + \ln(x) \quad (w, x > 0). \tag{6.8}$$

Proof To obtain identity (6.8), use the Chain Rule to calculate

$$\begin{aligned}\frac{d}{dx}\ln(wx) &= \frac{1}{wx}\cdot\frac{d}{dx}(wx)\\ &= \frac{1}{wx}\cdot w = \frac{1}{x}.\end{aligned}$$

In other words, for each fixed w, $\ln(wx)$ is an antiderivative of $1/x$. Because $\ln(wx)$ and $\ln(x)$ are antiderivatives of the same expression $1/x$, it follows that

$$\ln(wx) = \ln(x) + C \tag{6.9}$$

for some constant C. By setting $x = 1$ in equation (6.9), we see that $\ln(w) = 0 + C$. Substituting this value for C in equation (6.9) yields equation (6.8). ■

The natural logarithm was invented because there was a need for a function with property (6.8). In the late 16th century, multiplication was a tedious process known only to scholars. John Napier (1550–1617) invented logarithms to simplify the operation. With tables that convert a positive number u to $\ln(u)$ and back, equation (6.8) can be used to reduce the operation of multiplication to one of addition. The next example illustrates this idea.

Example 1 A table of logarithms contains the following entries:

x	$\ln(x)$
⋮	⋮
6975	8.850087607
⋮	⋮
7891	8.973478149
⋮	⋮
55039725	17.82356576
⋮	⋮

Use this information to find the product of 6975 and 7891.

Solution The table tells us that 8.850087607 and 8.973478149 are the natural logarithms of 6975 and 7891, respectively. The sum of these two logarithms is 17.82356576. The number 55,039,725 beside this sum is therefore the product of 6975 and 7891. (Logarithms were widely used in this way from their first publication in 1614 until the advent of handheld scientific calculators in the early 1970s.) ■

The next theorem gathers together some consequences of identity (6.8).

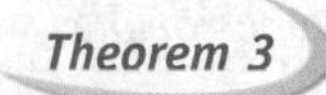

Theorem 3 If x is positive, then

$$\ln(x^{-1}) = -\ln(x). \tag{6.10}$$

If y is also positive, then

$$\ln\left(\frac{x}{y}\right) = \ln(x) - \ln(y). \tag{6.11}$$

If p is any number, then

$$\ln(x^p) = p \cdot \ln(x). \tag{6.12}$$

Proof Notice that

$$\ln(x) + \ln(x^{-1}) \overset{(6.8)}{=} \ln(x \cdot x^{-1}) = \ln(1) = 0,$$

which is equivalent to equation (6.10). To obtain equation (6.11), we write x/y as xy^{-1} and use equations (6.8) and (6.10) as follows:

$$\ln\left(\frac{x}{y}\right) = \ln(xy^{-1}) \overset{(6.8)}{=} \ln(x) + \ln(y^{-1}) \overset{(6.10)}{=} \ln(x) - \ln(y).$$

We can prove equation (6.12) in several stages. As you read the demonstration, notice how earlier steps are used to prove later steps. First, if $p = 0$, then $\ln(x^p) = \ln(1) = 0 = 0 \cdot \ln(x)$. Thus, equation (6.12) is true for $p = 0$. It is also clear that equation (6.12) holds for $p = 1$. Next, observe that

$$\ln(x^2) = \ln(x \cdot x) = \ln(x) + \ln(x) = 2 \cdot \ln(x).$$

Similarly,

$$\ln(x^3) = \ln(x^2 \cdot x) = \ln(x^2) + \ln(x) = 2 \cdot \ln(x) + \ln(x) = 3\ln(x).$$

This idea can be repeated, showing that equation (6.12) holds for every positive integer. If we now suppose that p is a negative integer, then $p = -|p|$ with $|p|$ a positive integer. By the positive integer case of equation (6.12), which has just been proved, we have

$$\ln(x^p) = \ln\left(x^{-|p|}\right) = \ln\left((x^{-1})^{|p|}\right) \overset{(6.12)}{=} |p|\ln(x^{-1}) \overset{(6.10)}{=} -|p|\ln(x) = p\ln(x),$$

which completes the proof that equation (6.12) is valid for all integers p. Now suppose that p is a rational number—that is, suppose there are two integers m and $n \neq 0$ such that $p = m/n$. By the integer case of equation (6.12),

$$m \cdot \ln(x) \overset{(6.12)}{=} \ln(x^m) = \ln(x^{p \cdot n}) = \ln((x^p)^n) \overset{(6.12)}{=} n \cdot \ln(x^p).$$

By dividing the first and last terms in this chain of equalities by n, we obtain $\ln(x^p) = (m/n) \cdot \ln(x) = p \cdot \ln(x)$. We have now proved that equation (6.12) holds for all rational values of p. In Section 6.4, we rigorously define irrational exponents in a way that ensures the validity of equation (6.12) for these values of p as well. ■

Example 2 Express $\ln(9/125)$ in terms of $\ln(3)$ and $\ln(5)$.

Solution We calculate that

$$\ln(9/125) = \ln(9) - \ln(125) = \ln(3^2) - \ln(5^3) = 2\ln(3) - 3\ln(5).$$ ■

Example 3 Express

$$A = \ln\left(\frac{\sqrt{a} \cdot b^2}{c^3 \cdot d}\right)$$

in terms of $\ln(a)$, $\ln(b)$, $\ln(c)$, and $\ln(d)$.

Solution We calculate as in Example 2. This time, however, we cite each use of Theorem 3:

$$\begin{aligned} \ln\left(\frac{\sqrt{a} \cdot b^2}{c^3 \cdot d}\right) &= \ln(a^{1/2} \cdot b^2) - \ln(c^3 \cdot d) && \text{Equation (6.11)} \\ &= \ln(a^{1/2}) + \ln(b^2) - [\ln(c^3) + \ln(d)] && \text{Equation (6.8)} \\ &= \frac{1}{2}\ln(a) + 2\ln(b) - 3\ln(c) - \ln(d). && \text{Equation (6.12)} \end{aligned}$$

■

Graphing the Natural Logarithm Function

Now we examine the graph of $y = \ln(x)$. We have already learned that the natural logarithm is an increasing function with x-intercept 1 (see Theorem 1b and 1e). Because

$$\frac{d^2}{dx^2}\ln(x) = \frac{d}{dx}\left(\frac{1}{x}\right) = -\frac{1}{x^2} < 0,$$

we infer that the graph of the natural logarithm is concave down. There are no relative maxima or minima since the derivative $1/x$ is never 0.

Let us now investigate the behavior of $\ln(x)$ as x tends to infinity. Since the natural logarithm is an increasing function, $\ln(x)$ either increases to a finite bound as x tends to infinity or $\ln(x)$ increases without bound. We may rule out the first possibility by considering the sequence $\{\ln(2^n)\}$. Because $2 > 1$, we deduce that $\ln(2) > \ln(1) = 0$. Therefore,

$$\ln(2^n) \overset{(6.12)}{=} n \cdot \ln(2) \to \infty, \quad \text{as } n \to \infty.$$

We conclude that

$$\lim_{x\to\infty} \ln(x) = \infty.$$

We can complete our sketch of $y = \ln(x)$ for $0 < x < \infty$ by investigating the behavior as $x \to 0^+$. In fact, a change of variable allows us to determine the behavior at 0^+ from the limit at infinity:

$$\lim_{x\to 0^+} \ln(x) \overset{u=1/x}{=} \lim_{u\to\infty} \ln\left(\frac{1}{u}\right) \overset{(6.10)}{=} -\lim_{u\to\infty} \ln(u) = -\infty.$$

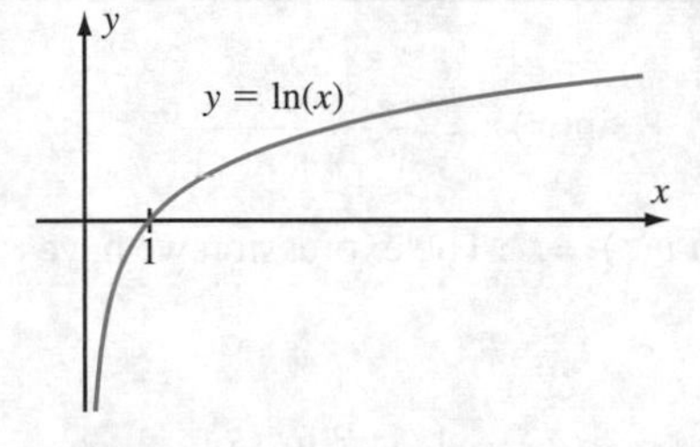

Figure 3

In other words, the y-axis is a vertical asymptote. (Figure 3 shows a sketch of the graph of $y = \ln(x)$.)

Since $x \mapsto \ln(x)$ is a continuous function that takes on arbitrarily large positive and negative numbers, the Intermediate Value Theorem (Section 2.3) tells us that the equation $\ln(x) = \gamma$ has a solution for each real number γ. Because the natural logarithm is an increasing function, the solution is in fact unique. We summarize our findings in the following theorem.

Theorem 4 The natural logarithm is an increasing function with domain equal to the set of positive real numbers. Its range is the set of all real numbers. The equation $\ln(x) = \gamma$ has a *unique* solution $x \in \mathbb{R}^+$ for every $\gamma \in \mathbb{R}$. The graph of the natural logarithm function is concave down. The y-axis is a vertical asymptote for the graph.

A LOOK BACK

Defining $\ln(x)$ to be the solution of the initial value problem $\frac{dy}{dx} = 1/x, y(0) = 1$ enables us to derive all the important properties of the natural logarithm. How do we know that this logarithm function is the same as the one defined in Section 3.6? The answer is that we have shown that the two logarithm functions satisfy the same initial value problem. *Therefore, they* must *be the same function since we know from Theorem 1 from Section 6.1 that the initial value problem has a unique solution.*

Differentiating Expressions Involving the Logarithm Function

The basic differentiation formula $\frac{d}{dx}\ln(x) = 1/x$ is often used in conjunction with the Chain Rule:

$$\frac{d}{dx}\ln(u) = \frac{1}{u}\cdot\frac{du}{dx} \tag{6.13}$$

and

$$\frac{d}{dx}v(\ln(x)) = v'(\ln(x))\cdot\frac{1}{x}. \tag{6.14}$$

Example 4 Calculate

$$\frac{d}{dx}\ln(x^2+\sin(x)), \quad \frac{d}{dx}\ln^3(x), \quad \text{and} \quad \frac{d}{dx}(\ln(x)\cdot\tan(x)).$$

Solution For the first problem, we use equation (6.13) with $u(x) = x^2 + \sin(x)$:

$$\frac{d}{dx}\ln(x^2+\sin(x)) \overset{(6.13)}{=} \frac{1}{x^2+\sin(x)}\,\frac{d}{dx}(x^2+\sin(x)) = \frac{2x+\cos(x)}{x^2+\sin(x)}.$$

For the second problem, we use equation (6.14) with $v(t) = t^3$. The expression we have to differentiate is $v(\ln(x))$.

$$\frac{d}{dx}\ln^3(x) \overset{(6.14)}{=} \frac{d}{dx}v(\ln(x)) = v'(\ln(x))\cdot\frac{1}{x} = 3\ln^2(x)\cdot\frac{1}{x} = \frac{3\ln^2(x)}{x}.$$

The last differentiation involves the Product Rule but not the Chain Rule since there is no composition of functions:

$$\begin{aligned}\frac{d}{dx}(\ln(x)\cdot\tan(x)) &= \left(\frac{d}{dx}\ln(x)\right)\cdot\tan(x) + \ln(x)\cdot\frac{d}{dx}\tan(x)\\ &= \frac{1}{x}\cdot\tan(x) + \ln(x)\cdot\sec^2(x).\end{aligned}$$

■

$y = \ln(|x|)$

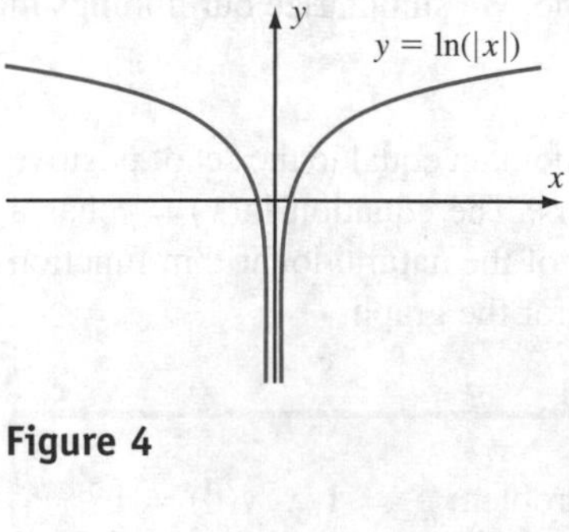

Figure 4

Because we have defined $\ln(x)$ only when $x > 0$, the graph can only be sketched to the right of the y-axis (as in Figure 3, previous page). However, it certainly makes sense to discuss $\ln(|x|)$ when $x \neq 0$. The graph of this *composite* function is shown in Figure 4. Now we will compute the derivative of $\ln(|x|)$.

Example 5 For $x \neq 0$, show that

$$\frac{d}{dx}\ln(|x|) = \frac{1}{x}. \tag{6.15}$$

Solution As usual with problems involving absolute values, we divide the computation into cases. If $x > 0$, then $|x| = x$. In this case,

$$\frac{d|x|}{dx} = 1 \quad \text{and} \quad \frac{d\ln(|x|)}{dx} \overset{(6.13)}{=} \frac{1}{|x|} \cdot 1 = \frac{1}{|x|} = \frac{1}{x}.$$

If $x < 0$, then $|x| = -x$. Therefore,

$$\frac{d|x|}{dx} = -1 \quad \text{and} \quad \frac{d\ln(|x|)}{dx} \overset{(6.13)}{=} \frac{1}{|x|} \cdot (-1) = \frac{1}{x}.$$

Thus, equation (6.15) is true, whatever the sign of x. ■

Integrals Involving the Natural Logarithm

Although the antiderivative formula

$$\int \frac{1}{x}\, dx = \ln(x) + C$$

is correct, we cannot use it to evaluate the definite integral

$$\int_{-2}^{-1} \frac{1}{x}\, dx.$$

If we attempted to do so, we would have

$$\int_{-2}^{-1} \frac{1}{x}\, dx = \ln(x)\Big|_{x=-2}^{x=-1} = \ln(-1) - \ln(-2).$$

In more advanced studies, we could define the logarithm for negative values in such a way that this integration is correct. Doing so, however, requires complex numbers and is quite sophisticated. Instead, we turn to equation (6.15) for a simple alternative. The equation tells us that $\ln(|x|)$ is an antiderivative of $1/x$ that is valid for negative as well as positive numbers. This gives us our basic integration formula:

$$\int \frac{1}{x}\, dx = \ln(|x|) + C. \tag{6.16}$$

Example 6 Calculate $\int_{-2}^{-1} 1/x\, dx$.

Solution We have

$$\int_{-2}^{-1} \frac{1}{x}\, dx = \ln(|x|)\Big|_{x=-2}^{x=-1} = \ln(1) - \ln(2) = 0 - \ln(2) = -\ln(2).$$ ■

Do not confuse the calculation in Example 6 with the following incorrect computation:

$$\int_{-2}^{+1} \frac{1}{x}\,dx \overset{\text{Wrong!}}{=} \ln(|x|)\Big|_{x=-2}^{x=+1} = \ln(1) - \ln(2) = -\ln(2).$$

Neither $1/x$ nor $\ln(|x|)$ is defined at the point $x = 0$ in the interval of integration $[-2, 1]$. Neither function can be extended to $x = 0$ in a continuous way. The application of the Fundamental Theorem of Calculus in the leftmost equation is not justified.

Example 7 Calculate $\int 1/(3+5x)\,dx$.

Solution By making the change of variable $u = 3 + 5x$, we have $\frac{du}{dx} = 5$ and $dx = du/5$:

$$\int \frac{1}{3+5x}\,dx = \int \frac{1}{u}\frac{du}{5} = \frac{1}{5}\ln(|u|) + C = \frac{1}{5}\ln(|3+5x|) + C.$$

Notice that it is appropriate to include the absolute value function even when calculating an indefinite integral. ■

Integrating Trigonometric Functions

Although we know how to differentiate each trigonometric function, we have, until now, only learned to integrate the sine and cosine functions. We can use equation (6.16) to obtain the antiderivatives of the four remaining trigonometric functions.

Example 8 Show that

$$\int \tan(x)\,dx = \ln(|\sec(x)|) + C. \tag{6.17}$$

Solution When we are presented with an integrand that is made up of trigonometric functions, it is sometimes helpful to rewrite the expression entirely in terms of sines and cosines. Thus, we have

$$\int \tan(x)\,dx = \int \frac{\sin(x)}{\cos(x)}\,dx.$$

The substitution $u = \cos(x)$, $du = -\sin(x)\,dx$ converts the last integral to

$$\int -\frac{1}{u}\,du = -\ln(|u|) + C.$$

Resubstituting $u = \cos(x)$ yields the solution

$$\int \tan(x)\,dx = -\ln(|\cos(x)|) + C.$$

We can convert this equality to formula (6.17) as follows:

$$\int \tan(x)\,dx = -\ln(|\cos(x)|) + C \overset{(6.10)}{=} \ln(|\cos(x)|^{-1}) + C = \ln(|\sec(x)|) + C.$$ ■

inSIGHT

As a rule of thumb, remember that the tangent and secant functions occur together in differentiation and integration formulas. For example, we have already learned from Section 3.4 that

$$\int \sec^2(x)\,dx = \tan(x) + C$$

and

$$\int \sec(x)\tan(x)\,dx = \sec(x) + C.$$

Equation (6.17) is another instance in which $\tan(x)$ and $\sec(x)$ appear together in an integral formula.

We obtain the integral of the cotangent function in a similar way:

$$\int \cot(x)\,dx = \int \frac{\cos(x)}{\sin(x)}\,dx = \ln(|\sin(x)|) + C = -\ln(|\csc(x)|) + C. \tag{6.18}$$

Formulas for the antiderivatives of the secant and cosecant are a bit more complicated:

$$\int \sec(x)\,dx = \ln(|\sec(x) + \tan(x)|) + C \tag{6.19}$$

and

$$\int \csc(x)\,dx = -\ln(|\csc(x) + \cot(x)|) + C.$$

Example 9 Derive formula (6.19).

Solution It is a routine matter to verify formula (6.19) by differentiating the right side. However, that solution sheds no light on how the right side can be found. The following direct approach is fast, but because it relies on a surprising algebraic manipulation, it does little to explain the discovery of the formula. The trick is to multiply and divide the integrand by the expression $\sec(x) + \tan(x)$:

$$\int \sec(x)\,dx = \int \sec(x) \cdot \frac{\sec(x) + \tan(x)}{\sec(x) + \tan(x)}\,dx = \int \frac{\sec(x)\tan(x) + \sec^2(x)}{\sec(x) + \tan(x)}\,dx.$$

Now we make the substitution $u = \sec(x) + \tan(x)$, $du = (\sec(x)\tan(x) + \sec^2(x))\,dx$. The last integral becomes $\int (1/u)\,du = \ln(|u|) + C$. When we resubstitute $u = \sec(x) + \tan(x)$, we obtain formula (6.19). The Genesis & Development for Chapter 7 provides a detailed historical discussion of the integral $\int \sec(x)\,dx$. ■

Summary of Integration Rules

$$\int \tan(x)\,dx = \ln(|\sec(x)|) + C \qquad \int \cot(x)\,dx = -\ln(|\csc(x)|) + C$$

$$\int \sec(x)\,dx = \ln(|\sec(x) + \tan(x)|) + C \qquad \int \csc(x)\,dx = -\ln(|\csc(x) + \cot(x)|) + C$$

quickquiz

1. What is the definition of the natural logarithm function?
2. What is the derivative of the function $\ln(1 + x^2)$?
3. Evaluate $\int 1/(2x + 1)\,dx$.
4. If $\ln(y) = 2\ln(a) - 3\ln(b)$, what is y?

EXERCISES

Problems for Practice

In Exercises 1–12, calculate the derivative with respect to x of the given expression.

1. $\ln(3x)$ **2.** $\ln(-x)$ **3.** $\ln(5x + 2)$
4. $\sqrt{2x} + \ln(7)$ **5.** $\ln(\sin(x))$ **6.** $\ln(\sqrt{1 - x})$
7. $\ln(1/x)$ **8.** $1/(4 + \ln(x))$ **9.** $\tan(2x) \cdot \ln(x)$
10. $\ln(\ln(x))$ **11.** $\ln^3(x)$ **12.** $\ln(2|x|)$

In Exercises 13–22, use the Method of Substitution to calculate the integral.

13. $\displaystyle\int \frac{1}{x+1}\,dx$ **14.** $\displaystyle\int \frac{1}{2x+1}\,dx$

15. $\displaystyle\int \frac{x}{x^2+4}\,dx$ **16.** $\displaystyle\int \frac{\ln(x)}{x}\,dx$

17. $\displaystyle\int \frac{1}{x \cdot \ln(x)}\,dx$ **18.** $\displaystyle\int \cot(x)\,dx$

19. $\displaystyle\int \frac{\sqrt{\ln(x)}}{x}\,dx$ **20.** $\displaystyle\int_0^1 \frac{x^3}{x^4+1}\,dx$

21. $\displaystyle\int \frac{\sec^2(x)}{\tan(x)}\,dx$ **22.** $\displaystyle\int \frac{\sin(x)}{8 - 3\cos(x)}\,dx$

In Exercises 23 and 24, find real numbers A, B, C, D, and E such that the expression has the form

$$A \cdot \ln(a) + B \cdot \ln(b) + C \cdot \ln(c) + D \cdot \ln(d) + E.$$

23. $\ln\left(\dfrac{a^3 b^{-2}}{c^2 d^3}\sqrt{2}\right)$ **24.** $\ln\left(\dfrac{21d}{a^{-2} b^3 c^2}\right)$

25. Simplify $\ln(\sec(x)) + \ln(\cos(x))$.

26. Simplify $\ln(\tan(x)) - \ln(\sin(x)) + 2\ln(\cos(x))$.

Further Theory and Practice

27. The mortality rates R associated with several cancers have been modeled by the equation

$$R(t) = B \cdot t^k$$

where $R(t)$ is the rate at age t and B and k are constants. This relationship is most often represented graphically by plotting $\ln(R(t))$ as a function of $\ln(t)$. Find a formula for $\ln(R(t))$ in terms of $\ln(t)$. Describe the plot.

28. **Logarithmic Differentiation** If f is a differentiable function that does not vanish, prove that

$$\frac{d}{dx}\ln(|f(x)|) = \frac{f'(x)}{f(x)}.$$

This simple fact can be used to calculate a number of otherwise unpleasant derivatives. In the following equations, calculate $f'(x)$ by taking the natural logarithm of both sides and differentiating that equation instead.

a. $f(x) = \sin(x) \cdot \cos^2(x) \cdot (x^2 - 5)^3$
b. $f(x) = (x + 1) \cdot (x + 2)^2 \cdot (x + 3)^3 \cdot (x + 4)^4$
c. $f(x) = (1 + x)^3 \tan^4(x)/(x^2 + 6)^2$

In Exercises 29–33, calculate the integral.

29. $\displaystyle\int_0^1 \frac{2x+3}{x+1}\,dx$ **30.** $\displaystyle\int \frac{\sin(x)\cos(x)}{1+\sin^2(x)}\,dx$

31. $\displaystyle\int \frac{1}{\sqrt{x}\cdot(2+\sqrt{x})}\,dx$ **32.** $\displaystyle\int \frac{\ln(\sqrt{x})}{x}\,dx$

33. $\displaystyle\int_1^e \left(\ln(x)+\ln\left(\frac{1}{x}\right)\right)dx$

In Exercises 34–37, state the domain of the function *f*. Determine where *f* is increasing and where *f* is decreasing. On what interval(s) is the graph of *f* concave up? Concave down? Identify any points of inflection. Identify any asymptotes. Sketch the graph of *f*.

34. $f(x)=\ln(x^2+1)$ **35.** $f(x)=x^2-2\ln(x)$

36. $f(x)=\ln^2(x)$ **37.** $f(x)=\ln(x/(x+1))$

38. The point (1, 1) is on the curve defined by the equation $2y\ln(xy)-x\sin(y-1)=0$. Find the slope of the tangent to this curve at the point (1, 1).

39. Find the slope of the tangent line to the curve defined by the equation $y\ln(x)-\ln(y+x)=0$ at the point (1, 0).

40. Express $\ln(2)+\ln(3)+\cdots+\ln(n)$ as a single logarithm.

41. If f is an odd function with values in the interval $(-A, A)$, show that $g(x)=\ln\left(\dfrac{A-f(x)}{A+f(x)}\right)$ is also an odd function.

42. It is known that $\int_1^e \ln(x)\,dx = 1$. If a is a positive constant, what is

$$\int_1^e \ln(ax)\,dx?$$

43. Suppose $g(x)=\ln(f(x))$. If $f(2.85)=-0.1$, $f(2.90)=0.1$, and $f(2.95)=0.5$, approximate $g'(2.90)$ by means of a central difference quotient.

44. Is $\int 1/|x|\,dx=\ln(|x|)+C$ a valid integration formula? Why or why not?

45. Use the methods you learned in Chapter 4 to find the minimum value m of $f(x)=x-\ln(x)$. Use the inequality $x-\ln(x)\ge m$ to deduce that $\ln(1+x)<x$ for $x>-1$. Prove that there is no real value of x such that $\ln(x)=x$.

46. Find $\frac{dy}{dx}$ if $y=\int_1^{\ln(x)} 1/t\,dt$ and $x>1$.

47. Let A and B be constants. Show that $y = A\cos(\ln(x))+B\sin(\ln(x))$ is a solution of the differential equation

$$x^2y''+xy'+y=0.$$

48. Let m be any positive number. Show by explicit construction that the graph of $y=\ln(x)$ has a tangent line with a slope of m.

49. Find constants A and B such that

$$\int \ln(x)\,dx = A\cdot x\ln(x)+B\cdot x+C$$

where C is an arbitrary constant.

50. Fix a positive h and graph $f(x)=1/(1+x)$ for $0\le x\le h$. By comparing the area under the graph of f with an inscribed rectangle and a circumscribed rectangle, show that

$$\frac{h}{1+h}<\ln(1+h)<h \quad (h>0).$$

51. In this exercise, you will demonstrate the inequality

$$\ln(t)\le\sqrt{t}-\frac{1}{\sqrt{t}} \quad (t\ge 1).$$

Start with the obvious inequality $0\le(\sqrt{t}-1)^2$. Deduce that

$$\frac{1}{t}\le\frac{1}{2}\frac{t+1}{t^{3/2}}.$$

Verify that this inequality amounts to

$$\frac{d}{dt}\ln(t)\le\frac{d}{dt}\left(\sqrt{t}-\frac{1}{\sqrt{t}}\right).$$

Use the Fundamental Theorem of Calculus to explain why this inequality implies the desired inequality.

52. In this exercise, you will demonstrate the inequality

$$2t\le(t+1)\ln(t)+2 \quad (t\ge 1).$$

Start by showing that $\ln(t)+1/t$ has a minimum value at $t=1$. Deduce that $1-1/t\le\ln(t)$. Deduce that

$$2\le\ln(t)+\frac{t+1}{t}.$$

Verify that this last inequality amounts to

$$\frac{d}{dt}(2t)\le\frac{d}{dt}((t+1)\ln(t)+2).$$

Use the Fundamental Theorem of Calculus to explain why this inequality implies the desired inequality.

53. Let f be a differentiable function defined on $\mathbb{R}^+$ and satisfying identity $f(xy)=f(x)+f(y)$. This exercise outlines a proof that

$$f(x)=f'(1)\cdot\ln(x).$$

a. Differentiate both sides of the equation with respect to x, treating y as constant, to show that $f'(xy)\cdot y=f'(x)$.

b. By setting $x = t$ and $y = 1/t$ in the equation obtained in part a, show that f is an antiderivative of the function $t \mapsto f'(1)/t$.

c. Use the result from part b to deduce that $f(t) = f'(1) \cdot \ln(t) + C$ where C is a constant. By evaluating $f(t)$ at $t = 1$, show that $C = f(1)$.

d. By considering the given property of f with $x = y = 1$, deduce that $f(1) = 0$.

54. Let $\kappa(x)$ be the curvature of the curve $y = \ln(x)$ at point $(x, \ln(x))$. Calculate

$$\lim_{x \to 0^+} \kappa(x) \quad \text{and} \quad \lim_{x \to \infty} \kappa(x).$$

What is the maximum value of κ and where does it occur?

55. The *logarithmic mean* $L(x, y)$ of two different positive numbers x and y is defined as

$$L(x, y) = \frac{y - x}{\ln(y) - \ln(x)}.$$

Notice that L is symmetric: $L(x, y) = L(y, x)$. We may therefore suppose $y > x$ without loss of generality. The arithmetic and geometric means of x and y are defined by

$$A(x, y) = \frac{x + y}{2} \quad \text{and} \quad G(x, y) = \sqrt{xy}.$$

Parts a and b that follow prove that

$$G(x, y) \leq L(x, y) \leq A(x, y).$$

a. Let $t = y/x$ in the inequality from Exercise 51 to deduce that $G(x, y) \leq L(x, y)$.

b. Let $t = y/x$ in the inequality from Exercise 52 to deduce that $L(x, y) \leq A(x, y)$.

Calculator/Computer Exercises

56. Use the Midpoint Rule with $N = 10$ to approximate $\ln(3)$.

57. Use Simpson's Rule with $N = 6$ to approximate $\ln(2)$.

58. Starting from the initial estimate $x_1 = 2.0$, use the Newton-Raphson Method to find successive approximations x_2, x_3, and x_4 of the solution of the equation $\ln(x) = 1$.

59. Investigate $\lim_{x \to 0^+} x \ln(x)$ graphically and numerically. What does this limit appear to be?

60. Which is larger, $a = 3^{100000}$ or $b = 100000^{9542}$? Do not use any numbers larger than 200,000 in your determination. Explain how you obtained your answer. Note that both numbers a and b are much larger than 10^{100}. Most scientific calculators cannot handle numbers of this magnitude. Nevertheless, you can use any scientific calculator to answer this exercise.

61. The formula $H(t) = 3.7 \ln(t) + 2t + 27.6$ has been used to model the height, in inches, of young children. Here t is measured in years and the formula is applied when $0.5 \leq x \leq 5.5$. Graph H in the window $[1/2, 5] \times [26, 44]$. At what age does the height of a typical child reach 40 inches?

62. Use the Newton-Raphson Method to find an approximation to the solution of the equation $\ln(x) = 3 - x$. How can you be sure that there is *exactly* one real solution?

63. The integral $\int_a^b 1/\ln(x)\,dx$ gives an approximation of the number of primes p in the interval $[a, b]$. For example, the 8000th and 9000th primes are 81,799 and 93,179, respectively. Thus, there are exactly 1000 primes in the interval [81800, 93179]. As you can see, the approximation $\int_{81800}^{93179} 1/\ln(x) = 1000.048848$ is quite a good one. Use Simpson's Rule with $N = 4$ to approximate the number of primes that lie in the interval [50000, 60000]. (Since the 5133th prime is 49,999 and the 6057th prime is 59,999, there are exactly $6057 - 5133 = 924$ primes in [50000, 60000].)

6.3 The Exponential Function

In Section 6.2, we learned that the equation $\ln(x) = y$ has a *unique* positive solution x for every real number y. In other words, the natural logarithm is an invertible function with domain $\mathbb{R}^+$ and image $\mathbb{R}$. In this section, we use that property to define and study the inverse of the natural logarithm. Our discussion relies on several general facts about inverse functions that we learned in Chapter 3. You might find it helpful to review Section 3.6 before continuing.

Definition The inverse of the natural logarithm is called the *exponential function* (or *natural exponential function*) and is written $x \mapsto \exp(x)$. The domain of the exponential function is the entire real line; the image of the exponential function is the set of positive real numbers.

Recall that, when we say the exponential function is the inverse function of the natural logarithm, we mean that a real number a and a positive number b are related by the equation

$$b = \exp(a)$$

if and only if

$$a = \ln(b).$$

We may also express the inverse relationship between the natural exponential and logarithm functions as follows:

$$\ln(\exp(a)) = a, \quad \text{for all real } a,$$

and

$$\exp(\ln(b)) = b, \quad \text{for all } b > 0.$$

Example 1 Using the definition of the exponential function, simplify the expressions

$$\exp(\ln(s) - \ln(t)) \quad \text{and} \quad \ln(5 \cdot \exp(a)).$$

Solution We have

$$\exp(\ln(s) - \ln(t)) = \exp\left(\ln\left(\frac{s}{t}\right)\right) = \frac{s}{t}$$

and

$$\ln(5 \cdot \exp(a)) = \ln(5) + \ln(\exp(a)) = \ln(5) + a.$$ ■

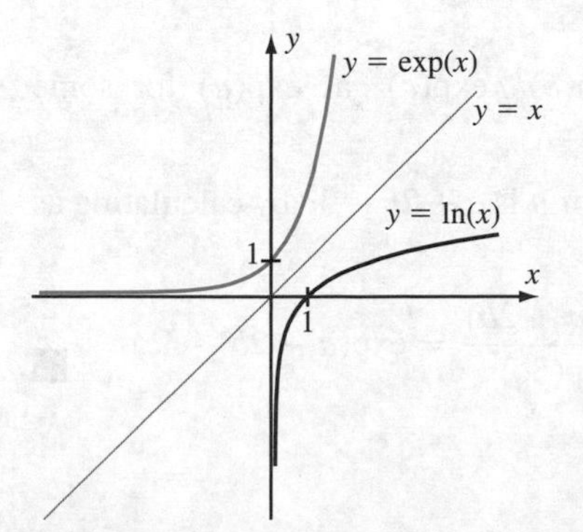

Figure 1

Properties of the Exponential Function

The graph of $y = \exp(x)$, which appears in Figure 1, is obtained by reflecting the graph of $y = \ln(x)$ through the line $y = x$. (Recall that this is the general procedure for finding the graph of an inverse function.) Because the graph of $y = \ln(x)$ lies below the line $y = x$ and is concave down and rising, we conclude that the graph of $y = \exp(x)$ lies above the line $y = x$ and is concave up and rising. In particular, the natural exponential is an increasing function. Since $\ln(1) = 0$, it follows that $\exp(0) = 1$. In other words, 1 is the y-intercept of the graph of $y = \exp(x)$. Because the exponential function only assumes positive values, there is no x-intercept.

Next we turn to some of the algebraic properties of the exponential function. The basic exponential law is

$$\exp(s+t) = \exp(s) \cdot \exp(t), \tag{6.20}$$

which holds for all real numbers s and t. This identity is the analogue of equation (6.8) for the logarithm, from which it can be derived. Indeed, notice that

$$\ln(\exp(s+t)) = s+t = \ln(\exp(s)) + \ln(\exp(t)) \overset{(6.8)}{=} \ln(\exp(s) \cdot \exp(t)).$$

Since the natural logarithm is a one-to-one function, it follows that $\exp(s+t)$ must equal $\exp(s) \cdot \exp(t)$, as identity (6.20) claims.

We can obtain a few other useful identities from equation (6.20). For instance, when s equals $-t$, equation (6.20) becomes $\exp(0) = \exp(-t)\exp(t)$, or

$$\exp(-t) = \frac{1}{\exp(t)}, \tag{6.21}$$

which is the analogue of logarithm identity (6.10). We can combine formulas (6.20) and (6.21) to produce

$$\exp(s-t) = \frac{\exp(s)}{\exp(t)}, \tag{6.22}$$

which is the analogue of logarithm identity (6.11). Finally, we obtain the exponential version of logarithm equation (6.12) by noting that

$$\ln(\exp(s)^t) \overset{(6.8)}{=} t\ln(\exp(s)) = st = \ln(\exp(st)).$$

Since the logarithm is one-to-one, we deduce that

$$(\exp(s))^t = \exp(st) \tag{6.23}$$

for all real s and t. Before continuing, observe that equations (6.20)–(6.23) are reminiscent of the following laws of exponents: $b^{s+t} = b^s b^t$, $b^{-t} = 1/b^t$, $b^{s-t} = b^s/b^t$, and $(b^s)^t = b^{st}$.

Example 2 Write the expression $\exp(a)\exp(b)^2/\exp(c)^3$ as $\exp(u)$ for some expression u.

Solution We find that the required expression for u is $a + 2b - 3c$ by calculating as follows:

$$\frac{\exp(a) \cdot \exp(b)^2}{\exp(c)^3} = \frac{\exp(a)\exp(2b)}{\exp(3c)} = \frac{\exp(a+2b)}{\exp(3c)} = \exp(a+2b-3c).$$

■

Derivatives and Integrals Involving the Exponential Function

Just as the algebraic properties of the exponential function follow from corresponding properties of the logarithm, so too can the calculus properties of the exponential function be deduced from their logarithm counterparts.

Theorem 1 The exponential function satisfies

$$\frac{d}{dx}\exp(x) = \exp(x) \quad \text{and} \quad \int \exp(x)\,dx = \exp(x) + C.$$

More generally,

$$\frac{d}{dx}\exp(u) = \exp(u)\frac{du}{dx} \quad \text{and} \quad \int \exp(u)\frac{du}{dx}\,dx = \exp(u) + C.$$

Proof When we studied inverse functions in Section 3.6, we learned that if g is the inverse function of f, if f is differentiable at $g(c)$, and if $f'(g(c)) \neq 0$, then g is differentiable at c and

$$g'(c) = \frac{1}{f'(g(c))}.$$

If $g(x) = \exp(x)$ and $f(x) = \ln(x)$, then

$$\left.\frac{d}{dx}\exp(x)\right|_{x=c} = \frac{1}{\left.\frac{d}{dx}\ln(x)\right|_{x=\exp(c)}} = \frac{1}{\left.\frac{1}{x}\right|_{x=\exp(c)}} = \exp(c).$$

In other words,

$$\frac{d}{dx}\exp(x) = \exp(x),$$

which is the first assertion. The other assertions follow from this one.

inSIGHT

Notice that the expression $\exp(x)$ is its own derivative. In fact, the only functions that satisfy the differential equation

$$\frac{dy}{dx} = y$$

are the functions $y(x) = C \cdot \exp(x)$ where C is a constant. We explore this assertion and some of its applications in Section 6.5.

Example 3 Compute the following derivatives:

$$\frac{d}{dx}(\exp(3x)) \quad \text{and} \quad \frac{d}{dx}\left(\frac{1}{1+\exp(x)}\right).$$

Solution For the first computation, we can use Theorem 1 with $u = 3x$:

$$\frac{d}{dx}\exp(3x) = \frac{d}{dx}\exp(u) = \exp(u)\frac{du}{dx} = \exp(3x)\cdot\frac{d}{dx}(3x) = 3\cdot\exp(3x).$$

To do the second differentiation, we use the Reciprocal Rule and Theorem 1:

$$\frac{d}{dx}\left(\frac{1}{1+\exp(x)}\right) = -\frac{\frac{d}{dx}(1+\exp(x))}{(1+\exp(x))^2} = -\frac{\exp(x)}{(1+\exp(x))^2}.$$

■

Example 4 Calculate the integrals

$$\int \exp(3x+5)\,dx \quad \text{and} \quad \int \exp(x)^2\,dx.$$

Solution We evaluate each integral by the Method of Substitution (Section 5.5). For the first, let $u = 3x + 5$. Thus, $du = 3\,dx$ and

$$\int \exp(3x+5)\,dx = \int \exp(u)\frac{1}{3}\,du = \frac{1}{3}\exp(u) + C = \frac{1}{3}\exp(3x+5) + C.$$

Similarly,

$$\int \exp(x)^2\,dx = \int \exp(2x)\,dx = \frac{1}{2}\exp(2x) + C.$$

■

Example 5 Evaluate the integral

$$\int \frac{\exp(x) - \exp(-x)}{\exp(x) + \exp(-x)}\,dx.$$

Solution We use the Method of Substitution with $u = \exp(x) + \exp(-x)$ because the expression $du = (\exp(x) - \exp(-x))\,dx$ is already present in the integral. Our integral is thus transformed to $\int 1/u\,du = \ln(|u|) + C$.

Resubstituting the expression involving x gives

$$\int \frac{\exp(x) - \exp(-x)}{\exp(x) + \exp(-x)}\,dx = \ln(|\exp(x) + \exp(-x)|) + C.$$

Since we know that the exponential function only assumes positive values, the absolute value in the argument of the logarithm is unnecessary:

$$\int \frac{\exp(x) - \exp(-x)}{\exp(x) + \exp(-x)}\,dx = \ln(\exp(x) + \exp(-x)) + C.$$

■

The Number *e*

We conclude this section by establishing the connection between the exponential function and the number

$$e = \lim_{n\to\infty}\left(1 + \frac{1}{n}\right)^n, \tag{6.24}$$

which was introduced in Section 2.6. Thanks to the Fundamental Theorem of Calculus and the theory developed in this chapter, we are finally able to demonstrate that this limit does exist. Indeed, we will now show that limit (6.24) equals $\exp(1)$. In other words,

$$e = \exp(1). \tag{6.25}$$

We begin with the definition of the derivative of $\ln(x)$ at $x = 1$:

$$\lim_{h\to 0}\frac{\ln(1+h) - \ln(1)}{h} = \frac{d}{dx}\ln(x)\bigg|_{x=1} = \frac{1}{x}\bigg|_{x=1} = 1.$$

In this equation, we let $h = 1/n$ where n tends to infinity through positive integers:

$$1 = \lim_{h\to 0} \frac{\ln(1+1/n) - \ln(1)}{1/n} = \lim_{n\to\infty} \frac{\ln(1+1/n)}{1/n} = \lim_{n\to\infty} n \ln\left(1+\frac{1}{n}\right) \overset{(6.12)}{=} \lim_{n\to\infty} \ln\left(\left(1+\frac{1}{n}\right)^n\right).$$

Now we exponentiate both sides of this equation and use the continuity of the exponential function to obtain

$$\exp(1) = \exp\left(\lim_{n\to\infty} \ln\left(\left(1+\frac{1}{n}\right)^n\right)\right) \overset{\text{Continuity}}{=} \lim_{n\to\infty} \exp\left(\ln\left(\left(1+\frac{1}{n}\right)^n\right)\right) = \lim_{n\to\infty}\left(1+\frac{1}{n}\right)^n.$$

This shows that limit (6.24) exists and equals exp(1), as claimed in equation (6.25). Figure 2a illustrates equation (6.25). Since $\exp(1) = e$, the inverse relationship of the natural logarithm and exponential functions tells us that

$$\ln(e) = 1. \tag{6.26}$$

This representation of e appears in Figure 2b. Finally, by noting that $\ln(e) = \int_1^e 1/x\,dx$, we see that e is the number such that the area under the graph of $y = 1/x$ and over the interval $[1, e]$ is 1 (see Figure 2c).

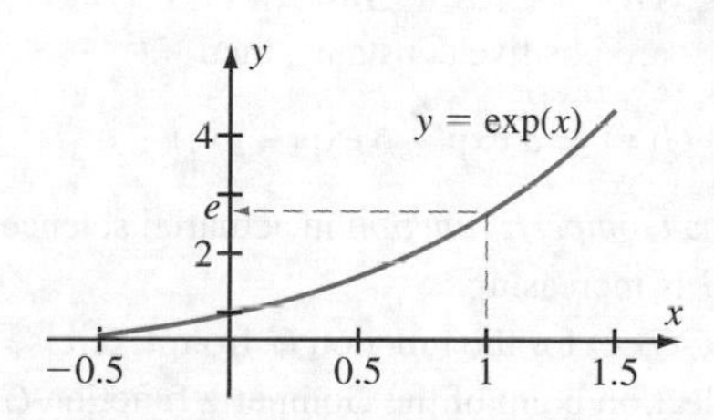

Figure 2a
$e = \exp(1)$

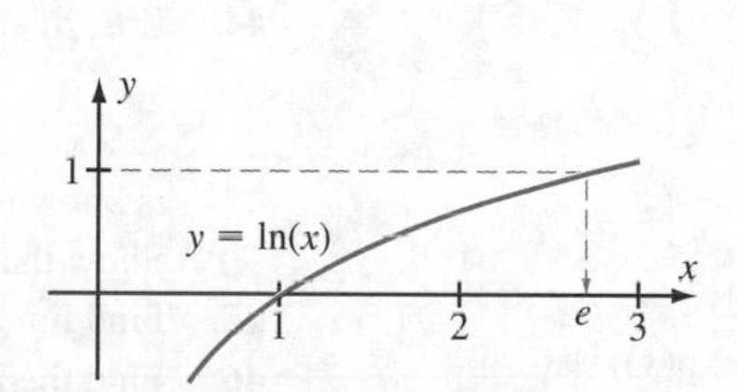

Figure 2b
$\ln(e) = 1$

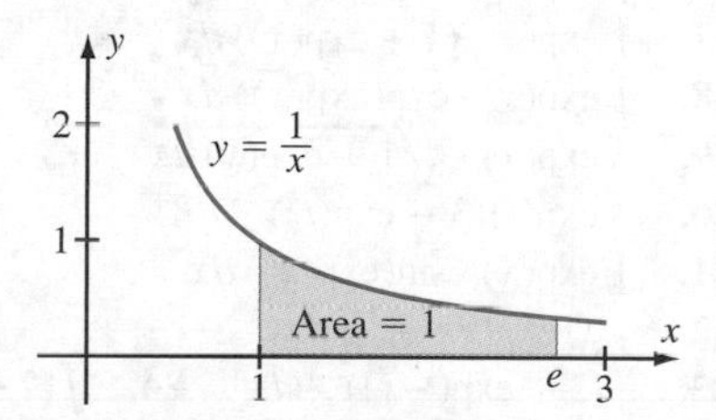

Figure 2c
$\int_1^e \frac{1}{x}\,dx = 1$

Example 6 Simplify the expression $\ln(e^3/7^4)$.

Solution We calculate that

$$\ln\left(\frac{e^3}{7^4}\right) = \ln(e^3) - \ln(7^4) = 3\ln(e) - 4\ln(7) \overset{(6.26)}{=} 3 - 4\ln(7).$$

■

quickquiz

1. What is the definition of the natural exponential function $\exp(x)$?
2. Calculate $\frac{d}{dx}(\exp(2x))$.
3. Calculate $\int x \cdot \exp(x^2)\,dx$.
4. How is the number e related to the natural exponential function?

EXERCISES

Problems for Practice

Calculate the derivatives in Exercises 1–12.

1. $\frac{d}{dx}\exp(-x)$ **2.** $\frac{d}{dx}\exp(3x)$

3. $\frac{d}{dx}(\sin(x)\exp(x))$ **4.** $\frac{d}{dx}\sin(\exp(x))$

5. $\frac{d}{dx}\exp(\sin(x))$ **6.** $\frac{d}{dx}(\exp(4x))^3$

7. $\frac{d}{dx}\ln(1+\exp(x))$ **8.** $\frac{d}{dx}\exp\left(\frac{x}{x+1}\right)$

9. $\frac{d}{dx}\exp(\exp(x))$ **10.** $\frac{d}{dx}\left(\frac{\exp(x)}{1+\exp(x)}\right)$

11. $\frac{d}{dx}\exp(1+\ln(x))$ **12.** $\frac{d^2}{dx^2}\exp(x^2-5x)$

Calculate the integrals in Exercises 13–24.

13. $\int \exp(-x)\,dx$

14. $\int \exp(1-2x)\,dx$

15. $\int (\exp(\pi x)-\exp(-\pi x))\,dx$

16. $\int \exp(\cos(x))\sin(x)\,dx$

17. $\int \exp(x)/(1+\exp(x))\,dx$

18. $\int \exp(x)\cdot\exp(\exp(x))\,dx$

19. $\int \exp(x)\cdot\sqrt{1+\exp(x)}\,dx$

20. $\int \exp(t)(3-\exp(t))^{-3}\,dt$

21. $\int \exp(x)\cdot\sin(\exp(x))\,dx$

22. $\int x(\exp(9x^2))^{3/2}\,dx$

23. $\int (3+\exp(-x))^{-1}\,dx$ **24.** $\int (2+\exp(s))^2\,ds$

In Exercises 25–30, use the properties of the exponential and natural logarithmic functions to simplify the expression.

25. $\exp(\ln(a^3)-\ln(b^2))$

26. $\ln(a\cdot b^2\exp(c-b))$

27. $\exp(\ln(ac)-4\ln(b^2c^{-5}))$

28. $\ln(\exp(a)\exp(b^2))$

29. $\ln\left(\frac{\exp(a)\exp(2b)}{\exp(c^{-2})}\right)$

30. $\exp(\ln(3ac)-\frac{1}{2}\ln(d^{-4}))$

In Exercises 31–34, simplify the expression.

31. $\ln(e^3\cdot 5)$ **32.** $\ln(e^4)-\ln(2e)$

33. $\ln(e/5)$ **34.** $\ln(4e^{-6})-2\ln(e^3)$

Further Theory and Practice

35. Suppose $f'(x)=f(x)$ for every x. Let $g(x)=f(x)/\exp(x)$. Compute $g'(x)$, making sure to simplify. Prove that $g(x)=f(0)$ for every x. (Cite a theorem from Section 4.1.) Deduce that $f(x)=f(0)\exp(x)$.

36. What is the domain of each function?

a. $\exp(\tan(x))$ **b.** $\ln(\exp(x)+1)$
c. $(4-\exp(x))^{1/2}$ **d.** $\tan(\exp(x))$
e. $\exp(1+\ln(x))$ **f.** $(4-\exp(x))^{1/3}$

37. Sketch the graph of each function.

a. $\exp(-3x)$ **b.** $\exp(1-x)$
c. $\exp(x)+x$ **d.** $2+\exp(-x)$
e. $3-\exp(-x)$

38. Find an antiderivative for $x\cdot\exp(x)$ by guessing that it has the form $A\exp(x)+Bx\cdot\exp(x)$ and then solving for A and B.

39. Calculate $\frac{d}{dx}(\exp(x))$ by differentiating the equation $\ln(\exp(x))=x$ implicitly.

In Exercises 40–43, evaluate the limit.

40. $\lim_{n\to\infty}((n+1)/n)^n$ **41.** $\lim_{n\to\infty}(n/(n+1))^n$

42. $\lim_{n\to\infty}(1+1/n)^{2n}$ **43.** $\lim_{n\to\infty}(1+1/n^2)^{n^2}$

44. If α, β, and γ are positive constants, then

$$G(x)=\alpha\exp(-\beta\exp(-\gamma x))$$

is known as a *Gompertz function* in actuarial science. Show that G is increasing.

45. Find $\lim_{x\to\infty}G(x)$ for the function G from Exercise 44.

46. Find the inflection point of the Gompertz function G from Exercise 44.

47. Find the slope of the tangent line to the curve

$$y\exp(2y-1)=2x\exp(3x)$$

at the point $(1, 2)$.

48. Describe the curve with parametric equations $x=\exp(t)$ and $y=\exp(2t)$ $(t>0)$.

49. Let $f(x)=x/\ln(x)$ for $x>1$. Calculate $f'(x)$ and $f''(x)$. Show that the graph of f is concave up. What is the minimum value of f?

50. Let $f(x)=\exp(\sqrt{x})$ for $x\geq 0$. Analyze the concavity of f and determine any points of inflection.

51. Determine the value of m such that the line $y=mx$ is tangent to the graph of $y=\ln(x)$.

52. Let k be a positive constant. Show that $f(x)=\exp(x)/x^k$ is decreasing for $0<x<k$ and increasing for $x>k$. Show also that

$$f''(x)=e^x\frac{(x-k)^2+k}{x^{k+2}}.$$

Deduce that the graph of f is concave up and that $\exp(k)/k^k$ is the minimum value of f. By taking $k = 2$, show that $\exp(x) > x^2$ for $x > 0$.

53. Use Exercise 52 to show that $\exp(\sqrt{x}) > x$ for $x > 0$ and that $\ln(x) < \sqrt{x}$ for $x > 0$.

54. Use the result from Exercise 53 to show that

$$\lim_{x\to\infty} \frac{\ln(x)}{x} = 0.$$

55. Show that if p is any positive number, then

$$\lim_{x\to\infty} \frac{\ln(x)}{x^p} = 0.$$

Hint: Replace x with x^p in Exercise 54.

56. Show that if r and q are any two positive numbers, then

$$\lim_{x\to\infty} \frac{(\ln(x))^r}{x^q} = 0.$$

Hint: Let $p = q/r$ in Exercise 55.

57. Show that if r is any positive number, then

$$\lim_{x\to\infty} \frac{x^r}{\exp(x)} = 0.$$

Hint: Replace q with 1 and x with $\exp(x)$ in Exercise 56.

58. Let r be any positive power. Use the result from Exercise 57 to show that there is a number N such that $\exp(x) > x^r$ for every $x > N$.

59. Notice that $1/u \to \infty$ as $u \to 0^+$. Replace x with $1/u$ in Exercise 55 to deduce that

$$\lim_{u\to 0^+} u^p \ln(u) = 0$$

for any positive constant p.

60. Exercise 50 from Section 6.2 outlines a proof of the inequalities

$$\frac{h}{1+h} < \ln(1+h) < h \quad (h > 0).$$

Use these inequalities with $h = 1/n$ to prove that

$$(1+1/n)^n < e < (1+1/n)^{n+1}$$

for any positive integer n.

61. The Arrhenius Equation The reaction rate of many chemical reactions involves a proportionality constant k that depends on temperature T (measured in degrees Kelvin) in the following way:

$$k = A\exp(-E_A/T))$$

where A is the Arrhenius factor and E_A is the activation energy. Describe the behavior of k as T approaches absolute zero. Compute $\frac{d}{dT}\exp(-E_A/T))$ for $T > 0$. Use Exercise 57 to show that

$$\lim_{T\to 0^+} \frac{d}{dT}\exp(-E_A/T)) = 0.$$

62. The function

$$f(x) = \exp(-x^2)\int_0^x \exp(t^2)\,dt,$$

defined for $0 \le x < \infty$, is known as Dawson's integral. Show that Dawson's integral satisfies the differential equation

$$f'(x) = 1 - 2xf(x).$$

Show that

$$f''(x) = (4x^2 - 2)f(x) - 2x.$$

There is a point c in the interval $[0.8, 1]$ at which $f'(c) = 0$. Show that a relative maximum occurs at this point.

63. The function

$$\nu \mapsto \frac{2h}{c^2}\frac{\nu^3}{e^{h\nu/(kT)} - 1}$$

gives the frequency density of Planck blackbody radiation at temperature T. The physical constants h, c, and k are positive. Determine the limit

$$\lim_{\nu\to 0^+} \frac{2h}{c^2}\frac{\nu^3}{e^{h\nu/(kT)} - 1}$$

by relating it to a known derivative.

64. Let $x_0 > 1$. For $n > 0$, let

$$x_{n+1} = \frac{x_n}{\ln(x_n)}.$$

Prove that $\lim_{n\to\infty} x_n$ exists. Evaluate the limit. *Hint:* Show that $e \le x_{n+1} \le x_n$ if $x_0 \ge e$ and $x_n \le x_{n+1} \le e$ if $1 < x_0 < e$.

65. Let f be continuous on $[a, b]$ and differentiable on (a, b). Suppose that $f(a) = f(b) = 0$. Show that there exists a point $c \in (a, b)$ such that $f'(c) = f(c)$. *Hint:* Apply Rolle's Theorem to $g(x) = \exp(-x)f(x)$.

Calculator/Computer Exercises

66. Graph $\ln(x)$ for $2.7 \le x \le 2.8$. Calculate the coordinates of the endpoints of the curve. Explain why you can conclude that $2.7 < e < 2.8$.

67. Use the Newton-Raphson Method to solve the equation $2\exp(-2x) = \exp(4x)$ to four decimal places.

68. Graph $f(x) = 4 + \exp(-x)\sin(x)$ in the window $[-\pi, \pi] \times [-3.5, 4.5]$. Approximate the minimum and maximum values of f. Approximate the points of inflection of f.

69. In 1956, Clair Patterson (1922–1995) devised the first method of accurately determining the age T of our solar system. His method (based on measuring uranium and lead isotopes in two meteorites) resulted in the equation

$$0.00725\frac{\exp(T/(1.015 \times 10^9)) - 1}{\exp(T/(6.45 \times 10^9)) - 1} = 0.641$$

where T is the age of the solar system in years. To three significant digits, approximately how old is the solar system?

70. The function $f(x) = x^3/(\exp(x) - 1)$ is important in physics and astronomy. Graph f for $0 < x \le 10$. Locate the point at which f has a maximum value. Locate the points of inflection.

71. An arrow is shot repeatedly at a very large round target. Suppose the probability that the arrow will land in the circle of radius r about the center of the target is

$$\frac{1}{\sqrt{\pi}}\int_{-r}^{r} \exp(-t^2)\, dt.$$

Estimate, to an accuracy of 10^{-3}, the probability that the arrow will land in the circle of radius 1. Do the same for the circle of radius 2. Of radius 5.

72. The ground state energy of a certain element is given by the integral

$$\int_2^4 \exp(-x^2)\, dx.$$

Estimate the value of this integral to an accuracy of 10^{-4}.

73. Approximate the point $c \in [0.8, 1]$ at which Dawson's integral has a maximum value. (Refer to Exercise 62.)

6.4 Logarithms and Powers with Arbitrary Bases

Let a be a positive number. For *rational* values of x, such as $x = 3$, $x = -2$, and $x = 1/3$, the expression a^x is defined by means of elementary concepts (as in Section 2.6). We can now use our knowledge of the logarithm and exponential functions to define a^x for *all* x. We begin with the observation that the expression $\exp(x \cdot \ln(a))$ is defined for all real values of x. Furthermore, when x is rational, we have the following simplification:

$$\exp(x \cdot \ln(a)) \overset{(6.12)}{=} \exp(\ln(a^x)) = a^x.$$

Because the expression $\exp(x \cdot \ln(a))$ equals a^x when x is rational and has a precise meaning even when x is irrational, we may use it to define a^x for all x.

Definition If a is any positive number, then we define the number a^x by

$$a^x = \exp(x \cdot \ln(a)) \quad (x \in \mathbb{R}). \tag{6.27}$$

The function $x \mapsto a^x$ is called the *exponential function a^x with base a*.

Bear in mind that our new notion of a^x serves only to *extend* the definition of exponentiation to irrational values of x. As an immediate consequence of this extension, we have

$$\ln(a^x) = \ln(\exp(x \cdot \ln(a))) = x \cdot \ln(a),$$

which gives meaning to formula (6.12) for irrational exponents. As the next example shows, it is easy to use definition (6.27) when working with exponents.

Example 1 Use formula (6.27) to derive the Power Rule of differentiation

$$\frac{d}{dt}t^p = p \cdot t^{p-1}.$$

Solution Since Chapter 3, we have used this basic differentiation rule without verification. Now we can prove it:

$$\begin{aligned}
\frac{d}{dt}t^p &= \frac{d}{dt}(\exp(p \cdot \ln(t))) && \text{Equation (6.27)}\\
&= \exp(p \cdot \ln(t)) \cdot \frac{d}{dt}(p \cdot \ln(t)) && \text{Chain Rule}\\
&= \exp(p \cdot \ln(t)) \cdot \left(p \cdot \frac{1}{t}\right) && \\
&= t^p \cdot \left(p \cdot \frac{1}{t}\right) && \text{Equation (6.27)}\\
&= p \cdot t^{p-1}. && \text{Algebraic simplification}
\end{aligned}$$

■

Notice that, when a is taken to be e, formula (6.27) simplifies to

$$e^x = \exp(x).$$

We may use this observation to rewrite equation (6.27) as

$$a^x = e^{x \ln(a)}. \tag{6.28}$$

Logarithms with Arbitrary Bases

Just as we have used the natural exponential function to define exponential functions with arbitrary bases, so too may we use the natural logarithm to define logarithms with arbitrary bases.

Definition If a is any positive number other than 1, then we define the *logarithm function with base a*, denoted $\log_a$, by

$$\log_a(x) = \frac{\ln(x)}{\ln(a)} \quad (x \in \mathbb{R}^+). \tag{6.29}$$

Notice that $\ln(a)$ makes sense in the previous two definitions because a is assumed to be positive. Furthermore, because we made the assumption $a \neq 1$, $\ln(a)$ is nonzero and the division in equation (6.29) is permitted. Observe that $\log_e(x)$ is the natural logarithm $\ln(x)$ since $\ln(a) = 1$ when $a = e$.

Theorem 1 For any fixed positive $a \neq 1$, the function $x \mapsto a^x$ has domain $\mathbb{R}$ and image $\mathbb{R}^+$. The function $x \mapsto \log_a(x)$ has domain $\mathbb{R}^+$ and image $\mathbb{R}$. These two functions satisfy the

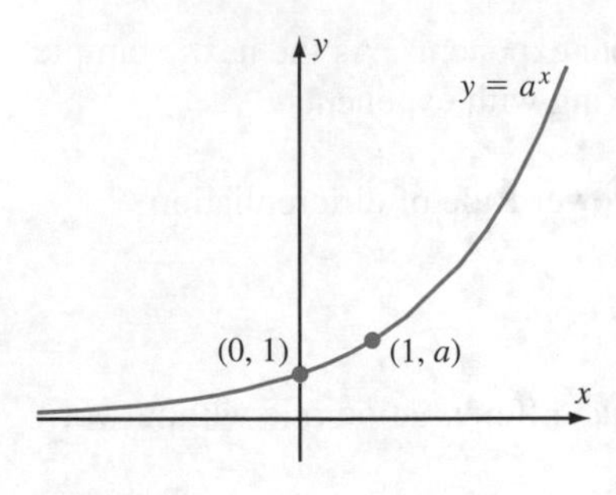

Figure 1
The graph of $y = a^x$, $a > 1$

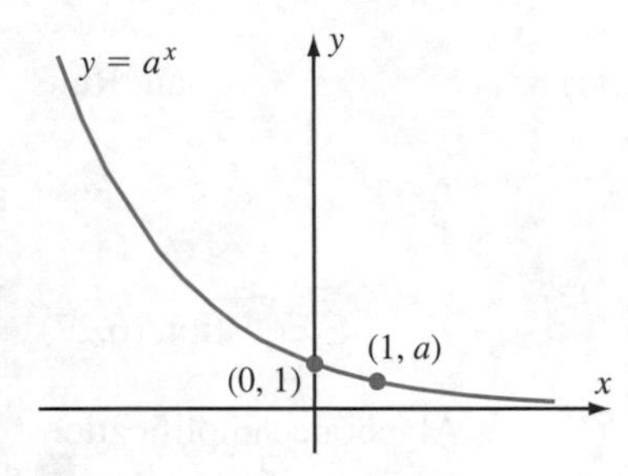

Figure 2
The graph of $y = a^x$, $0 < a < 1$

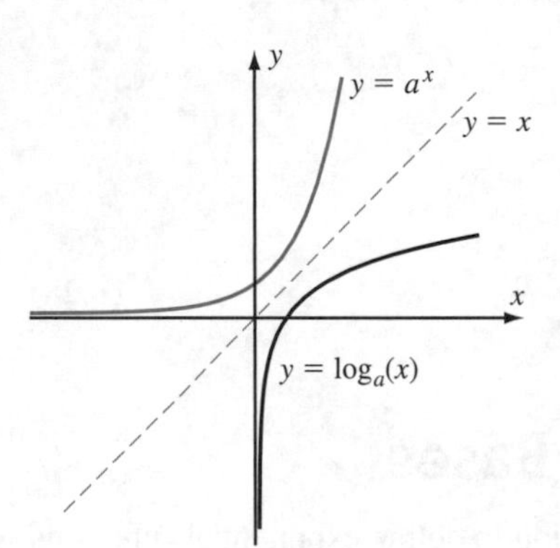

Figure 3

identities

$$a^{\log_a(x)} = x \quad (x \in \mathbb{R}^+) \qquad \text{and} \qquad \log_a(a^x) = x \quad (x \in \mathbb{R}).$$

In words, the functions $x \mapsto a^x$ and $x \mapsto \log_a(x)$ are *inverse* to each other.

Proof We leave verification of the domain and range assertions as an exercise. We obtain the inverse relationship between $x \mapsto a^x$ and $x \mapsto \log_a(x)$ from the inverse relationship between the natural logarithm and exponential functions. For $x > 0$, we have

$$a^{\log_a(x)} = a^{\ln(x)/\ln(a)} \overset{(6.27)}{=} \exp\left(\frac{\ln(x)}{\ln(a)} \cdot \ln(a)\right) = \exp(\ln(x)) = x,$$

whereas for any x, we have

$$\log_a(a^x) \overset{(6.29)}{=} \frac{\ln(a^x)}{\ln(a)} \overset{(6.27)}{=} \frac{\ln(\exp(x \cdot \ln(a)))}{\ln(a)} = \frac{x \cdot \ln(a)}{\ln(a)} = x.$$

■

Theorem 1 tells us that the graphs of $y = a^x$ and $y = \log_a(x)$ are reflections of each other through the line $y = x$ (see Figures 1, 2, and 3).

Example 2 Calculate $\log_3(81)$.

Solution We see that

$$\log_3(81) = \frac{\ln(81)}{\ln(3)} = \frac{\ln(3^4)}{\ln(3)} = \frac{4 \cdot \ln(3)}{\ln(3)} = 4.$$

We get the same result by thinking of the base 3 logarithm as the inverse function of the base 3 exponential function. That is, $\log_3(x)$ is the power to which we must raise 3 to obtain x. Since $3^4 = 81$, we infer that $\log_3(81) = 4$. ■

Elementary Properties of Exponentials and Logarithms with Arbitrary Bases

It is important to know that our new definition of raising a to the xth power, as in equation (6.27), works like an exponentiation. In other words, all the familiar power rules apply to a^x as defined by equation (6.27). For example,

$$a^{x+y} \overset{(6.27)}{=} \exp((x+y)\ln(a)) = \exp(x\ln(a) + y\ln(a)) \overset{(6.20)}{=} \exp(x\ln(a)) \cdot \exp(y\ln(a)) \overset{(6.27)}{=} a^x a^y.$$

This law and several others are listed in the next theorem.

Theorem 2 ***Laws of Exponents*** If $a, b > 0$ and $x, y \in \mathbb{R}$, then

a. $a^0 = 1$;
b. $a^1 = a$;
c. $a^{x+y} = a^x \cdot a^y$;

d. $a^{x-y} = a^x / a^y$;
e. $(a^x)^y = a^{x \cdot y}$;
f. $a^x = b$, if and only if $b^{1/x} = a$ (provided $x \neq 0$); and
g. $(a \cdot b)^x = a^x \cdot b^x$.

Example 3 Simplify the expression $e^{3\ln(5)-2\ln(7)}$.

Solution We write

$$e^{3\ln(5)-2\ln(7)} = e^{\ln(5^3)-\ln(7^2)} = e^{\ln(125)-\ln(49)} = \frac{e^{\ln(125)}}{e^{\ln(49)}} = \frac{125}{49}.$$ ■

Example 4 Simplify

$$\frac{3^{\pi+1} \cdot \pi^{3-\sqrt{2}}}{3^{\pi-1} \cdot \pi^{1-\sqrt{2}}} \quad \text{and} \quad \left(\sqrt[3]{\pi} \cdot 2^{1/6}\right)^3.$$

Solution We calculate

$$\frac{3^{\pi+1} \cdot \pi^{3-\sqrt{2}}}{3^{\pi-1} \cdot \pi^{1-\sqrt{2}}} = 3^{(\pi+1)-(\pi-1)} \cdot \pi^{(3-\sqrt{2})-(1-\sqrt{2})} = 3^2 \cdot \pi^2 = 9\pi^2$$

and

$$\left(\sqrt[3]{\pi} \cdot 2^{1/6}\right)^3 = (\pi^{1/3})^3 \cdot (2^{1/6})^3 = \pi\sqrt{2}.$$ ■

Now we introduce some useful properties of logarithm functions.

Theorem 3 Let a and b be positive bases. If $x > 0$ and $y > 0$, then

a. $\log_a(1) = 0$;
b. $\log_a(a) = 1$;
c. $\log_a(x \cdot y) = \log_a(x) + \log_a(y)$;
d. $\log_a(x/y) = \log_a(x) - \log_a(y)$;
e. For any exponent p, $\log_a(x^p) = p \cdot \log_a(x)$;
f. $\log_a(x) = \log_b(x)/\log_b(a)$; and
g. $\log_a(b) = 1/\log_b(a)$.

Proof The assertions of Theorem 3a and 3b are simply new ways of stating that $a^0 = 1$ and $a^1 = a$. We can derive the assertions of Theorem 3c, 3d, and 3e from corresponding properties of the natural logarithm. As a representative example, here is how we deduce the assertion of Theorem 3c from its natural logarithm analogue:

$$\log_a(x \cdot y) \overset{(6.29)}{=} \frac{\ln(x \cdot y)}{\ln(a)} \overset{(6.8)}{=} \frac{\ln(x) + \ln(y)}{\ln(a)} = \frac{\ln(x)}{\ln(a)} + \frac{\ln(y)}{\ln(a)} \overset{(6.29)}{=} \log_a(x) + \log_a(y).$$

We obtain Theorem 3f by using formula (6.29) twice:

$$\frac{\log_b(x)}{\log_b(a)} \overset{(6.29)}{=} \frac{\ln(x)/\ln(b)}{\ln(a)/\ln(b)} = \frac{\ln(x)}{\ln(a)} \overset{(6.29)}{=} \log_a(x).$$

The assertion of Theorem 3g is the special case of assertion f that results when x is taken to be b.

inSIGHT

Theorem 2c may be expressed in words as:

The exponential of a sum is the product of the exponentials.

Similarly, Theorem 3c states that

The logarithm of a product is the sum of the logarithms.

These two statements really say the same thing because applying the logarithm and applying the exponential are inverse operations.

Theorem 3f can be remembered by its similarity to the algebraic identity

$$\frac{x}{a} = \frac{x/b}{a/b}.$$

The following examples will familiarize you with logarithmic and exponential operations.

Example 5 Simplify the expression

$$\log_3(27) - 4 \cdot \log_2(4) - 3 \cdot \ln(e^2).$$

Solution The expression equals

$$\begin{aligned}\log_3(3^3) - 4 \cdot \log_2(2^2) - 3 \cdot \ln(e^2) &= 3 \cdot \log_3(3) - 4 \cdot (2 \cdot \log_2(2)) - 3 \cdot (2 \cdot \ln(e)) \\ &= 3 \cdot 1 - 4 \cdot 2 \cdot 1 - 3 \cdot 2 \cdot 1 \\ &= -11.\end{aligned}$$

■

Example 6 Solve the equation $2^x \cdot 3^{2x} = 5/4^x$ for the unknown x.

Solution We take the natural logarithm of both sides: $\ln(2^x \cdot 3^{2x}) = \ln(5/4^x)$. Applying the rules for logarithms from Theorem 3, we obtain

$$\ln(2^x) + \ln(3^{2x}) = \ln(5) - \ln(4^x),$$

or

$$x \cdot \ln(2) + 2x \cdot \ln(3) = \ln(5) - x \cdot \ln(4).$$

Gathering all the terms involving x yields

$$x \cdot (\ln(2) + 2 \cdot \ln(3) + \ln(4)) = \ln(5),$$

or

$$x \cdot \ln(2 \cdot 3^2 \cdot 4) = \ln(5).$$

Solving for x gives $x = \ln(5)/\ln(72) = 0.37633\ldots$. ■

Example 7 Find positive numbers a and b so that

$$Q = \frac{4 \cdot \log_3(5) - (1/2) \cdot \log_3(25)}{2 \cdot \log_3(6) + (1/3) \cdot \log_3(64)}$$

is equal to $\log_a(b)$.

Solution The numerator of Q equals

$$\log_3(5^4) - \log_3(25^{1/2}) = \log_3(625) - \log_3(5) = \log_3(625/5) = \log_3(125).$$

Similarly, the denominator can be rewritten as

$$\log_3(6^2) + \log_3(64^{1/3}) = \log_3(36) + \log_3(4) = \log_3(36 \cdot 4) = \log_3(144).$$

Thus, $Q = \log_a(b)$ for $a = 144$ and $b = 125$. Putting these two results together, and using Theorem 3f, we find that

$$Q = \frac{\log_3(125)}{\log_3(144)} = \log_{144}(125).$$

■

Differentiation and Integration of a^x and $\log_a(x)$

Differentiation formulas for a^x and $\log_a(x)$ are readily obtained from the derivatives of $\exp(x)$ and $\ln(x)$ and from the Chain Rule. For a fixed positive constant $a \neq 1$, we have

$$\frac{d}{dx} a^x = \ln(a) \cdot a^x \tag{6.30}$$

and

$$\frac{d}{dx} \log_a(x) = \frac{1}{x \cdot \ln(a)}. \tag{6.31}$$

We obtain formulas (6.30) and (6.31) as follows:

$$\frac{d}{dx}(a^x) = \frac{d}{dx} e^{x \cdot \ln(a)} = e^{x \cdot \ln(a)} \cdot \frac{d}{dx}(x \cdot \ln(a)) = e^{x \cdot \ln(a)} \cdot \ln(a) = a^x \cdot \ln(a)$$

and

$$\frac{d}{dx} \log_a(x) = \frac{d}{dx}\left(\frac{\ln(x)}{\ln(a)}\right) = \frac{1}{\ln(a)} \frac{d}{dx} \ln(x) = \frac{1}{\ln(a)} \cdot \frac{1}{x}.$$

Formula (6.30) is often combined with the Chain Rule and written as

$$\frac{d}{dx} a^u = \ln(a) \cdot a^u \cdot \frac{du}{dx}. \tag{6.32}$$

However, if you understand the use of the Chain Rule (and you should), then you need not memorize this last formula. Finally, it is useful to rewrite formula (6.30) in antiderivative form as

$$\int a^x \, dx = \frac{a^x}{\ln(a)} + C. \tag{6.33}$$

Example 8 Calculate

$$\frac{d}{dx}3^x, \quad \frac{d}{dx}5^{\sin(x)}, \quad \text{and} \quad \frac{d}{dx}\log_6(x).$$

Solution For the first problem, we use formula (6.30) with $a = 3$:

$$\frac{d}{dx}3^x = \ln(3) \cdot 3^x.$$

For the second problem, we can apply formula (6.32) with $a = 5$ and $u = \sin(x)$ to obtain

$$\frac{d}{dx}5^{\sin(x)} = \ln(5) \cdot 5^{\sin(x)} \cdot \frac{d}{dx}\sin(x) = \ln(5) \cdot 5^{\sin(x)} \cdot \cos(x).$$

We achieve the third of the required differentiations by using formula (6.31):

$$\frac{d}{dx}\log_6(x) = \frac{1}{x \cdot \ln(6)}.$$

■

Example 9 Use a substitution to evaluate $\int 4^{\tan(x)} \sec^2(x) \, dx$.

Solution If we set $u = \tan(x)$ and $du = \sec^2(x) \, dx$, then our integral becomes

$$\int 4^u \, du \overset{(6.33)}{=} \frac{1}{\ln(4)} \cdot 4^u + C.$$

Resubstituting the formula for u, we obtain

$$\int 4^{\tan(x)} \cdot \sec^2(x) \, dx = \frac{1}{\ln(4)} \cdot 4^{\tan(x)} + C.$$

■

The Magnitude of Exponential and Logarithmic Functions of a Large Argument

One of the key aspects of e^x, or any exponential function a^x with $a > 1$, is its rapid rate of growth. The next example gives you some feeling for exponential growth.

Example 10 The curve $y = e^x$ is to be graphed for $0 < x < 30$. The graph paper, which is scaled so that there are 10 units per inch, is 3 inches wide. How high does the graph paper have to be to accommodate the graph of f (assuming equal scaling on the x- and y-axes)?

Solution Since $e^{30} = 1.068647458152 \times 10^{13}$ to 12 decimal places, the graph paper will have to be about $(1.068647458152 \times 10^{13})/10 = 1.068647458152 \times 10^{12}$ inches high. Since there are $12 \times 5280 = 63360$ inches to the mile, the graph paper will have to be about $(1.068647458152 \times 10^{12})/63360 \simeq 16866279$ *miles* high. ■

Because the natural logarithm is the inverse function of the exponential, and because $\exp(x)$ tends to infinity very rapidly as x tends to infinity, we can deduce that $\ln(x)$ tends to infinity *very slowly* as x tends to infinity. For example, 10^{80} is often given as an order of magnitude estimate of the number of protons in the universe. Yet $\ln(10^{80})$ is less than 185.

Logarithmic Differentiation

We now show how to use the logarithm as an aid to differentiation. The key idea is that if f is a function taking positive values, then we can exploit the formula

$$\frac{d}{dx}\ln(f(x)) = \frac{f'(x)}{f(x)}. \tag{6.34}$$

Definition The derivative of the logarithm of f is called the *logarithmic derivative* of f. The process of differentiating $\ln(f(x))$ is called *logarithmic differentiation.*

As the right side of equation (6.34) indicates, the logarithmic derivative of f measures the *relative rate of change* of f. This quantity, which often conveys more useful information than f' itself, is frequently used in biology, pharmacology, medicine, and economics. Consider, for example, a product with a price that is increasing at the rate of \$100 per month. Is that a significant rate of change? The answer depends on the price of the product. Are we referring to a \$150,000 house or a \$1500 computer system? In the case of the house, the relative rate of change is only $1/1500$ per month, or 0.8% per year; for the computer system, the relative rate of change is $1/15$ per month, or 80% per year.

Logarithmic differentiation can also be useful in calculus. The idea is to use algebraic properties of the logarithm to simplify complicated products and quotients before differentiating. In addition, logarithmic differentiation can be an effective treatment for expressions in which both the base and the exponent vary, as the next two examples illustrate.

Example 11 Calculate the derivative of the function $f(x) = x^x$.

Solution First we take the natural logarithm of both sides:

$$\begin{aligned}\ln(f(x)) &= \ln(x^x)\\ &= x \cdot \ln(x).\end{aligned}$$

Then we differentiate each side, using the Product Rule to calculate the derivative on the right side:

$$\frac{d}{dx}\ln(f(x)) = 1 \cdot \ln(x) + x \cdot \frac{1}{x} = \ln(x) + 1.$$

Next we use formula (6.34) to simplify the left side:

$$\frac{f'(x)}{f(x)} = \ln(x) + 1.$$

Finally we solve for $f'(x)$ in the last equation to obtain

$$f'(x) = f(x) \cdot (\ln(x) + 1) = x^x \cdot (\ln(x) + 1).$$

A second way to do this problem is to use formula (6.28) with $a = x$:

$$x^x = e^{x \ln(x)}.$$

Thus,

$$\frac{d}{dx}x^x = \frac{d}{dx}e^{x\ln(x)} = e^{x\ln(x)}\frac{d}{dx}(x\ln(x)) = e^{x\ln(x)} \cdot (\ln(x) + 1) = x^x \cdot (\ln(x) + 1).$$

■

inSIGHT

Notice that we cannot calculate the derivative in Example 11 by using the Power Rule:

$$\frac{d}{dx}x^x \neq x \cdot x^{x-1} = x^x.$$

This incorrect computation treats the exponent x as if it were constant. *The Power Rule can be used only when the exponent is constant*. Nor can we use formula (6.30) with $a = x$:

$$\frac{d}{dx}x^x \neq \ln(x) \cdot x^x.$$

This incorrect computation treats the base x as if it were constant. *Formula (6.30) can be used only when the base is constant.*

Example 12 Calculate the derivative of

$$f(x) = \frac{x^3 \cdot e^x}{1 + x^2}.$$

Solution We can do this calculation in a straightforward but tedious fashion by means of the Quotient Rule and the Product Rule. Use of logarithmic differentiation, however, reduces the algebra that is necessary. To start,

$$\ln(f(x)) = \ln(x^3) + \ln(e^x) - \ln(1 + x^2) = 3\ln(x) + x - \ln(1 + x^2).$$

Therefore,

$$\frac{f'(x)}{f(x)} = \frac{3}{x} + 1 - \frac{2x}{1+x^2}.$$

Finally,

$$\begin{aligned} f'(x) &= f(x) \cdot \left(\frac{3}{x} + 1 - \frac{2x}{1+x^2}\right) \\ &= \frac{x^3 \cdot e^x}{1+x^2} \cdot \left(\frac{3}{x} + 1 - \frac{2x}{1+x^2}\right) \\ &= \frac{x^2 \cdot e^x \cdot (x^3 + x^2 + x + 3)}{(1+x^2)^2}. \end{aligned}$$

■

quickquiz

1. What does $\log_3(5)$ mean in terms of natural logarithms?
2. How is $3^{\sqrt{2}}$ defined in terms of the natural exponential function?
3. Differentiate $3^{\sin(2x)}$.
4. Differentiate $\log_9(x^2)$.
5. Evaluate $\int \log_6(x^3)\,dx$. (Simplify the integrand first.)

EXERCISES

Problems for Practice

In Exercises 1–12, calculate the derivative of the function.

1. $f(x) = 3^x$
2. $f(x) = 2^{3x}$
3. $f(x) = x^{-\sqrt{7}}$
4. $f(x) = \pi^{-x}$
5. $f(x) = (1+\sqrt{e})^x$
6. $f(x) = 8^x - x^8$
7. $f(x) = 4^{(x^2)}$
8. $f(x) = \sin(4^{2x})$
9. $f(x) = \ln(x)/(3^x + 1)$
10. $f(x) = \ln(4^x - x^4)$
11. $f(x) = (3^x + 4^x)/5^x$
12. $f(x) = 10^{2x} \cdot 2^{10x}$

In Exercises 13–24, calculate the integral.

13. $\int x^{-\sqrt{3}}\,dx$
14. $\int x^{\sqrt{8}}/x^{\sqrt{5}}\,dx$
15. $\int 5^x\,dx$
16. $\int 9^{3x-5}\,dx$
17. $\int \sec^2(x) \cdot 6^{\tan(x)}\,dx$
18. $\int 2^x \cdot \sin(2^x)\,dx$
19. $\int 10^x \cdot (10^x + 7)^{5/2}\,dx$
20. $\int \sqrt{3}^x/2^{2x}\,dx$
21. $\int x \cdot 3^{(x^2)}\,dx$
22. $\int 8^{-x}\sin(8^{-x})\,dx$
23. $\int 5^x/(1+5^x)\,dx$
24. $\int 3^{\sqrt{x}}/\sqrt{x}\,dx$

In Exercises 25–30, calculate the derivative of *f*.

25. $f(x) = \log_2(5x)$
26. $f(x) = \log_{10}(5x+3)$
27. $f(x) = \dfrac{1}{5 + \log_{10}(x)}$
28. $f(x) = x^3 \cdot \log_{10}(6-x)$
29. $f(x) = \dfrac{\log_3(x)}{\log_5(x) + \log_9(x)}$
30. $f(x) = \log_{10}(\log_2(x))$
31. Solve for x: $4^x \cdot 5^{2x} = 8 \cdot 6^{-x}$.
32. Solve for x: $7^x \cdot 3^{-x} \cdot 4^{2x} = 1$.

In Exercises 33–36, simplify the expression as far as you can. (Eliminate as many occurrences of exponentials and logarithms as possible.)

33. $\log_4(16^x) - \log_3(27) + 4^{\log_4(5)}$
34. $(e^3)^{\ln(4)} - (3^e)^{\log_3(5)}$
35. $\log_8(64 \cdot 4^{2x} \cdot 2^{-6})$
36. $\log_2(\log_2(4))$

Further Theory and Practice

37. Prior to January 1, 1981, only 87 cases of HIV infection had been diagnosed in the United States. By the end of 1983, that number had risen to 4603. If t represents

time measured in half-years with $t = 0$ corresponding to January 1, 1981 (so that $t = 1$ corresponds to July 1, 1981, etc.), then the function

$$P(t) = 87 \cdot 2^{t/1.048}$$

is a fairly good model of the total number of cases of HIV infection that had been diagnosed in the United States up until time t. The relative rate of growth of P at t is $P'(t)/P(t)$.

a. Show that the relative rate of change of $P(t)$ is constant.

b. What is the rate of change of P at time t? (State the units of measurement.)

c. How fast was the number of diagnosed HIV cases growing on January 1, 1982?

d. According to the model, about how many cases of HIV had been diagnosed by the end of 1993?

38. NASA estimates that the cost CE of estimating the cost CP of a deep space network project is given by

$$CE = k \cdot CP^{0.35}$$

where CE and CP are measured in thousands and millions of dollars, respectively, and where $k = 24$ for an order of magnitude estimate, $k = 60$ for a budget estimate, and $k = 115$ for a definitive estimate. Suppose these three relationships are illustrated graphically by plotting $\log_b(CE)$ as a function of $\log_b(CP)$ (a so-called log-log plot). What shapes do the three plots have? Compare the three plots. How would you describe their relationships to one another? What role does the particular choice of b play in the appearance of the plots?

In Exercises 39–44, use logarithmic differentiation or equation (6.28) to calculate the derivative of the function.

39. $f(x) = x^{3x}$

40. $g(x) = 5^{(3^x)}$

41. $f(x) = (\sqrt{x})^x$

42. $H(x) = (\sin(x))^{\cos(x)}$

43. $f(x) = \cos^{3^x}(x)$

44. $g(x) = \ln(x)^{\ln(x)}$

In Exercises 45–52, calculate the integral. For integrands involving mixed bases, first convert the exponentials or logarithms to a common base.

45. $\displaystyle\int \frac{\ln(x)}{\ln(x^5)}\,dx$

46. $\int 2^{3x} \cdot 3^{5x}\,dx$

47. $\displaystyle\int \frac{4^{7x}}{5^{2x}}\,dx$

48. $\int \sqrt{3^x} \cdot (5^{3x-7})^{1/3}\,dx$

49. $\displaystyle\int \frac{\log_2(5x)\log_7(3x)}{x}\,dx$

50. $\displaystyle\int \frac{\ln(x) \cdot \log_4(x)}{x \cdot \log_5(x)}\,dx$

51. $\displaystyle\int \frac{1}{x} \cdot \log_{10}(x)\,dx$

52. $\displaystyle\int \frac{\sin(\log_2(x))}{x}\,dx$

In Exercises 53–58, sketch the graph of the function.

53. $f(x) = -3^{2x}$

54. $g(x) = (1/3)^{-x}$

55. $f(x) = 4^{-5x}$

56. $g(x) = -6^{-4x}$

57. $f(x) = 7^{5x}$

58. $g(x) = -(1/4)^{6x}$

Exercises 59–61 concern the magnitude of earthquake energy on the Richter scale.

59. Let M be the energy of an earthquake measured in dyne-cm. The dimensionless number M_W, defined by

$$M_W = \frac{2}{3}\left(\log_{10}\left(\frac{M}{1 \text{ dyne-cm}}\right) - 16\right),$$

is said to be the magnitude of the earthquake on the Richter scale (a scale devised by the American seismologist Dr. Charles F. Richter in the 1930s). Solve for the energy M of an earthquake as a function of its magnitude M_W on the Richter scale.

60. Suppose M_0 and M_1 are the energies of two earthquakes measured in ergs. Suppose $(M_W)_0$ and $(M_W)_1$ are their magnitudes on the Richter scale. Finally, suppose that

$$(M_W)_1 = (M_W)_0 + 1.$$

There is a constant k that does not depend on M_0 such that

$$M_1 = k \cdot M_0.$$

State this property of the Richter scale in nontechnical language and determine the value of k.

61. Calculate $\frac{dM}{dM_W}$ from the formula for M obtained in Exercise 59. Using the differential approximation, estimate the energy of the 1964 Anchorage earthquake, which measured 8.4 on the Richter scale. For the base point, use the data of the 1906 San Francisco earthquake, a 2.82×10^{28} dyne-cm earthquake equivalent to 8.3 on the Richter scale.

62. Let $f(x) = x^x$ for $x > 0$. Where is f increasing? Decreasing? Concave up? What is the minimum value of f? Sketch the graph of f.

63. Which is greater: π^e or e^π? Answer this question without technological aids by determining where $f(x) = e^x/x^e$ is increasing and comparing the values $f(e)$ and $f(\pi)$.

64. Find the minimum value of $f(x) = x^{-1/x}$ $(x > 0)$.

65. Find the maximum value of $f(x) = x^{1/x^2}$ $(x > 0)$.

66. Find the maximum value of $f(x) = (1/x)^{\sqrt{x}}$ $(x > 0)$.

67. Evaluate $\lim_{x\to 0}(a^x - 1)/x$.

68. Suppose that a is positive and unequal to 1. Explain why it is true that

$$\log_a(x) = -\log_{1/a}(x) \quad (x > 0).$$

69. Verify that if $x = t^{1/(t-1)}$ and $y = t^{t/(t-1)}$ for positive t, then $x^y = y^x$. Thus, although the equation $x^y = y^x$ is symmetric in x and y, there are infinitely many points (x, y) on the curve $x^y = y^x$ with $x \neq y$. (It is a theorem of Euler that $x^y = y^x$ for positive x and y, $x \neq y$, if and only if $x = t^{1/(t-1)}$ and $y = t^{t/(t-1)}$ for positive t.)

70. Differentiate the equation $x^y = y^x$ implicitly with respect to x to find an equation for $\frac{dy}{dx}$ at a point (x, y) on the curve.

71. The German physician E. Heinz found that at time t after injection, the concentration of a drug in a patient's body is given by

$$c(t) = \frac{K}{b-a} \cdot (e^{-at} - e^{-bt})$$

where a, b, and K are positive constants with $b > a$. Find the value t_M at which c has a maximum value. Find the value t_I at which c has an inflection point. Show that $t_I = 2t_M$. What is $\lim_{t\to\infty} c(t)$? Sketch the graph of c.

72. In forestry, the function $E(T) = C \cdot e^{\alpha T/(T+\beta)}$ has been used to model water evaporation as a function of temperature T where α and β are constants with $\alpha > 2$ and $\beta > 0$. Show that

$$E'(T) = \frac{\alpha\beta}{(T+\beta)^2} E(T)$$

and

$$E''(T) = \frac{2((\alpha/2 - 1)\beta - T)}{(T+\beta)^4} E(T).$$

Deduce that E is an increasing function of T with a sigmoidal (or S-shaped) graph. What is the point of inflection of the graph of E?

Exercises 73–75 concern the *Chapman-Richards function,* which is used to model tree growth.

73. Suppose that M and C are positive constants. Suppose that q is a positive constant with $0 < q < 1$. The initial value equation

$$h'(t) = C \cdot h(t) \cdot \left(\left(\frac{M}{h(t)} \right)^q - 1 \right), \qquad h(0) = 0$$

is used in forest management to model tree growth. Here t represents time and h represents tree height (or some other growth indicator). The graph of the solution to this initial value equation has a sigmoidal shape somewhat like that of the logistic growth curve. Without solving for the solution, analyze the initial value equation and explain how it is consistent with a sigmoidally shaped (or S-shaped) solution curve.

74. The Chapman-Richards function is defined by

$$h(t) = M(1 - \exp(-rt))^p$$

where M, r, and p are three positive parameters.

a. If $p = 1/q$ and $r = q \cdot C$, show that h is a solution of the tree growth initial value problem given in Exercise 73.

b. Use the explicit form of $h(t)$ to explain why its graph has a sigmoidal shape.

c. What is the physical meaning of the parameter M? *Hint:* Let t tend to infinity.

75. There are three phases to the life of a tree: the juvenile stage, the mature stage, and the senescent stage. In the mature stage, $h(t)$ can be closely approximated by a linear function of t. In the senescent phase, the tree scarcely grows, and $h(t)$ is closely approximated by the constant M. Use the linear approximation to the exponential function at 0 to obtain a simple function that approximates tree growth in the juvenile stage.

Benjamin Gompertz was a self-educated scholar of wide-ranging interests. Applying calculus to actuarial questions in 1825, he formulated what is now known as *Gompertz's Law of Mortality*. This empirical law is considered in Exercise 76. A related growth function that also bears the name of Gompertz is considered in Exercise 77.

76. From any given (large) group, let $L(t)$ denote the number of individuals who will be alive at age t. The per capita instantaneous death rate, or *force of mortality* as it is known to actuaries, is then

$$\mu(t) = -\frac{L'(t)}{L(t)}.$$

a. Identify the force of mortality $\mu(t)$ as a logarithmic derivative of $L(t)$.

b. Gompertz showed that the force of mortality increases in a geometric progression:

$$\mu(t) = k \cdot r^t.$$

Use this relationship to identify the force of mortality $\mu(t)$ as the derivative of an exponential function.

c. Use parts a and b to find an explicit formula for $L(t)$.

d. What is the instantaneous death rate at time t?

e. When the force of mortality is plotted on a logarithmic scale, a straight line known as the *Gompertz function line* is obtained. What is the slope of the line?

77. The differential equation

$$P'(t) = \alpha \cdot e^{-\beta t} \cdot P(t)$$

with α and β as positive constants is known as the *Gompertz growth equation*. Use the Method of Separation of Variables to find an explicit formula for $P(t)$. Use your explicit solution to show that there is a number P_∞ (known as the *carrying capacity*) such that

$$\lim_{t\to\infty} P(t) = P_\infty.$$

Show that the Gompertz growth equation may be written in the form

$$P'(t) = k \cdot P(t) \cdot \ln\left(\frac{P_\infty}{P(t)}\right).$$

78. In this exercise, you will establish an inequality that is useful in analyzing motion with air resistance.

a. Show that 2 is the minimum value of $x + 1/x$ for $x > 0$. Use the methods from Chapter 4. Another way to establish this inequality is to notice that $(\sqrt{x} - 1/\sqrt{x})^2 \geq 0$.

b. Let c be a positive constant. Prove that $c^x + c^{-x} > 2$ if $x > 0$. Hence, $c^x - 1 > 1 - c^{-x}$.

c. Now suppose that $c > 1$. If a and b are positive constants and if

$$\int_0^a (c^x - 1)\,dx = \int_0^b (1 - c^{-x})\,dx,$$

what can we say about the relative magnitudes of a and b?

d. Show that if a and b are positive and if

$$e^a - e^{-b} = a + b,$$

then $b > a$.

Calculator/Computer Exercises

79. Graph $f(x) = e^x$ and $g(x) = (1 + x/10)^{10}$ in the same window $[0, 3] \times [0, 21]$. Now include the graph of $k(x) = (1 + x/100)^{100}$ in the window. Describe what you see and explain.

80. Graph $f(x) = e^x$ and $k(x) = (1 + x/100)^{100}$ in the same window $[0, 6] \times [0, 405]$. Now include the graph of $\ell(x) = (1 + x/10000)^{10000}$ in the window. Describe what you see and explain.

81. Although computer algebra systems can compute $\log_a(x)$ for any suitable a and x, most graphing calculators do not have the ability to do this directly.

a. Use the Newton-Raphson Method to approximate $\log_4(18)$. (If $x = \log_4(18) = 2.084962501$, then $4^x - 18 = 0$.)

b. All scientific calculators *do* have a key for the natural logarithm. Use the natural logarithm key to calculate $\log_4(18)$.

82. Analyze the graph of $f(x) = \log_x(\ln(x))$, $x > 1$.

83. Approximate the point on the curve $y = (1/3)^x$ at which the normal line passes through the origin.

84. Graph $f(x) = x^x$ and $g(x) = x^{1/x}$ in suitable windows. Use your graphs to determine the limits of $f(x)$ and $g(x)$ as $x \to 0^+$, if those limits exist. Obtain numerical evidence that substantiates any limit you have asserted.

85. Analyze the graph of $f(x) = x^{1/x}$ $(x > 0)$.

86. Analyze the graph of $f(x) = x^{x^{1/x}}$ $(x > 0)$.

6.5 Applications of the Exponential Function

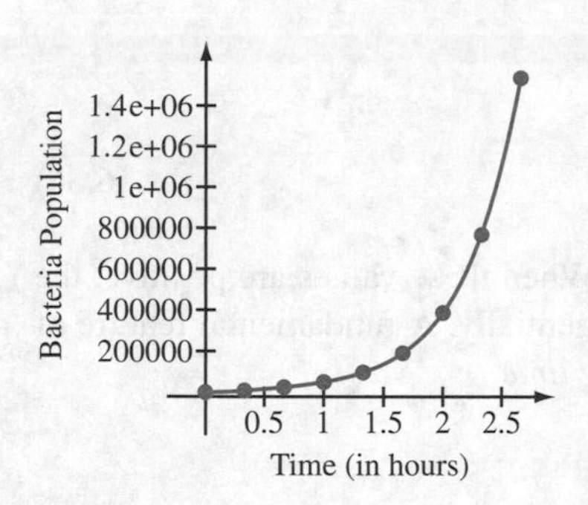

Figure 1
E. coli colony observed at 20-minute intervals

Many processes of nature involve logarithmic and exponential functions. For example, if we examine a population of bacteria, we notice that the rate at which the population grows is proportional to the number of bacteria present. A sketch of the bacteria population (Figure 1) shows that the growth is certainly not linear—indeed the shape of the curve appears to be of exponential form. In this case, we use a standard device of mathematical analysis: Even though the number of bacteria is always an integer, we represent the graph of the population of bacteria by a smooth curve, which enables us to apply the tools of calculus to the problem.

Exponential Growth

Evidence suggests that the number of bacteria $y(t)$ present in a given population at time t satisfies the differential equation

$$\frac{dy}{dt} = k \cdot y \tag{6.35}$$

where k is a positive constant of proportionality. In words, this differential equation states that the rate of change of the bacteria population at a particular time is proportional to the number of bacteria at that time. Equation (6.35) is separable and may be solved by the Method of Separation of Variables (Section 6.1):

$$\int \frac{1}{y} \cdot \frac{dy}{dt}\, dt = k \int dt,$$

or

$$\ln(y(t)) = kt + C$$

where C is a constant of integration. Exponentiating both sides of this equation gives

$$y(t) = e^{kt+C},$$

or

$$y(t) = e^C \cdot e^{kt} = A \cdot e^{kt}.$$

Notice that we have renamed the constant e^C with the simpler symbol A. It is also useful to observe that by evaluating $y(t)$ at $t = 0$, we can solve for A in terms of the initial population $A = y(0)$.

We have motivated our calculation by discussing bacteria, but in fact the calculation applies to any function that grows at a rate proportional to the size of the function. We summarize our results with the following theorem.

Theorem 1 If the rate of change of a function y is proportional to the value of y, that is, if $y'(t) = k \cdot y(t)$ for some constant k, then $y(t) = A \cdot e^{kt}$ where $A = y(0)$. In other words, the unique solution of the initial value problem

$$\frac{dy}{dt} = ky, \quad y(0) = y_0$$

is

$$y(t) = y_0 \cdot e^{kt}. \tag{6.36}$$

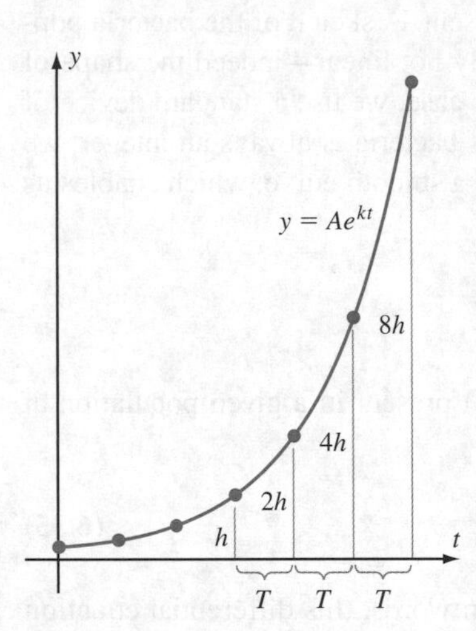

Figure 2
Constant doubling period T

Theorem 1 is valid for all values of k and y_0. When these values are positive, the function y defined by equation (6.36) grows exponentially. A fundamental feature of exponential growth is the existence of the *doubling time*

$$T = \frac{\ln(2)}{k}. \tag{6.37}$$

If the value of y is observed at any time t and if time T elapses and the value of y is observed again, then the second value is twice the first: $y(t + T) = 2 \cdot y(t)$. The proof follows easily from formula (6.36):

$$y(t+T) = y_0 e^{k(t+T)} = y_0 e^{kt} e^{kT} = y(t) e^{kT} = y(t) e^{k(\ln(2)/k)} = y(t) e^{\ln(2)} = 2 \cdot y(t)$$

(see Figure 2). If the doubling time T is known, it is often convenient to write the population as an exponential with base 2:

$$y(t) = y_0 \cdot 2^{t/T}. \tag{6.38}$$

To derive equation (6.38), we replace k with T in formula (6.35):

$$y(t) \overset{(6.36)}{=} y_0 e^{kt} \overset{(6.37)}{=} y_0 e^{\frac{\ln(2)}{T} t} = y_0 \cdot \left(e^{\ln(2)}\right)^{t/T} = y_0 \cdot 2^{t/T}.$$

Example 1 In a yeast-nutrient broth, the population of a colony of the bacterium *Escherichia coli* (*E. coli*) doubles every 20 minutes. If there are 6000 bacteria at 10:00 AM, then how many will there be at noon?

Solution To answer this question, let $y(t)$ be the number of bacteria at time t. For convenience, let $t = 0$ correspond to 10:00 AM so that $y(0) = 6000$. Suppose that time is measured in hours. Thus, noon corresponds to $t = 2$. We have been told that the doubling time T is 20 minutes. We convert this to the units with which we are working: $T = 1/3$ hour. Substituting this value of T and $A = 6000$ into equation (6.38) results in

$$y(t) = 6000 \cdot 2^{3t}.$$

We conclude that

$$y(2) = 6000 \cdot 2^6 = 384000.$$

■

Do not make the mistake of thinking that population problems can be done using only arithmetic. The reasoning "if the population doubles in 20 minutes, then it will quadruple in 40 minutes" is generally wrong because populations usually do not grow linearly.

Radioactive Decay

Another natural phenomenon that fits into our theoretical framework is *radioactive decay.* An unstable nucleus decays to stable form by emitting one or more particles. The rate at which the disintegrations take place follows the *Law of Radioactive Decay*. In the words of physicist Sir Ernest Rutherford and chemist Frederick Soddy, who first wrote about the phenomenon in 1903:

> [T]he law of radioactive change ... may be expressed in one statement—the *proportional* amount of radioactive matter that changes in unit time is a constant.... [This] constant λ may therefore be suitably called the "radioactive constant."

Expressed mathematically, the Law of Radioactive Decay takes the form

$$\frac{dR}{dt} = -\lambda R \tag{6.39}$$

where $R(t)$ is the amount at time t of a particular radioactive isotope (such as radioactive carbon C^{14}) and λ is a *positive* constant that is now known as the *decay* or *disintegration constant* of the radioactive substance. The unit of λ is $1/\text{time}$. A minus sign has been introduced in equation (6.39) because the amount of a radioactive isotope decreases. The derivative on the left side of equation (6.39) is therefore negative. The Law of Radioactive Decay is an *empirical law* (one accepted on the basis of observation). It may also be derived mathematically from a probability model for radioactive decay (as was proposed by the Austrian physicist Egon von Schweidler in 1905). It follows from Theorem 1 that

$$R(t) = R(0) \cdot e^{-\lambda t}. \tag{6.40}$$

The rate of decay is often described by a parameter τ that is defined by

$$\tau = \frac{\ln(2)}{\lambda}. \tag{6.41}$$

Notice that the unit of τ is the reciprocal of that of λ, namely, time. For any t and any value of $R(0)$, we have

$$R(t+\tau) = R(0) \cdot e^{-\lambda(t+\tau)} = R(0) \cdot e^{-\lambda t} e^{-\lambda \tau} = R(t) \cdot e^{-\lambda \tau} = R(t) \cdot e^{-\ln(2)} = \frac{1}{2} R(t).$$

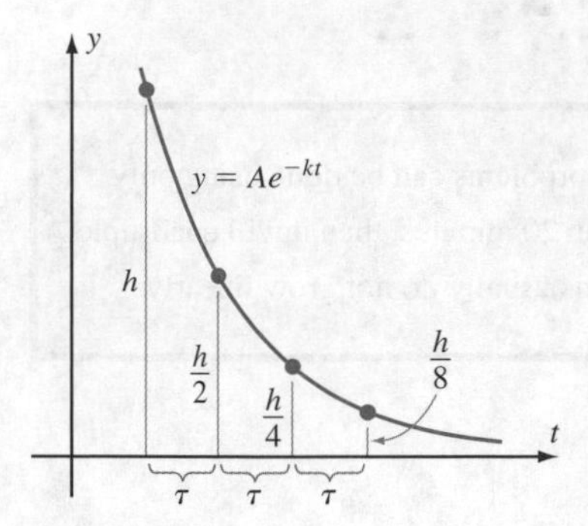

Figure 3
Constant half-life, τ, in exponential decay

In words, the amount of a radioactive substance at the end of a time period of length τ is half what it was at the start of that time period. The number τ defined by equation (6.41) is therefore called the *half-life* of the radioactive substance (refer to Figure 3). The exponential decay law can be expressed in terms of the half-life τ by

$$R(t) = R(0) \cdot \left(\frac{1}{2}\right)^{t/\tau}. \tag{6.42}$$

Example 2 In 100 years, 8 g of a certain radioactive isotope decay to 6 g. After how many more years will only 4 g remain?

Solution First note that the answer is *not* "we lose 2 grams every 100 years, so after 100 more years the isotope will have decayed from 6 grams to 4 grams." Instead, we let $R(t)$ denote the amount of radioactive material at time t. Theorem 1 guarantees that R has the form

$$R(t) = R(0) \cdot e^{\lambda t} = 8e^{\lambda t}.$$

Since $R(100) = 6$, we have

$$6 = R(100) = 8 \cdot (e^{\lambda})^{100}.$$

We conclude that

$$(e^{\lambda})^{100} = \frac{3}{4},$$

or

$$e^{\lambda} = \left(\frac{3}{4}\right)^{1/100}.$$

Thus, the formula for the amount of isotope present at time t is

$$R(t) = 8e^{\lambda t} = 8 \cdot (e^{\lambda})^t = 8 \cdot \left(\frac{3}{4}\right)^{t/100}.$$

We now have complete information about the function R, and we can answer the original question. There will be 4 g of material present when

$$4 = R(t) = 8 \cdot \left(\frac{3}{4}\right)^{t/100}, \quad \text{or} \quad \frac{1}{2} = \left(\frac{3}{4}\right)^{t/100}.$$

We solve for t by taking the natural logarithm of both sides:

$$\ln\left(\frac{1}{2}\right) = \ln\left(\left(\frac{3}{4}\right)^{t/100}\right) = \frac{t}{100} \cdot \ln\left(\frac{3}{4}\right).$$

We conclude that there are 4 g of radioactive material remaining when

$$t = 100 \cdot \frac{\ln(1/2)}{\ln(3/4)} \approx 240.942.$$

So, at about time $t = 240.942$—that is, about 140.942 years after the $t = 100$ observation, there will be 4 g of the isotope remaining. ■

A LOOK BACK

Compare the differential equations for bacterial growth (6.35) and radioactive decay (6.39). Each can be expressed as

$$f'(t) = \alpha \cdot f(t).$$

Since $f'(t)/f(t)$ is the *relative growth rate* of $f(t)$, each phenomenon can be characterized by its *constant relative growth rate*. For bacterial growth, the constant α is positive ($\alpha = k > 0$); for radioactive decay, the constant α is negative ($\alpha = -\lambda < 0$). Assuming that $f(t) > 0$ (as is the case when f represents a population or a mass), we see that $f'(t) = \alpha \cdot f(t)$ is positive when $\alpha > 0$ and negative when $\alpha < 0$. Thus, f increases if $\alpha > 0$ (*exponential growth*), and f decreases if $\alpha < 0$ (*exponential decay*).

Newton's Law of Heating and Cooling

We begin our discussion with a generalization of Theorem 1.

Theorem 2 Suppose that a and b are constants with $a \neq 0$. The initial value problem

$$\frac{du}{dt} = k \cdot u + b, \quad u(0) = u_0 \tag{6.43}$$

has the unique solution

$$u(t) = \left(u_0 + \frac{b}{k}\right) e^{kt} - \frac{b}{k}. \tag{6.44}$$

INSIGHT

The differential equation that is solved in the proof of Theorem 2 arises in a great many situations. Nevertheless, it is much more useful to understand the manner in which we solved the differential equation than it is to memorize the actual solution. In this case, we found a change of variable that transformed a "new" initial value problem into one for which we already had the solution. Many equations and formulas appear in calculus textbooks. You can simplify your study by mastering the basic ones and learning the techniques that allow you to handle variants.

Proof To exploit the similarity between this initial value problem and the one solved in Theorem 1, we make the change of variable $w(t) = u(t) + b/k$. Since $\frac{dw}{dt} = \frac{du}{dt}$, we have

$$\frac{dw}{dt} = \frac{du}{dt} = k \cdot u + b = k \cdot \left(w - \frac{b}{k}\right) + b = k \cdot w.$$

Also, $w(0) = u(0) + b/k = u_0 + b/k$. The initial value problem for w is, therefore,

$$\frac{dw}{dt} = k \cdot w, \quad w(0) = u_0 + \frac{b}{k}. \tag{6.45}$$

According to Theorem 1, $w(t) = w(0) \cdot e^{kt}$. On replacing $w(0)$ with the initial value found in equation (6.45) and substituting back for $u(t)$, we arrive at

$$u(t) + \frac{b}{k} = \left(u_0 + \frac{b}{k}\right) \cdot e^{kt}.$$

Thus, formula (6.44) follows. ■

Our first application of Theorem 2 concerns the transfer of heat. Suppose T is the temperature of an object surrounded by a body (e.g., water, air, etc.) at *constant* temperature T_∞—the reason for using this subscript will become apparent. Newton's

Law of Heating and Cooling states that the rate of change of T is proportional to the difference between T and T_∞:

$$\frac{dT}{dt} = K \cdot (T_\infty - T)$$

where the constant of proportionality K is *positive*. Note that if $T < T_\infty$, then $T' > 0$ and T increases. If instead $T > T_\infty$, then $T' < 0$ and T decreases. We can apply Theorem 2 with $u(t) = T(t)$, $k = -K$, $b = K \cdot T_\infty$, and $u_0 = T_0$. The solution is

$$T(t) = T_\infty + (T_0 - T_\infty)e^{-Kt}. \tag{6.46}$$

Notice that $e^{-Kt} \to 0$ as $t \to \infty$ and, therefore,

$$\lim_{t\to\infty} T(t) = T_\infty,$$

which explains why we have used the notation T_∞ for the ambient temperature.

Example 3 A thermometer is at room temperature (20.0°C). One minute after being placed in a patient's mouth, the thermometer reads 38.0°C. One minute later it reads 38.3°C. Is this second reading an accurate measure (to three significant digits) of the patient's temperature?

Solution Let $T(t)$ denote the temperature of the thermometer at time t with $t = 0$ corresponding to the moment it is placed in the patient's mouth. Let T_∞ denote the actual temperature of the patient. We then have

$$T(t) = T_\infty + (T_0 - T_\infty)e^{-Kt}$$

with $T_0 = 20$. Although in this example there are two unknowns (K and T_∞), we are given two observations with which to determine them:

$$38 = T(1) = T_\infty + (20 - T_\infty)e^{-K}$$

and

$$38.3 = T(2) = T_\infty + (20 - T_\infty)e^{-2K}.$$

We can write the first of these equations as

$$\frac{38 - T_\infty}{20 - T_\infty} = e^{-K}. \tag{6.47}$$

Since $e^{-2K} = (e^{-K})^2$, we may substitute the expression that we found for e^{-K} into the second equation:

$$38.3 = T_\infty + (20 - T_\infty)\left(\frac{38 - T_\infty}{20 - T_\infty}\right)^2.$$

This equation simplifies to a linear equation in the unknown T_∞, and we solve to find that $T_\infty = 38.305\ldots$. The reading at $t = 2$ is therefore an accurate measure of the patient's temperature to three significant digits. ■

inSIGHT

In Example 3, we can calculate the quantity $e^{-K} = 0.016662\ldots$ by using equation (6.47). When we substitute the values of T_0, T_∞, and e^{-K} into equation (6.46), we find that $T(t) = 38.305 - 18.305 \cdot (0.016662)^t$. The graph of $T(t)$ in the window $[0, 2] \times [30, 40]$ (see Figure 4a) indicates that $T(t)$ is very close to T_∞ after $t = 1.3$. A closer look using the window $[1.5, 2.5] \times [38.266, 38.305]$ reveals that $T(t)$ does continue to increase, but very slowly (see Figure 4b). This phenomenon is familiar to anyone who has watched the readout of a digital thermometer while waiting for its beep.

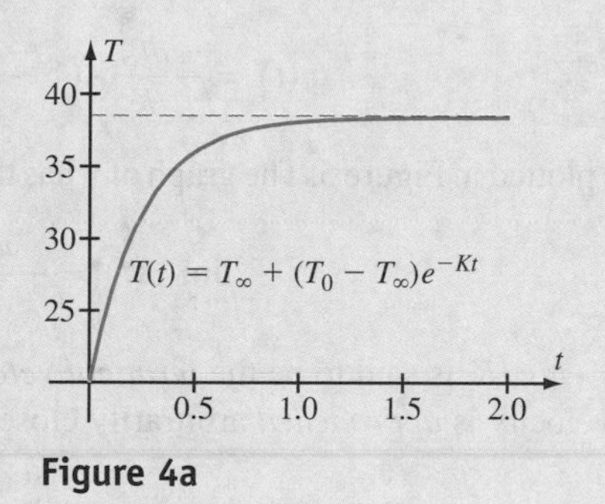

Figure 4a

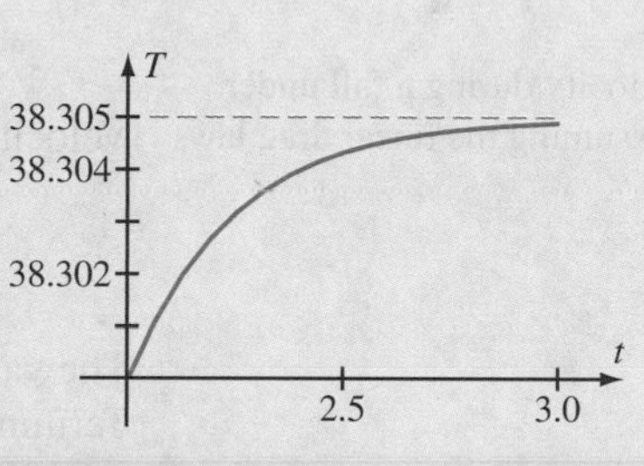

Figure 4b

The Linear Drag Law

We now consider vertical motion above Earth's surface. Let $y(t)$ denote the height of an object above Earth's surface at time t and let $v = \frac{dy}{dt}$ be the velocity. The force R of air resistance (or *air drag*) can often be approximated by a *linear drag law:*

$$R(v) = -K \cdot v$$

where K is a *positive* constant with units that are mass $\times$ time^{-1}. Notice that the presence of the minus sign in this formula ensures that $R(v)$ and v have opposite signs, which means the direction of air drag is opposite to the direction of motion. (In upward motion, y is increasing, $v = y'$ is positive, and $R(v) = -Kv$ is negative, hence a downward force. In downward motion, y is decreasing, $v = y'$ is negative, and $R(v) = -Kv$ is positive, hence an upward force.)

Example 4 Let $v(t)$ denote the velocity at time t of an object of mass m that is dropped from a great height. Assuming that air resistance is given by the formula $R(v) = -K \cdot v$, determine $v(t)$. What is the behavior of $v(t)$ for large t?

Solution Denoting the acceleration due to gravity by the positive constant g, we write the downward force of gravity as $-m \cdot g$. The equation of motion is therefore

$$\underbrace{\underbrace{m \cdot \frac{dv}{dt}}_{\text{mass} \times \text{acceleration}}}_{\text{Total force}} = \underbrace{-K \cdot v}_{\text{Resistive force}} + \underbrace{-m \cdot g}_{\text{Force of gravity}},$$

or

$$\frac{dv}{dt} = -\left(\frac{K}{m} \cdot v + g\right).$$

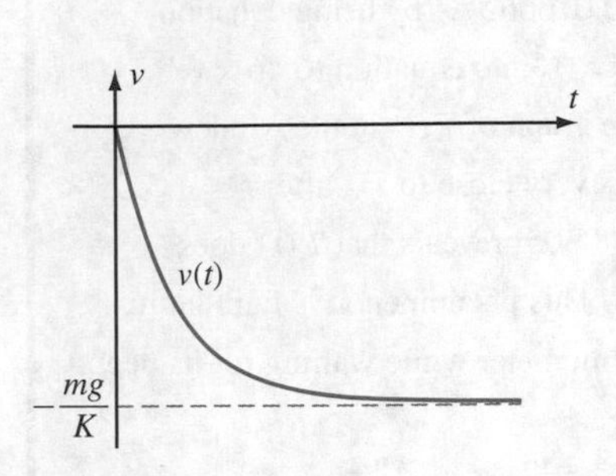

Figure 5
Plot of velocity during a fall under gravity, assuming the linear drag law

Although this example concerns only downward motion, the differential equation that we have just obtained for $v(t)$ is valid for both upward and downward motion. We may apply Theorem 2 with $k = -K/m$, $b = -g$, and $u_0 = 0$:

$$v(t) = \left(0 + \frac{-g}{-K/m}\right) \exp\left(-\frac{K}{m}t\right) - \left(\frac{-g}{-K/m}\right),$$

or

$$v(t) = -\frac{mg}{K} \cdot (1 - e^{-Kt/m}). \tag{6.48}$$

Velocity is plotted in Figure 5. The graph of v has the horizontal asymptote $v = -mg/K$:

$$\lim_{t\to\infty} v(t) = -\frac{mg}{K}.$$

The value $-mg/K$ is said to be the *terminal velocity,* which is somewhat misleading: Terminal velocity is *approached* arbitrarily closely but is *not* attained. ■

One reason that students from so many different backgrounds study calculus is its wide applicability. Mathematicians often state a theorem in an abstract form (such as the formulation of Theorem 2) to maximize the theorem's scope. For instance, we have seen that the differential equation studied in Theorem 2 applies to problems concerning heat transfer and motion. Here are a few more applications of the same differential equation:

- A drug delivered intravenously to a patient enters his bloodstream at a constant rate. The drug is broken down and eliminated from the patient's body at a rate proportional to the amount of the drug in the bloodstream. If $m(t)$ is the amount of the drug in the bloodstream, then

$$\frac{dm}{dt} = \alpha - \beta \cdot m \tag{6.49}$$

  for constants α and β.
- According to Fick's law, the diffusion of a solute through a cell membrane is described by

$$\frac{dc}{dt} = K(c_O - c(t)) \tag{6.50}$$

  where $c(t)$ is the concentration of solute in the cell, c_O is the concentration outside the cell, and K is a positive constant.
- A pacemaker attached to the heart creates a potassium current I that is determined by the differential equation

$$\frac{dI}{dt} = \alpha(1 - I) - \beta I \tag{6.51}$$

  involving positive constants α and β.

- In an evolving galaxy, the rate equation for the production and absorption of electromagnetic radiation N has the form

$$\frac{dN}{dt} = A \cdot (1 - K \cdot N) \qquad \textbf{(6.52)}$$

where A and K are positive constants.

Common Sense and Mathematical Modeling

Nearly all mathematical models involve some approximation. Many involve a compromise between accuracy and simplicity. A completely accurate model that is too difficult to solve may do us little good. By simplifying the model somewhat, we may be able to obtain a useful solution without sacrificing too much accuracy.

In fact, no model presented in this section is accurate for all time. Consider the examples presented in this section. Common sense tells us that bacteria populations do not grow exponentially forever. According to our model for radioactive decay, the amount of the radioactive substance will never be zero. Yet, at some point, there will be a last unstable isotope that disintegrates. In no uncertain terms, Newton's Law of Heating and Cooling indicates that a hot coal immersed in cold water will never quite reach the temperature of the surrounding water. Is this actually the case? We cannot know. As the model predicts, the temperature of the coal becomes so close to the temperature of the surrounding water that no measurement can detect any difference between the two. In using a mathematical model, it is important to anticipate any limitations that the model may have.

quickquiz

1. What is the relevance of exponential functions to bacterial growth and radioactive decay?
2. What form does the solution to a bacterial growth or radioactive decay problem have?
3. What is the relationship between the constant relative growth rate of a population and the doubling period of that population?
4. Describe the rate of change of the temperature of an object that is placed in surroundings that are kept at constant temperature.

EXERCISES

Problems for Practice

In Exercises 1–4, you are given certain information about a population of bacteria. Assuming exponential growth, find the number of bacteria at 8:00 PM.

1. 5000 bacteria at noon; 8000 bacteria at 3:00 PM
2. 4000 bacteria at 5:00 PM; population doubles every 2 hours
3. 6500 bacteria at 7:00 PM; 8000 bacteria at 9:00 PM
4. 7000 bacteria at 4:00 PM; population triples every 8 hours

In Exercises 5–8, you are given a certain amount of a radioactive isotope (in grams) and some information about its rate of decay. Assuming exponential decay, determine how much of the substance is present in the year 2000.

5. 5 g in 1945; 4 g in 1986

6. 8 g in 1984; half-life is 30 years
7. 12 g in 2018; 10 g in 2030
8. 15 g in 1990; 10 g in 2090

In Exercises 9–12, determine the half-life from the information about the rate of decay of a radioactive substance.

9. 12 g in 1945; 8 g in 1985
10. 10 g in 1950; 7 g in 2000
11. 14 g in 1880; 10 g in 1980
12. 20 g in 1900; 15 g in 1950

In Exercises 13 and 14, suppose that a very hot pan is placed in hot dishwater. The difference between the two temperatures is given at two times. Determine when the temperature difference is 100°F.

13. 400°F at 12:30 PM; 200°F at 12:35 PM
14. 300°F at 1:00 PM; 150°F at 1:06 PM

Exercises 15–17 concern vertical motion with a linear drag law $R(v) = -Kv$.

15. In terms of m (mass), l (length), and t (time), what are the dimensions of the drag coefficient K?
16. A 50 kg object is dropped. Its terminal velocity is 178 km/h. What is the drag coefficient K?
17. After falling for 10 seconds, a dropped object hits the ground at 99% of its terminal velocity. If the drag coefficient K is 2 kg/s, what is the mass of the object?
18. According to the 1950 census, the population of the United States was about 150 million. According to the 1980 census, it was about 225 million. Assuming that the rate of growth is exponential, what will be the U.S. population in the year 2005? In the year 2030?
19. When growing in a glucose medium, each cell of a population of the bacterium *E. coli* doubles in size and divides in half every 50 minutes. Under the stated conditions, what is the instantaneous rate of change of an *E. coli* population at the time when the population consists of 100,000 cells? (State the units.)
20. An *Escherichia coli* population doubles every 20 minutes in a yeast nutrient broth. What is the population size at the instant when the colony is growing at the rate of 13,863 bacteria per minute?

Further Theory and Practice

21. According to Newton's Law of Heating and Cooling, temperature $T(t)$ approaches ambient temperature T_∞ very slowly for large values of t. Use the differential equation that $T(t)$ satisfies to explain this phenomenon.
22. The *learning curve* for a class of gifted third-graders is exponential in form. After t weeks in the classroom, the class average (out of 100) on a standard aptitude test is

$$100(1 - e^{-Kt})$$

where K is a positive constant. If a particular class has an average score of 75 after 8 weeks, then what average score can we predict for the class after 16 weeks? When will the average score exceed 95?
23. The half-life of ^{226}Ra is 1620 years. If an area is contaminated with ^{226}Ra at a level that is ten times greater than the maximum safe level, then without a cleanup, how long will this area remain unsafe?
24. The *University of Southern California Chronicle* stated: "The Gompertz rule predicts that the risk of dying accelerates at a constant rate over time for all organisms, be they fruit flies or humans. For humans, for example, the chances of dying increase by approximately 9 percent each year after puberty—meaning risk of dying in any given year doubles with every eight years lived."

a. If $m(t)$ is the instantaneous rate of mortality at age t, then the assertions of the article may be translated as

$$m(t + 1) = 1.09 \cdot m(t)$$

and

$$m(t + 8) = 2 \cdot m(t).$$

Show that these growth laws are consistent (allowing for a small rounding inaccuracy).

b. Suppose that G is a function with the following property: There exists a $T > 0$ and an $r > 1$ such that

$$G(t + T) = r \cdot G(t)$$

for all $t > 0$. In terms of T and r, what is the doubling time for G? (Additional information about the Gompertz growth function may be found in Exercises 76 and 77 from Section 6.4.)

25. Suppose the rate of growth of a population P is exponential (for all $t \geq 0$) with rate constant K. Suppose the rate of growth of a subpopulation p of P is also exponential (for all $t \geq 0$) with a rate constant k.

a. Explain why $k \leq K$.

b. What differential equations do the populations p and P satisfy? Apply the Quotient Rule to p/P, then use the differential equations for p and P to find a differential equation that p/P satisfies. How could this equation have been deduced more simply from the formula for p/P?

26. *Glottochronology* refers to the application of statistics to determine the time at which two languages branched from a common ancestor. The fundamental premise is that the number of words $N(t)$ that the two languages have in common at time t after the branch point is $N(0) \cdot \exp(-kt)$. For two native American languages, the value of k is estimated to be 0.217 when time is measured in thousands of years. What is the rate of change of the number of common words when that number is 500? In how many years after that time will the two languages have only 100 words in common?

Radiocarbon Dating Living matter contains two isotopes of carbon, ^{12}C and ^{14}C, in a known fixed ratio. After death, carbon is no longer metabolized, and the amount of ^{14}C decreases due to radioactive decay. The mass of stable ^{12}C in a sample can be used to determine the mass m of ^{14}C that the sample had at the moment of death. Since the half-life of ^{14}C is known to be approximately 5700 years, the time of death can be calculated from the law of exponential decay. Willard Libby received the 1960 Nobel Prize in physics for discovering this method of dating. Exercises 27–33 concern radiocarbon dating. Where it appears, m represents the ^{14}C mass that the cited sample had at the moment of death.

27. The Shroud of Turin is a cloth believed by some to be the burial cloth of Jesus. In the 1980s, four strands of the cloth were given to four different institutions for dating. The results of the radiocarbon analysis were released on October 13, 1988. Each institution declared the Shroud of Turin to be a medieval forgery.

a. What percentage of m would the measured quantity of ^{14}C have been to authenticate the Shroud?

b. The Shroud was dated to around 1325 CE. Approximately what percentage of m was measured?

28. On September 19, 1991, a mummified human body was discovered in an Austrian glacier near the Tisenjoch Pass. Radiocarbon analysis has shown that the Iceman, as he has come to be called, died between 3350 and 3300 BCE. What percentage of m was measured?

29. In 1994, previously unknown cave art was discovered near Chauvet, France. Radiocarbon dating established that the oldest images were drawn about 32,410 years earlier. At the time of analysis what percentage of m was the ^{14}C content of the pigments used in the oldest images?

30. The Chauvet Cave contains evidence that points to the earliest occurrence of art restoration. When samples, all having equal ^{12}C content, were extracted from pigments coming from different areas of a cave drawing, one sample had ^{14}C content that was 1.837 times greater than the ^{14}C content found in the other samples. About how many years after the original work was the "touch-up" work done?

31. The Cro-Magnons migrated to Europe about 40,000 years ago. For a long time, scientists believed that the Neanderthals vanished soon afterward. However, recent radiocarbon dating of a Neanderthal fossil resulted in a ^{14}C value of $0.026m$. For about how long did the European Neanderthals and Cro-Magnons coexist (at the least)?

32. Spirit Cave Man, a partial mummy discovered in a Nevada cave in 1940, constitutes the oldest mummified remains that have been found in North America. Paleographic evidence led scholars to believe that Spirit Cave Man lived about 2000 years ago. In 1996, however, the Spirit Cave Man was scientifically dated for the first time. An analysis of hair samples revealed his age to be about 9400 years. In this instance, paleographic evidence had led to a 78.7% error. What percentage error in measuring the quantity of ^{14}C would have resulted in an underdating of 78.7%?

33. In 1941, the Dutch artist Han van Meegeren (1889–1947) painted *Christ and the Adultress*, which he sold as a Vermeer to German Reichsmarschall Hermann Göring. At that time, many of Meegeren's Vermeer and de Hooch forgeries had already been accepted as genuine. After World War II, to avoid imprisonment for selling artwork to the Nazis, Meegeren confessed to having forged the paintings he sold. He was sentenced to prison for forgery but died prior to serving his sentence. Presumably there are Meegeren forgeries yet to be found out. Given that Vermeer died in 1675, what is the maximum percentage of m that can be found in the pigments of one of his paintings? Given that Meegeren turned to forgery in the 1930s, what is the minimum percentage of m that his fakes could have?

34. John Graunt, a so-called collector and classifier of facts, was the inventor of modern scientific demography. By tabulating the births and deaths listed in the *Weekly Bills of Mortality for London,* Graunt determined that the population of London was then growing exponentially, with a doubling time of 64 years. (Graunt realized that the exponential law that he proposed could not be carried indefinitely backward, for then the city would be filled with "far more People, than are now in it.") Accepting that (1) Graunt's growth observations were valid throughout the years 1662–1664 and 1666–1700, inclusive; (2) the population of London declined by 100,000 during the Great Plague of 1665; and (3) the population of London at the end of 1700 was about 500,000, estimate the population of London at the start of 1662.

35. The rate of elimination of the excess amount A of a drug from the bloodstream is proportional to that excess amount:

$$\frac{dA}{dt} = -\frac{1}{K}A$$

where K is a time constant that depends on the drug and the individual. If K is $1/2$ hour for alcohol in a certain person, how long will it take for his blood alcohol content to reduce from 0.12% to 0.06%? (Since the normal blood alcohol content is 0, for the purposes of this problem, consider any blood alcohol content to be excess.)

36. A column of air with cross-sectional area 1 cm^2 extends indefinitely upward from sea level (that is, to infinity). The *atmospheric pressure* p at height y above sea level is the weight of that part of this infinite column of air that extends from y to infinity. Assume that the density of the air is proportional to the pressure. (This is an approximation that follows from Boyle's law, assuming constant temperature.) Show that there is a positive constant K such that

$$\frac{dp}{dy} = -Kp.$$

Under this model, what is the mathematical expression for atmospheric pressure? What physical dimensions does K carry?

37. A hot bar of iron with temperature 150°F is placed in a room with constant temperature 70°F. After 4 minutes, the temperature of the bar is 125°F. What is the temperature of the bar after 10 minutes? A hot rock (initial temperature 160°F) is placed in an environment with constant temperature 90°F. The rock cools to 150°F in 20 minutes. How much longer will it take the rock to cool to 100°F?

38. Suppose that you bring a mercury thermometer from a 72°F room to the outdoors. After 1 minute, the thermometer reads 50°F; after 90 seconds, the thermometer reads 40°F. What is the actual temperature outside?

39. In Hull's learning model, habit strength $H(r)$ depends on the number of reinforcements r according to $H(r) = \lambda(1 - e^{-\lambda r})$. Show that H satisfies differential equation (6.43). What are the constants k and b?

40. Several memory experiments have shown that when a pattern is shown to a subject and then withdrawn, the probability that the subject will forget the pattern in the time interval $[0, t]$ is

$$F(t) = p - c \cdot \lambda^t$$

for certain positive constants p, c, and λ with $c < p$ and $\lambda < 1$. Show that F satisfies differential equation (6.35). What are the constants k and b?

41. Suppose that during wartime a particular type of weapon is produced at a constant rate μ and destroyed in battle at a constant relative rate δ. If there were N_0 of these weapons at the outbreak of the war, then how many were there at time t? If the war is a lengthy one, about how many of these weapons will be on hand at time t when t is large?

42. Let $m(t)$ be the amount of a drug in the bloodstream. If the drug is delivered intravenously at a constant rate α and if it is eliminated from the patient's body at a constant relative rate β, then $m(t)$ satisfies equation (6.49). Use Theorem 2 to solve equation (6.49) for the amount $m(t)$ of a drug in the bloodstream. For a long-term patient, can an approximately constant level of the drug be maintained by this method of drug delivery?

43. Find the concentration of a solute inside a cell by using Theorem 2 to solve equation (6.50), which results from Fick's law for the diffusion of a solute through a cell membrane.

44. Use Theorem 2 to solve equation (6.51) to find a formula for the potassium current $I(t)$ induced by a pacemaker.

45. In an evolving galaxy, the rate equation for the production and absorption of electromagnetic radiation is given by differential equation (6.52). Use Theorem 2 to solve this differential equation.

46. A lake has a volume of 3×10^9 ft^3. The concentration of pollutants in the lake is 0.02%. An inflowing river brings clean water at the rate of 1 million ft^3/day. It flows out at the same rate. How long will it be before the pollution level is 0.001%?

47. A tank holds a saltwater solution. At time $t = 0$, there are L liters in the tank; dissolved in the L liters are

A kilograms of salt. Fresh water is added at a rate of r_{in} liters per hour and is mixed thoroughly with the brine solution. At the same time, the mixed solution is drained at the rate of r_{out} liters per hour. Let $m(t)$ denote the mass of the salt in the tank at time t.

a. Assuming perfect mixing prior to draining, show that $m(t)$ satisfies the differential equation

$$m'(t) = -\frac{r_{out}}{L + (r_{in} - r_{out})t} m(t).$$

b. What is $m(t)$ when $r_{in} = r_{out}$?

c. Suppose that $r_{in} \neq r_{out}$. Use the Method of Separation of Variables to solve for $m(t)$.

48. An object of mass m is dropped from height H. Let v denote its velocity at time t. Assume that air drag is linear: $R(v) = -Kv$. Use formula (6.48) to show that the height $y(t)$ of the object at time t is

$$y(t) = H - \frac{mg}{K}t + \frac{m^2 g}{K^2}(1 - e^{-Kt/m}).$$

49. Consider the object discussed in Exercise 48. Suppose that at time τ the value of the air drag coefficient changes from K to κ. (This happens, for example, when a sky diver opens his parachute.) Let $\eta(t)$ denote the object's height above Earth at time t and let $\sigma(t)$ denote the object's velocity at time t. For $0 \leq t \leq \tau$, $\sigma(t) = v(t)$ where $v(t)$ is given by formula (6.48) and $\eta(t) = y(t)$ where $y(t)$ is given by the displayed equation from Exercise 48. At time τ, the velocity of the object is

$$v_\tau = v(\tau) = -\frac{mg}{K}(1 - e^{-K\tau/m})$$

and its height above Earth is

$$y_\tau = y(\tau) = H - \frac{mg\tau}{K} + \frac{m^2 g}{K^2}(1 - e^{-K\tau/m}).$$

The initial value problems that describe the motion after this time τ are

$$m\frac{d\sigma}{dt} = -mg - \kappa\sigma, \quad \sigma(\tau) = v_\tau$$

and

$$\frac{d\eta}{dt} = \sigma, \quad \eta(\tau) = y_\tau.$$

Show that for $t \geq \tau$,

$$\sigma(t) = -\frac{mg}{\kappa} - e^{-\frac{1}{m}\kappa(t-\tau)}\left(\frac{mg}{\kappa} + v_\tau\right)$$

and

$$\eta(t) = y(\tau) + \int_\tau^t \sigma(u)\,du.$$

For later reference, obtain the explicit form of $\eta(t)$:

$$\eta(t) = H - \frac{mg\tau}{K} + \frac{m^2 g}{K^2}(1 - e^{-K\tau/m}) - \frac{mg}{\kappa}(t - \tau)$$
$$- \frac{m}{\kappa^2}(mg + v_\tau\kappa)\left(1 - e^{-(t-\tau)\kappa/m}\right).$$

Let K denote a positive constant. An object is thrown straight up from the ground with initial velocity v_0. Suppose that air resistance gives rise to the force $-K\frac{dv}{dt}$. Let T_u denote the time for the upward trajectory and T_d the time for the downward trajectory. Exercises 50–53 outline a derivation of the somewhat surprising inequality $T_u < T_d$.

50. Use Theorem 2 to show that

$$v(t) = \frac{mg}{K}\left(\left(1 + \frac{Kv_0}{mg}\right)e^{-Kt/m} - 1\right).$$

What is the value of $v(T_u)$? Show that

$$T_u = \frac{m}{K}\ln\left(1 + \frac{K}{mg}v_0\right).$$

51. Show that the height $y(t)$ of the projectile at time t is

$$y(t) = \frac{mg}{K^2}\left(\left(m + \frac{Kv_0}{g}\right)(1 - e^{-Kt/m}) - Kt\right)$$

for $0 \leq t \leq T_u + T_d$. In terms of the parameters of m, g, and K, what is the height H that the projectile reaches before beginning to fall? Use the formula for T_u to express $y(t)$ in the following form:

$$y(t) = \frac{mg}{K^2}(me^{KT_u/m}(1 - e^{-Kt/m}) - Kt).$$

52. Show that the time T_d for the downward trajectory is

$$T_d = -\frac{m}{K}\ln\left(1 + \frac{K}{mg}v_I\right)$$

where v_I is the velocity at the time of impact with the ground. Explain why the formula for T_d is consistent with the obvious physical requirement that $T_d > 0$.

53. Use the second formula for $y(t)$ found in Exercise 51, together with the equation $y(T_u + T_d) = 0$, to show that

$$(e^{KT_u/m} - e^{-KT_d/m}) = \frac{KT_u}{m} + \frac{KT_d}{m}.$$

Use Exercise 78 from Section 6.4 to deduce that $T_d > T_u$. In words, *it takes longer to come down than to go up*. (It may also be shown that $v_0 > v_I$.)

Since the 1920s, mathematical models have been developed to study running performance. These models rely on physical theory, empirical evidence, and statistical analysis. In 1973, J. B. Keller modified an existing mathematical theory of running to make it more appropriate for sprints. Exercises 54–56 concern the *Hill-Keller theory of sprinting*. In these exercises, $v(t)$ denotes the speed of a runner at time t, $p(t)$ denotes the horizontal component of the propulsive force *per unit mass* that is exerted by the runner at time t, and $R(v)$ denotes the total resistive force *per unit mass*.

54. Assume that $R(v) = v/\tau$ for some positive constant τ (the dimension of which is time). Suppose also that $p(t) = P$ for a positive constant P (the unit of which is distance/time^2). Obviously this assumption on $p(t)$ is not tenable for races longer than a sprint. Show that

$$\frac{dv}{dt} = P - \frac{v}{\tau}.$$

Use Theorem 2 to find an explicit solution for $v(t)$. (What initial condition must you use?)

55. Use the explicit solution for $v(t)$ found in Exercise 54 to show that

$$v_\infty = \lim_{t\to\infty} v(t)$$

exists. This terminal velocity is called *maximum velocity* in the track community.

56. Let $x(t)$ denote the distance a sprinter has run in time t. Show that

$$x(t) = v_\infty(t - \tau(1 - e^{-t/\tau})).$$

57. Bacterial populations grow by binary fission, a process in which a single bacterium divides into two bacteria. The time τ required for this to happen is the doubling time of the population, or the *generation time*. Let $N(t)$ be the number of bacteria at time t. The *Encyclopedia Britannica* states that two observations $N(t_0)$ and $N(t_0 + \Delta t)$, separated by *any* nonzero time increment Δt, determine τ by the formula

$$\tau = \frac{\Delta t}{3.3 \log_{10}(N(t_0 + \Delta t)/N(t_0))}.$$

Using this equation, show that $N(t)$ is given by

$$N(t) = N(0) \cdot 10^{t/(3.3\tau)} = N(0) \cdot \exp\left(\frac{\ln(10)}{3.3}\frac{t}{\tau}\right).$$

What value should be used instead of 3.3 if the *exact* value of the generation time is desired?

Calculator/Computer Exercises

58. Plot the population of London for the years 1662–1700 under the assumptions given in Exercise 34.

59. A 1 kg object is dropped. After 1 s, its (downward) speed is 9.4 m/s. Assuming that the drag is proportional to the velocity, what is the terminal velocity of the object?

60. A 0.5 kg object is dropped. Its terminal velocity is 49 m/s. Assuming that the drag is proportional to the velocity and that it was dropped from a sufficiently great height, at what time after being dropped is its speed 40 m/s?

61. On December 7, 1941, during the attack on Pearl Harbor, an 800 kg bomb was dropped from a Nakajima BN52 Kate bomber flying at an altitude of 3170 m. The bomb struck the battleship USS *Arizona*, igniting its black powder magazine, which in turn set off a series of catastrophic explosions. The ship sank in 9 minutes with a death toll of 1177. A 1997 analysis of the attack asserted that the flight of the bomb lasted 26 s. Let us accept that figure. Assume the linear drag law $R(v) = -Kv$.

a. What was the value of K? (Use Exercise 48. Include the units of K in your answer.)

b. What was the theoretical terminal velocity of the bomb?

c. What was the actual velocity of the bomb when it struck the USS *Arizona?* (State your answer in meters per second but also convert to miles per hour.)

d. Plot the height $y(t)$ of the bomb for $0 \le t \le 26$.

e. What would the time of fall have been had there been no air drag? The velocity on impact?

62. Suppose that for a sky diver the value of K in the linear drag law $R(v) = -Kv$ is 0.682 lb·s/ft. Let $P(t)$ denote the percentage of terminal velocity of a 120 lb sky diver t seconds into the jump. Plot the function P. How many seconds into the jump is it when the sky diver reaches 95% of terminal velocity? How many feet has the sky diver fallen by this time?

63. A 0.1 kg projectile is thrown up into the air with an initial speed of 20 m/s. Assume that $R(v) = -Kv$ where $K = 0.03\,\text{kg/s}^{-1}$. Approximate the maximum height H above ground that the object reaches, the impact velocity v_I with which the projectile strikes the ground, and the times T_u and T_d of the upward and downward trajectories. Refer to Exercises 50–52.

64. Let $y(t)$ and $v(t)$ denote the height and velocity of the object from Exercise 63. In the same coordinate plane, plot the kinetic energy $T(t) = mv(t)^2/2$, the potential energy $U(t) = mg \cdot y(t)$, and the total energy $E(t) = T(t) + U(t)$ of the object. Separately, plot $E'(t)$. The behavior of $E(t)$ and the range of values of $E'(t)$ reflect a property of resistive forces that is called *nonconservative*. Explain.

65. At the 1987 World Track and Field Championships in Rome, split times for the men's 100 meters competition show that Ben Johnson (Canada) and Carl Lewis (United States) each had maximum velocity v_∞ equal to 11.8 m/s. (Refer to the instructions for Exercises 54–56 for the Hill-Keller theory of sprinting.) Johnson's winning time was 9.83 s and Lewis's second place time was 9.93 s.

a. What were the approximate values of P and τ for these men? According to the Hill-Keller model, can Johnson's victory be attributed to greater propulsive force, lesser resistance to motion, or both? (At the 1988 Seoul Olympics, again running against Carl Lewis, Ben Johnson broke his own world record with a time of 9.79 s. However, Johnson was stripped of his Olympic Gold Medal and his world and Olympic records after he tested positive for steroids.)

b. In the same coordinate plane, plot $v(t)$ for both runners for $0 \le t \le 9.83$.

c. In meters, what was Johnson's winning margin?

66. As of this writing, former president George Herbert Bush is the only American president who has jumped from an airplane. During World War II, the future president parachuted from his airplane when it was shot down. On March 25, 1997, at the age of 72, the former president made a recreational skydiving/parachuting jump. In doing this problem, refer to Exercises 48 and 49 for notation and formulae. Assume that the mass of the president and his jumping gear was 7.03 lb·s^2/ft. Assume linear drag laws $R(v) = -Kv$ during the skydiving phase of the jump and $R(v) = -\kappa v$ after deployment of the parachute. Use the value $K = 1.4$ lb/s.

a. The president jumped from a height of 12,500 ft and did not open his parachute until he reached 4500 ft. Compute the duration τ of this sky dive. According to the model, what was the president's velocity $v(\tau)$ when he opened his parachute? Plot $v(t)$ for $0 \le t \le \tau$.

b. The total time of the president's dive and jump was about 9 min. In the absence of more precise data, use 540 s as the elapsed time between jumping from the plane and touching down on the ground. As in Exercise 49, use $\sigma(t)$ and $\eta(t)$ to denote President Bush's velocity and height above ground for $0 \le t \le 540$. From the equation $\eta(540) = 0$, determine κ to three significant digits.

c. According to this model, what was the terminal velocity of the president's parachute jump? With what velocity did he touch down?

d. Plot $\sigma(t)$ and $\eta(t)$ for $0 \le t \le 540$.

6.6 Inverse Trigonometric Functions

Figure 1, next page, shows the graphs of each of the six trigonometric functions. Notice that each graph has the property that some horizontal line intersects the graph at least twice. Therefore, not one of these functions is invertible. Another way of seeing this point is that each of the trigonometric functions repeats itself every 2π units: $f(x + 2\pi) = f(x)$; hence, each is *not* one-to-one.

We must restrict the domains of the trigonometric functions if we want to discuss inverses for them. In this section, we learn the standard methods for performing this restriction operation. As you read this section, you may find it useful to refer back to Section 1.6 to review the properties of the trigonometric functions.

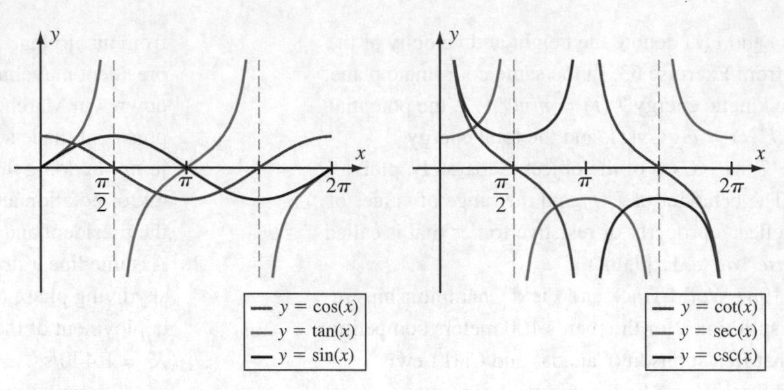

Figure 1

Inverse Sine and Cosine

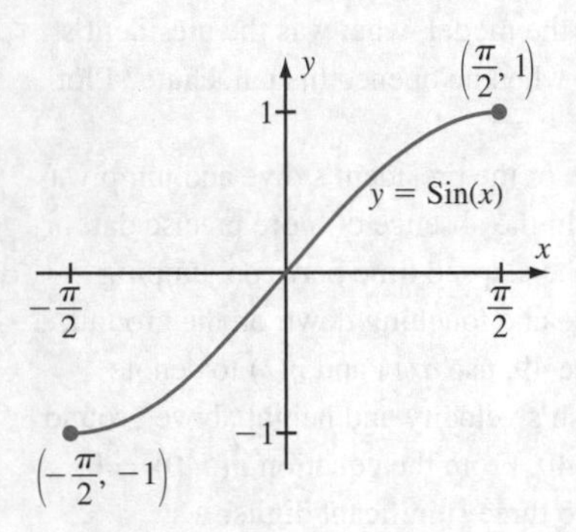

Figure 2

Consider the sine function with domain restricted to the interval $[-\pi/2, \pi/2]$. We denote this restricted function by $x \mapsto \mathrm{Sin}(x)$ (which is capitalized to emphasize that its domain differs from that of the sine function). The graph of $y = \mathrm{Sin}(x)$ in Figure 2 shows that $x \mapsto \mathrm{Sin}(x)$ is a one-to-one function from $[-\pi/2, \pi/2]$ *onto* $[-1, 1]$. Therefore, it is invertible. Its inverse function, $\mathrm{Sin}^{-1} : [-1, 1] \to [-\pi/2, \pi/2]$, is increasing, one-to-one, and onto. The expression $\mathrm{Sin}^{-1}(x)$ is read as "the inverse sine of x" or "the arcsine of x" and is often denoted by $\arcsin(x)$. We can obtain the graph of $y = \arcsin(x)$ by reflecting the graph of $y = \mathrm{Sin}(x)$ across the line $y = x$ (see Figure 3).

Example 1 Calculate $\arcsin(\sqrt{2}/2)$, $\arcsin(-\sqrt{3}/2)$, and $\arcsin(\sin(3\pi/2))$.

Solution To calculate $\arcsin(t)$, we look for the $s \in [-\pi/2, \pi/2]$ for which $t = \sin(s)$. Because $\sin(\pi/4) = \sqrt{2}/2$ and $\pi/4$ is in $[-\pi/2, \pi/2]$, we have $\arcsin(\sqrt{2}/2) = \pi/4$. Notice that even though $\sin(x)$ takes the value $\sqrt{2}/2$ at many different values of the variable x, $\mathrm{Sin}(x)$ equals $\sqrt{2}/2$ only at $x = \pi/4$ (see Figure 4). In a similar way, we calculate $\arcsin(-\sqrt{3}/2) = -\pi/3$. The last evaluation requires particular care:

$$\arcsin\left(\sin\left(\frac{3\pi}{2}\right)\right) = \arcsin(-1) = -\frac{\pi}{2}.$$

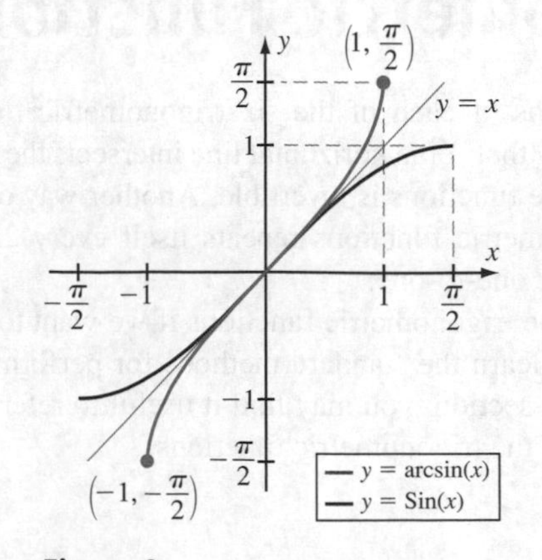

Figure 3

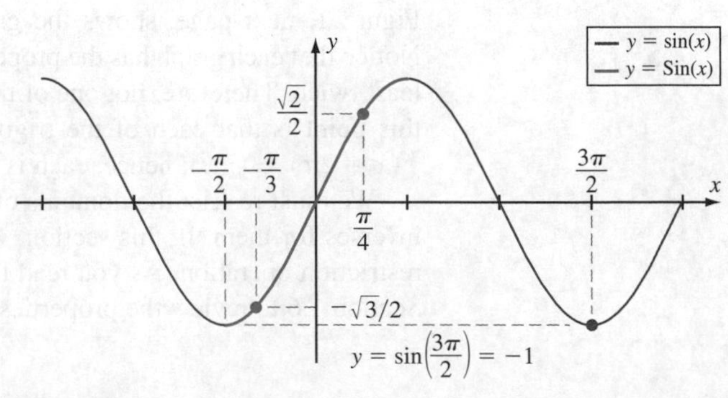

Figure 4

■

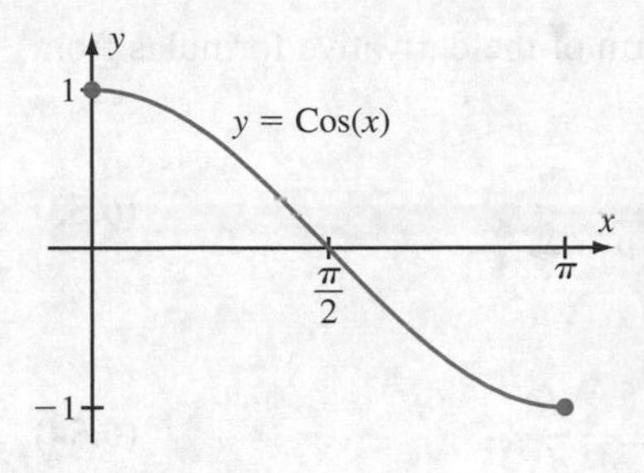

Figure 5

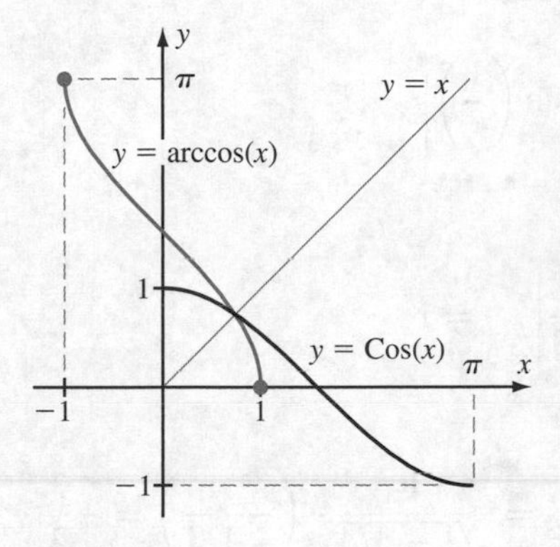

Figure 6

inSIGHT

> The function $x \mapsto \arcsin(x)$ is the inverse function of $x \mapsto \text{Sin}(x)$, not of $x \mapsto \sin(x)$. Otherwise we would have $\arcsin(\sin(3\pi/2)) = 3\pi/2$ instead of the correct value $-\pi/2$ obtained in Example 1. It is always true that $\sin(\arcsin(t)) = t$, but $\arcsin(\sin(s)) = s$ is true *only* when $s \in [-\pi/2, \pi/2]$.

The study of the inverse of cosine involves similar considerations, but we must select a different domain for our function. We define $\text{Cos}(x)$ to be the cosine function restricted to the interval $[0, \pi]$. Thus, as Figure 5 shows, $x \mapsto \text{Cos}(x)$ is a one-to-one function. Because it takes on all the values in the interval $[-1, 1]$, $\text{Cos} : [0, \pi] \to [-1, 1]$ is one-to-one and onto; therefore, it possesses an inverse. The inverse cosine $x \mapsto \text{Cos}^{-1}(x)$ is also called the *arccosine* and denoted by $\arccos(x)$. We reflect the graph of $y = \text{Cos}(x)$ across the line $y = x$ to obtain the graph of $y = \text{Cos}^{-1}(x)$, as is shown in Figure 6.

Example 2 Calculate $\arccos(-\sqrt{2}/2)$, $\arccos(0)$, and $\arccos(\sqrt{3}/2)$.

Solution We calculate as in Example 1 except that here we are looking for values in the interval $[0, \pi]$. We have $\arccos(-\sqrt{2}/2) = 3\pi/4$, $\arccos(0) = \pi/2$, and $\arccos(\sqrt{3}/2) = \pi/6$. ■

Theorem 1 The functions $t \mapsto \arcsin(t)$ and $t \mapsto \arccos(t)$ are differentiable on the open interval $(-1, 1)$ and

$$\frac{d}{dt}\arcsin(t) = \frac{1}{\sqrt{1-t^2}}$$

and

$$\frac{d}{dt}\arccos(t) = -\frac{1}{\sqrt{1-t^2}}.$$

Proof For t in the open interval $(-1, 1)$, let $s = \arcsin(t)$. Thus, s is in the open interval $(-\pi/2, \pi/2)$ and $\cos(s)$ is positive. Using this observation, together with the identity $\cos^2(s) = 1 - \sin^2(s)$, we calculate that

$$\cos(s) = \sqrt{\cos^2(s)} = \sqrt{1-\sin^2(s)} = \sqrt{1-(\sin(\arcsin(t)))^2} = \sqrt{1-t^2}.$$

We may now obtain the formula for the derivative of $\arcsin(t)$ by using the Inverse Function Derivative Rule that we learned in Section 3.6:

$$\frac{d}{dt}\arcsin(t) = \frac{1}{\frac{d}{ds}\sin(s)\Big|_{s=\arcsin(t)}} = \frac{1}{\cos(s)|_{s=\arcsin(t)}} = \frac{1}{\sqrt{1-t^2}}.$$

The derivative of $\arccos(t)$ is calculated in the same way. ■

Observe that $\arcsin'(t) > 0$ for all $t \in (-1, 1)$, because the inverse sine is an increasing function. Similarly, $\arccos'(t) < 0$ for all $t \in (-1, 1)$, because the inverse cosine is a decreasing function. The Chain Rule form of the derivative formulas from Theorem 1 are

$$\frac{d}{dx}\arcsin(u) = \frac{1}{\sqrt{1-u^2}}\frac{du}{dx} \tag{6.53}$$

and

$$\frac{d}{dx}\arccos(u) = -\frac{1}{\sqrt{1-u^2}}\frac{du}{dx}. \tag{6.54}$$

Example 3 Calculate the following derivatives:

$$\left.\frac{d}{dx}\arcsin(x)\right|_{x=0} \quad \text{and} \quad \left.\frac{d}{dx}\arcsin\left(\frac{1}{x}\right)\right|_{x=-2/\sqrt{3}}.$$

Solution We have

$$\left.\frac{d}{dx}\arcsin\right|_{x=0} = \left.\frac{1}{\sqrt{1-x^2}}\right|_{x=0} = 1$$

and

$$\left.\frac{d}{dx}\arcsin(1/x)\right|_{x=-2/\sqrt{3}} = \left.\frac{1}{\sqrt{1-(1/x)^2}}\cdot\left(-\frac{1}{x^2}\right)\right|_{x=-2/\sqrt{3}} = \frac{1}{\sqrt{1-3/4}}\cdot\left(-\frac{1}{4/3}\right) = -\frac{3}{2}.$$

■

Example 4 Calculate the following derivatives:

$$\left.\frac{d}{dx}\arccos(x)\right|_{x=1/2} \quad \text{and} \quad \left.\frac{d}{dx}\arccos\left(\sqrt{x}\right)\right|_{x=1/4}.$$

Solution We have

$$\left.\frac{d}{dx}\arccos(x)\right|_{x=1/2} = -\left.\frac{1}{\sqrt{1-x^2}}\right|_{x=1/2} = -\frac{1}{\sqrt{1-(1/2)^2}} = -\frac{2}{\sqrt{3}}$$

and

$$\left.\frac{d}{dx}\arccos\left(\sqrt{x}\right)\right|_{x=1/4} = -\left.\frac{1}{\sqrt{1-\left(\sqrt{x}\right)^2}}\cdot\left(\frac{1}{2}x^{-1/2}\right)\right|_{x=1/4} = -\frac{2}{\sqrt{3}}.$$

■

The Inverse Tangent Function

Define the function $x \mapsto \operatorname{Tan}(x)$ to be the restriction of $x \mapsto \tan(x)$ to the interval $(-\pi/2, \pi/2)$. Observe that the tangent function is undefined at the endpoints of this interval. The inequality

$$\frac{d}{dx}\operatorname{Tan}(x) = \sec^2(x) > 0$$

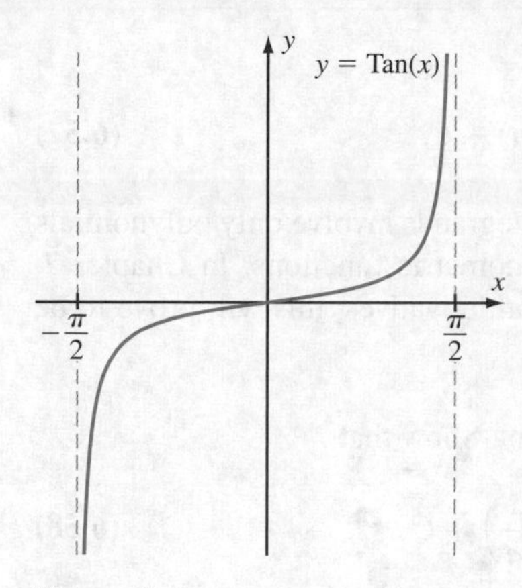

Figure 7

tells us that Tan(x) is increasing and, therefore, one-to-one (see Figure 7). Also Tan(x) takes arbitrarily large positive values when x is near to, but less than, $\pi/2$. In addition, Tan(x) takes negative values that are arbitrarily large in absolute value when x is near to, but greater than, $-\pi/2$. Therefore, Tan(x) takes all real values. Since Tan: $(-\pi/2, \pi/2) \to (-\infty, \infty)$ is one-to-one and onto, the inverse function $\text{Tan}^{-1}: (-\infty, \infty) \to (-\pi/2, \pi/2)$ exists. We read $\text{Tan}^{-1}(x)$ as "the inverse tangent of x" or "the arctangent of x." The notation arctan(x) is often used instead of $\text{Tan}^{-1}(x)$. The graph of the arctangent, obtained by reflecting the graph of $y = \text{Tan}(x)$ about the line $y = x$, is shown in Figure 8.

Example 5 Calculate $\arctan(1)$, $\arctan(\sqrt{3})$, and $\arctan(-1/\sqrt{3})$.

Solution We calculate $\arctan(t)$ by finding the unique s between $-\pi/2$ and $\pi/2$ for which $\tan(s) = t$. Thus, $\arctan(1) = \pi/4$, $\arctan(\sqrt{3}) = \pi/3$, and $\arctan(-1/\sqrt{3}) = -\pi/6$. ■

Theorem 2 The function $t \mapsto \arctan(t)$ is differentiable for each t and

$$\frac{d}{dt}\arctan(t) = \frac{1}{1+t^2} \quad (-\infty < t < \infty).$$

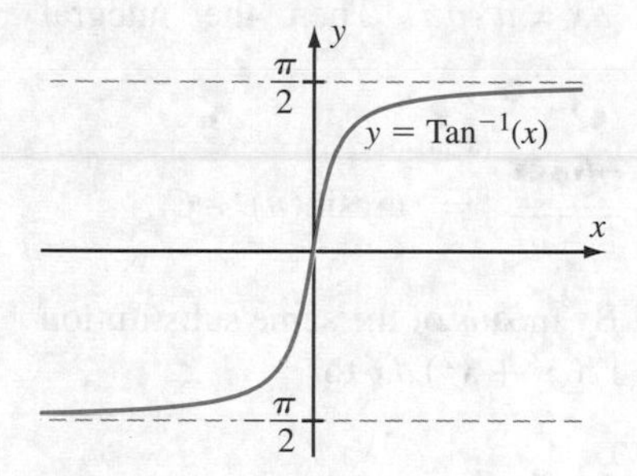

Figure 8

Proof As usual, we calculate the derivative of our new function using the method from Section 3.6:

$$\frac{d}{dt}\arctan(t) = \frac{1}{\frac{d}{ds}\text{Tan}(s)\Big|_{s=\arctan(t)}} = \frac{1}{\sec^2(s)|_{s=\arctan(t)}} = \frac{1}{(\sec(\arctan(t)))^2}.$$

Recall that $\sec^2(s) = 1 + \tan^2(s)$. Therefore,

$$\frac{d}{dt}\arctan(t) = \frac{1}{1+(\tan(\arctan(t)))^2} = \frac{1}{1+t^2}.$$

The Chain Rule form of the inverse tangent derivative formula is

$$\frac{d}{dx}\arctan(u) = \frac{1}{1+u^2}\frac{du}{dx}. \quad \textbf{(6.55)}$$

■

Example 6 Calculate $\frac{d}{dx}\arctan(e^x)|_{x=0}$.

Solution We use equation (6.55):

$$\frac{d}{dx}\arctan(e^x)\Big|_{x=0} = \frac{1}{1+(e^x)^2}\cdot e^x\Big|_{x=0} = \frac{1}{2}.$$ ■

Integrals in Which Inverse Trigonometric Functions Arise

We can write the differentiation formulas for inverse trigonometric functions in reverse as antidifferentiation formulas. We have

$$\int \frac{du}{\sqrt{1-u^2}} = \arcsin(u) + C \quad \textbf{(6.56)}$$

and

$$\int \frac{du}{1+u^2}\,du = \arctan(u) + C. \tag{6.57}$$

The important lesson here is that although the integrands involve only polynomials and roots, the antiderivatives involve inverse trigonometric functions. In Chapter 7, when we learn to analyze integrals and compute antiderivatives, this will prove to be important information.

Example 7 Suppose that a is a positive constant. Show that

$$\int \frac{dx}{\sqrt{a^2 - x^2}} = \arcsin\left(\frac{x}{a}\right) + C \tag{6.58}$$

and

$$\int \frac{dx}{a^2 + x^2} = \frac{1}{a}\arctan\left(\frac{x}{a}\right) + C. \tag{6.59}$$

Solution We use the substitution $x = a \cdot u$, $dx = a \cdot du$. Then the integral $\int 1/\sqrt{a^2 - x^2}\,dx$ becomes

$$\int \frac{a \cdot du}{\sqrt{a^2 - a^2u^2}} = \int \frac{a \cdot du}{\sqrt{a^2(1-u^2)}} = \int \frac{du}{\sqrt{1-u^2}} \overset{(6.56)}{=} \arcsin(u) + C.$$

We obtain equation (6.58) by substituting $u = x/a$. By means of the same substitution ($x = a \cdot u$, $dx = a \cdot du$), we convert the integral $\int 1/(a^2 + x^2)\,dx$ to

$$\int \frac{a \cdot du}{a^2 + a^2u^2} = \int \frac{a \cdot du}{a^2(1+u^2)} \overset{(6.57)}{=} \frac{1}{a}\arctan(u) + C.$$

We obtain equation (6.59) by substituting $u = x/a$. ■

Example 8 Use a substitution to evaluate

$$\int \frac{x}{\sqrt{1-x^4}}\,dx.$$

Solution We use the substitution $u = x^2$, $du = 2x\,dx$. The given integral becomes

$$\int \frac{1}{\sqrt{1-u^2}}\,\frac{1}{2}\,du = \frac{1}{2}\arcsin(u) + C.$$

Therefore,

$$\int \frac{x}{\sqrt{1-x^4}}\,dx = \frac{1}{2}\arcsin(x^2) + C.$$ ■

Motion in a Resisting Medium—Quadratic Drag Law

In Section 6.5, we studied vertical motion subject to a resistive force proportional to velocity. Sometimes the force R of air resistance is better approximated by

$$R(v) = \pm k \cdot v^2 \tag{6.60}$$

where the constant k is positive. The $\pm$ sign in equation (6.60) is chosen so that the direction of R is opposite to the direction of motion. Thus, $R(v) = k \cdot v^2$ for downward motion and $R(v) = -k \cdot v^2$ for upward motion. Equation (6.60) is called the quadratic drag law.

Example 9 A ball with mass 200 grams is thrown upward with initial velocity 28 meters per second. Let $v(t)$ denote the velocity of the ball at time t. Supposing that air resistance is approximated by the formula

$$R(v) = \pm \frac{1}{250} \cdot v^2$$

where the units of v are meters per second and the units of mass are grams, obtain a formula for $v(t)$. At what time, approximately, does the ball begin to fall?

Solution When the ball rises, $R(v)$ is a downward force. We therefore use the minus sign in the formula for $R(v)$. The equation of motion is

$$\underbrace{\underbrace{200 \cdot \frac{dv}{dt}}_{\text{mass} \times \text{acceleration}}}_{\text{Total force}} = \underbrace{-\frac{1}{250} \cdot v^2}_{\text{Resistive force}} + \underbrace{(-200 \cdot g)}_{\text{Force of gravity}},$$

or

$$\frac{dv}{dt} = -\left(\frac{1}{250 \cdot 200} \cdot v^2 + g\right).$$

Therefore, using the value 9.8 for g, we have

$$\frac{dv}{dt} = -\left(\frac{1}{50000} v^2 + 9.8\right) = -\frac{1}{50000}(v^2 + 490000) = -\frac{1}{50000}(v^2 + 700^2).$$

Separating variables, we obtain

$$\int \frac{1}{700^2 + v^2} \frac{dv}{dt}\, dt = -\int \frac{1}{50000}\, dt,$$

or, using integration formula (6.60), we have

$$\frac{1}{700} \arctan\left(\frac{v(t)}{700}\right) = -\frac{1}{50000} t + C.$$

This leads to

$$\arctan\left(\frac{v(t)}{700}\right) = -\frac{7}{500} t + A$$

where $A = 700C$. The initial condition, $v(0) = 28$, allows us to solve for A:

$$A = \arctan\left(\frac{28}{700}\right) = \arctan\left(\frac{1}{25}\right).$$

Therefore,

$$v(t) = 700\tan\left(-\frac{7}{500}t + \arctan\left(\frac{1}{25}\right)\right) = 700\cdot\tan\left(\arctan\left(\frac{1}{25}\right) - \frac{7}{500}t\right).$$

The ball begins to fall at the time when $v(t) = 0$. Using the solve utility of a calculator (or the Newton-Raphson Method from Chapter 4), we find the solution to this equation to be $t \approx 2.86$ seconds. ■

inSIGHT

Our analysis does not extend to the time when the ball falls back to Earth (during which the resistive force is *upward*). The differential equation that describes the descent is

$$\underbrace{\underbrace{200\cdot\frac{dv}{dt}}_{\text{mass}\times\text{acceleration}}}_{} = \underbrace{\underbrace{+\frac{1}{250}\cdot v^2}_{\text{Resistive force}} + \underbrace{(-200\cdot g)}_{\text{Force of gravity}}}_{\text{Total force}}.$$

Notice that for downward motion the resistive force is positive. We solve this differential equation in Section 6.7.

Other Inverse Trigonometric Functions

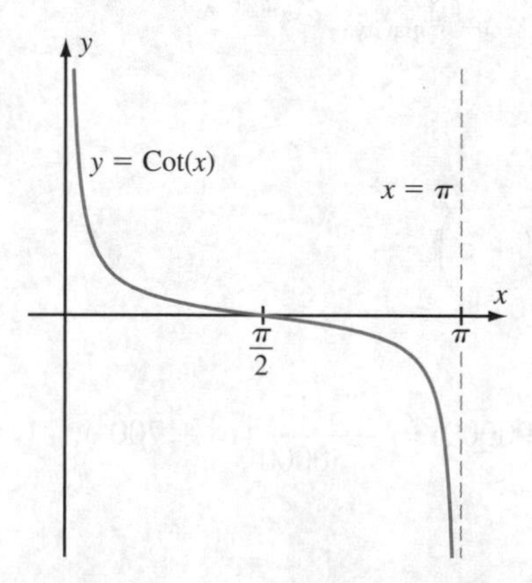

Figure 9

We now learn a few things about the inverse secant, the inverse cosecant, and the inverse cotangent, none of which arise frequently. We use $x \mapsto \mathrm{Cot}(x)$ to denote the restriction of the cotangent function to the interval $(0, \pi)$ (see Figure 9). Thus, $x \mapsto \mathrm{Cot}(x)$ is decreasing and takes on all real values. Therefore, the inverse,

$$\mathrm{Cot}^{-1} : (-\infty, \infty) \to (0, \pi),$$

is well defined. Refer to Figure 10 for the graph. We sometimes write $\mathrm{Cot}^{-1}(x)$ as $\mathrm{arccot}(x)$. A computation similar to the ones found in Theorems 1 and 2 shows that

$$\frac{d}{dx}\mathrm{Cot}^{-1}(x) = -\frac{1}{1+x^2}. \tag{6.61}$$

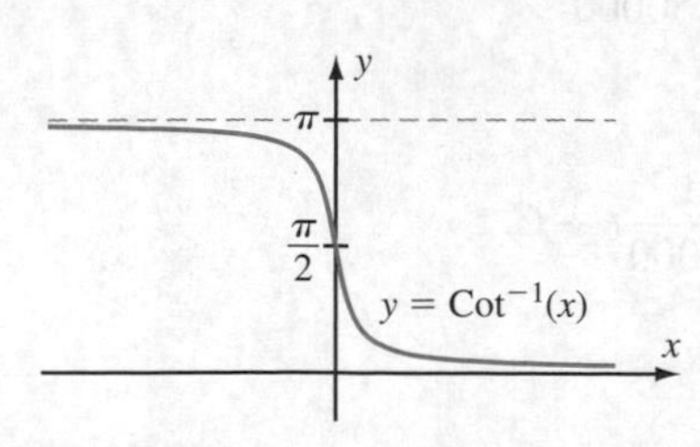

Figure 10

We define $x \mapsto \mathrm{Sec}(x)$ to be the restriction of the secant to the set $[0, \pi/2) \cup (\pi/2, \pi]$ (see Figure 11). Thus, $x \mapsto \mathrm{Sec}(x)$ is one-to-one. On the domain of $\mathrm{Sec}(x)$, the cosine function takes all values in the interval $[-1, 1]$ except for 0. Passing to the reciprocal, we see that secant takes all values greater than or equal to 1 and all values less than or equal to -1. The inverse function is

$$\mathrm{Sec}^{-1} : (-\infty, -1] \cup [1, \infty) \to [0, \pi/2) \cup (\pi/2, \pi]$$

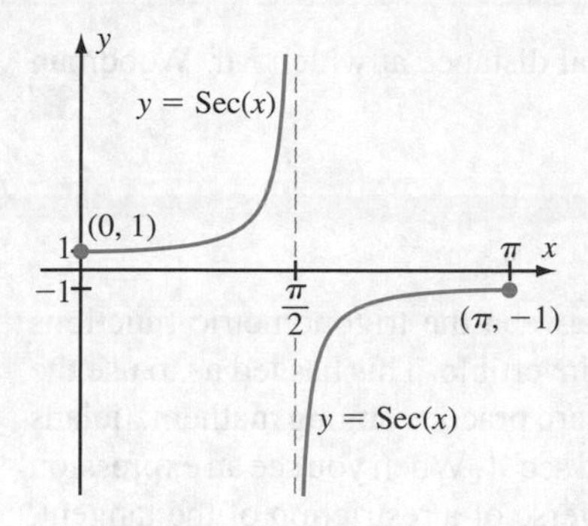

Figure 11

(see Figure 12). We sometimes write arcsec(x) for $\mathrm{Sec}^{-1}(x)$ and give its derivative by

$$\frac{d}{dx}\mathrm{Sec}^{-1}(x) = \frac{1}{|x| \cdot \sqrt{x^2 - 1}}, \ |x| > 1. \tag{6.62}$$

The function $x \mapsto \mathrm{Csc}(x)$ is defined as the restriction of the cosecant to the set $[-\pi/2, 0) \cup (0, \pi/2]$ (see the graph in Figure 13). We see that $x \mapsto \mathrm{Csc}(x)$ is one-to-one and, therefore, has an inverse function Csc^{-1} that maps the image of Csc to the domain of Csc. As x ranges over the values in the domain of Csc, $\sin(x)$ takes on all values in the interval $[-1, 1]$ except for 0. Therefore, the cosecant takes on all values greater than or equal to 1 and all values less than or equal to -1. Thus, Csc^{-1} has domain $(-\infty, -1] \cup [1, \infty)$ and range $[-\pi/2, 0) \cup (0, \pi/2]$, as seen in the plot of Csc^{-1} shown in Figure 14. The derivative of the inverse cosecant is given by

$$\frac{d}{dx}\mathrm{Csc}^{-1}(x) = -\frac{1}{|x| \cdot \sqrt{x^2 - 1}}, \ |x| > 1. \tag{6.63}$$

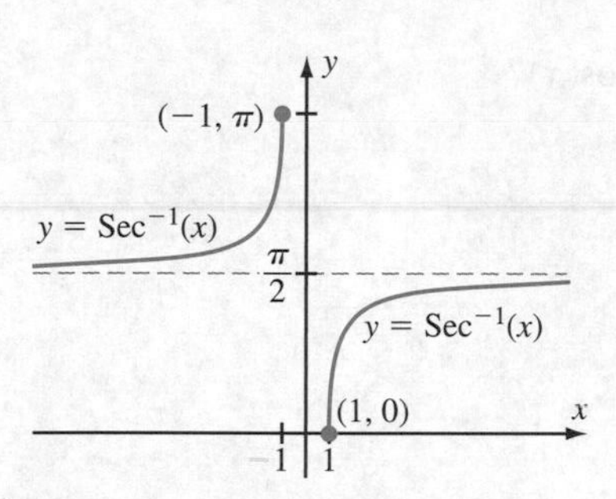

Figure 12

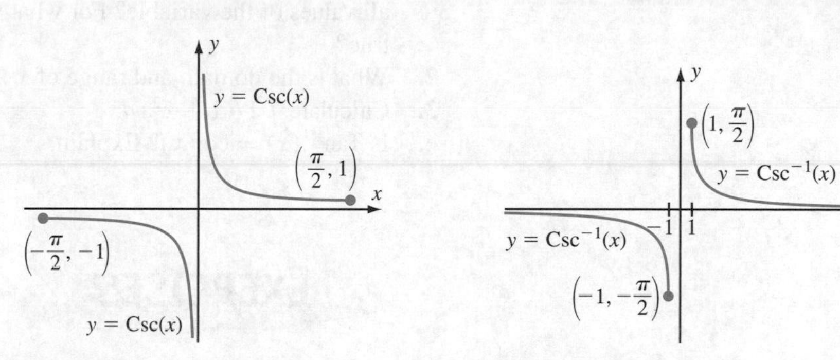

Figure 13 **Figure 14**

An Optimization Example Involving Inverse Trigonometric Functions

Example 10 Mr. Woodman is viewing a 12 ft long tapestry hung lengthwise on a wall. The bottom end of the tapestry is 3 ft above his eye level (see Figure 15). At what distance from the tapestry should Mr. Woodman stand to obtain the most favorable view? In other words, for what value of x is angle α maximized?

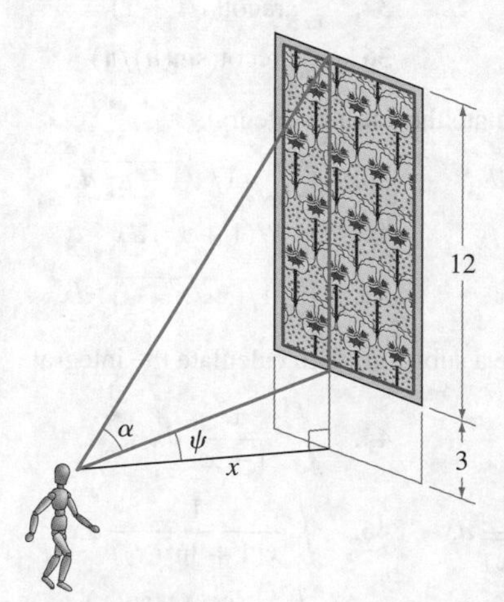

Figure 15
$\cot(\alpha + \psi) = x/15$ and $\cot(\psi) = x/3$

Solution From Figure 15, we see that $\cot(\alpha + \psi) = x/15$ and $\cot(\psi) = x/3$. Therefore, $\alpha + \psi = \mathrm{Cot}^{-1}(x/15)$ and $\psi = \mathrm{Cot}^{-1}(x/3)$. Thus,

$$\alpha = (\alpha + \psi) - \psi = \mathrm{Cot}^{-1}(x/15) - \mathrm{Cot}^{-1}(x/3).$$

Notice that as $x \to 0^+$ we have $\alpha \to 0$ and as $x \to \infty$ we also have $\alpha \to 0$. So angle α will have a maximum value at some positive value of x that can be found by differentiating α with respect to x, setting the derivative equal to 0, and solving for x. Using formula (6.61), we obtain

$$\frac{d\alpha}{dx} = -\frac{1}{15\left(1 + \frac{x^2}{225}\right)} + \frac{1}{3\left(1 + \frac{x^2}{9}\right)} = -12\frac{x^2 - 45}{(225 + x^2)(9 + x^2)}.$$

It follows that $x = \sqrt{45} = 3\sqrt{5}$ ft is the optimal distance at which Mr. Woodman should stand. ■

Notation

In this section, we have carefully distinguished between the trigonometric functions and their restrictions to domains on which they are invertible. This has led us to use the notations Sin^{-1}, Cos^{-1}, Tan^{-1}, and Sec^{-1}. It is standard practice among mathematicians and scientists to simply write $\sin^{-1}$, $\cos^{-1}$, $\tan^{-1}$, and $\sec^{-1}$. When you see an expression such as $\tan^{-1}(x)$, remember that it refers to the inverse of a restriction of the tangent. It is especially important not to confuse an inverse trigonometric function (such as $\mathrm{Tan}^{-1}(x)$) with the reciprocal of the corresponding trigonometric function (such as $(\mathrm{Tan}\,(x))^{-1} = 1/\mathrm{Tan}(x)$). The arc notation eliminates this danger.

quickquiz

1. Which of the equations $\sin(\arcsin(t)) = t$ and $\arcsin(\sin(s)) = s$ is true for all values of the variable? For what values of the variable is the other equation true?
2. What is the domain and range of $x \mapsto \arccos(x)$?
3. Calculate $\int 1/(1+x^2)\,dx$.
4. Is $\mathrm{Tan}^{-1}(x) = \cot(x)$? Explain.

EXERCISES

Problems for Practice

In Exercises 1–14, calculate the value of the inverse trigonometric function at the given point.

1. $\arcsin(1)$
2. $\arcsin(\sqrt{3}/2)$
3. $\arcsin(-1)$
4. $\arccos(-1)$
5. $\arccos(1/2)$
6. $\arccos(-\sqrt{3}/2)$
7. $\arccos(\sqrt{2}/2)$
8. $\arctan(-1)$
9. $\arctan(-\sqrt{3})$
10. $\mathrm{arcsec}(2)$
11. $\mathrm{arcsec}(\sqrt{2})$
12. $\mathrm{arccsc}(1)$
13. $\mathrm{arccsc}(-\sqrt{2})$
14. $\mathrm{arccot}(-\sqrt{3})$

In Exercises 15–36, calculate the derivative.

15. $\frac{d}{dx}\arcsin(3x)$
16. $\frac{d}{dx}\arcsin(\sqrt{x})$
17. $\frac{d}{dx}\arcsin(\ln(x))$
18. $\frac{d}{dt}(t\cdot\arcsin(t^{3/2}))$
19. $\frac{d}{dt}\arcsin(\sin^2(t))$
20. $\frac{d}{dx}\arcsin(x^3-x^2)$
21. $\frac{d}{dx}\arccos(x^4)$
22. $\frac{d}{dx}\arccos(e^{3x})$
23. $\frac{d}{dx}(\arcsin(x)\cdot\arccos(x))$
24. $\frac{d}{dx}(\ln(x)\cdot\arccos(x))$
25. $\frac{d}{dx}\arctan(e^x)$
26. $\frac{d}{dx}\arctan(3/x^3)$
27. $\frac{d}{dx}(\tan(x)\cdot\arctan(x))$
28. $\frac{d}{dx}\arctan(x/(x+1))$
29. $\frac{d}{dx}\cos(\arctan(x))$
30. $\frac{d}{dt}\mathrm{arcsec}(1/t)$
31. $\frac{d}{dt}\mathrm{arcsec}(2+\sin(t))$
32. $\frac{d}{dx}(\mathrm{arcsec}(x))^3$
33. $\frac{d}{dx}\mathrm{arccsc}(x^2)$
34. $\frac{d}{dt}\mathrm{arccot}(\sqrt{t+1})$
35. $\frac{d}{dt}\mathrm{arccot}(1/t^2)$
36. $\frac{d}{du}\mathrm{arccot}(\sin(u)/u)$

In Exercises 37–42, calculate the definite integral.

37. $\int_0^{1/2} 1/\sqrt{1-x^2}\,dx$
38. $\int_{-1/\sqrt{2}}^{0} 1/\sqrt{1-x^2}\,dx$
39. $\int_0^{\sqrt{3}} 1/\sqrt{4-x^2}\,dx$
40. $\int_0^1 1/(1+x^2)\,dx$
41. $\int_0^3 1/(9+x^2)\,dx$
42. $\int_{\sqrt{2}}^2 1/(x\sqrt{x^2-1})\,dx$

In Exercises 43–52, make a substitution to calculate the integral.

43. $\displaystyle\int_0^1 \frac{x}{1+x^4}\,dx$
44. $\displaystyle\int_0^1 \frac{x^2}{1+x^6}\,dx$
45. $\displaystyle\int \frac{1}{x\sqrt{1-\ln^2(x)}}\,dx$
46. $\displaystyle\int \frac{1}{x(1+\ln^2(x))}\,dx$
47. $\displaystyle\int_0^1 \frac{\arctan(x)}{1+x^2}\,dx$
48. $\displaystyle\int_0^{\pi/4} \frac{\sec(x)\tan(x)}{1+\sec^2(x)}\,dx$

49. $\displaystyle\int \frac{\sec^2(x)}{\sqrt{1-\tan^2(x)}}\,dx$ **50.** $\displaystyle\int \frac{1}{\sqrt{e^{2x}-1}}\,dx$

51. $\displaystyle\int \frac{e^x}{\sqrt{1-e^{2x}}}\,dx$ **52.** $\displaystyle\int_{2^{1/4}}^{\sqrt{2}} \frac{1}{x\sqrt{x^4-1}}\,dx$

Further Theory and Practice

53. Supply a proof of formula (6.61) for the derivative of $\mathrm{Cot}^{-1}(x)$.

54. Supply a proof of formula (6.62) for the derivative of $\mathrm{Sec}^{-1}(x)$.

55. Supply a proof of formula (6.63) for the derivative of $\mathrm{Csc}^{-1}(x)$.

56. Evaluate $\int (x^{3/2}+x^{1/2})^{-1}\,dx$

57. One of the gambling casinos in Las Vegas, Nevada, is famous for its glassed-in external elevator. It is easy to track its motion from nearby hotels. Suppose that the casino with the elevator is 300 ft tall and that the elevator is descending from the roof at a constant rate of 30 ft/s. You are observing from a horizontal distance of 150 ft and an elevation of 75 ft. At what height will the elevator appear to you to be moving most quickly?

58. Determine the values of x for which

$$\arcsin(x) = \frac{\pi}{2} - \arccos(x)$$

and prove the formula for those values of x.

59. Determine the values of x for which

$$\arctan(x) = \mathrm{arccot}\left(\frac{1}{x}\right),$$

and prove the formula for those values of x.

60. Determine the values of x for which

$$\arctan(x) = \mathrm{arcsec}\left(\sqrt{x^2+1}\right),$$

and prove the formula for those values of x.

In Exercises 61–64, evaluate the integral by completing the square and then making an appropriate substitution.

61. $\displaystyle\int \frac{1}{\sqrt{x-x^2}}\,dx$ **62.** $\displaystyle\int \frac{1}{\sqrt{8+2x-x^2}}\,dx$

63. $\displaystyle\int \frac{1}{2+2x+x^2}\,dx$

64. $\displaystyle\int \frac{1}{13+6x+x^2}\,dx$

65. Suppose c is a constant in the interval $(-1, 1)$. Find the linearization of $\arcsin(x)$ at $x = c$. What is this linearization for $c = 1/2$? For $c = -1/\sqrt{2}$?

66. For each real number c, find the linearization of $\arctan(x)$ at $x = c$. What is this linearization for $c = \sqrt{3}$? For $c = -1$?

The quadratic approximation to a function f at c is the degree two polynomial

$$T_2(x) = f(c) + f'(c)(x-c) + \frac{1}{2}f''(c)(x-c)^2.$$

In Exercises 67–70, find the quadratic approximation.

67. $f(x) = \arcsin(x)$, $c = 1/2$

68. $f(x) = \arctan(x)$, $c = 1$

69. $f(x) = \mathrm{arcsec}(x)$, $c = 2$

70. $f(x) = \tan(\pi x/3) + \arctan(x)$, $c = 1$

71. Suppose $a > 0$. Calculate $\int x^k/(a^2+x^2)\,dx$ for $k = 0$, 1, 2, and 3. What is the procedure for an integer greater than 3?

72. Show that $\cos^2(\arcsin(x))$ is the restriction to the interval $[-1, 1]$ of a polynomial in x.

73. Prove that there is a constant C such that

$$2\arctan(\sqrt{x}) = \arcsin\left(\frac{x-1}{x+1}\right) + C, x \geq 0.$$

What is the constant C ?

74. Show that

$$\arctan(x) + \arctan\left(\frac{1}{x}\right) = \frac{\pi}{2}, x > 0.$$

75. Use l'Hôpital's Rule to calculate the two limits

$$\lim_{x\to 0}\left(\frac{x-\arcsin(x)}{x^3}\right)$$

and

$$\lim_{x\to 0}\left(\frac{x\sqrt{1-x^2}-\arcsin(x)}{x^3}\right).$$

(Notice that these two computations show that the two limits $\lim_{x\to c}(f(x)k(x) - g(x))$ and $\lim_{x\to c}(f(x) - g(x))$ may exist and be unequal even though $\lim_{x\to c} k(x) = 1$.)

Calculator/Computer Exercises

In Exercises 76–79, use a central difference quotient to approximate $f'(c)$ for the given f and c. Also calculate $f'(c)$ exactly.

76. $f(x) = \arcsin(x)$, $c = 0.7$

77. $f(x) = \arcsin(x^2)$, $c = 0.8$

78. $f(x) = \arctan(x)$, $c = 1.3$

79. $f(x) = \arctan(\sqrt{x})$, $c = 0.65$

80. Verify that

$$F_1(x) = \frac{1}{2} \arctan\left(\frac{1}{2}\tan(x)\right)$$

and

$$F_2(x) = \frac{1}{2}\left(\arctan\left(\frac{1}{2}\tan(x)\right) + x - \arctan(\tan(x))\right)$$

are antiderivatives of $f(x) = (3\cos^2(x) + 1)^{-1}$ on $S = \{x \in \mathbb{R} : x \neq (k + 1/2)\pi \;\; k \in \mathbb{Z}\}$. To visualize the difference between F_1 and F_2, plot them over the interval $[-2\pi, 2\pi]$. What are the one-sided limits of F_1 and F_2 at the points not in their domain. Over what intervals $[a, b]$ is the formula

$$\int_a^b f(x)\,dx = F_1(b) - F_1(a)$$

valid? What if F_2 is used instead of F_1?

81. Analyze the graph of $f(x) = \arcsin(x/(1 + x^2))$.

82. Analyze the graph of $f(x) = \arctan(x) \cdot (1 - \arcsin(x))$ on the interval $[-1, 1]$.

83. Plot $y = x/\pi$ and $y = x^2/(x^2 + 1)$ for $0 \leq x \leq 0.38$. Find the abscissa b of the point of intersection. What is the area of the region between these two graphs and over the interval $[0, b]$?

84. Plot $y = 1/(1 + x^2)$ and $y = (3x + 4)/(3x + 6)$ in the window $[0, 0.6] \times [0, 1]$. Find the abscissa b of the point of intersection. Find the area between the two curves and over the interval $[0, b]$.

85. Plot $y = x^2/(x^2 + 1)$ and $y = (x + 3)/(x^2 + 1)$ in the window $[-1.6, 2.4] \times [0, 3.1]$. Find the abscissa a (and then b) of the point of intersection in the second (and then the first) quadrant. What is the area of the region enclosed by the two curves?

6.7 Enrichment: The Hyperbolic Functions*

In engineering, physics, and other applications, certain special functions, called *hyperbolic functions,* arise frequently. In this section, we define and briefly discuss these functions.

Definition For all real t, we define the *hyperbolic sine* of t to be

$$\sinh(t) = \frac{e^t - e^{-t}}{2},$$

and we define the *hyperbolic cosine* of t to be

$$\cosh(t) = \frac{e^t + e^{-t}}{2}.$$

The graphs of the hyperbolic sine and cosine are shown in Figures 1 and 2. Notice that the hyperbolic cosine is an even function, the hyperbolic sine is an odd function, and $\cosh(x)$ has an absolute minimum value of 1 that occurs at $x = 0$.

---

*Knowledge of this section is not required for subsequent sections.

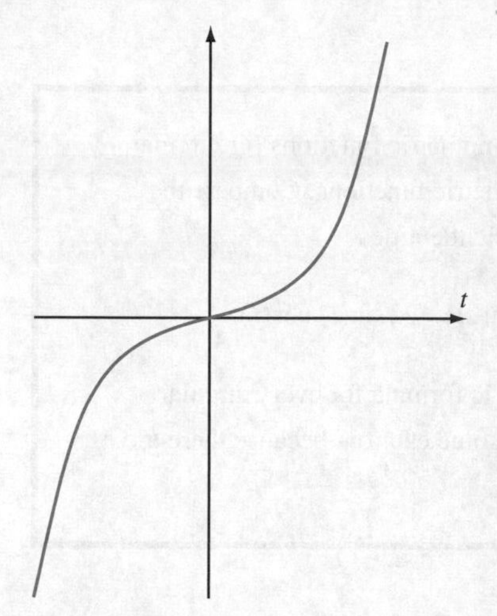

Figure 1
The graph of $\sinh(t)$

inSIGHT

Why is the term "hyperbolic" used? Recall that for classical trigonometry based on the unit circle, the numbers $x = \cos(t)$ and $y = \sin(t)$ satisfy

$$x^2 + y^2 = 1.$$

That is, the point $(\cos(t), \sin(t))$ lies on the unit circle.

By contrast, the numbers $x = \cosh(t)$ and $y = \sinh(t)$ satisfy

$$\begin{aligned} x^2 - y^2 &= \left(\frac{e^t + e^{-t}}{2}\right)^2 - \left(\frac{e^t - e^{-t}}{2}\right)^2 \\ &= \frac{e^{2t} + 2 + e^{-2t}}{4} - \frac{e^{2t} - 2 + e^{-2t}}{4} \\ &= \frac{4}{4} = 1. \end{aligned}$$

The hyperbolic cosine and hyperbolic sine therefore satisfy the identity

$$\cosh^2(t) - \sinh^2(t) = 1. \qquad \textbf{(6.64)}$$

In other words, the point $(\cosh(t), \sinh(t))$ lies on the unit hyperbola that has equation $x^2 - y^2 = 1$. In Example 2 (page 491), we will find a deeper reason for the terminology.

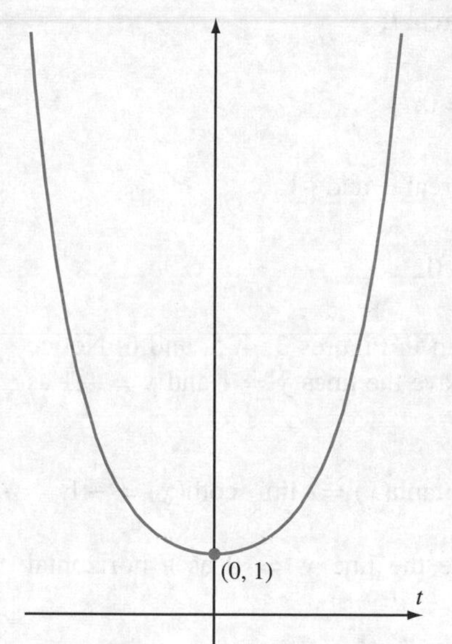

Figure 2
The graph of $\cosh(t)$

Example 1 Show that

$$\cosh(2t) = \cosh^2(t) + \sinh^2(t) \quad \text{and} \quad \sinh(2t) = 2\sinh(t)\cosh(t).$$

Solution The most basic method for establishing identities among the hyperbolic trigonometric functions is to convert them to exponential functions. Thus,

$$\begin{aligned} \cosh^2(t) + \sinh^2(t) &= \left(\frac{e^t + e^{-t}}{2}\right)^2 + \left(\frac{e^t - e^{-t}}{2}\right)^2 \\ &= \left(\frac{1}{4}e^{2t} + \frac{1}{2} + \frac{1}{4e^{2t}}\right) + \left(\frac{1}{4}e^{2t} - \frac{1}{2} + \frac{1}{4e^{2t}}\right). \end{aligned}$$

On simplification, we obtain

$$\cosh^2(t) + \sinh^2(t) = \frac{1}{2}e^{2t} + \frac{1}{2}e^{-2t} = \cosh(2t).$$

Similarly,

$$\begin{aligned} 2\sinh(t)\cosh(t) &= 2\left(\frac{e^t - e^{-t}}{2}\right)\left(\frac{e^t + e^{-t}}{2}\right) \\ &= \frac{1}{2}(e^{2t} - e^{-2t}) = \sinh(2t). \end{aligned}$$

■

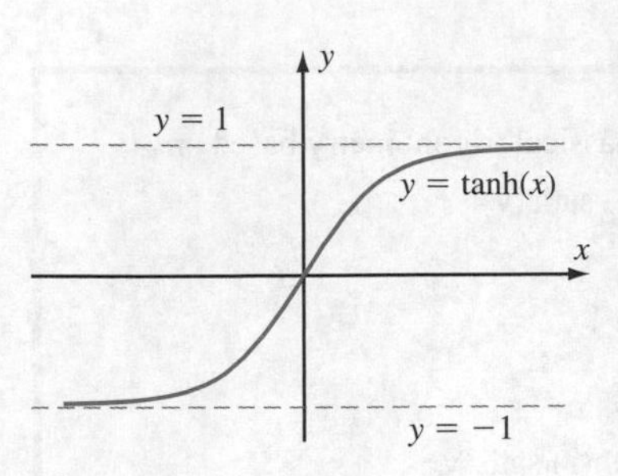

Figure 3

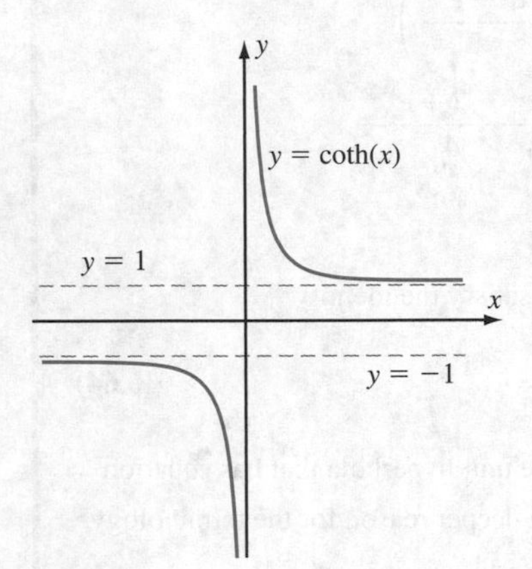

Figure 4

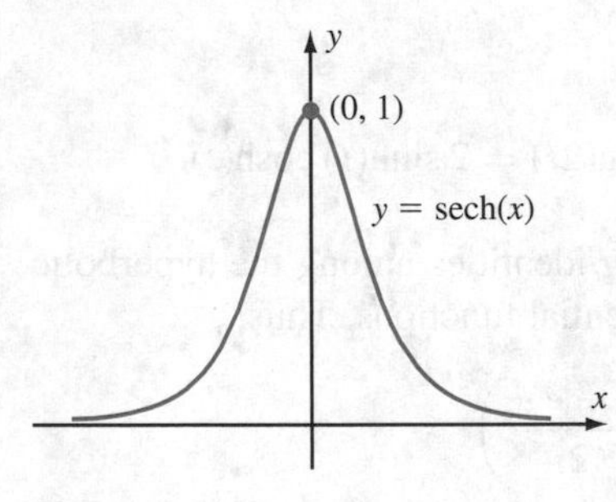

Figure 5

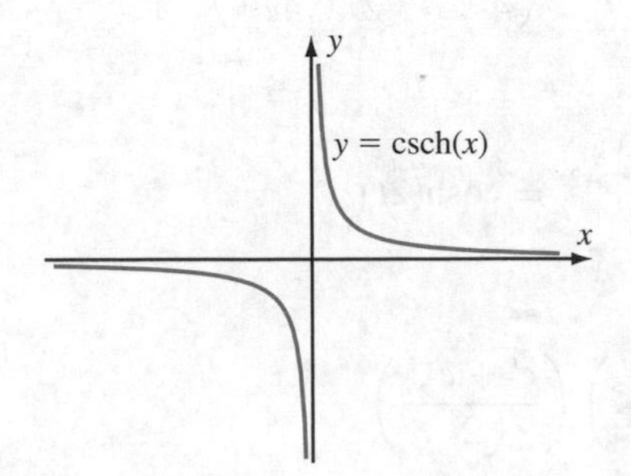

Figure 6

INSIGHT

There are many analogies between the standard trigonometric functions (or circular trigonometric functions) and the hyperbolic trigonometric functions. Compare the identities in Example 1 with the circular trigonometric identities

$$\cos(2t) = \cos^2(t) - \sin^2(t) \quad \text{and} \quad \sin(2t) = 2\sin(t)\cos(t).$$

As a rule of thumb, there is a hyperbolic trigonometric formula for every circular trigonometric formula. However, you must exercise some caution, because there are some differences in the occurrence of minus signs.

Just as we can define the tangent, the cotangent, the secant, and the cosecant as quotients involving the sine and cosine, so too can we define the hyperbolic tangent (tanh), the hyperbolic cotangent (coth), the hyperbolic secant (sech), and the hyperbolic cosecant (csch):

$$\tanh(t) = \frac{\sinh(t)}{\cosh(t)}, \quad \text{all real } t;$$

$$\coth(t) = \frac{\cosh(t)}{\sinh(t)}, \quad t \neq 0;$$

$$\operatorname{sech}(t) = \frac{1}{\cosh(t)}, \quad \text{all real } t; \text{ and}$$

$$\operatorname{csch}(t) = \frac{1}{\sinh(t)}, \quad t \neq 0.$$

The graphs of these hyperbolic functions are given in Figures 3, 4, 5, and 6. Notice that the graphs of $y = \tanh(x)$ and $y = \coth(x)$ have the lines $y = 1$ and $y = -1$ as horizontal asymptotes:

$$\lim_{x\to\infty} \tanh(x) = \lim_{x\to\infty} \coth(x) = 1 \quad \text{and} \quad \lim_{x\to-\infty} \tanh(x) = \lim_{x\to-\infty} \coth(x) = -1.$$

The graphs of $y = \operatorname{sech}(x)$ and $y = \operatorname{csch}(x)$ have the line $y = 0$ as a horizontal asymptote:

$$\lim_{x\to\infty} \operatorname{sech}(x) = \lim_{x\to-\infty} \operatorname{sech}(x) = 0 \quad \text{and} \quad \lim_{x\to\infty} \operatorname{csch}(x) = \lim_{x\to-\infty} \operatorname{csch}(x) = 0.$$

The graphs of $y = \coth(x)$ and $y = \operatorname{csch}(x)$ have the line $x = 0$ as a vertical asymptote:

$$\lim_{x\to 0^+} \operatorname{csch}(x) = \lim_{x\to 0^+} \coth(x) = \infty \quad \text{and} \quad \lim_{x\to 0^-} \operatorname{csch}(x) = \lim_{x\to 0^-} \coth(x) = -\infty.$$

Derivatives of the Hyperbolic Functions

It is easy to calculate that

$$\frac{d}{dt}\cosh(t) = \frac{d}{dt}\left(\frac{e^t + e^{-t}}{2}\right) = \frac{e^t - e^{-t}}{2} = \sinh(t).$$

Similarly,

$$\frac{d}{dt}\sinh(t) = \frac{d}{dt}\left(\frac{e^t - e^{-t}}{2}\right) = \frac{e^t + e^{-t}}{2} = \cosh(t).$$

The calculation of the derivatives of the remaining four hyperbolic functions follow from these formulas and from the Quotient Rule. For example,

$$\begin{aligned}\frac{d}{dt}\tanh(t) &= \frac{d}{dt}\left(\frac{\sinh(t)}{\cosh(t)}\right)\\ &= \frac{\cosh(t)\sinh'(t) - \sinh(t)\cosh'(t)}{\cosh^2(t)}\\ &= \frac{\cosh^2(t) - \sinh^2(t)}{\cosh^2(t)}\\ &= \frac{1}{\cosh^2(t)}\\ &= \operatorname{sech}^2(t).\end{aligned}$$

As usual, differentiation formulas allow us to analytically determine features of functions that appear in their graphs. For example, the hyperbolic sine and hyperbolic tangent are increasing functions since each has an everywhere positive first derivative.

Differentiation Formulas for the Hyperbolic Trigonometric Functions

$$\frac{d}{dt}\sinh(t) = \cosh(t) \qquad \frac{d}{dt}\coth(t) = -\operatorname{csch}^2(t)$$

$$\frac{d}{dt}\cosh(t) = \sinh(t) \qquad \frac{d}{dt}\operatorname{sech}(t) = -\operatorname{sech}(t)\cdot\tanh(t)$$

$$\frac{d}{dt}\tanh(t) = \operatorname{sech}^2(t) \qquad \frac{d}{dt}\operatorname{csch}(t) = -\operatorname{csch}(t)\cdot\coth(t)$$

Example 2 The arc of the unit circle $x^2 + y^2$ between $(1, 0)$ and $(\cos(t), \sin(t))$ subtends a sector of area $t/2$ (see Figure 7). Show that the arc of the unit hyperbola $x^2 - y^2 = 1$ between $(1, 0)$ and $(\cosh(t), \sinh(t))$ also subtends a sector of area $t/2$. That is, show that the shaded area $A(t)$ in Figure 8, next page, is $t/2$.

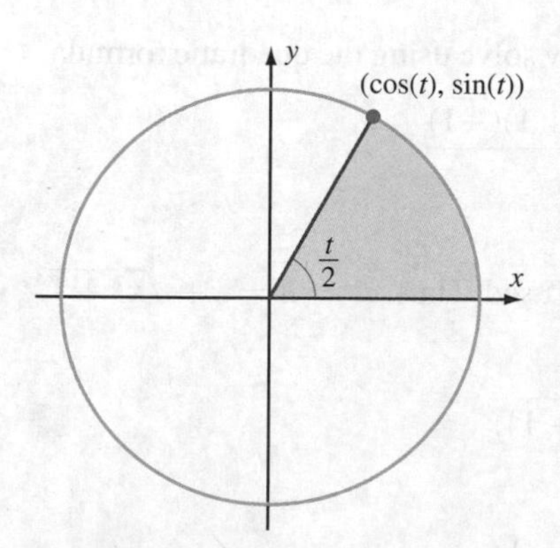

Figure 7

Solution We obtain a formula for $A(t)$ by subtracting the area under the arc of the hyperbola, from $(1, 0)$ to $(\cosh(t), \sinh(t))$, from the area of the right triangle with base $\cosh(t)$ and height $\sinh(t)$:

$$A(t) = \frac{1}{2}\cosh(t)\sinh(t) - \int_1^{\cosh(t)} \sqrt{x^2 - 1}\,dx.$$

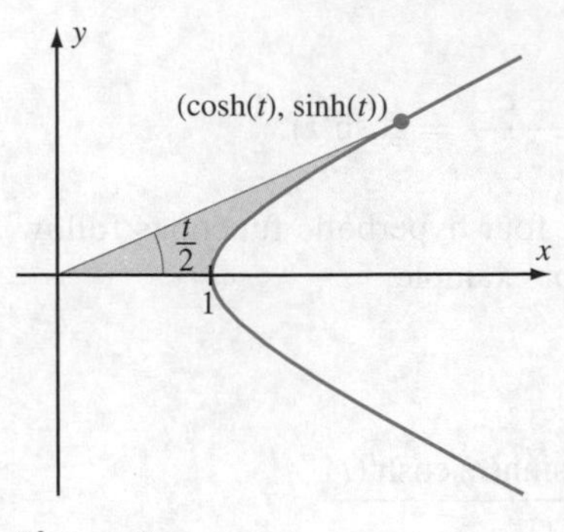

Figure 8

The Chain Rule form of the Fundamental Theorem of Calculus tells us that $A(t)$ is differentiable and that

$$A'(t) = \frac{1}{2}\frac{d}{dt}(\cosh(t)\sinh(t)) - \sqrt{\cosh^2(t) - 1}\,\frac{d}{dt}\cosh(t).$$

Using the Product Rule, we have

$$A'(t) = \frac{1}{2}(\cosh^2(t) + \sinh^2(t)) - \sinh(t)\,\sinh(t) = \frac{1}{2}(\cosh^2(t) - \sinh^2(t)) = \frac{1}{2}.$$

Clearly, $A(0) = 0$. Therefore,

$$A(t) = A(t) - A(0) \overset{\text{Fundamental Theorem of Calculus}}{=} \int_0^t A'(u)\,du = \int_0^t \frac{1}{2}\,du = \frac{t}{2}.$$

■

The Inverse Hyperbolic Functions

Since the hyperbolic sine is increasing on the entire real line, it is one-to-one. We may therefore discuss its inverse, $\sinh^{-1}(x)$. The other hyperbolic trigonometric functions are either not one-to one or not onto $\mathbb{R}$. If their domains and ranges are suitably restricted, however, then their inverses may be defined. We may explicitly calculate these inverses in terms of functions that we already understand.

Example 3 Show that

$$\sinh^{-1}(x) = \ln\left(x + \sqrt{x^2+1}\right).$$

Solution If $y = \sinh^{-1}(x)$, then $\sinh(y) = x$, or

$$e^y - e^{-y} = 2x.$$

Because the expression e^y occurs frequently in our calculation, we denote it by u. The last equation becomes $u - u^{-1} = 2x$, or

$$u^2 - 2x \cdot u - 1 = 0.$$

This is simply a quadratic equation in u that we may solve using the quadratic formula:

$$\begin{aligned} u &= \frac{2x \pm \sqrt{(-2x)^2 - 4(1)(-1)}}{2 \cdot 1} \\ &= x \pm \sqrt{x^2+1}. \end{aligned}$$

Since $u = e^y$ cannot be negative, the positive sign is used. Thus, $e^y = u = x + \sqrt{x^2+1}$. From this, we conclude that

$$y = \ln\left(x + \sqrt{x^2+1}\right).$$

■

Formulas for the Inverses of the Hyperbolic Functions

$$\sinh^{-1}(x) = \ln\left(x + \sqrt{x^2+1}\right), \quad \text{all real } x \qquad \coth^{-1}(x) = \frac{1}{2}\ln\left(\frac{x+1}{x-1}\right), \quad |x| > 1$$

$$\cosh^{-1}(x) = \ln\left(x + \sqrt{x^2-1}\right), \quad x \geq 1 \qquad \operatorname{sech}^{-1}(x) = \ln\left(\frac{1+\sqrt{1-x^2}}{x}\right), \quad 0 < x \leq 1$$

$$\tanh^{-1}(x) = \frac{1}{2}\ln\left(\frac{1+x}{1-x}\right), \quad -1 < x < 1 \qquad \operatorname{csch}^{-1}(x) = \ln\left(\frac{1}{x} + \frac{\sqrt{1+x^2}}{|x|}\right), \quad x \neq 0$$

Example 4 Show that

$$\frac{d}{dx}\sinh^{-1}(x) = \frac{1}{\sqrt{1+x^2}}, \qquad \text{all real } x;$$

$$\frac{d}{dx}\tanh^{-1}(x) = \frac{1}{1-x^2}, \qquad -1 < x < 1; \text{ and}$$

$$\frac{d}{dx}\operatorname{sech}^{-1}(x) = -\frac{1}{x\sqrt{1-x^2}}, \qquad 0 < x < 1.$$

Solution We can use the explicit formulas for the inverse hyperbolic functions. For example,

$$\frac{d}{dx}\sinh^{-1}(x) = \frac{d}{dx}\ln\left(x+\sqrt{x^2+1}\right) = \frac{\frac{d}{dx}\left(x+\sqrt{x^2+1}\right)}{x+\sqrt{x^2+1}}$$

$$= \frac{1+\frac{x}{\sqrt{x^2+1}}}{x+\sqrt{x^2+1}} = \frac{\frac{\sqrt{x^2+1}+x}{\sqrt{x^2+1}}}{x+\sqrt{x^2+1}} = \frac{1}{\sqrt{x^2+1}}$$

and

$$\frac{d}{dx}\tanh^{-1}(x) = \frac{1}{2}\frac{d}{dx}(\ln(1+x) - \ln(1-x))$$

$$= \frac{1}{2}\left(\frac{1}{1+x} + \frac{1}{1-x}\right) = \frac{1}{1-x^2}.$$

We can also calculate these derivatives using the Inverse Function Derivative Rule (see Section 3.6). For practice, let us use that rule to differentiate the inverse hyperbolic secant. We first need to use the identity

$$1 - \tanh^2(t) = \operatorname{sech}^2(t), \tag{6.65}$$

which is obtained by dividing each term of identity (6.64) by $\cosh^2(t)$. Let x be positive (so that it is in the domain of sech), and set $y = \operatorname{sech}^{-1}(x)$. Notice that y is positive; therefore, $\tanh(y)$ is also positive. From this observation and from identity (6.65), it follows that $\tanh(y) = \sqrt{1-\operatorname{sech}^2(y)}$. Since $y = \operatorname{sech}^{-1}(x)$, we have $x = \operatorname{sech}(y)$

and

$$\frac{d}{dx}\operatorname{sech}^{-1}(x) = \left.\frac{1}{\frac{d}{dy}\operatorname{sech}(y)}\right|_{y=\operatorname{sech}^{-1}(x)} = -\left.\frac{1}{\operatorname{sech}(y)\tanh(y)}\right|_{y=\operatorname{sech}^{-1}(x)}$$

$$= -\left.\frac{1}{\operatorname{sech}(y)\sqrt{1-\operatorname{sech}^2(y)}}\right|_{y=\operatorname{sech}^{-1}(x)} = -\frac{1}{x\sqrt{1-x^2}}.$$

■

inSIGHT

It is certainly possible to do without the hyperbolic trigonometric functions. They are merely arithmetic combinations of functions that we already know and understand. However, knowledge of these new functions enables us to recognize certain patterns and expressions that frequently arise in applications. They also considerably simplify many formulas. It is handy to express the differentiation formulas that we calculated in Example 4 in antiderivative form:

$$\int \frac{dx}{\sqrt{1+x^2}} = \sinh^{-1}(x) + C, \quad \int \frac{dx}{1-x^2} = \tanh^{-1}(x) + C, \quad \text{and}$$

$$\int \frac{dx}{x\sqrt{1-x^2}} = -\operatorname{sech}^{-1}(x) + C.$$

In Exercises 40–42, you are asked to verify the following generalizations:

$$\int \frac{dx}{\sqrt{a^2+x^2}} = \sinh^{-1}\left(\frac{x}{a}\right) + C, \tag{6.66}$$

$$\int \frac{dx}{a^2-x^2} = \frac{1}{a}\tanh^{-1}\left(\frac{x}{a}\right) + C, \tag{6.67}$$

and

$$\int \frac{dx}{x\sqrt{a^2-x^2}} = -\frac{1}{a}\operatorname{sech}^{-1}\left(\frac{x}{a}\right) + C. \tag{6.68}$$

Motion in a Resisting Medium—The Quadratic Drag Law

In Section 6.6, we studied upward motion under the quadratic drag law:

$$R(v) = \pm k \cdot v^2$$

where k is a positive constant and the sign is chosen so that the direction of $R(v)$ is opposite the direction of motion. In the next example, we use our knowledge of hyperbolic functions to study downward motion under the quadratic drag law.

Example 5 Let $v(t)$ denote the velocity at time t of an object that is dropped from a height at time $t = 0$. Assume that throughout the object's fall, air resistance can be approximated by the formula

$$R(v) = +k \cdot v^2.$$

If the mass of the object is m, show that

$$v(t) = \sqrt{\frac{mg}{k}} \tanh\left(-t\sqrt{\frac{kg}{m}}\right) \tag{6.69}$$

until impact.

Solution Because the object is falling, $R(v)$ is an upward force. We therefore use the positive sign in the quadratic drag law. The equation of motion is

$$\underbrace{\underbrace{m \cdot \frac{dv}{dt}}_{\text{mass} \times \text{acceleration}}}_{\text{Total force}} = \underbrace{+k \cdot v^2}_{\substack{\text{Resistive force} \\ \text{(upward)}}} + \underbrace{(-m \cdot g)}_{\substack{\text{Force of gravity} \\ \text{(downward)}}},$$

or

$$\frac{m}{k}\frac{dv}{dt} = v^2 - \frac{mg}{k}.$$

Let $a = \sqrt{mg/k}$:

$$\int \frac{1}{a^2 - v^2}\frac{dv}{dt}\, dt = -\int \frac{k}{m}\, dt.$$

Using integration formula (6.67), we obtain

$$\frac{1}{a}\tanh^{-1}\left(\frac{v(t)}{a}\right) = -\frac{k}{m} \cdot t + C.$$

To solve for the constant of integration C, we evaluate each side at $t = 0$, using the initial value $v(0) = 0$ (since the object is *dropped*):

$$\frac{1}{a}\tanh^{-1}\left(\frac{0}{a}\right) = -\frac{k}{m} \cdot 0 + C,$$

or $C = 0$. Thus,

$$\frac{1}{a}\tanh^{-1}\left(\frac{v(t)}{a}\right) = -\frac{k}{m} \cdot t,$$

or

$$v(t) = a \cdot \tanh\left(-\frac{ak}{m} \cdot t\right).$$

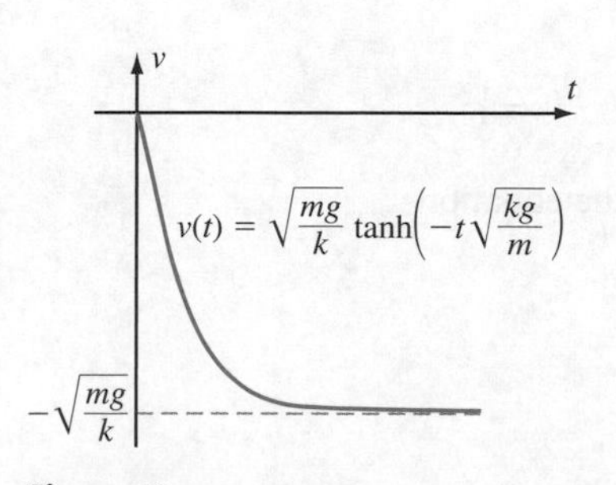

Figure 9

We obtain formula (6.69) by substituting $a = \sqrt{mg/k}$. Figure 9 contains a plot of the velocity. Notice that

$$\lim_{t\to\infty} v(t) = \lim_{t\to\infty} \sqrt{\frac{mg}{k}} \tanh\left(-t\sqrt{\frac{kg}{m}}\right) = -\sqrt{\frac{mg}{k}},$$

which is *terminal velocity* under the quadratic drag law. ■

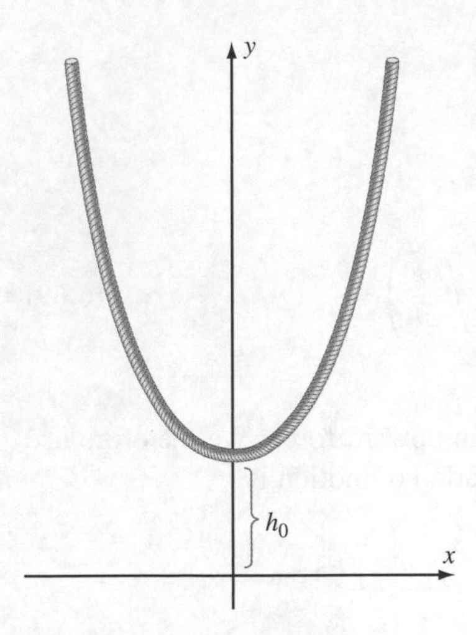

Figure 10a
A hanging cable

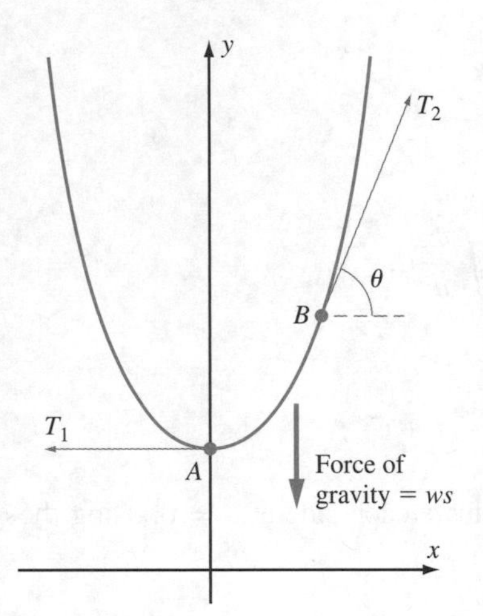

Figure 10b

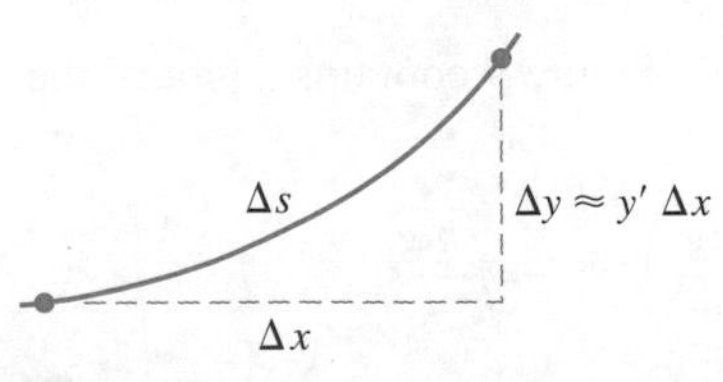

Figure 11

The Hanging Cable

We conclude by applying the hyperbolic functions to an engineering problem. Imagine a heavy cable of uniform thickness and density with both ends clamped down. The cable hangs under the influence of its own weight alone, as shown in Figure 10a. We let T_1 be the horizontal tension at A; T_2, the component of tension *tangent* to the cable at B; and w, the weight of the cable per unit of length. Our task is to determine the shape that the cable will assume. Put another way, the hanging cable can be represented by the graph of a function y. We seek to determine the mathematical form of $y(x)$.

If s is the length of the cable between A and B, then $w \cdot s$ is the downward force of gravity on this portion of the cable (as shown in Figure 10b). We use the symbol θ to denote the angle that the tangent to the cable at B makes with the horizontal. By Newton's First Law, we may equate horizontal components of force to obtain

$$T_1 = T_2 \cos(\theta).$$

Likewise, we equate vertical components of force to obtain

$$w \cdot s = T_2 \sin(\theta).$$

It follows that

$$\frac{w \cdot s}{T_1} = \frac{T_2 \sin(\theta)}{T_2 \cos(\theta)} = \tan(\theta).$$

Since y' at B equals $\tan(\theta)$, we may rewrite this equation as

$$\frac{dy}{dx} = \frac{w \cdot s}{T_1}.$$

We set $q = y'$ so that

$$q(x) = \frac{w}{T_1} \cdot s(x). \tag{6.70}$$

If Δx is an increment of x, then $\Delta q = q(x + \Delta x) - q(x)$ is the corresponding increment of q and $\Delta s = s(x + \Delta x) - s(x)$ is the increment in s. As Figure 11 indicates, Δs is approximated by

$$\Delta s \approx ((\Delta x)^2 + (y'\Delta x)^2)^{1/2} = (1 + (y')^2)^{1/2}\Delta x = (1 + q^2)^{1/2}\Delta x.$$

Thus, from equation (6.70), we have

$$\Delta q = \frac{w}{T_1}\Delta s \approx \frac{w}{T_1}(1 + q^2)^{1/2}\Delta x.$$

Dividing by Δx and letting Δx tend to zero gives the equation

$$\frac{dq}{dx} = \frac{w}{T_1}(1 + q^2)^{1/2},$$

which may be rewritten as

$$\int \frac{1}{(1 + q^2)^{1/2}} \cdot \frac{dq}{dx}\, dx = \frac{w}{T_1} \int dx.$$

Using integration formula (6.66), we obtain

$$\sinh^{-1}(q(x)) = \frac{w}{T_1}x + C.$$

We know that the cable has a horizontal tangent when $x = 0$ (which corresponds to point A in Figure 10b). Thus, $q(0) = y'(0) = 0$. Substituting this information into the last equation gives $C = 0$. Thus, our solution is $\sinh^{-1}(q(x)) = wx/T_1$, or $q(x) = \sinh(wx/T_1)$. However, because $q = y'$, we have

$$\frac{dy}{dx} = \sinh\left(\frac{w}{T_1}x\right).$$

Finally, we integrate this last equation to obtain

$$y(x) = \frac{T_1}{w}\cosh\left(\frac{w}{T_1}x\right) + D$$

where D is a constant of integration. The constant D can be determined from the height h_0 of the point A from the x-axis (Figure 10b):

$$h_0 = y(0) = \frac{T_1}{w}\cosh(0) + D;$$

hence,

$$D = h_0 - \frac{T_1}{w}.$$

Our hanging cable is completely described by the equation

$$y(x) = \frac{T_1}{w}\cosh\left(\frac{w}{T_1}x\right) + h_0 - \frac{T_1}{w}. \tag{6.71}$$

A curve with an equation of the form

$$y = h + a \cdot \cosh\left(\frac{x}{a}\right)$$

for positive constants a and h is called a *catenary*. As we have seen, a hanging cable is a catenary. The architectural plans of the Gateway Arch in St. Louis, Missouri, which is very nearly an inverted catenary, specify the equation

$$y = 693.8597 - 68.7672 \cdot \cosh\left(\frac{3.0022}{299.2239}x\right)$$

for the centroids of its cross sections (when x and y are measured in feet).

quickquiz

1. How can the hyperbolic functions be used to parameterize the curve $x^2 - y^2 = 1$?
2. What is the definition of the hyperbolic tangent function? What is its derivative?
3. What is $\int \sinh(x)\,dx$?
4. Evaluate $\int (1 - x^2)^{-1}\,dx$.

EXERCISES

Problems for Practice

In Exercises 1–10, differentiate the given function.

1. $f(x) = \sinh(3x)$ **2.** $g(x) = \sinh(\ln(x))$
3. $f(x) = \ln(\sinh(x))$ **4.** $g(x) = \sinh(x)\cosh(x)$
5. $f(x) = \operatorname{sech}(\sqrt{x})$ **6.** $g(x) = \coth(x^2)$
7. $f(x) = \cosh(x^2 - 5x)$ **8.** $g(x) = \sinh(\cosh(x))$
9. $f(x) = \ln(\tanh(x))$ **10.** $g(x) = \tanh(\tan(x))$

In Exercises 11–20, perform the integration.

11. $\int \cosh(3x)\,dx$ **12.** $\displaystyle\int \frac{\cosh(x)}{\sinh(x)}\,dx$
13. $\displaystyle\int \frac{\cosh(2x)}{\sqrt{\sinh(2x)}}\,dx$ **14.** $\int x\tanh(x^2)\,dx$
15. $\int x^2 \operatorname{sech}(x^3)\tanh(x^3)\,dx$
16. $\int \sinh(4x)\cosh(4x)\,dx$
17. $\int \tanh^3(x)\operatorname{sech}^2(x)\,dx$ **18.** $\int e^{-3x}\cosh(3x)\,dx$
19. $\int \cosh(2x+5)\,dx$ **20.** $\displaystyle\int \frac{e^x - e^{-x}}{1+\cosh(x)}\,dx$

In Exercises 21–26, calculate the derivatives.

21. $f(x) = \cosh^{-1}(x)$ **22.** $g(x) = \coth^{-1}(x)$
23. $h(x) = \operatorname{csch}^{-1}(x)$ **24.** $f(x) = \dfrac{\sinh^{-1}(x)}{x + \tanh^{-1}(x)}$
25. $g(x) = \operatorname{sech}^{-1}(x)\tanh^{-1}(2x)$
26. $h(x) = \ln(\tanh^{-1}(x))$

Further Theory and Practice

27. Calculate the derivatives of $\tanh(x)$, $\coth(x)$, $\operatorname{sech}(x)$, and $\operatorname{csch}(x)$ using only the Quotient Rule and the rule for the derivative of the exponential function.

In Exercises 28–31, use the graphing techniques developed in Chapter 4 to justify the graph of the hyperbolic function.

28. $\tanh(x)$ **29.** $\coth(x)$
30. $\operatorname{sech}(x)$ **31.** $\operatorname{csch}(x)$

32. Verify the identity $\cosh(x+y) = \cosh(x)\cosh(y) + \sinh(x)\sinh(y)$.

33. Verify the identity $\sinh(x+y) = \sinh(x)\cosh(y) + \cosh(x)\sinh(y)$.

34. Verify the identity $\cosh(2x) = 2\cosh^2(x) - 1$.

35. Verify the identity $\cosh(2x) = 2\sinh^2(x) + 1$.

36. Verify the identity

$$\tanh(x+y) = \frac{\tanh(x) + \tanh(y)}{1 + \tanh(x)\tanh(y)}.$$

37. Show that $y = \sinh(\omega x)$ and $y = \cosh(\omega x)$ are solutions of the differential equation $y'' - \omega^2 y = 0$.

38. Suppose that $y(x)$ is a solution of $y'' - \omega^2 y = 0$. By differentiating, show that $\omega y(x)\cosh(\omega x) - y'(x)\sinh(\omega x)$ is a constant A. Similarly, show that $-\omega y(x)\sinh(\omega x) + y'(x)\cosh(\omega x)$ is a constant B. Use the equations for A and B to eliminate $y'(x)$ and deduce that y must have the form

$$y(x) = A\cosh(\omega x) + B\sinh(\omega x)$$

where A and B are constants.

39. Show that

$$(\cosh(x) + \sinh(x))^n = \cosh(nx) + \sinh(nx)$$

for any n, integer or not.

40. Verify equation (6.66):

$$\int \frac{1}{\sqrt{a^2 + x^2}}\,dx = \sinh^{-1}\left(\frac{x}{a}\right) + C.$$

41. Verify equation (6.67):

$$\int \frac{1}{a^2 - x^2}\,dx = \frac{1}{a}\tanh^{-1}\left(\frac{x}{a}\right) + C.$$

42. Verify equation (6.68):

$$\int \frac{1}{x\sqrt{a^2 - x^2}}\,dx = -\frac{1}{a}\operatorname{sech}^{-1}\left(\frac{x}{a}\right) + C.$$

43. The relationship between the wavelength λ and the velocity v of water waves is

$$v = \sqrt{\frac{g}{2\pi}\lambda\tanh\left(\frac{2\pi d}{\lambda}\right)}$$

where d is the depth of the water. Use the linear approximation of $\tanh(x)$ at $x = 0$ to show that $v \approx \sqrt{gd}$ when d is small compared with λ. By considering the horizontal asymptote of the hyperbolic tangent, show that $v \approx \sqrt{(g/2\pi)\lambda}$ when d is large compared with λ.

Let a be a positive constant and set

$$T(x) = a \cdot \operatorname{sech}^{-1}(x/a) - \sqrt{a^2 - x^2}.$$

Exercises 44–47 pertain to the curve $y = T(x)$, which is called a tractrix (a particular type of pursuit curve, as Exercise 46 illustrates).

44. Show that

$$T'(x) = -\frac{\sqrt{a^2 - x^2}}{x}.$$

45. Use the result of Exercise 44 and the Fundamental Theorem of Calculus to show that $y = a \cdot T(x)$ is the unique solution of the initial value problem

$$\frac{dy}{dx} = -\frac{\sqrt{a^2 - x^2}}{x}, \quad y(0) = 0.$$

46. A chipmunk is at the base of a tree. An owl is on the ground a units away from the chipmunk. The chipmunk begins to run up the tree at the instant the owl begins to fly directly toward the chipmunk. The owl continuously adjusts its flight direction so that at each instant of time it is headed directly toward the chipmunk. The speeds of the animals are such that the distance between the chipmunk and the owl remains a units at all times. Use Exercise 45 to show that the path the owl flies along is a tractrix.

47. A boat is pulled by a tether of length a (see Figure 12). Show that the tip of the boat moves along a tractrix.

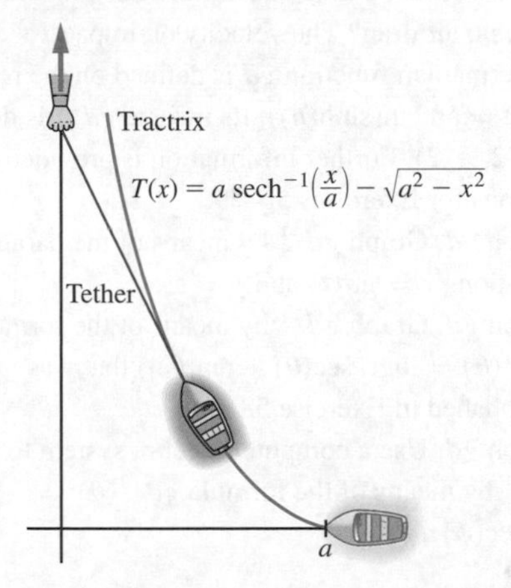

Figure 12

48. The following mathematical model of war was proposed by F.W. Lanchester (1868–1946): Let x denote the number of combatants on one side X, and let z denote the number of forces of the opposing side Z. Each side has a certain lethalness: ξ for the X forces, ζ for the Z forces. This means that the instantaneous rate of decrease of z is proportional to x, with ξ being the proportionality constant. Likewise, the rate at which x decreases is ζz.

a. Relate x, z, and t by two *coupled* first order differential equations. The differentiation is with respect to t. (By "coupled," we mean that each equation involves both x and z.)

b. Show that x and z satisfy the same second order differential equation.

c. Use Exercise 38 to find explicit expressions for $x(t)$ and $z(t)$. These expressions will each involve two constants.

d. Let x_0 and z_0 denote the number of X and Z forces at time $t = 0$. What are $x'(0)$ and $z'(0)$? Use x_0, z_0, $x'(0)$, and $z'(0)$ to calculate the four constants from part c.

49. A hanging cable weighs 3 pounds per linear foot. It hangs between two poles of equal height so that the horizontal tension at the lowest point is 12 foot-pounds. Assuming that the lowest point has y-coordinate 1, write the equation describing the shape of the cable.

50. Show that $\int \tanh(t)\, dt = \ln(\cosh(t)) + C$.

51. Use Exercise 50 to show that the position $y(t)$ of an object of mass m dropped from a height H is

$$y(t) = H - \frac{m}{K} \ln\left(\cosh\left(\sqrt{\frac{Kg}{m}}\,t\right)\right),$$

assuming the quadratic drag law with proportionality constant K.

52. Refer to Exercise 51. Show that the moment of impact is at

$$\sqrt{\frac{m}{Kg}}\, \ln\left(e^{HK/m} + \sqrt{e^{2HK/m} - 1}\right).$$

53. Refer to Exercises 51 and 52. Show that the speed at impact is

$$v_I(H) = -\sqrt{\frac{mg}{K}}\left(1 - \frac{1}{e^{2HK/m} + e^{HK/m}\sqrt{e^{2HK/m} - 1}}\right).$$

54. Refer to Exercises 51 and 53. Show that the terminal velocity v_∞ satisfies $v_\infty = \lim_{H\to\infty} v_I(H)$.

55. Show that

$$\frac{d}{dx}\sinh^{-1}(x) = \frac{d}{dx}\sinh^{-1}\left(x\sqrt{1+y^2} + y\sqrt{1+x^2}\right).$$

Verify the identity

$$\sinh^{-1}(x) + \sinh^{-1}(y) = \sinh^{-1}\left(x\sqrt{1+y^2} + y\sqrt{1+x^2}\right).$$

There is a direct connection between the usual trigonometric (or circular) functions and the hyperbolic functions. The link is the *gudermanian function gd*, which is defined on the real line by $gd(u) = \arctan(\sinh(u))$. Exercises 56–59 concern the gudermanian.

56. Show that

$$\begin{aligned} \cosh(u) &= \sec(gd(u)) & \operatorname{sech}(u) &= \cos(gd(u)) \\ \tanh(u) &= \sin(gd(u)) & \operatorname{csch}(u) &= \cot(gd(u)). \\ \coth(u) &= \csc(gd(u)) \end{aligned}$$

57. Show that $gd'(u) = \operatorname{sech}(u)$.

58. Use the Inverse Function Derivative Rule to derive

$$\frac{d}{d\theta} gd^{-1}(\theta) = \sec(\theta).$$

Deduce that $\int_0^\theta \sec(\phi)\, d\phi = gd^{-1}(\theta)$. Conclude that

$$gd^{-1}(\theta) = \ln(|\sec(\theta) + \tan(\theta)|).$$

59. The *loxodrome* is a curve on the sphere that cuts each meridian in the same angle. (A *meridian* is a semicircle with the north and south poles as endpoints.) To be specific, suppose that a curve starts at a point O on the equator. Let N be the point at the North Pole. Let $x(P)$ and $y(P)$ be the latitude and longitude coordinates of a point P on the loxodrome that makes an angle α with meridians: If $M(P)$ is the intersection of the meridian through P with the equator, then $x(P)$ is the arc length of the equator from O to $M(P)$, and $y(P)$ is the arc length of the meridian from $M(P)$ to P. From spherical trigonometry, we can determine that

$$\frac{dy}{dx} = \cos(y)\tan(\alpha).$$

Deduce that $y(x) = gd(x\tan(\alpha))$.

Calculator/Computer Exercises

In Exercises 60–63, analyze the graph of the function f. Locate all local extrema and points of inflection.

60. $f(x) = \tanh(x^2 - 1)$

61. $f(x) = \sinh\left(\dfrac{x-1}{x^2+1}\right)$

62. $f(x) = \dfrac{2\sinh(x) + 1}{2\cosh(x) - 1}$

63. $f(x) = \tanh(x) - \operatorname{sech}(x)$

64. On the same coordinate axes, plot x and $\tanh(x)$ for $0 \le x \le 0.3$. Next, for $1.5 \le x \le 10$, plot x and $x\tanh(x)$ on the same coordinate axes. Divide the positive x-axis into three subintervals. Find a continuous, piecewise-defined function that is linear or quadratic on each of the three subintervals and that is a reasonable approximation of $x\tanh(x)$.

65. Graph the six hyperbolic functions in the rectangle $[0, 1.5] \times [0, 2]$. Identify the seven points of intersection in the first quadrant.

66. The attack on the battleship USS *Arizona* was briefly described in Exercise 61 in Section 6.5. In summary, an 800 kg bomb was dropped from a height of 3170 m. A 1997 analysis of the attack asserted that the flight of the bomb lasted 26 s, a figure that we will accept for the purposes of this exercise. Assume the quadratic drag law $R(v) = Kv^2$.

a. What was the value of K? (Use the result from Exercise 51. Include the units of K in your answer.)

b. What was the theoretical terminal velocity of the bomb?

c. What was the actual velocity of the bomb when it struck the USS *Arizona?* (State your answer in meters per second but also convert to miles per hour.)

d. Plot the height $y(t)$ of the bomb for $0 \le t \le 26$.

e. What would the time of fall have been had there been no air drag? The velocity on impact?

67. The gudermanian function gd is defined on the real line by $gd(u) = \arctan(\sinh(u))$. Its inverse gd^{-1} is defined on $(-\pi/2, \pi/2)$. Further information is provided in the instructions for Exercises 56–59.

a. Graph gd. Graph gd^{-1} by means of the parametric equations $x = gd(t)$ and $y = t$.

b. Graph gd. Graph gd^{-1} by means of the formula $gd^{-1}(\theta) = \log(|\sec(\theta) + \tan(\theta)|)$ that was established in Exercise 58.

c. Graph gd. Use a computer algebra system to graph gd^{-1} by means of the formula $gd^{-1}(\theta) = \int_0^\theta \sec(\phi)\, d\phi$.

Summary of Key Topics

Method of Separation of Variables (Section 6.1)

The differential equation

$$\frac{dy}{dx} = g(x)h(y)$$

is said to be *separable*. It is solved by separating y and x onto different sides of the equation and integrating each side:

$$\int \frac{1}{h(y)} \frac{dy}{dx}\, dx = \int g(x)\, dx.$$

The Natural Logarithm (Section 6.2)

The natural logarithm function is defined as

$$\ln(x) = \int_1^x \frac{1}{t}\, dt, \quad x > 0.$$

The basic property of the logarithm is

$$\ln(ab) = \ln(a) + \ln(b) \quad (a, b > 0).$$

We can apply the Fundamental Theorem of Calculus to the definition of the logarithm to see that

$$\frac{d}{dx} \ln(x) = \frac{1}{x}.$$

More generally,

$$\frac{d}{dx} \ln(|u|) = \frac{1}{u} \frac{du}{dx}$$

and

$$\int \frac{1}{u} du = \ln(|u|) + C.$$

Logarithms to Other Bases (Section 6.4)

If a and b are positive numbers with $a \neq 1$, then we define

$$\log_a(b) = \frac{\ln(b)}{\ln(a)}.$$

The Exponential Function and the Number e (Section 6.3)

The natural logarithm has domain the positive reals and range the entire real line. It is one-to-one. Its inverse function is called $\exp(x)$. The function exp satisfies

$$\frac{d}{dx} \exp(x) = \exp(x).$$

The exponential function has the basic property

$$\exp(a+b) = \exp(a) \cdot \exp(b).$$

The number $\exp(1)$ is denoted by the symbol e.

Other Exponential Functions (Section 6.4)

If a is positive, we define a^x by

$$a^x = e^{x\ln(a)}.$$

The derivative and integral of this exponential function are given by

$$\frac{d}{dx}a^x = a^x\ln(a) \quad \text{and} \quad \int a^x\,dx = \frac{1}{\ln(a)}a^x + C \quad (a \neq 1).$$

Differential Equations with Exponential Solutions (Section 6.5)

The unique solution of the initial value problem

$$\frac{dy}{dt} = ky, \quad y(0) = A$$

is $y(t) = Ae^{kt}$. This equation describes exponential growth when $k > 0$ and exponential decay when $k < 0$.

For appropriate values of the parameters k and b, the differential equation

$$\frac{du}{dt} = k \cdot u + b$$

can be used to model the temperature of an object. It also describes the velocity of an object that is subjected to the sum of a constant force and linear drag. This equation can be solved by making the change of variable $w = u + b/k$.

The Inverse Trigonometric Functions (Section 6.6)

By restricting the domains of the trigonometric functions, we obtain invertible functions Sin, Cos, Tan, Cot, Sec, and Csc:

$\mathrm{Sin}(x) = \sin(x), -\frac{\pi}{2} \le x \le \frac{\pi}{2}$ $\qquad$ $\mathrm{Cos}(x) = \cos(x), 0 \le x \le \pi$

$\mathrm{Tan}(x) = \tan(x), -\frac{\pi}{2} < x < \frac{\pi}{2}$ $\qquad$ $\mathrm{Cot}(x) = \cot(x), 0 < x < \pi$

$\mathrm{Sec}(x) = \sec(x), x \in \left[0, \frac{\pi}{2}\right) \cup \left(\frac{\pi}{2}, \pi\right]$ $\qquad$ $\mathrm{Csc}(x) = \csc(x), x \in \left[-\frac{\pi}{2}, 0\right) \cup \left(0, \frac{\pi}{2}\right]$

The derivatives of these functions are given by:

$$\frac{d}{dx}\mathrm{Sin}^{-1}(x) = \frac{1}{\sqrt{1-x^2}}, -1 < x < 1 \qquad \frac{d}{dx}\mathrm{Cos}^{-1}(x) = -\frac{1}{\sqrt{1-x^2}}, -1 < x < 1$$

$$\frac{d}{dx}\mathrm{Tan}^{-1}(x) = \frac{1}{1+x^2}, -\infty < x < \infty \qquad \frac{d}{dx}\mathrm{Cot}^{-1}(x) = -\frac{1}{1+x^2}, -\infty < x < \infty$$

$$\frac{d}{dx}\mathrm{Sec}^{-1}(x) = \frac{1}{|x|\cdot\sqrt{x^2-1}}, |x| > 1 \qquad \frac{d}{dx}\mathrm{Csc}^{-1}(x) = -\frac{1}{|x|\cdot\sqrt{x^2-1}}, |x| > 1$$

$$\int \frac{du}{\sqrt{1-u^2}} = \mathrm{Sin}^{-1}(u) + C \qquad \int \frac{du}{\sqrt{1-u^2}} = -\mathrm{Cos}^{-1}(u) + C$$

$$\int \frac{du}{1+u^2}du = \mathrm{Tan}^{-1}(u) + C \qquad \int \frac{du}{1+u^2}du = -\mathrm{Cot}^{-1}(u) + C$$

$$\int \frac{du}{|u|\cdot\sqrt{u^2-1}} = \mathrm{Sec}^{-1}(u) + C \qquad \int \frac{du}{|u|\cdot\sqrt{u^2-1}} = -\mathrm{Csc}^{-1}(u) + C$$

The Hyperbolic Trigonometric Functions (Section 6.7)

The hyperbolic sine and hyperbolic cosine are defined by

$$\sinh(t) = \frac{e^t - e^{-t}}{2}$$

and

$$\cosh(t) = \frac{e^t + e^{-t}}{2},$$

respectively. The other hyperbolic trigonometric functions are defined in terms of

$$\tanh(t) = \frac{\sinh(t)}{\cosh(t)}, \quad \coth(t) = \frac{\cosh(t)}{\sinh(t)},$$

$$\mathrm{sech}(t) = \frac{1}{\cosh(t)}, \quad \mathrm{csch}(t) = \frac{1}{\sinh(t)}.$$

These functions have algebraic and calculus properties that are analogous to properties of the usual trigonometric functions. For example, $\cosh^2(t) - \sinh^2(t) = 1$ is the analogue of $\cos^2(t) + \sin^2(t) = 1$. Here are the derivatives of the hyperbolic trigonometric functions:

$$\frac{d}{dt}\sinh(t) = \cosh(t) \qquad \frac{d}{dt}\coth(t) = -\mathrm{csch}^2(t)$$

$$\frac{d}{dt}\cosh(t) = \sinh(t) \qquad \frac{d}{dt}\mathrm{sech}(t) = -\mathrm{sech}(t)\cdot\tanh(t)$$

$$\frac{d}{dt}\tanh(t) = \mathrm{sech}^2(t) \qquad \frac{d}{dt}\mathrm{csch}(t) = -\mathrm{csch}(t)\cdot\coth(t)$$

The formulas for the inverses of the hyperbolic functions are

$$\sinh^{-1}(x) = \ln\left(x + \sqrt{x^2+1}\right), \text{ all real } x;$$
$$\cosh^{-1}(x) = \ln\left(x + \sqrt{x^2-1}\right), x \geq 1;$$
$$\tanh^{-1}(x) = \frac{1}{2}\ln\left(\frac{1+x}{1-x}\right), -1 < x < 1;$$
$$\coth^{-1}(x) = \frac{1}{2}\ln\left(\frac{x+1}{x-1}\right), |x| > 1;$$
$$\operatorname{sech}^{-1}(x) = \ln\left(\frac{1+\sqrt{1-x^2}}{x}\right), 0 < x \leq 1; \text{ and}$$
$$\operatorname{csch}^{-1}(x) = \ln\left(\frac{1}{x} + \frac{\sqrt{1+x^2}}{|x|}\right), x \neq 0.$$

The basic derivative formulas for the inverse hyperbolic functions are

$$\frac{d}{dx}\sinh^{-1}(x) = \frac{1}{\sqrt{1+x^2}}, \quad \frac{d}{dx}\tanh^{-1}(x) = \frac{1}{1-x^2}, \text{ and}$$
$$\frac{d}{dx}\operatorname{sech}^{-1}(x) = -\frac{1}{x\sqrt{1-x^2}}.$$

The following integral formulas are often useful:

$$\int \frac{dx}{\sqrt{a^2+x^2}} = \sinh^{-1}\left(\frac{x}{a}\right) + C, \quad \int \frac{dx}{a^2-x^2} = \frac{1}{a}\tanh^{-1}\left(\frac{x}{a}\right) + C,$$
$$\int \frac{dx}{x\sqrt{a^2-x^2}} = -\frac{1}{a}\operatorname{sech}^{-1}\left(\frac{x}{a}\right) + C.$$

genesis & DEVELOPMENT

By the end of the 16th century, the needs of commerce, surveying, map-making, navigation, and astronomy created an unprecedented demand for efficient mathematical calculation. Yet even simple arithmetic required an advanced education. Any new computational shortcut was therefore eagerly greeted. For example, trigonometric identities such as

$$\cos(A)\cos(B) = \frac{1}{2}(\cos(A+B) + \cos(A-B))$$

were prized because they convert the product of two numbers into a sum. In the early 1590s, John Napier (1550–1617) set about to find a more direct procedure for transforming a multiplication problem into one of addition. After 20 years of research and computation, he published his work as *The Description of the Wonderful Canon of Logarithms* (1614). He coined the name *logarithm* because it means "ratio number."

The Napierian Logarithm

Napier was a titled landowner by profession. He first attained public renown as a scholar and theologian. His inventiveness demonstrated itself in a number of ways. Napier conducted controlled agricultural experiments with manure long before such trials became common. He invented and secured a patent for hydraulic equipment that was used in mining. In defense of his country, Napier devised four weapons, one of which was a "musket-proof armoured chariot" that he described more ominously as a "moving mouth of metal."

Napier conceived the logarithm by imagining two points moving in straight lines. In his construction, one point moves along the segment $(0, r]$ of the x-axis, starting from $x = r$ with speed v_0 and traveling to the left with decreasing speed proportionate to the remaining distance to the origin; its position at time t is denoted by $x(t)$. The other point, representing the logarithm of x *as Napier defined it*, starts at the origin of a parallel axis and moves to the right with constant velocity v_0 along that axis; its position at time t is denoted by $y(t)$. Refer to Figure 1. We write $\mathrm{nl}_r(x)$ for Napier's logarithm of x to distinguish it from the

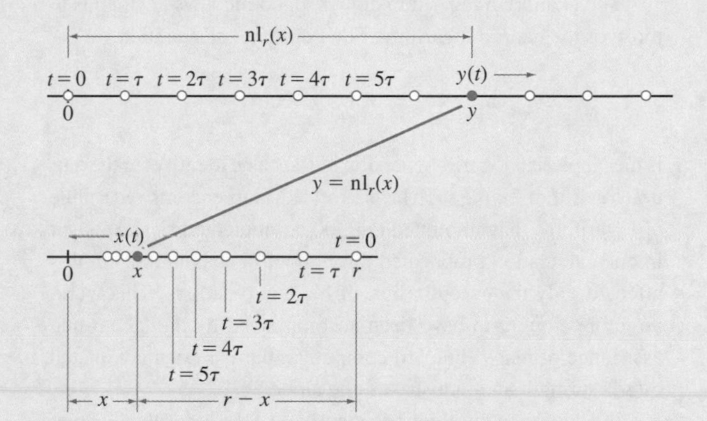

Figure 1
In the figure, τ represents an arbitrary positive increment of time. On the upper axis, a particle moves from the origin to the right at constant speed. At the same time, another particle moves to the left toward the origin of the lower axis and decelerates so that its speed is proportional to its distance to the origin. The particle on the upper axis has traveled $\mathrm{nl}_r(x)$ units when the particle on the lower axis has traveled $r - x$ units.

natural logarithm, $\ln(x)$, from which it differs slightly. Observe that by setting $x(t)|_{t=0} = r$ and $y(t)|_{t=0} = 0$, Napier adopted the normalization $\mathrm{nl}_r(r) = y(x)|_{x=r} = 0$.

With the benefit of differential calculus, which was not yet invented in Napier's time, we can now express the point movements that Napier envisioned by means of the equations

$$\frac{dy}{dt} = v_0, \quad \frac{dx}{dt} = -\frac{x}{r}v_0, \quad y = \mathrm{nl}_r(x), \quad \text{and} \quad \mathrm{nl}_r(r) = 0.$$

Using the Chain Rule, we can recast these equations as the initial value problem

$$\frac{dy}{dx} = -\frac{r}{x}, \quad y(r) = 0.$$

Napier used kinematical considerations to solve this problem. Nowadays we solve it by integration, obtaining the relationship between Napier's logarithm and its modern replacement:

$$\begin{aligned}\mathrm{nl}_r(x) &= \int_r^x \left(-\frac{r}{u}\right) du = -r(\ln(x) - \ln(r)) \\ &= r\ln\left(\frac{r}{x}\right) = r\log_{e^{-1}}\left(\frac{x}{r}\right).\end{aligned}$$

Napier chose $r = 10^7$, but other large integers would have suited his purposes as well.

The Napierian logarithm enjoys algebraic laws analogous to those of the natural logarithm. For example, the equation

$$\mathrm{nl}_r(rxy) = \mathrm{nl}_r(rx) + \mathrm{nl}_r(ry)$$

is the Napierian logarithm version of the basic identity for the natural logarithm. Using such laws as an aid, Napier compiled a table of logarithms that amounted to 9300 computed entries. An error in one calculation propagated throughout his tabulation, but this affected only the seventh digit of his computations. This work, he wrote, "ought to have been accomplished by the labour and assistance of many [human] computers, but has been completed by the strength and industry of one alone."

In addition to the algebraic identities of the logarithm, Napier also discovered the inequalities

$$\frac{r}{x+h} < \frac{\mathrm{nl}_r(x) - \mathrm{nl}_r(x+h)}{h} < \frac{r}{x}.$$

Had the concept of *derivative* existed in his time, Napier would have been able to calculate $\mathrm{nl}_r'(x) = -r/x$ easily, since this derivative formula is an immediate consequence of his inequalities.

Other Early Logarithms

We know that, prior to the publication of his tables, Napier wished to modify his logarithm so that it would have the form $10^9 \log_{10}(x)$ (in modern notation). In particular, he was aware of the chief defect of his logarithm: $\mathrm{nl}_r(1) \neq 0$. However, painful gout and lawsuits inconvenienced Napier for the last few years of his life, and he was not up to the task of such a revision. He died in Merchiston Castle in Scotland, where he had been born. Still standing, the castle has been incorporated into Napier Technical College, with drab modern buildings adjoining it.

The importance of Napier's wonderful canon of logarithms was recognized immediately. Henry Briggs (1561–1630) traveled to Edinburgh in the summers of 1615 and 1616 to meet with Napier and to calculate a system of base 10 logarithms. Tables of several other types of logarithms were soon published. Kepler learned of the Napierian logarithm in 1617 and used them in the work that led to his third law of planetary motion. Kepler wrote enthusiastically to his astronomy teacher Michael Maestlin, who advised Kepler that "it is unseemly for a professor of mathematics to be so childishly pleased about any shortening of calculation." Kepler was not deterred, and his investigations added much to the theory of logarithms. For example, Kepler discovered the inequality

$$\frac{\ln(x) - \ln(y)}{x - y} < \frac{1}{\sqrt{xy}}, \quad x, y > 0.$$

He even devised and calculated tables of his own logarithm, which he defined by

$$\mathrm{kl}(x) = \lim_{n\to\infty} 10^5 \cdot 2^n \cdot \left(1 - \sqrt[2^n]{\frac{x}{10^5}}\right).$$

The limit on the right of this equation can be calculated using l'Hôpital's Rule. It turns out that the Keplerian logarithm is given more simply by the formula

$$\mathrm{kl}(x) = 10^5 \ln\left(\frac{10^5}{x}\right).$$

Thus, despite its rather strange formulation, the Keplerian logarithm differs from the Napierian logarithm only in the choice of r.

Exponential Functions in Calculus

As calculus developed in the 17th century, the important roles of the natural logarithm and its inverse, the natural exponential, soon became apparent. In 1649, even before Newton and Leibniz came on the scene, Alphons Anton de Sarasa (1618–1667) discovered the basic integration formula $\ln(x) = \int_1^x 1/t\, dt$. In 1661, Christiaan Huygens studied curves that satisfy the differential equation $\frac{dy}{dx} = y/s$ where s is a positive constant. These are the curves that have a constant subtangent (Figure 2). Huygens

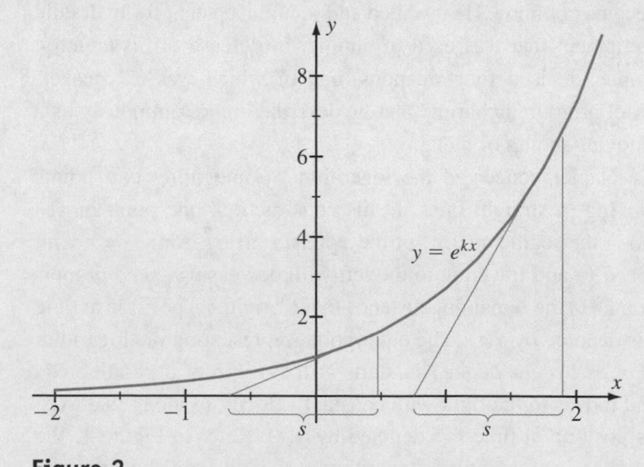

Figure 2
The graph of the exponential function $y = e^{kx}$ has subtangents of constant length.

discovered that curves of the form $y = Ce^{x/s}$ have this property. Several years later Leibniz proved that the differential equation $\frac{dy}{dx} = y/s$ has no other solution, as we learned in Section 6.5.

By the end of the 17th century, Leibniz was using the logarithm function to define and study general exponential functions. For example, in a letter of 1694, he used the formula $\log_a(y) = x \log_a(x)$ to define the curve $y = x^x$. Leibniz was aware that the inverse of the natural logarithm is an exponential function—that is, a function of the form $x \mapsto a^x$. He used the symbol b to denote the base of the natural exponential function, but he did not study the number b, which we now denote by e, any further. Indeed, considering the importance of e, it is remarkable that so much time elapsed before it was computed. There was a near miss in 1624 when Briggs recorded the number $0.43429448190325\ldots$, which is $\log_{10}(e)$. Briggs does not appear, however, to have calculated e itself. The first calculation of e seems to have been by Roger Cotes (1682–1716). In a posthumous collection of his papers that appeared in 1722, Cotes computed the base of the natural logarithm to be 2.7182818. Also appearing in this work is a famous relationship between the exponential function of a complex argument and the sine and cosine functions:

$$i\theta = \ln(\cos(\theta) + i \sin(\theta)).$$

This formula was rediscovered by Abraham de Moivre in 1730 and then by Euler in 1748, who published the more familiar form

$$e^{i\theta} = \cos(\theta) + i \cdot \sin(\theta).$$

Euler calculated e to 23 decimal places and derived the limit formula

$$\lim_{n\to\infty} \left(1 + \frac{x}{n}\right)^n = e^x.$$

Euler introduced the symbol e, and it has been in use ever since.

If you take $\theta = \pi$ in Euler's formula $e^{i\theta} = \cos(\theta) + i \cdot \sin(\theta)$, then you obtain the equation $e^{i\pi} + 1 = 0$ that relates the five most basic constants. There are a number of other interesting relationships between e and π. For example, Srinivasa Ramanujan (1887–1920) discovered that many numbers of the form $e^{\pi\sqrt{m}}$ (m prime) are "near integers." The transcendental number $e^{\pi\sqrt{163}}$, for instance, has decimal expansion

$$e^{\pi\sqrt{163}} = 262537412640768743.99999999999992\ldots.$$

The number e itself was shown to be transcendental by Charles Hermite (1822–1901) in 1874.

The Catenary

The shape of a cable hanging freely under gravity was first determined in 1690 by Leibniz, who coined the name catenary, and, independently, by Jakob Bernoulli. The derivation of the equation and its solution that is given in Section 6.7 is a standard one that is typical of the treatment given in engineering statics texts. There is another approach that is interesting because it relies on a "variational principle" similar to Fermat's Principle of Least Time (Genesis & Development, Chapter 4). Among all curves of fixed length l joining two points P and Q, the catenary arc of length l with endpoints P and Q has lowest center of gravity, or, equivalently, the least potential energy.

7

Techniques of Integration

PREVIEW

The purpose of this chapter is to develop some facility with calculating integrals. Until now, the integrals we have calculated have been of two types. Either we have immediately recognized an antiderivative, or we have found an antiderivative by making a direct substitution. In fact, an array of diverse procedures can be brought to bear on the calculation of integrals. This chapter covers the most important of these procedures.

Tables of integrals and computer algebra systems are valuable resources for calculating integrals. However, neither renders obsolete the need for a basic working knowledge of integration theory. In this chapter, we learn fundamental algebraic and trigonometric manipulations that are the basis for evaluating integrals that frequently arise in mathematical applications. However, the methods in this chapter have a significance that transcends integration theory. Indeed, these techniques are needed for the development of many important topics in engineering, science, and mathematics.

7.1 Integration by Parts

The integral of the sum of two functions is the sum of their integrals. This simple observation enables us to break complicated integration problems into several simple ones. But what if we want to integrate the *product* of two functions? A technique called *integration by parts* allows us to evaluate many integrals involving a product of two functions.

The Product Rule in Reverse

If u and v are two differentiable functions, then we know that

$$\frac{d}{dx}(u \cdot v) = \frac{du}{dx} \cdot v + \frac{dv}{dx} \cdot u.$$

We rewrite this equation as

$$u \cdot \frac{dv}{dx} = \frac{d}{dx}(u \cdot v) - v \cdot \frac{du}{dx}.$$

Integrating both sides gives

$$\int u \cdot \frac{dv}{dx}\, dx = \int \frac{d}{dx}(u \cdot v)\, dx - \int v \cdot \frac{du}{dx}\, dx.$$

Using the Fundamental Theorem of Calculus, we obtain the *Integration by Parts Formula:*

$$\int u \cdot \frac{dv}{dx}\, dx = u \cdot v - \int v \cdot \frac{du}{dx}\, dx. \tag{7.1}$$

inSIGHT

We often use differential notation and summarize the Integration by Parts Formula as

$$\int u\, dv = uv - \int v\, du \tag{7.2}$$

where dv is shorthand for $\frac{dv}{dx}\, dx$ and du is shorthand for $\frac{du}{dx}\, dx$. This compact form of the Integration by Parts Formula is the easiest to remember. Bear in mind, however, that it conceals the original variable of integration (x in our discussion). We may use a version of formula (7.2) for definite integrals, provided that we use appropriate limits of integration:

$$\int_{x=a}^{x=b} u\, dv = u \cdot v \Big|_{x=a}^{x=b} - \int_{x=a}^{x=b} v\, du. \tag{7.3}$$

Example 1 Calculate $\int x \cdot \cos(x)\,dx$.

Solution We integrate by parts. Our first job is to decide which function will play the role of u and which will play the role of v. If we take $u = x$ and $dv = \frac{dv}{dx}\,dx = \cos(x)\,dx$, then $\int x \cdot \cos(x)\,dx = \int u\,dv$. Therefore, according to formula (7.2),

$$\int x \cdot \cos(x)\,dx = u \cdot v - \int v\,du. \tag{7.4}$$

Our task is to work out the right side of the equation. We already know what u is. We must determine what v is. Since $dv = \cos(x)\,dx$, we may take $v = \sin(x)$. Likewise, since $u = x$, it follows that $du = \frac{du}{dx} \cdot dx = 1 \cdot dx$. If we substitute for u, v, and du on the right side of equation (7.4), then we obtain

$$\int x \cdot \cos(x)\,dx = x \cdot \sin(x) - \int \sin(x) \cdot 1\,dx.$$

We have converted our original integration problem into a much easier one. Since $\int \sin(x)\,dx = -\cos(x) + C$, we conclude that

$$\int x \cdot \cos(x)\,dx = x \cdot \sin(x) - (-\cos(x) + C) = x \cdot \sin(x) + \cos(x) - C.$$

If you prefer, you may replace the arbitrary constant C with $-C$ so that the answer takes on the more familiar form $x \cdot \sin(x) + \cos(x) + C$. As usual with indefinite integrals, we can check our work by differentiating our answer and verifying that the result equals the integrand:

$$\frac{d}{dx}(x \cdot \sin(x) + \cos(x) - C) = (x\cos(x) + \sin(x)) + (-\sin(x)) - 0 = x\cos(x).$$

■

INSIGHT

The purpose of integration by parts is to convert a difficult integral into another, easier-to-solve integral. Bear this notion in mind as you choose u and v. In Example 1, if we had selected $u = \cos(x)$ and $dv = x\,dx$, then the new integral

$$\int v\,du = -\int \frac{x^2}{2} \sin(x)\,dx$$

would have been *harder,* not easier, than the one we started with.

When a power of x is present in the integrand, it is *often* a good rule to choose u to be the power of x. Then, when integration by parts is applied, the power of x will be decreased. In this way, we guarantee in advance that *one* factor becomes simpler. Of course, we must also consider the role of dv. As the next example shows, that consideration may be paramount.

Example 2 Calculate $\int_1^4 2x \cdot \ln(x)\,dx$.

Solution In this example, it would be foolish to take $dv = \ln(x)\,dx$, because we do not know an antiderivative for $\ln(x)$. So instead we take $dv = 2x\,dx$ and $u = \ln(x)$.

We then have

$$u = \ln(x), \qquad dv = 2x\,dx,$$
$$du = \frac{1}{x}\,dx, \qquad v = x^2.$$

Thus,

$$\int_1^4 2x \cdot \ln(x)\,dx = \int_1^4 \ln(x) 2x\,dx = \int_1^4 u\,dv = uv\Big|_1^4 - \int_1^4 v\,du = \ln(x) \cdot x^2\Big|_1^4 - \int_1^4 x^2 \cdot \frac{1}{x}\,dx.$$

Once again, we have converted a difficult integral into a new integral that is much simpler. We continue calculating:

$$\int_1^4 2x \cdot \ln(x)\,dx = x^2 \ln(x)\Big|_1^4 - \frac{x^2}{2}\Big|_1^4 = 16 \cdot \ln(4) - \left(8 - \frac{1}{2}\right) = 16 \cdot \ln(4) - \frac{15}{2}.$$

Example 3 Calculate $\int \ln(x)\,dx$.

Solution This is a new type of problem for us. On the one hand, we do not know an antiderivative for $\ln(x)$. On the other hand, the integrand does not appear to be a product because we see only one function—$\ln(x)$. However, we can rewrite the integral as

$$\int \ln(x) \cdot 1\,dx.$$

We take $u = \ln(x)$ and $dv = 1\,dx$ to obtain

$$u = \ln(x), \qquad dv = 1\,dx,$$
$$du = \frac{1}{x}\,dx, \qquad v = x.$$

Thus,

$$\int \ln(x) \cdot 1\,dx = \int u\,dv = u \cdot v - \int v\,du = \ln(x) \cdot x - \int x \cdot \frac{1}{x}\,dx = x\ln(x) - \int 1\,dx = x\ln(x) - x + C.$$

Advanced Applications of Integration by Parts

Sometimes applying integration by parts just once is not sufficient to solve the problem at hand. Further applications of the formula might be needed, as the next two examples illustrate.

Example 4 In atomic measurements, it is sometimes convenient to use the Bohr radius $\alpha_0 = 0.52917725 \times 10^{-10}$ m as a unit of distance. Let ρ denote any (unitless) positive number. At a given instant of time, the probability of finding the hydrogen electron at a distance no greater than $\rho \cdot \alpha_0$ from the atomic nucleus is

$$\int_0^\rho 4r^2 e^{-2r}\,dr.$$

What is the probability p that the hydrogen electron is less than twice the Bohr radius from the nucleus?

Solution The answer is given by

$$p = \int_0^2 4r^2 e^{-2r}\, dr = 4 \cdot \int_0^2 r^2 e^{-2r}\, dr = 4 \cdot J \quad \text{where } J = \int_0^2 r^2 e^{-2r}\, dr.$$

To calculate this last integral, we take $u = r^2$ and $dv = \exp(-2r)\, dr$ to obtain

$$u = r^2, \qquad dv = \exp(-2r)\, dr,$$
$$du = 2r\, dr, \qquad v = -\frac{1}{2}\exp(-2r).$$

We therefore have

$$J = \int_0^2 r^2 \cdot e^{-2r}\, dr = \int_{r=0}^{r=2} u\, dv = u \cdot v\Big|_{r=0}^{r=2} - \int_{r=0}^{r=2} v\, du = r^2 \cdot \left(-\frac{e^{-2r}}{2}\right)\Big|_{r=0}^{r=2} - \int_0^2 \left(-\frac{e^{-2r}}{2}\right) \cdot 2r\, dr.$$

Working with the expression on the right, we obtain

$$J = -\frac{1}{2}r^2 \cdot e^{-2r}\Big|_{r=0}^{r=2} + \int_0^2 r \cdot e^{-2r}\, dr = \int_0^2 r \cdot e^{-2r}\, dr - 2e^{-4}.$$

Thus,

$$J = K - 2e^{-4} \quad \text{where } K = \int_0^2 r \cdot e^{-2r}\, dr.$$

We see that our new integral K is *simpler* than the one we started with (the power of r is lower), but it requires more work. We apply the Integration by Parts Formula to K, taking $u = r$ and $dv = e^{-2r}\, dr$. We now have

$$u = r, \qquad dv = e^{-2r}\, dr,$$
$$du = 1\, dr, \qquad v = -\frac{e^{-2r}}{2}.$$

Therefore,

$$\begin{aligned} K &= \int_0^2 r \cdot e^{-2r}\, dr \\ &= \int_{r=0}^{r=2} u\, dv \\ &= u \cdot v\Big|_{r=0}^{r=2} - \int_{r=0}^{r=2} v\, du \\ &= r\left(\frac{e^{-2r}}{-2}\right)\Big|_{r=0}^{r=2} - \int_0^2 \left(\frac{e^{-2r}}{-2}\right) dr \\ &= \frac{1}{2}\int_0^2 e^{-2r}\, dr - e^{-4} \\ &= \frac{1}{2}\left(\frac{e^{-2r}}{-2}\right)\Big|_{r=0}^{r=2} - e^{-4} = \frac{1}{4} - \frac{5}{4}e^{-4}. \end{aligned}$$

inSIGHT

If we organize our work properly and keep repeated applications of the Integration by Parts Formula separate, then there is no harm in reusing the symbols u and v each time. Also, notice that we can use the Integration by Parts Formula with a variable of integration other than x.

Finally, the required probability p is given by

$$p = 4 \cdot J = 4 \cdot (K - 2e^{-4}) = 4\left(\left(\frac{1}{4} - \frac{5}{4}e^{-4}\right) - 2e^{-4}\right) = 1 - 13e^{-4} \approx 0.7618967.$$

■

We conclude this section with an example in which we indirectly evaluate an integral. This method is particularly useful for integrals involving trigonometric or exponential functions.

Example 5 Calculate the integral

$$S = \int e^x \cos(x)\,dx.$$

Solution We take $u = e^x$ and $dv = \cos(x)\,dx$ to obtain

$$\begin{aligned} u &= e^x, & dv &= \cos(x)\,dx, \\ du &= e^x\,dx, & v &= \sin(x). \end{aligned}$$

Therefore,

$$S = \int u\,dv = u \cdot v - \int v\,du = e^x \sin(x) - \int \sin(x) e^x\,dx.$$

We have converted the original integral problem S to a new integration problem, but the new one does not look any simpler than the original—indeed, all we have done is replace $\cos(x)$ with $\sin(x)$. Nevertheless, we integrate by parts again, taking $u = e^x$ and $dv = \sin(x)\,dx$. As a result,

$$\begin{aligned} u &= e^x, & dv &= \sin(x)\,dx, \\ du &= e^x\,dx, & v &= -\cos(x). \end{aligned}$$

We have

$$S = e^x \sin(x) - \int \sin(x) e^x\,dx = e^x \sin(x) - \left(-e^x \cos(x) + \int \cos(x) e^x\,dx\right),$$

or

$$S = e^x(\sin(x) + \cos(x)) - \int e^x \cos(x)\,dx.$$

The integral we calculated in Example 5 arises naturally in signal processing and may be found in many engineering texts. A remarkable feature of the method used in the example is that we never calculate the integral in the usual way. Instead, we express the integral (which we think of as the unknown S) in terms of itself and then obtain the solution through algebra.

By observing that the term on the far right is just S, we can rewrite this last equation as

$$S = e^x(\sin(x) + \cos(x)) - S.$$

We can solve this equation for S to obtain

$$S = \frac{1}{2}(e^x \sin(x) + e^x \cos(x)).$$

Finally, observe that because S is an indefinite integral, a constant of integration should be added to get the general solution

$$S = \frac{1}{2}(e^x \sin(x) + e^x \cos(x)) + C.$$

■

We can also apply the method from Example 5 to integrals of expressions of the form $\sin(ax)\cos(bx)$, $\sin(ax)\sin(bx)$, and $\cos(ax)\cos(bx)$. Because these integrals are sometimes needed in applications, we record them here for reference (*not* for memorization):

$$\int \sin(ax)\cos(bx)\,dx = -\frac{1}{2}\frac{\cos((a-b)x)}{a-b} - \frac{1}{2}\frac{\cos((a+b)x)}{a+b} + C, \quad a \neq b, \quad (7.5)$$

$$\int \sin(ax)\sin(bx)\,dx = \frac{1}{2}\frac{\sin((a-b)x)}{a-b} - \frac{1}{2}\frac{\sin((a+b)x)}{a+b} + C, \quad a \neq b, \quad (7.6)$$

$$\int \cos(ax)\cos(bx)\,dx = \frac{1}{2}\frac{\sin((a-b)x)}{a-b} + \frac{1}{2}\frac{\sin((a+b)x)}{a+b} + C, \quad a \neq b. \quad (7.7)$$

quickquiz

1. State the Integration by Parts Formula.
2. The Integration by Parts Formula is another way of looking at what formula for the derivative?
3. What is the result of applying the Integration by Parts Formula to $\int x^n v'(x)\,dx$?
4. Calculate $\int \ln(2x)\,dx$.

EXERCISES

Problems for Practice

In Exercises 1–6, use the Integration by Parts Formula to evaluate the indefinite integral.

1. $\int x \cdot e^x\,dx$
2. $\int x \cdot e^{-x}\,dx$
3. $\int x \cdot \sin(x)\,dx$
4. $\int x \cdot \sin(2x)\,dx$
5. $\int x \cdot \cos(2x)\,dx$
6. $\int x^3 \cdot \ln(2x)\,dx$

In Exercises 7–12, use the Integration by Parts Formula to evaluate the definite integral.

7. $\int_0^1 x \cdot e^x\,dx$
8. $\int_0^\pi x \cdot \cos(x)\,dx$
9. $\int_0^{\pi/4} x \cdot \cos(2x)\,dx$
10. $\int_1^e x \cdot \ln(x)\,dx$
11. $\int_1^{e/3} x \cdot \ln(3x)\,dx$
12. $\int_0^1 x \cdot 3^x\,dx$

In Exercises 13–18, use the Integration by Parts Formula successively to evaluate the definite integral.

13. $\int x^2 \cdot e^x\,dx$
14. $\int x^2 \cdot \cos(x)\,dx$
15. $\int x^2 \cdot \sin(3x)\,dx$
16. $\int \ln^3(x)\,dx$
17. $\int x^3 \cdot e^{-2x}\,dx$
18. $\int x^3 \cdot \sin(x)\,dx$

In Exercises 19–22, simplify the integrand before applying the Integration by Parts Formula.

19. $\int \ln(\sqrt{x})\,dx$
20. $\int x\ln(x^5)\,dx$
21. $\int x^3\ln(1/x)\,dx$
22. $\int x\sin(x)\cos(x)\,dx$

In Exercises 23–26, calculate the integral.

23. $\int_0^1 \arctan(x)\,dx$
24. $\int_0^1 \arcsin(x)\,dx$
25. $\int_1^{\sqrt{2}} x\,\mathrm{arcsec}(x)\,dx$
26. $\int_1^{\sqrt{3}} x\arctan(x)\,dx$
27. What is the area of the first quadrant region bounded by $y = \pi x/2$ and $y = \arcsin(x)$?
28. What is the area of the first quadrant region bounded by $y = \pi x/4$ and $y = \arctan(x)$?

Further Theory and Practice

In Exercises 29–40, evaluate the integral.

29. $\int x \cdot 2^x\,dx$
30. $\int \ln(x)/x^2\,dx$
31. $\int \ln(x)/\sqrt{x}\,dx$
32. $\int \sqrt{x}\ln(x)\,dx$
33. $\int x\sec(x)\tan(x)\,dx$
34. $\int x\sec^2(x)\,dx$
35. $\int \ln(1+x^2)\,dx$
36. $\int (e^x - x)^2\,dx$
37. $\int (x-2)^2\ln(x)\,dx$
38. $\int x \cdot \sin(3x+4)\,dx$
39. $\int_{-1}^1 x/(x+3)^{1/2}\,dx$
40. $\int x \cdot (x-1)^{1/2}\,dx$

In Exercises 41–46, make an appropriate substitution before applying the Integration by Parts Formula.

41. $\int \sin(\sqrt{x})\,dx$
42. $\int e^{\sqrt{x}}\,dx$
43. $\int x^5 \cdot e^{x^2}\,dx$
44. $\int e^{2x}\sin(e^x)\,dx$

45. $\int x^5 \sin(x^3)\,dx$ **46.** $\int x^3 \sin(x^2)\cos(x^2)\,dx$

In Exercises 47–50, calculate the integral by following the procedure used in Example 5.

47. $\int 2^x e^x\,dx$ **48.** $\int \cos(\ln(x))\,dx$

49. $\int \cos(x)\sin(3x)\,dx$ **50.** $\int e^{-3x}\sin(2x)\,dx$

51. Derive formula (7.5). **52.** Derive formula (7.6).

53. Derive formula (7.7).

54. Show that

$$\int \sec^3(x)\,dx = \frac{1}{2}\sec(x)\tan(x) + \frac{1}{2}\ln(|\sec(x)+\tan(x)|) + C.$$

Use the Integration by Parts Formula with $u = \sec(x)$. You will obtain an integral with an integrand that contains the factor $\tan^2(x)$. Use the identity $\tan^2(x) = \sec^2(x) - 1$ and the method from Example 5.

55. Let f be a k times differentiable function with all derivatives continuous. Calculate

$$\int_0^1 x^k \cdot f^{(k)}(x)\,dx.$$

56. Derive the formula

$$\int e^{ax}\cos(bx)\,dx = \frac{e^{ax}(a\cos(bx) + b\sin(bx))}{a^2+b^2}.$$

57. Derive the formula

$$\int e^{ax}\sin(bx)\,dx = \frac{e^{ax}(a\sin(bx) - b\cos(bx))}{a^2+b^2}.$$

58. The sine integral $\mathrm{Si}(x)$ is defined by

$$\mathrm{Si}(x) = \int_0^x \frac{\sin(t)}{t}\,dt.$$

a. Explain why there is no problem beginning the integration at 0 even though the integrand is not defined there.

b. Integrate by parts to calculate $\int \mathrm{Si}(x)\,dx$. (Your answer will involve $\mathrm{Si}(x)$ and elementary functions.)

c. Integrate by parts to calculate $\int x\mathrm{Si}(x)\,dx$. (Your answer will involve $\mathrm{Si}(x)$ and elementary functions.)

59. The Fresnel sine and Fresnel cosine functions, denoted by FresnelS and FresnelC, are important in the theory of optics. They are defined by

$$\mathrm{FresnelS}(x) = \int_0^x \sin\left(\frac{\pi}{2}t^2\right)dt \quad \text{and}$$

$$\mathrm{FresnelC}(x) = \int_0^x \cos\left(\frac{\pi}{2}t^2\right)dt.$$

a. Integrate by parts to calculate $\int \mathrm{FresnelS}(x)\,dx$. (Your answer will involve $\mathrm{FresnelS}(x)$ and elementary functions.)

b. Integrate by parts to calculate $\int \mathrm{FresnelC}(x)\,dx$. (Your answer will involve $\mathrm{FresnelC}(x)$ and elementary functions.)

c. Integrate by parts to calculate $\int x\mathrm{FresnelS}(x)\,dx$ and $\int x\mathrm{FresnelC}(x)\,dx$. (Each answer will involve $\mathrm{FresnelS}(x)$, $\mathrm{FresnelC}(x)$, and elementary functions.)

60. The error function (denoted by erf), which is central to the subjects of probability and statistics, is defined by

$$\mathrm{erf}(x) = \frac{2}{\sqrt{\pi}}\int_0^x \exp(-t^2)\,dt.$$

a. Integrate by parts to calculate $\int \mathrm{erf}(x)\,dx$. (Your answer will involve $\mathrm{erf}(x)$ and elementary functions.)

b. Integrate by parts to calculate $\int x^2\exp(-x^2)\,dx$. (Take $u = x$, $dv = x\exp(-x^2)\,dx$. Your answer will involve $\mathrm{erf}(x)$ and elementary functions.)

c. Integrate by parts to calculate $\int x^2\,\mathrm{erf}(x)\,dx$. (You will need the result from part b. Your answer will involve $\mathrm{erf}(x)$ and elementary functions.)

Calculator/Computer Exercises

61. Refer to Example 4. For approximately what value ρ is the probability of finding the hydrogen electron within a distance $\rho \cdot \alpha_0$ of the nucleus equal to 1/2?

62. Refer to Example 4. If the probability of finding the hydrogen electron within a distance $\rho \cdot \alpha_0$ of the hydrogen nucleus is exactly 2/3 the probability of finding the hydrogen electron within a distance $2\rho \cdot \alpha_0$ of the hydrogen nucleus, then what is ρ?

63. Graph $y = x\ln(x)$ for $0 < x < e$. Use your graph to determine

$$\lim_{x\to 0^+} x\ln(x).$$

(Notice that $x\ln(x)$ is not defined for $x = 0$.) Add the line $y = x$ to your graph. The region bounded above by $y = x$ and below by $y = x\ln(x)$ will appear to have a well-defined area. Use the Integration by Parts Formula to determine that area. (Use the limit that you just determined to calculate $\lim_{x\to 0^+} x^2\ln(x)$.)

64. Find the area of the region that is bounded above by $y = 1 + 2x$ and below by $y = xe^x$.

65. Plot $\arctan(x) - 1$ and $\ln(x)/x$ for $0.7 \leq x \leq 5$. Find the two points of intersection. Calculate the area of the region enclosed by the two curves.

66. Suppose $f(x)$ is an odd function: $f(-x) = -f(x)$. If we let

$$b_n = \frac{1}{\pi}\int_{-\pi}^{\pi} f(x)\sin(nx)\,dx \qquad (n \geq 1),$$

then

$$\sum_{n=1}^{N} b_n \sin(nx)$$

is said to be the order N Fourier sine-polynomial approximation of f. For $N = 3$ and 5, plot the order N Fourier sine-polynomial approximation of $f(x) = x$. Use the viewing window $[-2, 2] \times [-0.2, 4]$.

67. Suppose $f(x)$ is an even function: $f(-x) = f(x)$. If we let

$$a_n = \frac{1}{\pi}\int_{-\pi}^{\pi} f(x)\cos(nx)\,dx \qquad (n \geq 0),$$

then

$$\frac{a_0}{2} + \sum_{n=1}^{N} a_n \cos(nx)$$

is the order N Fourier cosine-polynomial approximation of f. For $N = 3$ and 5, plot the order N Fourier cosine-polynomial approximation of $f(x) = x^2$. Use the viewing window $[-2, 2] \times [-0.2, 4]$.

68. Let the Fourier coefficients a_n $(n \geq 0)$ and b_n $(n \geq 1)$ of a function f be defined as in Exercises 66 and 67. If f is a differentiable function on the interval $(-\pi, \pi)$, then the order N Fourier polynomial

$$\frac{a_0}{2} + \sum_{n=1}^{N} (a_n \cos(nx) + b_n \sin(nx))$$

converges to $f(x)$ as N tends to infinity. Calculate the orders 2 and 3 Fourier polynomials of $f(x) = \exp(x)$. Plot these Fourier polynomials and the function f in the viewing window $[-2.6, 2.6] \times [-2, 13]$. (With these small values of N, you can only begin to see the Fourier polynomials taking on the shape of f. Figures 1a and 1b show the plot of f and the approximating Fourier polynomials of order 5 and 20, respectively.)

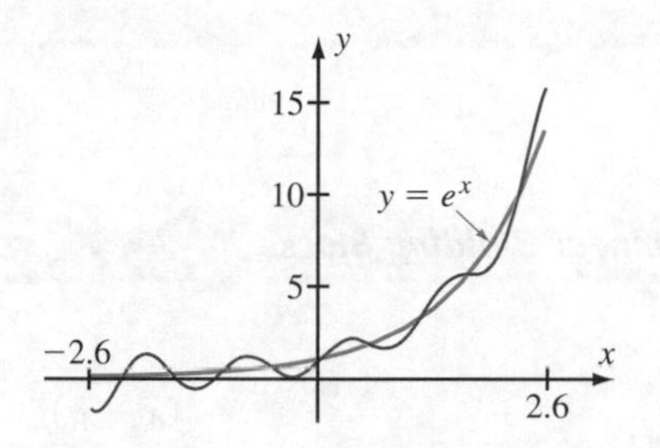

Figure 1a
The order 5 Fourier polynomial of the exponential function

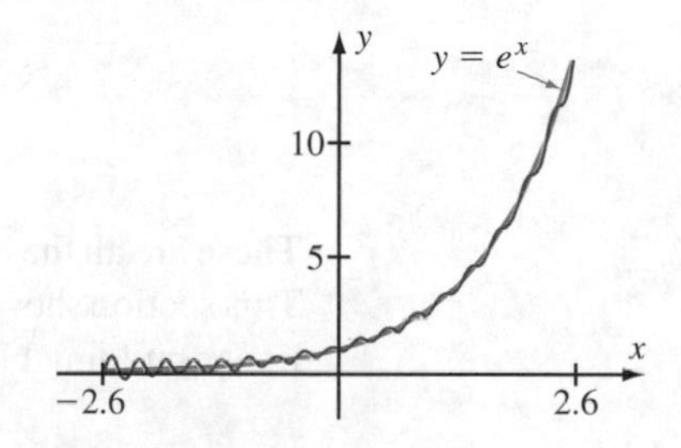

Figure 1b
The order 20 Fourier polynomial of the exponential function

7.2 Partial Fractions—Linear Factors

In this section, we begin to learn how to integrate quotients of polynomials; such quotients are called *rational functions*. The basic scheme, called *partial fractions*, consists of writing any rational function as a sum of building blocks that we already know how to integrate. Let us now describe some of these building blocks.

Simple Linear Building Blocks

$$\frac{A}{x-a}$$

where A and a are constants. The integral of this building block is

$$A \cdot \ln(|x-a|).$$

Repeated Linear Building Blocks

$$\frac{A}{(x-a)^m}$$

where m is an integer greater than 1. The integral of such a building block is

$$-\frac{A}{m-1} \cdot \frac{1}{(x-a)^{m-1}}.$$

These are all the linear building blocks, and you can see that they are easy to understand. This section shows how a great many rational functions can be written as a sum of simple linear building blocks and repeated linear building blocks.

The Method of Partial Fractions for Linear Factors

Example 1 Integrate

$$\int \frac{3}{(x-1)(x+2)}\, dx.$$

Solution The scheme is to rewrite the integrand as

$$\frac{3}{(x-1)(x+2)} = \frac{A}{x-1} + \frac{B}{x+2}$$

where A and B are constants that we must determine. We put the terms on the right over a common denominator:

$$\frac{3}{(x-1)(x+2)} = \frac{A(x+2)+B(x-1)}{(x-1)(x+2)}.$$

Both sides of this equation have the same denominator. Thus, the only way that equality can hold is if the numerators are equal:

$$3 = A(x+2) + B(x-1). \tag{7.8}$$

We can reorganize equation (7.8) to more clearly reflect its status as an identity of polynomials in x:

$$0x + 3 = (A+B)x + (2A - B).$$

For two polynomials to be equal, their constant terms must be equal, their linear terms must be equal, and so on. Thus,

$$3 = 2A - B \qquad \text{and} \qquad 0 = A + B.$$

We solve these two linear equations simultaneously. The second equation tells us that $B = -A$; substituting this value for B in the first equation results in $3 = 3A$. Thus, $A = 1$ and $B = -1$. We have discovered that

$$\frac{3}{(x-1)(x+2)} = \frac{1}{x-1} + \frac{-1}{x+2}.$$

Therefore,

$$\begin{aligned}\int \frac{3}{(x-1)(x+2)}\,dx &= \int \frac{1}{x-1}\,dx - \int \frac{1}{x+2}\,dx \\ &= \ln(|x-1|) - \ln(|x+2|) + C \\ &= \ln\left(\left|\frac{x-1}{x+2}\right|\right) + C.\end{aligned}$$

■

inSIGHT

It is important to understand that we do not solve equation (7.8) for x. Indeed, the procedure is to find values for A and B that convert equation (7.8) into an *identity* in x, or *an equation that is valid for every allowable* x. For each value of x, equation (7.8) becomes an equation in the variables A and B. In a sense, identity (7.8) gives us infinitely many equations with which we could work. Actually, we need only two equations for determining the two unknowns A and B. This leads to a very simple procedure: In equation (7.8), substitute values of x that isolate the unknowns. Letting x be 1 in equation (7.8) results in $3 = 3A$, or $A = 1$. Letting x be -2 results in $3 = -3B$, or $B = -1$. Notice that there is no need to solve simultaneous equations with this technique.

The process used in Example 1 is a special case of the following general technique.

The Method of Partial Fractions for Distinct Linear Factors

To integrate a function of the form

$$\frac{p(x)}{(x-a_1)(x-a_2)\cdots(x-a_K)}$$

where $p(x)$ is a polynomial and the a_j are distinct real numbers, you must first perform the following two algebraic steps:

1. Be sure that the degree of p is less than the degree of the denominator; if it is not, *divide the denominator into the numerator.*
2. Decompose the function into the form

$$\frac{A_1}{x-a_1}+\frac{A_2}{x-a_2}+\cdots+\frac{A_K}{x-a_K}$$

and solve for the numerators $A_1, A_2, \ldots, A_K$.

Example 2 Calculate the integral

$$\int \frac{2x^3-9x^2-5x+7}{x^2-4x-5}\,dx.$$

Solution We first notice that the numerator has degree exceeding that of the denominator, so we must divide. We find that

$$\frac{2x^3-9x^2-5x+7}{x^2-4x-5} = 2x-1+\frac{x+2}{x^2-4x-5} = 2x-1+\frac{x+2}{(x+1)(x-5)}.$$

We now need to decompose

$$\frac{x+2}{(x+1)(x-5)} = \frac{A}{x+1}+\frac{B}{x-5}.$$

Putting the right side over a common denominator and equating numerators leads to

$$x+2 = A(x-5)+B(x+1) \qquad \textbf{(7.9)}$$

We continue with the simple method described in the Insight on page 519. We substitute $x=-1$ and $x=5$. The first substitution results in $-1+2=A(-1-5)+B\cdot 0$, or $A=-1/6$. The substitution $x=5$ results in $5+2=A\cdot 0+B(5+1)$, or $B=7/6$. In conclusion,

$$\int \frac{2x^3-9x^2-5x+7}{x^2-4x-5}\,dx = \int (2x-1)\,dx + \int \frac{-1/6}{x+1}\,dx + \int \frac{7/6}{x-5}\,dx$$

$$= x^2-x-\frac{1}{6}\ln(|x+1|)+\frac{7}{6}\ln(|x-5|)+C. \quad \blacksquare$$

inSIGHT

Notice how we choose the values of x in Example 2. We choose $x=-1$ so that the coefficient of B will be 0. The result is an equation that involves only A. Similarly, the choice $x=5$ eliminates the unknown A, leaving us with an equation in B alone.

An Application—Logistic Growth

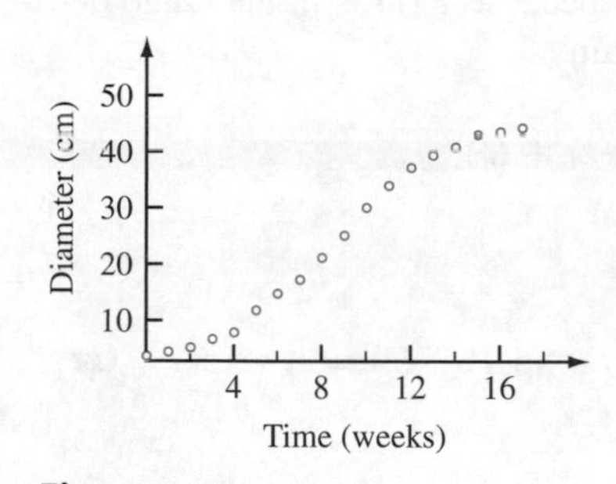

Figure 1
Pumpkin growth

Figure 1 shows the growth of a pumpkin in a controlled agricultural experiment. At first, the growth of the pumpkin appears to be exponential. However, the growth plot eventually becomes concave down and can no longer be modeled by exponential growth. To obtain a more accurate model, scientists often make the following growth assumptions:

1. There is a size P_∞ that cannot be exceeded. In population models, the constant P_∞ is called the *carrying capacity*.
2. $P'(t) \to 0$ as $P(t) \to P_\infty$, or in words, as $P(t)$ approaches the carrying capacity P_∞, its rate of growth decreases to zero.

One way to incorporate these assumptions into a growth model is to assume that the rate of population increase is jointly proportional to the population $P(t)$ and to $P_\infty - P(t)$. The resulting differential equation,

$$\frac{dP}{dt} = k \cdot P(t) \cdot (P_\infty - P(t)), \qquad P(0) = P_0, \tag{7.10}$$

is called the *logistic growth equation*. In this differential equation, k, P_0, and P_∞ represent fixed positive constants. Using the Method of Separation of Variables, we rewrite equation (7.10) as

$$\int \frac{1}{P \cdot (P_\infty - P)} \frac{dP}{dt}\, dt = \int k \cdot dt. \tag{7.11}$$

Example 3 Use the Method of Partial Fractions to solve equation (7.11).

Solution We apply partial fractions and write

$$\frac{1}{P \cdot (P_\infty - P)} = \frac{A}{P} + \frac{B}{P_\infty - P}.$$

Before we go any further, it is a good idea to make sure that we understand the role of each quantity in this equation: P_∞ is a fixed constant and P is a variable. We must find values for A and B so that the equation is valid for all values of P. Putting the right side over a common denominator and equating numerators leads to

$$1 = A \cdot (P_\infty - P) + B \cdot P.$$

If we let P be 0, we find that $A = 1/P_\infty$. Letting P be P_∞ results in $B = 1/P_\infty$. With these values, equation (7.11) becomes

$$\int \frac{1}{P_\infty} \cdot \left(\frac{1}{P} + \frac{1}{P_\infty - P} \right) \frac{dP}{dt}\, dt = \int k\, dt,$$

or

$$\frac{1}{P_\infty} \cdot (\ln(P(t)) - \ln(P_\infty - P(t))) = kt + C.$$

We have omitted absolute values in the logarithms because $P(t)$ is in the range $0 < P(t) < P_\infty$. We can combine the logarithms to obtain

$$\ln\left(\frac{P(t)}{P_\infty - P(t)}\right) = P_\infty \cdot (kt + C),$$

or, on exponentiating and letting $A = \exp(P_\infty \cdot C)$,

$$\frac{P(t)}{P_\infty - P(t)} = \exp(P_\infty \cdot kt + P_\infty \cdot C) = \exp(P_\infty \cdot kt)\exp(P_\infty \cdot C) = A \cdot \exp(P_\infty \cdot kt).$$

We solve for the constant A by substituting $t = 0$:

$$\frac{P_0}{P_\infty - P_0} = \frac{P(0)}{P_\infty - P(0)} = A.$$

Therefore,

$$\frac{P(t)}{P_\infty - P(t)} = \frac{P_0}{P_\infty - P_0} \cdot \exp(P_\infty \cdot kt).$$

Isolating $P(t)$, which is elementary but somewhat tedious, results in

$$P(t) = \frac{P_0 \cdot P_\infty}{P_0 + (P_\infty - P_0)\exp(-k \cdot P_\infty \cdot t)}. \tag{7.12}$$

■

The function P defined by equation (7.12) is called a *logistic growth function*. Biology also uses the term *sigmoidal growth*. Notice that there are three positive constants that determine logistic growth: the proportionality constant k, the initial value P_0 of P, and the carrying capacity P_∞. The graph of $P(t)$

1. is everywhere rising,
2. is concave up for $P_0 \le P < \frac{1}{2}P_\infty$ (see Exercise 44),
3. is concave down for $\frac{1}{2}P_\infty < P < P_\infty$ (see Exercise 44), and
4. has $y = P_\infty$ as a horizontal asymptote.

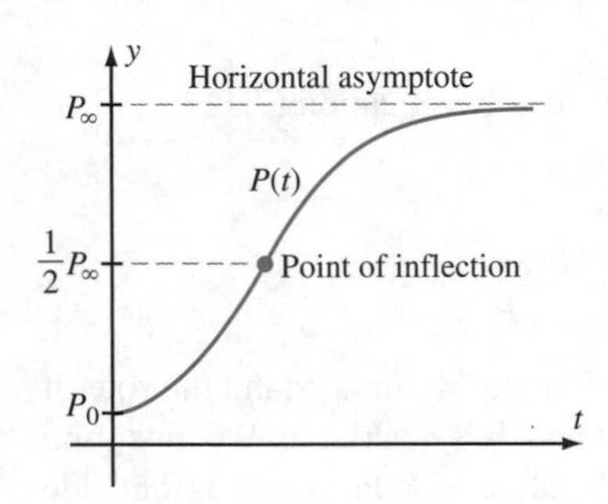

Figure 2
Logistic growth curve

Property 4 can be expressed as

$$\lim_{t\to\infty} P(t) = P_\infty,$$

which explains the subscript ∞. Figure 2 shows a typical logistic curve. In Figure 3, we replotted our pumpkin growth data, adding the graph of the logistic function determined by $P_0 = 3.6$, $P_\infty = 46$, and $k = 0.0065$.

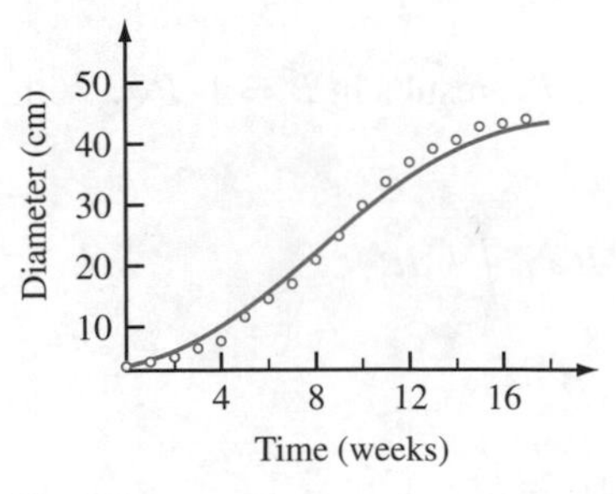

Figure 3
Pumpkin growth

Example 4 Suppose that 100 children attend a day-care center. Let $P(t)$ be the fraction of the day-care population that has a viral infection at time t (measured in days). Suppose that, at the start of Monday morning ($t = 0$), one child is infected. Suppose that, at the start of Tuesday morning ($t = 1$), there are three infected children (including the original child). Assume that the rate of change of P is jointly proportional to P (since this represents the fraction of the population capable of spreading the infection) and $1 - P$ (since this represents the fraction of the population that can still be infected). By what day will the infection have spread to half the day-care population?

Solution The initial fraction of the population that is infected is given to be 1 child out of 100 children; that is, $P(0) = 1/100$. The size P_∞ that the fraction P cannot exceed is 1. We obtain the differential equation that models the spread of the infection by substituting these values into equation (7.10):

$$\frac{dP}{dt} = k \cdot P(t) \cdot (1 - P(t)), \qquad P(0) = \frac{1}{100}.$$

The solution of this differential equation is given by formula (7.12):

$$P(t) = \frac{1/100}{1/100 + (1 - 1/100)\exp(-k \cdot 1 \cdot t)} = \frac{1}{1 + 99e^{-kt}}.$$

We may now use the observation $P(1) = 3/100$ to solve for k:

$$\frac{3}{100} = P(1) = \frac{1}{1+99e^{-k}}, \quad \text{or} \quad 1 + 99e^{-k} = \frac{100}{3}, \quad \text{or} \quad e^{-k} = \frac{1}{99}\left(\frac{100}{3} - 1\right) = \frac{97}{297}.$$

By applying the natural logarithm to each side of the last equation, we obtain $-k = \ln(97/297)$. Our final expression for P is

$$P(t) = \frac{1}{1 + 99e^{t\ln(97/297)}} = \frac{1}{1 + 99\exp(\ln((97/297)^t))} = \frac{1}{1 + 99 \cdot (97/297)^t}.$$

The solution to our problem is found by solving the equation

$$\frac{1}{1 + 99 \cdot (97/297)^t} = \frac{1}{2}$$

for t. After some simplification, we obtain $99 = (297/97)^t$ or

$$t = \frac{\ln(99)}{\ln(297/97)} \approx 4.1.$$

According to our model, the 50th child will be infected sometime on Friday morning. ■

Heaviside's Method

In Examples 1 and 2, we learned a shortcut for determining the partial fractions decomposition of

$$\frac{p(x)}{(x-a_1)(x-a_2)\cdots(x-a_K)} = \frac{A_1}{x-a_1} + \frac{A_2}{x-a_2} + \cdots + \frac{A_K}{x-a_K} \qquad \textbf{(7.13)}$$

where $a_1, a_2, \ldots, a_K$ are distinct real numbers. (It is worth highlighting this method, which is named after the electrical engineer Oliver Heaviside (1850–1925) who systematically used it in his work.) Let $q_j(x)$ denote what is left of the denominator $(x-a_1)(x-a_2)\cdots(x-a_K)$ after the factor $(x-a_j)$ is removed. To determine the coefficient A_j, simply multiply both sides of equation (7.13) by $x - a_j$. This results in

$$\frac{p(x)}{q_j(x)} = A_j + (x - a_j) \cdot (\text{terms that involve the other coefficients}).$$

After a little practice, you will be able to carry out Heaviside's method with ease. If you had looked at the final decomposition from Example 5 without seeing the calculation, then you might have thought tedious algebra was required to obtain such coefficients as 1/6, 29/15, and 9/10. As we have seen, however, Heaviside's method is as easy to apply when the coefficients are messy as it is when the coefficients are simple integers. Remember, though, that the procedure described in Example 5 works only when the denominator of the integrand factors into distinct linear factors.

Now set $x = a_k$. We immediately see that

$$A_j = \frac{p(a_j)}{q_j(a_j)}.$$

Example 5 Use Heaviside's method to find the partial fractions decomposition for

$$\frac{3x^2 + x - 1}{x(x-3)(x+2)}.$$

Solution Write

$$\frac{3x^2 + x - 1}{x(x-3)(x+2)} = \frac{A}{x} + \frac{B}{x-3} + \frac{C}{x+2}. \tag{7.14}$$

To determine A, multiply both sides by x. The result is

$$\frac{3x^2 + x - 1}{(x-3)(x+2)} = A + x \cdot \left(\frac{B}{x-3} + \frac{C}{x+2} \right).$$

Now we substitute $x = 0$ into this equation to find that $A = -1/(-6) = 1/6$. To find B, multiply equation (7.14) through by $x - 3$. The result is

$$\frac{3x^2 + x - 1}{x(x+2)} = B + (x-3) \cdot \left(\frac{A}{x} + \frac{C}{x+2} \right).$$

Substituting in $x = 3$ yields $B = 29/15$. A similar computation shows that $C = 9/10$. Try it! ■

Repeated Linear Factors

When one or more of the linear factors in the denominator is repeated, we must use a slightly different partial fractions scheme.

Partial Fractions When Repeated Linear Factors Are Present

Consider the rational function

$$\frac{p(x)}{(x-a_1)^{m_1}(x-a_2)^{m_2} \cdots (x-a_K)^{m_K}}$$

where $p(x)$ is a polynomial, the a_js are distinct real numbers, and the m_js are positive integers (some of them exceeding 1). To integrate a function of this type, perform the following two steps.

1. Be sure that the degree of p is less than the degree of the denominator; if it is not, *divide the denominator into the numerator.*
2. For each of the factors $(x - a_j)^{m_j}$ in the denominator of the rational function being considered, the partial fractions decomposition must contain terms of the form

$$\frac{A_1}{(x-a_j)^1} + \frac{A_2}{(x-a_j)^2} + \cdots + \frac{A_{m_j}}{(x-a_j)^{m_j}}.$$

Example 6 Evaluate the integral

$$\int \frac{5x^2 + 18x - 1}{(x+4)^2(x-3)}\,dx.$$

Solution We must decompose the integrand into the form

$$\frac{5x^2 + 18x - 1}{(x+4)^2(x-3)} = \frac{A_1}{x+4} + \frac{A_2}{(x+4)^2} + \frac{B}{x-3}.$$

Notice that because $(x+4)$ is a repeated factor in the denominator of the integrand, our partial fractions decomposition must contain *all powers* of $(x+4)$, up to and including the highest power occurring in the denominator of the integrand.

Putting the right side over a common denominator (and equating numerators) leads to

$$5x^2 + 18x - 1 = A_1(x+4)(x-3) + A_2(x-3) + B(x+4)^2.$$

Equating the coefficients of like powers of x gives the system

$$\begin{aligned} A_1 \quad\quad\;\; + \;\; B &= 5 \\ A_1 + A_2 + 8B &= 18 \\ -12A_1 - 3A_2 + 16B &= -1. \end{aligned}$$

The first equation allows us to eliminate $B = 5 - A_1$ from the remaining two equations, which, in turn, become

$$-7A_1 + A_2 = -22 \quad\text{and}\quad -28A_1 - 3A_2 = -81.$$

We solve these equations in standard fashion to obtain $A_1 = 3$ and $A_2 = -1$. Finally, $B = 5 - A_1 = 2$. Thus,

$$\begin{aligned} \int \frac{5x^2 + 18x - 1}{(x+4)^2(x-3)}\,dx &= \int \frac{3}{x+4}\,dx + \int \frac{(-1)}{(x+4)^2}\,dx + \int \frac{2}{x-3}\,dx \\ &= 3\ln(|x+4|) + \frac{1}{(x+4)} + 2\ln(|x-3|) + C. \end{aligned}$$

■

inSIGHT

In Example 6, you might be tempted to leave out one of the terms $A_1/(x+4)$ or $A_2/(x+4)^2$ on the right side. Try it. You will find that you *cannot* solve the resulting algebraic system for the coefficients. The more complicated partial fractions decomposition *is* necessary. Theory tells us that the coefficients of a partial fractions expansion are unique; thus, you cannot succeed when you start out with the wrong form—unless an omitted coefficient just happens to be zero.

Summary of Basic Partial Fraction Forms

We conclude by highlighting the techniques of this section with several quick instances of the different forms of partial fractions that can arise:

$$\frac{3x-1}{(x-4)(x+7)} = \frac{A}{x-4} + \frac{B}{x+7}$$

$$\frac{x^2+x+1}{x(x^2-1)} = \frac{A}{x} + \frac{B}{x+1} + \frac{C}{x-1}$$

$$\frac{x+3}{(x-2)^2(x+1)} = \frac{A}{x-2} + \frac{B}{(x-2)^2} + \frac{C}{x+1}$$

$$\frac{x^4-2x-1}{(x+8)^3(x-1)^2} = \frac{A}{x+8} + \frac{B}{(x+8)^2} + \frac{C}{(x+8)^3} + \frac{D}{x-1} + \frac{E}{(x-1)^2}$$

As you do the exercises, refer to these examples to determine which form of the partial fractions technique you should use. Notice this unifying theme: The number of unknown coefficients in any partial fractions problem is equal to the degree of the denominator.

quickquiz

1. If $(5x-7)/((x-3)(x+2)) = A/(x-3) + B/(x+2)$, what are the simultaneous equations that A and B satisfy?
2. If $A(x+1)(x+2) + Bx(x+2) + Cx(x+1) = 2x^2 + 6x + 2$, what are A, B, and C?
3. What is the form of the partial fractions expansion of $(2x^2+1)/(x^3(x+1))$?
4. What is the form of the partial fractions expansion of $(x^2+3)/(x(x^2-4))$?

EXERCISES

Problems for Practice

In Exercises 1–10, write the form of the partial fractions decomposition of the rational function. Do *not* explicitly calculate the coefficients.

1. $\dfrac{x+2}{x(x+1)}$
2. $\dfrac{2x+3}{x^2-1}$
3. $\dfrac{3x+1}{x^2-4}$
4. $\dfrac{21}{(x-2)(x+5)}$
5. $\dfrac{x^3}{(2x-5)(x+4)}$
6. $\dfrac{x^2+2x+2}{(x-4)^2(x+2)}$
7. $\dfrac{x^3+7}{x^2(x+1)(x-2)}$
8. $\dfrac{x^5+1}{x^2(x+1)^2}$
9. $\dfrac{x^7+x^6-17}{(x-5)^2(x+3)^2}$
10. $\dfrac{2x^4}{(x-7)^3(x+1)}$

In Exercises 11–20, decompose the fraction into partial fractions. Calculate the coefficients.

11. $\dfrac{1}{x(x+1)}$
12. $\dfrac{2}{x^2-1}$
13. $\dfrac{x+6}{x^2-4}$
14. $\dfrac{21}{(x-2)(x+5)}$
15. $\dfrac{26x}{(2x-5)(x+4)}$
16. $\dfrac{36}{(x-4)^2(x+2)}$
17. $\dfrac{2}{x^2(x+1)(x-2)}$
18. $\dfrac{1}{x^2(x+1)^2}$
19. $\dfrac{2x^3-19x^2+8x+313}{(x-5)^2(x+3)^2}$
20. $\dfrac{6x^2-96x+410}{(x-7)^3(x+1)}$

In Exercises 21–30, use the Method of Partial Fractions to calculate the integral.

21. $\displaystyle\int \frac{3x+1}{x^2-1}\,dx$
22. $\displaystyle\int \frac{3x+4}{x^2+x-6}\,dx$
23. $\displaystyle\int \frac{9x+18}{(x-3)(x+6)}\,dx$
24. $\displaystyle\int \frac{x^3-2x^2-2x-2}{x^2-x}\,dx$
25. $\displaystyle\int \frac{x^3-4x^2+6}{x^2-3x-4}\,dx$
26. $\displaystyle\int \frac{3x^2}{x^2+2x}\,dx$

27. $\displaystyle\int \frac{9x}{(x-2)^2(x+1)}\,dx$ 28. $\displaystyle\int \frac{3x+2}{(x-2)^2(x+2)}\,dx$

29. $\displaystyle\int \frac{x^2+1}{x^3+x^2}\,dx$ 30. $\displaystyle\int \frac{25}{(x+1)(x-4)^2}\,dx$

In Exercises 31–34, use the Method of Partial Fractions to calculate the integral.

31. $\displaystyle\int_3^4 \frac{x^2+3x+6}{(x-2)(x+5)}\,dx$ 32. $\displaystyle\int_0^1 \frac{x^2}{(x+1)^2(x+2)^2}\,dx$

33. $\displaystyle\int_1^4 \frac{x+2}{(2x-1)(3x+4)}\,dx$

34. $\displaystyle\int_4^9 \frac{64x}{(x+1)(x-3)^3}\,dx$

Further Theory and Practice

In Exercises 35–39, make a substitution before applying the Method of Partial Fractions to calculate the integral.

35. $\displaystyle\int \frac{e^t}{e^{2t}-1}\,dt$

36. $\displaystyle\int \frac{\cos(x)}{\sin^2(x)-5\sin(x)+6}\,dx$

37. $\displaystyle\int \frac{16\cos(x)}{(3+\sin(x))^2(1-\sin(x))}\,dx$

38. $\displaystyle\int \frac{dx}{x^{3/2}-x}$ (Try $u = x^{1/2}$.)

39. $\displaystyle\int \frac{dx}{x^{4/3}-x^{2/3}}$ (Try $u = x^{1/3}$.)

40. What happens if you attempt a partial fractions decomposition of $1/(x-3)^4$ into

$$\frac{A}{x-3} + \frac{B}{(x-3)^2} + \frac{C}{(x-3)^3} + \frac{D}{(x-3)^4}?$$

Explain your answer.

41. Mr. Woodman set up the partial fractions decomposition

$$\frac{5x^2-4x+2}{x^2(x-1)^2} = \frac{A}{x^2} + \frac{B}{(x-1)^2}$$

and correctly solved $A = 2$ and $B = 3$.

a. What is wrong with this procedure?

b. Why was he able, *in this case*, to get the right answer even though he started out with the wrong form of the partial fractions decomposition?

c. Show that if the numerator is $5x^2 - 4x + 1$ instead of $5x^2 - 4x + 2$, then the procedure used will fail. What is the correct partial fractions decomposition of $(5x^2 - 4x + 1)/(x^2(x-1)^2)$?

42. Assume that a is a positive constant and that $|x| < a$. Verify the formula

$$\int \frac{dx}{a^2-x^2} = \frac{1}{2a}\ln\left(\frac{a+x}{a-x}\right) + C$$

by differentiation. Use the Method of Partial Fractions to derive this formula.

43. A population P satisfies the differential equation

$$P'(t) = 10^{-5} \cdot P(t) \cdot (15000 - P(t)).$$

Answer the following true/false questions and explain your answers.

a. If $P(0) = 25000$, then the population size will increase.

b. If $P(0) = 5000$, then the population size will reach 35000.

c. The graph of $P(t)$ has a horizontal asymptote.

d. The greatest initial growth rate occurs if $P(0) = 7500$.

44. The logistic growth differential equation (7.12) gives $\frac{dP}{dt}$ as a *quadratic* expression of $P(t)$.

a. Complete the square of this expression.

b. Use the completed square to analyze the sign of $P'(t)$.

c. Use the completed square to show that the solution $P(t)$ of the logistic growth equation has exactly one inflection point, which occurs when $P(t) = P_\infty/2$.

45. The solution $P(t)$ to the logistic differential growth equation, as given by equation (7.12), has the inflection point $(t_I, P(t_I))$ where $P(t_I) = P_\infty/2$. Show that the time t_I at which this occurs is given by

$$t_I = \frac{1}{kP_\infty}\ln\left(\frac{P_\infty - P_0}{P_0}\right).$$

46. Let $P(t)$ be the logistic growth function defined by equation (7.12). Let t_I be the time at which $P(t)$ has an inflection point (see Exercise 45 for a formula for t_I). Show that $P(t)$ can be expressed in terms of the parameters k, P_∞, and t_I by means of the formula

$$P(t) = \frac{P_\infty}{1 + \exp(-kP_\infty(t - t_I))}.$$

47. The plot of the pumpkin data in Figure 1 on page 521 has an inflection point at (9, 25). The initial diameter $P(0)$ is 3.6 cm. Use Exercise 46 to find k. Approximate the rate of change of $P(t)$ at $t = 9$ in the following two ways.

a. Use the logistic differential equation.

b. Use the values $P(8) = 21.2$ and $P(10) = 29.8$ to obtain a central difference quotient approximation.

48. If a population $P(t)$ follows the logistic growth model according to equation (7.12), then there are three parameters to solve for: the initial population P_0, the growth constant k, and the carrying capacity P_∞. It is easiest to solve for these constants if a time interval τ is chosen, if a particular time T is chosen, and if measurements of $P(t)$ are taken at the equally spaced times $T-\tau$, T, and $T+\tau$. Choose the origin of the time axis so that $t=0$ corresponds to T. Because $P_0 = P(0)$ is the measured value of P at time T, it is therefore known. Also known are $A = P(-\tau)$ and $B = P(\tau)$.

a. Use the equation $B = P(\tau)$ to solve for $\exp(-k \cdot P_\infty \cdot \tau)$ in terms of the known values B and P_0 and the unknown P_∞.

b. Use the equation $A = P(-\tau)$ to solve for $\exp(k \cdot P_\infty \cdot \tau)$ in terms of the known values A and P_0 and the unknown P_∞.

c. The product of $\exp(-k \cdot P_\infty \cdot \tau)$ and $\exp(k \cdot P_\infty \cdot \tau)$ is 1. Use this observation, together with parts a and b, to obtain a quadratic equation for P_∞ in terms of A, B, and P_0. (Once this equation is solved, the remaining variable k can be solved for.)

d. A town's population (in thousands) was 36 in 1970, 60 in 1980, and 90 in 1990. Assuming that the growth of the population of this town is logistic, what will the population be in the year 2010?

Calculator/Computer Exercises

49. Find the area of the region in the first quadrant bounded on the side by the y-axis, above by $y = (2x^2+5)/((x+2)(x+3)(x+5))$, and below by $y = (x^3+2)/((x+2)(x+3)(x+5))$.

50. Calculate $\int_8^{10} (x+3)/(x^3-8x^2+2)\,dx$ by finding a partial fractions decomposition of the integrand.

51. Plot $y = \ln(x)$ and $y = x^3/(x^2+3x+2)$ in the viewing window $[1.35, 2.25] \times [0.30, 0.83]$. Find the two points of intersection. Then use a partial fractions decomposition to find the area enclosed by the two curves.

52. Plot $y = x^4 - 5x^3 + 6x - 1$ for $-2 \le x \le 5$. Locate the four real roots. Plot $y = (2x^3+x+2)/(x^4-5x^3+6x-1)$ and $y = 4+2x-x^2$ in the viewing window $[0.24, 1] \times [1, 5]$. Find the two points of intersection. Then use a partial fractions decomposition to find the area enclosed by the two curves.

53. Computer algebra systems can calculate the partial fractions decomposition of a rational function, provided that the arithmetic can be done with rational numbers. For example,

```
convert(R,parfrac,x);
```

is the Maple command for obtaining the partial fractions decomposition of a rational expression R in the variable x. Use a computer algebra system to find the partial fractions decomposition of the following rational functions.

a. $\dfrac{x^3+5x^2+7x}{x^4+10x^3+35x^2+50x+24}$

b. $\dfrac{2x^7+17x^6+45x^5+2x^4-245x^3-595x^2-279x+582}{x^6+7x^5+5x^4-55x^3-90x^2+108x+216}$

c. $\dfrac{5x^6+25x^5-6x^4-233x^3-520x^2-444x-123}{x^7+6x^6+10x^5-4x^4-23x^3-10x^2+12x+8}$

d. $\dfrac{11x^6+24x^5-26x^4-107x^3-64x^2-49x+3}{x^7+x^6-3x^5-3x^4+3x^3+3x^2-x-1}$

54. The return of wildlife has been a measurable benefit of pollution reduction. Figure 4 contains a plot of the number P of double-crested cormorant nests in the Great Lakes region for 1979–1993 (estimates based on data obtained from Environment Canada). A logistic growth curve has been superimposed on the plot. In this exercise, you will obtain the formula for the logistic curve

$$P(t) = \frac{P_0 \cdot P_\infty}{P_0 + (P_\infty - P_0)\exp(-k \cdot P_\infty \cdot (t-1979))}.$$

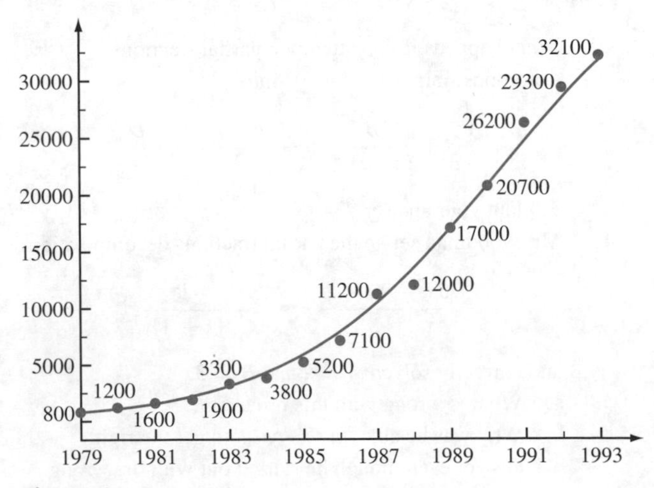

Figure 4
Cormorant nests (1979–1993)

Use $P_\infty = 44000$ for the carrying capacity and $P_0 = 800$ for the value of the initial population. (The formula for $P(t)$ differs from that of equation (7.12) in that we use $t - 1979$ here so that $P_0 = P(1979)$.)

a. For $t = 1984, 1985, \ldots, 1988$ use a central difference quotient to approximate $P'(t)$. For example,

$$P'(1980) \approx \frac{P(1981) - P(1979)}{1981 - 1979} = \frac{1600 - 800}{2}.$$

The logistic differential equation $P'(t) = kP(t)(P_\infty - P(t))$ then permits an estimate for k. Average the five estimates to get a working value of k.

b. Calculate $P(t)$ for $t = 1990, \ldots, 1993$. (Although the observed values for these years appear in the plot, they were not used to obtain the model. Comparing the predicted values with the observed values is a reality check for the model.)

c. Plot the logistic growth curve for $1979 \le t \le 1993$.

d. According to your logistic model, about when did the number of nests reach 43,000?

7.3 Powers and Products of Trigonometric Functions

This section contains a collection of devices for treating integrals that involve powers of sines and cosines. The techniques are a combination of trigonometric identities and substitutions. The problems are grouped by type so that you can easily study the technique for each type of problem. There are no formulas to memorize in this section. Rather, to master this material, you must learn how and when to apply each technique.

Squares of Sine or Cosine

To integrate the expressions $\sin^2(\theta)$ and $\cos^2(\theta)$, we use the equations

$$\sin^2(\theta) = \frac{1 - \cos(2\theta)}{2} \tag{7.15}$$

and

$$\cos^2(\theta) = \frac{1 + \cos(2\theta)}{2}, \tag{7.16}$$

which are obtained by replacing θ with 2θ in Half-Angle Formulas (1.14) and (1.15). Notice that the right sides of identities (7.15) and (7.16) differ only in the sign that precedes the term $\cos(2\theta)$. In particular, the trigonometric expression $\cos(2\theta)$ appears in the numerator of both.

Example 1 Show that

$$\int \sin^2(t)\,dt = \frac{1}{2}t - \frac{1}{4}\sin(2t) + C \tag{7.17}$$

and

$$\int \cos^2(t)\,dt = \frac{1}{2}t + \frac{1}{4}\sin(2t) + C. \tag{7.18}$$

Solution Identity (7.15) is used to derive formula (7.17), and identity (7.16) can be used to obtain equation (7.18). Let us illustrate with the second formula:

$$\begin{aligned}\int \cos^2(t)\,dt &= \int \frac{1+\cos(2t)}{2}\,dt \\ &= \int \frac{1}{2}\,dt + \int \frac{\cos(2t)}{2}\,dt \\ &= \frac{1}{2}t + \frac{1}{4}\sin(2t) + C.\end{aligned}$$

■

inSIGHT

In the next subsection, we learn that it is sometimes best to express the antiderivatives of $\sin^2(t)$ and $\cos^2(t)$ in terms of trigonometric functions of the argument t (instead of $2t$). Using the identity $\sin(2t) = 2\sin(t)\cos(t)$, we may rewrite formulas (7.17) and (7.18) as

$$\int \sin^2(t)\,dt = \frac{1}{2}(t - \sin(t)\cos(t)) + C \qquad \textbf{(7.19)}$$

and

$$\int \cos^2(t)\,dt = \frac{1}{2}(t + \sin(t)\cos(t)) + C. \qquad \textbf{(7.20)}$$

We can then use equations (7.19) and (7.20) to evaluate some definite integrals that frequently arise:

$$\int_0^{\pi} \cos^2(t)\,dt = \int_0^{\pi} \sin^2(t)\,dt = \frac{\pi}{2} \quad \text{and} \quad \int_0^{2\pi} \cos^2(t)\,dt = \int_0^{2\pi} \sin^2(t)\,dt = \pi. \qquad \textbf{(7.21)}$$

Working with Reduction Formulas

In Example 4 from Section 7.1, we evaluated the integral $\int r^2 e^{-2r}\,dr$ by applying the Integration by Parts Formula twice. The first application resulted in an integral of the form $\int re^{-2r}\,dr$, which is the same as the original integral except that the power of r is reduced by one. When we applied integration by parts on this second integral, we reduced it to $\int r^0 e^{-2r}\,dr$, or $\int e^{-2r}\,dr$, which is a standard integral that does not require further reduction. In summary, applying integration by parts to $\int r^n e^{-2r}\,dr$ results in a formula involving the integral $\int r^{n-1}e^{-2r}\,dr$, which is just like the original except for the lower power. Such a formula is known as a *reduction formula*. For reference, we record several reduction formulas involving trigonometric functions:

$$\int \sin^n(x)\,dx = -\frac{1}{n}\sin^{n-1}(x)\cos(x) + \frac{n-1}{n}\int \sin^{n-2}(x)\,dx \qquad (n \neq 0), \qquad \textbf{(7.22)}$$

$$\int \cos^n(x)\,dx = \frac{1}{n}\sin(x)\cos^{n-1}(x) + \frac{n-1}{n}\int \cos^{n-2}(x)\,dx \qquad (n \neq 0), \qquad \textbf{(7.23)}$$

$$\int \sec^n(x)\,dx = \frac{1}{n-1}\sec^{n-2}(x)\tan(x) + \frac{n-2}{n-1}\int \sec^{n-2}(x)\,dx \qquad (n \neq 1), \tag{7.24}$$

$$\int \tan^n(x)\,dx = \frac{1}{n-1}\tan^{n-1}(x) - \int \tan^{n-2}(x)\,dx \qquad (n \neq 1). \tag{7.25}$$

Example 2 Derive identity (7.22).

Solution Let $I_n = \int \sin^n(x)\,dx$ denote the required integral. We apply the Integration by Parts Formula as follows:

$$I_n = \int \underbrace{\sin^{n-1}(x)}_{u}\,\underbrace{\sin(x)\,dx}_{dv} = \underbrace{\sin^{n-1}(x)}_{u}\cdot\underbrace{(-\cos(x))}_{v} - \int \underbrace{(-\cos(x))}_{v}\underbrace{(n-1)\sin^{n-2}(x)\cos(x)\,dx}_{du},$$

which simplifies to

$$I_n = -\sin^{(n-1)}(x)\cos(x) + (n-1)\int \sin^{(n-2)}\cos^2(x)\,dx.$$

We replace $\cos^2(x)$ with $1 - \sin^2(x)$ and expand the integral:

$$\begin{aligned}
I_n &= -\sin^{(n-1)}(x)\cos(x) + (n-1)\int \sin^{(n-2)}(1-\sin^2(x))\,dx \\
&= -\sin^{(n-1)}(x)\cos(x) + (n-1)\int \sin^{(n-2)}\,dx - (n-1)\int \sin^n(x)\,dx \\
&= -\sin^{(n-1)}(x)\cos(x) + (n-1)\int \sin^{(n-2)}\,dx - (n-1)I_n.
\end{aligned}$$

We now solve this equation in I_n by bringing the unknown I_n to one side of the equation and isolating it:

$$I_n + (n-1)I_n = -\sin^{(n-1)}(x)\cos(x) + (n-1)\int \sin^{(n-2)}\,dx.$$

Since the left side of this equation simplifies to $n \cdot I_n$, dividing through by n results in formula (7.22). ■

A reduction formula is used successively until the power is sufficiently reduced (usually to 0 or 1) that further reduction is not required, as the examples that follow illustrate.

Example 3 Evaluate $\int \sec^3(x)\,dx$.

Solution Reduction formula (7.24) with $n = 3$ asserts that

$$\int \sec^3(x)\,dx = \frac{1}{3-1}\sec^{3-2}(x)\tan(x) + \frac{3-2}{3-1}\int \sec^{3-2}(x)\,dx = \frac{1}{2}\sec(x)\tan(x) + \frac{1}{2}\int \sec(x)\,dx.$$

This last integral is one we learned in Chapter 6:

$$\int \sec(x)\,dx = \ln(|\sec(x) + \tan(x)|) + C.$$

Thus,

$$\int \sec^3(x)\,dx = \frac{1}{2}\sec(x)\tan(x) + \frac{1}{2}\ln(|\sec(x) + \tan(x)|) + C.$$

■

Example 4 Evaluate $\int \cos^6(x)\,dx$.

Solution After one application of formula (7.23), we obtain

$$\int \cos^6(x)\,dx = \frac{1}{6}\sin(x)\cos^5(x) + \frac{5}{6}\int \cos^4(x)\,dx.$$

When we apply formula (7.23) to $\int \cos^4(x)\,dx$, we obtain

$$\int \cos^6(x)\,dx = \frac{1}{6}\sin(x)\cos^5(x) + \frac{5}{6}\left(\frac{1}{4}\sin(x)\cos^3(x) + \frac{3}{4}\int \cos^2(x)\,dx\right),$$

or

$$\int \cos^6(x)\,dx = \frac{1}{6}\sin(x)\cos^5(x) + \frac{5}{24}\sin(x)\cos^3(x) + \frac{5}{8}\int \cos^2(x)\,dx.$$

At this point, we can exploit equation (7.20). Alternatively, we can apply reduction formula (7.23) to $\int \cos^2(x)\,dx$:

$$\int \cos^2(x)\,dx = \frac{1}{2}\sin(x)\cos(x) + \frac{1}{2}\int \cos^0(x)\,dx = \frac{1}{2}\sin(x)\cos(x) + \frac{1}{2}x.$$

Thus,

$$\int \cos^6(x)\,dx = \frac{1}{6}\sin(x)\cos^5(x) + \frac{5}{24}\sin(x)\cos^3(x) + \frac{5}{16}\sin(x)\cos(x) + \frac{5}{16}x.$$

■

Odd Powers of Sine or Cosine

Although we can use reduction formulas (7.22) and (7.23) to evaluate $\int \sin^n(x)\,dx$ and $\int \cos^n(x)\,dx$ when n is any positive integer, there is a simpler, less tedious method available when n is odd. As an example, we use the identities

$$\cos^2(x) = 1 - \sin^2(x) \qquad \text{and} \qquad \sin^2(x) = 1 - \cos^2(x).$$

Suppose that k and ℓ are positive integers. We integrate an odd power $\sin^{2k+1}(x)$ of $\sin(x)$ by writing

$$\sin^{2k+1}(x) = (\sin^2(x))^k \sin(x) = (1 - \cos^2(x))^k \sin(x). \qquad \textbf{(7.26)}$$

We integrate an odd power $\cos^{2\ell+1}(x)$ of $\cos(x)$ by writing

$$\cos^{2\ell+1}(x) = (\cos^2(x))^\ell \cos(x) = (1 - \sin^2(x))^\ell \cos(x). \qquad \textbf{(7.27)}$$

The next two examples illustrate how we use these formulas.

Example 5 Use formula (7.27) to calculate $\int \cos^3(x)\,dx$.

Solution Formula (7.27) tells us that

$$\begin{aligned}\int \cos^3(x)\,dx &= \int (\cos^2(x)) \cdot \cos(x)\,dx \\ &= \int (1 - \sin^2(x)) \cdot \cos(x)\,dx.\end{aligned}$$

We make the substitution $u = \sin(x)$, $du = \cos(x)\,dx$ to obtain

$$\int \cos^3(x)\,dx = \int (1 - u^2)\,du = u - \frac{u^3}{3} + C \overset{\text{Resubstituting } x}{=} \sin(x) - \frac{\sin^3(x)}{3} + C.$$

You can check this answer by differentiating. ■

The following is a slightly more complicated example.

Example 6 Calculate $\int \sin^5(x)\,dx$.

Solution Using formula (7.26), we write

$$\int \sin^5(x)\,dx = \int (\sin^2(x))^2 \sin(x)\,dx = \int (1 - \cos^2(x))^2 \sin(x)\,dx.$$

We use the substitution $u = \cos(x)$, $du = -\sin(x)\,dx$ to rewrite this equation as

$$-\int (1 - u^2)^2\,du = \int (-1 + 2u^2 - u^4)\,du = -u + \frac{2u^3}{3} - \frac{u^5}{5} + C.$$

Resubstituting the expressions in x yields the final solution:

$$\int \sin^5(x)\,dx = -\cos(x) + \frac{2\cos^3(x)}{3} - \frac{\cos^5(x)}{5} + C.$$ ■

Integrals Involving Both Sine and Cosine

So far in this section, we have learned how to integrate positive powers of $\sin(x)$ and $\cos(x)$. We now consider integrands that involve *both* of these functions:

$$\int \sin^m(x) \cos^n(x)\,dx$$

where m and n are positive integers. There are two possibilities for this equation: Either at least one of the numbers m and n is odd or both of them are even.

If *at least* one of m or n is odd, then we apply the technique of the preceding subsection on odd powers of sine or cosine. (If both are odd, then apply formula (7.26) or (7.27) to the *smaller* of the two.)

If *neither* m nor n is odd, then both are even. We use the identity $\cos^2(x) + \sin^2(x) = 1$ to convert the integrand to a sum of even powers of sine or of cosine. Then we apply the appropriate reduction formula.

Our next two examples illustrate these techniques. Other options are explored in Exercises 31–35.

Example 7 Evaluate the integral $\int \cos^3(x) \sin^4(x)\,dx$.

Solution Since cosine is raised to an odd power, we apply the technique associated with formula (7.27). Our integral equals

$$\int \cos(x) \cos^2(x) \sin^4(x)\,dx = \int \cos(x) \cdot (1 - \sin^2(x)) \cdot \sin^4(x)\,dx.$$

We use the substitution $u = \sin(x),\ du = \cos(x)\,dx$ to get

$$\begin{aligned}\int \cos^3(x)\sin^4(x)\,dx &= \int (1-u^2)\cdot u^4\,du\\ &= \int u^4 - u^6\,du\\ &= \frac{u^5}{5} - \frac{u^7}{7} + C\\ &= \frac{\sin^5(x)}{5} - \frac{\sin^7(x)}{7} + C.\end{aligned}$$

Example 8 Convert $\int \sin^6(\theta)\cos^4(\theta)\,d\theta$ to a sum of integrals that can be evaluated by a reduction formula.

Solution We have

$$\begin{aligned}\int \sin^6(\theta)\cos^4(\theta)\,d\theta &= \int \sin^6(\theta)(\cos^2(\theta))^2\,d\theta\\ &= \int \sin^6(\theta)(1-\sin^2(\theta))^2\,d\theta\\ &= \int \sin^6(\theta)(1-2\sin^2(\theta)+\sin^4(\theta))\,d\theta\\ &= \int \sin^6(\theta)\,d\theta - 2\int \sin^8(\theta)\,d\theta + \int \sin^{10}(\theta)\,d\theta.\end{aligned}$$

At this point in the solution, we have obtained the required conversion because each summand can be handled by reduction formula (7.22). However, we will carry the computation for a few more steps to indicate how it can be continued in an efficient fashion. We apply the reduction formula to the integral with the *highest* power of sine. Thus,

$$\int \sin^6(\theta)\cos^4(\theta)\,d\theta = \int \sin^6(\theta)\,d\theta - 2\int \sin^8(\theta)\,d\theta + \left(-\frac{1}{10}\sin^9(\theta)\cos(\theta) + \frac{9}{10}\int \sin^8(\theta)\,d\theta\right).$$

Next we group the terms that involve the same integrals. Then we apply the reduction formula again:

$$\begin{aligned}\int \sin^6(\theta)\cos^4(\theta)\,d\theta &= -\frac{1}{10}\sin^9(\theta)\cos(\theta) + \int \sin^6(\theta)\,d\theta - \frac{11}{10}\int \sin^8(\theta)\,d\theta\\ &= -\frac{1}{10}\sin^9(\theta)\cos(\theta) + \int \sin^6(\theta)\,d\theta - \frac{11}{10}\left(-\frac{1}{8}\sin^7(\theta)\cos(\theta) + \frac{7}{8}\int \sin^6(\theta)\,d\theta\right)\\ &= -\frac{1}{10}\sin^9(\theta)\cos(\theta) + \frac{11}{80}\sin^7(\theta)\cos(\theta) + \left(1 - \frac{77}{80}\right)\int \sin^6(\theta)\,d\theta\\ &= -\frac{1}{10}\sin^9(\theta)\cos(\theta) + \frac{11}{80}\sin^7(\theta)\cos(\theta) + \frac{3}{80}\left(-\frac{1}{6}\sin^5(\theta)\cos(\theta) + \frac{5}{6}\int \sin^4(\theta)\,d\theta\right)\\ &= -\frac{1}{10}\sin^9(\theta)\cos(\theta) + \frac{11}{80}\sin^7(\theta)\cos(\theta) - \frac{1}{160}\sin^5(\theta)\cos(\theta)\\ &\quad + \frac{5}{160}\left(-\frac{1}{4}\sin^3(\theta)\cos(\theta) + \frac{3}{4}\int \sin^2(\theta)\,d\theta\right).\end{aligned}$$

We can now use formula (7.19) to finish off the calculation.

Sometimes, when other trigonometric functions are converted to sines and cosines, the methods that have been discussed in this section can be brought to bear, as the next example illustrates.

Example 9 Evaluate $\int \tan^5(\theta)\sec^3(\theta)\, d\theta$.

Solution We have

$$\tan^5(\theta)\sec^3(\theta) = \frac{\sin^5(\theta)}{\cos^5(\theta)}\sec^3(\theta) = \cos^{-8}(\theta)(\sin^2(\theta))^2\sin(\theta) = \cos^{-8}(\theta)(1-\cos^2(\theta))^2\sin(\theta).$$

The substitution $u = \cos(\theta)$, $du = -\sin(\theta)\, d\theta$ converts the given integral to

$$-\int u^{-8}(1-u^2)^2\, du = -\int (u^{-8} - 2u^{-6} + u^{-4})\, du = \frac{1}{7u^7} - \frac{2}{5u^5} + \frac{1}{3u^3} + C.$$

We resubstitute $u = \cos(\theta)$ to obtain

$$\int \tan^5(\theta)\sec^3(\theta)\, d\theta = \frac{1}{7\cos^7(\theta)} - \frac{2}{5\cos^5(\theta)} + \frac{1}{3\cos^3(\theta)} + C.$$ ■

quickquiz

1. Evaluate $\int \cos^2(x)\, dx$.
2. Express $\int \sin^{12}(x)\, dx$ in terms of $\int \sin^{10}(x)\, dx$.
3. Evaluate $\int \sin^3(x)\cos^5(x)\, dx$.
4. List the different possible combinations of sine and cosine that can occur under an integral and describe the techniques for treating those combinations.

EXERCISES

Problems for Practice

In Exercises 1–4, evaluate the integral.

1. $\int \cos^2(4\theta)\, d\theta$
2. $\int \sin^2(3\theta)\, d\theta$
3. $\int (\cos^2(t) + \sin^2(t))\, dt$
4. $\int (\cos^2(t) + \sec^2(t))\, dt$

In Exercises 5–8, evaluate the integral.

5. $\int \sin^3(2x)\, dx$
6. $\int \cos^5(x)\, dx$
7. $\int \sin^3(x+2)\, dx$
8. $\int \cos^7(x)\, dx$

In Exercises 9–12, evaluate the integral.

9. $\int_0^{\pi/2} \sin(t)\cos^4(t)\, dt$
10. $\int_0^{\pi} \sin^3(w)\cos^2(w)\, dw$
11. $\int_0^{\pi} \cos^4(\theta)\sin^3(\theta)\, d\theta$
12. $\int_{-\pi/2}^{\pi/2} \sin^2(x)\cos^5(x)\, dx$

In Exercises 13–20, use a reduction formula to evaluate the integral.

13. $\int \cos^4(x)\, dx$
14. $\int_0^1 \cos^4(\pi x)\, dx$
15. $\int_{\pi}^{2\pi} \sin^4(x)\, dx$
16. $\int \cos^6(x/2)\, dx$

In Exercises 17–20, use a reduction formula to evaluate the integral.

17. $\int_0^{\pi/4} \cos^2(t)\sin^2(t)\, dt$
18. $\int_0^{\pi} \cos^4(t)\sin^2(t)\, dt$
19. $\int_{\pi/6}^{\pi/3} \cos^2(x)\sin^4(x)\, dx$
20. $\int_0^{\pi/4} \sec^4(t)\tan^2(t)\, dt$

In Exercises 21–25, evaluate the integral by converting the integrand to an expression in sines and cosines.

21. $\int \tan(t)\sec^3(t)\, dt$
22. $\int \tan(t)\sec^4(t)\, dt$
23. $\int \tan^3(t)\, dt$
24. $\int \tan^3(t)\sec(t)\, dt$
25. $\int \tan(t)\sin(t)\, dt$

In Exercises 26–30, the integrand is of the form $a \pm b$. Multiply numerator and denominator by $a \pm b$ to obtain a difference of squares in the denominator. Then use an appropriate trigonometric identity before integrating.

26. $\int \frac{1}{1-\sin(\theta)}\, d\theta$
27. $\int \frac{1}{1+\cos(\theta)}\, d\theta$
28. $\int \frac{\cos(\theta)}{1-\cos(\theta)}\, d\theta$

29. $\displaystyle\int \frac{1}{\sec(\theta)-1}\,d\theta$ **30.** $\displaystyle\int \frac{\tan(\theta)}{\sec(\theta)+1}\,d\theta$

Further Theory and Practice

In Exercises 31–34, the integrand involves even powers of $\sin(t)$ and $\cos(t)$. Use identities (7.15) and (7.16) to halve each power. Repeat until the nonconstant summands of the integrands involve only second or odd powers of $\cos(2t), \cos(4t), \ldots$. Use formula (7.18) to evaluate the integral of the square of a cosine function. Use the method from Example 5 to treat any odd powers greater than 1. For example,

$$\begin{aligned}\int \cos^4(t)\,dt &= \int (\cos^2(t))^2\,dt\\ &= \frac{1}{4}\int (1+\cos(2t))^2\,dt\\ &= \frac{1}{4}\int (1+2\cos(2t)+\cos^2(2t))\,dt.\end{aligned}$$

The first two summands are elementary, and the last may be treated with formula (7.18). The method outlined above, which is an alternative method for handling integrals that involve even powers of sines and cosines, relies on a reduction technique but not on a reduction formula.

31. $\int \cos^6(t)\,dt$ **32.** $\int \sin^6(t)\,dt$

33. $\int \cos^2(t)\sin^4(t)\,dt$ **34.** $\int \cos^4(t)\sin^2(t)\,dt$

35. When $\sin(x)$ and $\cos(x)$ are both raised to the same power in an integrand, the identity $\sin(2x) = 2\sin(x)\cos(x)$ may be used to simplify the integral. Use this observation as an aid in evaluating the following.

a. $\int \cos^2(x)\sin^2(x)\,dx$ **b.** $\int \cos^4(x)\sin^4(x)\,dx$

36. Let m and n be positive integers. Without evaluating the integral, explain why $\int_{-\pi}^{\pi} \cos^{2m}(x)\sin^{2n+1}(x)\,dx = 0$.

37. Let m and n be positive integers. Without evaluating the integral, explain why $\int_{-\pi}^{\pi} \cos^{2m}(x)\sin^{2n}(x)\,dx \neq 0$.

38. Sketch the graphs of $y = \sin^2(x)$ and $y = \cos^2(x)$ for $0 \le x \le 2\pi$. Use the identity $\sin^2(x)+\cos^2(x) = 1$ to explain why

$$\int_0^{2\pi} \cos^2(x)\,dx = \int_0^{2\pi} \sin^2(x)\,dx = \pi$$

(without performing any "integrations"). State a few subintervals $[a, b]$ of $[0, 2\pi]$ for which

$$\int_a^b \cos^2(x)\,dx = \int_a^b \sin^2(x)\,dx = \frac{1}{2}(b-a).$$

39. Explain why the integrations

$$\int \sin(x)\cos(x)\,dx = \frac{1}{2}\sin^2(x) + C$$

and

$$\int \sin(x)\cos(x)\,dx = -\frac{1}{2}\cos^2(x) + C$$

and

$$\int \sin(x)\cos(x)\,dx = -\frac{1}{4}\cos(2x) + C$$

are all complete and correct. Explain why your answer is not contradictory.

40. Calculate $\int_0^{\pi/2} (\cos(x)+\sin(x))^2\,dx$.

41. Calculate $\int_0^{\pi/2} x\cos^2(x)\,dx$.

42. Find the area of the region bounded by $y = \sin^2(x)$ and $y = \sin^3(x)$ for $0 \le x \le \pi/2$.

43. Evaluate $\int_0^{\pi/4} \sin^2(x)\tan(x)\,dx$.

44. Evaluate $\int_0^{\pi/4} \sin^3(x)/\sqrt{\cos(x)}\,dx$.

45. Find a reduction formula for $\int \ln^n(x)\,dx$.

46. Let n be a nonnegative integer. Derive the following reduction formula:

$$\int x^n e^{ax}\,dx = \frac{1}{a}x^n e^{ax} - \frac{n}{a}\int x^{n-1}e^{ax}\,dx.$$

47. Calculate $\int \tan^2(x)\,dx$.

48. Derive reduction formula (7.23).

49. Derive reduction formula (7.24).

50. Derive reduction formula (7.25).

51. Let m and n be integers such that $m \neq -n$. Prove the reduction formula

$$\int \sin^n(x)\cos^m(x)\,dx = -\frac{\sin^{n-1}(x)\cos^{m+1}(x)}{m+n} + \frac{n-1}{m+n}\int \sin^{n-2}(x)\cos^m(x)\,dx.$$

52. Find a reduction formula for integrals of the form

$$\int \tan^{2k}(\theta)\sec^{2j+1}(\theta)\,d\theta$$

where j and k are nonnegative integers.

53. Use the Integration by Parts Formula to show that

$$\int \frac{1}{(x^2+a^2)^{n+1}}\,dx = \frac{1}{2a^2 n}\cdot\frac{x}{(x^2+a^2)^n} + \frac{(2n-1)}{2a^2 n}\int \frac{1}{(x^2+a^2)^n}\,dx.$$

54. For each natural number n, let

$$W_n = \int_0^{\pi/2} \sin^n(x)\,dx.$$

a. Use formula (7.21) to show that

$$W_n = \frac{n-1}{n}W_{n-2} \qquad (n \ge 2).$$

b. By calculating W_0 and using part a, show that

$$W_{2k} = \frac{1 \cdot 3 \cdot 5 \cdots (2k-1)}{2 \cdot 4 \cdot 6 \cdots (2k)} \cdot \frac{\pi}{2}.$$

c. By calculating W_1 and using part a, show that

$$W_{2k+1} = \frac{2 \cdot 4 \cdot 6 \cdots (2k)}{3 \cdot 5 \cdot 7 \cdots (2k+1)}.$$

d. By comparing integrands, deduce that $W_{2k+1} \leq W_{2k} \leq W_{2k-1}$.

e. Use the results from parts b, c, and d to deduce that

$$\frac{2^2 \cdot 4^2 \cdots (2k)^2}{3^2 \cdot 5^2 \cdots (2k-1)^2} \frac{1}{(2k+1)} \leq \frac{\pi}{2}, \text{ and}$$

$$\frac{\pi}{2} \leq \frac{2^2 \cdot 4^2 \cdots (2k)^2}{3^2 \cdot 5^2 \cdots (2k-1)^2} \frac{1}{2k}.$$

f. Deduce that there is a number θ_k in $[0, 1]$ such that

$$\frac{\pi}{2} = \frac{2^2 \cdot 4^2 \cdot 6^2 \cdots (2k)^2}{3^2 \cdot 5^2 \cdot 7^2 \cdots (2k-1)^2} \cdot \frac{1}{2k+\theta_k}.$$

g. Deduce the following formula for π, which John Wallis published in 1655:

$$\pi = \lim_{k\to\infty} \left(\frac{2^2 \cdot 4^2 \cdot 6^2 \cdots (2k)^2}{3^2 \cdot 5^2 \cdot 7^2 \cdots (2k-1)^2} \cdot \frac{1}{k} \right).$$

Note: The *Wallis limit formula* is beautiful but quite useless from the point of view of approximating π. Although the quotient of products in Wallis's formula is computationally expensive, we can remove that difficulty at the price of storage. If we define Q_N recursively by $Q_1 = 4$ and

$$Q_k = \frac{4k(k-1)}{(2k-1)^2} \cdot Q_{k-1} \quad (k \geq 2),$$

then Wallis's formula becomes $\pi = \lim_{k\to\infty} Q_k$. By storing each value of Q_{k-1}, starting with Q_1, we can easily get the next value Q_k. The real problem, however, is that convergence to the limit is extremely slow. For example $Q_{100000} = 3.141604\ldots$; only three decimal places are correct. By comparison, at the time of this writing, more than 51 billion decimal digits of π are known.

55. Suppose A and B are constants, not both of which are 0. Notice that

$$A\cos(x) + B\sin(x) = \sqrt{A^2+B^2}(\cos(x)\cdot\xi + \sin(x)\cdot\eta)$$

where $\xi = A/\sqrt{A^2+B^2}$ and $\eta = B/\sqrt{A^2+B^2}$. Observing that (ξ, η) is a point on the unit circle, deduce that $A\cos(x) + B\sin(x) = \sqrt{A^2+B^2}\cos(x+\phi)$ for some ϕ. Evaluate

$$\int \frac{1}{(A\cos(x) + B\sin(x))^k}\, dx$$

for $k = 1$ and $k = 2$.

Calculator/Computer Exercises

56. Plot $y = 10\sin^3(x) + 2$ and $y = \sec^3(x)$ in the viewing window $[-0.45, 1.1] \times [0, 9]$. Find the area between the two curves.

57. Plot $y = \sin^2(x)\cos^3(x)$ and $y = \cos(x)^2\sin(x)^5$ in the viewing window $[0, 1] \times [0, 0.2]$. Find the area between the two curves.

58. The line $y = 0.6x$ intersects the curve $y = \sin(x)^4$ at the origin and at two points with positive coordinates. Two regions in the first quadrant are bounded by these curves. Find the area of each region.

59. Let x_0 be the least positive value of x such that the tangent line to $y = \sin(x)^4$ at $x = x_0$ passes through the origin. Find the area enclosed by this tangent line and $y = \sin(x)^4$ for $0 \leq x \leq x_0$.

7.4 Integrals Involving Quadratic Expressions

In this section, we learn to treat integrands that involve expressions of the form

$$Ax^2 + Bx + C, \quad A \neq 0.$$

We have already encountered a few isolated instances of integrals that contain quadratic expressions. For example, if $A = -1$ or 1, $B = 0$, and $C = a^2 > 0$, then Ax^2+Bx+C

becomes $a^2 - x^2$ or $a^2 + x^2$. Formulas (6.58) and (6.59) from Section 6.6 tell us that

$$\int \frac{dx}{\sqrt{a^2 - x^2}} = \arcsin\left(\frac{x}{a}\right) + C \quad \text{and} \quad \int \frac{dx}{a^2 + x^2} = \frac{1}{a}\arctan\left(\frac{x}{a}\right) + C.$$

As these two equations show, inverse trigonometric functions can appear in the antiderivatives of functions that involve the expression $Ax^2 + Bx + C$.

Trigonometric Substitution

Consider the three quadratic polynomials $a^2 - x^2$, $a^2 + x^2$, and $x^2 - a^2$ where a is a positive constant. By making an appropriate substitution in each expression, we may exploit a trigonometric identity to reduce these sums to a single square:

1. After substituting $x = a\sin(\theta)$, the expression $a^2 - x^2$ becomes $a^2 - a^2\sin^2(\theta) = a^2(1 - \sin^2(\theta)) = a^2\cos^2(\theta)$.
2. After substituting $x = a\tan(\theta)$, the expression $a^2 + x^2$ becomes $a^2 + a^2\tan^2(\theta) = a^2(1 + \tan^2(\theta)) = a^2\sec^2(\theta)$.
3. After substituting $x = a\sec(\theta)$, the expression $x^2 - a^2$ becomes $a^2\sec^2(\theta) - a^2 = a^2(\sec^2(\theta) - 1) = a^2\tan^2(\theta)$.

These calculations suggest a new substitution technique. To integrate a function that contains the expression $a^2 - x^2$, consider making the change of variable $x = a\sin(\theta)$, $dx = a\cos(\theta)\,d\theta$. Similarly, the expression $a^2 + x^2$ in an integral suggests the change of variable $x = a\tan(\theta)$, $dx = a\sec^2(\theta)\,d\theta$. Finally, the expression $x^2 - a^2$ points to the substitution $x = a\sec(\theta)$, $dx = a\sec(\theta)\tan(\theta)\,d\theta$. These substitutions are examples of a type of change of variable known as the *Method of Inverse (or Indirect) Substitution*. Table 1 provides a summary of the trigonometric substitutions we use.

Expression	**Substitution**	**Result of Substitution**	**After Simplification**
$a^2 - x^2$	$x = a \cdot \sin(\theta)$, $dx = a \cdot \cos(\theta)\,d\theta$	$a^2 \cdot (1 - \sin^2(\theta))$	$a^2 \cdot \cos^2(\theta)$
$a^2 + x^2$	$x = a \cdot \tan(\theta)$, $dx = a \cdot \sec^2(\theta)\,d\theta$	$a^2 \cdot (1 + \tan^2(\theta))$	$a^2 \cdot \sec^2(\theta)$
$x^2 - a^2$	$x = a \cdot \sec(\theta)$, $dx = a \cdot \sec(\theta)\tan(\theta)\,d\theta$	$a^2 \cdot (\sec^2(\theta) - 1)$	$a^2 \cdot \tan^2(\theta)$

Table 1

If you already know the fundamental trigonometric identities that are the basis of these three substitutions, then Table 1 requires no memorization. Instead, work enough practice problems so that you recognize when a particular trigonometric substitution is called for.

Example 1 Calculate

$$\int \frac{x^2}{\sqrt{1 - x^2}}\,dx.$$

Solution The expression $1 - x^2$ suggests the substitution $x = 1 \cdot \sin(\theta)$, $dx = \cos(\theta)\,d\theta$. With this change of variable, the integral becomes

$$\int \frac{\sin^2(\theta)}{\sqrt{1 - \sin^2(\theta)}}\cos(\theta)\,d\theta = \int \frac{\sin^2(\theta)}{\sqrt{\cos^2(\theta)}}\cos(\theta)\,d\theta = \int \frac{\sin^2(\theta)}{|\cos(\theta)|}\cos(\theta)\,d\theta.$$

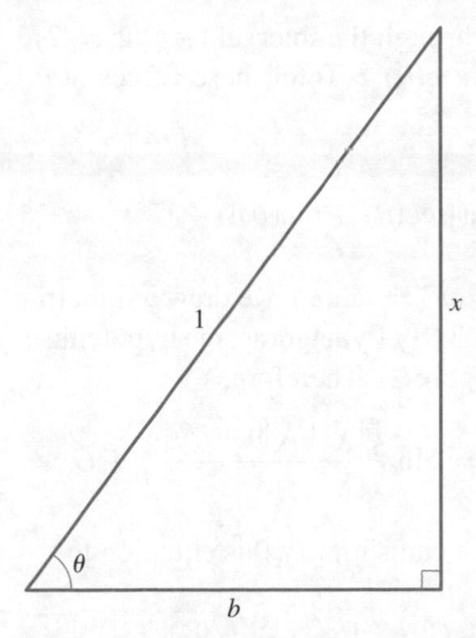

Figure 1
$\frac{x}{1} = \sin(\theta) \Rightarrow \cos(\theta) = \frac{b}{1} = \sqrt{1-x^2}$

Notice that the variable x must lie in the open interval $(-1, 1)$. Because $\sin(\theta)$ assumes every value between -1 and 1 as θ ranges from $-\pi/2$ to $\pi/2$, we may assume, without loss of generality, that θ belongs to the interval $(-\pi/2, \pi/2)$. For these values of θ, we have $\cos(\theta) > 0$. Therefore, $|\cos(\theta)| = \cos(\theta)$, and, using equation (7.19) from Section 7.3, we obtain

$$\int \frac{\sin^2(\theta)}{|\cos(\theta)|} \cos(\theta)\, d\theta = \int \sin^2(\theta)\, d\theta = \frac{1}{2}(\theta - \sin(\theta)\cos(\theta)) + C.$$

We finish the calculation by resubstituting. Since $x/1 = \sin(\theta)$, we draw a right triangle with angle θ, opposite side x, and hypotenuse 1 (see Figure 1). By Pythagoras, the adjacent side b equals $\sqrt{1-x^2}$ and $\cos(\theta) = \sqrt{1-x^2}/1$. We therefore resubstitute $\theta = \arcsin(x)$, $x = \sin(\theta)$, and $\cos(\theta) = \sqrt{1-x^2}$ to obtain

$$\int \frac{x^2}{\sqrt{1-x^2}}\, dx = \frac{1}{2}\left(\arcsin(x) - x\sqrt{1-x^2}\right) + C. \quad \blacksquare$$

Example 2 Calculate

$$\int_{2\sqrt{2}}^{4} \frac{1}{x\sqrt{x^2-4}}\, dx.$$

Solution The expression $x^2 - 2^2$ suggests the substitution $x = 2\sec(\theta)$, $dx = 2 \cdot \sec(\theta)\tan(\theta)\, d\theta$. The limits of integration $x = 2\sqrt{2}$ and $x = 4$ correspond to $\sec(\theta) = \sqrt{2}$ and $\sec(\theta) = 2$, or $\theta = \pi/4$ and $\theta = \pi/3$, respectively. Therefore,

$$\int_{2\sqrt{2}}^{4} \frac{1}{x\sqrt{x^2-4}}\, dx = \int_{\pi/4}^{\pi/3} \frac{2\sec(\theta)\tan(\theta)}{2\sec(\theta)\sqrt{4\sec^2(\theta)-4}}\, d\theta$$

$$= \int_{\pi/4}^{\pi/3} \frac{\tan(\theta)}{2\sqrt{\sec^2(\theta)-1}}\, d\theta = \int_{\pi/4}^{\pi/3} \frac{\tan(\theta)}{2\sqrt{\tan^2(\theta)}}\, d\theta.$$

For $\pi/4 \le \theta \le \pi/3$, we have $\tan(\theta) > 0$ and therefore $\sqrt{\tan^2(\theta)} = |\tan(\theta)| = \tan(\theta)$. We conclude that

$$\int_{2\sqrt{2}}^{4} \frac{1}{x\sqrt{x^2-4}}\, dx = \frac{1}{2}\int_{\pi/4}^{\pi/3} 1\, d\theta = \frac{1}{2}\theta\Big|_{\pi/4}^{\pi/3} = \frac{1}{2}\left(\frac{\pi}{3} - \frac{\pi}{4}\right) = \frac{\pi}{24}. \quad \blacksquare$$

inSIGHT

> Examples 1 and 2 concern expressions of the form $\sqrt{a^2-x^2}$ and $\sqrt{x^2-a^2}$. Because $|\sin(\theta)| \le 1$, the substitution $x = a\sin(\theta)$ forces $|x| \le a$, which is appropriate for the expression $\sqrt{a^2-x^2}$. Similarly, because $|\sec(\theta)| \ge 1$, the substitution $x = a\sec(\theta)$ forces $|x| \ge a$, which is appropriate for the expression $\sqrt{x^2-a^2}$. For the same reason, an interval of integration can also dictate which of the two substitutions should be applied to a difference of squares. Exercise 75 illustrates this point.

Example 3 Calculate the integrals

$$\int \frac{2}{\sqrt{9+x^2}}\, dx \quad \text{and} \quad \int \frac{2x}{\sqrt{9+x^2}}\, dx.$$

Solution Both integrands involve the expression $3^2 + x^2$. Therefore, the substitution $x = 3\tan(\theta)$, $dx = 3\sec^2(\theta)\, d\theta$ is indicated. With this substitution, the first integral becomes

$$\int \frac{2}{\sqrt{9+9\tan^2(\theta)}} 3\sec^2(\theta)\, d\theta = 2\int \frac{\sec^2(\theta)}{\sqrt{1+\tan^2(\theta)}}\, d\theta$$

$$= 2\int \frac{\sec^2(\theta)}{\sqrt{\sec^2(\theta)}}\, d\theta = 2\int \frac{\sec^2(\theta)}{|\sec(\theta)|}\, d\theta.$$

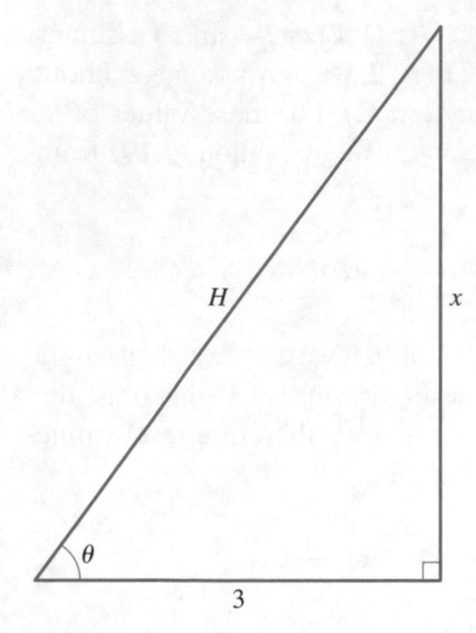

Figure 2
$\frac{x}{3} = \tan(\theta) \Rightarrow \sec(\theta) = \frac{H}{3} = \frac{\sqrt{3^2+x^2}}{3}$

Because $\tan(\theta)$ takes on all real values as θ ranges through the interval $(-\pi/2, \pi/2)$, we may assume that θ lies in this interval. Because $\sec(\theta) > 0$ for these values of θ, we have $|\sec(\theta)| = \sec(\theta)$ and

$$2\int \frac{\sec^2(\theta)}{|\sec(\theta)|}\,d\theta = 2\int \sec(\theta)\,d\theta \overset{(6.19)}{=} 2\ln(|\sec(\theta)+\tan(\theta)|) + C.$$

We complete the integration by resubstituting. Since $x/3 = \tan(\theta)$, we draw a right triangle with angle θ, opposite side x, and adjacent side 3. By Pythagoras, the hypotenuse H equals $\sqrt{9+x^2}$ and $\sec(\theta) = \sqrt{9+x^2}/3$ (see Figure 2). Therefore,

$$\int \frac{2}{\sqrt{9+x^2}}\,dx = 2\ln\left(\left|\frac{\sqrt{9+x^2}}{3} + \frac{x}{3}\right|\right) + C = 2\ln\left(\left|\frac{x+\sqrt{9+x^2}}{3}\right|\right) + C.$$

Using the logarithm law $\ln(u/v) = \ln(u) - \ln(v)$, we can simplify this equation to

$$\int \frac{2}{\sqrt{9+x^2}}\,dx = 2\ln\left(\left|x+\sqrt{9+x^2}\right|\right) - 2\ln(3) + C = 2\ln\left(\left|x+\sqrt{9+x^2}\right|\right) + \widetilde{C}$$

where we have written the (arbitrary) constant $C - 2\ln(3)$ as $\widetilde{C}$.

It is possible to calculate the second integral by making the same substitution. However, we notice that the derivative $2x$ of $9+x^2$ is present in the integral. The direct substitution $u = 9+x^2$, $du = 2x\,dx$ is therefore more efficient. With this substitution, the integral $\int 2x/\sqrt{9+x^2}\,dx$ becomes $\int u^{-1/2}\,du = 2\sqrt{u} + C$. Therefore,

$$\int \frac{2x}{\sqrt{9+x^2}}\,dx = 2\sqrt{9+x^2} + C.$$

■

A LOOK BACK

In the Method of Substitution (Section 5.5), we select an expression that is already present in a given integral and replace that expression with a new variable. For example, we treat the integral

$$\int \frac{2x}{\sqrt{9+x^2}}\,dx$$

by making the (direct) substitution $u = 9 + x^2$ for the expression $9 + x^2$ that is already in the integral. In the Method of Inverse Substitution, we replace the original variable of integration with a function that is nowhere to be found in the original integral. For example, we treat the integral

$$\int \frac{2}{\sqrt{9+x^2}}\,dx$$

by making the substitution $x = 3\tan(\theta)$, even though the term $\tan(\theta)$ does not have any apparent connection to the integrand. In theory, the two methods of substitution are distinct. *In practice, however, the two methods are carried out in exactly the same manner*.

General Quadratic Expressions That Appear under a Radical

Since a general quadratic expression $Ax^2 + Bx + C$ can be converted (by completing the square) to a sum or difference of squares, the trigonometric substitutions we have just learned can be applied to integrals that involve these general quadratic expressions. The following two examples show how this works in practice.

Example 4 Calculate the integral

$$\int \frac{dt}{\sqrt{t^2 - 8t + 25}}.$$

Solution The first step is to complete the square under the radical sign:

$$t^2 - 8t + 25 = \left(t^2 - 8t + \left(\frac{8}{2}\right)^2\right) + 25 - \left(\frac{8}{2}\right)^2 = (t-4)^2 + 9.$$

To evaluate the integral

$$\int \frac{dt}{\sqrt{(t-4)^2 + 9}},$$

we first make the direct substitution $x = t - 4$, $dx = dt$, which results in the integral

$$\int \frac{dx}{\sqrt{x^2 + 9}}.$$

The substitution $x = 3\tan(\theta)$, $dx = 3\sec^2(\theta)\, d\theta$ is appropriate. In fact, in Example 3, we calculated this integral to be $\ln(|x + \sqrt{9 + x^2}|) + C$. Resubstituting $x = t - 4$, we obtain

$$\begin{aligned}\int \frac{dt}{\sqrt{t^2 - 8t + 25}} &= \ln\left(\left|t - 4 + \sqrt{(t-4)^2 + 9}\right|\right) + C \\ &= \ln\left(\left|t - 4 + \sqrt{t^2 - 8t + 25}\right|\right) + C.\end{aligned}$$

■

Example 5 Calculate $\int \sqrt{21 + 8x - 4x^2}\, dx$.

Solution First we complete the square:

$$\begin{aligned}21 + 8x - 4x^2 &= 21 - 4(x^2 - 2x) = 21 + 4 - 4(x^2 - 2x + 1) \\ &= 25 - 4(x-1)^2 = 4\left(\left(\frac{5}{2}\right)^2 - (x-1)^2\right).\end{aligned}$$

Therefore,

$$\int \sqrt{21 + 8x - 4x^2}\, dx = 2\int \sqrt{\left(\frac{5}{2}\right)^2 - (x-1)^2}\, dx.$$

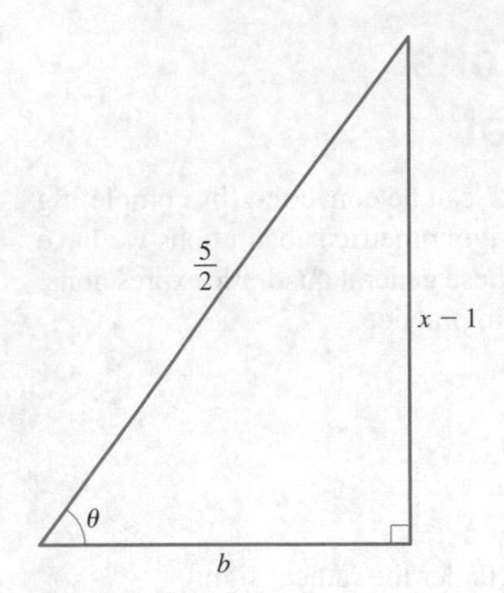

Figure 3
$(x-1)/(5/2) = \sin(\theta) \Rightarrow$
$\cos(\theta) = b/(5/2) = \frac{\sqrt{(5/2)^2-(x-1)^2}}{5/2}$

Next we make the substitution $x - 1 = (5/2)\sin(\theta)$, $dx = (5/2)\cos(\theta)\,d\theta$. The last integral becomes

$$2\int\sqrt{\left(\frac{5}{2}\right)^2 - \left(\frac{5}{2}\right)^2 \sin^2(\theta)}\,\frac{5}{2}\cos(\theta)\,d\theta = 2\cdot\frac{5}{2}\cdot\frac{5}{2}\int\sqrt{1-\sin^2(\theta)}\cos(\theta)\,d\theta$$
$$= \frac{25}{2}\int \cos^2(\theta)\,d\theta,$$

which equals

$$\frac{25}{4}(\theta + \sin(\theta)\cos(\theta)) + C$$

by equation (7.20). The triangle associated with the change of variable,

$$\frac{x-1}{5/2} = \sin(\theta),$$

is shown in Figure 3. By calculating the length of the adjacent side, we find that $\cos(\theta) = \sqrt{(5/2)^2 - (x-1)^2}/(5/2)$. Therefore,

$$\int\sqrt{21+8x-4x^2}\,dx = \frac{25}{4}\left(\arcsin\left(\frac{x-1}{5/2}\right) + \frac{(x-1)}{(5/2)}\cdot\frac{\sqrt{(5/2)^2-(x-1)^2}}{(5/2)}\right) + C.$$

■

Quadratic Expressions Not under a Radical Sign

Suppose $P(x)$ is a polynomial. To calculate

$$\int \frac{P(x)}{Ax^2+Bx+C}\,dx,$$

it is best not to attempt trigonometric substitutions immediately. Instead, follow these three steps:

1. If the degree of P is less than or equal to 1, then proceed directly to step 2. Otherwise, *divide* the denominator into the numerator, which will result in

$$Q(x) + \frac{R(x)}{Ax^2+Bx+C}$$

where $Q(x)$ is a polynomial and $R(x)$ has degree less than or equal to 1. The polynomial $Q(x)$ is easy to integrate. Continue with step 2 applied to the new fraction $R(x)/(Ax^2+Bx+C)$.

2. If the degree of $P(x)$ is 0, or if you have divided in step 1 and the degree of $R(x)$ is 0, then proceed to step 3. If $P(x)$ or $R(x)$ has degree 1, then separate the

Quadratic Expressions Not under a Radical Sign (continued)

numerator into a multiple of $(2Ax + B)$ plus a constant K. Integrate the expression

$$\frac{2Ax + B}{Ax^2 + Bx + C}$$

by making the substitution $u = Ax^2 + Bx + C$, $du = (2Ax + B)\,dx$.

3. Integrate the expression

$$\frac{K}{Ax^2 + Bx + C}$$

by completing the square in the denominator.

The only way to understand these steps is to see them used in practice.

Example 6 Integrate

$$\int \frac{2x^3 - 8x^2 + 20x - 5}{x^2 - 4x + 8}\,dx.$$

Solution The degree of the numerator is not 0 or 1, so we start with step 1. First we divide:

$$\begin{array}{r} 2x \\ x^2 - 4x + 8\;\overline{)\,2x^3 - 8x^2 + 20x - 5} \\ -(2x^3 - 8x^2 + 16x) \\ \hline 4x - 5 \end{array}$$

After division, we obtain

$$\frac{2x^3 - 8x^2 + 20x - 5}{x^2 - 4x + 8} = 2x + \frac{4x - 5}{x^2 - 4x + 8}.$$

Therefore,

$$\int \frac{2x^3 - 8x^2 + 20x - 5}{x^2 - 4x + 8}\,dx = x^2 + \int \frac{4x - 5}{x^2 - 4x + 8}\,dx.$$

We rewrite the integral on the right as

$$\int \frac{2(2x - 4) + 3}{x^2 - 4x + 8}\,dx = 2 \cdot \int \frac{(2x - 4)}{x^2 - 4x + 8}\,dx + \int \frac{3}{x^2 - 4x + 8}\,dx. \qquad \textbf{(7.28)}$$

We can now see the motivation for this step: With the u-substitution $u = x^2 - 4x + 8$, $du = 2x - 4\,dx$, the first integral on the right side of equation (7.28) is

$$2\int \frac{du}{u} = 2 \cdot \ln|u| + C = 2\ln(|x^2 - 4x + 8|) + C. \qquad \textbf{(7.29)}$$

According to step 3, we write the second integral on the right side of equation (7.28) as

$$\int \frac{3}{x^2-4x+8}\,dx = \int \frac{3}{(x-2)^2+4}\,dx$$

by completing the square. We next make the conventional u-substitution $u=x-2$, $du = dx$ to obtain

$$\int \frac{3}{u^2+4}\,du \overset{(6.59)}{=} \frac{3}{2}\arctan\left(\frac{u}{2}\right)+C.$$

We then resubstitute x:

$$\int \frac{3}{x^2-4x+8}\,dx = \frac{3}{2}\arctan\left(\frac{x-2}{2}\right)+C. \qquad \textbf{(7.30)}$$

Substituting the information from lines (7.29) and (7.30) into equation (7.28) gives

$$\int \frac{4x-5}{x^2-4x+8}\,dx = 2\ln(|x^2-4x+8|) + \frac{3}{2}\arctan\left(\frac{x-2}{2}\right)+C.$$

Therefore,

$$\int \frac{2x^3-8x^2+20x-5}{x^2-4x+8}\,dx = x^2 + 2\ln(|x^2-4x+8|) + \frac{3}{2}\arctan\left(\frac{x-2}{2}\right)+C.$$

■

quickquiz

1. What indirect substitution is appropriate for $\int 1/\sqrt{1-4x^2}\,dx$?
2. What indirect substitution is appropriate for $\int 1/\sqrt{10+2x+x^2}\,dx$?
3. What indirect substitution is appropriate for $\int 1/\sqrt{x^2-4}\,dx$?
4. Evaluate $\int (2+x^2)/(1+x^2)\,dx$.

EXERCISES

Problems for Practice

In Exercises 1–4, evaluate the integral.

1. $\int \frac{x}{1-x^2}\,dx$ **2.** $\int \frac{x}{\sqrt{1-x^2}}\,dx$

3. $\int \frac{4}{\sqrt{4-x^2}}\,dx$ **4.** $\int \frac{\sqrt{2}}{\sqrt{1-4x^2}}\,dx$

In Exercises 5–8, evaluate the integral.

5. $\int \frac{x}{x^2+1}\,dx$ **6.** $\int \frac{x+1}{x^2+2x}\,dx$

7. $\int \frac{2}{9x^2+1}\,dx$ **8.** $\int \frac{\sqrt{2}}{x^2+4}\,dx$

In Exercises 9–12, evaluate the integral.

9. $\int_2^3 \frac{1}{x\sqrt{x^2-1}}\,dx$ **10.** $\int_{-3}^{-2} \frac{1}{x\sqrt{x^2-1}}\,dx$

11. $\int_{\sqrt{5}}^{2\sqrt{2}} \frac{1}{x\sqrt{x^2-4}}\,dx$ **12.** $\int \frac{1}{|x|\sqrt{4x^2-1}}\,dx$

In Exercises 13–16, evaluate the integral.

13. $\int \frac{2x+3}{\sqrt{1-x^2}}\,dx$ **14.** $\int \frac{2x+3}{x^2+1}\,dx$

15. $\int \frac{x^2+2}{x^2+1}\,dx$ **16.** $\int_2^3 \frac{x+1}{x\sqrt{x^2-1}}\,dx$

In Exercises 17–20, evaluate the integral.

17. $\int \frac{x^2}{\sqrt{1-x^2}}\,dx$ **18.** $\int \sqrt{9-x^2}\,dx$

19. $\int \frac{x^2}{(1-x^2)^{3/2}}\,dx$ **20.** $\int (4-x^2)^{3/2}\,dx$

In Exercises 21–24, evaluate the integral. Recall that the integral of $\sec^3(x)$ was computed in Example 3 from Section 7.3.

21. $\int \frac{1}{(1+x^2)^{3/2}}\,dx$ **22.** $\int \frac{1}{\sqrt{1+x^2}}\,dx$

23. $\int \frac{x^2}{\sqrt{1+x^2}}\,dx$ **24.** $\int \sqrt{1+x^2}\,dx$

In Exercises 25–28, evaluate the integral.

25. $\int \frac{x}{\sqrt{x^2-1}}\,dx$ **26.** $\int \frac{1}{\sqrt{x^2-1}}\,dx$

27. $\int \frac{1}{(x^2-1)^{3/2}}\,dx$ **28.** $\int \sqrt{x^2-1}\,dx$

In Exercises 29–33, evaluate the integral.

29. $\int_0^2 \frac{1}{7x^2+8}\,dx$ **30.** $\int_0^1 \sqrt{25-16x^2}\,dx$

31. $\int_0^1 \frac{1}{\sqrt{16+9x^2}}\,dx$ **32.** $\int_1^{\sqrt{20}} \frac{1}{\sqrt{25x^2-16}}\,dx$

33. $\int_0^1 \frac{x^2}{\sqrt{9-4x^2}}\,dx$

In Exercises 34–37, evaluate the integral. Integrate by parts as the first step.

34. $\int_1^{\sqrt{2}} \operatorname{arcsec}(x)\,dx$ **35.** $\int_0^{1/\sqrt{2}} x\arcsin(x)\,dx$

36. $\int_0^{1/2} x^2\arcsin(x)\,dx$ **37.** $\int_1^{\sqrt{2}} x^2\operatorname{arcsec}(x)\,dx$

38. Calculate the area under $y = 1-x^2$ and over $y = (x+1)\sqrt{1-x^2}$ for $0 \le x \le 1$.

39. Calculate the area under the graph of $y = (2-x)\sqrt{x^2-1}$, $1 \le x \le 2$.

40. Calculate the area under the graph of $y = x^2\sqrt{1-x^2}$, $0 \le x \le 1$.

Further Theory and Practice

In Exercises 41–46, evaluate the integral.

41. $\int \frac{1}{(x^2+1)^2}\,dx$ **42.** $\int \frac{x^2-1}{(x^2+1)^2}\,dx$

43. $\int \frac{\sqrt{1+x^2}}{x}\,dx$ **44.** $\int \frac{x^2}{\sqrt{x^2-1}}\,dx$

45. $\int \frac{\sqrt{1+x^2}}{x^2}\,dx$ **46.** $\int \frac{x}{\sqrt{x^4+1}}\,dx$

In Exercises 47–62, evaluate the integral.

47. $\int \frac{1}{\sqrt{2x-x^2}}\,dx$ **48.** $\int \sqrt{x^2+2x+2}\,dx$

49. $\int \frac{x}{\sqrt{x^2+2x+2}}\,dx$ **50.** $\int \sqrt{x^2-x-1}\,dx$

51. $\int \frac{1}{(x^2-2x+2)^2}\,dx$ **52.** $\int \sqrt{x^2-2x-3}\,dx$

53. $\int \frac{x}{x^2-14x+58}\,dx$ **54.** $\int \frac{x^2}{x^2+x+1}\,dx$

55. $\int \frac{x^2-6x+8}{x^2+4}\,dx$

56. $\int \frac{2x^3+13x^2+16x-60}{x^2+8x+20}\,dx$

57. $\int \frac{3x}{2x^2+6x+5}\,dx$ **58.** $\int \frac{5x^2+20x-6}{x^2+4x+6}\,dx$

59. $\int \frac{x}{(x^2+2x+2)^3}\,dx$ **60.** $\int \frac{x}{\sqrt{x^2+4x-5}}\,dx$

61. $\int \frac{(x+1)}{\sqrt{10x-x^2}}\,dx$ **62.** $\int \frac{2x}{\sqrt{16x-x^2-48}}\,dx$

A nonnegative function f that satisfies $\int_a^b f(x)\,dx = 1$ is said to be a *probability density function* on the interval $[a, b]$. If a nonnegative function g satisfies

$$c = \int_a^b g(x)\,dx > 0,$$

then $f(x) = g(x)/c$ is a probability density function. One important two-parameter family of probability density functions on the interval $[0, 1]$ arises by taking $g(x) = x^{\alpha-1}(1-x)^{\beta-1}$ for α and β positive. The associated probability density functions are called *beta densities*. In Exercises 63 and 64, make an indirect trigonometric substitution to find the value of c that corresponds to the values of α and β.

63. $\alpha = 2, \beta = 5/2$ **64.** $\alpha = 3/2, \beta = 3$

In each of Exercises 65–68, divide before integrating.

65. $\int \frac{1+x^2}{4+x^2}\,dx$ **66.** $\int \frac{2x^3+x^2}{9+x^2}\,dx$

67. $\int \frac{1+x^2}{5+4x+x^2}\,dx$ **68.** $\int \frac{x^3+1}{32-8x+x^2}\,dx$

In Exercises 69–74, make an indirect substitution $x = \phi(u)$ to evaluate the given integral.

69. $\int \frac{1}{1+\sqrt{x}}\,dx$ **70.** $\int \sqrt{1+\sqrt{x}}\,dx$

71. $\int \frac{1-\sqrt{x}}{1+\sqrt{x}}\,dx$ **72.** $\int \frac{x^{1/4}+1}{x^{1/2}+1}\,dx$

73. $\int \frac{\sqrt{1+x}}{x}\,dx$ **74.** $\int \frac{1}{1+x^{1/3}}\,dx$

75. Show that the substitution $x = a\sin(\theta)$ leads to the formula

$$\int \frac{dx}{a^2 - x^2} = \frac{1}{a}\ln\left(\frac{x+a}{\sqrt{a^2-x^2}}\right).$$

Show that the substitution $x = a\sec(\theta)$ leads to the formula

$$\int \frac{dx}{a^2 - x^2} = \frac{1}{a}\ln\left(\frac{x+a}{\sqrt{x^2-a^2}}\right).$$

Evaluate

$$\int_0^{1/2} \frac{dx}{1-x^2}$$

and

$$\int_{\sqrt{2}}^{2} \frac{dx}{1-x^2}.$$

Also, calculate both definite integrals using the Method of Partial Fractions, as discussed in Section 7.2.

Calculator/Computer Exercises

76. Calculate the two definite integrals from Exercise 75 using formula (6.67).

77. Plot $y = 2x^2/\sqrt{4-x^2}$ and $y = x$ for $0 \le x \le 1$. Find the area enclosed by the two curves.

78. Find the area of the region that is bounded above by $y = (1+x^2)^{-3/2}$, below by $y = x^2/\sqrt{1-x^2}$, and on the left by the y-axis.

79. Plot $y = 3 - x/\sqrt{1-x^2}$ and $y = \sqrt{1-x^2}/x$ for $0.32 \le x \le 0.96$. Find the points of intersection of the two graphs. Then find the area of the region enclosed by the two graphs.

80. Plot $y = 2x - x^2$ and $y = x^2/(2+2x+x^2)$ for $0 \le x \le 2$. Find the area enclosed by the two curves.

81. Plot $y = (x^3 - 2x + 1)/(x^2 - 2x + 2)$ for $0.6 \le x \le 1$. Find the x-intercept $a < 1$. Calculate the area enclosed by this curve and the x-axis.

7.5 Partial Fractions—Irreducible Quadratic Factors

In Section 7.2, we learned how to evaluate integrals of the form

$$\int \frac{p(x)}{(x-a_1)^{m_1}(x-a_2)^{m_2}\cdots(x-a_K)^{m_K}}\,dx \tag{7.31}$$

where $p(x)$ is a polynomial and $a_1, a_2, \ldots, a_K$ are real.

To evaluate this type of integral, we decompose the integrand into a sum of simpler expressions using the Method of Partial Fractions. The purpose of the present section is to extend that method to denominators that include irreducible quadratic factors. Since the Fundamental Theorem of Algebra tells us that any polynomial can be split into a product of linear and irreducible quadratic factors, our study of the Method of Partial Fractions will then be complete. As a result, we will be able to integrate *any* rational function $p(x)/q(x)$.

Rational Functions with Quadratic Terms in the Denominator

In evaluating integrals of the form (7.31), we made use of two simple algebraic expressions: the simple linear building block $A/(x-a)$ and the repeated linear building block $A/(x-a)^m$. To find a partial fractions decomposition of the rational function $p(x)/q(x)$ when $q(x)$ has one or more irreducible quadratic factors, we need two new building blocks.

Simple Quadratic Building Blocks

$$\frac{Bx + C}{x^2 + bx + c} \tag{7.32}$$

where we assume that the denominator $x^2 + bx + c$ cannot be factored; otherwise, we would be in the situation of the linear building blocks studied in Section 7.2. (Recall that we learned how to integrate expressions of the form (7.32) in Section 7.4.)

Repeated Quadratic Building Blocks

$$\frac{Bx + C}{(x^2 + bx + c)^n}$$

where n is an integer greater than 1. We studied integrands of this form in Section 7.4.

We can now state our completed partial fractions rule. You will find that this rule simply builds upon steps with which you are already familiar. The presence of irreducible quadratic terms in the denominator does not change the way we treat the linear terms.

The Method of Partial Fractions

To integrate a rational function of the form

$$\frac{p(x)}{(x - a_1)^{m_1} \cdots (x - a_K)^{m_K} \cdot (x^2 + b_1 x + c_1)^{n_1} \cdots (x^2 + b_L x + c_L)^{n_L}},$$

perform the following four steps.

1. Be sure that the degree of p is less than the degree of the denominator; if it is not, *divide the denominator into the numerator.*
2. Factor the denominator into linear and irreducible quadratic factors. *Be sure that the quadratic factors $x^2 + b_j x + c_j$ cannot be factored into linear factors with real coefficients.* (Do this by verifying that $b_j^2 - 4c_j < 0$; if $b_j^2 - 4c_j \geq 0$ for any quadratic term $x^2 + b_j x + c_j$, then that expression must be factored into a product of two linear terms.)
3. For each of the factors $(x - a_j)^{m_j}$ in the denominator of the rational function being considered, the partial fractions decomposition must contain terms

The Method of Partial Fractions (continued)

of the form

$$\frac{A_1}{(x-a_j)^1} + \frac{A_2}{(x-a_j)^2} + \cdots + \frac{A_{m_j}}{(x-a_j)^{m_j}}.$$

4. For each of the factors $(x^2 + b_j x + c_j)^{n_j}$ in the denominator of the integrand being considered, the partial fractions decomposition must contain terms of the form

$$\frac{B_1 x + C_1}{x^2 + b_j x + c_j} + \frac{B_2 x + C_2}{(x^2 + b_j x + c_j)^2} + \cdots + \frac{B_{n_j} x + C_{n_j}}{(x^2 + b_j x + c_j)^{n_j}}.$$

The summary at the end of this section provides several illustrative partial fractions expansions, which will give you a more concrete idea of how to set up decompositions into partial fractions (without the distraction of determining the unknown coefficients or performing subsequent integrations). It is a good idea to study the summary at this time.

Checking for Irreducibility

It is very important to begin with the right form of a partial fractions decomposition before attempting to solve for the unknown coefficients. In most cases, it is impossible to solve for the unknown coefficients when the partial fractions decomposition has not been set up correctly.

Example 1 Find the correct form of the partial fractions decomposition for the rational expression

$$\frac{3x^2 + 8x + 1}{(x+1)(x^2 + 7x + 10)}.$$

Solution Since the degree of the numerator is 2 and the degree of the denominator is 3, a preliminary division is not necessary. Notice that the correct partial fractions decomposition is *not*

$$\frac{3x^2 + 8x + 1}{(x+1)(x^2 + 7x + 10)} = \frac{A}{x+1} + \frac{Bx + C}{x^2 + 7x + 10}.$$

The reason is that we have not factored the denominator as far as possible. The quadratic polynomial $x^2 + 7x + 10$ factors as

$$x^2 + 7x + 10 = (x+2)(x+5).$$

Thus, the given rational function really falls under the case of distinct linear factors, which we studied in Section 7.2. The correct partial fractions decomposition is

$$\frac{A}{x+1} + \frac{B}{x+2} + \frac{C}{x+5}.$$

You may want to practice the two methods of determining the unknown coefficients studied in Section 7.2 to find that $A = -1$, $B = 1$, and $C = 3$. ■

inSIGHT

How can we tell in advance whether a quadratic factor $x^2 + bx + c$ is irreducible? The quadratic formula tells us. If the discriminant $b^2 - 4c$ is negative, then the polynomial is irreducible; otherwise, it can be factored into linear terms. A quick calculation of the discriminant of $x^2 + 7x + 10$ from Example 1 results in the positive number $7^2 - 4 \cdot 10$, which alerts us to the reducibility of the quadratic factor.

It is important to realize that the Fundamental Theorem of Algebra guarantees that any real polynomial can be factored into linear and quadratic factors. In the theory of partial fractions, we never have to use building blocks that involve irreducible cubic, quartic, or higher order factors because they do not exist.

Example 2 Find the correct form of the partial fractions decomposition of $3/(x^3+1)$.

Solution This expression may appear to have an irreducible denominator, but it does not. Notice that -1 is a root of the polynomial x^3+1. Therefore, $(x+1)$ divides the polynomial. Thus, we have

$$x^3+1=(x+1)(x^2-x+1).$$

The quadratic polynomial has a negative discriminant, so it is irreducible. Thus, we see that the correct partial fractions decomposition for our rational expression is

$$\frac{3}{x^3+1}=\frac{A}{x+1}+\frac{Bx+C}{x^2-x+1}. \tag{7.33}$$

■

Calculating the Coefficients of a Partial Fractions Decomposition

Now that we have considered the forms that decompositions into partial fractions can take, let us see how the unknown coefficients can be computed in practice.

Example 3 Find the partial fractions decomposition of the rational function $3/(x^3+1)$.

Solution In Example 2, we determined the correct form of the partial fractions decomposition of $3/(x^3+1)$. (See equation (7.33).) The right side of equation (7.33) can be rewritten as

$$\frac{A(x^2-x+1)+(Bx+C)(x+1)}{(x+1)(x^2-x+1)},$$

or, after expanding both the numerator and denominator,

$$\frac{(A+B)x^2+(-A+B+C)x+(A+C)}{x^3+1}.$$

The numerator of this expression must equal the numerator of the left side of equation (7.33):

$$0x^2+0x+3=(A+B)x^2+(-A+B+C)x+(A+C).$$

Equating the coefficients of like powers of x on each side of this equation leads to the system

$$A+B=0, \quad -A+B+C=0, \quad A+C=3.$$

The first equation allows us to replace B with $-A$ in the second equation to obtain $-2A+C=0$. Thus, $C=2A$. The third equation then tells us that $3A=3$, or $A=1$; hence, $B=-1$ and $C=2$. In conclusion,

$$\frac{3}{x^3+1}=\frac{1}{x+1}+\frac{-x+2}{x^2-x+1}.$$

■

Example 4 Calculate

$$\int \frac{x^3 - x^2 - 3x + 3}{(x^2 + 4x + 5)(x^2 + 9)}\, dx.$$

Solution The degree of the denominator is 4, which is 1 greater than the degree of the numerator. Division is therefore not necessary. Because the discriminants of $x^2 + 4x + 5$ and $x^2 + 9$ are the negative numbers $(4)^2 - (4)(5) = -4$ and $0^2 - (4)(9) = -36$, respectively, we can be sure that further factoring of the denominator is unnecessary. We may therefore proceed directly to the partial fractions decomposition:

$$\frac{x^3 - x^2 - 3x + 3}{(x^2 + 4x + 5)(x^2 + 9)} = \frac{Ax + B}{x^2 + 4x + 5} + \frac{Cx + D}{x^2 + 9}.$$

We combine the two summands on the right to obtain

$$\begin{aligned} \frac{x^3 - x^2 - 3x + 3}{(x^2 + 4x + 5)(x^2 + 9)} &= \frac{(Ax + B)(x^2 + 9) + (Cx + D)(x^2 + 4x + 5)}{(x^2 + 4x + 5)(x^2 + 9)} \\ &= \frac{(A + C)x^3 + (B + 4C + D)x^2 + (9A + 5C + 4D)x + 9B + 5D}{(x^2 + 4x + 5)(x^2 + 9)}. \end{aligned}$$

We may now equate the numerators on the left with the numerator on the right:

$$x^3 - x^2 - 3x + 3 = (A + C)x^3 + (B + 4C + D)x^2 + (9A + 5C + 4D)x + 9B + 5D.$$

By equating coefficients of like powers of x, we obtain the four equations

$$A + C = 1, \quad B + 4C + D = -1, \quad 9A + 5C + 4D = -3, \quad 9B + 5D = 3.$$

We may use the first and second of these equations to solve for A and B in terms of C and D: $A = 1 - C$ and $B = -1 - 4C - D$. By eliminating A and B in the third and fourth equations, we obtain two equations in the two unknowns C and D:

$$9(1 - C) + 5C + 4D = -3, \quad \text{or} \quad -4C + 4D = -12,$$

and

$$9(-1 - 4C - D) + 5D = 3, \quad \text{or} \quad -36C - 4D = 12.$$

We solve these last two equations in C and D, obtaining $C = 0$ and $D = -3$. It follows that $A = 1 - C = 1 - 0 = 1$ and $B = -1 - 4C - D = -1 - 4(0) - (-3) = 2$. Thus,

$$\frac{x^3 - x^2 - 3x + 3}{(x^2 + 4x + 5)(x^2 + 9)} = \frac{x + 2}{x^2 + 4x + 5} - \frac{3}{x^2 + 9}$$

and

$$\int \frac{x^3 - x^2 - 3x + 3}{(x^2 + 4x + 5)(x^2 + 9)}\, dx = \int \frac{x + 2}{x^2 + 4x + 5}\, dx - \int \frac{3}{x^2 + 9}\, dx. \qquad \textbf{(7.34)}$$

We complete the square $x^2 + 4x + 5 = (x+2)^2 + 1$ and make the u-substitution $u = x + 2, du = dx$ to convert the first integral on the right side of equation (7.34) to

$$\int \frac{u}{u^2+1}\,du.$$

By means of the substitution $w = u^2 + 1,\ dw = 2u\,du$, this last integral is converted to

$$\frac{1}{2}\int \frac{1}{w}\,dw = \frac{1}{2}\ln(|w|) + C.$$

We resubstitute, observing that $w = u^2 + 1 = x^2 + 4x + 5$ is positive, to obtain

$$\int \frac{x+2}{x^2+4x+5}\,dx = \frac{1}{2}\ln(x^2+4x+5) + C.$$

Finally, we use formula (6.59) to evaluate the right-most integral of equation (7.34):

$$\int \frac{x^3 - x^2 - 3x + 3}{(x^2+4x+5)(x^2+9)}\,dx = \frac{1}{2}\ln(x^2+4x+5) - \arctan\left(\frac{x}{3}\right) + C. \quad \blacksquare$$

Example 5 Calculate the integral

$$\int \frac{x-1}{x(x^2+1)^2}\,dx.$$

Solution The degree of the numerator is less than that of the denominator, so we proceed directly with the partial fractions decomposition:

$$\frac{x-1}{x(x^2+1)^2} = \frac{A}{x} + \frac{Bx+C}{x^2+1} + \frac{Dx+E}{(x^2+1)^2}.$$

Putting the right side over a common denominator leads to

$$\begin{aligned} x - 1 &= A(x^2+1)^2 + (Bx+C)(x^2+1)x + (Dx+E)x \\ &= A(x^4+2x^2+1) + (Bx+C)(x^3+x) + Dx^2 + Ex \\ &= x^4(A+B) + x^3(C) + x^2(2A+B+D) + x(C+E) + A. \end{aligned}$$

Equating like coefficients of x leads to the system

$$A + B = 0, \quad C = 0, \quad 2A + B + D = 0, \quad C + E = 1, \quad A = -1.$$

We conclude that $A = -1$, $B = 1$, $C = 0$, $D = 1$, and $E = 1$. Therefore,

$$\begin{aligned} \int \frac{x-1}{x(x^2+1)^2}\,dx &= \int \frac{-1}{x}\,dx + \int \frac{x}{x^2+1}\,dx + \int \frac{x+1}{(x^2+1)^2}\,dx \\ &= -\ln(|x|) + \frac{1}{2}\int \frac{2x}{x^2+1}\,dx + \frac{1}{2}\int \frac{2x}{(x^2+1)^2}\,dx + \int \frac{1}{(x^2+1)^2}\,dx. \end{aligned}$$

The first and second integrals are handled with the u-substitution $u = x^2 + 1$, $du = 2x\,dx$. We treat the last with the substitution $x = \tan(\theta)$, $dx = \sec^2(\theta)\,d\theta$ (see

Section 7.4). This results in the integral

$$\int \frac{\sec^2(\theta)}{(\tan^2(\theta)+1)^2}\,d\theta = \int \frac{\sec^2(\theta)}{\sec^4(\theta)}\,d\theta = \int \cos^2(\theta)\,d\theta \overset{(7.20)}{=} \frac{1}{2}\theta + \frac{1}{2}\sin(\theta)\cos(\theta) + C.$$

After resubstituting for x, we have

$$\int \frac{1}{(x^2+1)^2}\,dx = \frac{1}{2}\arctan(x) + \frac{1}{2}\frac{x}{x^2+1} + C.$$

Assembling all pieces, we conclude that our original integral is given by

$$\int \frac{x-1}{x(x^2+1)^2}\,dx = -\ln(|x|) + \frac{1}{2}\ln(x^2+1) - \frac{1}{2}\cdot\frac{1}{x^2+1} + \frac{x}{2(1+x^2)} + \frac{1}{2}\arctan(x) + C.$$

■

Summary of Basic Partial Fractions Forms

We conclude by summarizing the techniques of this section with several quick instances of the different partial fractions forms that can arise:

$$\frac{3}{(x-4)(x+7)} = \frac{A}{x-4} + \frac{B}{x+7}$$

$$\frac{1}{(x-5)(x^2+6)} = \frac{A}{x-5} + \frac{Cx+D}{x^2+6}$$

$$\frac{x+3}{(x-2)^2(x+1)} = \frac{A}{x-2} + \frac{B}{(x-2)^2} + \frac{C}{x+1}$$

$$\frac{x}{(x+8)^3(x-1)^2} = \frac{A}{x+8} + \frac{B}{(x+8)^2} + \frac{C}{(x+8)^3} + \frac{D}{x-1} + \frac{E}{(x-1)^2}$$

$$\frac{1}{(x^2+x+1)^2(x-1)^2} = \frac{Ax+B}{x^2+x+1} + \frac{Cx+D}{(x^2+x+1)^2} + \frac{E}{x-1} + \frac{F}{(x-1)^2}$$

When you do the exercises for this section, refer to these examples to determine which form of the partial fractions technique you should use. Notice this unifying theme: The number of unknown coefficients in any partial fractions problem is equal to the degree of the denominator.

quickquiz

1. How do the partial fractions decompositions of $(x+1)/((x-1)x^2)$ and $(x+1)/((x-1)(x^2+1))$ differ in form?
2. How do the partial fractions decompositions of $(x+1)/(x^2-3x-4)^2$ and $(x+1)/(x^2-3x+4)^2$ differ in form?
3. What is the form of the partial fractions expansion of $1/((x^2+4)(x+1))$?
4. What is the form of the partial fractions expansion of $1/((x^2+4)^2(x+1))$?

EXERCISES

Problems for Practice

In Exercises 1–10, write the *form* of the partial fractions decomposition of the rational function. *Do not explicitly calculate the coefficients.*

1. $\dfrac{2x^3+x+1}{(x^2+1)(x^2+4)}$
2. $\dfrac{2x^4}{(x^2+2x+2)(2x^2+5x+3)}$
3. $\dfrac{2x^3+x+1}{(x^2+1)(x^2+x+1)^2}$
4. $\dfrac{2x^4}{(x^2+1)(17x^2+4x+1)}$
5. $\dfrac{2x+1}{(x^2+x+3)(x-4)}$
6. $\dfrac{3x^3+x+1}{x^2(x+1)^2}$
7. $\dfrac{2x^6}{(x^2+4)^3(x-2)}$
8. $\dfrac{3x^5+x+1}{x^2(x^2-1)^2}$
9. $\dfrac{3x^5+x+1}{x^2(x^2-1)(x^2+1)}$
10. $\dfrac{7x^8}{(x-3)^3(x^2+6x+10)^3}$

In Exercises 11–20, explicitly calculate the partial fractions decomposition of the rational function.

11. $\dfrac{3x^2-5x+4}{(x-1)(x^2+1)}$
12. $\dfrac{x^2+2}{(x^2+1)x^2}$
13. $\dfrac{7x^3+9x-3x^2-6}{(x^2+2)(x^2+1)}$
14. $\dfrac{x^2+2x}{(x^2+1)^2}$
15. $\dfrac{x^3-x}{(x^2+1)^2}$
16. $\dfrac{2x^2}{(x+1)^2(x^2+1)}$
17. $\dfrac{x^3+12x^2-9x+48}{(x-3)(x^2+4)}$
18. $\dfrac{2x^2+4x+2}{(x^2+1)^3}$
19. $\dfrac{3x^3-5x^2+10x-19}{(x^2+4)(x^2+3)}$
20. $\dfrac{2x^4+15x^2+30}{(x^2+4)(x^2+3)^2}$

In Exercises 21–30, use the Method of Partial Fractions to decompose the integrand. Then evaluate the given integral.

21. $\displaystyle\int \frac{3x^2-5x+4}{(x-1)(x^2+1)}\,dx$
22. $\displaystyle\int \frac{x^2+2}{(x^2+1)x^2}\,dx$
23. $\displaystyle\int \frac{7x^3+9x-3x^2-6}{(x^2+2)(x^2+1)}\,dx$
24. $\displaystyle\int \frac{x^2+2x}{(x^2+1)^2}\,dx$
25. $\displaystyle\int \frac{x^3-x}{(x^2+1)^2}\,dx$
26. $\displaystyle\int \frac{2x^2}{(x+1)^2(x^2+1)}\,dx$
27. $\displaystyle\int \frac{x^3+12x^2-9x+48}{(x-3)(x^2+4)}\,dx$
28. $\displaystyle\int \frac{2x^2+4x+2}{(x^2+1)^3}\,dx$
29. $\displaystyle\int \frac{3x^3-5x^2+10x-19}{(x^2+4)(x^2+3)}\,dx$
30. $\displaystyle\int \frac{2x^4+15x^2+30}{(x^2+4)(x^2+3)^2}\,dx$

Further Theory and Practice

31. What happens if you attempt a partial fractions decomposition of

$$\frac{1}{(x^2+3)^4}$$

into

$$\frac{A_1x+B_1}{x^2+3}+\frac{A_2x+B_2}{(x^2+3)^2}+\frac{A_3x+B_3}{(x^2+3)^3}+\frac{A_4x+B_4}{(x^2+3)^4}?$$

Explain.

In Exercises 32–35, use the Method of Partial Fractions to calculate the integral.

32. $\displaystyle\int \frac{2x^2+4x+9}{x^3-1}\,dx$
33. $\displaystyle\int \frac{5x^2-2x+2}{x^3+1}\,dx$
34. $\displaystyle\int \frac{8x}{x^5-x^4-x+1}\,dx$
35. $\displaystyle\int \frac{48}{(x^2+1)^4}\,dx$

36. At normal temperatures, conduction and convection are the primary means of heat transfer. Newton's Law of Heat Change is reasonably accurate under these circumstances. At very high temperatures, radiation is the predominant method of heat transfer. *Stefan's Law of Radiation* states that the temperature T (in degrees Kelvin) of an object placed in an environment at temperature T_0 (in degrees Kelvin) satisfies

$$\frac{dT}{dt}=k\left(T_0^4-T^4\right)$$

for some positive constant k. By separating variables, solve this differential equation. (You will not obtain T explicitly as a function of t. Instead, you will obtain an equation that relates the two variables.)

Calculator/Computer Exercises

37. Plot $y = x^4 + 2$ and $y = (5x^3 + 3x + 2)/(x^3 + x^2 + x + 1)$ for $0 \le x \le 0.8$. Find the abscissa b of the point of intersection. Calculate the area enclosed by the two curves.

38. Plot $y = x^2$ and $y = (3x^3 + 4x)/(x^4 + 3x^2 + 2)$ in the viewing window $[0, 1.1] \times [0, 1.2]$. Find the area of the region enclosed by the two curves.

In Exercises 39 and 40, find all roots of the denominator of the integrand. Find a partial fractions decomposition and evaluate the given integral.

39. $$\int_1^2 \frac{2x+4}{x^5 - 7x^4 + 4x^3 - 23x^2 + 20}\, dx$$

40. $$\int_2^3 \frac{3x+1}{x^5 - 2x^3 - x^2 - 3x + 1}\, dx$$

41. Computer algebra systems can quickly decompose many a fearsome rational function into its partial fractions expansion. Figure 1 shows an example of this, created with Maple. Find the partial fractions decomposition of the following rational functions.

a. $$\frac{2x^5 + 13x^4 - 28x^3 + 30x^2 - 37x - 8}{x^6 - x^5 + 2x^4 - x^3 + 2x^2 - x + 1}$$

b. $$\frac{17x^5 + 39x^4 + 140x^3 + 140x^2 + 199x + 33}{x^6 + 3x^5 + 12x^4 + 19x^3 + 36x^2 + 27x + 27}$$

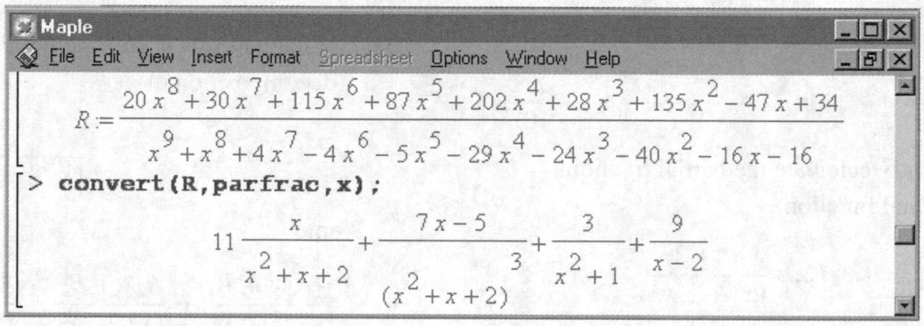

Maple

File Edit View Insert Format Spreadsheet Options Window Help

$$R := \frac{20x^8 + 30x^7 + 115x^6 + 87x^5 + 202x^4 + 28x^3 + 135x^2 - 47x + 34}{x^9 + x^8 + 4x^7 - 4x^6 - 5x^5 - 29x^4 - 24x^3 - 40x^2 - 16x - 16}$$

```
> convert(R,parfrac,x);
```

$$11\frac{x}{x^2+x+2} + \frac{7x-5}{(x^2+x+2)^3} + \frac{3}{x^2+1} + \frac{9}{x-2}$$

Figure 1

Summary of Key Topics

Integration by Parts (Section 7.1)

If u and v are continuously differentiable functions, then

$$\int u\frac{dv}{dx}\, dx = uv - \int v\frac{du}{dx}\, dx,$$

or, more simply,

$$\int u\, dv = uv - \int v\, du.$$

This formula is often used to treat integrals of products of functions.

Powers and Products of Sines and Cosines (Section 7.3)

It is possible to evaluate integrals of the form

$$\int \sin^m(x)\cos^n(x)\, dx$$

for m and n nonnegative integers by using one of three procedures. If m is odd, then $m = 2k+1$ for some nonnegative integer k. We then rewrite the integral as

$$\int \sin^m(x)\cos^n(x)\,dx = \int (\sin^2(x))^k \cos^n(x)\sin(x)\,dx = \int (1-\cos^2(x))^k \cos^n(x)\sin(x)\,dx.$$

The u-substitution $u = \cos(x)$, $du = -\sin(x)\,dx$ can be applied to this last integral. If $n = 2\ell + 1$ is odd, we proceed in a similar manner. We write

$$\int \sin^m(x)\cos^n(x)\,dx = \int (\cos^2(x))^\ell \sin^m(x)\cos(x)\,dx = \int (1-\sin^2(x))^\ell \sin^m(x)\cos(x)\,dx$$

and make the u-substitution $u = \sin(x)$, $du = \cos(x)\,dx$. If both powers are even, then we write the integral as

$$\int \sin^m(x)\cos^{2\ell}(x)\,dx = \int \sin^m(x)(1-\sin^2(x))^\ell\,dx,$$

expand the integrand, and integrate the resulting terms using the reduction formula

$$\int \sin^n(x)\,dx = -\frac{1}{n}\sin^{n-1}(x)\cos(x) + \frac{n-1}{n}\int \sin^{n-2}(x)\,dx.$$

Inverse Trigonometric Functions (Section 7.4)

By restricting the domain of $\sin(\theta)$ to $[-\pi/2, \pi/2]$, the domain of $\tan(\theta)$ to $(-\pi/2, \pi/2)$, and the domain of $\sec(\theta)$ to $[0, \pi/2) \cup (\pi/2, \pi]$, we obtain the inverse trigonometric functions $\arcsin(x)$, $\arctan(x)$, and $\operatorname{arcsec}(x)$. The derivatives of these functions are

$$\frac{d}{dx}\arcsin(x) = \frac{1}{\sqrt{1-x^2}}, \quad \frac{d}{dx}\arctan(x) = \frac{1}{1+x^2}, \quad \text{and} \quad \frac{d}{dx}\operatorname{arcsec}(x) = \frac{1}{|x|\sqrt{x^2-1}}.$$

Integrals Involving Quadratic Expressions (Section 7.4)

If a is a positive constant, then we may treat an integrand involving an expression of the form $a^2 - x^2$, $a^2 + x^2$, and $x^2 - a^2$ by making the indirect substitution indicated in the table.

Expression	**Substitution**	**Result of Substitution**	**After Simplification**
$a^2 - x^2$	$x = a \cdot \sin(\theta)$, $dx = a \cdot \cos(\theta)\,d\theta$	$a^2 \cdot (1 - \sin^2(\theta))$	$a^2 \cdot \cos^2(\theta)$
$a^2 + x^2$	$x = a \cdot \tan(\theta)$, $dx = a \cdot \sec^2(\theta)\,d\theta$	$a^2 \cdot (1 + \tan^2(\theta))$	$a^2 \cdot \sec^2(\theta)$
$x^2 - a^2$	$x = a \cdot \sec(\theta)$, $dx = a \cdot \sec(\theta)\tan(\theta)\,d\theta$	$a^2 \cdot (\sec^2(\theta) - 1)$	$a^2 \cdot \tan^2(\theta)$

To treat an integral that involves the expression $\sqrt{\alpha x^2 + \beta x + \gamma}$, we complete the square under the radical to rewrite

$$\alpha x^2 + \beta x + \gamma = \alpha\left(\left(x + \frac{\beta}{2\alpha}\right)^2 + \left(\frac{\gamma}{\alpha} - \frac{\beta^2}{4\alpha^2}\right)\right)$$

If we let $a = \sqrt{|\gamma/\alpha - \beta^2/4\alpha^2|}$ and make the u-substitution $u = x + \beta/2\alpha$, $du = dx$, then we have $\alpha x^2 + \beta x + \gamma = \alpha \cdot (u^2 \pm a^2)$. We may apply one of the three trigonometric substitutions listed in the table to this last expression.

Partial Fractions (Sections 7.2 and 7.5)

To integrate a rational function of the form

$$\frac{p(x)}{(x-a_1)^{n_1}\cdots(x-a_k)^{n_k}\cdot(\alpha_1x^2+\beta_1x+\gamma_1)^{m_1}\cdots(\alpha_\ell x^2+\beta_\ell x+\gamma_\ell)^{m_\ell}},$$

perform the following three steps:

1. Be sure that the degree of p is less than the degree of the denominator; if it is not, *divide the denominator into the numerator.*
2. Factor the denominator into linear and *irreducible* quadratic factors.
3. For each of the factors $(x-a_j)^{n_j}$ in the denominator of the rational function being considered, the partial fractions decomposition must contain terms of the form

$$\frac{A_1}{(x-a_j)^1}+\frac{A_2}{(x-a_j)^2}+\cdots+\frac{A_{n_j}}{(x-a_j)^{n_j}}.$$

4. For each of the factors $(\alpha_jx^2+\beta_jx+\gamma_j)^{m_j}$ in the denominator of the integrand being considered, the partial fractions decomposition must contain terms of the form

$$\frac{B_1x+C_1}{\alpha_jx^2+\beta_jx+\gamma_j}+\frac{B_2x+C_2}{(\alpha_jx^2+\beta_jx+\gamma_j)^2}+\cdots+\frac{B_{m_j}x+C_{m_j}}{(\alpha_jx^2+\beta_jx+\gamma_j)^{m_j}}.$$

genesis & DEVELOPMENT

Each technique of integration in Chapter 7 was discovered early in the history of calculus. The Integration by Parts Formula, to cite one example, appeared in Leibniz's first development of calculus (1673). By that time, many particular integration formulas were already known. What distinguishes the work of Leibniz and Newton is that these men searched for *general* techniques that would apply to a large class of functions. Before their work, integration formulas were developed only as needed. For example, Kepler discovered the equation

$$\int_0^\theta \sin(\phi)\,d\phi = 1 - \cos(\theta)$$

because it arose in the course of his investigation of the orbit of Mars. Similarly, the formula

$$\int_0^\theta \sec(\phi)\,d\phi = \ln(|\sec(\theta) + \tan(\theta)|) \qquad \textbf{(7.35)}$$

became known because of a problem of maritime navigation.

In 1569, Gerardus Mercator (or Gerhard Kremer, 1512–1594) introduced a world map based on a new projection. Prior to Mercator's projection, nautical maps were unreliable at best. The great advantage of the Mercator map is that a line of constant bearing intersects meridians at constant angles. There is, however, a catch: Navigation by the Mercator map depends on the evaluation of the secant integral. That obstacle was overcome by two English mathematicians who put to sea on separate naval expeditions late in the 1500s.

When the English Navy drew up plans in 1591 to raid the Spanish fleet in the Azores, skilled navigators were needed. Queen Elizabeth I granted Edward Wright (1561–1615) a sabbatical from Cambridge so that he could serve as a navigator on board a man-of-war. Although the campaign went badly for the English, Wright survived the fighting. In 1599, he organized his navigational work into a book titled *Certaine Errors of Navigation Corrected*. Wright described one computation as "the perpetual addition of the secantes answerable to the latitudes of each point or parallel into the summe compounded of all former secantes." Translated into the language of calculus, Wright approximated the integral $\int_0^\theta \sec(\phi)\,d\phi$ by using Riemann sums in which the increment $\Delta\phi$ was taken to be $1/360$ of a degree. Until the 1960s, historians thought that this was the first calculation of the secant integral. Then a paper of Thomas Harriot (1560–1621) came to light.

In 1585, Sir Walter Raleigh appointed Harriot to be a mathematical handyman on an expedition to Virginia. Harriot's duties included navigation, cartography, and surveying. After his return to England, Harriot wrote a series of papers devoted to the technical aspects of his service. In one of these papers, which went unnoticed for centuries, Harriot described his calculation of the secant integral $\int_0^\theta \sec(\phi)\,d\phi$.

With his days of exploration behind him, Harriot did not settle into the calm routine of a scholar. Instead, he assumed the intellectual leadership of the circle that assembled around Sir Walter Raleigh. When Shakespeare wrote

Oh paradox! Black is the badge of hell
The hue of dungeons and the school of night

in *Love's Labor's Lost* (4.3.250), he was referring to the evening sessions at Raleigh's country estate during which Harriot presided over discussions of astronomy, geography, chemistry, and philosophy. It was a dangerous circle. One of its members, the poet and playwright Christopher Marlowe, died in a knife fight over a tavern bill. Raleigh was sentenced for sedition in 1603, locked up in the Tower of London, and hanged in 1618. For his part in a subsequent conspiracy to blow up the English houses of parliament, Harriot's patron, the Earl of Northumberland, was imprisoned for 16 years.

Harriot's scientific reputation suffered less from his own brief incarceration than from the careless disposition of his last effects. After his death, Harriot's scientific papers were split among two heirs. One parcel was soon misplaced. By the time it turned up in 1784, mixed in among stable accounts, priority for Harriot's discoveries had already been claimed by others. Although a selection of papers from the second parcel was published in 1631, the larger part of them remain lost. Today, Harriot's most obvious legacies are the symbols that we use to express inequalities.

The first analytic proof of equation (7.35) was given by James Gregory in 1688. Edmond Halley assessed Gregory's derivation as, "A long train of Consequences and Complication of Proportions whereby the evidence of the Demonstration is in a great measure lost and the Reader wearied before he attain it." In the 1660s Isaac Barrow, Newton's predecessor at Cambridge, published a new evaluation of the secant integral:

$$\begin{aligned}\int_0^\theta \sec(\phi)\,d\phi &= \int_0^\theta \frac{\cos(\phi)}{\cos^2(\phi)}\,d\phi = \int_0^\theta \frac{\cos(\phi)}{1-\sin^2(\phi)}\,d\phi \\ &= \frac{1}{2}\int_0^\theta \frac{\cos(\phi)}{1-\sin(\phi)}\,d\phi + \frac{1}{2}\int_0^\theta \frac{\cos(\phi)}{1+\sin(\phi)}\,d\phi \\ &= \frac{1}{2}(-\ln(|1-\sin\theta|)+\ln(|1+\sin\theta|)) \\ &= \frac{1}{2}\ln\left(\left|\frac{1+\sin(\theta)}{1-\sin(\theta)}\right|\right).\end{aligned}$$

As an exercise, you should verify that Barrow's formula is equivalent to formula (7.35).

Notice that Barrow's proof uses the Method of Partial Fractions, a technique that Leibniz systematically used later in the 17th century. The theoretical basis for the Method of Partial Fractions is the Fundamental Theorem of Algebra, which states that every nonconstant polynomial with complex coefficients has a complex root. Such a result was conjectured by Harriot at the beginning of the 17th century. It was finally proved in 1797 by Carl Friedrich Gauss (1777–1855), who was at that time only 20 years old.

"Archimedes, Newton, and Gauss, these three, are in a class by themselves among the great mathematicians, and it is not for ordinary mortals to attempt to range them in order of merit." So wrote E. T. Bell, one of the most widely read mathematical expositors. We know little of the early life of Archimedes. We do know enough about Newton's childhood to be certain that he was no mathematical prodigy. Gauss, on the other hand, was astonishingly precocious. Anecdotal evidence indicates that Gauss taught himself to calculate before he learned to talk. According to one story, the young Gauss corrected one of his father's wage computations at the age of three. When Gauss was still in elementary school, his arithmetic teacher assigned his class the task of calculating $1+2+3+\cdots+100$. Gauss arrived at the correct answer nearly instantaneously, having already determined the rule for the sum of arithmetic progressions.

At the age of 14, Gauss discovered that, given any two nonnegative numbers a and b, there is a number $M(a,b)$ (called the *arithmetic-geometric mean,* or *AGM,* of a and b) that is the common limit of the two sequences defined inductively by

$$a_n = \begin{cases} a, & n=0 \\ \frac{1}{2}(a_{n-1}+b_{n-1}), & n>0 \end{cases} \quad \text{and} \quad b_n = \begin{cases} b, & n=0 \\ \sqrt{a_{n-1}b_{n-1}}, & n>0 \end{cases}.$$

This was a fine feat for a young teenager (even though the result was not new). Eight years later, Gauss found a remarkable application of the arithmetic-geometric mean. Ever since the 1660s, mathematicians had searched in vain for an effective method of calculating

$$\int_0^\pi \frac{d\phi}{\sqrt{1-x^2\sin^2(\phi)}}.$$

This *elliptic integral,* as it is known, is important in several branches of mathematics and physics. In 1799, Gauss discovered an unsuspected formula that permits the evaluation of elliptic integrals by means of the AGM:

$$\int_0^\pi \frac{d\phi}{\sqrt{1-x^2\sin^2(\phi)}} = \frac{\pi}{M(1+x,1-x)} \qquad 0 \le x < 1. \tag{7.36}$$

In 1801, Gauss made a contribution to celestial mechanics that brought him great fame in scientific circles. Table 1 lists the seven planets that were known at that time and the distance (in astronomical units) between the sun and each planet.

In the 1700s, astronomers observed that the number

$$B_m = 0.4 + 2^{m-2}\cdot 0.3 \tag{7.37}$$

is a good approximation to the distance in astronomical units (au) of the mth planet from the sun. In particular, it was noticed that *Bode's Law,* as formula (7.37) is known, would work well through Uranus *were there a still undiscovered planet between Mars and Jupiter approximately* 2.8 au *from the sun.*

	Mercury	Venus	Earth	Mars	*	Jupiter	Saturn	Uranus
Order from sun	1	2	3	4	*	5(?)	6(?)	7(?)
m	$-\infty$	2	3	4	5	6	7	8
Distance to sun (au)	0.39	0.72	1.00	1.52	*	5.20	9.54	19.18
B_m	0.4	0.7	1.0	1.6	2.8	5.2	10.0	19.6

Table 1 Bode's Law and the Seven Known Planets of 1801

On the first night of the 19th century, Giuseppe Piazzi (1746–1826) noticed a faint body that had not been previously catalogued. He continued to watch it. Hopeful that the object might turn out to be the missing planet, Piazzi wrote: "It has occurred to me several times that it might be something better than a comet. But I have been careful not to advance this supposition to the public. I will try to calculate its elements when I have made more observations." The object, however, faded from his sight on February 11, 1801. Piazzi had tracked it for 41 nights over an arc of only 3 deg. In June 1801, he published the observations that he had made. They did not, however, help other astronomers relocate the object.

Where astronomers with telescopes failed, Gauss succeeded, using mathematics alone. In particular, Gauss applied a mathematical tool of his own invention—the Method of Least Squares. On December 7, 1801, the first clear viewing night after Piazzi's object emerged from the shadow of the sun, it was spotted exactly where Gauss had predicted it would be. As Gauss remarked, "[T]he fugitive was restored to observation."

Many descendants of Gauss are scattered across the United States. The two youngest of Gauss's four sons, Eugen and Wilhelm, emigrated to the United States. Writing from New York City in 1831, Eugen informed his father, "Your name is well known even here in this wilderness." Eugen's army enlistment took him to Fort Snelling, Minnesota, where he served under the command of future president General Zachary Taylor. He settled in St. Charles, Missouri, across the river from St. Louis. The house that he built there still stands.

8

Applications of the Integral

PREVIEW

There is a common philosophical thread running throughout this chapter. Namely, each section introduces certain physical quantities. We approximate these quantities in a plausible way by a mathematical procedure that leads to a Riemann sum. By letting the mesh of the sum tend to 0 in each case, we are led to an integral. We then *declare* that integral to represent the physical quantity that we started out to analyze. What gives us the right to do this? There is a good practical answer.

In each instance discussed in this chapter, the mathematical model for the physical quantity being discussed gives *the same measurement in practical examples as one or more independent methods*. For instance, we develop several mathematical formulas for volumes of revolution. When applied to calculate the volume inside a sphere, the formula generates the same answer as obtained by immersing a sphere in water and measuring the displacement. We also develop a formula for the length of a curve. In practice, this formula gives the same answers as those obtained by measuring the length of the same curves with a tape measure. Similar comments apply to the other physical quantities—work and center of mass—discussed in this chapter.

When applying mathematics, it is always important to have objective methods for testing the validity of mathematical models. Scientists are constantly challenging, checking, and revising their methods for *modeling*—a crucial step in the advancement of scientific knowledge.

8.1 Volumes

In this section, we think about volume in much the same way that we learned to think about area in Chapter 5. That is, we calculate the volume of a solid by cutting up the solid into elementary pieces.

Volumes by Slicing—The Disk Method

You may have already learned that

$$V = \frac{1}{3}\pi r^2 h \tag{8.1}$$

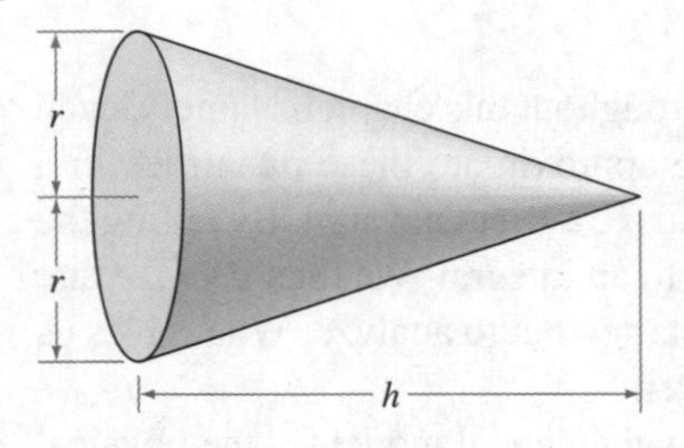

Figure 1

is the volume of a right circular cone of height h and radius r (Figure 1). Let us see for ourselves how we can deduce this formula. We begin by slicing the cone into N equally thick pieces, as in Figure 2. The volume of each slice is approximately equal to the volume of a corresponding disk—see Figure 2. If the jth disk has radius r_j and thickness Δx, then its volume is the product $(\pi r_j^2) \cdot \Delta x$ of its cross-sectional area πr_j^2 and its thickness Δx. To obtain an approximation to the volume of the cone, we add up the volumes of the disks:

$$\sum_{j=1}^{N} \pi r_j^2 \, \Delta x. \tag{8.2}$$

To get a better approximation, we make the disks thinner by increasing the value of N. To get the precise volume V, we let the thickness of the disks tend to 0 by letting N tend to infinity:

$$V = \lim_{N\to\infty} \sum_{j=1}^{N} \pi r_j^2 \, \Delta x. \tag{8.3}$$

To compute this limit, we identify expression (8.2) as a Riemann sum. Formula (8.3) expresses the volume of the cone as a limit of Riemann sums—that is, an integral (as we learned in Chapter 5). The following example provides the details of the technique just described.

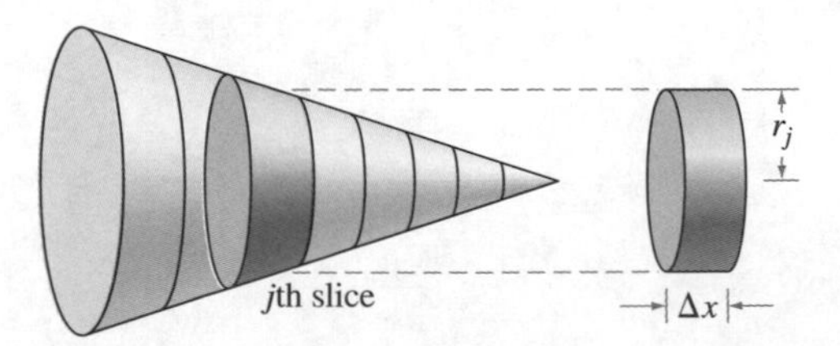

Figure 2
A disk is used to approximate a slice of the cone.

Example 1 Calculate the volume V of a right circular cone that has height 11 and base of radius 5.

Solution Figure 3 depicts the xy-plane and the cone in question. The cone is positioned so that the x-axis is its axis of symmetry and the origin is the center of its base.

The line segment formed by the intersection of the edge of the cone and the xy-plane has slope $-5/11$ and y-intercept $(0, 5)$. Its equation is therefore $y = -5x/11 + 5$.

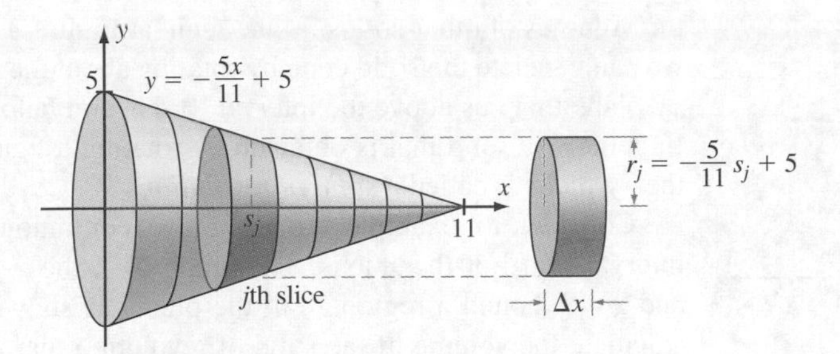

Figure 3

In Figure 3, we partitioned the interval $[0, 11]$ into N subintervals of equal length Δx. For each j from 1 to N, a point s_j is chosen in the jth subinterval. (Since the particular value of s_j will not matter, a formula for s_j is not needed.) As Figure 3 indicates, we approximate the jth slice by means of a disk of thickness Δx and radius r_j equal to the vertical distance from the point $(s_j, 0)$ to the edge of the cone. The approximating sum (8.2) becomes

$$\sum_{j=1}^{N} \pi \left(\frac{-5s_j}{11} + 5 \right)^2 \cdot \Delta x. \tag{8.4}$$

As N increases and Δx becomes smaller, sum (8.4) gives a more accurate approximation to the desired volume; however, expression (8.4) is also a Riemann sum for the integral

$$\int_0^{11} \pi \left(\frac{-5x}{11} + 5 \right)^2 dx. \tag{8.5}$$

Thus, both the volume V of the cone and integral (8.5) equal the limit of sum (8.4) as N tends to infinity. We conclude that V is equal to

$$\int_0^{11} \pi \left(\frac{-5x}{11} + 5 \right)^2 dx = \pi \left(\frac{5}{11} \right)^2 \int_0^{11} (-x+11)^2 \, dx = \pi \left(\frac{5}{11} \right)^2 \left. \frac{(-x+11)^3}{-3} \right|_0^{11}$$

$$= \pi \left(\frac{5}{11} \right)^2 \left(\frac{0}{-3} - \frac{11^3}{-3} \right) = \frac{1}{3} \pi \cdot 5^2 \cdot 11.$$

■

inSIGHT

In Example 1, if we replace each occurrence of 5 with r and each 11 with h, then we find that the volume of the right circular cone with base r and height h is

$$V = \pi \left(\frac{r}{h} \right)^2 \int_0^h (-x+h)^2 \, dx = -\frac{1}{3} \pi \left(\frac{r}{h} \right)^2 (-x+h)^3 \Big|_{x=0}^{x=h} = \frac{1}{3} \pi r^2 h,$$

as given in formula (8.1).

Solids of Revolution

With little modification, we can use the method that we developed for the cone to find the volumes of other solids. Look again at Figure 3. The key step is to recognize that we can generate the solid cone by rotating about the x-axis the triangular region of the xy-plane that lies above the interval [0, 11] and below the graph of $y = -5x/11 + 5$. In general, a solid that is obtained by rotating a figure in the xy-plane about a line in the xy-plane is called a *solid of revolution*.

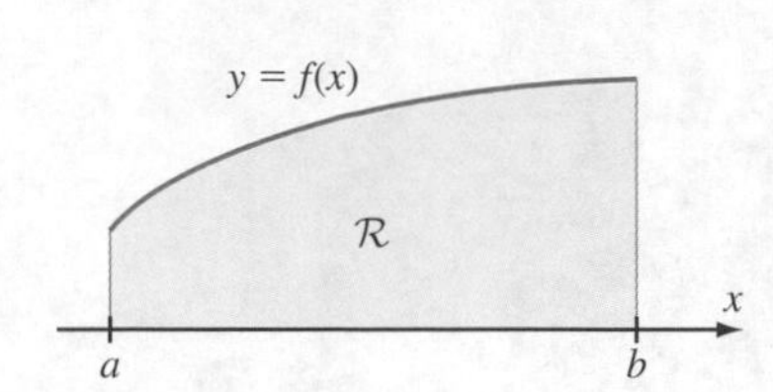

Figure 4

Consider, for example, a nonnegative, continuous function f that is defined on an interval $[a, b]$ of the x-axis. The graph of f, the x-axis, and the vertical lines $x = a$ and $x = b$ bound a region $\mathcal{R}$ in the plane, as shown in Figure 4. Let us see how to calculate the volume of the solid of revolution that is generated by rotating region $\mathcal{R}$ about the x-axis.

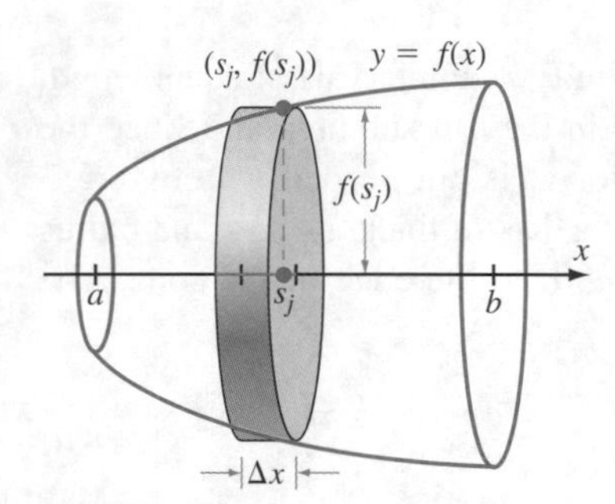

Figure 5

As before, we partition the interval $[a, b]$ into N subintervals of equal length Δx. For each j from 1 to N, a point s_j is chosen in the jth subinterval. Figure 5 indicates that we approximate the jth slice of the solid of revolution with a disk of radius $f(s_j)$ and thickness Δx. The contribution to volume from this disk is $(\pi f(s_j)^2) \cdot \Delta x$. The sum of the volumes of the N disks is

$$\sum_{j=1}^{N} \pi f(s_j)^2 \cdot \Delta x.$$

As we let N tend to infinity, the union of the approximating disks becomes closer to the solid of revolution. We therefore *define* the volume V of the solid of revolution to be

$$V = \lim_{N\to\infty} \sum_{j=1}^{N} \pi f(s_j)^2 \cdot \Delta x.$$

Notice that the expression in this limit is a Riemann sum for the integral

$$\int_a^b \pi \cdot f(x)^2 \, dx.$$

Since

$$\int_a^b \pi \cdot f(x)^2 \, dx = \lim_{N\to\infty} \sum_{j=1}^{N} \pi f(s_j)^2 \cdot \Delta x,$$

we see that the volume of the solid of revolution is represented by a Riemann integral. We may state our conclusion as a theorem.

Theorem 1

Method of Disks (Rotation about the x-axis) Suppose that f is a nonnegative, continuous function on the interval $[a, b]$. Let $\mathcal{R}$ denote the region of the xy-plane that is bounded by the graph of f, the x-axis, and the vertical lines $x = a$ and $x = b$. The volume V of the solid obtained by rotating $\mathcal{R}$ about the x-axis is given by

$$V = \pi \int_a^b f(x)^2 \, dx.$$

Example 2 Calculate the volume of the solid of revolution generated by rotating about the x-axis the region of the xy-plane that is bounded by $y = x^2$, $y = 0$, $x = 1$, and $x = 3$.

Solution Figure 6 shows both the region to be revolved as well as the solid of revolution. According to Theorem 1, the volume of this solid is

$$V = \pi \int_1^3 (x^2)^2\,dx = \pi \int_1^3 x^4\,dx = \pi \left(\frac{x^5}{5} \Bigg|_{x=1}^{x=3} \right) = \pi \left(\frac{3^5 - 1}{5} \right) = \frac{242}{5}\pi.$$

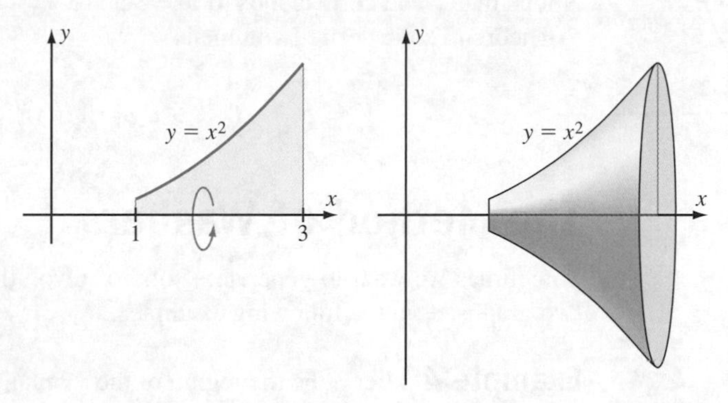

Figure 6

■

In some problems, it is useful to think of the curve as $x = g(y)$ and to rotate about the y-axis (see Figure 7). Reasoning analogous to that which led to Theorem 1 gives the following result.

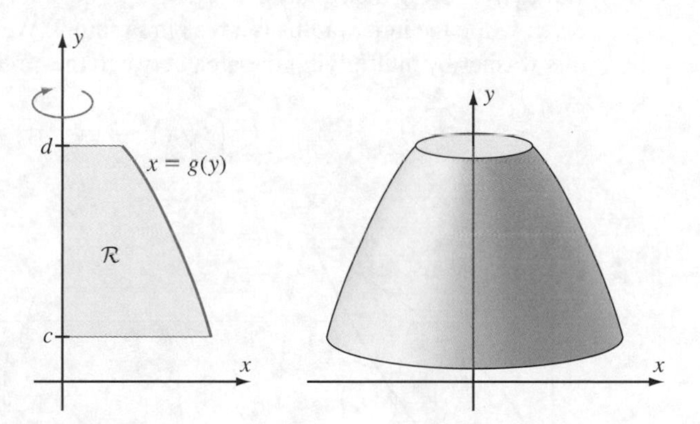

Figure 7
A planar region $\mathcal{R}$ and the solid that $\mathcal{R}$ generates when rotated about the y-axis

Theorem 2 ***Method of Disks (Rotation about the y-axis)*** Suppose that $x = g(y)$ is a nonnegative, continuous function on the interval $c \le y \le d$. Let $\mathcal{R}$ denote the region of the xy-plane bounded by the graph of $x = g(y)$, the y-axis, and the horizontal lines $y = c$ and $y = d$. The volume V of the solid obtained by rotating $\mathcal{R}$ about the y-axis is given by

$$V = \pi \int_c^d g(y)^2\,dy.$$

Example 3 Calculate the volume enclosed when the graph of $y = x^3, 2 \le x \le 4$, is rotated about the y-axis.

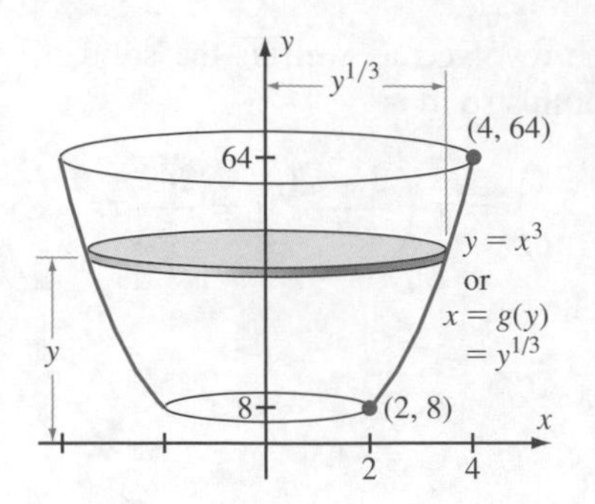

Figure 8

Solution Refer to Figure 8. Since we are rotating about the y-axis, we want to use Theorem 2. Therefore, we express x as a function of y:

$$x = g(y) = y^{1/3}.$$

Notice that $x = 2$ corresponds to $y = 8$ and $x = 4$ corresponds to $y = 64$. According to Theorem 2, the desired volume is

$$V = \pi \int_8^{64} (y^{1/3})^2 \, dy = \frac{3}{5}\pi y^{5/3} \Big|_{y=8}^{y=64} = \frac{2976}{5}\pi.$$

■

The Method of Washers

Sometimes we want to generate a solid of revolution by rotating the region *between* two graphs, as in the following example.

Example 4 Let $\mathcal{D}$ be the region of the xy-plane bounded above by $y = 8\sqrt{x}$ and below by $y = x^2$. Calculate the volume of the solid of revolution generated when $\mathcal{D}$ is rotated about the x-axis.

Solution When we decompose this solid of revolution into thin slices, we obtain pieces that are approximately washers, not disks (see Figure 9). To approximate the solid, we use washers that correspond to values of x from $x = 0$ to $x = 4$. We find these values by solving the equation $8\sqrt{x} = x^2$. The outer radius of the washer located at x is $8\sqrt{x}$, while the inner radius is x^2, as in Figure 9. We obtain the volume contribution of this washer by multiplying the area between the circles by the thickness of the washer:

$$\left(\pi \left(8\sqrt{x}\right)^2 - \pi (x^2)^2\right) \cdot \Delta x.$$

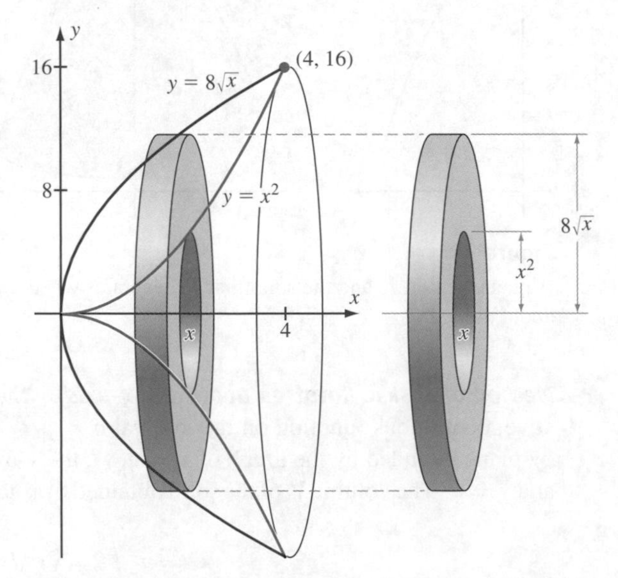

Figure 9

Therefore, the volume of the solid is

$$\pi \int_0^4 \left(\left(8\sqrt{x}\right)^2 - (x^2)^2\right) dx = \pi \int_0^4 (64x - x^4)\, dx = \pi \left(32x^2 - \frac{x^5}{5}\right)\Bigg|_{x=0}^{x=4} = \frac{1536}{5}\pi.$$

■

inSIGHT

In Example 4, it would be an error to use Theorem 1 with the function $f(x) = 8\sqrt{x} - x^2$, because we find the area between two circles by *subtracting the areas of the circles*. We do not find it by subtracting the radii and squaring the difference.

The reasoning from Example 4 may be used to derive the following theorem.

Theorem 3 ***Method of Washers*** Suppose that U and L are nonnegative, continuous functions on the interval $[a, b]$ with $L(x) \le U(x)$ for each x in this interval. Let $\mathcal{R}$ denote the region of the xy-plane that is bounded above by the graph of U, below by the graph of L, and on the sides by the vertical lines $x = a$ and $x = b$. The volume V of the solid obtained by rotating $\mathcal{R}$ about the x-axis is given by

$$V = \pi \int_a^b \left(U(x)^2 - L(x)^2\right) dx. \tag{8.6}$$

inSIGHT

We may write equation (8.6) as

$$V = \pi \int_a^b U(x)^2\, dx - \pi \int_a^b L(x)^2\, dx.$$

This formula is to be expected because we can obtain the solid from Theorem 3 by first rotating the region under the graph of $y = U(x)$ and then removing the solid that is obtained by rotating the region under the graph of $y = L(x)$.

Example 5 Let $\mathcal{R}$ be the region of the xy-plane that is bounded above by $y = e^x$, below by $y = \sqrt{x}e^{x^2}$, on the left by the vertical line $x = 0$, and on the right by the vertical line $x = 1$. Calculate the volume of the solid of revolution that is generated when $\mathcal{R}$ is rotated about the x-axis.

Solution Figure 10 shows the plots of $U(x) = e^x, 0 \le x \le 1$, and $y = L(x) = \sqrt{x}e^{x^2}$. To the right of these plots are the surfaces they generate when rotated about the x-axis. According to Theorem 3, the volume of the solid bounded by these surfaces is

$$\pi \int_0^1 \left(U(x)^2 - L(x)^2\right) dx = \pi \int_0^1 \left(e^{2x} - xe^{2x^2}\right) dx = \pi \left(\frac{e^{2x}}{2} - \frac{e^{2x^2}}{4}\right)\Bigg|_{x=0}^{x=1} = \pi\, \frac{e^2 - 1}{4}.$$

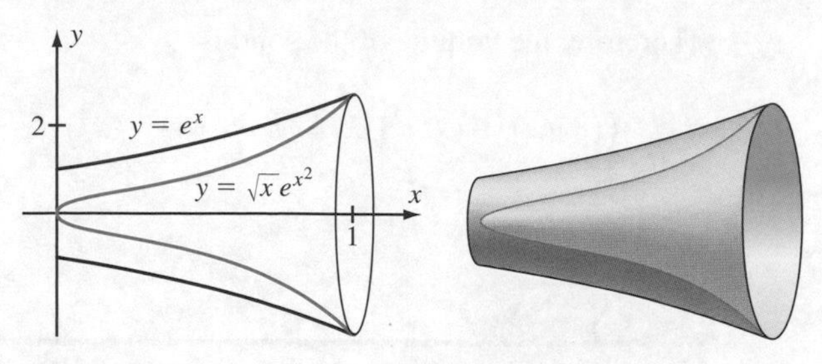

Figure 10

Rotation about a Line That Is Not a Coordinate Axis

With only a small modification, we can apply the analysis leading to Theorems 1 and 2 to calculate the volume of a solid of revolution about a line that is parallel to (but not equal to) one of the coordinate axes, as in the next example.

Example 6 Rotate the parallelogram bounded by $y = 3$, $y = 4$, $y = x$, and $y = x - 1$ about the line $x = 1$ and find the resulting volume.

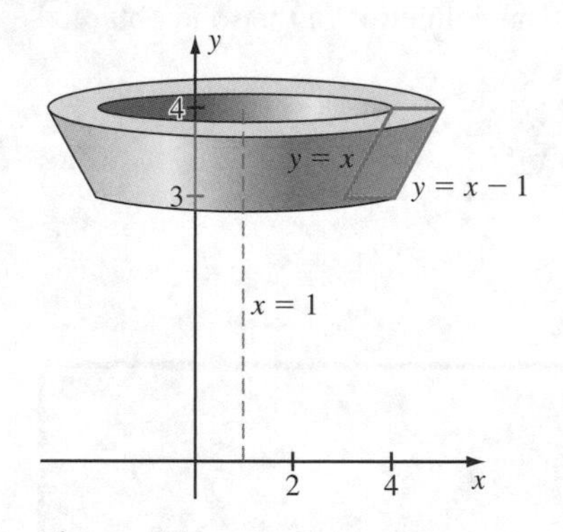

Figure 11

Solution Figure 11 illustrates the parallelogram and solid of revolution. We approximate the solid by a union of washers corresponding to values of y between 3 and 4. (A typical washer at height y is shown in Figure 12.) Since y will be the variable of integration, we must express the volume of this washer in terms of y. Refer to Figure 12 and note that the distance of the outer edge of the washer to the y-axis is the x-coordinate of point P. Because point P is on the line $y = x - 1$, that coordinate, expressed in terms of y, is $y + 1$. From this expression we must subtract 1 to find the distance of P to the axis of rotation $x = 1$. We conclude that the outer radius of the washer is $(y + 1) - 1$, or y. Similarly, the inner radius of the washer is 1 unit less than the x-coordinate of point Q. Expressed in terms of y, the inner radius is $y - 1$. The volume contribution of this washer is therefore

$$(\pi y^2 - \pi (y-1)^2) \cdot \Delta y.$$

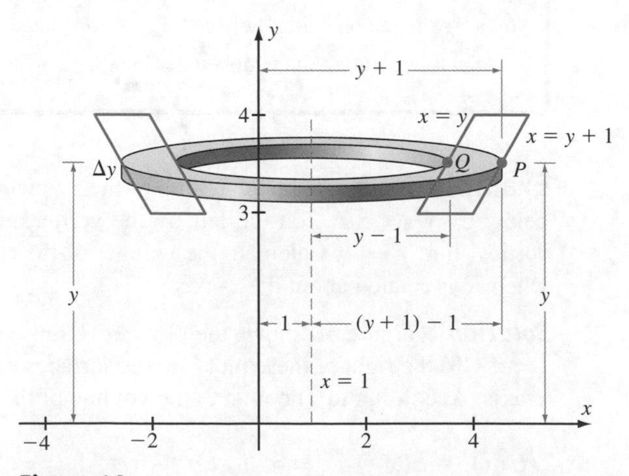

Figure 12
Washer at height y

As before, we sum these volumes and let Δy tend to 0. It follows that the volume we seek is

$$\int_3^4 (\pi y^2 - \pi(y-1)^2)\,dy = \pi \int_3^4 (y^2 - (y^2 - 2y + 1))\,dy = \pi \int_3^4 (2y - 1)\,dy = \pi(y^2 - y)\Big|_{y=3}^{y=4} = 6\pi.$$

Now that we have seen several instances of calculating volume by the method of slicing, it is a good idea to summarize the basic steps that constitute the method.

Steps for Calculating Volume by the Method of Slicing

1. Identify the shape of each slice.
2. Identify the independent variable that gives the position of each slice.
3. Write an expression, in terms of the independent variable, that describes the cross-sectional area of each slice.
4. Identify the interval $[a, b]$ over which the independent variable ranges.
5. With respect to the independent variable from step 2, integrate the expression for the cross-sectional area from step 3 over the interval $[a, b]$ from step 4.

In the next example, we explicitly follow these five steps. Notice that if each step can be completed, then the method of slicing can be applied even if the slices are neither disks nor washers.

Example 7 ***A Solid That Is Not Generated by Rotation*** If V is the volume of a solid pyramid that has height h and rectangular base of area A, then

$$V = \frac{1}{3}Ah.$$

Verify this formula for a solid pyramid of height 5 if the width and depth of the base are 2 and 3, respectively.

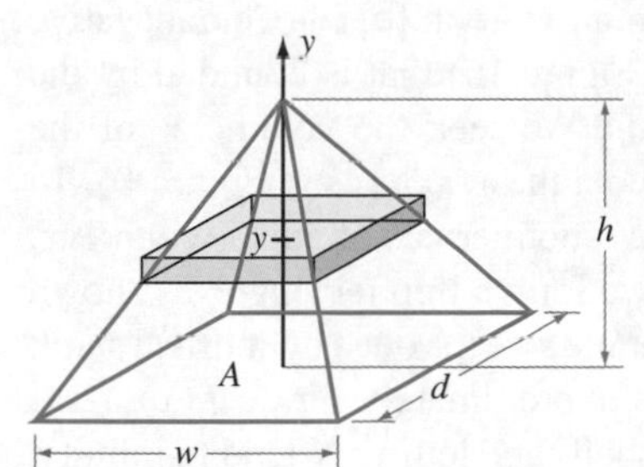

Figure 13

Solution

1. The cross section of each slice will be a rectangle (see Figure 13).
2. Since the slices are perpendicular to the y-axis, each slice is located by its position on the y-axis. So we make y the independent variable.
3. Let $w(y)$ and $d(y)$ denote the width and depth of the rectangular cross section of the slice at height y. An argument with similar triangles reveals that $w(y)/2 = (5 - y)/5$, or $w(y) = 2 \cdot (5 - y)/5 = 2(1 - y/5)$ (see Figure 14). Likewise, $d(y)/3 = (5 - y)/5$, or $d(y) = 3 \cdot (5 - y)/5 = 3(1 - y/5)$. Therefore, the cross-sectional area of the slice is
$$w(y)d(y) = 2 \cdot 3 \cdot \left(1 - \frac{y}{5}\right)^2.$$
4. The range of the independent variable y is determined by the extreme values, 0 and 5, of y: $y \in [0, 5]$.

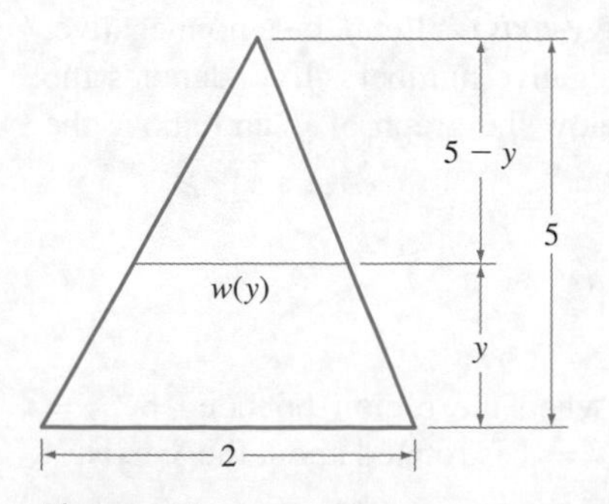

Figure 14

5. The volume is

$$\int_0^5 6\cdot\left(1-\frac{y}{5}\right)^2 dy = -\frac{1}{3}6\cdot 5\cdot\left(1-\frac{y}{5}\right)^3\Bigg|_{y=0}^{y=5} = \frac{1}{3}6\cdot 5 = \frac{1}{3}A\cdot h.$$ ■

The Method of Cylindrical Shells

For certain solids of revolution, the Method of Disks or Washers is either difficult or not feasible. To handle such cases, we will now develop an alternative method for calculating the volumes of solids of revolution. We call this the *Method of Cylindrical Shells* because it uses pipelike objects rather than disks or washers as basic building blocks (see Figure 15). This new method rests on a simple observation: The volume of a *thin* cylindrical shell of radius r, height h, and thickness t is approximately $(2\pi r)\cdot h\cdot t$. (Imagine cutting straight down the side of the cylindrical shell and flattening it out. The result is approximately a parallelepiped with side lengths $2\pi r$, h, and t, as in Figure 15.)

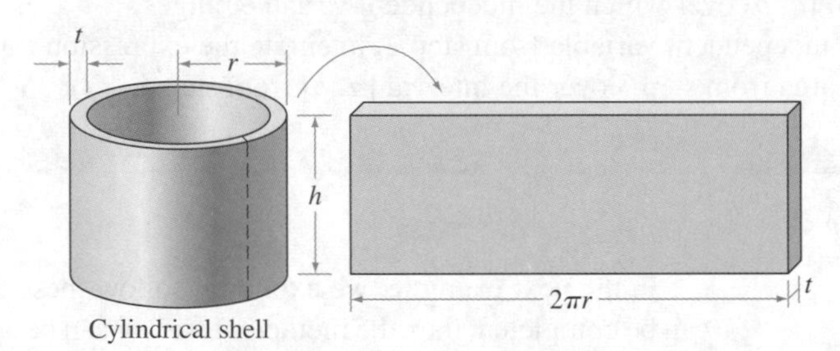

Figure 15

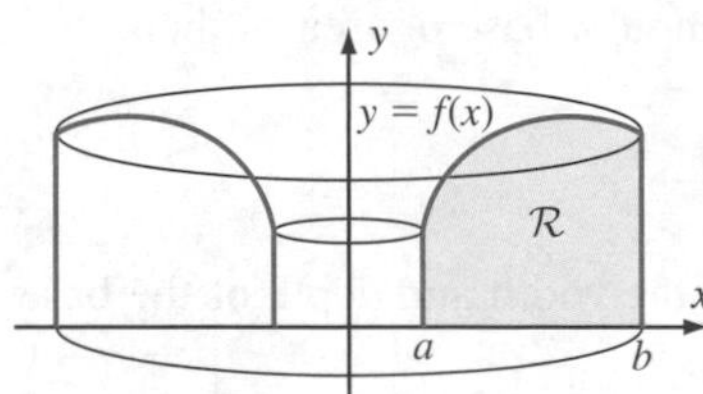

Figure 16

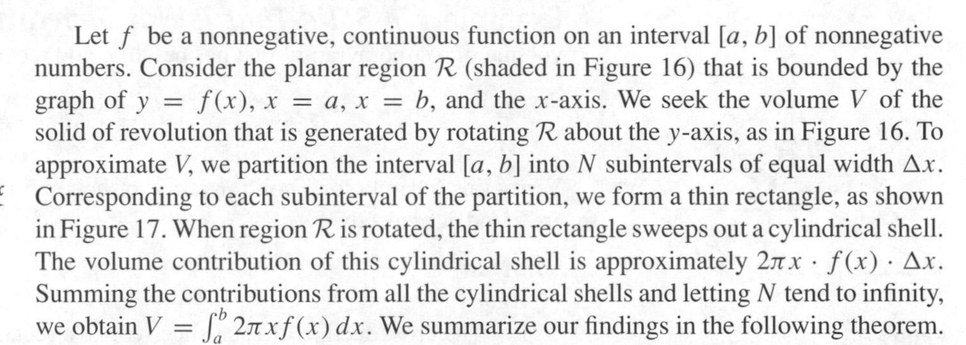

Let f be a nonnegative, continuous function on an interval $[a, b]$ of nonnegative numbers. Consider the planar region $\mathcal{R}$ (shaded in Figure 16) that is bounded by the graph of $y = f(x)$, $x = a$, $x = b$, and the x-axis. We seek the volume V of the solid of revolution that is generated by rotating $\mathcal{R}$ about the y-axis, as in Figure 16. To approximate V, we partition the interval $[a, b]$ into N subintervals of equal width Δx. Corresponding to each subinterval of the partition, we form a thin rectangle, as shown in Figure 17. When region $\mathcal{R}$ is rotated, the thin rectangle sweeps out a cylindrical shell. The volume contribution of this cylindrical shell is approximately $2\pi x\cdot f(x)\cdot\Delta x$. Summing the contributions from all the cylindrical shells and letting N tend to infinity, we obtain $V = \int_a^b 2\pi x f(x)\,dx$. We summarize our findings in the following theorem.

Theorem 4 ***Method of Cylindrical Shells (Rotation about the y-axis)*** Let f be a nonnegative, continuous function on an interval $[a, b]$ of nonnegative numbers. If V denotes the volume of the solid generated when the region below the graph of f and above the interval $[a, b]$ is rotated about the y-axis, then

$$V = 2\pi\int_a^b x f(x)\,dx.$$

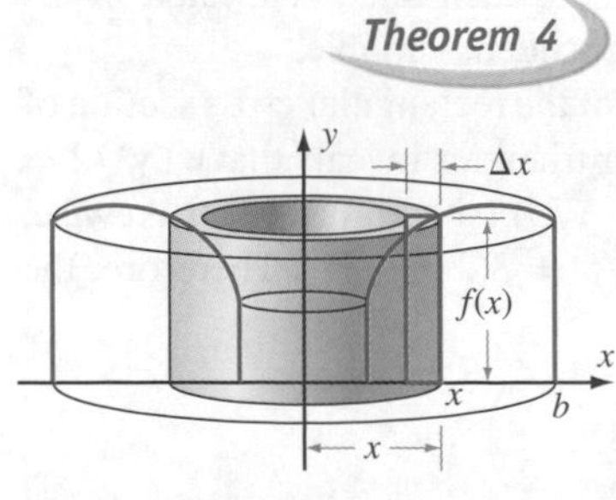

Figure 17

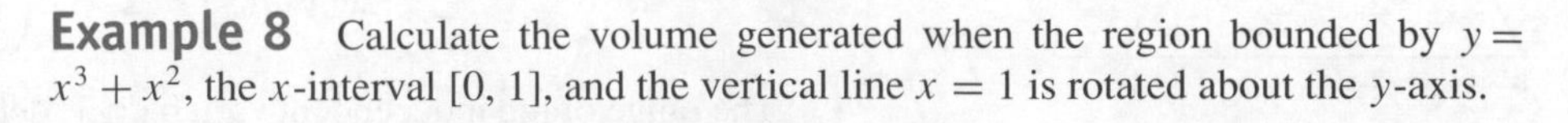

Example 8 Calculate the volume generated when the region bounded by $y = x^3 + x^2$, the x-interval $[0, 1]$, and the vertical line $x = 1$ is rotated about the y-axis.

Solution Theorem 3 applies with $f(x) = x^3 + x^2$, $a = 0$ and $b = 1$:

$$V = 2\pi \int_0^1 x \cdot (x^3 + x^2)\, dx = 2\pi \int_0^1 (x^4 + x^3)\, dx = 2\pi \left(\frac{x^5}{5} + \frac{x^4}{4} \right)\Bigg|_{x=0}^{x=1} = \frac{9}{10}\pi. \quad \blacksquare$$

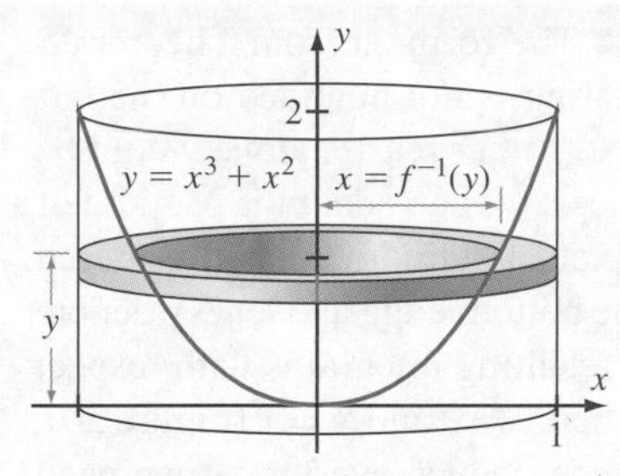

Figure 18

inSIGHT

In principle, we could have done the problem in Example 8 by the Method of Washers (see Figure 18). The formula would have been

$$V = \pi \int_0^2 (1^2 - f^{-1}(y)^2)\, dy.$$

However, finding a formula for $f^{-1}(y)$ would entail solving the cubic equation $y = x^3 + x^2$ for x. Doing so is no easy feat, and the resulting formula is extremely complicated. We have been able to avoid these difficulties by using the Method of Cylindrical Shells.

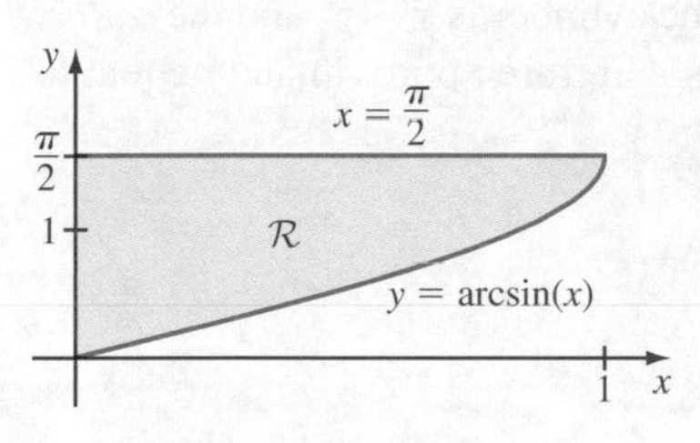

Figure 19a

Similar reasoning applies to the situation in which we rotate a region between $x = g(y)$ and the y-axis about the x-axis, as in the following theorem.

Theorem 5 ***Method of Cylindrical Shells (Rotation about the x-axis)*** Let g be a nonnegative, continuous function on an interval $[c, d]$ of nonnegative numbers. The volume of the solid generated when the region bounded by $x = g(y)$, the y-axis, $y = c$, and $y = d$ is rotated about the x-axis is

$$2\pi \int_c^d y g(y)\, dy.$$

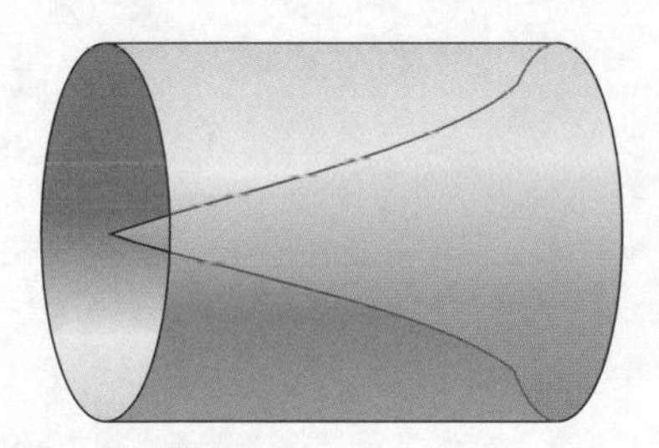

Figure 19b

Example 9 Let $\mathcal{R}$ be the region bounded above by the horizontal line $y = \pi/2$, below by the curve $y = \arcsin(x)$, $0 \le x \le 1$, and on the left by the y-axis. Let V denote the volume of the solid obtained when $\mathcal{R}$ is rotated about the x-axis. Use the Method of Cylindrical Shells to calculate V.

Solution Region $\mathcal{R}$ is shown in Figure 19a. When we rotate $\mathcal{R}$ about the x-axis, we obtain the solid that lies between the cylinder and the "sorcerer's hat" shown in Figure 19b. In Figure 19c, we shift our perspective a bit and cut away part of the solid's outer boundary to reveal one of the cylindrical shells that we can use to calculate the volume. According to Theorem 5, with $c = 0$, $d = \pi/2$, and $g(y) = \sin(y)$, the volume V is $2\pi \int_0^{\pi/2} y \sin(y)\, dy$. We apply the Method of Integration by Parts with $u = y$ and $dv = \sin(y)\, dy$ to obtain $du = dy$, $v = -\cos(y)$, and

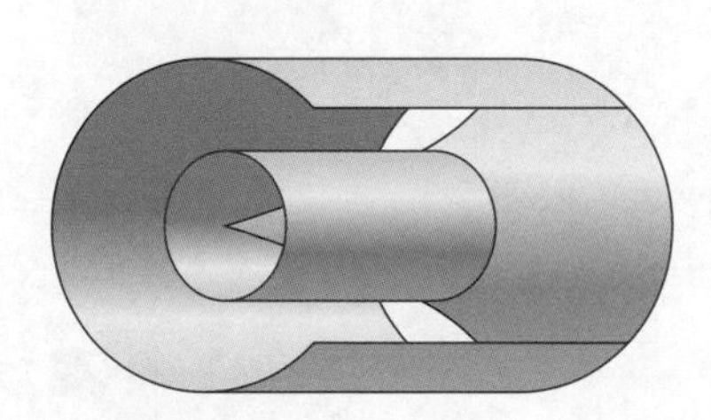

Figure 19c

$$V = 2\pi \left(-y\cos(y)\Big|_0^{\pi/2} + \int_0^{\pi/2} \cos(y)\, dy \right) = 2\pi \left(\sin(y) - y\cos(y) \right)\Big|_0^{\pi/2} = 2\pi. \quad \blacksquare$$

As we have already seen in the case of washers, the methods of this section apply to rotation about lines other than the coordinate axes. The following is an example in the context of cylindrical shells.

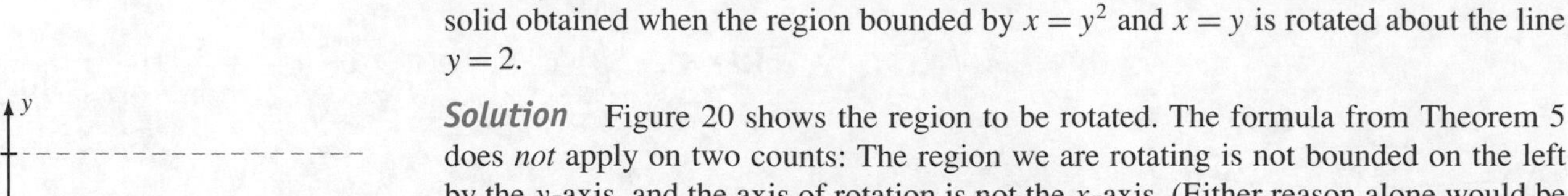

Example 10 Use the Method of Cylindrical Shells to calculate the volume of the solid obtained when the region bounded by $x = y^2$ and $x = y$ is rotated about the line $y = 2$.

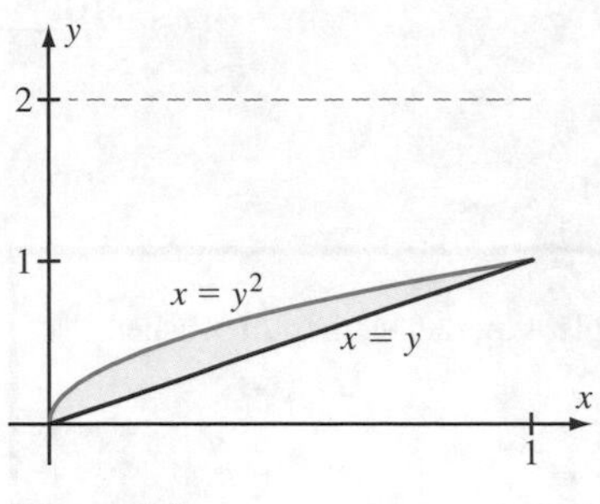

Figure 20

Solution Figure 20 shows the region to be rotated. The formula from Theorem 5 does *not* apply on two counts: The region we are rotating is not bounded on the left by the y-axis, and the axis of rotation is not the x-axis. (Either reason alone would be enough to disqualify the use of Theorem 5.) Nevertheless, we can adapt the Method of Cylindrical Shells to the solid at hand. Figure 21 depicts a cylindrical shell centered about the axis of rotation. Let y denote the height of the bottom edge and let Δy denote the thickness of the shell. We let Δy tend to 0, so the resulting integral is with respect to the variable y. We notice that the curves intersect when $y = 0$ and $y = 1$ (Figure 20), so these values will be our limits of integration. We can deduce everything we need from the coordinates (which we express in terms of the variable of integration) of the points labeled P and Q in Figure 21. The *height* of the cylinder is $y - y^2$ and the *radius* is $2 - y$. The volume V of the solid of revolution is therefore approximately equal to a sum of terms of the form

$$2\pi(2 - y)(y - y^2)\Delta y.$$

By letting Δy tend to 0, we obtain

$$V = \int_0^1 2\pi(2-y)\cdot(y-y^2)\,dy = 2\pi\int_0^1 (2y-3y^2+y^3)\,dy = 2\pi\left(y^2 - y^3 + \frac{y^4}{4}\right)\Bigg|_{y=0}^{y=1} = \frac{1}{2}\pi.$$

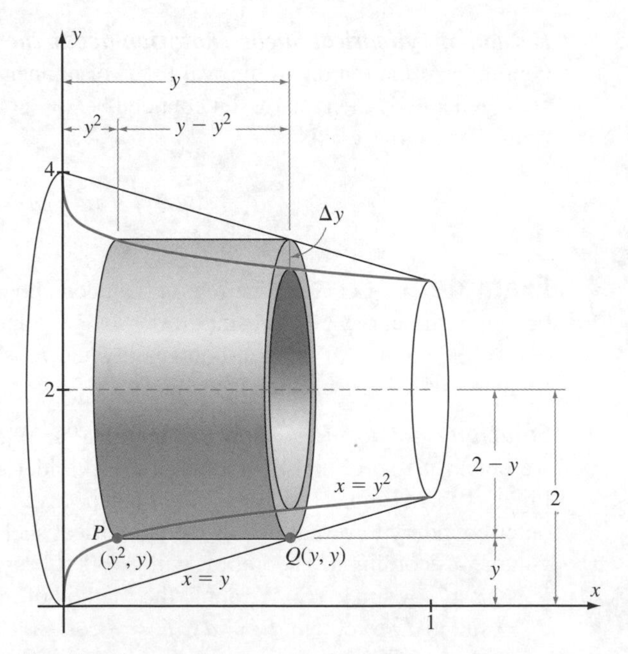

Figure 21

■

A Final Remark

It is best not to memorize the theorems in this section. Instead, it is better to reason about disks, washers, and cylinders each time you calculate a volume. That way, you will be sure to take into account the position of the axis of rotation and the relative positions of the graphs involved.

quickquiz

1. A region in the first quadrant of the xy-plane is rotated about the y-axis. If the Method of Disks is used to calculate the volume of the resulting solid of revolution, what is the variable of integration? Explain.
2. A region in the first quadrant of the xy-plane is rotated about the x-axis. If the Method of Cylindrical Shells is used to calculate the volume of the resulting solid of revolution, what is the variable of integration? Explain.
3. Use the Method of Cylindrical Shells to express the volume of the cone in Figure 3 as an integral.
4. Calculate the volume of the solid that results when the square $\{(x, y): |x - 2| < 1, |y - 3| < 1\}$ is rotated about the x-axis.

EXERCISES

Problems for Practice

In Exercises 1–4, use the Method of Disks to calculate the volume obtained by rotating the planar region $\mathcal{R}$ about the x-axis.

1. $\mathcal{R}$ is the region below the graph of $y = 3x^{1/3}$, above the x-axis, and between $x = 1$ and $x = 8$.
2. $\mathcal{R}$ is the region between the x-axis and the parabola $y = 8 - x^2$.
3. $\mathcal{R}$ is the region between the x-axis and $y = \sqrt{\sin(x)}$, $0 \le x \le \pi$.
4. $\mathcal{R}$ is the region between the x-axis and $y = \sec(x)$, $0 \le x \le \pi/4$.

In Exercises 5–8, use the Method of Disks to calculate the volume obtained by rotating the planar region $\mathcal{R}$ about the y-axis.

5. $\mathcal{R}$ is the region between the y-axis and the graph of $x = y^3$, $0 \le y \le 1$.
6. $\mathcal{R}$ is the region in the first quadrant bounded by the y-axis and the curve $x = 2y - y^2$.
7. $\mathcal{R}$ is the region between the y-axis and the curve $y = \ln(x)$, $1 \le x \le e$.
8. $\mathcal{R}$ is the region between the y-axis and the curve $y = x^2$, $1 \le x \le 2$.

In Exercises 9–12, use the Method of Washers to calculate the volume obtained by rotating the planar region $\mathcal{R}$ about the x-axis.

9. $\mathcal{R}$ is the region between the curves $y = x$ and $y = x^2$, $0 \le x \le 1$.
10. $\mathcal{R}$ is the region between the curves $y = x^2$ and $y = x^4$, $0 \le x \le 1$.
11. $\mathcal{R}$ is the region bounded above by the curve $y = 4 - x^2$ and below by the line $y = x + 2$.
12. $\mathcal{R}$ is the region bounded above by the curve $y = 9 - x^2$ and below by the line $y = x + 7$.

In Exercises 13–16, use the Method of Washers to calculate the volume obtained by rotating the planar region $\mathcal{R}$ about the y-axis.

13. $\mathcal{R}$ is the region between the graphs of $x = 2y$ and $x = y^2$, $0 \le y \le 2$.
14. $\mathcal{R}$ is the region between the curves $x = y^2$ and $x = y^3$, $0 \le y \le 1$.
15. $\mathcal{R}$ is the region bounded on the left by the curve $x = y^2 + 2$ and on the right by the curve $x = 4 - y$.
16. $\mathcal{R}$ is the region between the curves $x = y^2 + y + 1$, $x = 8y$, $y = 2$, and $y = 3$.

In Exercises 17–20, use the Method of Cylindrical Shells to calculate the volume obtained by rotating the planar region $\mathcal{R}$ about the y-axis.

17. $\mathcal{R}$ is the region below the graph of $y = x^2$, above the x-axis, and between $x = 0$ and $x = 1$.
18. $\mathcal{R}$ is the region bounded above by the curve $y = 2x - x^2$ and below by the x-axis.
19. $\mathcal{R}$ is the region below the graph of $y = x^2 + 1$, above the x-axis, and between $x = 2$ and $x = 4$.
20. $\mathcal{R}$ is the region below the graph of $y = 16 - x^3$, $0 \le x \le 2$, and above the x-axis.

In Exercises 21–24, use the Method of Cylindrical Shells to calculate the volume obtained by rotating the planar region $\mathcal{R}$ about the x-axis.

21. $\mathcal{R}$ is the region between the y-axis and the curve $x = \sqrt{y}, 0 \le y \le 4$.
22. $\mathcal{R}$ is the region between the y-axis, the curve $y = x^3$, and the lines $y = 1$ and $y = 2$.
23. $\mathcal{R}$ is the region between the y-axis and the curve $x = 2 + y + y^2, 0 \le y \le 2$.
24. $\mathcal{R}$ is the region between the y-axis and the curve $x = \exp(y^2), 0 \le y \le 1$.

In Exercises 25–28, use the Method of Cylindrical Shells to calculate the volume obtained by rotating the planar region $\mathcal{R}$ about the y-axis.

25. $\mathcal{R}$ is the region in the first quadrant bounded by $y = x$ and $y = x^2$.
26. $\mathcal{R}$ is the region bounded by $y = x$, $y = x^2$, $x = 2$, and $x = 3$.
27. $\mathcal{R}$ is the region below the graph of $y = x^2 + 1$, above the graph of $y = x$, and between $x = 2$ and $x = 4$.
28. $\mathcal{R}$ is the region below the graph of $y = 1 - x^3$ and above the graph of $y = 1 - x$.

In Exercises 29–32, use the Method of Cylindrical Shells to calculate the volume obtained by rotating the planar region $\mathcal{R}$ about the x-axis.

29. $\mathcal{R}$ is the region bounded by the x-axis, the curve $y = \sqrt{x}$, and the line $x = 4$.
30. $\mathcal{R}$ is the region bounded by the curve $y = x^2$ and the lines $y = 1$ and $x = 2$.
31. $\mathcal{R}$ is the region bounded by the curve $y = 2\sqrt{x}$ and the lines $y = 1$ and $y = 3x - 1$.
32. $\mathcal{R}$ is the region bounded by the x-axis, the curve $y = \sqrt{2x - 1}$, and the line $y = x$.

Further Theory and Practice

In Exercises 33–36, use the Method of Disks to calculate the volume obtained by rotating the planar region $\mathcal{R}$ about the x-axis.

33. $\mathcal{R}$ is the region below the graph of $y = \sin(x)$, $0 \le x \le \pi$.
34. $\mathcal{R}$ is the region below the graph of $y = \cos(x)$, $-\pi/6 \le x \le \pi/3$.
35. $\mathcal{R}$ is the region between the x-axis, the curve $y = \sqrt{x} \cdot \exp(x)$, and the line $x = 1$.
36. $\mathcal{R}$ is the region between the x-axis and the curve $y = \sqrt{x \sin(x)}, 0 \le x \le \pi$.

In Exercises 37–40, use the Method of Disks to calculate the volume obtained by rotating the planar region $\mathcal{R}$ about the y-axis.

37. $\mathcal{R}$ is the region between the curves $y = \exp(x)$, the y-axis, and $y = e$.
38. $\mathcal{R}$ is the region between the curve $y = \sin(x^2)$ and the lines $x = 0$ and $y = 1$.
39. $\mathcal{R}$ is the region between the curve $y = x^2/(1 - x^2)$ and the lines $x = 0$ and $y = 1/3$.
40. $\mathcal{R}$ is the region in the first quadrant bounded by the coordinate axes and the curve $x = (1 - y^2)^{3/4}$.

In Exercises 41–46, use the Method of Cylindrical Shells to calculate the volume obtained by rotating the planar region $\mathcal{R}$ about the given line ℓ.

41. $\mathcal{R}$ is the region between the curves $y = x^2 - 4x - 5$ and $y = -x^2 + 4x - 3$; ℓ is the line $x = -3$.
42. $\mathcal{R}$ is the region between the graphs of $y = \exp(x)$, $y = x\exp(x)$, and the y-axis; ℓ is the line $x = 0$.
43. $\mathcal{R}$ is the region between the curves $y = x^2 - x - 5$ and $y = -x^2 + x + 7$; ℓ is the line $x = 6$.
44. $\mathcal{R}$ is the region between the curves $x = y^3$ and $x = -y^2$; ℓ is the line $y = 3$.
45. $\mathcal{R}$ is the region between the curves $x = y^4$ and $x = y^{1/2}$; ℓ is the line $y = -1$.
46. $\mathcal{R}$ is the region between the curves $y = 2\sin(x)$, $\pi/6 \le x \le 5\pi/6$, and $y = 1$; ℓ is the line $y = -1$.

In Exercises 47–54, choose disks, washers, or cylindrical shells, whichever is best, to calculate the volume of the solid obtained when the region $\mathcal{R}$ is rotated about the line ℓ.

47. $\mathcal{R}$ is the region between $y = -x^2 + 6$ and $y = -x$; ℓ is the line $x = -4$.
48. $\mathcal{R}$ is the region between the curves $y = x^2$ and $y = x^{1/2}$; ℓ is the y-axis.
49. $\mathcal{R}$ is the region between the curve $y = \cos(x)$ and the x-axis, $\pi/2 \le x \le 3\pi/2$; ℓ is the line $x = 3\pi$.
50. $\mathcal{R}$ is the region between the curve $x = \tan(y)$ and the y-axis, $0 \le y \le \pi/4$; ℓ is the line $x = 4$.
51. $\mathcal{R}$ is the region bounded by the curve $x = \sin(y)$, the y-axis, and the line $y = \pi/2$; ℓ is the line $y = 4$.
52. $\mathcal{R}$ is the region between the curves $y = \sin(x)$ and $y = \cos(x)$, $0 \le x \le \pi/4$; ℓ is the line $y = 4$.
53. $\mathcal{R}$ is the region between the curves $y = x^3 + x$, $y = 0$, and $x = 1$; ℓ is the line $x = 0$.
54. $\mathcal{R}$ is the region between the curves $y = 1/(1 + x^2)$, $y = 1/2$, and $x = 0$; ℓ is the line $x = 0$.
55. Suppose $R > r > 0$. Calculate the volume of the solid obtained when the disk $\{(x, y) : x^2 + y^2 \le r^2\}$ is rotated about the line $x = R$. (This solid is called a *torus*.)

56. Calculate the volume of the solid obtained when the triangle with vertices (2, 5), (6, 1), and (4, 4) is rotated about the line $x = -3$.

57. Calculate the volume obtained when the region outside the square $\{(x, y) : |x| < 1, |y| < 1\}$ and inside the circle $\{(x, y) : x^2 + y^2 \leq 4\}$ is rotated about the line $y = 7$.

58. A solid has the ellipse $x^2 + 4y^2 = 16$ as its base. The vertical slices *parallel* to the line $y = 2x$ are equilateral triangles. Find the volume.

59. The base of a solid $\mathcal{S}$ is the disk $x^2 + y^2 \leq 25$. For each $k \in [-5, 5]$, the plane through the line $x = k$ and perpendicular to the xy-plane intersects $\mathcal{S}$ in a square. Find the volume of $\mathcal{S}$.

60. A solid has as its base the region bounded by the parabola $x - y^2 = -8$ and the left branch of the hyperbola $x^2 - y^2 - 4 = 0$. The vertical slices perpendicular to the x-axis are squares. Find the volume of the solid.

61. Old Boniface he took his cheer,
Then he drilled a hole in a solid sphere
Clear through the center straight and strong,
And the hole was just ten inches long.
Now tell us, when the end was gained,
What volume in the sphere remained.
Sounds like you've not been told enough,
But that's all you need; it's not too tough.

62. An open cylindrical beaker with circular base has height L and radius r. It is partially filled with a volume V of a fluid. Consider the parameters L, r, and V to be constant. The axis of symmetry of the beaker is along the positive y-axis, and one diameter of its base is along the x-axis. When the tank is revolved about the y-axis with angular speed ω, the surface of the fluid assumes a shape that is the paraboloid of revolution that results when the curve

$$y = h + \frac{\omega^2 x^2}{2g}, \, 0 \leq x \leq r,$$

is revolved about the y-axis. This formula is valid for angular speeds at which the surface of the fluid has not yet touched the base or the mouth of the beaker. The number $h = h(\omega)$ is in the interval $[0, V/(\pi r^2)]$ and depends on ω. (When $\omega = 0$, $h = V/(\pi r^2)$. As ω increases, h decreases.)

a. Find a formula for $h(\omega)$.

b. At what value ω_S of ω does spilling begin, assuming that $h(\omega) > 0$ for $\omega < \omega_S$?

c. At what value ω_B of ω does the surface touch the bottom of the beaker, assuming that spilling does not occur for $\omega < \omega_B$?

d. As ω increases, does the surface of the fluid touch the bottom of the beaker or the mouth of the beaker first?

Calculator/Computer Exercises

63. The region between the graphs of $y = x\exp(x)$ and $y = \sqrt{x}$ is rotated about the y-axis. Use Simpson's Rule to calculate the resulting volume to four decimal places.

64. The region below the graph of $y = \exp(-x^2)$, $-1 \leq x \leq 1$, is rotated about the x-axis. Use Simpson's Rule to calculate the resulting volume to four decimal places.

65. A flashlight reflector is made of an aluminum alloy that has mass density 3.743 g/cm^3. The reflector occupies the solid region obtained when the region bounded by $y = 2.530\sqrt{x}$, $y = 2.530\sqrt{x} + 0.300$, $x = 0$, and $x = 2.5$ cm is rotated about the x-axis. Sketch the cross section of the solid in the xy-plane. What is its mass?

66. The equation of the Gateway Arch in St. Louis, Missouri, is

$$y = 693.8597 - 34.38365(\exp(kx) + \exp(-kx))$$

for $k = 0.0100333$ and $-299.2239 < x < 299.2239$ where both x and y are measured in feet. Rotate this curve about its vertical axis of symmetry and compute the resulting volume.

8.2 Arc Length and Surface Area

Just as we have approximated area by a sum of areas of rectangles, so can we approximate the length of a curve by a sum of lengths of line segments. This process leads to an integral that is used to calculate the length of a graph of a function. We can apply analogous ideas to calculate the area of a surface of revolution.

The Basic Method for Calculating Arc Length

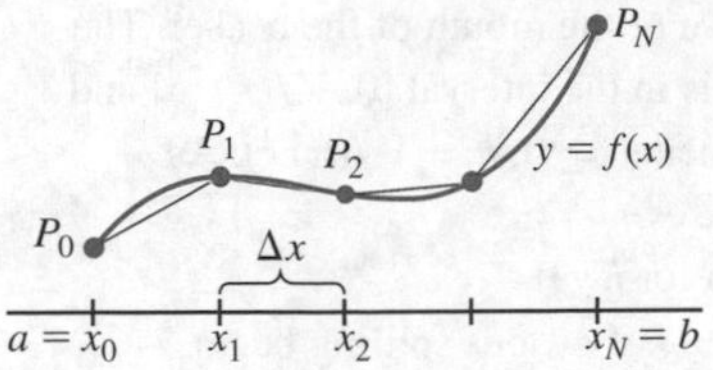

Suppose that f is a function with continuous derivative on a domain that contains the interval $[a, b]$. Let us calculate the length L of the graph of f over this interval. Fix a positive integer N and let $a = x_0 < x_1 < x_2 < \cdots < x_{N-1} < x_N = b$ be a uniform partition of the interval $[a, b]$. As Figure 1 suggests, we use a piecewise linear curve to approximate the graph of f. To be specific, we use the line segment with endpoints $P_{j-1} = (x_{j-1}, f(x_{j-1}))$ and $P_j = (x_j, f(x_j))$ to approximate the part of the graph of f that lies over the jth subinterval $[x_{j-1}, x_j]$ (see Figure 2). The length ℓ_j of this line segment then approximates the length of the arc of the graph between P_{j-1} and P_j. We sum the lengths ℓ_j to obtain an approximate length for the curve:

Figure 1

$$L \approx \sum_{j=1}^{N} \ell_j.$$

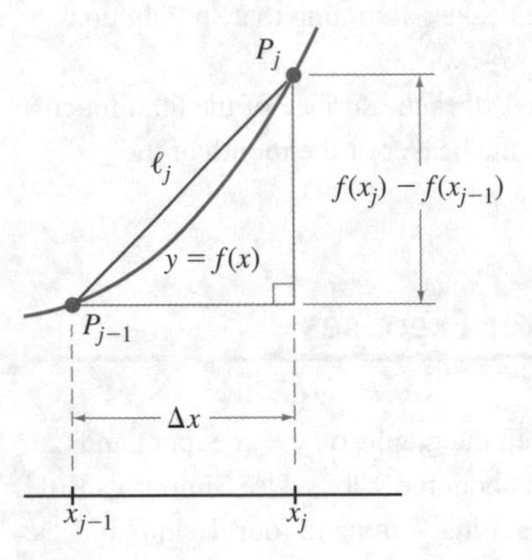

Figure 2

The accuracy of this approximation is improved by increasing the number N of subintervals (Figure 3). As N increases to infinity and Δx tends to 0, these approximating sums tend to what we think of as the length of the curve:

$$L \overset{\text{Intuition}}{=} \lim_{N\to\infty} \sum_{j=1}^{N} \ell_j. \tag{8.7}$$

We therefore use equation (8.7) to *define* the length L of the graph of f.

Refer again to Figure 2. Notice that the length ℓ_j is given by the usual planar distance formula:

$$\ell_j = ((x_j - x_{j-1})^2 + (f(x_j) - f(x_{j-1}))^2)^{1/2}.$$

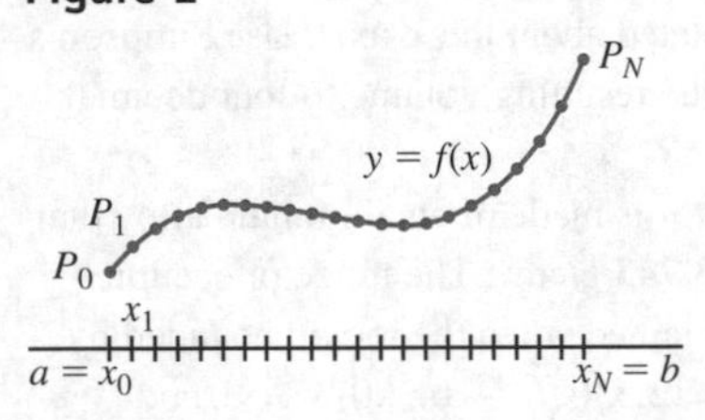

Figure 3

We denote the quantity $x_j - x_{j-1}$ by Δx and apply the Mean Value Theorem to the expression $f(x_j) - f(x_{j-1})$ to obtain $f(x_j) - f(x_{j-1}) = f'(\xi_j)\Delta x$ for some ξ_j between x_{j-1} and x_j. We may now rewrite the formula for ℓ_j as

$$\ell_j = ((\Delta x)^2 + (f'(\xi_j)\Delta x)^2)^{1/2} = (1 + f'(\xi_j)^2)^{1/2}\Delta x.$$

If we use this formula to substitute for ℓ_j in equation (8.7), then we have

$$L = \lim_{N\to\infty} \sum_{j=1}^{N} (1 + f'(\xi_j)^2)^{1/2}\Delta x.$$

Notice that these sums are also Riemann sums for the integral $\int_a^b (1 + f'(x)^2)^{1/2}\,dx$. We conclude that this integral represents the length of the curve:

Definition

If f has a continuous derivative on an interval containing $[a, b]$, then the *arc length* L of the graph of f over the interval $[a, b]$ is given by

$$L = \int_a^b (1 + f'(x)^2)^{1/2}\,dx.$$

The formula for arc length leads to integrals that are often difficult or impossible to evaluate. In such cases, we may use the techniques of numerical integration that we studied in Section 5.7 to obtain accurate approximations. In this section, we concentrate on simple examples in which the integrals work out neatly.

Example 1 Let us calculate the arc length L of the graph of $f(x) = 2x^{3/2}$ over the interval $[0, 7]$.

Solution By definition, the length is

$$L = \int_0^7 (1 + f'(x)^2)^{1/2}\, dx = \int_0^7 (1 + (3x^{1/2})^2)^{1/2}\, dx = \int_0^7 (1 + 9x)^{1/2}\, dx.$$

In the last integral, we make the change of variable $u = 1 + 9x$, $du = 9dx$ and note that u varies from 1 to 64 as x varies from 0 to 7:

$$L = \int_1^{64} u^{1/2}\frac{1}{9}\, du = \frac{1}{9} \cdot \frac{2}{3} u^{3/2}\Big|_{u=1}^{u=64} = \frac{2}{27}(512 - 1) = \frac{1022}{27}.$$ ■

Example 2 Calculate the length L of the graph of the function $f(x) = (e^x + e^{-x})/2$ over the interval $[1, \ln(8)]$.

Solution We first calculate that $f'(x) = (e^x - e^{-x})/2$ and

$$1 + f'(x)^2 = 1 + \frac{(e^x - e^{-x})^2}{4} = \frac{4 + e^{2x} - 2 + e^{-2x}}{4} = \frac{e^{2x} + 2 + e^{-2x}}{4} = \left(\frac{e^x + e^{-x}}{2}\right)^2.$$

Thus,

$$L = \int_1^{\ln(8)} (1 + f'(x)^2)^{1/2} dx = \int_1^{\ln(8)} \left(\frac{e^x + e^{-x}}{2}\right) dx = \frac{e^x - e^{-x}}{2}\Big|_{x=1}^{x=\ln(8)}$$

$$= \frac{1}{2}\left(\left(8 - \frac{1}{8}\right) - \left(e - \frac{1}{e}\right)\right).$$ ■

Sometimes an arc length problem is more conveniently solved if we think of the curve as being the graph of $x = g(y)$.

Definition

If g' is continuous, then the arc length L of the graph of $x = g(y)$ for $c \leq y \leq d$ is given by

$$L = \int_c^d (1 + g'(y)^2)^{1/2}\, dy. \qquad \textbf{(8.8)}$$

This is only a restatement of the definition of arc length and does not require additional analysis.

Example 3 Calculate the length L of that portion of the graph of the curve $9x^2 = 4y^3$ between the points $(0, 0)$ and $(2/3, 1)$.

Solution If we express the curve as $x = (2/3)y^{3/2}$ and set $g(y) = (2/3)y^{3/2}$, then $g'(y) = y^{1/2}$. Formula (8.8) becomes

$$L = \int_0^1 (1 + (y^{1/2})^2)^{1/2}\, dy = \int_0^1 (1 + y)^{1/2}\, dy.$$

We can easily evaluate this integral by means of the substitution $u = 1 + y$, $du = dy$:

$$L = \int_1^2 u^{1/2}\, du = \frac{2}{3}u^{3/2}\Big|_{u=1}^{u=2} = \frac{2}{3}(2\sqrt{2} - 1).$$ ■

In Example 3, if we had expressed the curve as $y = f(x) = \sqrt[3]{9/4}x^{2/3}$, then L would be given by the formula

$$L = \int_0^{2/3} \left(1 + \sqrt[3]{\frac{4}{9}}x^{-2/3}\right)^{1/2} dx.$$

This integral would have been considerably more difficult to evaluate. (If you wish to try your hand at it, consider the indirect substitution $x = (2/3)\tan^3(\theta)$.) In general, there is no method for predicting whether it is advantageous to set up an arc length integral in a particular way. The fact is, it is usually impossible to calculate arc length *exactly,* whichever variable of integration is chosen.

Surface Area

Let f be a nonnegative function with continuous derivative on the interval $[a, b]$. Imagine rotating the graph of f about the x-axis. This procedure generates a *surface of revolution,* as shown in Figure 4. We will now develop a procedure for determining the area of such a surface. As with previously developed methods that have led to integral formulas, we first divide the geometric object into a number of pieces that can be approximated by a simple shape. The basic shape that we use here, a *frustum,* is shown in Figure 5 and results when a conical surface is cut by two planes that are perpendicular to the axis of symmetry. The formula for the surface area S of a frustum is given by

$$S = 2\pi\left(\frac{r+R}{2}\right)s \tag{8.9}$$

where r and R are the radii of the circular edges and s is the *slant height,* as indicated in Figure 5. Notice that S also represents the surface area of a cylinder with height equal to the slant height of the frustum and with radius equal to the average radius of the cross sections of the frustum. Formula (8.9) is elementary in the sense that its derivation (as outlined in Exercise 35) does not require calculus.

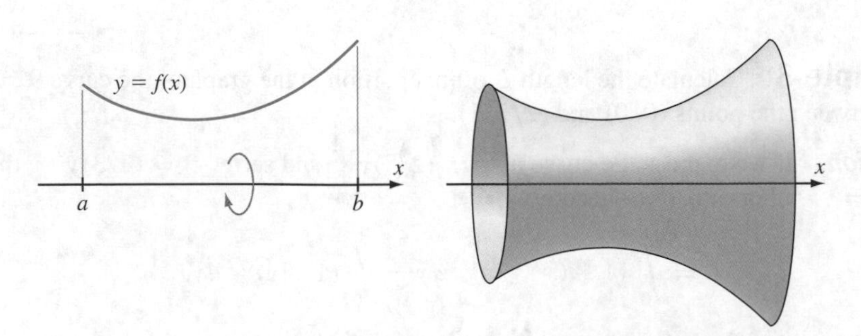

Figure 4
The graph of a function and the surface it generates when rotated about the x-axis

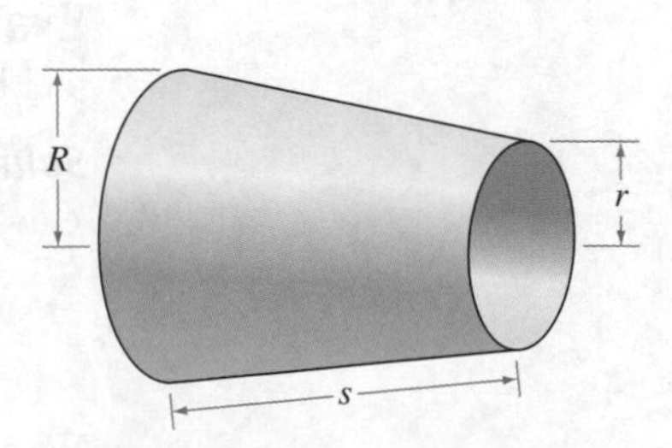

Figure 5
Surface area of frustum $= 2\pi(\frac{r+R}{2})s$

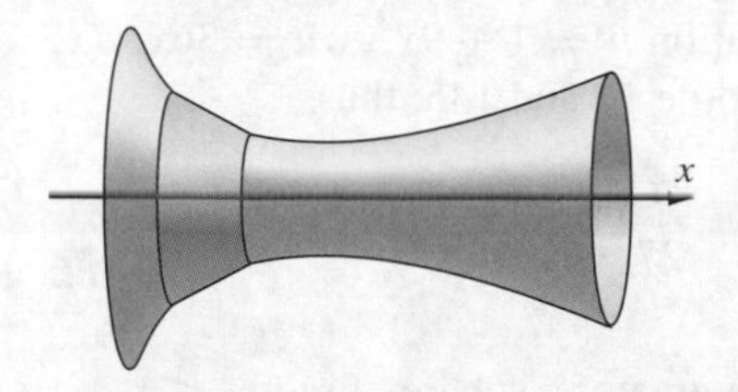

Figure 6
Slice of surface approximated by a frustum of a cone

For each positive integer N, we form a uniform partition $a = x_0 < x_1 < x_2 < \cdots < x_{N-1} < x_N = b$ of the interval $[a, b]$. Let Δx denote the common length of the N subintervals. When the arc of the curve that lies over the jth subinterval is rotated about the x-axis, we obtain a surface that we can approximate by a frustum (as in Figure 6). The slant height of this frustum can be approximated by the arc length ℓ_j of the graph of f from $(x_{j-1}, f(x_{j-1}))$ to $(x_j, f(x_j))$. In the discussion of arc length earlier in this section, we reasoned that there is a point ξ_j in the jth subinterval such that

$$\ell_j \approx (1 + f'(\xi_j)^2)^{1/2} \Delta x.$$

We may use the value $f(\xi_j)$ at this interior point ξ_j to approximate the average value $(f(x_{j-1}) + f(x_j))/2$ of f at the endpoints. It follows from formula (8.8) that the area contribution of the jth increment of our surface is approximately

$$2\pi \cdot f(\xi_j)(1 + f'(\xi_j)^2)^{1/2} \Delta x.$$

If we sum up the area contribution from each subinterval of the partition, we find that the area of our surface of revolution is approximately

$$\sum_{j=1}^{N} 2\pi \cdot f(\xi_j)(1 + f'(\xi_j)^2)^{1/2} \Delta x.$$

The accuracy of this approximation is improved by increasing the number N of subintervals. As N increases to infinity and Δx tends to 0, these approximating sums tend to what we think of as surface area S of the surface of revolution:

$$S = \lim_{N \to \infty} \sum_{j=1}^{N} 2\pi \cdot f(\xi_j)(1 + f'(\xi_j)^2)^{1/2} \Delta x.$$

These sums are also Riemann sums for the integral

$$2\pi \int_a^b f(x)(1 + f'(x)^2)^{1/2}\, dx.$$

We conclude that this integral represents the surface area of the surface of revolution.

Definition If a nonnegative function f has a continuous derivative on an interval containing $[a, b]$, then the *surface area* of the surface of revolution obtained when the graph of f over $[a, b]$ is rotated about the x-axis is given by

$$2\pi \int_a^b f(x)(1 + f'(x)^2)^{1/2}\, dx.$$

Example 4 Let $f(x) = x^3$. For $1 \le x \le 2$, we rotate the graph of f about the x-axis. Calculate the resulting surface area S.

Solution We have

$$S = 2\pi \int_1^2 f(x)(1 + f'(x)^2)^{1/2}\, dx = 2\pi \int_1^2 x^3(1 + (3x^2)^2)^{1/2}\, dx$$

$$= \frac{\pi}{27} \int_1^2 \frac{3}{2}(1 + 9x^4)^{1/2}(36x^3)\, dx.$$

This integral is easily calculated using the substitution $u = 1 + 9x^4$, $du = 36x^3\,dx$. With this substitution, the limits of integration become 10 and 145; thus,

$$S = \frac{\pi}{27}\int_{10}^{145} \frac{3}{2}u^{1/2}\,du = \frac{\pi}{27}u^{3/2}\Big|_{10}^{145} = \frac{\pi}{27}(145^{3/2} - 10^{3/2}). \quad \blacksquare$$

Example 5 Show that $S = 4\pi r^2$ is the surface area of a sphere of radius r.

Solution It is convenient to think of the sphere as the surface of revolution generated by rotating the graph of the function $f(x) = \sqrt{r^2 - x^2}$, $-r \le x \le r$, about the x-axis. Then

$$\sqrt{1 + f'(x)^2} = \sqrt{1 + \left(\frac{-x}{\sqrt{r^2 - x^2}}\right)^2} = \sqrt{1 + \left(\frac{x^2}{r^2 - x^2}\right)}$$

$$= \sqrt{\frac{(r^2 - x^2) + x^2}{r^2 - x^2}} = \frac{r}{\sqrt{r^2 - x^2}}$$

and

$$S = 2\pi\int_{-r}^{r} f(x)\sqrt{1 + f'(x)^2}\,dx = 2\pi\int_{-r}^{r}\sqrt{r^2 - x^2}\frac{r}{\sqrt{r^2 - x^2}}\,dx = 2\pi\int_{-r}^{r} r\,dx = 4\pi r^2. \quad \blacksquare$$

We may also consider the area of a surface obtained by rotating the graph of a function about the y-axis. We do so by using y as the independent variable, as in the next example.

Example 6 Set up, but do not evaluate, the integral for finding the area of the surface obtained when the graph of $f(x) = x^4$, $1 \le x \le 5$, is rotated about the y-axis (see Figure 7).

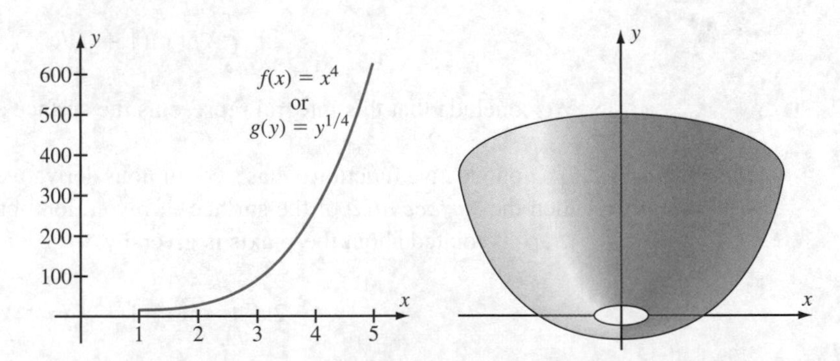

Figure 7
The surface generated by rotating the graph of f about the y-axis. The xy-plane is tilted to reveal the opening at the bottom of the surface.

Solution We think of the curve as the graph of $g(y) = y^{1/4}$, $1 \le y \le 625$. The formula for surface area is

$$2\pi\int_{1}^{625} g(y)(1 + g'(y)^2)^{1/2}\,dy.$$

Calculating $g'(y)$ and substituting, we find that the desired surface area is the value of the integral

$$2\pi \int_1^{625} y^{1/2}\left(1+\left(\frac{1}{4}y^{-3/4}\right)^2\right)^{1/2} dy. \qquad \blacksquare$$

quickquiz

1. What piecewise linear approximation do we use when estimating the length of a curve?
2. Write the integral that represents the length of a semicircle and evaluate it.
3. What shape is used to approximate the area of a surface of revolution?
4. Write the integral that represents the surface area generated by rotating $y = mx$, $0 \le x \le h$, about the x-axis. Evaluate it.

EXERCISES

Problems for Practice

In Exercises 1–4, use the integral formula to calculate the arc length of the graph of the function $f(x)$ over the interval I. Verify that the arc length is the distance between the two endpoints.

1. $f(x) = 6x$, $I = [3, 8]$
2. $f(x) = 2x + 1$, $I = [0, 4]$
3. $f(x) = 3 - x$, $I = [-1, 2]$
4. $f(x) = 10 - 2x$, $I = [0, 5]$

In Exercises 5–8, calculate the arc length of the graph of the function $f(x)$ over the interval I.

5. $f(x) = 2 + x^{3/2}$, $I = [1, 4]$
6. $f(x) = (x-1)^{3/2}$, $I = [2, 5]$
7. $f(x) = 3 + (2x+1)^{3/2}$, $I = [0, 4]$
8. $f(x) = (3x-5)^{3/2}$, $I = [2, 10]$

In Exercises 9–16, calculate the area of the surface obtained when the function $f(x)$ is rotated about the x-axis over the interval I.

9. $f(x) = 2x^3$, $I = [-2, 3]$
10. $f(x) = 3x^{1/2}$, $I = [4, 9]$
11. $f(x) = x^3/3 + 1/(4x)$, $I = [1, 3]$
12. $f(x) = e^x/2 + e^{-x}/2$, $I = [-1, 1]$
13. $f(x) = 3x - 1$, $I = [1, 4]$
14. $f(x) = (x+1)^{1/2}$, $I = [3, 8]$
15. $f(x) = x^4 + x^{-2}/32$, $I = [-1, -1/2]$
16. $f(x) = x^3/2 + x^{-1}/6$, $I = [1, 2]$

In Exercises 17–20, calculate the area of the surface obtained when the given function is rotated about the y-axis over the interval I.

17. $f(x) = x^{1/3}$, $I = [8, 27]$
18. $f(x) = \ln\left(x + \sqrt{x^2 - 1}\right)$, $I = [3, 5]$
19. $f(x) = x^2$, $I = [-6, -2]$
20. $f(x) = x/4 - 2$, $I = [-8, -4]$

Further Theory and Practice

In Exercises 21–25, calculate the arc length of the graph of the given function over the interval I. (In these exercises, the functions have been contrived to permit a simplification of the radical in the arc length formula.)

21. $f(x) = (1/3)(x^2+2)^{3/2}$, $I = [0, 1]$
22. $f(x) = (1/6)x^3 + (1/2)x^{-1}$, $I = [1, 2]$
23. $f(x) = e^x/2 + e^{-x}/2 + 7$, $I = [0, 1]$
24. $f(x) = (2^x + 2^{-x})/(2\ln(2))$, $I = [0, 2]$
25. $f(x) = 2x^4 + 1/(64x^2)$, $I = [-2, -1]$
26. Find the arc length of the graph of $y^4 - 16x + 8/y^2 = 0$ between $P = (9/8, 2)$ and $Q = (9/16, 1)$.
27. Find the arc length of the graph of $30yx^3 - x^8 = 15$ between $P = (0.1, 500.0\ldots)$ and $Q = (1, 8/15)$.
28. Consider a curve that is the graph of a function f joining the points $(a, f(a))$ and $(b, f(b))$. If we translate this curve horizontally by an amount h and vertically by an amount v so that the translated curve joins the points $(a+h, f(a)+v)$ and

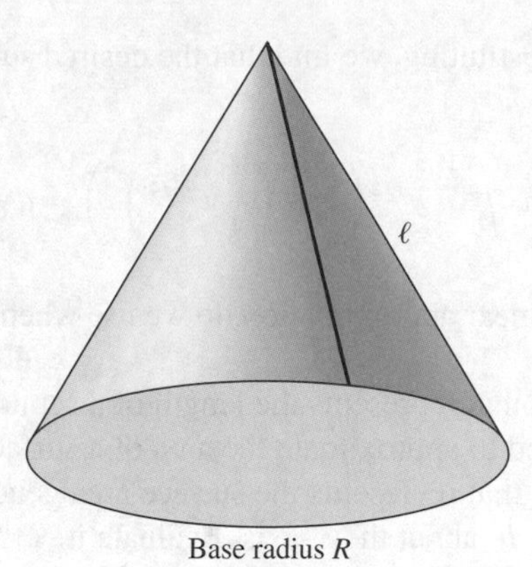

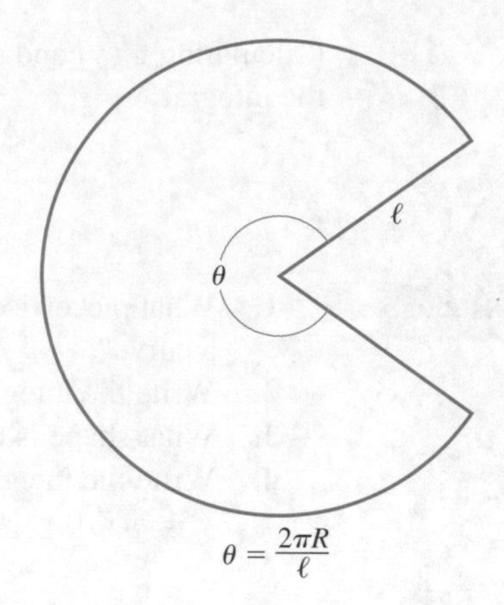

Figure 8

$(b+h, f(b)+v)$, then the translated curve should have the same arc length as the original. Prove this.

29. Suppose that $a > b > 0$. Let $\varepsilon = \sqrt{a^2 - b^2}/a$. Show that for any $0 < \xi < a$,

$$\int_0^{\xi} \sqrt{\frac{a^2 - \varepsilon^2 x^2}{a^2 - x^2}}\, dx$$

is the length of the arc of the ellipse $x^2/a^2 + y^2/b^2 = 1$ that lies above the interval $[0, \xi]$ of the x-axis. Make a change of variable to write this integral in terms of the so-called *elliptic integral* $\int_0^{\eta} \sqrt{1 - \varepsilon^2 \sin^2(t)}\, dt$ where η depends on a and ξ. Elliptic integrals cannot be evaluated using elementary functions, so they are tabulated.

Suppose each of the functions $t \mapsto x(t)$ and $t \mapsto y(t)$ has a continuous derivative. The curve that is defined parametrically by the equations $x = x(t)$ and $y = y(t)$ $(\alpha \le t \le \beta)$ has arc length

$$\int_{\alpha}^{\beta} \sqrt{x'(t)^2 + y'(t)^2}\, dt.$$

Although this formula can be obtained using the methods learned in this section, we will instead derive it in a more general context in Chapter 12. For now, use the formula to calculate the length of the parametric curves in Exercises 30–33.

30. $x = t\cos(t),\ y = t\sin(t),\ 0 \le t \le 3\pi$

31. $x = \exp(t)\cos(t),\ y = \exp(t)\sin(t),\ 0 \le t \le 1$

32. $x = t - \sin(t),\ y = 1 - \cos(t),\ 0 \le t \le 2\pi$

33. $x = 2\arctan(t),\ y = \ln(1 + t^2),\ 0 \le t \le 1$

34. Observe that we can parameterize the ellipse $x^2/a^2 + y^2/b^2 = 1$ by the equations $x = a\cos(t)$ and $y = b\sin(t)$ $(0 \le t \le 2\pi)$. Let $E(s)$ denote the arc length of that part of the ellipse given by $0 \le t \le s$. Using the instructions to Exercises 30–33, express $E(s)$ by means of an integral. (Do not attempt to evaluate the integral.)

35. When the right circular cone of slant height ℓ and base radius R is cut along a lateral edge and flattened, it becomes a sector of a circle of radius ℓ and central angle θ, as in Figure 8. Prove that $\theta = 2\pi R/\ell$. Use this observation to derive formula (8.9) for the surface area of a frustum of a cone.

36. By means of integration, calculate the surface area of a right circular cone of radius r and height h. Verify that your answer coincides with the area given by formula (8.9). (Think about why this is not really a new method for obtaining the formula.)

37. Find the area of a surface obtained by rotating the graph of $f(x) = \sin(x)$, $\pi/4 \le x \le 3\pi/4$, about the x-axis.

38. Find the area of a surface obtained by rotating the graph of $f(x) = \cos(x)$, $0 \le x \le \pi/3$, about the x-axis.

39. Find the area of a surface obtained by rotating the graph of $f(x) = e^x$, $0 \le x \le 2$, about the x-axis.

40. Let T denote the solid of revolution that is enclosed when the curve $y = 1/x$, $1 \le x < \infty$, is rotated about the x-axis. This solid is called *Torricelli's infinitely long solid*. Let V_N and S_N denote the volume and surface area of that part for which $1 \le x \le N$. Calculate the finite number $\lim_{N\to\infty} V_N$. Show that $S_N > \ln(N)$ and therefore $\lim_{N\to\infty} S_N = \infty$.

Calculator/Computer Exercises

41. Use Simpson's Rule to calculate the arc length of the ellipse

$$\frac{x^2}{4} + \frac{y^2}{16} = 1$$

to two decimal places of accuracy.

42. Use Simpson's Rule to calculate the arc length of the graph of $y = \sin(x)$, $0 \le x \le 2\pi$, to four decimal places of accuracy.

43. Plot $T(x) = 4x^3 - 3x$ for $-1 \le x \le 1$. Notice that the plot is contained in the square $[-1, 1] \times [-1, 1]$. Of all degree three polynomials that have this containment property, T has the longest arc length. Use Simpson's Rule to calculate the arc length of the graph of $y = T(x)$, $-1 \le x \le 1$, to four decimal places of accuracy.

44. The surface of a flashlight reflector is obtained when $y = 2.530\sqrt{x/(x+1)}$, $0 \le x \le 2.5$ cm, is rotated about the x-axis. Calculate its surface area to three decimal places.

45. The equation of the Gateway Arch in St. Louis, Missouri, is

$$y = 693.8597 - 34.38365(\exp(kx) + \exp(-kx))$$

for $k = 0.0100333$ and $-299.2239 < x < 299.2239$ where x and y are measured in feet. Find an equation of a parabola with the same vertex and feet. Plot both curves on the same set of coordinate axes. Compute the length of each.

46. Calculate the surface area that results when the curve in Exercise 45 is rotated about the y-axis.

8.3 The Average Value of a Function

If we took five temperature readings during the day, and if those measurements were 50, 58, 68, 70, and 54 (in degrees Fahrenheit), then the *average value* (or *mean value*) of our readings would be

$$\frac{50 + 58 + 68 + 70 + 54}{5} = \frac{300}{5} = 60.$$

We often need to average infinitely many values. For instance, consider the plot of temperature given in Figure 1. To speak of an average temperature *during a day,* we must average the temperature at infinitely many instants of time. The method for averaging infinitely many values, which is the subject of this section, is to use an integral.

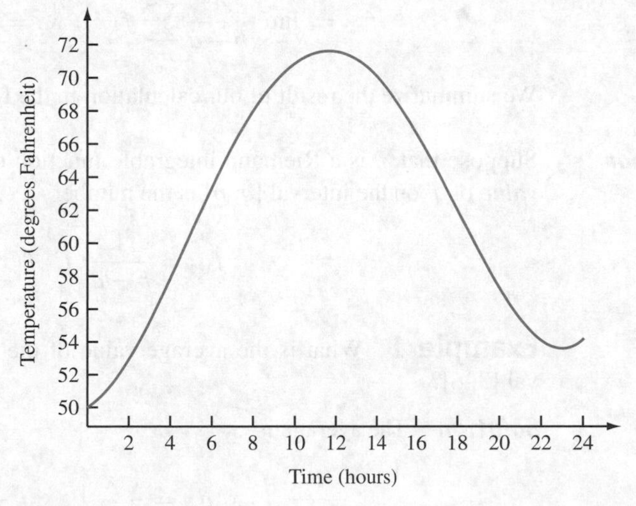

Figure 1

The Basic Technique

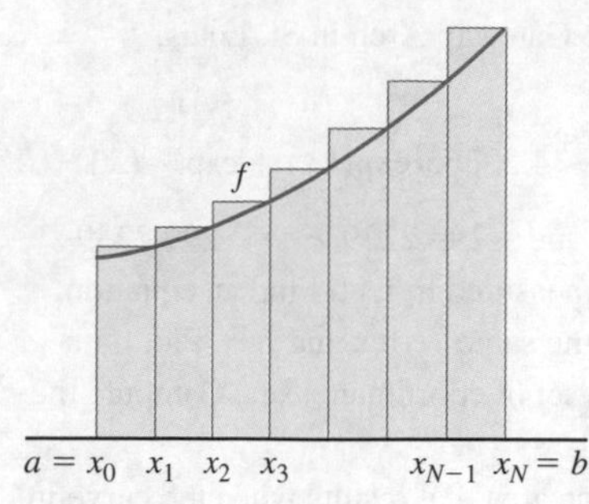

Figure 2

Suppose f is a continuous function with a domain that contains the interval $[a, b]$. To estimate the average value of f on this interval, we take a uniform partition of the interval:

$$a = x_0 < x_1 < x_2 < \cdots < x_{N-1} < x_N = b.$$

As usual, let $\Delta x = (b - a)/N$ denote the common length of the subintervals formed by the partition. If we add up the values of f at the right endpoints of the partition (see Figure 2) and divide by the number of points, then the resulting expression is an approximation to the average value f_{ave} of f:

$$f_{\text{ave}} \approx \frac{f(x_1) + f(x_2) + \cdots + f(x_{N-1}) + f(x_N)}{N}.$$

As we let N increase to infinity, our approximation uses more and more points from the interval $[a, b]$ and becomes more accurate. We therefore define the average value f_{ave} of $f(x)$ for x in $[a, b]$ to be

$$f_{\text{ave}} = \lim_{N\to\infty} \frac{f(x_1) + f(x_2) + \cdots + f(x_{N-1}) + f(x_N)}{N}.$$

We may rewrite the expression in the limit as

$$\begin{aligned}\frac{f(x_1) + f(x_2) + \cdots + f(x_{N-1}) + f(x_N)}{N} &= \frac{1}{N}\sum_{j=1}^{N} f(x_j) = \frac{1}{N\,\Delta x}\sum_{j=1}^{N} f(x_j)\,\Delta x \\ &= \frac{1}{b-a}\sum_{j=1}^{N} f(x_j)\,\Delta x.\end{aligned}$$

Since $\sum_{j=1}^{N} f(x_j)\,\Delta x$ is a Riemann sum for the integral $\int_a^b f(x)\,dx$, we conclude that

$$f_{\text{ave}} = \lim_{N\to\infty} \frac{1}{b-a}\sum_{j=1}^{N} f(x_j)\,\Delta x = \frac{1}{b-a}\int_a^b f(x)\,dx.$$

We summarize the result of our calculation in the following definition.

Definition Suppose that f is a Riemann integrable function on the interval $[a, b]$. The *average value* of f on the interval $[a, b]$ is the number

$$f_{\text{ave}} = \frac{1}{b-a}\int_a^b f(x)\,dx. \qquad \textbf{(8.10)}$$

Example 1 What is the average value of the function $f(x) = x^2$ on the interval $[3, 6]$?

Solution The average is

$$\frac{1}{6-3}\int_3^6 x^2\,dx = \frac{1}{3}\,\frac{x^3}{3}\bigg|_{x=3}^{x=6} = \frac{1}{9}(6^3 - 3^3) = 21. \qquad \blacksquare$$

Example 2 A rod of length 9 cm has temperature distribution $(2x - 6x^{1/2})$°C for $0 \le x \le 9$. This means that, at position x on the rod, the temperature is $(2x - 6x^{1/2})$°C. Calculate the average temperature of the rod.

Solution The computation

$$\frac{1}{9-0}\int_0^9 (2x - 6x^{1/2})\,dx = \frac{1}{9}(x^2 - 4x^{3/2})\Big|_{x=0}^{x=9} = \frac{1}{9}((81 - 4\cdot 27) - (0 - 4\cdot 0)) = -3$$

shows that the average temperature of the rod is 3°C below zero. ■

inSIGHT

The expected cost of heating a building during the winter is often calculated using the concept of *degree days*. A simple definition of degree day for a given day is 68°F, less the average outdoor temperature for that day. For example, if the cost of heating the weather station in Example 3 is 35 cents per degree day, then we would expect the heating bill for the day in question to be about $\$0.35 \cdot (68 - 62.205687) \approx \$2.03$.

Example 3 Suppose that, at a weather station, the temperature f in degrees Fahrenheit is recorded to be

$$f(x) = 0.00103265x^4 - 0.04678932x^3 + 0.50831530x^2 + 0.64927850x + 50$$

for $0 \le x \le 24$ hours. What is the average temperature for that day?

Solution The graph of f is given in Figure 1 (see page 583). By scanning the values taken by f, we might estimate the average value to be between 60 and 64. At the beginning of this section, we used five values—$f(0) = 50$, $f(4) = 58$, $f(8) = 68$, $f(14) = 70$, and $f(22) = 54$—to obtain a sample average of 60. Now we can calculate the average over *all* values of f:

$$f_{\text{ave}} = \frac{1}{24}\int_0^{24} f(x)\,dx.$$

Using a computer algebra system or calculator to do the arithmetic calculations, we obtain $f_{\text{ave}} = 62.21$ (to two decimal places). ■

Example 4 Let f_{ave} denote the average value of f over the interval $[a, b]$. Show that the function f and the constant function $g(x) = f_{\text{ave}}$ have the same integral over $[a, b]$.

Solution We obtain the required equality as follows:

$$\int_a^b g(x)\,dx = \int_a^b f_{\text{ave}}\,dx = (b-a)\cdot f_{\text{ave}} = (b-a)\cdot\left(\frac{1}{(b-a)}\int_a^b f(x)\,dx\right) = \int_a^b f(x)\,dx.$$

■

inSIGHT

The result in Example 4 is the continuous analogue of a well-known fact about discrete averages. Namely, if $\overline{y}$ is the average value of the N numbers $y_1, y_2, \ldots, y_N$, then

$$\sum_{j=1}^{N} \overline{y} = N\cdot\overline{y} = N\cdot\left(\frac{1}{N}\sum_{j=1}^{N} y_j\right) = \sum_{j=1}^{N} y_j.$$

It is often the case that a theorem about finite sums has a continuous analogue in which the sums are replaced by integrals.

A LOOK BACK

In Section 5.3, we learned the Mean Value Theorem for Integrals, which asserts that if f is continous on an interval $[a, b]$, then there is a value c in the interval such that

$$f(c) = \frac{1}{b-a}\int_a^b f(x)\,dx. \tag{8.11}$$

Since the right side of equation (8.11) is f_{ave}, we may interpret the Mean Value Theorem for Integrals in this way: A continuous function on a closed bounded interval *assumes* its average value. This is a property that discrete samples do not necessarily have. The average of the five temperatures in Example 3 is 60, which does not equal any of the five sample temperatures.

Example 5 At what point (or points) is the temperature of the metal rod from Example 2 equal to the average temperature of the rod?

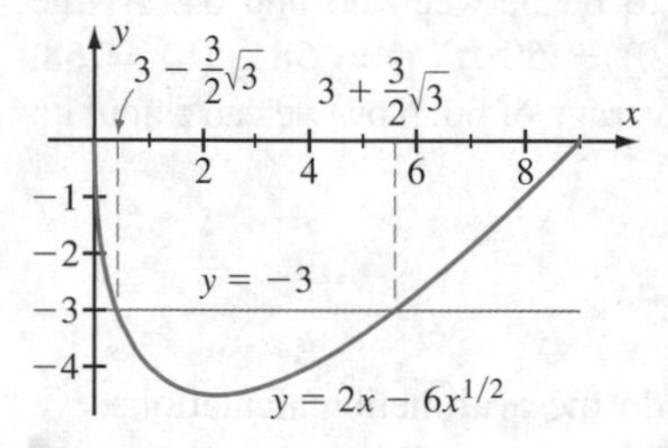

Figure 3

Solution Since the temperature of the rod, $2x - 6x^{1/2}$, is continuous, we can be sure that there is at least one point at which the temperature is equal to the average temperature $-3°$C (a value that we calculated in Example 2). Figure 3 shows that there are actually two points at which the average temperature is attained. To find these two points, we solve the equation $2x - 6x^{1/2} = -3$. We rewrite this as $2x + 3 = 6x^{1/2}$. If we square each side, then we obtain the quadratic equation $(2x+3)^2 = 36x$, or $4x^2 - 24x + 9 = 0$. According to the quadratic formula, the roots are $3 - 3\sqrt{3}/2 \approx 0.40192$ and $3 + 3\sqrt{3}/2 \approx 5.59808$. As expected, both roots are in the interval $[0, 9]$. ■

Random Variables

Many important variables assume their values in a "random" manner. For example, a company may be able to say quite a bit *in general* about the longevity of the computer hard drives it produces. The manufacturer cannot, however, say anything definite about the lifetime X of the particular hard drive in *your* computer. We say that X is a *random variable*. Similarly, a paleontologist may know what range of values the heights of velociraptors lie in. What they cannot specify for certain is the height of the next specimen to be discovered: It is a random variable.

The theory of probability is largely concerned with the analysis of random variables. The field of statistics uses probability theory to make estimates and inferences about random variables based on observed values. In this subsection, we learn how calculus plays a role in determining the average value of a random variable.

Suppose X is a random variable with values that all lie in an interval $I = [a, b]$. If $[\alpha, \beta]$ is a subinterval of I, then we write $P(\alpha \le X \le \beta)$ to denote the probability that the random variable X takes on a value in the subinterval $[\alpha, \beta]$. Just as in everyday informal language, a probability in mathematics is a number between 0 and 1. For example, we have $P(a \le X \le b) = 1$ because it is certain, according to our assumptions, that X takes a value in this range.

Definition Suppose that X is a random variable with values that all lie in an interval I. If there is a function f such that

$$P(\alpha \leq X \leq \beta) = \int_{\alpha}^{\beta} f(x)\,dx$$

for every subinterval $[\alpha, \beta]$ of I, then we say that f is a *probability density function* of X. The abbreviation *p.d.f.* is commonly used.

A continuous probability density function f on $I = [a, b]$ is nonnegative. In other words, $f(x) \geq 0$ for all $x \in I$ (see Exercise 33). Also, a p.d.f. on I satisfies $\int_a^b f(x)\,dx = 1$, as discussed earlier. In probability theory, it is shown that any continuous function that satisfies both of these properties is the p.d.f. of some random variable.

Example 6 Let X denote the time to failure (in hours) of an integrated circuit that has been placed into service. If $f(x) = (1/20000)e^{-x/20000}$ is the p.d.f. of X, then what is the probability that the circuit will fail within its first 10,000 hours?

Solution The required probability is given by

$$P(0 \leq X \leq 10000) = \int_0^{10000} \frac{1}{20000} e^{-x/20000}\,dx = -e^{-x/20000}\Big|_{x=0}^{x=10000} = 1 - e^{-1/2} \approx 0.3935. \quad \blacksquare$$

Average Values in Probability Theory

Suppose that X is a random variable with range contained in the interval $I = [a, b]$. Suppose that f is a continuous function that is the probability density function of X. Our goal is to determine the average value μ of X. We cannot simply use our previous formula and integrate X because we do not have a deterministic formula for X. We can calculate the average value of f, but f_{ave} does not give us the average value μ of X.

To calculate μ, we again rely on a discrete model. If we made a large number N of observations $X_1, \ldots, X_N$ of X, then we would expect

$$\mu \approx \frac{X_1 + X_2 + \cdots + X_N}{N} \tag{8.12}$$

to be a good approximation. Form a partition $a = x_0 < x_1 < \cdots < x_M = b$ of the interval I into M subintervals of equal length Δx. For each j, we approximate the contribution of the terms in the numerator of formula (8.12) that lie in the jth subinterval $I_j = [x_{j-1}, x_j]$. First we estimate the number of these observations. To do so, we note that the probability that an observation of X will lie in the subinterval I_j is

$$P(x_{j-1} \leq X \leq x_j) = \int_{x_{j-1}}^{x_j} f(x)\,dx = f(c_j)\Delta x$$

where $f(c_j)$ is the average value of f in the subinterval I_j—refer to formula (8.11). It follows that approximately $N \cdot (f(c_j)\Delta x)$ of the observations will lie in I_j. Each of these observations will be between x_{j-1} and x_j. Since c_j is in this small interval, we may use c_j as a representative value of the observations in I_j. The contribution of the terms in I_j is therefore approximately $c_j \cdot (N \cdot f(c_j)\Delta x)$. By grouping the summands of $X_1 + X_2 + \cdots + X_N$ according to the subintervals in which they lie, we may rewrite our approximation of μ as

$$\mu \approx \frac{1}{N} \sum_{j=1}^{M} c_j \cdot (N \cdot f(c_j)\Delta x) = \sum_{j=1}^{M} c_j \cdot f(c_j)\Delta x.$$

Since $\sum_{j=1}^{M} c_j \cdot f(c_j)\Delta x$ is a Riemann sum for the integral $\int_I xf(x)\,dx$, we are led to the following definition.

Definition If f is the probability density function of a random variable X that takes values in an interval $I = [a, b]$, and if $xf(x)$ is Riemann integrable over I, then the *average* (or *mean*) μ of X is defined as

$$\mu = \int_a^b xf(x)\,dx.$$

This value is also said to be the *expectation* of X. Other notations for the expectation of X are $\overline{X}$ and $E(X)$.

Example 7 Let X denote the fraction of total impurities that are filtered out in a particular purification process. Suppose X has probability density function $f(x) = 20x^3(1-x)$ for $0 \le x \le 1$. What is the average of X?

Solution According to our definition, the average is

$$\int_0^1 x \cdot 20x^3(1-x)\,dx = \int_0^1 (20x^4 - 20x^5)\,dx = 20\left(\frac{x^4}{5} - \frac{x^6}{6}\right)\Bigg|_{x=0}^{x=1} = \frac{2}{3}. \quad \blacksquare$$

quickquiz

1. What does it mean to calculate the average of a continuous function over a closed bounded interval?
2. What is the average of the function $\sin(2x)$ over the interval $[0, \pi]$? Does your answer have a geometric interpretation?
3. Suppose that a random variable X has range $[a, b]$. If f is the probability density function of X, then what is the value of $\int_a^b f(x)\,dx$?
4. Suppose that a random variable X has p.d.f. $f(x) = x/2$ for $0 \le x \le 2$. What is the mean of X?

EXERCISES

Problems for Practice

In Exercises 1–8, calculate the average value of the function $f(x)$ on the interval I.

1. $f(x) = \cos(x)$, $I = [0, \pi/2]$
2. $f(x) = x^2$, $I = [3, 7]$
3. $f(x) = 1/x$, $I = [1, 4]$
4. $f(x) = 3x^2 - 6x + 1$, $I = [3, 7]$
5. $f(x) = \sin(x)$, $I = [0, \pi]$
6. $f(x) = 1 + \cos(x)$, $I = [0, 2\pi]$
7. $f(x) = (x-1)^{1/2}$, $I = [2, 5]$
8. $f(x) = x^{1/2} - x^{1/3}$, $I = [1, 64]$

In Exercises 9–14, the probability density function f of a random variable X is given. Calculate $P(\alpha \le X \le \beta)$ for the interval $J = [\alpha, \beta]$.

9. $f(x) = 3x^2/8, 0 \le x \le 2, J = [1, 2]$
10. $f(x) = 12x^2(1-x), 0 \le x \le 1, J = [0, 1/2]$
11. $f(x) = \sin(x)/2, 0 \le x \le \pi, J = [\pi/4, \pi/2]$
12. $f(x) = 3(1+x)/(8\sqrt{x}), 0 \le x \le 1, J = [1/4, 1/2]$
13. $f(x) = e^{1-x}/(e-1), 0 \le x \le 1, J = [0, 1/2]$
14. $f(x) = \ln(x), 1 \le x \le e, J = [3/2, 2]$

In Exercises 15–20, calculate the mean of the random variable with the given probability density function.

15. $f(x) = 3x^2, I = [0, 1]$
16. $f(x) = 2(1-x), I = [0, 1]$
17. $f(x) = 1/(3\sqrt{x}), I = [1/4, 4]$
18. $f(x) = 6x(1-x), I = [0, 1]$
19. $f(x) = x^2/3, I = [-1, 2]$
20. $f(x) = 3/(\pi(1+x^2)), I = [0, \sqrt{3}]$
21. A patient with a fever has temperature at time t (measured in hours) given by

$$T(t) = 99.6 - t + 0.8t^2 \text{ degrees}$$

where temperature is measured in degrees Fahrenheit. Find the patient's average temperature as t ranges from 0 to 3.
22. The temperature distribution on a uniform rod of length 4 m is $T(s) = 18 + 6s^2 - 2s + \sqrt{s}, 0 \le s \le 4$. What is the average temperature of the rod?

Further Theory and Practice

In Exercises 23–28, calculate the average of the given expression over the given interval.

23. $x\sin(x^2), 0 \le x \le \pi$
24. $\cos(x)\sqrt{1+\sin(x)}, 0 \le x \le \pi$
25. $\tan(x), 0 \le x \le \pi/4$
26. $\sec(x), 0 \le x \le \pi/4$
27. $x\sin(x), 0 \le x \le \pi$
28. $\ln(x), 1 \le x \le e$
29. For what value of c is $-1/2$ the average value of $(x-c)\sin(x)$ over the interval $[0, \pi/3]$?
30. Is the average value of $\cos(x)$ for $0 \le x \le \pi/4$ equal to the reciprocal of the average value of $1/\cos(x)$ over the same x-interval?
31. The outdoor temperature in a certain town is given by

$$T(t) = 40 - \frac{(t-45)^2}{200}$$

for t ranging from 0 to 72 hours. Use the definition of degree days given in Example 3 to calculate the degree days for each of the three successive 24-hour periods. If fuel costs a certain homeowner 30 cents per degree day, what does it cost her to heat her house during these three days?
32. The temperature of an aluminum rod at point x is given by

$$T(x) = \begin{cases} x^3 - x^2 + 32, & 0 \le t \le 2 \\ 36 - 2t + t^2, & 2 < t \le 8 \end{cases}.$$

Find the average temperature of the rod.
33. Suppose f is a continuous p.d.f. of a random variable X on an interval I. Suppose that ξ is a point in I. For each n sufficiently large, choose a subinterval $I_n = [\alpha_n, \alpha_n + 1/n]$ of I that contains ξ. Show that there is a point c_n in I_n such that

$$f(c_n) = n\int_{\alpha_n}^{\alpha_n+1/n} f(x)\,dx.$$

Show that $c_n \to \xi$ and deduce that $f(\xi) \ge 0$. In other words, *a continuous probability density function must be nonnegative.*

Suppose that g is a nonnegative, continuous function on an interval I. In courses in probability theory, it is shown that if the integral of g over I is a positive real number c, then there is a random variable that takes values in I and that has $f(x) = (1/c)g(x)$ as its probability density function. Exercises 34–39 involve an interval I and a nonnegative function g on I. Find a value c such that $f(x) = (1/c)g(x)$ is a probability density function.

34. $g(x) = \ln(x), I = [e, e^2]$
35. $g(x) = x^3(1-x)^2, I = [0, 1]$
36. $g(x) = x^2(1-x^3)^{-1/2}, I = [0, 1]$
37. $g(x) = (9+x^2)^{-1/2}, I = [0, 4]$
38. $g(x) = \arcsin(x), I = [0, 1]$
39. $g(x) = x(1-x)^{-1/2}, I = [0, 1]$
40. For what a, $0 \le a \le \pi$, does the function $f(x) = \sin(x) - \cos(x)$ have the greatest average over the interval $[a, a+\pi]$ of length π?
41. Given $c > 0$, for what value b, $0 < b < c$, is $\exp(b)$ equal to the average of $\exp(x)$ for $0 \le x \le c$?

In Exercises 42–45, calculate the expectation of a random variable with the given probability density function.

42. $(1+\cos(x))/(2\pi), 0 \le x \le 2\pi$
43. $\ln(x), 1 \le x \le e$
44. $e^{1-x}/(e-1), 0 \le x \le 1$
45. $(\exp(1-x/e) - 1)/(e^2 - 2e), 0 \le x \le e$
46. Let p be a positive constant. Let X be a random variable with p.d.f. $f(x) = (p+1)x^p$ for $0 \le x \le 1$. Show that $\overline{X} \ne f_{\text{ave}}$. (In general, the average of a

random variable is *not* equal to the average of its probability density function.)

We say that m is a median of a random variable X if $P(X \le m) = P(X \ge m) = 1/2$. In Exercises 47–50, calculate a median of a random variable with the given probability density function.

47. $\cos(x), 0 \le x \le \pi/2$

48. $3x^2/26, 1 \le x \le 3$

49. $e^{1-x}/(e-1), 0 \le x \le 1$

50. $4x(1-x^2), 0 \le x \le 1$

51. Let $g(u) = 1 + u - 2^u$ for $0 \le u \le 1$. Calculate $g'(u)$. Show that $g(0) = g(1) = 0$ but that $g(u) > 0$ for $0 < u < 1$. Let p be a positive constant. By setting $u = 1/(p+1)$ deduce that $(p+2)/(p+1) > 2^{1/(p+1)}$. Let X be a random variable with p.d.f. $f(x) = (p+1)x^p$ for $0 \le x \le 1$. Show that $\overline{X}$ is not equal to the median of X. (In general, the mean of a random variable is *not* equal to its median.)

52. Suppose that f is a continuous positive function on the unbounded interval $[a, \infty)$. Is it appropriate to make the definition

$$f_{\text{ave}} = \lim_{N \to \infty} \frac{1}{N-a} \int_a^N f(x)\,dx?$$

Discuss why or why not.

Calculator/Computer Exercises

In Exercises 53 and 54, choose a small positive value of h. For the function f, define $F(x)$ to be the average of f over the interval $[x-h, x+h]$. Plot $f(x)$ and $F(x)$ for $-1/2 \le x \le 1/2$. The smoothness gained by the averaging process should be evident.

53. $f(x) = |x|$

54. $f(x) = \begin{cases} 0 & -2 \le x < 0 \\ 1 & 0 \le x \le 2 \end{cases}$

In Exercises 55 and 56, perform the following.

a. Plot the function f over the indicated interval.
b. Calculate the average f_{ave} of f over the interval.
c. Find a value c in the interval such that $f(c) = f_{\text{ave}}$.
d. Plot a rectangle with area equal to the area under the graph of f.

55. $f(x) = \exp(-x^2), 0 \le x \le 1$

56. $f(x) = \sin(\pi(x^2 - x^3)), 0 \le x \le 1$

Let μ be any real number. Suppose that σ is a positive real number. In a sense that will be made precise in Section 8.7, the function

$$f(x) = \frac{1}{\sqrt{2\pi}\sigma} \exp\left(-\frac{1}{2}\left(\frac{x-\mu}{\sigma}\right)^2\right),$$

$-\infty < x < \infty$, is a probability density function of a random variable X that takes values in the unbounded interval $(-\infty, \infty)$. We call f a *normal* probability density and X a *normal* or *Gaussian* random variable. In the case that $\mu = 0$ and $\sigma = 1$, we use the term *standard normal*. The parameter σ is called the *standard deviation* of X, and μ is the mean of X. Exercises 57 and 58 concern a normal random variable X.

57. For most practical purposes, $[\mu - 5\sigma, \mu + 5\sigma]$ may be used as the range of values taken by a random variable X with a normal probability density function. This is true because it is nearly certain that X takes a value that is within five standard deviations of its mean. In fact,

$$P(\mu - 5\sigma \le X \le \mu + 5\sigma) > 0.999999.$$

Choose specific values for μ and σ and verify this inequality for your choices.

58. Suppose X is a normal random variable with mean μ and standard deviation σ. Let p be the probability that X takes a value that is within one standard deviation of its mean. It is not hard to show that p does not depend on the specific values of μ and σ. Choose a value for each of these two parameters and determine p.

8.4 Center of Mass

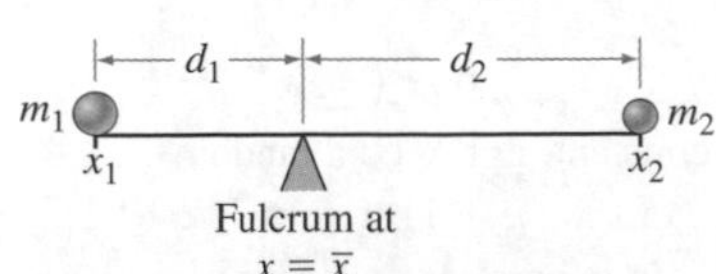

Figure 1

When two children of different masses m_1 and m_2 play on a seesaw, they find that they can balance the seesaw if they position themselves appropriately. Let us set up a simple model of the seesaw. Suppose that two point masses m_1 and m_2 are situated at endpoints of an interval $[x_1, x_2]$ on the x-axis, as in Figure 1. If a fulcrum could be positioned underneath the axis and if the interval could pivot about the fulcrum, then

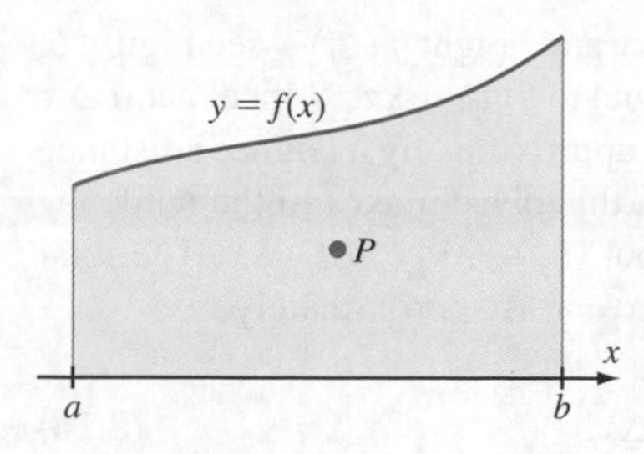

Figure 2
The shaded region balances at point *P*.

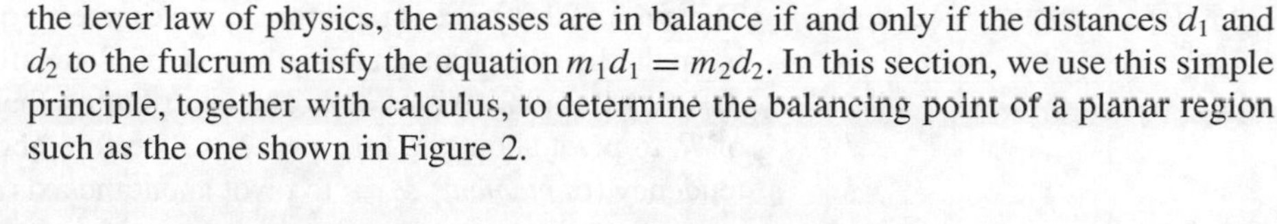

at what coordinate $\overline{x}$ should we place the fulcrum to achieve a balance? According to the lever law of physics, the masses are in balance if and only if the distances d_1 and d_2 to the fulcrum satisfy the equation $m_1 d_1 = m_2 d_2$. In this section, we use this simple principle, together with calculus, to determine the balancing point of a planar region such as the one shown in Figure 2.

Moments of Two-Point Systems

Before turning our attention to planar figures, it is useful to study the seesaw in Figure 1 in greater detail. The point $\overline{x}$ is said to be the *center of mass* of the system. We may rewrite the equation $m_1 d_1 = m_2 d_2$ in terms of the center of mass as follows: $m_1 \cdot (\overline{x} - x_1) = m_2 \cdot (x_2 - \overline{x})$, or

$$m_1 \cdot (x_1 - \overline{x}) + m_2 \cdot (x_2 - \overline{x}) = 0. \tag{8.13}$$

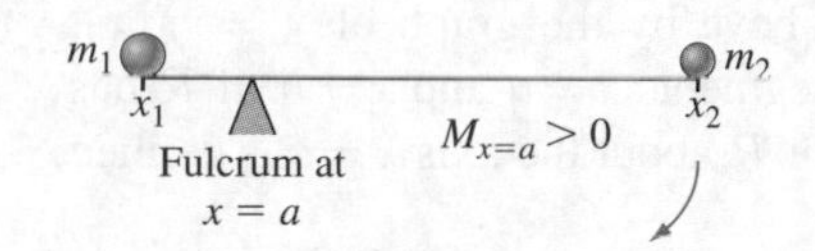

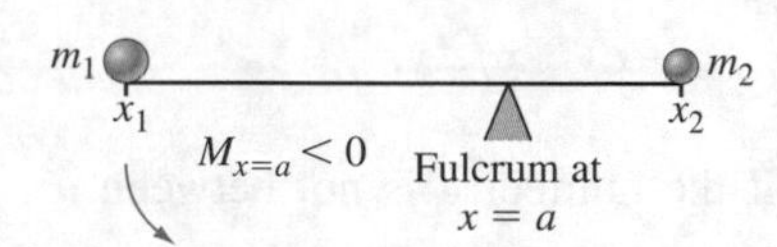

Figure 3

The left side of this equation is called the *moment* about the axis $x = \overline{x}$. The lever law, as expressed by equation (8.13), tells us that our system will balance if and only if the *moment* about the axis $x = \overline{x}$ is zero. In general, if we position the fulcrum at any point $x = c$, then we define the *moment* $M_{x=c}$ about the axis $x = c$ to be the expression

$$M_{x=c} = m_1 \cdot (x_1 - c) + m_2 \cdot (x_2 - c). \tag{8.14}$$

The moment about the y-axis,

$$M_{x=0} = m_1 \cdot x_1 + m_2 \cdot x_2, \tag{8.15}$$

is an important special case. Moments may be positive, negative, or zero. In the first two cases, the axis will swing about the fulcrum as shown in Figure 3.

Equation (8.13) provides us with a method for determining the center of mass $\overline{x}$ of our two-point system. We simply solve for $\overline{x}$ in the equation $M_{x=\overline{x}} = 0$: $m_1 \cdot x_1 + m_2 \cdot x_2 - \overline{x} \cdot (m_1 + m_2) = 0$, or

$$\overline{x} = \frac{m_1 \cdot x_1 + m_2 \cdot x_2}{m_1 + m_2}.$$

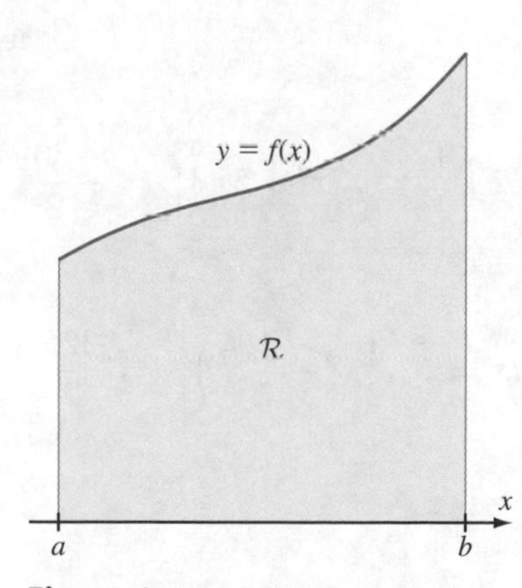

Figure 4

If we let $M = m_1 + m_2$ denote the total mass of the system, then we can rewrite our formula for $\overline{x}$ in the form

$$\overline{x} = \frac{M_{x=0}}{M}.$$

Moments

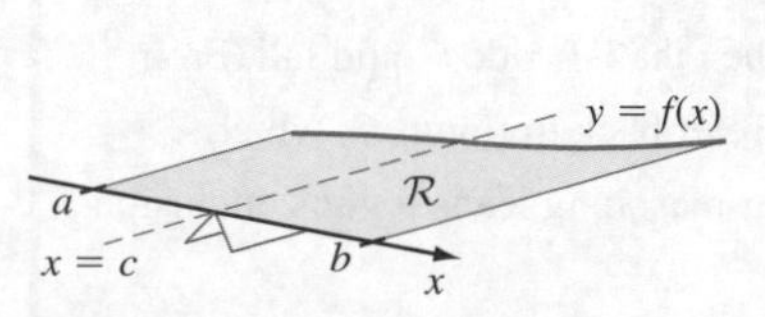

Figure 5

Let us now consider the region $\mathcal{R}$ shown in Figure 4. We suppose that the region has uniform mass density δ. By this, we mean that the mass of any subset of $\mathcal{R}$ is δ times the area of the subset. Imagine the xy-plane to be horizontal and position a fulcrum underneath the axis $x = c$ (see Figure 5). What must c be so that region $\mathcal{R}$ does not swing to one side or the other? To answer this question, we form a partition $a = x_0 < x_1 < \cdots < x_N = b$ of the interval $[a, b]$. Let Δx denote the common length of the subintervals, and let ξ_j be the midpoint of the jth subinterval I_j. We divide region $\mathcal{R}$ into the N pieces that lie over the subintervals $I_1, I_2, \ldots, I_N$. The piece that

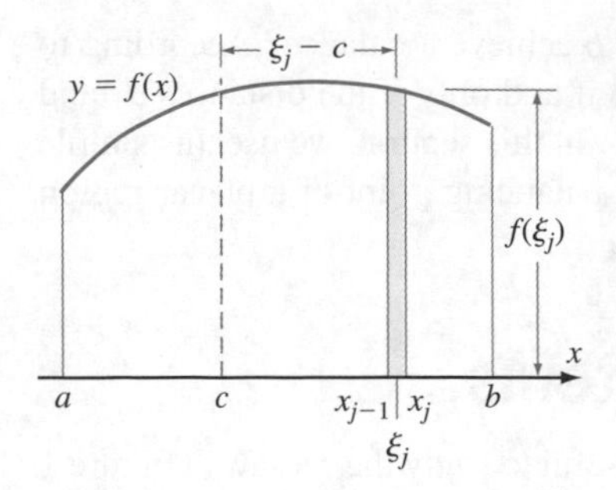

Figure 6

lies over I_j is approximately a rectangle with base Δx and height $f(\xi_j)$—see Figure 6. The mass of this piece is therefore approximately equal to $\delta f(\xi_j)\Delta x$. Notice that if Δx is small, then the points in this jth piece of $\mathcal{R}$ are all approximately a (signed) distance $(\xi_j - c)$ from the axis $x = c$. The contribution that this piece makes to the tendency of $\mathcal{R}$ to pivot about the line $x = c$ is therefore about $(\xi_j - c) \cdot \delta f(\xi_j)\Delta x$. The total tendency (or *moment*) $M_{x=c}$ to pivot about the axis $x = c$ is approximately

$$\sum_{j=1}^{N} (\xi_j - c) \cdot \delta f(\xi_j)\Delta x. \tag{8.16}$$

As N increases, Δx decreases and expression (8.16) becomes a better approximation to the tendency of $\mathcal{R}$ to rotate about the line $x = c$. Since expression (8.16) is a Riemann sum for the integral $\int_a^b (x - c)\delta f(x)\, dx$, we are led to the following definition.

Definition Let c be any real number. Suppose that f is continuous and nonnegative on the interval $[a, b]$. Let $\mathcal{R}$ denote the planar region bounded above by the graph of $y = f(x)$, below by the x-axis, and on the sides by the line segments $x = a$ and $x = b$. If $\mathcal{R}$ has a uniform mass density δ, then the moment $M_{x=c}$ of $\mathcal{R}$ about the axis $x = c$ is defined by the equation

$$M_{x=c} = \int_a^b (x - c)\delta f(x)\, dx = \delta \int_a^b (x - c) f(x)\, dx.$$

Notice that the moment $M_{x=c}$ is defined even if the number c is not between a and b.

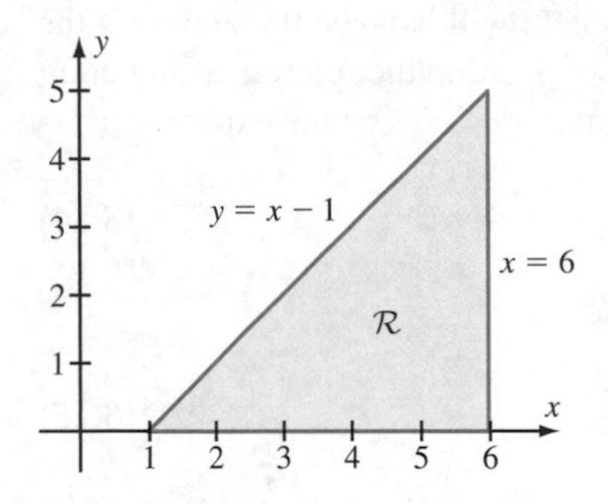

Figure 7

Example 1 Let $\mathcal{R}$ be the region bounded by $y = x - 1$, $y = 0$, and $x = 6$ (as shown in Figure 7). Suppose $\mathcal{R}$ has uniform mass density $\delta = 2$. Calculate the moments about the axes $x = 5$ and $x = 0$.

Solution The required moments are

$$M_{x=5} = 2\int_1^6 (x - 5) \cdot (x - 1)\, dx = 2\int_1^6 (x^2 - 6x + 5)\, dx = 2\left(\frac{1}{3}x^3 - 3x^2 + 5x\right)\Bigg|_1^6 = -\frac{50}{3}$$

and

$$M_{x=0} = 2\int_1^6 (x - 0) \cdot (x - 1)\, dx = 2\int_1^6 (x^2 - x)\, dx = 2\left(\frac{1}{3}x^3 - \frac{1}{2}x^2\right)\Bigg|_1^6 = \frac{325}{3}.$$

■

inSIGHT

In Example 1, the signs of $M_{x=5}$ and $M_{x=0}$ could have been determined simply by inspecting Figure 7. Imagine a horizontal plate in the shape of $\mathcal{R}$. Our experience tells us that if it were allowed to pivot about the axis $x = 5$, then the left side would fall. From this we conclude that $M_{x=5} < 0$. (It may be useful to refer back to Figure 3, which illustrates the analogous behavior of a seesaw.) Similar reasoning leads to the conclusion that $M_{x=0}$ must be positive.

Consider again the region $\mathcal{R}$ shown in Figure 4. We would like to define the moment $M_{y=d}$ of $\mathcal{R}$ about a horizontal axis $y = d$. Unfortunately, we cannot do so in a way that is analogous to the definition of moments about vertical axes. The reason can be seen in Figure 6. The key idea behind the definition of the moment $M_{x=c}$ is that, if the rectangle is thin, then the point masses within the rectangle are nearly equidistant from the vertical axis $x = c$. Clearly that cannot be true for any horizontal axis.

A thorough solution to the problem of defining $M_{y=d}$ requires multivariable calculus. For now, let us consider only the moment $M_{y=0}$, or the moment about the x-axis. Look again at Figure 6. The average distance to the x-axis of the points in the rectangle is $f(\xi_j)/2$. Therefore, it seems plausible to use

$$\underbrace{\frac{1}{2}f(\xi_j)}_{\text{Average distance}} \cdot \underbrace{\delta f(\xi_j)\Delta x}_{\text{Approximate mass}}$$

as an approximation of the moment of the rectangle about the x-axis. The resulting approximation to the total moment of $\mathcal{R}$ about the x-axis is then

$$\sum_{j=1}^{N} \frac{1}{2} f(\xi_j) \cdot \delta f(\xi_j)\Delta x.$$

Since this is a Riemann sum of the integral $\int_a^b (1/2)\delta f(x)^2\,dx$, we are led to define $M_{y=0}$ by the formula

$$M_{y=0} = \frac{1}{2}\delta \int_a^b f(x)^2\,dx. \tag{8.17}$$

Multivariable calculus provides the tools to rigorously justify this formula.

Example 2 Let $\mathcal{R}$ be the region bounded by $y = x - 1$, $y = 0$, and $x = 6$ (as in Example 1). Suppose $\mathcal{R}$ has uniform mass density $\delta = 2$. Calculate the moment $M_{y=0}$.

Solution The required moment is

$$M_{y=0} = \frac{1}{2}(2)\int_1^6 (x-1)^2\,dx = \left.\frac{(x-1)^3}{3}\right|_1^6 = \frac{125}{3}. \quad \blacksquare$$

Center of Mass

The center of mass $(\overline{x}, \overline{y})$ of a region $\mathcal{R}$ is the point at which the region balances (as in Figure 2). To find the center of mass, we must translate this physical property into mathematical equations.

Definition Let $\mathcal{R}$ be a region as shown in Figure 4. The x-coordinate $\overline{x}$ of the center of mass of $\mathcal{R}$ is the real number such that $M_{x=\overline{x}} = 0$.

Theorem 1 Let f be a continuous, nonnegative function on the interval $[a, b]$. Let $\mathcal{R}$ denote the region bounded above by the graph of $y = f(x)$, below by the x-axis, and on the sides by the line segments $x = a$ and $x = b$. Let M denote the mass of $\mathcal{R}$. If $\mathcal{R}$ has a uniform mass density δ, then the x-coordinate $\overline{x}$ of the center of mass of $\mathcal{R}$ is given by

$$\overline{x} = \frac{M_{x=0}}{M} = \frac{\int_a^b xf(x)\,dx}{\int_a^b f(x)\,dx}. \tag{8.18}$$

The y-coordinate $\overline{y}$ of the center of mass is given by

$$\overline{y} = \frac{M_{y=0}}{M} = \frac{(1/2)\int_a^b f(x)^2\,dx}{\int_a^b f(x)\,dx}. \tag{8.19}$$

Proof By definition, $M_{x=\overline{x}} = 0$. To determine $\overline{x}$, we solve for $\overline{x}$ in the equation $\delta \int_a^b (x - \overline{x}) f(x)\,dx = 0$. After expanding the integrand and expressing the resulting integral as a sum, we obtain

$$\delta \int_a^b xf(x)\,dx - \overline{x} \cdot \delta \int_a^b f(x)\,dx = 0.$$

It follows that

$$\overline{x} = \frac{\delta \int_a^b xf(x)\,dx}{\delta \int_a^b f(x)\,dx} = \frac{M_{x=0}}{M}.$$

(Notice that the uniform mass density δ cancels and does not appear in the final formula for $\overline{x}$.) We obtain the formula for $\overline{y}$ by replacing x with y in the equation $\overline{x} = M_{x=0}/M$ and then using equation (8.17). ■

Example 3 Let $\mathcal{R}$ be the region bounded by $y = x - 1$, $y = 0$, and $x = 6$ (as in Example 1). Suppose that $\mathcal{R}$ has uniform mass density $\delta = 2$. Calculate the center of mass of $\mathcal{R}$.

Solution In Example 1, we calculated $M_{x=0} = 325/3$. The total mass M of $\mathcal{R}$ is computed as follows:

$$M = 2\int_1^6 (x - 1)\,dx = 2\left(\frac{1}{2}x^2 - x\right)\Big|_1^6 = 25.$$

Therefore,

$$\overline{x} = \frac{325/3}{25} = \frac{13}{3}.$$

In Example 2, we showed that $M_{y=0} = 125/3$. Therefore,

$$\overline{y} = \frac{125/3}{25} = \frac{5}{3}.$$

■

quickquiz

1. Let $\mathcal{R}$ denote the triangle with vertices $(0, 0)$, $(0, 2)$, and $(4, 0)$. Without any computation, what can you say about $M_{x=2}$?
2. Let $\mathcal{R}$ denote the part of the unit circle that lies in the first and second quadrants. For what value c is $M_{x=c} = 0$?
3. Let $\mathcal{R}$ denote the part of the unit circle that lies in the first quadrant. Without any computation, what can you say about the coordinates of the center of mass of $\mathcal{R}$?

EXERCISES

Problems for Practice

In Exercises 1–8, find the moment of the region $\mathcal{R}$ about the axis. Assume that $\mathcal{R}$ has uniform unit mass density.

1. $\mathcal{R}$ is the triangular region with vertices $(0, 0)$, $(0, 2)$, and $(6, 0)$; about $x = 3$.
2. $\mathcal{R}$ is the triangular region with vertices $(0, 0)$, $(0, 2)$, and $(6, 0)$; about $x = -1$.
3. $\mathcal{R}$ is the triangular region with vertices $(0, 0)$, $(0, 2)$, and $(6, 0)$; about $y = 0$.
4. $\mathcal{R}$ is the first quadrant region bounded by $y = 4x - x^3$ and the x-axis; about $x = 1$.
5. $\mathcal{R}$ is the first quadrant region bounded by $y = 4x - x^3$ and the x-axis; about $x = 3$.
6. $\mathcal{R}$ is the first quadrant region bounded by $y = 4x - x^3$ and the x-axis; about $y = 0$.
7. $\mathcal{R}$ is the region bounded above by $y = 1/x$, below by the x-axis, and on the sides by the vertical lines $x = 1$ and $x = 2$; about $x = -3$.
8. $\mathcal{R}$ is the region bounded above by $y = 1/x$, below by the x-axis, and on the sides by the vertical lines $x = 1$ and $x = 2$; about $y = 0$.

In Exercises 9–20, find the center of mass of the region $\mathcal{R}$, assuming that it has uniform mass density.

9. $\mathcal{R}$ is the triangular region with vertices $(0, 0)$, $(0, 2)$, and $(6, 0)$.
10. $\mathcal{R}$ is the region bounded by $y = 4 - x^2$ and the x-axis.
11. $\mathcal{R}$ is the region bounded by $y = \sqrt{a^2 - x^2}$ and the x-axis.
12. $\mathcal{R}$ is the region bounded by $y = \cos(x)$, $-\pi/2 \leq x \leq \pi/2$, and the x-axis.
13. $\mathcal{R}$ is the region bounded by $y = 8 - 2x - x^2$ and the x-axis.
14. $\mathcal{R}$ is the region bounded by $y = x - x^3$, $0 \leq x \leq 1$, and the x-axis.
15. $\mathcal{R}$ is the region bounded by $y = 1/x$, $1 \leq x \leq 2$, and the x-axis.
16. $\mathcal{R}$ is the region bounded by $y = \sqrt{x}$, $x = 4$, $x = 9$, and the x-axis.
17. $\mathcal{R}$ is the region bounded by $y = x^2 - 4x + 5$, $x = 1$, $x = 4$, and the x-axis.
18. $\mathcal{R}$ is the region bounded by $y = 1/\sqrt{x}$, $4 \leq x \leq 9$, and the x-axis.
19. $\mathcal{R}$ is the region bounded by $y = (9 - x^2)^{3/2}$ and the x-axis. (The area of $\mathcal{R}$ is $243\pi/8$.)
20. $\mathcal{R}$ is the region bounded by $y = x + 1/x$, $x = 1/2$, $x = 2$, and the x-axis.

Further Theory and Practice

In Exercises 21–30, find the center of mass of the region $\mathcal{R}$, assuming that it has uniform mass density.

21. $\mathcal{R}$ is the region bounded by $y = \cos(x)$, $0 \leq x \leq \pi/2$, and the x-axis.
22. $\mathcal{R}$ is the region bounded by $y = \exp(x)$, $x = -1$, $x = 1$, and the x-axis.
23. $\mathcal{R}$ is the region bounded by $y = x$, $y = 2 - x^2$, and the x-axis.
24. $\mathcal{R}$ is the region bounded by $y = 1/(1 + x)$, $1 \leq x \leq 2$, and the x-axis.
25. $\mathcal{R}$ is the region bounded by $y = 1/\sqrt{1 + x^2}$, $x = 0$, $x = 1$, and the x-axis.
26. $\mathcal{R}$ is the region bounded above by $y = x/\sqrt{1 + x^2}$, $x = 1$, $x = 3$, and the x-axis.
27. $\mathcal{R}$ is the region bounded above by $y = \ln(x)$, $x = 1$, $x = e$, and the x-axis.
28. $\mathcal{R}$ is the region bounded above by $y = 3x$ $(0 \leq x \leq 1)$, $y = 4 - x^2$ $(1 \leq x \leq 2)$, and the x-axis.
29. $\mathcal{R}$ is the region bounded above by
$$y = \begin{cases} x^2 & 0 \leq x \leq 2 \\ 4(x-3)^2 & 2 \leq x \leq 3 \end{cases}$$
and the x-axis.
30. $\mathcal{R}$ is the region bounded above by $y = |x|$, $-1 \leq x \leq 2$, and the x-axis.
31. Suppose that $f(x) \geq g(x)$ for all x in the interval $[a, b]$. Let $\mathcal{R}$ denote the region bounded above and below by the graphs of f and g and laterally by the vertical lines $x = a$ and $x = b$. Assume that $\mathcal{R}$ has uniform mass density. Show that the center of mass $(\overline{x}, \overline{y})$ of $\mathcal{R}$ is given by
$$\overline{x} = \frac{1}{M}\int_a^b x(f(x) - g(x))\, dx$$
and
$$\overline{y} = \frac{1}{2M}\int_a^b (f(x)^2 - g(x)^2)\, dx$$
where $M = \int_a^b (f(x) - g(x))\, dx$. Notice that in the formula for $\overline{y}$ the integrand is *not* $(f(x) - g(x))^2$.

In Exercises 32–35, use the formulas from Exercise 31 to calculate the center of mass of the region $\mathcal{R}$.

32. $\mathcal{R}$ is the region bounded above by $y = 4x$ $(1 \le x \le 3)$, below by $y = x$ $(1 \le x \le 3)$, and laterally by the vertical lines $x = 1$ and $x = 3$.

33. $\mathcal{R}$ is the region bounded above by $y = 2 + x$ $(1 \le x \le 2)$, below by $y = x^2$ $(1 \le x \le 2)$, and laterally by the vertical line $x = 1$.

34. $\mathcal{R}$ is the region bounded above by $y = 4x$ $(0 \le x \le 2)$ and below by $y = 2x^2$ $(0 \le x \le 2)$.

35. $\mathcal{R}$ is the region bounded above by $y = 5 - x^2$, below by $y = (x-1)^2$, and laterally by the vertical lines $x = 0$ and $x = 1$.

36. Suppose that a nonnegative function f defined on an interval $[a, b]$ is the p.d.f. of a random variable X. Let $\mathcal{R}$ denote the region bounded by the graph of f, $x = a$, $x = b$, and the x-axis. What is the relationship between X and the abscissa of the center of mass of $\mathcal{R}$?

The moments we have defined in this section are, to be precise, *first moments*. By using higher powers, we obtain *higher moments*. For example, if f is a continuous function on $[a, b]$, then the *second moment* of f about the vertical axis $x = c$ is defined as $\int_a^b (x-c)^2 f(x)\,dx$. Second moments are used in moment of inertia calculations in physics as well as throughout probability and statistics. To be specific, if f is the p.d.f. of a random variable X with mean μ_X, then the second moment of f about the axis $x = \mu_X$ is called the *variance* of X and is denoted by $\mathrm{Var}(X)$, or σ_X^2. The quantity σ_X is called the *standard deviation* of X. Exercises 37–40 concern these quantities.

37. If X has p.d.f $f(x) = 3x^2$ on $[0, 1]$, calculate $\mathrm{Var}(X)$.

38. If X has p.d.f $f(x) = 1/(b-a)$ on $[a, b]$, calculate the standard deviation of X.

39. Suppose that a random variable X has p.d.f. $f(x) = cx^p(1-x)$ for $0 \le x \le 1$ where c and p are positive constants. Find a formula for c in terms of p. Then calculate the variance of X in terms of p.

40. If X is a random variable, then X^2 is as well. It may be shown that $E(X^2) = \int_a^b x^2 f(x)\,dx$. Use this fact to deduce that $\mathrm{Var}(X) = E(X^2) - E(X)^2$.

Calculator/Computer Exercises

41. Find the center of mass of the region that lies above the interval $[0, \sqrt{\pi}]$ and below the graph of $y = \sin(\pi - x^2)$.

42. The function $f(x) = 6x/\exp(x) - 1$ has a root a in the interval $[0, 2]$ and a root b in the interval $[2, 4]$. Find the center of mass of the region that lies above the interval $[a, b]$ and below the graph of f.

43. Let (c, c) be the point of intersection of $y = x$ and $y = 5 - \exp(x)$. Calculate the center of mass of the region that lies above the x-axis and that is below the graph of $y = x$ for $0 \le x \le c$ and below the graph of $y = 5 - \exp(x)$ for $x \ge c$.

44. The equation of the Gateway Arch in St. Louis, Missouri, is

$$y = 693.8597 - 34.38365(\exp(kx) + \exp(-kx))$$

for $k = 0.0100333$ and $-299.2239 < x < 299.2239$. Find the center of mass for the region that lies above the x-axis and below this curve.

45. For each of the three choices—0.2, 1, and 2—of σ, plot

$$f(x) = \frac{1}{\sqrt{2\pi}\sigma}\exp\left(-\frac{1}{2}\left(\frac{x}{\sigma}\right)^2\right), \quad -10 \le x \le 10.$$

Let $g(\sigma)$ be the y-coordinate of the center of mass for the region under the graph of f. Calculate $g(\sigma)$ for each of the three specified values of σ. Plot the three points $(\ln(\sigma), \ln(g(\sigma)))$. What formula for g is suggested by these plotted points? Choose a fourth value of σ between 0 and 2 and use it to test your conjecture.

8.5 Work

Suppose that a body is moved a distance d along a straight line while being acted on by a force of *constant* magnitude F *in the direction of motion* (Figure 1). By definition, the *work* performed in this move is force times distance:

$$W = F \cdot d. \tag{8.20}$$

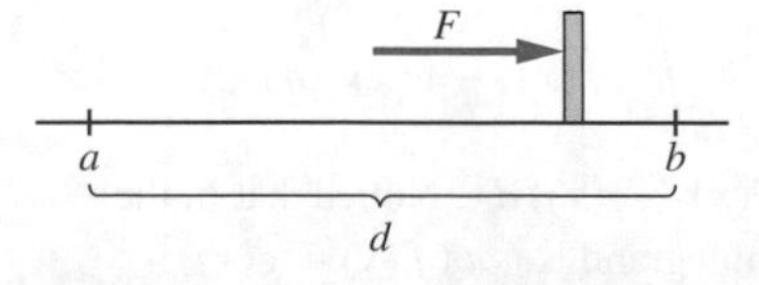

Figure 1

Bear in mind that this is a technical definition of work that differs somewhat from everyday usage. For example, a person has to exert a force of 100 lb to *hold up* a 100 lb

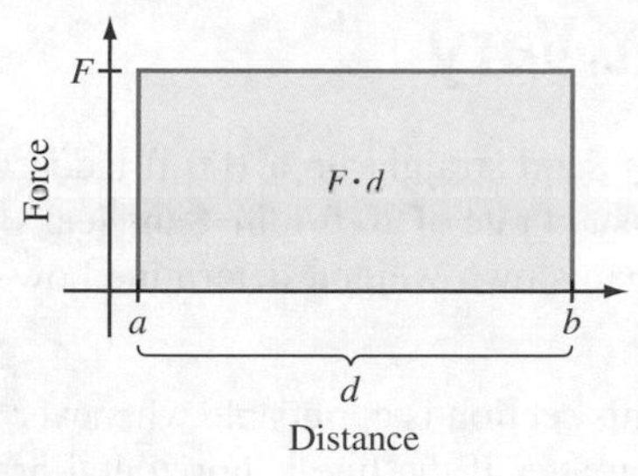

Figure 2

weight. Although this may be a strain and the person may tire from the exertion, if the weight does not move, then no work is done.

If we graph the constant function F as in Figure 2, then we see that work is related to area. When we think of area under a graph, however, the distance from 0 to 1 on each coordinate axis represents a unit length. The unit used to express area must represent $(\text{length})^2$. The square foot and square meter are examples. In contrast, it is evident from formula (8.20) that a unit of work must represent the product of a force and a length. In Figure 2, the unit of measurement along the vertical axis is a unit of force, not length.

Any combination of units that represents force × distance or, breaking it down into basic components, mass × distance^2/time^2 can be used to represent work. However, the unit of work can be expressed most simply by consistently using units from one of three systems of measurement. In the British engineering system, the standard unit of force is the *pound* (lb). This leads to the *foot-pound* (ft-lb) as the unit of measure for work: It is the amount of work done when a constant force of 1 lb is applied for 1 ft. In the *mks system* of measurement, force is measured in newtons (N) and distance is measured in meters (m). The unit of work is the *newton-meter,* which is also called the *joule* (J): 1 J is the amount of work done when a constant force of 1 N is applied for 1 m. In the *cgs system,* 1 *erg* is the amount of work done by 1 dyne acting over 1 centimeter. One joule is equal to 10^7 ergs and approximately 0.73756 ft-lb.

Using Integrals to Calculate Work

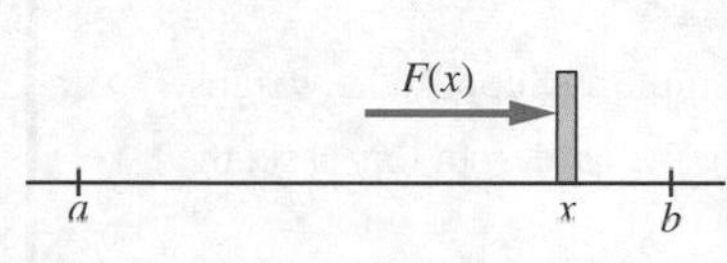

Figure 3

Suppose that a body is being moved from $x = a$ to $x = b$ along the x-axis and that the force applied in the direction of motion at the point x is $F(x)$ (see Figure 3). How much work is done altogether? If F is not a constant function, then formula (8.20) cannot be applied. Instead, we examine this problem by partitioning the interval into N subintervals of equal length Δx. If Δx is sufficiently small, then $F(x)$ will not vary greatly over each subinterval of the partition. In other words, if s_j is any point in the jth subinterval, then the force applied throughout the jth subinterval may be approximated by the constant force $F(s_j)$. The work done in the incremental move over the jth subinterval is therefore about $F(s_j) \cdot \Delta x$ (see Figure 4). Notice that if the unit of measurement on the vertical axis were length rather than force, our approximation would represent the area of the rectangle over the jth subinterval. Summing over j, we obtain an approximation to the total work:

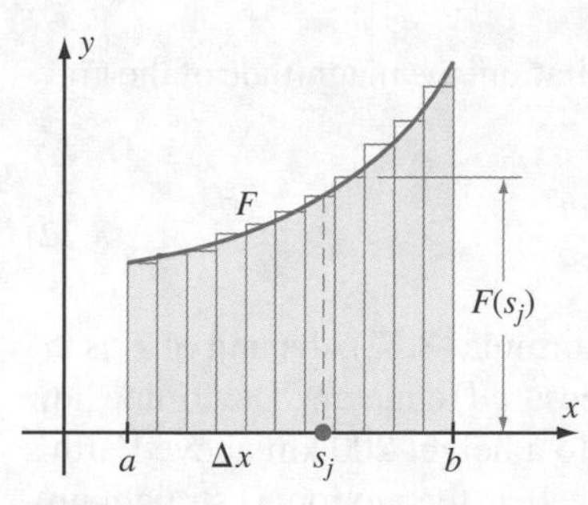

Figure 4

$$W \approx \sum_{j=1}^{N} F(s_j)\Delta x. \qquad \textbf{(8.21)}$$

As N increases and Δx becomes smaller, approximation (8.21) becomes more accurate. Notice that these approximating sums are also Riemann sums for the integral $\int_a^b F(x)\,dx$. From Chapter 5, we know that these Riemann sums tend to the integral as the number N of subintervals tends to infinity. We conclude that $\int_a^b F(x)\,dx$ equals the total work. Observe that there is again an analogy between work and area under the force curve. We summarize this in the next definition.

Definition

Suppose a body is moved linearly from $x = a$ to $x = b$ by a force in the direction of motion. If the magnitude of the force at each point $x \in [a, b]$ is $F(x)$, then the work performed is

$$\int_a^b F(x)\,dx.$$

Examples with Weights That Vary

Example 1 A man carries a leaky 50 lb sack of sand straight up a 100 ft ladder that runs up the side of a building. He climbs at a constant rate of 20 ft/min. Sand leaks out of the sack at a rate of 4 lb/min. Ignoring the man's own weight, determine how much work he performs on this trip.

Solution The definition of work given earlier in this section is applicable whenever a force is applied along a straight line—it does not matter if the line is horizontal or vertical (or any other direction). We must calculate the weight $F(y)$ of the sack of sand at height y, because that is the force that must be applied. The weight w of the sack at time t is given by $w = 50 - 4t$ lb when t is measured in minutes. Since the height y of the man is given by $y = 20t$ ft when t is measured in minutes, we may eliminate t to obtain $w = 50 - 4y/20$ lb, or $F(y) = 50 - y/5$ lb. The total work that is performed therefore equals

$$W = \int_0^{100} \left(50 - \frac{y}{5}\right) dy = 50y - \frac{y^2}{10}\Bigg|_{y=0}^{y=100} = 4000 \text{ ft-lb.}$$

■

Suppose that we did not ignore the man's weight in Example 1. Suppose he weighs 200 lb. He also has to lift his own weight when climbing the ladder. In this case, the work performed is

$$W = \int_0^{100} \left(\left(50 - \frac{y}{5}\right) + 200\right) dy = 24000 \text{ ft-lb.}$$

Example 2 According to Newton's Law of Gravitation, the magnitude of the gravitational force F exerted by Earth on a mass m is

$$F = 3.98621 \times 10^{14} \frac{m}{r^2} \tag{8.22}$$

when the mass is a distance r from Earth's center. In formula (8.22), the unit of F is the newton when m is measured in kilograms and r is measured in meters. Determine how much work is performed in lifting a 5000 kg rocket to a height 200 km above Earth's surface (assuming that the direction of the force, as well as the motion, is straight up). Use 6375.58 km for Earth's radius.

Solution We convert all units to the mks system. By doing so, we arrive at a value of work that is measured in joules. Since $r = 6375580$ m at Earth's surface and $r = (6375580 + 200000)$ m $= 6575580$ m when the rocket is 200 km above Earth's surface,

> **inSIGHT**
>
> At Earth's surface, r equals 6375580 m and equation (8.22) becomes $F = mg$ with $g = 9.80665$ m/s^2. *Near* Earth's surface, we may continue to use the equation $F = mg$ as an *approximation* to the weight-mass relationship. However, as a rocket rises to great heights, its weight decreases significantly according to equation (8.22).

we have

$$\begin{aligned} W &= \int_{6375580}^{6575580} F(r)\,dr \\ &= \int_{6375580}^{6575580} 3.98621 \times 10^{14} \frac{5000}{r^2}\,dr \\ &= -3.98621 \times 10^{14} \frac{5000}{r} \Bigg|_{r=6375580}^{r=6575580} \\ &= -3.98621 \times 10^{14} \cdot 5000 \left(\frac{1}{6575580} - \frac{1}{6375580} \right) \\ &= 9.50838 \times 10^{9}\ \text{J}. \end{aligned}$$

■

An Example Involving a Spring

Consider a spring attached to a rigid support with one free end (as in Figure 5a). According to Hooke's law, the force F that the spring exerts is proportional to the amount that it has been stretched or compressed. Although Hooke's law is not an exact physical law, it is a very good approximation for small deformations. When working problems with springs, it is usually convenient to set up a coordinate axis with origin at the free end of the relaxed spring, a point sometimes called the *equilibrium position*. If we let x denote the coordinate of the free end of the spring, then x represents the amount of extension when $x > 0$ and the amount of compression when $x < 0$. Hooke's law may be written as

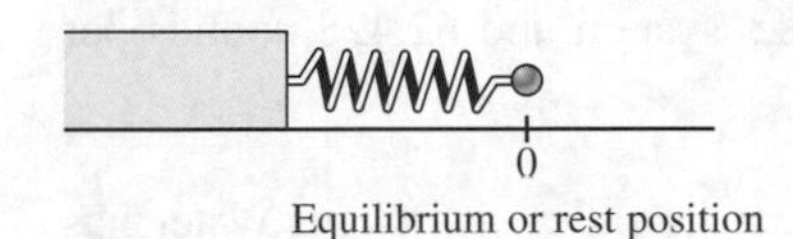

Figure 5a
Spring at rest

$$F = -kx$$

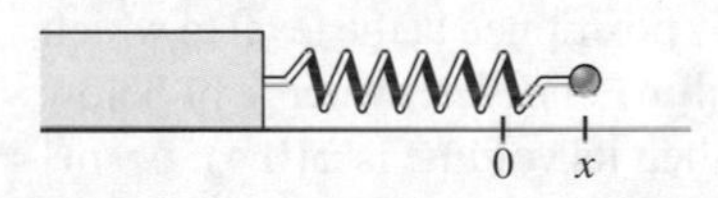

Figure 5b
Stretched spring ($x > 0$)

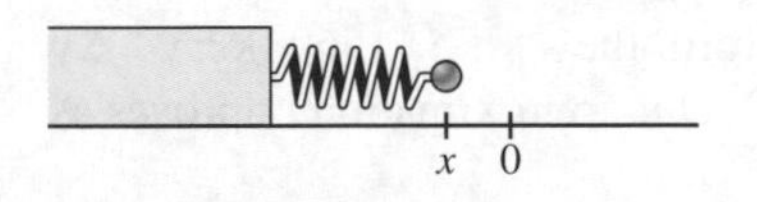

Figure 5c
Compressed spring ($x < 0$)

for a positive constant k (known as the *spring constant*). The negative sign in Hooke's law is required to reflect the direction of the force. The force of the spring is a *restorative* force, which means it is always directed toward the equilibrium position. Thus, when the spring is extended and $x > 0$, the force is directed toward the negative x-axis (Figure 5b). When the spring is compressed and $x < 0$, the force is directed toward the positive x-axis (Figure 5c). Notice that the constant k carries units that express force/length. The constant therefore represents a stiffness density.

To stretch or compress a spring, a force must be exerted that is equal in magnitude to the restorative spring force, but opposite in direction. Suppose that $0 \leq a \leq b$. From the definition of work, it follows that the work done in stretching a spring from a to b is

$$W = \int_a^b kx\,dx.$$

Example 3 If 5 J of work is done in extending a spring 0.2 m beyond its equilibrium position, then how much extra work is required to extend it an additional 0.2 m?

Solution We first use the given information to solve for the spring constant:

$$5\ \text{J} = \int_{0\ \text{m}}^{0.2\ \text{m}} kx\,dx = \frac{1}{2}kx^2 \Bigg|_{x=0\ \text{m}}^{x=0.2\ \text{m}} = 0.02k\ \text{m}^2,$$

or

$$k = \frac{5}{0.02}\ \mathrm{J/m^2} = 250\ \mathrm{Nm/m^2} = 250\ \mathrm{N/m}.$$

Finally, the required work W is calculated as follows:

$$W = \int_{0.2\ \mathrm{m}}^{0.4\ \mathrm{m}} kx\,dx = \frac{1}{2}kx^2\Big|_{x=0.2\ \mathrm{m}}^{x=0.4\ \mathrm{m}} = \frac{1}{2}k(0.4^2\ \mathrm{m}^2 - 0.2^2\ \mathrm{m}^2) = 0.06k\ \mathrm{m}^2.$$

We obtain our final answer by substituting in the value for the spring constant:

$$W = (0.06)(250\ \mathrm{N/m})\ \mathrm{m}^2 = 15\ \mathrm{Nm} = 15\ \mathrm{J}.$$ ■

Examples That Involve Pumping a Fluid from a Reservoir

The analysis that results in the definition of work as an integral can be used to solve a variety of problems. Until now, we have considered examples that concern a variable force. In our next application, pumping a fluid to the top of a reservoir, it is the distance that varies because the fluid at each horizontal level must be raised a distance equal to its depth. In the two examples that follow, we need the weight density of water: 9806.65 newtons per cubic meter (N/m^3) in the mks system and 62.428 pounds per cubic foot (lb/ft^3) in the British engineering system.

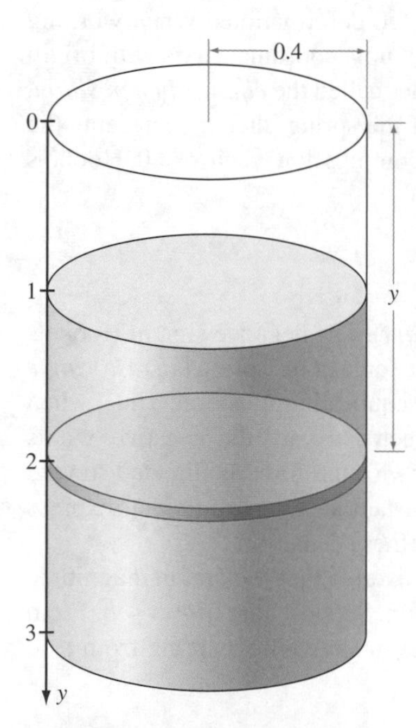

Figure 6

Example 4 A cylindrical sump pit is 3 m deep; its radius is 0.4 m. Water has accumulated in the pit to a depth of 2 m. A sump pump floats on the water's surface and pumps water to the top of the pit. How much work is done in pumping out half the water?

Solution In this type of problem, it is usually convenient to orient the positive y-axis downward, as indicated in Figure 6. The origin is usually positioned at the level to which the fluid must be pumped. Figure 6 also illustrates a thin "slice" of water y m below the top of the pit. If the thickness of the slice is Δy m, then its volume is $\pi(0.4)^2\Delta y$ m^3 and its weight is $\pi(0.4)^2\Delta y\ \mathrm{m}^3 \times 9806.65\ \mathrm{N/m}^3 = 1569.06\pi\,\Delta y$ N. The work done to pump this slice y m to the top is therefore *about* $1569.06\pi\,\Delta y \cdot y$ J, by equation (8.20). The reason this expression is not exact is that the volume pumped does have some thickness; consequently, not all points are exactly y m from the top. To approximate the total work done in pumping out half the pit, we form the sum $\sum 1569.06\pi y \cdot \Delta y$ of the contributions of the slices from $y = 1$ to $y = 2$. Our approximation improves as Δy tends to 0. In passing to the limit, we obtain

$$W = \int_1^2 1569.06\pi y\,dy = 1569.06\pi\frac{y^2}{2}\Big|_{y=1}^{y=2} = 2353.59\pi\ \mathrm{N} \approx 7394\ \mathrm{N}.$$ ■

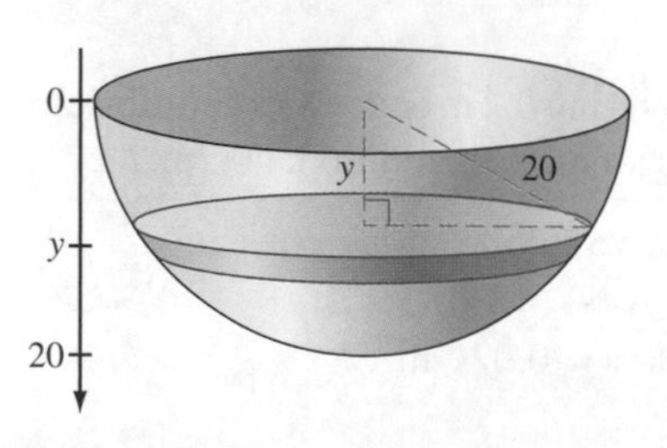

Figure 7

Example 5 A pond full of water is in the shape of a hemisphere of radius 20 ft. A pump floats on the water's surface and pumps the water from the surface to the level of the edge of the pond, where the water runs off. How much work is done in emptying the pond?

Solution The pond is sketched in Figure 7. As in Example 4, we have positioned a downward oriented vertical axis with origin at the top of the reservoir. Consider a thin horizontal "slice" at depth y. By solving for the base of the right triangle indicated in

Figure 7, we find that the cross section of the slice is a circle of radius $\sqrt{20^2 - y^2}$. If the thickness of the slice is Δy, then the volume of the slice is $\pi(\sqrt{20^2 - y^2})^2 \Delta y$ ft^3. It follows that the weight of the slice is $62.428 \cdot \pi(20^2 - y^2)\Delta y$ lb, and the work done in raising this to the surface is about $62.428 \cdot \pi(20^2 - y^2)\Delta y \cdot y$ ft-lb. Passing to an integral, as in Example 4, we see that the total work is

$$\begin{aligned}\int_0^{20} 62.428\pi(20^2 - y^2)y\,dy &= 62.428\pi \int_0^{20} (400y - y^3)\,dy \\ &= 62.428\pi \left(200y^2 - \frac{y^4}{4}\right)\Bigg|_{y=0}^{y=20} \\ &= 62.428\pi \left(200 \cdot 20^2 - \frac{20^4}{4}\right),\end{aligned}$$

which comes to about 7.84493×10^6 ft-lb. ■

quickquiz

1. How does *work* differ from *force?*
2. Explain the analogy between work and area.
3. If the amount of work in stretching a spring 0.01 m beyond its equilibrium position is 1/2 J, what force is necessary to maintain the spring at that position?
4. A cube of side length 1 m is filled with a fluid that weighs 1 N/m^3. What work is done in pumping the fluid to the surface?

EXERCISES

Problems for Practice

1. A steam shovel lifts a 500 lb load of gravel from the ground to a point 80 ft above the ground. However, the gravel is fine, and it leaks from the shovel at the rate of 1 lb/s. If it takes the steam shovel 1 min to lift its load at a constant rate, then how much work is performed?
2. How much work is done in lifting an 800 lb satellite from Earth's surface to a height of 200 mi?
3. A rocket is climbing straight up at the constant rate of 1000 mi/h. Its total weight, including fuel, is 7000 lb. If fuel is consumed at the constant rate of 30 lb/mi, how much work is performed in lifting the rocket the first 20 mi into space?
4. A man stands at the top of a tall building and pulls a chain up the side of the building. The chain is 50 ft long and weighs 3 lb per linear foot. How much work does the man do in pulling the chain to the top?
5. A crane lifts a large boat out of its berth to place it in dry dock. The vessel, including its contents, weighs 80 tons. However, as the boat is lifted, it releases water from its holds at the constant rate of 50 ft^3/min. The crane raises the boat a distance of 50 ft at the constant rate of 5 ft/min. How much work is performed in lifting the boat?
6. A heavy uniform cable is used to lift a 300 lb load from ground level to the top of a 100 ft tall building. If the cable weighs 20 lb per linear foot, how much work is done?
7. With regard to the cable and load from Exercise 6, how much work is done in lifting the load from ground level to a height 30 ft above ground level?
8. With regard to the cable and load from Exercise 6, if the load is lifted to the top of the building, how much work is done in lifting it the final (uppermost) 30 ft?
9. If a spring with spring constant 8 lb/in. is stretched 7 in. beyond its equilibrium position, then how much work is done?
10. A spring is stretched 2 in. beyond its equilibrium position. If the force required to maintain it in its

stretched position is 60 lb, how much work has been done?

11. A force of 280 lb compresses a spring 4 in. from its natural length of 16 in. How much work is done compressing it an additional 4 in.?
12. If 30 ft-lb of work are done in stretching a spring 3 in. beyond its natural length, how much work is done stretching it 1 in. more?
13. A spring is stretched a certain length beyond its natural length. What percentage of the total amount of work is expended by the first half of the stretch?
14. If 2/3 ft-lb of work is done in extending a spring at rest 2 in. beyond its equilibrium position, how much force is required to maintain the spring in that position?
15. A swimming pool full of water has square base of side 15 ft. It is 10 ft deep and is being pumped dry. The pump floats on the surface of the water and pumps the water to the top of the pool, at which point the water runs off. How much work does the pump perform in emptying the pool?
16. A swimming pool full of water has rectangular base with side lengths of 16 and 30 ft. It is 10 ft deep but not completely filled—the depth of the water it contains is 8 ft. A pump floats on the surface of the water and pumps the water to the top of the pool, at which point the water runs off. How much work does the pump perform in emptying the pool?
17. Referring to the swimming pool in Exercise 16, if power is interrupted when half the water has been pumped, how much work has been done?
18. A semicylindrical tank filled with water is 16 ft long. It has rectangular horizontal cross sections and vertical cross sections that are semicircles of radius 6 ft. A pump that floats on the surface of the tank pumps water to the top, at which point the water runs off. How much work is done in pumping the water?
19. A filled reservoir is in the shape of an inverted cone; the radius of the base of the cone and the depth at the center are both 100 ft. A pump that floats on the surface of the reservoir pumps water to the top, at which point the water runs off. How much work is done in pumping the water in the reservoir to a depth of 50 ft?
20. A tank filled with water is in the shape of an inverted pyramid with square base. The base of the pyramid measures 25 ft on a side, and the height of the pyramid is 40 ft. A pump floats on the surface of the water and pumps water to the upper edge, at which point the water runs off. If the tank is to be emptied so that the remaining water has a depth of 15 ft, what is the amount of work performed by the pump?

Further Theory and Practice

21. A stonemason carries a 50 lb sack of mortar 120 ft straight up a ladder at the rate of 20 ft/min. The sack leaks at the rate of 2 lb/min. When the mason is 60 ft off the ground, he shifts the sack and accidentally enlarges the hole. Now the sack leaks at the rate of 3 lb/min, but he continues up at the same rate. If the mason weighs 180 lb, how much work does he perform climbing the ladder while carrying the sack?
22. Suppose a cylinder of cross-sectional area A in.^{2} has one fixed end (the cylinder head). Suppose that a movable piston closes the other end a variable distance x in. away from the cylinder head. The pressure (force per area) p of gas confined in the cylinder is a continuous function of volume: $p = p(v)$. If $0 < x_1 < x_2$, show that the work W done by the piston in compressing the gas from a volume Ax_2 to Ax_1 is
$$W = A\int_{x_1}^{x_2} p(Ax)\,dx.$$
23. Suppose a cylinder with cross-sectional area 2 in.^{2} contains 20 in.^{3} of a gas under 50 lb/in.^{2} pressure. If the expression $pv^{1.4}$ is constant as the piston moves, then how much work does a piston do in compressing the gas to 4 in.^{3}? (Refer to Exercise 22.)
24. Refer to Exercise 23. For $0 < a < 10$, let $W(a)$ denote the work done in compressing the gas from 20 in.^{3} to $2a$ in.^{3}. Investigate $\lim_{a\to 0^+} W(a)$.

Suppose a tank has the shape of a paraboloid of revolution that results from rotating the curve $y = 2x^2 - 18$, $-3 \le x \le 3$, about the y-axis (x and y measured in feet). In Exercises 25–28, a pump floats on the surface of water and pumps the water to the top of the tank. Calculate the work done in performing the task described.

25. The tank, initially filled, is pumped until the remaining water is 3 ft deep at the center.
26. The tank, initially filled, is pumped until the remaining water has half the original volume.
27. The tank, initially filled to a depth of 4 ft at the center, is pumped until it is empty.
28. The tank, initially filled to a depth of 4 ft at the center, is pumped until the remaining water is 2 ft deep at the center.
29. The introduction of a force $F(x) > 0$ causes a mass m, initially at rest at $x = 0$, to move along the x-axis with velocity $v(x)$. Let $W(b)$ be the work done in moving the body from $x = 0$ to $x = b$. Show that $W(b) = mv(b)^2/2$. In other words, the work done is equal

to the gain in kinetic energy. *Hint:* Start from Newton's Law, which involves the derivative of velocity with respect to time. Use the Chain Rule to calculate the derivative of velocity with respect to x.

30. The gas tank of a tractor is in the shape of a cylinder lying on its side. The tank is positioned on the tractor so that its central axis is 5 ft above the ground. The radius of the tank is 1.5 ft and its length is 5 ft. For winter storage of the tractor, gas is pumped straight up out of the tank from the surface of the gasoline into a holding tank with opening 15 ft from ground level. If gasoline weighs 35 lb/ft^3, how much work is done in emptying a full tank of gas on the tractor into the holding tank?

31. The magnitude of a force that stretches a spring is given by $F(x) = kx$ when the spring is extended a distance x beyond its equilibrium position. Let $W(x)$ denote the work done in stretching the spring a distance x from its equilibrium position. Describe the curve that results from plotting the parametric equations $F = F(x)$, $W = W(x)$ in the FW-plane.

32. An object sits on a flat surface that presents an irregular frictional force. The object is pushed forward. In pushing it x units from its rest position, the work done is $x + \arctan(x)$. Describe the pushing force as a function of x.

Calculator/Computer Exercises

Exercises 33–36 concern a filled reservoir in the shape $y = 10(1/e^2 - \exp(-x^2/2))$ feet for $-2 \le x \le 2$ (x measured in feet). A pump that floats on the surface of the reservoir pumps water to the top, at which point the water runs off.

33. How much work is done pumping out a volume of water that leaves the remaining water 4 ft deep at the center?

34. How much work is done in pumping out half the water in the reservoir?

35. How much work is done in pumping out all the water in the reservoir?

36. In pumping out all the water in the reservoir, how deep is the remaining water at the center when half the work is done?

8.6 Improper Integrals—Unbounded Integrands

The theory of the integral that we learned in Chapter 5 enables us to integrate a continuous function $f(x)$ on a closed, bounded interval $[a, b]$. However, it is often necessary to integrate an unbounded function, or a function that is defined on an unbounded interval. In this section and the next, we learn to do so, and we see some applications of this new technique.

Integrals with Infinite Integrands

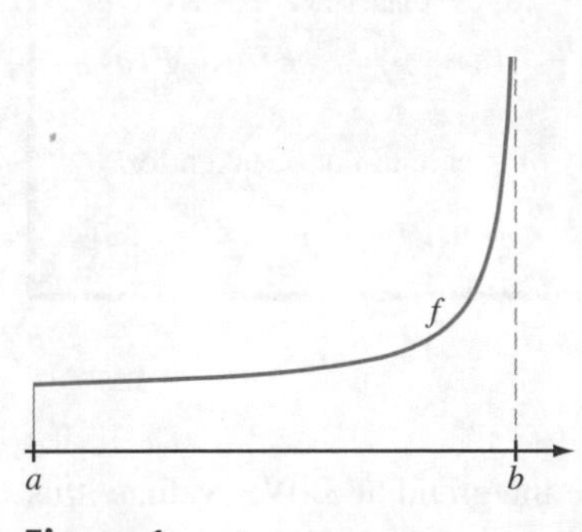

Figure 1

Let f be a continuous function on the interval $[a, b)$. Suppose f is unbounded as $x \to b^-$. (Refer to Figure 1). The integral $\int_a^b f(x)\,dx$ is called an *improper integral* with infinite integrand at b. The next definition tells us how to evaluate such an integral.

Definition If $\int_a^b f(x)\,dx$ is an improper integral with infinite integrand at b, then the value of the integral is defined to be

$$\lim_{\epsilon \to 0^+} \int_a^{b-\epsilon} f(x)\,dx,$$

provided that this limit exists and is finite. In this case, the integral is said to *converge*. Otherwise the integral is said to *diverge*. Figure 2 illustrates the method by which we evaluate this type of improper integral.

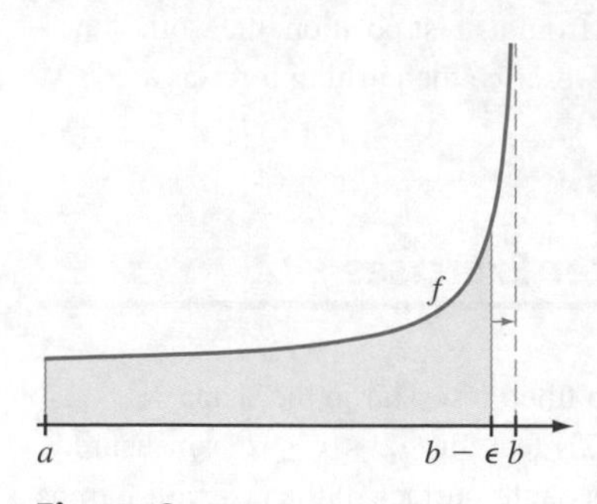

Figure 2
The improper integral $\int_a^b f(x)\,dx$ is defined to be the limit of the area of the shaded region as $b-\epsilon$ approaches b.

inSIGHT

The idea behind the method for evaluating improper integrals is to integrate *almost up to* the singularity—to within ϵ—and then let ϵ tend to 0. Refer again to Figure 2.

Example 1 Evaluate the integral $\int_0^8 (8-x)^{-1/3}\,dx$.

Solution The given integral is improper with infinite integrand at 8. We calculate

$$\lim_{\epsilon\to 0^+}\int_0^{8-\epsilon}(8-x)^{-1/3}\,dx = \lim_{\epsilon\to 0^+}\left(-\frac{(8-x)^{2/3}}{2/3}\bigg|_0^{8-\epsilon}\right) = -\frac{3}{2}\lim_{\epsilon\to 0^+}(\epsilon^{2/3}-8^{2/3})$$

$$= -\frac{3}{2}\lim_{\epsilon\to 0^+}(0-4) = 6.$$

We conclude that the given integral is convergent and its value is 6. ■

inSIGHT

In Example 1, we are justified in applying the Fundamental Theorem of Calculus to $f(x) = (8-x)^{-1/3}$ on the closed interval $[0, 8-\epsilon]$ because f is continuous on that interval. However, because f is not continuous on $[0, 8]$, we may not casually apply the Fundamental Theorem of Calculus directly to the given improper integral. Nevertheless, in this case, as in many others, a correct answer *does* result from such an application, because the antiderivative $F(x) = -3(8-x)^{2/3}/2$ of $f(x)$ is continuous at $x = 8$. Thus, if F is an antiderivative of f on $[a, b)$ that is continuous on $[a, b]$, then

$$\int_a^b f(x)\,dx = \lim_{\epsilon\to 0^+}\int_a^{b-\epsilon} f(x)\,dx = \lim_{\epsilon\to 0^+}(F(b-\epsilon)-F(a)) \overset{\text{Continuity}}{=} F(b)-F(a).$$

Of course, the continuity of F at an endpoint singularity of f should not be taken for granted.

Example 2 Analyze the integral $\int_1^3 (x-3)^{-2}\,dx$.

Solution This is an improper integral with infinite integrand at 3. We evaluate this integral by considering

$$\lim_{\epsilon\to 0^+}\int_1^{3-\epsilon}\frac{1}{(x-3)^2}\,dx = \lim_{\epsilon\to 0^+} -(x-3)^{-1}\Big|_1^{3-\epsilon}$$

$$= \lim_{\epsilon\to 0^+}[\epsilon^{-1}-2^{-1}]$$

$$= +\infty.$$

We conclude that the given improper integral diverges. ■

An improper integral with an integrand that is infinite at the left endpoint of integration is handled in a manner similar to the right endpoint case.

Definition If $f(x)$ is continuous on $(a, b]$ and unbounded as $x \to a^+$, then the value of the improper integral $\int_a^b f(x)\,dx$ is defined to be

$$\lim_{\epsilon \to 0^+} \int_{a+\epsilon}^{b} f(x)\,dx,$$

provided that this limit exists and is finite. In this case, the integral is said to *converge*. Otherwise, the integral is said to *diverge*. Figure 3 illustrates this technique.

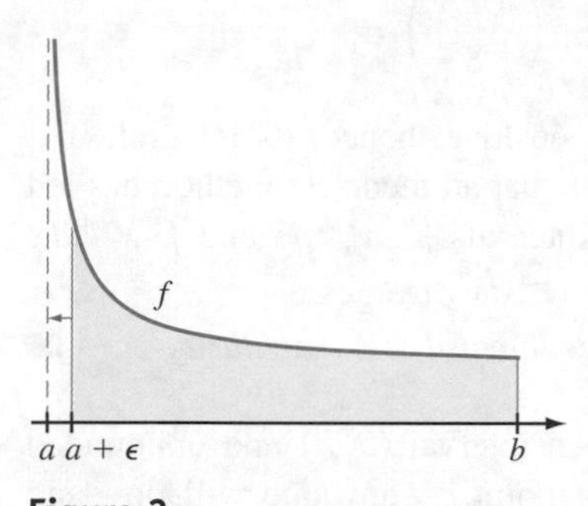

Figure 3
The improper integral $\int_a^b f(x)\,dx$ is defined to be the limit of the area of the shaded region as $a+\epsilon$ approaches a.

Example 3 Evaluate $\int_0^9 x^{-1/2}\,dx$.

Solution This integral is improper with infinite integrand at 0. The value of the integral is defined as

$$\int_0^9 x^{-1/2}\,dx = \lim_{\epsilon \to 0^+} \int_{0+\epsilon}^{9} x^{-1/2}\,dx = \lim_{\epsilon \to 0^+} 2x^{1/2}\Big|_{\epsilon}^{9} = \lim_{\epsilon \to 0^+} \left(2\sqrt{9} - 2\sqrt{\epsilon}\right) = 6. \quad ■$$

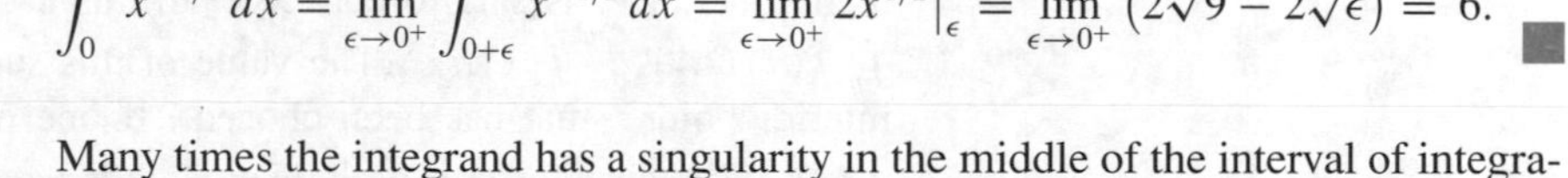

Many times the integrand has a singularity in the middle of the interval of integration. In these circumstances, we divide the interval of integration into two subintervals, one on each side of the singularity. We then integrate over each subinterval separately. If both of these integrals converge, then the original integral is said to converge. Otherwise, we say that the original integral diverges. The following is an example of this technique.

Example 4 Evaluate the improper integral $\int_{-3}^{2} 8(x+1)^{-1/5}\,dx$.

Solution The integrand is unbounded as x tends to -1. Therefore, we separately evaluate the two improper integrals $\int_{-3}^{-1} 8(x+1)^{-1/5}\,dx$ and $\int_{-1}^{2} 8(x+1)^{-1/5}\,dx$. For the first of these, we have

$$\begin{aligned}\int_{-3}^{-1} 8(x+1)^{-1/5}\,dx &= \lim_{\epsilon \to 0^+} \int_{-3}^{-1-\epsilon} 8(x+1)^{-1/5}\,dx = \lim_{\epsilon \to 0^+} \frac{8(x+1)^{4/5}}{4/5}\Bigg|_{-3}^{-1-\epsilon} \\ &= 10 \lim_{\epsilon \to 0^+} \left((-\epsilon)^{4/5} - (-2)^{4/5}\right) = -10 \cdot 2^{4/5}.\end{aligned}$$

The second integral is evaluated in a similar way:

$$\begin{aligned}\int_{-1}^{2} 8(x+1)^{-1/5}\,dx &= \lim_{\epsilon \to 0^+} \int_{-1+\epsilon}^{2} 8(x+1)^{-1/5}\,dx = \lim_{\epsilon \to 0^+} \left(\frac{8(x+1)^{4/5}}{4/5}\Bigg|_{-1+\epsilon}^{2}\right) \\ &= 10 \lim_{\epsilon \to 0^+} \left(3^{4/5} - \epsilon^{4/5}\right) = 10 \cdot 3^{4/5}.\end{aligned}$$

We conclude that the original integral converges and that

$$\begin{aligned}\int_{-3}^{2} 8(x+1)^{-1/5}\,dx &= \int_{-3}^{-1} 8(x+1)^{-1/5}\,dx + \int_{-1}^{2} 8(x+1)^{-1/5}\,dx \\ &= -10 \cdot 2^{4/5} + 10 \cdot 3^{4/5} = 10(3^{4/5} - 2^{4/5}). \quad ■\end{aligned}$$

It is dangerous to try to save work by not dividing the integral at the singularity. The next example illustrates what can go wrong.

Example 5 Evaluate the improper integral $\int_{-2}^{2} x^{-4}\,dx$.

Solution What we *should* do is divide this into the two integrals $\int_{-2}^{0} x^{-4}\,dx$ and $\int_{0}^{2} x^{-4}\,dx$. Suppose that, instead, we try to save work by antidifferentiating:

$$\int_{-2}^{2} x^{-4}\,dx = -\frac{1}{3}x^{-3}\Big|_{-2}^{2} = -\frac{1}{3}\left(\frac{1}{8} - \left(-\frac{1}{8}\right)\right) = -\frac{1}{12}.$$

Clearly something is wrong. The function x^{-4} is positive; hence, its integral, if it exists, should be positive too. What has happened is that an incorrect method has led to an incorrect negative answer. In fact, each of the integrals $\int_{-2}^{0} x^{-4}\,dx$ and $\int_{0}^{2} x^{-4}\,dx$ diverges, so *by definition,* the improper integral $\int_{-2}^{2} x^{-4}\,dx$ diverges.

The moral of this example is always to divide the integral at a singularity. ■

When an integrand f is continuous on an open interval (a, b) and unbounded at both endpoints a and b, we choose an interior point c—any one will do—and investigate the two improper integrals $\int_a^c f(x)\,dx$ and $\int_c^b f(x)\,dx$. If *both* converge, then $\int_a^b f(x)\,dx$ is said to converge. In this case, $\int_a^b f(x)\,dx$ is defined as the sum $\int_a^c f(x)\,dx + \int_c^b f(x)\,dx$. (The value of this sum does not depend on the particular interior point c that has been chosen.) If one of the two integrals, $\int_a^c f(x)\,dx$ and $\int_c^b f(x)\,dx$, diverges or if both diverge, then we say that $\int_a^b f(x)\,dx$ also diverges.

INSIGHT

In Example 6, there is no need to consider the second component integral $K = \int_{1/2}^{1} 1/(x(1-x)^{1/3})\,dx$. The divergence of just one of the two component integrals is enough to ensure that the full integral diverges. The matter would have been quite different had we started our investigation with K, which is in fact convergent (see Exercise 40). We would have had to proceed with an examination of the other component integral J to determine the behavior of I: As Example 6 shows, the convergence of just one component integral does not ensure that the full integral converges.

Example 6 Determine whether the improper integral $I = \int_0^1 1/(x(1-x)^{1/3})\,dx$ converges or diverges.

Solution The integrand of I is singular at both endpoints 0 and 1:

$$\lim_{x\to 0^+} \frac{1}{x(1-x)^{1/3}} = \lim_{x\to 1^-} \frac{1}{x(1-x)^{1/3}} = \infty.$$

Accordingly, we choose $c = 1/2$ (or any other point between the limits of integration) and separately analyze the two improper integrals $J = \int_0^{1/2} 1/(x(1-x)^{1/3})\,dx$ and $K = \int_{1/2}^{1} 1/(x(1-x)^{1/3})\,dx$. In considering J, notice that

$$\frac{1}{x(1-x)^{1/3}} \geq \frac{1}{x\cdot(1)^{1/3}} = \frac{1}{x}$$

for $0 < \epsilon \leq x \leq 1/2$. It follows that

$$\int_0^{1/2} \frac{1}{x(1-x)^{1/3}}\,dx = \lim_{\epsilon\to 0^+} \int_\epsilon^{1/2} \frac{1}{x(1-x)^{1/3}}\,dx \geq \lim_{\epsilon\to 0^+} \int_\epsilon^{1/2} \frac{1}{x}\,dx$$
$$= \lim_{\epsilon\to 0^+} \left(\ln\left(\frac{1}{2}\right) - \ln(\epsilon)\right).$$

Since $\lim_{\epsilon\to 0^+} \ln(\epsilon) = -\infty$, we conclude that J diverges. Because one of the component integrals of I diverges, we conclude that I also diverges. ■

An Application to Area and Volume

Suppose that f is a nonnegative, continuous function on the interval $(a, b]$, which is unbounded as $x \to a^+$. Let $\mathcal{R}$ denote the region bounded above by the graph of f, below by the interval $(a, b]$, and laterally by $x = a$ and $x = b$. If the improper integral

$\int_a^b f(x)\,dx$ converges, then it is natural to define the area of $\mathcal{R}$ as the value of this improper integral. Similarly, the volume of the solid $\mathcal{S}$ that is generated by rotating $\mathcal{R}$ about the x-axis is defined as the improper integral $\int_a^b \pi f(x)^2\,dx$, if it converges. In addition, the surface of $\mathcal{S}$ has surface area $\int_a^b 2\pi f(x)(1+f'(x)^2)^{1/2}\,dx$, if this improper integral converges.

Example 7 Calculate the area A above the x-axis and under the curve $y = 1/(x\ln^2(x))$ for $0 < x \le 1/2$.

Solution According to the preceding discussion,

$$A = \int_0^{1/2} \frac{1}{x\ln^2(x)}\,dx = \lim_{\epsilon\to 0^+} \int_\epsilon^{1/2} \frac{1}{x\ln^2(x)}\,dx.$$

We simplify the last integral by making the substitution $u = \ln(x)$, $du = (1/x)\,dx$:

$$A = \lim_{\epsilon\to 0^+} \int_{\ln(\epsilon)}^{\ln(1/2)} \frac{1}{u^2}\,du = \lim_{\epsilon\to 0^+} \left(-\frac{1}{u}\Bigg|_{\ln(\epsilon)}^{\ln(1/2)} \right) = \lim_{\epsilon\to 0^+} \left(-\frac{1}{\ln(1/2)} + \frac{1}{\ln(\epsilon)} \right).$$

Now as $\epsilon \to 0$, we have $\ln(\epsilon) \to -\infty$ and, therefore, $1/\ln(\epsilon) \to 0$. We conclude that our improper integral converges and $A = -1/\ln(1/2) = 1/\ln(2)$. ■

Example 8 Calculate the volume V enclosed by the surface obtained when $y = x^{-1/3}$, $0 < x \le 1$, is rotated about the x-axis.

Solution We have

$$V = \int_0^1 \pi(x^{-1/3})^2\,dx = \pi \lim_{\epsilon\to 0^+} \int_\epsilon^1 x^{-2/3}\,dx = \pi \lim_{\epsilon\to 0^+} \left(\frac{x^{1/3}}{1/3}\Bigg|_\epsilon^1 \right) = 3\pi \lim_{\epsilon\to 0^+} (1-\epsilon^{1/3}) = 3\pi.$$

■

Example 9 The graph of $y = x^{-1/3}$, $0 < x \le 1$, is rotated about the x-axis. Show that the area of the resulting surface of revolution is infinite. (Note: This is the same surface that encloses the finite volume in Example 8.)

Solution If we let $f(x) = x^{-1/3}$, then $f'(x) = -x^{-4/3}/3$ and the surface area S is given by

$$S = \int_0^1 2\pi(x^{-1/3}) \left(1 + \left(-\frac{x^{-4/3}}{3} \right)^2 \right)^{1/2} dx = \frac{2\pi}{3} \lim_{\epsilon\to 0^+} \int_\epsilon^1 x^{-1/3}(9+x^{-8/3})^{1/2}\,dx.$$

An estimate suffices to show that this surface area is infinite. Notice that $x^{-1/3}(9 + x^{-8/3})^{1/2} > x^{-1/3}(x^{-8/3})^{1/2} = x^{-5/3}$. We conclude that

$$S \ge \frac{2\pi}{3} \lim_{\epsilon\to 0^+} \int_\epsilon^1 x^{-5/3}\,dx = \frac{2\pi}{3} \lim_{\epsilon\to 0^+} \left(\frac{x^{-2/3}}{-2/3} \right)\Bigg|_\epsilon^1 = \pi \lim_{\epsilon\to 0^+} (\epsilon^{-2/3} - 1) = \infty.$$

Therefore, the area of the surface of revolution is infinite. ■

quickquiz

1. What is an improper integral with infinite integrand?
2. How do we evaluate an improper integral with infinite integrand?
3. Calculate $\int_{-2}^{3} (2x)^{-1/3}\,dx$ as an improper integral.
4. Discuss $\int_{-2}^{3} (2x)^{-4/3}\,dx$ as an improper integral.

EXERCISES

Problems for Practice

In Exercises 1–8, determine whether the improper integral is convergent or divergent. If it converges, evaluate it.

1. $\int_1^5 (x-5)^{-4/3}\,dx$
2. $\int_{-3}^{-2} 1/(x+2)\,dx$
3. $\int_2^4 (4-x)^{-0.9}\,dx$
4. $\int_1^2 (x-2)^{-1/5}\,dx$
5. $\int_0^{\pi/2} \tan(x)\,dx$
6. $\int_0^{\pi/2} \sec^2(x)\,dx$
7. $\int_0^1 x/(1-x^2)^{1/4}\,dx$
8. $\int_0^1 1/\sqrt{1-x^2}\,dx$

In Exercises 9–16, determine whether the improper integral is convergent or divergent. If it converges, evaluate it.

9. $\int_{-3}^2 (x+3)^{-1.1}\,dx$
10. $\int_{-4}^0 (x+4)^{-0.1}\,dx$
11. $\int_0^8 x^{-1/3}\,dx$
12. $\int_{-1}^3 (x+1)^{-3}\,dx$
13. $\int_0^1 \ln(x)/x\,dx$
14. $\int_4^{13} 1/\sqrt{x-4}\,dx$
15. $\int_0^3 x^{-1/2}(1+x)\,dx$
16. $\int_1^2 1/(x\cdot\ln(x)^{1/3})\,dx$

In Exercises 17–24, determine the point where the integrand is singular, then divide the integral into two pieces and calculate the improper integral. If it diverges, then say so; if it converges, then give its value.

17. $\int_0^2 1/(x-1)\,dx$
18. $\int_{-2}^3 x^{-1/7}\,dx$
19. $\int_{-5}^{-1} 1/(x+3)^{2/5}\,dx$
20. $\int_3^6 (x-4)^{-2}\,dx$
21. $\int_{-2}^4 (x+1)^{-2/3}\,dx$
22. $\int_{1/e}^e \ln^{-2/7}(x)/x\,dx$
23. $\int_0^3 x/(x^2-2)\,dx$
24. $\int_{-3}^3 x^{-1/3}(x+1)\,dx$

In Exercises 25–28, calculate the volume of the solid obtained by rotating about the x-axis the region below the graph of the function f over the given interval.

25. $f(x)=x^{-1/3}$, $x\in(0,2]$
26. $f(x)=1/(3-x)^{2/5}$, $x\in[1,3)$
27. $f(x)=1/(1-x^2)^{1/4}$, $x\in[0,1)$
28. $f(x)=x^3/(128-x^7)^{1/7}$, $x\in[0,2)$

Further Theory and Practice

In Exercises 29–36, determine whether the improper integral is convergent or divergent. If it converges, evaluate it.

29. $\int_{-5}^2 \ln(x+5)\,dx$
30. $\displaystyle\int_0^2 \frac{1}{\sqrt{4-x^2}}\,dx$
31. $\displaystyle\int_{-2}^{-3/2} \frac{1}{(x+2)\ln(x+2)}\,dx$
32. $\int_0^1 \ln(x)^2\,dx$
33. $\int_0^7 x^{-1/2}\ln(x)\,dx$
34. $\int_0^{10} x^{-3/2}\ln(x)\,dx$
35. $\int_{-1}^3 \ln|x|\,dx$
36. $\displaystyle\int_1^6 \frac{\ln^2(x-1)}{x-1}\,dx$
37. Suppose p and b are positive numbers. Show that $\int_0^b 1/x^p\,dx$ is convergent if and only if $p<1$.
38. Suppose p is positive and $0<a<b$. Show that $\int_a^b 1/(b-x)^p\,dx$ is convergent if and only if $p<1$.
39. For what positive values of p is the improper integral $\int_1^e 1/x(\ln^p(x))\,dx$ convergent?

Suppose $f(x)$ and $g(x)$ are continuous on $[a,b)$ and unbounded as $x\to b^-$. Suppose also that $0\le f(x)\le g(x)$ for all $a<x<b$. The *Comparison Theorem* states that (i) if $\int_a^b g(x)\,dx$ is convergent, then $\int_a^b f(x)\,dx$ is also convergent; and (ii) if $\int_a^b f(x)\,dx$ is divergent, then $\int_a^b g(x)\,dx$ is also divergent. Similar conclusions hold if $f(x)$ and $g(x)$ are continuous on $(a,b]$ and unbounded as $x\to a^+$. In Exercises 40–48, use the Comparison Theorem to determine whether the improper integral is convergent or divergent. In some cases, you may have to break up the integration before applying the Comparison Theorem.

40. $\displaystyle\int_{1/2}^1 \frac{1}{x(1-x)^{1/3}}\,dx$
41. $\displaystyle\int_1^2 \frac{1}{x(2-x)^{4/3}}\,dx$
42. $\int_0^1 x^{-1/2}(1-x)^{-3/4}\,dx$
43. $\displaystyle\int_0^4 \frac{1}{\sqrt{4x-x^2}}\,dx$
44. $\displaystyle\int_0^5 \frac{1}{x^2(x-5)}\,dx$
45. $\displaystyle\int_0^3 \frac{1}{x\sqrt{3-x}}\,dx$
46. $\int_{-1}^1 (1-x^2)^{-1/2}\,dx$
47. $\displaystyle\int_0^1 \frac{e^x}{\sqrt{x}}\,dx$
48. $\displaystyle\int_0^{1/4} \frac{\sec(\pi x)}{1-4x}\,dx$

Calculator/Computer Exercises

In Exercises 49–52, the integrand f of the integral $\int_0^1 f(x)\,dx$ has a singularity at 0. Determine a function $g(x) = cx^p$ such that (i) $0 \le f(x) \le g(x)$ for each x in $(0, 1]$ and (ii) $\int_0^1 g(x)\,dx$ is convergent. (This shows that the given improper integral is convergent—see the instructions to Exercises 40–48.) Plot f and g in a suitable viewing window. Evaluate $\int_0^1 g(x)\,dx$ and approximate $\int_0^1 f(x)\,dx$ to one decimal place.

49. $\int_0^1 x^{\;1/2}\cos(x^2)\,dx$

50. $\int_0^1 \sqrt{1 + 1/x}\,dx$

51. $\int_0^1 \exp(2x)\sin^2(x)x^{-7/3}\,dx$

52. $\int_0^1 \sqrt{x}\csc(x)\,dx$

8.7 Improper Integrals—Unbounded Intervals

Suppose we want to calculate the integral of a continuous function $f(x)$ over an unbounded interval of the form $[A, +\infty)$ or $(-\infty, B]$. Like the integrals with unbounded integrand that we studied in Section 8.6, these integrals are said to be improper. The theory of the integral that we learned in Chapter 5 does not cover these improper integrals; thus, some new concepts are needed.

The Integral on an Infinite Interval

Definition Let f be a continuous function on the interval $[A, \infty)$. The value of the improper integral $\int_A^\infty f(x)\,dx$ is defined to be

$$\lim_{N\to+\infty} \int_A^N f(x)\,dx,$$

provided that the limit exists and is finite. When the limit exists, the integral is said to *converge*. Otherwise, it is said to *diverge* (see Figure 1a). Similarly, if g is a continuous function on the interval $(-\infty, B]$, then the value of the improper integral $\int_{-\infty}^B g(x)\,dx$ is defined as

$$\lim_{M\to-\infty} \int_M^B g(x)\,dx,$$

provided that the limit exists and is finite. When the limit exists, the integral is said to *converge*. Otherwise, it is said to *diverge* (see Figure 1b).

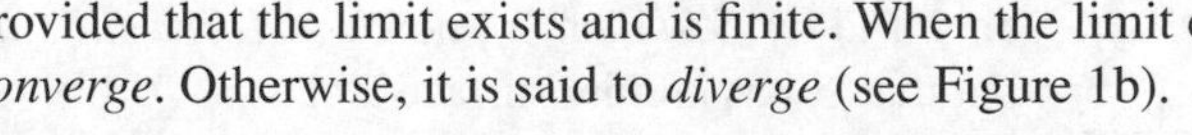

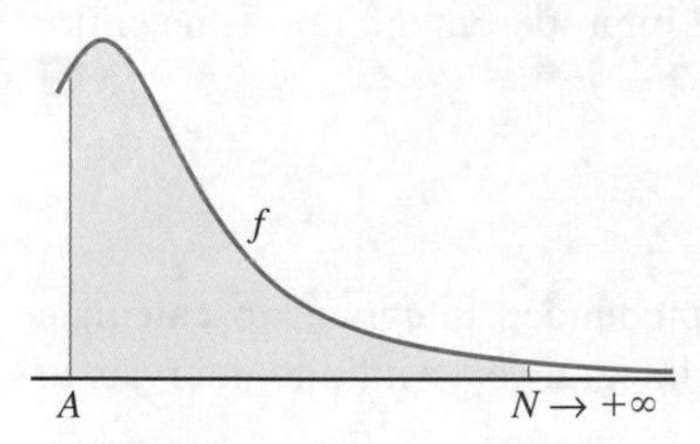

Figure 1a
The improper integral $\int_A^\infty f(x)\,dx$ is defined to be $\lim_{N\to+\infty} \int_A^N f(x)\,dx$.

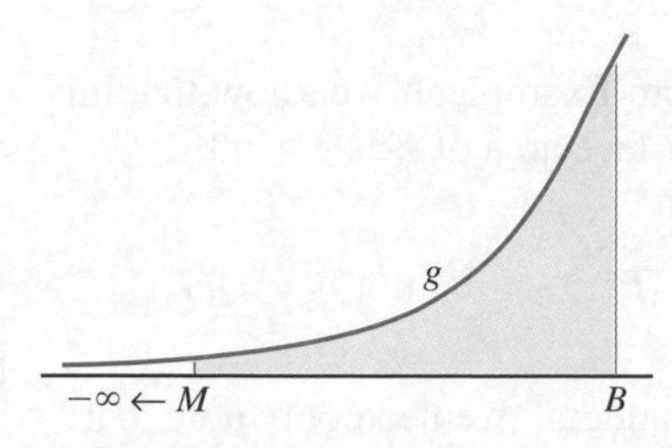

Figure 1b
The improper integral $\int_{-\infty}^B g(x)\,dx$ is defined to be $\lim_{M\to-\infty} \int_M^B g(x)\,dx$.

Example 1 Calculate the improper integral $\int_1^\infty x^{-2}\,dx$.

Solution We have

$$\int_1^\infty x^{-2}\,dx = \lim_{N\to\infty} \int_1^N x^{-2}\,dx = \lim_{N\to\infty} \left(-\frac{1}{x}\bigg|_1^N \right) = \lim_{N\to\infty} \left(-\frac{1}{N} - \left(-\frac{1}{1}\right)\right) = 1.$$

We conclude that the integral converges and has value 1. ■

Example 2 Determine whether the improper integral $\int_{-\infty}^{-8} x^{-1/3}\,dx$ converges or diverges.

Solution We do this by evaluating the limit

$$\lim_{M\to-\infty}\int_M^{-8} x^{-1/3}\,dx = \lim_{M\to-\infty}\left(\frac{x^{2/3}}{2/3}\bigg|_M^{-8}\right) = \frac{3}{2}\lim_{M\to-\infty}\left((-8)^{2/3} - M^{2/3}\right) = -\infty.$$

We conclude that the integral diverges. ■

Sometimes we must integrate over the entire real line. We do so by breaking the integral $\int_{-\infty}^{\infty} f(x)\,dx$ into two separate improper integrals $\int_{-\infty}^{c} f(x)\,dx$ and $\int_{c}^{\infty} f(x)\,dx$. The original integral is said to converge precisely when both component integrals converge. In this case, $\int_{-\infty}^{\infty} f(x)\,dx$ is defined as $\int_{-\infty}^{c} f(x)\,dx + \int_{c}^{\infty} f(x)\,dx$. An elementary computation shows that this sum does not depend on the point c that is chosen to break up the integration.

Example 3 Evaluate the improper integral $\int_{-\infty}^{\infty} 1/(1+x^2)\,dx$.

Solution To evaluate this integral, we break the interval into two pieces: $(-\infty,\infty) = (-\infty,0]\cup[0,\infty)$. The choice of zero as a place to break the interval is not important; any other point would do in this example. Thus, we evaluate the integrals $\int_0^{\infty} 1/(1+x^2)\,dx$ and $\int_{-\infty}^{0} 1/(1+x^2)\,dx$ separately. For the first of these two improper integrals, we have

$$\int_0^{\infty}\frac{1}{1+x^2}\,dx = \lim_{N\to\infty}\int_0^N \frac{1}{1+x^2}\,dx = \lim_{N\to\infty}\arctan(x)\Big|_0^N = \lim_{N\to\infty}(\arctan(N) - 0) = \frac{\pi}{2}.$$

The second integral evaluates to $\pi/2$ in a similar manner. Since each integral on the half line is convergent, we conclude that the original improper integral over the entire real line is convergent and that its value is $\pi/2 + \pi/2 = \pi$. ■

Geometric Applications

In Section 8.6, we used improper integrals with unbounded integrand to calculate surface area and volume. We can also use improper integrals over infinite intervals in those calculations.

Example 4 *Surface Area and Volume* Let $\mathcal{S}$ be the surface generated by rotating $y = 1/x$, $1 \le x < \infty$, about the x-axis (see Figure 2). Show that the solid enclosed by $\mathcal{S}$, which is known as *Torricelli's infinitely long solid,* has finite volume but that $\mathcal{S}$ has infinite surface area.

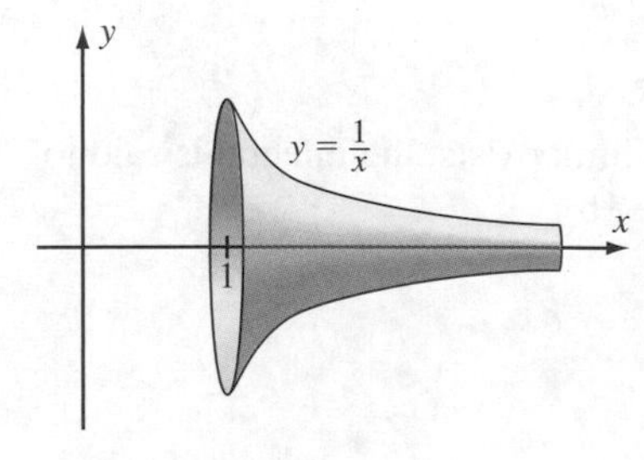

Figure 2
A section of Torricelli's infinitely long solid

Solution The volume equals $\int_1^{\infty} \pi(1/x)^2\,dx$. From Example 1, we know that this improper integral converges and has value π. The surface area of $\mathcal{S}$ is given by

$$\int_1^{\infty} 2\pi\cdot\frac{1}{x}\cdot\sqrt{1+\left(\frac{d}{dx}\left(\frac{1}{x}\right)\right)^2}\,dx = \lim_{N\to\infty}\int_1^N 2\pi\cdot\frac{1}{x}\cdot\sqrt{1+\frac{1}{x^4}}\,dx.$$

Rather than trying to find an exact value for the last integral, we need only note that

$$\int_1^N 2\pi\cdot\frac{1}{x}\cdot\sqrt{1+\frac{1}{x^4}}\,dx > \int_1^N 2\pi\cdot\frac{1}{x}\,dx = 2\pi\ln(N).$$

Since $\ln(N)$ tends to infinity as N tends to infinity, we conclude that the surface area of S is infinite. ■

Applications to Finance

If an amount A of money is allowed to compound continuously at an annual interest rate of r, then it will grow to Ae^{rt} in t years (as we learned in Section 2.6). In other words, the growth factor is e^{rt}. In particular, the amount Ae^{-rt} invested now will return A in t years. We say that Ae^{-rt} is the *present value* of an amount A that is to be received t years from now.

Instead of a single payment in the future, consider a future *income stream*. We assume that the income will begin to come in T_1 years in the future and that it will terminate at time $T_2 > T_1$. We assume that the income at time t is $f(t)$ dollars for $0 \leq T_1 \leq t \leq T_2 < \infty$. Let us figure out the present value of the income stream. Divide the interval $[T_1, T_2]$ into small subintervals using the uniform partition $T_1 = t_0 < t_1 < \cdots < t_N = T_2$. Over the time increment $[t_{j-1}, t_j]$, the money earned is about $f(t_j) \cdot \Delta t$ dollars. The present value of those earnings is about $(f(t_j) \cdot \Delta t) \cdot e^{-rt_j}$. Thus, the present value of the entire income stream is about

$$\sum_{j=1}^{N} f(t_j) \cdot e^{-rt_j} \Delta t.$$

This quantity is a Riemann sum for the integral $\int_{T_1}^{T_2} f(t)e^{-rt}\,dt$ that results when $N \to \infty$ and $\Delta t \to 0$. We have found that if r is the current annual interest rate, then

$$\int_{T_1}^{T_2} f(t)e^{-rt}\,dt \text{ is the present value of the income stream } f(t) \text{ for } T_1 \leq t \leq T_2.$$

If an income stream continues in perpetuity, then we let T_2 tend to infinity. In other words, the improper integral $\int_{T_1}^{\infty} f(t)e^{-rt}\,dt$ is the present value of an income stream that begins T_1 years in the future and continues in perpetuity.

Example 5 ***Present Value of a Perpetuity*** Suppose a trust is established that pays $2t + 50$ dollars per year for every year in perpetuity where t is time measured in years (here the present corresponds to time $t = 0$). Assume a constant interest rate of 6%. What is the total value, in today's dollars, of all the money that will ever be earned by this trust account?

Solution The present value of the perpetuity is given by

$$\int_0^{\infty} (2t+50)e^{-0.06t}\,dt = \lim_{N\to\infty} \int_0^N \underbrace{(2t+50)}_{u}\ \underbrace{e^{-0.06t}\,dt}_{dv}$$

$$= \lim_{N\to\infty} \left(\underbrace{(2t+50)}_{u}\ \underbrace{\frac{e^{-0.06t}}{-0.06}}_{v} \Bigg|_0^N - \int_0^N \underbrace{\frac{e^{-0.06t}}{-0.06}}_{v}\ \underbrace{2\,dt}_{du} \right).$$

Thus, the present value of the perpetuity is equal to

$$\lim_{N\to\infty}\left((2t+50)\frac{e^{-0.06t}}{-0.06}-2\frac{e^{-0.06t}}{(-0.06)^2}\bigg|_0^N\right)=\lim_{N\to\infty}\left(\left((2N+50)\frac{e^{-0.06N}}{-0.06}-2\frac{e^{-0.06N}}{(-0.06)^2}\right)-\left((50)\frac{e^{-0}}{-0.06}-2\frac{e^{-0}}{(-0.06)^2}\right)\right).$$

The second group of terms in the last limit evaluates to $-1388.888\ldots$. We finish our computation by using l'Hôpital's Rule to show that $\lim_{N\to\infty} Ne^{-0.06N}=0$. It follows that the present value of the perpetuity is \$1388.89, to the nearest penny. ■

Random Variables

In Section 8.3, we considered random variables that take values in a finite interval. It turns out that many important random variables are not of that type. Consider, for example, the time X that an unstable isotope takes to decay to stable form. Because we cannot say for certain that the decay will take place in any fixed finite time interval, it is appropriate to use $[0,\infty)$ as the range of X. Using improper integrals, we are able to define probability density functions and expectations of random variables such as X.

Suppose, then, that X is a random variable that takes values in an interval I. The interval I is not assumed to be finite. It may be $[0,\infty)$ or $(-\infty,\infty)$, for example. We say that f is a probability density function of X if $P(\alpha < X < \beta)=\int_\alpha^\beta f(x)\,dx$ for any subinterval (α,β) of I. This integral is improper if the interval (α,β) is not finite. The mean (or expectation) μ of X is given by the same formula as the one we derived in Section 8.3:

$$\mu=\int_I xf(x)\,dx$$

where I is the interval of integration. If I is not finite, then this is an improper integral.

Example 6 Let X denote the time until decay of a certain unstable isotope. Suppose that X has probability density function given by $f(x)=\exp(-x/82)/82$ for positive x (measured in years). What is the probability that the isotope will still be unstable after 500 years? What is the mean of X?

Solution Saying that the isotope will still be unstable after 500 years means that $500 < X < \infty$. The required probability is

$$\begin{aligned}P(500 < X < \infty)&=\frac{1}{82}\int_{500}^{\infty}e^{-x/82}\,dx=\lim_{N\to\infty}\frac{1}{82}\int_{500}^{N}e^{-x/82}\,dx\\&=\lim_{N\to\infty}(-e^{-x/82})\big|_{500}^{N}=e^{-500/82}-\lim_{N\to\infty}e^{-N/82}.\end{aligned}$$

Since $\lim_{N\to\infty}e^{-N/82}=0$, we conclude that $P(500 < X < \infty)=e^{-500/82}\approx 0.00225$.

The mean μ_X of X is calculated in a similar manner but requires integration by parts:

$$\begin{aligned}\frac{1}{82}\int\underbrace{x}_{u}\,\underbrace{e^{-x/82}\,dx}_{dv}&=\frac{1}{82}\left(\underbrace{x}_{u}\,\underbrace{(-82e^{-x/82})}_{v}-\int\underbrace{(-82e^{-x/82})}_{v}\,\underbrace{dx}_{du}\right)\\&=-xe^{-x/82}-82e^{-x/82}.\end{aligned}$$

Thus,

$$\begin{aligned}\mu_X &= \frac{1}{82}\int_0^\infty x e^{-x/82}\,dx \\ &= \lim_{N\to\infty} \frac{1}{82}\int_0^N x e^{-x/82}\,dx \\ &= \lim_{N\to\infty}\left(-x\exp\left(-\frac{x}{82}\right) - 82\exp\left(-\frac{x}{82}\right)\right)\Bigg|_0^N \\ &= \lim_{N\to\infty}\left(-N\exp\left(-\frac{N}{82}\right) - 82\exp\left(-\frac{N}{82}\right)\right) + 82.\end{aligned}$$

The value of $\lim_{N\to\infty}\exp(-N/82)$ is clearly 0 and the value of $\lim_{N\to\infty} N\exp(-N/82)$ is also 0 by an application of l'Hôpital's Rule. It follows that $\mu_X = 82$. ■

An especially important probability density function f defined on the entire real line is given by

$$f(x) = \frac{1}{\sqrt{2\pi}\sigma}\exp\left(-\frac{1}{2}\left(\frac{x-\mu}{\sigma}\right)^2\right), \quad -\infty < x < \infty. \tag{8.23}$$

In this formula, μ is any real number and σ is any positive number. The function f is called the *normal* probability density function (p.d.f.) with mean μ and standard deviation σ. A random variable X that has f as its p.d.f. is called a *normal* or *Gaussian* random variable. If $\mu = 0$ and $\sigma = 1$, then

$$f(x) = \frac{1}{\sqrt{2\pi}}\exp\left(-\frac{1}{2}x^2\right), \quad -\infty < x < \infty. \tag{8.24}$$

This function is called the *standard normal* probability density function.

Example 7 The duration X of a human pregnancy is often assumed to be a normal random variable with mean 268 days and standard deviation 16 days. Using this model, express as an improper integral the probability that a pregnancy will last at least 300 days.

Solution We set $\mu = 268$ and $\sigma = 16$ in formula (8.23). The required probability is

$$P(300 \le X < \infty) = \frac{1}{16\sqrt{2\pi}}\int_{300}^\infty \exp\left(-\frac{1}{2}\left(\frac{x-268}{16}\right)^2\right) dx.$$ ■

INSIGHT

Whenever we have a normal p.d.f. as an integrand, we can make a change of variable that results in the standard normal p.d.f. as the integrand. The advantage is that tables of such integrals are readily available. In Example 7, we make the substitution $u = (x - 268)/16$, $du = dx/16$ to obtain

$$P(300 \le X < \infty) = \frac{1}{\sqrt{2\pi}}\int_2^\infty e^{-u^2/2}\,du$$

The value of this integral can be looked up in a table (or calculated) to be about 0.02275.

quickquiz

1. What is an improper integral defined on an infinite interval?
2. How do we evaluate an improper integral defined on an infinite interval?
3. Evaluate $\int_1^\infty (1+x)^{-3}\,dx$.
4. Discuss $\int_1^\infty (1+x)^{-1}\,dx$.

EXERCISES

Problems for Practice

In Exercises 1–16, determine whether the improper integral converges or diverges. If it converges, evaluate it.

1. $\int_3^\infty x^{-3/2}\,dx$ **2.** $\int_9^\infty 1/\sqrt{x}\,dx$

3. $\int_{-1}^\infty 1/(3+x)^{3/2}\,dx$ **4.** $\int_4^\infty 1/(1+x)\,dx$

5. $\int_0^\infty 1/(1+x^2)\,dx$ **6.** $\int_0^\infty x/(1+x^2)\,dx$

7. $\int_0^\infty x/(1+x^2)^2\,dx$ **8.** $\int_2^\infty e^{-3x}\,dx$

9. $\int_0^\infty e^x/(e^x+1)\,dx$ **10.** $\int_0^\infty e^x/(e^x+1)^3\,dx$

11. $\int_1^\infty xe^{-3x^2}\,dx$ **12.** $\int_1^\infty xe^{-2x}\,dx$

13. $\int_e^\infty 1/(x\ln(x))\,dx$ **14.** $\int_1^\infty 1/(x\ln(x)^2)\,dx$

15. $\int_1^\infty \arctan(x)/(1+x^2)\,dx$

16. $\int_1^\infty (2/3)^x\,dx$

In Exercises 17–24, determine whether the improper integral converges or diverges. If it converges, evaluate it.

17. $\int_{-\infty}^{-2} x^{-1/3}\,dx$ **18.** $\int_{-\infty}^{-2} x^{-3}\,dx$

19. $\int_{-\infty}^{-100} 1/\sqrt{|x|}\,dx$ **20.** $\int_{-\infty}^{0} \sin(x)\,dx$

21. $\int_{-\infty}^{4} e^{x/3}\,dx$ **22.** $\int_{-\infty}^{4} e^{-x}\,dx$

23. $\int_{-\infty}^{-5} e^x/(e^{2x}+1)\,dx$ **24.** $\int_{-\infty}^{-3} \ln(|x|)/x\,dx$

In Exercises 25–30, determine whether the given improper integral converges or diverges. If it converges, evaluate it.

25. $\int_{-\infty}^{\infty} e^{-x}\,dx$ **26.** $\int_{-\infty}^{\infty} xe^{-x^2}\,dx$

27. $\int_{-\infty}^{\infty} 1/(1+x^2)\,dx$ **28.** $\int_{-\infty}^{\infty} x/(1+x^2)\,dx$

29. $\int_{-\infty}^{\infty} x/(1+x^2)^2\,dx$ **30.** $\int_{-\infty}^{\infty} 1/(x^2+2x+10)\,dx$

Further Theory and Practice

In Exercises 31–34, determine whether the improper integral converges or diverges. If it converges, evaluate it.

31. $\int_1^\infty x^{-2}\ln(x)\,dx$ **32.** $\int_{-\infty}^{4} x2^x\,dx$

33. $\int_{-\infty}^{0} x^2e^{x+1}\,dx$ **34.** $\int_1^\infty e^{-x}\cos(x)\,dx$

In Exercises 35–38, the integrand is not defined at the left endpoint of integration. In addition, the interval of integration is infinite. Determine whether the given improper integral converges or diverges. If it converges, evaluate the integral or show that it diverges.

35. $\int_0^\infty (e^{-1/x})/x^2\,dx$ **36.** $\int_0^\infty (e^{-\sqrt{x}})/\sqrt{x}\,dx$

37. $\int_1^\infty 1/(x\ln^2(x))\,dx$ **38.** $\int_1^\infty 1/(x\ln^{1/3}(x))\,dx$

In Exercises 39–42, calculate the volume contained inside the surface of revolution obtained by rotating the graph of f about the x-axis.

39. $f(x) = 1/(x\sqrt{x}),\ 1 \le x < \infty$

40. $f(x) = e^{-x},\ 0 \le x < \infty$

41. $f(x) = 1/(1+x),\ 0 \le x < \infty$

42. $f(x) = 1/\sqrt{1+x^2},\ 0 \le x < \infty$

43. Show that there is no value of p for which $\int_0^\infty 1/x^p\,dx$ is convergent.

44. For which values of p is $\int_e^\infty 1/(x\ln^p(x))\,dx$ convergent?

Suppose that $f(x)$ and $g(x)$ are continuous on the unbounded interval $[a, \infty)$. Suppose also that $0 \le f(x) \le g(x)$ for all $a \le x < \infty$. The Comparison Theorem states that (i) if $\int_a^\infty g(x)\,dx$ is convergent, then $\int_a^\infty f(x)\,dx$ is also convergent; and (ii) if $\int_a^\infty f(x)\,dx$ is divergent, then $\int_a^\infty g(x)\,dx$ is also divergent. Similar conclusions hold if $f(x)$ and $g(x)$ are continuous on $(-\infty, b]$. In Exercises 45–50, use the Comparison Theorem to establish that the improper integral $\int_a^\infty f(x)\,dx$ is convergent.

45. $\int_1^\infty x/(1+x^3)\,dx$

46. $\int_1^\infty (2+\sin(x))/x^2\,dx$

47. $\int_1^\infty 1/\sqrt{1+x^{5/2}}\,dx$

48. $\int_1^\infty x/(1+e^{2x})\,dx$

49. $\int_1^\infty e^{-x}/\sqrt{x}\,dx$

50. $\int_1^\infty (1+x+2x^2)/(1+2x+3x^2+4x^3\sqrt{x})\,dx$

Read the instructions for Exercises 45–50. In Exercises 51–56, use the Comparison Theorem to establish that the improper integral $\int_a^\infty g(x)\,dx$ is divergent.

51. $\int_1^\infty 1/\sqrt{1+x^2}\,dx$ **52.** $\int_{10}^\infty 1/\ln(x)\,dx$

53. $\int_1^\infty (\sin^2(x)+x)/x^{3/2}\,dx$

54. $\int_1^\infty (3-\sin(x))/(x^{1/6}-1/2)\,dx$

55. $\int_1^\infty \exp(x)/(x\exp(x)-1)\,dx$

56. $\int_1^\infty \arctan(x)/\sqrt{1+x}\,dx$

57. What is the present value of an annuity that will pay \$10,000 once per year, at the end of the year, in perpetuity? What is the present value of a continuous income stream that will pay \$10,000 per year in perpetuity?

58. A growing business projects $4t+500$ dollars profit per annum t years into the future. Assuming an annual interest rate of 8%, what is the present value of the total profits in perpetuity of this business?

59. A company's book value is \$25 per share. (The *book value* of a company is the sum of the fair market valuations of all tangible assets—equipment, real estate, investments, accounts receivable, and so on—less the company's debt, accounts payable, other obligations.) Over the next 3 years, the company is expected to produce a per share profit stream of $2+0.35t$ dollars per year at time t. Assuming that the operating profits are the only changes to the book value of the company and that the annual interest rate is 8%, what is the present value of one share's worth 3 years hence? If the company could sustain its expected growth rate in perpetuity, what would be the present value of one share?

60. The lifetime X of an electronic component is a random variable with p.d.f. $f(x)=\lambda e^{-\lambda x}$ for $x>0$. What is λ if the mean lifetime of this component is 20,000 hours? What is the probability that the component will last more than 10,000 hours?

Suppose that X is a random variable with normal probability density function f with mean μ and standard deviation σ. Exercises 61–65 pertain to X and f.

61. Determine where the graph of f is concave up and concave down. What is the maximum value that f assumes?

62. The parameter μ is called the mean because it *is* the mean in the technical sense given in this section. Prove it.

63. Suppose that X_0 is a random variable that has a standard normal probability density function. Show that $P(\mu-\lambda\sigma<X<\mu+\lambda\sigma)=P(-\lambda<X_0<\lambda)$.

64. Let X_0 be a standard normal random variable. What does $P(-\infty<X_0<\infty)$ equal? What are the values of $\int_{-\infty}^\infty \exp(-x^2/2)\,dx$ and $\int_{-\infty}^\infty \exp(-x^2)\,dx$?

65. The *error function* erf is defined by

$$\operatorname{erf}(x)=\frac{2}{\sqrt{\pi}}\int_0^x \exp(-t^2)\,dt.$$

Show that $\lim_{x\to\infty}\operatorname{erf}(x)=1$. Use the error function to express $P(X<\mu+\lambda\sigma)$.

66. A part is to have length 1 cm but will be acceptable if its length is between 0.9 and 1.1 cm. Suppose that due to machine error, the length of a machined part is a normal random variable with mean 1 cm and standard deviation 0.05 cm. Write an expression that approximates the percentage of unacceptable parts that are produced in a large production run.

67. The improper integral $\Gamma(s)=\int_0^\infty x^{s-1}e^{-x}\,dx$ is convergent for $s>0$ and is called the *gamma function*.

a. Integrate by parts to show that $\Gamma(s+1)=s\Gamma(s)$.

b. Calculate $\Gamma(1)$. Use this value, together with the equation of part a, to calculate $\Gamma(2)$. In a similar way, calculate $\Gamma(3)$ and then $\Gamma(4)$.

c. Deduce that $\Gamma(n+1)=n!$ for each positive integer n.

d. Suppose that λ and s are positive constants. For what value of c is $f(x)=cx^{s-1}e^{-\lambda x}$ a p.d.f. on $[0,\infty)$?

68. Suppose f is a continuous, nonnegative function such that $\int_0^\infty f(x)\,dx$ is convergent. At first, one might expect that $\lim_{x\to\infty} f(x)=0$, but, in fact, this limit does not have to hold. For each $n\in\mathbb{Z}^+$, let the point $V_n=(n,1)$ be the vertex of an isosceles triangle with base points $P_n=(n-1/2^n,0)$ and $Q_n=(n+1/2^n,0)$. Let O denote the origin. Finally, let f be the nonnegative, continuous function defined on $[0,\infty)$ so that its graph consists of the line segments $\overline{OP_1}, \overline{P_1V_1}, \overline{V_1Q_1}, \overline{Q_1P_2}, \overline{P_2V_2}, \overline{V_2Q_2}, \overline{Q_2P_3}, \ldots$ (ad infinitum). Show that $\lim_{x\to\infty} f(x)$ does not exist. Show that $\lim_{N\to\infty}\int_0^N f(x)\,dx$ does exist and evaluate $\int_0^\infty f(x)\,dx$.

Calculator/Computer Exercises

In Exercises 69–72, approximate the improper integral to two decimal places. Use Simpson's Rule after first applying the change of variable $x=1/\xi$ to convert the given integral to an improper integral over a finite interval.

69. $\int_1^\infty 1/\sqrt{1+x^3}\,dx$ **70.** $\int_{1/2}^\infty \exp(-x^2)\,dx$

71. $\int_1^\infty 1/(x\sqrt{1+x})\,dx$ **72.** $\int_2^\infty (1+x)/(1+x^3)\,dx$

Summary of Key Topics

Volumes by Slicing (Section 8.1)

If the cross sections, perpendicular to the x-axis, of a certain solid have area $A(x)$ at $x, a \leq x \leq b$, then the volume of the solid is

$$\int_a^b A(x)\,dx.$$

Solids of Revolution: The Method of Washers (Section 8.1)

Let $y = f(x)$ be a continuous function on the interval $[a, b]$ and assume that the line $y = \beta$ does not cross the graph of f. The volume V of the solid enclosed when the graph of f is rotated about the line $y = \beta$ is given by

$$V = \int_a^b \pi(f(x) - \beta)^2\,dx.$$

Let $x = g(y)$ be a continuous function on the interval $y \in [c, d]$ and assume that the line $x = \alpha$ does not cross the graph of g. The volume V of the solid enclosed when the graph of g is rotated about the line $x = \alpha$ is given by

$$V = \int_c^d \pi(g(y) - \alpha)^2\,dy.$$

Solids of Revolution: The Method of Cylindrical Shells (Section 8.1)

Let $y = f(x)$ be a nonnegative, continuous function on the interval $[a, b]$. Assume that the line $x = \alpha$ does not cross the graph of f. The volume V of the solid generated when the region under the graph of f and above the x-axis is rotated about the line $x = \alpha$ is given by

$$V = \int_a^b 2\pi|x - \alpha| f(x)\,dx.$$

Let $x = g(y)$ be a nonnegative, continuous function on the interval $[c, d]$ and assume that the line $y = \beta$ does not cross the graph of g. The volume of the solid generated when the region between the graph of g and the y-axis is rotated about the line $y = \beta$ is given by

$$V = \int_c^d 2\pi|y - \beta| g(y)\,dy.$$

Arc Length (Section 8.2)

If f is a continuously differentiable function on the interval $[a, b]$, then the arc length of the graph of f is

$$\int_a^b \sqrt{1 + f'(x)^2}\,dx.$$

Surface Area (Section 8.2)

If f is a nonnegative, continuously differentiable function on the interval $[a, b]$, then the surface area of the surface obtained when the graph of f is rotated about the x-axis is

$$\int_a^b 2\pi f(x)\sqrt{1 + f'(x)^2}\, dx.$$

Average Value of a Function (Section 8.3)

If $f(x)$ is a continuous function on an interval $[a, b]$, then the average value of f on $[a, b]$ is

$$f_{\text{ave}} = \frac{1}{b-a}\int_a^b f(x)\, dx.$$

If X is a random variable that takes values in an interval $[a, b]$ and if f is a nonnegative function on $[a, b]$ such that the probability that X takes a value in any subinterval $[\alpha, \beta]$ is $\int_\alpha^\beta f(x)\, dx$, then the mean μ_X of X (or expectation of X) is

$$\mu_X = \int_a^b x f(x)\, dx.$$

Center of Mass (Section 8.4)

Suppose that $f(x)$ is a nonnegative, continuous function on an interval $[a, b]$. Let $\mathcal{R}$ be the region that is bounded above by the graph of $y = f(x)$, below by the x-axis, and on the sides by the line segments $x = a$ and $x = b$. If $\mathcal{R}$ has a uniform mass density δ and if c is any real number, then the moment $M_{x=c}$ of $\mathcal{R}$ about the axis $x = c$ is defined by the equation

$$M_{x=c} = \delta\int_a^b (x - c) f(x)\, dx.$$

The center of mass $(\overline{x}, \overline{y})$ of $\mathcal{R}$ is given by the equations

$$\overline{x} = \frac{M_{x=0}}{M}$$

and

$$\overline{y} = \int_a^b \frac{f(x)^2}{2M}\, dx$$

where $M = \delta\int_a^b f(x)\, dx$ is the mass of $\mathcal{R}$.

Work (Section 8.5)

If a body travels a linear path with position given by x (measured in feet), $a \le x \le b$, under the influence of a force in the direction of motion that at position x is $F(x)$ pounds, then the work performed in traveling from a to b is

$$\int_a^b F(x)\, dx \text{ ft-lb.}$$

Improper Integrals (Sections 8.6 and 8.7)

If $f(x)$ is a continuous function on $[a, b)$, which is unbounded as $x \to b^-$, then the integral $\int_a^b f(x)\,dx$ is defined as

$$\lim_{\epsilon \to 0^+} \int_a^{b-\epsilon} f(x)\,dx,$$

provided that this limit exists. A similar definition applies for singularities on the left.

If the singularity occurs in the middle of the interval of integration, then the integral should be broken up at the singularity to reduce to the two previous cases.

If f is a continuous function on the interval $[A, +\infty)$, then the integral $\int_A^{+\infty} f(x)\,dx$ is defined as

$$\lim_{N \to +\infty} \int_A^N f(x)\,dx,$$

provided that this limit exists. A similar definition applies to integrals on an interval of the form $(-\infty, B]$.

Improper integrals may be applied to calculate areas, volumes, probabilities, and a variety of other physical quantities.

genesis & DEVELOPMENT

Archimedes and the Sphere

Arc lengths, surface areas, and volumes—these concepts were all investigated by the geometers of ancient Greece. Archimedes, in particular, became occupied with their computation. In the best known of his calculations, Archimedes demonstrated that the volume of a sphere is two thirds the volume of the circumscribing cylinder and that the surface area of a sphere is (also) two-thirds that of the circumscribing cylinder (including its circular ends). Stated more explicitly, he proved that the volume and surface area of a sphere of radius r are $4\pi r^3/3$ and $4\pi r^2$, respectively.

These results must have pleased Archimedes for he requested that his tomb be decorated with a sphere and cylinder and that the ratio giving their relationship be inscribed. His wishes were carried out, as we learn from Cicero, who in 75 BCE rediscovered and repaired the overgrown, forgotten site:

> *I remembered some verses which I had been informed were engraved on his monument, and these set forth that on the top of the tomb there was placed a sphere with a cylinder.... I observed a small column standing out a little above the briers with the figure of a sphere and a cylinder upon it.... When we could get at it ... I found the inscription, though the latter parts of all the verses were almost half effaced. Thus, one of the noblest cities of Greece, and one which at one time likewise had been very celebrated for learning, had known nothing of the monument of its greatest genius.*

The restored grave of Archimedes eventually vanished again. However, in modern times, workers excavating the site of a new hotel in Syracuse reported coming upon a stone tablet inscribed with a sphere and cylinder.

Kepler and Solids of Revolution

In 1612, Johannes Kepler (1571–1630) took up residence in Linz. Because the wine harvest was good that year, the Danube River banks were filled with wine casks. "This being so," in the words of Kepler's biographer, Max Caspar, "as husband and good *paterfamilias,* he [Kepler] considered it his duty to provide his home with the necessary drink." Accordingly, Kepler had some wine barrels installed in his house. Seeing the wine merchant simply insert a measuring stick into the cask to determine its volume, without regard to its exact shape, Kepler set himself the problem of developing a more scientific system of measurement.

Inspired by these wine barrels, Kepler derived formulas for the volumes of several solids of revolution (not all of which were suitable for wine storage). Among other results, he rediscovered Pappus's formula, $2\pi b \cdot \pi a^2$, for the volume of the *torus,* an inner-tube shaped solid that is generated when a disk of radius a is revolved about an axis a distance $b > a$ from its center (Figure 1). The rigorous deductions of Archimedes were somewhat beyond Kepler, who resorted to the very free use of infinitesimal arguments. In Kepler's own words: "[W]e could obtain absolute and perfect proofs from the books of Archimedes if we did not shrink from the thorny reading of his writings." Kepler published the results of his investigations in 1615. Despite his best efforts to raise funds, he was forced to underwrite the publication costs himself. With the exception of sales to one nobleman, two mathematicians, and the Königsberg Library, Kepler received no revenue from which to recoup his outlay.

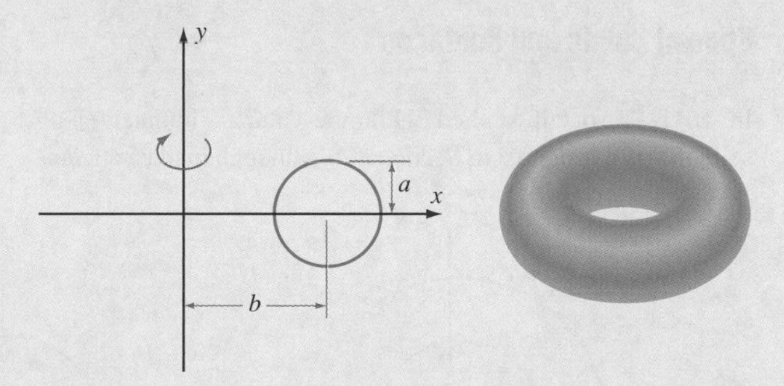

Figure 1

Arc Length

The arc length problem (known then as *rectification*) was also revived by mathematicians in the 1600s. Some leading philosophers believed rectification to be insoluble. Descartes, for one, was emphatic on this point in his *Géométrie:* "The ratios between straight and curved lines are not known, and I believe cannot be discovered by human minds." The idea that it might be possible to compare the arc lengths of two different curves seems to have originated with another philosopher, Thomas Hobbes (1588–1679), who communicated his idea to Roberval in 1642. By the end of that year, Roberval was able to produce equal-length arcs of the Archimedean spiral $\sqrt{x^2+y^2} = \arctan(y/x)$ (Figure 2a) and of the parabola $y = x^2/2$. Three years later, Torricelli calculated the arc length of the logarithmic spiral shown in Figure 2b.

After a 12-year hiatus, a number of arc length calculations were published in quick succession. In 1657, William Neile (1637–1670) rectified the *semicubical parabola* $y^3 = ax^2$. One year later, Christopher Wren (1632–1723), better known as the architect of St. Paul's Cathedral, computed the arc length of the cycloid. The general arc length integral, $\int \sqrt{1+(dy/dx)^2}\,dx$, was first published by Hendrik van Heuraet (b. 1633) in 1659. Fermat also solved the rectification problem more or less simultaneously.

The *characteristic triangle* or *differential triangle*—the right "triangle" with dx and dy for legs and ds for "hypotenuse" (Figure 3)—appears first in the work of Blaise Pascal in 1658, who became interested in rectification after hearing of Wren's result. In 1672, Huygens introduced Leibniz to the work of Pascal. In the hands of Leibniz, the differential triangle became the basis of a method of quadrature that supplemented its role in rectification. Indeed, it led Leibniz to the formula for integration by parts and provided him with his first recognition of the relationship between tangents (derivatives) and areas (integrals).

Special Solids and Surfaces

In 1641, Torricelli studied the finite-volume, infinitely long solid that is considered in Section 8.7. Although mathematicians accepted Torricelli's *Solidum Hyperbolicum Acutum,* as it was called, some philosophers, notably Thomas Hobbes, remained unconvinced. A spirited exchange between Hobbes and John Wallis ensued. In Wallis's words, understanding Torricelli's infinitely long solid "requires more Geometry and Logic than Mr. Hobs is Master of." The reply of Hobbes was equally pointed: "I think Dr. Wallis does wrong, ... for, to understand this for sense, it is not required that a man should be a geometrician or a logician but that he should be mad."

Another surface of revolution that has historical interest was introduced by Euler. Suppose that $P = (a, \alpha)$ and $Q = (b, \beta)$ are points in the plane that lie above the x-axis. Suppose that they are joined by a curve $\mathcal{C}$ that is an arc of the graph of a function f. Rotating $\mathcal{C}$ about the x-axis generates a surface of revolution. Euler posed the question: Among all such surfaces of revolution, which has minimal surface area? Analytically, the question may be asked in this way: Which function f satisfies $f(a) = \alpha$ and $f(b) = \beta$ and minimizes the integral

$$I(f) = 2\pi \int_a^b f(x)\sqrt{1+(f'(x)^2)}\,dx?$$

The task of finding a function that minimizes an integral such as $I(f)$ is substantially more complicated than finding an extremum of a numerical function. This kind of problem is studied in a branch of mathematics called the *calculus of variations*. Euler solved his own problem, proving that $I(f)$ is minimized when $\mathcal{C}$ is the arc of a catenary. The surface of revolution, which is called a catenoid, arises naturally—indeed, you can generate a catenoid with some wire and soapy water. Form wires in the shape of two circles, bring them together in a soap solution, and draw them apart so that they are circles of revolution about the same axis. The resulting soap film stretching between the two circular wires will assume the shape of a catenoid. In general, soap films give minimal surfaces. This fact can be proved from their surface tension equations.

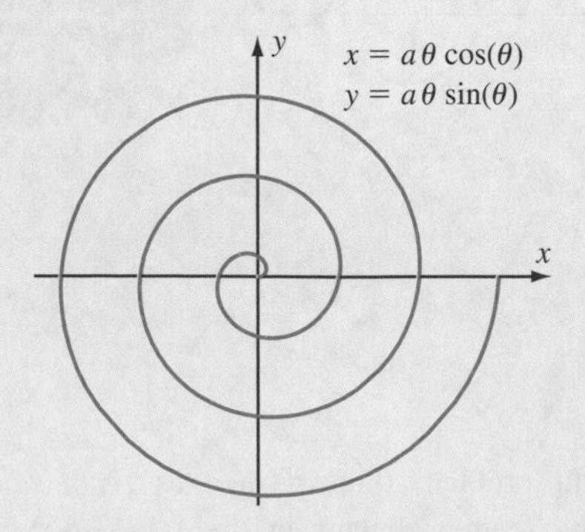

Figure 2a
Archimedean spiral

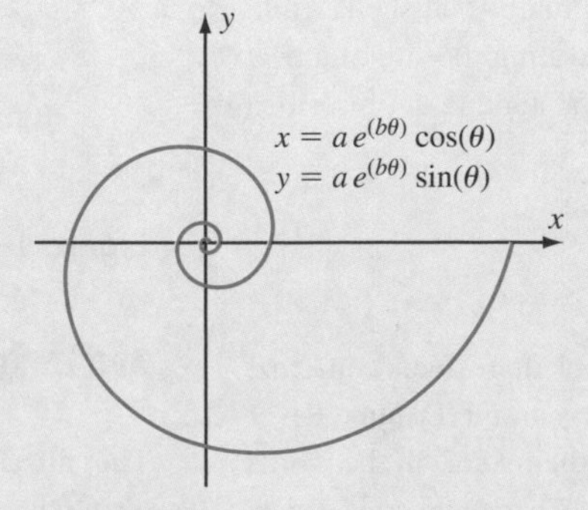

Figure 2b
Logarithmic spiral

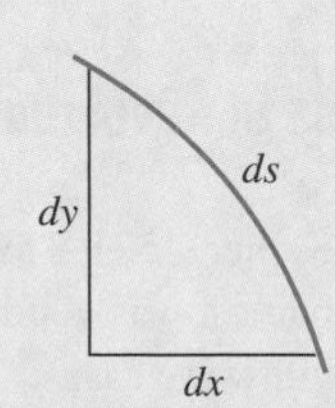

Figure 3
Leibniz's characteristic triangle

9

Infinite Series

PREVIEW

When we learn to write fractions in decimal form, we soon become aware that some numbers result in a nonterminating sequence of digits. For example, when we decimalize $1/3$, we apply the long division algorithm. No matter how many times we divide, we are left with a remainder. If we terminate the procedure after any division, we are left with $0.33\ldots33$, which is a little too small. If we round up to $0.33\ldots34$, then the result is a number that is a little too great. Like it or not, we must accept that the decimalization of $1/3$ is $0.333\ldots$ without end.

Of course, in grade school, we do not pause long to reflect on the infinite process symbolized by the trailing dots. Intuitively, we understand that the notation must stand for the "infinite" sum

$$\frac{3}{10}+\frac{3}{100}+\frac{3}{1000}+\frac{3}{10000}+\cdots,$$

even if we have only a fuzzy idea of what that might mean.

Now that we have studied some calculus, we are prepared to understand *exactly* what such sums signify. The concept of "limit" that is needed for the infinite processes of calculus is just what we need to understand the infinite sums that arise in grade school and elsewhere. In fact, our approach is nothing more than a framework for turning our intuition into something precise. If you think of the sequence

$$0.3=\frac{3}{10},\ 0.33=\frac{3}{10}+\frac{3}{100},\ 0.333=\frac{3}{10}+\frac{3}{100}+\frac{3}{1000},\ \ldots$$

as a sequence of approximations of $1/3$ that become ever more accurate, then you already have the key idea of understanding an "infinite" sum as a limit of its "partial sums."

Infinite sums are important in calculus, but they are not confined to mathematical disciplines. If you study scientific subjects such as biology or thermodynamics, then you are likely to encounter infinite series. If you learn about the money supply in economics or annuities in finance, you will see that infinite series are needed. This chapter provides a good basis for your understanding of infinite sums, wherever they arise.

9.1 Series

The idea of adding together finitely many numbers is a simple and familiar one. We are so used to the properties of finite sums that we tend not to think about them. Now we will learn about sums of *infinitely many* numbers, so we will have to be a bit more careful. As the following example shows, contradictions can arise if we do not establish certain ground rules.

Example 1 Group the terms of $-1+1-1+1-1+1-1+\cdots$ in two ways to produce different results.

Solution If we group the terms as $(-1+1)+(-1+1)+(-1+1)+\cdots$, the sum seems to be $0+0+0+\cdots$, which ought to be 0. On the other hand, if we group the terms as $-1+(1-1)+(1-1)+(1-1)+\cdots$, the sum seems to be $-1+0+0+0+\cdots$, which ought to be -1. Apparently, the associative law of addition fails for this infinite sum. ■

Example 1 points to the need for a precise definition that establishes what it means to "add" infinitely many numbers.

The Definition of an Infinite Series

Recall that an *infinite sequence* $\{a_n\}_{n=1}^{\infty}$ (or simply $\{a_n\}$) is a list of terms indexed by the positive integers: $a_1, a_2, a_3, \ldots$. For example, $1, 1/2, 1/4, 1/8, 1/16, \ldots$ is the infinite sequence $\{1/2^{n-1}\}_{n=1}^{\infty}$ of nonnegative powers of 1/2. If $a_1, a_2, a_3, \ldots$ is an infinite sequence, then the (formal) expression $a_1+a_2+a_3+\cdots$ is called an *infinite series*. It is convenient to extend the sigma notation

$$\sum_{n=M}^{N} a_n = a_M + a_{M+1} + \cdots + a_{N-1} + a_N$$

for finite sums introduced in Section 5.1. To signify an infinite series, we simply use the symbol ∞ for the upper limit of summation. Thus, we let $\sum_{n=1}^{\infty} a_n$ denote the infinite series $a_1+a_2+a_3+\cdots$. For instance, if $a_n = 1/2^{n-1}$, then $\sum_{n=1}^{\infty} a_n$ denotes the infinite series $1+1/2+1/4+1/8+1/16+\cdots$.

Example 1 suggests that some care is needed if we are to assign a value to the infinite series $\sum_{n=1}^{\infty} a_n$. Nevertheless, there is a quite natural way to define and interpret infinite series. Consider the decimalization of 1/3, namely 0.3333... (ad infinitum). In reality, the decimal number 0.3333... symbolizes the infinite series $\sum_{n=1}^{\infty} a_n$ with $a_n = 3/10^n$. If we need a decimal expansion accurate to two decimal places, then we use $a_1+a_2=0.33$. Similarly, if we require five decimal digits of 1/3, then we use $a_1+a_2+a_3+a_4+a_5=0.33333$, and so on. By taking "partial" sums of the infinite series $\sum_{n=1}^{\infty} a_n$, we obtain approximations that become closer and closer to the exact value of $\sum_{n=1}^{\infty} a_n$. This familiar example suggests the method by which we define the "sum" of an infinite series in general.

Definition If $a_1, a_2, a_3, \ldots$ is an infinite sequence, then we say that the expression $S_N = \sum_{n=1}^{N} a_n$ is a *partial sum* of the infinite series $\sum_{n=1}^{\infty} a_n$. The sequence $\{S_N\}_{N=1}^{\infty}$ is called the *sequence of partial sums* of the infinite series $\sum_{n=1}^{\infty} a_n$.

Example 2 Compute the sequence of partial sums of the infinite series $\sum_{n=1}^{\infty}(-1)^n$.

Solution According to the definition, $S_1 = \sum_{n=1}^{1}(-1)^n = -1$, $S_2 = \sum_{n=1}^{2}(-1)^n = (-1)^1 + (-1)^2 = -1 + 1 = 0$, $S_3 = \sum_{n=1}^{3}(-1)^n = (-1)^1 + (-1)^2 + (-1)^3 = -1$, and so on. The sequence of partial sums of $\sum_{n=1}^{\infty}(-1)^n$ is $-1, 0, -1, 0, -1, 0, \ldots$. ■

Example 3 Find a formula for the partial sum $\sum_{n=1}^{N} 1/2^{n-1}$ of the infinite series $\sum_{n=1}^{\infty} 1/2^{n-1}$.

Solution To calculate the partial sum

$$\sum_{n=1}^{N}\frac{1}{2^{n-1}} = 1 + \frac{1}{2} + \frac{1}{2^2} + \frac{1}{2^3} + \cdots + \frac{1}{2^{N-1}},$$

we use the equation

$$1 + r + r^2 + \cdots + r^{N-1} = \frac{r^N - 1}{r - 1} \quad (r \neq 1) \tag{9.1}$$

that we derived in Theorem 3 from Section 2.5. If we specify $r = 1/2$ in equation (9.1), we see that

$$\begin{aligned}\sum_{n=1}^{N}\left(\frac{1}{2}\right)^{n-1} &= 1 + \frac{1}{2} + \frac{1}{2^2} + \frac{1}{2^3} + \cdots + \frac{1}{2^{N-1}} \\ &= \frac{(1/2^N - 1)}{(1/2 - 1)} \\ &= 2\left(1 - \frac{1}{2^N}\right) = 2 - \frac{1}{2^{N-1}}.\end{aligned}$$

■

Convergence of Infinite Series

The meaning that we give to the "sum" of an infinite series $\sum_{n=1}^{\infty} a_n$ is determined by the behavior of its sequence of partial sums $\{S_N\}$ as N tends to infinity. In Section 2.5, we learned that an infinite sequence $\{S_N\}$ has a limit ℓ as N tends to infinity if the terms S_N become and remain arbitrarily close to ℓ for N sufficiently large. For example, the terms of the sequence $\{2 - 1/2^{N-1}\}_{N=1}^{\infty}$ become and remain arbitrarily close to 2, so we write $\lim_{N\to\infty}(2 - 1/2^{N-1}) = 2$. On the other hand, the terms of the sequence $\{(-1)^N\}_{N=1}^{\infty}$ do not become *and remain* close to one number. Instead, they alternate between $+1$ and -1. In this case, we say that $\lim_{N\to\infty}(-1)^N$ does not exist.

Definition

Let $\{S_N\}$ be the sequence of partial sums of an infinite series $\sum_{n=1}^{\infty} a_n$. If $\lim_{N\to\infty} S_N = \ell$, then we say that the infinite series $\sum_{n=1}^{\infty} a_n$ *converges* to ℓ. In this case, when we refer to $\sum_{n=1}^{\infty} a_n$, we mean the number ℓ, and we write $\sum_{n=1}^{\infty} a_n = \ell$. We call ℓ the *sum* of the infinite series. If the sequence $\{S_N\}$ does not converge, then we say the series *diverges*. A series that converges is said to be *convergent;* otherwise, we say that it is *divergent*.

inSIGHT

The terms *convergent* and *divergent* suggest an analogy between infinite series $\sum_{n=1}^{\infty} a_n$ and improper integrals $\int_1^{\infty} f(x)\,dx$. Indeed, the approaches to the two concepts are similar. To calculate $\int_1^{\infty} f(x)\,dx$, we compute $\int_1^{N} f(x)\,dx$ for large values of N and take the limit of $\int_1^{N} f(x)\,dx$ as N tends to infinity. Analogously, we can calculate the partial sum $a_1 + a_2 + a_3 + \cdots + a_N$ for any large integer N. If this sum has a limit ℓ as the number N of summands tends to infinity, then we define $\sum_{n=1}^{\infty} a_n$ to be this limit ℓ. Although we call ℓ the "sum" of the series, *it is not a sum* in the conventional sense: By definition, it is a *limit*.

inSIGHT

Example 5 sheds some light on the apparent contradiction encountered in Example 1. When we first investigated the series $\sum_{n=1}^{\infty} (-1)^n = -1 + 1 - 1 + 1 - 1 + \cdots$, we formed two different groupings of its summands. Those groupings suggested different values, -1 and 0, for the series. Now we understand that -1 is the limit of the odd partial sums and 0 is the limit of the even partial sums. The partial sums taken together do not have a limit, and therefore the series does not converge. The lesson is that we should not attempt to perform arithmetic operations on series that do not converge.

Example 4 Does the series $\sum_{n=1}^{\infty} 1/2^{n-1}$ converge? If so, to what limit?

Solution In Example 3, we calculated the Nth partial sum $S_N = 2 - 1/2^{N-1}$ of this infinite series. We may now pass to the limit:

$$\lim_{N\to\infty} S_N = \lim_{N\to\infty} \left(2 - \frac{1}{2^{N-1}}\right) = 2.$$

By definition, we say that the series $\sum_{n=1}^{\infty} 1/2^{n-1}$ converges to 2. ■

Example 5 Discuss convergence for the series $\sum_{n=1}^{\infty} (-1)^n = -1 + 1 - 1 + 1 - 1 + \cdots$.

Solution If N is even, then the partial sum $S_N = \sum_{n=1}^{N} (-1)^n = -1 + 1 - \cdots - 1 + 1$ comprises an equal number of $+1$s and -1s. Therefore, $S_N = 0$ for N even. If N is odd, then there is one more -1 than $+1$ in the sum $S_N = \sum_{n=1}^{N} (-1)^n$, and $S_N = -1$. Therefore, the sequence $\{S_N\}_{N=1}^{\infty}$ of partial sums is simply the sequence $-1, 0, -1, 0, -1, 0, \ldots$, which does not converge. We conclude that the series itself does not converge. ■

A LOOK BACK

It is important to understand the difference between a sequence and a series. Remember, a *sequence* is a *list* of numbers, whereas a *series* is a *sum* (or a *limit of a sum*) of numbers. We use a sequence $\{a_n\}$ to create an infinite series $\sum_{n=1}^{\infty} a_n$. Then we form the sequence $\{S_N\}_{N=1}^{\infty}$ of partial sums $S_N = \sum_{n=1}^{N} a_n$ of the infinite series. We determine whether we can assign a value to the infinite series by looking at the limiting behavior of its sequence of partial sums.

A Telescoping Series

It is relatively rare to find an explicit formula for the partial sums of an infinite series. In the next example, we employ a useful algebraic "trick" to evaluate the partial sums.

Example 6 Discuss convergence for the series $\sum_{n=1}^{\infty} 1/(n(n+1))$.

Solution If you use a calculator to compute some partial sums, then you will probably be convinced that the series converges. We can justify this conclusion by writing each summand in its partial fractions decomposition:

$$\frac{1}{n(n+1)} = \left(\frac{1}{n} - \frac{1}{n+1}\right).$$

Then

$$\begin{aligned} S_N &= \frac{1}{1\cdot 2} + \frac{1}{2\cdot 3} + \frac{1}{3\cdot 4} + \cdots + \frac{1}{N\cdot(N+1)} \\ &= \left(\frac{1}{1} - \frac{1}{2}\right) + \left(\frac{1}{2} - \frac{1}{3}\right) + \left(\frac{1}{3} - \frac{1}{4}\right) + \cdots + \left(\frac{1}{N-1} - \frac{1}{N}\right) + \left(\frac{1}{N} - \frac{1}{N+1}\right). \end{aligned}$$

Almost everything cancels, and we have $S_N = 1 - 1/(N+1)$. Clearly $\lim_{N\to\infty} S_N = 1$. Therefore, the series $\sum_{n=1}^{\infty} 1/(n(n+1))$ converges to 1. ■

INSIGHT

A series that can be written in the form $\sum_{n=1}^{\infty} (f(n) - f(n+1))$ is called a *collapsing* or *telescoping series,* because each partial sum collapses to two summands:

$$\begin{aligned} \sum_{n=1}^{N} (f(n) - f(n+1)) &= (f(1) - f(2)) + (f(2) - f(3)) + (f(3) - f(4)) + \cdots \\ &\quad + (f(N-1) - f(N)) + (f(N) - f(N+1)) \\ &= f(1) - f(N+1). \end{aligned}$$

Such a series converges if and only if $\lim_{N\to\infty} f(N) = 0$. Example 6 illustrates a convergent telescoping series.

The Harmonic Series

The infinite series $\sum_{n=1}^{\infty} 1/n$ comes up often in mathematics and physics. It is called the *harmonic series* because of its role in the theory of acoustics: If the natural frequencies at which a given body vibrates are $\rho_1, \rho_2, \ldots$ and if $\rho_n/\rho_{n+1} = 1/(n+1)$ for all n, then the frequencies are said to form a sequence of *harmonics*. The harmonic series is divergent; however, if you calculate a large number of partial sums of the harmonic series using a computer, then you might not suspect it. The divergence takes place so slowly that it is difficult to be certain what is actually happening. For example, if we sum the first million terms, we obtain $S_{1000000} = 14.3927$ (rounded to four decimal places). If we add the next thousand terms to this, then we obtain $S_{1001000} = 14.3937$—the change is barely noticeable despite the additional thousand terms. Add 10^{43} terms of the harmonic series, and you still have a sum that is less than 100. To be certain of the divergence, we need a mathematical proof.

Theorem 1 The harmonic series $\sum_{n=1}^{\infty} 1/n$ is divergent.

Proof Notice that

$$\begin{aligned} S_1 &= 1 = \frac{2}{2} \\ S_2 &= 1 + \frac{1}{2} = \frac{3}{2} \\ S_4 &= 1 + \frac{1}{2} + \left(\frac{1}{3} + \frac{1}{4}\right) \\ &> 1 + \frac{1}{2} + \left(\frac{1}{4} + \frac{1}{4}\right) = 1 + \frac{1}{2} + \frac{1}{2} = \frac{4}{2} \\ S_8 &= 1 + \frac{1}{2} + \left(\frac{1}{3} + \frac{1}{4}\right) + \left(\frac{1}{5} + \frac{1}{6} + \frac{1}{7} + \frac{1}{8}\right) \\ &> 1 + \frac{1}{2} + \left(\frac{1}{4} + \frac{1}{4}\right) + \left(\frac{1}{8} + \frac{1}{8} + \frac{1}{8} + \frac{1}{8}\right) = \frac{5}{2}. \end{aligned}$$

Now let us examine S_{16}. The first eight terms sum to S_8, which we have observed to be greater than $5/2$. Each of the remaining eight terms, $1/9, 1/10, \ldots, 1/16$, is greater than or equal to $1/16$. The sum of these eight terms is therefore at least $8 \cdot (1/16)$, or $1/2$. We conclude that $S_{16} > 5/2 + 1/2 = 6/2$.

In general, this argument shows that $S_{2^k} \geq (k+2)/2$. The sequence S_Ns is increasing because the series contains only positive terms. The fact that the partial sums $S_1, S_2, S_4, S_8, \ldots$ increase without bound shows that the entire sequence of partial sums must increase without bound. We conclude that the series diverges. ■

Series of Powers—Geometric Series

We now consider series $\sum_{n=0}^{\infty} r^n = 1 + r + r^2 + r^3 + \cdots$ of all nonnegative integer powers of a fixed number r. These are called *geometric series*. (Up until now we have always used $n = 1$ as the first index value of our infinite series. Starting with $n = 0$ does not affect the theory at all. See Exercise 48.)

Theorem 2 If $|r| < 1$, then the series $\sum_{n=0}^{\infty} r^n$ converges to $1/(1-r)$. That is,

$$\sum_{n=0}^{\infty} r^n = \frac{1}{1-r} \quad (|r| < 1). \tag{9.2}$$

If $|r| \geq 1$, then the series $\sum_{n=0}^{\infty} r^n$ diverges.

Proof If $r \geq 1$, then it is clear that the partial sum $S_N = \sum_{n=0}^{N} r^n$ of the geometric series $\sum_{n=0}^{\infty} r^n$ satisfies $S_N = \sum_{n=0}^{N} r^n \geq \sum_{n=0}^{N} 1 = N + 1$. It follows that the sequence $\{S_N\}$ does not converge to any real number ℓ. In other words, the series $\sum_{n=0}^{\infty} r^n$ is divergent for $r \geq 1$. We leave the case $r \leq -1$ as an exercise for the interested reader. (In Section 9.2, we will see how to prove this divergence without analyzing partial sums.)

Now suppose that $|r| < 1$. Let S_N denote the Nth partial sum. Using formula (9.1), we have

$$S_N = \sum_{n=0}^{N-1} r^n = (1 + r + r^2 + \cdots + r^{N-1}) = \frac{r^N - 1}{r - 1}.$$

Since $|r| < 1$, we have $\lim_{N\to\infty} r^N = 0$ and $\lim_{N\to\infty} S_N = (0-1)/(r-1) = 1/(1-r)$. ■

Example 7 Evaluate the repeating decimal 1.11111 . . . .

Solution The infinite decimal expansion 1.11111 . . . represents the geometric series $\sum_{n=0}^{\infty} 1/10^n$. According to Theorem 2 with $r = 1/10$, the series $\sum_{n=0}^{\infty} 1/10^n$ converges to $1/(1 - 1/10) = 10/9$. ■

Basic Properties of Series

We now present some elementary properties of series that are similar to the properties of sequences that we learned in Chapter 2. Because they are so familiar in nature, we shall omit their proofs.

Theorem 3 Suppose that $\sum_{n=1}^{\infty} a_n$ converges to A and $\sum_{n=1}^{\infty} b_n$ converges to B. Then

a. $\sum_{n=1}^{\infty} (a_n + b_n)$ converges to $A + B$ and $\sum_{n=1}^{\infty} (a_n - b_n)$ converges to $A - B$; and

b. $\sum_{n=1}^{\infty} (\lambda a_n)$ converges to λA, where λ is any real constant.

Theorem 4 If $\sum_{n=1}^{\infty} a_n$ diverges and λ is any nonzero constant, then $\sum_{n=1}^{\infty} (\lambda a_n)$ also diverges.

Example 8 Discuss convergence of the series $\sum_{n=1}^{\infty} (2^{-n+7} - 10^{-n+3})$.

Solution Since $2^{-n+7} = 2^6/2^{n-1} = 64/2^{n-1}$ and $10^{-n+3} = 10^2/10^{n-1} = 100/10^{n-1}$, we may rewrite the series as

$$\sum_{n=1}^{\infty} \left(64 \left(\frac{1}{2}\right)^{n-1} - 100 \left(\frac{1}{10}\right)^{n-1} \right).$$

Example 4, combined with Theorem 3b, tells us that $\sum_{n=1}^{\infty} 64/2^{n-1}$ converges to 128. Similarly, Example 7 and Theorem 3b tell us that $\sum_{n=1}^{\infty} 100/10^{n-1}$ converges to 1000/9. Hence, by using Theorem 3a, we may conclude that the given series converges to $128 - 1000/9 = 152/9$. ■

Example 9 Let $a_n = (2^n - n)/(n2^n)$. Does the series $\sum_{n=1}^{\infty} a_n$ converge?

Solution We may write the summand a_n as

$$a_n = \frac{2^n - n}{n2^n} = \frac{2^n}{n2^n} - \frac{n}{n2^n} = \frac{1}{n} - \frac{1}{2^n}.$$

Let $b_n = 1/2^n$. By Example 4, we know that the series $\sum_{n=1}^{\infty} b_n$ converges. If the series $\sum_{n=1}^{\infty} a_n$ were to converge, then so would the series $\sum_{n=1}^{\infty} (a_n + b_n)$, by Theorem 3. But $a_n + b_n = 1/n$, and we know that the harmonic series diverges (Theorem 1). Therefore, $\sum_{n=1}^{\infty} a_n$ must diverge. ■

Example 10 At a certain aluminum recycling plant, the recycling process turns n lb of used aluminum into $9n/10$ lb of new aluminum. Including the initial quantity, how much usable aluminum will 100 lb of virgin aluminum ultimately yield, if we assume that it is continually returned to the same recycling plant?

Solution We begin with 100 lb of aluminum. When it is returned to the plant as scrap and recycled, $0.9 \cdot 100$ lb of new aluminum results. When that aluminum is returned to the plant as scrap and recycled, $0.9 \cdot (0.9 \cdot 100) = 0.9^2 \cdot 100$ lb of new aluminum results. This process continues forever. Writing the initial quantity of aluminum as $0.9^0 \cdot 100$, and using Theorem 3b and equation (9.2), we see that the total amount of aluminum created is

$$0.9^0 \cdot 100 + 0.9 \cdot 100 + 0.9^2 \cdot 100 + \cdots = \sum_{n=0}^{\infty} (0.9^n \cdot 100) = 100 \sum_{n=0}^{\infty} 0.9^n$$
$$= 100 \cdot \frac{1}{1-0.9} = 1000.$$

Therefore, 1000 lb of usable aluminum is ultimately generated by an initial 100 lb of virgin aluminum. ■

quickquiz

1. What is an infinite series?
2. What does it mean for an infinite series to converge?
3. What is the difference between a sequence and a series?
4. Discuss convergence for the series $\sum_{n=1}^{\infty} 1/(n+1)$.

EXERCISES

Problems for Practice

In Exercises 1–8, a series $\sum_{n=1}^{\infty} a_n$ is given. Calculate the first five partial sums of the series. That is, calculate $S_N = \sum_{n=1}^{N} a_n$ for $N = 1, 2, 3, 4, 5$.

1. $\sum_{n=1}^{\infty} 1/n$ **2.** $\sum_{n=1}^{\infty} 2^n/n!$
3. $\sum_{n=1}^{\infty} (4^{-n}+1)$ **4.** $\sum_{n=1}^{\infty} (1/n - 1/(n+1))$
5. $\sum_{n=1}^{\infty} 2^n/3^n$ **6.** $\sum_{n=1}^{\infty} 1/n^2$
7. $\sum_{n=1}^{\infty} (-1)^{n+1}/n^2$ **8.** $\sum_{n=1}^{\infty} n^2/10^n$

In Exercises 9–16, find the sum of the series.

9. $\sum_{n=1}^{\infty} 8^{-n}$ **10.** $\sum_{n=1}^{\infty} (3/7)^n$
11. $\sum_{n=1}^{\infty} (2/3)^{2n}$ **12.** $\sum_{n=0}^{\infty} (-2/3)^n$
13. $\sum_{n=1}^{\infty} 7^{-n/3}$ **14.** $\sum_{n=3}^{\infty} (2^{-n}+3^{-n})$
15. $\sum_{n=1}^{\infty} (2^{-n} \cdot 3^{-n} + 7^{-n})$ **16.** $\sum_{n=4}^{\infty} (5^{-n} \cdot 3^{-n} \cdot 4^n)$

In Exercises 17–20, write the repeating decimal as a constant multiplied by a geometric series (the geometric series will contain powers of 0.1). Use the formula for the sum of a geometric series to express the repeating decimal as a rational number.

17. 0.8888888 ... **18.** 0.131313131313 ...
19. 0.017017017 ... **20.** 0.983983983 ...

Further Theory and Practice

In Exercises 21–24, prove that the series diverges by showing that the Nth partial sum satisfies $S_N \geq k \cdot N$ for some positive constant k.

21. $\sum_{n=1}^{\infty} n/(n+1)$ **22.** $\sum_{n=1}^{\infty} n/(2n+3)$
23. $\sum_{n=1}^{\infty} n/\sqrt{n^2+1}$ **24.** $\sum_{n=1}^{\infty} (1.01)^n$

In Exercises 25–30, calculate the Nth partial sum S_N of the series in closed form. Sum the series by finding $\lim_{N\to\infty} S_N$.

25. $\displaystyle\sum_{n=1}^{\infty} \left(\frac{1}{n} - \frac{1}{n+1}\right)$

26. $\displaystyle\sum_{n=1}^{\infty} \left(\frac{n+1}{n+2} - \frac{n}{n+1}\right)$

27. $\displaystyle\sum_{n=1}^{\infty} \left(\frac{2n-2}{n^3} - \frac{2n}{(n+1)^3}\right)$

28. $\displaystyle\sum_{n=1}^{\infty} (n^{-2} - (n+1)^{-2})$

29. $\displaystyle\sum_{n=1}^{\infty} \left(\frac{1}{\sqrt{n}} - \frac{1}{\sqrt{n+1}}\right)$

30. $\displaystyle\sum_{n=1}^{\infty} (\arctan(n+1) - \arctan(n))$

In Exercises 31–34, use partial fractions to calculate the Nth partial sum S_N of the series in closed form. Sum the series by finding $\lim_{N\to\infty} S_N$.

31. $\displaystyle\sum_{n=1}^{\infty} \frac{1}{n(n+2)}$ **32.** $\displaystyle\sum_{n=1}^{\infty} \frac{1}{n(n+4)}$

33. $\displaystyle\sum_{n=1}^{\infty} \frac{1}{(2n+1)(2n+3)}$ **34.** $\displaystyle\sum_{n=1}^{\infty} \frac{2n+1}{(n^2+n)^2}$

35. Calculate the Nth partial sum of $\sum_{n=1}^{\infty} \ln(n/(n+1))$. Show that the series is divergent.

36. Show that

$$\sum_{n=2}^{N} \ln\left(1 - \frac{1}{n^2}\right) = \ln((N-1)!) + \ln((N+1)!) - 2\ln(N) - \ln(2)$$

and use this formula to sum $\sum_{n=2}^{\infty} \ln(1 - 1/n^2)$.

37. Suppose that r is a constant greater than 1. Calculate the Nth partial sum of

$$\sum_{n=1}^{\infty} \frac{r^n}{(r^n - 1)(r^{n+1} - 1)}.$$

Use your formula to sum the series.

38. The Fibonacci sequence $\{f_n\}$ is recursively defined by $f_1 = f_2 = 1$ and $f_{n+2} = f_n + f_{n+1}$ for $n \geq 0$. Show that

$$\frac{1}{f_n f_{n+2}} = \frac{1}{f_n f_{n+1}} - \frac{1}{f_{n+1} f_{n+2}}.$$

Use this formula to sum $\sum_{n=1}^{\infty} 1/(f_n f_{n+2})$.

39. It is known that $\sum_{n=1}^{\infty} 1/n^2 = \pi^2/6$. What are the values of the convergent series

$$\frac{1}{2^2} + \frac{1}{4^2} + \frac{1}{6^2} + \frac{1}{8^2} + \cdots$$

and

$$1 + \frac{1}{3^2} + \frac{1}{5^2} + \frac{1}{7^2} + \cdots?$$

40. Use the inequality $\ln(1 + x) > x/2$ for $x \in (0, 1)$ to prove that $\sum_{n=1}^{\infty} \ln(1 + 1/n)$ diverges.

41. Prove that $\sum_{n=1}^{\infty} \ln(1/n)$ diverges.

42. A tournament Ping-Pong ball bounces to 2/3 its original height when it is dropped from a height of 12 in. or less onto a hard surface. How far will the ball travel if dropped from a height of 10 in. and allowed to bounce forever?

43. Suppose that every dollar we spend generates (through wages, profits, etc.) 90 cents for someone else to spend. That 90 cents will generate a further 81 cents for spending, and so on. How much spending will result from the purchase of a \$16,000 automobile? (This phenomenon is known as the *multiplier effect*.)

44. Read the description of the multiplier effect in Exercise 43. How much spending will be generated by a \$100 billion tax cut (that is, $\$10^{11}$) if the national savings rate is 5%?

45. A heart patient must take a 0.5 mg daily dose of a medication. Each day, his body eliminates 90% of the medication present. The amount S_N of the medicine that is present after N days is a partial sum of which infinite series? Approximately what amount of the medicine is maintained in the patient's body after a long period of treatment?

46. Suppose that $r > 0$. A factory produces a pollutant that breaks down in water in the following way: If a mass M is introduced into water, then after time T the mass of the pollutant is reduced to Me^{-rT}. Suppose that at intervals $0, T, 2T, 3T, \ldots$ the plant discharges an amount M into a holding tank containing water. After a long period, the mixture from this tank is fed into a river, but the quantity of pollutant is required to be no greater than Q. In terms of r, T, and Q, how large can M be if the factory is compliant?

47. Sketch the graphs of $y = x^n$, $n = 1, 2, 3, \ldots$, in one viewing window $[0, 1] \times [0, 1]$. Find the area between two consecutive graphs $y = x^{n-1}$ and $y = x^n$. Use your calculation to evaluate $\sum_{n=1}^{\infty} 1/(n \cdot (n+1))$ by a geometric argument.

48. Any integer M may be used as the first index of an infinite series. To see why, suppose that a sequence $b_M, b_{M+1}, b_{M+2}, \ldots$ begins with M as its first index. Let S_N denote the sum of the first N terms of the sequence $\{b_j\}_{j=M}^{\infty}$. That is, $S_N = \sum_{m=M}^{M+N-1} b_m$. Use the partial sums $\{S_N\}_{N=1}^{\infty}$ to define $\sum_{m=M}^{\infty} b_m$. Show that there is a sequence $\{a_n\}_{n=1}^{\infty}$ beginning with index 1 such that $\sum_{n=1}^{\infty} a_n$ has the same sequence of partial sums as $\sum_{m=M}^{\infty} b_m$. Consequently, $\sum_{m=M}^{\infty} b_m$ is convergent if and only if $\sum_{n=1}^{\infty} a_n$ is convergent.

49. Mr. Woodman has an unlimited supply of 1 in. long, 3/16 in. thick dominoes. He stacks N dominoes so that, for each $2 \leq n \leq N$, the nth domino, counting from the *bottom* of the stack, protrudes $1/(2(N - n + 1))$ in. beyond the right end of the $(n-1)$th domino. Show that the center of mass of Mr. Woodman's tower of dominoes lies over the bottom domino (and so the stack will not fall). Deduce that by using a sufficiently large number N, Mr. Woodman can make his stack span, from left to right, any given distance. About how many dominoes would it take to span the 10 ft length of his playroom. How high would the tower be?

50. The infinite series $\sum_{n=1}^{\infty} nr^{n-1}$ arises in genetics and other fields. Observe that the partial $S_N = \sum_{n=1}^{N} nr^{n-1}$ is equal to $\frac{d}{dr}\sum_{n=1}^{N} r^n$. Use this to derive a closed form expression for S_N. Show that $\sum_{n=1}^{\infty} nr^{n-1}$ converges for $|r| < 1$, and determine its value.

Calculator/Computer Exercises

In Exercises 51–54, plot the partial sums S_N of the series. From your plot, does it appear that the given series converges? Is so, then approximately what number does it converge to?

51. $\sum_{n=1}^{\infty} e^{-n}/n$

52. $\sum_{n=1}^{\infty} \sqrt{n}/n!$

53. $\sum_{n=1}^{\infty} (1.1/n)^n$

54. $\sum_{n=1}^{\infty} \sin(n)/n^3$

55. The geometric series

$$\sum_{n=0}^{\infty}\left(\frac{1}{1+10^{-50}}\right)^n$$

converges, but its sum S is very large. By about how much does the millionth partial sum $S_{1000000}$ differ from the full sum S? How large must N be so that S_N is within 0.1 of S?

9.2 Determining Convergence

In Section 9.1, we learned that the convergence of a series $\sum_{n=1}^{\infty} a_n$ depends on the behavior of the Nth partial sum $S_N = \sum_{n=1}^{N} a_n$ as N tends to infinity. The problem is that it is often difficult or impossible to find an explicit formula for S_N. Fortunately, in many cases, we do not need such an explicit formula. Beginning in this section, we will develop tests that permit us to decide whether an infinite series converges or diverges without having to analyze the sequence of partial sums.

The Divergence Test

Even when we cannot find an explicit expression for $S_N = a_1 + a_2 + \cdots + a_{N-1} + a_N$, we may always write S_N in terms of the preceding partial sum, S_{N-1}:

$$S_N = (a_1 + a_2 + \cdots + a_{N-1}) + a_N = S_{N-1} + a_N.$$

This equation may also be written as $a_N = S_N - S_{N-1}$. Notice that if we have $\lim_{N\to\infty} S_N = \ell$, then we also have $\lim_{N\to\infty} S_{N-1} = \ell$. Thus, if the infinite series $\sum_{n=1}^{\infty} a_n$ converges to ℓ, then it follows that

$$\lim_{N\to\infty} a_N = \lim_{N\to\infty}(S_N - S_{N-1}) = \ell - \ell = 0.$$

We can use this observation as follows: If we inspect the terms of a series and see that they either do not converge or converge to a nonzero number, then we can conclude that the infinite series diverges. We record this idea as a theorem.

Theorem 1 ***Divergence Test*** If the summands a_n of an infinite series $\sum_{n=1}^{\infty} a_n$ do not tend to 0, then the series $\sum_{n=1}^{\infty} a_n$ diverges.

Notice that the Divergence Test tells us that certain infinite series diverge, but *it never tells us that a particular infinite series converges*. The next two examples show how the Divergence Test can be used and how it should not be used.

Example 1 What does the Divergence Test tell us about the geometric series $\sum_{n=0}^{\infty} r^n$?

Solution Let $a_n = r^n$. If $|r| < 1$, then $\lim_{n\to\infty} a_n = 0$. In this case, the hypothesis of the Divergence Test is not satisfied; the test therefore yields no information about the series $\sum_{n=0}^{\infty} r^n$ when $|r| < 1$. Of course, we do know from Section 9.1 that the series $\sum_{n=0}^{\infty} r^n$ converges for $|r| < 1$. On the other hand, if $|r| \geq 1$, then $|r^n| \geq 1$ and $\lim_{n\to\infty} a_n \neq 0$. In this case, the hypothesis of the Divergence Test *is* satisfied; the test tells us that $\sum_{n=0}^{\infty} r^n$ diverges when $|r| \geq 1$. (This application of the Divergence Test completes the proof of Theorem 2 from Section 9.1.) ■

inSIGHT

It is a common error to conclude that an infinite series converges because its terms tend to 0. Such an inference is *false:* The harmonic series shows that a series $\sum_{n=1}^{\infty} a_n$ may be divergent even though $\lim_{n\to\infty} a_n = 0$. In the theory of infinite series, it often happens that one particular test is inconclusive, which only means that we must resort to a different line of reasoning.

Example 2 What does the Divergence Test tell us about the two series $\sum_{n=1}^{\infty} 1/n$ and $\sum_{n=1}^{\infty} 1/(n^2+n)$?

Solution The answer is "Not a thing!" The Divergence Test can only be applied to infinite series $\sum_{n=1}^{\infty} a_n$ for which $\lim_{n\to\infty} a_n$ does not exist or exists but is nonzero. Neither of these hypotheses holds for the series in this example. We do know from Section 9.1 that $\sum_{n=1}^{\infty} 1/n$ diverges, but the Divergence Test does not provide us with an alternative demonstration. The second series $\sum_{n=1}^{\infty} 1/(n^2+n) = \sum_{n=1}^{\infty} 1/(n(n+1))$ has been proved convergent in Example 6 from Section 9.1, but the Divergence Test does not provide us with an alternative demonstration—the test can *never* be used to conclude that a series converges. ■

Series with Nonnegative Terms

Some series have both positive and negative terms and converge because the terms tend to cancel each other out. The situation is more straightforward when a series has summands that are all nonnegative. Suppose $a_n \geq 0$ for each n. Let $S_N = \sum_{n=1}^{N} a_n$. Notice that $S_N = S_{N-1} + a_N$ and, therefore, $S_N \geq S_{N-1}$. Thus, the sequence of partial sums is *increasing:* $0 \leq S_1 \leq S_2 \leq S_3 \leq \ldots$. There are now only two possibilities: Either the S_Ns increase without bound, or they do not. If they increase without bound, then the series diverges. On the other hand, if the S_Ns are bounded above, then by the Monotone Convergence Property (Section 2.6), the S_Ns converge to a limit ℓ. In this latter case, $\sum_{n=1}^{\infty} a_n = \ell$. We state this observation as a theorem.

Theorem 2 Suppose that $a_n \geq 0$ for each n. Let $S_N = \sum_{n=1}^{N} a_n$. If there is a real number U such that $S_N \leq U$ for all N, then the infinite series $\sum_{n=1}^{\infty} a_n$ converges.

Theorem 2 does not tell us the value of a convergent series $\sum_{n=1}^{\infty} a_n$, but it is extremely useful nonetheless. Up until now, we have had to find an explicit formula for the partial sum S_N in order to determine the convergence of $\sum_{n=1}^{\infty} a_n$. Theorem 2 allows us to determine convergence with less precise information about the partial sums. The following is an example of Theorem 2 in practice.

Example 3 Discuss convergence for the series $\sum_{n=1}^{\infty} 1/n^n$.

Solution The summands $1/n^n$ are all positive. Theorem 2 tells us that we may conclude that $\sum_{n=1}^{\infty} 1/n^n$ is convergent if we can find a real number U such that

$S_N \le U$ for all N. In fact, $U = 3/2$ will do:

$$\begin{aligned} S_N &= 1 + \frac{1}{2^2} + \frac{1}{3^3} + \frac{1}{4^4} + \cdots + \frac{1}{N^N} \\ &\le 1 + \frac{1}{2^2} + \frac{1}{2^3} + \frac{1}{2^4} + \cdots + \frac{1}{2^N} \\ &= \left(1 + \frac{1}{2^1} + \frac{1}{2^2} + \frac{1}{2^3} + \frac{1}{2^4} + \cdots + \frac{1}{2^N}\right) - \frac{1}{2^1} \\ &< \sum_{n=0}^{\infty} \left(\frac{1}{2}\right)^n - \frac{1}{2} \\ &= 2 - \frac{1}{2}. \qquad \text{By equation (9.2) with } r = 1/2 \end{aligned}$$

To recapitulate, the infinite series $\sum_{n=1}^{\infty} 1/n^n$ is convergent because its terms are positive and its partial sums are bounded above. ■

insight

In Example 3, we can conclude that the infinite series $\sum_{n=1}^{\infty} n^{-n}$ converges *without* actually finding its sum ℓ. Because our analysis shows that $S_N \le 3/2$ for all N, we deduce that $\sum_{n=1}^{\infty} n^{-n} = \lim_{N\to\infty} S_N \le 3/2$. If we need a more precise estimate, then we must calculate partial sums. Using a computer, we calculate $S_9 = 1.291285997$ (to ten significant digits) and $S_{10} = 1.291285997$. Indeed, with a little more calculation, we see that adding more terms changes only those digits that are beyond the ninth decimal place.

Example 4 Give an example of a divergent infinite series $\sum_{n=1}^{\infty} a_n$ with bounded partial sums $S_N = \sum_{n=1}^{N} a_n$.

Solution According to Theorem 2, the summands a_n cannot all be nonnegative; otherwise the series would converge. In fact, we have already encountered an infinite series that has the properties we seek. We have observed that the partial sums of the series $\sum_{n=1}^{\infty} (-1)^n$ are all either -1 or 0 (Section 9.1, Example 2). Moreover, this series diverges (Section 9.1, Example 5). The lesson is that the hypothesis $a_n \ge 0$ cannot be omitted from Theorem 2. ■

The Tail End of a Series

Whether a given series converges does not depend on the first million or so terms. Suppose that we are given the series $\sum_{n=1}^{\infty} a_n$ and that we are able to ascertain that $\sum_{n=10^6+1}^{\infty} a_n$ converges. This means that there is a real number such that $\lim_{N\to\infty} \sum_{n=10^6+1}^{N} a_n = \ell$. This limit allows us to see that the partial sums of the full series converge:

$$\lim_{N\to\infty} \sum_{n=1}^{N} a_n = \lim_{N\to\infty} \left(\sum_{n=1}^{10^6} a_n + \sum_{n=10^6+1}^{N} a_n \right) = \sum_{n=1}^{10^6} a_n + \ell.$$

Of course, there is nothing special about the number 10^6. If M is any positive integer, then $\sum_{n=1}^{\infty} a_n$ is convergent if and only if $\sum_{n=M+1}^{\infty} a_n$ is convergent. In this case, the sum of the two infinite series $\sum_{n=1}^{\infty} a_n$ and $\sum_{n=M+1}^{\infty} a_n$ differ only by the finite number $\sum_{n=1}^{M} a_n$.

Example 5 Does the series $\sum_{n=75}^{\infty} (1.1)^n$ converge?

Solution The series must diverge. If it were convergent, then it would follow that $\sum_{n=0}^{\infty} (1.1)^n$ converges, which is false since $|1.1| > 1$ (Section 9.1, Theorem 2). ■

Example 6 Does the series $\sum_{n=20}^{\infty} (0.9)^n$ converge?

Solution Yes. Because $|0.9| < 1$, the geometric series $\sum_{n=0}^{\infty} (0.9)^n$ converges (Section 9.1, Theorem 2). The difference between this series and the given one is simply the finite sum $0.9^1 + 0.9^2 + \cdots + 0.9^{19}$. ■

Example 7 Sum the series $\sum_{n=20}^{\infty} (0.9)^n$ explicitly.

Solution Using equation (9.2) to evaluate $\sum_{n=0}^{\infty} (0.9)^n$ and formula (9.1) to evaluate $\sum_{n=0}^{19} (0.9)^n$, we have

$$\sum_{n=20}^{\infty} (0.9)^n = \sum_{n=0}^{\infty} (0.9)^n - \sum_{n=0}^{19} (0.9)^n = \frac{1}{1-0.9} - \frac{(0.9)^{20}-1}{(0.9-1)} = \frac{(0.9)^{20}}{1-0.9} = 10 \cdot (0.9)^{20}.$$ ■

inSIGHT

Because geometric series often arise with an initial summation index M that is different from 1, it is convenient to generalize the result of Example 7 as

$$\sum_{n=M}^{\infty} r^n = \frac{r^M}{1-r} \qquad (M \in \mathbb{Z}, |r| < 1). \tag{9.3}$$

A proof of this formula is outlined in Exercise 51.

quickquiz

1. True or false: $\sum_{n=1}^{\infty} 1/\sqrt{n}$ converges because $\lim_{n\to\infty}(1/\sqrt{n}) = 0$.
2. True or false: The Divergence Test cannot be applied to the series $\sum_{n=1}^{\infty} 1/n^2$.
3. True or false: If $\sum_{n=1}^{\infty} a_n$ diverges, then $\lim_{n\to\infty} a_n \neq 0$.
4. True or false: A series with bounded partial sums is convergent.

EXERCISES

Problems for Practice

In Exercises 1–20, state what conclusion, if any, may be drawn from the Divergence Test.

1. $\sum_{n=1}^{\infty} ne^{-n}$ **2.** $\sum_{n=1}^{\infty} 1/\sqrt{1+\sqrt{n}}$

3. $\sum_{n=1}^{\infty} n^2/(n^2+1)$ **4.** $\sum_{n=1}^{\infty} 3^n/(3^n+4)$

5. $\sum_{n=1}^{\infty} 3^n/(4^n+3)$ **6.** $\sum_{n=1}^{\infty} \ln(n)/n$

7. $\sum_{n=1}^{\infty} 1/(1+1/n)$ **8.** $\sum_{n=1}^{\infty} 1/(1+\ln(n))$

9. $\sum_{n=1}^{\infty} (3^n+5^n)/8^n$ **10.** $\sum_{n=1}^{\infty} 5^n/(2^n+3^n)$

11. $\sum_{n=1}^{\infty} (2^n+1)/(2^n+n^2)$

12. $\sum_{n=1}^{\infty} n^{1/4}/\sqrt{1+\sqrt{n}}$

13. $\sum_{n=1}^{\infty} 2n/(3n^2+1)$ **14.** $\sum_{n=1}^{\infty} 2n^2/(3n^2+n+1)$

15. $\sum_{n=1}^{\infty} n^{1/n}$ **16.** $\sum_{n=1}^{\infty} n!/n^n$

17. $\sum_{n=1}^{\infty} (\pi/2 - \arctan(n))$ **18.** $\sum_{n=1}^{\infty} \cos(1/n)$

19. $\sum_{n=1}^{\infty} \sin(1/n)$

20. $\sum_{n=1}^{\infty} (n/(n+1) - 1/(n+2))$

In Exercises 21–26, calculate the sum of the infinite series.

21. $\sum_{n=0}^{\infty} (0.1)^n$ **22.** $\sum_{n=4}^{\infty} (0.2)^n$

23. $\sum_{n=3}^{\infty} (0.1)^n/(0.2)^{n+2}$ **24.** $\sum_{n=2}^{\infty} 1/2^{n/2}$

25. $\sum_{n=-1}^{\infty} (2/3)^{2n+1}$ **26.** $\sum_{n=-3}^{\infty} (1/5)^n$

In Exercises 27–30, use either *may* or *must* to fill in the blank so that the completed sentence is correct. Explain your answer by referring to a theorem or example.

27. A series with summands tending to 0 ____ converge.

28. A series that converges ____ have summands that tend to zero.

29. If a series diverges, then the Divergence Test ____ succeed in proving the divergence.

30. If the partial sums of an infinite series are bounded, then the series ____ converge.

Further Theory and Practice

In Exercises 31–44, state what conclusion, if any, may be drawn from the Divergence Test.

31. $\sum_{n=1}^{\infty} (1/n)^{1/n}$

32. $\sum_{n=1}^{\infty} (\sin(1/n) - 1/n)$

33. $\sum_{n=1}^{\infty} (1 + 1/n)^n$

34. $\sum_{n=1}^{\infty} (\cos(1/n) - \sec(1/n))$

35. $\sum_{n=1}^{\infty} \ln(n^3)/n$

36. $\sum_{n=1}^{\infty} (\sqrt{n^2 + 3} - n)$

37. $\sum_{n=1}^{\infty} ((1 + 1/n)^n - n)$

38. $\sum_{n=1}^{\infty} \sin(n)/n$

39. $\sum_{n=1}^{\infty} n \sin(1/n)$

40. $\sum_{n=1}^{\infty} (n!)^2/(2n)!$

41. $\sum_{n=1}^{\infty} 1/\arctan(n)$

42. $\sum_{n=1}^{\infty} \tan(\pi/3 - 1/n)$

43. $\sum_{n=1}^{\infty} (\arctan(n) - \pi/2)$

44. $\sum_{n=1}^{\infty} ((n - 1)/n)^n$

45. Show that $\sum_{n=1}^{\infty} (1/2)^{n^2}$ is convergent.

46. Prove that the infinite series $\sum_{n=1}^{\infty} 1/n^2$ is convergent by showing that its partial sums are bounded by 2. *Hint:* In Section 9.1, we proved that $\sum_{n=1}^{\infty} 1/(n(n+1)) = 1$.

47. Show that the series $\sum_{n=1}^{\infty} \exp(-2n\ln(n) + 7)$ is convergent.

48. Show that the series $\sum_{n=1}^{\infty} (1/\ln(1+n))^n$ is convergent.

49. Prove that $\sqrt{n}/4^n < 1/2^n$ for positive integers n. Show that the infinite series $\sum_{n=1}^{\infty} \sqrt{n}/4^n$ is convergent by showing that its partial sums are bounded by 1.

50. Notice that

$$n! = n \cdot (n-1) \ldots 4 \cdot 3 \cdot 2 \cdot 1 > 2 \cdot 2 \ldots 2 \cdot 2 \cdot 2 \cdot 1$$

for $n \geq 3$. Use this observation to find a number U that is an upper bound of the infinite series $\sum_{n=1}^{\infty} 1/n!$ (thereby showing that this series is convergent).

51. Suppose that $|r| < 1$. Observe that for any integers N and M with $N \geq M$,

$$\sum_{n=M}^{N} r^n = r^M \sum_{n=0}^{N-M} r^n.$$

Use formula (9.1) to deduce that

$$\sum_{n=M}^{N} r^n = r^M \frac{(r^{N-M+1} - 1)}{r - 1}.$$

Use this last equation to prove formula (9.3).

Calculator/Computer Exercises

In Exercises 52–57, use a plot of the terms of the series to decide if the Divergence Test may be applied.

52. $\sum_{n=1}^{\infty} (1/n)^{1/n}$

53. $\sum_{n=1}^{\infty} \sin(1/n)\csc(2/n)$

54. $\sum_{n=1}^{\infty} n((1 + 1/n)^n - e)$

55. $\sum_{n=1}^{\infty} (\cos(1/n) - \sec(1/n))$

56. $\sum_{n=1}^{\infty} (n!)^2/(2n)!$

57. $\sum_{n=1}^{\infty} (\sqrt{n^2 + 3n} - n)$

9.3 Series with Nonnegative Terms—The Integral Test

In Section 9.2, we observed that an infinite series $\sum_{n=1}^{\infty} a_n$ with nonnegative terms gives rise to an *increasing* sequence $\{S_N\}$ of partial sums. In view of the Monotone Convergence Property of the real number system, the convergence of the sequence $\{S_N\}$ (and hence of the infinite series $\sum_{n=1}^{\infty} a_n$) comes down to whether the sequence $\{S_N\}$ is bounded. In this section, we study a specific way of determining that a sequence of partial sums is bounded. Because it is critical that $\{S_N\}$ be increasing, we limit our discussion in this section to *series of nonnegative terms*.

The Integral Test

We have already noted the analogy between infinite series and improper integrals over an infinite interval. We develop this idea further to obtain a convergence test for infinite series of positive terms.

Theorem 1 ***Integral Test*** Let f be a positive, continuous, decreasing function on the interval $[1, \infty)$. The infinite series $\sum_{n=1}^{\infty} f(n)$ converges if and only if the improper integral $\int_1^{\infty} f(x)\,dx$ converges. In this case, we have

$$\int_1^{\infty} f(x)\,dx \leq \sum_{n=1}^{\infty} f(n) \leq f(1) + \int_1^{\infty} f(x)\,dx. \tag{9.4}$$

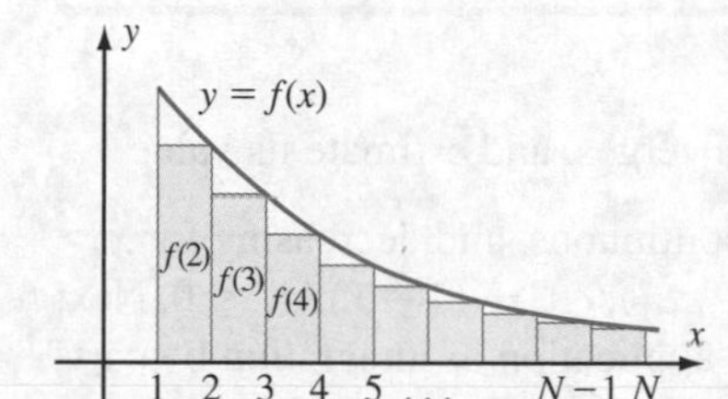

Figure 1
The area of the shaded rectangles is the series $\sum_{n=2}^{N} f(n)$.

Proof Look at Figure 1. Because the base of each inscribed rectangle has length 1, the area of each rectangle is equal to its height. Thus, the area of the first rectangle is $f(2)$, the area of the second is $f(3)$, and so on. *All the rectangles lie under the graph.* Therefore,

$$\sum_{n=2}^{N} f(n) < \int_1^{N} f(x)\,dx. \tag{9.5}$$

If the improper integral $\int_1^{\infty} f(x)\,dx$ is convergent, then the numbers $\int_1^{N} f(x)\,dx$ increase to a (finite) limit as N tends to infinity. This limit is denoted by $\int_1^{\infty} f(x)\,dx$. Using inequality (9.5), we deduce that

$$\sum_{n=1}^{N} f(n) = f(1) + \sum_{n=2}^{N} f(n) < f(1) + \int_1^{N} f(x)\,dx \leq f(1) + \int_1^{\infty} f(x)\,dx.$$

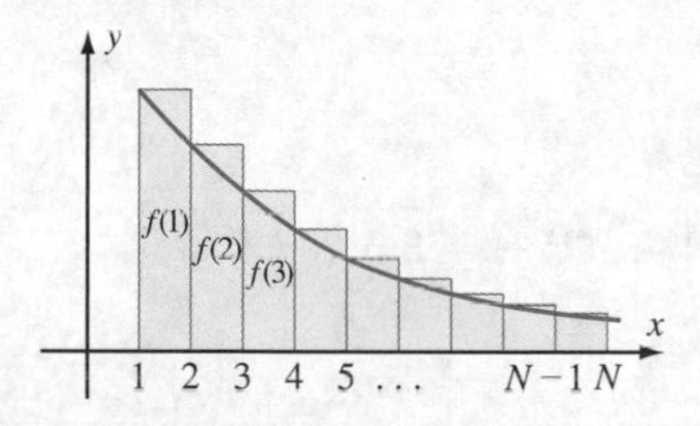

Figure 2
The area of the shaded rectangles is the series $\sum_{n=1}^{N-1} f(n)$.

The partial sums of $\sum_{n=1}^{\infty} f(n)$ are therefore bounded, and, as a result, the series $\sum_{n=1}^{\infty} f(n)$ is convergent.

Now look at the superscribed rectangles in Figure 2. Clearly

$$\int_1^{N} f(x)\,dx < \sum_{n-1}^{N-1} f(n). \tag{9.6}$$

If $\sum_{n=1}^{\infty} f(n)$ is convergent, then inequality (9.6) tells us that the numbers $\int_1^{N} f(x)\,dx$ are bounded above by $\sum_{n=1}^{\infty} f(n)$; thus, $\int_1^{\infty} f(x)\,dx$ is convergent.

Finally, we note that the inequalities in line (9.4) are consequences of our analysis. ■

The crux of the proof of the Integral Test is in the geometry. Remember the pictures, and you will remember why the test works. Here are several examples.

Example 1 Does the series $\sum_{n=1}^{\infty} 1/(1+n^2)$ converge?

Solution We apply the Integral Test to the continuous, positive, decreasing function $f(x) = 1/(1+x^2)$. Because

$$\begin{aligned}\int_1^{\infty} f(x)\,dx &= \int_1^{\infty} \frac{1}{1+x^2}\,dx = \lim_{N\to\infty} \arctan(x)\Big|_{x=1}^{x=N} \\ &= \lim_{N\to\infty}\left(\arctan(N) - \frac{\pi}{4}\right) = \frac{\pi}{2} - \frac{\pi}{4} = \frac{\pi}{4},\end{aligned}$$

we conclude that $\int_1^{\infty} f(x)\,dx$ is convergent. According to the Integral Test, the series $\sum_{n=1}^{\infty} f(n) = \sum_{n=1}^{\infty} 1/(1+n^2)$ is also convergent. ■

inSIGHT

The Integral Test not only shows us that $\sum_{n=1}^{\infty} 1/(1+n^2)$ converges, it also provides an estimate of the series. Because $\int_1^{\infty} 1/(1+x^2)\,dx = \pi/4$, line (9.4) tells us that

$$0.785398 \approx \frac{\pi}{4} = \int_1^{\infty} \frac{1}{1+x^2}\,dx \le \sum_{n=1}^{\infty} \frac{1}{1+n^2} \le \frac{1}{1+1^2} + \int_1^{\infty} \frac{1}{1+x^2}\,dx = \frac{1}{2} + \frac{\pi}{4} \approx 1.285398.$$

In fact, a computer calculation shows that $\sum_{n=1}^{\infty} 1/(1+n^2) = 1.07667\ldots$.

Example 2 Show that the series $\sum_{n=1}^{\infty} n/e^n$ converges, and estimate its value.

Solution The function $f(x) = x/e^x$ is positive, continuous, and decreasing for $x \ge 1$, since for these values of x we have $f'(x) = (e^x - xe^x)/e^{2x} = (1-x)/e^x \le 0$. Next we investigate the improper integral of $f(x)$. By an application of integration by parts with $u = x$ and $dv = e^{-x}\,dx$, we have

$$\int_1^{\infty} f(x)\,dx = \lim_{N\to\infty} \int_1^N xe^{-x}\,dx = \lim_{N\to\infty} \left((-xe^{-x})\Big|_1^N + \int_1^N e^{-x}\,dx \right)$$
$$= \lim_{N\to\infty} (-xe^{-x} - e^{-x})\Big|_1^N .$$

Therefore,

$$\int_1^{\infty} f(x)\,dx = \lim_{N\to\infty} (-Ne^{-N} - e^{-N}) - (-e^{-1} - e^{-1}) = -\lim_{N\to\infty} Ne^{-N} + \frac{2}{e}.$$

An application of l'Hôpital's Rule shows that

$$\lim_{x\to\infty} xe^{-x} = \lim_{x\to\infty} \frac{x}{e^x} = \lim_{x\to\infty} \frac{\frac{d}{dx}x}{\frac{d}{dx}e^x} = \lim_{x\to\infty} \frac{1}{e^x} = 0.$$

In particular, we have $\lim_{N\to\infty} Ne^{-N} = 0$. We conclude that $\int_1^{\infty} f(x)\,dx$ converges and equals $2/e$. It follows from the Integral Test that $\sum_{n=1}^{\infty} n/e^n$ also converges. Line (9.4) tells us that

$$0.7358 \approx \frac{2}{e} = \int_1^{\infty} \frac{x}{e^x}\,dx \le \sum_{n=1}^{\infty} \frac{n}{e^n} \le \frac{1}{e^1} + \int_1^{\infty} \frac{x}{e^x}\,dx = \frac{1}{e} + \frac{2}{e} = \frac{3}{e} \approx 1.1036.$$

Using more advanced techniques, we can show that the value of the given series is $e/(e-1)^2 = 0.92067\ldots$, which indeed lies between our lower and upper estimates. ■

p-Series

If p is a fixed number, then the infinite series $\sum_{n=1}^{\infty} 1/n^p$ is called a *p-series*. The next theorem completely describes how the convergence of the p-series depends on the value of p.

Theorem 2 Fix a real number p. The series $\sum_{n=1}^{\infty} 1/n^p$ converges if $p > 1$; it diverges if $p \leq 1$.

Proof If p is not positive, then the terms of the series do not tend to 0. According to the Divergence Test, the series must diverge. If p is positive, then $f(x) = 1/x^p$ is positive, continuous, and decreasing on the interval $[1, \infty)$. We may therefore apply the Integral Test. There are three cases to consider. If $0 < p < 1$, then

$$\int_1^{\infty} \frac{1}{x^p}\, dx = \lim_{N\to\infty} \frac{1}{1-p} \cdot x^{1-p}\Big|_1^N = \infty.$$

The Integral Test tells us that the series diverges. If $p = 1$, then the series $\sum_{n=1}^{\infty} 1/n^p$ is the harmonic series, which diverges. Finally, if $p > 1$, then

$$\int_1^{\infty} \frac{1}{x^p}\, dx = \lim_{N\to\infty} \frac{x^{1-p}}{1-p}\Big|_1^N = \frac{1}{1-p} \lim_{N\to\infty} \left(\frac{1}{N^{p-1}} - 1\right) = \frac{1}{p-1}.$$

Thus, the Integral Test tells us that the series converges. ■

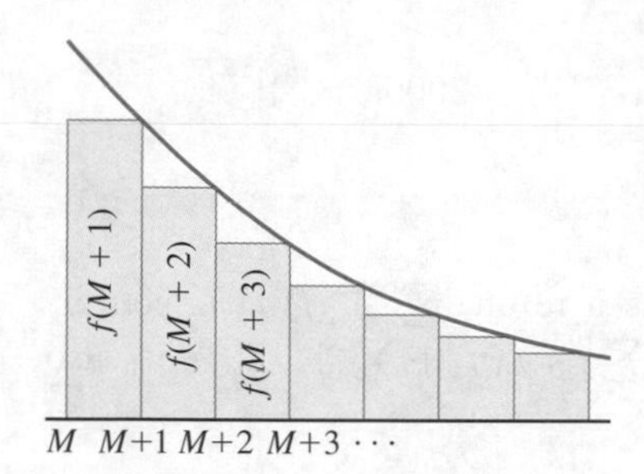

Example 3 Determine whether $\sum_{n=1}^{\infty} (n+2)/n^{3/2}$ is convergent.

Solution The series $\sum_{n=1}^{\infty} 1/n^{3/2}$ is a p-series with $p = 3/2 > 1$. Therefore, it is convergent. If the given series were convergent, then Theorem 3 from Section 9.1 (with $a_n = (n+2)/n^{3/2}$, $b_n = 1/n^{3/2}$, and $\lambda = -2$) would tell us that

$$\sum_{n=1}^{\infty} (a_n + 2b_n) = \sum_{n=1}^{\infty} \left(\frac{n+2}{n^{3/2}} - 2\frac{1}{n^{3/2}}\right) = \sum_{n=1}^{\infty} \frac{n}{n^{3/2}} = \sum_{n=1}^{\infty} \frac{1}{n^{1/2}}$$

is convergent. However, since $\sum_{n=1}^{\infty} 1/n^{1/2}$ is a p-series with $p = 1/2 \leq 1$, we know it is divergent. We conclude that the given series must be divergent; otherwise we would have a contradiction. ■

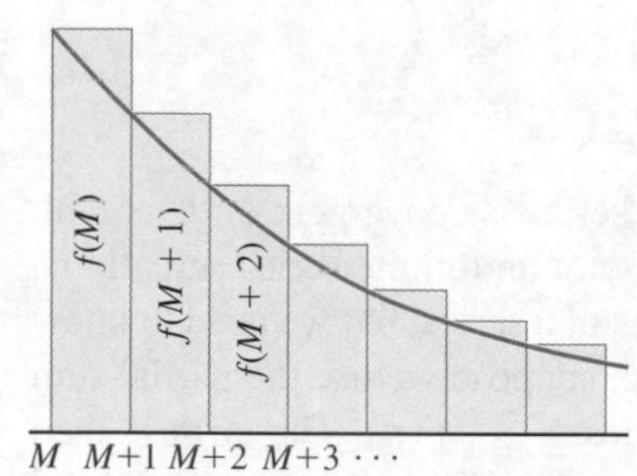

Figure 3

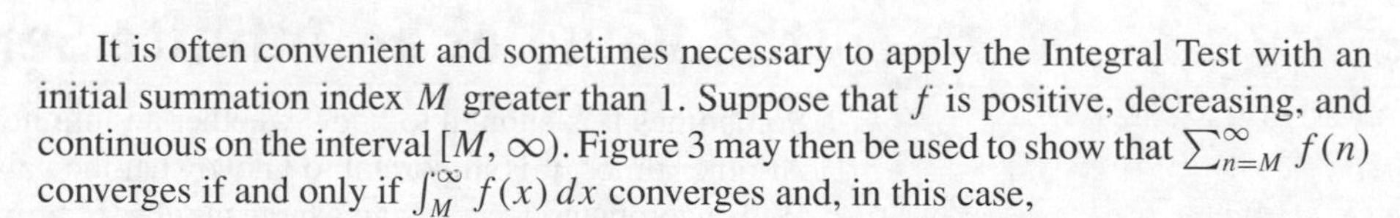

It is often convenient and sometimes necessary to apply the Integral Test with an initial summation index M greater than 1. Suppose that f is positive, decreasing, and continuous on the interval $[M, \infty)$. Figure 3 may then be used to show that $\sum_{n=M}^{\infty} f(n)$ converges if and only if $\int_M^{\infty} f(x)\, dx$ converges and, in this case,

$$\int_M^{\infty} f(x)\, dx \leq \sum_{n=M}^{\infty} f(n) \leq f(M) + \int_M^{\infty} f(x)\, dx. \tag{9.7}$$

(The analysis does not differ from that of Theorem 1 and need not be repeated.)

Example 4 Determine whether the series $\sum_{n=2}^{\infty} 1/(n \ln(n))$ converges.

Solution This series begins with $n = 2$; notice that the summand is not even defined for $n = 1$. Because x and $\ln(x)$ are increasing functions of x, so is the product $x \ln(x)$. The reciprocal $f(x) = 1/(x \ln(x))$ of this product is therefore decreasing. Because f is also positive and continuous on the interval $[2, \infty)$, we may apply the Integral Test. We calculate

$$\int_2^{\infty} f(x)\, dx = \int_2^{\infty} \frac{1}{x \cdot \ln(x)}\, dx = \lim_{N\to\infty} (\ln(\ln(x)))\Big|_{x=2}^{x=N} = \infty.$$

The divergence of the improper integral $\int_2^{\infty} f(x)\, dx$ implies the divergence of the infinite series $\sum_{n=2}^{\infty} f(n) = \sum_{n=2}^{\infty} 1/(n \ln(n))$. ■

We can use inequalities (9.7) to establish convergence or divergence of the tail series $\sum_{n=M}^{\infty} f(n)$, and, as we discussed in Section 9.2, the convergence of a full series $\sum_{n=1}^{\infty} f(n)$ is determined by that of any tail series $\sum_{n=M}^{\infty} f(n)$.

Example 5 Use the Integral Test to verify that the series $\sum_{n=1}^{\infty} n^2 e^{-n/10}$ converges.

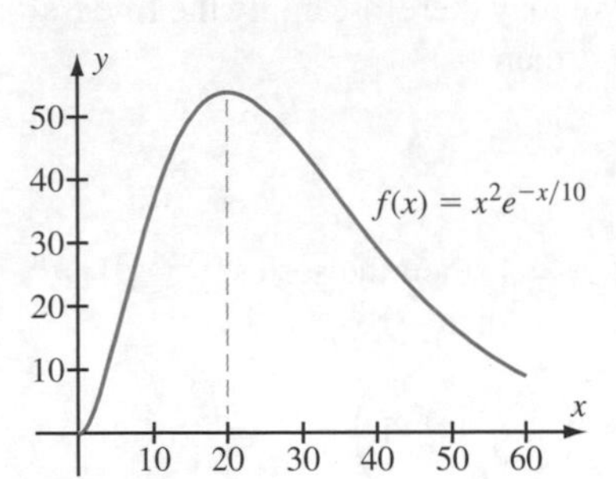

Figure 4

Solution The given series is equal to $\sum_{n=1}^{\infty} f(n)$ for $f(x) = x^2 e^{-x/10}$. In this case, f has an increasing factor, namely x^2. To apply the Integral Test, we must be sure that f itself is decreasing. The plot of f in Figure 4 suggests that $f(x)$ increases for $0 < x < 20$ and decreases for $x > 20$. The Derivative Test confirms this observation:

$$f'(x) = 2xe^{-x/10} + x^2\left(-\frac{1}{10}\right)e^{-x/10} = \frac{1}{10}xe^{-x/10}(20 - x) < 0, \quad \text{for } x > 20.$$

We may therefore apply the Integral Test over the interval $[20, \infty)$. By performing integration by parts twice, we calculate:

$$\begin{aligned}\int_{20}^{\infty} f(x)\,dx &= \lim_{N\to\infty} \int_{20}^{N} x^2 e^{-x/10}\,dx \\ &= \lim_{N\to\infty} \left(-10x^2 e^{-x/10} - 200xe^{-x/10} - 2000e^{-x/10}\right)\Big|_{x=20}^{x=N} \\ &= \frac{10000}{e^2}.\end{aligned}$$

It follows that $\int_{20}^{\infty} f(x)\,dx$ is convergent, and as a result, $\sum_{n=20}^{\infty} f(n)$ is convergent. Therefore, $\sum_{n=1}^{\infty} n^2 e^{-n/10} = \sum_{n=1}^{\infty} f(n) = \sum_{n=1}^{19} f(n) + \sum_{n=20}^{\infty} f(n)$ is also convergent. ■

Enrichment: Approximating the Value of an Infinite Series*

Sometimes it is enough to know whether an infinite series is convergent or divergent. At other times, it is important to know what the value of an infinite series actually is. With a computer, we can sum a large number of terms of a series, but we need to know when it is safe to stop the summation. To be specific, suppose we use the partial sum $\sum_{n=1}^{M-1} f(n)$ as an initial estimate of the infinite series $\sum_{n=1}^{\infty} f(n)$. The error is then the tail series $f(M) + f(M+1) + f(M+2) + \cdots$. The difficulty is that it is hard to know what this error amounts to. The inequalities of line (9.7) allow us to "correct" the initial estimate $\sum_{n=1}^{M-1} f(n)$ so that the error becomes less than the first term $f(M)$ that is omitted from the partial sum.

Theorem 3 Let f be a positive, continuous, decreasing function on the interval $[1, \infty)$. If the infinite series $\sum_{n=1}^{\infty} f(n)$ converges, then

$$\sum_{n=1}^{M-1} f(n) + \int_{M}^{\infty} f(x)\,dx$$

underestimates the series $\sum_{n=1}^{\infty} f(n)$ by an amount that is at most $f(M)$.

*Knowledge of this topic is not required for subsequent sections.

We use Theorem 3 in the following way: To evaluate $\sum_{n=1}^{\infty} f(n)$ with an error no greater than ϵ, we find the first index M for which $f(M) < \epsilon$. We then use $\sum_{n=1}^{M-1} f(n) + \int_M^{\infty} f(x)\,dx$ as our estimate.

Example 6 Calculate $\sum_{n=1}^{\infty} 1/n^4$ with error no greater than 0.001.

Solution The given series is a convergent p-series with $p = 4$. Since $5^4 = 625$ and $6^4 = 1296$, we see that 6 is the first value of n for which $1/n^4 < 1/1000 = 0.001$. Therefore,

$$\sum_{n=1}^{\infty} \frac{1}{n^4} \approx \sum_{n=1}^{5} \frac{1}{n^4} + \int_6^{\infty} \frac{1}{x^4}\,dx \approx 1.0804 + \frac{1}{3} 6^{-3} = 1.0819\ldots.$$

We can be sure that our error is no greater than $1/1296 = 0.0007\ldots < 0.001$. (The exact value of $\sum_{n=1}^{\infty} 1/n^4$ is known to be $\pi^4/90 = 1.0823\ldots$, which shows that our error is actually about 0.0004.) ■

A LOOK BACK

The Integral Test works by comparing the partial sums of a series with the area under a curve. It is often natural to compare the partial sums of one series with those of a series with known convergence. Section 9.4 is devoted to this powerful technique.

A LOOK FORWARD

quickquiz

1. State the hypotheses of the Integral Test.
2. Does the Integral Test establish only convergence, only divergence, or both?
3. What does the Integral Test tell us about geometric series with ratio r?
4. True or false: If α is any fixed number for which $\sum_{n=1}^{\infty} n^{-\alpha}$ converges, then $\sum_{n=1}^{\infty} n^{-p}$ converges for all $p > \alpha$.

EXERCISES

Problems for Practice

In Exercises 1–20, use the Integral Test to determine whether the series converges or diverges. Before you apply the test, be sure that the hypotheses are satisfied.

1. $\sum_{n=1}^{\infty} e^{-n}$ **2.** $\sum_{n=1}^{\infty} n/(n^2+1)$
3. $\sum_{n=1}^{\infty} 1/(n^2+4)$ **4.** $\sum_{n=1}^{\infty} 1/(n(n+2))$
5. $\sum_{n=1}^{\infty} 1/(n+3)$ **6.** $\sum_{n=1}^{\infty} n^2 e^{-n^3}$
7. $\sum_{n=1}^{\infty} 2n^2/(n^3+4)$ **8.** $\sum_{n=2}^{\infty} 1/(n\ln(n))$
9. $\sum_{n=1}^{\infty} \ln(n)/n$ **10.** $\sum_{n=3}^{\infty} 1/(n^2-n)$
11. $\sum_{n=8}^{\infty} e^n/(1+e^n)^2$ **12.** $\sum_{n=1}^{\infty} n/10^n$
13. $\sum_{n=2}^{\infty} 3/(n^2+n)$ **14.** $\sum_{n=4}^{\infty} 1/(n\ln^4(n))$
15. $\sum_{n=1}^{\infty} 1/(n+3)^{5/4}$ **16.** $\sum_{n=1}^{\infty} 1/(n^2+2n+1)$
17. $\sum_{n=1}^{\infty} ne^{-2n}$ **18.** $\sum_{n=1}^{\infty} n2^{-n}$
19. $\sum_{n=2}^{\infty} 1/(n\sqrt{n^2-1})$ **20.** $\sum_{n=3}^{\infty} 1/(n\ln^{3/2}(n))$

In Exercises 21–28, use known facts about p-series to determine whether the series converges or diverges.

21. $\sum_{n=1}^{\infty} 1/\sqrt{n}$ **22.** $\sum_{n=1}^{\infty} 1/n^{(\pi-e)}$
23. $\sum_{n=1}^{\infty} \sqrt{3}/n^{\sqrt{2}}$ **24.** $\sum_{n=1}^{\infty} (7-\sqrt{2})/(4n^2)$
25. $\sum_{n=1}^{\infty} (\sqrt{n}+5)/n^2$ **26.** $\sum_{n=1}^{\infty} (n^3+1)/n^4$

27. $\sum_{n=1}^{\infty} (n^2 + 11)/n^4$
28. $\sum_{n=1}^{\infty} (2n^{1/8} + 3n^{1/16} + 1)/n^{5/4}$

Further Theory and Practice

29. Define $f(x) = \sin^2(\pi x) + 1/x^2$ for $x \geq 1$. Notice that f is positive and continuous. Show that $\int_1^{\infty} f(x)\,dx$ diverges, yet $\sum_{n=1}^{\infty} f(n)$ converges. Explain why the conclusion of the Integral Test is not valid.
30. For which values of a and b does the series

$$\sum_{n=2}^{\infty} \frac{1}{n^a \cdot \ln^b(n)}$$

converge? For which values does it diverge?
31. For which values of a is

$$\sum_{n=1}^{\infty} \left(\frac{n}{n^2+1}\right)^a$$

convergent?
32. For which values of a and b is the series

$$\sum_{n=1}^{\infty} \frac{n^a}{1+n^b}$$

convergent?
33. Use the Integral Test to determine which series converges and which diverges.

a. $\displaystyle\sum_{n=1}^{\infty} \frac{n^4}{2^n}$ b. $\displaystyle\sum_{n=1}^{\infty} \frac{\ln(n)}{n^2}$

c. $\displaystyle\sum_{n=1}^{\infty} \frac{1}{1+n^{1/2}}$ d. $\displaystyle\sum_{n=1}^{\infty} \ln\left(\frac{n+3}{n}\right)$

e. $\displaystyle\sum_{n=1}^{\infty} \ln\left(\frac{n^2+1}{n^2}\right)$ f. $\displaystyle\sum_{n=1}^{\infty} n^3 e^{-2n}$

34. Determine which values of the constant c will make the series converge and which values will make the series diverge.

a. $\sum_{n=1}^{\infty} n^{-3c}$ b. $\sum_{n=1}^{\infty} (c)^{3n}$

c. $\sum_{n=1}^{\infty} (2n+5)^c$ d. $\sum_{n=1}^{\infty} c^n/n^2$

35. Prove that if $a_n \geq 0$ and $\sum_{n=1}^{\infty} a_n$ converges, then there exist numbers $b_n > a_n$ such that $\sum_{n=1}^{\infty} b_n$ still converges. Prove that if $c_n > 0$ and $\sum_{n=1}^{\infty} c_n$ diverges, then there exist numbers d_n satisfying $0 < d_n < c_n$ and such that $\sum_{n=1}^{\infty} d_n$ diverges. This exercise shows that there is no largest convergent series and no smallest divergent series.
36. Determine whether $\sum_{n=2}^{\infty} \int_n^{2n} x^{-2}\,dx$ converges.
37. Determine whether $\sum_{n=2}^{\infty} \int_n^{2n} e^{-x}\,dx$ converges.
38. Use the Integral Test to prove that the series

$$\sum_{n=20}^{\infty} \frac{1}{n\ln(n)\ln(\ln(n))}$$

diverges.
39. Use the Integral Test to prove that the series

$$\sum_{n=20}^{\infty} \frac{1}{n\ln(n)(\ln(\ln(n)))^2}$$

converges.

Let $H_N = \sum_{n=1}^{N} 1/n$ denote the partial sums of the harmonic series. Let $A_N = H_N - \ln(N)$. Exercises 40–43 establish that $\lim_{N\to\infty} A_N$ exists. This important number, usually denoted γ, is called the *Euler-Mascheroni constant*. Its value is 0.5772156649 to ten decimal places.

40. Sketch $y = 1/x$. Use the ideas of the Integral Test to illustrate the geometric significance of A_N. Draw a staircase-shaped region with area H_N. Shade the region that comprises $N-1$ "triangular" regions and one rectangle and that has area A_N.
41. Let a_n denote the area of the "triangular" region above $y = 1/x$, below $y = 1/n$, and between $x = n$ and $x = n+1$. Your sketch should show that

$$A_N = \sum_{n=1}^{N-1} a_n + \frac{1}{N}.$$

By comparison with a circumscribed rectangle, show that

$$0 < a_n < \left(\frac{1}{n} - \frac{1}{n+1}\right).$$

42. Show that the partial sums of the series $\sum_{n=1}^{\infty} a_n$ are bounded by the value of the convergent series

$$\sum_{n=1}^{\infty} \left(\frac{1}{n} - \frac{1}{n+1}\right) = 1.$$

Deduce that the limit $\gamma = \lim_{N\to\infty} A_N$ exists and is a number between 1/2 and 1.
43. Here is an analytic alternative approach to γ. Let $f(x) = x + \ln(1-x)$ for $0 \leq x < 1$. Show that $f'(x) < 0$ for $0 < x < 1$. Deduce that $f(x) < 0$ for $0 < x < 1$. Show that $A_{N+1} - A_N = f(1/(N+1))$. Deduce that $\{A_N\}$ is a positive decreasing sequence, and hence it is convergent.

Enrichment Exercises

In Exercises 44–47, a convergent infinite series $\sum_{n=1}^{\infty} f(n)$ is given together with its value to four decimal places. Find an integer M such that $A_M = \sum_{n=1}^{M-1} f(n) + \int_M^{\infty} f(x)\,dx$ approximates $\sum_{n=1}^{\infty} f(n)$ to within 0.005. Calculate the approximation A_M.

44. $\sum_{n=1}^{\infty} 1/n^6 = 1.0173\ldots$

45. $\sum_{n=1}^{\infty} 10^{-n} = 0.1111\ldots$

46. $\sum_{n=1}^{\infty} 1/(n+4)^4 = 0.0035\ldots$

47. $\sum_{n=1}^{\infty} n/(n^2+1)^3 = 0.1453\ldots$

48. Let f satisfy the hypotheses of Theorem 1. Show that

$$\int_M^{\infty} f(x)\,dx \le \sum_{n=M}^{\infty} f(n) \le \int_{M-1}^{\infty} f(x)\,dx.$$

In other words, if the partial sum $\sum_{n=L}^{M-1} f(n)$ is used to estimate the series $\sum_{n=L}^{\infty} f(n)$, then the error is at least $\int_M^{\infty} f(x)\,dx$ but no greater than $\int_{M-1}^{\infty} f(x)\,dx$.

49. Let f satisfy the hypotheses of Theorem 1. Show that

$$f(M) + \int_L^M f(x)\,dx \le \sum_{n=L}^{M} f(n) \le f(L) + \int_L^M f(x)\,dx.$$

50. Suppose that $0 < r < 1$. What (exactly) is the error that results when $\sum_{n=1}^{M-1} r^n$ is used as an approximation to $\sum_{n=1}^{\infty} r^n$? Is it smaller than the first omitted term? Show that

$$r^M\left(\frac{1}{\ln(r)} + \frac{1}{1-r}\right)$$

is the error that results when

$$\sum_{n=1}^{M-1} r^n + \int_M^{\infty} r^x\,dx$$

is used as an approximation to $\sum_{n=1}^{\infty} r^n$. Theory tells us that $1/\ln(r) + 1/(1-r)$ must be between 0 and 1 for $0 < r < 1$. For a challenge, prove this directly.

51. It is known that both the series $\sum_{n=1}^{\infty} n^{-2}$ and $1 + \sum_{n=1}^{\infty} n^{-2}(n+1)^{-1}$ converge to $\pi^2/6$. For what M does

$$\sum_{n=1}^{M-1} \frac{1}{n^2} + \int_M^{\infty} x^{-2}\,dx$$

approximate $\pi^2/6$ to within 0.0001? For what M does

$$1 + \sum_{n=1}^{M-1} n^{-2}(n+1)^{-1} + \int_M^{\infty} x^{-2}(x+1)^{-1}\,dx$$

approximate $\pi^2/6$ to within 0.0001? What value does this last integral contribute to the estimate?

Calculator/Computer Exercises

In Exercises 52–55 a positive decreasing function f is given. Calculate the value of the convergent infinite series $\sum_{n=1}^{\infty} f(n)$ to two decimal places. Evaluate $\int_1^{\infty} f(x)\,dx$ and verify the two inequalities of line (9.4).

52. $f(n) = \exp(n)/(1+\exp(2n))$

53. $f(n) = \exp(n)/(1+\exp(n))^3$

54. $f(n) = n/(1+n^2)^{5/2}$

55. $f(n) = n^2/(1+n^3)^3$

Enrichment Exercises

In Exercises 56–59, a convergent infinite series $\sum_{n=1}^{\infty} f(n)$ is given together with its value to six decimal places. Find an integer M such that $A_M = \sum_{n=1}^{M-1} f(n) + \int_M^{\infty} f(x)\,dx$ approximates $\sum_{n=1}^{\infty} f(n)$ to within 0.0005. Calculate the approximation A_M.

56. $\sum_{n=1}^{\infty} e^{-n^2} = 0.386318\ldots$

57. $\sum_{n=1}^{\infty} \dfrac{n}{2^{n^2}} = 0.630920\ldots$

58. $\sum_{n=1}^{\infty} \dfrac{1}{(1+n)\ln^{10}(1+n)^2} = 1.921177\ldots$

59. $\sum_{n=1}^{\infty}(1+n+n^3)^{-1} = 0.4947249\ldots$

60. Let $a_n = (n\ln(n)(\ln(\ln(n)))^2)^{-1}$. The series $\sum_{n=20}^{\infty} a_n$ converges (Exercise 39). Use Exercise 48 to determine the smallest value of M for which $\sum_{n=20}^{M-1} a_n$ approximates $\sum_{n=20}^{\infty} a_n$ to within 0.001 accuracy.

61. The infinite series

$$\sum_{n=20}^{\infty} \frac{1}{n\ln(n)\ln(\ln(n))}$$

diverges (Exercise 38). Although the partial sums S_N must increase without bound, it turns out that even when $N = 10^{(10^{100})}$ (a *googolplex*), S_N is still less than 6. Use Exercise 49 to prove this. Determine a value of N such that $\sum_{n=20}^{N}(n\ln(n)\ln(\ln(n)))^{-1} > 6$.

9.4 Series with Nonnegative Terms—The Comparison Test

In Section 9.3, we studied the convergence of a series $\sum_{n=1}^{\infty} a_n$ of nonnegative terms by comparing the series with an improper integral. The idea was to show the boundedness of the increasing sequence of partial sums. In this section, we see that the same goal can often be reached by comparing the given infinite series with another series.

The Comparison Test for Convergence

Suppose that $\sum_{n=1}^{\infty} b_n$ is a convergent series of nonnegative terms with $\sum_{n=1}^{\infty} b_n = \ell$. Because the partial sums $\sum_{n=1}^{N} b_n$ form an increasing sequence, these partial sums must increase to ℓ. In particular, each partial sum satisfies $\sum_{n=1}^{N} b_n \leq \ell$. If $\sum_{n=1}^{\infty} a_n$ is another series satisfying $0 \leq a_n \leq b_n$ for every n, then the partial sums for this series satisfy

$$\sum_{n=1}^{N} a_n \leq \sum_{n=1}^{N} b_n \leq \ell.$$

Thus, the partial sums of the series $\sum_{n=1}^{\infty} a_n$ are increasing (since the a_ns are nonnegative) and bounded above by ℓ. By the Monotone Convergence Property (Section 2.6), we conclude that the sequence of partial sums of $\sum_{n=1}^{\infty} a_n$ converge.

We summarize in the following theorem.

Theorem 1 ***Comparison Test for Convergence*** Let $0 \leq a_n \leq b_n$ for every n. If the series $\sum_{n=1}^{\infty} b_n$ converges, then the series $\sum_{n=1}^{\infty} a_n$ also converges.

Example 1 Show that the series $\sum_{n=1}^{\infty} 1/(n(n^2+1))$ converges.

Solution Observe that $n(n^2+1) = n^3 + n \geq n^3$. We conclude that $1/(n(n^2+1)) \leq 1/n^3$ for every n. Also, $\sum_{n=1}^{\infty} 1/n^3$ is a p-series that is convergent (because $p = 3 > 1$). By the Comparison Test for Convergence, the given series $\sum_{n=1}^{\infty} 1/(n(n^2+1))$ converges. ■

inSIGHT

When we apply the Comparison Test for Convergence to a series $\sum_{n=1}^{\infty} a_n$, we must produce a *second* series $\sum_{n=1}^{\infty} b_n$ for comparison. How do we find such a series? In Example 1, look at the denominator $n(n^2+1) = n^3 + n$ of the general term. Compare the contribution of each summand to the total. When n is one hundred, for instance, n^3 is one million. Observe that for large n, the summand n^3 dominates the expression $n^3 + n$. Thus, for large n, the quantity $1/(n(n^2+1))$ is similar in size to the simpler expression $1/n^3$, which suggests the comparison used in Example 1.

inSIGHT

It is often possible to come to an intuitive decision about the convergence of a series without doing a precise comparison. In Example 2, we see that for large n, the summand $1/(3n-2)^2$ is about the same size as $1/(3n)^2$. Such a rough comparison is enough for us to predict the convergence of $\sum_{n=1}^{\infty} 1/(3n-2)^2$ by comparison with $(1/9)\sum_{n=1}^{\infty} 1/n^2$. The Limit Comparison Test, which appears later in this section, will make this idea precise.

Example 2 Discuss convergence for the series $\sum_{n=1}^{\infty} 1/(3n-2)^2$.

Solution Notice that $3n-2 = n+2(n-1) \geq n$ for all $n \geq 1$. Therefore, $(3n-2)^2 \geq n^2$ so that $1/(3n-2)^2 \leq 1/n^2$. Finally, $\sum_{n=1}^{\infty} 1/n^2$ is a convergent p-series (with $p = 2 > 1$). By the Comparison Test for Convergence, we conclude that $\sum_{n=1}^{\infty} 1/(3n-2)^2$ converges. ■

Example 3 Discuss convergence for the series $\sum_{n=1}^{\infty} \sin^2(2n+5)/(n^4+8n+6)$.

Solution Observe that

$$0 \leq \frac{\sin^2(2n+5)}{n^4+8n+6} \leq \frac{1}{n^4+8n+6} \leq \frac{1}{n^4}.$$

Also, $\sum_{n=1}^{\infty} 1/n^4$ is a p-series that is convergent (because $p=4>1$). By the Comparison Test for Convergence, the given series converges. ■

An Extension of the Comparison Test for Convergence

Because the convergence or divergence properties of a series do not depend on the first several terms, we could state the Comparison Test for Convergence by requiring that $0 \leq a_n \leq b_n$ *for all sufficiently large n*. Let us look at an example to see how this might work in practice.

Example 4 Determine whether the series $\sum_{n=1}^{\infty} (n^2+100n+101)/n^5$ converges.

Solution Let $a_n = (n^2+100n+101)/n^5$. When n is large, the summand n^2 is the dominant term in the numerator of a_n. For example, when n is one million, n^2 is ten thousand times larger than $100n$. We therefore expect that $n^2 + (100n+101) \leq n^2 + n^2 = 2n^2$ for large n. We establish this inequality by noting that $2n^2 - (n^2+100n+101) = n^2 - 100n - 101 = (n+1)(n-101)$, which is nonnegative for $n \geq 101$. Therefore, $n^2 + 100n + 101 \leq 2n^2$ for $n \geq 101$. Thus, for $n \geq 101$, $a_n \leq (2n^2)/n^5 = 2/n^3$. We compare $\sum_{n=101}^{\infty} a_n$ with the series $\sum_{n=101}^{\infty} (2/n^3) = 2\sum_{n=101}^{\infty} 1/n^3$, which is convergent. We conclude that the given series converges as well. ■

The Comparison Test for Divergence

We can now reverse our reasoning to obtain a comparison test for divergence. Namely, suppose that $0 \leq b_n \leq a_n$ and that the series $\sum_{n=1}^{\infty} b_n$ diverges. The series $\sum_{n=1}^{\infty} a_n$ would then have to diverge also. (If the latter series converged, then the Comparison Test for Convergence would imply that $\sum_{n=1}^{\infty} b_n$ converges, which would be a contradiction.) We summarize this in the following theorem.

Theorem 2 ***Comparison Test for Divergence*** Let $0 \leq b_n \leq a_n$ for all sufficiently large n. If the series $\sum_{n=1}^{\infty} b_n$ diverges, then the series $\sum_{n=1}^{\infty} a_n$ also diverges.

Example 5 Show that the series $\sum_{n=1}^{\infty} \ln(n+4)/n$ diverges.

Solution Observe that $\ln(n+4) \geq \ln(1+4) > 1$ for every $n \geq 1$. It follows that $\ln(n+4)/n > 1/n$ for $n \geq 1$. Because the series $\sum_{n=1}^{\infty} 1/n$ diverges, the Comparison Test for Divergence allows us to conclude that the given series diverges as well. ■

Example 6 Analyze the series $\sum_{n=2}^{\infty}(1/(\sqrt{n}\cdot\sqrt{\ln(n)}))$.

Solution We notice that $n\ln(n) > 1$ for each $n \geq 2$. Therefore, $n\ln(n) > \sqrt{n\ln(n)}$ and $1/(\sqrt{n}\cdot\sqrt{\ln(n)}) > 1/(n\ln(n))$. In Example 4 from Section 9.3, we proved that the series $\sum_{n=2}^{\infty} 1/(n\ln(n))$ diverges. By the Comparison Theorem for Divergence, the series in this example also must diverge. ■

Example 7 Use the Comparison Theorem for Divergence to study the series

$$\sum_{n=1}^{\infty}\frac{1}{\sqrt{\sqrt{n}+3}}.$$

Solution A rough analysis tells us that we are dealing with terms that are comparable to $1/n^{1/4}$ for large n. Because the p-series with $p = 1/4$ is divergent, we predict that the given series diverges. We can make this informal observation precise by noting that for $n \geq 9$ we have $\sqrt{n} \geq 3$, which implies $2\sqrt{n} = \sqrt{n}+\sqrt{n} \geq \sqrt{n}+3$. This last inequality is equivalent to $\sqrt{2\sqrt{n}} \geq \sqrt{\sqrt{n}+3}$, or

$$\frac{1}{\sqrt{\sqrt{n}+3}} \geq \frac{1}{\sqrt{2}}\cdot\frac{1}{n^{1/4}}.$$

The series $(1/\sqrt{2})\sum_{n=1}^{\infty} 1/n^{1/4}$ diverges, and therefore so does $\sum_{n=1}^{\infty} 1/\sqrt{\sqrt{n}+3}$ by the Comparison Test for Divergence. ■

The next example is more sophisticated than the preceding ones. As you read the solution, pay particular attention to the way the logarithm is handled. The basic principle to keep in mind is that logarithmic growth is very slow. In fact,

$$\lim_{x\to\infty}\frac{\ln(x)}{x^q} = 0 \tag{9.8}$$

for *any* positive number q. To derive limit formula (9.8), we apply l'Hôpital's Rule as follows:

$$\lim_{x\to\infty}\frac{\frac{d}{dx}\ln(x)}{\frac{d}{dx}x^q} = \frac{1}{q}\lim_{x\to\infty}\frac{1/x}{x^{q-1}} = \frac{1}{q}\lim_{x\to\infty}\frac{1}{x^q} = 0.$$

Example 8 Does the series $\sum_{n=3}^{\infty}\ln(n)/n^2$ converge?

Solution This series of nonnegative summands termwise exceeds the series $\sum_{n=3}^{\infty} 1/n^2$. Yet this last series *converges*. Thus, the Comparison Test for Divergence *does not tell us anything* about the convergence or divergence of the series $\sum_{n=3}^{\infty}\ln(n)/n^2$ based on the comparison we have made. We must try another approach. This time we analyze the size of the summand $\ln(n)/n^2$ more carefully. Notice that we can use some of the growth in the denominator to counteract the logarithmic growth in the numerator. To this end, we write

$$\frac{\ln(n)}{n^2} = \frac{\ln(n)}{n^{1/2}}\cdot\frac{1}{n^{3/2}}.$$

Taking $q = 1/2$ in formula (9.8), we infer that there is some integer M such that $\ln(x)/x^{1/2} \leq 1$ for $x \geq M$. Thus, for $n \geq M$ we have $\ln(n)/n^2 \leq 1/n^{3/2}$. Because the infinite series $\sum_{n=M}^{\infty} 1/n^{3/2}$ is a convergent p-series, it follows from the Comparison Test for Convergence that the series $\sum_{n=M}^{\infty} \ln(n)/n^2$ converges. Since the given series $\sum_{n=3}^{\infty} \ln(n)/n^2$ has a convergent tail, it must also be convergent. ■

Example 8 reminds us that the comparison tests are *not* "if and only if" statements. If we find a *divergent* series $\sum_{n=1}^{\infty} b_n$ such that $a_n \le b_n$, then the Comparison Test for Convergence tells us nothing about $\sum_{n=1}^{\infty} a_n$ because not all the hypotheses of the test are satisfied. Likewise, if we find a *convergent* series $\sum_{n=1}^{\infty} b_n$ such that $a_n \ge b_n$, then the Comparison Test for Divergence tells us nothing about $\sum_{n=1}^{\infty} a_n$.

The Limit Comparison Test

We close this section with a variant of the comparison tests called the Limit Comparison Test. In practice, you may find this new test easier to apply than the comparison tests from Theorems 1 and 2. The idea behind this new test is also easy to understand: If $\sum_{n=1}^{\infty} a_n$ and $\sum_{n=1}^{\infty} b_n$ are series of positive terms and if

$$\ell = \lim_{n\to\infty} \frac{a_n}{b_n} \text{ exists} \quad \text{and} \quad 0 < \ell < \infty,$$

then for large n we have the approximation $a_n \approx \ell \cdot b_n$. Since the terms of the tail series of $\sum_{n=1}^{\infty} a_n$ and $\sum_{n=1}^{\infty} b_n$ are approximately proportional, we expect the two series to have the same convergence properties. We state this as a theorem.

Theorem 3 Let $\sum_{n=1}^{\infty} a_n$ and $\sum_{n=1}^{\infty} b_n$ be series of positive terms. If $\lim_{n\to\infty} (a_n/b_n)$ exists and is a (finite) positive number, then $\sum_{n=1}^{\infty} a_n$ converges if and only if $\sum_{n=1}^{\infty} b_n$ converges.

Example 9 Does the series $\sum_{n=1}^{\infty} n/(2n^3 - 4n + 6)$ converge?

Solution The nth summand a_n is about $n/(2n^3) = 1/(2n^2)$ for large values of n. This suggests that we apply the Limit Comparison Test with the series $\sum_{n=1}^{\infty} b_n = \sum_{n=1}^{\infty} 1/(2n^2)$. We calculate

$$\lim_{n\to\infty} \frac{a_n}{b_n} = \lim_{n\to\infty} \frac{n/(2n^3 - 4n + 6)}{1/(2n^2)} = \lim_{n\to\infty} \frac{2n^3}{2n^3 - 4n + 6}$$
$$= \lim_{n\to\infty} \frac{2}{2 - 4n/n^2 + 6/n^3} = 1.$$

Because the limit is positive and finite and because the simpler series $\sum_{n=1}^{\infty} 1/(2n^2)$ converges, we conclude that the original series converges as well. ■

inSIGHT

The Limit Comparison Test allows us to compare series without having to do a term-by-term comparison. In Example 9, we are able to compare the series $\sum_{n=1}^{\infty} n/(2n^3 - 4n + 6)$ with the series $\sum_{n=1}^{\infty} 1/(2n^2)$. In doing so, we do not have to worry about proving the inequality $n/(2n^3 - 4n + 6) \leq 1/(2n^2)$ as we would with the Comparison Test for Convergence. (In fact, for $n > 1$, this inequality is false.)

Example 10 Show that the series $\sum_{n=1}^{\infty} (2^n + 1)/(n2^n + 1)$ diverges.

Solution The nth summand $a_n = (2^n + 1)/(n2^n + 1)$ may be written as $a_n = (1+2^{-n})/(n+2^{-n})$. It follows that $a_n \approx 1/n$ for large n. We apply the Limit Comparison Test using the infinite series $\sum_{n=1}^{\infty} b_n = \sum_{n=1}^{\infty} 1/n$ for comparison:

$$\lim_{n\to\infty} \frac{a_n}{b_n} = \lim_{n\to\infty} \frac{(2^n + 1)/(n2^n + 1)}{1/n} = \lim_{n\to\infty} \frac{n2^n + n}{n2^n + 1} = \lim_{n\to\infty} \frac{1 + 1/2^n}{1 + 1/(n2^n)} = 1.$$

Since this limit is finite and positive and since $\sum_{n=1}^{\infty} 1/n$ diverges, we conclude that the given series diverges. ■

quickquiz

1. State the Comparison Test for Convergence. Be sure to include all hypotheses.
2. State the Comparison Test for Divergence. Be sure to include all hypotheses.
3. Analyze the series $\sum_{n=1}^{\infty} (n+1)/(n^3+2n^2)$ using one of the comparison tests.
4. Analyze the series $\sum_{n=1}^{\infty} (n+1)/(n^2+2n)$ using one of the comparison tests.

EXERCISES

Problems for Practice

In Exercises 1–16, use the Comparison Test for Convergence to show that the series converges.

1. $\sum_{n=1}^{\infty} n/(n^3 + 1)$
2. $\sum_{n=1}^{\infty} (5n - 4)/n^{9/4}$
3. $\sum_{n=1}^{\infty} (2 + \sin(n))/n^4$
4. $\sum_{n=1}^{\infty} (2n - 1)e^{-n}/n$
5. $\sum_{n=1}^{\infty} (n - 2)/(2n^{5/2} + 3)$
6. $\sum_{n=1}^{\infty} (n^2 + 2n + 10)/(2n^4)$
7. $\sum_{n=1}^{\infty} 2^n/(n3^n)$
8. $\sum_{n=1}^{\infty} (1/3)^{n^2}$
9. $\sum_{n=1}^{\infty} 1/n!$
10. $\sum_{n=1}^{\infty} 1/n^n$
11. $\sum_{n=1}^{\infty} 1/(2n\sqrt{n} - 1)$
12. $\sum_{n=1}^{\infty} \sqrt{n}/(n^2 + 1)$
13. $\sum_{n=1}^{\infty} n^3/(n^{4.01} + 1)$
14. $\sum_{n=1}^{\infty} (2^n + 3^n)/(7^n + 5^n)$
15. $\sum_{n=2}^{\infty} \sqrt{n^2 + 1}/(n\ln(n))^2$
16. $\sum_{n=2}^{\infty} ((1 + \ln(n^2))/\ln(n^4))^n$

In Exercises 17–24, use the Comparison Test for Divergence to show that the series diverges.

17. $\sum_{n=2}^{\infty} 1/\ln(n)$
18. $\sum_{n=1}^{\infty} \ln(n)/(n + 10)$
19. $\sum_{n=1}^{\infty} 1/\sqrt{10 + n^2}$
20. $\sum_{n=1}^{\infty} (2n + 11)/(n^2 + 1)$
21. $\sum_{n=1}^{\infty} (n + 2)/(2n^{3/2} + 3)$
22. $\sum_{n=1}^{\infty} (2^n + 1)/(2^n + n^2)$

23. $\sum_{n=1}^{\infty} (3^n + n)/(\sqrt{n}3^n + 1)$
24. $\sum_{n=1}^{\infty} (2 + \sin(n))/\sqrt{n}$

In Exercises 25–32, use the Limit Comparison Test to determine whether the series converges.

25. $\sum_{n=1}^{\infty} (2^n + 11)/(3^n - 1)$
26. $\sum_{n=1}^{\infty} (\sqrt{n} + 1)/(2n^2 - n)$
27. $\sum_{n=1}^{\infty} (2^n + n^3)/(3^n + n^2)$
28. $\sum_{n=1}^{\infty} (10n + 3)/(100n^3 - 99)$
29. $\sum_{n=1}^{\infty} (n + \ln(n))/n^3$
30. $\sum_{n=1}^{\infty} 7^n/(10^n + n^{10})$
31. $\sum_{n=1}^{\infty} (n^2 + 2)/(n^2 + 1)^2$
32. $\sum_{n=1}^{\infty} ((2n + 5)/(3n + 7))^n$

Further Theory and Practice

In Exercises 33–46, use a comparison test to determine whether the series converges or diverges.

33. $\sum_{n=1}^{\infty} \ln(n^3)/n^3$
34. $\sum_{n=2}^{\infty} n/\ln(n)^3$
35. $\sum_{n=1}^{\infty} (2 + \sin(n))/\sqrt{n}$
36. $\sum_{n=1}^{\infty} (2n + 1)/n^2$
37. $\sum_{n=1}^{\infty} (n + 2)/(2n^{3/2} + 3)$
38. $\sum_{n=1}^{\infty} 100^n/n^{n/2}$
39. $\sum_{n=1}^{\infty} \sin(1/n)$
40. $\sum_{n=1}^{\infty} (1 - \cos(1/n))$
41. $\sum_{n=1}^{\infty} (1 + 1/n)/n$
42. $\sum_{n=1}^{\infty} n^{1/n-1/2}$
43. $\sum_{n=1}^{\infty} \tan(1/n)/(1 + n)$
44. $\sum_{n=1}^{\infty} (1 + 1/n)^{-n^2}$
45. $\sum_{n=1}^{\infty} (\arctan(n))/n$
46. $\sum_{n=1}^{\infty} \sqrt{\sin(1/n^3)}$
47. Let $\sum_{n=1}^{\infty} a_n$ and $\sum_{n=1}^{\infty} b_n$ be series of positive terms. Suppose that $\lim_{n\to\infty} a_n/b_n = 0$. Show that if $\sum_{n=1}^{\infty} b_n$ converges, then $\sum_{n=1}^{\infty} a_n$ also converges.
48. Let $\sum_{n=1}^{\infty} a_n$ and $\sum_{n=1}^{\infty} b_n$ be series of positive terms. Suppose that $\lim_{n\to\infty} a_n/b_n = \infty$. Show that if $\sum_{n=1}^{\infty} b_n$ diverges, then $\sum_{n=1}^{\infty} a_n$ also diverges.
49. Suppose that $\{a_n\}$ is a sequence of positive numbers. Show that if $\sum_{n=1}^{\infty} a_n$ converges, then $\sum_{n=1}^{\infty} a_n^2$ also converges.
50. Suppose that $\{a_n\}$ and $\{b_n\}$ are sequences of positive numbers. Suppose that $b_n \to \ell$ where ℓ is a real number. Show that if $\sum_{n=1}^{\infty} a_n$ converges, then $\sum_{n=1}^{\infty} a_n b_n$ also converges.
51. Let p be any number. Show that $\sum_{n=1}^{\infty} n^p r^n$ is convergent if $0 < r < 1$.
52. Let q be any number. Show that $\sum_{n=1}^{\infty} \ln^q(n)/n^p$ is convergent if $p > 1$.
53. Show that $\sum_{n=1}^{\infty} n!/n^n$ is convergent.

Calculator/Computer Exercises

54. As a consequence of limit formula (9.8), there is a positive integer M such that $\ln(n)/n^{1/100} < 1/2$ for $n \geq M$. Find such an M and illustrate the inequality with an appropriate plot. Prove that $\sum_{n=1}^{\infty} (\ln(n)/n^{1/100})^n$ is convergent.
55. Find a positive integer M such that $n^{100} \cdot 0.99^n < 0.995^n$ for all $n \geq M$. Illustrate this inequality with an appropriate plot. Prove that the series $\sum_{n=1}^{\infty} n^{100} \cdot 0.99^n$ is convergent.

9.5 Alternating Series

We know that the terms of *any* convergent series tend to 0. We also know that the converse is not true: The terms of a series may become small and yet the series may diverge. Recall, for example, the harmonic series—its terms tend to 0, but its sequence of partial sums is unbounded. When a sequence of positive terms converges, we understand that it does so because the terms become small *sufficiently rapidly*.

With a general series, the matter is less simple. When a series with both positive and negative terms converges, the convergence may take place due to cancellation.

Because the partial sums of such series do not form an increasing sequence, the tests of the preceding two sections do not apply. In this section, we study a different approach.

The Alternating Series Test

The idea of cancellation motivates the next convergence test. We assume that the terms of our series have alternating signs: $\ldots, +, -, +, -, \ldots$. Such series are called *alternating series*. In an alternating series, the terms with odd indices all have the same sign and the terms with even indices all have the same sign. When the odd terms have positive sign, we account for the sign changes in the series by writing it as $\sum_{n=1}^{\infty}(-1)^{n+1}a_n = a_1 - a_2 + a_3 - \cdots$ where $a_n > 0$. If the signs of the even terms are positive, then we write the series as $\sum_{n=1}^{\infty}(-1)^n a_n = -a_1 + a_2 - a_3 + \cdots$ where $a_n > 0$. Because this last series is simply the negative of $\sum_{n=1}^{\infty}(-1)^{n+1}a_n$, each statement that we make about one of the series will hold for the other.

Theorem 1 ***Alternating Series Test*** If $\{a_n\}$ is a sequence of nonnegative numbers that satisfies

a. $a_1 \geq a_2 \geq a_3 \geq \cdots$ and
b. $\lim_{n\to\infty} a_n = 0$,

then the series $\sum_{n=1}^{\infty}(-1)^{n+1}a_n$ converges. Furthermore, the limit $\ell = \sum_{n=1}^{\infty}(-1)^{n+1}a_n$ lies between each pair of consecutive partial sums $S_N = \sum_{n=1}^{N}(-1)^{n+1}a_n$ and $S_{N+1} = S_N + (-1)^N a_{N+1}$. In particular, the limit ℓ satisfies the inequality

$$|\ell - S_N| \leq a_{N+1}. \tag{9.9}$$

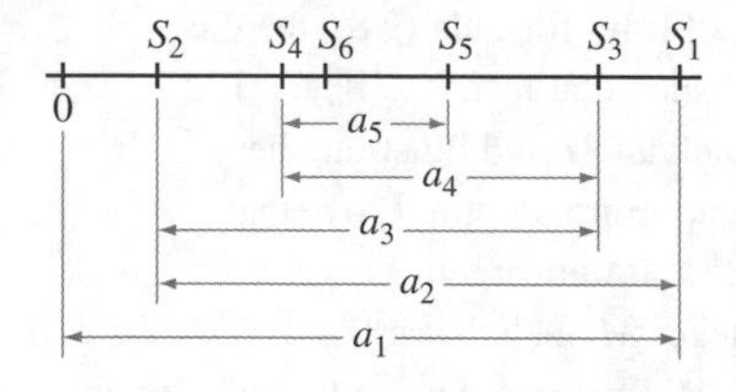

Figure 1

Proof First, observe that $-a_2 + a_3 \leq 0$, by hypothesis. Therefore, we have $S_1 = a_1 \geq a_1 + (-a_2 + a_3) = S_3$. Similarly, $-a_4 + a_5 \leq 0$, and so $S_3 \geq S_3 + (-a_4 + a_5) = S_5$. In general, this argument shows that the odd partial sums form a decreasing sequence. In a similar way, we see that the even partial sums form an increasing sequence. The sequence of even partial sums lies to the left of the sequence of odd partial sums (see Figure 1). Therefore, the two monotone sequences are bounded and hence are convergent. Let ℓ denote the limit of the odd partial sums. For any positive integer N, $S_{2N} - S_{2N-1} = a_{2N} \to 0$. Thus, the even and odd partial sums tend to the same limit ℓ, which means the series converges. Since ℓ lies between each pair of consecutive partial sums S_N and $S_N + a_{N+1}$, inequality (9.9) follows. ■

The Alternating Series Test applies to series in which the terms (i) have alternating signs and (ii) decrease in absolute value to 0. Always be sure to verify these hypotheses when applying the test.

Example 1 Apply the Alternating Series Test to the series $\sum_{n=1}^{\infty}(-1)^{n+1}/n$.

Solution If we set $a_n = 1/n$, then $a_1 \geq a_2 \geq a_3 \geq \cdots$ and $a_n \to 0$. Thus, the series has the form of an alternating series, as in Theorem 1. The Alternating Series Test applies, and $\sum_{n=1}^{\infty}(-1)^{n+1}/n$ converges. ■

inSIGHT

Notice that if we remove the minus signs from the series in Example 1, then we obtain the harmonic series, which is divergent. The minus signs cause enough cancellation to occur so that the series converges. Theorem 1 does not tell us the value ℓ of $\sum_{n=1}^{\infty} (-1)^{n+1}/n$, but it does allow us to make an estimate. If we want to evaluate ℓ with an error no greater than 0.01, then we find the first index n for which $1/n < 0.01$. This value is 101. By inequality (9.9), we can be sure that

$$\left| \ell - \sum_{n=1}^{100} \frac{(-1)^{n+1}}{n} \right| \leq \frac{1}{101} < 0.01.$$

With a computer, we can calculate $\sum_{n=1}^{100} (-1)^{n+1}/n \approx 0.6882$, which is our approximation. In fact, it is known that $\sum_{n=1}^{\infty} (-1)^{n+1}/n$ is *exactly* equal to $\ln(2) = 0.693\ldots.$

Example 2 Analyze the series $\sum_{n=1}^{\infty} (-1)^n/\sqrt{n}$.

Solution Because $1/\sqrt{n}$ decreases to 0, we may apply the Alternating Series Test to the series $\sum_{n=1}^{\infty} (-1)^{n+1}/\sqrt{n}$, which is the negative of the given series. Therefore, the original series also converges. ■

Example 3 Does the series

$$\sum_{n=1}^{\infty} (-1)^n \frac{1}{1+1/n}$$

converge?

Solution The series is indeed alternating. However, the terms do *not* tend to 0. Therefore, the Alternating Series Test *does not apply:* We may draw no conclusion from the test. On the other hand, the Divergence Test does apply; it tells us that the series must diverge because the terms do not tend to 0. ■

Because the first terms of a series do not affect convergence, we may be able to deduce that a series converges by applying the Alternating Series Test to an appropriate tail, as the next example illustrates.

Example 4 Show that the series

$$\sum_{n=1}^{\infty} (-1)^{n+1} \frac{\ln(n^4)}{\sqrt{n}}$$

converges.

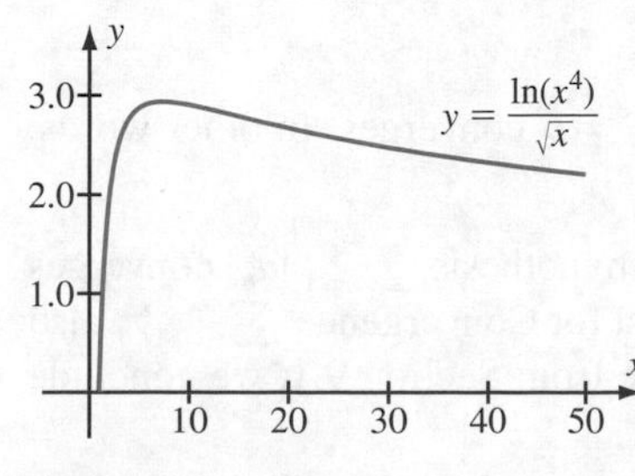

Figure 2

Solution A plot of $\ln(x^4)/\sqrt{x}$ (shown in Figure 2) indicates that this expression is not decreasing for all positive x, but it does appear to decrease from some point forward. To be sure, we apply the Derivative Test:

$$\frac{d}{dx}\left(\frac{\ln(x^4)}{\sqrt{x}}\right) = 2 \cdot \frac{2 - \ln(x)}{x^{3/2}}.$$

This expression is negative for $\ln(x) > 2$—that is, for $x > e^2 \approx 7.39$. It follows that the sequence $a_n = \ln(n^4)/\sqrt{n}$ decreases for $n \geq 8$. Also, notice that

$$\lim_{n\to\infty} a_n = \lim_{n\to\infty} \frac{\ln(n^4)}{\sqrt{n}} = 4 \lim_{n\to\infty} \frac{\ln(n)}{n^{1/2}} \overset{(9.8)}{=} 0.$$

Thus, the Alternating Series Test does apply to the tail series $\sum_{n=8}^{\infty} (-1)^{n+1} \ln(n^4)/\sqrt{n}$. Therefore, the full series also converges. ■

Absolute Convergence

It is useful to distinguish between series that converge because of cancellation and series that converge because of size. We can learn something by comparing the two. These considerations motivate the following definition.

Definition Let $\sum_{n=1}^{\infty} a_n$ be a series, possibly containing both positive and negative terms. If the series $\sum_{n=1}^{\infty} |a_n|$ of absolute values converges, then we say that the series $\sum_{n=1}^{\infty} a_n$ converges *absolutely*.

Example 5 Does the series $\sum_{n=1}^{\infty} (-1)^{n+1}/n$ converge absolutely?

Solution The given series converges by the Alternating Series Test, as noted in Example 1. However, the series $\sum_{n=1}^{\infty} |(-1)^{n+1}/n| = \sum_{n=1}^{\infty} 1/n$ diverges: It is the harmonic series. Therefore, the series $\sum_{n=1}^{\infty} (-1)^{n+1}/n$ does *not* converge absolutely. ■

Example 6 Does the series $\sum_{n=1}^{\infty} (-1/2)^n$ converge? Does it converge absolutely?

Solution The given series is a geometric series with ratio $-1/2$, which is less than 1 in absolute value. Therefore, the given series is convergent. For the same reason, the series of absolute values, $\sum_{n=1}^{\infty} |(-1/2)^n| = \sum_{n=1}^{\infty} (1/2)^n$, converges. Therefore, the given series converges absolutely. ■

INSIGHT

Our intuition tells us that absolute convergence is "better" than ordinary convergence. Absolute convergence cannot happen by virtue of cancellation. Example 5 suggests that it is harder for a series to converge absolutely than it is for a series to simply converge. The next theorem makes this intuition substantive.

Theorem 2 If the series $\sum_{n=1}^{\infty} |a_n|$ converges, then the series $\sum_{n=1}^{\infty} a_n$ converges. In other words, absolute convergence implies convergence.

Proof If we let $b_n = a_n + |a_n|$, then $0 \leq b_n \leq 2|a_n|$. By hypothesis, $\sum_{n=1}^{\infty} |a_n|$ converges so that $\sum_{n=1}^{\infty} 2|a_n|$ converges. By the Comparison Test for Convergence, $\sum_{n=1}^{\infty} b_n$ also converges. However, $a_n = b_n - |a_n|$. By Theorem 3a from Section 9.1, we conclude that $\sum_{n=1}^{\infty} a_n$ converges, as required.

Example 7 Test the series $\sum_{n=1}^{\infty} \sin(n)/2^n$ for convergence.

Solution Notice that $\sin(n)$ is sometimes positive and sometimes negative, *but it does not alternate in sign.* The pattern of signs begins $+, +, +, -, -, -, +, +, +, -, -, - \ldots$, but the pattern is less simple than these initial terms suggest: $\sin(22)$, $\sin(23)$, $\sin(24)$, $\sin(25)$ produce four consecutive negative terms. Therefore, the Alternating Series Test cannot be used. The comparison and integral tests do not apply because the series has both positive and negative terms. In short, *we cannot use any of the tests for convergence that we have learned in this chapter.* However, $\sum_{n=1}^{\infty} |\sin(n)/2^n|$ converges by the Comparison Test for Convergence because it is a sequence of positive terms that satisfies $0 \leq |\sin(n)/2^n| \leq 1/2^n$. Comparison with the convergent geometric series $\sum_{n=1}^{\infty} 1/2^n$ tells us that $\sum_{n=1}^{\infty} |\sin(n)/2^n|$ converges. Theorem 2 then tells us that the original series $\sum_{n=1}^{\infty} \sin(n)/2^n$ converges. ■

The sum of absolutely convergent series remains absolutely convergent. To see why, suppose that $\sum_{n=0}^{\infty} a_n$ and $\sum_{n=0}^{\infty} b_n$ are absolutely convergent. Notice that $|a_n + b_n| \leq |a_n| + |b_n|$ for each index n by the Triangle Inequality. Therefore,

$$\sum_{n=1}^{N} |a_n + b_n| \leq \sum_{n=1}^{N} |a_n| + \sum_{n=1}^{N} |b_n| \leq \sum_{n=1}^{\infty} |a_n| + \sum_{n=1}^{\infty} |b_n| < \infty.$$

It follows that the partial sums of the series $\sum_{n=1}^{\infty} |a_n + b_n|$ are bounded and that the series is convergent. We record this and some other simple facts as a theorem.

Theorem 3 Let $\sum_{n=0}^{\infty} a_n$ and $\sum_{n=0}^{\infty} b_n$ be absolutely convergent series.

a. $\sum_{n=0}^{\infty} (a_n + b_n)$ and $\sum_{n=0}^{\infty} (a_n - b_n)$ are absolutely convergent.
b. $\sum_{n=0}^{\infty} \lambda a_n$ converges absolutely for any real constant λ.

Conditional Convergence

We have seen that the alternating harmonic series $\sum_{n=1}^{\infty} (-1)^{n+1}/n$ is convergent (Example 1) but not absolutely convergent (Example 5). This suggests that there is an interesting class of series that converge but do not converge absolutely.

Definition If a series $\sum_{n=1}^{\infty} a_n$ converges, but does not converge absolutely, then we say that the series converges *conditionally.*

Example 8 Determine which positive values of p make the alternating p-series $\sum_{n=1}^{\infty} (-1)^{n+1}/n^p$ conditionally convergent.

Solution For $p > 1$, the series $\sum_{n=1}^{\infty} |(-1)^{n+1}/n^p|$ is just the convergent p-series $\sum_{n=1}^{\infty} 1/n^p$. Therefore, $\sum_{n=1}^{\infty} (-1)^{n+1}/n^p$ is absolutely convergent for $p > 1$. By definition, absolutely convergent series cannot be conditionally convergent.

Now suppose that $0 < p \leq 1$. The series $\sum_{n=1}^{\infty} |(-1)^{n+1}/n^p|$ is the divergent p-series $\sum_{n=1}^{\infty} 1/n^p$. Thus, $\sum_{n=1}^{\infty} (-1)^{n+1}/n^p$ is *not* absolutely convergent. Since $a_n = 1/n^p$ decreases to 0 as n tends to infinity, the series $\sum_{n=1}^{\infty} (-1)^{n+1}/n^p$ *is* convergent by the Alternating Series Test. We conclude that $\sum_{n=1}^{\infty} (-1)^{n+1}/n^p$ is conditionally convergent for $0 < p \leq 1$ and for no other positive values of p. ■

quickquiz

1. What is the Alternating Series Test? Include all hypotheses.
2. What is an absolutely convergent series?
3. What is a conditionally convergent series?
4. True or false: Every series is either absolutely convergent or conditionally convergent or divergent.

EXERCISES

Problems for Practice

In Exercises 1–12, the series may be shown to converge by using the Alternating Series Test. Show that the hypotheses of the Alternating Series Test are satisfied.

1. $\sum_{n=1}^{\infty}(-1)^n/(n^3+1)$
2. $\sum_{n=1}^{\infty}(-1)^n/\sqrt{n}$
3. $\sum_{n=1}^{\infty}(-1)^n/\sqrt{n^2+1}$
4. $\sum_{n=2}^{\infty}(-1)^n/\ln(n)$
5. $\sum_{n=1}^{\infty}(-1)^{n+1}(2/3)^n$
6. $\sum_{n=1}^{\infty}(-1)^{n+1}/n!$
7. $\sum_{n=1}^{\infty}(-4/5)^n/(n+2)$
8. $\sum_{n=1}^{\infty}(-1)^{n+1}n/(n^2+1)$
9. $\sum_{n=2}^{\infty}(-1)^{n+1}n^2/e^n$
10. $\sum_{n=1}^{\infty}(-1)^n(1+1/n)\ln(1+1/n)$
11. $\sum_{n=2}^{\infty}\cos(\pi n)\sin(\pi/n)$
12. $\sum_{n=1}^{\infty}(-1)^n(\pi/2-\arctan(n))$

In Exercises 13–24, determine if the series converges absolutely, converges conditionally, or diverges.

13. $\sum_{n=1}^{\infty}(-1)^n/(n^3+1)$
14. $\sum_{n=1}^{\infty}(-1)^n/\sqrt{n}$
15. $\sum_{n=1}^{\infty}(-1)^n n/(n^2+1)$
16. $\sum_{n=1}^{\infty}(-1)^n n/\sqrt{n^2+1}$
17. $\sum_{n=2}^{\infty}(-1)^n/\ln(n)$
18. $\sum_{n=1}^{\infty}(-1)^n(1+1/n)$
19. $\sum_{n=1}^{\infty}(-1)^n(1-1/n)^n$
20. $\sum_{n=1}^{\infty}(-1)^{n+1}(2/3)^n$
21. $\sum_{n=1}^{\infty}(-1)^{n+1}(n+\ln(n))/n^{3/2}$
22. $\sum_{n=2}^{\infty}(-1)^n n/\ln^3(n)$
23. $\sum_{n=1}^{\infty}\sin(n)e^{-n}$
24. $\sum_{n=1}^{\infty}(-1)^n\arctan(n)/(2n^4-n)$

In Exercises 25–30, a convergent alternating series $\sum_{n=1}^{\infty}(-1)^n a_n$ is given. Find a value of M such that the partial sum $\sum_{n=1}^{M}(-1)^n a_n$ approximates the infinite series to within 0.01.

25. $\sum_{n=1}^{\infty}(-1)^n(-1/n)$
26. $\sum_{n=1}^{\infty}(-1)^n/n^3$
27. $\sum_{n=1}^{\infty}(-1)^n(4/(1-2n))$
28. $\sum_{n=1}^{\infty}(-1)^n/n!$
29. $\sum_{n=1}^{\infty}(-1)^n/(n^2+15n)$
30. $\sum_{n=1}^{\infty}(-1)^n(1/\ln(n+1))$

Further Theory and Practice

In Exercises 31–34, find a value of M for which the Alternating Series Test may be applied to the tail $\sum_{n=M}^{\infty}(-1)^n a_n$ of the given series $\sum_{n=1}^{\infty}(-1)^n a_n$.

31. $\sum_{n=1}^{\infty}(-1)^n/(n^2-20n+101)$
32. $\sum_{n=1}^{\infty}(-1)^n/(n^4-32n^2+400)$
33. $\sum_{n=1}^{\infty}(-1)^n n^5/2^n$
34. $\sum_{n=1}^{\infty}(-1)^n\ln(n^{10})/\sqrt{n}$

In Exercises 35–38, calculate the alternating series to three decimal places.

35. $\sum_{n=1}^{\infty}(-1)^n/n!$
36. $\sum_{n=1}^{\infty}(-1)^n((1+1/n)/10)^n$
37. $\sum_{n=1}^{\infty}(-1)^n n!/(2n)!$
38. $\sum_{n=1}^{\infty}(-1)^n n^{-2n}$

Calculator/Computer Exercises

In Exercises 39–44, the given equation expresses the value of an alternating series. Illustrate the equation by calculating several partial sums S_N and plotting the points (N, S_N). To your plot, add the horizontal line that is the asymptote of the plotted points.

39. $\sum_{n=1}^{\infty}(-1)^{n+1}/n=\ln(2)$
40. $\sum_{n=1}^{\infty}(-1)^{n+1}4/(2n-1)=\pi$
41. $\sum_{n=1}^{\infty}(-1)^{n+1}/n!=1-1/e$

42. $\sum_{n=1}^{\infty} (-1)^{n+1}/(2n-1)! = \sin(1)$

43. $\sum_{n=0}^{\infty} (-1)^n/(2n)! = \cos(1)$

44. $\sum_{n=0}^{\infty} (-1)^n(\pi/3)^{2n}/(2n)! = 1/2$

In Exercises 45–48, find a value of M for which the Alternating Series Test may be applied to the tail $\sum_{n=M}^{\infty} (-1)^n a_n$ of the given series $\sum_{n=1}^{\infty} (-1)^n a_n$.

45. $\sum_{n=1}^{\infty}(-1)^n(9n^2+13)/(n^3+55n+60)$

46. $\sum_{n=1}^{\infty} (-1)^n n^4 e^{-n} \ln(n)$

47. $\sum_{n=1}^{\infty} (-1)^n(100n^{9/4}+n)/(150+n^{5/2})$

48. $\sum_{n=1}^{\infty} (-1)^n n^3/\exp(\sqrt{n/100})$

9.6 The Ratio and Root Tests

There are two powerful and easy-to-use tests that tell when a series converges absolutely: the Ratio Test and the Root Test. Bear in mind, however, that these tests are of no use in determining whether a series converges conditionally.

The Ratio Test

Suppose that $\sum_{n=1}^{\infty} a_n$ is a series of positive terms for which each term a_{n+1} is approximately equal to a fixed multiple L of its predecessor: $a_{n+1} \approx L \cdot a_n$. Thus, $a_2 \approx L \cdot a_1$ and $a_3 \approx L \cdot a_2$. Notice that we can combine these two approximations to conclude that $a_3 \approx L^2 \cdot a_1$. Similarly, $a_4 \approx L \cdot a_3$, and we can use our preceding estimate to conclude that $a_4 \approx L^3 \cdot a_1$. If this process is continued, then we estimate $a_n \approx L^{n-1} \cdot a_1$ for each positive integer n.

Now, if the approximate ratio L is less than 1, then the series $\sum_{n=1}^{\infty} a_n$ will converge by comparison with the convergent geometric series $a_1 \sum_{n=1}^{\infty} L^{n-1}$. On the other hand, if the approximate ratio L is greater than 1, then $a_{n+1} > a_n$ and the terms of the series do not tend to 0. The series diverges by the Divergence Test. These arguments are the ideas behind the following theorem.

Theorem 1 ***Ratio Test*** Let $\sum_{n=1}^{\infty} a_n$ be a series. Suppose that $\lim_{n\to\infty} |a_{n+1}/a_n| = L$.

a. If $L < 1$, then the series converges absolutely.
b. If $L > 1$, then the series diverges.
c. If $L = 1$, then the test gives no information.

Take particular notice that the Ratio Test yields no information when $L = 1$. This point will be made clearer in the examples. Also, remember that the limit of the ratio $|a_{n+1}/a_n|$ might not even exist, in which case we could not use Theorem 1. Finally, notice that the test only depends on the *limit* of the ratio $|a_{n+1}/a_n|$. The outcome does *not* depend on the first million or so terms of the series, and that is the way it should be: The convergence or divergence of a series depends only on its tail.

Now we look at some examples that illustrate when the Ratio Test gives convergence and when it gives divergence.

Example 1 Use the Ratio Test to analyze the series $\sum_{n=1}^{\infty} 2^n/n^{10}$.

Solution We set $a_n = 2^n/n^{10}$. As $n \to \infty$,

$$L = \lim_{n\to\infty}\left|\frac{a_{n+1}}{a_n}\right| = \lim_{n\to\infty}\frac{2^{n+1}/(n+1)^{10}}{2^n/n^{10}} = \lim_{n\to\infty} 2\left(\frac{n}{n+1}\right)^{10} = 2\cdot 1^{10} = 2.$$

Because $L > 1$, the Ratio Test says that the series diverges. (For practice, use the Divergence Test to reach the same conclusion.) ■

Example 2 Discuss the convergence or divergence of the series $\sum_{n=1}^{\infty}\sqrt{n+1}/2^n$.

Solution In this example, we let $a_n = \sqrt{n+1}/2^n$. We have

$$L = \lim_{n\to\infty}\left|\frac{a_{n+1}}{a_n}\right| = \lim_{n\to\infty}\frac{\sqrt{n+2}/2^{n+1}}{\sqrt{n+1}/2^n} = \lim_{n\to\infty}\frac{1}{2}\sqrt{\frac{n+2}{n+1}} = \frac{1}{2}\lim_{n\to\infty}\sqrt{\frac{1+2/n}{1+1/n}} = \frac{1}{2}.$$

Part a of the Ratio Test tells us that the series converges. (As a matter of style, the adjective *absolute* is usually not used in discussions about the convergence of a series of positive terms a_n. Doing so would add no information, since $\sum_{n=1}^{\infty}|a_n| = \sum_{n=1}^{\infty}a_n$.) ■

Example 3 Use the Ratio Test to show that the series $\sum_{n=0}^{\infty}x^n/n!$ converges for every real number x.

Solution Recall that $0! = 1$, $1! = 1$, and $n! = n\cdot(n-1)\cdot(n-2)\cdots 3\cdot 2\cdot 1$. When using the Ratio Test, it is easiest to spot cancellation when a larger factorial is written in terms of a smaller one. Thus,

$$\begin{aligned}(n+1)! &= (n+1)\cdot n\cdot(n-1)\cdot(n-2)\cdots 3\cdot 2\cdot 1\\ &= (n+1)\cdot n!.\end{aligned}$$

Let $a_n = x^n/n!$ so that

$$\left|\frac{a_{n+1}}{a_n}\right| = \frac{(|x|^{n+1})/(n+1)!}{|x|^n/n!} = \frac{|x|\cdot n!}{(n+1)!} = \frac{|x|}{n+1}.$$

As $n \to \infty$, the limit of this expression is 0 for every x. Therefore, in this example, the limit L exists and is less than 1. By part a of the Ratio Test, we may conclude that the series converges. ■

inSIGHT

The Ratio Test often works well when the general term a_n of a series factors into powers and factorials. In such cases, the ratio $|a_{n+1}/a_n|$ can simplify greatly due to cancellation, as Example 3 illustrates.

Example 4 Test the series $\sum_{n=1}^{\infty}n^n/n!$ for convergence.

Solution If we set $a_n = n^n/n!$, then

$$\begin{aligned}\left|\frac{a_{n+1}}{a_n}\right| &= \frac{(n+1)^{n+1}/(n+1)!}{n^n/n!}\\ &= \left(\frac{n+1}{n}\right)^n\\ &= \left(1+\frac{1}{n}\right)^n.\end{aligned}$$

This last expression converges to the number e when $n \to \infty$, as we know from Section 2.6. Since $e = 2.718\ldots > 1$, the series must diverge. Exercise 52 contains some additional remarks about this example. ■

Example 5 Apply the Ratio Test to the series $\sum_{n=1}^{\infty} 1/n$.

Solution Setting $a_n = 1/n$, we have

$$\lim_{n\to\infty}\left|\frac{a_{n+1}}{a_n}\right| = \lim_{n\to\infty}\frac{1/(n+1)}{1/n} = \lim_{n\to\infty}\frac{n}{n+1} = 1.$$

Therefore, the Ratio Test gives no information whatsoever. Of course, we already know from independent reasoning that this series diverges. ■

inSIGHT

Examples 5 and 6 together explain what we mean when we say that the Ratio Test gives no information when the limit L is equal to 1. Under these circumstances, the series could diverge (Example 5) or the series could converge (Example 6). There is no way to tell when $L = 1$: *You must use another convergence test to find out.*

Example 6 Apply the Ratio Test to the series $\sum_{n=1}^{\infty} 1/n^3$.

Solution If we set $a_n = 1/n^3$, then

$$\lim_{n\to\infty}\left|\frac{a_{n+1}}{a_n}\right| = \lim_{n\to\infty}\frac{1/(n+1)^3}{1/n^3} = \lim_{n\to\infty}\frac{n^3}{(n+1)^3} = \lim_{n\to\infty}\left(\frac{n}{n+1}\right)^3 = 1^3 = 1.$$

Therefore, the Ratio Test gives no information. Of course, we already know from independent reasoning that this series is a convergent p-series. ■

The Root Test

The next test that we will learn, the Root Test, is similar to the Ratio Test. However, sometimes it is easier to apply the Ratio Test than it is to apply the Root Test (and vice versa). Thus, even though the tests look rather similar, you should be well-versed in using both of them.

Theorem 2 Let $\sum_{n=1}^{\infty} a_n$ be a series. Suppose that $\lim_{n\to\infty} |a_n|^{1/n} = L$.

a. If $L < 1$, then the series converges absolutely.
b. If $L > 1$, then the series diverges.
c. If $L = 1$, then the test gives no information.

Once again, take particular note that when $L = 1$, the Root Test gives no information whatsoever. Also, the limit L might not even exist, in which case we could not use Theorem 2. Finally, the Root Test does not depend on the first million or so terms of the series—only on the tail.

The limit formula

$$\lim_{x\to\infty} x^{1/x} = 1 \tag{9.10}$$

is often useful when applying the Root Test. To evaluate this limit, we use the basic formula $\lim_{x\to\infty} \ln(x)/x = 0$, which is obtained from equation (9.8) by setting $q = 1$. It then follows that

$$\lim_{x\to\infty} x^{1/x} = \lim_{x\to\infty} \exp\left(\frac{\ln(x)}{x}\right) = e^0 = 1.$$

We now look at some examples that illustrate when the Root Test gives convergence, when it gives divergence, and when it gives no information.

inSIGHT

Notice that

$$\frac{n}{(n^2+6)^n} \leq \frac{n}{n^{2n}} \leq \frac{n}{n^4} = \frac{1}{n^3},$$

provided that $n \geq 2$. Thus, the series from Example 7 is termwise dominated by the p-series $\sum_{n=1}^{\infty} 1/n^3$. We can therefore deduce that the given series converges by applying the Comparison Test for Convergence. The Root Test, however, proves to be more straightforward in this example.

Example 7 Analyze the series $\sum_{n=1}^{\infty} n/(n^2+6)^n$.

Solution If we apply the Root Test with $a_n = n/(n^2+6)^n$, then we see that $|a_n|^{1/n} = n^{1/n}/(n^2+6)$. The numerator of this expression tends to 1 by equation (9.10). Since the denominator becomes unbounded, $L = \lim_{n\to\infty} |a_n|^{1/n} = 0$. Because this limit is less than 1, the Root Test tells us that the series converges. ■

Example 8 Apply the Root Test to the series $\sum_{n=3}^{\infty} 1/\ln^n(n)$.

Solution When $a_n = 1/\ln^n(n)$, we have $|a_n|^{1/n} = 1/\ln(n)$. This expression tends to 0 as $n \to \infty$. Therefore, $L = 0 < 1$, and the Root Test says that the series converges. ■

inSIGHT

When applied to the series from Example 8, both the Ratio Test and the Integral Test result in a difficult calculation. By contrast, the Root Test is an obvious choice for Example 8 since a_n is an nth power. The Comparison Test for Convergence might also be used. For example, $\ln(n)$ is greater than 2 for $n \geq 8$. Therefore, $1/\ln^n(n) < 1/2^n$ for $n \geq 8$. We deduce that the given series is convergent by comparison with the convergent geometric series $\sum_{n=8}^{\infty} 1/2^n$.

Example 9 Apply the Root Test to the series $\sum_{n=1}^{\infty} (-2)^n/n^{10}$.

Solution With $a_n = (-2)^n/n^{10}$, we have

$$\lim_{n\to\infty} |a_n|^{1/n} = \lim_{n\to\infty} \frac{2}{n^{10/n}} = \frac{2}{(\lim_{n\to\infty} n^{1/n})^{10}} \overset{(9.10)}{=} \frac{2}{1} > 1.$$

We conclude, using the Root Test, that the series diverges. ■

Example 10 Let p be positive. Apply the Root Test to the p-series $\sum_{n=1}^{\infty} 1/n^p$.

Solution If $a_n = 1/n^p$, then

$$L = \lim_{n\to\infty} \left|a_n^{1/n}\right| = \lim_{n\to\infty} \frac{1}{n^{p/n}} = \frac{1}{(\lim_{n\to\infty} n^{1/n})^p} \overset{(9.10)}{=} 1.$$

Whatever the value of p, the Root Test is inconclusive. ■

inSIGHT

Example 10 explains what we mean when we say that the Root Test gives no information when the limit L is equal to 1. Under these circumstances, the series could diverge (a p-series with $p \leq 1$) or the series could converge (a p-series with $p > 1$). There is no way to tell from the Root Test: *You must use another test to find out.*

quickquiz

1. State the Ratio Test. Be sure to distinguish three cases.
2. True or false: The Ratio Test always fails when applied to a conditionally convergent series.
3. Apply the Ratio Test to the series $\sum_{n=1}^{\infty} 4^n/(n!)$.
4. Apply the Root Test to the series $\sum_{n=1}^{\infty} 4^n/7^{2n}$.

EXERCISES

Problems for Practice

In Exercises 1–10, use the Ratio Test to determine the convergence or divergence of the series.

1. $\sum_{n=1}^{\infty} n/e^n$
2. $\sum_{n=1}^{\infty} n^2/2^n$
3. $\sum_{n=1}^{\infty} 2^n/n^3$
4. $\sum_{n=1}^{\infty} 10^n/n!$
5. $\sum_{n=1}^{\infty} n^{100}/n!$
6. $\sum_{n=1}^{\infty} n!/(1000^n(n+12)^{13})$
7. $\sum_{n=1}^{\infty} n!/(n3^n)$
8. $\sum_{n=1}^{\infty} \sqrt{n!}/(n^5 \cdot 7^n)$
9. $\sum_{n=1}^{\infty} 2^n\sqrt{n}/3^n$
10. $\sum_{n=1}^{\infty} (3n+n^2)/3^n$

In Exercises 11–16, verify that the Ratio Test yields no information about the convergence of the series. Use other methods to determine whether the series converges absolutely, converges conditionally, or diverges.

11. $\sum_{n=2}^{\infty} (-1)^n/\ln(n^2)$
12. $\sum_{n=1}^{\infty} (-1)^n(n+1)/(n^3+1)$
13. $\sum_{n=1}^{\infty} (-3)^n/(n^3+3^n)$
14. $\sum_{n=1}^{\infty} (-1)^n\sqrt{n}/(n+4)$
15. $\sum_{n=1}^{\infty} (-1)^n \ln(n)/n^2$
16. $\sum_{n=1}^{\infty} (-1)^n \sin(1/n)$

In Exercises 17–22, use the Root Test to determine the convergence or divergence of the series.

17. $\sum_{n=1}^{\infty} (-1)^n 10^n/n^{10}$
18. $\sum_{n=1}^{\infty} (-1)^n n^{100}/(1+n)^n$
19. $\sum_{n=1}^{\infty} (-37/n)^n$
20. $\sum_{n=1}^{\infty} (-1)^n n^7/\ln^n(n)$
21. $\sum_{n=1}^{\infty} ((n^2+7n+13)/(2n^2+1))^n$
22. $\sum_{n=1}^{\infty} (1-1/n)^{n^2}$

Further Theory and Practice

In Exercises 23–28, use the Ratio Test to determine the convergence or divergence of the series.

23. $\sum_{n=1}^{\infty} (2n)!/(3n)!$
24. $\sum_{n-1}^{\infty} (n!)^2/(2n)!$
25. $\sum_{n=1}^{\infty} (2n)!/(n! \cdot 2^n)$
26. $\sum_{n=1}^{\infty} n!/2^{n^n}$
27. $\sum_{n=1}^{\infty} (n+3^n)/(n^3+2^n)$
28. $\sum_{n=1}^{\infty} 2^n/(1+\ln^n(n))$

In Exercises 29–32, use the Root Test to determine the convergence or divergence of the series.

29. $\sum_{n=1}^{\infty} 4^n/(n^{1/n}+2)^n$
30. $\sum_{n=1}^{\infty} (n^{1/n}+1/2)^n$
31. $\sum_{n=1}^{\infty} 2^n/(1+\ln^n(n))$
32. $\sum_{n=1}^{\infty} \exp(n)/(1+\ln^n(n))$

In Exercises 33–48, determine whether the series converges absolutely, converges conditionally, or diverges. Because the tests developed in this section are not the most appropriate for some of these series, you may use any test that has been discussed in this chapter.

33. $\sum_{n=1}^{\infty} (-1)^n/\sqrt{n+10}$
34. $\sum_{n=1}^{\infty} (-1)^n/(n\sqrt{n+10})$
35. $\sum_{n=1}^{\infty} (-1)^n n!/(3^n)$
36. $\sum_{n=1}^{\infty} (-3/4)^n n$
37. $\sum_{n=1}^{\infty} (-1)^n n/(n^2-11)$
38. $\sum_{n=2}^{\infty} (-1)^n/(n\ln^3(n))$

39. $\sum_{n=2}^{\infty}(-1)^n n^{1/n}/\ln(n)$
40. $\sum_{n=1}^{\infty}(-1)^n \ln(n^{1/n})$
41. $\sum_{n=1}^{\infty}(-1)^n n^{1/n}/(1+1/n)$
42. $\sum_{n=1}^{\infty}(-1)^n(n^2+2n+2)/(3n^3+7)$
43. $\sum_{n=1}^{\infty}(-1)^n e^{1/n}/n^e$
44. $\sum_{n=1}^{\infty}(-1)^n \ln(2+1/n)$
45. $\sum_{n=1}^{\infty}(-1)^n \ln(1+1/n)$
46. $\sum_{n=1}^{\infty}(-1)^n n 2^{-n^2}$
47. $\sum_{n=1}^{\infty}(-1)^n/(1+1/n)^n$
48. $\sum_{n=1}^{\infty}(-1)^n \ln(n)/\sqrt{n}$
49. Let p be a fixed real number (possibly negative). Show that the Ratio Test results in the same conclusion when applied to both $\sum_{n=1}^{\infty} n^p a_n$ and $\sum_{n=1}^{\infty} a_n$. Explain how this observation can save work in the application of the Ratio Test. How can you see instantly that the Ratio Test is inconclusive when applied to all p-series?
50. If p is a polynomial, then show that the Ratio Test results in the same conclusion when applied to both $\sum_{n=1}^{\infty} p(n)a_n$ and $\sum_{n=1}^{\infty} a_n$.
51. Let p be a fixed real number (possibly negative). Show that the Ratio Test results in the same conclusion when applied to both $\sum_{n=2}^{\infty} \ln^p(n)a_n$ and $\sum_{n=2}^{\infty} a_n$.
52. In Example 4, the Ratio Test is used to show that $\sum_{n=1}^{\infty} n^n/n!$ diverges. Use the Divergence Test to reach the same conclusion. The work done in Example 4 need not be wasted. To wit, show that $\sum_{n=1}^{\infty} n!/n^n$ converges.

Calculator/Computer Exercises

Let $\{a_n\}$ be a sequence of positive numbers. In Chapter 10, we will learn that if the two limits $\lim_{n\to\infty} a_{n+1}/a_n$ and $\lim_{n\to\infty} a_n^{1/n}$ exist, then they are equal. In Exercises 53–56, produce a plot that illustrates the equality of these two limits. Your plot should include a horizontal line that is the asymptote of the points $\{(n, a_{n+1}/a_n)\}$ and $\{(n, a_n^{1/n})\}$.

53. $a_n = 1/n^2$
54. $a_n = n/2^n$
55. $a_n = e^n/\ln(1+n)$
56. $a_n = n!/n^n$

Summary of Key Topics

Series (Section 9.1)

An infinite sum $\sum_{n=1}^{\infty} a_n$ is called a series. (Sometimes it is convenient to begin a series at an index other than $n = 1$.) The expression $S_N = \sum_{n=1}^{N} a_n$ is called the Nth partial sum of the series. The series is said to converge to a limit ℓ if $S_N \to \ell$ as $N \to \infty$.

Some Special Series (Sections 9.1, 9.2, and 9.3)

The harmonic series $\sum_{n=1}^{\infty} 1/n$ diverges.

The geometric series $\sum_{n=M}^{\infty} ar^n$ with ratio r converges if and only if $|r| < 1$. When $|r| < 1$, the series sums to $ar^M/(1-r)$.

The p-series $\sum_{n=1}^{\infty} 1/n^p$ converges if and only if $p > 1$.

Properties of Series (Section 9.1)

If $\sum_{n=1}^{\infty} a_n$ converges to A and $\sum_{n=1}^{\infty} b_n$ converges to B, then

a. $\sum_{n=1}^{\infty}(a_n+b_n) = A + B$ and $\sum_{n=1}^{\infty}(a_n-b_n) = A - B$; and
b. $\sum_{n=1}^{\infty} \lambda a_n = \lambda A$ for any real constant λ.

The Divergence Test (Section 9.2)

If $\sum_{n=1}^{\infty} a_n$ converges, then $a_n \to 0$. Equivalently, if the summands of a series do not tend to 0, then the series cannot converge.

The Integral Test (Section 9.3)

If f is a positive, continuous, decreasing function with domain $\{x : 1 \le x < \infty\}$, then $\int_1^\infty f(x)\,dx$ is finite if and only if $\sum_{n=1}^\infty f(n)$ converges.

The Comparison Tests (Section 9.4)

If $0 \le a_n \le b_n$ and $\sum_{n=1}^\infty b_n$ converges, then $\sum_{n=1}^\infty a_n$ converges.

If $0 \le b_n \le a_n$ and $\sum_{n=1}^\infty b_n$ diverges, then $\sum_{n=1}^\infty a_n$ diverges.

Let $\sum_{n=1}^\infty a_n$ and $\sum_{n=1}^\infty b_n$ be series of positive terms. If $\lim_{n\to\infty} a_n/b_n$ exists as a (finite) positive number, then $\sum_{n=1}^\infty a_n$ converges if and only if $\sum_{n=1}^\infty b_n$ converges.

Alternating Series Test (Section 9.5)

If $a_1 \ge a_2 \ge a_3 \ge \cdots$ and if $\lim_{n\to\infty} a_n = 0$, then $\sum_{n=1}^\infty (-1)^{n+1} a_n$ converges.

Series with Positive and Negative Terms (Section 9.5)

A series $\sum_{n=1}^\infty a_n$ is said to *converge absolutely* if $\sum_{n=1}^\infty |a_n|$ converges.

If a series converges absolutely, then it converges.

A series that is convergent but not absolutely convergent is called *conditionally convergent*.

The Ratio and Root Tests (Section 9.6)

The Ratio Test Let $\sum_{n=1}^\infty a_n$ satisfy $\lim_{n\to\infty} |a_{n+1}/a_n| = \ell$. If $\ell < 1$, then the series converges absolutely. If $\ell > 1$, then the series diverges. If $\ell = 1$, then no conclusion is possible *from the Ratio Test.*

The Root Test Let $\sum_{n=1}^\infty a_n$ satisfy $\lim_{n\to\infty} |a_n|^{1/n} = \ell$. If $\ell < 1$, then the series converges absolutely. If $\ell > 1$, then the series diverges. If $\ell = 1$, then no conclusion is possible *from the Root Test.*

genesis & DEVELOPMENT

The concept of infinite series, like many of the mathematical ideas that we study today, originated in ancient Greece. As we have seen, one of Zeno's paradoxes arises from decomposing the number 1 into infinitely many summands: $1 = \sum_{n=1}^{\infty} 1/2^n$. It is often stated that Zeno's paradoxes instilled in Greek mathematicians a paralyzing fear of the infinite. In fact, the lesson that Greek mathematicians brought away from Zeno's paradoxes is that especially careful argumentation is needed in the presence of the infinite. Archimedes, for example, knew the formula

$$\sum_{n=0}^{N} r^n = \frac{1 - r^{N+1}}{1 - r}$$

for a finite geometric series with $r \neq 1$. He also had a perfect understanding of the formula

$$\sum_{n=0}^{\infty} r^n = \frac{1}{1 - r} \quad (0 \leq r < 1),$$

although the mathematical style of the time did not permit such statements. Nevertheless, using cumbersome techniques designed to avoid any skepticism over his handling of the infinite, Archimedes used precisely this formula to compute the area of a region bounded by a parabola and a chord.

The study of infinite series—and much else—was put on hold when the school of classical Greek mathematics declined. No further progress was made until the 14th century. Nevertheless, after a millennium of neglect, spectacular results were obtained on two fronts.

Nicole Oresme was one of the greatest of the medieval scholastic philosophers. Trained in theology, Oresme abandoned an academic career to rise through the ranks of the clergy, obtaining a bishop's chair in 1377. Along the way, he served as financial advisor to France's King Charles V (Charles the Wise). In response to a request from Charles, Oresme translated several of Aristotle's works from Latin into French. These translations are considered to have had an important influence on the development of the French language.

Oresme's own writings were both extensive and varied. He is commonly held to be the leading economist of the Middle Ages; one can make a case for a similar status in physics. However, it was in mathematics that Oresme was at his most original. Among his mathematical achievements is the near introduction of analytic geometry 200 years before its implementation by Descartes and Fermat. It was Oresme who first demonstrated that an infinite series can be divergent even though its terms converge to 0. Indeed, the proof of the divergence of the harmonic series given in Section 9.1 is Oresme's own argument. On the flip side of the convergence coin, Oresme obtained one of the first nontrivial evaluations of an infinite series when he demonstrated that $\sum_{n=1}^{\infty} n/2^{n-1} = 4$.

Oresme turned out to be something of an isolated genius in medieval mathematics. He had no immediate followers to develop his theories and extend his mathematical work. Although not all of Oresme's work sank into oblivion, his work on infinite series certainly did. More than 300 years later, prominent mathematicians were quite pleased with themselves when they "discovered" his theorem on the harmonic series.

Even more astonishing is the singular appearance of Mādhavan (1350–1425). Like his near-contemporary Oresme, Mādhavan discovered remarkable facts about infinite series, facts that would not take root in European mathematics for several centuries. Like the results of Oresme, Mādhavan's work on infinite series would become known through their "discovery" hundreds of years later. Indeed, Mādhavan's work remained largely unknown until its recognition midway through the 20th century. Two examples will suffice to show Mādhavan's depth and originality:

$$\frac{\pi}{\sqrt{12}} = \sum_{n=0}^{\infty} (-1)^n \frac{1}{(2n+1)3^n} = 1 - \frac{1}{3 \times 3} + \frac{1}{5 \times 3^2} - \frac{1}{7 \times 3^3} + \cdots$$

and

$$\frac{\pi}{4} = \sum_{n=0}^{\infty} (-1)^n \frac{1}{(2n+1)} = 1 - \frac{1}{3} + \frac{1}{5} - \frac{1}{7} + \cdots.$$

The simplicity of this second formula is striking. Think of the geometric meaning of π. Does it not seem remarkable that π admits such a representation in terms of the reciprocals of the odd integers? In Chapter 10, we will learn how to derive these and other formulas of Mādhavan. What we do not know is exactly how Mādhavan himself obtained these formulas. Although the statements of his main theorems were recorded by later mathematicians, nothing of his own mathematical writing has survived.

The study of infinite series began to flourish in the 17th century. Pietro Mengoli (1626–1686) found the value of the alternating harmonic series,

$$\ln(2) = \sum_{n=1}^{\infty} (-1)^{n+1}\frac{1}{n} = 1 - \frac{1}{2} + \frac{1}{3} + \frac{1}{4} - \frac{1}{5} + \cdots,$$

as did several other mathematicians. Mengoli also reproved Oresme's forgotten theorem concerning the divergence of the harmonic series. Mādhavan's work did not become known in the West until modern times, but his infinite series for $\pi/4$ was rederived several times in the second half of the 17th century. In a letter of 1676, Leibniz communicated to Newton Mādhavan's formula $\pi/4 = 1 - 1/3 + 1/5 - 1/7 + \cdots$ (now usually called Leibniz's series). Newton, who had obtained the same result a decade before Leibniz, responded with an astonishing example of scientific one-upmanship. Providing only the sketchiest of hints, Newton sent Leibniz a similar but much less transparent series:

$$\frac{\pi}{2\sqrt{2}} = \frac{1}{1} + \frac{1}{3} - \frac{1}{5} - \frac{1}{7} + \frac{1}{9} + \frac{1}{11} - \frac{1}{13} - \frac{1}{15} + \frac{1}{17} + \frac{1}{19} - \frac{1}{21} - \frac{1}{23} + \cdots.$$

There is no evidence that Leibniz ever came to grips with Newton's remarkable equation.

Whether Newton could have stumped Euler, we can only guess. A contemporary of Euler once remarked that Euler calculated as effortlessly as "men breathe, as eagles sustain themselves in the air." Euler was a master of both convergent and divergent infinite series. One of his earliest successes was the evaluation of a series that eluded his predecessors:

$$\frac{\pi^2}{6} = \sum_{n=1}^{\infty} \frac{1}{n^2} = 1 + \frac{1}{4} + \frac{1}{9} + \frac{1}{16} + \frac{1}{25} + \cdots.$$

In fact, Euler discovered a method to evaluate all p-series $\zeta(p) = \sum_{n=1}^{\infty} n^{-p}$ for which p is a positive even integer. More precisely, Euler was able to determine rational numbers q_k such that $\zeta(2k) = q_k\pi^{2k}$ for each positive integer k. For example, $q_1 = 1/6$, $\zeta(2) = \pi^2/6$; $q_2 = 1/90$, $\zeta(4) = \pi^4/90$; and $q_3 = 1/945$, $\zeta(6) = \pi^6/945$. The numerators of q_1 through q_5 are all 1, but that pattern ends with q_6. Euler calculated explicitly as far as $q_{13} = 1315862/11094481976030578125$.

Given our knowledge of $\zeta(2)$, $\zeta(4)$, $\zeta(6)$, and so on, it is natural to wonder about $\zeta(3)$. Ever since the time of Euler, the p-series $\sum_{n=1}^{\infty} 1/n^{2k+1}$ with odd powers has defied nearly all efforts to unlock its secrets. In 1978, Roger Apéry (1916–1994) created a sensation when he proved that $\zeta(3)$ is irrational. Despite a great deal of effort since then, we do not know the exact value of $\zeta(3)$, and we know still less about other odd-exponent p-series.

Although the formulas that appear in this section can be used to calculate the digits of π, they are not efficient. If you were to sum 1000 terms of Leibniz's series, you would not even get the third decimal place of π for your effort. (The grade school approximation $22/7$ is very nearly as accurate.) We close our discussion with remarkable infinite series for π and $1/\pi$. The first was discovered by Srinivasa Aiyangar Ramanujan (1887–1920) early in the 20th century:

$$\frac{1}{\pi} = \frac{\sqrt{8}}{9801} \sum_{n=0}^{\infty} \frac{(4n)!(1103 + 26390n)}{(n!)^4(396)^{4n}}.$$

It is an astonishing formula, but that was Ramanujan's stock-in-trade. At his death, he left notebooks filled with mysterious equations. Mathematicians have been at work ever since analyzing and proving them.

The last infinite series that we record was discovered late in the 20th century. In 1995, David Bailey, Peter Borwein, and Simon Plouffe announced the formula

$$\pi = \sum_{n=0}^{\infty} \left(\frac{4}{8n+1} - \frac{2}{8n+4} - \frac{1}{8n+5} - \frac{1}{8n+6}\right)\left(\frac{1}{16}\right)^n.$$

Is this just another complicated formula? Actually, there is more to it than that. Bailey, Borwein, and Plouffe used this formula to compute the *ten billionth* hexadecimal digit of π. What makes this equation special is that it is a digit extractor. It can be used to calculate digits of π without having to calculate the preceding digits.

10

Taylor Series

PREVIEW

In Chapter 9, we studied infinite series $\sum_{n=0}^{\infty} u_n$ in which each term u_n is a constant. It is very useful, however, to allow the terms of an infinite series to be functions of a variable x. In practice, the particular choice of function will depend on the application. For example, many basic differential equations of physics can be solved by infinite series with terms of the form $u_n = a_n \cos(nx) + b_n \sin(nx)$. In this chapter, we study a simpler class of series in which $u_n = a_n(x - c)^n$ where c and a_n are constants. The resulting expression, $\sum_{n=0}^{\infty} a_n(x - c)^n$, is said to be a *power series*. It is a function of x that converges for x lying in an interval centered at c.

Power series have many applications in mathematics and science. For example, several important differential equations from physics and engineering are best solved by power series methods. Functions defined by power series have become fundamental tools in the study of such diverse subjects as civil engineering, aeronautics, structural geology, and celestial mechanics.

Power series are not used only to create new functions. In many cases it is important to be able to express the standard elementary functions by means of power series. The question becomes: Given a function f and a point c, can we find coefficients $\{a_n\}$ so that $f(x) = \sum_{n=0}^{\infty} a_n(x - c)^n$? The answer lies in the theory of Taylor series, which is the main focus of this chapter. Because the partial sums $\sum_{n=0}^{N} a_n(x - c)^n$ of a power series are polynomials, our study of Taylor series will lead to powerful techniques for approximating functions by polynomials.

It is easy to overlook the importance of approximation when calculators and software packages speedily evaluate all the familiar functions of mathematics. Bear in mind that the computations that go on behind the scenes must involve only the basic operations of arithmetic. The algorithms that are employed are often similar in spirit to the Taylor polynomial approximations that we will learn in this chapter.

10.1 Introduction to Power Series

The simplest functions that we know are those that are of the form $x \mapsto x^n$ where n is a nonnegative integer, such as the constant function $f(x) = x^0 = 1$ and the functions $g(x) = x$, $h(x) = x^2$, and so on. These functions are easy to understand and simple to compute. We use them to form polynomials $a_0 + a_1 \cdot x + \cdots + a_N \cdot x^N$, which are also straightforward to handle. Now that we understand the concept of infinite series, it is natural to create functions by summing infinitely many powers of x.

Definition An expression of the form $a_0 + a_1 x + a_2 x^2 + a_3 x^3 + \cdots$, where the a_ns are constants, is called a *power series* in x. It is convenient to denote the constant term a_0 by $a_0 x^0$ so that we can use the sigma notation $\sum_{n=0}^{\infty} a_n x^n$ to denote the power series $a_0 + a_1 x + a_2 x^2 + a_3 x^3 + \cdots$.

Example 1 Are the infinite series $\sum_{n=0}^{\infty} (\sqrt{x})^n$, $\sum_{n=3}^{\infty} (2x)^n$, and $\sum_{n=0}^{\infty} (x^2)^n$ power series in x?

Solution In the definition of a power series, each power of x that appears is a *nonnegative integer*. Therefore, the series $\sum_{n=0}^{\infty} (\sqrt{x})^n = \sum_{n=0}^{\infty} x^{n/2} = 1 + \sqrt{x} + x + x\sqrt{x} + \cdots$ is *not* a power series.

The series $\sum_{n=3}^{\infty} (2x)^n = \sum_{n=3}^{\infty} 2^n x^n$ may be written as $0 + 0 \cdot x + 0 \cdot x^2 + 2^3 \cdot x^3 + 2^4 \cdot x^4 + \cdots$. Therefore $\sum_{n=3}^{\infty} (2x)^n$ *is* a power series, even though it begins with the index $n = 3$. Similarly,

$$\sum_{n=0}^{\infty} (x^2)^n = 1 + x^2 + x^4 + \cdots = 1 + 0 \cdot x + x^2 + 0 \cdot x^3 + 1 \cdot x^4 + \cdots$$

is a power series. ■

Notice that the power series $a_0 + a_1 x + a_2 x^2 + \cdots$ converges for $x = 0$ because it reduces to its initial term a_0 when $x = 0$. Thus, the set on which the series $\sum_{n=0}^{\infty} a_n x^n$ converges is not empty: It contains the point 0 at the least. We may therefore regard $\sum_{n=0}^{\infty} a_n x^n$ as a function of x. The domain of this function is the (nonempty) set of x for which the series converges.

Example 2 For what values of x does the power series $f(x) = \sum_{n=0}^{\infty} x^n$ converge?

Solution Observe that $f(x)$ is a geometric series with ratio x, so we know that it converges when $|x| < 1$ and diverges when $|x| \geq 1$. Thus, the function f is defined for $x \in (-1, 1)$. From our study of geometric series, we know that $f(x) = 1/(1 - x)$ for these values of x. ■

Radius and Interval of Convergence

The set of values at which a power series converges always has a special form. Suppose, for example, that the power series $\sum_{n=0}^{\infty} a_n x^n$ converges at some $x = t$ and that $|s| < |t|$. Then $\lim_{n\to\infty} |a_n t^n| = 0$, by the Divergence Test (Section 9.2). In particular, we have $|a_n t^n| \leq 1$ for sufficiently large n. For such n, we see that if $r = |s|/|t| < 1$ then $|a_n s^n| = r^n \cdot |a_n \cdot t^n| \leq r^n$. We conclude that $\sum_{n=0}^{\infty} |a_n s^n|$ converges by comparison with the convergent geometric series $\sum_{n=0}^{\infty} r^n$. In summary, we have proved that if a power series $\sum_{n=0}^{\infty} a_n x^n$ converges at some value of x, then it converges absolutely at all points that have a smaller absolute value. It follows that the set of points where the series converges is an *interval* centered at the origin. This interval can have infinite length or (finite) positive length or zero length. Refer to Figure 1. We state this as the following theorem.

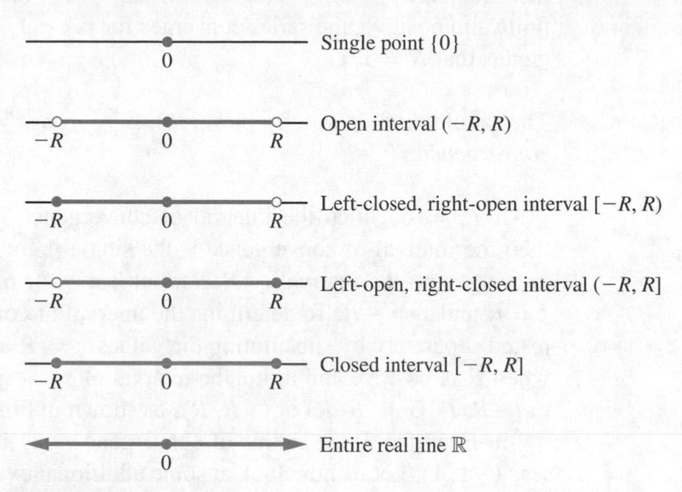

Figure 1
Intervals of convergence

Theorem 1 If we let $\sum_{n=0}^{\infty} a_n x^n$ be a power series, then precisely one of the following statements holds.

a. The series converges absolutely for every real x.
b. There is a positive number R such that the series converges absolutely for $|x| < R$ and diverges for $|x| > R$.
c. The series converges only at $x = 0$.

Whichever the case, there is a (possibly infinite) nonnegative number R such that the power series $\sum_{n=0}^{\infty} a_n x^n$ converges absolutely for $|x| < R$ and diverges for $|x| > R$. In case a of Theorem 1, $R = \infty$, and in case c, $R = 0$. In all cases, the number R is called the *radius of convergence* of the power series. We now learn how to compute this quantity.

Theorem 2 Suppose that the limit $\ell = \lim_{n\to\infty} |a_n|^{1/n}$ exists (as a nonnegative real number or ∞). Let R denote the radius of convergence of the power series $\sum_{n=0}^{\infty} a_n x^n$. If $0 < \ell < \infty$, then $R = 1/\ell$. If $\ell = 0$, then $R = \infty$. If $\ell = \infty$, then $R = 0$.

Proof We apply the Root Test (Section 9.6) to the power series for $x \neq 0$. Thus, we have convergence when $L = \lim_{n\to\infty} |a_n x^n|^{1/n} < 1$ and divergence when $L = \lim_{n\to\infty} |a_n x^n|^{1/n} > 1$. Notice that $L = |x| \cdot \lim_{n\to\infty} |a_n|^{1/n} = |x| \cdot \ell$. Thus, the series converges for

$$|x| \cdot \ell < 1 \qquad \textbf{(10.1)}$$

and diverges for

$$|x| \cdot \ell > 1. \tag{10.2}$$

If $\ell = 0$, then inequality (10.1) is satisfied for all x. Consequently, $R = \infty$. If $\ell = \infty$, then inequality (10.2) is satisfied for all $x \neq 0$; therefore, $R = 0$. Otherwise, if ℓ is finite and positive, the series converges for $|x| < 1/\ell$ and diverges for $|x| > 1/\ell$. This means that $R = 1/\ell$. ■

Definition The set of points at which a power series $\sum_{n=1}^{\infty} a_j x^n$ converges is called the *interval of convergence*.

If $R = \infty$, then the interval of convergence is the entire real line. If $R = 0$, then the interval of convergence is the single point $\{0\}$. When $0 < R < \infty$, the series converges on the interval $(-R, R)$, but it may or may not converge at the endpoints $x = R$ and $x = -R$. To determine the interval of convergence, each endpoint must be tested separately by substituting the values $x = R$ and $x = -R$ into the series. Thus, when R is positive and finite, the interval of convergence will have the form $[-R, R]$ or $(-R, R]$ or $[-R, R)$ or $(-R, R)$, as shown in Figure 1, page 665.

In Example 2, the radius of convergence was 1, and the interval of convergence was $(-1, 1)$. Let us now look at some additional examples.

Example 3 Find the interval of convergence for the power series $\sum_{n=0}^{\infty} x^n/3^n$.

Solution We have $a_n = 1/3^n$, and the radius of convergence R is given by

$$R = \left(\lim_{n\to\infty} |a_n|^{1/n}\right)^{-1} = \left(\lim_{n\to\infty} \left|\frac{1}{3^n}\right|^{1/n}\right)^{-1} = \left(\lim_{n\to\infty} \frac{1}{3}\right)^{-1} = 3.$$

This tells us that the series converges absolutely when $|x| < 3$ and diverges when $|x| > 3$. To completely determine the interval of convergence, we must investigate each endpoint. When $x = 3$, the series becomes $\sum_{n=0}^{\infty} 1$, which diverges. When $x = -3$, the series becomes $\sum_{n=0}^{\infty} (-1)^n$, which also diverges. The interval of convergence is therefore $(-3, 3)$. As a function of x, the domain of the power series $\sum_{n=0}^{\infty} x^n/3^n$ is the open interval $(-3, 3)$. ■

Example 4 Find the interval of convergence for the series $\sum_{n=1}^{\infty} x^n/n$.

Solution We set $a_n = 1/n$ and calculate that the radius of convergence R is given by

$$R = \left(\lim_{n\to\infty} |a_n|^{1/n}\right)^{-1} = \left(\lim_{n\to\infty} \left|\frac{1}{n}\right|^{1/n}\right)^{-1} = \left(\lim_{n\to\infty} \frac{1}{n^{1/n}}\right)^{-1} = \lim_{n\to\infty} n^{1/n} \overset{(9.10)}{=} 1.$$

The series converges absolutely for $|x| < 1$ and diverges when $|x| > 1$. When $x = 1$, the series becomes $\sum_{n=1}^{\infty} 1/n$; this is the harmonic series, which diverges. When $x = -1$, the series becomes $\sum_{n=1}^{\infty} (-1)^n/n$; this is the alternating harmonic series, which is convergent by the Alternating Series Test. We conclude that the interval of convergence is $[-1, 1)$. The power series $\sum_{n=1}^{\infty} x^n/n$ defines a function of x with domain $[-1, 1)$. ■

Example 5 Find the radius and interval of convergence for the power series $\sum_{n=1}^{\infty} x^n/(5^n \cdot n^2)$.

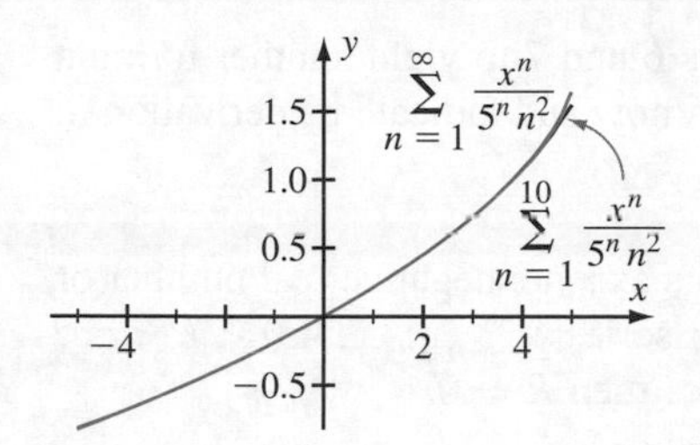

Figure 2

Solution We calculate the radius of convergence:

$$R \overset{(9.10)}{=} \left(\lim_{n\to\infty} \left| \frac{1}{5^n \cdot n^2} \right|^{1/n} \right)^{-1} = \lim_{n\to\infty} 5 \cdot (n^{1/n})^2 = 5.$$

Thus, the series converges absolutely when $|x| < 5$ and diverges when $|x| > 5$. When $x = 5$, the series becomes $\sum_{n=1}^{\infty} 1/n^2$, which is a convergent p-series. When $x = -5$, the series becomes $\sum_{n=1}^{\infty} (-1)^n/n^2$, which, by the Alternating Series Test, also converges. We conclude that the interval of convergence is $[-5, 5]$. The power series $\sum_{n=1}^{\infty} x^n/(5^n n^2)$ defines a function with domain $[-5, 5]$ (see Figure 2). The graph of an approximating partial sum, $\sum_{n=1}^{10} x^n/(5^n n^2)$, is also shown in Figure 2. Figure 3 suggests the divergence of the power series $\sum_{n=1}^{\infty} x^n/(5^n n^2)$ for $|x| > 5$. In Figure 3a, we plotted the partial sum $\sum_{n=1}^{50} x^n/(5^n n^2)$ over the interval $[-6, 6]$; and in Figure 3b, the partial sum $\sum_{n=1}^{100} x^n/(5^n n^2)$. Notice the scales on the vertical axes. ■

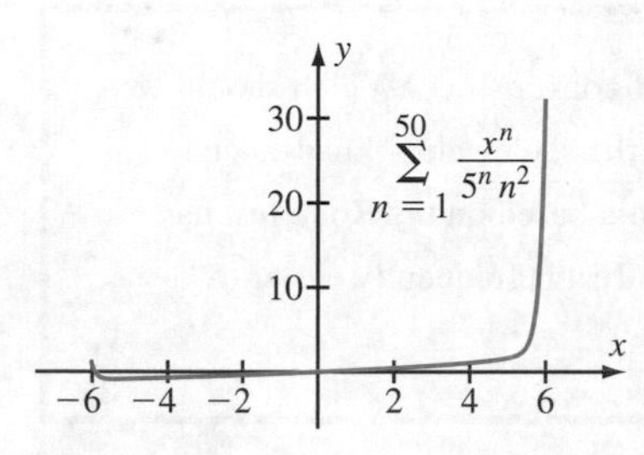

Figure 3a

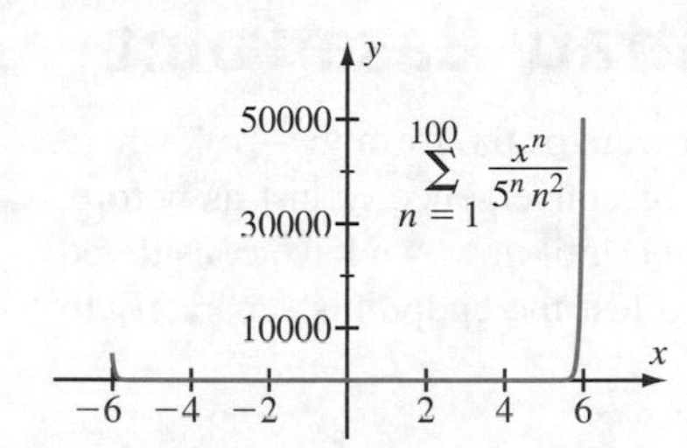

Figure 3b

Sometimes it is convenient to use the Ratio Test when calculating the radius of convergence.

Example 6 Find the interval of convergence for the power series $\sum_{n=0}^{\infty} x^n/n!$ by using the Ratio Test.

Solution Let $u_n = x^n/n!$. According to the Ratio Test (Section 9.6), the series $\sum_{n=1}^{\infty} u_n$ converges absolutely if $\lim_{n\to\infty} |u_{n+1}/u_n| < 1$ and diverges if $\lim_{n\to\infty} |u_{n+1}/u_n| > 1$. We have, for every x,

$$\lim_{n\to\infty} \left| \frac{u_{n+1}}{u_n} \right| = \lim_{n\to\infty} \left| \frac{x^{n+1}/(n+1)!}{x^n/n!} \right| = \lim_{n\to\infty} \left| \frac{n!}{(n+1)!} \cdot \frac{x^{n+1}}{x^n} \right|$$

$$= \lim_{n\to\infty} \left| \frac{n!}{(n+1) \cdot n!} \cdot x \right| = \lim_{n\to\infty} \left| \frac{x}{n+1} \right| = 0.$$

In particular, the limit is *always* less than 1. Hence, the Ratio Test tells us that the series converges absolutely for every real x. In other words, $R = \infty$ and the interval of convergence is $(-\infty, \infty)$. ■

We can simplify a ratio of factorials by peeling off factors from the larger factorial until both numerator and denominator involve the same factorial, which is then cancelled. Example 6 illustrates the technique, as does the next example.

Example 7 Find the radius and interval of convergence for the series $\sum_{n=1}^{\infty} (2n)!\, x^n/n!$.

Solution For $u_n = (2n)!\, x^n/n!$, we have

$$\lim_{n\to\infty} \left| \frac{u_{n+1}}{u_n} \right| = \lim_{n\to\infty} \left| \frac{((2n+2)!/(n+1)!)x^{n+1}}{((2n)!/n!)x^n} \right| = \lim_{n\to\infty} \left| \frac{(2n+2)! \cdot n!}{(2n)! \cdot (n+1)!} x \right|.$$

Since $(n+1)! = (n+1) \cdot n!$ and $(2n+2)! = (2n+2)(2n+1) \cdot (2n)!$, we have

$$\lim_{n\to\infty} \left| \frac{u_{n+1}}{u_n} \right| = \lim_{n\to\infty} \left| \frac{(2n+2)(2n+1) \cdot (2n)! \cdot n!}{(2n)! \cdot (n+1) \cdot n!} x \right|$$

$$= \lim_{n\to\infty} \left| \frac{(2n+2)(2n+1)}{(n+1)} x \right| = \lim_{n\to\infty} |2(2n+1)x|.$$

This limit is less than 1 only when $x = 0$. It exceeds 1 (in fact, the limit is infinite) for any nonzero value of x. Therefore, $R = 0$ and the interval of convergence is the set $\{0\}$ that contains only the origin. ■

We can generalize the calculations in Examples 6 and 7 to yield another formula for the radius of convergence. We state this formula now and indicate its derivation in Exercise 59.

Theorem 3 Suppose that the limit $\ell = \lim_{n\to\infty} |a_{n+1}|/|a_n|$ exists as a nonnegative real number or ∞. Let R be the radius of convergence of the power series $\sum_{n=0}^{\infty} a_n x^n$. If $0 < \ell < \infty$, then $R = 1/\ell$. If $\ell = 0$, then $R = \infty$, and if $\ell = \infty$, then $R = 0$.

Theorems 2 and 3 each give a formula for the radius of convergence. Which should we use? Naturally, when both limits exist, they must have the same value. In advanced calculus courses, it is shown that, of the two, the formula based on the Root Test has wider scope. However, the formula based on the Ratio Test is frequently easier to calculate. Therefore, the Ratio Test is often tried first.

Power Series About an Arbitrary Base Point

It is often useful to consider series that consist of powers of the form $(x - c)^n$ where c is *any* fixed constant. These series have a radius of convergence R just as before, but now *the interval of convergence is centered at* c. (Until now, we have considered only power series centered at zero.) We still have to test the endpoints separately to determine the exact interval of convergence.

Definition An expression of the form

$$S = a_0 + a_1(x - c) + a_2(x - c)^2 + a_3(x - c)^3 + \cdots$$

where c and the a_ns are constants is called a *power series in x with base point (or center) c*. It is convenient to denote the constant term a_0 by $a_0(x - c)^0$ so that we may use the sigma notation $\sum_{n=0}^{\infty} a_n(x - c)^n$ to denote S.

Notice that when the base point c has value 0, this definition of power series reduces to the one with which we already have been working. The next theorem expresses the content of Theorems 1, 2, and 3 for an arbitrary base point.

Theorem 4 Let $\sum_{n=0}^{\infty} a_n(x - c)^n$ be a power series for which the limit $\ell = \lim_{n\to\infty} |a_n|^{1/n}$ exists as a nonnegative real number or ∞. Set $R = 1/\ell$ if ℓ is a positive real number. If $\ell = 0$, then set $R = \infty$. If $\ell = \infty$, then set $R = 0$.

a. If the limit $\lim_{n\to\infty} |a_{n+1}|/|a_n|$ also exists as a nonnegative real number or ∞, then $\ell = \lim_{n\to\infty} |a_{n+1}|/|a_n|$.
b. The series converges absolutely for $|x - c| < R$ and diverges for $|x - c| > R$. In particular, if $R = \infty$, then the series converges absolutely for every real x. If $R = 0$, then the series converges only for $x = c$.

As in the case $c = 0$, we call R the *radius of convergence* of the power series. The set on which the series converges is called the *interval of convergence*. If $R = \infty$, then the interval of convergence is the entire real line. If $R = 0$, then the interval of convergence is the single point $\{c\}$. When $0 < R < \infty$, the series may or may not converge at the endpoints $x = c + R$ and $x = c - R$. To determine the interval of convergence, we must test each endpoint separately by substituting the values $x = c + R$ and $x = c - R$ into the series. Thus, when R is positive and finite, the interval of convergence will have the form $[c - R, c + R]$ or $(c - R, c + R]$ or $[c - R, c + R)$ or $(c - R, c + R)$. Figure 4 illustrates the possibilities.

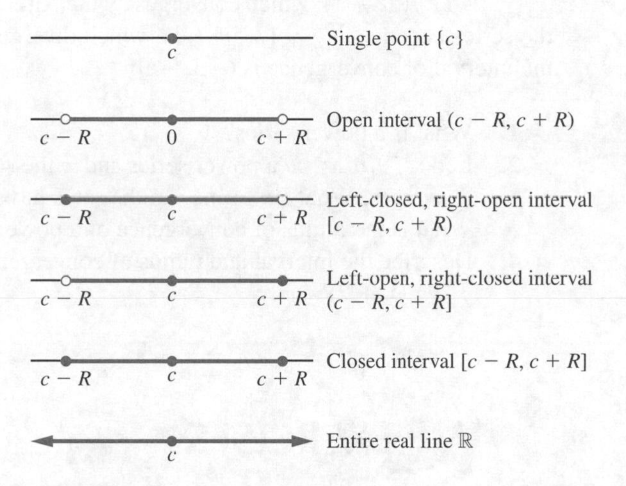

Figure 4
Intervals of convergence centered at c

Example 8 Calculate the interval of convergence for the power series $\sum_{n=0}^{\infty} (x - 2)^n$.

Solution Since $\sum_{n=0}^{\infty} (x-2)^n = \sum_{n=0}^{\infty} a_n(x-2)^n$ with $a_n = 1$ for all n, we calculate $\ell = \lim_{n\to\infty} |a_n|^{1/n} = \lim_{n\to\infty} 1^{1/n} = 1$. We set $R = 1/\ell = 1$. Theorem 4 tells us that the series converges absolutely for $|x - 2| < 1$ and diverges for $|x - 2| > 1$. The points $x = 2 + 1 = 3$ and $x = 2 - 1 = 1$ must be tested separately. When $x = 3$, the series becomes $\sum_{n=0}^{\infty} 1^n$, which diverges. When $x = 1$, the series becomes $\sum_{n=0}^{\infty} (-1)^n$, which also diverges. Thus, the interval of convergence is the set $(1, 3)$. In conclusion, the given power series $\sum_{n=0}^{\infty} (x - 2)^n$ defines a function with the interval $(1, 3)$ as its domain. ■

The importance of Theorem 4a is that it provides us with an alternative method of calculating ℓ, and thereby R, when the limit $\lim_{n\to\infty} |a_n|^{1/n}$ is difficult to evaluate.

Example 9 Determine the interval of convergence of the series $\sum_{n=0}^{\infty} (-1)^n(2x + 5)^n/(5n + 1)$.

Solution The first step is to write this power series in the standard form $\sum_{n=0}^{\infty} a_n(x - c)^n$:

$$\sum_{n=0}^{\infty} (-1)^n \frac{(2x+5)^n}{5n+1} = \sum_{n=0}^{\infty} \frac{(-1)^n 2^n}{5n+1}\left(x + \frac{5}{2}\right)^n = \sum_{n=0}^{\infty} \frac{(-2)^n}{5n+1}\left(x - \left(-\frac{5}{2}\right)\right)^n.$$

Now we can identify the base point $c = -5/2$ and the coefficients $a_n = (-2)^n/(5n+1)$. We calculate that

$$\ell = \lim_{n\to\infty} \frac{|a_{n+1}|}{|a_n|} = \lim_{n\to\infty} \frac{|(-2)^{n+1}/(5(n+1)+1)|}{|(-2)^n/(5n+1)|} = \lim_{n\to\infty} \frac{(5n+1)\cdot 2^{n+1}}{(5n+6)\cdot 2^n}$$
$$= 2\lim_{n\to\infty} \frac{5n+1}{5n+6} = 2\lim_{n\to\infty} \frac{5+1/n}{5+6/n} = 2.$$

Therefore, $R = 1/\ell = 1/2$. The endpoints of the interval of convergence are $c - R = -5/2 - 1/2 = -3$ and $c + R = -5/2 + 1/2 = -2$. When $x = -2$, the series becomes $\sum_{n=0}^{\infty} (-1)^n/(5n+1)$, which converges by the Alternating Series Test. When $x = -3$, the series becomes $\sum_{n=0}^{\infty} 1/(5n+1)$, which diverges by the Integral Test. Therefore, the interval of convergence is $(-3, -2]$. ■

quickquiz

1. What is a power series?
2. Let $\sum_{n=0}^{\infty} a_n x^n$ be a power series and $\mathcal{I}$ the set of x for which the power series converges. What six forms can the set $\mathcal{I}$ have?
3. What is the radius of convergence of a power series?
4. Describe the interval and radius of convergence of the power series $\sum_{n=1}^{\infty} (x+2)^n/n^2$.

EXERCISES

Problems for Practice

In Exercises 1–10, use Theorem 2 to calculate the radius of convergence. Determine the interval of convergence by checking the endpoints.

1. $\sum_{n=0}^{\infty} (2x)^n$
2. $\sum_{n=1}^{\infty} nx^n$
3. $\sum_{n=0}^{\infty} n^3x^n$
4. $\sum_{n=1}^{\infty} (4x)^n/n$
5. $\sum_{n=0}^{\infty} (-x/3)^n$
6. $\sum_{n=0}^{\infty} (2nx/(3n+1))^n$
7. $\sum_{n=1}^{\infty} n^nx^n$
8. $\sum_{n=1}^{\infty} (-x/n)^n$
9. $\sum_{n=1}^{\infty} (1+1/n)^nx^n$
10. $\sum_{n=0}^{\infty} n^{3n}x^n/(2n^3+3n^2+1)^n$

In Exercises 11–20, use Theorem 3 to calculate the radius of convergence. Determine the interval of convergence by checking the endpoints.

11. $\sum_{n=0}^{\infty} \exp(n)x^n$
12. $\sum_{n=0}^{\infty} (x/17)^n$
13. $\sum_{n=3}^{\infty} (3n+5)x^n/(n-2)$
14. $\sum_{n=0}^{\infty} n^2x^n/(n^3+1)$
15. $\sum_{n=1}^{\infty} n^3(x/3)^n$
16. $\sum_{n=0}^{\infty} (1+2^n)x^n/(1+3^n)$
17. $\sum_{n=0}^{\infty} (-1)^n nx^n/\sqrt{n+1}$
18. $\sum_{n=2}^{\infty} \ln(n)x^n$
19. $\sum_{n=0}^{\infty} (2x)^n/n!$
20. $\sum_{n=0}^{\infty} n!\,x^n/(2n)!$

In Exercises 21–28, use Theorem 4 to calculate the radius of convergence. Determine the interval of convergence by checking the endpoints.

21. $\sum_{n=0}^{\infty} (x+1)^n/(0.1)^n$
22. $\sum_{n=1}^{\infty} (\pi - x)^n/\sqrt{n}$
23. $\sum_{n=0}^{\infty} (x/\sqrt{2}-6)^n$
24. $\sum_{n=0}^{\infty} (x+4)^n/(n^5+1)$
25. $\sum_{n=0}^{\infty} (x-3)^n/(2n+3)$
26. $\sum_{n=1}^{\infty} (x+6)^n/\sqrt{n}$
27. $\sum_{n=0}^{\infty} (-1)^n n^2(x+5)^n/(3n+1)$
28. $\sum_{n=1}^{\infty} (x+1)^n/n^n$

Further Theory and Practice

We know that $\sum_{n=0}^{\infty} x^n = 1/(1-x)$, provided that $-1 < x < 1$. Use this fact, together with some algebra, to find a power series representation for each function in Exercises 29–36.

29. $1/(1+x)$
30. $1/(1-x^2)$
31. $1/(1-x^4)$
32. $x^3/(1-x)$
33. $x^2/(1+x^2)$
34. $x/(1+2x)$
35. $3/(3-x)$
36. $(2+x)/(1+x)$

In Exercises 37–42, find the radius of convergence of the power series.

37. $\sum_{n=0}^{\infty} n(n+1)(n+2)(2x+3)^n$
38. $\sum_{n=0}^{\infty} (1-1/2^n)(x+1)^n$
39. $\sum_{n=1}^{\infty} (n!)^2 x^n/(2n)!$
40. $\sum_{n=0}^{\infty} (n!)^2 (x+1)^n/(n+2)!$
41. $\sum_{n=1}^{\infty} n!(x-2)^n/(n^2)!$
42. $\sum_{n=0}^{\infty} 10^{2^n} x^n/(2^n)^{10}$

Each power series in Exercises 43–46 comprises only even powers $(x-c)^{2n}$. Make the substitution $t = (x-c)^2$, find the interval of convergence of the series in t, and use it to find the interval of convergence of the original series.

43. $\sum_{n=0}^{\infty} (-1)^n (x+1)^{2n}/9^n$
44. $\sum_{n=1}^{\infty} 4^n (x+3)^{2n}/n^2$
45. $\sum_{n=0}^{\infty} 4^n x^{2n}/n!$
46. $\sum_{n=1}^{\infty} n(3x)^{2n}/2^n$

Each power series in Exercises 47–50 comprises only odd powers $(x-c)^{2n+1}$. Factor out $(x-c)^1$ to produce a series consisting of even powers. Follow the instructions for Exercises 43–46 to find the interval of convergence of the given series.

47. $\sum_{n=1}^{\infty} (2x)^{2n+1}/(n+1)$
48. $\sum_{n=1}^{\infty} n(x-1)^{2n+1}/25^n$
49. $\sum_{n=0}^{\infty} (x+e)^{2n+1}/3^n$
50. $\sum_{n=0}^{\infty} 2^n (x+\sqrt{2})^{2n+1}$

Suppose that α and β are positive real numbers with $\alpha < \beta$. In Exercises 51–54, find a power series with an interval of convergence that is precisely the given interval.

51. (α, β)
52. $(\alpha, \beta]$
53. $[\alpha, \beta)$
54. $[\alpha, \beta]$

55. Is it possible for a power series to have interval of convergence $(-\infty, 1)$? Could it have interval of convergence $(0, \infty)$? Explain your answers.

56. What is the radius of convergence of $\sum_{n=1}^{\infty} n!\,(x/n)^n$?

57. Let $a_n = \sum_{k=1}^{n} 1/k^2$. What is the radius of convergence of $\sum_{n=1}^{\infty} a_n x^n$?

58. Let $b_n = \sum_{k=1}^{n} 1/\sqrt{k}$. What is the radius of convergence of $\sum_{n=1}^{\infty} b_n x^n$?

59. Suppose that $a_n \neq 0$ for $n \geq 0$. Show that if $\ell = \lim_{n\to\infty} |a_{n+1}|/|a_n|$ exists (as a nonnegative number or ∞), then $R = 1/\ell$ is the radius of convergence of $\sum_{n=0}^{\infty} a_n x^n$. *Hint:* Follow Examples 6 and 7 and apply the Ratio Test to the series $\sum_{n=0}^{\infty} u_n$ where $u_n = a_n x^n$.

60. Let $a_1 = 1$, $a_2 = 0$, $a_3 = 3^{-3}$, $a_4 = a_5 = 0$, $a_6 = 6^{-6}$, $a_7 = a_7 = a_9 = 0$, $a_{10} = 10^{-10}$, and in general, $a_N = N^{-N}$ if $N = n(n+1)/2$ for positive integers n and $a_N = 0$ otherwise. Calculate the radius of convergence of $\sum_{n=0}^{\infty} a_n x^n$.

61. For each positive integer k, let $a_{2k-1} = 2k-1$. For each positive integer k, let $a_{2k} = 2a_{2k-1}$. Calculate the radius of convergence of $\sum_{n=0}^{\infty} a_n x^n$.

Calculator/Computer Exercises

62. We know that $\sum_{n=0}^{\infty} (-1)^n x^n = 1/(1+x)$ for $-1 < x < 1$. In this exercise, we explore this equation by graphing partial sums of the series.

a. For $N = 2$, plot $S_N(x) = \sum_{n=0}^{N} (-1)^n x^n$ and $f(x) = 1/(1+x)$ for $-0.5 < x < 0.5$.
b. Repeat for $N = 3$, $N = 4$, and $N = 10$.
c. Repeat parts a and b, but plot over the interval $0.5 < x < 0.8$. Notice that the convergence is less rapid away from zero.
d. Repeat parts a and b, but plot over the interval $-0.8 < x < -0.5$. Notice that the convergence is less rapid away from zero.

63. We know that $\sum_{n=0}^{\infty} x^n = 1/(1-x)$ for $-1 < x < 1$. In this exercise, we explore this equation by graphing partial sums of the series.

a. For $N = 2$, plot $S_N(x) = \sum_{n=0}^{N} x^n$ and $f(x) = 1/(1-x)$ for $-0.5 < x < 0.5$.
b. Repeat for $N = 3$, $N = 4$, and $N = 10$.
c. Repeat parts a and b, but plot over the interval $0.5 < x < 0.8$. Notice that the convergence is less rapid away from zero.
d. Repeat parts a and b, but plot over the interval $-0.8 < x < -0.5$. Notice that the convergence is less rapid away from zero.

64. We know that $\sum_{n=0}^{\infty} nx^n = x/(x-1)^2$ for $-1 < x < 1$.

a. For $N = 3$, plot $S_N(x) = \sum_{n=0}^{N} nx^n$ and $f(x) = x/(x-1)^2$ for $-0.5 < x < 0.5$.
b. Repeat part a for $N = 4$ and $N = 8$.

c. Repeat parts a and b, but plot over the interval $0.5 < x < 0.8$. Notice that the convergence is less rapid away from zero.

d. Repeat parts a and b, but plot over the interval $-0.8 < x < -0.4$. Notice that the convergence is less rapid away from zero.

65. We know that $\sum_{n=0}^{\infty} (-1)^n t^n = 1/(1+t)$ for $-1 < t < 1$. Therefore,

$$\ln(1+x) = \int_0^x \frac{1}{1+t}\, dt = \int_0^x \sum_{n=0}^{\infty} (-1)^n t^n \, dt.$$

In this exercise, we investigate graphically the approximation that results when an infinite series is truncated and then integrated term by term.

a. For $N = 5$, plot $\ln(1+x)$ and $\int_0^x \sum_{n=0}^{N} (-1)^n t^n \, dt$ for $0 < x < 1$.

b. Repeat part a for $N = 10$.

10.2 Operations on Power Series

Each power series defines a function on its interval of convergence. We would like to have simple formulas that tell us how to add two such functions, how to differentiate a function defined by a power series, or how to integrate a function defined by a power series. In this section, we learn about such results.

Addition and Scalar Multiplication

The next theorem tells us how to add or subtract two power series *with the same base point* and how to multiply a power series by a constant. Each assertion of the theorem follows from a straightforward application of Theorem 3 from Section 9.5.

Theorem 1 Let λ be any real number and let k be a positive integer. Suppose that $\sum_{n=0}^{\infty} a_n(x-c)^n$ and $\sum_{n=0}^{\infty} b_n(x-c)^n$ are power series that converge absolutely on an interval I.

a. The power series $\sum_{n=0}^{\infty} (a_n + b_n)(x-c)^n$ and $\sum_{n=0}^{\infty} (a_n - b_n)(x-c)^n$ also converge absolutely on I; moreover,

$$\sum_{n=0}^{\infty} a_n(x-c)^n + \sum_{n=0}^{\infty} b_n(x-c)^n = \sum_{n=0}^{\infty} (a_n + b_n)(x-c)^n$$

and

$$\sum_{n=0}^{\infty} a_n(x-c)^n - \sum_{n=0}^{\infty} b_n(x-c)^n = \sum_{n=0}^{\infty} (a_n - b_n)(x-c)^n.$$

b. The power series $\sum_{n=0}^{\infty} \lambda a_n(x-c)^n$ converges absolutely on I and

$$\lambda \sum_{n=0}^{\infty} a_n(x-c)^n = \sum_{n=0}^{\infty} \lambda a_n(x-c)^n.$$

c. The power series $\sum_{n=0}^{\infty} a_n(x-c)^{n+k}$ converges absolutely on I and

$$\sum_{n=0}^{\infty} a_n(x-c)^{n+k} = (x-c)^k \sum_{n=0}^{\infty} a_n(x-c)^n.$$

Example 1 Discuss the convergence of the series $\sum_{n=0}^{\infty}(5-(1/3)^n)(x-2)^n$.

Solution The series $\sum_{n=0}^{\infty}(x-2)^n$ converges absolutely when $x \in (1, 3)$ (Example 8, Section 10.1). Therefore, so does the series $\sum_{n=0}^{\infty} 5(x-2)^n$ by Theorem 1b. The series $\sum_{n=0}^{\infty}(1/3)^n(x-2)^n$ converges absolutely when $x \in (-1, 5)$, because $R = 1/\lim_{n\to\infty}|(1/3)^n|^{1/n} = 1/\lim_{n\to\infty}(1/3) = 3$. By Theorem 1a, the difference $\sum_{n=0}^{\infty}(5-(1/3)^n)(x-2)^n$ of the two series converges absolutely on the intersection $(1, 3)$ of the two intervals. ■

Example 2 Express $x^3/(4+x^2)$ by means of a power series with base point 0.

Solution We begin with the identity

$$\frac{1}{4+x^2} = \frac{1}{4(1-(-x^2/4))} = \frac{1}{4}\frac{1}{(1-u)}$$

where $u = -x^2/4$. Since $1/(1-u) = \sum_{n=0}^{\infty} u^n$ for $|u| < 1$, we deduce that

$$\frac{1}{4+x^2} = \frac{1}{4}\sum_{n=0}^{\infty}\left(-\frac{x^2}{4}\right)^n \quad \text{for } \left|\frac{x^2}{4}\right| < 1.$$

Using Theorem 1b and 1c, we conclude that

$$\frac{x^3}{4+x^2} = \frac{x^3}{4}\sum_{n=0}^{\infty}\left(-\frac{x^2}{4}\right)^n = \sum_{n=0}^{\infty}(-1)^n\left(\frac{1}{4}\right)^{n+1} x^{2n+3} \quad \text{for } |x| < 2.$$

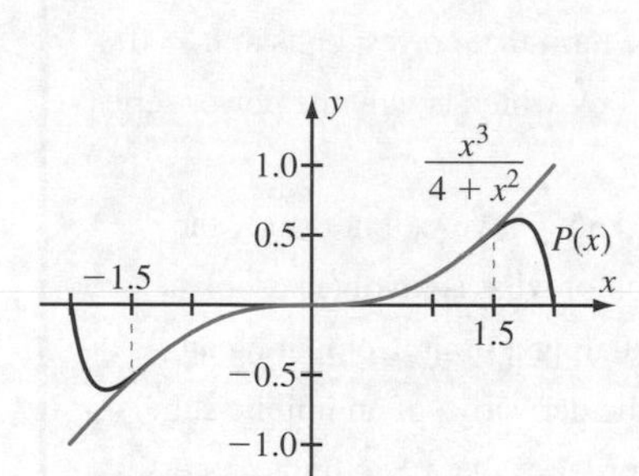

Figure 1

If we write out several terms, then this equation becomes

$$\frac{x^3}{4+x^2} = \underbrace{\frac{1}{4}x^3 - \frac{1}{16}x^5 + \frac{1}{64}x^7 - \frac{1}{256}x^9 + \frac{1}{1024}x^{11} - \frac{1}{4096}x^{13}}_{P(x)} + \cdots \quad (-2 < x < 2).$$

Figure 1 shows that the degree 13 polynomial $P(x)$ that we obtain by truncating the infinite series is quite a good approximation to $x^3/(4+x^2)$ for x not too far away from the center of the interval of convergence. In the figure, the plots of $x^3/(4+x^2)$ and $P(x)$ are barely distinguishable for $-1.5 < x < 1.5$. ■

Differentiation and Antidifferentiation of Power Series

Next we learn how to perform calculus operations on power series.

Theorem 2 Let I denote the interval $(c-R, c+R)$ where R is the radius of convergence of $f(x) = \sum_{n=0}^{\infty} a_n(x-c)^n$.

a. The function f is infinitely differentiable on I and

$$f'(x) = \sum_{n=1}^{\infty} n a_n (x-c)^{n-1}.$$

The power series for f' also converges absolutely on the interval I.

b. The power series

$$F(x) = \sum_{n=0}^{\infty} \frac{a_n}{n+1}(x-c)^{n+1}$$

converges absolutely on I. The function F satisfies the equation $F(x) = \int_c^x f(t)\,dt$ for all x in I. In particular, $F'(x) = f(x)$ on I. The indefinite integral of $f(x)$ is given by $\int f(x)\,dx = F(x) + C$ where C is an arbitrary constant.

inSIGHT

Theorem 2 says that a power series may be differentiated termwise and integrated termwise. Notice that in the power series for $f(x)$, the term that corresponds to $n = 0$ is the constant a_0. It vanishes when we differentiate $f(x)$, which is why the power series for $f'(x)$ begins with $n = 1$.

Exercise 48 outlines a proof of Theorem 2a. Even though we will not study the proof of Theorem 2, we should have some appreciation of what is involved. Look at Theorem 2a. Computing a derivative amounts to calculating a limit. Computing an infinite sum also involves taking a limit. Saying that the derivative of an infinite sum is the sum of the derivatives involves interchanging the order of these two limit processes. This is a subtle matter and is rather difficult to justify. Similar comments apply to Theorem 2b.

Example 3 Calculate the derivative and the indefinite integral of the power series $f(x) = \sum_{n=1}^{\infty} \sqrt{n} \cdot x^n$ for x in the interval of convergence $I = (-1, 1)$.

Solution By Theorem 2, $f'(x) = \sum_{n=1}^{\infty} n\sqrt{n}\, x^{n-1}$ for x in I. Although this formula correctly represents $f'(x)$ as a power series, it is often desirable to express the powers more simply. If we change the index of summation by setting $m = n - 1$, then as n ranges over the integers from 1 to infinity, m ranges over the integers from 0 to infinity and $f'(x) = \sum_{m=0}^{\infty} (m+1)\sqrt{m+1}\, x^m$.

Also, according to Theorem 2, the indefinite integral of $f(x)$ is

$$\int f(x)\,dx = \sum_{n=1}^{\infty} \left(\frac{\sqrt{n}}{n+1}\right) x^{n+1} + C.$$

To write this power series in standard form, we change the index of summation to $m = n + 1$. As n ranges over the integers from 1 to infinity, m ranges over the integers from 2 to infinity and

$$\int f(x)\,dx = C + \sum_{m=2}^{\infty} \left(\frac{\sqrt{m-1}}{m}\right) x^m.$$

■

inSIGHT

If we explicitly write out several terms of $f(x)$, $f'(x)$, and $\int f(x)\,dx$, then the results of our term-by-term differentiation and integration become more evident:

$$f(x) = x + \sqrt{2}x^2 + \sqrt{3}x^3 + \cdots,$$
$$f'(x) = 1 + 2\sqrt{2}x + 3\sqrt{3}x^2 + \cdots, \text{ and}$$
$$\int f(x)\,dx = C + \frac{1}{2}x^2 + \frac{\sqrt{2}}{3}x^3 + \frac{\sqrt{3}}{4}x^3 + \cdots.$$

Frequently, Theorem 2 can help us find an explicit formula for a power series, as the next example demonstrates.

Example 4 Suppose that $-1 < x < 1$. Sum the series $\sum_{n=1}^{\infty} nx^n = x + 2x^2 + 3x^3 + 4x^4 + \cdots$ explicitly.

Solution The key is to notice that $nx^n = x \cdot nx^{n-1} = x \cdot \frac{d}{dx}x^n$. Thus, using Theorem 2a to guarantee that we may differentiate a series term by term, we have

$$\sum_{n=1}^{\infty} nx^n = x \cdot \sum_{n=1}^{\infty} \frac{d}{dx}x^n = x \cdot \frac{d}{dx}\left(\sum_{n=1}^{\infty} x^n\right) \overset{(9.3)}{=} x \cdot \frac{d}{dx}\left(\frac{x}{1-x}\right) = x\left(\frac{(1-x) - x(-1)}{(1-x)^2}\right) = \frac{x}{(1-x)^2}.$$

(In applying equation (9.3), we have set the parameter M of the formula equal to 1.) Thus,

$$\frac{x}{(1-x)^2} = x + 2x^2 + 3x^3 + 4x^4 + \cdots \quad (-1 < x < 1).$$

■

Power Series Expansions of Some Standard Functions

In Example 3, we found an explicit form for the power series $\sum_{n=1}^{\infty} nx^n$. We cannot do the same for the (apparently) similar power series $\sum_{n=1}^{\infty} \sqrt{n}x^n$ from Example 2. A natural question to ask is, "If a power series defines a function, why can't we always say what function it is?" The answer is that most functions do not have names. Open the financial section of a newspaper and look at the graphs of the gross national product or unemployment. These are certainly functions, but they are not $\sin(x)$, $\cos(x)$, or any of our other old friends. Nevertheless, as several of the next examples show, some of our old friends do indeed arise as power series.

Example 5 Find a power series representation for the function $\ln(1 + x)$.

Solution Let $F(x) = \ln(1 + x)$. Theorem 2b tells us that if we know the power series of $f(x) = F'(x)$, then we can integrate this series term by term to find the power series of $F(x)$. Our starting point is therefore

$$F'(x) = \frac{d}{dx}\ln(1+x) = \frac{1}{1+x} = \frac{1}{1-(-x)} = \frac{1}{1-u},$$

where $u = -x$. Notice that x lies in the interval $(-1, 1)$ precisely when u does. For these values of u, the expression $1/(1-u)$ is the sum of the geometric series $\sum_{n=0}^{\infty} u^n$. Thus, for $-1 < x < 1$, we have

$$F'(x) = \sum_{n=0}^{\infty} u^n = \sum_{n=0}^{\infty} (-x)^n = \sum_{n=0}^{\infty} (-1)^n x^n.$$

Next, we integrate both sides of this equation. As mentioned, Theorem 2 guarantees that the right side may be integrated termwise. The result is

$$F(x) = \sum_{n=0}^{\infty} (-1)^n \frac{x^{n+1}}{n+1} + C \quad (-1 < x < 1)$$

where C is a constant of integration. By setting $x = 0$ in this equation, we obtain $F(0) = C$. Of course, it is also true that $F(0) = \ln(1+0) = 0$. We conclude that $C = 0$. Thus, we have derived the formula

$$\ln(1+x) = \sum_{n=0}^{\infty} (-1)^n \frac{x^{n+1}}{n+1} \quad (-1 < x < 1).$$

If we change the index of summation by setting $m = n + 1$, then our formula for $\ln(1+x)$ simplifies to

$$\ln(1+x) = \sum_{m=1}^{\infty} \frac{(-1)^{m-1}}{m} x^m \quad (-1 < x < 1). \tag{10.3}$$

■

inSIGHT

It is usually a good idea to write out the first several terms of a new series as a visual aid. For instance, equation (10.3) becomes

$$\ln(1+x) = x - \frac{1}{2}x^2 + \frac{1}{3}x^3 - \frac{1}{4}x^4 + \frac{1}{5}x^5 - \frac{1}{6}x^6 + \frac{1}{7}x^7 - \cdots \quad (-1 < x < 1).$$

Example 6 Use formula (10.3) to calculate $\ln(1.1)$ to within 0.0001.

Solution Formula (10.3) tells us that

$$\ln(1+0.1) = 0.1 - \frac{1}{2}(0.1)^2 + \frac{1}{3}(0.1)^3 - \frac{1}{4}(0.1)^4 + \cdots.$$

We can estimate this quantity by truncating the alternating infinite series. According to inequality (9.9) of the Alternating Series Test, the error that results is less than the first term that is omitted. Therefore,

$$\ln(1+0.1) = 0.1 - \frac{1}{2}(0.1)^2 + \frac{1}{3}(0.1)^3 + \epsilon$$

where ϵ is an error that is less than $(0.1)^4/4 = 0.000025$ in absolute value. The required approximation is

$$\ln(1.1) \approx 0.1 - \frac{1}{2}(0.1)^2 + \frac{1}{3}(0.1)^3 = 0.09533\ldots.$$

A check with a calculator will verify the indicated accuracy. ■

Example 7 Find a power series expansion for $\arctan(x)$, $-1 < x < 1$.

Solution Let $F(x) = \arctan(x)$. As in Example 3, we identify the power series of $F'(x)$ and use Theorem 2b to integrate this series term by term. Our starting point is therefore

$$F'(x) = \frac{d}{dx}\arctan(x) = \frac{1}{1+x^2} = \frac{1}{1-u}$$

where $u = -x^2$. Notice that $-1 < u \leq 0$ for $-1 < x < 1$. For these values of u, the geometric series $\sum_{n=0}^{\infty} u^n$ converges to $1/(1-u)$. Therefore,

$$F'(x) = \sum_{n=0}^{\infty} u^n = \sum_{n=0}^{\infty} (-x^2)^n = \sum_{n=0}^{\infty} (-1)^n x^{2n} \quad (-1 < x < 1).$$

Now we find antiderivatives of both sides. The result is

$$\arctan(x) = \sum_{n=0}^{\infty} \frac{(-1)^n x^{2n+1}}{2n+1} + C \quad (-1 < x < 1).$$

Setting $x = 0$ shows that $C = \arctan(0) = 0$. Therefore,

$$\arctan(x) = \sum_{n=0}^{\infty} \frac{(-1)^n x^{2n+1}}{2n+1} \quad (-1 < x < 1), \tag{10.4}$$

or, in a form that is more easily recognized and remembered,

$$\arctan(x) = x - \frac{1}{3}x^3 + \frac{1}{5}x^5 - \frac{1}{7}x^7 + \frac{1}{9}x^9 - \cdots \quad (-1 < x < 1).$$ ■

Look again at equations (10.3) and (10.4). Our theory guarantees that the power series representations for $\ln(1+x)$ and $\arctan(x)$ are valid for $-1 < x < 1$. Yet the Alternating Series Test tells us that the series in these formulas converge when $x = 1$ as well. Since each side of equations (10.3) and (10.4) makes sense for $x = 1$, it is plausible to think that these equations remain valid for $x = 1$. That is,

$$\ln(2) = 1 - \frac{1}{2} + \frac{1}{3} - \frac{1}{4} + \frac{1}{5} - \frac{1}{6} + \frac{1}{7} - \cdots \tag{10.5}$$

and, since $\arctan(1) = \pi/4$,

$$\frac{\pi}{4} = 1 - \frac{1}{3} + \frac{1}{5} - \frac{1}{7} + \frac{1}{9} - \cdots. \tag{10.6}$$

In fact, both of these formulas are true. Proving them requires the sort of delicate interchange of limits that has been previously discussed. For example, if we accept the equality indicated by the arrow in the middle of the next line, then we have

$$\ln(2) = \lim_{x\to 1^-} \ln(1+x) = \lim_{x\to 1^-} \sum_{m=1}^{\infty} \frac{(-1)^{m-1}}{m} x^m \overset{\downarrow}{=} \sum_{m=1}^{\infty} \left(\lim_{x\to 1^-} \frac{(-1)^{m-1}}{m} x^m \right) = \sum_{m=1}^{\infty} \frac{(-1)^{m-1}}{m} = 1 - \frac{1}{2} + \frac{1}{3} - \frac{1}{4} + \frac{1}{5} - \cdots.$$

Using advanced techniques, we can justify the highlighted step. The formulas that result are certainly interesting, but from a computational viewpoint, they are not very efficient. For instance, after summing the first 100 terms that appear on the right side of formula (10.6), we get

$$\frac{\pi}{4} \approx 1 - \frac{1}{3} + \frac{1}{5} - \frac{1}{7} + \frac{1}{9} - \frac{1}{11} + \cdots - \frac{1}{199} = 0.7828982\ldots.$$

This results in the approximation $\pi \approx 4 \times 0.7828982\ldots = 3.131\ldots$, which is correct to only one decimal place.

Example 8 Expand $1/x$ and $1/x^2$ in powers of $(x-1)$ for $|x-1|<1$.

Solution First notice that

$$\frac{1}{x} = \frac{1}{1-(1-x)} = \sum_{n=0}^{\infty}(1-x)^n = \sum_{n=0}^{\infty}(-1)^n(x-1)^n \quad \text{for } |x-1|<1.$$

We can differentiate both sides of the last equation by using Theorem 2a to differentiate the right side term by term. We obtain $-1/x^2 = \sum_{n=1}^{\infty} n(-1)^n(x-1)^{n-1}$, or

$$\frac{1}{x^2} = \sum_{n=1}^{\infty}(-1)^{n-1}n(x-1)^{n-1} \quad \text{for } |x-1|<1.$$

There is nothing wrong with leaving the answer in this form. Nevertheless, it is common to simplify the exponent. To that end, we set $m = n-1$. As n runs through the integers from 1 to infinity, the index m runs through the integers from 0 to infinity. Therefore,

$$\frac{1}{x^2} = \sum_{m=0}^{\infty}(-1)^m(m+1)(x-1)^m \quad \text{for } |x-1|<1.$$

If we write out several terms, then we obtain

$$\frac{1}{x^2} = 1 - 2(x-1) + 3(x-1)^2 - 4(x-1)^3 + \cdots \quad \text{for } |x-1|<1.$$

■

A Uniqueness Theorem with Applications to Differential Equations

We conclude this section with an important uniqueness theorem for power series.

Theorem 3 If $\sum_{n=0}^{\infty} a_n(x-c)^n = \sum_{n=0}^{\infty} b_n(x-c)^n$ for all x in an open interval containing the base point c, then $a_n = b_n$ for every n. In other words, if we fix a base point, then the power series expansion of a function is unique.

Proof Let $f(x) = \sum_{n=0}^{\infty} a_n(x-c)^n$ and $g(x) = \sum_{n=0}^{\infty} b_n(x-c)^n$. Substitute $x = c$ to obtain $a_0 = f(c) = g(c) = b_0$. Now differentiate the identity $f(x) = g(x)$ to obtain

$$\sum_{n=1}^{\infty} na_n(x-c)^{n-1} = f'(x) = g'(x) = \sum_{n=1}^{\infty} nb_n(x-c)^{n-1}.$$

Substitute $x = c$ to obtain

$$1 \cdot a_1 + 0 + 0 + \cdots = f'(c) = g'(c) = 1 \cdot b_1 + 0 + 0 + \cdots,$$

or $a_1 = f'(c)/1! = g'(c)/1! = b_1$. Differentiating again yields

$$\sum_{n=2}^{\infty} n(n-1)a_n(x-c)^{n-2} = f''(x) = g''(x) = \sum_{n=2}^{\infty} n(n-1)b_n(x-c)^{n-2}.$$

Setting $x = c$ results in

$$2 \cdot 1 \cdot a_2 + 0 + 0 + \cdots = f''(c) = g''(c) = 2 \cdot 1 \cdot b_2 + 0 + 0 + \cdots,$$

or $a_2 = f''(c)/2! = g''(c)/2! = b_2$. Continuing in this fashion, we find that $a_3 = f^{(3)}(c)/3! = g^{(3)}(c)/3! = b_3$, and so on. In other words, $a_n = f^{(n)}(c)/n! = g^{(n)}(c)/n! = b_n$ for every nonnegative integer n. ■

Theorem 3 tells us that if we use one base point to obtain power series expansions of a function in two different fashions, then the same series will result. The next example shows how to use this theorem to solve a differential equation.

Example 9 Find a power series solution of the initial value problem $\frac{dy}{dx} = y - x$, $y(0) = 2$.

Solution Suppose that $y(x) = \sum_{n=0}^{\infty} a_n x^n$. Using Theorem 2, we obtain $\frac{dy}{dx} = \sum_{n=1}^{\infty} n a_n x^{n-1}$. For y to satisfy the given differential equation, we must have

$$\underbrace{\sum_{n=1}^{\infty} n a_n x^{n-1}}_{y'} = \underbrace{\sum_{n=0}^{\infty} a_n x^n}_{y} - x.$$

It is convenient to make a change of summation index for the series on the left side: $m = n - 1$. Since n runs from 1 to infinity, we see that m runs from 0 to infinity. We get

$$\sum_{m=0}^{\infty} (m+1)a_{m+1}x^m = \sum_{n=0}^{\infty} a_n x^n - x,$$

or, by renaming the index summation of the series on the left,

$$\sum_{n=0}^{\infty} (n+1)a_{n+1}x^n = \sum_{n=0}^{\infty} a_n x^n - x.$$

We may write this equation as

$$a_1 + 2a_2 x + \sum_{n=2}^{\infty} (n+1)a_{n+1}x^n = a_0 + (a_1 - 1)x + \sum_{n=2}^{\infty} a_n x^n$$

and use the Uniqueness Theorem to deduce that $a_1 = a_0$, $2a_2 = a_1 - 1$, and $(n+1)a_{n+1} = a_n$ for $n \geq 2$. The initial condition $y(0) = 2$ tells us that $a_0 = 2$. It follows that $a_1 = 2$ and $a_2 = (a_1 - 1)/2 = (2-1)/2 = 1/2$. Substituting $n = 2$ in the equation $(n+1)a_{n+1} = a_n$, we obtain $a_3 = a_2/3 = 1/3!$. Substituting $n = 3$ in the equation $(n+1)a_{n+1} = a_n$, we obtain $a_4 = a_3/4 = 1/4!$. Continuing in this way, we obtain $a_n = 1/n!$ for $n \geq 2$. It follows that on its interval of convergence,

$$y(x) = 2 + 2x + \sum_{n=2}^{\infty} \frac{x^n}{n!}$$

is a solution of the given differential equation. Example 6 from Section 10.1 shows that this power series converges on the entire real line. In Section 10.5, we will learn what its limit is, thereby finding a simple expression for $y(x)$. ■

The Relationship between the Coefficients and Derivatives of a Power Series

The next theorem, which is a consequence of the proof of Theorem 3, tells us the relationship between the coefficients of a power series and the function that the series represents.

Theorem 4 If $f(x) = \sum_{n=0}^{\infty} a_n(x-c)^n$ converges on an open interval centered at c, then

$$a_n = \frac{f^{(n)}(c)}{n!}, \qquad n = 0, 1, 2, 3, \ldots.$$

(As usual, $f^{(0)}(c)$ is understood to be $f(c)$, and, for $n > 1$, $f^{(n)}(c)$ is the nth derivative of f at c.)

Example 10 Let $a_n = n \cdot 2^n$ for $0 \le n \le 3$ and $a_n = 0$ for $n > 3$. Verify Theorem 4 for the series $f(x) = \sum_{n=0}^{\infty} a_n(x+1)^n$.

Solution Since $(x+1)^n = (x-(-1))^n$, we use Theorem 4 with $c = -1$. Our first task is to sum the series and find $f(x)$:

$$f(x) = \sum_{n=0}^{\infty} a_n(x+1)^n = \sum_{n=0}^{3} n2^n(x+1)^n = 2(x+1) + 8(x+1)^2 + 24(x+1)^3,$$

or, after some algebraic simplification, $f(x) = 24x^3 + 80x^2 + 90x + 34$. The following table includes all the information we need for the verification of Theorem 4.

n	$f^{(n)}(x)$	$f^{(n)}(-1)$	$f^{(n)}(-1)/n!$	$a_n = n \cdot 2^n$
0	$24x^3 + 80x^2 + 90x + 34$	0	0	$0 \cdot 2^0$
1	$72x^2 + 160x + 90$	2	2	$1 \cdot 2^1$
2	$144x + 160$	16	8	$2 \cdot 2^2$
3	144	144	24	$3 \cdot 2^3$
$4, 5, \ldots$	0	0	0	0

Observe that the last two columns are identical, which verifies Theorem 4. ■

Example 11 The *error function* $x \mapsto \operatorname{erf}(x)$ is defined by

$$\operatorname{erf}(x) = \frac{2}{\sqrt{\pi}} \int_0^x e^{-t^2}\, dt \quad (-\infty < x < \infty).$$

As its name suggests, the error function arises in the analysis of observational errors. It also plays an important role in the theory of refraction and the theory of heat conduction. It is known that the error function has a convergent power series representation $\sum_{n=0}^{\infty} a_n x^n$. Calculate the partial sum $\sum_{n=0}^{3} a_n x^n$.

Solution Let $f(x) = \text{erf}(x)$. By the Fundamental Theorem of Calculus, we have $f'(x) = (2/\sqrt{\pi})e^{-x^2}$. We calculate $f^{(2)}(x) = -(4/\sqrt{\pi})xe^{-x^2}$ and $f^{(3)}(x) = -(4/\sqrt{\pi})e^{-x^2} + (8/\sqrt{\pi})x^2 e^{-x^2}$. It follows that $f(0) = 0$, $f^{(1)}(0) = 2/\sqrt{\pi}$, $f^{(2)}(0) = 0$, and $f^{(3)}(0) = -4/\sqrt{\pi}$. According to Theorem 4, the power series representation $\sum_{n=0}^{\infty} a_n x^n$ of $f(x)$ has $a_n = f^{(n)}(0)/n!$ for each n. Therefore,

$$\sum_{n=0}^{3} \frac{f^{(n)}(0)}{n!}(x-0)^n = 0 + \frac{(2/\sqrt{\pi})}{1!}x + \frac{0}{2!}x^2 + \frac{(-4/\sqrt{\pi})}{3!}x^3 = \frac{2}{\sqrt{\pi}}x - \frac{2}{3\sqrt{\pi}}x^3$$

is the required partial sum. Figure 2 reveals that even though this partial sum has only two nonzero terms, it is still a very good approximation to the error function for values of x that are close to the base point 0. ■

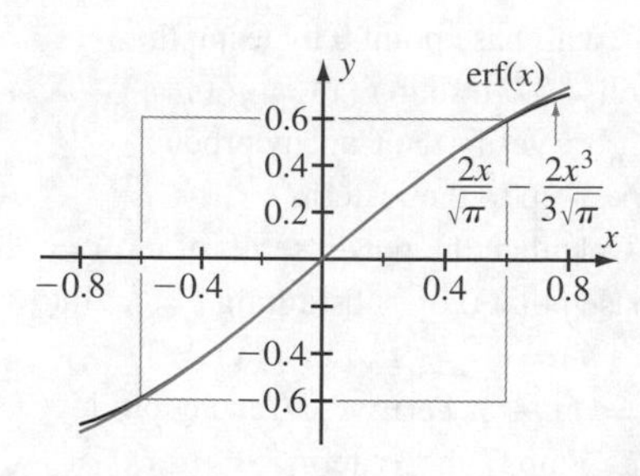

Figure 2

Given a differentiable function f, Theorem 4 tells us that the power series $\sum_{n=0}^{\infty} ((f^{(n)}(c))/n!)(x-c)^n$ is the only power series with base point c that could converge to $f(x)$ for all x in an interval about c. Whether it does or not is the subject of the next three sections.

A LOOK FORWARD

quickquiz

1. What arithmetic operations are respected by power series?
2. When are we allowed to differentiate a power series term by term?
3. How do we integrate a power series?
4. State the uniqueness theorem for power series.

EXERCISES

Problems for Practice

In Exercises 1–10, express the given function as a power series in x with base point 0. Calculate the radius of convergence.

1. $1/(1-2x)$ **2.** $x/(1+x)$
3. $1/(2-x)$ **4.** $1/(1+x^2)$
5. $1/(4-x)$ **6.** $1/(9-x^2)$
7. $(x^2+1)/(1-x^2)$ **8.** $x/(1+x^2)$
9. $x^3/(1+x^4)$ **10.** $1/(16-x^4)$

In Exercises 11–16, use equation (10.3), together with the operations on power series that you learned in this section, to express the given function as a power series in x with base point 0. Calculate the radius of convergence.

11. $\ln(1+4x)$ **12.** $\ln(5+2x)$
13. $\ln(1+x^2)$ **14.** $x\ln(1+x^2)$
15. $\int_0^x \ln(1+t^2)\,dt$ **16.** $\int_0^x t\ln(1+t)\,dt$

In Exercises 17–20, use equation (10.4), together with the operations on power series that you learned in this section, to express the given function as a power series in x with base point 0. Calculate the radius of convergence.

17. $\arctan(x^2)$ **18.** $\int_0^x \arctan(t)\,dt$
19. $\int_0^x \arctan(t^2)\,dt$ **20.** $\int_0^x t\arctan(t)\,dt$

In Exercises 21–26, use the equation

$$\frac{1}{1-(t-c)} = \sum_{j=0}^{\infty}(t-c)^j, \ |t-c| < 1,$$

together with some algebra, to express the given function as a power series in $(x - c)$. In each exercise, find the radius of convergence.

21. $1/(x+1)$, $c = -3$
22. $1/(x-4)$, $c = 1$
23. $1/(2x+5)$, $c = -1$
24. $1/x$, $c = 2$
25. $\ln(x)$, $c = 1$
26. $\ln(2x+5)$, $c = -1/2$

Further Theory and Practice

By noticing that

$$\frac{d}{dx}\left(\frac{1}{1-x}\right) = \frac{1}{(1-x)^2},$$

find power series for each function in Exercises 27–30.

27. $1/(1-x)^2$
28. $1/(1-2x)^2$
29. $(2+x)/(1-x)^2$
30. $1/(1-x)^3$

In Exercises 31–36, find the sum of the series in closed form. Also, calculate the radius of convergence.

31. $\sum_{n=0}^{\infty} x^{n+2}$
32. $\sum_{n=1}^{\infty} nx^{n+1}$
33. $\sum_{n=1}^{\infty} n(n+1)x^n$
34. $\sum_{n=2}^{\infty} n(n-1)x^{n-2}$
35. $\sum_{n=1}^{\infty} nx^{2n+1}$
36. $\sum_{n=0}^{\infty} (x+2)^{n+3}$

37. Find all solutions of the differential equation

$$\frac{d^5}{dx^5} f(x) = 0$$

by guessing that $f(x) = \sum_{n=0}^{\infty} a_n x^n$, substituting this expression into the equation, and solving for the coefficients. Explain how the Uniqueness Theorem is used.

In Exercises 38–43, use the Uniqueness Theorem to determine the coefficients $\{a_n\}$ of the solution $y(x) = \sum_{n=0}^{\infty} a_n x^n$ of the initial value problem.

38. $\dfrac{dy}{dx} = y$, $y(0) = 1$
39. $\dfrac{dy}{dx} = 2y$, $y(0) = 3$
40. $\dfrac{dy}{dx} = x + y$, $y(0) = 1$
41. $\dfrac{dy}{dx} = 2x - y$, $y(0) = 1$
42. $\dfrac{dy}{dx} = 1 + x + y$, $y(0) = 0$
43. $\dfrac{dy}{dx} = x + xy$, $y(0) = 0$

If $f(x) = \sum_{n=0}^{\infty} a_n x^n$ and $g(x) = \sum_{n=0}^{\infty} b_n x^n$ converge on $(-R, R)$, then it is possible to formally multiply the series as though they were polynomials. That is, if $h(x) = f(x)g(x)$, then

$$h(x) = \sum_{n=0}^{\infty}\left(\sum_{k=0}^{n} a_k b_{n-k}\right) x^n.$$

The product series, which is called the *Cauchy product,* also converges on $(-R, R)$. Exercises 44–47 concern the Cauchy product.

44. Suppose $|x| < 1$. Calculate the power series of $h(x) = 1/(1-x)^2$ with base point 0 by using the method from Exercise 27. Using $f(x) = g(x) = 1/(1-x) = \sum_{n=0}^{\infty} x^n$, verify the Cauchy product formula for $h = f \cdot g$ up to the x^6 term.

45. Suppose $|x| < 1$. Calculate the power series of $h(x) = 1/(1-x^2)$ with base point 0 by substituting $t = x^2$ into the equation $1/(1-t) = \sum_{n=0}^{\infty} t^n$. Let $f(x) = 1/(1-x)$ and $g(x) = f(-x)$. Verify the Cauchy product formula for $h = f \cdot g$ up to the x^8 term.

46. The secant function has a known power series expansion that begins

$$\sec(x) = 1 + \frac{1}{2}x^2 + \frac{5}{24}x^4 + \frac{61}{720}x^6 \cdots.$$

The sine function has a known power series expansion that begins

$$\sin(x) = x - \frac{x^3}{3!} + \frac{x^5}{5!} - \frac{x^7}{7!} \cdots.$$

The tangent function has a known power series expansion that begins

$$\tan(x) = x + \frac{1}{3}x^3 + \frac{2}{15}x^5 + \frac{17}{315}x^7 + \cdots.$$

Verify the Cauchy product formula for $\tan(x) = \sin(x)\sec(x)$ up to the x^7 term.

47. If we suppose that the series $\sum_{n=0}^{\infty} a_n x^n$ converges to a function $f(x)$ on $(-R, R)$ and that $|f(x)| \geq c > 0$ for some positive constant c, then $1/f(x)$ has a power series expansion $\sum_{n=0}^{\infty} b_n x^n$ that converges on $(-R, R)$. Use the equation $f(x) \cdot g(x) = 1$ to solve for the b_ns in terms of the a_ns.

48. Complete the details of the following outline to see why Theorem 2a is true.

a. If $f(x) = \sum_{n=1}^{\infty} a_n (x-c)^n$ converges for $|x - c| < R$, then so does $g(x) = \sum_{n=1}^{\infty} na_n (x-c)^{n-1}$. Use the Root Test to prove this assertion.

b. Fix an $x \in (c - R, c + R)$. Apply the Mean Value Theorem to the nth summand on the right side of

$$\left|\frac{f(x+h) - f(x)}{h} - g(x)\right|$$
$$= \left|\sum_{n=1}^{\infty} a_n \cdot \frac{(x+h-c)^n - (x-c)^n}{h} - g(x)\right|$$

to see that the summand equals $a_n \cdot n \cdot (t_n - c)^{n-1}$ for some t_n between x and $x + h$.

c. We now have

$$\left|\frac{f(x+h)-f(x)}{h} - g(x)\right| = \left|\sum_{n=1}^{\infty} n \cdot a_n \cdot ((t_n - c)^{n-1} - (x - c)^{n-1})\right|.$$

Apply the Mean Value Theorem to each summand on the right to obtain

$$\left|\frac{f(x+h)-f(x)}{h} - g(x)\right| = \left|\sum_{n=2}^{\infty} na_n(t_n - x)(n-1)(s_n - c)^{n-2}\right|$$

for some s_n between x and t_n.

d. Estimate

$$\left|\frac{f(x+h)-f(x)}{h} - g(x)\right| \le |h| \cdot \sum_{n=2}^{\infty} |n(n-1)| \cdot |a_n| \cdot |h|^{n-2}.$$

e. If $|h| < R/2$, then there is a constant C_0, independent of h, such that

$$\sum_{n=2}^{\infty} |n \cdot (n-1)| \cdot |a_n| \cdot |h|^{n-2} \le C_0.$$

Hence,

$$\left|\frac{f(x+h)-f(x)}{h} - g(x)\right| \le C_0|h|.$$

f. Conclude that

$$\lim_{h\to 0} \frac{f(x+h)-f(x)}{h}$$

exists and equals $g(x)$.

49. Suppose that P is a degree N polynomial. Let c be any real number. Show that

$$P(x) = \sum_{n=0}^{N} \frac{P^{(n)}(c)}{n!}(x-c)^n.$$

Hint: First show that $P(x) = \sum_{n=0}^{N} b_n(x-c)^n$ for some set of coefficients $\{b_n\}$.

Calculator/Computer Exercises

50. Let $f(x) = (x+1)^4/(x^4+1)$. It is known that f has a power series expansion of the form

$$f(x) = 1 + 4x + 6x^2 + 4x^3 - 4x^5 - 6x^6 - 4x^7 + 4x^9 + \cdots.$$

a. Plot the central difference quotient approximation $D_0 f(x, 10^{-5})$ of $f'(x)$ for $-0.5 < x < 0.5$.

b. Use the given power series to find a degree eight polynomial approximation of $f'(x)$. Add the plot of this polynomial to the viewing rectangle of part a.

c. Repeat parts a and b, but plot for $-3/4 < x < 3/4$. Notice that the approximation is less accurate away from zero, the base point of the given series.

51. Let $f(x) = \sqrt{5 + x^2}$. It is known that f has a power series expansion of the form

$$f(x) = 3 + \frac{2}{3}(x-2) + \frac{5}{54}(x-2)^2 - \frac{5}{243}(x-2)^3 + \frac{55}{17496}(x-2)^4 + \cdots.$$

a. Plot the central difference quotient approximation $D_0 f(x, 10^{-5})$ of $f'(x)$ for $0 < x < 4$.

b. Use the given power series to find a degree three polynomial approximation of $f'(x)$. Add the plot of this polynomial to the viewing rectangle of part a. Notice that the approximation is less accurate away from two, the base point of the given series.

52. The unique solution to the initial value problem

$$\frac{dy}{dx} = x^2 + y, \quad y(0) = 1$$

is $y(x) = 3e^x - x^2 - 2x - 2$.

a. Calculate the power series expansion $y(x) = \sum_{n=0}^{\infty} a_n x^n$ of the solution up to the x^5 term.

b. Plot the exact solution and $S_3(x) = \sum_{n=0}^{3} a_n x^n$ in the same viewing window with $-1 < x < 1$.

c. Repeat part b in the viewing window with $-3 < x < 3$.

d. To judge the accuracy of the approximation, plot the exact solution and $S_5(x) = \sum_{n=0}^{5} a_n x^n$ in the same viewing window with $-3 < x < 3$.

53. The unique solution to the initial value problem

$$\frac{dy}{dx} = 2 - x - y, \quad y(0) = 1$$

is $y(x) = 3 - x - 2e^{-x}$.

a. Calculate the power series expansion $y(x) = \sum_{n=0}^{\infty} a_n x^n$ of the solution up to the x^5 term.

b. Plot the exact solution and $S_3(x) = \sum_{n=0}^{3} a_n x^n$ in the same viewing window with $-1 < x < 1$.

c. Repeat part b in the viewing window with $-2 < x < 2$.

d. To judge the accuracy of the approximation, plot the exact solution and $S_5(x) = \sum_{n=0}^{5} a_n x^n$ in the same viewing window with $-2 < x < 2$.

10.3 Taylor Polynomials

In Section 10.2, we saw how some functions, such as $\ln(1 + x)$ and $\arctan(x)$, can be expanded in power series. As we noted in Example 6 of that section, a power series expansion can be a useful computational tool. Unfortunately, the method used in Section 10.2 for obtaining power series involved some luck: We differentiated or integrated a known power series and performed some algebraic manipulations to obtain a limited number of new power series expansions. We would like a systematic technique for associating to any reasonable function a power series expansion, which is the subject of this section and the next two. The key to these matters is Taylor's expansion. Our first step toward understanding the Taylor series expansion is to develop a family of approximating polynomials.

Generalizing the Tangent Line Approximation

When we first studied the derivative in Chapter 3, we learned how to use it to approximate a function. Specifically, if f is differentiable at c, then the graph of

$$P_1(x) = f(c) + f'(c) \cdot (x - c)$$

is the tangent line of the graph of f at $(c, f(c))$. Refer to Figure 1. Notice that P_1 is a degree one polynomial such that

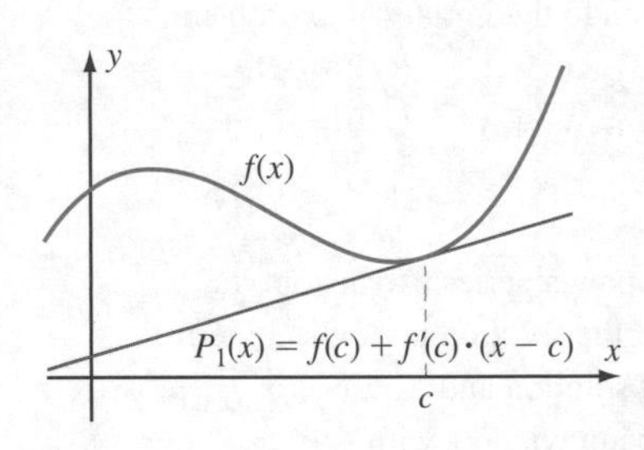

Figure 1

$$P_1(c) = f(c) \quad \text{and} \quad P_1'(c) = f'(c). \tag{10.7}$$

To better approximate the graph of f near $(c, f(c))$, we must use a curve that has the same concavity as the graph of f. As we know from Chapter 4, it is the second derivative f'' that captures concavity information. With that in mind, let us find a degree two polynomial P_2 with derivatives at c that agree with those of f up to order 2:

$$P_2(c) = f(c), \quad P_2'(c) = f'(c), \quad \text{and} \quad P_2''(c) = f''(c). \tag{10.8}$$

We may write $P_2(x) = f(c) + f'(c) \cdot (x - c) + \alpha \cdot (x - c)^2$ and determine the required value of α. Notice that $P_2(c) = f(c)$, $P_2'(c) = f'(c) \cdot 1 + 2\alpha \cdot (x - c)|_{x=c} = f'(c)$, and $P_2''(c) = 2\alpha$. To ensure that the three equations of line (10.8) are satisfied, we need only set $\alpha = f''(c)/2$. Thus,

$$P_2(x) = f(c) + f'(c) \cdot (x - c) + \frac{f''(c)}{2} \cdot (x - c)^2 \tag{10.9}$$

is the required degree two polynomial.

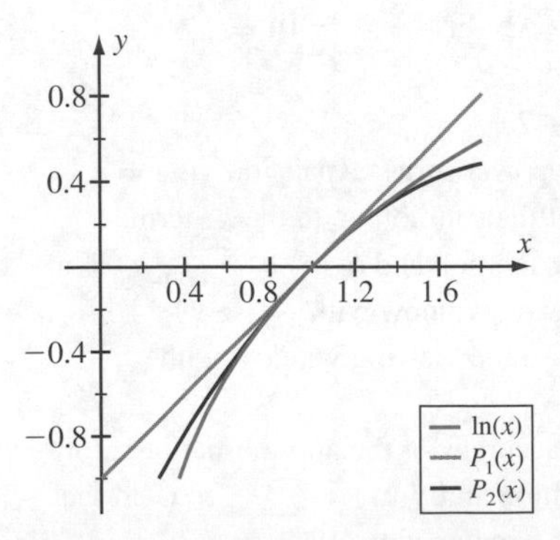

Figure 2

Example 1 Let $f(x) = \ln(x)$. Find a degree two polynomial P_2 such that $P_2(1) = f(1)$, $P_2'(1) = f'(1)$, and $P_2''(1) = f''(1)$.

Solution We calculate $f'(x) = 1/x$ and $f''(x) = -1/x^2$, and we note that $f(1) = 0$, $f'(1) = 1$, and $f''(1) = -1$. According to formula (10.9) with $c = 1$, the required polynomial is $P_2(x) = 0 + 1 \cdot (x - 1) + (-1/2) \cdot (x - 1)^2$, or $P_2(x) = (x - 1) - (x - 1)^2/2$. The plots of f, its tangent line at $(1, 0)$, and P_2 are shown in Figure 2. ■

inSIGHT

Recall from equation (10.3) that $\ln(1+x) = x - x^2/2 + x^3/3 - \cdots$. Observe that if we replace x with $x-1$ in this identity, then we obtain $\ln(x) = (x-1) - (x-1)^2/2 + (x-1)^3/3 - \cdots$. It is no accident that the polynomial P_2 that we found in Example 1 agrees with the first two terms of the power series expansion of $\ln(x)$ with base point 1. The purpose of this section is to develop this relationship between approximating polynomials and power series expansions.

Higher-Order Approximating Polynomials

We now suppose that f is N times continuously differentiable (as defined in Section 3.7). The next theorem tells us that we can explicitly construct a degree N polynomial P_N that matches the derivatives of f at c up to order N. The formula for P_N can be more conveniently expressed by using $f^{(0)}$ to denote f. In other words, the zeroth derivative of f is just f itself.

Theorem 1 Let N be a nonnegative integer. If f is an N times continuously differentiable function, then the polynomial

$$P_N(x) = \frac{f^{(0)}(c)}{0!}\cdot(x-c)^0 + \frac{f^{(1)}(c)}{1!}\cdot(x-c)^1 + \frac{f^{(2)}(c)}{2!}\cdot(x-c)^2 + \frac{f^{(3)}(c)}{3!}\cdot(x-c)^3 + \cdots + \frac{f^{(N)}(c)}{N!}\cdot(x-c)^N,$$

which can be expressed more compactly as

$$P_N(x) = \sum_{n=0}^{N} \frac{f^{(n)}(c)}{n!}(x-c)^n, \tag{10.10}$$

is the unique degree N polynomial such that

$$P_N(c) = f(c), \quad P_N'(c) = f'(c), \quad P_N''(c) = f''(c), \quad P_N^{(3)}(c) = f^{(3)}(c), \quad \ldots \quad P_N^{(N)}(c) = f^{(N)}(c).$$

Proof Let us fix an integer k between 0 and N. Observe that

$$\frac{d^k}{dx^k}(x-c)^n = \begin{cases} 0 & \text{if } n < k \\ k! & \text{if } n = k \\ n(n-1)\cdots(n-k+1)\cdot(x-c)^{n-k} & \text{if } n > k \end{cases}.$$

Notice that for $n > k$,

$$\left.\frac{d^k}{dx^k}(x-c)^n\right|_{x=c} = n\cdot(n-1)\cdots(n-k+1)\cdot(c-c)^{n-k} = 0.$$

By dividing P_N into three parts,

$$P_N(x) = \underbrace{\sum_{n=0}^{k-1}\frac{f^{(n)}(c)}{n!}(x-c)^n}_{n<k} + \underbrace{\frac{f^{(k)}(c)}{k!}(x-c)^k}_{n=k} + \underbrace{\sum_{n=k+1}^{N}\frac{f^{(n)}(c)}{n!}(x-c)^n}_{n>k}, \tag{10.11}$$

we see that

$$\left.\frac{d^k}{dx^k} P_N(x)\right|_{x=c} = 0 + \frac{f^{(k)}(c)}{k!} \cdot k! + 0 = f^{(k)}(c).$$

Thus, $P_N^{(k)}(c) = f^{(k)}(c)$. The assertion about uniqueness is left as Exercise 43. ■

Definition We call the polynomial P_N of Theorem 1 the *Taylor polynomial of degree N and base point c for the function f.*

inSIGHT

Because $0! = 1$, $f^{(0)}(c) = f(c)$, and $(x - c)^0 = 1$, the first term

$$\frac{f^{(0)}(c)}{0!} \cdot (x - c)^0$$

of a Taylor polynomial always simplifies to $f(c)$. Although it is not necessary to use the longer, unsimplified notation, this notation does emphasize that the first term fits into the pattern of all the other terms.

Example 2 Compute the Taylor polynomials of degree one, two, and three for the function $f(x) = e^{2x}$ expanded about the point $c = 0$.

Solution First,

$$\begin{aligned} f(0) &= 1 \\ f^{(1)}(0) &= 2 \cdot e^{2x}|_{x=0} = 2 \\ f^{(2)}(0) &= 4 \cdot e^{2x}|_{x=0} = 4 \\ f^{(3)}(0) &= 8 \cdot e^{2x}|_{x=0} = 8. \end{aligned}$$

Therefore, according to Theorem 1,

$$P_1(x) = \frac{f(0)}{0!} \cdot (x - 0)^0 + \frac{f^{(1)}(0)}{1!} \cdot (x - 0)^1 = 1 + 2x,$$

$$P_2(x) = \frac{f(0)}{0!} \cdot (x - 0)^0 + \frac{f^{(1)}(0)}{1!} \cdot (x - 0)^1 + \frac{f^{(2)}(0)}{2!} \cdot (x - 0)^2 = 1 + 2x + 2x^2,$$

and

$$\begin{aligned} P_3(x) &= \frac{f(0)}{0!} \cdot (x - 0)^0 + \frac{f^{(1)}(0)}{1!} \cdot (x - 0)^1 \\ &\quad + \frac{f^{(2)}(0)}{2!} \cdot (x - 0)^2 + \frac{f^{(3)}(0)}{3!} \cdot (x - 0)^3 \\ &= 1 + 2x + 2x^2 + \frac{4}{3}x^3. \end{aligned}$$

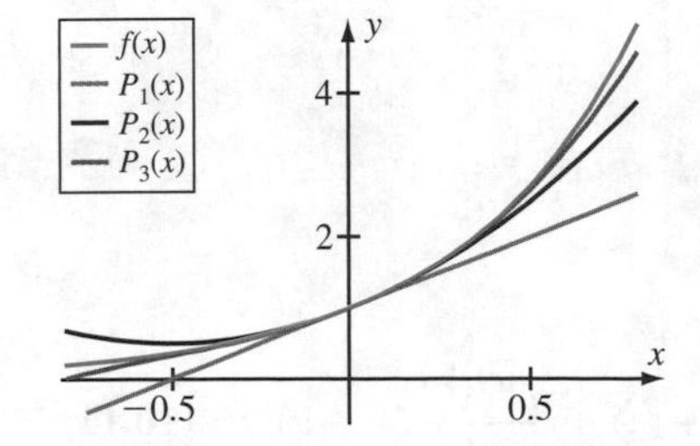

Figure 3

See Figure 3. ■

> **inSIGHT**
>
> Each Taylor polynomial is obtained by adding one term to the Taylor polynomial of one degree less. If we already have computed $P_N(x)$, then we need not start from scratch in computing the next Taylor polynomial $P_{N+1}(x)$; we only have to add the next term:
>
> $$P_{N+1}(x) = P_N(x) + \frac{f^{(N+1)}(c)}{(N+1)!}(x-c)^{N+1}. \tag{10.12}$$
>
> Thus, in Example 2, $P_2(x) = P_1(x) + 2x^2$ and $P_3(x) = P_2(x) + (4/3)x^3$.

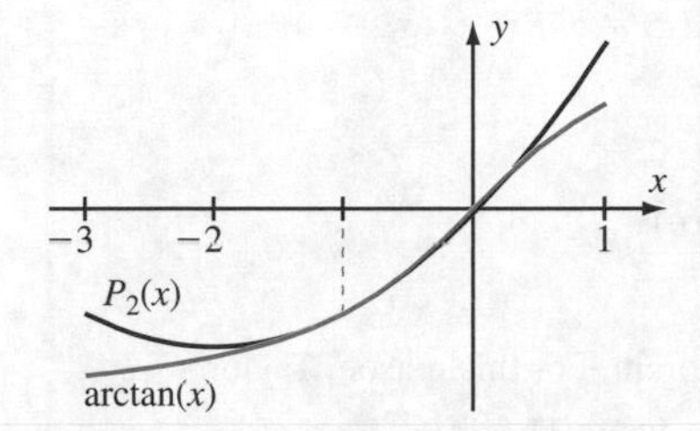

Figure 4

Example 3 Find the Taylor polynomial of degree two for the function $f(x) = \arctan(x)$ about the point $c = -1$.

Solution We begin by calculating

$$f(-1) = -\frac{\pi}{4}$$
$$f^{(1)}(-1) = \left.\frac{1}{1+x^2}\right|_{x=-1} = \frac{1}{2}$$
$$f^{(2)}(-1) = \left.\frac{-2x}{(1+x^2)^2}\right|_{x=-1} = \frac{1}{2}.$$

Therefore,

$$P_2(x) = \frac{-\pi/4}{0!}(x+1)^0 + \frac{1/2}{1!}(x+1)^1 + \frac{1/2}{2!}(x+1)^2 = -\frac{\pi}{4} + \frac{1}{2}(x+1) + \frac{1}{4}(x+1)^2.$$

The plots of f and P_2 are shown in Figure 4. ■

> **inSIGHT**
>
> Notice that our Taylor expansion for arctan(x) differs from the power series expansion for arctan(x) in formula (10.4) from Section 10.2. The reason is that the point $c = 0$ is the base point of expansion (10.4), whereas the point $c = -1$ is the base point used in Example 3.

Taylor's Theorem

Figures 2, 3, and 4 show that Taylor polynomials can be very effective approximations near the center point c. As those plots indicate, the accuracy of a Taylor polynomial approximation at x will decrease as the distance between x and c increases, which is to be expected. In constructing Taylor polynomials, we use only data observed at point c. The next theorem is important because it allows us to estimate the error that results when using a Taylor polynomial $P_N(x)$ to approximate $f(x)$.

Theorem 2 ***Taylor's Theorem*** Suppose that f is $N+1$ times continuously differentiable on an open interval I centered at c. If $x_0 \in I$, then there is a number ξ between c and x_0 such that

$$f(x_0) = P_N(x_0) + R_N(x_0) \tag{10.13}$$

where

$$R_N(x_0) = \frac{f^{(N+1)}(\xi)}{(N+1)!} \cdot (x_0 - c)^{N+1}. \tag{10.14}$$

Equation (10.13) expresses $f(x_0)$ as a degree N polynomial $P_N(x_0)$ plus a *remainder term,* or *error term,* $R_N(x_0)$. The remainder $R_N(x_0)$ is similar to the term that must be added to $P_N(x_0)$ to obtain the next Taylor polynomial $P_{N+1}(x_0)$, as you can see by reviewing equations (10.12) and (10.14). The difference is that the derivative $f^{(N+1)}$ is not evaluated at the base point c but at some other, unspecified point ξ.

For $N = 0$, Taylor's Theorem tells us that there is a point ξ between c and x_0 such that

$$f(x_0) = P_0(x_0) + \frac{f^{(1)}(\xi)}{1!} \cdot (x_0 - c)^1 = f(c) + f^{(1)}(\xi) \cdot (x_0 - c),$$

or

$$f^{(1)}(\xi) = \frac{f(x_0) - f(c)}{x_0 - c}.$$

This is precisely the assertion of the Mean Value Theorem. For this reason, Taylor's Theorem may be seen as a generalization of the Mean Value Theorem. A proof is outlined in Exercise 36 for $N = 1$ and in Exercise 42 for the general case.

We call equation (10.12) an Nth order Taylor expansion. In this section, we concentrate on learning how to compute P_N and R_N. In Section 10.4, we will study the significance of the error term.

Example 4 Apply Taylor's Theorem with $N = 7$ to the function $f(x) = \sin(x)$ about the point $c = 0$.

Solution Fix x_0. We first calculate

$$\begin{aligned} f(0) &= \sin(0) = 0 & f^{(1)}(0) &= \cos(x)|_{x=0} = 1 \\ f^{(2)}(0) &= -\sin(x)|_{x=0} = 0 & f^{(3)}(0) &= -\cos(x)|_{x=0} = -1 \\ f^{(4)}(0) &= \sin(x)|_{x=0} = 0 & f^{(5)}(0) &= \cos(x)|_{x=0} = 1 \\ f^{(6)}(0) &= -\sin(x)|_{x=0} = 0 & f^{(7)}(0) &= -\cos(x)|_{x=0} = -1 \\ & & f^{(8)}(x) &= \sin(x). \end{aligned}$$

Therefore,

$$\begin{aligned} P_7(x_0) &= \frac{0}{0!} \cdot (x_0 - 0)^0 + \frac{1}{1!} \cdot (x_0 - 0)^1 + \frac{0}{2!} \cdot (x_0 - 0)^2 + \frac{(-1)}{3!} \cdot (x_0 - 0)^3 \\ &\quad + \frac{0}{4!} \cdot (x_0 - 0)^4 + \frac{1}{5!} \cdot (x_0 - 0)^5 + \frac{0}{6!} \cdot (x_0 - 0)^6 + \frac{(-1)}{7!} \cdot (x_0 - 0)^7 \\ &= x_0 - \frac{x_0^3}{3!} + \frac{x_0^5}{5!} - \frac{x_0^7}{7!}. \end{aligned}$$

The remainder term is given by

$$R_7(x_0) = \frac{f^{(8)}(\xi)}{8!} \cdot (x_0 - 0)^8 = \frac{\sin(\xi)}{8!} \cdot x_0^8.$$

Thus,

$$\sin(x_0) = x_0 - \frac{x_0^3}{3!} + \frac{x_0^5}{5!} - \frac{x_0^7}{7!} + \frac{\sin(\xi)}{8!} \cdot x_0^8$$

for some undetermined value of ξ between 0 and x_0. ■

> **inSIGHT**
>
> Example 4 already reveals the value of Taylor's Theorem. Because $|\sin(\xi)|$ can be no greater than 1 and $1/8! < 0.000025$, we can make the estimate
>
> $$\left| \frac{\sin(\xi)}{8!} \cdot x_0^8 \right| < 0.000025 \quad \text{for} \quad |x_0| \leq 1.$$
>
> Therefore, $\sin(x_0)$ and the polynomial $x_0 - x_0^3/3! + x_0^5/5! - x_0^7/7!$ agree to four decimal places when $|x_0| \leq 1$. As Figure 5 shows, $x_0 - x_0^3/3! + x_0^5/5! - x_0^7/7!$ is a good approximation of $\sin(x_0)$ on an even larger interval.

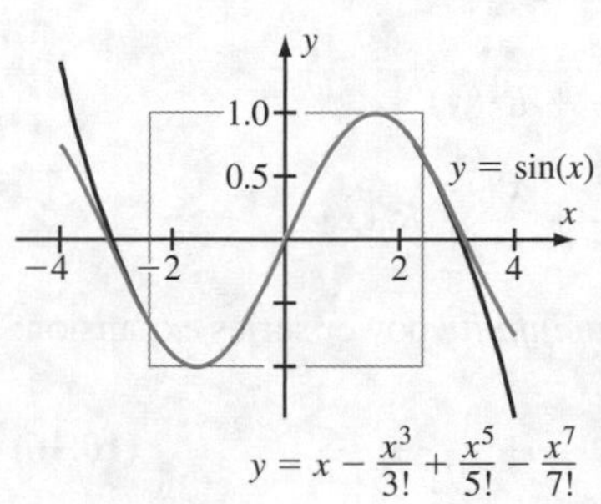

Figure 5

Example 5 Apply Taylor's Theorem with $N = 4$ to the function $f(x_0) = \sin(x_0)$ about the point $c = \pi/4$.

Solution Using the derivatives calculated in Example 4, we have $f(\pi/4) = 1/\sqrt{2}$, $f^{(1)}(\pi/4) = \cos(\pi/4) = 1/\sqrt{2}$, $f^{(2)}(\pi/4) = -\sin(\pi/4) = -1/\sqrt{2}$, $f^{(3)}(\pi/4) = -\cos(\pi/4) = -1/\sqrt{2}$, $f^{(4)}(\pi/4) = \sin(\pi/4) = 1/\sqrt{2}$, and $f^{(5)}(x) = \cos(x)$. Thus we see that

$$P_4(x_0) = \frac{1/\sqrt{2}}{0!}\left(x_0 - \frac{\pi}{4}\right)^0 + \frac{1/\sqrt{2}}{1!}\left(x_0 - \frac{\pi}{4}\right)^1 - \frac{1/\sqrt{2}}{2!}\left(x_0 - \frac{\pi}{4}\right)^2 - \frac{1/\sqrt{2}}{3!}\left(x_0 - \frac{\pi}{4}\right)^3 + \frac{1/\sqrt{2}}{4!}\left(x_0 - \frac{\pi}{4}\right)^4$$

and

$$R_4(x_0) = \frac{f^{(5)}(\xi)}{5!}\left(x_0 - \frac{\pi}{4}\right)^5 = \frac{\cos(\xi)}{5!}\left(x_0 - \frac{\pi}{4}\right)^5$$

where ξ is between $\pi/4$ and x_0. ■

> **inSIGHT**
>
> The Taylor polynomials of a function may look quite different when calculated at different base points. When $c = 0$ is used as the base point, the Taylor polynomials of $\sin(x)$ involve only odd powers of x (Example 4), which is not the case when $c = \pi/4$ is used as the base point (Example 5).

Concluding Remarks about Convergence and the Size of the Error Term

It is natural to wonder why we bother with the error term. We never seem to do much with it, and in any event it involves a value ξ that is not specified. For instance, because

$$\left.\frac{d^j}{dx^j} e^x \right|_{x=0} = e^x|_{x=0} = 1$$

for every nonnegative integer j, we can quickly use Theorem 2 to calculate that

$$e^x = \underbrace{1 + x + \frac{x^2}{2!} + \frac{x^3}{3!} + \frac{x^4}{4!} + \cdots + \frac{x^N}{N!}}_{P_N(x)} + R_N(x). \qquad \textbf{(10.15)}$$

In other words, we have

$$\begin{aligned} e^x &= 1 + x + \frac{x^2}{2!} + \frac{x^3}{3!} + \frac{x^4}{4!} + R_4(x) \\ &= 1 + x + \frac{x^2}{2!} + \frac{x^3}{3!} + \frac{x^4}{4!} + \frac{x^5}{5!} + R_5(x) \\ &= 1 + x + \frac{x^2}{2!} + \frac{x^3}{3!} + \frac{x^4}{4!} + \frac{x^5}{5!} + \frac{x^6}{6!} + R_6(x) \end{aligned}$$

and so on. Why not discard the error term and write an *infinite* power series expansion:

$$e^x = \sum_{n=0}^{\infty} \frac{x^n}{n!} = 1 + x + \frac{x^2}{2!} + \frac{x^3}{3!} + \frac{x^4}{4!} + \cdots + \frac{x^N}{N!} + \cdots \qquad \textbf{(10.16)}$$

for each x? The reason we do not do this is that discarding the error term and writing an infinite Taylor expansion is wishful thinking, *not* mathematics. Here is a more careful analysis of the situation: Equation (10.16) asserts that the partial sums $P_N(x) = \sum_{n=0}^{N} x^n/n!$ converge to e^x as N tends to infinity. Another way to state this is that equation (10.16) is true if and only if $(e^x - P_N(x)) \to 0$. By equation (10.15), this statement is equivalent to the assertion that $R_N(x) \to 0$. *These considerations contain the following crucial point about the error term.*

> The formal infinite Taylor expansion for a function f actually converges to $f(x)$ at a point x if and only if the error term $R_N(x)$ tends to zero as N tends to infinity.

The next two sections are devoted to studying the use of the error term.

quickquiz

1. What is the connection between Taylor polynomials with base point c and the tangent line to the graph of a function f at $(c, f(c))$?
2. What is the degree two Taylor polynomial of $\cos(x)$ when 0 is used as the base point?
3. If $3x^2 - x + 1$ is a Taylor polynomial of f with base point 0, then what is the equation of the tangent line to the graph of f at $(0, f(0))$?
4. If $3x^2 - x + 1$ is a Taylor polynomial of f with base point 0, then what is $f^{(2)}(0)$?

EXERCISES

Problems for Practice

In Exercises 1–16, compute the Taylor polynomial of the function f about the point c to the order N.

1. $f(x) = \cos(x)$, $N = 4$, $c = \pi/3$
2. $f(x) = \sin(x)$, $N = 5$, $c = \pi/2$
3. $f(x) = \cos(2x + \pi)$, $N = 5$, $c = 0$
4. $f(x) = e^x$, $N = 6$, $c = 1$
5. $f(x) = 5x^2 + e^x$, $N = 5$, $c = 0$
6. $f(x) = \sin(3x)$, $N = 5$, $c = 0$
7. $f(x) = 1/x^3$, $N = 5$, $c = 2$
8. $f(x) = x^2 + 1 + 1/x^2$, $N = 5$, $c = -1$
9. $f(x) = \ln(x + 3)$, $N = 5$, $c = 0$
10. $f(x) = \ln(x)$, $N = 4$, $c = e$
11. $f(x) = \ln(3x + 7)$, $N = 4$, $c = -2$
12. $f(x) = (x - 2)/(x^2 + 4)$, $N = 2$, $c = -1$
13. $f(x) = \exp(x + 2)/x$, $N = 2$, $c = -2$
14. $f(x) = \cos(x) + \sin(x)$, $N = 2$, $c = \pi/2$
15. $f(x) = \tan(x)$, $N = 3$, $c = 0$
16. $f(x) = \sec(x)$, $N = 4$, $c = 0$

A polynomial of degree N is its own Taylor polynomial of order N. You can see this for yourself by computing P_N and R_N in Exercises 17–22. (Although it may not be obvious that $f(x) = P_N(x)$, it must be so if $R_N(x) = 0$.)

17. $f(x) = 2 - x + x^2$, $N = 2$, $c = 3$
18. $f(x) = 2 - x + x^3 + 4x^4$, $N = 4$, $c = 1$
19. $f(x) = x + 10x^4$, $N = 4$, $c = 1$
20. $f(x) = x^5$, $N = 5$, $c = 1$
21. $f(x) = 1 - 2(x - 2) + 3(x - 2)^2 - 4(x - 2)^3$, $N = 3$, $c = 0$
22. $f(x) = (x - 2)^2 - (x + 2)^3$, $N = 3$, $c = -1$

Further Theory and Practice

In Exercises 23–30, compute the Taylor polynomial of the function f about the point c to the order N.

23. $f(x) = x/(x^2 + 1)$, $N = 4$, $c = 1$
24. $f(x) = x \cdot e^x$, $N = 3$, $c = 0$
25. $f(x) = (x^2 + 1) \cdot e^{-x}$, $N = 3$, $c = 0$
26. $f(x) = (x - 3)/\sqrt{x^2 + 16}$, $N = 2$, $c = 3$
27. $f(x) = x \cdot e^x$, $N = 3$, $c = 0$
28. $f(x) = x(x + 5)^{1/3}$, $N = 3$, $c = 3$
29. $f(x) = (1 + x^3)^{1/2}$, $N = 3$, $c = 2$
30. $f(x) = \exp(x)\cos(x)$, $N = 3$, $c = 0$
31. Fix a base point c. How are the coefficients of the Taylor polynomials of f' related to those of f? Illustrate with an example.
32. Let $f(x) = x^4$ and $c = 0$. Verify that for $N = 1$, $N = 2$, and $N = 3$ the Taylor polynomial $P_N(x) = 0$ for all x. For each of these three values of N, what is the value of ξ (as a function of x) that is used in the formula for $R_N(x)$?
33. Let $f(x) = x^4$ and $c = 1$. Calculate the Taylor polynomial $P_N(x)$ for each of the three values $N = 1$, $N = 2$, and $N = 3$. For each value, find the value of ξ (as a function of x) that is used in the formula for $R_N(x)$.
34. Expand $\sin(t + x)$ by using the Addition Formula. Replace $\cos(x)$ and $\sin(x)$ in this equation with their degree ten Taylor polynomials using $c = 0$. Now calculate the degree ten Taylor polynomial of $x \mapsto \sin(t + x)$ using $c = 0$ and compare.
35. Calculate the degree six Taylor polynomial of $x \mapsto \cos^2(x)$ using $c = 0$. Use a trigonometric formula to obtain the same result from the Taylor polynomial of $x \mapsto \cos(2x)$.
36. Suppose that f is twice continuously differentiable in an open interval I that is centered at c. Let x be any fixed point in I. Consider the function

$$\varphi(t) = f(x) - f(t) - (x - t)f'(t) - \left(\frac{x - t}{x - c}\right)^2 (f(x) - P_1(x))$$

for $c < t < x$. Apply Rolle's Theorem to φ to deduce that there is a ξ between c and x for which $\varphi^{(1)}(\xi) = 0$. Conclude that

$$f(x) = P_1(x) + \frac{f^{(2)}(\xi)}{2}(x - c)^2$$

for some ξ between c and x.

37. Consider the function

$$f(x) = \begin{cases} -x^4 & \text{if } x \le 0 \\ x^4 & \text{if } x > 0 \end{cases}.$$

Verify that f is three times continuously differentiable. Compute the Taylor polynomial of degree three for f about $c = 0$. What can you say about the remainder here?

38. **Application to Extrema** If $f'(c) = f''(c) = \cdots = f^{(N-1)}(c) = 0$ and $f^{(N)}(c) \neq 0$ for an integer N that is greater than 2, then c is a critical point of f at which the Second Derivative Test fails. Use $P_N(x)$ expanded about c to discuss the behavior of f near c.
39. Let N be a positive integer. Calculate the degree N Taylor series of $x \mapsto (a+x)^N$ using $c = 0$ as the base point. Deduce the binomial formula, $\sum_{k=0}^{N} \binom{N}{k} a^{N-k} x^k$, from your calculation.
40. Let $f(t) = \tan(t/3) + 2\sin(t/3)$. Snell's inequality, discovered in 1621, states that $t \leq f(t)$ for $0 < t < 3\pi/2$. The approximation $t \approx f(t)$ is remarkably good for $0 < t < \pi/2$. Calculate the Taylor polynomial of degree five for f about $c = 0$. Use it to explain the accuracy of the approximation.
41. The inequality
$$\frac{3\sin(t)}{2+\cos(t)} < t \quad \text{(for all } t > 0)$$
was discovered by Nicholas Cusa in 1458. Calculate the Taylor polynomial of degree five for
$$t \mapsto \frac{3\sin(t)}{2+\cos(t)}$$
about $c = 0$. Use it to explain Cusa's inequality for small positive values of t.
42. Suppose f is N times continuously differentiable in an open interval I centered at c. Let x be any fixed point in I. Define the function
$$\rho_N(t) = f(x) - \sum_{n=0}^{N} \frac{f^{(n)}(t)}{n!}(x-t)^n$$
for $c < t < x$. Let
$$\varphi(t) = \rho_N(t) - \left(\frac{x-t}{x-c}\right)^{N+1} \rho_N(c).$$
Note that $\varphi(c) = \varphi(x) = 0$. Apply Rolle's Theorem to φ to deduce that there is a ξ between c and x for which $\varphi^{(1)}(\xi) = 0$. Deduce Taylor's Theorem from this fact.
43. Use Exercise 49 from Section 10.2 to prove the uniqueness assertion of Theorem 1.

Calculator/Computer Exercises

In Exercises 44–47, plot the function f and its Taylor polynomials $P_1(x)$, $P_2(x)$, $P_5(x)$ in one viewing window with $-1 < x < 1$. Use $c = 0$. Notice that the higher the degree, the more closely the Taylor polynomial approximates f. Repeat the plots in a viewing window with $-2 < x < 2$. Notice that the further x is from the base point 0, the poorer the Taylor polynomial approximation tends to be.

44. $f(x) = e^x$
45. $f(x) = \sin(x + \pi/4)$
46. $f(x) = x + 5\sin(x)$
47. $f(x) = x^3 + \cos(2x)$

Each of Exercises 48–51 corresponds to one of the preceding four exercises. In each exercise, a different base point is specified to obtain a better approximation over the given interval.

48. For $f(x) = e^x$ and base point $c = -1$, plot $f(x)$ and $P_5(x)$ in the viewing window with $-2 < x < 0$. Compare the plot with that of Exercise 44.
49. For $f(x) = \sin(x + \pi/4)$ and base point $c = 1$, plot $f(x)$ and $P_5(x)$ in the viewing window with $0 < x < 2$. Compare the plot with that of Exercise 45.
50. For $f(x) = x + 5\sin(x)$ and base point $c = 3/2$, plot $f(x)$ and $P_5(x)$ in the viewing window with $1 < x < 2$. Compare the plot with that of Exercise 46.
51. For $f(x) = x^3 + \cos(2x)$ and base point $c = 1$, plot $f(x)$ and $P_5(x)$ in the viewing window with $0 < x < 2$. Compare the plot with that of Exercise 47.

Fix a c. Let $P_N(x)$ be the degree N Taylor polynomial of f expanded about c. Define
$$r_N(x,t) = \frac{f^{(N+1)}(t)}{(N+1)!} \cdot (x-c)^{N+1}.$$

In Exercises 52–55, a function f, a base point c, a positive integer N, and a point x_0 are given. In the ty-plane, plot $y = f(x_0) - P_N(x_0)$ and $y = r_N(x_0, t)$ for t between c and x_0. Explain how your plot illustrates Taylor's Theorem. Determine the value of ξ for which $f(x_0) = P_N(x_0) + r_N(x_0, \xi)$.

52. $f(x) = e^x$, $c = 0$, $N = 3$, $x_0 = -1$
53. $f(x) = \cos(x)$, $c = 0$, $N = 7$, $x_0 = 3\pi/2$
54. $f(x) = x^6$, $c = 1$, $N = 3$, $x_0 = 2$
55. $f(x) = x^6$, $c = 1$, $N = 4$, $x_0 = 2$

10.4 Estimating the Error Term—The Rate of Convergence of Taylor's Expansion

In Section 10.3, we learned that we may approximate a differentiable function $f(x)$ by means of a Taylor polynomial $P_N(x)$. This approximation is not of much use unless we have a way of telling how large the absolute error $|R_N(x)| = |f(x) - P_N(x)|$ is. The following theorem is the key to estimating the error term.

Theorem 1 Let f be a function that is $N+1$ times continuously differentiable on an open interval I centered at c. For each x_0 in I, let J denote the closed interval with endpoints x_0 and c. Thus, $J = [c, x_0]$ if $c \leq x_0$ and $J = [x_0, c]$ if $x_0 < c$. The error term $R_N(x_0)$ in the Nth order Taylor expansion of f about the point c satisfies

$$|R_N(x_0)| \leq M_{N+1} \cdot \frac{|x_0 - c|^{N+1}}{(N+1)!} \quad (x \in I) \qquad \textbf{(10.17)}$$

where

$$M_{N+1} = \max_{x \in J} \left| f^{(N+1)}(x) \right|. \qquad \textbf{(10.18)}$$

Proof The idea of this theorem is simple. The remainder $R_N(x_0)$ is expressed in terms of a derivative $f^{(N+1)}(\xi)$ that is evaluated at an undetermined point. We bound the remainder by using the largest possible value for this derivative. Thus, by formula (10.14), we have, for some ξ between c and x_0,

$$|R_N(x_0)| = \left| \frac{f^{(N+1)}(\xi)}{(N+1)!} \cdot (x_0 - c)^{N+1} \right| = \frac{\left| f^{(N+1)}(\xi) \right|}{(N+1)!} \cdot |x_0 - c|^{N+1} \leq M_{N+1} \frac{|x_0 - c|^{N+1}}{(N+1)!}.$$

■

Now we learn to make use of the important estimate contained in Theorem 1.

Example 1 Use the degree seven Taylor polynomial expanded about 0 to approximate $\sin(0.1)$. Use Theorem 1 to estimate the error. Repeat for $\sin(1.0)$.

Solution Here $N = 7$, $c = 0$, and $f(x) = \sin(x)$. In Example 4 from Section 10.3, we calculated $P_7(x) = x - x^3/3! + x^5/5! - x^7/7!$ and $f^{(8)}(x) = \sin(x)$. When we take $x_0 = 0.1$, we obtain $P_7(0.1) = 0.0998334166468\ldots$ for our approximation of $\sin(0.1)$. Next we calculate

$$M_8 = \max_{0 \leq x \leq 0.1} \left| f^{(8)}(x) \right| = \max_{0 \leq x \leq 0.1} \left| \sin(x) \right|.$$

Because $\sin(x)$ is increasing for $0 \leq x \leq 0.1$, we conclude that $M_8 = \sin(0.1)$, which is precisely the unknown quantity that we are trying to approximate! No matter—we may safely say that $M_8 = \sin(0.1) \leq 1$. According to Theorem 1, the absolute error $|R_7(0.1)|$ satisfies

$$|R_7(0.1)| \leq M_8 \cdot \frac{|0.1 - 0|^8}{8!} \leq \frac{|0.1 - 0|^8}{8!} < 2.5 \times 10^{-13}.$$

Our error estimate assures us that our approximation is correct to 12 decimal places.

inSIGHT

Why can the accuracy of the approximation $P_7(1.0) \approx \sin(1.0)$ be so much worse than that for $P_7(0.1) \approx \sin(0.1)$? The answer is simple, and it is important for us to keep in mind. Look at the bound for the error R_N given in Theorem 1. The closer x_0 is to c, the smaller the factor $|x_0 - c|^{N+1}$ is in the estimate for $R_N(x_0)$.

Turning to the second value, $\sin(1.0)$, our approximation is $P_7(1.0) = 0.841468\ldots$. The absolute error $|R_7(1.0)|$ satisfies

$$|R_7(1.0)| \le M_8 \cdot \frac{|1.0 - 0|^8}{8!} \le \frac{|1.0 - 0|^8}{8!} < 2.5 \times 10^{-5}.$$

Our approximation of $\sin(1.0)$ is therefore correct to four decimal places. ■

Example 2 Use the third order Taylor expansion of e^x expanded about the point $c = 0$ to find an approximate value for $e^{-0.1}$. Estimate your accuracy.

Solution Equation (10.15) from Section 10.3 tells us that the degree three Taylor polynomial of e^x about $c = 0$ is $P_3(x) = 1 + x + x^2/2! + x^3/3!$. Our approximation is therefore $P_3(-0.1) = 0.90483\ldots$. Next, to estimate the accuracy of this approximation, we calculate

$$M_4 = \max_{x_0 \le x \le c} \left| \frac{d^4}{dx^4} e^x \right| = \max_{-0.1 \le x \le 0} e^x = 1.$$

Therefore,

$$|R_3(-0.1)| \le 1 \cdot \frac{|-0.1 - 0|^4}{4!} < 5 \times 10^{-6}.$$

Our error estimate guarantees that our approximation is accurate to five decimal places. A check with a calculator shows this to be correct. ■

inSIGHT

As in Example 2, when $|f^{(N+1)}(x)|$ is a monotonic function on the interval J between c and x_0, the maximum $M_{N+1} = \max_{x \in J} |f^{(N+1)}(x)|$ occurs at an endpoint of J.

Achieving a Desired Degree of Accuracy

As we have learned, one way to force the error term to be small is to work with points x_0 and c that are close to each other. Choosing a large value of N is a second strategy for producing a small error term. This strategy often works because of division by $(N+1)!$ in the estimate for $R_N(x_0)$. Furthermore, if $|x_0 - c| < 1$, then $|x_0 - c|^{N+1}$ decreases to zero as N becomes larger. The correct choice of N will vary from case to case. Studying several examples is the best way to understand the method of choosing the proper c and large enough N to obtain a particular degree of accuracy.

Example 3 Compute $\ln(1.2)$ to an accuracy of four decimal places.

Solution Let $f(x) = \ln(x)$ and $x_0 = 1.2$. Saying that we require four decimal places of accuracy means that we want the error term to be less than 5×10^{-5}. Thus, we require $|R_N(x_0)| < 5 \times 10^{-5}$. We can arrange for this estimate to be true by choosing c sensibly and by making N sufficiently large. We take $c = 1$ (which is reasonably close to $x_0 = 1.2$) and note that for $f(x) = \ln(x)$, we have $f^{(1)}(x) = x^{-1}$, $f^{(2)}(x) = -1 \cdot x^{-2}$, $f^{(3)}(x) = 2 \cdot x^{-3}$, $f^{(4)}(x) = -3 \cdot 2 \cdot x^{-4} = -3!x^{-4}$, $f^{(5)}(x) = 4!x^{-5}$, and, in general,

$$f^{(n)}(x) = (-1)^{n-1}(n-1)!x^{-n} \quad (n = 1, 2, 3, \ldots). \tag{10.19}$$

Thus, $f^{(n)}(1)/n! = (-1)^{n-1}/n$ and the degree N Taylor polynomial of f with base point 1 is

$$P_N(x_0) = \ln(c) + \sum_{n=1}^{N} \frac{f^{(n)}(c)}{n!}(x_0 - c)^n = \ln(1) + \sum_{n=1}^{N} \frac{(-1)^{n-1}}{n}(1.2 - 1)^n = \sum_{n=1}^{N} \frac{(-1)^{n-1}(0.2)^n}{n}. \tag{10.20}$$

To estimate the error, we note that the interval J from Theorem 1 is [1, 1.2]. Also, $f^{(N+1)}(x) = (-1)^N N! x^{-(N+1)}$ by formula (10.19). Therefore,

$$M_{N+1} = \max_{1\leq x\leq 1.2} \left|f^{(N+1)}(x)\right| = \max_{1\leq x\leq 1.2} \left|(-1)^{N+2} N! x^{-(N+1)}\right| = N! \max_{1\leq x\leq 1.2} \frac{1}{x^{N+1}} = N!.$$

Thus,

$$|R_N(1.2)| \leq M_{N+1} \cdot \frac{|1.2-1|^{N+1}}{(N+1)!} = N! \cdot \frac{(0.2)^{N+1}}{(N+1)!} = \frac{(0.2)^{N+1}}{(N+1)}.$$

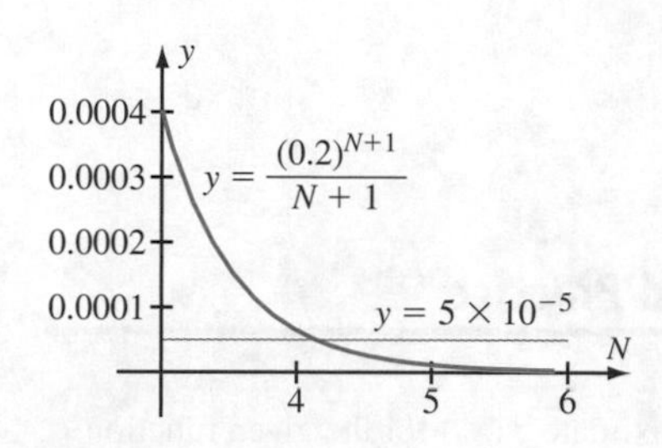

Figure 1

To ensure that the error is less than 5×10^{-5}, it suffices to choose N large enough so that $(0.2)^{N+1}/(N+1) < 5 \times 10^{-5}$. Substituting in a few values for N, we see that the inequality will first be true when $N = 5$. We could also determine this value of N by plotting $(0.2)^{N+1}/(N+1)$, as in Figure 1. Notice that we use the smallest N that gives the required accuracy. Using a larger value of N would result in greater accuracy but more calculation.

In conclusion, the formula $\ln(1.2) = P_5(1.2) + R_5(1.2)$ and the inequality $|R_5(1.2)| < 5 \times 10^{-5}$ assure us that $\ln(1.2)$ and $P_5(1.2)$ agree to four decimal places. Substituting $N = 5$ in line (10.20), we calculate

$$\ln(1.2) \approx P_5(1.2) = \sum_{n=1}^{5} \frac{(-1)^{n-1}(0.2)^n}{n} = 0.18233\ldots.$$

inSIGHT

The problem posed in Example 3 is different from the preceding ones because the accuracy is mandated in advance, which is what would usually happen in real life. One knows in advance what degree of accuracy is required for a given problem. For the parts on a cyclotron, perhaps ten decimal places of accuracy are required; for building a fence in your backyard, one decimal place of accuracy is more than sufficient; a surveyor laying off acreage for the National Forestry Service cannot achieve, and does not require, one decimal place of accuracy.

Example 4 In Example 1 from Section 3.9, we used the differential approximation to calculate three decimal places of $\sqrt{4.1}$. Now use a second order Taylor polynomial to estimate $\sqrt{4.1}$. How many decimal places of accuracy are obtained?

Solution Let $f(x) = \sqrt{x}$, $x_0 = 4.1$, and $c = 4$. We calculate $f^{(1)}(x) = (1/2)x^{-1/2}$, $f^{(2)}(x) = (-1/4)x^{-3/2}$, and $f^{(3)}(x) = (3/8)x^{-5/2}$. Using the first two of these derivatives, we obtain

$$\begin{aligned} P_2(x_0) &= \frac{f^{(0)}(4)}{0!}(x_0-4)^0 + \frac{f^{(1)}(4)}{1!}(x_0-4)^1 + \frac{f^{(2)}(4)}{2!}(x_0-4)^2 \\ &= 2 + \frac{(x_0-4)}{4} - \frac{(x_0-4)^2}{64}, \end{aligned}$$

or

$$P_2(4.1) = 2 + \frac{(0.1)}{4} - \frac{(0.1)^2}{64} = 2.02484375.$$

Using the last calculated derivative, $f^{(3)}(x) = (3/8)x^{-5/2}$, we have

$$M_3 = \max_{4\leq x\leq 4.1} \left|f^{(3)}(x)\right| = \max_{4\leq x\leq 4.1} \left|\frac{3}{8}x^{-5/2}\right| = \frac{3}{8}4^{-5/2} = \frac{3}{256}.$$

Theorem 1 tells us that

$$|R_2(4.1)| \leq M_3 \cdot \frac{(4.1-4)^3}{3!} = \frac{3}{256} \cdot \frac{(0.1)^3}{3!} = 1.953125 \times 10^{-6} < 5 \times 10^{-6}.$$

Thus, $P_2(4.1) = 2.02484375$ approximates $\sqrt{4.1}$ with an error less than 5×10^{-6}. In other words, our approximation is accurate to five decimal places. ■

quickquiz

1. If the Taylor polynomial $P_N(x)$ is used to approximate $f(x)$, then what order derivative of f is involved in the error estimate?
2. What does it mean for an estimate to be accurate to k decimal places?
3. Sketch the steps for approximating the value of a function f at a point x to k decimal places of accuracy.
4. Use a Taylor polynomial to estimate $\sqrt{1.01}$ to five decimal places of accuracy.

EXERCISES

Problems for Practice

In Exercises 1–12,

a. Approximate the value $f(x_0)$ by using a Taylor polynomial of order N and base point c.
b. Estimate the error term using Theorem 1.

1. $f(x) = e^x$, $x_0 = 0.1$, $c = 0$, $N = 4$
2. $f(x) = e^{2x}$, $x_0 = -0.01$, $c = 0$, $N = 3$
3. $f(x) = \sin(x)$, $x_0 = 0.3$, $c = 0$, $N = 5$
4. $f(x) = \sin(x)$, $x_0 = 1.6$, $c = \pi/2$, $N = 5$
5. $f(x) = \cos(x)$, $x_0 = -0.2$, $c = 0$, $N = 4$
6. $f(x) = \cos(x)$, $x_0 = -1.56$, $c = -\pi/2$, $N = 4$
7. $f(x) = \ln(x)$, $x_0 = 3$, $c = e$, $N = 5$
8. $f(x) = \ln(x)$, $x_0 = 1.2$, $c = 1$, $N = 5$
9. $f(x) = \arctan(2x)$, $x_0 = 0.2$, $c = 0$, $N = 2$
10. $f(x) = 1/\sqrt{x}$, $x_0 = 3.9$, $c = 4$, $N = 3$
11. $f(x) = \sqrt{1+x}$, $x_0 = 8.2$, $c = 8$, $N = 3$
12. $f(x) = \sqrt{x}$, $x_0 = 24.4$, $c = 25$, $N = 3$

In Exercises 13–22, a function f, a point x_0, a base point c, and an order N are given. Show that the maximum value of $|f^{(N+1)}(x)|$ for x between c and x_0 occurs at an endpoint. Calculate the value of M_{N+1} from Theorem 1 for the given data, and state the estimate of the remainder term $R_N(x_0)$.

13. $f(x) = \sqrt{1+x}$, $x_0 = 0.4$, $c = 0$, $N = 4$
14. $f(x) = x + \sin(x)$, $x_0 = \pi/6$, $c = \pi/4$, $N = 3$
15. $f(x) = x\sin(x)$, $x_0 = \pi/6$, $c = 0$, $N = 2$
16. $f(x) = x + \exp(x)$, $x_0 = 0.8$, $c = 0$, $N = 3$
17. $f(x) = x^2 + \exp(x)$, $x_0 = -1.1$, $c = -2$, $N = 3$
18. $f(x) = x/(1+x)$, $x_0 = 0.5$, $c = 0$, $N = 2$
19. $f(x) = \ln(x)$, $x_0 = 2.5$, $c = e$, $N = 5$
20. $f(x) = \ln(1+x^2)$, $x_0 = \sqrt{2} - 1$, $c = 0$, $N = 2$
21. $f(x) = \arctan(x)$, $x_0 = 0.6$, $c = 0$, $N = 2$
22. $f(x) = 1/(1+x^2)$, $x_0 = 0.2$, $c = 0$, $N = 2$

Further Theory and Practice

In Exercises 23–30, approximate $f(x_0)$ for the given function $f(x)$ and point x_0. Use a Taylor polynomial $P_N(x_0)$ with base point c. The order N is not specified. Estimate the error term $R_N(x_0)$ for a general value of N. Using your estimate, determine the smallest order N that is guaranteed to yield the given number of decimal places of accuracy. Finally, calculate the approximation.

23. $f(x) = \cos(x)$, $x_0 = 0.2$, $c = 0$, 6 places
24. $f(x) = \ln(1+x)$, $x_0 = -0.1$, $c = 0$, 4 places
25. $f(x) = \sin(x)$, $x_0 = 1.5$, $c = \pi/2$, 8 places
26. $f(x) = e^x$, $x_0 = 0.5$, $c = 0$, 4 places
27. $f(x) = e^{-x}$, $x_0 = 0.1$, $c = 0$, 4 places
28. $f(x) = 1/\sqrt{1+x}$, $x_0 = 0.3$, $c = 0$, 4 places
29. $f(x) = \ln(x)$, $x_0 = 1.2$, $c = 1$, 4 places
30. $f(x) = \ln(2x)$, $x_0 = 1.5$, $c = e/2$, 5 places
31. Compute $\ln(50)$ to an accuracy of four decimal places by finding a Taylor polynomial approximation with an appropriate base point c. Do not use a calculator to evaluate any value of $\ln(x)$, but you may use a calculator for arithmetic with the number e and its powers.
32. Use a Taylor polynomial with an appropriate base point to compute $\sin(2)$ to an accuracy of six decimal places.
33. By writing $\sin(x) = P_3(x) + R_3(x)$, using $c = 0$ as base point, prove that
$$\lim_{x\to 0} \frac{\sin(x) - x}{x^3} = -\frac{1}{6}.$$
34. By writing $\cos(x) = P_2(x) + R_2(x)$, using $c = 0$ as base point, prove that
$$\lim_{x\to 0} \frac{1-\cos(x)}{x^2} = \frac{1}{2}.$$

In Exercises 35–40, calculate the limit $\lim_{x\to 0} f(x)/x^N$ by expressing $f(x) = P_N(x) + R_N(x)$ and showing that $\lim_{x\to 0} R_N(x)/x^N = 0$.

35. $\lim_{x\to 0} \dfrac{\ln(1+x) - x}{x^2}$

36. $\lim_{x\to 0} \dfrac{2 - 2\cos(x) - x^2}{x^4}$

37. $\lim_{x\to 0} \dfrac{\exp(x) + \exp(-x) - 2}{x^2}$

38. $\lim_{x\to 0} \dfrac{\exp(3x) - \exp(-3x) - 6x}{x^3}$

39. $\lim_{x\to 0} \dfrac{\arctan(x) - x}{x^3}$

40. $\lim_{x\to 0} \dfrac{\sin(x) - x\cos(x)}{x^3}$

41. Evaluate

$$\lim_{x\to 0} \frac{\sin(x) - x}{(1 - \cos(x)) \cdot \ln(1+x)}$$

by approximating each of the functions $\sin(x)$, $\cos(x)$, and $\ln(1+x)$ with its degree three Taylor polynomial $P_3(x)$ using base point $c = 0$.

42. If f'' is continuous on an open interval centered at c, show that

$$\lim_{h\to 0} \frac{f(c+h) - 2f(c) + f(c-h)}{h^2} = f''(c).$$

Calculator/Computer Exercises

In Exercises 43 50, a function f, a point x_0, a base point c, and an order N are given. Plot $|f^{(N+1)}(x)|$ for x between c and x_0. Use your plot to estimate the quantity M_{N+1} from Theorem 1. Use your value of M_{N+1} to obtain an upper bound $R_N(x_0)$ for $|f(x_0) - P_N(x_0)|$, the absolute error that results when $f(x_0)$ is approximated by the degree N Taylor polynomial with base point c. Verify that $|f(x_0) - P_N(x_0)| \le R_N(x_0)$.

43. $f(x) = x^2\sin(x^2)$, $x_0 = \pi/4$, $c = \pi/12$, $N = 2$

44. $f(x) = x^2/e^x$, $x_0 = 2.1$, $c = 1.7$, $N = 2$

45. $f(x) = x/(1+x^2)$, $x_0 = 4$, $c = 3.5$, $N = 3$

46. $f(x) = x\exp(-x^2)$, $x_0 = 1.8$, $c = 1.2$, $N = 3$

47. $f(x) = 1/x + 5\sin(2x)$, $x_0 = 5$, $c = 4.4$, $N = 4$

48. $f(x) = \cos(x) + \sin(x)$, $x_0 = 5.8$, $c = 5$, $N = 4$

49. $f(x) = x/(1+\sqrt{x})$, $x_0 = 1.4$, $c = \sqrt{2}$, $N = 2$

50. $f(x) = \exp(\sin(x))$, $x_0 = 1.5$, $c = \pi/2$, $N = 3$

10.5 Taylor Series

As was mentioned at the end of Section 10.3, the best and simplest possible situation occurs when we can actually express a given function as a convergent power series. If this can be done, then the role of the error term fades into the background—the sum of the series *actually equals* the original function. In this situation, we have a powerful and useful formula.

Definition Suppose that for every nonnegative integer N the function f is N times continuously differentiable on an open interval I centered at c. The power series

$$T(x) = \sum_{n=0}^{\infty} \frac{f^{(n)}(c)}{n!}(x-c)^n$$

is called the *Taylor series* of f with *base point* c. We say that the Taylor series is *expanded about* c (or *centered* at c). A Taylor series with base point 0 is also called a *Maclaurin series*.

Notice that for any function f that has derivatives of all orders at c, we can form a Taylor series $T(x)$ expanded about c. The generated Taylor series $T(x)$ enjoys all the properties of power series discussed in Sections 10.1 and 10.2. In particular, $T(x)$ has an interval of convergence. However, a Taylor series does not necessarily converge to the function f that generates it. That is, $T(x)$ may not equal $f(x)$ even where $T(x)$ converges. Exercise 55 discusses one such instance. Fortunately, such surprises do not occur with the functions commonly used in mathematics, science, and engineering.

To understand when a Taylor series of f converges to f, we turn to Taylor's Theorem (Theorem 2 from Section 10.3), which tells us that for every N we can write

$$f(x) = P_N(x) + R_N(x) \tag{10.21}$$

where $P_N(x) = \sum_{n=0}^{N} f^{(n)}(c)(x-c)^n/n!$ is the degree N Taylor polynomial of f with base point c and $R_N(x)$ is a remainder term for which we have an estimate. When we let N tend to infinity in equation (10.21), we obtain

$$f(x) = \lim_{N\to\infty}\left(\sum_{n=0}^{N} \frac{f^{(n)}(c)}{n!}(x-c)^n + R_N(x)\right).$$

From this equation, we see that

$$T(x) = \lim_{N\to\infty}\sum_{n=0}^{N} \frac{f^{(n)}(c)}{n!}(x-c)^n$$

exists if and only if $\lim_{N\to\infty} R_N(x)$ exists, and $f(x) = T(x)$ if and only if $\lim_{N\to\infty} R_N(x) = 0$. We record the most important of these observations as a theorem.

Theorem 1 Suppose that for every nonnegative integer N the function f is N times continuously differentiable on an open interval I centered at c. Let

$$R_N(x) = f(x) - \sum_{n=0}^{N} \frac{f^{(n)}(c)}{n!}(x-c)^n$$

be the remainder in the Nth order Taylor expansion about c. If

$$\lim_{N\to\infty} R_N(x) = 0,$$

then $f(x)$ equals the Taylor series of f expanded about c:

$$f(x) = \sum_{n=0}^{\infty} \frac{f^{(n)}(c)}{n!}(x-c)^n.$$

Power Series Expansions of the Common Transcendental Functions

In Section 10.2, we calculated power series expansions of $\ln(1+x)$ and $\arctan(x)$. Now we will do the same for some other basic functions. Before we attack some concrete examples, however, we must first compute the limit of a certain sequence that will occur in most of our examples.

Theorem 2 If b is a fixed positive constant, then

$$\lim_{k\to\infty} \frac{b^k}{k!} = 0.$$

Proof Since b is fixed, we may choose an integer ℓ that exceeds b. Notice that b^k and $k!$ each have k factors. For $k > \ell$, we have

$$\frac{b^k}{k!} = \frac{b}{k} \cdot \underbrace{\left(\frac{b}{(k-1)} \cdot \frac{b}{(k-2)} \cdots \frac{b}{\ell} \right)}_{\text{Number less than 1}} \cdot \underbrace{\left(\frac{b}{(\ell-1)} \cdot \frac{b}{(\ell-2)} \cdots \frac{b}{2} \cdot \frac{b}{1} \right)}_{\text{Fixed number}}.$$

Since the first factor on the right tends to 0 as k tends to infinity, the theorem follows. ■

INSIGHT

Reread Example 1 to identify the main steps. You will notice that we use the Taylor expansion of f to generate a power series that we *hope* converges to f. We check that it *does* converge to f by verifying that the remainder term tends to 0. It is not enough simply to verify that the Taylor series converges for each x. *We want to know that the series converges back to the function f that we* began with. To check this, we *must* use the error term.

Example 1 *Power Series Expansion of the Exponential Function*

Show that e^x is represented by its Maclaurin series on the entire real line:

$$e^x = \sum_{n=0}^{\infty} \frac{x^n}{n!} \quad (-\infty < x < \infty). \tag{10.22}$$

Solution By observing that $\frac{d^n}{dx^n} e^x = e^x$ and $\frac{d^n}{dx^n} e^x|_{x=0} = e^x|_{x=0} = 1$ for every n, we see that the Taylor polynomial $P_N(x)$ with base point 0 is $\sum_{j=0}^{N} x^n/n!$, and the Taylor series is $\sum_{n=0}^{\infty} x^n/n!$. In Example 3 from Section 9.6, we used the Ratio Test to show that this Taylor series is convergent for every real x. Now fix an x_0. Let J denote the interval between 0 and x_0. Theorem 1 from Section 10.4 tells us that the remainder $R_N(x_0)$ satisfies

$$|R_N(x_0)| \le \left(\max_{x \in J} \left| \frac{d^{N+1}}{dx^{N+1}} e^x \right| \right) \cdot \frac{|x_0|^{N+1}}{(N+1)!} = \max_{x \in J}(e^x) \frac{|x_0|^{N+1}}{(N+1)!} \le e^{|x_0|} \cdot \frac{|x_0|^{N+1}}{(N+1)!}.$$

Because x_0 is fixed, Theorem 2 says that the right side of this inequality tends to zero as N tends to infinity. We conclude that $\lim_{N \to \infty} R_N(x_0) = 0$, which is precisely what is needed to show that e^x is equal to its Maclaurin series for every value of x. ■

A LOOK BACK

In Example 9 from Section 10.2, we proved that $y(x) = 2 + 2x + \sum_{n=2}^{\infty} x^n/n!$ is the solution of the initial value problem $\frac{dy}{dx} = y - x$, $y(0) = 2$. We may now simplify the expression for $y(x)$ as follows:

$$\begin{aligned} y(x) = 2 + 2x + \sum_{n=2}^{\infty} \frac{x^n}{n!} &= 1 + x + \left(1 + x + \sum_{n=2}^{\infty} \frac{x^n}{n!} \right) \\ &= 1 + x + \left(\sum_{n=0}^{\infty} \frac{x^n}{n!} \right) = 1 + x + e^x. \end{aligned}$$

Substitute this expression for $y(x)$ into the differential equation $\frac{dy}{dx} = y - x$ to verify that it is indeed a solution.

Example 2 *Power Series Expansion of the Sine Function*

Show that $\sin(x)$ is represented by its Maclaurin series on the entire real line:

$$\sin(x) = x - \frac{x^3}{3!} + \frac{x^5}{5!} - \frac{x^7}{7!} + \frac{x^9}{9!} - \frac{x^{11}}{11!} + \cdots \quad (-\infty < x < \infty). \tag{10.23}$$

Solution The calculations of Example 4 from Section 10.3 show that the right side of equation (10.23) is indeed the Maclaurin series of $\sin(x)$. Let x_0 be fixed. We apply

Theorem 1 from Section 10.4 to $R_N(x_0)$. In this application, $M_{N+1} = \max_{x \in J} |f^{(N+1)}(x)|$ where J is the interval between c and x_0. Since $f^{(N+1)}(x)$ is $\pm\sin(x)$ or $\pm\cos(x)$, we may say that $M_{N+1} \leq 1$. Therefore,

$$|R_N(x_0)| \leq \frac{|x_0|^{N+1}}{(N+1)!}.$$

Using Theorem 2, we conclude that $\lim_{N\to\infty} R_N(x_0) = 0$, which is exactly what is required for equation (10.23) to hold. ■

inSIGHT

Not every property of a function is quickly detected from its power series. However, because odd powers of x appear in the Maclaurin series of $\sin(x)$, we can easily see that $\sin(x)$ is an odd function: $\sin(-x) = -\sin(x)$. Because we know that $\cos(x)$ is an even function, that is, $\cos(-x) = \cos(x)$ for all x, we expect the Maclaurin series of $\cos(x)$ to involve only even powers of x. Indeed, a calculation that is analogous to that from Example 2 shows that

$$\cos(x) = 1 - \frac{x^2}{2!} + \frac{x^4}{4!} - \frac{x^6}{6!} + \frac{x^8}{8!} - \frac{x^{10}}{10!} + \cdots \quad (-\infty < x < \infty). \tag{10.24}$$

The Binomial Series

Fix a nonzero real number α. Let us expand the function $f(x) = (1+x)^\alpha$ about the point $c = 0$. We calculate $f^{(1)}(x) = \alpha \cdot (1+x)^{\alpha-1}$, $f^{(2)}(x) = \alpha \cdot (\alpha - 1) \cdot (1+x)^{\alpha-2}$, $f^{(3)}(x) = \alpha \cdot (\alpha-1) \cdot (\alpha-2) \cdot (1+x)^{\alpha-3}$, and so on. In general, $f^{(n)}(x) = \alpha \cdot (\alpha - 1) \cdot (\alpha - 2) \cdots (\alpha - (n-1)) \cdot (1+x)^{\alpha-n}$ for $n = 0, 1, 2, \ldots$. Substituting $x = 0$, we obtain

$$f^{(n)}(0) = \alpha \cdot (\alpha - 1) \cdot (\alpha - 2) \cdots (\alpha - (n-2)) \cdot (\alpha - (n-1)).$$

The Maclaurin series of f is therefore

$$\sum_{n=0}^{\infty} \frac{\alpha \cdot (\alpha-1) \cdot (\alpha-2) \cdots (\alpha - (j-2)) \cdot (\alpha - (n-1))}{n!} x^n.$$

There is a useful notation that we can use to simplify the unwieldy appearance of this Taylor series. For any real number α and nonnegative integer n, we define the expression $\binom{\alpha}{n}$ by $\binom{\alpha}{0} = 1$ when $n = 0$ and, for $n \in \mathbb{Z}^+$,

$$\binom{\alpha}{n} = \frac{\alpha \cdot (\alpha-1) \cdot (\alpha-2) \cdots (\alpha-(n-2)) \cdot (\alpha-(n-1))}{n!}. \tag{10.25}$$

Notice that when α is a positive integer, equation (10.25) is the same as the usual definition of the binomial coefficient $\binom{\alpha}{n}$. Using this notation, the Maclaurin series associated to the function $f(x) = (1+x)^\alpha$ can be written as $\sum_{n=0}^{\infty} \binom{\alpha}{n} x^n$. This series is known as *Newton's binomial series,* and its coefficients are known as *binomial coefficients,* even when α is not a positive integer.

A simple check with the Ratio Test shows that, no matter what the value of α, the series converges for $|x| < 1$. For these values of x, the Maclaurin series converges *to the function* f:

$$(1+x)^{\alpha} = \sum_{n=0}^{\infty} \binom{\alpha}{n} x^n \quad (-1 < x < 1). \tag{10.26}$$

Exercise 47 explores a proof of this assertion for general α and $x \in (-1, 1)$.

Example 3 Calculate the binomial series of $\sqrt{1+x}$ up until the x^4 term. Observe that $\sqrt{1+x} = (1+x)^{\alpha}$ with $\alpha = 1/2$.

Solution We have

$$\binom{1/2}{0} = 1$$
$$\binom{1/2}{1} = \frac{1/2}{1!} = \frac{1}{2}$$
$$\binom{1/2}{2} = \frac{(1/2)\cdot(-1/2)}{2!} = -\frac{1}{8}$$
$$\binom{1/2}{3} = \frac{(1/2)\cdot(-1/2)\cdot(-3/2)}{3!} = \frac{1}{16}$$
$$\binom{1/2}{4} = \frac{(1/2)\cdot(-1/2)\cdot(-3/2)\cdot(-5/2)}{4!} = -\frac{5}{128}$$
$$\ldots.$$

We conclude that

$$\sqrt{1+x} = 1 + \frac{1}{2}x - \frac{1}{8}x^2 + \frac{1}{16}x^3 - \frac{5}{128}x^4 + \cdots \quad (-1 < x < 1). \tag{10.27}$$

■

Using Taylor Series to Approximate

Section 10.4 was devoted to an analysis of the remainder term of the Taylor expansion with a view to approximation. It should not be overlooked that if a convergent Taylor series is alternating, then inequality (9.9) of the Alternating Series Test provides an estimate of the remainder. The next two examples illustrate this point.

Example 4 Approximate $\sqrt{1.02}$ to five decimal places.

Solution We may use the binomial series expansion given by equation (10.27):

$$(1+0.02)^{1/2} = 1 + \frac{0.02}{2} - \frac{(0.02)^2}{8} + \frac{(0.02)^3}{16} - \frac{(0.02)^4}{128} + \cdots.$$

Because this is an alternating series, the truncation error is less than the first term omitted. The first term that is less than 5×10^{-6}, the maximum allowable error, is $(0.02)^3/16 = 5.0 \times 10^{-7}$. Our approximation is therefore $1 + 0.02/2 - (0.02)^2/8 = 1.00995$. A calculator evaluation yields $\sqrt{1.02} = 1.0099504\ldots$, which confirms our analysis.

■

Example 5 Calculate $\int_0^{0.3} \cos(u^2)\, du$ to six decimal places.

Solution By substituting $x = u^2$ in equation (10.24), we obtain

$$\cos(u^2) = 1 - \frac{u^4}{2!} + \frac{u^8}{4!} - \frac{u^{12}}{6!} + \frac{u^{16}}{8!} - \frac{u^{20}}{10!} + \cdots.$$

Recalling that we may integrate a power series term by term in its interval of convergence, we have

$$\int_0^{0.3} \cos(u^2)\,du = \left(u - \frac{u^5}{(2!)\cdot 5} + \frac{u^9}{(4!)\cdot 9} - \frac{u^{13}}{(6!)\cdot 13} + \cdots \right)\Bigg|_{u=0}^{u=0.3}.$$

It follows that

$$\int_0^{0.3} \cos(u^2)\,du = 0.3 - \frac{(0.3)^5}{(2!)\cdot 5} + \frac{(0.3)^9}{(4!)\cdot 9} - \frac{(0.3)^{13}}{(6!)\cdot 13} + \cdots.$$

The series on the right is alternating. Since the third term, $(0.3)^9/(4!\cdot 9) = 9.1125 \times 10^{-8}$, is less than 5×10^{-7}, we have

$$\int_0^{0.3} \cos(u^2)\,du = 0.3 - \frac{(0.3)^5}{(2!)\cdot 5} = 0.299757,$$

which is correct to six decimal places. Software for numerical integration gives us $\int_0^{0.3} \cos(u^2)\,du = 0.29975709\ldots$, confirming our estimate. ■

quickquiz

1. How do we tell whether the Taylor series of a function converges back to the function?
2. Can $x \mapsto |x|$ have a power series expansion about 0? About π?
3. What are the Maclaurin series for $\exp(x)$, $\sin(x)$, and $\cos(x)$?
4. What is the Newton binomial series associated to $\alpha \in \mathbb{R}$? What function does it converge to?

EXERCISES

Problems for Practice

In Exercises 1–12, derive the Maclaurin series of the function by using a known Maclaurin series. (The functions of Exercises 10–12 are extended by continuity at 0.)

1. $\sin(2x)$
2. $x\cos(x)$
3. $\exp(1-x)$
4. $\exp(-x^2)$
5. $x^3 + \cos(x^2)$
6. $x^2/(1-x)$
7. $x/(1+x^2)$
8. $\sin^2(x)$
9. $\sqrt{1+x^3}$
10. $\sin(x)/x$
11. $\ln(1+x^2)/x$
12. $(1-\cos(x))/x^2$

In Exercises 13–20, compute the Taylor series expansion for the given function about the point c on the given interval I.

13. $f(x) = \sqrt{x}, c = 1, I = (0, 2)$
14. $f(x) = e^{3x}, c = 0, I = (-\infty, \infty)$
15. $f(x) = x^2 \cdot e^x, c = 0, I = (-\infty, \infty)$
16. $f(x) = 1/x, c = 2, I = (1, 3)$
17. $f(x) = \ln(x), c = 1, I = (0, 2)$
18. $f(x) = \ln(x), c = 3, I = (2, 4)$
19. $f(x) = \sin(x)\cos(x), c = 0, I = (-\infty, \infty)$
20. $f(x) = x \cdot \ln(x), c = 3, I = (2, 4)$

In Exercises 21–24, use Taylor's Theorem to approximate the expression to four decimal places.

21. $\cos(0.2)$
22. e
23. $\exp(-0.2)$
24. $\arctan(0.15)$

In Exercises 25–30, write out Newton's binomial series for the expression up to and including the x^3 term.

25. $(1+x)^{3/4}$ **26.** $(1+x)^4$
27. $1/\sqrt{1+x}$ **28.** $1/(1+x)$
29. $1/(1+x)^{3/2}$ **30.** $1/(1+x)^2$

Further Theory and Practice

In Exercises 31–36, evaluate the series (exactly).

31. $\frac{\pi}{4} - \frac{\pi^3}{4^3 \cdot 3!} + \frac{\pi^5}{4^5 \cdot 5!} - \frac{\pi^7}{4^7 \cdot 7!} + \cdots$

32. $1 - \frac{\pi^2}{3^2 \cdot 2!} + \frac{\pi^4}{3^4 \cdot 4!} - \frac{\pi^6}{3^6 \cdot 6!} + \frac{\pi^8}{3^8 \cdot 8!} - \cdots$

33. $1 - \frac{\pi^2}{2^2 \cdot 3!} + \frac{\pi^4}{2^4 \cdot 5!} - \frac{\pi^6}{2^6 \cdot 7!} + \cdots$

34. $1 + \frac{1}{1!} + \frac{1}{2!} + \frac{1}{3!} + \frac{1}{4!} + \cdots$

35. $1 - \frac{1}{1!} + \frac{1}{2!} - \frac{1}{3!} + \frac{1}{4!} - \cdots$

36. $\frac{2}{1!} + \frac{4}{2!} + \frac{6}{3!} + \frac{8}{4!} + \frac{10}{5!} + \frac{12}{6!} + \cdots$

37. Find the Maclaurin series of

$$\cosh(x) = \frac{\exp(x) + \exp(-x)}{2}.$$

38. Find the Maclaurin series of

$$\sinh(x) = \frac{\exp(x) - \exp(-x)}{2}.$$

In Exercises 39 42, use a Taylor polynomial to calculate the integral to five decimal places.

39. $\int_0^{1/2} e^{-x^2}\, dx$ **40.** $\int_0^{\pi/4} \sin(x)/x\, dx$
41. $\int_0^{1/3} 1/(1+x^5)\, dx$ **42.** $\int_0^{1/2} \sqrt{1+x^3}\, dx$

43. Let c be a critical point of a function f that is equal to its Taylor series with base point c on an open interval centered at c. Use the Taylor series of f to derive the Second Derivative Test for a local extremum at c.

44. Use the Maclaurin series of $\cos(x)$, namely $1+u$ where $u = \sum_{n=1}^{\infty}(-1)^n x^{2n}/(2n)!$, and the Maclaurin series $\sum_{m=1}^{\infty}(-1)^{m+1}u^m/m$ of $\ln(1+u)$ to show that the Maclaurin series of $\ln(\cos(x))$ is $-x^2/2 - x^4/12 - x^6/45 - \cdots$. Use this result to obtain the Maclaurin series of $\tan(x)$, up to and including the x^5 term.

45. Demonstrate the Maclaurin series expansion

$$\ln\left(\frac{1+x}{1-x}\right) = 2x + \frac{2x^3}{3} + \frac{2x^5}{5} + \frac{2x^7}{7} + \cdots,$$

which is valid for $-1 < x < 1$. Use this expansion to estimate $\ln(2)$ to five decimal place accuracy. The series used here is more rapidly convergent than the more familiar series

$$\ln(2) = 1 - \frac{1}{2} + \frac{1}{3} - \frac{1}{4} + \frac{1}{5} - + \cdots.$$

46. Show that

$$\binom{-1/2}{n} = \frac{(-1)^n(2n)!}{2^{2n}(n!)^2}.$$

Using this formula, state the Maclaurin series of $(1+u)^{-1/2}$ and derive the Maclaurin series of $\arcsin(x)$.

47. Fix a nonzero real number α. Define $f(x) = (1+x)^\alpha$ and

$$g(x) = \sum_{n=0}^{\infty} \binom{\alpha}{n} x^n.$$

Your goal is to show that $f(x) = g(x)$ for $x \in (-1, 1)$. You know from the text that the series certainly converges on that interval. Now complete the following steps.

a. Prove that

$$(n+1) \cdot \binom{\alpha}{n+1} + n \cdot \binom{\alpha}{n} = \alpha \cdot \binom{\alpha}{n}.$$

b. Prove that $(1+x) \cdot g'(x) = \alpha \cdot g(x)$, $x \in (-1, 1)$.

c. Prove that $(1+x) \cdot f'(x) = \alpha \cdot f(x)$, $x \in (-1, 1)$.

d. Notice that $f(0) = g(0) = 1$.

e. Solve the equation $(1+x)y' = \alpha \cdot y$ by rewriting it as

$$\frac{y'}{y} = \frac{\alpha}{1+x}.$$

f. Find the unique solution to the equation $(1+x)y' = \alpha \cdot y$ satisfying $y(0) = 1$. Conclude that $f = g$.

48. Prove that for any real numbers θ and x, we have

$$e^{x \cdot \cos(\theta)} \cdot \cos(x \cdot \sin(\theta)) = \sum_{n=0}^{\infty} \cos(n\theta) \cdot \frac{x^n}{n!}.$$

49. Suppose f is a function with derivatives of all orders. Prove that if f satisfies

$$(1-x^2)f^{(n+2)}(x) - (2n+1)xf^{(n+1)}(x) - n^2 f^{(n)}(x) = 0 \quad \textbf{(10.28)}$$

for a particular nonnegative integer n, then it also satisfies equation (10.28) for the next integer $n+1$. In other words, if (10.28) is true for a particular value of n, then the equation obtained from (10.28) by replacing n with $n+1$ is also true. Show that for $f(x) = \arcsin(x)$, equation (10.28) is true for $n = 0$. Deduce that for $f(x) = \arcsin(x)$, equation (10.28) is true for each nonnegative integer n. Use (10.28) to express $f^{(n+2)}(0)$ in terms of $f^{(n)}(0)$. Hence, show that

$$x + \frac{1^2}{3!}x^3 + \frac{1^2 \cdot 3^3}{5!}x^5 + \frac{1^2 \cdot 3^3 \cdot 5^2}{7!}x^7 + \cdots$$

is the Maclaurin series of $\arcsin(x)$.

50. Suppose f is a function with derivatives of all orders. Prove that if f satisfies

$$(1 - x - x^2)f^{(n+2)}(x) - (n+2)(1+2x)f^{(n+1)}(x) - (n+2)(n+1)f^{(n)}(x) = 0 \qquad \textbf{(10.29)}$$

for a particular nonnegative integer n, then it also satisfies equation (10.29) for the next integer $n+1$. In other words, if (10.29) is true for a particular value of n, then the equation obtained from (10.29) by replacing n with $n+1$ is also true. Show that for $f(x) = x/(1 - x - x^2)$, equation (10.29) is true for $n = 0$. Deduce that for $f(x) = x/(1 - x - x^2)$, equation (10.29) is true for each nonnegative integer n. Use (10.29) to express $f^{(n+2)}(0)$ in terms of $f^{(n+1)}(0)$ and $f^{(n)}(0)$. Hence, show that the Maclaurin series of $x/(1 - x - x^2)$ is $\sum_{n=0}^{\infty} a_n x^n$ where $a_0 = 0$, $a_1 = 1$, and $a_n = a_{n-1} + a_{n-2}$ for $n \geq 2$.

Calculator/Computer Exercises

51. Let $f(x) = x - 3\sin(x)/(2 + \cos(x))$. Its Taylor series is

$$f(x) = \frac{x^5}{180} + \frac{x^7}{1512} + \frac{x^9}{25920} + \cdots.$$

Try plotting $f(x)/x^5$ for $0.01 < x < 0.02$. What should the plot look like based on the Taylor series of f?

52. Plot

$$f(x) = \frac{\sin(\tan(x)) - \tan(\sin(x))}{\arcsin(\arctan(x)) - \arctan(\arcsin(x))},$$

$0.01 < x < 0.02$. If your plot did not come out correctly, then the following considerations may help. The numerator of $f(x)$ has Maclaurin series

$$-\frac{x^7}{30} - \frac{29x^9}{756} + \cdots$$

and the denominator has Maclaurin series

$$-\frac{x^7}{30} + \frac{13x^9}{756} + \cdots.$$

Use these two series to plot f in the given interval.

53. Verify that

$$J_0(x) = \sum_{n=0}^{\infty} (-1)^n \frac{x^{2n}}{2^{2n}(n!)^2}$$

is a solution of the differential equation

$$x^2\frac{d^2y}{dx^2} + x\frac{dy}{dx} + x^2 y = 0.$$

The function $J_0(x)$ is called *Bessel's function of order 0*. Its plot looks like that of $\cos(x)$ but with diminishing peaks as x tends to infinity. To get an idea, plot $\sum_{n=0}^{9}(-1)^n x^{2n}/(2^{2n}(n!)^2)$ for $0 \leq x \leq 8$. The plot is a quite accurate representation of $J_0(x)$ up to $x = 7$.

54. The error function is denoted by erf and is defined by

$$\operatorname{erf}(x) = \frac{2}{\sqrt{\pi}} \int_0^x e^{-t^2}\, dt.$$

By expanding $\exp(-t^2)$ in a Maclaurin series and integrating term-by-term, obtain a Maclaurin series of $\operatorname{erf}(x)$. By truncating this series at a suitable degree, plot the error function for $-1.5 < x < 1.5$.

55. Define

$$f(x) = \begin{cases} 0 & \text{if } x \leq 0 \\ e^{1-1/x^2} & \text{if } x > 0 \end{cases}.$$

It may be shown that f has derivatives of all orders at every point (including 0) and that $f^{(n)}(0) = 0$ for every nonnegative integer n. The Taylor series of f with base point 0 is therefore everywhere convergent to zero; in particular, it converges to $f(x)$ only at the base point. This exercise provides some visual evidence for this phenomenon. Plot x^2 for $-1 \leq x \leq 1$. One at a time, add the plots of x^4, x^6, and x^8 in the viewing window $[-1, 1] \times [0, 1]$. Describe the behavior of the plots of x^{2n} as n increases. Add the plot of f to the viewing window. Now plot x^8 and $f(x)$ in the viewing window $[-0.3, 0.3] \times [0, 0.00002]$. Continue by plotting x^{14} and $f(x)$ in the viewing window $[0.02, 0.03] \times [0, 5 \times 10^{-22}]$. The plots suggest that as x tends to 0 the expression $f(x)$ approaches 0 more rapidly than any positive power of x. Notice that $\exp(-1/x^2)/x^{2n} = u^n/e^u$ where $u = 1/x^2$ approaches infinity as x approaches zero. Use l'Hôpital's Rule to show that $\lim_{x\to 0} \exp(-1/x^2)/x^{2n} = 0$ for every positive integer n.

Summary of Key Topics

Power Series and Radius of Convergence (Section 10.1)

An expression

$$\sum_{n=0}^{\infty} a_n(x-c)^n$$

is called a *power series with base point* c (or *centered* at c). We set $\ell = \lim_{n\to\infty} |a_n|^{1/n}$, provided that this limit exists as a finite or infinite number. If it is convenient, then the formula $\ell = \lim_{n\to\infty} |a_{n+1}|/|a_n|$ may be used instead. The number $R = 1/\ell$ is called the *radius of convergence* of the power series (with the understanding that $R=0$ when $\ell=\infty$ and $R=\infty$ when $\ell=0$). The power series will converge absolutely for $|x-c|<R$ and diverge for $|x-c|>R$. The *interval of convergence* of a power series is the set of points at which the series is convergent. If R is a positive finite number, then the interval of convergence will be either $(c-R, c+R)$, $[c-R, c+R)$, $(c-R, c+R]$, or $[c-R, c+R]$. If R is infinite, then the interval of convergence is the entire real line. If $R=0$, then the interval of convergence is just the point c.

Basic Properties of Power Series (Section 10.2)

In the interval of convergence,

$$\sum_{n=0}^{\infty}(a_n+b_n)x^n = \sum_{n=0}^{\infty} a_n + \sum_{n=0}^{\infty} b_n, \quad \sum_{n=0}^{\infty}(a_n-b_n)x^n = \sum_{n=0}^{\infty} a_n - \sum_{n=0}^{\infty} b_n, \quad \text{and}$$

$$\sum_{n=0}^{\infty} \lambda a_n x^n = \lambda \sum_{n=0}^{\infty} a_n x^n \quad \text{for every } \lambda \in \mathbb{R}.$$

Power series may be differentiated and integrated term by term in the interior of the interval of convergence.

These theorems may be used to generate convergent power series expansions for several functions.

Expanding Functions in Power Series (Section 10.3)

If f is $N+1$ times continuously differentiable on an interval I centered at c, then for any $x_0 \in I$ we have

$$f(x_0) = \underbrace{\sum_{n=0}^{N} \frac{f^{(n)}(c)}{n!}\cdot(x_0-c)^n}_{P_N(x_0)} + \underbrace{\frac{f^{(N+1)}(\xi)}{(N+1)!}\cdot(x_0-c)^{N+1}}_{R_N(x_0)}$$

for some ξ between c and x_0. We call P_N the Taylor polynomial of degree N for the function f expanded about the point c. We call R_N the remainder term of order N.

Estimating the Remainder Term (Section 10.4)

The remainder term R_N satisfies the estimate

$$|R_N(x_0)| \le M_{N+1}\cdot\frac{|x_0-c|^{N+1}}{(N+1)!}$$

where

$$M_{N+1} = \max_{x \in J} \left| f^{(N+1)}(x) \right|$$

and J is the interval $[c, x_0]$ if $c \leq x_0$ or $[x_0, c]$ if $x_0 < c$. The power series

$$\sum_{n=0}^{\infty} \frac{f^{(n)}(c)}{n!}(x-c)^n$$

converges to $f(x)$ if and only if $R_N(x) \to 0$ as $N \to \infty$.

Theorem 2 from Section 10.3 (Taylor's Theorem) and Theorem 1 from Section 10.4 (the error estimate) may be used to estimate the values of many transcendental functions to any desired degree of accuracy.

Power Series Representations of Functions (Section 10.5)

Many familiar functions have convergent power series representations. Indeed,

$$\sin(x) = \sum_{n=0}^{\infty} \frac{(-1)^n}{(2n+1)!} x^{2n+1} \quad (-\infty < x < \infty),$$

$$\cos(x) = \sum_{n=0}^{\infty} \frac{(-1)^n}{(2n)!} x^{2n} \quad (-\infty < x < \infty),$$

$$\ln(1+x) = \sum_{n=1}^{\infty} \frac{(-1)^{n-1}}{n} x^n \quad (-1 < x \leq 1),$$

$$e^x = \sum_{n=0}^{\infty} \frac{1}{n!} x^n \quad (-\infty < x < \infty).$$

The function $f(x) = (1+x)^\alpha$ has a particularly interesting power series representation called the binomial series:

$$(1+x)^\alpha = \sum_{n=0}^{\infty} \binom{\alpha}{n} x^n \quad \text{for } |x| < 1$$

where

$$\binom{\alpha}{n} = \frac{\alpha \cdot (\alpha - 1) \cdots (\alpha - n + 1)}{n!}.$$

genesis & DEVELOPMENT

Calculus is usually considered to be a product of the rise of scientific thought in 17th century Europe. For the most part, it is. However, one notable exception is the discovery of power series in India nearly 300 years before their introduction in the West. Although many details are not yet known, it is certain that Mādhavan and Nilakantha (?–1545) had the basic trigonometric power series, including the Maclaurin series for $\sin(x)$, $\cos(x)$, and $\arctan(x)$, in their possession. It is believed that the need for an accurate value of π in astronomical computations was the motivation for the discovery of these identities. For example, by substituting $x = 1$ in the Maclaurin series for the arctangent, we obtain the series

$$\frac{\pi}{4} = 1 - \frac{1}{3} + \frac{1}{5} - \frac{1}{7} + \frac{1}{9} - \frac{1}{11} + \cdots,$$

now known as the *Leibniz series,* or sometimes as the *Gregory-Leibniz series,* even though it had been explicitly recorded in India centuries earlier. Because of the slow convergence of this series, however, it is useless for the calculation of π. However, Indian mathematicians derived many other *more rapidly convergent* series. One example is

$$\frac{\pi}{16} = \sum_{n=1}^{\infty} \frac{(-1)^{n-1}}{(2n-1)^5 + 4(2n-1)},$$

the first ten terms of which give six decimal places of accuracy. Whether any of these remarkable results spread to other parts of India and beyond is not known at the time of this writing.

Early Power Series Developments

Before the general theory of Taylor series was expounded, many particular power series expansions were discovered and rediscovered. For example, in 1624 Henry Briggs (1561–1631) derived the power series formula

$$(1+x)^{1/2} = 1 + \frac{1}{2}x - \frac{1}{8}x^2 + \frac{1}{16}x^3 - \frac{5}{128}x^4 + \cdots$$

in the course of calculating logarithms. Briggs's formula is a special case of the binomial series. The Maclaurin series of $\ln(1+x)$,

$$\ln(1+x) = x - \frac{1}{2}x^2 + \frac{1}{3}x^3 - \frac{1}{4}x^4 + \frac{1}{5}x^5 - \frac{1}{6}x^6 + \frac{1}{7}x^7 - \cdots,$$

was independently discovered by Johann Hudde (1628–1704) in 1656, by Sir Isaac Newton in 1665, and by Nicolaus Mercator in 1668. By the time Mercator published his series for the logarithm, James Gregory and Sir Isaac Newton had developed quite general techniques for calculating power series expansions.

James Gregory and Sir Isaac Newton

Historical research in the 20th century has revealed that both James Gregory and Sir Isaac Newton anticipated Taylor in the discovery of Taylor series. Of the two, Newton's successes with power series came earlier, but Gregory found Taylor series first.

In 1665, when he was "in the prime of his age for invention," Newton hit upon the binomial series, as well as several other power series. In fact, power series played an important role in his conception of calculus. Early in the 1670s, Newton began to correspond with James Gregory, with John Collins acting as intermediary. Through Collins, Newton sent Gregory examples of his method of power series, and Gregory responded in kind. By 1671, Gregory was in possession of the Maclaurin series for $\arcsin(x)$, $\arctan(x)$, $\tan(x)$, and $\sec(x)$. During the 1930s, an analysis of Gregory's private papers proved that Gregory derived the power series expansions of $\tan(x)$ and $\sec(x)$ *by means of Taylor series,* which means that James Gregory knew the Taylor series formula 14 years before Taylor was born!

From his letters to Collins, it appears that Gregory believed that he had merely rediscovered the method that Newton used. As he wrote to Collins, "As for Mr. Newton's universal method, I imagine I have some knowledge of it." It seems likely that Gregory withheld disclosure of his important discovery in order to cede the privilege of first publication to Newton. This is ironic, for

Newton did not discover Taylor series until 1691, and that discovery, hidden in his private papers, did not come to light until 1976!

It is often written that the binomial series is engraved on the monument to Newton at Westminster Abbey. This myth was flatly rejected by De Morgan in the mid-1800s. When the Dean of Westminster Abbey was asked to verify De Morgan's assertion near the end of the 19th century, he wrote, "In front of the half-recumbent figure of Sir Isaac Newton are two winged youths holding a small scroll.... I fear that the figures on the small marble scroll are quite obliterated.... Time, I fear, and London atmosphere have done their sad work."

Brook Taylor

Brook Taylor (1685–1731) was born into a well-to-do family with connections to the nobility. Although he studied law at Cambridge, his scientific career was already under way by the time he graduated. In 1712, when Taylor was still preparing for a legal career, he was selected by the Royal Society to be a member of a committee that was charged with the investigation of the Newton-Leibniz calculus dispute. In this capacity, Taylor was introduced to many of England's greatest mathematicians. Encouraged by these contacts to develop and communicate his discoveries, Taylor wrote a letter in 1712, privately disclosing the series that now bears his name. He published a number of influential scientific articles and one major book, *The Method of Increments,* which first appeared in 1715. It was in this book that Taylor made public his discovery of the formula

$$f(x+h) = f(x) + \frac{f'(x)}{1!}h + \frac{f''(x)}{2!}h^2 + \frac{f'''(x)}{3!}h^3 + \cdots + \frac{f^{(n)}(x)}{n!}h^n + \cdots,$$

written here in modern notation.

The terminology *Taylor series* has been traced to 1786; it quickly became established. By the mid-1800s, one writer commented, "A single analytical formula in the *Method of Increments* has conferred a celebrity on its author, which the most voluminous works have not often been able to bestow." Later in the 19th century, however, the attribution came into question. Although the unpublished work of Gregory and Newton was not yet known, there were other anticipations that were published: Johann Bernoulli's *series universalissima* that he published in 1694 and a general series expansion that Abraham de Moivre published in 1708. Like many historical questions, this one has no clear-cut resolution. However, the contemporaries of Bernoulli, de Moivre, and Taylor did not deny Taylor's claim to priority. Moreover, Taylor's immediate successors, including Maclaurin and Euler, credit him not only for introducing the series that now bear his name but also for demonstrating the uses to which they can be put. Modern opinion has accepted this verdict. Nevertheless, it should be noted that the error term in Taylor's Theorem was introduced by Lagrange later in the 18th century.

Taylor's scientific activity was, like his life, relatively brief. Most of his mathematical work was concentrated in the years 1712–1719. In addition to the theorem for which he is best known, he wrote an influential study of the vibrating string. He was probably led to that problem through his interest in music, for he was an accomplished harpsichordist and even wrote a book on music. Taylor was also a talented painter whose canvases decorated the walls of the family estate. The last decade of Taylor's life, however, was not occupied by the happy pursuit of music and art. Rather, it was marked by misfortune and emotional distress. Taylor's marriage to a woman without means led to an estrangement from his father. Two years later, Taylor's wife died in childbirth. He remarried, but his second wife also died in childbirth. This last tragedy precipitated a rapid decline in Taylor's health, which led to his own premature death soon afterward.

Colin Maclaurin

Colin Maclaurin (1698–1746) was born a minister's son. Orphaned at the age of 9, he became the ward of an uncle, who was also a minister. The precocious boy entered the University of Glasgow just two years later. The plan was that he too would become a minister. At the age of 12, however, Maclaurin discovered the *Elements* of Euclid and was led astray. Within a few days, he mastered the first six books of Euclid. His interest in mathematics did not wane thereafter. Three years later, at the age of 15, he defended his thesis on gravity and graduated from university.

The young Maclaurin had to bide his time for a few years, but he was still only 19 when he was appointed professor of mathematics at the University of Aberdeen in 1717. During the next two years, he forged a warm friendship with Sir Isaac Newton, a bond that Maclaurin described as his "greatest honour and happiness." On Newton's recommendation, the University of Edinburgh offered Maclaurin a professorship in 1725. It came in the nick of time—Aberdeen was about to dismiss Maclaurin for being absent without leave for nearly three years. Maclaurin's teaching at Edinburgh had happier results—one student wrote that "He made mathematics a fashionable study."

Maclaurin's research also received the highest praise. Lagrange described Maclaurin's study of tides as "a masterpiece of geometry, comparable to the most beautiful and ingenious results that Archimedes left to us." Maclaurin's *Treatise of Fluxions,* which appeared in 1742, is considered his major work. He conceived it as an answer to the criticisms of Bishop Berkeley that were discussed in the Genesis & Development for Chapters 2 and 3. It is ironic that Maclaurin is now primarily known for the series that bear his name. Although *Maclaurin series* are to be found in the *Treatise of Fluxions,* Maclaurin himself attributed them to Taylor. Thus, the terminology *Maclaurin series* is not historically justified. Nonetheless, it is convenient, as it saves us from having to say "Taylor series expanded about 0." Furthermore, it is appropriate to remember this great mathematician in some way because ideas such as the Integral Test, which Maclaurin did introduce in his *Treatise,* have not been named for him.

APPENDIX: ANSWERS TO SELECTED EXERCISES

Section 1.1

1. 213/100 **3.** 23/99 **5.** 4996/999 **7.** 0.025

9. $1.\bar{6}$ (The overline identifies the repeating block.)

11. 0.72 **13.** $[1, 3]$ **15.** $(1, \infty)$ **17.** $[-14, 6]$

19.

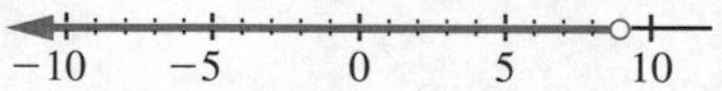

−10 −5 0 5 10

21.

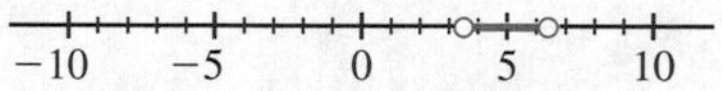

−10 −5 0 5 10

23.

−10 −5 0 5 10

25. $\{x : |x - 1| \leq 2\}$

27. $\{x : |x - 1| < \pi + 1\}$

31. Both approximations are rational numbers, but π is not.

33. $[34.155, 34.845]$; $\{x : |x - 34.5| \leq 0.345\}$

35.

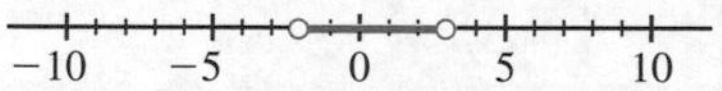

−10 −5 0 5 10

37.

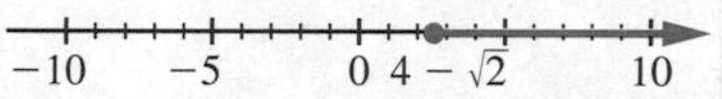

−10 −5 0 $4 - \sqrt{2}$ 10

39.

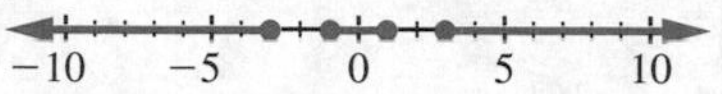

−10 −5 0 5 10

41.

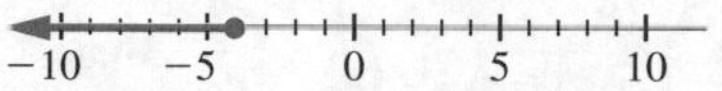

−10 −5 0 5 10

43.

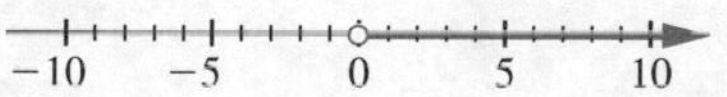

−10 −5 0 5 10

45. $\{-3\}$

47. $\{x : 1 < x \leq 2\sqrt{2}\}$

49. $\{s : -13 < s < -5/3\}$

51. $\{w : |w + 1|/(w + 1) = 1\}$

59. $[0.4485, 0.4495]$

61. $[-0.00049001, 0.00050999]$

63. 4.001

65. The following table records the calculation of $\alpha_0 \cdot \beta_0$ using $n = 12, 14, 16, 18$, and 20 digits:

n	$\alpha_0 \cdot \beta_0$
12	34562960.8
14	−345629.6
16	0.000000
18	0.000000
20	1.0369

67.

x	Relative Error
10^{10}	0.5
10^{12}	9.901×10^{-3}
10^{14}	9.999×10^{-5}
10^{16}	9.99999×10^{-7}
10^{18}	9.9999999×10^{-9}

x	Relative Error
-10^{12}	0.0101
-10^{14}	1.0001×10^{-4}
-10^{16}	1.000001×10^{-6}
-10^{18}	$1.00000001 \times 10^{-8}$
-10^{20}	$1.0000000001 \times 10^{-10}$

69.

x	Relative Error
10^{3}	1.031
10^{6}	0.0345
10^{9}	0.00003
10^{12}	0.3×10^{-7}
10^{15}	0.3×10^{-10}

x	Relative Error
-10^{3}	0.9709
-10^{6}	0.0322
-10^{9}	0.00003
-10^{12}	0.3×10^{-7}
-10^{15}	0.3×10^{-10}

Section 1.2

1. $(7, 3)$, $(\sqrt{3}, \sqrt{6})$, $(\pi^2, \sqrt{\pi})$, $(\pi + 3, \pi - 3)$, $\left(-\frac{8}{3}, -\frac{2}{\pi}\right)$, $(2, -4)$

3. $(5, 7)$, $(4, 6)$, $(3, 5)$, $(2, 4)$, $(1, 3)$, $(0, 2)$, $(-1, 1)$, $(-2, 0)$, $(-4, -2)$, $(-5, -3)$

5. $|\overline{AB}| = 2\sqrt{13}$, $|\overline{AC}| = \sqrt{130}$, $|\overline{BC}| = \sqrt{170}$

7. Center: $(1, 3)$; radius: 3

9. Center: $(-2, 1)$; radius: $3/\sqrt{2}$

11. Center: $(0, -5)$; radius: $\sqrt{2}$

13. Center: $(-2, 1)$; radius: $\sqrt{17/3}$

15. Center: $(0, 1/2)$; radius: $1/2$

17. $(x + 3)^2 + (y - 5)^2 = 36$

19. $(x + 4)^2 + (y - \pi)^2 = 25$

21. Vertex: $(0, -3)$; axis of symmetry: $x = 0$

23. Vertex: $(1, 1)$; axis of symmetry: $x = 1$

25.

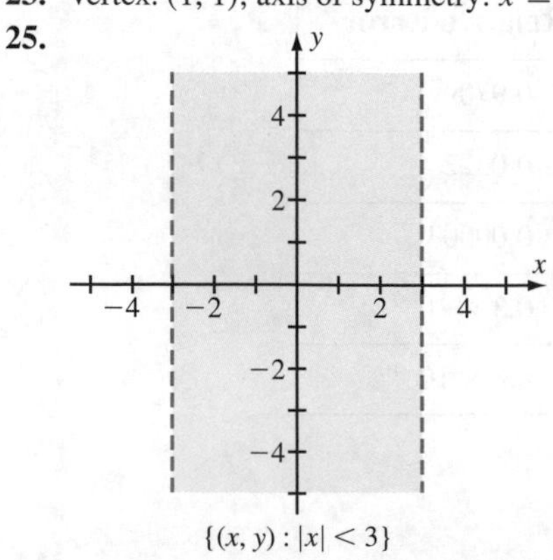

$\{(x, y) : |x| < 3\}$

27.

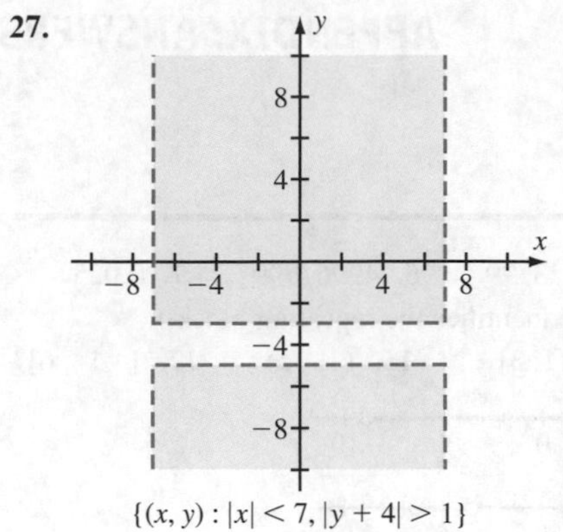

$\{(x, y) : |x| < 7, |y + 4| > 1\}$

29.

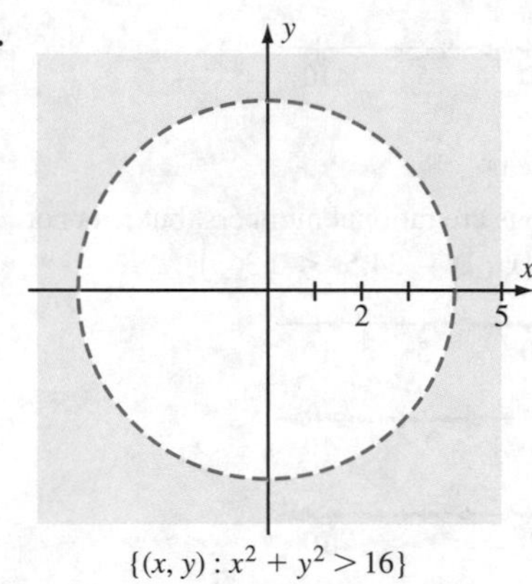

$\{(x, y) : x^2 + y^2 > 16\}$

31.

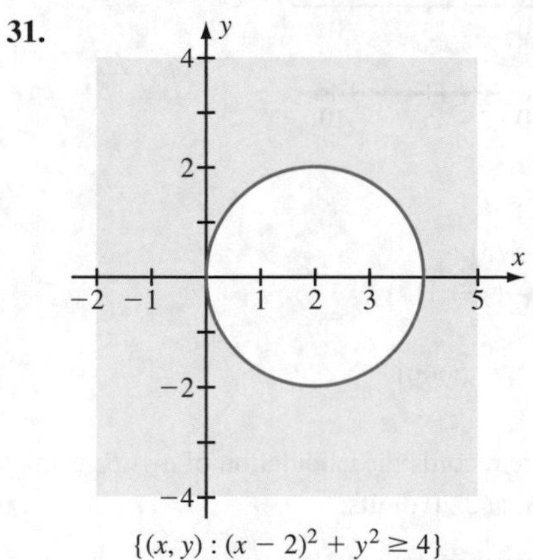

$\{(x, y) : (x - 2)^2 + y^2 \geq 4\}$

33.

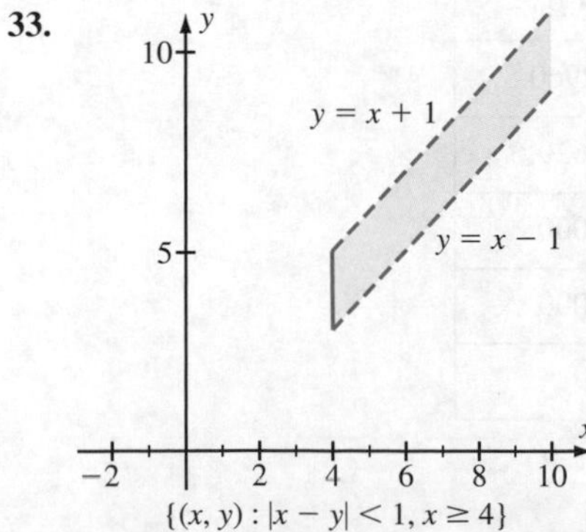

$\{(x, y) : |x - y| < 1, x \geq 4\}$

35.

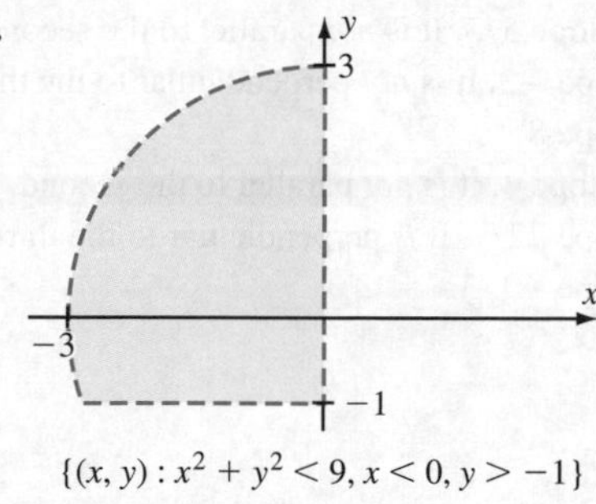

$\{(x, y) : x^2 + y^2 < 9, x < 0, y > -1\}$

37. $(447/82, 427/82)$

39. $(0, 0)$

41. Yes, it is a circle centered at $(1/5, -3/10)$ with radius $\sqrt{133}/10$.

43. No, it is the equation of a hyperbola.

45.

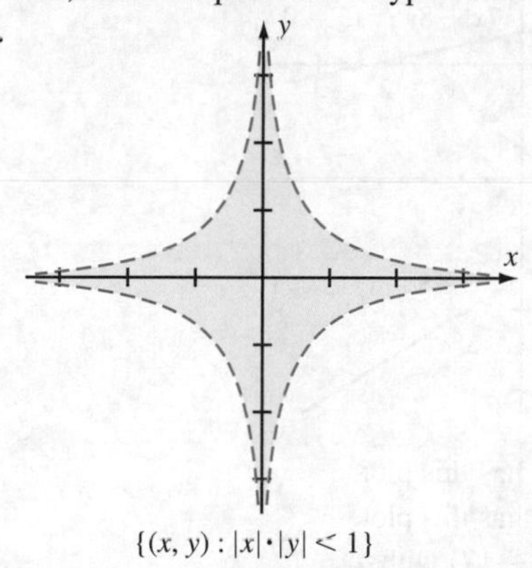

$\{(x, y) : |x| \cdot |y| < 1\}$

47.

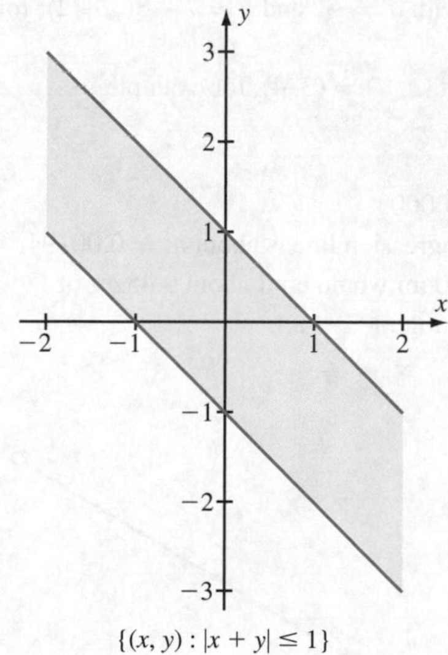

$\{(x, y) : |x + y| \leq 1\}$

49.

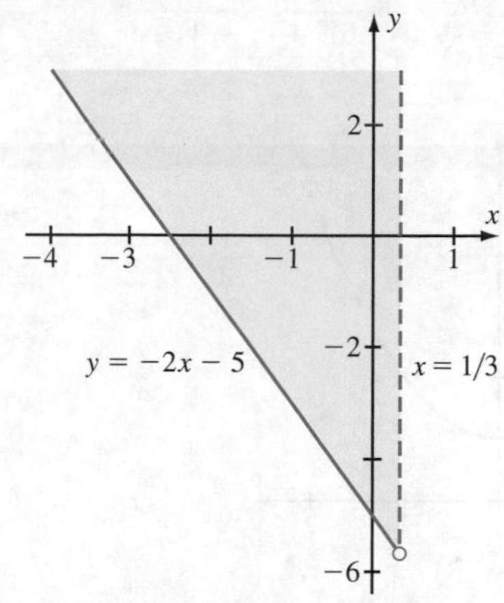

$\{(x, y) : 3x - 1 < 0 \leq y + 2x + 5\}$

51.

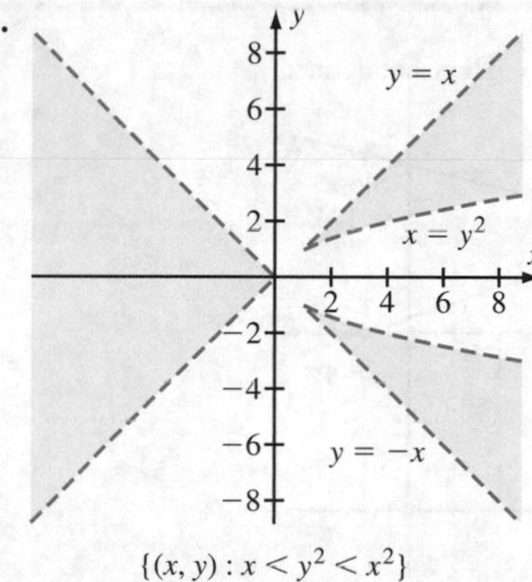

$\{(x, y) : x < y^2 < x^2\}$

53.

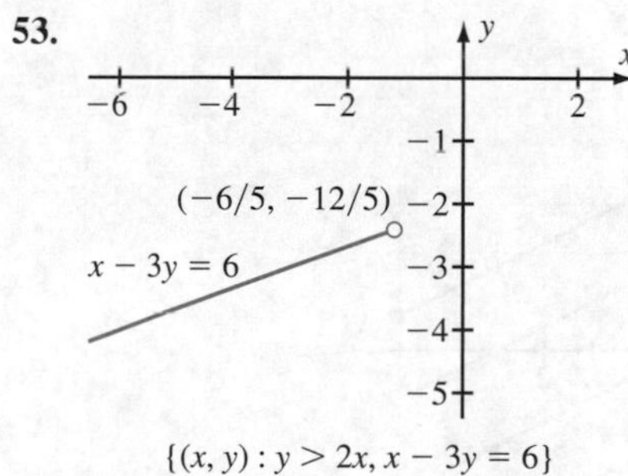

$\{(x, y) : y > 2x, x - 3y = 6\}$

55. A point $(u, 2u)$ on the circle $(x - h)^2 + (y - k)^2 = r^2$ satisfies the quadratic equation $(u - h)^2 + (2u - k)^2 = r^2$ in u. A quadratic equation has zero, one, or two solutions.

59. The equation is $y = \sqrt{(x-0)^2 + (y-1)^2}$, or $y = 1/2 + x^2/2$.

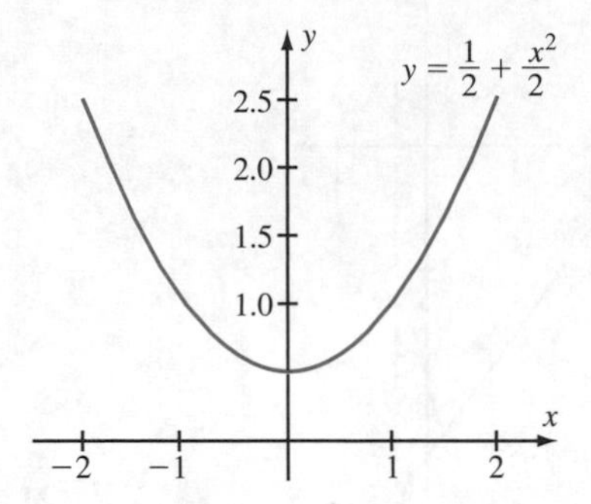

Section 1.3

1.

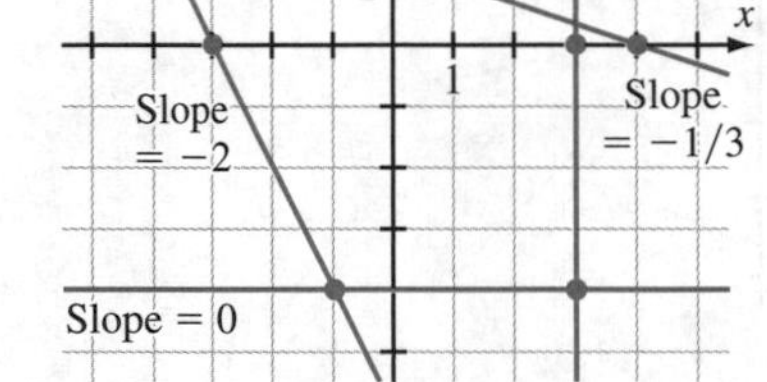

3.

5. $y = 5(x+3) + 7$

7. $y = -4x + 9$

9. $7x/9 + 5y/18 = 1$

11. $y = (147/220)(x - 18) + 14$

13. $y = 3(x+4)$

15. $y = (-1/3)x - 5$

17. $-2x + 4y + 8 = 0$

19. $y = (-1/3)x - 1/8$

21. The first line has slope $5/3$. It is *not* parallel to the second line, which has slope -3. It is *not* perpendicular to the third line, which has slope 8.

23. The first line has slope 3. It is *not* parallel to the second line, which has slope $22/7$. It *is* perpendicular to the third line, which has slope $-1/3$.

25.

27.

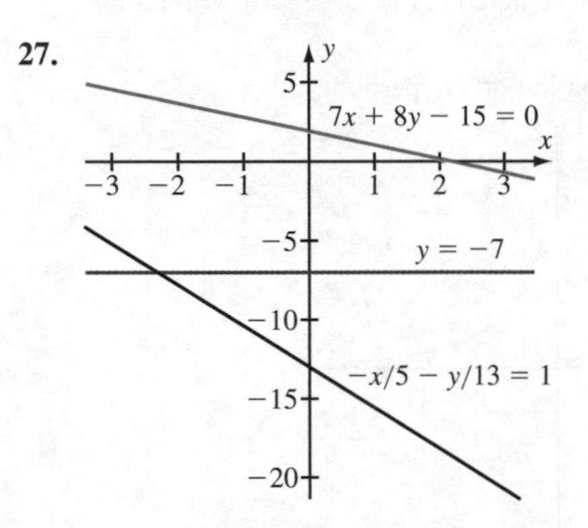

29. The figure for 27 has this plot.

31. The figure for 27 has this plot.

35. All points $(a, 2a - 17)$ with $a \neq 5$

37. All points (a, b) with $a \neq -2$ and $b = 7 - 3(a+2)$; for example, $(1, -2)$

39. $(a, b) = (1, 0)$ and $(c, d) = (3, 4)$, for example

41. $(332/73, 88/73)$

43. $(2, 1)$; $x + y = 3$

47. -1999997 and 2000003

49. The slope of the regression line is about $m = 0.00194$. A car with 100,000 mi would emit about 0.452 g of hydrocarbons per mile.

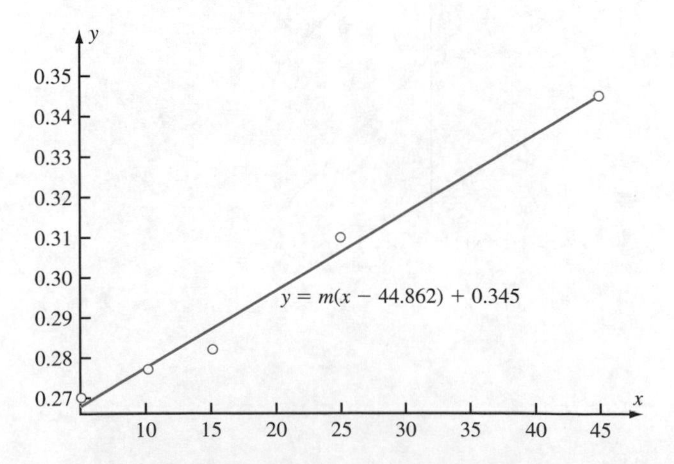

51. Equation of budget line: $x \cdot p_X + y \cdot p_Y = C$, $x \geq 0$ and $y \geq 0$; x-intercept: C/p_X, y-intercept: C/p_Y; slope: $-p_X/p_Y$; the budget line L' that corresponds to C' is parallel to the budget line L that corresponds to C. L' lies above L if $C' > C$; L' lies below L if $C' < C$.

57. As an example, zoom in to the window $[0.999, 1.001] \times [0.999^2, 1.001^2]$. The slope is $(1.002001 - 0.998001)/(1.001 - 0.999)$, or 2.0.

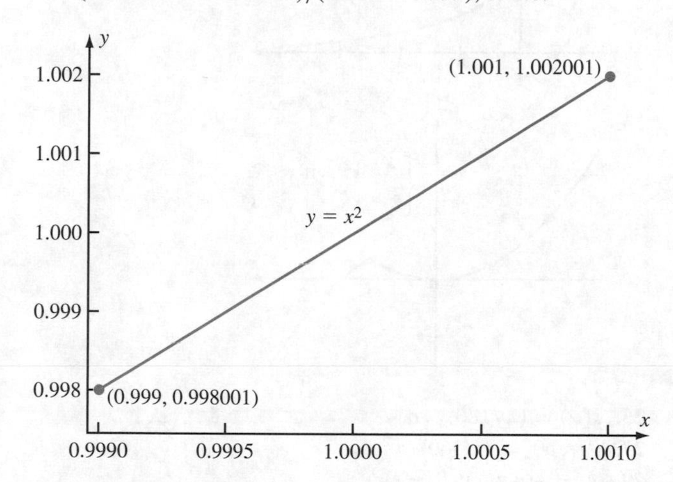

59. As an example, zoom in to the window $[-0.001, 0.001] \times [-0.001999998, 0.001999998]$. The slope is $(0.001999998 - (-0.001999998))/(0.001 - (-0.001))$, or 1.999998.

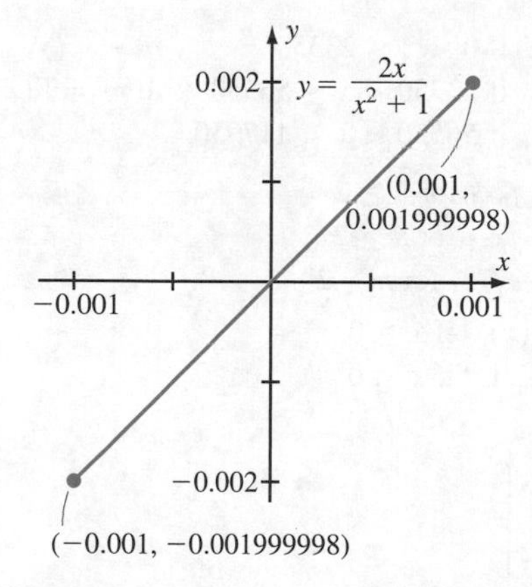

61. As an example, zoom in to the window $[1 - 10^{-8}, 1 + 10^{-8}] \times [0, 2]$. The slope is 0.

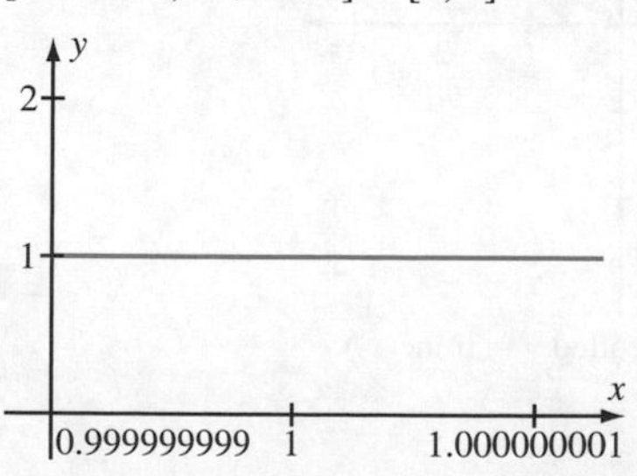

63. $Q = (3/4, \sqrt{3}/2)$; line ℓ has equation $y = (1/\sqrt{3})(x - 3/4) + \sqrt{3}/2$.

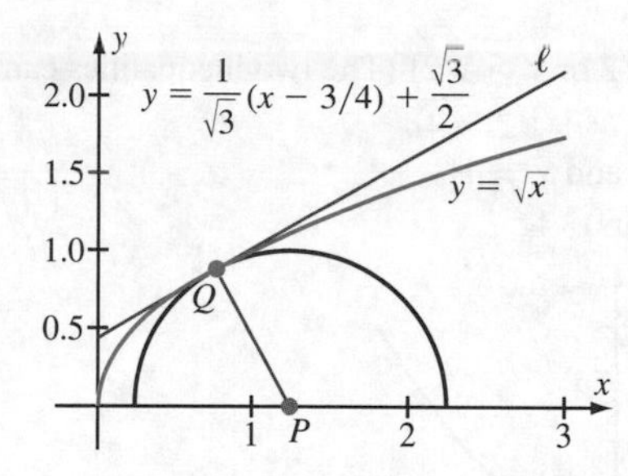

65. $m \approx 4.76923$ minimizes the sum of the squared errors; $m \approx 4.6666667$ minimizes the sum of the absolute errors.

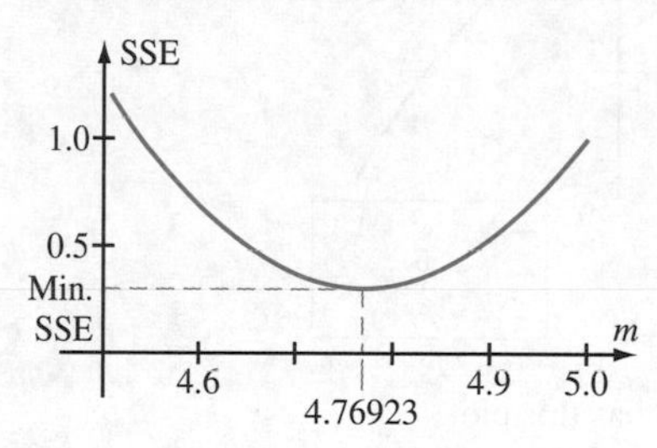

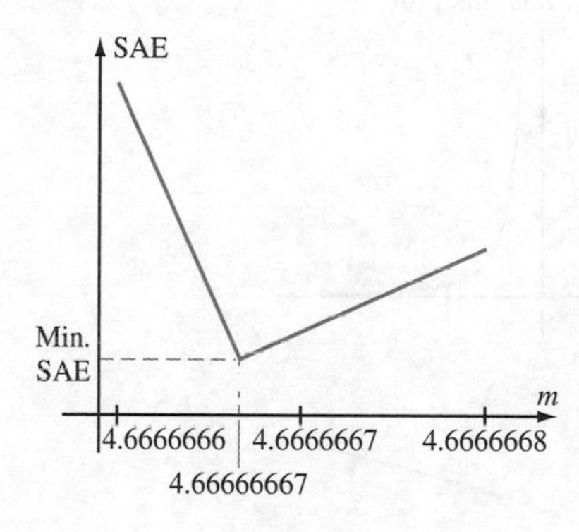

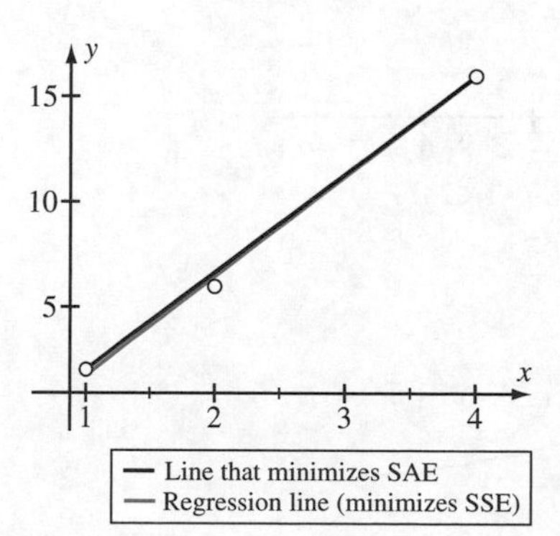

Section 1.4

1. $\{x \in \mathbb{R} : x \neq -1\}$

3. $\{x \in \mathbb{R} : x \leq -\sqrt{2} \text{ or } x \geq \sqrt{2}\}$ (The two inequalities can also be expressed as $|x| \geq \sqrt{2}$.)

5. $\{x \in \mathbb{R} : x \neq -1 \text{ and } x \neq 1\}$

7. $\mathbb{R}$ (all real numbers)

9.

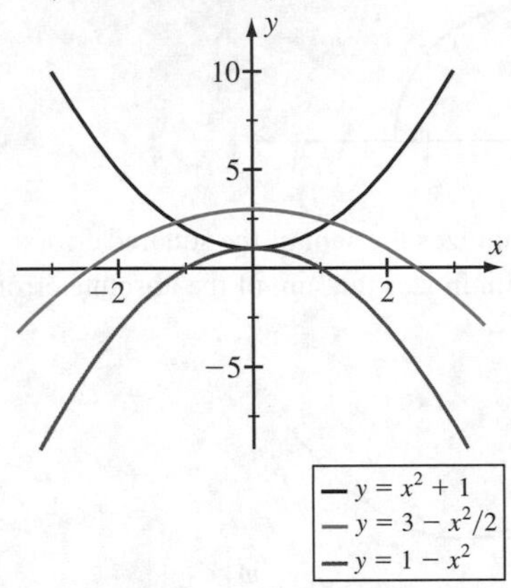

11. The figure for 9 has this plot.

13. The figure for 9 has this plot.

15.

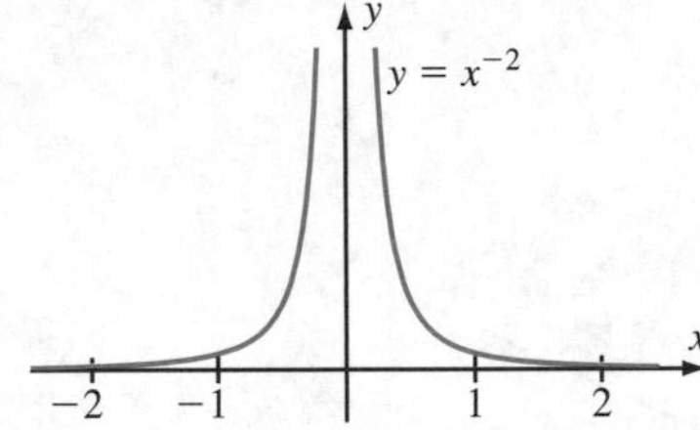

17.

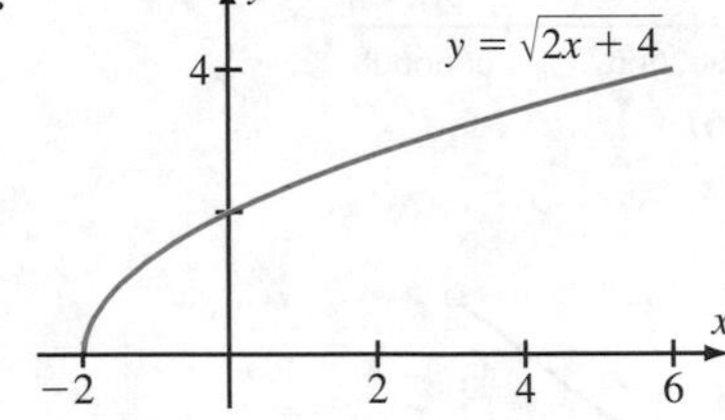

19.

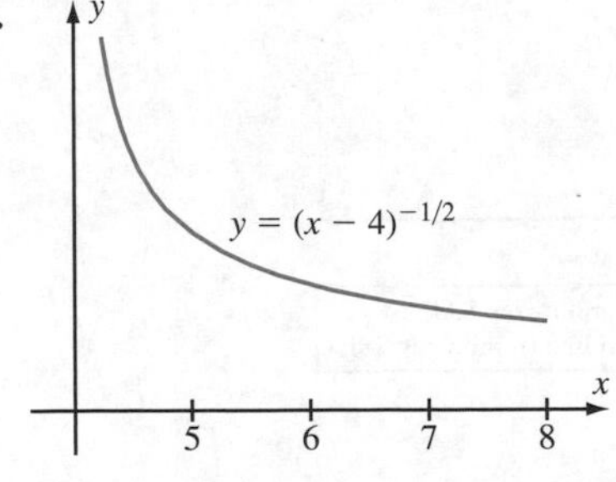

21.

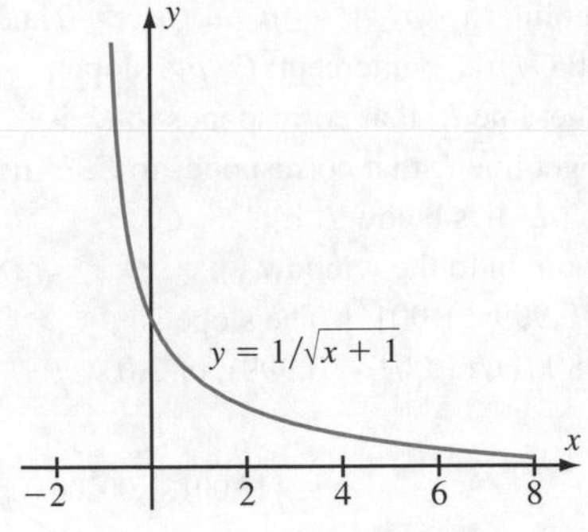

23.

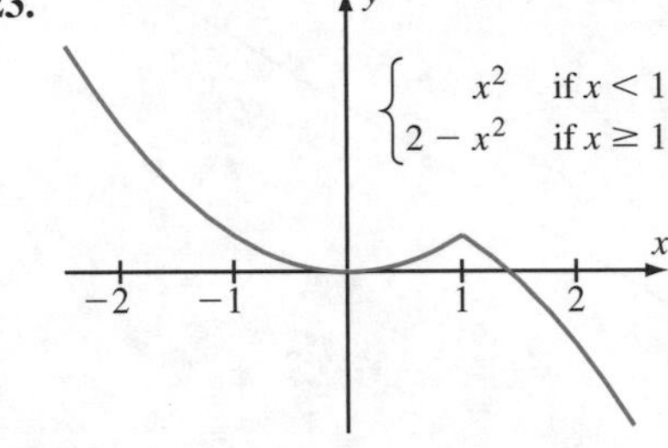

25. $s(\alpha) = (\pi/180) \cdot r\alpha$

27. $A(r, \alpha) = (\pi/360) \cdot r^2 \cdot \alpha$

29. $A = -(r + s)$; $B = rs$

31. $f(x) = \begin{cases} 1 + 2x & \text{if } 0 \leq x \leq 1 \\ 3(3 - x)/2 & \text{if } 1 < x \leq 3 \\ x - 3 & \text{if } 3 < x \leq 4 \\ 1 & \text{if } 4 < x \leq 5 \end{cases}$

33. $m(x) = \begin{cases} 0.15 & \text{if } 0 < x \leq 23350 \\ 0.28 & \text{if } 23350 < x \leq 56550 \\ 0.31 & \text{if } 56550 < x < 117950 \end{cases}$; $A(x) = T(x)$

35. $f_1 = 2$, $f_n = 2 \cdot f_{n-1}$ for $n \geq 2$

37. $f_1 = 1$, $f_n = n \cdot f_{n-1}$ for $n \geq 2$

39. $f_1 = 1$, $f_n = n + f_{n-1}$ for $n \geq 2$

41. $\lfloor x \rfloor = \begin{cases} \text{Int}(x) & \text{if } x \geq 0 \\ \text{Int}(x) - 1 & \text{if } x < 0 \end{cases}$

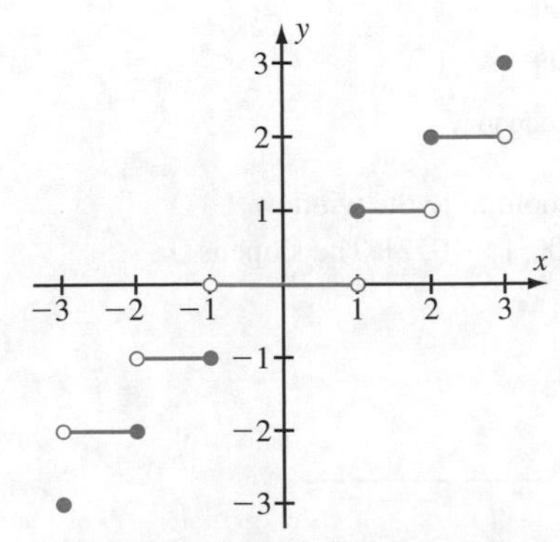

$y = \text{Int}(x)$ (also called $y = \text{trunc}(x)$)

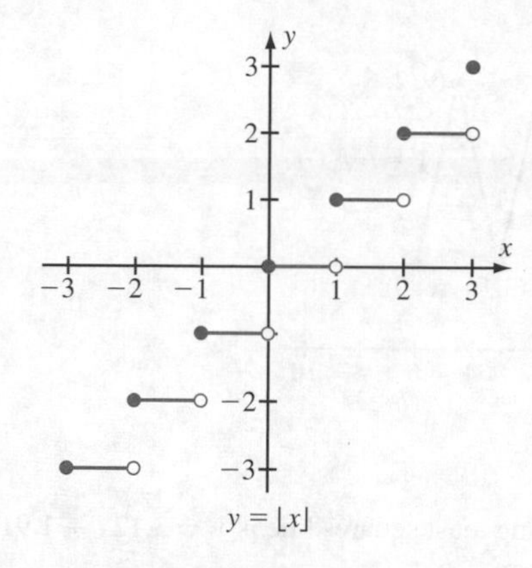

43. $I(P, m, n) = 12 \cdot m \cdot n - P$

45. 1, 1, 2, 5, 14, 42, 132, 429, 1430; using the first recursive definition, we see that c_0 and $c_1 = c_0c_0 = 1$ are integers. It follows that $c_2 = c_0c_1 + c_1c_0$ is an integer. In general, if $c_0, c_1, \ldots, c_n$ are integers, then $c_{n+1} = c_0c_n + c_1c_{n-1} + \cdots + c_{n-1}c_1 + c_nc_0$ is also an integer.

47. $H(x) = (x \geq 0)$

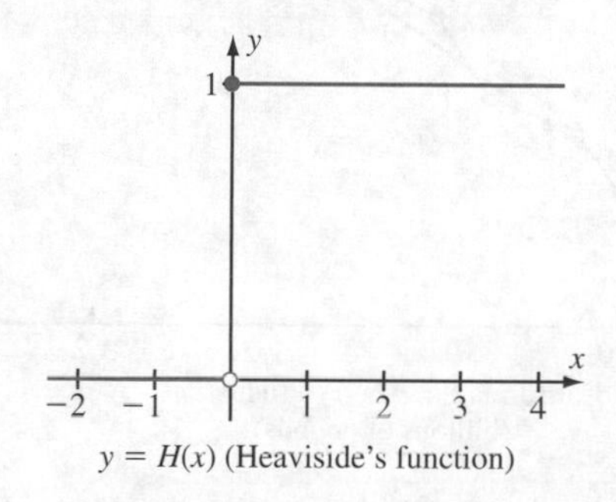

49. $b(x) = x(0 < x)(x \leq 1) + (2 - x)(1 < x)(x \leq 2)$

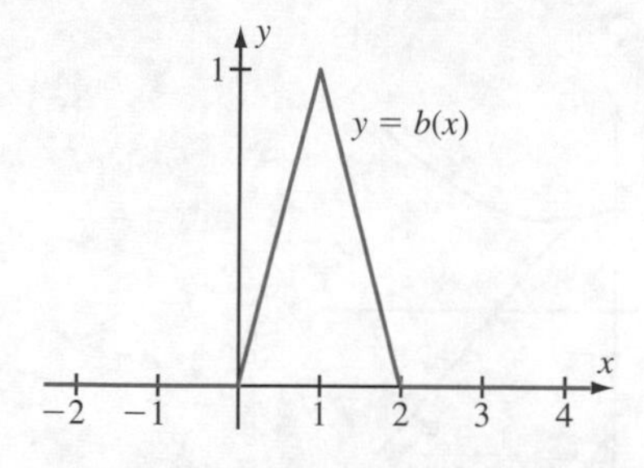

51. If the incidence of a particular disease is small, then the probability that a positive result is a true positive is small. Routine screening (without regard to the presence of risk factors) is not advisable.

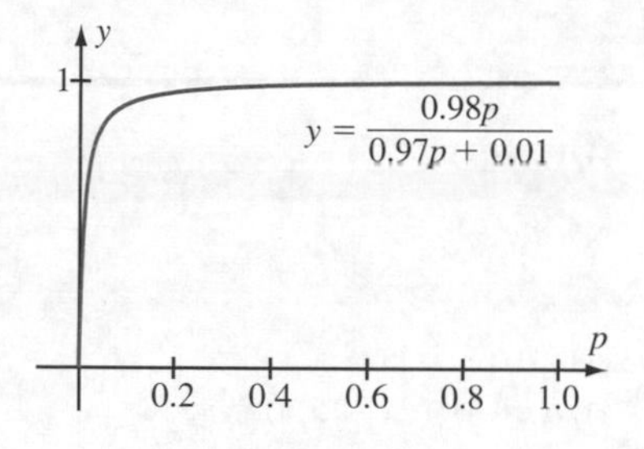

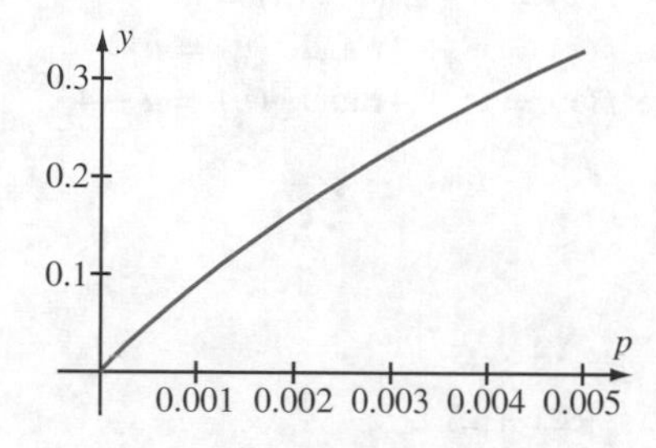

53. The maximum of $f(x)$ occurs when $x = 4$. Set $x_0 = 4$. Let $x_1 = 3$. (The choice of a different value in the interval $(2, x_0)$ will lead to similar results.) We have:

	$h = 10^{-4}$	$h = 10^{-5}$	$h = 10^{-6}$	$h = 10^{-8}$
$F(h, x_1)$	0.72e − 2	0.72e − 3	0.72e − 4	0.72e − 6
$F(h, x_0)$	0.72e − 6	0.72e − 8	0.72e − 10	0.72e − 12

At the point x_0 at which f attains a maximum value, the backward differences are orders of magnitude smaller than the backward differences at other points x_1.

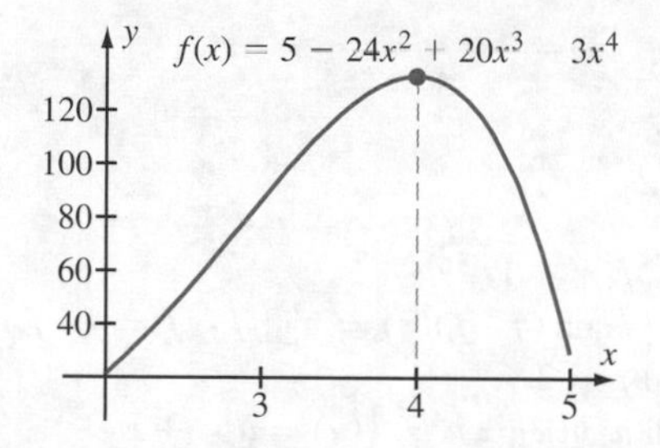

57. $p_1 = 2.828427124$, $p_2 = 3.061467458$, $p_3 = 3.121445152$, $p_4 = 3.136548492$, $p_5 = 3.140331158$, $p_6 = 3.141277254$, $p_7 = 3.141513803$, $p_8 = 3.141572945$

Section 1.5

1. $(x^3 - x^2 + 6x - 4)/(x - 1)$
3. $(x - 2)/(x - 3)$
5. $(2x - 5)(x + 1)/(x - 1)$
7. $(x - 2)^2/(x - 3)^2 + 5$
9. $2(x^2 + 5)(x^2 - 3x + 3)/(x - 1)$
11. $(x^3 - x^2 + 6x - 4)/((x - 1)(2x - 5)^2)$
13. $(g \circ f)(x)$ where $f(x) = x - 2$ and $g(u) = u^2$
15. $(g \circ f)(x)$ where $f(x) = x^3 + 3x$ and $g(u) = u^4$
17. $(g \circ f)(x)$ where $f(x) = x^2 + 4x$ and $g(u) = u + 4$
19. $g(u) = 3u^2 + 1$
21. $g(u) = u/(u^2 + 2)$
23. 3
25. 1/2
27. $(x + 5)(x - 1)$
29. $(x - 2)(x + 2)(x^2 + 2x + 2)$
31. $g(x) = f(x + 2)$; the plot of g is obtained by shifting the plot of f by 2 units to the left.
33. $g(x) = f(x - 3)$; the plot of g is obtained by shifting the plot of f by 3 units to the right.
35. $g(x) = f(x + 1) + 4$; the plot of g is obtained by shifting the plot of f by 1 unit to the left and 4 units up.
37. $g(x) = f(x - 1) + 1$; the plot of g is obtained by shifting the plot of f by 1 unit to the right and 1 unit up.
39. $y = 3x - 7$
41. The horizontal ray that has $(1, 3)$ as a left endpoint
43. $y = 1/x, 0 < x \leq 1$
45. $x \mapsto x$ and, for any constant k, $x \mapsto -x + k$
47. $(3x^2 + 2x + 3) - (8x^2 + 3x + 3)/(x^3 - x + 3)$
49. $g(u) = u^2 + 2$
51. $g(u) = 2u + 18$
53. $f(x) = x - 4$
55. $f(x) = (x^2 - 1)^{1/3}$
57. $f(x) = (x - 3)^2$
59. $f(x) = (1 - x^3)/(x^2 + 1)$
61. $(f \circ f)(x) = x^{p^2}$ and $(f \cdot f)(x) = x^{2p}$; $f \circ f = f \cdot f$ when $p = 0$ and $p = 2$
63. $g(u) = \pi$ for all u; $h(u) = u^{1/5}$; $k(u) = u^{1/5} + \pi$
65.

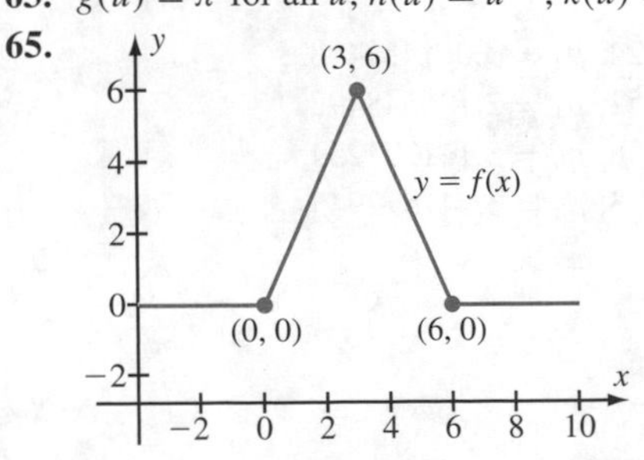

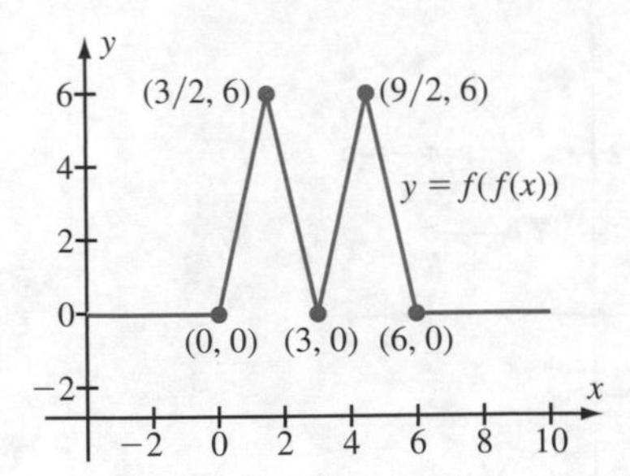

69. $x = y^{2/3}$
71. $y = 3 \cdot 2^x$
73. The approximating least squares line is $y = 3.12x + 1.91$.

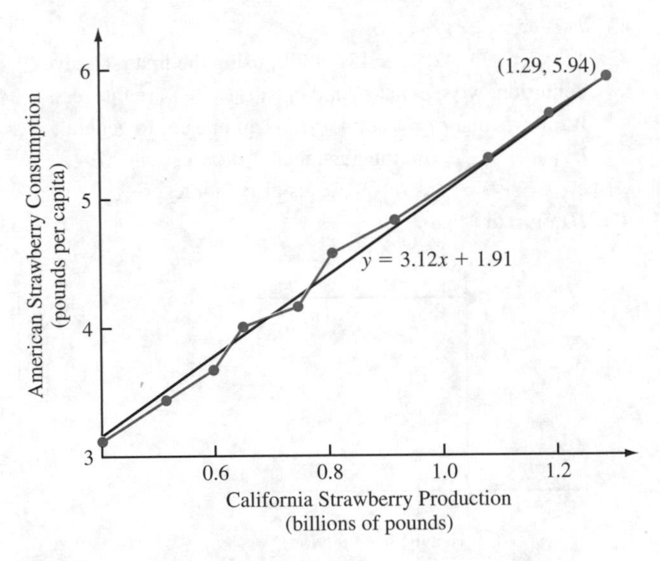

79. C' is the graph of $y = (x - 3)^2$.
81. C' is obtained by replacing x with $-x$ in the equation that defines C.

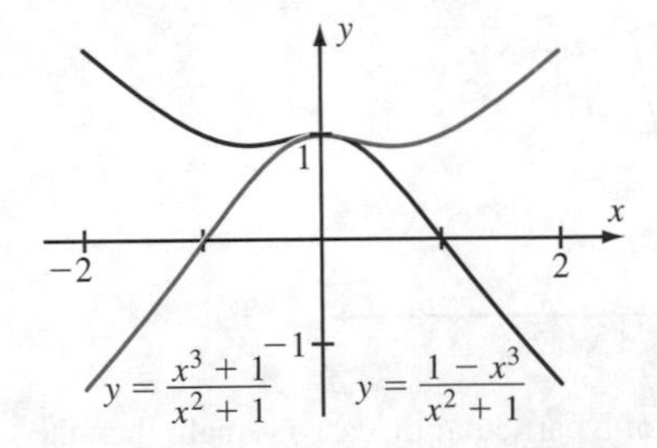

85. $r(3) = 0.288675$; The line is $y = 0.288675(x - 3) + \sqrt{3}$.

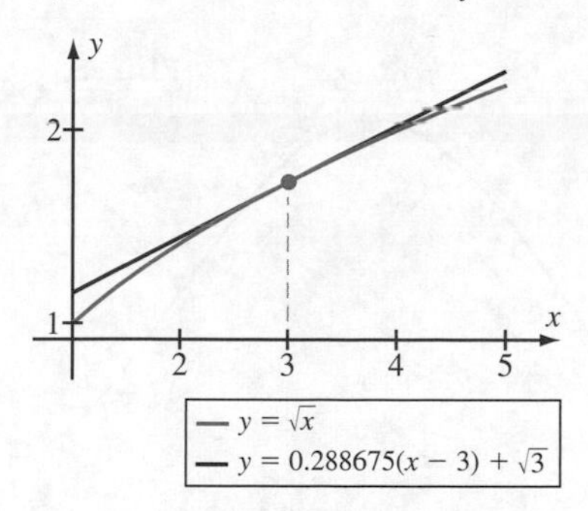

87. $r(2) = 1.64645$; The line is $y = 1.64645(x - 2) + (4 - \sqrt{2})$.

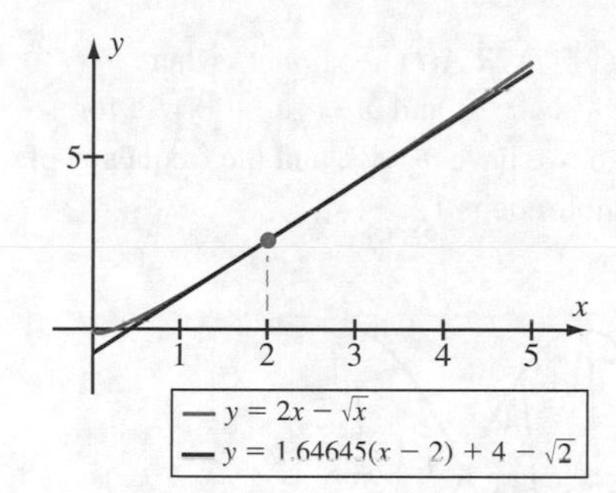

89.

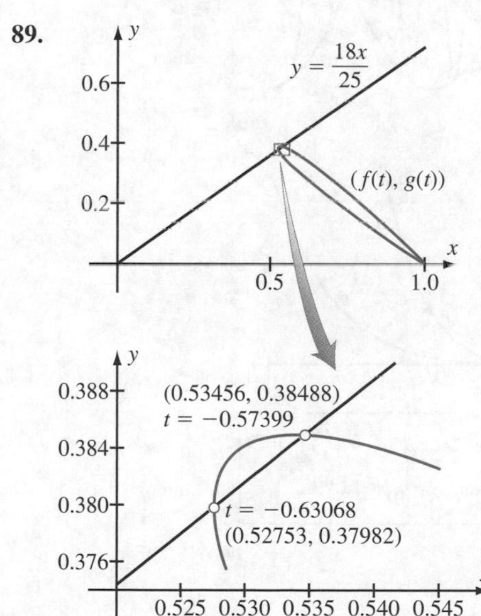

Section 1.6

1. $\cos(\pi/6) = \sqrt{3}/2$, $\sin(\pi/6) = 1/2$, $\tan(\pi/6) = \sqrt{3}/3$, $\cot(\pi/6) = \sqrt{3}$, $\sec(\pi/6) = 2\sqrt{3}/3$, $\csc(\pi/6) = 2$

3. $\cos(2\pi/3) = -1/2$, $\sin(2\pi/3) = \sqrt{3}/2$, $\tan(2\pi/3) = -\sqrt{3}$, $\cot(2\pi/3) = -\sqrt{3}/3$, $\sec(2\pi/3) = -2$, $\csc(2\pi/3) = 2\sqrt{3}/3$

5. $\sqrt{3}/4$ **7.** $(\sqrt{3}+1)/2$ **9.** 3

11. 1 **13.** -1 **15.** $2\sqrt{2}/3$

17. 24/25

19. $(1/2)\sqrt{(12+\sqrt{119})/6}$

21. All six

23. Tangent, cotangent

25.

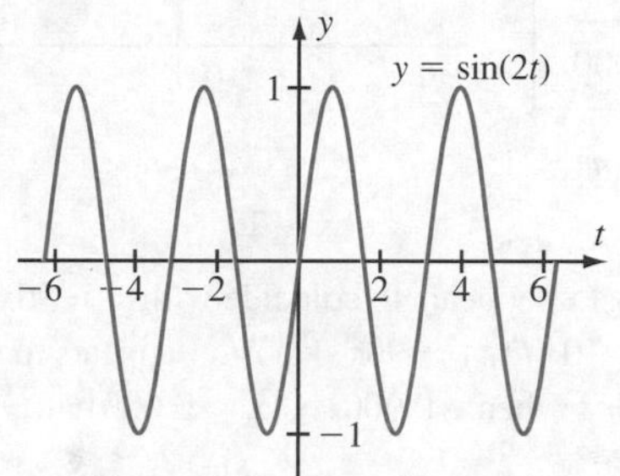

27. $y = \sin\left(t - \frac{\pi}{6}\right)$

29. $x = \cos(t)$, $y = \sin(t)$

31. $x = \sin(t)$, $y = \cos(t)$

33. $s(r, \theta) = r\theta$

35. The first equation holds for some values of its variable. The solutions are $n\pi/2$ where n is an integer. The second equation is an *identity*. It holds for all values of its variable.

45. $\cos(7\pi/12) = -(\sqrt{2-\sqrt{3}})/2$, $\sin(7\pi/12) = (\sqrt{2+\sqrt{3}})/2$, $\tan(7\pi/12) = -2-\sqrt{3}$, $\cot(7\pi/12) = -2+\sqrt{3}$, $\sec(7\pi/12) = -2/\sqrt{2-\sqrt{3}}$, $\csc(7\pi/12) = 2/\sqrt{2+\sqrt{3}}$

51. $C = \sqrt{A^2+B^2}$ and ϕ satisfies $\sin(\phi) = A/C$ and $\cos(\phi) = B/C$

55. A line segment joining $(a, 0)$ to $(0, b)$; from $(0, b)$, it doubles back to $(a, 0)$ and then repeats, ending at $(a, 0)$.

57. The hyperbola $y = ab/x$, $x > 0$

59. 2π **61.** π **63.** 2π

65. $T(t) = 27.5 + 23.5\sin(\pi t/6 - \pi/2)$; In this model, the low occurs at the end of December ($t = 12$), and the high occurs at the beginning of July ($t = 6$).

69.

n	$n \cdot \sin(1/n)$
10	0.9983342
10^2	0.9999833
10^3	0.9999998
10^4	1.0000000
10^5	1.0000000
10^6	1.0000000

$\lim_{n\to\infty} n \cdot \sin(1/n) = 1$;
$\lim_{x\to 0} f(x) = 1$

71. $f(x) = 90x + \cos(x)$ appears to coincide with $y = 90x$ when $-10 \le x \le 10$. $f(x) = 90x + \cos(x)$ appears to coincide with $y = 1$ when $-0.0001 \le x \le 0.0001$.

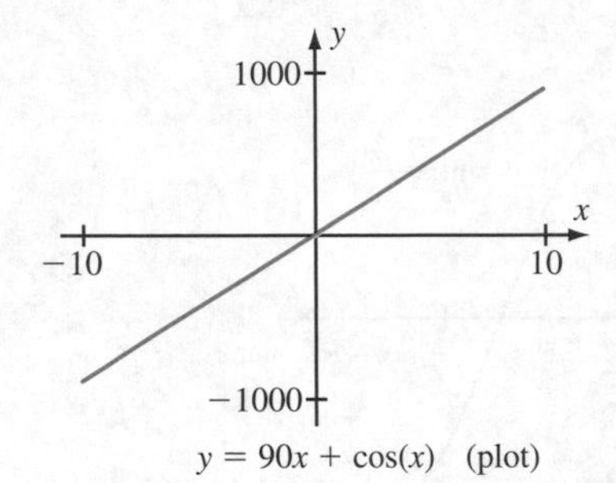

$y = 90x + \cos(x)$ (plot)

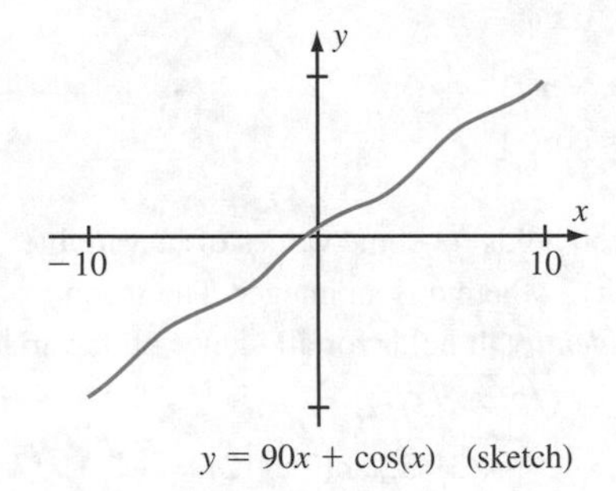

$y = 90x + \cos(x)$ (sketch)

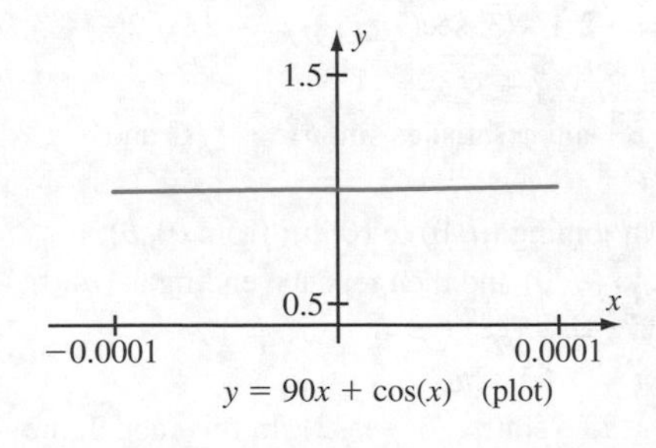

$y = 90x + \cos(x)$ (plot)

73. $T(t) = 59 + 19\sin(\pi t/6 - 7\pi/12)$

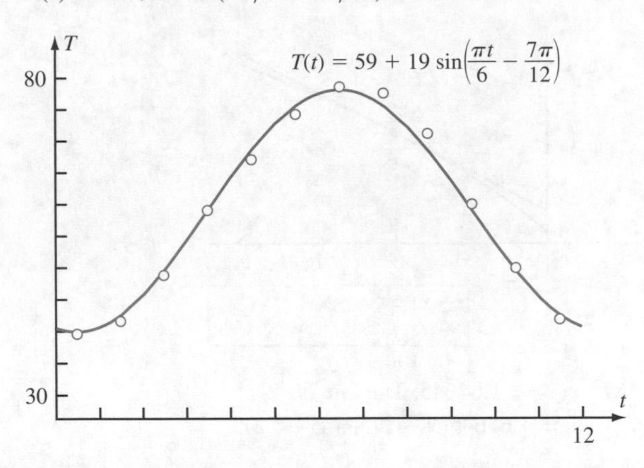

75. $\cos(\nu_1 \cdot t) + \cos(\nu_2 \cdot t) = A(t) \cdot \cos(\omega \cdot t)$ where $A(t) = 2\cos((\nu_1 - \nu_2)t/2)$ and $\omega = (\nu_1 + \nu_2)/2$; for $\nu_1 = 8$ and $\nu_2 = 6$, we have $\omega = 7$, and the frequency of the modulated amplitude is 1.

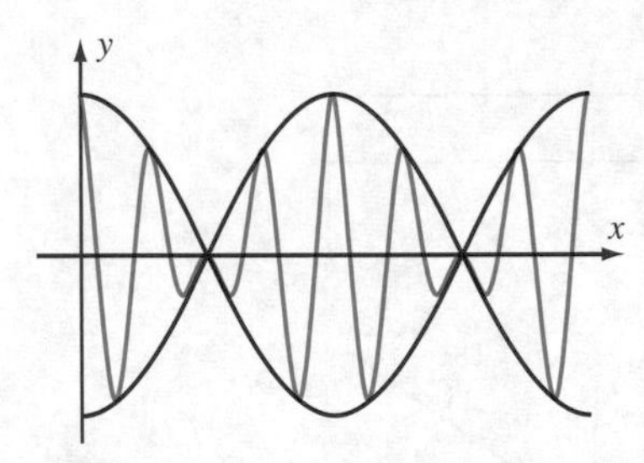

77.

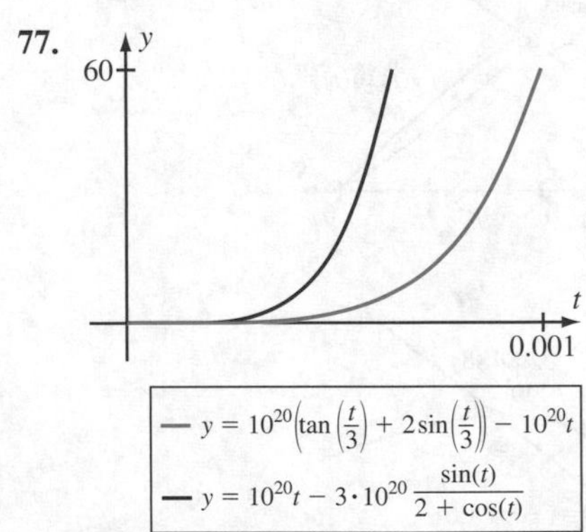

Section 2.1

1. 5 **3.** 57 **5.** −1
7. −4 **9.** 10 **11.** −1/14
13. Does not have a limit. **15.** $\lim_{x\to 2} f(x) = -2$

17. $\lim_{x\to 5} f(x) = 7$

19. 0.01 **21.** 0.002

25. **a.** The domains of g, f, h, and k are, respectively, $\{x \in \mathbb{R} : x \neq 1\}$, $\mathbb{R}$, $\mathbb{R}$, and $\mathbb{R}$. **b.** The functions f and h are the same because they have the same domain and $f(x) = h(x)$ for each x in their domain. **c.** When $x \neq 1$, we have $(x^2 - 1)/(x - 1) = ((x+1)(x-1))/(x-1) = x + 1$. Therefore, $g(x) = f(x) = h(x) = k(x)$ for $x \neq 1$. It follows that $\lim_{x\to 1} g(x) = \lim_{x\to 1} f(x) = \lim_{x\to 1} h(x) = \lim_{x\to 1} k(x) = 2$.

27. **a.** The domains of f, g, k, and h are, respectively, $\mathbb{R}$, $\mathbb{R}$, $\mathbb{R}$, and $\{x \in \mathbb{R} : x \neq 0\}$. **b.** The functions f and k are the same because they have the same domain and $f(x) = k(x)$ for each x in their domain (including $x = 0$); h differs from f and k because it has a different domain; g differs from f and k because $1 = g(0) \neq f(0) = k(0) = 0$. **c.** When $x \neq 0$, we have $g(x) = f(x) = h(x) = k(x)$. It follows that $\lim_{x\to 0} g(x) = \lim_{x\to 0} f(x) = \lim_{x\to 0} h(x) = \lim_{x\to 0} k(x) = 0$.

29. $\lim_{x\to 0^+} H(x) = 1$; $\lim_{x\to 0^-} H(x) = 0$; $\lim_{x\to 0} H(x)$ does not exist.

31. 3.28×10^{-3}

33. **a.** $\lim_{x\to 0}\lfloor x \rfloor$ does not exist. **b.** $\lim_{x\to 1/2}\lfloor x \rfloor = 0$ **c.** $\lim_{x\to 1^+} 1/\lfloor x \rfloor = 1$; however, $1/\lfloor x \rfloor$ is not defined just to the left of 1, so we do not talk about $\lim_{x\to 1^-} 1/\lfloor x \rfloor$. **d.** $\lim_{x\to -1/2} 1/\lfloor x \rfloor = 1/(-1) = -1$

35. See the figure for the answer to Exercise 41, Section 1.4; the graph shows that $\lim_{x\to n^+}\lfloor x \rfloor = (\lim_{x\to n^-}\lfloor x \rfloor) + 1$; hence, $\lim_{x\to n^+}\lfloor x \rfloor \neq \lim_{x\to n^-}\lfloor x \rfloor$.

37. **a.** 3 **b.** 2.5 **c.** 2.1 **d.** $2 + h$ **e.** 2

39. **a.** 65 **b.** 46.125 **c.** 34.481 **d.** $32 + 24h + 8h^2 + h^3$ **e.** 32

41. **a.** 0 **b.** $1/\sqrt{2}$

43. $|x - 2| < 0.0095$

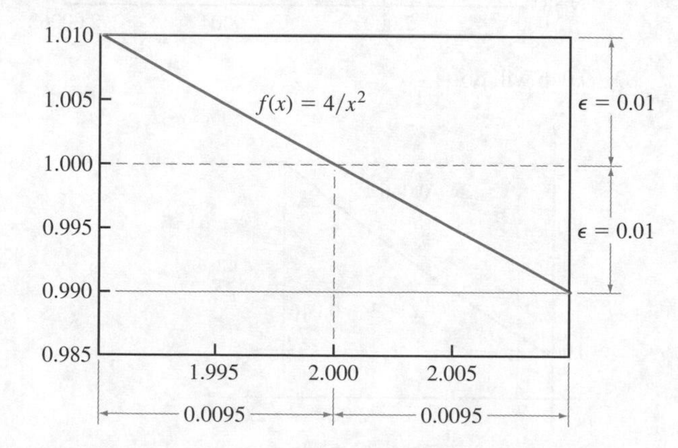

45. $|x - 10| < 0.00046$

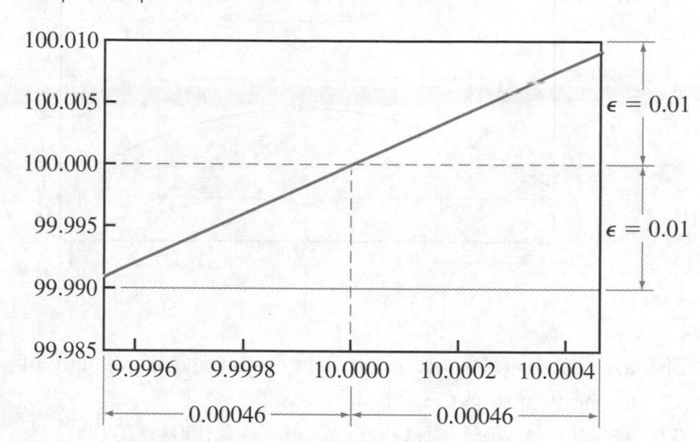

49. 7.82×10^{-4} mm

51. **a.**

x	$f(x)$
$\pi/2 - 1/10$	−0.99833
$\pi/2 - 1/100$	−0.99998
$\pi/2 - 1/1000$	−0.99999

b. −1

c. $\delta = 0.24$ will do.

d.

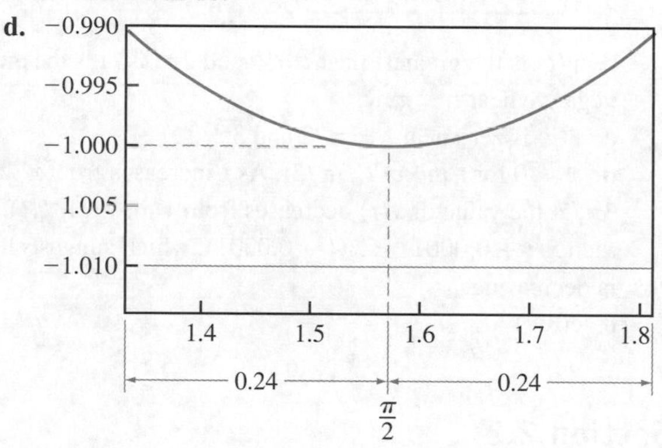

53. **a.**

x	$f(x)$
1/10	0.16658
1/100	0.16666
1/1000	0.16666

b. 1/6

c. $\delta = 1.05$ will do.

d.

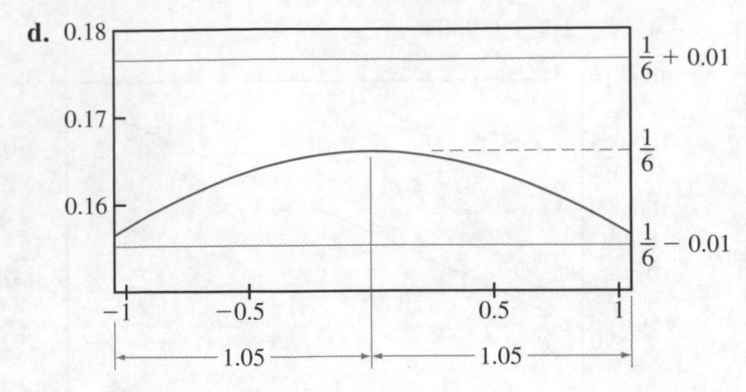

55. a. 0.1 **b.** 0.01 **c.** ϵ **d.** The instantaneous velocity of the body at time $t = 2$ is 4.

57. a. $\bar{v}(t) = (\sin(t + 0.0001) - \sin(t - 0.0001))/0.0002$

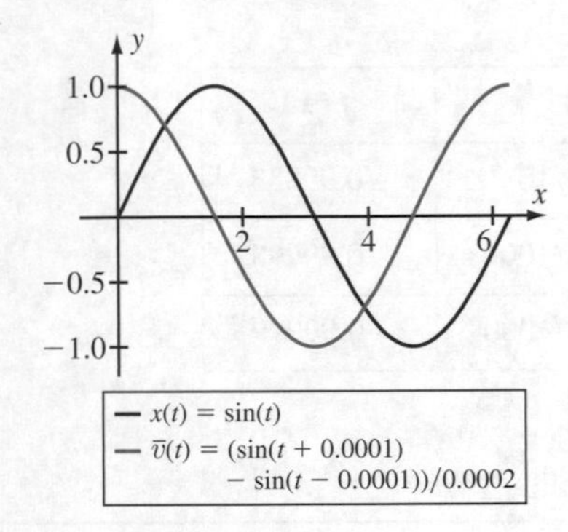

b. $\bar{v}(t)$ is the greatest near $t = 0$ and 2π; $\bar{v}(t)$ is the most negative near $t = \pi$.
c. $\bar{v}(t)$ is zero near $t = \pi/2$ and $3\pi/2$.
d. $\bar{v} < 0$ for t in $[\pi/2, 3\pi/2]$. As t increases from $\pi/2$ to $3\pi/2$, the value of $x(t)$ decreases from 1 to −1. If $\bar{v}(t) < 0$, then $x(t + 0.0001) < x(t - 0.0001)$, which suggests that x is decreasing.
e. $\cos(t)$

Section 2.2

1. 6 **3.** 0
9. $\lim_{x\to 4^-} f(x) = 17$; $\lim_{x\to 4^+} f(x) = -7$
11. $\lim_{x\to 4^-} f(x) = 64$; $\lim_{x\to 4^+} f(x) = 64$
13. 14
15. −5
17. 2
19. −4/5
21. 1/2
23. 6
25. −2
27. $(\pi - \sqrt[3]{\pi})/(\sqrt{\pi} - 3)$
29. 0
31. 1
33. 3
35. Yes, $\lim_{x\to 2} f(x) = 2$
37. Insufficient information
39. $\lim_{x\to 1^+} f(x) = 0$ and $\lim_{x\to 2^-} f(x) = 0$
41. $\lim_{x\to (-2)^+} f(x) = 2$ and $\lim_{x\to 2^-} f(x) = 0$
43. 1
45. −14
47. 36
51. 0
53. 3
55. 2
57. 0
59. $\pi/180$
61. 6
63. 4
65. 1
67. −1/8
69. 3/2
71. −1
73. 3/2
75. 1
85. 0.00062 will do.

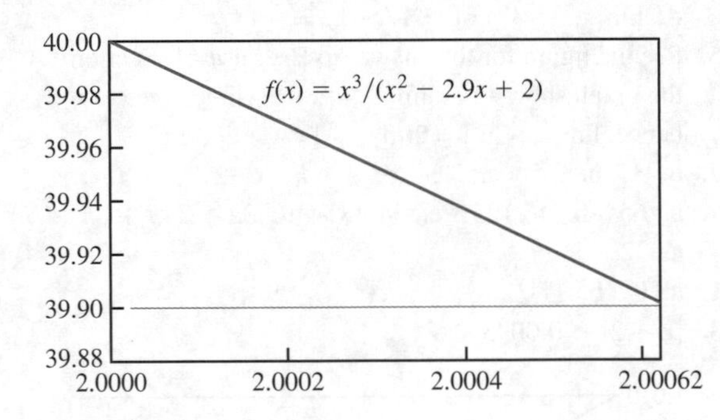

87. 0.046 will do.

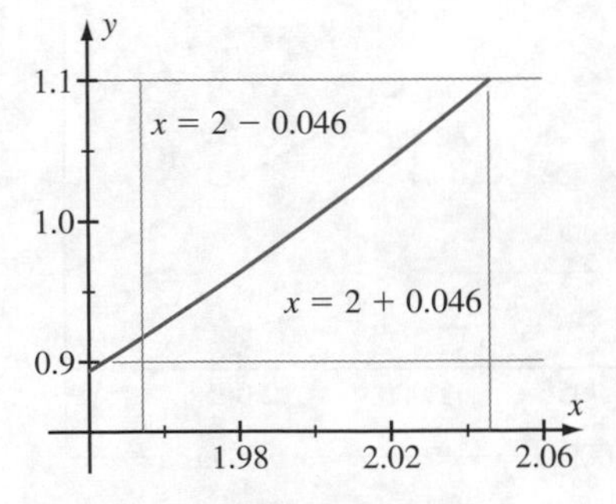

89.

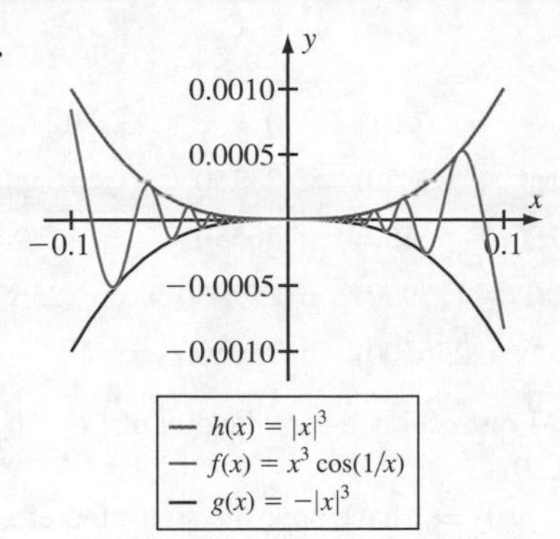

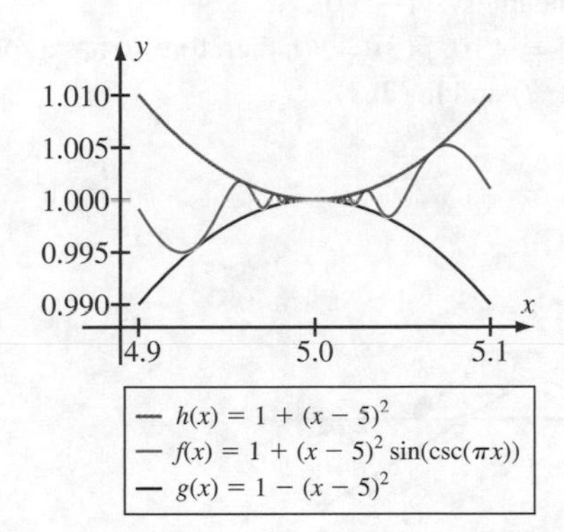

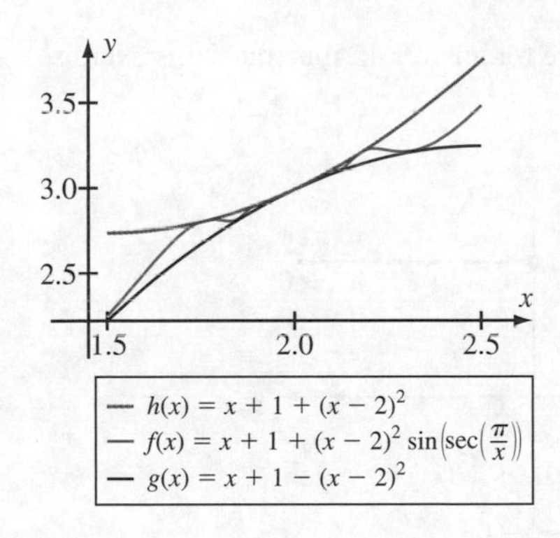

93. $\lim_{x\to c} F(x) = 3$; $G(x) \le F(x) \le H(x)$ for x in $[1.99, 2.01]$

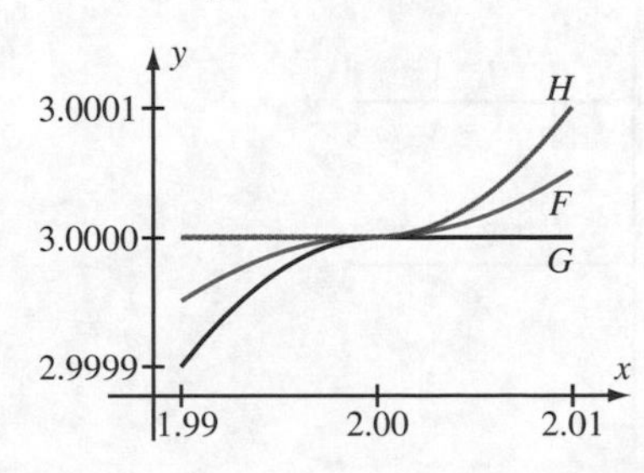

95. $\lim_{x\to c} F(x) = 8$; $G(x) \le F(x) \le H(x)$ for x in $[2.95, 3.05]$

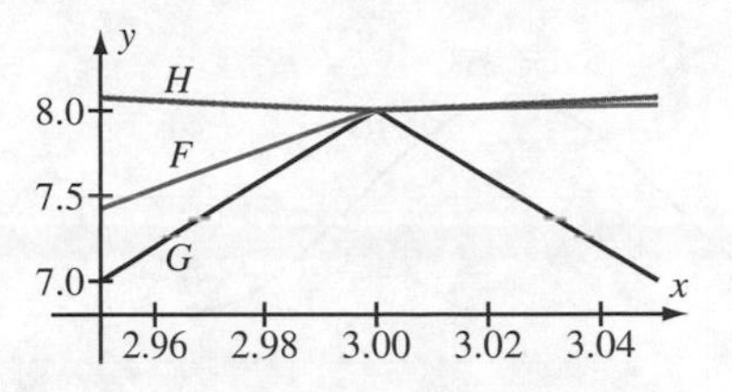

Section 2.3

1. f is continuous at all values of its domain, which is $\mathbb{R}$.
3. g is continuous at all values of its domain, which is $\{x \in \mathbb{R} : x \neq -1\}$.
5. f is continuous at all values of its domain, which is $\mathbb{R}$.
7. H is continuous at all values of its domain, which is $\{x \in \mathbb{R} : x \neq 3, x \neq -4\}$.
9. The domain of g is $\mathbb{R}$; g is continuous at each point of $\{x \in \mathbb{R} : x \neq 3\}$.
11. H is continuous at all values of its domain, which is $\mathbb{R}$.
13. g is continuous at all values of its domain, which is $\mathbb{R}$.
15. $F(2) = 11$
17. $F(2) = 4$
19. Left continuous on $(-\infty, \infty)$; right continuous on $\{x \in \mathbb{R} : x \neq 5\}$; continuous on $\{x \in \mathbb{R} : x \neq 5\}$
21. Left continuous on $\{x \in \mathbb{R} : x \neq 3\}$; right continuous on $(-\infty, \infty)$; continuous on $\{x \in \mathbb{R} : x \neq 3\}$
23.

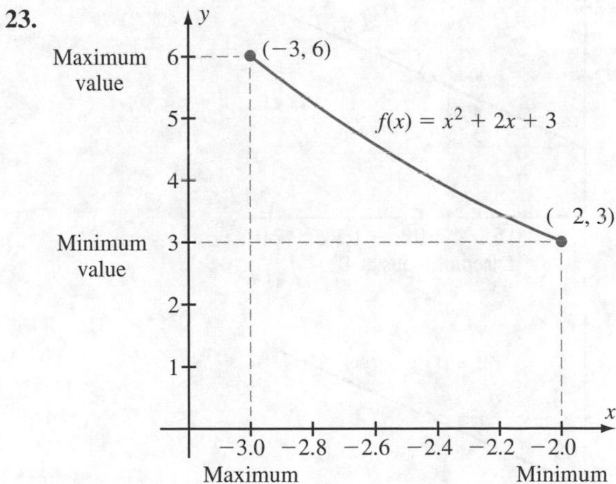

25.

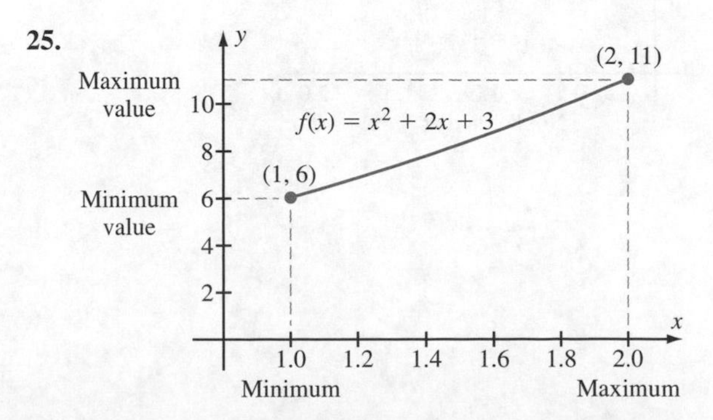

27.

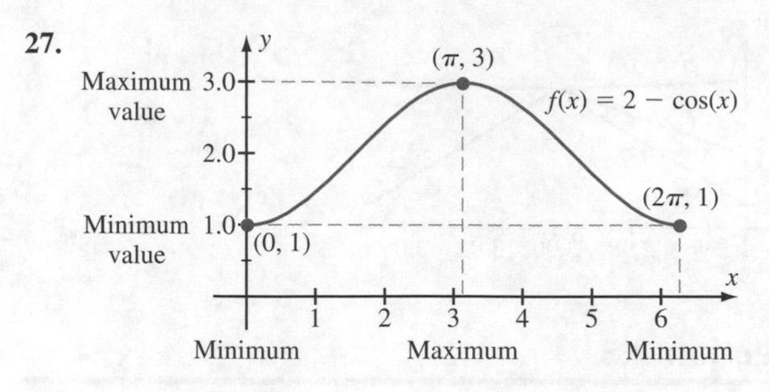

29. **a.** $f(x) = 26 - 9x$ **b.** $f(x) = (25x - 21)/6$
c. $f(x) = (9 - 7x)/2$

31. $\lim_{x\to 1} f(x)$, $\lim_{x\to 1} g(x)$, $\lim_{x\to 1} h(x)$, and $\lim_{x\to 1} k(x)$ all exist and equal 2; g and h are continuous at 1, and k is discontinuous at 1; g and h are identical.

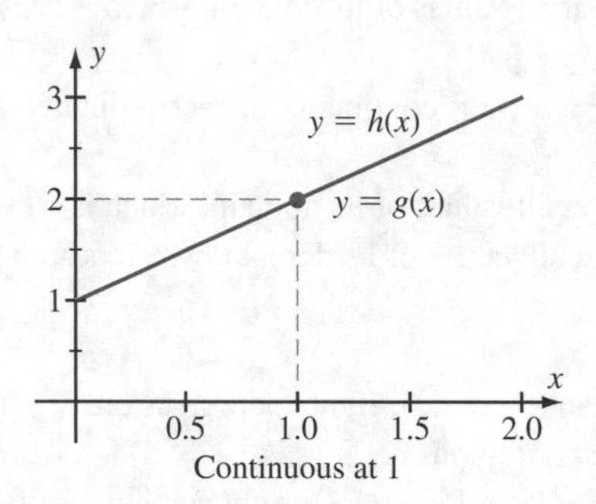

Continuous at 1

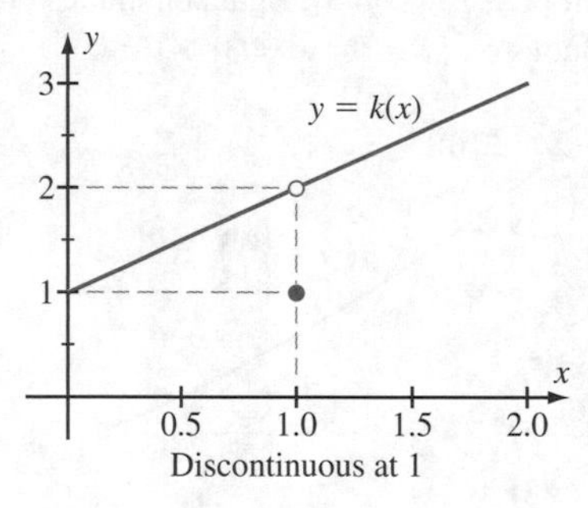

Discontinuous at 1

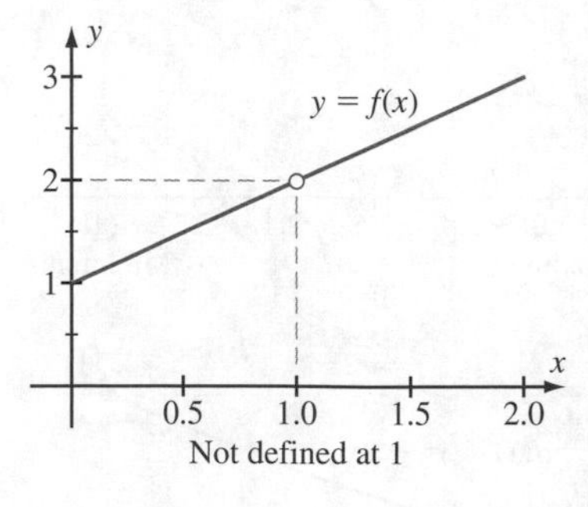

Not defined at 1

33.

$$T(x) = \begin{cases} 0.10x & \text{if } 0 < x \le 23350 \\ 2335 + 0.15(x - 23350) & \text{if } 23350 < x \le 56550 \\ 7315 + 0.25(x - 56550) & \text{if } 56550 < x \le 117950 \\ 22665 + 0.30(x - 117950) & \text{if } 117950 < x \le 256500 \\ 64230 + 0.35(x - 256500) & \text{if } 256500 < x \end{cases}$$

35. **a.** See Exercise 41 of Section 1.4 for the plot of $\lfloor x \rfloor$ **b.** 2 **c.** −1 **d.** 1 **e.** 0

41. $p(-1) = -1$ and $p(0) = 1$ have opposite signs; therefore, p has a root in the interval $(-1, 0)$.

43. $p(1) = 14$, $p(2) = -10$, $p(3) = 90$; therefore, p has a root in the interval $(1, 2)$ and in $(2, 3)$.

45. False

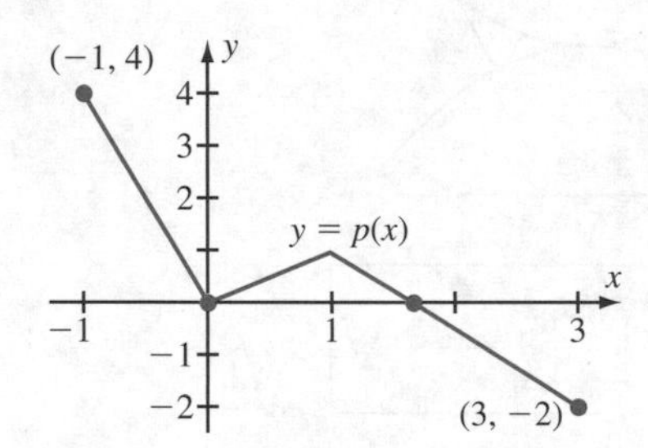

47. False (The figure for answer 45 illustrates this exercise as well.)

49. False

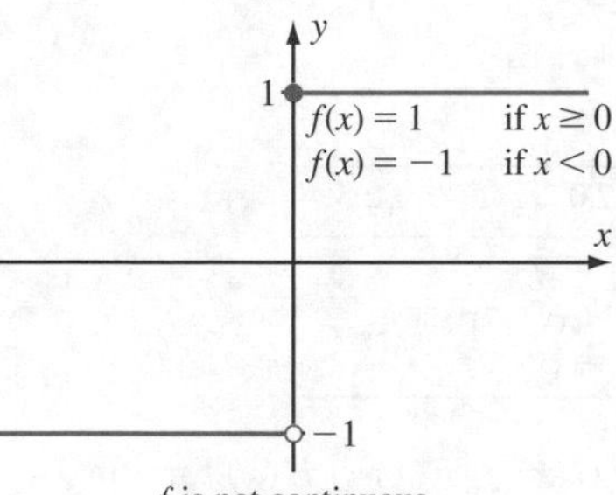

f is not continuous

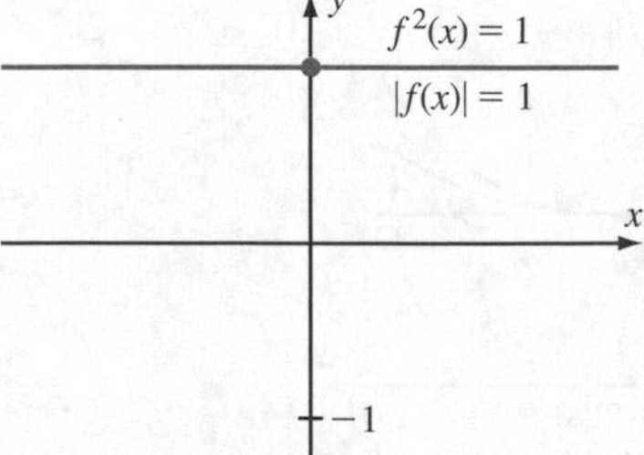

f^2 and $|f|$ are continuous

51. False (The figure for answer 49 illustrates this exercise as well.)

53. True

61. $\lim_{x\to 0} f(x) = f(0) = -1/1000000$

Function	Reason for Continuity
$F(x) = \sqrt[1001]{x}$	Theorem 4, Section 2.2
$G(x) = (\pi/2) \cdot F(x)$	Theorem 1a, Section 2.3
$H(x) = \cos(x)$	Theorem 7, Section 2.2
$J(x) = H(G(x))$	Theorem 2, Section 2.3
$K(x) = x^3$	Theorem 3, Section 2.2
$L(x) = K(J(x))$	Theorem 2, Section 2.3
$M(x) = L(x)/1000000$	Theorem 1a, Section 2.3
$N(x) = x^2$	Theorem 3, Section 2.2
$f(x) = N(x) - M(x)$	Theorem 1a, Section 2.3

63. 4.301695685

65. No continuous extension

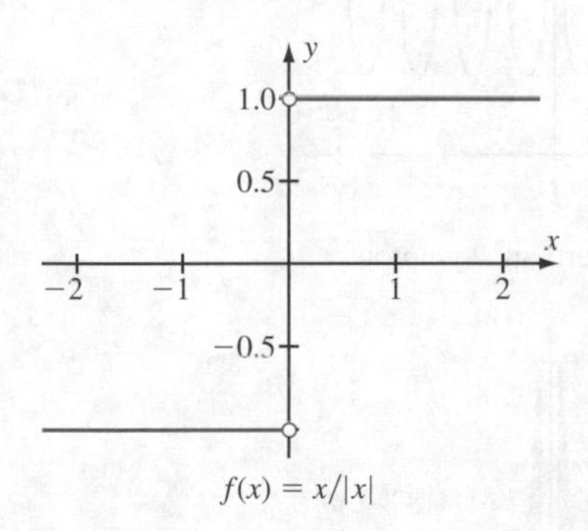

67. A continuous extension F is obtained by setting $F(2) = -8$.

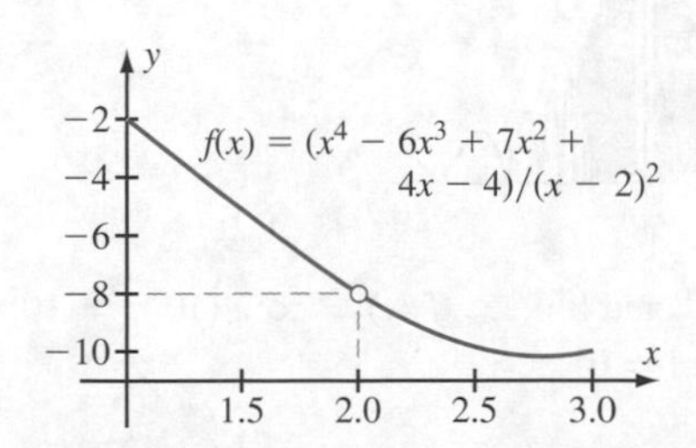

69. For each $\gamma \geq 10$, the equation $p(x) = \gamma$ has at least one solution x; the equation has no solution when $\gamma < 10$. When $\gamma = 20$, the solutions are 0.7267 and 4.1132.

Section 2.4

1. ∞ **3.** 0 **5.** 1 **7.** 1

9. ∞ **11.** 1/3 **13.** ∞ **15.** ∞

17. $-\infty$

19. $-\infty$

21. Vertical asymptote: $x = 7$; horizontal asymptote: $y = 1$

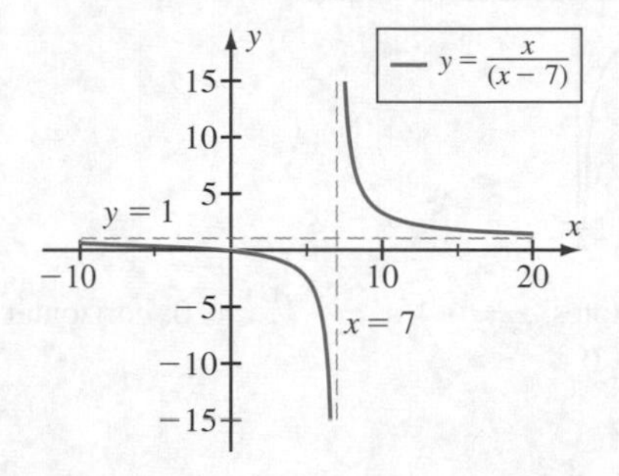

23. Vertical asymptote: $x = 0$; horizontal asymptote: $y = 0$

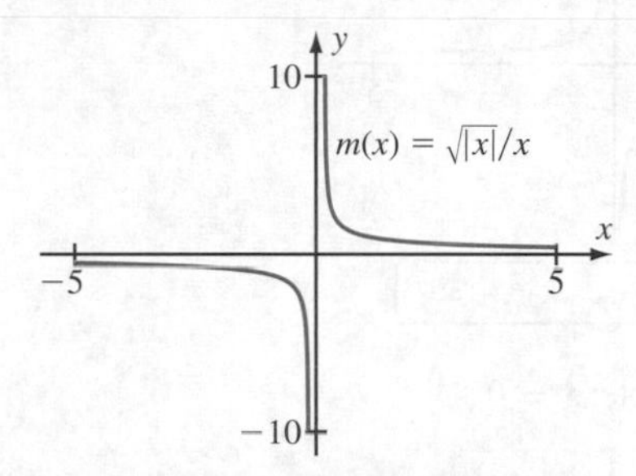

25. Horizontal asymptote: $y = 1$

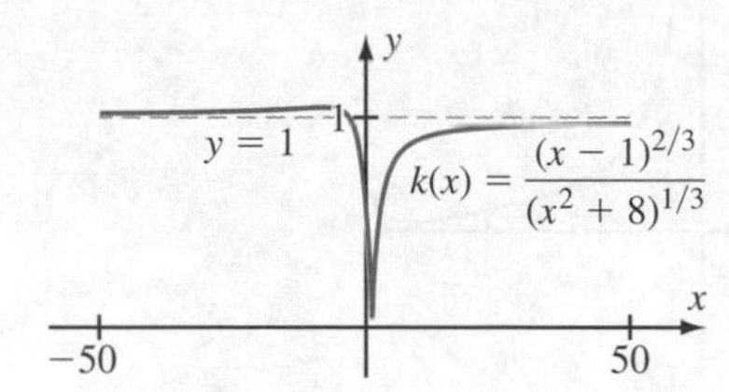

27. Horizontal asymptote: $y = 0$

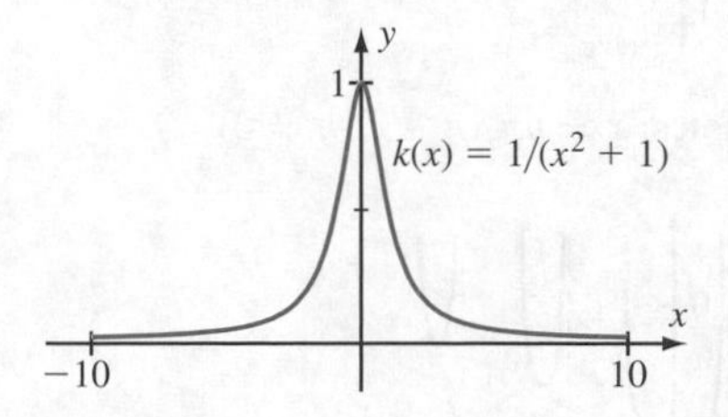

29. Vertical asymptote: $x = 0$; horizontal asymptote: $y = 0$

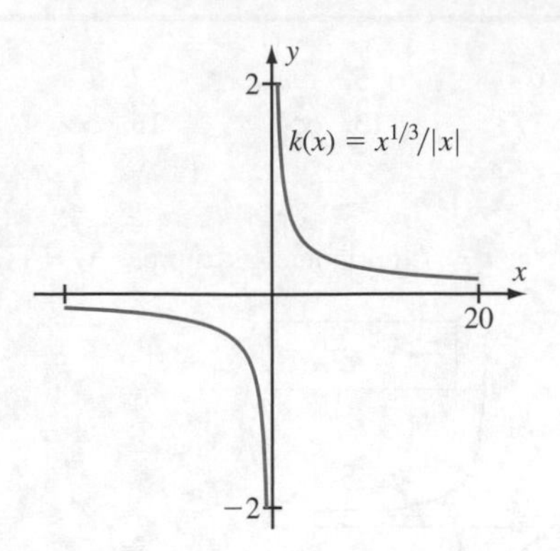

31. Vertical asymptotes: $x = -4$, $x = -1$, $x = 0$; horizontal asymptote: $y = 0$

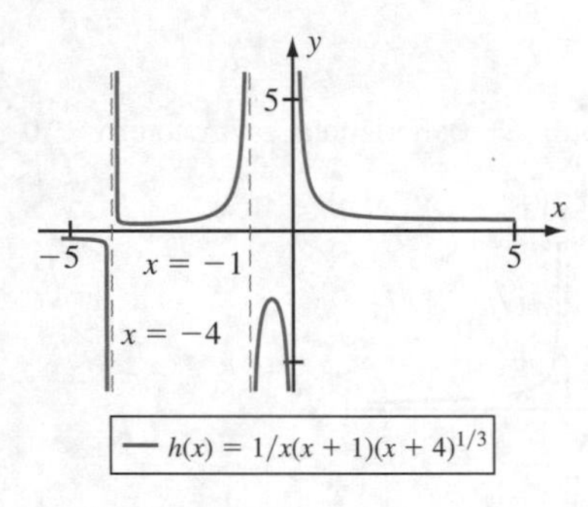

33. Vertical asymptote: $x = 0$

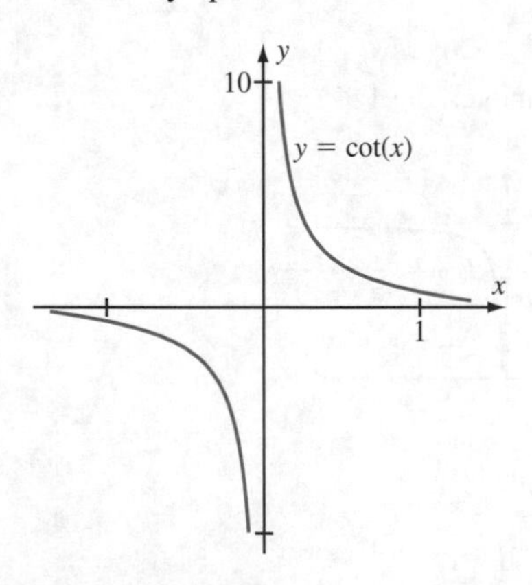

41. Vertical asymptotes: $x = n\pi$ ($n \in \mathbb{Z}, n \neq 0$)

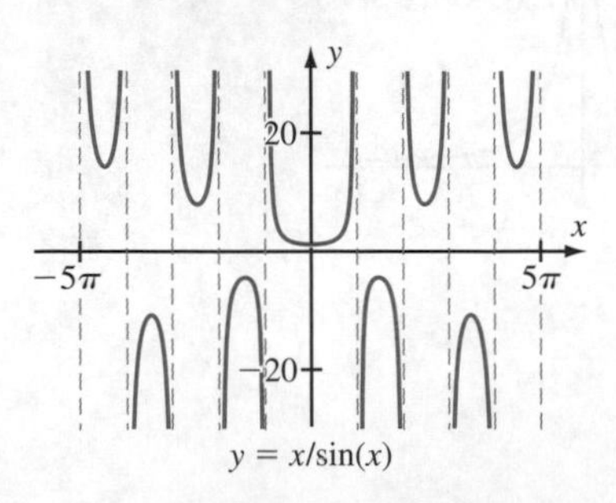

43. Vertical asymptote: $x = 0$; horizontal asymptote: $y = 0$

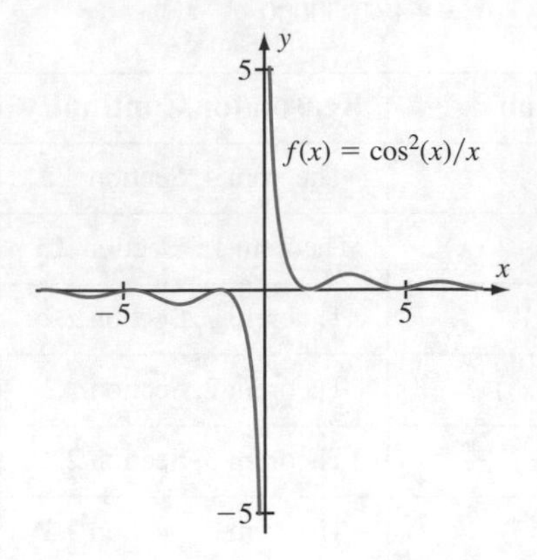

45. Vertical asymptotes: $x = n\pi$ ($n \in \mathbb{Z}$)

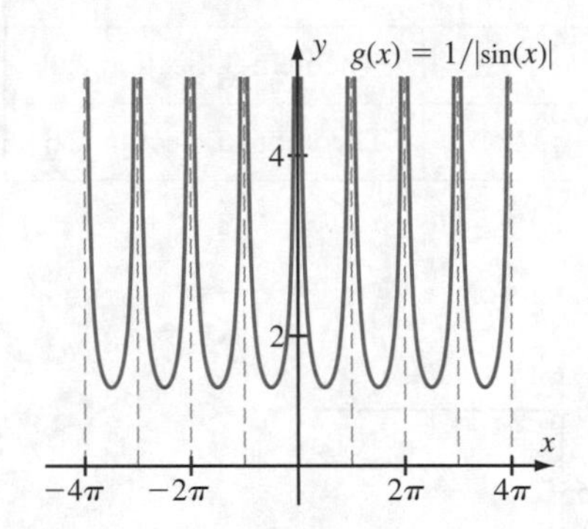

47. $x = 0$ is not a vertical asymptote.

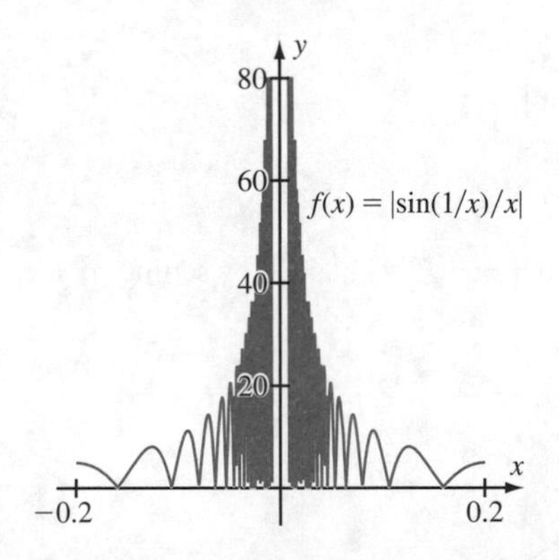

49. $f(10^{80}) = 10^{-20}$ and $\lim_{x\to\infty} f(x) = \infty$; $g(10^{80}) = 10^{20}$ and $\lim_{x\to\infty} g(x) = 0$

51.

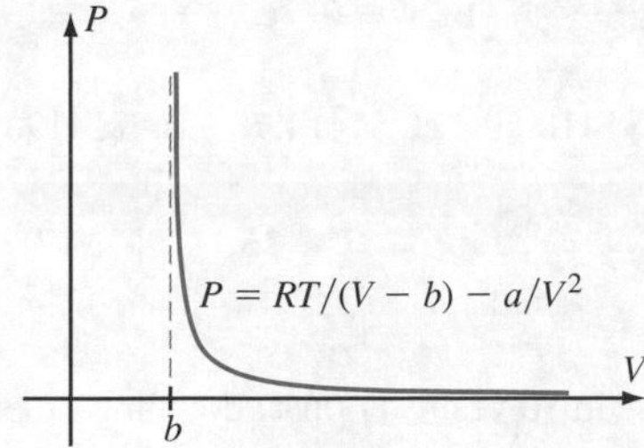

57. $x^2 - 3x + 9$ is a parabolic asymptote of $F(x)$; $x^2 - 5x + 25$ is a parabolic asymptote of $G(x)$; $x^2 + 1$ is a parabolic asymptote of $H(x)$.

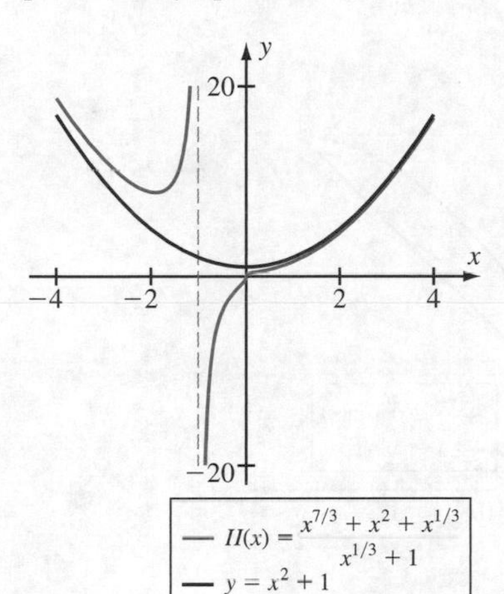

59. $x > (500 + \sqrt{249999})^2 \approx 999997.999999$

61.

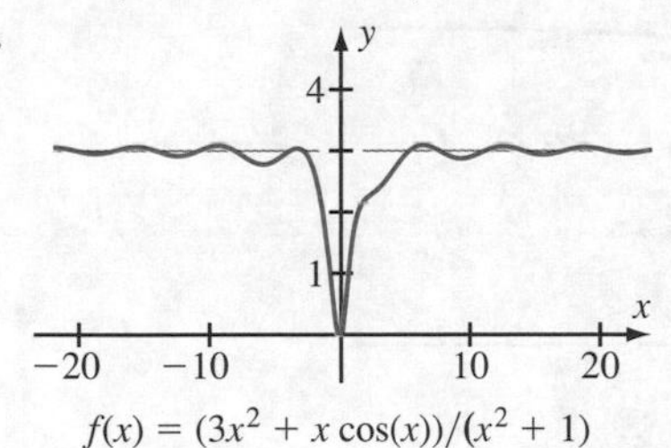

$f(x) = (3x^2 + x\cos(x))/(x^2 + 1)$

Graphed in the viewing window $[-25, 25] \times [0, 4]$

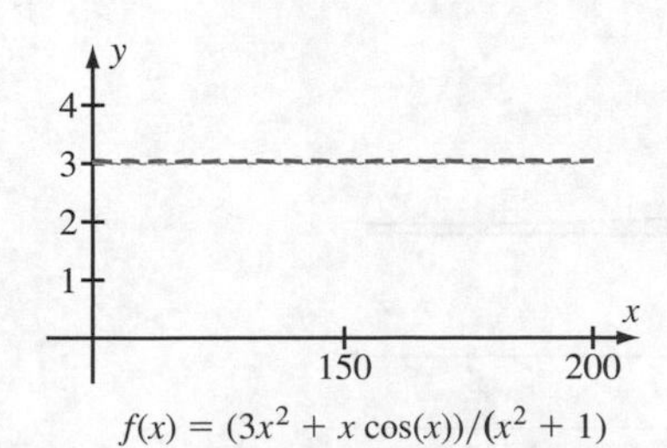

$f(x) = (3x^2 + x\cos(x))/(x^2 + 1)$

Graphed in the viewing window $[100, 200] \times [0, 4]$

63.

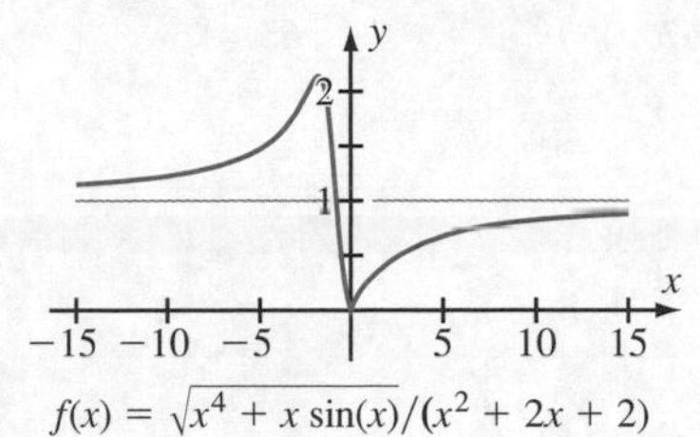

$f(x) = \sqrt{x^4 + x\sin(x)}/(x^2 + 2x + 2)$

Graphed in the viewing window $[-15, 15] \times [0, 2]$

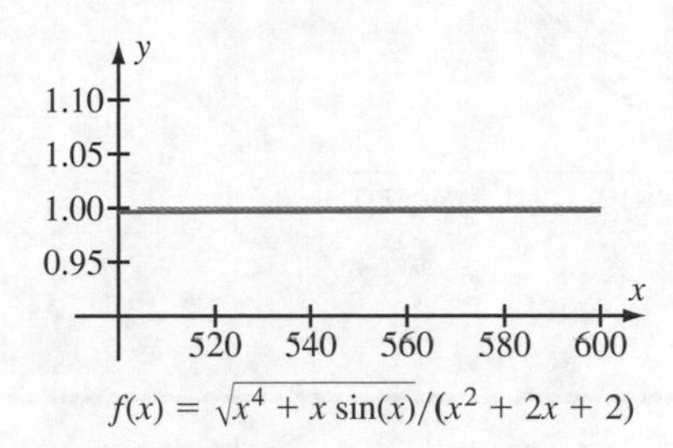

$f(x) = \sqrt{x^4 + x\sin(x)}/(x^2 + 2x + 2)$

Graphed in the viewing window $[500, 600] \times [0.9, 1.1]$

65.

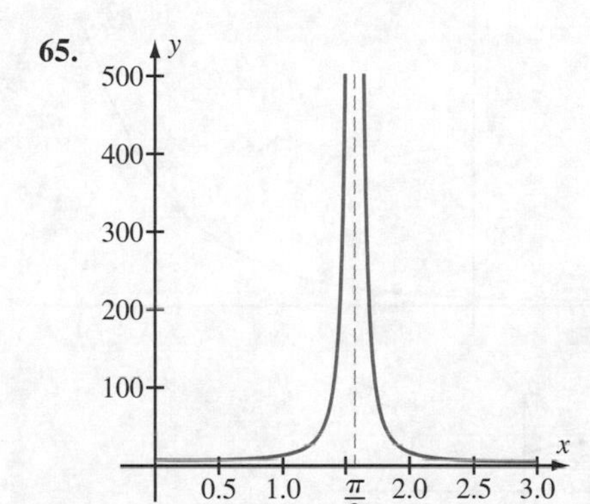

$f(x) = (4 + \cos(x))/(\sin(x) + \cos(2x))$

Section 2.5

1. 0 **3.** 1 **5.** 0

7. Not convergent

9. 0 **11.** ∞ **13.** 4 **15.** 1

17. 3 **19.** 2 **21.** 0 **23.** 0

25. 0 **27.** 1 **29.** 0 **31.** 2

33. −1 **35.** 4 **37.** 3/4 **39.** 10/9

41. 0 **43.** 2 **45.** 1/2 **47.** 16

49. 608911/4950

51. $a_n = (-1)^n$

59. $\lim_{n\to\infty} \lim_{m\to\infty} a_{n,m} = 0$, but $\lim_{m\to\infty} \lim_{n\to\infty} a_{n,m} = 1$

61. 1 **63.** 1 **65.** 1

67. $\lim_{j\to\infty} j^{1/j} = 1$

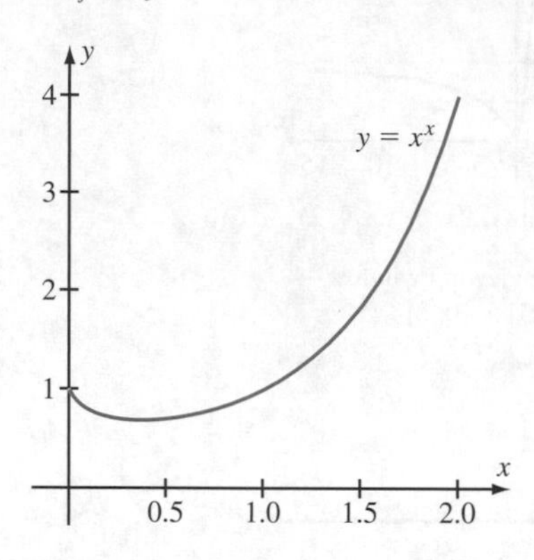

Section 2.6

1. $2^{\sqrt{3}}$ **3.** 2^{π} **5.** 11

7.

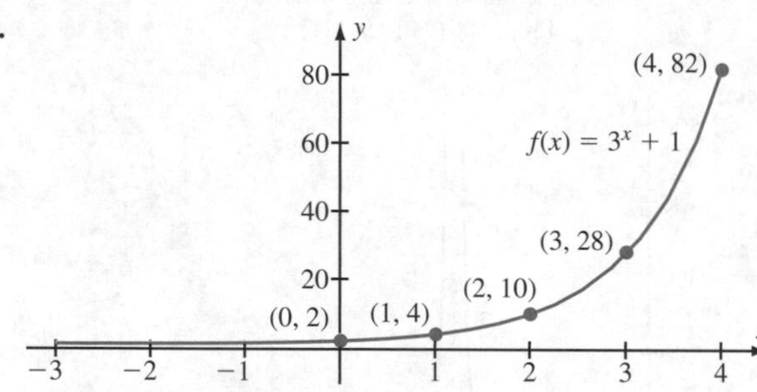

9.

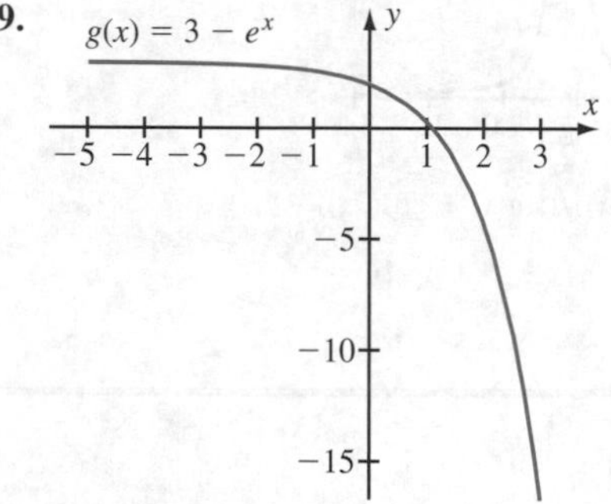

11.

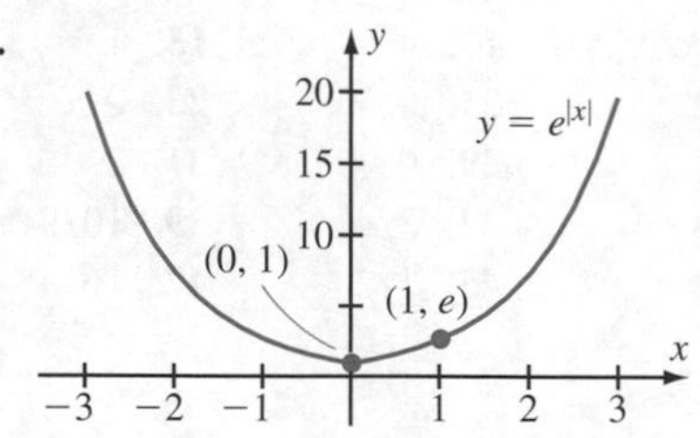

13. 1

15. ∞

17. a. $x = 0, y = 1, y = -1$ **b.** $y = 0$ **c.** $x = e, y = 3$ **d.** $x = 1, y = 1$

19. a. \$1402.55 **b.** \$1410.60 **c.** \$1414.78 **d.** \$1419.02 **e.** \$1419.07

21. e^2 **23.** e^{-1} **25.** $(3+\sqrt{17})/2$

27. 2

29. $e^{k\tau} = 2$

31. Percentage increase in 10 years: 41.06%; doubling time: 20.15 yr

33. e^6

37. α

39. $\ell = (1+\sqrt{5})/2$

43. $\lim_{n\to\infty} x_n = \pi/2$

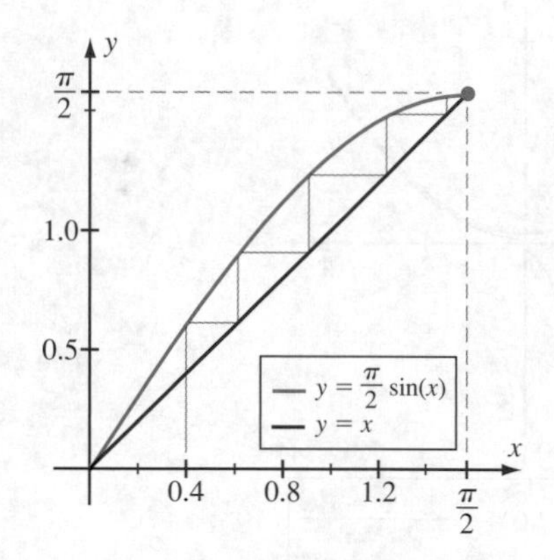

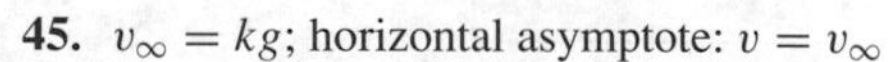

45. $v_\infty = kg$; horizontal asymptote: $v = v_\infty$

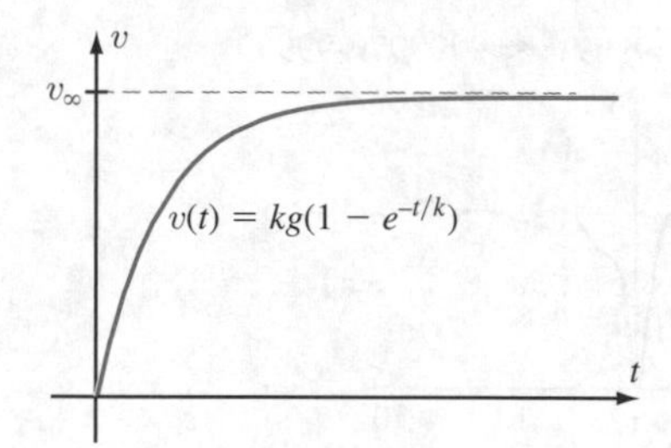

47. $\lim_{t\to\infty} c(t) = C$; horizontal asymptote: $y = C$

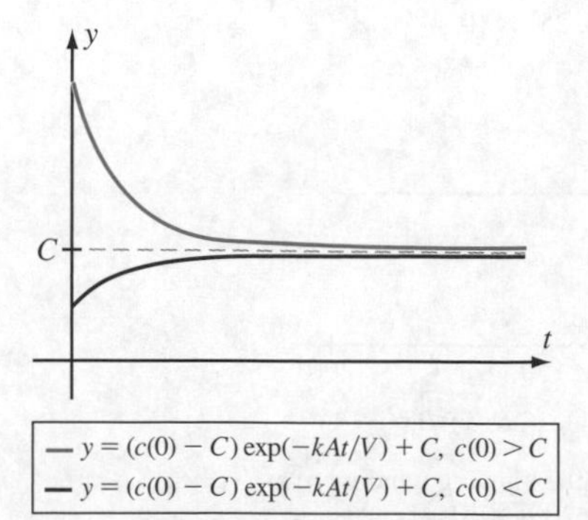

49. **a.** e **b.** e^4

51. **a.** $v(t) = e^{-0.045t}$

b.

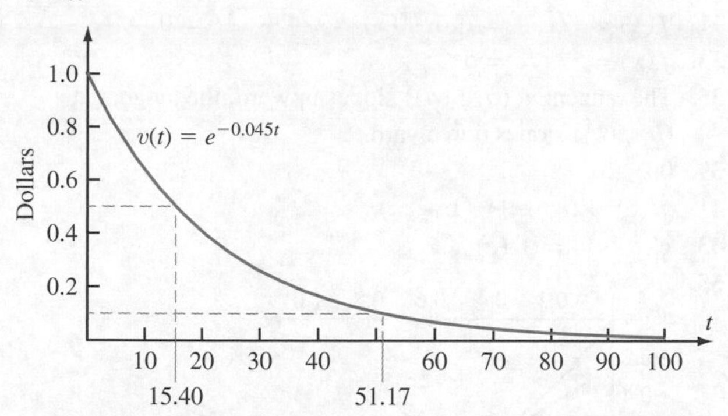

c. 15.40 yr; 51.17 yr

53. Nearly 120 months

55. $m_1 = 1/2, m_2 = 3/4, m_3 = 5/8, m_4 = 11/16,$ $m_5 = 23/32, m_6 = 45/64, m_7 = 89/128, m_8 = 177/256,$ $m_9 = 355/512, m_{10} = 709/1024 \approx 0.6924$; the length of I_n is $1/2^n$. $|\gamma - m_n| < 1/2^n$, so $|\gamma - 709/1024| < 1/2^{10} \approx 0.00098$.

57. 3761.52 yr

59. 61.6%

61. 20048 yr

63. 0.32

Section 3.1

1. 0

3. −10

5. −4

7. Forward

9. Backward

11. 6

13. −6

15. $y = 20(x - 5) + 50$

17. $y = 12(x + 2) - 7$

19. $y = (-1/20)(x - 5) + 50$

21. $y = (-1/12)(x + 2) - 7$

23. 72 per hour

25. 2043 per month

27. 64π in.^{2}/in.

29. Positive for $t > -1/2$; negative for $t < -1/2$

31. A: 10; B: 3; C: −1; D: −3; E: 0; F: 20

33. **a.** 18 ft **b.** $13.8 - 32t$ **c.** Maximum height 21.0 ft at $t = 13.8/32$ s **d.** 6.96 ft/s **e.** [0, 13.8/16] **f.** 1.58 s **g.** −36.64 ft/s **h.** −12.05 ft/s **i.** 0.808 s

37. $y = -6(x + 1) + 3$ and $y = 18(x - 3) + 27$

39. 6

41. $f(x) = \begin{cases} -2x + 6 & \text{if } -\infty < x < 1 \\ x + 3 & \text{if } 1 \le x < 3 \\ -x + 9 & \text{if } 3 \le x < \infty \end{cases}$

45. **a.** **i.** −8 **ii.** 6.4 **iii.** 7.84 **iv.** 7.984 **b.** 8

47.

n	0	1	2	3	4
Ave. Velocity over [0, 10^{-n}]	0.079401	1.6209	1.7219	1.7310	1.7320

Guess for instantaneous velocity: $v = \sqrt{3}$

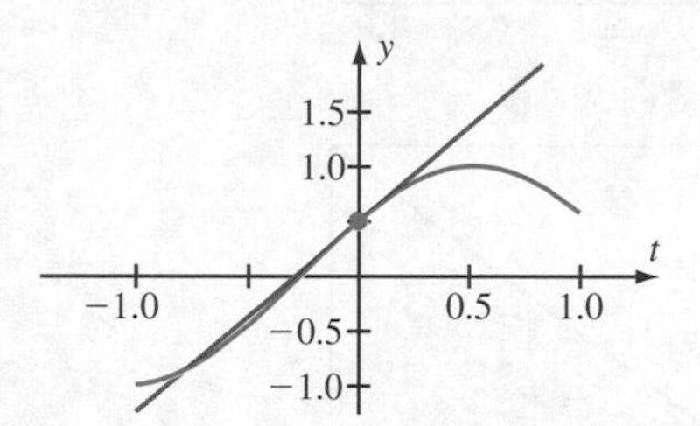

49.

n	1	2	3	4
Ave. Velocity over [2, 2 + $10^{(-n)}$]	7.7711	7.4261	7.3928	7.3894
Ave. Velocity over [3, 3 + $10^{(-n)}$]	21.124	20.186	20.096	20.087

For $t = 2$ and $t = 3$, the instantaneous velocity at t and the position $p(t)$ are (apparently) equal.

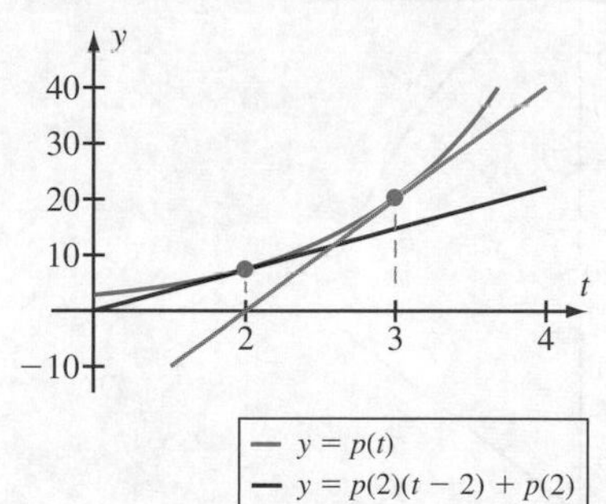

51. Zooming in to the viewing window $[\pi/4 - h, \pi/4 + h] \times [\tan(\pi/4 - h), \tan(\pi/4 + h)]$ results in a slope of $(\tan(\pi/4 + h) - \tan(\pi/4 - h))/(2h) \approx 2.0006$. For $Q(\Delta x) = (\tan(\pi/4 + \Delta x) - \tan(\pi/4))/\Delta x$,

Δx	0.01	0.001	0.0001	0.0001
$Q(\Delta x)$	2.02027	2.002003	2.0002	2.0000

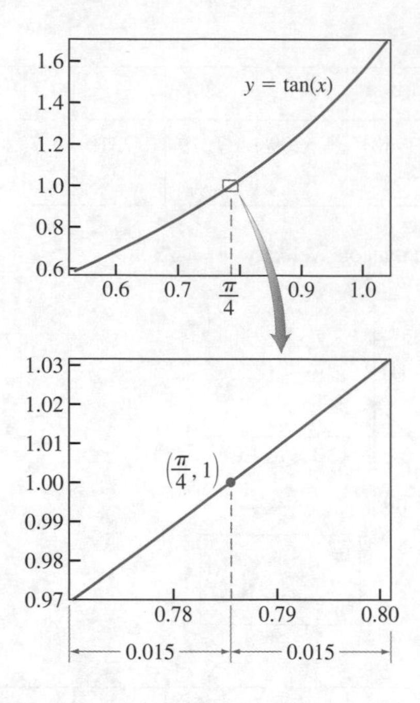

Section 3.2

1. -6
3. 5
5. 592
7. $-\pi$
9. $12x^2 + 12x$
11. 3
13. $-\sqrt{2}$ and $\sqrt{2}$
15.

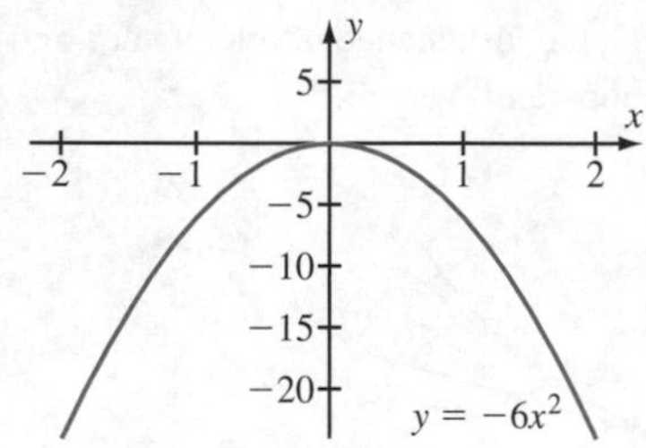

17.

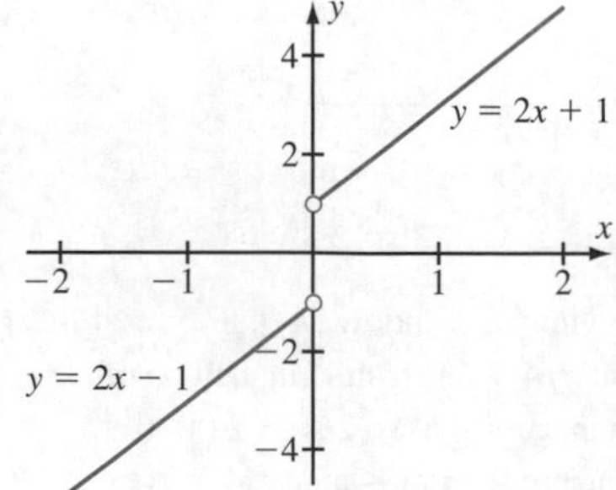

19. 0
21. $1/6$
23. $H = K', G = H', F = G'$
25. $p(x) = -5(x - 2) + 6$
27. Not differentiable at $x = 0$
29. Not differentiable at $x = 0$
31. $f(x) = \sqrt{x}, c = 4$, or $f(x) = \sqrt{4 + x}, c = 0$
33. $f(x) = 1/x, c = 5$
35. The tangent at $(a, f(a))$ slopes upward; the tangent at $(b, f(b))$ slopes downward.
39. 0
41. $f'(c) = (c^2 - 1)/(1 + c^2)^2$
43. $g'(c - k) = f'(c)$
55.

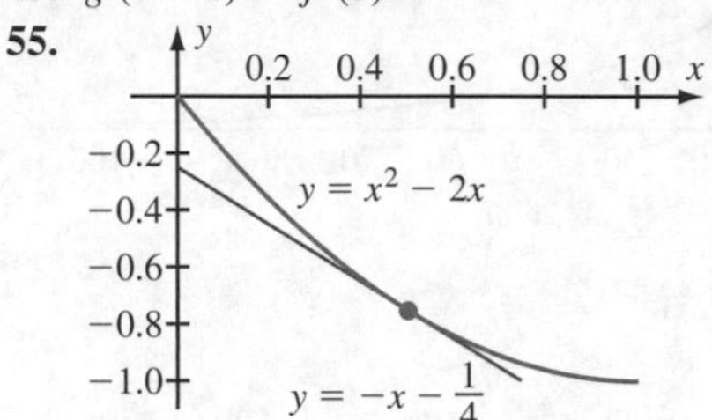

57.

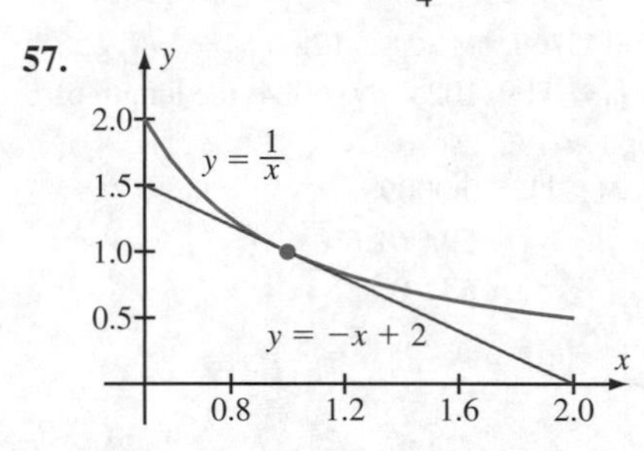

59. $f'(c) \approx 0.2272727275$ (slope of secant line through $(\varsigma, f(\varsigma))$)

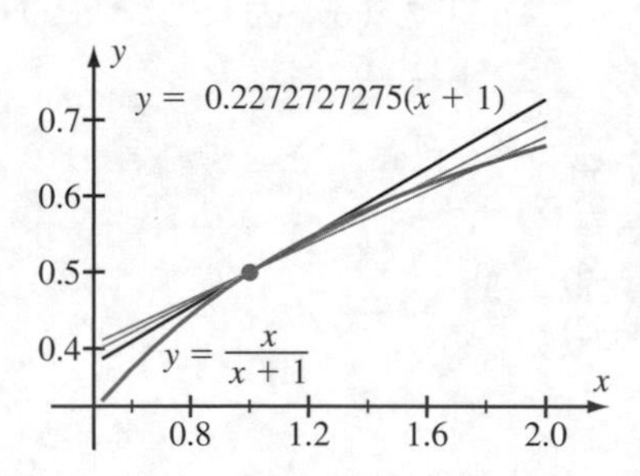

61. $f'(c) \approx 0.7019192346$ (slope of secant line through $(\varsigma, f(\varsigma))$)

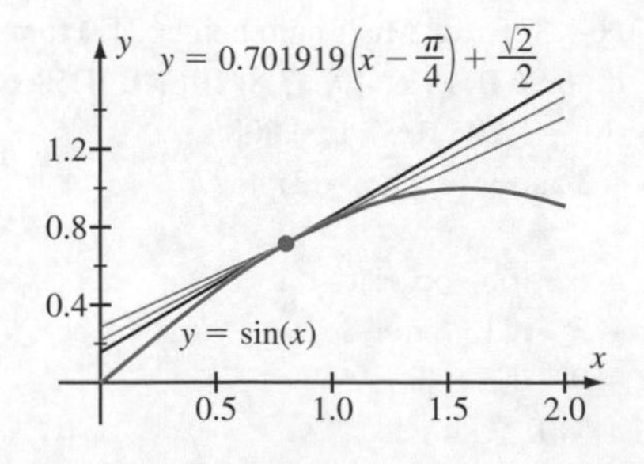

67. $f'(1)$ does not exist.

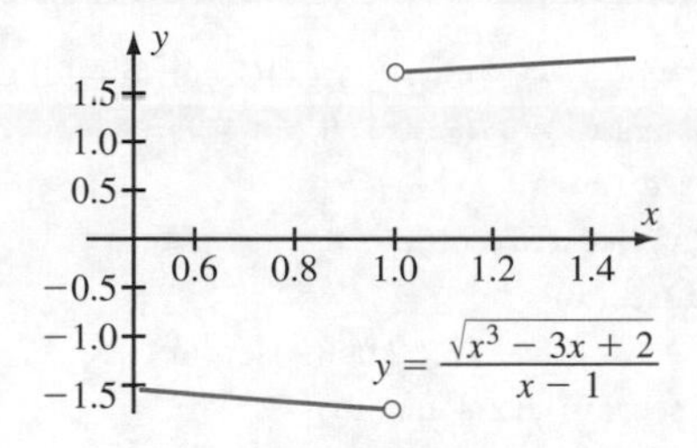

69. $f'(\pi/2)$ does not exist.

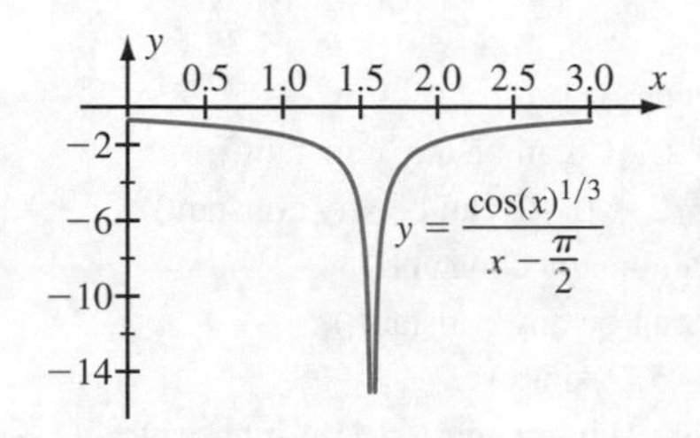

71. $f'(0) = 0$

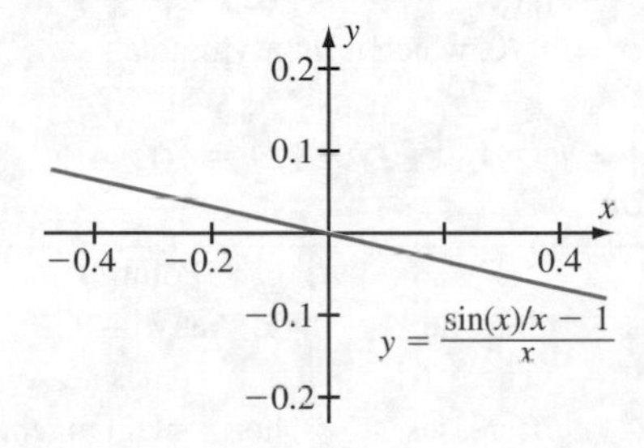

73.

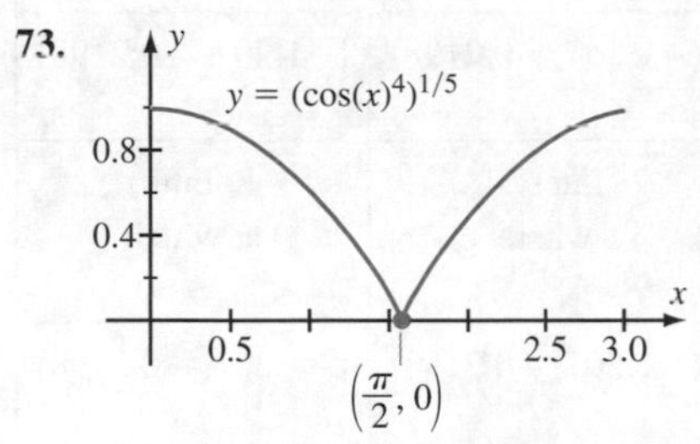

75.

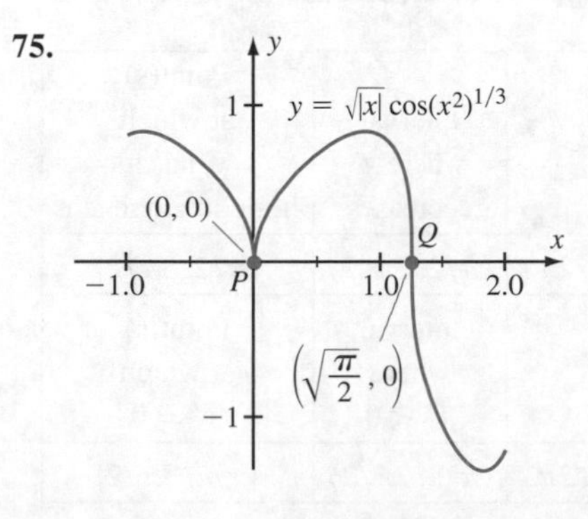

77. $f'(x) = 2\cos(2x)$

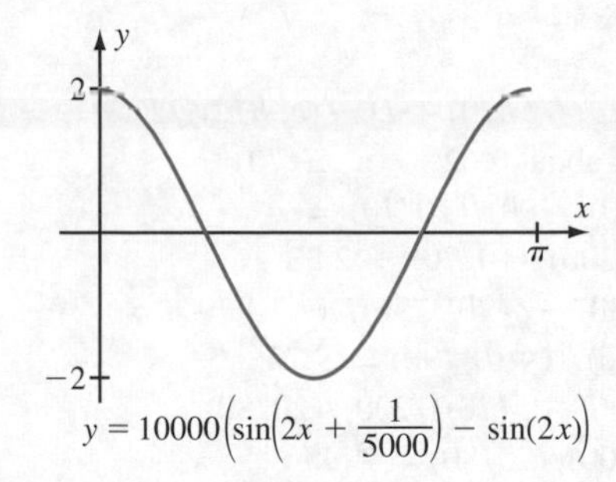

Section 3.3

1. $12x^2 + 6x$

3. $(2x - 5)(4x^3 + x^2) + (x^2 - 5x)(12x^2 + 2x)$, or $20x^4 - 76x^3 - 15x^2$

5. $6x\sin(x) - \cos(x)$

7. $(\sin(x) - x\cos(x))/\sin^2(x)$

9. $-(3x^2 + 2x)/(x^3 + x^2 + 1)^2$

11. $2x(x^2 + 2)^{-2}$

13. $3(1 - x^2)/(x^2 + 1)^2$

15. $(x^2\cos(x) + \sin^2(x))/(x + \sin(x))^2$

17. $2(x^3 - 4x^2 + 5x + 8)/(x - 1)^2$

19. $6x^5 + 12x^3 + 4x$

21. $\sin(x)(2 + \tan^2(x))$

23. $y = (32/25)(x - 1) - 2/5$

25. $y = x$

27. 5.0

29. 20.0

31. $10x^9$

33. $\sec^2(x)$

35. $2\cos(2x)$

37. $(1/f)'(3) = -3/8$; $(6f)'(3) = 36$

39. $(f \cdot g^2)'(4) = -41$; $(g \cdot f^2)'(4) = -56$; $(1/f^2)'(4) = 5/4$; $(g/f)'(4) = -13/4$

41. $-1, 3/2, 3$

43. **a.** The domain of f is $\{x \neq -1\}$. Two points, x_1 and x_2, with $x_1 < -1 < x_2$ appear to have the required property. **b.** The slope of the tangent line to the graph of f at $(c, f(c))$ is $2/(c + 1)^2$. The common property of the tangents to the graph of f is that they rise; all tangents have positive slope, which is consistent with the formula $f'(c) = 2/(c + 1)^2$. **c.** $y = (2/(c + 1)^2)(x - c) + 2c/(c + 1)$ **d.** $c_1 = -3 - 2\sqrt{2}$ and $c_2 = -3 + 2\sqrt{2}$ are the abscissas of the required points.

45. If $p(x) = (x - \alpha)^3 q(x)$, then $p'(x) = (x - \alpha)^2 Q(x)$ where $Q(x) = (x - \alpha)q'(x) + 3q(x)$.

47. $\alpha \geq 1/2$

49. The expression $x = \underbrace{1 + \cdots + 1}_{x \text{ summands}}$ is only meaningful when x is a positive integer. Therefore, the function

$\mathbb{N} \ni x \mapsto \underbrace{1 + \cdots + 1}_{x \text{ summands}}$ cannot even be differentiated with respect to x.

51. $D_0 f(c, h) = (D_+ f(c, h/2) + D_- f(c, h/2))/2$

55. b. A decrease of about 0.62% **c.** $E(70) \approx -((q(80) - q(70))/300)/(10/70) = -((200 - 300)/300)/(10/70) = 2.33$; $E(90) \approx -((q(80) - q(90))/96)/((-10)/90) = -((200 - 96)/96)/((-10)/90) = 9.75$; $E(80) \approx -((q(90) - q(70))/200)/(20/80) = -((96 - 300)/200)/(20/80) = 4.08$

57.

n	Forward	Backward	Central
1	0.48943	−0.48943	0.00000
2	0.04934	−0.04934	0.00000
3	0.00493	−0.00493	0.00000
4	0.00049	−0.00049	0.00000
5	0.00005	−0.00005	0.00000

59.

n	Forward	Backward	Central
1	0.24279	0.24401	0.24342
2	0.24337	0.24349	0.24343
3	0.24342	0.24343	0.24343
4	0.24344	0.24343	0.24344
5	0.24340	0.24340	0.24350

61. 0.86603; $h = 1/1000$ suffices

63. −1.41421; $h = 1/1000$ suffices

65.

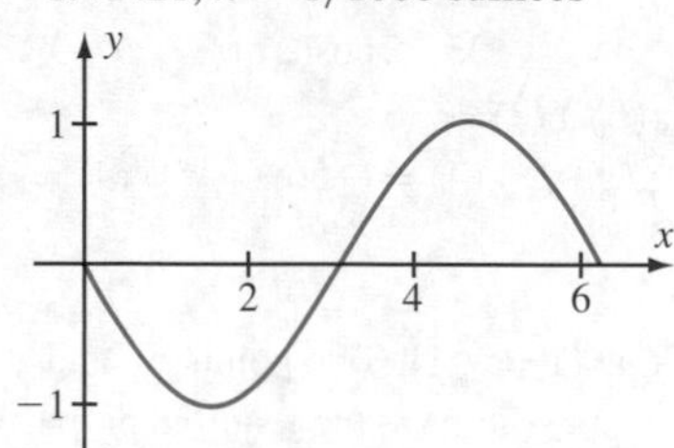

Plots of $D_0 f(x, 10^{-5})$ and $f'(x) = -\sin(x)$ (The two curves cannot be distinguished in this figure.)

67. $L(2.70) \approx 0.99322$ and $L(2.73) \approx 1.00435$; $a_0 \approx 2.718$ and $\frac{d}{dx} a_0^x \big|_{x=0} \approx L(2.718) = 1$

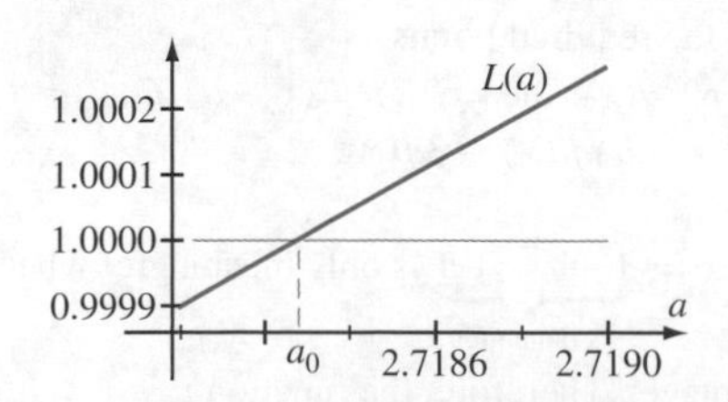

Section 3.4

1. $80x^9 + 30x^{-6}$ **3.** $-\csc(x)(2\cot^2(x) + 1)$

5. $\sec(x)((\sqrt{x} + 1)\tan(x) - 1/(2\sqrt{x}))/(\sqrt{x} + 1)^2$

7. $\sec(x)\tan(x) - \sec^2(x)$

9. $\cot(x) - x\csc^2(x) + \csc(x)\cot(x)$

11. $\sec^3(x) + \tan^2(x)\sec(x)$

13. $x^{-6}(x - 5)e^x$ **15.** $2\tan(x)\sec^2(x)$

17. $(2 + \tan(x) - x\sec^2(x))/(2 + \tan(x))^2$

19. $x((2 - x)e^x + 2\sin(x) - x\cos(x))/(e^x + \sin(x))^2$

21. $y = -(x - \pi)/\pi^3$ **23.** $y = 42(x - \pi/3) + 6\sqrt{3}$

25. $y = \pi(x - \pi)$ **27.** $2 + e^x$

29. $-e^{-x}(\cos(x) + \sin(x))$

31. $x^7 - 2x^2 + 6x + C$ (C can be any constant.)

33. $x^9/9 + x^6 - 3x^2/2 + C$ (C can be any constant.)

35. $\sin(x) + C$ (C can be any constant.)

37. $8\cot(x) + C$ (C can be any constant.)

39. $(9/2)\cos(x) + (-3/2)\sin(x)$

41. $y = e^t(x - t) + e^t$; x-intercept: $t - 1$; y-intercept: $e^t(1 - t)$

45. −1 **47.** 2

49. $f_n'(x) = n\, f_{n-1}(x) + f_n(x)$

51. $M(t) = \cos^2(t)$; $x = \pi/2$, which is an asymptote.

53. 1

57. $\deg((p \cdot q)') = n + m - 1$; $\deg((p \circ q)') = nm - 1$; $n + m - 1$ and $2m$

67.

Interval where f increases	Interval where f decreases	Point(s) at which f has a horizontal tangent
$(-2, -1.207)$, $(-0.2198, 2)$	$(-1.207, -0.2198)$	$\{-1.207, -0.2198\}$
Interval where $f' > 0$	Interval where $f' < 0$	Point(s) at which $f' = 0$
$(-2, -1.207)$, $(-0.2198, 2)$	$(-1.207, -0.2198)$	$\{-1.207, -0.2198\}$

69.

Interval where f increases	Interval where f decreases	Point(s) at which f has a horizontal tangent
$(0, \pi/2)$, $(3\pi/2, 2\pi)$	$(\pi/2, 3\pi/2)$	$\{\pi/2, 3\pi/2\}$
Interval where $f' > 0$	Interval where $f' < 0$	Point(s) at which $f' = 0$
$(0, \pi/2)$, $(3\pi/2, 2\pi)$	$(\pi/2, 3\pi/2)$	$\{\pi/2, 3\pi/2\}$

71. $g_k'(0) = k$

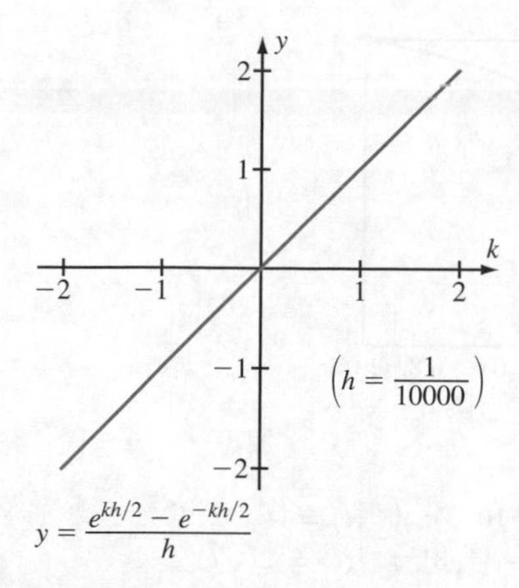

73. e^x grows faster than x^{20} on $(88.083\ldots, \infty)$; $e^x = x^{20}$ for $x \approx 89.9951$.

75. a. $(\mu^4\tau_0 p_0 - p_0\tau_0 + 2\mu^2\Delta)/(p + p_0\mu^2)^2$

e.

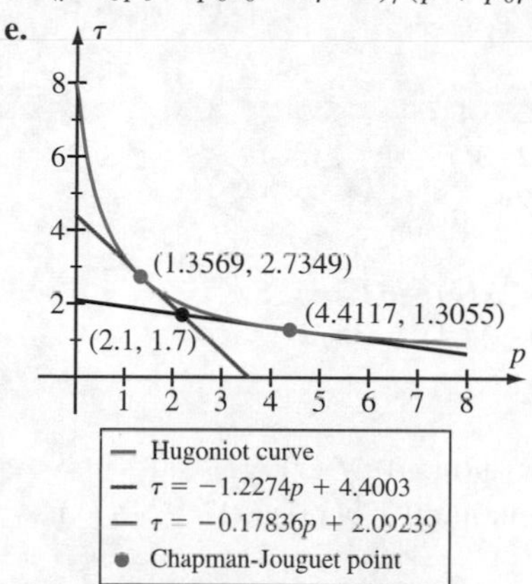

Section 3.5

1. $10(x^2 + 3x)^9(2x + 3)$
3. $\sec^2(\sin(x))\cos(x)$
5. $-\sec(\cos(x))\tan(\cos(x))\sin(x)$
7. $(8x^7 + 2x)\cos(x^8 + x^2)$
9. $6(\sin x + x^2)^5(\cos(x) + 2x)$
11. $3x^2\sec^2(x^3)$
13. $2x\exp(x^2 + 1)$
15. $-\sin(\sqrt{x})/(2\sqrt{x})$
17. $-7\sec^2(1 - 7x)$
19. $(f \circ g)'(x) = 2x + 7$; $(g \circ f)'(x) = (2x^3 + 7x)/\sqrt{x^4 + 7x^2}$
21. $(f \circ g)'(x) = 3x^2/(x^3 + 1)^2$; $(g \circ f)'(x) = 3x^2/(x + 1)^4$
23. 16
25. 0
27. $(-1/49)(1 + 2\sin(2/7))\cos(2/7)$
29. 3992365
35. $m'(t) = -0.121286\exp(-0.00121286t)$; $m'(t) = -0.00121286 \cdot m(t)$; $\lim_{t\to\infty} m(t) = \lim_{t\to\infty} m'(t) = 0$
37. $8\sqrt{2}$ in.^{2}/min
39. $g'(t) = 2tf'(t^2)$; $g'(\sqrt{t}) = 2\sqrt{t}f'(t)$
41. $35\sec^5(7x)\tan(7x)$
43. $-2\sin(2x)\exp(\cos(2x))$
45. $-2x\sin(\sqrt{2x^2 + 3})/\sqrt{2x^2 + 3}$
47. 0
49. $(5\sec^2(5x) + 1/\sqrt{2x + 1})/(2\sqrt{\tan(5x) + \sqrt{2x + 1}})$
51. Yes; $c = 1/3$
55. $v_\infty = \sqrt{g/\kappa}$; $v'(t) = 4g\exp(2t\sqrt{g\kappa})/(\exp(2t\sqrt{g\kappa}) + 1)^2$; $\lim_{t\to\infty} v'(t) = 0$
57. 2.33577
61. -1.921×10^{-11} g/yr
63. a.

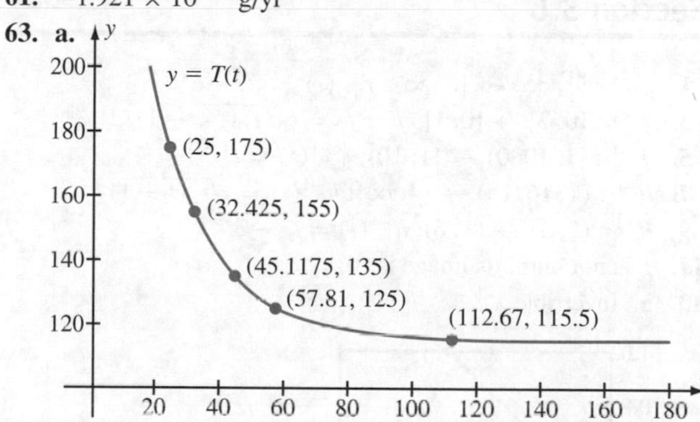

b.

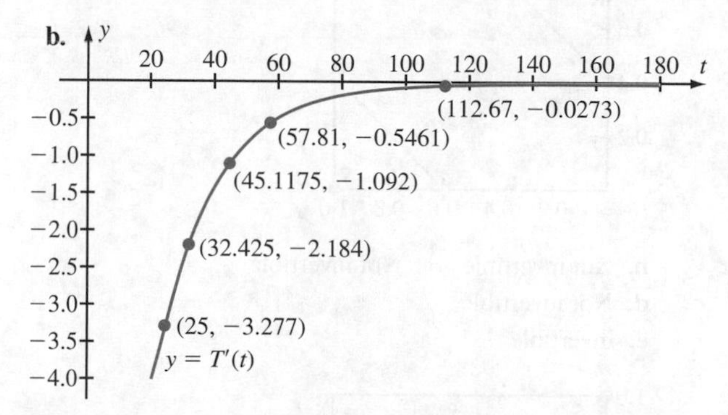

65. v is everywhere continuous and differentiable at all points except $t = 8$; $v'(t)$ is continuous and differentiable on its domain, which does not include the point $t = 8$.

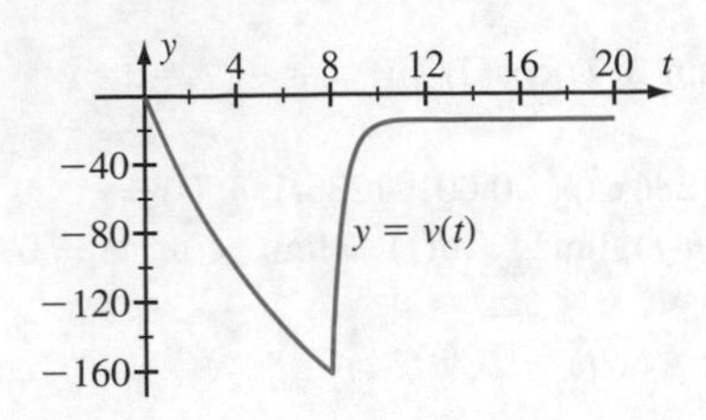

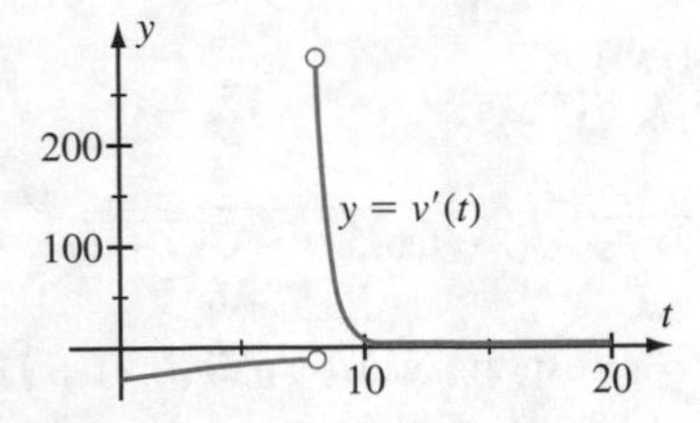

Section 3.6

1. $f^{-1}: [1, \infty) \to [0, \infty)$; $f^{-1}(t) = \sqrt{t-1}$
3. $f^{-1}: [0, 2] \to [0, 1]$; $f^{-1}(t) = (\sqrt{1+4t} - 1)/2$
5. $f^{-1}: [1, 1000) \to [1, 10)$; $f^{-1}(t) = t^{1/3}$
7. $h^{-1}: (1, 16/15) \to (4, \infty)$; $h^{-1}(t) = \sqrt{t/(t-1)}$
9. $h^{-1}: (2, 3) \to (1, 6)$; $h^{-1}(t) = t^2 - 3$
11. g is not onto; its image is $(0, 1/2]$.
13. **a.** Invertible

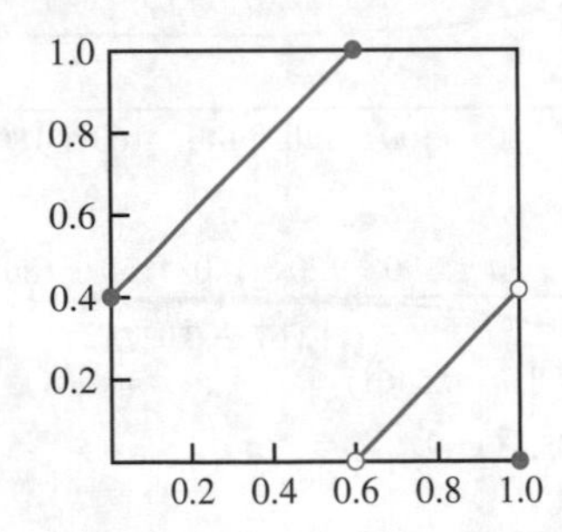

b. Not invertible **c.** Not invertible
d. Not invertible
e. Invertible

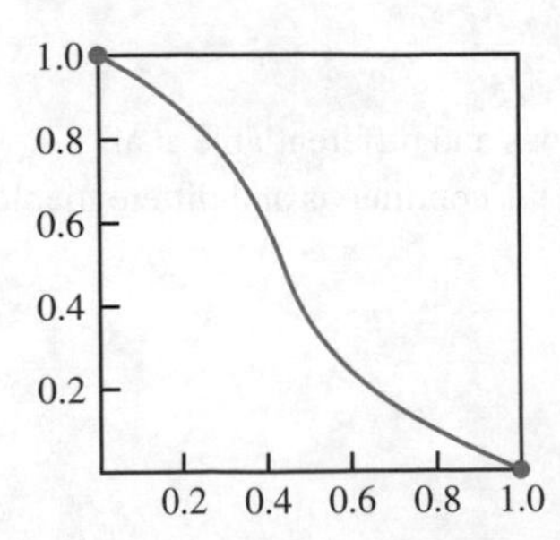

f. Invertible

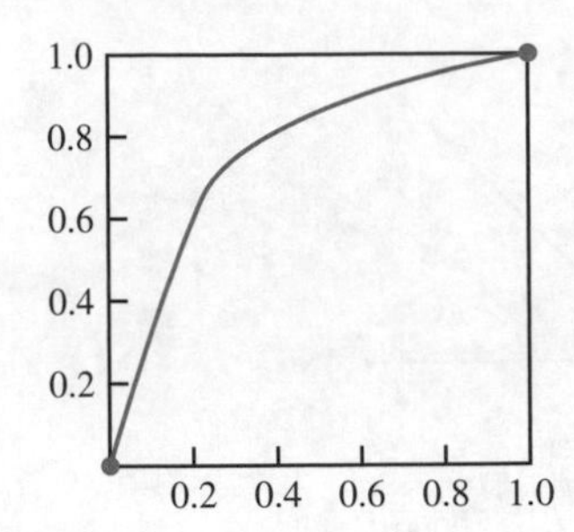

15. $1/2$
17. $1/3$
19. $f^{-1}: (2, 245) \to (0, 3)$; $f^{-1}(t) = (t-2)^{1/5}$
21. $f^{-1}: [1/64, 1] \to [1, 8]$; $f^{-1}(t) = 1/\sqrt{t}$
23. 3
25. $1/8$
27. $e^x/(1+e^x)$
29. $2^{\ln(x)} \ln(2)/x$
31. $1/(2x\sqrt{\ln(x)})$
33. $-\sin(\ln(x))/x$
35. $2 \cdot 3^x \ln(3) \cos(2 \cdot 3^x)$
37. $\pi \cdot 2^{x^\pi} x^{\pi-1} \ln(2)$
39. $2/x$
45. $f^{-1}(3) = 0$; $(f^{-1})'(3) = 1/2$
47. $f^{-1}(\pi) = 0$; $(f^{-1})'(\pi) = 3\pi/2$
49. $k = (1/\tau)\ln(2)$
51. $1/k$
53. $f(x) = \exp(v(x)\ln(u(x)))$; $f'(x) = v'(x) \cdot u(x)^{v(x)} \cdot \ln(u(x)) + v(x) \cdot u(x)^{v(x)-1} \cdot u'(x)$
57. $\ln(a)$
61. $\sqrt{g/\kappa}(1 - \exp(2t\sqrt{g\kappa}))/(1 + \exp(2t\sqrt{g\kappa}))$
67. -1.3078 g/yr; 2.052 yr
69. 0.3065
71. 0.02701
73. $c = 0.56714$

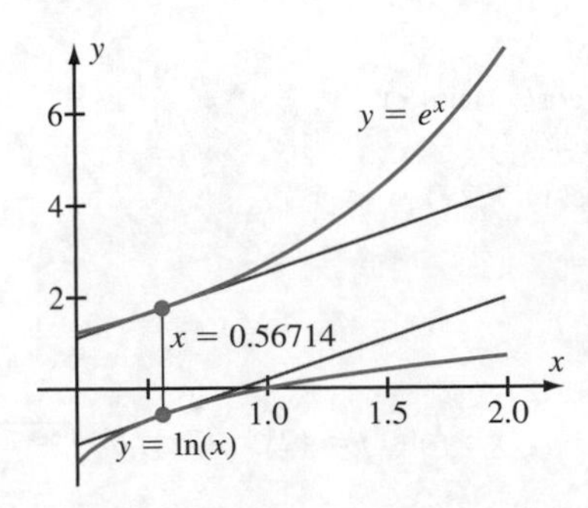

75. $\dfrac{d}{dx}\ln(x) = 1/x$

77. Macaques: 0.1440; Humans: 0.1230

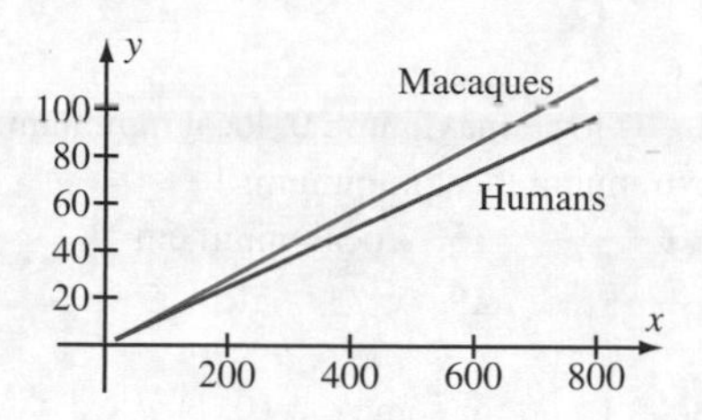

79.

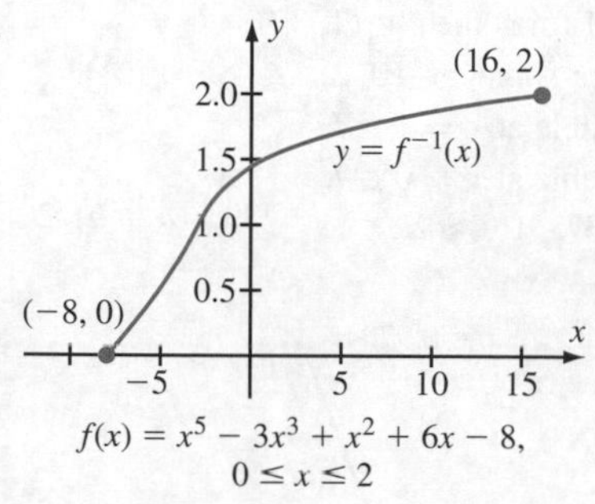

$f(x) = x^5 - 3x^3 + x^2 + 6x - 8,$
$0 \le x \le 2$

81.

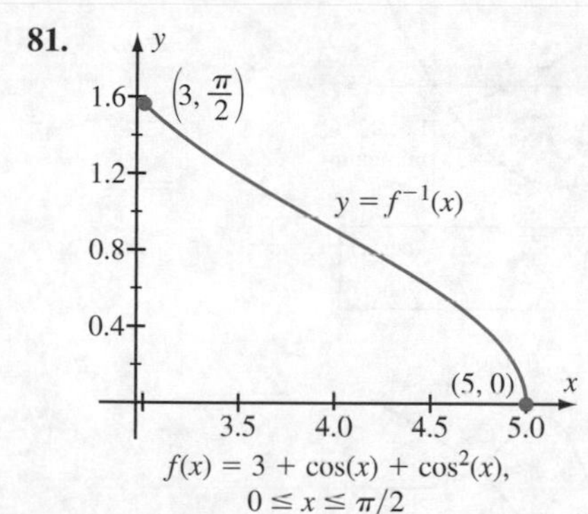

$f(x) = 3 + \cos(x) + \cos^2(x),$
$0 \le x \le \pi/2$

Section 3.7

1. $f'(x) = 12x^2 + 35x^{-6} + 5x^{3/2}$;
$f''(x) = 24x - 210x^{-7} + (15/2)\sqrt{x}$;
$f'''(x) = 24 + 1470x^{-8} + 15/(4\sqrt{x})$
3. $g'(t) = \sec^2(t)$; $g''(t) = 2\sec^2(t)\tan(t)$;
$g'''(t) = 2\sec^2(t)(2\tan^2(t) + \sec^2(t))$
5. $k'(x) = -2/(x-1)^2$; $k''(x) = 4/(x-1)^3$;
$k'''(x) = -12/(x-1)^4$
7. $k'(x) = (3x^2 - 3)\cos(x^3 - 3x)$;
$k''(x) = -(3x^2-3)^2\sin(x^3 - 3x) + 6x\cos(x^3 - 3x)$;
$k'''(x) = (6 - (3x^2 - 3)^3)\cos(x^3 - 3x) -$
$18x(3x^2 - 3)\sin(x^3 - 3x)$
9. $H'(t) = \tan(t) + t\sec^2(t)$; $H''(t) = 2\sec^2(t)(1 + t\tan(t))$;
$H'''(t) = 2\sec^2(t)(3\tan(t) + 3t\tan^2(t) + t)$
11. $\cos(x)$
13. $6(234t^2 - 390t + 175)(3t^2 - 5t)^{-8}$
15. $-2880x^2 + 840x$
17. $-16\cos(4x + 3)$
19. $\sin(x)\sin(\sin(x)) - \cos^2(x)\cos(\sin(x))$
21. 1 ft/s^2; 7 ft/s^2
23. 38 m/min^2; 74 m/min^2
29. $x^3 - 4x^2 + 2x + 2$
35. $(6x^4 + 9x^2 + 2)/(x^2 + 1)^{3/2}$
37. $6x\cos(x) - 6x^2\sin(x) - x^3\cos(x)$
39. $P_1(x) = x$; $P_2(x) = (3x^2 - 1)/2$; $P_3(x) = (5x^3 - 3x)/2$;
$P_4(x) = (35x^4 - 30x^2 + 3)/8$
43. 0

Section 3.8

1. $-8/5$
3. $-1/16$
5. $1/8$
7. $y = 8 - x$
9. $y = x$
11. $y = (20 - x)/9$
13. $y = 2\pi - x$
15. 4; $-4/3$
17. 4; -54
19. -1; 0
21. $y = x/4 - 1$
23. $y = 2$
25. $10\text{ in.}^2/\text{min}$
27. $92/\sqrt{97}$ units/s
29. $28/5$ ft/s
31. $-3/101$ radians/min
39. $4(\ln(2) - 1)/(2\ln(2) - 1)$
41. $s = a^2$ for every tangent T
43. $-Ay^3(2b + 3x)\exp(b/x)/(2x^{7/2}(y - 2))$
45. **a.** $q'(6) = -0.2893$; **b.** $E(6) = 0.5025$
47. Decreasing at the rate of $2\text{ in.}^2/\text{min}$
49. $1/3\text{ ft}^2/\text{min}$
51. -0.99751
53. -1.0453
55. $y = -1.0002x - 0.66397$
57. $y = -0.096965x - 10.861$
59.

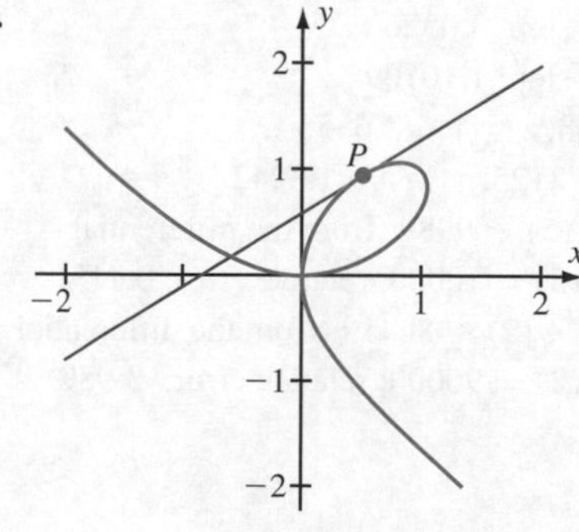

61.

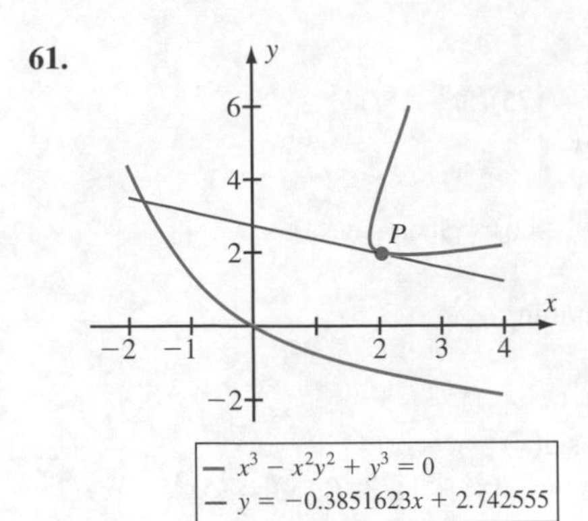

63. $x = -1.22074$ and $x = 0.724492$

65. $q'(4) = -0.43170\ldots$;
$q'(4) \approx (q(4.2) - q(3.8))/(4.2 - 3.8) = -0.43217\ldots$

Section 3.9

1. 1.975 **3.** 0.37024 **5.** 1

7. 0.1 **9.** 1.8403 **11.** 0.1152

13. 0.71227

15. $f(x) = \sqrt{x}$; $c = 25$; $f(24) \approx 4.9$; error $= -0.001\ldots$

17. $f(x) = \sqrt{1 + \sqrt{x}}$; $c = 9$; $f(9.1) \approx 2.00417$; error $= -0.158\ldots$

19. $f(x) = \cos(\pi x/180)$; $c = 60$; $f(59) \approx 0.5151$; error $= -0.00007\ldots$

21. $df = g(h(c)) + g'(h(c))h'(c)\,dx$

23. $L(x) = f(0)$

27. 2.6

29. 3/2

31. Slope $= -0.2612$; about 3235 units would be sold at \$6.80; about 3287 units would be sold at \$6.60.

33. 41001

35. 1-step: 3.03333; 3-step: 3.03263

37. 1-step: 1.10364; 2-step: 1.10109

39. $y_1 = 2.6$; $z_1 = 2.68$; $y(x_1) = 2.6856\ldots$

41. $y_1 = 1$; $z_1 = 1.2578125$; $y(x_1) = 1.8242\ldots$

43. $q(5.10) = 9612$; $q(5) \approx 9989$ (from the differential approximation); $q(5) = 10000$; relative error: 0.11%

45. $q(1.80) = 111111$; $q(2) \approx 86419$ (from the differential approximation); $q(2) = 90000$; relative error: 3.98%

Section 4.1

1. Local minimum: 2

3. Local maximum: 0

5. Local maxima: $\{(4k+1)\pi/8 : k \in \mathbb{Z}\}$; local minima: $\{(4k+3)\pi/8 : k \in \mathbb{Z}\}$

7. Local minimum: 6

9. Local minimum: -1; local maximum: 0; local minimum: 1

11. -5, not a local extremum; local minimum: 1

13. Local minimum: 1 **15.** Local minimum: 0

17. $5x + C$ **19.** $x^3/3 + \pi x + C$

21. $\sin(x) + C$

23. $4c^3 + 21c^2 - 18c + 1 = 0$ for some $c \in (0, 1)$

25. $(1/5)c^{-4/5} = 1/31$ for some $c \in (1, 32)$

27. $-4(c-1)^3 = 0$ for some $c \in (0, 2)$

31. f is not differentiable at $x = 1 \in I$.

33. f is not differentiable at $x = 0 \in I$.

35. $1/\sqrt{3}$ **37.** $1 - \sqrt{2}$ **39.** $(a+b)/2$

41. $\sqrt{2}$

71.

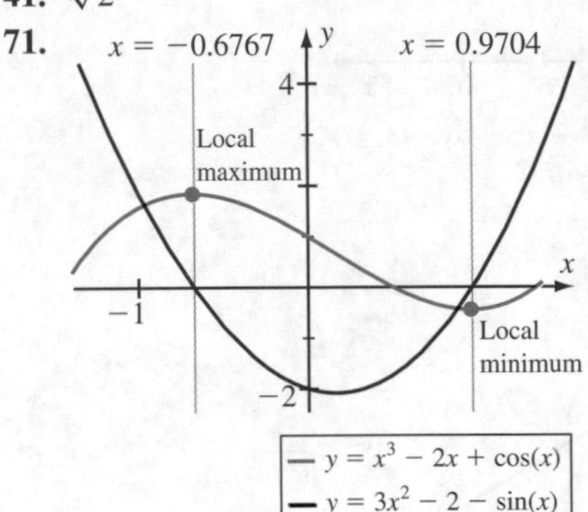

73.

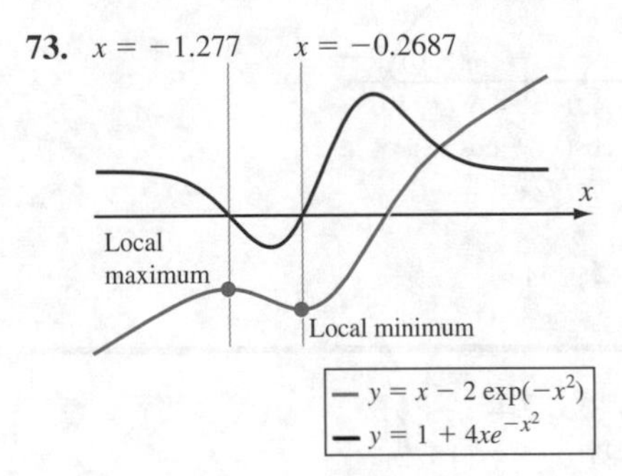

75. 1.3124

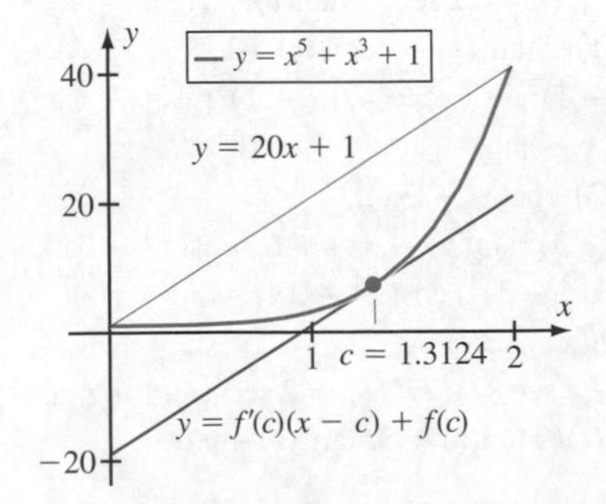

77. 1.826

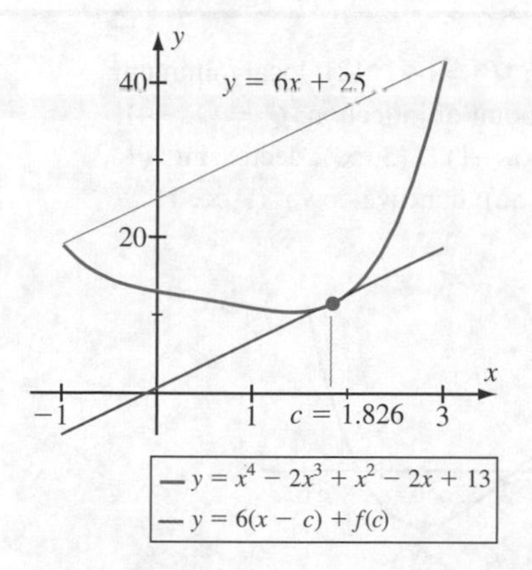

Section 4.2

1. Increasing: $(-\infty, -5) \cup (3, \infty)$; decreasing: $(-5, 3)$
3. Increasing: $(-\infty, -\sqrt{5}) \cup (\sqrt{5}, \infty)$; decreasing: $(-\sqrt{5}, \sqrt{5})$
5. Decreasing: $(-\infty, 1) \cup (1, \infty)$
7. Increasing: $(-2, -3/2) \cup (-1, \infty)$; decreasing: $(-\infty, -2) \cup (-3/2, -1)$
9. Increasing: $(-\infty, \infty)$
11. Increasing: $(0, \infty)$; decreasing: $(-\infty, 0)$
13. Local minimum: $-1/2$
15. Local maximum: 0; local minimum: 4
17. Local maximum: 0
19. Local minimum: 0; local maximum: 2
21. Local minimum: $1/2$; 7, not a local extremum
23. Local minima: $n\pi$ $(n \in \mathbb{Z})$; local maxima: $(2n+1)\pi/2$ $(n \in \mathbb{Z})$
25. Local maximum: -2; local minimum: $34/13$; 4, not a local extremum
27. 0, not a local extremum; local maximum: $\sqrt{3}/9$
29. Increasing: $(-3/2, -1) \cup (0, 3/2) \cup (2, 5/2)$; decreasing: $(-1, 0) \cup (3/2, 2)$
39. Forward: $t < 1/2$ and $t > 2/3$; reverse: $1/2 < t < 2/3$
53. Increasing: $(1.3479, \infty)$; decreasing: $(-\infty, 1.3479)$
55. Increasing: $(-\infty, -1.90757) \cup (1.06832, \infty)$; decreasing: $(-1.90757, 1.06832)$
57. Increasing: $(-\infty, 0.69725)$; decreasing: $(0.69725, \infty)$
59. Increasing: $(0, 1.7632)$; decreasing: $(1.7632, \infty)$
61. Increasing: $(-0.90456, 0) \cup (0.90456, \pi/2)$; decreasing: $(-\pi/2, -0.90456) \cup (0, 0.90456)$; critical points and local minima: -0.90456, 0.90456; critical point and local maximum: 0
63. Increasing: $(0, 0.27777)$; decreasing: $(0.27777, 1)$; critical point and local maximum: 0.27777

Section 4.3

1. 25 ft (each side)
3. $(2^{-1/3}, 2^{1/6})$
5. $10/3$ in. (each side)
7. -2
9. $\sqrt{3}/3$
11. Height = width = $2 + 4\sqrt{5}$ in.
13. Base: $4\sqrt{3}/3$; height: $8/3$
15. Side length of base: $2 \cdot 9^{1/3}$ m
17. Triangle side length: $20/\sqrt{6-\sqrt{3}}$ m; other side length of rectangle: $5\sqrt{6-\sqrt{3}} - 5\sqrt{(6+\sqrt{3})/11}$ m
19. $5\pi/3$ (radius 1 and cylinder height 1)
21. $\pi/3$
23. $(b + \sqrt{a^2+b^2})/a$
25. Circumference of circle: $\pi L/(\pi+4)$
27. Maximum area: $L^2/(4\pi)$; minimum area: $\pi L^2/(6\pi^2 + 8\pi + 8)$
29. $(b/a)^{1/4}$
31. $(2b)^{1/6}$
35. $(r/R)^{1/4}$
37. 500 m^2
39. $8/\pi$
41. $\ell/6$
45. 0.78622
47. $10(2 - 2^{2/3} + 2^{1/3})/3$ from the stronger source
49. $\cos(2\alpha) = gh/(v^2 + gh)$
51. 0.51282

Section 4.4

1. Concave up: $(-3, \infty)$; concave down: $(-\infty, -3)$; critical points: $\{-7, 1\}$; point of inflection: -3; local maximum: -7; local minimum: 1
3. Concave up: $(1/2, \infty)$; concave down: $(-\infty, 1/2)$; critical points: $\{-1, 2\}$; point of inflection: $1/2$; local maximum: -1; local minimum: 2
5. Concave up: $(-\infty, 0) \cup (7/2, \infty)$; concave down: $(0, 7/2)$; critical points: $\{0, 21/4\}$; points of inflection: $\{0, 7/2\}$; local maximum: $21/4$
7. Concave up: $(-3, -3/2) \cup (0, \infty)$; concave down: $(-\infty, -3) \cup (-3/2, 0)$; critical points: $\{-3, 0\}$; points of inflection: $\{-3, -3/2, 0\}$; no local extrema
9. Concave up: $(-\infty, 0)$; concave down: $(0, \infty)$; no critical points; point of inflection: 0; no local extrema
11. Concave up: $(-\infty, -5/2) \cup (0, \infty)$; concave down: $(-5/2, 0)$; critical points: $\{0, 5/4\}$; points of inflection: $\{-5/2, 0\}$; local minimum: $5/4$

13. Concave up: $(0, \infty)$; concave down: $(-\infty, 0)$; critical points: $-1, 1$; no point of inflection; local maximum: -1; local minimum: 1

15. Concave up: the intervals $((2n-1)\pi, 2n\pi)$; concave down: the intervals $(2n\pi, (2n+1)\pi)$; critical points $\{m\pi/2 : m \text{ an odd integer}\}$; points of inflection: $\{n\pi : n \text{ an integer}\}$; no local extrema

17. Concave down: $(0, \infty)$; no critical points, no points of inflection, no local extrema

19. Concave up: $(0, \infty)$; critical point: $1/e$; no points of inflection; local minimum at $1/e$

21. Local maximum: 0; local minimum: 1

23. Local maximum: -1; local minimum: 1

25. Local maximum: -1; local minimum: 1

27. 0, 1

29. $-1, 1$

31. $-1, 1$

33. Local minimum: $1/2$

35. Local maximum: 4; local minimum: 0

37. Points of inflection: $-2a/\sqrt{3}$ and $2a/\sqrt{3}$; concave up: $(-\infty, -2a/\sqrt{3}) \cup (2a/\sqrt{3}, \infty)$; concave down: $(-2a/\sqrt{3}, 2a/\sqrt{3})$

39. Concave up

47. $6|x|/(9x^4 + 12x^2 + 5)^{3/2}$

49. $2/(4x+1)^{3/2}$

51. $1/\sqrt{2}$

53. $(2n+1)\pi/2, n \in \mathbb{Z}$

61. Critical point: 1.279 (local maximum); points of inflection: -2.399, 2.399

63. Critical point: -0.808 (local minimum); points of inflection: -0.344, 0.533

65. Critical points: -1 (local maximum), 1 (local minimum); points of inflection: -1.532, -0.347

67.

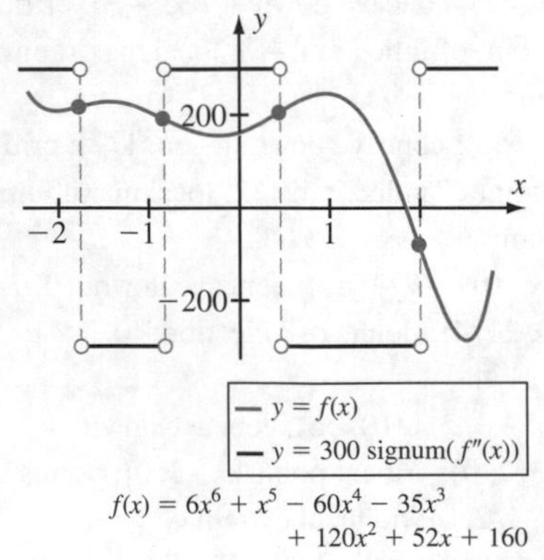

Section 4.5

1. Local maximum: $P = (-1, 12)$; local minimum: $R = (3, -20)$; point of inflection: $Q = (1, -4)$; increasing: $(-\infty, -1) \cup (3, \infty)$; decreasing: $(1, 3)$; concave up: $(1, \infty)$; concave down: $(-\infty, 1)$

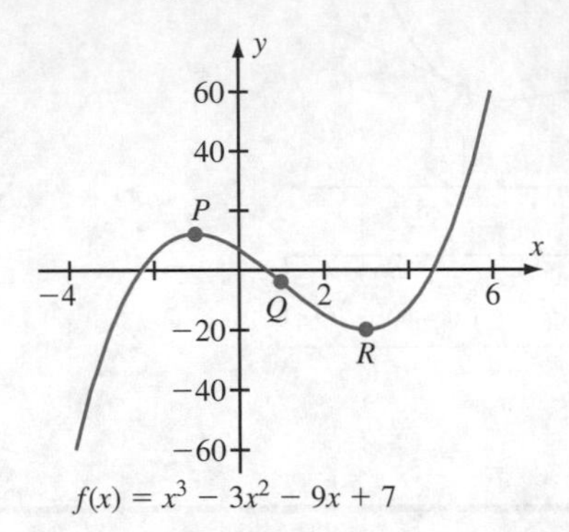

3. Local maximum: $Q = (-27/8, 4/27)$; point of inflection: $P = (-(9/5)^3, 100/729)$; increasing: $(-\infty, -27/8)$; decreasing: $(-27/8, 0) \cup (0, \infty)$; concave up: $(-\infty, -(9/5)^3) \cup (0, \infty)$; concave down: $(-(9/5)^3, 0)$; vertical asymptote: $x = 0$; horizontal asymptote: $y = 0$

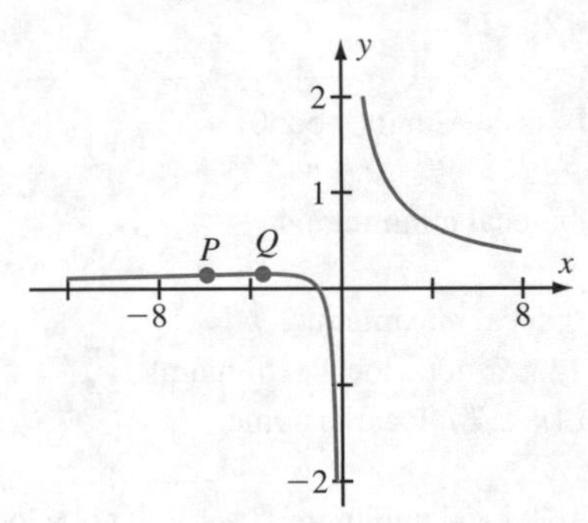

5. Local minimum: $P = (1/2, -3 \cdot 2^{-4/3})$; points of inflection: $Q = (0, 0)$ and $R = (-1, 3)$; increasing: $(1/2, \infty)$; decreasing: $(-\infty, 1/2)$; concave up: $(-\infty, -1) \cup (0, \infty)$; concave down: $(-1, 0)$

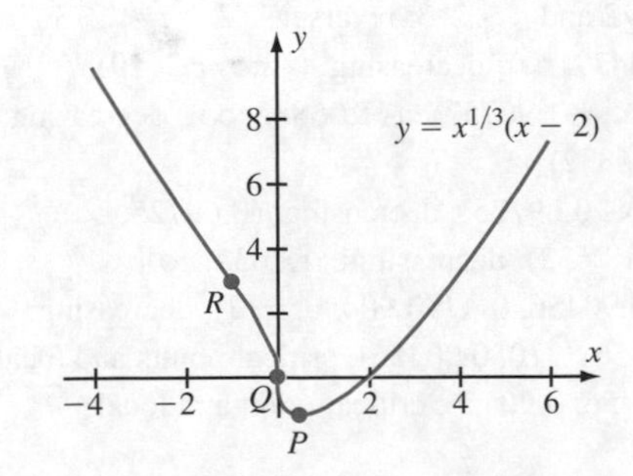

7. Local maximum: $P = (0, 4)$; local minimum: $R = (2, 0)$; point of inflection: $Q = (1, 2)$; increasing: $(-\infty, 0) \cup (2, \infty)$; decreasing: $(0, 2)$; concave up: $(1, \infty)$; concave down: $(-\infty, 1)$

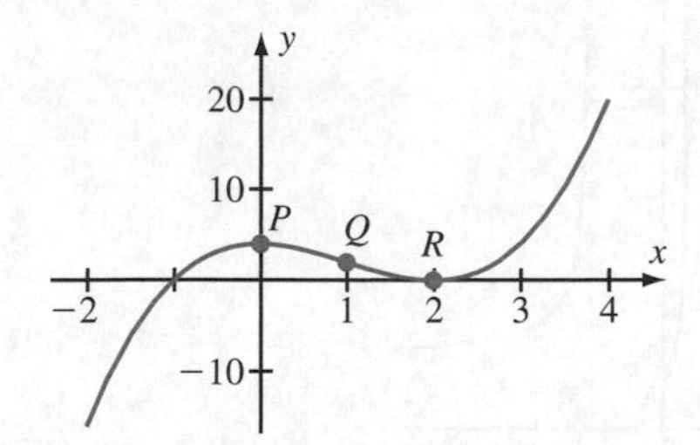

9. Points of inflection: $\{(n\pi, -n\pi) : n \in \mathbb{Z}\}$; concave down: intervals of the form $(2k\pi, (2k+1)\pi)$, $k \in \mathbb{Z}$; concave up: intervals of the form $((2k-1)\pi, 2k\pi)$, $k \in \mathbb{Z}$

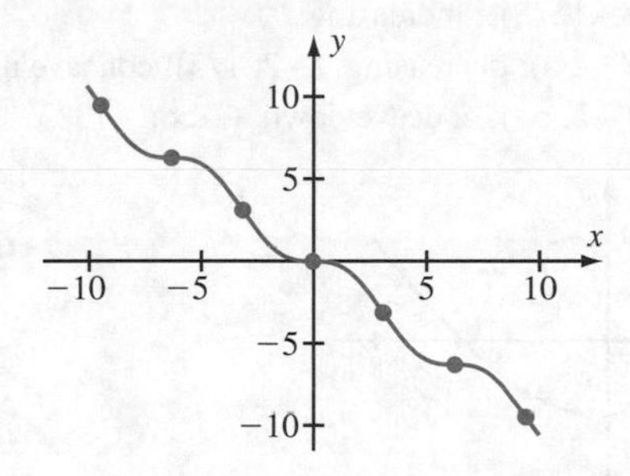

11. Local maximum: $S = (2, 1/4)$; local minimum: $Q = (-2, -1/4)$; points of inflection: $P = (-2\sqrt{3}, -\sqrt{3}/8)$, $R = (0, 0)$, and $T = (2\sqrt{3}, \sqrt{3}/8)$; increasing: $(-2, 2)$; decreasing: $(-\infty, -2) \cup (2, \infty)$; concave up: $(-2\sqrt{3}, 0) \cup (2\sqrt{3}, \infty)$; concave down: $(-\infty, -2\sqrt{3}) \cup (0, 2\sqrt{3})$; horizontal asymptote: $y = 0$

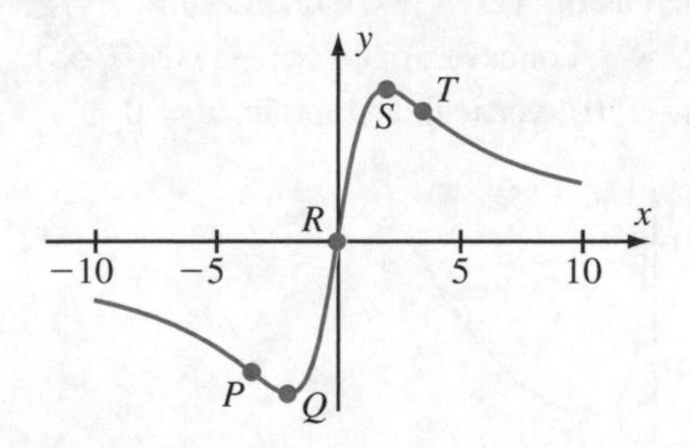

13. Local maximum: $P = (0, -1)$; increasing: $(-\infty - 2) \cup (-2, 0)$; decreasing: $(0, 2) \cup (2, \infty)$; concave up: $(-\infty, 2) \cup (2, \infty)$; concave down: $(-2, 2)$; vertical asymptotes $x = -2$ and $x = 2$; horizontal asymptote: $y = 1$

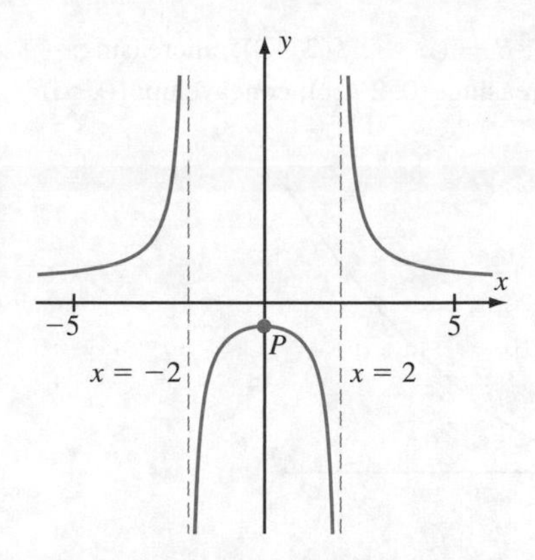

15. Local maximum: $Q = (3/2, f(3/2))$; points of inflection: $P = (3/2 - 9\sqrt{5}/10, f(3/2 - 9\sqrt{5}/10))$, $O = (0, 0)$, and $R = (3/2 + 9\sqrt{5}/10, f(3/2 + 9\sqrt{5}/10))$; increasing: $(-\infty, -3) \cup (-3, 3/2)$; decreasing: $(3/2, \infty)$; concave up: $(-\infty, -3) \cup (3/2 - 9\sqrt{5}/10, 0) \cup (3/2 + 9\sqrt{5}/10, \infty)$; concave down: $(-3, 3/2 - 9\sqrt{5}) \cup (0, 3/2 + 9\sqrt{5})$; vertical asymptote $x = -3$; horizontal asymptote: $y = 0$

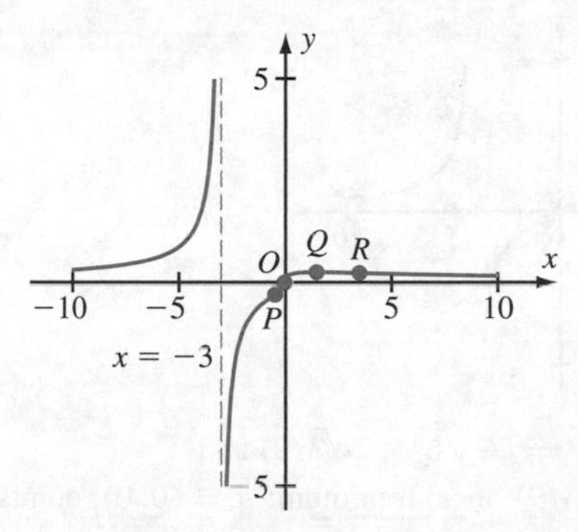

17. Local minimum: $P = (-5/6, f(-5/6))$; points of inflection: $Q = (-5/3, f(-5/3))$ and $R = (-1, 0)$; increasing: $(-5/3, \infty)$; decreasing: $(-\infty, -5/3)$; concave up: $(-\infty, -5/3) \cup (-1, \infty)$; concave down: $(-5/3, -1)$

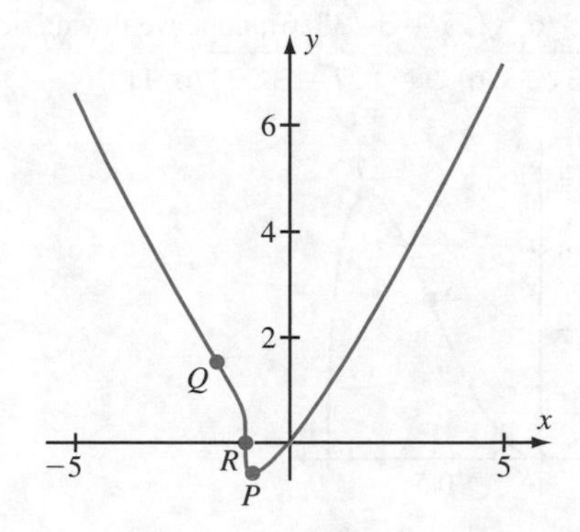

19. Local minimum: $P = (2^{-4/3}, f(2^{-4/3}))$; increasing: $(2^{-4/3}, \infty)$; decreasing: $(0, 2^{-4/3})$; concave up: $(0, \infty)$

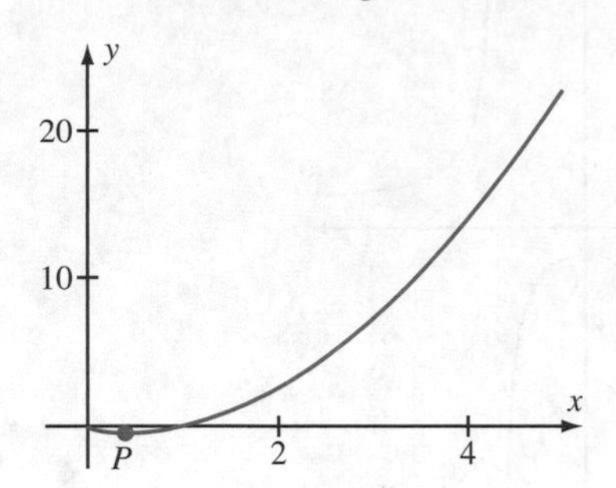

21. Local maximum: $Q = (-2, f(-2))$; points of inflection: $P = (-2 - 6\sqrt{5}/5, f(-2 - 6\sqrt{5}/5))$, $R = (0, 0)$, $S = (-2 + 6\sqrt{5}/5, f(-2 + 6\sqrt{5}/5))$; increasing: $(-\infty, -2)$; decreasing: $(-2, 4) \cup (4, \infty)$; concave up: $(-\infty, -2 - 6\sqrt{5}/5) \cup (0, -2 + 6\sqrt{5}/5) \cup (4, \infty)$; concave down: $(-2 - 6\sqrt{5}/5, 0) \cup (-2 + 6\sqrt{5}/5, 4)$; vertical asymptote: $x = 4$; horizontal asymptote: $y = 0$

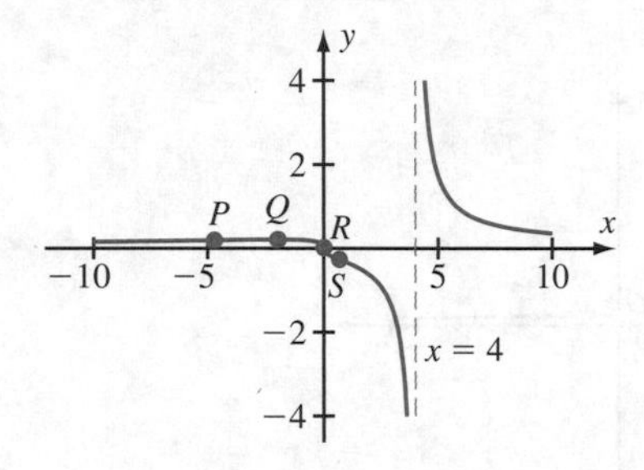

23. Local maxima: $P = (-\sqrt{6}/3, 2\sqrt{3}/9)$ and $T = (\sqrt{6}/3, 2\sqrt{3}/9)$; local minimum: $R = (0, 0)$; points of inflection: $Q = (-\sqrt{27 - 3\sqrt{33}}/6, f(-\sqrt{27 - 3\sqrt{33}}))$ and $S = (\sqrt{27 - 3\sqrt{33}}/6, f(\sqrt{27 - 3\sqrt{33}}))$; increasing: $[-1, -\sqrt{6}/3) \cup (0, \sqrt{6}/3)$; decreasing: $(-\sqrt{6}/3, 0) \cup (\sqrt{6}/3, 1]$; concave up: $(-\sqrt{27 - 3\sqrt{33}}/6, \sqrt{27 - 3\sqrt{33}}/6)$; concave down: $[-1, -\sqrt{27 - 3\sqrt{33}}/6) \cup (\sqrt{27 - 3\sqrt{33}}/6, 1]$

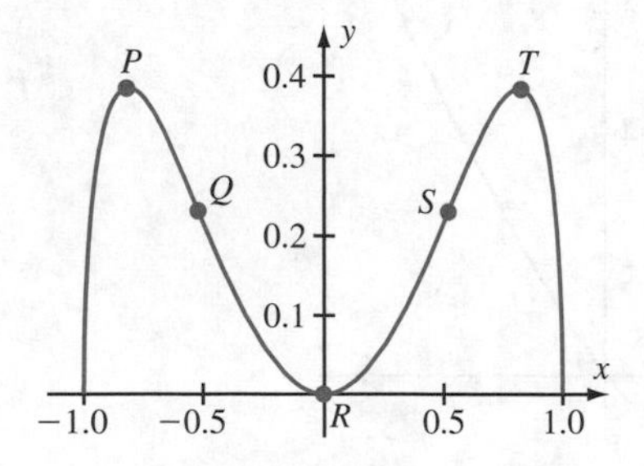

25. Local minima: $P = (-2, 0)$ and $Q = (1, 0)$; point of inflection: $R = (0, \sqrt{6}/3)$; decreasing: $(-\infty, -2] \cup (-1, 1] \cup (3, \infty)$; concave up: $(-1, 0) \cup (3, \infty)$; concave down: $(-\infty, -2) \cup (0, 1)$; vertical asymptotes: $x = -1$ and $x = 3$; horizontal asymptote: $y = 1$

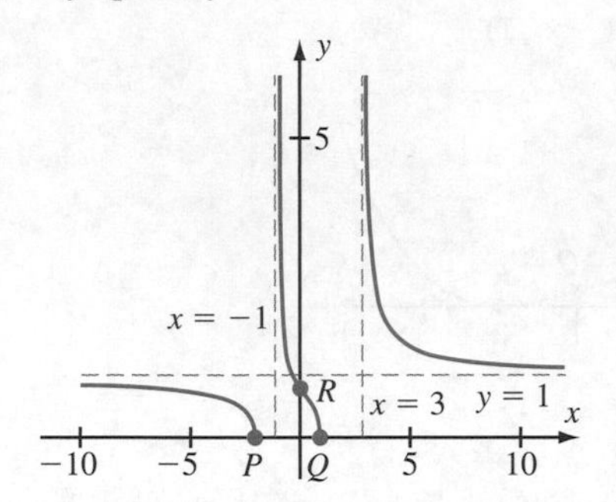

27. Local minimum: $Q = (-2, 0)$; local maximum: $R = (1/3, f(1/3))$; point of inflection: $P = (-13/3, f(-13/3))$; increasing: $(-\infty, -2) \cup (1/3, \infty)$; decreasing: $(-2, 1/3)$; concave up: $(-13/3, -2) \cup (-2, \infty)$; concave down: $(-\infty, -13/3)$

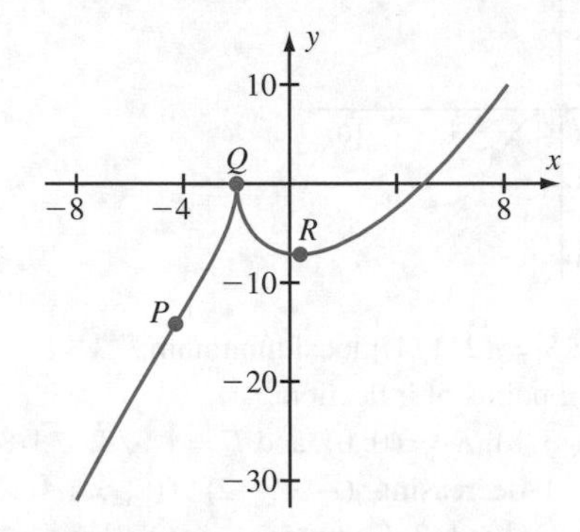

29. Local minimum: $Q = (2^{-1/3}, f(2^{-1/3}))$; point of inflection: $P = (-1, 1)$; increasing: $(2^{-1/3}, \infty)$; decreasing: $(-\infty, 0) \cup (0, 2^{-1/3})$; concave up: $(-\infty, -1) \cup (0, \infty)$; concave down: $(-1, 0)$; vertical asymptote: $x = 0$

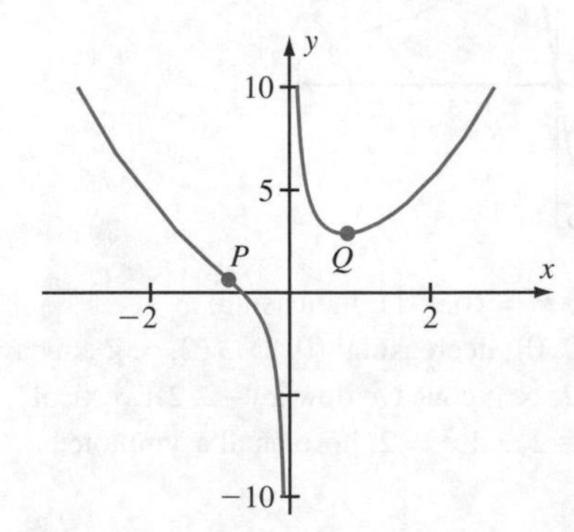

31. No local extrema; no points of inflection; increasing: on its entire domain; concave up: intervals of the form $((2k - 1)\pi/2, (2k + 1)\pi/2)$ $(k \in \mathbb{Z})$; concave down:

intervals of the form $((2k+1)\pi/2, (2k+3)\pi/2)$ $(k \in \mathbb{Z})$; vertical asymptotes: lines of the form $x = (2k+1)\pi/2$ $(k \in \mathbb{Z})$

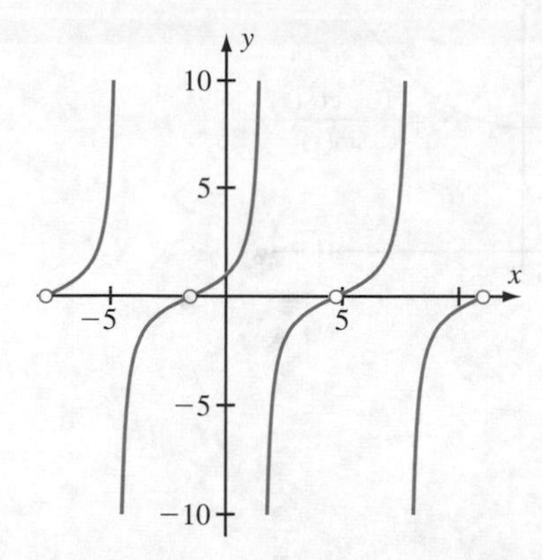

33. Local minima: $P = (-1, 0)$ and $R = (0, 0)$; local maximum: $Q = (-1/3, 4/27)$; points of inflection: $S = (-2/3, 2/27)$ and $R = (0, 0)$; increasing: $(-1, -1/3) \cup (0, \infty)$; decreasing: $(-\infty, -1) \cup (-1/3, 0)$; concave up: $(-\infty, -2/3) \cup (0, \infty)$; concave down: $(-2/3, 0)$

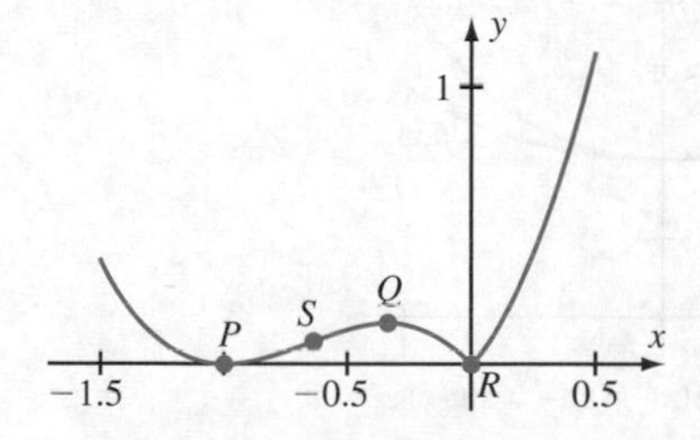

35. Local minimum: $Q = (-14/13 - 3\sqrt{3}/13, f(-14/13 - 3\sqrt{3}/13))$; local maximum: $R = (-14/13 + 3\sqrt{3}/13, f(-14/13 + 3\sqrt{3}/13))$; point of inflection: $P = (-2, -4)$; concave up: $(-2, -1) \cup (1, \infty)$; concave down: $(-\infty, -2) \cup (-1, 1)$; vertical asymptotes: $x = -1$ and $x = 1$

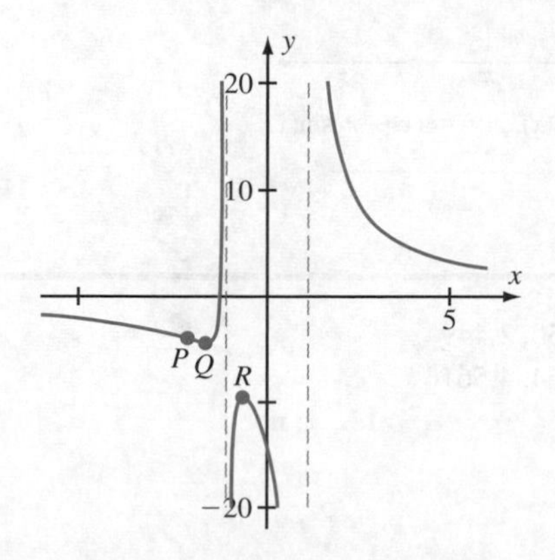

37. Increasing: $(-\infty, 0) \cup (0, \infty)$; concave up: $(-\infty, 0)$; concave down: $(0, \infty)$; vertical asymptote: $x = 0$; skew asymptote: $y = x/2$

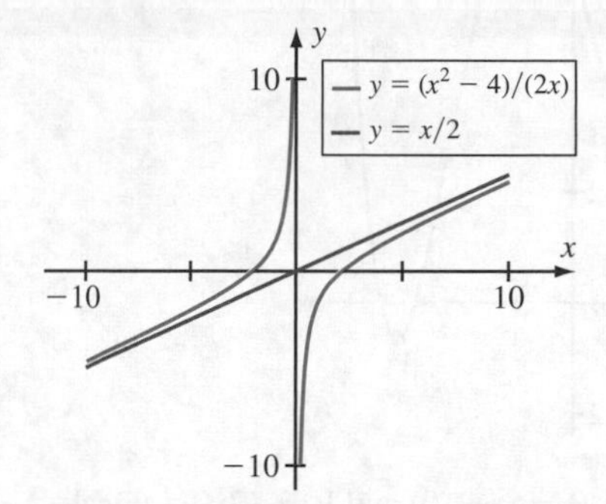

41. Global minimum: $P = (-0.60583, f(-0.60583))$; increasing: $(-0.60583, \infty)$; decreasing: $(-\infty, -0.60583)$; concave up: $(-\infty, \infty)$

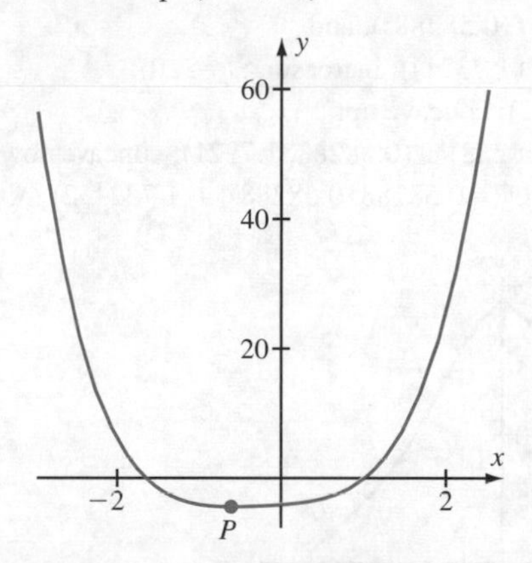

43. Local maxima: $P = (-1.6012, f(-1.6012))$ and $T = (0.39915, f(0.39915))$; local minima: $R = (-1.1839, f(-1.1839))$ and $V = (1.5859, f(1.5859))$; points of inflection: $Q = (-1.4117, f(-1.4117))$, $S = (-0.31451, f(-0.31451))$, and $U = (1.1262, f(1.1262))$; increasing: $(-\infty, -1.6012) \cup (-1.1839, 0.39915) \cup (1.5859, \infty)$; decreasing: $(-1.6012, -1.1839) \cup (0.39915, 1.5859)$; concave up: $(-1.4117, -0.31451) \cup (1.1262, \infty)$; concave down: $(-\infty, -1.4117) \cup (-0.31451, 1.1262)$

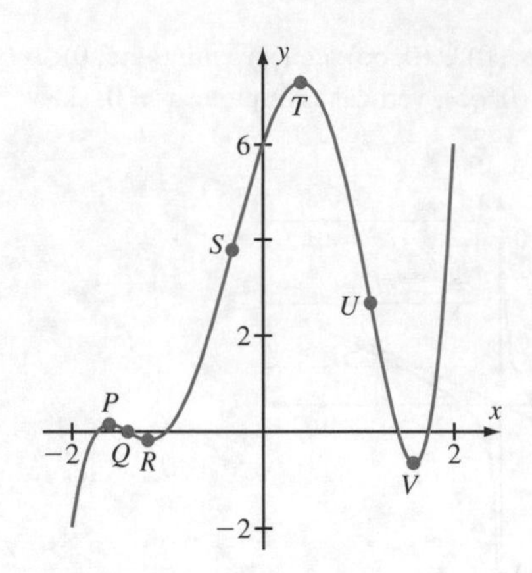

47. Global minima: $P = (-2, 0)$ and $V = (2, 0)$; global maxima: $S = (0, 2)$; points of inflection: $Q = (-1.7321, f(-1.7321))$, $R = (-0.58288, f(-0.58288))$, $T = (0.58288, f(0.58288))$, and $U = (1.7321, f(1.7321))$; increasing: $(-2, 0)$; decreasing: $(0, 2)$; concave up: $(-1.7321, -0.58288) \cup (0.58288, 1.7321)$; concave down: $(-2, -1.7321) \cup (-0.58288, 0.58288) \cup (1.7321, 2)$

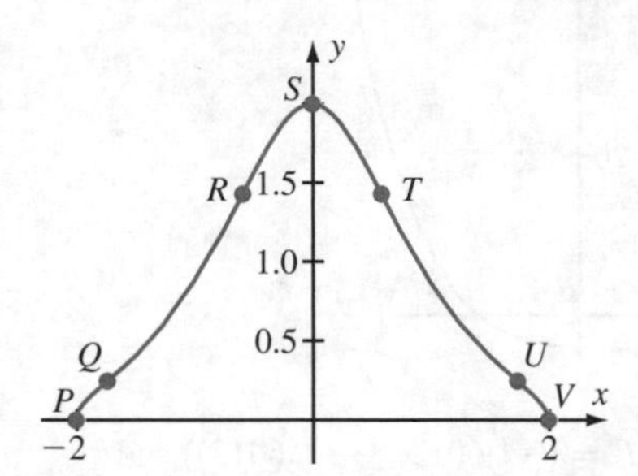

Section 4.6

1. −1 **3.** 1/5 **5.** −1/8
7. 1/4 **9.** Does not exist **11.** 2
13. 1/3 **15.** 1 **17.** 0
19. 1 **21.** $4/\pi^2$ **23.** Does not exist
25. 0 **27.** 0 **29.** 1
31. 0 **33.** 1 **35.** 1
37. 1 **39.** 1 **41.** Does not exist
43. 0 **45.** Does not exist **47.** 0
49. −1 **51.** −1/2 **53.** e^x
55. 1/2
57. **a.** −1 **b.** −2 **c.** 1/3 **d.** 1 **e.** 0 **f.** 0

63. The limit is 1/2.

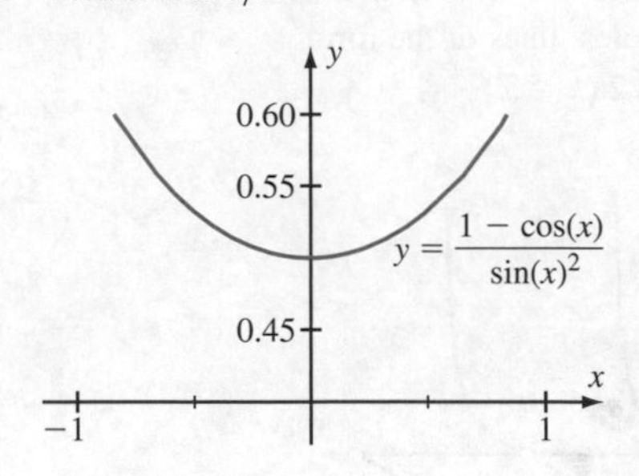

65. The limit is 1.

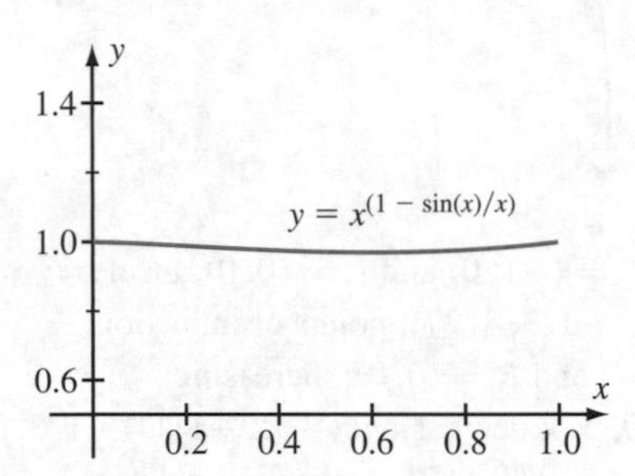

67.

$f(x) = 1 - \cos(x)$, $g(x) = x \sin(x)$

69.

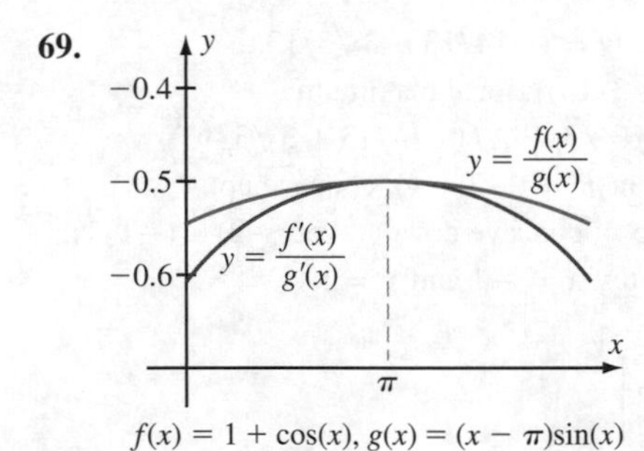

$f(x) = 1 + \cos(x)$, $g(x) = (x - \pi)\sin(x)$

Section 4.7

1. 2.0, 2.500, 2.450, 2.449
3. 4.0, 4.667, 4.564, 4.5616
5. 2.0, 1.9875
13. 1; no

25. 1.08891683

27. 1.8

29. 0.9955988

31. 0.6290

33. $x_{12} = 1.00034$; $N = 12$

34. $x_{11} = -1.99966$; $N = 11$

35. $x_{11} = 0.50044$; $N = 11$

36. $x_{10} = 2.00049$; $N = 10$

37. $x_4 = 1.00000$; $x_4 = -2.00000$; $x_3 = 0.49994$; $x_4 = 1.99999$; Compare $M \leq 4$ with $N \geq 10$

39. 1.080

41. **a.** 7.75% **b.** 5.82%

43. The second; it has an effective yield of 6.50%, whereas the first has an effective yield of 6.25%.

45. The second; it has an effective yield of 7.37%, whereas the first has an effective yield of 7.00%.

47. 1.8030059

49. 1.0288, occuring at $x = 0.51282$

Section 4.8

1. $(1/3)x^3 - (5/2)x^2 + C$

3. $e^x + C$

5. $(2/3)(x+2)^{3/2} + C$

7. $-1/x - (1/6)x^{-6} + C$

9. $(1/3)(x+1)^3 + C$

11. $(1/e)\exp(ex) + C$

13. $(-3/4)x^{-4/3} - 12x^{1/3} + C$

15. $(3/4)\sin(4x) + x^2 + C$

17. $(1/8)\tan(8x) + C$

19. $(1/12)(3x-2)^4 + C$

21. **a.** 1 **b.** $-3/2$ **c.** -4 **d.** $3e - 4$

23. $4\sqrt{10}$ m/s

25. $625/4$ ft; $-20\sqrt{21}$ ft/s

27. $121/54$ ft/s^2

29. $5\sqrt{15}/4$ s

31. $-(1/4)\cos(2x) + C$ $\tan(x) - x + C$

33. $\tan(x) - x + C$

35. $(1/2)(x + (1/2)\sin(2x)) + C$

37. $G(f(x)) + C$

39. **a.** $(x^2+1)^{101}/202 + C$ **b.** $\sin^2(x)/2 + C$ **c.** $\sin^3(x)/3 + C$ **d.** $e^{x^2}/2 + C$ **e.** $\sin(\sin(x)) + C$

45. $27075/64$ ft; 95 ft/s

47. $290377/8 - 493\sqrt{337705}/8 \approx 485.4$ ft

49. 120 s

51. $3825/4$ ft; $8 + (25/8 + 15\sqrt{17}/8)$ s; $-60\sqrt{17}$ ft/s

53. $\ln(2)/k$

55. $H - kgt + k^2 g(1 - e^{-t/k})$

57. 1839.6 yr

59. $\lim_{t\to\infty} T(t) = 40$; 25.54 s elapse as the object cools from 50° C to 46° C; 20.27 s elapse as the object cools from 46° C to 44° C; 14.39 s elapse as the object cools from 44° C to 43° C.

61. $v(t) = 8.855(1 + e^{4.427t})^{-1} - 4.4275$; $v_\infty = -4.4275$; $v(t) = 0.999v_\infty$ when $t = 1.7168$

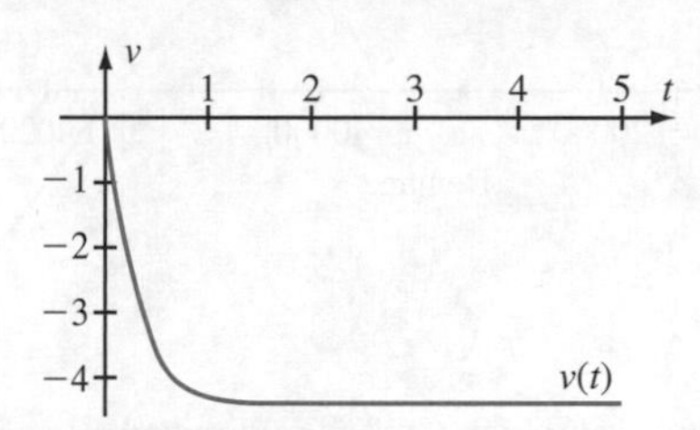

Section 4.9

1. 2

3. 19

5. **a.** 2000 **b.** 3 **c.** 3 **d.** 23

7. **a.** 1200 **b.** 449.25 **c.** 448.50 **d.** 606

9. 100

11. 25

13. 49

15. 89

17. 2 times per year

19. 3 times per year

21. $\lfloor 1/k \rfloor$ or $\lfloor 1/k \rfloor + 1$

23. 5657

25. 400

27. $E(p) = \sqrt{p}/(8 - 2\sqrt{p})$; inelastic demand: $p < 64/9$; elastic demand: $p > 64/9$

29. 16

31. 40

33. 10% decrease

35. 10% increase

37. **a.** $x = 3500 - 400p$ **b.** \$150 **c.** $E(3.75) = 0.75$ and $E(4.50) = 1.06$ **d.** Raising the price further will decrease revenue. **e.** 1520

49. 400

51. 159

71. 38958

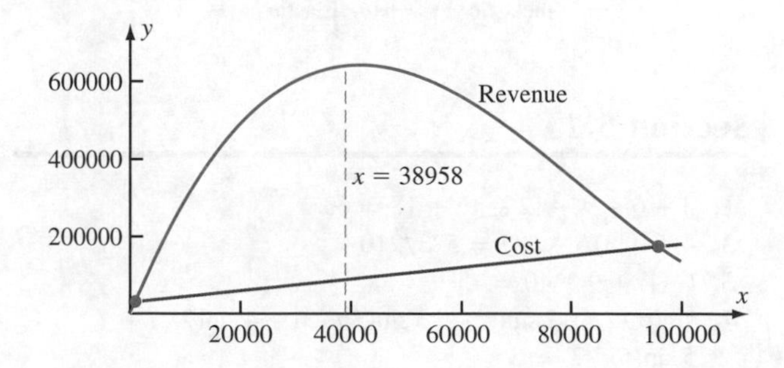

73. 25039

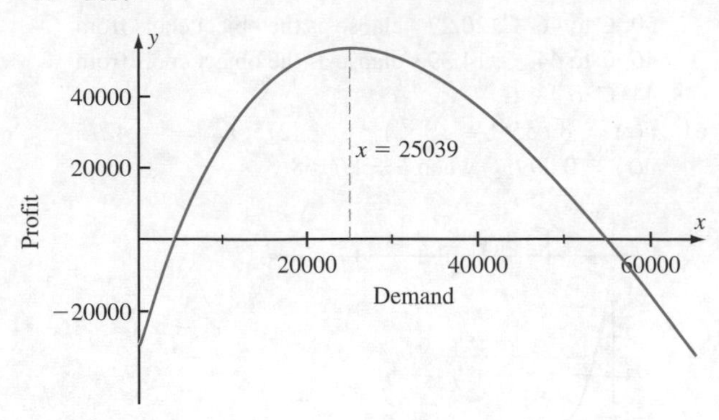

75. 24495; the demand x at which $\bar{C}(x)$ is minimized

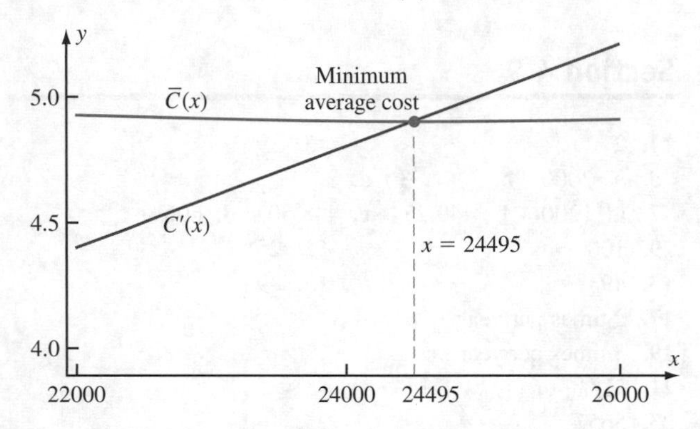

77. Inelastic: $p < 15.41$; elastic: $p > 15.41$; maximum revenue: $p = 15.41$

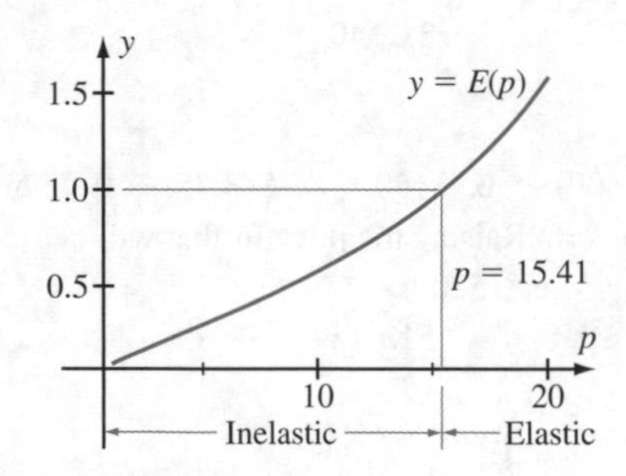

Section 5.1

1. $3 + 6 + 9 + 12 + 15 + 18 = 63$

3. $4/5 + 5/6 + 6/7 = 523/210$

5. $(-4) + 9 + 40 = 45$

7. $\sin(\pi/2) + 2\sin(\pi) + 3\sin(3\pi/2) + 4\sin(2\pi) + 5\sin(5\pi/2) = 3$

9. $\sum_{j=2}^{6} j$

11. $\sum_{j=1}^{6}(5 + 4j)$

13. $\sum_{j=4}^{8}(1/j)$

15. 52.875

17. 3π

19. 4.71

21. $\pi^2/2$

23. 1496

25. 89357

27. Right endpoint approximation: 60; trapezoidal approximation: 32

29. Right endpoint approximation: 5/4; trapezoidal approximation: 3/2

31. 16/3

33. $p = 5050$; $q = 100!$

35. $N(2N - 1)(2N + 1)/3$

49. $ab - c$

53. $N = 25$: 0.165...; $N = 50$: 0.166...

55. $N = 25$: 1.997...; $N = 50$: 1.999...

57. $N = 25$: 1.945...; $N = 50$: 1.948...

59. $N = 25$: 0.306...; $N = 50$: 0.307...

Section 5.2

1. **a.** 72 **b.** 72

3. **a.** 72 **b.** 224/3

5. **a.** 378 **b.** 696

7. **a.** 42 **b.** 37

9. **a.** $2\exp(-2) + 2 + 2\exp(2)$ **b.** $\exp(2) - \exp(-4)$

11. **a.** $\pi/3$ **b.** $-\sqrt{3}/2$

13. $2\sqrt{2}$

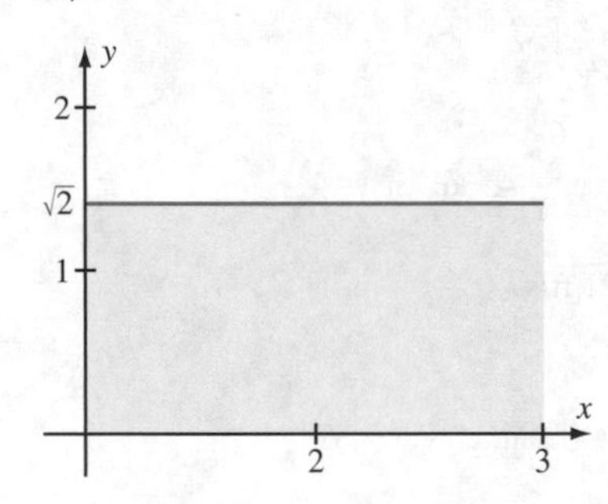

15. 5

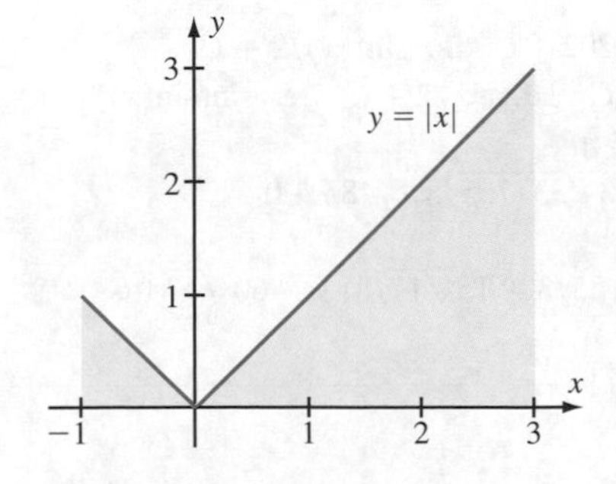

17. 3/2

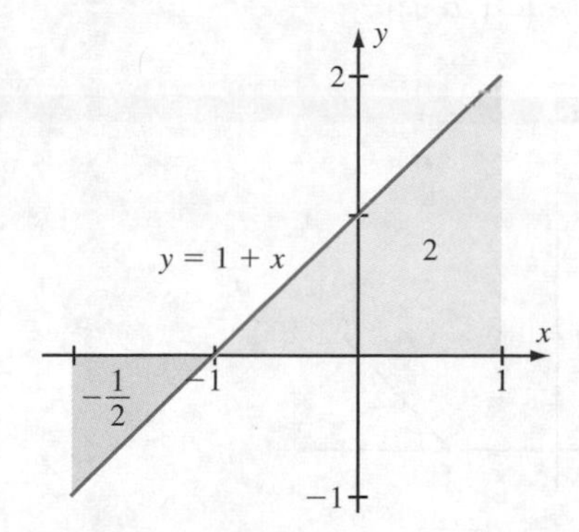

19. 13/2

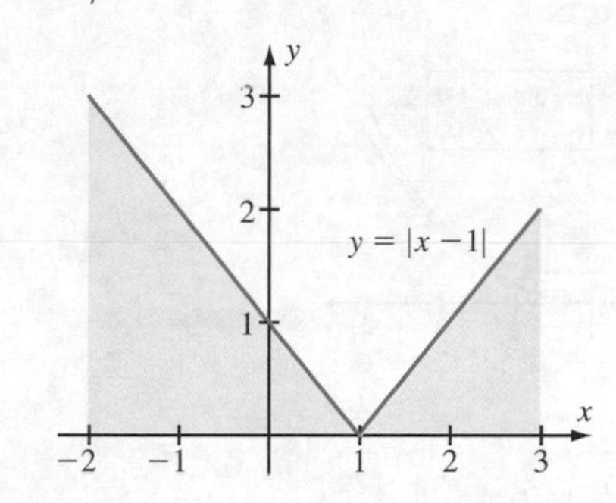

21. 11 **23.** 14/3
25. 1 **27.** 1
29. 195/4 **31.** 1
33. **a.** e **b.** 1 **c.** $1 \le \sqrt{e} \le e$
35. **a.** 5/6 **b.** 7/12 **c.** $7/12 \le 24/35 \le 5/6$
37. **a.** $U = \sqrt{2} + 3 + 2\sqrt{7}$ **b.** $L = 4 + \sqrt{2}$
c. $L \le \sqrt{2}(3 + \sqrt{7}\sqrt{5} + \sqrt{7}\sqrt{19})/4 \le U$
39. Upper Riemann sum: $\pi(\sqrt{3}+1)/3$; lower Riemann sum: $\pi\sqrt{3}/6$; $\int_0^\pi \sin(x)\,dx = 2 \in (\pi\sqrt{3}/6, \pi(\sqrt{3}+1)/3)$
41. Upper Riemann sum: $e^2 + 2e$; lower Riemann sum: $e + 2$; $\int_0^\pi e^{|x|}\,dx = e^2 + e - 2 \in (e+2, e^2+2e)$
43. Upper Riemann sum: 22; lower Riemann sum: 11; $\int_0^3 f(x)\,dx = 33/2 \in (11, 22)$
45. Yes; no, except when $f(u) = f(t)$
49. $s_1 = \sqrt{3}/3$; $s_2 = \sqrt{21}/3$
53. $(1 - N + \sqrt{N^2 + 1/3})/2$
57. 0.375
59. 3.14195
61. Upper Riemann sum: 0.55563; lower Riemann sum: 0.55469
63. Upper Riemann sum: 0.40977; lower Riemann sum: 0.40549

Section 5.3

1. 4 **3.** 8 **5.** 10 **7.** 5
9. −64 **11.** −18 **13.** −5
15. $f_{\text{ave}} = 4$; $c = 3$
17. $f_{\text{ave}} = 14/9$; $c = 196/81$
19. $f_{\text{ave}} = 89$; $c = \sqrt{7}$
21. 0 and −1, respectively
23. −17/13 and −50/13, respectively
25. $6 - 5\ln(3)$ **27.** $3 - 1/e$ **29.** $3 - 6/e$
31. $\int_a^b f(x)\,dx = 0$ or $b = a + 1$
33. $A = 1$; $B = 3$ **35.** $A = 3$; $B = 3 \cdot 5^{1/3}$
37. $\pi/2$ **43.** −4/15
45. $\ell = 6/17$; $u = 2(1.42032)$
47. $\ell = (\pi/2)$; $u = (\pi/2)(3.337)$
49. $f_{\text{ave}} = 2/\pi$; $c = 0.6901$

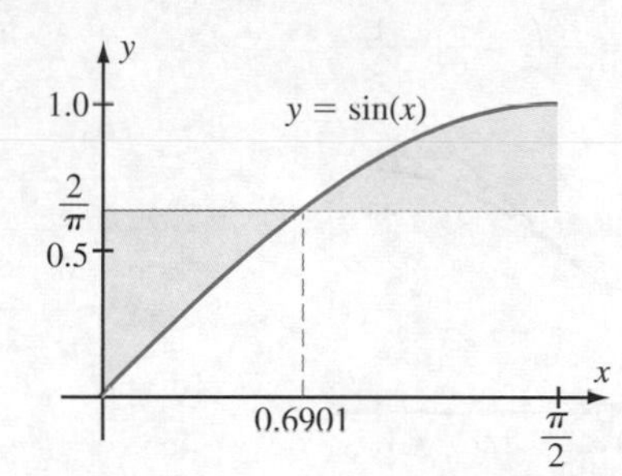

51. $f_{\text{ave}} = 49/12$; $c = 2.19645$

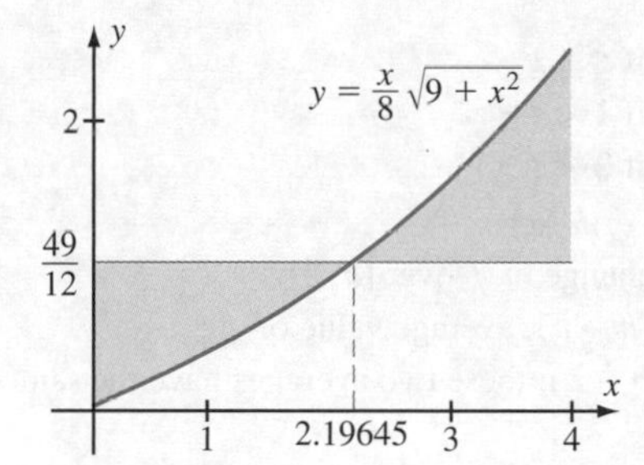

Section 5.4

1. 52/5 **3.** −3 **5.** 1370/7 **7.** $e^2 - e^{-9}$
9. 1 **11.** 324 **13.** 332/21 **15.** 15/2
17. $e - e^{-1}$ **19.** $x^3 + x + 2$
21. $(3/4)x^{4/3} - 12$ **23.** $(e^x - e^{-x})/2$
25. $\sec(x) - 1$ **27.** $\tan(x^2)$
29. $3\cot(3x)$ **31.** $2\sqrt{1 + 4x^2}$

33. $F'(x) = x(x-1)$; $F''(x) = 2x - 1$; F increases on $(-\infty, 0) \cup (1, \infty)$ and decreases on $(0, 1)$; F is concave down on $(-\infty, 1/2)$ and concave up on $(1/2, \infty)$.

35. $F'(x) = x\ln(x)$; $F''(x) = \ln(x) + 1$; F decreases on $(0, 1)$ and increases on $(1, \infty)$; F is concave down on $(0, 1/e)$ and concave up on $(1/e, \infty)$.

37. $3(27x^3 + 3x)^{1/4} - 2(8x^3 + 2x)^{1/4}$

39. $1/x$

41. $F'(x) \equiv 0$

45.

$y = \frac{x}{e}$

$y = \ln(x)$

47.

$y = \frac{\sqrt{3}}{2}\left(x - \frac{\pi}{6}\right) + \frac{1}{2}$

$y = \sin(x)$

49. $f(t) = \begin{cases} 2 & \text{if } t \leq 1 \\ 0 & \text{if } 1 < t \leq 2 \\ -1 & \text{if } 2 < t \leq 3 \end{cases}$

53. $f(g(h(x)))g'(h(x))h'(x)$

55. Average rate of change of f over $[a, b]$: $(f(b) - f(a))/(b - a)$; average value of f': $(\int_a^b f'(x)\,dx)/(b - a)$; these two averages have the same value.

59. $1 - 2xF(x)$

65. $A_f(x) = \begin{cases} -x & \text{if } x \leq -1/10 \\ 5x^2 + 1/20 & \text{if } -1/10 < x \leq 1/10 \\ x & \text{if } 1/10 < x \end{cases}$

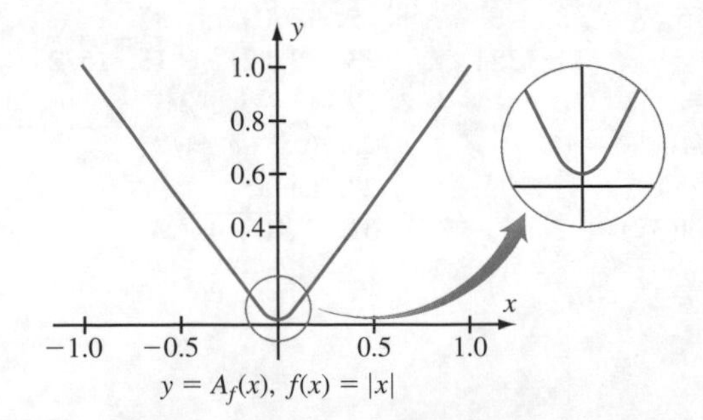

$y = A_f(x),\ f(x) = |x|$

67. Continuous: $(-2, 2)$; differentiable: $\{x \in (-2, 2) : x \notin \{-1, 0, 1\}\}$

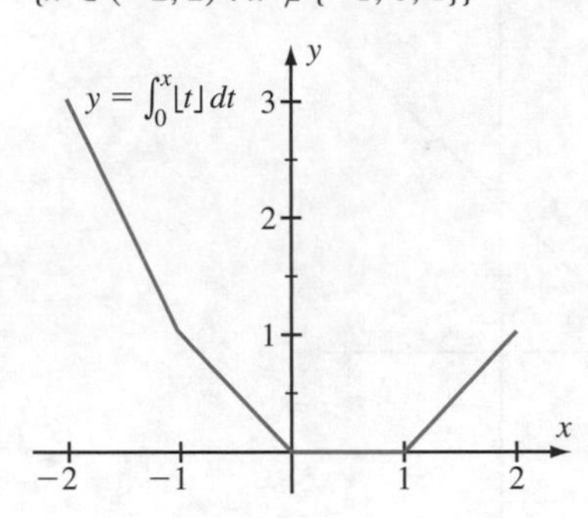

69. Decreasing: $(-\infty, -0.71)$; increasing: $(-0.71, \infty)$; concave up: $(-\infty, \infty)$

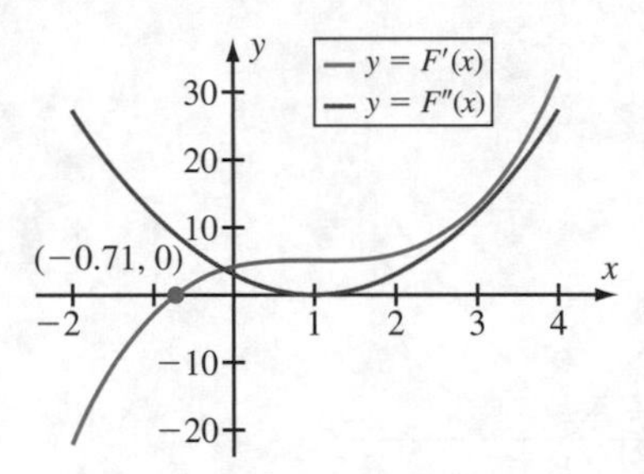

$F(x) = \int_{-3}^x (t^3 - 3t^2 + 3t + 4)\,dt$

71. Decreasing: $(-\infty, -1) \cup (0, 1)$; increasing: $(-1, 0) \cup (1, \infty)$; concave down: $(-\infty, -3.2143) \cup (-0.46081, 0.67513)$; concave up: $(-3.2143, -0.46081) \cup (0.67513, \infty)$

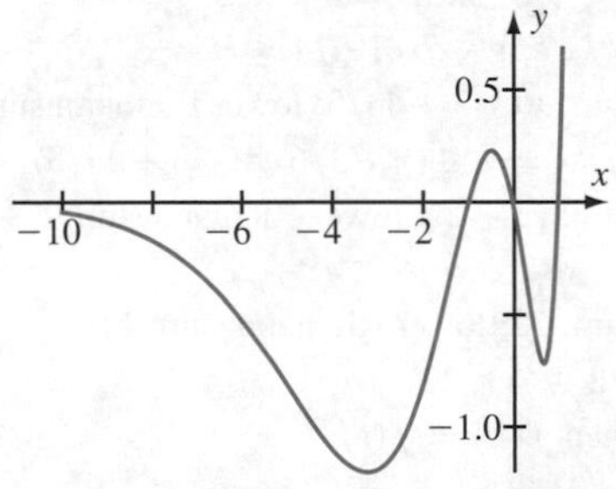

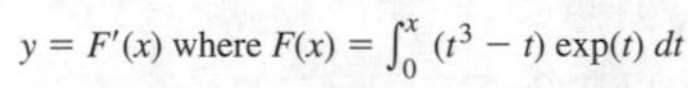

$y = F'(x)$ where $F(x) = \int_0^x (t^3 - t)\exp(t)\,dt$

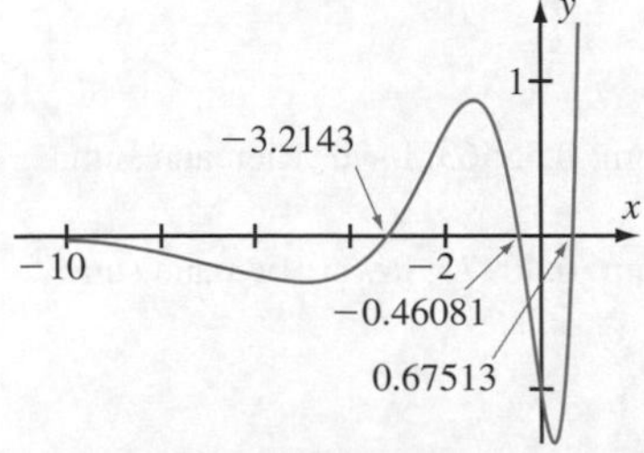

$y = F''(x)$ where $F(x) = \int_0^x (t^3 - t)\exp(t)\,dt$

73. Approximation: 0.99990; $F'(0) = 1$

75. Approximation: 0.70710677; $F'(0) = 1/\sqrt{2} \approx 0.70710678$

79. $365 - N(37) \approx 6$ represents the number of days on which the temperature reached 37°C or higher.

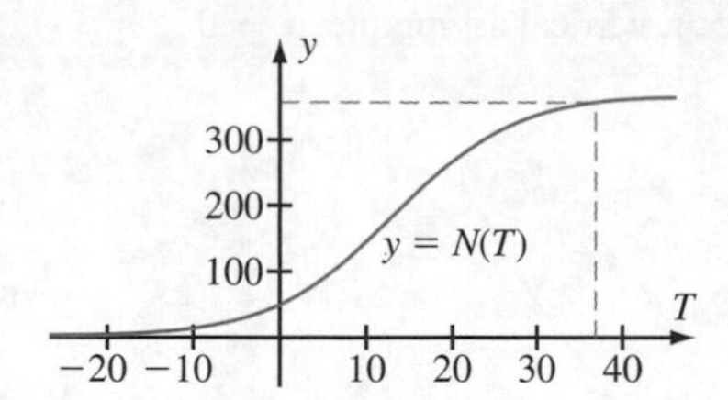

81. $a = 4.49341$; $b = 7.72525$

Section 5.5

1. $u = 3x, du = 3\,dx$; $-(1/3)\cos(3x) + C$

3. $u = x^8 + 1, du = 8x^7 dx$; $-(1/32)(x^8 + 1)^{-4} + C$

5. $u = x^3 - 5, du = 3x^2\,dx$; $(2/5)(x^3 - 5)^{5/2} + C$

7. $u = \cos(s), du = -\sin(s)\,ds$; $-(1/5)\cos^5(s) + C$

9. $u = \pi/t, du = -(\pi/t^2)\,dt$; $(8/\pi)\cos(\pi/t) + C$

11. $(1/8)(x^2 + 1)^8 + C$

13. $(1/3)(1 + x^2)^{3/2} + C$

15. $1/\cos(x) + C$

17. $(1/12)\sin^6(2x) + C$

19. $5\sqrt{1 + x^2} + C$

21. $-2\cos(\sqrt{x}) + C$

23. $32/3$ **25.** $9 - (5/3)\sqrt{5}$

27. 0 **29.** $-5/64$

31. -2 **33.** $(2/3)\sqrt{2} - 1/3$

35. $(2 - \sqrt{3})/(\sqrt{3} - 1)$

39. $(1/10)(2x + 3)^{5/2} - (1/2)(2x + 3)^{3/2} + C$

41. $(2/3)(x + 3)^{3/2} - 6(x + 3)^{1/2} + C$

43. $(6 + s^2)^{1/2}(s^4 - 8s^2 + 96)/5 + C$

45. $(2/5)(s - 5)^{5/2} + (14/3)(s - 5)^{3/2} + C$

47. $\sqrt{x^4 + x^2 + 1} + C$

49. 1 **51.** $9/28$

63. $A = 2$; $B = -1$; $\ln(t^2 + 1) + 1/(t + 1) + C$

65. 2.323745 **67.** 1.510129

Section 5.6

1. $2 - \sqrt{3}/2$ **3.** $776/3$ **5.** $4\sqrt{2} + 2$

7. $101/384$ **9.** $1/\pi$ **11.** $5/6$

13. $95/6$ **15.** $3 - \sqrt{2}$ **17.** $125/2$

19. 36 **21.** $32/3$ **23.** $1/2$

25. $608/5$ **27.** 4 **29.** $1/2$

31. $\ln(2)/2 - 1/4$

33. 4 **35.** $64/3$ **37.** $1/12$

39. $\int_0^{\pi} \sqrt{2}\,dy$

41. $\int_1^{e} \ln(y)\,dy$

43. $\int_0^2 (2y - y^2)\,dy$

45. $\int_0^3 (y - (-y))\,dy$

47. $\int_1^2 (x - 1)\,dx + \int_2^3 (3 - x)\,dx$

49. $\int_{-1}^0 (2 - x^2 + x)\,dx + \int_0^1 (2 - x^2 - x)\,dx$

51. $\int_0^1 ((3 - y) - (1 + y))\,dy$

53. $\int_0^1 (y - (-y))\,dy + \int_1^2 (\sqrt{2 - y} - (-\sqrt{2 - y}))\,dy$

55. $\sqrt{5}/6$ **57.** 0.32219

59. 5.1097 **61.** 0.69765

63. $b \approx 1.3644$; the area is approximately 0.769.

Section 5.7

1. Midpoint Rule: $135/16$; Trapezoidal Rule: $81/8$; Simpson's Rule: 9

3. Midpoint Rule: $-\pi(\sqrt{2 - \sqrt{2}} + \sqrt{2 + \sqrt{2}})/4$; Trapezoidal Rule: $-\pi(\sqrt{2} + 1)/4$; Simpson's Rule: $-\pi(2\sqrt{2} + 1)/6$

5. Midpoint Rule: 12.6721; Trapezoidal Rule: 12.6558; Simpson's Rule: 12.6665

7. 9; Midpoint Rule error bound: $9/16$; Trapezoidal Rule error bound: $9/8$; Simpson's Rule error bound: 0

9. -2; Midpoint Rule error bound: $\pi^3/384$; Trapezoidal Rule error bound: $\pi^3/192$; Simpson's Rule error bound: $\pi^5/46080$

11. $38/3$; Midpoint Rule error bound: $125/12288$; Trapezoidal Rule error bound: $125/6144$; Simpson's Rule error bound: $3125/393216$

13. 0.36 **15.** 0.383

17. Trapezoidal Rule: 255; Simpson's Rule: 18

19. 6.86 L/min **27.** $103/250$

29. Simpson's Rule: 652868 km^2 (0.45% error); Trapezoidal Rule: 655988 km^2 (0.93% error); Midpoint Rule: 646628 km^2 (0.51% error)

33. $\mathcal{S}_2 = 2(4\sqrt{2} + 2)/3 \approx 5.1$; $\mathcal{M}_1 = 4\sqrt{2} = 5.66$; $\mathcal{T}_1 = 4$

35. $\mathcal{S}_4 = \pi(2\sqrt{2} + 1)/6 \approx 2.005$; $\mathcal{M}_2 = \pi\sqrt{2}/2 \approx 2.22$; $\mathcal{T}_2 = \pi/2 \approx 1.57$

39. 115

41. 18

45. 1; Simpson's Rule: 1.00081; $c = 1.57195$

47. 2; Simpson's Rule: 2.000036; $c = 10.7956$

49. 0.6828 (which is accurate to three decimal places)

51. **a.** $F(46) \approx 364.998$ (using $N = 50$); exact value: 365 **b.** About 8 days **c.** 14 (75% increase)

Section 6.1

9. $y(x) = x^2 + 2$
11. $y(x) = \sin(x) + 2$
13. $y(x) = \sqrt{x^2+1}$
15. $y(x) = 2/(2\cos(x) - 1)$
17. $2(1+y^2)^{3/2} = 3x^2 + 4\sqrt{2}$
19. $y(x) = \sqrt{10e^{2x} - 1}$
21. $y(x) = 3\exp(\cos(1/x))$
23. $y(x) = 2\exp(4x^3/3)$
25. $y(x) = g(f(x)) + 4$
27. $y(x) = e^{-x}(x + C)$
29. $2\ln(y) - x^2/y^2 = C$
31. $P(T) = 2\sqrt{C/17}T^{17/4}$
33. $P(r) = 2G\pi\rho^2(R^2 - r^2)/3$
35. $25\sqrt{2}/6$ s
37. 19273.9
39. $[N_2O_5] = 2.32\exp(-0.00061194t)$; k is the negative of the slope of the line $\ln([N_2O_5]) = 0.84157 - 0.00061194t$.

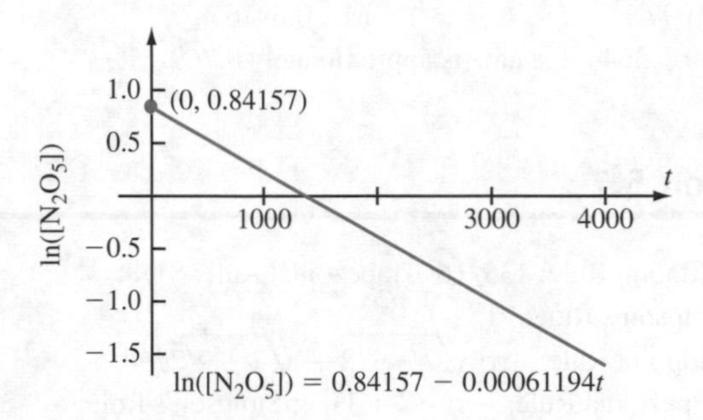

41. $y(x) = (1/\sqrt{2\pi})\exp(-x^2/2)$
43. $y(t) = A + C \cdot \exp(-\lambda t^p/p^2)$; $\lim_{t\to\infty} y(t) = A$
45. $L(x) = C/\exp(Bg^x/\ln(g))$
47. $T(t) = \tau \cdot \exp(-2k\sqrt{t})$
49. $a(x) = C(x + x_0)^{-\gamma}$; $y = 23.29(x + 3.346)^{-0.7}$

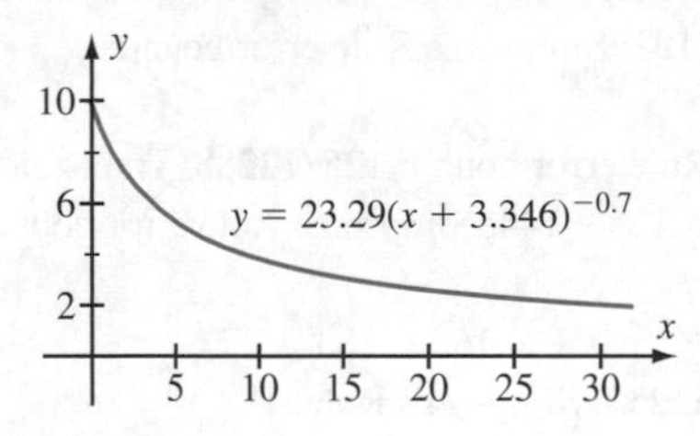

53. $y^5(x)/5 + y(x) = x + x^3/3$; $y(2) = 1.713385$
55. $y^2(x)/2 - y(x) + \ln(1 + y(x)) = \ln(x+1)$; $y(2) = 2.0$
57. 4.9337 weeks

Section 6.2

1. $1/x$
3. $5/(5x+2)$
5. $\cot(x)$
7. $-1/x$
9. $2\sec^2(2x)\ln(x) + \tan(2x)/x$
11. $\ln^2(x)/x$
13. $\ln(x+1) + C$
15. $(1/2)\ln(x^2+4) + C$
17. $\ln(\ln(x)) + C$
19. $2(\ln(x))^{3/2}/3 + C$
21. $\ln(\tan(x)) + C$
23. $A = 3$, $B = C = -2$, $D = -3$, $E = (1/2)\ln(2)$
25. 0
27. $\ln(R(t)) = \ln(B) + k \cdot \ln(t)$; A straight line with slope k
29. $2 + \ln(2)$
31. $2\ln(2 + \sqrt{x}) + C$
33. 0
35. Domain: $(0, \infty)$; increasing: $(1, \infty)$; decreasing: $(0, 1)$; concave up: $(0, \infty)$; vertical asymptote: $x = 0$

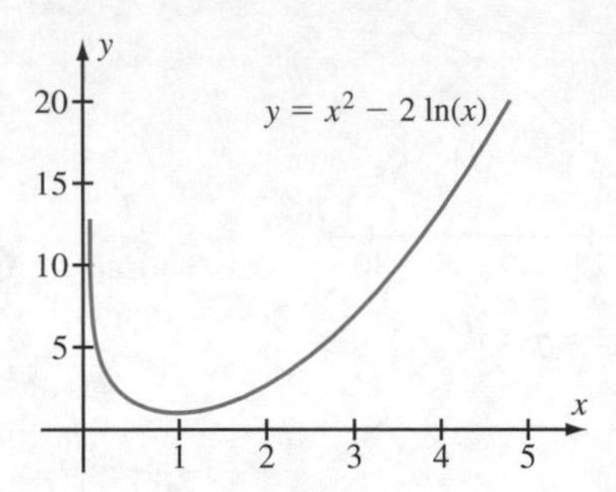

37. Domain: $(-\infty, -1) \cup (0, \infty)$; increasing: $(-\infty, -1) \cup (0, \infty)$; concave up: $(-\infty, -1)$; concave down: $(0, \infty)$; vertical asymptotes: $x = -1$, $x = 0$

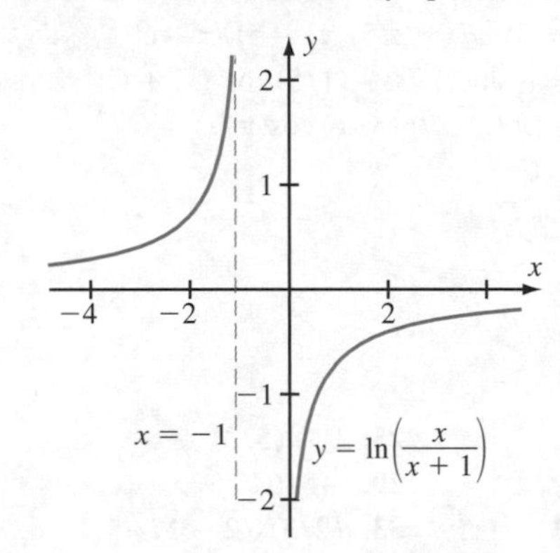

39. −1
43. 60
49. $A = 1$, $B = -1$
57. $\ln(2) = \int_1^2 1/x\,dx \approx 0.69317$
59. 0
61. 3.75 yr
63. 916

Section 6.3

1. $-\exp(-x)$
3. $\exp(x)(\cos(x) + \sin(x))$
5. $\cos(x)\exp(\sin(x))$
7. $\exp(x)/(1 + \exp(x))$
9. $\exp(x) \cdot \exp(\exp(x))$
11. e
13. $-\exp(-x) + C$
15. $(\exp(\pi x) + \exp(-\pi x))/\pi + C$
17. $\ln(1 + \exp(x)) + C$
19. $2(1 + \exp(x))^{3/2}/3 + C$
21. $-\cos(\exp(x)) + C$
23. $(\ln(3 + \exp(-x)) + x)/3 + C$
25. a^3/b^2
27. ac^{21}/b^8

29. $(ac^2 + 2bc^2 - 1)/c^2$ **31.** $3 + \ln(5)$
33. $1 - \ln(5)$
37.

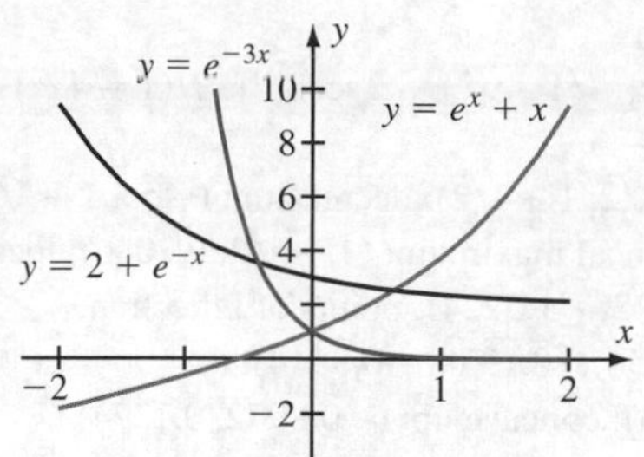

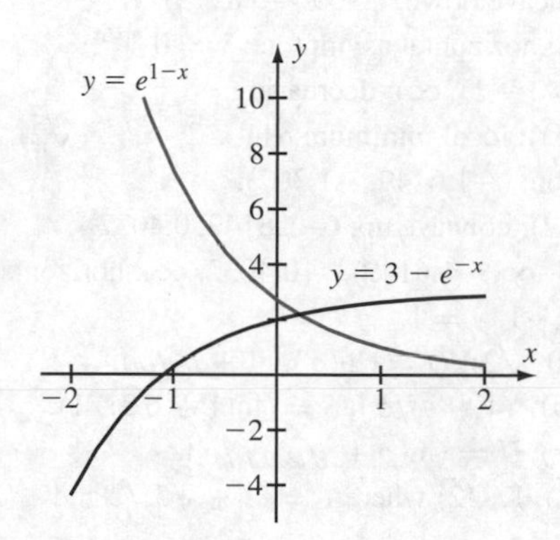

41. $1/e$ **43.** e **45.** α **47.** $8/5$
49. Minimum value: e occuring at $x = e$
51. $1/e$ **63.** 0 **67.** 0.1155
69. 4.6007×10^9 yr
71. 0.8427; 0.9953; 1.0000
73. 0.92413887

Section 6.4

1. $3^x \ln(3)$ **3.** $-\sqrt{7}x^{-\sqrt{7}-1}$
5. $(1 + \sqrt{e})^x \ln(1 + \sqrt{e})$ **7.** $2\ln(4) \cdot x4^{x^2}$
9. $(3^x + 1 - x3^x \ln(x)\ln(3))/(x(3^x + 1)^2)$
11. $(3/5)^x \ln(3/5) + (4/5)^x \ln(4/5)$
13. $x^{-\sqrt{3}+1}/(-\sqrt{3} + 1) + C$
15. $\int 5^x\,dx = 5^x/\ln(5) + C$
17. $6^{\tan(x)}/\ln(6) + C$
19. $2(10^x + 7)^{7/2}/(7\ln(10)) + C$
21. $3^{(x^2)}/(2\ln(3)) + C$ **23.** $\ln(1 + 5^x)/\ln(5) + C$
25. $1/(x\ln(2))$
27. $-1/((5 + \log_{10}(x))^2 x \ln(10))$
29. 0 **31.** $\ln(8)/(\ln(6) + \ln(100))$
33. $2x + 2$ **35.** $4x/3$
39. $x^{3x}(3\ln(x) + 3)$ **41.** $(\sqrt{x}^x/2)(\ln(x) + 1)$
43. $3^x \cos^{3x}(x)(\ln(3)\ln(\cos(x)) - \tan(x))$
45. $x/5 + C$
47. $(4^7/5^2)^x/(7\ln(4) - 2\ln(5)) + C$

49. $\log_2(5)\log_7(3)\ln(x) + (1/2)(\ln(5) + \ln(3))\log_2(x)\log_7(x) + (1/2)\log_2(x)\log_7(x)\ln(x) + C$
51. $(\ln(10)/2)(\log_{10}(x))^2 + C$
59. $10^{16} \cdot 10^{(3/2)M_W}$ dyne cm
61. $M'(M_W) = 15 \cdot 10^{15} \cdot 10^{(3/2)M_W} \ln(10)$; $M(8.4) \approx 3.79 \times 10^{28}$
65. Maximum: $f(e^{1/2}) = \exp(1/(2e))$
67. $\ln(a)$
71. $t_M = (\ln(a) - \ln(b))/(a - b)$; $t_I = (\ln(a^2) - \ln(b^2))/(a - b)$; $\lim_{t\to\infty} c(t) = 0$
83. $(0.4285, (1/3)^{0.4285})$
85. Local maximum: $(e, e^{1/e})$; points of inflection: $(0.5819327, f(0.5819327))$ and $(4.36777, f(4.36777))$; increasing: $(0, e)$; decreasing: (e, ∞); concave up: $(0, 0.58193) \cup (4.36777, \infty)$; concave down: $(0.58193, 4.36777)$; horizontal asymptote: $y = 1$

Section 6.5

1. 17510 **3.** 7211
5. 3.7066 g **7.** 15.7744 g
9. $40\ln(2)/\ln(3/2)$ **11.** $100\ln(2)/\ln(7/5)$
13. 12:40 PM **15.** m/t
17. 4.343 kg **19.** 1386.3 cells/min
23. 5381.5 yr **27.** **a.** 79% **b.** 92%
29. 1.94% **31.** About 10000 yr
33. Maximum: about 96%; minimum: about 99%
35. 0.35 h **37.** 101.35°; 232.5 min
39. $k = -\lambda$; $b = \lambda^2$
41. $(N_0 - \mu/\delta)\exp(-\delta t) + \mu/\delta$; μ/δ
43. $(c(0) - c_O)\exp(-Kt) + c_O$
45. $(N(0) - 1/K)\exp(-A \cdot K \cdot t) + 1/K$
47. **b.** $A\exp(-r_{out}t/L)$ kg
c. $m(t) = A \cdot (L/(L + (r_{in} - r_{out})t))^{r_{out}/(r_{in} - r_{out})}$
55. $v_\infty = P\tau$ **57.** $\ln(10)/\ln(2)$
59. 11.4 m/s (downward)
61. **a.** 4.1977 kg/s **b.** 1870 m/s **c.** 238 m/s (about 533 m/h)
d.

$$y(t) = 359477.6 - 445.38 \times (4.1977te^{\alpha t} + 800)e^{-\alpha t}$$
$$\alpha = 0.0052471$$

e. Time of fall: 25.422 s; velocity at impact: 249.4 m/s

63. $H = 14.6$ m; $v_I = -14.16$ m/s; $T_u = 1.59$ s; $T_d = 1.89$ s
65. **a.** Johnson: $\tau = 1.356$, $P = 8.702$; Lewis: $\tau = 1.457$, $P = 8.099$ **c.** 1.18 m

Section 6.6

1. $\pi/2$ **3.** $-\pi/2$ **5.** $\pi/3$
7. $\pi/4$ **9.** $-\pi/3$ **11.** $\pi/4$
13. $-\pi/4$ **15.** $3/\sqrt{1-9x^2}$
17. $1/x\sqrt{1-\ln^2(x)}$ **19.** $2\sin(t)\cos(t)/\sqrt{1-\sin^4(t)}$
21. $-4x^3/\sqrt{1-x^8}$
23. $(\arccos(x)-\arcsin(x))/\sqrt{1-x^2}$
25. $e^x/(1+e^{2x})$
27. $\sec^2(x)\arctan(x)+\tan(x)/(1+x^2)$
29. $-x/(1+x^2)^{3/2}$
31. $\cos(t)/((2+\sin(t))\sqrt{3+4\sin(t)+\sin^2(t)})$
33. $-2/(x\sqrt{x^4-1})$ **35.** $2t/(1+t^4)$
37. $\pi/6$ **39.** $\pi/3$ **41.** $\pi/12$ **43.** $\pi/8$
45. $\arcsin(\ln(x))+C$ **47.** $\pi^2/32$
49. $\arcsin(\tan(x))+C$ **50.** $\text{arcsec}(e^x)+C$
51. $\arcsin(e^x)+C$ **57.** 75 ft
61. $\arcsin(2x-1)+C$ **63.** $\arctan(x+1)+C$
65. $\arcsin(c)+(x-c)/\sqrt{1-c^2}$; $\pi/6+(\sqrt{3}/3)(2x-1)$; $1-\pi/4+\sqrt{2x}$
67. $\pi/6+(2\sqrt{3}/3)(x-1/2)+(2\sqrt{3}/9)(x-1/2)^2$
69. $\pi/3+(\sqrt{3}/6)(x-2)-(7\sqrt{3}/72)(x-2)^2$
71. $k=0$: $(1/a)\arctan(x/a)+C$; $k=1$: $(1/2)\ln(a^2+x^2)+C$; $k=2$: $x-a\arctan(x/a)+C$; $k=3$: $x^2/2-(a^2/2)\ln(a^2+x^2)+C$
73. $C=\pi/2$ **75.** $-1/6$; $-2/3$
77. 2.082316825 **79.** 0.3758628321
81. Domain: $(-\infty,\infty)$; local minimum: $(-1,-\pi/6)$; local maximum: $(1,\pi/6)$; increasing: $(-1,1)$; decreasing: $(-\infty,-1)\cup(1,\infty)$; points of inflection: $(-1.690141555, -0.4536496788)$, $(0,0)$, and $(1.690141555, 0.4536496788)$; concave up: $(-1.690141555, 0)\cup(1.690141555, 0, \infty)$; concave down: $(-\infty,-1.690141555)\cup(0, 1.690141555)$; horizontal asymptote: $y=0$; symmetry: about the origin
83. $b=0.3594330037$; area: 0.006182100655
85. $a=-1.302775638$; $b=2.302775638$; area: 5.127796048

Section 6.7

1. $3\cosh(3x)$ **3.** $\coth(x)$
5. $-\text{sech}(\sqrt{x})\tanh(\sqrt{x})/(2\sqrt{x})$
7. $(2x-5)\sinh(x^2-5x)$ **9.** $\coth(x)-\tanh(x)$
11. $(1/3)\sinh(3x)+C$ **13.** $\sqrt{\sinh(2x)}+C$
15. $(1/6)\tanh^2(x^3)+C$ **17.** $(1/4)\tanh^4(x)+C$
19. $(1/2)\sinh(2x+5)+C$ **21.** $1/\sqrt{x^2-1}$
23. $-1/(|x|\sqrt{1+x^2})$
25. $-\text{arctanh}(2x)/(|x|^2\sqrt{1-x^2})+2\,\text{sech}^{-1}(x)/(1-4x^2)$
49. $y=4\cosh(x/4)-3$
61. Increasing: $(1-\sqrt{2}, 1+\sqrt{2})$; decreasing: $(-\infty, 1-\sqrt{2})\cup(1+\sqrt{2},\infty)$; local maximum: $(1+\sqrt{2}, 0.2086)$; local minimum: $(1-\sqrt{2}, -1.5224)$; points of inflection: $(-0.8972, -1.2556)$, $(0.1374, -0.9514)$, $(-0.8814, 0.1843)$; concave up: $(-0.8972, 0.1374)\cup(3.7211,\infty)$; concave down: $(-\infty,-0.8972)\cup(0.1374, 3.7211)$; horizontal asymptote: $y=0$
63. Increasing: $(\ln(\sqrt{2}-1),\infty)$; decreasing: $(-\infty, \ln(\sqrt{2}-1))$; local minimum: $(\ln(\sqrt{2}-1), -\sqrt{2})$; points of inflection: $(-1.6149, -1.307)$, $(0.4032, -0.5412)$; concave up: $(-1.6149, 0.4032)$; concave down: $(-\infty,-1.6149)\cup(0.4032,\infty)$; horizontal asymptotes: $y=-1$, $y=1$
65. $P=(\ln(1+\sqrt{2}), \sqrt{2})$, $Q=(\ln(\sigma), (1+\rho)/\sigma)$, $R=(\ln(\rho+\sqrt{\rho}), (1+\rho)/\sigma)$, $S=(\ln(1+\sqrt{2}), 1)$, $T=(\ln(\sigma), \rho/\sigma)$, $U=(\ln(\rho+\sqrt{\rho}), \rho/\sigma)$, $V=(\ln(1+\sqrt{2}), 1/\sqrt{2})$ where $\rho=(1+\sqrt{5})/2$ and $\sigma=\sqrt{1+2\rho}$

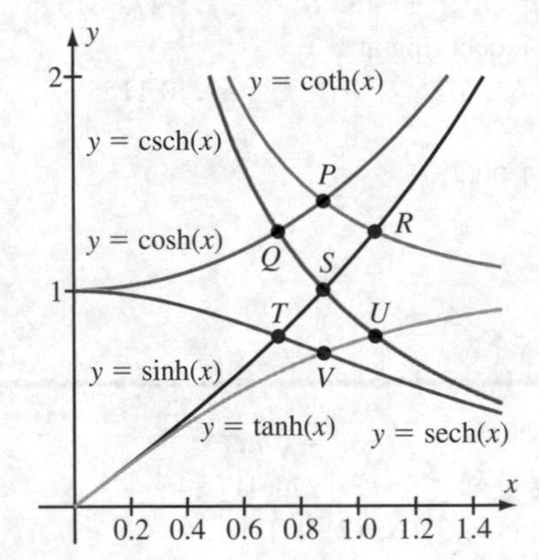

67.

$y = \frac{\pi}{2}$, $y = -\frac{\pi}{2}$, $x = -\frac{\pi}{2}$, $x = \frac{\pi}{2}$

— $y = gd(x)$
— $y = gd^{-1}(x)$

Section 7.1

1. $xe^x - e^x + C$ **3.** $\sin(x) - x\cos(x) + C$
5. $\cos(2x)/4 + x\sin(2x)/2 + C$
7. 1 **9.** $\pi/8 - 1/4$
11. $e^2/36 - (1/2)\ln(3) + 1/4$ **13.** $(x^2 - 2x + 2)e^x + C$
15. $-(1/3)x^2\cos(3x) + (2/27)\cos(3x) + (2/9)x\sin(3x) + C$
17. $-(x^3 + (3/2)x^2 + (3/2)x + 3/4)e^{-2x}/2 + C$
19. $x(\ln(x) - 1)/2 + C$
21. $(1/4)x^4\ln(1/x) + x^4/16 + C$
23. $\pi/4 - \ln(2)/2$ **25.** $\pi/4 - 1/2$
27. $1 - \pi/4$
29. $-2^x(1/\ln(2) - x)/\ln(2) + C$
31. $2\sqrt{x}(\ln(x) - 2) + C$
33. $x\sec(x) - \ln(|\sec(x) + \tan(x)|) + C$
35. $x\ln(x^2 + 1) - 2x + 2\arctan(x) + C$
37. $(1/3)x^3\ln(x) - (1/9)x^3 - 2x^2\ln(x) + x^2 + 4x\ln(x) - 4x + C$
39. $2(7\sqrt{2} - 10)/3$
41. $2\sin(\sqrt{x}) - 2\sqrt{x}\cos(\sqrt{x}) + C$
43. $(1/2)x^4e^{x^2} - x^2e^{x^2} + e^{x^2} + C$
45. $(1/3)(\sin(x^3) - x^3\cos(x^3)) + C$
47. $(2e)^x/(1 + \ln(2)) + C$
49. $-(3/8)\cos(x)\cos(3x) - (1/8)\sin(x)\sin(3x)$
55. $f^{(k-1)}(1) - kf^{(k-2)}(1) + k(k-1)f^{(k-3)}(1) - \cdots + (-1)^{k-1}k!f^{(0)}(1) + (-1)^k k!\int_0^1 f(x)\,dx$
59. **a.** $x\text{FresnelS}(x) + (1/\pi)\cos(\pi x^2/2) + C$
b. $x\text{FresnelC}(x) - (1/\pi)\sin(\pi x^2/2) + C$
c. $x^2\text{FresnelS}(x)/2 + x\cos(\pi x^2/2)/(2\pi) - \text{FresnelC}(x)/(2\pi)$; $x^2\text{FresnelC}(x)/2 - x\sin(\pi x^2/2)/(2\pi) + \text{FresnelS}(x)/(2\pi)$
61. $1.337 \cdot \alpha_0$ **63.** 3/4 **65.** 0.5452
67.

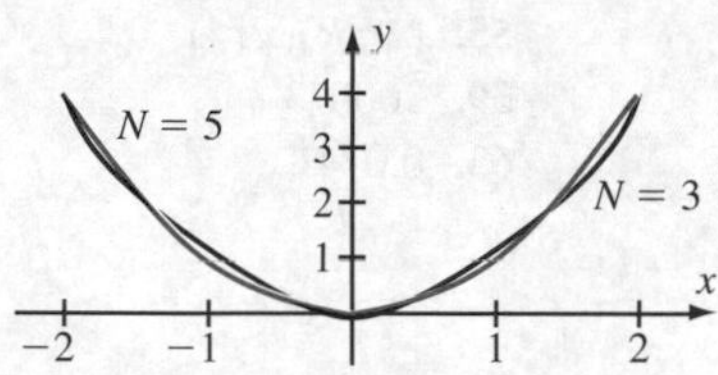

Fourier cosine-polynomial approximations of $f(x) = x^2$.

Section 7.2

1. $A/x + B/(x+1)$ **3.** $A/(x-2) + B/(x+2)$
5. $(Ax + B) + C/(2x - 5) + D/(x + 4)$
7. $A/x + B/x^2 + C/(x+1) + D/(x-2)$
9. $Ax^3 + Bx^2 + Cx + D + E/(x-5) + F/(x-5)^2 + G/(x+3) + H/(x+3)^2$
11. $1/x - 1/(x+1)$ **13.** $2/(x-2) - 1/(x+2)$
15. $10/(2x-5) + 8/(x+4)$
17. $-1/x^2 + 1/(2x) - (2/3)/(x+1) + (1/6)/(x-2)$
19. $2/(x-5)^2 - 1/(x-5) + 1/(x+3)^2 + 3/(x+3)$
21. $2\ln(|x-1|) + \ln(|x+1|) + C$
23. $5\ln(|x-3|) + 4\ln(|x+6|) + C$
25. $(1/2)x^2 - x - (1/5)\ln(|x+1|) + (6/5)\ln(|x-4|) + C$
27. $-6/(x-2) + \ln(|x-2|) - \ln(|x+1|) + C$
29. $-1/x - \ln(|x|) + 2\ln(|x+1|) + C$
31. $1 + (16/7)(\ln(2) + \ln(8) - \ln(9))$
33. $(19/66)\ln(7) - (2/33)\ln(16)$
35. $(1/2)\ln(|e^t - 1|/(e^t + 1)) + C$
37. $-4/(3 + \sin(x)) + \ln(3 + \sin(x)) - \ln(1 - \sin(x)) + C$
39. $(3/2)\ln(|x^{1/3} - 1|/|x^{1/3} + 1|) + C$
41. **c.** $1/x^2 - 2/x + 2/(x-1) + 2/(x-1)^2$
43. **a.** False; $P'(t) < 0$ **b.** False; $P(t) < 15000$ for all t
c. True; $\lim_{t\to\infty} P(t) = 15000$
d. True; $P'(0) = 10^{-5}(7500^2 - (P(0) - 7500)^2)$
47. $k = 5.6808 \times 10^{-3}$; **a.** $P'(9) \approx 3.55$ **b.** $P'(9) \approx 4.3$
49. $25.887 - (19/3)\ln(2) + 24\ln(3) - 89/3\ln(5)$
51. 0.02079

Section 7.3

1. $\theta/2 + \sin(4\theta)\cos(4\theta)/8 + C$
3. $t + C$
5. $-(1/3)\cos(2x)(1 + \sin^2(2x)/2) + C$
7. $-(1/3)\cos(x+2)(2 + \sin^2(x+2)) + C$
9. 1/5 **11.** 4/35
13. $3x/8 + 3\cos(x)\sin(x)/8 + \sin(x)\cos^3(x)/4 + C$
15. $3\pi/8$ **17.** $\pi/32$
19. $\pi/96 + \sqrt{3}/64$ **21.** $(1/3)(\cos(t))^{-3} + C$
23. $(1/2)(\cos(t))^{-2} + \ln(|\cos(t)|) + C$
25. $-\sin(t) + \ln(|\sec(t) + \tan(t)|) + C$
27. $-\cot(\theta) + \csc(\theta) + C$
29. $-\csc(\theta) - \cot(\theta) - \theta + C$
31. $(5/16)t + (11/48)\sin(2t) + (3/64)\sin(4t) + (1/48)\sin(2t)\cos^2(2t) + C$
33. $(1/16)t - (1/48)\sin(2t) - (1/64)\sin(4t) + (1/48)\sin(2t)\cos^2(2t) + C$
41. $\pi^2/16 - 1/4$ **43.** $\ln(\sqrt{2}) - 1/4$
45. $\int \ln^n(x)\,dx = x\ln^n(x) - n\int \ln^{n-1}(x)\,dx$
57. 0.07428
59. $x_0 = 1.393249$; Area $= 0.238668$

Section 7.4

1. $-(1/2)\ln(|1 - x^2|) + C$ **3.** $4\arcsin(x/2) + C$
5. $(1/2)\ln(x^2 + 1) + C$ **7.** $(2/3)\arctan(3x) + C$

9. $\operatorname{arcsec}(3) - \pi/3$ **11.** $\pi/8 - (1/2)\operatorname{arcsec}(\sqrt{5}/2)$
13. $-2\sqrt{1-x^2} + 3\arcsin(x) + C$
15. $x + \arctan(x) + C$
17. $(-x\sqrt{1-x^2} + \arcsin(x))/2 + C$
19. $x/\sqrt{1-x^2} - \arcsin(x) + C$
21. $x/\sqrt{x^2+1} + C$
23. $(x\sqrt{x^2+1} - \ln(x + \sqrt{x^2+1}))/2 + C$
25. $\sqrt{x^2-1} + C$
27. $-x/\sqrt{x^2-1} + C$
29. $(\sqrt{14}/28)\arctan(\sqrt{14}/2)$
31. $\ln(2)/3$
33. $(9/16)\arcsin(2/3) - \sqrt{5}/8$
35. $1/8$
37. $((\pi - 1)\sqrt{2} - \ln(\sqrt{2}+1))/6$
39. $\sqrt{3} - \ln(2 + \sqrt{3})$
41. $(x/(x^2+1) + \arctan(x))/2 + C$
43. $\sqrt{x^2+1} - \ln((1 + \sqrt{x^2+1})/|x|) + C$
45. $\ln(x + \sqrt{x^2+1}) - \sqrt{x^2+1}/x + C$
47. $\arcsin(x-1) + C$
49. $\sqrt{x^2+2x+2} - \ln(\sqrt{x^2+2x+2} + x + 1) + C$
51. $((x-1)/(x^2-2x+2) + \arctan(x-1))/2 + C$
53. $(1/2)\ln(|x^2 - 14x + 58|) + (7/3)\arctan((x-7)/3) + C$
55. $x - 3\ln(x^2+4) + 2\arctan(x/2) + C$
57. $(3/4)\ln(2x^2 + 6x + 5) - (9/2)\arctan(2x+3) + C$
59. $-(1/4)(x+2)/(x^2+2x+2)^2 - (3/8)(x+1)/(x^2+2x+2) - (3/8)\arctan(x+1) + C$
61. $-\sqrt{10x - x^2} + 6\arcsin(x/5 - 1) + C$
63. $c = 4/35$
65. $x - (3/2)\arctan(x/2) + C$
67. $x - 2\ln(5 + 4x + x^2) + 4\arctan(x+2) + C$
69. $2\sqrt{x} + \ln(|2\sqrt{x} - x - 1|/(x-1)^2) + C$
71. $-x + 4\sqrt{x} + 2\ln(|2\sqrt{x} - x - 1|/(x-1)^2) + C$
73. $2\sqrt{x+1} + \ln(|(2\sqrt{x+1} - x - 2)/x|) + C$
75. $\int_0^{1/2}(1/(1-x^2))\,dx = (1/2)\ln(3)$; $\int_{\sqrt{2}}^{2}(1/(1-x^2))\,dx = \ln(\sqrt{3}) - \ln(\sqrt{2}+1)$
77. 0.1454 **79.** 0.87803 **81.** 0.02164

Section 7.5

1. $(Ax+B)/(x^2+1) + (Cx+D)/(x^2+4)$
3. $(Ax+B)/(x^2+1) + (Cx+D)/(x^2+x+1) + (Ex+F)/(x^2+x+1)^2$
5. $A/(x-4) + (Cx+D)/(x^2+x+3)$
7. $A/(x-2) + (Bx+C)/(x^2+4) + (Dx+E)/(x^2+4)^2 + (Fx+G)/(x^2+4)^3$
9. $A/x + B/x^2 + C/(x-1) + D/(x+1) + (Ex+F)/(x^2+1)$
11. $1/(x-1) + (2x-3)/(x^2+1)$
13. $5x/(x^2+2) + (2x-3)/(x^2+1)$
15. $x/(x^2+1) - 2x/(x^2+1)^2$
17. $1 + 12/(x-3) + (3x-4)/(x^2+4)$
19. $(2x-1)/(x^2+4) + (x-4)/(x^2+3)$
21. $\ln(|x-1|) + \ln(x^2+1) - 3\arctan(x) + C$
23. $(5/2)\ln(x^2+2) + \ln(x^2+1) - 3\arctan(x) + C$
25. $1/(x^2+1) + (1/2)\ln(x^2+1) + C$
27. $x + 12\ln(x-3) + (3/2)\ln(x^2+4) - 2\arctan(x/2) + C$
29. $\ln(x^2+4) - (1/2)\arctan(x/2) + (1/2)\ln(x^2+3) - (4/\sqrt{3})\arctan(x/\sqrt{3}) + C$
33. $3\ln(|x+1|) + \ln(x^2 - x + 1) + C$
35. $8x/(x^2+1)^3 + 10x/(x^2+1)^2 + 15x/(x^2+1) + 15\arctan(x) + C$
37. $b = 0.7585$; area $= 0.07744$
39. Roots: $-0.80061, 0.90579, 6.8948$; integral: -0.24263

Section 8.1

1. $837\pi/5$ **3.** 2π
5. $\pi/7$ **7.** $\pi(e^2-1)/2$
9. $2\pi/15$ **11.** $108\pi/5$
13. $64\pi/15$ **15.** $162\pi/5$
17. $\pi/2$ **19.** 132π
21. $128\pi/5$ **23.** $64\pi/3$
25. $\pi/6$ **27.** $284\pi/3$
29. 8π **31.** $49\pi/72$
33. $\pi^2/2$ **35.** $\pi(e^2+1)/4$
37. $\pi(e-2)$ **39.** $\pi(1/3 - \ln(4/3))$
41. $400\pi\sqrt{5}/3$ **43.** $1375\pi/3$
45. $7\pi/5$ **47.** $375\pi/2$
49. $8\pi^2$ **51.** 6π
53. $16\pi/15$ **55.** $(2\pi R)(\pi r^2)$
57. $56\pi(\pi-1)$ **59.** $2000/3$
61. $500\pi/3$ **63.** 0.0741
65. 49.685 g

Section 8.2

1. $5\sqrt{37}$ **3.** $3\sqrt{2}$
5. $(80\sqrt{10} - 13\sqrt{13})/27$
7. $\sqrt{2}(82\sqrt{41} - 10\sqrt{5})/27$
9. $2\pi(2917\sqrt{2917} - 577\sqrt{577})/108$
11. $1505\pi/18$ **13.** $39\pi\sqrt{10}$
15. $1179\pi/1024$
17. $2\pi(365\sqrt{730}/27 - 145\sqrt{145}/54)$
19. $\pi(145\sqrt{145} - 17\sqrt{17})/6$
21. $4/3$ **23.** $(e - e^{-1})/2$
25. $7683/256$ **27.** 499.53

31. $\sqrt{2}(e-1)$
33. $2\ln(1+\sqrt{2})$
37. $\pi(\sqrt{3}-\ln(-1+\sqrt{3})+\ln(1+\sqrt{3}))$
39. $\pi(e^2\sqrt{1+e^4}+\ln(e^2+\sqrt{1+e^4})-\sqrt{2}-\ln(1+\sqrt{2}))$
41. 19.38
43. 6.5186
45. Length of arch: 1480.28 ft; length of parabola: 1438.54 ft

Section 8.3

1. $2/\pi$
3. $(1/3)\ln(4)2/\pi$
7. $14/9$
9. $7/8$
11. $\sqrt{2}/4$
13. $\sqrt{e}/(\sqrt{e}+1)$
15. $3/4$
17. $7/4$
19. $5/4$
21. 100.5°F
23. $(1/(2\pi))(1-\cos(\pi^2))$
25. $(2/\pi)\ln(2)$
27. 1
29. $\sqrt{3}$
31. Total cost: \$27.51
35. $1/60$
37. $\ln(3)$
39. $4/3$
41. $\ln((e^c-1)/c)$
43. $(1/4)(e^2+1)$
45. $e(2e-5)/(2(e-2))$
47. $\pi/6$
49. $1-\ln((e+1)/2)$
53.

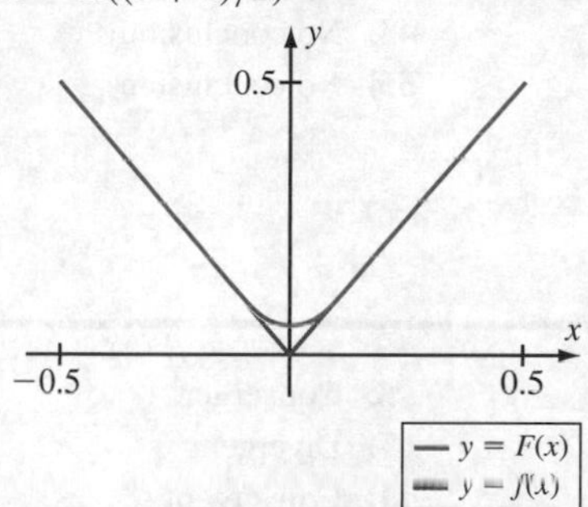

55. b. $f_{\text{ave}}=\sqrt{\pi}\,\text{erf}(1)/2\approx 0.7468$
c. $c=\sqrt{\ln(2/(\sqrt{\pi}\,\text{erf}(1)))}\approx 0.5403$
d.

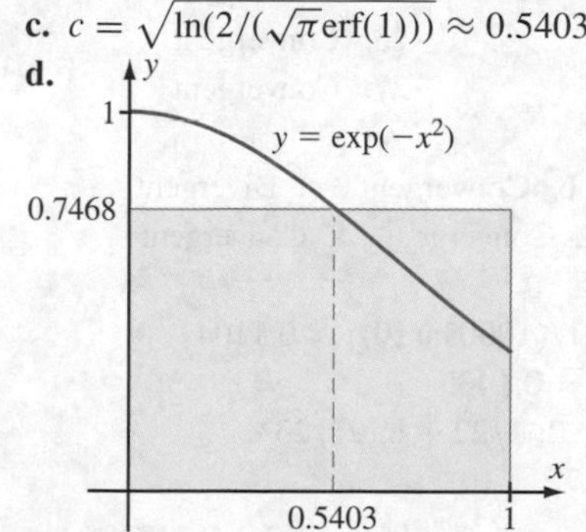

Section 8.4

1. -6
3. 4
5. $-116/15$
7. $1+3\ln(2)$
9. $\bar{x}=2, \bar{y}=2/3$
11. $\bar{x}=0, \bar{y}=4a/(3\pi)$
13. $\bar{x}=-1, \bar{y}=18/5$
15. $\bar{x}=1/\ln(2), \bar{y}=1/(4\ln(2))$
17. $\bar{x}=23/48, \bar{y}=13/60$
19. $\bar{x}=0, \bar{y}=1152/(35\pi)$
21. $\bar{x}=\pi/2-1, \bar{y}=\pi/8$
23. $\bar{x}=7/(16\sqrt{2}-14), \bar{y}=2(16\sqrt{2}-19)/(5(8\sqrt{2}-7))$
25. $\bar{x}=(\sqrt{2}-1)/\ln(1+\sqrt{2}), \bar{y}=\pi/(8\ln(1+\sqrt{2}))$
27. $\bar{x}=(e^2+1)/4, \bar{y}=e/2-1$
29. $\bar{x}=7/4, \bar{y}=6/5$
33. $\bar{x}=19/14, \bar{y}=92/35$
35. $\bar{x}=1/2, \bar{y}=5/2$
37. $3/80$
39. $c=(p+1)(p+2)$; $\text{Var}(X)=2(p+1)/((p+3)^2(p+4))$
41. $\bar{x}\approx 1.1175, \bar{y}\approx 0.37431$
43. $\bar{x}=0.9765796, \bar{y}=0.4376893$
45. $g(0.2)=0.70524, g(1)=0.14105, g(2.0)=0.070524$

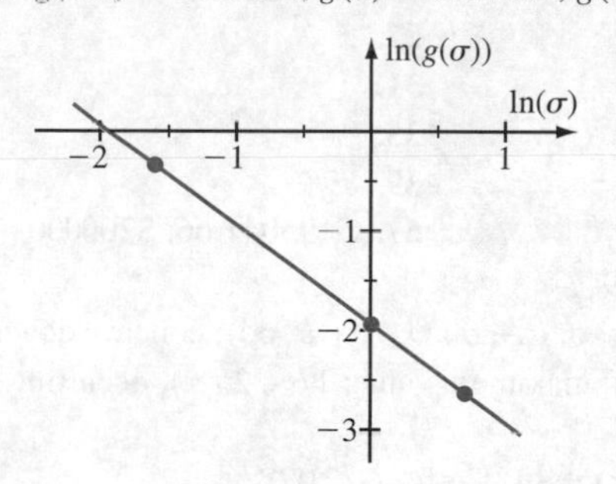

Section 8.5

1. 37600 ft lb
3. 134000 ft-lb
5. 7.21965×10^6 ft-lb
7. 60000 ft-lb
9. $49/3$ ft-lb
11. 140 ft-lb
13. 25%
15. 7.02315×10^5 ft-lb
17. 2.3972×10^5 ft-lb
19. 5.7203×10^7 ft-lb
21. 26790 ft-lb
23. 188.2 ft-lb
25. 88255.5 ft-lb
27. 12028.9 ft-lb
31. The right half of the parabola $W=F^2/(2k)$
33. 3582 ft-lb
35. 5754.9 ft-lb

Section 8.6

1. Divergent
3. $10\cdot 2^{1/10}$
5. Divergent
7. $2/3$
9. Divergent
11. 6
13. Divergent
15. $4\sqrt{3}$
17. Divergent
19. $10\cdot 2^{3/5}/3$
21. $3\sqrt[3]{5}+3$
23. Divergent
25. $3\cdot 2^{1/3}\cdot\pi$
27. $\pi^2/2$

29. $7(\ln(7)-1)$
31. Divergent
33. $2\sqrt{7}(\ln(7)-2)$
35. $3\ln(3)-4$
39. $p<1$
41. Divergent
43. Convergent
45. Divergent
47. Convergent
49. $g(x)=x^{-1/2}$; 1.9
51. $g(x)=e^2x^{-1/3}$; 3.5

Section 8.7

1. $2/\sqrt{3}$
3. $\sqrt{2}$
5. $\pi/2$
7. $1/2$
9. Divergent
11. $e^{-3}/6$
13. Divergent
15. $3\pi^2/32$
17. Divergent
19. Divergent
21. $3e^{4/3}$
23. $\arctan(e^{-5})$
25. Divergent
27. π
29. 0
31. 1
33. $2e$
35. 1
37. Divergent
39. $\pi/2$
41. π
57. \$195041.66; \$200000
59. \$31.68; \$104.69
61. Concave up: $(-\infty,\mu-\sigma)\cup(\mu+\sigma,\infty)$; concave down: $(\mu-\sigma,\mu+\sigma)$; maximum value: $1/(\sqrt{2\pi}\sigma)$, occurring at $x=\mu$
65. $P(X<\mu+\lambda\sigma)=(1+\operatorname{erf}(\lambda/\sqrt{2}))/2$
67. **b.** $\Gamma(1)=1, \Gamma(2)=1, \Gamma(3)=2, \Gamma(4)=6$
d. $c=\lambda^s/\Gamma(s)$
69. 1.89
71. 1.76

Section 9.1

1. 1, 3/2, 11/6, 25/12, 137/60
3. 5/4, 37/16, 213/64, 1109/256, 5461/1024
5. 2/3, 10/9, 38/27, 130/81, 422/243
7. 1, 3/4, 31/36, 115/144, 3019/3600
9. 1/7
11. 4/5
13. $1/(7^{1/3}-1)$
15. 11/30
17. $8\sum_{n=1}^{\infty}(1/10)^n=8/9$
19. $17\sum_{n=1}^{\infty}(1/1000)^n=17/999$
21. $S_N\geq(1/2)\cdot$
23. $S_N\geq(1/\sqrt{2})\cdot N$
25. $S_N=1-1/(N+1)$; 1
27. $S_N=-2N/(N+1)^3$; 0
29. $S_N=1-1/\sqrt{N+1}$; 1
31. $N(3N+5)/(4(N+2)(N+1))$; 3/4
33. $N/(3(2N+3))$; 1/6
35. $S_N=-\ln(N+1)$
37. $S_N=r(r^N-1)/((r^{N+1}-1)(r-1)^2)$; $1/(r-1)^2$
39. $\pi^2/24$; $\pi^2/8$
43. \$144000
45. $\sum_{n=2}^{\infty}5/10^n$; 1/18 mg
49. About 1.2924243×10^{103} dominoes; about 6.5×10^{84} light-years high
51. 0.458675
53. 1.458
55. 1000001; 1.17432×10^{52}

Section 9.2

1. No conclusion
3. Series diverges
5. No conclusion
7. Series diverges
9. No conclusion
11. Series diverges
13. No conclusion
15. Series diverges
17. No conclusion
19. No conclusion
21. 10/9
23. 25/16
25. 27/10
27. May
29. May
31. Series diverges
33. Series diverges
35. No conclusion
37. Series diverges
39. Series diverges
41. Series diverges
43. No conclusion
53. Series diverges
55. No conclusion
57. Series diverges

Section 9.3

1. Convergent
3. Convergent
5. Divergent
7. Divergent
9. Divergent
11. Convergent
13. Convergent
15. Convergent
17. Convergent
19. Convergent
21. Divergent
23. Convergent
25. Convergent
27. Convergent
31. $a>1$
33. **a.** Convergent **b.** Convergent **c.** Divergent **d.** Divergent **e.** Convergent **f.** Convergent
37. Convergent
45. $A_3=11/100+1/(1000\ln(10))\approx0.1104$
47. $A_3=287/2000=0.1435$
51. $M=101$; $M=22$; $1/22+\ln(22/23)$
53. $\sum_{n=1}^{\infty}f(n)=6.79\times10^{-2}$; $\int_1^{\infty}\exp(x)/(1+\exp(x))^3\,dx=1/(2(1+e)^2)\approx 3.62\times10^{-2}$; $f(1)=e/(1+e)^3\approx5.29\times10^{-2}$; inequality (9.4) states $3.62\times10^{-2}<6.79\times10^{-2}<5.29\times10^{-2}+3.62\times10^{-2}$, which is true.

55. $\sum_{n=1}^{\infty} f(n) = 0.13$; $\int_1^{\infty} \frac{n^2}{(1+n^3)^3}\, dn = 1/24 \approx 0.04$; $f(1) = 1/8 = 0.125$; inequality (9.4) states $0.04 < 0.13 < 0.125 + 0.04$, which is true.
57. $A_4 = 0.63087\ldots$
59. $A_{13} = 0.49449\ldots$
61. exp(exp(exp(6.1)))

Section 9.4

1. Compare with $\sum_{n=1}^{\infty} 1/n^2$
3. Compare with $\sum_{n=1}^{\infty} 3/n^4$
5. Compare with $\sum_{n=1}^{\infty} 1/n^{3/2}$
7. Compare with $\sum_{n=1}^{\infty} (2/3)^n$
9. Compare with $\sum_{n=1}^{\infty} 2(1/2)^n$
11. Compare with $\sum_{n=1}^{\infty} 1/n^{3/2}$
13. Compare with $\sum_{n=1}^{\infty} 1/n^{1.01}$
15. Compare with $\sum_{n=2}^{\infty} 2/(n \ln^2(n))$
17. Compare with $\sum_{n=2}^{\infty} 1/n$
19. Compare with $\sum_{n=1}^{\infty} 1/(4n)$
21. Compare with $\sum_{n=1}^{\infty} 1/(2n^{1/2})$
23. Compare with $\sum_{n=1}^{\infty} 1/\sqrt{n}$
25. Convergent
27. Convergent
29. Convergent
31. Convergent
33. Convergent
37. Divergent
39. Divergent
41. Divergent
43. Convergent
45. Divergent
55. $M = 246435$;
$\sum_{n=M}^{\infty} n^{100} \cdot 0.99^n < \sum_{n=M}^{\infty} 0.995^n = 200(0.995)^M$

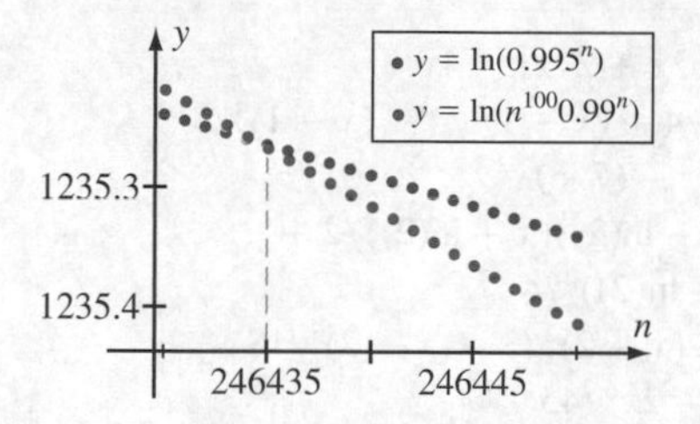

Section 9.5

13. Converges absolutely
15. Converges conditionally
17. Converges conditionally
19. Diverges
21. Converges conditionally
23. Converges absolutely
25. 100
27. 200
29. 5
31. 10
33. 8
35. −0.632
37. −0.424
39.

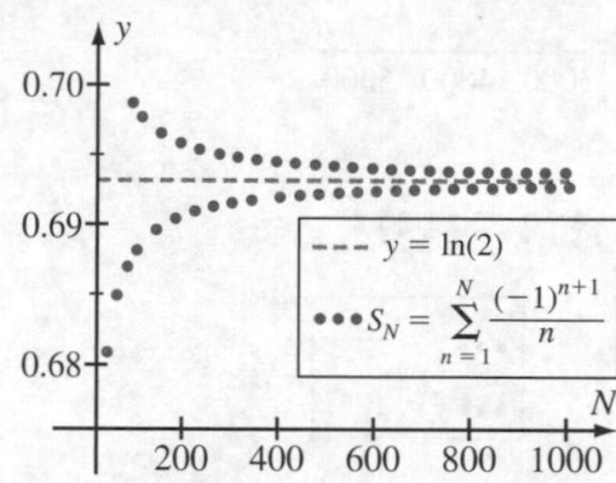

41.

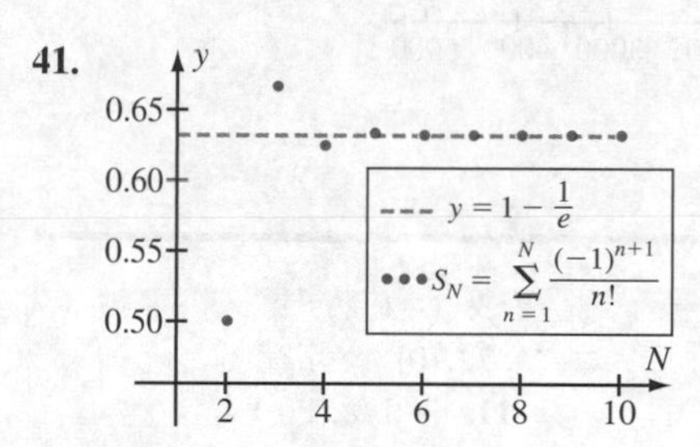

45. 9
47. 18

Section 9.6

1. Convergent
3. Divergent
5. Convergent
7. Divergent
9. Convergent
11. Converges conditionally
13. Diverges
15. Converges absolutely
17. Divergent
19. (Absolutely) convergent
21. (Absolutely) convergent
23. Convergent
25. Divergent
27. Divergent
29. Divergent
31. Convergent
33. Converges conditionally
35. Diverges
37. Converges conditionally
39. Converges conditionally
41. Diverges
43. Converges absolutely
45. Converges conditionally
47. Diverges

53.

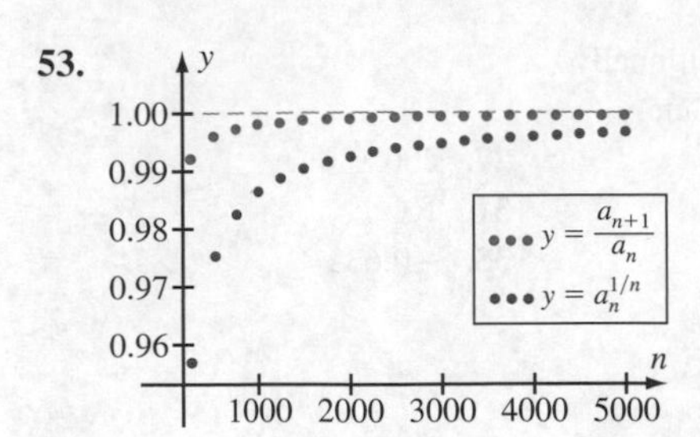

55.

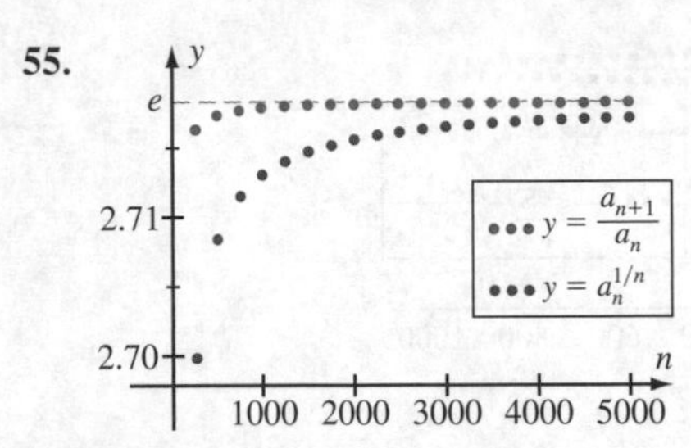

Section 10.1

1. $(-1/2, 1/2)$ **3.** $(-1, 1)$
5. $(-3, 3)$ **7.** $\{0\}$
9. $(-1, 1)$ **11.** $(-1/e, 1/e)$
13. $(-1, 1)$ **15.** $(-3, 3)$
17. $(-1, 1)$ **19.** $(-\infty, \infty)$
21. $(-1.1, -0.9)$ **23.** $(5\sqrt{2}, 7\sqrt{2})$
25. $[2, 4)$ **27.** $(-6, -4)$
29. $\sum_{n=0}^{\infty}(-1)^n x^n$ **31.** $\sum_{n=0}^{\infty} x^{4n}$
33. $\sum_{m=1}^{\infty}(-1)^{m-1}x^{2m}$ **35.** $\sum_{n=0}^{\infty} 3^{-n}x^n$
37. $1/2$ **39.** 4
41. ∞ **43.** $(-4, 2)$
45. $[-1/2, 1/2]$ **47.** $(-1/2, 1/2)$
49. $(-e - \sqrt{3}, -e + \sqrt{3})$ **57.** 1
61. 1

Section 10.2

1. $\sum_{n=0}^{\infty}(2x)^n$; radius of convergence: $1/2$
3. $\sum_{n=0}^{\infty}(1/2)^{n+1}x^n$; radius of convergence: 2
5. $\sum_{n=0}^{\infty}(-1)^n(1/4)^{n+1}x^n$; radius of convergence: 4
7. $1 + \sum_{n=1}^{\infty} 2x^{2n}$; radius of convergence: 1
9. $\sum_{n=0}^{\infty}(-1)^n x^{4n+3}$; radius of convergence: 1
11. $\sum_{m=1}^{\infty}(-1)^{m-1}4^m x^m/m$; radius of convergence: $1/4$
13. $\sum_{m=1}^{\infty}(-1)^{m-1}x^{2m}/m$; radius of convergence: 1
15. $\sum_{m=1}^{\infty}(-1)^{m-1}x^{2m+1}/(m(2m+1))$; radius of convergence: 1
17. $\sum_{n=1}^{\infty}(-1)^{n+1}x^{4n-2}/(2n-1)$; radius of convergence: 1
19. $\sum_{n=1}^{\infty}(-1)^{n+1}x^{4n-1}/((2n-1)(4n-1))$; radius of convergence: 1
21. $\sum_{n=0}^{\infty}(-1)2^{-(n+1)}(x+3)^n$; radius of convergence: 2
23. $\sum_{n=0}^{\infty}(-1)^n(2^n/3^{n+1})(x+1)^n$; radius of convergence: $3/2$
25. $\sum_{m=1}^{\infty}(-1)^{m-1}(x-1)^m/m$; radius of convergence: 1
27. $\sum_{m=0}^{\infty}(m+1)x^m$
29. $\sum_{j=0}^{\infty}(3j+2)\cdot x^j$
31. $x^2/(1-x)$; radius of convergence: 1
33. $2x/(1-x)^3$; radius of convergence: 1
35. $x^3/(x^2-1)^2$; radius of convergence: 1
39. $a_n = 3 \cdot 2^n/n!$
41. $a_0 = 1, a_1 = -1, a_n = 3(-1)^n/n!\,, 2 \le n < \infty$
43. $a_0 = 0, a_{2k+1} = 0, a_{2k} = 1/(2^k k!), 1 \le k < \infty$

Section 10.3

1. $1/2 - (\sqrt{3}/2)(x - \pi/3) - (x - \pi/3)^2/4 + (\sqrt{3}/12)(x - \pi/3)^3 + (x - \pi/3)^4/48$
3. $-1 + 2x^2 - 2x^4/3$
5. $1 + x + (11/2)x^2 + (1/6)x^3 + (1/24)x^4 + (1/120)x^5$
7. $1/8 - (3/16)(x-2) + (3/16)(x-2)^2 - (5/32)(x-2)^3 + (15/128)(x-2)^4 - (21/256)(x-2)^5$
9. $\ln(3) + x/3 - x^2/18 + x^3/81 - x^4/324 + x^5/1215$
11. $3(x+2) - (9/2)(x+2)^2 + 9(x+2)^3 - (81/4)(x+2)^4$
13. $1 - (x+2)/2 - (x+2)^2/8$
15. $x + x^3/3$
17. $P_2(x) = 8 + 5(x-3) + (x-3)^2$
19. $P_4(x) = 11 + 41(x-1) + 60(x-1)^2 + 40(x-1)^3 + 10(x-1)^4$
21. $P_3(x) = 49 - 62x + 27x^2 - 4x^3$
23. $1/2 - (x-1)^2/4 + (x-1)^3/4 - (x-1)^4/8$
25. $1 - x + (3/2)x^2 - (7/6)x^3$
27. $-\ln(2)/2 + (1 - \ln(2))(x + \ln(2))/2 + (2 - \ln(2))(x + \ln(2))^2/4$
29. $3 + 2(x-2) + (x-2)^2/3 - (x-2)^3/18$
35. $1 - x^2 + x^4/3 - 2x^6/45$
51. $\xi \approx -0.18741$ **53.** $\xi \approx 1.21106$

Section 10.4

1. Approximation: 1.10517083; error estimate: 0.92×10^{-7}
3. Approximation: 0.29552025; error estimate: 0.299×10^{-6}
5. Approximation: 0.9800666667; error estimate: 0.53×10^{-6}
7. Approximation: 1.098612478; error estimate: 0.207×10^{-6}
9. Approximation: 0.4; error estimate: 0.0213

11. Approximation: 3.033150206; error estimate: 0.29×10^{-7}

13. $M_5 = 105/32$ and $R_4(0.4) = 0.00028$

15. $M_3 = 3\sin(\pi/6) + (\pi/6)\cos(\pi/6)$ and $R_2(\pi/6) = 0.0467355$

17. $M_4 = \exp(-1.1)$ and $R_3(-1.1) = 9.099863251 \times 10^{-3}$

19. $M_6 = 120/(2.5)^6$ and $R_5(2.5) = 7.384382694 \times 10^{-8}$

21. $M_3 = 2$ and $R_2(0.6) = 0.072$

23. $P_5(0.2) = 0.980067$

25. $P_5(1.5) = 0.99749499$

27. $P_3(0.1) = 0.9048$

29. $P_5(1.2) = 0.1823$

35. $-1/2$

37. 1

39. $-1/3$

41. $-1/3$

43. $M_3 \approx 10.45$; $|R_2(\pi/4)| \leq 0.25$

45. $M_4 \approx 0.007215$; $|R_3(4)| \leq 1.88 \times 10^{-5}$

47. $M_5 \approx 160.02$; $|R_4(5)| \leq 0.103693$

49. $M_3 \approx 0.071075$; $|R_2(1.4)| \leq 3.4 \times 10^{-8}$

Section 10.5

1. $\sum_{n=1}^{\infty}(-1)^{n+1}(2x)^{2n-1}/(2n-1)!$

3. $\sum_{n=0}^{\infty}(-1)^n e \cdot x^n/n!$

5. $1 + x^3 + \sum_{n=1}^{\infty}(-1)^n x^{4n}/(2n)!$

7. $\sum_{n=0}^{\infty}(-1)^n x^{2n+1}$, $-1 < x < 1$

9. $\sum_{n=0}^{\infty}\binom{1/2}{n}x^{3n}$, $-1 < x < 1$

11. $\sum_{n=1}^{\infty}(-1)^{n+1}x^{2n-1}/n$, $-1 < x < 1$

13. $\sum_{n=0}^{\infty}\binom{1/2}{n}(x-1)^n$

15. $\sum_{m=2}^{\infty}x^m/(m-2)!$

17. $\sum_{n=1}^{\infty}(-1)^{n+1}(x-1)^n/n$

19. $\sum_{n=1}^{\infty}(-1)^{n+1}2^{2(n-1)}x^{2n-1}/(2n-1)!$

21. 0.9800

23. 0.8187

25. $(1+x)^{3/4} = 1 + 3x/4 - 3x^2/32 + 5x^3/128 + \cdots$

27. $1/\sqrt{1+x} = 1 - x/2 + 3x^2/8 - 5x^3/16 + \cdots$

29. $1/(1+x)^{3/2} = 1 - 3x/2 + 15x^2/8 - 35x^3/16 + \cdots$

31. $1/\sqrt{2}$

33. $2/\pi$

35. $1/e$

37. $\sum_{k=0}^{\infty}x^{2k}/(2k)!$

39. 0.46128

41. 0.33310

Index

A

abscissa, 12
absolute convergence, 650–651
absolute error, 7, 28, 29
absolute maximum or minimum, defined, 251
  See also maxima and minima
absolute value
  definition of, 4
  speed as, 156
  symbol for, 4
absolute value function, 35
absolute zero, 85
acceleration
  antidifferentiation of, 319–322
  approximation of, 218–219
  definition of, 216
  description of, 152
  as higher derivative, 216–217
  linear drag law, 469–470
  quadratic drag law, 483–484, 494–495
  unit of measure for, 319
accuracy, degree of. *See* degree of accuracy
Achilles (Zeno), 72
addition. *See* sums (or differences) of functions
Addition Formulas, trigonometric identities, 62
affine functions, 54
air drag (resistance), 469, 483–484, 494–495
d'Alembert, Jean le Rond, 147–148
algebraic numbers, 71
Algebra (Wallis), 344
allometric growth, 429–430
alternating series, 647–650
Alternating Series Test, 647–650
Ampère, André-Marie, 344
amplitude, 67
amplitude modulation, 68
Amthor, A., 71
Analyst, The (Berkeley), 147
analytic geometry
  of circles, 12–13
  of conic sections, 15–17
  elements of, 11–13
  history of, 72
  of lines, 21–26
  translations and, 49–50
angle of elevation, 66
antidifferentiation, 315–322
  definition of, 315–316
  notation for, 315–316
  of powers of x, 316–318
  rules of, 317
  of trigonometric functions, 318–319
  of velocity and acceleration, 319–322
  See also indefinite integrals
Apéry, Roger, 661
Apollonius, 245
Approximate Derivative Rule, 180–182
approximation, 7–8
  of area, 350–352, 357
  definition of, 7
  of derivative values, 180–182
  explicit equation solutions. *See* Newton-Raphson Method
  of extreme values, 269
  historical perspectives, 415–416
  of infinite series, 638–639
  of integrals, 399–407, 413
  of limits, 171
  mathematical modeling and, 471
  by Method of Increments. *See* Increments Method
  notation of, 234, 239
  of second derivatives, 217–219
  Taylor polynomials and. *See* Taylor polynomials
  with Taylor series, 701–702
arccosine, 479–480, 486
Archimedes
  E. T. Bell on, 558
  and infinite process, 72, 415
  and infinite series, 660
  and irrational numbers, 71
  Kepler on, 619
  Method of Exhaustion and, 414–415
  and the sphere, 619
  tomb of, 619
arc length
  basic method for, 576–578
  definitions of, 576, 577
  historical perspectives, 619–620
  summarized, 616
arcsine, 478–480, 481–482, 486, 555
arctangent, 480–484, 486, 555, 677–678, 687
area
  accuracy and, 357
  approximation of, 350–352, 357
  definition of, 352–353
  Fundamental Theorem of Calculus and, 365, 377
  historical perspectives, 414–415
  improper integrals and, 606–607
  as limit, 76
  reversing the roles of the axes and, 396–397
  Riemann sums and, 358–359
  rules for calculating, 393–394
  by subdivision into rectangles, 348–349
  summary of, 412–413
  surface, 578–581, 610–611, 617
  of a triangle, 353
  between two curves, 395–397
  work compared to, 597
area function, 377, 410
argument of a function, 33
Aristotle, 71, 72
Arithmetica Infinitorum (Wallis), 416
arithmetic-geometric mean (AGM), 558
Arithmetic-Geometric Mean Inequality, 143
Arithmetic-Geometric Mean Method for π, 45
Arrhenius equation, 449
arrow notation, 33
Arrow (Zeno), 72
Assayer, The (Galileo), 73
associated principal angle, 60–61
astronomical unit (AU), 44
asymptotes
  horizontal, 116–118
  limits at infinity and, 115–116
  skew, 120, 292–293
  vertical, 114–115, 117–118
atmospheric pressure, 474
atomic packing factor, 314
average cost, 331–332
average (mean) value of a function, 583
  basic technique, 584–586
  geometric interpretation of, 374
  Mean Value Theorem for Integrals and, 586
  in probability theory, 587–588
  random variables and, 586–587
  summarized, 617
average rate of change, 55
axes, 11
axis of symmetry, 15

B

backward difference, 45
backward difference quotient, 180
Bailey, David, 661
bandwidth, 68
Barrow, Isaac, 415, 558
baseband signal, 68
Baye's law, 44
Beardon, Alan, 265
Beaugrand, Jean de, 343
Bell, E. T., 558
Berkeley, George, 147, 247, 248, 709
Bernoulli, Daniel, 148
Bernoulli, Jakob, 507

Bernoulli, Johann, 708
Bernoulli differential equation, 427
Bernoulli's inequality, 257
Bessel horn, 430
Bessel's function, 704
best linear approximation, 236–238, 239, 244
beta densities, 545
binomial coefficients, 221, 700
binomial series, 700–701, 707–708
biology applications, 162, 199, 213, 215, 233, 275, 429–430, 442, 461–462, 472–473, 474, 476, 527–529
Bode's Law, 558–559
Bohr radius, 512
Boltzmann's constant, 214
Bolzano, Bernard, 148
bond valuation, 311–312
Boolean functions, 43–44
Borwein, Peter, 661
bounded intervals, 5
bounded sequences, 131–133
Boyle, Robert, 73
Boyle's law, 474
Briggs, Henry, 506, 507, 707
British unit system, 597
budget line, 32, 184
business. *See* economics, business, and finance applications

C

calculators
  computer algebra system capabilities on, 9
  inverse functions evaluated on, 211
  limitations of, correction for, 293–296
  loss of significance and, 8
  See also computers
calculus genesis and development
  analytic geometry, 72
  the catenary, 507
  celestial mechanics, 558–559
  completeness property, 73
  correct presentation of foundations, 148–149
  differentiation, 247–248
  exponential functions, 506–507
  extrema, 342
  Fermat's Principle of Least Time, 343–344
  infinite process, 72, 415
  infinite series, 660–661
  infinitesimals, 147–148
  inflection points, 342–343
  integral calculus, 415–417, 557–558
  least upper bound property, 148
  logarithms, 505–506
  Mean Value Theorem, 344
  Method of Exhaustion, 414–415
  Method of Partial Fractions, 558
  Newton-Raphson Method, 344–345
  optics controversy, 343–344
  power series, 707–709
  real numbers, 71–72, 73
  Rolle's Theorem, 344
  scientific revolution, 72–73
  secant integral, 557–558
  tangent problem, 245–247
  Theory of Indivisibles, 415
calculus of variations, 620
Cantor, Georg, 71–72
capital gain or loss, 312
Caratheodory, Constantine, 248
cardiac output, 406
carrier signal, 68
carrying capacity, 522
Cartesian equation, 12
Cartesian plane, 11–12
  displacement on, 12
  regions in, 18–19
  symbol for, 11
Cascades, Method of, 344
Catalan number, 43
catastrophic cancellation, 8
catenary, 496–497
cattle problem, 71
Cauchy, Augustin-Louis, 148
Cauchy's Mean Value Theorem, 305
Cavalieri, Bonaventura, 344, 415
center of mass, 590–591, 593–594, 617
central difference quotient, 180
Certaine Errors of Navigation Corrected (Wright), 557
cgs system, 597
Chain Rule, 193–196
  evaluation issues, 195, 210
  historical perspectives, 416
  multiple compositions and, 197–198
  powers and, 196
  proof of, 248
  Substitution Method and, 389–390
  summarized, 243
change of variable. *See* Substitution Method
Chapman-Jouguet points, 193
Chapman-Richards function, 461
characteristic triangle, 620
chemistry applications, 83, 120, 199, 226–228, 274, 276, 314, 325, 425, 428, 449, 474–475, 629
Chevilliet's form of the error in Simpson's Rule, 408
Cicero, 619
circle(s)
  standard form of equation for, 13, 15
  tangent lines and, 158, 222
circumference, defined, 76
Clerselier, Claude, 343, 344
closed intervals, 5
closed regions, 18–19
closeness, degree of. *See* degree of accuracy
coefficient of inequality, 403
collapsing (telescoping) series, 624–625
collapsing (telescoping) sum, 355
combining functions, 45
  arithmetic operations, 46
  composition. *See* composition
  linear combinations, 177
  parameterization, 51–53
  polynomial functions, 46–47
  summarized, 70
  translations, 49–50
commodity space, 32
Comparison Test for Convergence, 642–646
Comparison Test for Divergence, 643–645
Comparison Theorem, 608
completeness property, 73
  Extreme Value Theorem, 105–106, 132
  Intermediate Value Theorem, 105–106, 132
  monotone convergence property and, 132
  nested interval property, 73
Completing the Square, Method of, 14–15
composite integers, 37
composition
  continuity and, 105, 128–129
  differentiating. *See* Chain Rule
  of functions, 47–49
compound interest, 136–139
computers
  algebra systems on, 9
  Newton-Raphson Method on, 310–311, 312
  partial fractions on, 528
  right endpoint approximation on, 357
  See also calculators
concavity
  definition of, 277
  increasing/decreasing functions and, 278
  inflection points and, 279–280
  maxima/minima and, 281
  parabolas and, 17, 32
  Second Derivative Test for, 278
  Simpson's Rule and, 404
  slope and, 250, 277
  summarized, 339
  up or down, 32, 277–278
  zero slope and, 250
conditional convergence, 651
cone, volume of, 41, 562–563
Conics (Apollonius), 245
conic sections, Cartesian equations of, 15–17
constant functions
  differentiation of, 175
  Mean Value Theorem and, 255
constant relative growth rate, 467
Constant Rule, 175, 182
constants
  Boltzmann's constant, 214
  decay, 139, 465
  definite integrals and arbitrary, 379
  differential equations and, 421
  Euler-Mascheroni constant, 640
  growth, 139, 464, 467
  Hubble's constant, 40
  of integration, 316, 317
  integration of, 369
  linear combinations, derivative of, 177
  product of function plus, derivative of, 176
  of proportionality, 42
  radioactive, 465
  See also e; pi
constructive definitions, 38
consumer equilibrium, 184
continued fraction expansion, 130

continuity
  analogous description of, 98
  on closed interval, 105–106
  composition and, 105, 128–129
  continuous extensions, 101–102
  definition of, 98–99
  differentiability and, 167–169, 208–209
  equivalent formulation of, 99–101
  Extreme Value Theorem and, 108
  graphing and, 102, 106
  historical perspectives, 148
  Horizontal Chord Theorem and, 111
  Intermediate Value Theorem and, 106–108
  inverse functions and, 208–209
  jump discontinuity, 103–104
  one-sided, 103–104
  as predictability, 167
  properties of real numbers and, 105
  recursive sequences and, 133
  theorems about, 104–108
  uniform, 361
continuous analogues, 585
continuous compounding, 136–139
continuous extensions, 101–102
convergence, 122–123
  definition of, 122–123
  fixed point of *f*, 133
  improper integrals and, 603, 605, 606, 608, 609
  of infinite series. *See* convergence of infinite series
  limit rules for, 125–126
  monotonic convergence property, 131–133
  numerical investigation of, 123–125
  Riemann sums and, 360, 363
  of sequences, 122–123, 131–133
  tail of a sequence and, 123
  See also divergence
convergence of infinite series
  absolute convergence, 650–651
  Alternating Series Test, 647–650
  arithmetic operations and, 624
  comparison test for, 642–646
  conditional convergence, 651
  definition of, 623
  Divergence Test, 630–631
  geometric series, 626–627
  harmonic series, 625–626, 631, 649
  Integral Test for, 635–636, 709
  nonnegative terms and, 631–632
  power series, 664–670, 698, 699, 701
  p-series, 636–638
  Ratio Test, 653–655
  Root Test, 655–656
  summarized, 658–659
  tail end of a series and, 632–633
  Taylor expansion and, 689–690, 693–695
  Taylor series and, 698, 699
  telescoping series, 624–625
convergent sequences, 122–123, 131–133
corners, 160, 168
cosecant function
  antiderivatives of, 439–440
  defined, 61
  derivative of, 187
  graphs of, 61
  inverse, 485
  as periodic, 291
  values of, 63
cosecant function, hyperbolic. *See* hyperbolic functions
cosine function
  definition of, 59
  derivatives of, 170–171
  graphs of, 61
  identities of, 62–64
  integration of even powers of, 533, 534
  integration of odd powers of, 532–534
  integration by parts and, 511, 514–515
  integration by reduction formulas, 530–532
  integration of squares of, 529–530
  inverse (arccosine), 479–480, 486
  limit theorems for, 93–95
  as periodic, 291
  power series expansion of, 700
  properties of, 60
  superposition of, 67–68
  triangle definition, 60–61
  values of, 63
cosine function, hyperbolic. *See* hyperbolic functions
cotangent function
  defined, 61
  derivative of, 187
  graphs of, 61
  identities of, 62
  inverse (arccot), 484, 485–486
  as periodic, 291
  values of, 63
cotangent function, hyperbolic. *See* hyperbolic functions
Cotes, Roger, 507
countable, 71
coupon rate, 311
Cours d'Analyse (Cauchy), 148
critical points, 261–262, 266, 279
Cureau de la Chambre, Marin, 343
curvature, 282–284
curve(s), 12
  arc length. *See* arc length
  area between two, 395–397
  catenary, 496–497
  cycloid, 416
  demand, 214
  Folium of Descartes, 232, 246, 247
  Hugoniot, 192–193
  indifference, 184, 285
  Lorenz, 403
  loxodrome, 500
  normal lines to, 160, 241, 245–246
  parametric. *See* parametric curves
  Phillip's, 163
  solution, 420
  tangent lines of, 166–167
Cusa, Nicholas, 68
cusps, 161, 168
cycling, 310
cycloid, 416
cylindrical shells, method of, 570–572, 616

D

Darboux, Jean Gaston, 265
Darboux's Theorem, 265
data, functions from. *See* mathematical models
Dawson's integral, 384, 430, 449
Debye function, 424
decay, modeling, 139, 206–208, 465–467
decay constant, 139, 465
decimal expansions
  antiderivatives of, 439–440
  floating point decimal representation, 8
  integer part of a real number, 43
  midpoint approximation compared to, 402
  nonterminating, 3
  nonterminating, nonrepeating, 3
  terminating, 2
decreasing functions. *See* increasing and decreasing functions
decreasing sequences, 131
Dedekind, Julius Richard, 73
definite integrals, 361–363
  distinguished from indefinite integrals, 362, 379
  existence of, theorem for, 362–363
  improper. *See* improper integrals
  Integration by Parts Formula for, 510
  limits of integration, 362, 371–372, 389
  Mean Value Theorem for, 373–374
  natural logarithms, 437–438
  notation for, 361–362
  order properties of, 372–373
  properties of, 369, 412
  summarized, 411
  See also definite integrals, applications of; integration techniques; Riemann sums
definite integrals, applications of
  arc length. *See* arc length
  area. *See* area
  average value. *See* average (mean) value of a function
  center of mass, 590–591, 593–594, 617
  moments, 591–593, 596
  volume. *See* volume
  water pressure, 214, 233, 600–603
  work. *See* work
degree of accuracy
  coinciding, 171
  definition of limit and, 87–88
  graphical methods of finding, 81–82
  Method of Increments and, 235
  Midpoint Rule and, 401–402, 403
  Newton-Raphson Method and, 306, 308–310
  refinement of, 88
  rule of thumb in calculating, 133
  Simpson's Rule and, 405–406, 408
  specified degrees, 80–81
  Taylor expansion and, 694–695
  Trapezoidal Rule and, 403
  See also error
degree days, 585
degree of a polynomial, 47
demand curve, 214

demand equation, 326
demand function, 326
Democritus, 414
demodulation, 68
De Moivre, Abraham, 507, 708
De Morgan, Augustus, 708
denumerable, 71
dependent variable, 39
derivatives
  applications. *See* acceleration; differential equations; maxima and minima; velocity
  calculating. *See* differentiation
  definition of, 165–166
  domain of, 168–169
  as function, 167, 215–216
  higher. *See* higher derivatives
  increasing or decreasing functions determined by, 259–260
  maxima and minima and. *See* maxima and minima
  notation for, 165, 166–167, 242
Desargues, Gérard, 246
Descartes, René
  analytic geometry development and, 72
  and arc length problem, 619–620
  life of, 72
  and optics controversy, 343, 344
  and scientific revolution, 73
  and tangent problem, 245–247
Description of the Wonderful Canon of Logarithms, The (Napier), 505
Dichotomy (Zeno), 72
difference quotients
  approximation of second derivatives and, 217–219
  numeric differentiation and, 180–182
Difference Rule, 182
differences of functions. *See* sums (or differences) of functions
differentiability
  continuity and, 167–169
  defined, 165, 167
  graphs determining, 169–170
differential approximation, 236–238, 239, 244
differential equations
  Bernoulli, 427
  coupled, 499
  defined, 420
  first order, 420
  graphs of, 420
  homogeneous of degree 0, 427
  implicit solutions, 423
  initial value problem (IVP) and, 421
  linear, 427
  logistic, 521
  Method of Separation of Variables for, 421–423
  power series and, 678–680, 699
  separable, 421–422
  solving, 420–426
  See also differential equations, applications of
differential equations, applications of
  allometric growth, 429–430
  animal growth, 429
  athletics, 476, 477
  blood serum surface tension, 429
  brass instruments, 430
  chemical kinetics, 425, 428
  disease propagation, 522–523
  drainage (Torricelli's Law), 424
  electromagnetic radiation in a galaxy, 471
  exponential growth, 463–465
  hanging cable (the catenary), 496–497
  heat transfer (Newton's Law), 467–469
  logistic growth, 521
  motion in a resisting medium, 469–470, 475, 476, 477, 483–484, 494–495, 499, 500
  motion under diminishing gravitational force, 425–426
  pharmacology, 470, 474
  predator-prey ecosystems, 429, 430
  psychology, 474
  pursuit (the tactrix), 429, 499
  radiative heat transfer, 553
  radioactive decay, 465–467
  radiocarbon dating, 473
  solute concentration, 470, 474–475
  stellar physics, 428
  tree growth, 461
  warfare, 499
  water evaporation, 461
differentials, 166, 238–239
differential triangle, 620
differentiation
  of constant function, 175
  of exponential functions, 208, 455
  of expressions, 171–172
  of higher derivatives. *See* higher derivatives
  of hyperbolic functions, 490–492
  implicit. *See* Implicit Differentiation Method
  of inverse functions, 208–212
  of inverse hyperbolic functions, 493–494
  of inverse trigonometric functions, 479–481, 484–485
  logarithmic method of, 457–459
  of logarithms with arbitrary bases, 455
  natural exponentials and, 188–190, 445
  of natural logarithms, 211–212, 436–437
  Newton-Raphson Method. *See* Newton-Raphson Method
  numeric, 180–182
  of power series, 673–675
  of powers of *x*, 186–187
  as process of calculation, 165
  of products of functions, 177–178, 219
  reversal of. *See* antidifferentiation
  rules for, 175–182, 186–187, 239
  of sums (or differences) of functions, 176–177
  of trigonometric functions, 170–171, 187–188, 190
dimensional analysis, 198
Dioptrique, La (Descartes), 343
discontinuity, point of, 99
  See also continuity
discrete dynamical systems, 257
disks, method of, 562–566, 569
displacement, 12
distance
  definition of, 4
  formula for, 12–13
distance formula, 12–13
divergence
  defined, 123
  improper integrals and, 603, 605, 606, 608, 609
  of infinite series, 623–624, 643–645
  Newton-Raphson Method and, 310
  See also convergence
Divergence Test, 630–631
divergent sequences, 123–125
domain, 33
  consisting of real numbers, 33–34
  of derived function, 168–169
  of inverse functions, 201
  restriction of, 205, 477–478, 479, 486, 492
Doppler effect, 221
Double Angle Formulas, trigonometric identities, 6
doubling time, 140, 464, 476

E

e, 137–138
  and exponential function, 446–447
  historical perspectives, 507
  as limit of a continuous variable, 138
  notation for, 137
  See also natural exponential functions; natural logarithms
economic lot size, 329–330
economic order quantity, 337
economics, business, and finance
    applications, 110, 136–139, 141, 143, 144, 162, 163, 173–174, 184–185, 192, 214, 220, 232, 236, 240, 262, 264, 273, 285–286, 311–312, 315, 326–334, 335–338, 460, 611–613, 615, 629
economies of scale, 174
effective yield, 312
Einstein's Relativistic Mass Law, 120
elasticity of demand, 173, 184, 333–334
elements of a set, 3
ellipse(s), 16
elliptic integral, 558
empty set, 5
endpoint(s), 5
  maxima/minima and, 265–266, 269–271
  one or none, 6
  one-sided limits and. *See* one-sided limits
engineering applications, 82, 86, 143, 200, 217, 232, 233, 235, 266–267, 269–270, 272, 273, 274, 276, 627–628
Epitome Astronomiae Copernicanae (Kepler), 416
equilibrium point, 257
equilibrium position, 599

error
absolute, 7, 28, 29
area and, 348
discontinuity and, 99
relative, 7
roundoff, 8
total, 181
truncation, 7, 181
See also degree of accuracy
error function, 516, 680–681, 704
error term of Taylor expansion, 687–688, 689–690, 693–695, 699
Euclid, 38, 708
Eudoxus, 414
Euler, Leonard, 148, 507, 620, 661, 708
Euler-Mascheroni constant, 640
Euler's Method, 240, 241
even functions. *See* odd and even functions
Exhaustion, Method of, 414–415
existence theorems, 107–108
expectation of a real number, 588
exponential functions
as continuous, 135
derivatives of, 208
e written as base of, 208
graphs of, 135
irrational exponents, 134–135, 450–451
modeling with, 139–140, 463–471
natural. *See* natural exponential functions
See also exponential functions, applications of; exponential functions with arbitrary bases
exponential functions, applications of
compound interest, 136–139
decay, 139, 206–208, 465–467
growth, 139–140, 463–465, 467
linear drag law, 469–470
Newton's Laws of Heating and Cooling, 467–469
exponential functions with arbitrary bases
definition of, 450–451
differentiation of, 455–456
integration of, 455–456
magnitude of, with large arguments, 456–457
properties of, 452–455
exponents, irrational, 134–135, 450–451
expressions, derivatives of, 171–172
Extreme Value Theorem, 108, 145
extremum (extrema), 108, 251
See also maxima and minima

F

Faà di Bruno, Francesco, 222
face value, 311
factorials, simplifying ratios of, 667
falling bodies, applications, 142, 163, 164, 199–200, 214, 232, 254, 324–325, 425–426, 428, 472, 476, 477–478, 499, 500, 629
Fermat, Pierre de
analytic geometry development by, 72
arc length problem and, 620
and extrema, 342
and inflection points, 342–343
and integral calculus, 415
and Law of Refraction, 343–344
life of, 342
tangent problem and, 246–247
Fermat's Principle of Least Time, 343–344
Fermat's Theorem, 252–253
Feynman, Richard, 344
Fibonacci sequence, defined, 38
Fick's law, 214, 470
finance. *See* economics, business, and finance applications
first coordinate, 11
First Derivative Test, 260–262, 282
first moments, 596
fixed costs, 326
fixed point, 133
floating point decimal representation, 8
fluents, 247
fluxions, 147, 247
Folium of Descartes, 232, 246, 247
foot-pound (ft-lb), 597
force
unit of measure for, 597
work and. *See* work
force of mortality, 174, 429, 461–462
forward difference quotient, 180
Fourier, Joseph, 417
Fourier cosine-polynomial approximation, 517
Fourier polynomial approximation, 517
Fourier sine-polynomial approximation, 517
frequency, 67
Fresnel cosine function, 516
Fresnel function, 424
Fresnel sine function, 516
Fresnel sine integral, 384
frustum, surface area of, 578–579
functions, 33
absolute value. *See* absolute value
affine, 54
approximation of. *See* Increments Method
area, 377, 410
argument of, 33
average value of. *See* average (mean) value of a function
average value of, 374
Bessel's, 704
Boolean, 43–44
combining. *See* combining functions
constant, 175, 255
continuous. *See* continuity
derivatives as, 167, 215–216
domain of. *See* domain
error, 516, 680–681, 704
even. *See* odd and even functions
exponential. *See* exponential functions
gamma, 615
greatest integer, 43
gudermanian, 500
Heaviside, 44, 85
hyperbolic. *See* hyperbolic functions
image of, 33
increasing. *See* increasing and decreasing functions
inside and outside, 194–195, 197
inverse. *See* inverse functions
Lambert's *W* function, 315, 430
logarithms. *See* natural logarithms
logistic, 521–523
as mathematical model, generally, 38–41
multivariable, 41–42
odd. *See* odd and even functions
periodic, 290–292
period of, 60, 67
piecewise-defined (multicase), 34–35, 43–44
polynomial. *See* polynomial functions
power. *See* power functions
products of. *See* products of functions
range of, 33
rational. *See* rational functions
real-valued, 33
of a real variable, 33–34
restriction, 34
roots of. *See* roots (zeroes) of a function
rule of, 33
signum (sign), 35–37
sums of. *See* sums (or differences) of functions
trigonometric. *See* trigonometric functions
Fundamental Theorem of Algebra, 47
history of, 558
in Partial Fractions Method, 546, 549
Fundamental Theorem of Calculus
definite integrals evaluated with, 364–365, 379–382
statement of, 378
differentiation/integration as inverse operations, 378–379
geometric interpretation of, 377–378
historical perspectives, 417
minus sign as pitfall in, 365
notation for, 364
proof of, 363–364

G

Galileo, 73, 416
gamma function, 615
Gauss, Carl Friedrich, 558–559
Gaussian random variable, 590, 613
generation time, 476
See also doubling time
Geometria Indivisibilibus Continuorum (Cavalieri), 344
geometric series, 126–128
convergence of, 626–627
definition of, 127
sigma notation, 350
Géométrie, La (Descartes), 72, 245, 343, 416, 620
Géostatique (Beaugrand), 343
Gini coefficient, 403
global (absolute) maximum or minimum, defined, 251
See also maxima and minima
glottochronology, 473
Gompertz, Benjamin, 429, 461
Gompertz function line, 462
Gompertz growth equation, 462
Gompertz's Law of Mortality, 461

graphing, 11–12
  continuity and, 102, 106
  interpolation, 52
  planar, notation for, 11–12
  sketching, steps for, 286–290
  symmetry and, 50–51
graphing calculators. *See* calculators
graphs, 35–37
  concavity of. *See* concavity
  curvature of, 282–284
  definition of, 35
  degrees of accuracy using, 81–82
  differentiability determined from, 169–170
  of solutions of differential equations, 420
  of exponential functions, 135
  of hyperbolic functions, 488–489, 490
  of inverse functions, 206
  of inverse trigonometric functions, 478, 479, 481, 484–485
  maxima and minima of. *See* maxima and minima
  of natural exponential functions, 443
  of natural logarithms, 207, 431, 435
  scatter plots, 38–39, 40–41
  of sequences, 122
  slope of tangent line as equal to instantaneous rate of change, 159
  smoothness, tangent lines as measure of, 161
  symmetry and. *See* symmetry
  time-series, 27
  of trigonometric functions, 61–62
  vertical line test of, 35
Graunt, John, 473–474
greatest integer function, 43
Gregory, James, 557–558, 707
Gregory-Leibniz series, 707
growth constant (k), 139
  decay constant contrasted to, 467
  doubling time and, 464
growth modeling
  exponential, 139–140, 463–465, 467
  logistic, 142, 521–523
gudermanian function, 500

H

Hail Stone Sequence, 142
Half-Angle Formulas, trigonometric identities, 6
half closed intervals, 5
half-life, 208, 466
half open intervals, 5
Halley, Edmond, 147, 557–558
Hamilton, Sir William Rowan, 248
harmonic series, 625–626, 631, 649, 661
Harriot, Thomas, 557, 558
Heaviside, Oliver, 523
Heaviside function, 44, 85
Heaviside's method, 523–524
Heiberg, Johan Ludvig, 415
Heinz, E., 461
Hermite, Charles, 507
Heuraet, Hendrik van, 620
higher derivatives
  acceleration and, 216–217
  approximation of, 217–219
  definition of, 216
  as functions, 215–216
  Leibniz's Rule for computing, 219
  Method of Implicit Differentiation and, 229–230
  notation for, 216, 217
  physical significance of, 217
  summarized, 243–244
higher moments, 596
Hilbert, David, 149
Hill-Keller theory of sprinting, 476
Hipparchus, 63
Hobbes, Thomas, 620
homogeneous of degree 0, 427
Hooke, Robert, 73
Hooke's law, 599
horizontal asymptotes, 116–118
Horizontal Chord Theorem, 111
horizontal line(s), slope as zero for, 22, 250
Horizontal Line Test, 201, 205
horizontal translations, 49
Hubble, Edwin, 38–40
Hubble's constant, 40
Hudde, Johann, 707
Hugoniot curve, 192–193
Hull's learning model, 474
Huntington, E. V., 275
Huxley allometry equation, 430
Huygens, Christiaan, 73, 344, 506–507, 620
hydrostatic pressure applications, 214, 233, 600–603
hyperbola(s)
  Cartesian equation of, 16
  definition of, 16
hyperbolic functions
  antiderivatives of, 494
  catenary and, 496–497
  definitions of, 488–489, 490
  derivatives of, 490–492
  graphs of, 488–489, 490
  inverse, 492–494
  quadratic drag law and, 494–495
  standard trigonometric functions compared to, 490
  summarized, 503–504

I

identities of rational functions, 519
identity, trigonometric, 62–64
image of a function, 33
Implicit Differentiation Method, 222–226
  compared to Method of Separation of Variables, 423
  form of equation and, 225
  higher derivatives and, 229–230
  inapplicability of, 224–225
  related rates and, 226–229
  summarized, 244
  value of y specification and, 226
improper integrals
  area and, 606–607
  finance applications, 611–612
  geometric applications, 610–611
  with infinite integrands, 603–606
  on an infinite interval, 609–610
  random variables and, 612–613
  summarized, 618
  volume and, 607
Improved Euler's Method, 241
Impulse Injection Method, 406
income stream, 611
increasing and decreasing functions
  concavity and, 278
  definition of, 258
  derivative and indentification of, 259–260
  and invertibility, 206–207
  as one-to-one, 206
  summarized, 339
increasing sequences, 131–133
increment, maximum value and, 44–45
incremental costs, 326
Increments Method, 233–235
  differentials, 238–239
  in economics, 236
  Euler's Method, 240
  Improved Euler's Method, 241
  linearization, 236–238
indefinite integrals
  constant of integration, 316, 317
  distinguished from definite integrals, 362, 379
  of exponentials with arbitrary bases, 456
  hyperbolic functions, 494
  of inverse trigonometric functions, 481–482
  of logarithms with arbitrary bases, 456
  notation for, 316
  of power series, 674–675
  of powers of x, 317
  of trigonometric functions, 438–440
  two solutions for, 390–391
  See also antidifferentiation; integration techniques
independent variable, 39
indeterminate form
  defined, 298
  l'Hôpital's Rule for, 298–301, 303
  natural logarithm and, 301–303
index of summation, 349
indifference curve, 184, 285
inductively defined sequences, 38
inelastic demand, 333
inequalities
  Arithmetic-Geometric Mean, 143
  Bernoulli's, 257
  intervals of, 5, 6
  preserved under multiplication, 5
  reversed under multiplication, 5
  as set, sketching, 3–4
  Triangle, 6–7, 373
  trigonometric, 68
infinite integrands. *See* improper integrals
infinite process, 72, 415
infinite sequences, 37
infinite series
  alternating, 647–650
  approximating value of, 638–639
  as concept, 621

convergence of. *See* convergence of infinite series
definition of, 622–623
distinguished from sequences, 624
divergence of, defined, 623
estimation of, 636
geometric. *See* geometric series
harmonic, 625–626, 631, 649, 661
historical perspectives, 660–661
as limit, 624
with nonnegative terms, 631–632
notation for, 622
partial sums of, defined, 622–623
power. *See* power series
properties of, 627–628
p-series, 636–638, 661
summarized, 658–659
telescoping (collapsing), 624–625
infinitesimals, 147–148, 238, 247–248, 344
infinite sums, clarification of, 128
infinite-valued limits
definition of, 112–113
vertical asymptotes and, 114–115
infinity, limits at, 115–116
inflection points, 279–280
distinguished from critical points, 279
historical perspectives, 342–343
initialization, 38
initial value problems (IVP), differential equations and, 421
inside function, 194–195, 197
instantaneous rate of change
definition of, 156
pattern of, for degree of function, 157, 160
slope of tangent line equivalent to, 159
specific uses of, 156
instantaneous velocity
calculation of, 154–155
defined intuitively, 76
definition of, 154
negative values for, meaning of, 156
as particular example of instantaneous rate of change, 156
integers, 2
Catalan numbers, 43
composite, 37
greatest integer in *x*, 43
least positive, 71
near, 507
prime, 37–38
sum of consecutive positive, 350
integral calculus, historical perspectives, 415–417
integrals, improper. *See* improper integrals
integral sign, 362
Integral Test for convergence, 635–636, 709
integrand, 362
integration, as term, 362
integration by parts
choice of variable and, 511, 513
compact form of Formula, 510
exponential functions and, 512–513, 514
Formula for, 510
historical perspectives, 557
indirect evaluation with, 514
logarithms and, 511–512
reduction formulas and, 530
repeated applications of, 512–513
summarized, 554
trigonometric functions and, 511, 514–515
Integration by Parts Formula, 510
integration techniques
approximation, 373, 399–407, 413
Fundamental Theorem of Calculus and, 364–365, 379–382
Inverse (or Indirect) Substitution Method, 541–545
Midpoint Rule, 400–402, 403, 404
partial fractions. *See* Partial Fractions Method
by parts. *See* integration by parts
for powers of trigonometric functions, 529–535
for products, 510–515
for quadratics not under a radical, 542–544
for quadratics under a radical, 541–542
for rational functions, 518–520, 523–526, 542–544, 546–552
reversing the roles of the axes, 396–397
rules of, 369–374
Simpson's Rule, 404–407, 408
Substitution Method, 386–391
Trapezoidal Rule, 402–404
trigonometric substitution, 537–544
See also definite integrals; indefinite integrals
intercept form of a line, 26
interest, compound, 136–139
Intermediate Value Theorem, 106–108
historical perspectives, 148
interpolation, 52
intervals, 5–6
bounded, 5
of convergence, 664–669, 705
endpoints, 5, 6
infinite, improper integrals and, 609–610
midpoints, 5
nested, 73, 105
notation for, 5, 6
unbounded, 6
intervals of convergence, 664–669, 705
Inverse Function Derivative Rule, 210
inverse functions
definition of, 201
differentiation of, 208–211
distinguished from reciprocals, 203
graphs of, 206
Horizontal Line Test and, 201, 205
hyperbolic, 492–494
increasing functions and, 206–207
in integrals of quadratic functions, 538
notation for, 201, 203
recognition of, 204–205
rule for finding, 202–204
summarized, 243
trigonometric. *See* inverse trigonometric functions
Inverse (or Indirect) Substitution Method
general quadratic expressions under a radical, 541–542
quadratic expressions not under a radical, 542–544
summarized, 555
technique, 538–540
inverse trigonometric functions
defined, 478–479, 480–481, 484–485
derivatives of, 555
distinguished from reciprocal, 486
domain restriction and, 477–478, 479, 486
integrals involving, 481–482
notation and, 486
optimization example involving, 485–486
power series and, 677–678
quadratic drag law and, 483–484
summarized, 502–503
invertible, definition of, 201
irrational numbers, 2
as exponents, 134–135
historical perspectives, 71–72
monotone convergence property of, 131–132
nonterminating, nonrepeating decimals as, 3
irreducible polynomials, 47, 548

J

Janoschek growth function, 429
jerk, 217
jointly proportional variables, 42
joule (J), 597
jump discontinuity, 103–104

K

Keller, J. B., 476
Kepler, Johannes
on extrema, 44–45, 342
integral calculus and, 416, 557
life of, 619
and logarithms, 506
and planetary orbits, 73
and solids of revolution, 619
Kepler's equation, 309–310
Kepler's Third Law, 73
Klee, V. L., 275
Kremer, Gerhard (Gerardus Mercator), 557

L

La Géométrie, (Descartes), 72, 245, 343, 416, 620
Lagrange, Joseph-Louis, 344, 708
Lambert, Johann Heinrich, 71
Lambert's *W* function, 315, 430
Lanchester, F. W., 499
lattice point, 32
Law of Charles and Gay-Lussac, 85
law of demand, 214
Law of Exponents, 452–453
Law of Radioactive Decay, 465
Law of Reflection, 343
Law of Refraction, historical perspectives, 343–344
leading coefficient, 47
leading term, 11

least positive integer solution, 71
Least Squares Method, 27–29
  historical perspectives, 559
  scatter plots and, 40–41
least upper bound property, 148
Lecomte Du Nouy, Pierre, 429
Legendre, Adrien-Marie, 71
Legendre, A. M., 221
Legendre polynomials, 221
Leibniz, Gottfried Wilhelm von
  and the catenary, 507
  and infinite series, 661
  and infinitesimals, 247–248
  and integration by parts, 557, 620
  and logarithms, 507
  and Method of Partial Fractions, 558
  notation development by, 166, 416–417
Leibniz notation, 166–167
  Chain Rule and, 194, 197
  historical perspectives, 416–417
  notation of differentials and, 239
Leibniz series, 707
Leibniz's Rule for Second Derivatives, 219
Leibniz's series, 661
Letters of Descartes (Clerselier), 343, 344
lever law, 591
l'Hôpital's Rule, 298–301, 303
Libby, Willard, 473
Limit Comparison Test, 645–646
limit(s)
  algebraic manipulations to evaluate, 79–80
  algebraic properties of, 90
  applications of, 82–83
  basic theorems of, 90–91
  defined informally, 77
  definition of ($\varepsilon - \delta$), 87–88
  degrees of accuracy, 80–82
  geometrical explanation of, 75–76
  geometric series, 126–128
  historical perspectives, 147–149
  improper integrals, 609
  indeterminate. *See* indeterminate form
  infinite series as, 624
  infinite-valued, 112–113, 114–115
  at infinity, 115–116
  of integration. *See* limits of integration
  nonexistence of, 92, 145
  one-sided. *See* one-sided limits
  Pinching Theorem. *See* Pinching Theorem
  of recursive sequences, 133
  of sequences. *See* sequences
  summarized, 144
  trigonometric, 93–95
  uniqueness of, 90, 145
limits at infinity, 115–116
limits of integration, 362
  reversal of, 371–372, 389
Limit Uniqueness Theorem, 90, 145
Lindemann, Carl Louis Ferdinand von, 72
linear building blocks. *See* Partial Fractions Method
linear combinations, 177
linear differential equations, 427
linear drag law, 469–470
linear equations
  approximation of, 236–238, 239, 244
  basic rule for obtaining, 24
  intercept form, 26
  point-slope form, 25–26
  slope-intercept form, 26
  two-point form, 31
linear factors. *See* Partial Fractions Method
linearization of f at c, 236–238
linear polynomials, 47
line of best fit, 27
  Least Squares Method, 27–29
line(s)
  equations for. *See* linear equations
  geometrical descriptions of, 23
  horizontal, slope as zero for, 22, 250
  least squares (regression) line, 27–29
  parallel, slope of, 23
  perpendicular. *See* perpendicular lines
  secant, 158
  tangent. *See* tangent line(s)
  vertical. *See* vertical line(s)
Liouville, Joseph, 71
localized information, 55
local maximum or minimum, defined, 250–251
  See also maxima and minima
locus, 12
logarithmic differentiation, 457–459
logarithmic mean, 442
logarithms, historical perspectives, 505–506
logarithms, natural. *See* natural logarithms
logarithms with arbitrary bases
  definition of, 451–452
  differentiation of, 455–456
  integration of, 455–456
  magnitude of, with large argument, 457
  properties of, 453–455
logistic growth equation, 521
logistic growth formula, 142
logistic growth function, 521–523
Lorenz curve, 403
Lorenz function, 403
loss of significance, 8, 181
Lotka, A. J., 429
Lotka-Volterra equation, 429
lower bound, 131
lower-order terms, 11
lower Riemann sum, 360
loxodrome, 500

M

Machin, John, 64
Maclaurin, Colin, 708–709
Maclaurin series, 697, 699–701, 707–709
Mādhavan, 660–661, 707
Maestlin, Michael, 506
Makeham, William, 429
Makeham's Law, 429
marginal cost, 156, 162, 236, 326
marginal product of labor, 264
marginal product of labor (MPL), 236
marginal profit, 236
marginal rate of substitution, 285
marginal revenue, 236
mass, center of, 590–591, 593–594, 617
mathematical models
  approximation and, 471
  exponential functions in, 139–140, 463–471
  functions as, generally, 38–41
  with trigonometry, 64–65
  validity of, checking, 561
mature, 311
maxima and minima, 250–252
  approximation of, 269
  closed bounded intervals and, 266–267
  concavity and, 281
  critical points and, 261–262, 266
  definition of, 108
  definitions of local, 250–251
  endpoints and, 265–266, 269–271
  Extreme Value Theorem, 108
  Fermat's Theorem and, 252–253
  First Derivative Test for, 260–262, 282
  historical perspectives, 342, 620
  horizontal tangents and, 250, 252, 254
  identification of, 258, 260–262, 280–282
  increments and, 44–45
  locating candidates for, 252–253
  Mean Value Theorem and, 254–255
  Rolle's Theorem and, 253–254
  Second Derivative Test for, 280–282
  slope at zero and, 250
  terminology and, 250–252
  vertex of a parabola and, 17
maximum profit principle, 328–329
maximum revenue principle, 334
maximum velocity (sprinting), 476
mean, 410
  logarithmic, 442
  of a random variable, 588
Mean Value Theorem, 254–255, 256
  historical perspectives, 344
  Taylor's Theorem as generalization of, 688
Mean Value Theorem for Integrals, 373–374, 586
medical applications, 142, 200, 214, 228–229, 274, 461, 468, 470, 474, 522–523, 629
Mengoli, Pietro, 661
Mercator, Gerardus (Gerhard Kremer), 557
Mercator, Nicolaus, 707
meridian, 500
Mersenne, Marin, 343
Method of Cascades, 344
Method of Completing the Square, 14–15
Method Concerning Mechanical Theorems, dedicated to Eratosthenes (Archimedes), 415
Method of Exhaustion, 414–415
Method of Implicit Differentiation. *See* Implicit Differentiation Method
Method of Increments. *See* Increments Method
Method of Increments, The (Taylor), 708
Method of Inverse Substitution. *See* Inverse (or Indirect) Substitution Method
Method of Least Squares. *See* Least Squares Method

Method of Partial Fractions. *See* Partial Fractions Method
Method of Separation of Variables, 422–423
Method of Substitution. *See* Substitution Method
Methodus Fluxionum (Newton), 247
midpoint approximation, 400
Midpoint Rule, 400–402, 403, 404
midpoint(s), 5
minima. *See* maxima and minima
minimum average cost principle, 332
mks system, 597
models. *See* mathematical models
modulation index, 68
moments, 591–593, 596
moments of fluxions, 247
monotone convergence property of real numbers, 131–133
monotonic sequences, 131
Moore's law, 143
Moulton, Forrest, 310
moving average, 384–385
multicase functions. *See* piecewise-defined (multicase) functions
multiplication. *See* products of functions
multiplicative inverse, 203
multiplier effect, 629
multivariable functions, 41–42

N

Napier, John, 432, 505–506
Napierian logarithm, 505–506
natural exponential functions
- definition of, 443
- derivative and, 188–190, 445
- *e* and, 446–447
- general techniques sought by, 557
- graphs of, 443
- integrals and, 445–446
- integration by parts and, 512–513, 514
- as inverse of natural logarithms, 442–443
- inverse of. *See* natural logarithms
- magnitude of, with large argument, 456–457
- notation for, 188
- power series expansion of, 699
- properties of, 443–444
- Taylor expansion and, 694
- *See also* exponential functions

natural logarithms
- definition of, 431
- differentiation and, 211–212, 436–437
- differentiation method using, 457–459
- graphs of, 207, 431, 435
- historical perspectives, 505–506
- for indeterminate form, 301–303
- integrals involving, 437–438
- integration by parts and, 511–521
- as inverse of exponential function, 207
- inverse of. *See* natural exponential functions
- magnitude of, with large argument, 457
- mean defined by, 442
- power series and, 675–677
- properties of, 207–208, 431–434
- summarized, 501
- Taylor expansion and, 694–695
- *See also* logarithms with arbitrary bases

natural numbers, 2
near integers, 507
negative reciprocals, slope of normal line as, 160
Neile, William, 620
nested intervals, 73, 105
Newton, Sir Isaac
- and Fundamental Theorem of Calculus, 416
- and infinite series, 661
- and infinitesimals, 147, 247, 248
- life of, 558, 708
- Maclaurin and, 708
- and Newton-Raphson Method, 344–345
- notation of, 167, 217
- and pi, 71
- and power series, 707–708
- and scientific revolution, 73

newton-meter, 597
newton (N), 597
Newton notation, 167
- for higher derivatives, 217

Newton-Raphson Method, 305–306
- accuracy of, 306, 308–310
- calculating with, 307–308
- computer implementation of, 310–311, 312
- economics application of, 311–312
- geometry of, 306–307
- historical perspectives, 344–345
- pitfalls of, 310
- variant for roots of higher multiplicity, 314

Newton's binomial series, 700–701, 707–708
Newton's Law of Gravitation, 425–426, 598
Newton's Law of Heating and Cooling, 142, 467–469, 471
Newton's Method. *See* Newton-Raphson Method
Nilakantha, 707
nonconservative property of resistive forces, 477
nondenumerable, 71
nonrepeating decimals, 3
nonterminating, nonrepeating decimals, 3
nonterminating decimals, 3
normal distribution, 410
normal (perpendicular) lines to curves, 160, 241, 245–246
normal probability density, 590
normal random variable, 590, 613
notation and symbols
- absolute value, 4
- antiderivatives, 315–319
- approximation ($\approx$), 234, 239
- arrow, 33
- average (mean) of a random variable, 588
- average value of a function, 584
- Cartesian plane ($\mathbb{R}^2$), 11
- composition of functions, 48
- degrees of accuracy, 80
- derivatives, 165, 166–167, 242
- differentials, 238–239
- divergent sequences, 124
- doubling time, 140
- *e*, 137
- element of a set ($\in$), 3
- empty set ($\emptyset$), 5
- ε and δ, origins of, 148
- exponentials with arbitrary bases, 450
- function, 33
- Fundamental Theorem of Calculus, 364
- greatest integer ($\lfloor x \rfloor$), 43
- growth (or decay) constant, 139
- higher derivatives, 216
- historical perspectives, 416–417
- indefinite integrals, 316
- infinite series, 622
- infinite value (∞), 6
- integers ($\mathbb{Z}$), 2
- inverse functions, 201, 203
- inverse trigonometric functions, 478, 486
- limits, 77
- logarithms with arbitrary bases, 451
- maximum value, 108
- minimum value, 108
- multivariable functions, 41–42
- natural exponential function, 188
- natural logarithms, 207
- natural numbers ($\mathbb{N}$), 2
- ordered pair, 11
- planar coordinates, 11–12
- positive integers ($\mathbb{Z}^+$), 2
- probability density function, 587
- radian measure, 58
- rational numbers ($\mathbb{Q}$), 2
- real numbers ($\mathbb{R}$), 2
- reciprocals, 203
- regions, graphing of, 18
- Riemann integrals, 361–362
- Riemann sums, 358
- sequences, 37
- sets, 3
- sigma, 349, 622
- summation, 349
- Taylor polynomials, 686

*n*th roots
- as continuous, 99
- Intermediate Value Theorem and, 107
- limit theorem of, 91

*n*th term, 37
number lines, real numbers represented on, 3
number systems
- integers. *See* integers
- natural numbers, 2
- rational. *See* rational numbers
- real numbers. *See* real numbers

numeric differentiation, 180–182

O

oblique asymptotes. *See* skew asymptotes
"October 1666 Tract on Fluxions, The" (Newton), 416
odd and even functions
- derivatives of, 173
- integration of trigonometric, 532–534
- symmetry of, 50–51

one-sided continuity, 103–104

one-sided limits, 79
  corners and, 160–161
  definition of left limit, 89
  definition of right limit, 89
  endpoint investigation and, 88
  infinite-valued, 112
  notation for, 79
  summarized, 144
  vertical asymptotes and, 114
one-to-one property
  definition of, 204
  increasing and decreasing functions and, 206
  inverse functions as, 204–205
onto property
  definition of, 204
  inverse functions as, 204–205
open intervals, 5
optics controversy, 343–344
ordered field, 73
ordered pair(s), 11–12
ordinate, 12
Oresme, Nicole, 72, 660
origin, 11
  symmetry with respect to, 297
outside function, 194–195, 197
overshoot, 310

P

palimpsest, 415
Pappus's formula, 619
parabola(s)
  Cartesian equation of, 16–17
  concave up, 17, 32
  definition of, 16
  vertex of, 17
parallel lines, slope of, 23
parameter, 51
parameterization, 51
parametric curves, 51–53
  definition of, 51
  eliminating the parameter, 53
parametric equations, 51
Partial Fractions Method, 517
  coefficient calculation, 549–552
  Heaviside's method, 523–524
  irreducibility, checking for, 548–549
  linear building blocks, 518
  for linear factors, 518–520
  logistic growth and, 521–523
  quadratic building blocks, 547
  for quadratic factors, 546–548, 552
  for repeated linear factors, 524–525
  summarized, 526, 552, 556
partial sums, definition of, 622
Pascal, Blaise, 620
path, 53
Patterson, Clair, 450
Pearl, Raymond, 250
Pell equations, 71
period of a function, 60, 67
periodic functions
  definition of, 290–291
  sketching of, 291–292
  trigonometric functions as, 291
perpendicular lines
  to a curve (normal line), 160, 241, 245–246
  slope of, 22
phase shift, 67–68
Phillip's curve, 163
physics applications, 86, 120, 142, 143, 199, 220–221, 231, 257, 264, 274, 275, 276, 324, 325, 424–426, 428, 449, 450, 468, 471, 498–499, 516, 553, 575, 590–596, 596–603
pi (π)
  Arithmetic-Geometric Mean Method for, 45
  defined, 2
  digits, computation of, 71–72, 661
  historical perspectives, 71–72, 507, 660–661, 707
  as limit of a sequence, 131
  power series and, 707
  sequence defined for, 38
  Viète's algorithm for, 45
Piazzi, Giuseppe, 559
piecewise-defined (multicase) functions, 34–35
  Boolean functions and, 43–44
piecewise linear curve, arc length and, 576
Pinching Theorem, 92–93
  applied to Riemann sums, 360, 363
  for sequences, 125
Planck blackbody radiation, 449
Plouffe, Simon, 661
point(s)
  critical, 261–262, 266, 279
  endpoints, 5, 6, 88, 265–266, 269–271
  of inflection. *See* inflection points
  lattice, 32
  midpoints, 5
  ordered sets, 11–12
point-slope form of a line, 25–26
polygon(s), regular, 67
polynomial functions, 46–47
  as continuous, 99
  degree of, 47
  differentiation of, degree and, 187
  factoring of, 47
  fifth-degree and higher, solving explicitly, 305
  irreducible, 47
  leading coefficient of, 47
  limit theorem for, 91
  quotients of. *See* rational functions
polynomials
  leading term of, 11
  Legendre, 221
  lower-order terms of, 11
  Taylor. *See* Taylor polynomials
positive integers, 2, 71, 350
pound (lb), 597
power functions
  antidifferentiation of, 316–318
  Chain Rule applied to, 196
  as continuous, 135
Power Rule, 186–187, 190
  constant exponent required for, 458
  proof of, 451
power series
  about an arbitrary base point, 668–670
  arithmetic operations on, 672–673
  coefficients of, and function represented by, 680–681
  convergence of, 664–670, 698, 699, 701
  definition of, 664
  derivatives of, 673–675
  expansions of standard functions, 675–678, 698–700, 705–706
  explicit formula for, 675
  historical perspectives, 707–709
  integrals of, 674–675
  interval of convergence, 664–669, 705
  Maclaurin series, 697
  radius of convergence, 665–669, 705
  summarized, 705–706
  Taylor series, 697–702, 707–709
  uniqueness theorem for, 678–680
present value, 141, 611
prime numbers, 37–38, 442
principal, 136
principal associated angle, 60–61
Principia (Newton), 247
probability density function, 545, 587, 589, 590
probability theory, 586
  average values (μ) in, 587–588
  random variables, 586–587, 590, 612–613
product of labor, 236, 264
Product Rule, 178
  historical perspectives on, 247–248
  proof of, 248
  in reverse. *See* Integration by Parts Formula
products, scalar, derivative of constant times a function, 176–177, 182
products of functions
  antiderivatives of, 317
  combining operation, 46
  definite integral properties and, 369
  derivatives of, 177–178
  higher derivatives of, Leibniz's Rule for, 219
  instantaneous rate of change of, 157–158
  integration by parts for. *See* integration by parts
  limits of, 90–91
profit maximization, 326–329
proportionality
  defined, 42
  jointly proportional variables, 42
proportionality constant, 42
p-series, 636–638, 661
Ptolemy, 63
pumping fluids, 600–601
pyramid, volume of, 569–570
Pythagoras, 71

Q

quadrants, 11
quadratic drag law, 483–484, 494–495
quadratic equations, solving
  continued fraction expansion, 130
  Newton-Raphson Method, 305–312

quadratic expressions, integrals involving. *See* Inverse (or Indirect) Substitution Method
quadratic factors. *See* Partial Fractions Method
quotient of polynomials. *See* rational functions
Quotient Rule, 179–180
quotients, derivatives of, 177, 178–180
  See also indeterminate form

R

radian measure, 58–59
  limit theorem and, 94–95
radioactive decay, 144, 206–208, 465–466
radiocarbon dating, 144, 473
radius of convergence, 665–669, 705
Raleigh, Sir Walter, 557
Ramanujan, Srinivasa Aiyangar, 507, 661
random variables, 586–587, 590, 612–613
range, 33
  of real numbers, 33
  See also domain
Raphson, Joseph, 345
rational functions, 54
  arithmetic combination of, 46
  asymptote determination for, 117–118
  as continuous, 99
  integration of. *See* Partial Fractions Method
  limits at infinity of, 116
  limits of, 91
rationalist methodology, 72–73
rational numbers, 2
  nonterminating decimals as, 3
  as ordered field, 73
  symbol for, 2
  terminating decimals as, 2
  theory of exponents and sequences of, 134–135
Ratio Test, 653–655
real numbers, 2–3
  algebraic, 71
  completeness property of. *See* completeness property
  historical perspectives, 71–72, 73
  monotone convergence property of, 131–133
  as ordered field, 73
  symbol for, 2
  transcendental, 71–72, 507
real-valued function, 33
recession velocity, 38–39
Reciprocal Rule, 178–179
reciprocals, distinguished from inverses, 203, 486
rectification, 619–620
recursively defined sequences, 38
recursive sequences, limits of, 133
redeem, 311
reduction formulas, 530–532
regions in the plane, 18–19
regression (least squares) line, 27–29
regular polygons, 67
related rates, 226–229
relative error, 7
  polynomials and, 11
relative growth rate, 467
relative (local) maximum or minimum, defined, 250–251
  See also maxima and minima
relative rate of change, 457
remainder term. *See* error term of Taylor expansion
restricting the domain, 205, 477–478, 479, 486, 492
restriction, 34
Richter, Charles F., 460
Richter scale, 460
Riemann, Georg Friedrich Bernhard, 417
Riemann sums, 358–361
  continuous analogues for, 585
  limit definition for, 361
  lower and upper, 359–361
  midpoint approximation, 400
  moment, 592, 593
  negative, meaning of, 359, 393
  notation for, 358
  point selection for, 358–359
  work approximation using, 597
  See also definite integrals
right endpoint approximation, 351, 357
rise, 22
rise over run, 21–22
risk-reward relationship, 52
Roberval, Giles, 247, 342–343, 415, 416, 620
Rodrigues, Benjamin Olinde, 221
Roe, P. L., 275
Rolle, Michel, 254, 344
Rolle's Theorem, 253–254
  historical perspectives, 344
roots (zeroes) of a function, 47
  approximation by Newton-Raphson Method, 307
  fixed point, 133
Root Test, 655–656
rounding
  agreement to q decimal places and, 8
  as not transitive, 11
roundoff error, 8, 181
rule of a function, 33
run, 21
Russell, Bertrand, 72
Rutherford, Sir Ernest, 465

S

Saha's equation, 264
Sarasa, Alphons Anton de, 506
Scalar Multiplication Rule, 176, 182
scatter plot (scatter diagram), 38–39
  Least Squares Method and, 40–41
Schweidler, Egon von, 465
scientific revolution, 72–73
secant function
  antiderivatives of, 439–440
  defined, 61
  graphs of, 61
  historical perspectives on integrals of, 557–558
  integrals involving, 535
  inverse (arcsec), 484–485, 486, 555
  as periodic, 291
  values of, 63
secant function, hyperbolic. *See* hyperbolic functions
secant line(s), slope of tangent line and, 158
second central difference quotient, 218
second coordinate, 11
second derivative, 216
  See also higher derivatives
Second Derivative Test for Concavity, 278–279
second moments, 596
second-order processes, 309
semicubical parabola, 620
sensitivity, 156
Separation of Variables Method, 422–423
sequence of partial sums, definition of, 622
sequences, 37–38
  bounded, 131–133
  composition and, 128–129
  convergent, 122–123, 131–133
  distinguished from series, 624
  divergent, 123–125
  graphing of, 122
  increasing and decreasing, 131
  index variables for, 122
  infinity, tending to, 123–125
  initialization of, 38
  irrational exponents and, 135
  limit theorems for, 125–126
  monotone convergence property and, 131–133
  monotonic, 131
  recursive, limits of, 133
  tail of, 123
  See also infinite series
series, geometric. *See* geometric series
series, infinite. *See* infinite series
serpentine of Newton, 21
sets
  intervals. *See* intervals
  notation for, 3
  sketching, 3–4
  summarized, 69
sidebands, 68
sigma notation, 349, 622
sigmoidal growth. *See* logistic growth function
significance, loss of, 8, 181
significant digits, 8, 181–182
signum (sign) function, 35, 36–37
simple interest, 136, 139
Simpson's Rule, 404–407, 408
sine function
  antiderivative of, 318–319
  definite integrals of, 380
  definition of, 59
  derivatives of, 170–171
  graphs of, 61
  identities of, 62–64
  integration of even powers of, 533, 534
  integration of odd powers of, 532–534
  integration by parts and, 511, 514–515
  integration by reduction formulas, 530–532

sine function (*continued*)
integration of squares of, 529–530
inverse (arcsine), 478–480, 481–482, 486, 555
limit theorems for, 93–95
modeling with, 64–65
as periodic, 291
power series expansion of, 699–700
properties of, 60
Taylor expansion and, 688–689, 693–694
triangle definition, 60–61
values of, 63
sine function, hyperbolic. *See* hyperbolic functions
sine integral, 384, 516
sketching
graphs, 286–290
periodic functions, 291–292
sets, 3–4
skew (slant, oblique) asymptotes, 120, 292–293
slant asymptotes. *See* skew asymptotes
slant height, 578
slicing. *See* volume by method of slicing
slope
and basic rule for obtaining linear equations, 24–25
intercept form, 26
negative, 22
of parallel lines, 23
of perpendicular lines, 22
point-slope form, 25–26
positive, 22
as rise over run, 21–22
of secant line, 158
slope-intercept form, 26
summarized, 69–70
of tangent line, 158, 159
undefined for vertical line, 22
zero, of horizontal line, 22, 250
slope-intercept form of a line, 26
smoothness
differentiability as degree of, 167
presence of tangent lines as measure of, 161
Snell's inequality, 68
Snell's Law, 344
Snel(l) van Roijen, Willebrord van, 73, 344
Soddy, Frederick, 465
solid(s)
historical perspectives, 619, 620
not generated by rotation, 569–570
of revolution, defined, 564
rotated about a line not a coordinate axis, 568–569, 572–573
volume of. *See* volume
Solidum Hyperbolicum Acutum (Torricelli), 620
solution curves, 420
speed, 156
See also velocity
springs, 599–600
square errors, sum of, 29
square root, absolute value and, 4
square root formula, 330–331
square roots, algorithms for extraction, 45
squares, sum of consecutive, 350
stable equilibrium point, 257
Stade (Zeno), 72
standard deviation, 410, 590, 596
standard normal, 590, 613
statistics, probability theory and, 586
Stefan's Law of Radiation, 553
Stirling's form, of error in Simpson's Rule, 408
Substitution Method
arbitrary constants and, 390–391
Chain Rule and, 389–390
definite integrals and, 388–389
historical perspectives, 416
indefinite integration by, 387–388, 390–391
steps for, 386–387
See also Inverse (or Indirect) Substitution Method
subtraction. *See* sums (or differences) of functions
sum of consecutive positive integers, 350
sum of consecutive squares, 350
summation
index of, 349
notation for, 349
properties of, 349
summation notation, 349
Sum Rule, 157–158, 182
sums, integrals as limit of, 362
sums (or differences) of functions
antiderivatives of, 317
combining operation, 46
definite integral properties and, 369
derivatives of, 176–177
instantaneous rate of change for, 157–158
limits of, 90
superposition of waves, 67–68
surface area, 578–581, 610–611, 617
surface of revolution, defined, 578
surge, 217
symbols. *See* notation and symbols
symmetry, 50
axis of, 15
even and odd functions, 50–51
of function and its inverse, 202
with respect to the origin, 297
with respect to *y*-axis, 297

T

tail end of a series, 632–633
tangent function
antiderivatives of, 438–439, 440
defined, 61
derivative of, 187–188
graphs of, 61
identities of, 62
integrals involving, 535
inverse (arctangent), 480–484, 486, 555, 677–678, 687
as periodic, 291
values of, 63
tangent function, hyperbolic. *See* hyperbolic functions
tangent line approximation, 236–238, 239, 244
tangent line(s)
approximation of, 236–238, 239, 244, 684
to a circle, 158, 222
of curves that are not graphs of functions, 166–167
definition of, 158
differentiability defined in terms of, 165, 167–168
historical perspectives on, 245–247
implicit differentiation and. *See* Implicit Differentiation Method
maxima and minima and, 250, 252, 254
as measure of smoothness, 161
Newton-Raphson Method and, 306–307
nonexistence of, 160–161
normal lines and, 160, 245–246
slope of, 158, 159
summarized, 241
vertical, discussed, 161
Taylor, Brook, 708, 709
Taylor polynomials
computation of, 687–689
convergence and, 689–690, 693–695
definition of, 686
degree of accuracy desired for, 694–695
error term, 687–688, 689–690, 693–695, 699
estimation of error term, 693–695
higher-order approximating polynomials, 685–687
*N*th order Taylor expansion, 687, 688
tangent line approximation generalized, 684–685
Taylor series, 697–702, 707–709
Taylor's Theorem, 687–689
telescoping (collapsing) series, 624–625
telescoping (collapsing) sum, 355
terminal velocity, 142, 470, 495
terminating decimals, 2
terms
leading, 11
lower-order, 11
*n*th, 37
testing concepts, 353
Theory of Indivisibles, 415
third derivative, 216
See also higher derivatives
time-series data, 27
time-series graph, 27
Torricelli, Evangelista, 73, 247, 415, 620
Torricelli's infinitely long solid, 582, 610–611, 620
Torricelli's law, 424
torus, volume of, 619
total error, 181
tractrix, 429
Traité d'Algèbre (Rolle), 344
Traité de la Lumière (Cureau), 343
trajectory, 53
transcendental equations, solving, 305–306
transcendental numbers, 71–72, 507
translations, 49–50
trapezoidal approximation, 402
Trapezoidal Rule, 402–404

Treatise of Fluxions (Maclaurin), 708–709
Triangle Inequality, 6–7, 373
trident of Newton, 21
trigonometric functions, 58–65
  antidifferentiation of, 318–319
  associated principal angle, 60–61
  definite integrals of, 380
  definitions of, 59, 61
  derivatives of, 170–171, 187–188, 190
  fundamental properties of, 60
  identities, 62–64
  inequalities, 68
  integration of converted functions, 535
  integration of even powers, 533, 534
  integration of odd powers, 532–534
  integration by parts and, 511, 514–515
  integration by reduction formulas, 530–532
  integration of squares, 529–530
  inverse. *See* inverse trigonometric functions
  modeling with, 64–65
  as periodic, 291–292
  period of, 60
  power series expansion of, 699–700
  radian measure, 58–59
  summarized, 70
  superposition of waves, 67–68
  Taylor expansion and, 688–689, 693–694
  See also hyperbolic functions
trigonometric limits, 93–95
trigonometric substitution, 537–544
truncation error, 7, 181
two-point form of a line, 31
two-point systems, moments of, 591

U

unbounded integrands. *See* improper integrals
unbounded intervals, 6
uncountable, 71
uniform continuity, 361
uniform partition, 350–351
unit circle, 13
  radian measure and, 58
unit costs, 174
unit elasticity, 333
units of measure
  for acceleration, 319
  for velocity, 319
  for work, 597
upper bound, 131
utility (consumer satisfaction), 285

V

Van der Waals's equation, 120
variables
  dependent, 39
  independent, 39
  Leibniz notation and, 166
  Newton notation and, 167
  random, 586–587, 590, 612–613
  separable, 421–423
variance, 596
velocity
  antidifferentiation of, 319–322
  average, 153
  higher derivatives of, 216–217
  instantaneous. *See* instantaneous velocity
  linear drag law, 469–470
  Mean Value Theorem and, 254–255
  negative values for, meaning of, 156
  quadratic drag law, 483–484, 494–495
  terminal, 142, 470, 495
  unit of measure for, 319
vertex (vertices)
  definition of, 17
  minimum and maximum values of, 17
vertical asymptotes, 114–115, 117–118
vertical line(s)
  form of equation for, 22
  slope as undefined for, 22
  tangents, discussed, 161
Vertical Line Test, 35
vertical translations, 49–50
Viète, François, 72, 342
Viète's algorithm for π, 45
viewing rectangle, 19
viewing window, 19
Volterra, Vito, 429
volume
  finding by cylindrical shells, 570–572, 616
  finding by slicing. *See* volume by method of slicing
  improper integrals and, 607
  instantaneous rate of change of, 157
  memorization of formulas not recommended, 572–573
  of rectangular box, 42
  of right circular cone, 41, 562–563
  of solid not generated by rotation, 569–570
  of solid rotated about a line not a coordinate axis, 568–569, 571–572
volume by method of slicing
  disk method, 562–566, 569
  for a solid not of revolution, 569–570
  summarized, 569, 616
  washer method, 566–569, 571, 616

W

Wallis, John, 344, 415, 416, 537, 620
Wallis limit formula, 537
washers, method of, 566–569, 571, 616
water pressure applications, 214, 233, 600–603
Weierstrass, Karl, 73, 148–149
work
  area compared to, 597
  definition of, 596–597
  integrals to calculate, 597
  pumping fluid, 600–601
  springs, 599–600
  summarized, 617
  unit of measure for, 597
  and weights that vary, 598–599
Wren, Christopher, 620
Wright, Edward, 557

X

x-axis, 11
x-coordinate, 12
x-intercept, 24
xy-plane, 11

Y

y-axis, 11
  symmetry with respect to, 297
y-coordinate, 12
y-intercept, 24

Z

Zeno of Elea, 72
Zeno's paradoxes, 72, 660
zero
  absolute, 85
  as derivative of a constant, 175
  numerator and denominator. *See* indeterminate form
  as slope of horizontal line, 22, 250
zeroes. *See* roots (zeroes) of a function
zeroth derivative, 216